# *Coastal Geomorphology of Great Britain*

THE GEOLOGICAL CONSERVATION REVIEW SERIES

The comparatively small land area of Great Britain contains an unrivalled sequence of rocks, mineral and fossil deposits, and a variety of landforms that record much of the Earth's long history. Well-documented ancient volcanic episodes, famous fossil sites, and sedimentary rock sections used internationally as comparative standards have given these islands an importance out of all proportion to their size. The long sequences of strata and their organic and inorganic contents have been studied by generations of leading geologists, thus giving Britain a unique status in the development of the science. Many of the divisions of geological time used throughout the world are named after British sites or areas; for instance, the Cambrian, Ordovician and Devonian systems, the Ludlow Series and the Kimmeridgian and Portlandian stages.

The Geological Conservation Review (GCR) was initiated by the Nature Conservancy Council in 1977 to assess and document the most important parts of this rich heritage. The GCR records the current state of knowledge of the key Earth science sites in Great Britain and provides a firm basis upon which site conservation can be founded in years to come. Each GCR title in the 42-volume series describes networks of sites of national or international importance in the context of a portion of the geological column, or a geological, palaeontological or mineralogical topic.

Within each volume, every GCR site is described in detail in a self-contained account, consisting of an introduction (with a concise history of previous work), a description, an interpretation (assessing the fundamentals of the site's scientific interest and importance), and a conclusion (written in simpler terms for the non-specialist). Each site report is a justification of the particular scientific interest in a locality, of its importance in a British or international setting, and ultimately of its worthiness for conservation.

The aim of the Geological Conservation Review Series is to provide a public record of the features of interest in sites that have been notified as – or are being considered for notification as – Sites of Special Scientific Interest (SSSIs). It is written to the highest scientific standards but in such a way that the assessment and conservation value of the site is clear. It is a public statement of the value placed on our geological and geomorphological heritage by the Earth science community and it will be used by the Joint Nature Conservation Committee, the Countryside Council for Wales, English Nature and Scottish Natural Heritage in carrying out their conservation functions. The three country agencies are also active in helping to establish sites of local and regional importance. Regionally Important Geological/Geomorphological Sites (RIGS) augment the SSSI coverage, with local groups identifying and conserving sites that have educational, historical, research or aesthetic value, enhancing the wider Earth heritage conservation perspective.

All the sites in this volume have been proposed for notification as SSSIs; the final decision to notify, or re-notify, sites lies with the governing councils of the appropriate country conservation agency.

Information about the GCR publication programme may be obtained from:

GCR Unit,
Joint Nature Conservation Committee,
Monkstone House,
City Road,
Peterborough PE1 1JY.

www.jncc.gov.uk

Copies of published volumes can be purchased from:

NHBS Ltd,
2–3 Wills Road,
Totnes,
Devon TQ9 5XN.

www.nhbs.com

Published titles in the GCR Series

1. **An Introduction to the Geological Conservation Review**
   N.V. Ellis (ed.), D.Q. Bowen, S. Campbell, J.L. Knill, A.P. McKirdy,
   C.D. Prosser, M.A. Vincent and R.C.L. Wilson

2. **Quaternary of Wales**
   S. Campbell and D.Q. Bowen

3. **Caledonian Structures in Britain
   South of the Midland Valley**
   Edited by J.E. Treagus

4. **British Tertiary Volcanic Province**
   C.H. Emeleus and M.C. Gyopari

5. **Igneous Rocks of South-West England**
   P.A. Floyd, C.S. Exley and M.T. Styles

6. **Quaternary of Scotland**
   Edited by J.E. Gordon and D.G. Sutherland

7. **Quaternary of the Thames**
   D.R. Bridgland

8. **Marine Permian of England**
   D.B. Smith

9. **Palaeozoic Palaeobotany of Great Britain**
   C.J. Cleal and B.A. Thomas

10. **Fossil Reptiles of Great Britain**
    M.J. Benton and P.S. Spencer

11. **British Upper Carboniferous Stratigraphy**
    C.J. Cleal and B.A. Thomas

12. **Karst and Caves of Great Britain**
    A.C. Waltham, M.J. Simms, A.R. Farrant and H.S. Goldie

13. **Fluvial Geomorphology of Great Britain**
    Edited by K.J. Gregory

14. **Quaternary of South-West England**
    S. Campbell, C.O. Hunt, J.D. Scourse, D.H. Keen and N. Stephens

15. **British Tertiary Stratigraphy**
    B. Daley and P. Balson

16. **Fossil Fishes of Great Britain**
    D.L. Dineley and S.J. Metcalf

17. **Caledonian Igneous Rocks of Great Britain**
    D. Stephenson, R.E. Bevins, D. Millward, A.J. Highton, I. Parsons, P. Stone and W.J. Wadsworth

18. **British Cambrian to Ordovician Stratigraphy**
    A.W.A. Rushton, A.W. Owen, R.M. Owens and J.K. Prigmore

19. **British Silurian Stratigraphy**
    R.J. Aldridge, David J. Siveter, Derek J. Siveter, P.D. Lane, D. Palmer and N.H. Woodcock

20. **Precambrian Rocks of England and Wales**
    J.N. Carney, J.M. Horák, T.C. Pharaoh, W. Gibbons, D. Wilson, W.J. Barclay, R.E. Bevins,
    J.C.W. Cope and T.D. Ford

Published GCR volumes

21. **British Upper Jurassic Stratigraphy**
    Oxfordian to Kimmeridgian
    J.K. Wright and B.M. Cox

22. **Mesozoic and Tertiary Palaeobotany of Great Britain**
    C.J. Cleal, B.A. Thomas, D.J. Batten and M.E. Collinson

23. **British Upper Cretaceous Stratigraphy**
    R.N. Mortimore, C.J. Wood and R.W. Gallois

24. **Permian and Triassic Red Beds and the Penarth Group of Great Britain**
    M.J. Benton, E. Cook and P. Turner

25. **Quaternary of Northern England**
    D. Huddart and N.F. Glasser

26. **British Middle Jurassic Stratigraphy**
    B.M. Cox and M.G. Sumbler

27. **Carboniferous and Permian Igneous Rocks of Great Britain**
    North of the Variscan Front
    D. Stephenson, S.C. Loughlin, D. Millward, C.N. Waters and I.T. Williamson

28. **Coastal Geomorphology of Great Britain**
    V.J. May and J.D. Hansom

Coastal Geomorphology of Great Britain

V.J. May

School of Conservation Sciences,
Bournemouth University, UK

and

J.D. Hansom

Department of Geography and Geomatics,
University of Glasgow, UK

GCR Editors: **K.M. Clayton** and **E.C.F. Bird**

Published by the Joint Nature Conservation Committee, Monkstone House, City Road, Peterborough, PE1 1JY, UK

First edition 2003

© 2003 Joint Nature Conservation Committee

Typeset in 10/12pt Garamond ITC by JNCC
Printed in Great Britain by CLE Print Limited on Huntsman Velvet 100 gsm
ISBN 1 86107 4840.

Apart from any fair dealing for the purposes of research or private study, or criticism or review, as permitted under the UK Copyright Designs and Patents Act, 1988, this publication may not be reproduced, stored, or transmitted, in any form or by any means, without the prior permission in writing of the publishers, or in the case of reprographic reproduction only in accordance with the terms of the licences issued by the Copyright Licensing Agency in the UK, or in accordance with the terms and licences issued by the appropriate Reproduction Rights Organization outside the UK. Enquiries concerning reproduction outside the terms stated here should be sent to the GCR Team, JNCC.

The publisher makes no representation, express or implied, with regard to the accuracy of the information contained in this book and cannot accept any legal responsibility or liability for any errors or omissions that may be made.

British Geological Survey Copyright protected materials

1. The copyright of materials derived from the British Geological Survey's work is vested in the Natural Environment Research Council (NERC). No part of these materials (geological maps, charts, plans, diagrams, graphs, cross-sections, figures, sketch maps, tables, photographs) may be reproduced or transmitted in any form or by any means, or stored in a retrieval system of any nature, without the written permission of the copyright holder, in advance.

2. To ensure that copyright infringements do not arise, permission has to be obtained from the copyright owner. In the case of BGS maps this includes **both BGS and the Ordnance Survey**. Most BGS geological maps make use of Ordnance Survey topography (Crown Copyright), and this is acknowledged on BGS maps. Reproduction of Ordnance Survey materials may be permitted independently by the licences issued by Ordnance Survey to users of their materials. Users who do not have an Ordnance Survey licence to reproduce the topography must make their own arrangements with the Ordnance Survey, Copyright Branch, Romsey Road, Southampton SO9 4DH (Tel. 0230 879 2913).

3. Permission to reproduce BGS materials must be sought in writing from the Intellectual Property Rights Manager, British Geological Survey, Kingsley Dunham Centre, Keyworth, Nottingham NG12 5GG (Tel. 0115 936 3100).

A catalogue record for this book is available from the British Library.

Recommended example citations

May, V.J. and Hansom, J.D. (2003) *Coastal Geomorphology of Great Britain*, Geological Conservation Review Series, No. 28, Joint Nature Conservation Committee, Peterborough, 737 pp.

May, V.J. (2003) Carmarthen Bay. In *Coastal Geomorphology of Great Britain*, Geological Conservation Review Series, No. 28, (V.J. May and J.D. Hansom), Joint Nature Conservation Committee, Peterborough, pp. 611–24.

Contents

Acknowledgements — xi
Access to the countryside — xiii
Preface *V.J. May, J.D. Hansom, K.M. Clayton and E.C.F. Bird* — xv

1 An introduction to the coastal geomorphology of Great Britain — 1

Introduction — 3
Organization of this volume — 3
Coastal research in Britain *K.M. Clayton* — 3
The geological background *K.M. Clayton* — 5
The coastal marine environment: tides, waves, surges and currents
 K.M. Clayton — 10
Coastal sediment supply and sediments cells *K.M. Clayton* — 13
Sea-level history *K.M. Clayton* — 15
Coastal management and coastal engineering *K.M. Clayton* — 18
Further reading — 20
GCR site selection guidelines *V.J. May and N.V. Ellis* — 21
Anthropogenic influences and the GCR *V.J. May and N.V. Ellis* — 23
Legal protection of the GCR sites *V.J. May and N.V. Ellis* — 28
GCR site selection in conclusion — 30

2 The geomorphology of the coastal cliffs of Great Britain — 31
K.M. Clayton, E.C.F. Bird and J.D. Hansom

Introduction — 33
Inland and coastal cliffs — 35
Coastal slope processes — 36
Processes of marine erosion of cliffs and shore platforms — 38
The form of coastal cliffs — 40
 Geological controls on cliff form — 40
 Characteristic medium-scale features of cliffs — 41
 Influence of inland topography on cliff form — 44
Shore platforms — 45
Evidence of inheritance — 48
Sea cliffs as biological SSSIs and Special Areas of Conservation (SACs)
 N.V. Ellis and C.R. McLeod — 51

Contents

3 Hard-rock cliffs – GCR site reports   **55**

Introduction *J.D. Hansom*   57
St Kilda, Western Isles *J.D. Hansom*   60
Villians of Hamnavoe, Shetland *J.D. Hansom*   68
Papa Stour, Shetland *J.D. Hansom*   72
Foula, Shetland *J.D. Hansom*   76
West Coast of Orkney *J.D. Hansom*   81
Duncansby to Skirza Head, Caithness *J.D. Hansom*   86
Tarbat Ness, Ross and Cromarty *J.D. Hansom*   90
Loch Maddy–Sound of Harris Coastline, Western Isles *J.D. Hansom*   94
Northern Islay, Argyll and Bute (Potential GCR site) *J.D. Hansom*   98
Bullers of Buchan, Aberdeenshire *J.D. Hansom*   103
Dunbar, East Lothian *J.D. Hansom*   107
St Abb's Head, Berwickshire *J.D. Hansom*   110
Tintagel, Cornwall *V.J. May*   113
South Pembroke Cliffs, Pembrokeshire *V.J. May*   117
Hartland Quay, Devon *V.J. May*   121
Solfach, Pembrokeshire *V.J. May*   126

4 Soft-rock cliffs – GCR site reports   **129**

Introduction *V.J. May and K.M. Clayton*   131
Ladram Bay, Devon *V.J. May*   138
Robin Hood's Bay, Yorkshire *V.J. May*   141
Blue Anchor–Watchet–Lilstock, Somerset *V.J. May*   145
Nash Point, Glamorgan *V.J. May*   148
Lyme Regis to Golden Cap, Dorset *V.J. May*   151
South-west Isle of Wight *V.J. May*   158
Kingsdown to Dover, Kent *V.J. May*   165
Beachy Head to Seaford Head, East Sussex *V.J. May*   170
Ballard Down, Dorset *V.J. May*   176
Flamborough Head, Yorkshire *V.J. May*   181
Joss Bay (GCR Name: Foreness Point), Kent *V.J. May*   187
Porth Neigwl, Gwynedd *V.J. May*   191
Holderness (Potential GCR Site), Yorkshire *K.M. Clayton*   195

5 Beaches, spits, barriers and dunes – an introduction   **201**
E.C.F. Bird, K.M. Clayton and J.D. Hansom

Introduction   203
Provenance of beach sediments   204
Coastal sediment movements   209
Beach plan   212
Beach profiles   213
Beach states   214
Bars and troughs   214
Lateral grading   215
Prograding beaches   216
Beach ridges   217
Spits, tombolos and cuspate forelands   217
Coastal barriers   219
Coastal dunes   221
Causes of beach and dune erosion   226

Contents

6 Gravel and 'shingle' beaches – GCR site reports   **229**

    Introduction *V.J. May*   231
    Westward Ho! Cobble Beach, Devon *V.J. May*   238
    Loe Bar, Cornwall *V.J. May*   241
    Slapton Sands and Hallsands, Devon *V.J. May*   244
       Slapton Sands *V.J. May*   246
       Hallsands *V.J. May*   248
    Budleigh Salterton Beach, Devon *V.J. May*   251
    Chesil Beach, Dorset *V.J. May*   254
    Porlock, Somerset *J. Orford*   266
    Hurst Castle Spit, Hampshire *V.J. May*   271
    Pagham Harbour, West Sussex *V.J. May*   278
    The Ayres of Swinister, Shetland *J.D. Hansom*   281
    Whiteness Head, Moray *J.D. Hansom*   285
    Spey Bay, Moray *J.D. Hansom*   290
    The West Coast of Jura, Argyll and Bute *J.D. Hansom*   297
    Benacre Ness, Suffolk *V.J. May*   301
    Orfordness and Shingle Street, Suffolk *V.J. May*   304
    Dungeness and Rye Harbour *V.J. May*   310
       Rye Harbour, East Sussex *V.J. May*   314
       Dungeness, Kent *V.J. May*   315

7 Sandy beaches and dunes – GCR site reports   **327**

    Introduction *V.J. May*   329
    Marsden Bay, County Durham *V.J. May*   337
    South Haven Peninsula, Dorset *V.J. May*   340
    Upton and Gwithian Towans, Cornwall *V.J. May*   345
    Braunton Burrows, Devon *V.J. May*   348
    Oxwich Bay, Glamorgan *V.J. May*   354
    Tywyn Aberffraw, Anglesey *V.J. May*   356
    Ainsdale, Lancashire *V.J. May*   359
    Luce Sands, Dumfries and Galloway *J.D. Hansom*   364
    Sandwood Bay, Sutherland *J.D. Hansom*   370
    Torrisdale Bay and Invernaver, Sutherland *J.D. Hansom*   375
    Dunnet Bay, Caithness *J.D. Hansom*   380
    Balta Island, Shetland *J.D. Hansom*   384
    Strathbeg, Aberdeenshire *J.D. Hansom*   387
    Forvie, Aberdeenshire *J.D. Hansom*   393
    Barry Links, Angus *J.D. Hansom*   400
    Tentsmuir, Fife *J.D. Hansom*   407

8 Sand spits and tombolos – GCR site reports   **415**

    Introduction *V.J. May*   417
    Pwll-ddu, Glamorgan *V.J. May*   422
    Ynyslas, Ceredigion *V.J. May*   424
    East Head (Chichester Harbour), West Sussex *V.J. May*   426
    Spurn Head, Yorkshire *V.J. May*   430
    Dawlish Warren, Devon *V.J. May*   435
    Gibraltar Point, Lincolnshire *C.A.M King and V.J. May*   439
    Walney Island, Lancashire *V.J. May*   443
    Winterton Ness, Norfolk *V.J. May*   446

Contents

Morfa Harlech, Gwynedd *V.J. May*	449
Morfa Dyffryn, Gwynedd *V.J. May*	453
St Ninian's Tombolo, Shetland *J.D. Hansom*	458
Coast of the Isles of Scilly *V.J. May*	462
Central Sanday, Orkney *J.D. Hansom*	465

9 Machair *J.D. Hansom* — 471

Introduction	473
Machir Bay, Islay, Argyll and Bute	478
Eoligarry, Barra, Western Isles	482
Ardivachar to Stoneybridge, South Uist, Western Isles	487
Hornish and Lingay Strands (Machairs Robach and Newton), North Uist, Western Isles	492
Pabbay, Harris, Western Isles	495
Luskentyre and Corran Seilebost, Harris, Western Isles	499
Mangersta, Lewis, Western Isles	503
Tràigh na Berie, Lewis, Western Isles	506
Balnakeil, Sutherland	510

10 Saltmarshes — 515

Introduction *E.C.F. Bird*	517
Culbin, Moray *J.D. Hansom* (see also Chapter 11)	529
Morrich More, Ross and Cromarty *J.D. Hansom* (see also Chapter 11)	530
St Osyth Marsh, Essex *V.J. May*	531
Dengie Marsh, Essex *V.J. May*	534
Keyhaven Marsh, Hurst Castle, Hampshire *V.J. May*	538
Solway Firth saltmarshes *J.D. Hansom*	539
Solway Firth (north shore), Dumfries and Galloway *J.D. Hansom*	541
Upper Solway flats and marshes (south shore), Cumbria, *J.D. Hansom*	548
Cree Estuary, Outer Solway Firth, Dumfries and Galloway *J.D. Hansom*	552
Loch Gruinart, Islay, Argyll and Bute *J.D. Hansom*	556

11 Coastal assemblage GCR sites — 563

Introduction *V.J. May*	565
Culbin, Moray *J.D. Hansom*	567
Morrich More, Ross and Cromarty *J.D. Hansom*	576
Carmarthen Bay, Carmarthenshire *V.J. May*	583
The Coast of Caernarfon Bay (Newborough Warren and Morfa Dinlle) *V.J. May*	593
Newborough Warren, Isle of Anglesey *V.J. May*	595
Morfa Dinlle, Gwynedd *V.J. May*	600
Holy Island, Northumberland *V.J. May*	604
North Norfolk Coast *V.J. May*	611
The Dorset Coast: Peveril Point to Furzy Cliff *V.J. May*	624

References	643
Glossary	703
Index	717

Acknowledgements

Work began on selecting British coastal geomorphology sites for the Geological Conservation Review in the early 1980s with a widespread consultation exercise co-ordinated by Drs J.E. Gordon and A.S. Mather, and Professors V.J. May and W. Ritchie, under the auspices of the then-Head of the Geological Conservation Review Unit of the former Nature Conservancy Council (NCC), Dr W.A. Wimbledon. Grateful acknowledgement is given to all those who contributed to this consultation and site-selection exercise, particularly the late R.W.G. Carter, E.C.F. Bird, T. Sunamura, the late A. Guilcher, the late J.A. Steers, D. Brunsden, A.P. Carr, G. de Boer, C.A.M. King and H. Caldwell.

Supporting documentation for the selected sites was compiled by Professor V.J. May (England and Wales) and Professor W. Ritchie (Scotland) during the site selection programme, and this material, including maps, has been invaluable in the preparation of the present volume.

In 1988, Dr W.A. Wimbledon invited Professor V.J. May to prepare a text for publication as a GCR Series volume, detailing the English and Welsh sites, and this was completed for the newly formed JNCC in 1992. Thanks are especially due to Dr Wimbledon for guidance and encouragement. However, in reviewing its plans for the publication of the GCR Series, a decision was made to produce a single volume covering the coastal geomorphology of the whole of Britain, rather than just England and Wales. Therefore Dr J.D. Hansom was invited by JNCC to prepare text for publication for the Scottish sites, and Professor V.J. May updated the already drafted reports for England and Wales.

Within this volume, the descriptions and interpretations of individual sites lean heavily on the observations and research of many individuals. Although published source material is referenced, the authors of the volume have also contributed their own personal knowledge of many of the sites. This text is thus a synthesis of understanding where the credits reach far wider than the names attributed to each part of the book. Grateful acknowledgement is therefore accorded to Professor W. Ritchie, Dr L. Pierce, Dr S. Gemmell, Dr A. Dawson, Dr S. Angus, Dr G. Lees and Dr F. Mactaggart.

The authors are grateful to the editors, Professor K.M. Clayton and Professor E.C.F. Bird, for their help in the development of the site descriptions and for providing some textual overviews.

The various topographical maps and diagrams have been compiled from numerous sources, and have inevitably extracted information from the many high

Acknowledgements

quality maps produced for this country by the British Geological Survey and the Ordnance Survey. Thanks go to P. and A. Macdonald, Glyn Satterley, Ken Crossan and Lorne Gill, whose photographs of the Scottish coast enliven the text and who kindly gave permission to use their work. Les Hill digitally prepared the Scottish photographs. Thanks are also due to all those who agreed to the use of their material in this volume, in particular, Cambridge University for the use of aerial photographs from their extensive collection. Diagrams were drafted by Ian Foulis and Associates. Each figure is accompanied by an acknowledgement of the source unless it is an original diagram prepared by the authors especially for the present volume. Thanks also go to Linda Cannings for her painstaking help in compiling the reference list for sites reports for England and Wales, and to Rebecca Cook for her help in bringing it up-to-date.

Thanks are also due to the GCR Publication Production Team: Neil Ellis (Publications Manager), Anita Carter, Emma Durham and Colin McLeod (Production Editors) for their unfailing patience and support in sustaining the preparation of this volume, and also to Neil Cousins, Nicholas D.W. Davey, and Val Wyld, former GCR Editorial Officers with NCC/JNCC.

Access to the countryside

This volume is not intended for use as a field guide. The description or mention of any site should not be taken as an indication that access to a site is open. Most sites described are in private ownership, and their inclusion herein is solely for the purpose of justifying their conservation. Their description or appearance on a map in this work should not be construed as an invitation to visit. Prior consent for visits should always be obtained from the landowner and/or occupier.

Information on conservation matters, including site ownership, relating to Sites of Special Scientific Interest (SSSIs) or National Nature Reserves (NNRs) in particular counties or districts may be obtained from the relevant country conservation agency headquarters listed below:

> Countryside Council for Wales,
> Maes-y-Ffynnon,
> Penrhosgarnedd,
> Bangor,
> Gwynedd LL57 2DN.
>
> English Nature,
> Northminster House,
> Peterborough PE1 1UA.
>
> Scottish Natural Heritage,
> 12 Hope Terrace,
> Edinburgh EH9 2AS.

Preface

Few countries can boast a coastline as geomorphologically diverse as that of Great Britain. From the island and fjord coastline of north-west Scotland to the ephemeral sand and mud coastlines of eastern England, it is a landscape of contrasts. The spectacular sheer cliffs of St Kilda have hardly changed in centuries, but in areas such as Holderness in eastern England, erosion has been so rapid that land and homes have been lost to the sea at dramatic rates. Areas of extensive coastal urbanization are in stark contrast to other coasts that are surprisingly untouched by development. There are classic 'textbook' examples of typical coastal geomorphological features cited the world over, such as Scolt Head Island, Lulworth Cove, Chesil Beach and St Ninian's Tombolo, and yet others such as the machair of the west coast of Scotland that are unique to the British Isles. These features can be visually spectacular, and many have earned international renown scientifically and aesthetically.

The broad outline of the coastline owes much to the variety of rocks and large-scale geological structures (such as the Great Glen Fault in Scotland), which have different levels of resistance to erosion; the pattern of the coastline of northern and western Britain can be largely attributed to the differential resistance to erosion of the rocks over many millennia and several glaciations. Ice-Age glaciation and fluvial and marine processes have superimposed drainage networks and carved an intricate pattern of coastal landforms including headlands, bays and estuaries. Glaciers have also deposited vast stores of sediment offshore, much of which has been brought ashore subsequently by marine transgression in the Holocene Epoch to form large, depositional, coastal landforms.

Relative to land, sea levels have varied so much during the last 20 000 years that coastal landforms include surfaces now many metres above the sea, as well as submerged features formed when sea level was relatively low compared to present-day. If present-day sea level continues to rise, many of the beach sites will be affected further by erosion, part of the continuing evolution of the British coastline. Many of the sites in the southern part of Britain provide excellent opportunities for the monitoring and modelling of the effects of sea-level change because they have lengthy records of shoreline change and beach profiles that can serve as baselines. They may help us understand future coastal changes associated with sea-level rise (or its absence) resulting from climate change.

Whereas small-scale features may reach equilibrium within a single tidal cycle, beaches and saltmarshes may take several centuries to reach equilibrium and the larger-scale configuration of the coastline may require several thousands of years

Preface

to adjust to the Holocene rise in sea level and isostatic rebound.

In many locations, the coastline also owes its present characteristics to such human activities as land 'reclamation' (land-claim, an activity that has been going on since Roman times), gravel and sand extraction, flood defences, protection against erosion, and harbour construction. The rarity of saltmarsh along much of the British coast reflects the role of catchment management affecting fine-grained sediment transport and the effects of land-claim.

The coast has, as a result, been the focus of considerable political attention, arising from major engineering programmes including the Thames Barrage (the development of which was stimulated by the 1953 floods), and the construction and servicing of the North Sea oil and gas resources. The coast has also been the centre of debate about the quality of the environment, especially beach quality. The coast is also important for recreation and as the location of many of Britain's resorts. The 19th century growth of the seaside resorts meant that many areas of previously rural coastal land were developed. Since then, protection measures have had significant impacts on coastal processes. In many cases, a policy of non-intervention might prove more efficient in the longer term, not least where the beaches themselves provide a natural form of protection that can continue to transgress with a rising sea level. Such a policy is geomorphologically sound, but may be politically insensitive where communities live close to sea level. Nevertheless it may prove the sounder economic policy if vast sums are not to be devoted to sea defences that will need to be rebuilt regularly.

Many of Britain's coastal features are of worldwide significance, reflecting not only their intrinsic nature, but also the substantial record of scholarship and research devoted to their description and understanding. There is still much more to learn; the sites continue to play a part in future research. The sites selected for the Geological Conservation Review represent not only this international reputation, but also provide a nationwide framework of landforms and processes within which research and education can continue to develop. They have been chosen to represent the rich variety of coastal landforms and to provide a network of sites that reflect the different results of rock-type, structure, sediments, wave and tidal conditions, and climate. In an environment as dynamic as the coast, it has been taken as axiomatic that some sites should be included despite significant levels of human intervention in natural processes. This should help to ensure that sites that are of international importance can be protected and that the influence of human activities can be integrated with studies of the natural processes.

Britain's coast is also the home of many rare species and the location of fragile habitats. Much of the coast is noted for nesting and roosting birds; for example, Little Terns on shingle beaches and Guillemots on hard-rock cliffs. Therefore, the stability or dynamism of the geomorphological features and processes is intrinsically linked to the future of coastal wildlife and habitat sites. Although 'biological' conservation sites are not described in this volume *per se*, key sites are also important as wildlife and habitat Sites of Special Scientific Interest; internationally important sites are protected through further designations – all, however, depend ultimately on the relationships between the geomorphological and oceanographic processes affecting their ecosystems.

This volume deals with the state of knowledge of the sites available at the time of writing, in 1995–2002, and must be seen in this context. Geomorphology, like any other science, is an ever-developing pursuit with new discoveries being made, and existing models are subject to continual testing and modification as new data come to light. Increased or hitherto unrecognized significance may be seen in new sites, and it is possible that additional sites worthy of conservation will be identified in future years. Effects of coastal processes and development means that the GCR site list must, like the sites themselves, be dynamic.

Preface

This volume is not a field guide to the sites, nor does it cover the practical problems of their future conservation. Its remit is to put on record the scientific justification for conserving the sites. It will be invaluable as an essential reference book and, it is hoped, will provide a stimulus for further scientific research. The conservation value of the sites is mostly based on a specialist understanding of the features present and is, therefore, of a technical nature. The account of each site in this book ends, however, with a brief summary of the geomorphological interest, framed in less technical language, in order to help the non-specialist. The first chapter of the volume, used in conjunction with the glossary, is also aimed at a less specialized audience.

The educational significance of the coast, and the interest stimulated by appropriate information, is difficult to assess, but the number of people whose first interest in geology, geomorphology and environmental processes was awakened on a visit to the coast is incalculable.

There is still much more to learn and the sites described in this volume are as important today as they have ever been in increasing our knowledge and understanding of the geological history of Britain. This account clearly demonstrates the value of these sites for research, and their important place in Britain's scientific and natural heritage. This, after all, is the main objective of the GCR Series of publications.

V.J. May, J.D. Hansom, K.M. Clayton and E.C.F. Bird, January 2003

Chapter 1

An introduction to the coastal geomorphology of Great Britain

Coastal research in Britain

INTRODUCTION

This volume summarizes the results of the site evaluation and selection programme of Britain's coastal regions that was undertaken between 1980 and 1990 as part of the Geological Conservation Review (GCR), although the site evaluation phase of the Review is ongoing. With the aim of representing the highlights of Britain's coastal geomorphology, 99 sites (see Figure 1.2) were selected eventually for this part of the GCR, to be considered for long-term conservation under British law.

The descriptions of the GCR sites in this book are intended not only to justify why it is important to conserve geomorphologically important parts of our coast, but also to demonstrate the major contribution that British sites have made, and continue to make, in developing our understanding of coastal geomorphological processes and the features that they create.

ORGANIZATION OF THIS VOLUME

In the following sections of Chapter 1, an overview of coastal research is followed by an introduction to the coastal geomorphology of Great Britain – discussing geomorphological processes and their controls (a glossary of geomorphological and technical terms is provided at the back of this book). There is a brief discussion of the impact of coastal management and engineering on the environment, and the chapter concludes with details of the rationale and methods of selection of the coastal geomorphology GCR sites.

Most of the GCR sites described in the present volume are dominated by one coastal landform, especially in terms of their associated research significance, and this is the basis of their arrangement into chapters. This has the advantage that each set of landforms can be introduced in terms of our understanding of coastal geomorphological processes and the linked landforms. However, while cliff site descriptions are divided into 'hard-rock' (Chapter 3) and 'soft-rock' (Chapter 4), one chapter (Chapter 2) serves to introduce them. Similarly, whereas it is convenient to separate out into successive chapters groups of sites representing gravel ('shingle') beaches, sandy beaches, and spits and tombolos, it is simpler to cover all three chapters (6–8) by a single introduction (Chapter 5). There follow chapters on machair (9), and saltmarshes (10), each with their own introduction. Finally the selected GCR sites include a number that are complex in their assemblage of linked geomorphological forms, and so they have been grouped into a final chapter (11) with the title 'Coastal Assemblages'.

COASTAL RESEARCH IN BRITAIN

K.M. Clayton

With its geologically varied coastline and high-energy coastal regime, much coastal research has been stimulated in Britain over the past two hundred years. Books describing the landforms and scenery of the British Isles have paid considerable attention to the coast, and descriptions of many British sites have found their way into textbooks about coastal forms and processes as archetypal exemplars. Steers has published two very valuable descriptions of the entire British coast, one on England and Wales (Steers, 1946a) and the second on Scotland (Steers, 1973). Perhaps it is because of the diversity of British coastal forms – as well as the long history of study – that British authors have contributed several key texts on coastal geomorphology to the literature at a variety of levels of technical complexity (e.g. Pethick, 1984; Carter, 1985, 1988; Hansom, 1988; Bird, 2000, Haslett, 2000).

There is considerable bias in the published literature, however. In geographical terms, the north and west of Britain are poorly represented; in topical terms, soft-rock coasts have had more attention than hard-rock, and low coasts have had far more than cliffs. Coasts that change form more rapidly have also received greater attention, whereas hard-rock cliffs, which seem to change very little in a lifetime, have received much less.

A computer-database bibliography of some 9000 items has been accumulated covering a large proportion of all papers and books published on the geomorphology of the British Isles since 1960, together with a selection of papers published before that date (based on Clayton 1964, together with *Geomorphological Abstracts*, *Geographical Abstracts* and *Geo Abstracts*). Of these, 1400 are classified under the heading 'Coasts'. Table 1.1 gives a breakdown by date (the list is not comprehensive until after 1960, so the two columns are not comparable, but the fall-off in rate of publication of coastal papers since 1989 seems quite remark-

An introduction to British coastal geomorphology

Table 1.1 Number of items in the computerized bibliography of geomorphology of Britain that are classified as 'Coasts' (total 1400), by year of publication.

Year	Items
1830–1859	5
1860–1899	15
1900–1909	1
1910–1919	10
1920–1929	24
1930–1939	36
1940–1949	28
1950–1954	68
1955–1959	73
1960–1964	102
1965–1969	86
1970–1974	121
1975–1979	197
1980–1984	229
1985–1989	209
1990–1994	68
1995–1999	102

Table 1.2 Number of items under selected keywords (some items appear more than once as several keywords are allocated to each).

Beach	284
Erosion	267
Sea level	186
Cliffs	126
Saltmarsh	104
Sand dunes	90
Gravel/Shingle	86
Littoral/Longshore drift	79
Coastal protection	74
Spit	66
Coastal platform	42
Accretion	27
Sediment cell	3

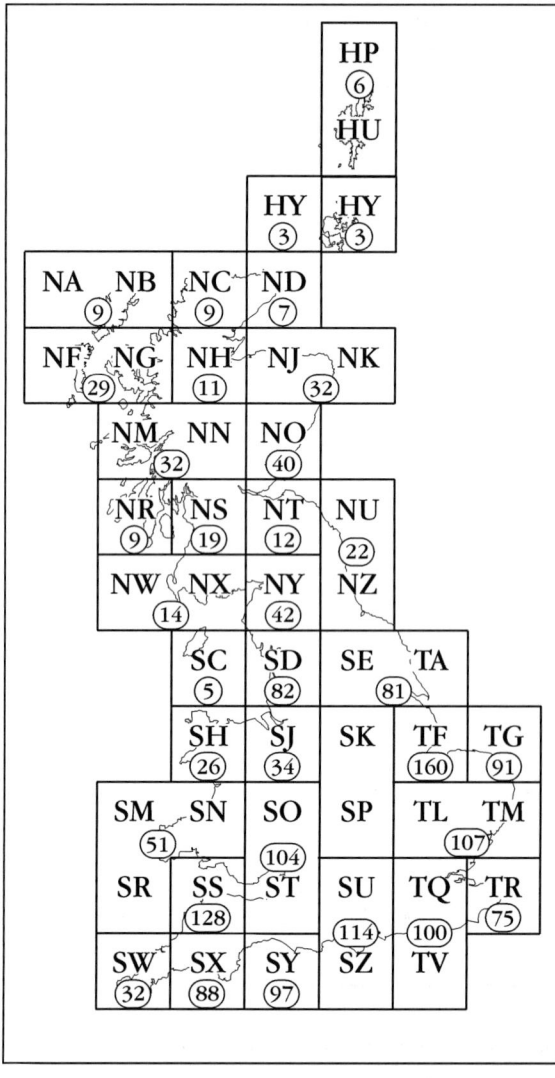

Figure 1.1 Geographical distribution of UK coastal research, based on a comprehensive computerized bibliography of books and papers on the geomorphology of the British Isles, containing some 9000 entries, compiled by K.M. Clayton. Of the 9000 entries, some 1400 are classified as dealing with coasts. These in turn are indexed under the 100 km squares of the National Grid and the number of published articles is shown (encircled number) for each relevant National Grid square, or combination of National Grid squares. As the map shows, they are strongly biased to the southern half of Britain. Because some articles cover the coast in more than one grid square, the total number of entries on this map is 1671.

able, and presumably is matched by a growth in other geomorphological topics, no doubt Quaternary geomorphology in particular). Table 1.2 gives the number of published papers and books indexed under particular keywords (the categories are not exclusive), and gives a good idea of the relative cumulative interest in various aspects of coastal geomorphology. Figure 1.1 is a map that displays the number of items in the bibliography dealing with coasts, allocated to the 100 km squares of the National Grid to which they refer.

As Figure 1.1 shows, the geographical distribution of published coastal research is very uneven, with a clear bias to the south and east. The largest numbers are for TF (Lincolnshire, The Wash and North Norfolk) and SS (north and southern shores of the Bristol Channel).

Of course the length of coastline within a grid

The geological background

Table 1.3 Geographical analysis of the British coastal literature, using selected grid squares only.

Grid square	Estimated length of coastline	Number of publications	Coastline length per number of publications
SY (Dorset)	110 km	97	1.13 km
TM (Suffolk/Essex)	120 km	95	1.26 km
SD (Lancashire/S. Cumbria)	150 km	82	1.83 km
SN (Fishguard to Aberdovey)	95 km	35	2.71 km
NJ (south side of Moray Firth)	100 km	22	4.55 km
NZ (Durham/North Yorkshire)	130 km	18	7.22 km
NC (Sutherland)	150 km	9	16.67 km

square varies considerably, but if we take the following examples of grid squares that have a generally linear coast without long indentations, we find the pattern set out in Table 1.3. Clearly in this analysis, such coasts as Dorset, Suffolk/Essex and Lancashire/south Cumbria are among the most studied, whereas Sutherland is among the least. Therefore, Figure 1.1 helps to highlight those areas of the British coast that are better understood geomorphologically, and, perhaps, identifies those areas where further study may help us to gain a more complete understanding of the coastal geomorphology of Britain.

THE GEOLOGICAL BACKGROUND

K.M. Clayton

The pattern of geological outcrops along the British coast (Figure 1.2) has a fundamental control on the nature of the coastline. This is for several reasons, outlined below.

- **Coastal topography**: Underlying geology has influenced the topography of the land, and the detailed outline of the coast in large part reflects the relief of the littoral zone. Rocks that are susceptible to erosion tend to form bays and inlets, whereas erosion-resistant rocks form headlands. Local differences in the level to which outcrops adjacent to the coast have been lowered by subaerial erosion – and the impact of such differences on the coastal form – are seen best along the English Channel coast, such as along the coast of Dorset, or demonstrated in such contrasting situations as Beachy Head and the adjoining Pevensey Levels.

- **Dissection in rocks of different strengths**: Where relatively erosion-resistant rocks have been deeply dissected by erosion, the coastal outline is complex, such as in western Scotland. Where weaker rocks have been dissected, former headlands and bays may have been truncated by marine erosion, such as the Seven Sisters in Sussex.

- **Geological control on cliff profile**: Rocks of all strengths can be cut back by erosion to form cliffs, but weaker rocks generally fail more readily and so form sloping cliffs with angles from 20° to 40°, whereas erosion-resistant rocks are more likely to form near-vertical cliffs, such as at Duncansby, Caithness. In the more resistant rocks, the details of bedding and jointing commonly influence cliff form, both in plan and profile; thus seaward-dipping rocks are likely to suffer slide failure as basal erosion persists, leading to gentler slopes than on horizontal or landward-dipping strata.

- **Lithological control of landsliding**: Where weak rocks underlie stronger ones, landslides are likely to occur; good examples are Folkestone Warren, now largely controlled by drainage and 'toe loading', and Ventnor–Shanklin on the Isle of Wight, where seaward-

Overleaf:
Figure 1.2 Geological map of Great Britain, also showing the locations of the Coastal Geomorphology GCR Sites. The map shows sedimentary rocks classified according to their age of deposition and igneous rocks according to their mode of origin. The numbers in the key indicate age in millions of years (Ma). (Permit number IPR/26-45C British Geological Survey © NERC. All rights reserved.)

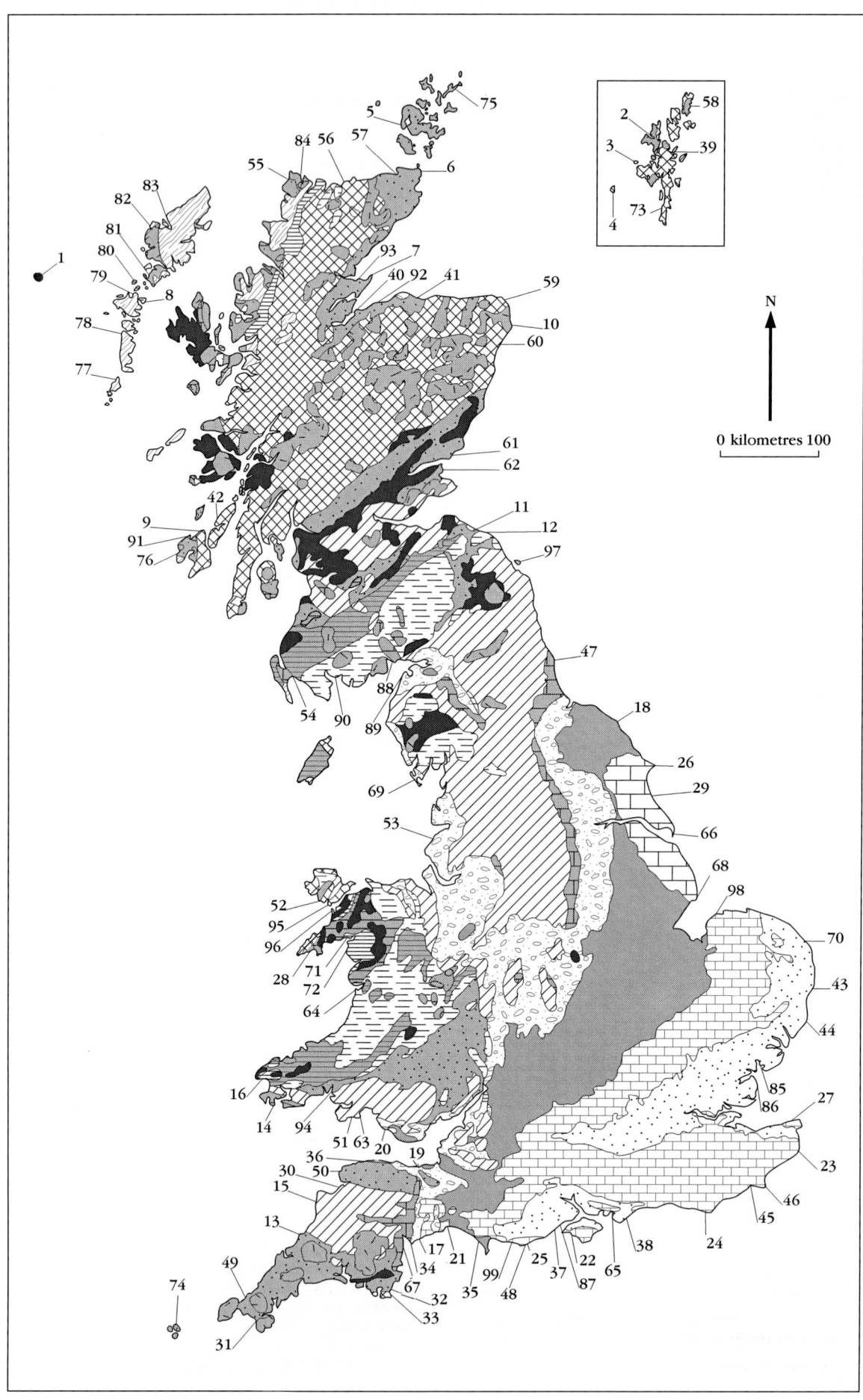

SEDIMENTARY ROCKS   Age (Ma)
CAINOZOIC

Tertiary and marine Pleistocene
Mainly clays and sands. Pleistocene glacial drift not shown   up to 65

MESOZOIC

Cretaceous Mainly chalk, clays and sand   65–140

Jurassic Mainly limestones and clays   140–195

Triassic Marls, sandstones and conglomerates   195–230

PALAEOZOIC

Permian Mainly magnesian limestones, marls and sandstones   230–280

Carboniferous Limestones, sandstones, shales and coal seams   280–345

Devonian Sandstones, shales, conglomerates, (Old Red Sandstone) slates and limestones   345–395

Silurian Shales, mudstones, greywacke, some limestones   395–445

Ordovician Mainly shales and mudstones, limestone in Scotland   445–510

Cambrian Mainly shales, slate and sandstones, limestone in Scotland   510–570

UPPER PROTEROZOIC

Late Precambrian Mainly sandstones, conglomerates and slitstones   600–1000

METAMORPHIC ROCKS

Lower Palaeozoic and Proterozoic
Mainly schists and gneisses   500–1000

Early Precambrian (Lewisian)
Mainly gneisses   1500–300

IGNEOUS ROCKS

Intrusive: Mainly granite, granodiorite, gabbro and dolerite

Volcanic: Mainly basalt, rhyolite, andesite and tuffs

Hard-rock cliffs (Chapter 3)
1. St Kilda Archipelago, Western Isles
2. Villians of Hamnavoe, Shetland
3. Papa Stour, Shetland
4. Foula, Shetland
5. West Coast of Orkney
6. Duncansby to Skirza Head, Caithness
7. Tarbat Ness, Easter Ross
8. Loch Maddy–Sound of Harris coastline, Western Isles
9. Northern Islay, Argyll and Bute
10. Bullers of Buchan, Aberdeenshire
11. Dunbar, East Lothian
12. St Abb's Head, Berwickshire
13. Tintagel, Cornwall
14. South Pembroke Cliffs, Pembrokeshire
15. Hartland Quay, Devon
16. Solfach, Pembrokeshire

Soft-rock cliffs (Chapter 4)
17. Ladram Bay, Devon
18. Robin Hood's Bay, Yorkshire
19. Blue Anchor–Watchet–Lilstock, Somerset
20. Nash Point, Glamorgan
21. Lyme Regis to Golden Cap, Dorset
22. South-west Isle of Wight
23. Kingsdown to Dover, Kent
24. Beachy Head to Seaford Head, East Sussex
25. Ballard Down, Dorset
26. Flamborough Head, Yorkshire
27. Joss Bay (GCR Name: Foreness Point), Kent
28. Porth Neigwl, Caernarfonshire
29. Holderness, Yorkshire

Gravel and 'shingle' beaches (Chapter 6)
30. Westward Ho! Cobble Beach, Devon
31. Loe Bar, Cornwall
32. Slapton Sands, Devon
33. Hallsands, Devon
34. Budleigh Salterton Beach, Devon
35. Chesil Beach, Dorset
36. Porlock, Somerset
37. Hurst Castle Spit, Hampshire
38. Pagham Harbour, West Sussex
39. The Ayres of Swinister, Shetland
40. Whiteness Head, Nairnshire
41. Spey Bay, Morayshire
42. West Coast of Jura
43. Benacre Ness, Suffolk
44. Orfordness and Shingle Street, Suffolk
45. Rye Harbour, East Sussex
46. Dungeness, Kent

Sandy beaches and dunes (Chapter 7)
47. Marsden Bay, County Durham
48. South Haven Peninsula, Dorset
49. Upton and Gwithian Towans, Cornwall
50. Braunton Burrows, Devon
51. Oxwich Bay, Glamorgan
52. Tywyn Aberffraw, Angelsey
53. Ainsdale, Lancashire
54. Luce Sands, Dumfries and Galloway
55. Sandwood Bay, Sutherland
56. Torrisdale Bay and Invernaver, Sutherland
57. Dunnet Bay, Caithness
58. Balta Island, Shetland
59. Strathbeg, Aberdeenshire
60. Forvie, Aberdeenshire
61. Barry Links, Angus
62. Tentsmuir, Fife

Sand spits and tombolos (Chapter 8)
63. Pwll-ddu, Glamorgan
64. Ynyslas, Ceredigion
65. East Head, West Sussex
66. Spurn Head, Yorkshire
67. Dawlish Warren, Devon
68. Gibraltar Point, Lincolnshire
69. Walney Island, Lancashire
70. Winterton Ness, Norfolk
71. Morfa Harlech, Gwynedd
72. Morfa Dyffryn, Gwynedd
73. St Ninian's Tombolo, Shetland
74. Isles of Scilly
75. Central Sanday, Orkney

Machair (Chapter 9)
76. Machir Bay, Islay
77. Eoligarry, Barra, Western Isles
78. Ardivacher to Stoneybridge, South Uist, Western Isles
79. Hornish and Lingay Strands (Machairs Robach and Newton), North Uist, Western Isles
80. Pabbay, Harris, Western Isles
81. Luskentyre and Corran Seilebost, Harris, Western Isles
82. Mangersta, Lewis, Western Isles
83. Tràigh na Berie, Lewis, Western Isles
84. Balnakeil, Sutherland

Saltmarshes (Chapter 10)
85. St Osyth Marsh, Essex
86. Dengie Marsh, Essex
87. Keyhaven Marsh, Hurst Castle, Hampshire
88. Solway Firth (north shore), Dumfries and Galloway
89. Solway Firth: Upper Solway flats and marshes (south shore)
90. Solway Firth: Cree Estuary, Dumfries and Galloway
91. Loch Gruinart, Islay, Argyll and Bute

Coastal assemblages (Chapter 11)
92. Culbin, Moray
93. Morrich More, Ross and Cromarty
94. Carmarthen Bay, Carmarthenshire
95. Newborough Warren, Anglesey
96. Morfa Dinlle, Gwynedd
97. Holy Island, Northumberland
98. North Norfolk Coast
99. The Dorset Coast: Peveril Point to Furzy Cliff

An introduction to British coastal geomorphology

dipping Cretaceous strata are still mobile. Where weaker rocks overlie more resistant strata, 'slope-over-wall cliffs' occur (see Chapter 2); examples include the south-west England peninsula (where the upper cliff retains a periglacial facet) and the till-capped Jurassic cliffs of North Yorkshire.

- **Control by interfluve level**: Rock resistance influences the general level of interfluves, and as cliffs cut back farther inland by wave action, cliff-height may increase as progressively higher interfluves are encountered; maximum cliff height would be attained at the point where the cliffline crosses the crest of the highest interfluves. However, in many areas cliffs are appreciably lower than this potential maximum, suggesting that the rocks are resistant enough to have prevented the landward movement necessary to attain the maximum potential cliff height (or perhaps that the period of stability since sea level reached its present position at that site has been insufficient to achieve much cliff recession).

It is clear from the foregoing that development of a map of relative rock resistance to erosion (Figure 1.3) is particularly useful for interpreting coastal form. Whereas geological maps stress rock age and differentiation by lithology within the major age groups, coastal form reflects markedly – in both planform and elevation – contrasts in rock strength, and it is the local contrasts that lead to the detail and diversity of our coasts. The outcrop pattern is intimately linked to rock strength (compare Figures 1.2 and 1.3; see also Table 2.1).

By studying the relative elevation and dissection of rocks in Britain (using a database for the kilometre squares of the Ordnance Survey National Grid) a relative order of resistance to erosion was established for the 71 most common (outcrops >500 km²) rocks in Britain (Clayton and Shamoon, 1998). While an exact fit to the resistance of rocks to erosion-rates measured at the coast is unlikely to be achieved, the general order of resistance is likely to be similar, and this may aid comparison between sites in future investigations. With six categories, the pattern shown in Table 1.4 emerges.

Therefore, geological maps of Britain are important documents in understanding coastal geomorphology. Although Figure 1.2 is at a small scale, broad connections between outcrop pattern and the outline of the coast can be

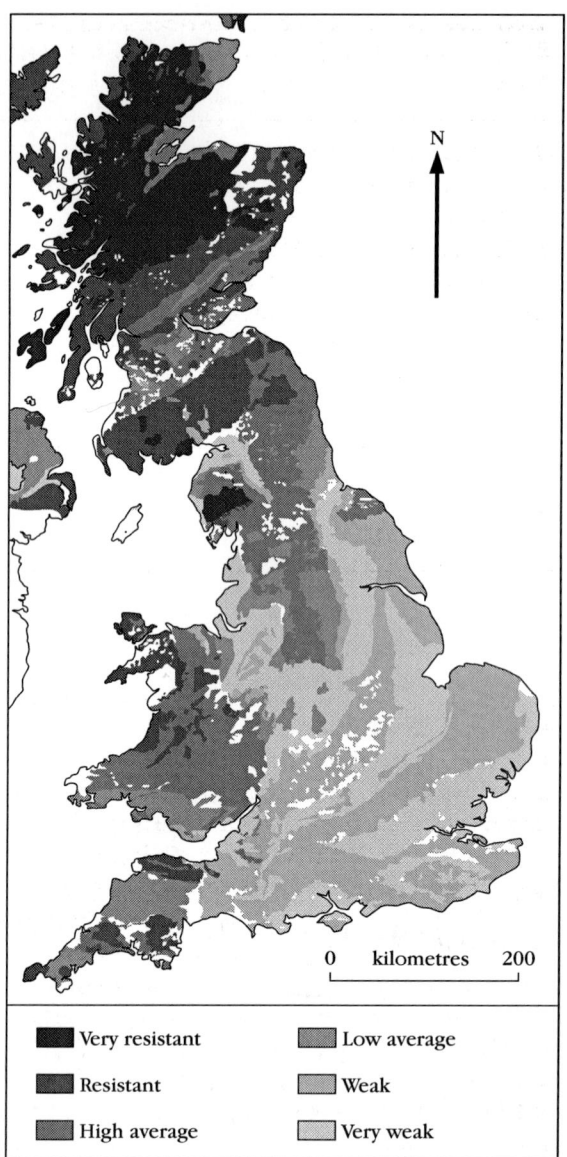

Figure 1.3 Relative rock resistance for 71 different outcrops (divided by lithology and age) was established through computer analysis of data on altitude, dissection, and geology for a grid of kilometre squares covering Great Britain and the surrounding continental shelf. Six consistent classes were established using up to 19 variables in various combinations. White areas are unclassified. (From Clayton and Shamoon, 1998, fig. 1).

traced; larger-scale maps, showing more detailed outcrop patterns, emphasize the geological control of coastal form at local levels still further.

Thus while there are differences between the older and generally far more resistant rocks of

The geological background

Table 1.4 General order of resistance to erosion of British rock types (from Clayton and Shamoon, 1998).

Very Resistant: Precambrian metamorphosed sediments, Cambrian quartzite and sandstone, Ordovician tuff.
Resistant: Old Red Sandstone, Lower Palaeozoic slates, Palaeozoic basalt and andesite.
High Average: Skiddaw slate, Millstone Grit, Carboniferous limestone, Yoredale series.
Low Average: Palaeozoic shale, Coal Measures, Devonian greywackes, Tertiary basalts.
Weak: Magnesian (Permian) limestone, Jurassic limestone, Hastings Beds, Chalk.
Very Weak: Mesozoic and Cainozoic mudrocks, Thanet sand.

northern and western Britain and the younger and weaker rocks found in east and southern England, within each of these zones local contrasts dominate the coastal geomorphology. From Flamborough Head in Yorkshire southwards and westwards to the Exe estuary in Devon, the Chalk and sandstones that form the cuestas of the scarpland and vale landscape also form the major coastal headlands (Flamborough Head, North Foreland in Kent, Beachy Head in Sussex, and the Needles on the Isle of Wight, for example, all on Chalk) and between them on the intervening clays or on till-covered littoral plateaux, wide bays, locally fronted by saltmarshes and sand dunes, alternate with low cliffs cut into the low till-capped plateaux of Holderness, Norfolk and Suffolk.

Geological influence on sediment supply into the coastal system

A further influence of geology on coastal geomorphology is in the provision of sediment that can be incorporated into beaches. Beaches around Britain vary considerably in their texture (from fine-grained sand to boulders) and in their lithology (from shelly sands to flint cobbles) and reflect the local supply of material of appropriate dimensions. Some coarse sediments are still brought to the coast by rivers, especially in Scotland and Wales, where gradients are steep and coarse-grained material is readily transported by floods. In contrast, very little sediment other than mud (clay and silt) is now brought down the rivers of lowland Britain to the coast. Thus, especially in areas with more gentle inland relief, it is the delivery of sediment from offshore as well as from retreating cliffs that has provided most of the material for the local beaches. Boulders and coarse gravel are derived from erosion of resistant rocks in areas such as Scotland and parts of the Welsh coast, their initial size depending on rock-joint spacing. Locally along the English coast, quartzites are the source of coarse gravel (e.g. at Budleigh Salterton); flints form the commonest pebbles and cobbles on beaches in the south of England.

Many 'shingle' (gravel) beaches (such as Slapton Sands in Devon, Chesil Beach in Dorset and Blakeney Point in Norfolk) have been built from offshore gravels, swept ashore as sea level rose during the Holocene marine transgression, and former sea-floor sediment has contributed to many beaches elsewhere (see Figure 6.2). In places, flints are derived from erosion of the Chalk in which they occur. Indeed, Chalk cliffs are generally associated with flint beaches because eroded Chalk debris is quickly broken down by wave action so that Chalk cobbles form a minor part of the beach material. Most flint in English beaches is secondarily derived from quite a wide range of intermediate sources. These include the Pebble Beds of the Tertiary succession of south-east England, where the 'Pebbles' are either derived directly from local erosion, or through their incorporation into river gravels, such as the sequence of Early Pleistocene river terraces – attributed in part to the River Thames – cropping out in Essex, Suffolk and Norfolk. Thus at Dunwich, the cliff contributes flints from gravels at the top of the exposure that were deposited by the ancient River Thames. In contrast, with no local landward source, the flints that dominate Slapton Sands must have been brought ashore from an offshore source. The former river concerned flowed down the English Channel, no doubt fed by such tributaries as the present-day Seine, Rhine and Thames at a time when much of the present-day area of the North Sea was occupied by an ice sheet.

Farther north in England, while we cannot rule out such offshore sources, a large proportion of the gravel has been eroded from glacial gravels and till cropping out along the coast or

An introduction to British coastal geomorphology

offshore. In Briton's Lane Pit, within the Cromer Ridge of north Norfolk, a 30 m-thick sequence of coarse flint gravels demonstrates the power of the ice and its associated meltwater in eroding nearby Chalk (for there are few quartzites or other erratics in this section) and so incorporating flints into its deposits. Normally, glacial gravels and tills include a range of erosion-resistant lithologies alongside flint, notably quartzites (which in part at least have been derived from Triassic strata) and metamorphic rocks high in quartz content, as well as well-cemented sandstones perhaps of Carboniferous or Jurassic age. Less chemically stable rocks, such as granites and limestones, are uncommon as clasts and may be absent. Thus it is this glacial assemblage of resistant gravel and (quartz) sand that dominates beaches along most of the coast north of the former glacial limit. Indeed flints are relatively common even on the Irish Sea coasts, for Chalk crops out on the floor of the Irish Sea, as well as lying beneath the lavas of Antrim.

In Scotland and Wales, by far the greatest source of gravel and sediment has been derived from glaciogenic sources, deposited both inland and on the adjacent shelf by either glaciers or glacial meltwater. As a result, the sediments are as varied as the rocks that were originally eroded by ice.

In the north and west of Scotland, erosion of Torridonian sandstones has yielded even older gravels onto beaches, which also commonly contain quartzite, metamorphic schist and gneiss clasts in addition to sandstones. In the northern and western isles a wide, nearshore shelf is the source of biogenic shell sand that has been swept onto beaches, locally reaching almost 100% of the beach materials (e.g. the 'Coral' beaches at Dunvegan, Skye). In the east of Scotland, igneous- and Old Red Sandstone-derived gravels dominate the extensive beaches sourced both from steep-fast flowing rivers and from offshore deposits over much of the Holocene Epoch. In south-east Scotland, Old Red Sandstone and Carboniferous sandstones and shales form the bulk of the beach material and produce beaches with a high proportion of quartz sand.

In Wales, sediment and gravel, other than that of glaciogenic origin, have predominantly been derived from Lower Palaeozoic and Precambrian shales and slates in the north and west, and mainly Carboniferous limestones in the south.

THE COASTAL MARINE ENVIRONMENT: TIDES, WAVES, SURGES AND CURRENTS

K.M. Clayton

In global terms, the British Isles have unusually high tides and unusually stormy conditions; thus they have a very dynamic coast, one of the reasons why British coastal research has made such an important contribution to the world literature. However, each of these influences also varies greatly around Great Britain. Tidal range is highest at the head of inlets such as the Bristol Channel, and lowest on the English Channel coast between Start Point and Portsmouth, on the East Anglian coast within the North Sea, Cardigan Bay in Wales, and in Shetland in Scotland (Figure 1.4). Wave energy is highest on coasts exposed to the strong winds of western Britain and the North Atlantic swell; it is lower in such relatively sheltered areas as the Irish Sea and the North Sea. Given its western exposure to the Atlantic Ocean, the English Channel coast tends to fall into an intermediate category (Figure 1.5).

Both tidal movements and the advance of waves into shallow water create currents and these move sediment in the nearshore zone, shaping sandbanks, which in turn can affect the local conditions, for example, by forcing larger waves to break and so lose energy as they touch bottom on submerged banks, or even to break against them at low tide. In constricted bedrock channels, perhaps the best example is the Pentland Firth, the tides create extremely strong currents, and where they are channelled between sandbanks as in the Thames estuary or off Great Yarmouth, the patterns of ebb and flow (often dominating different channels) run much faster than in the open sea.

Tidal range has an effect on coastal landforms. Barrier beaches, behind which saltmarshes form, tend to be restricted to areas with relatively low tidal range, such as the north Norfolk coast. Such areas also have spits, which in some cases grow to many kilometres in length, for example, Blakeney Point and Orfordness. This is because wave processes are more efficiently focused on a narrow vertical range and so waves and wave-generated currents assume greater relative importance in shaping coastal form in these areas than tides.

High tidal-ranges can occur towards the head

The coastal marine environment

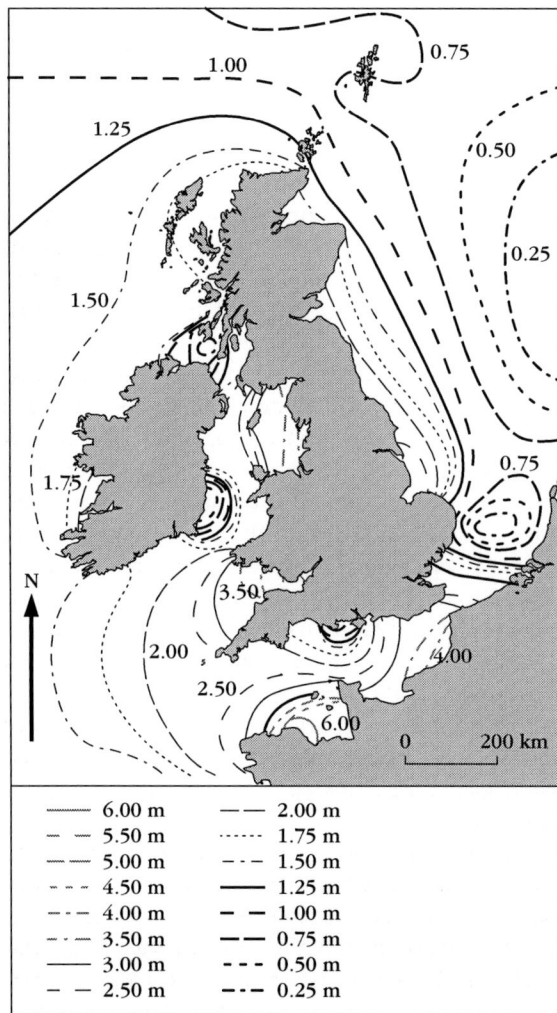

Figure 1.4 Spring tidal amplitude (measured in metres) around the British coast. Elevations should be doubled to give spring tidal range. (UKDMAP 1998, © NERC.)

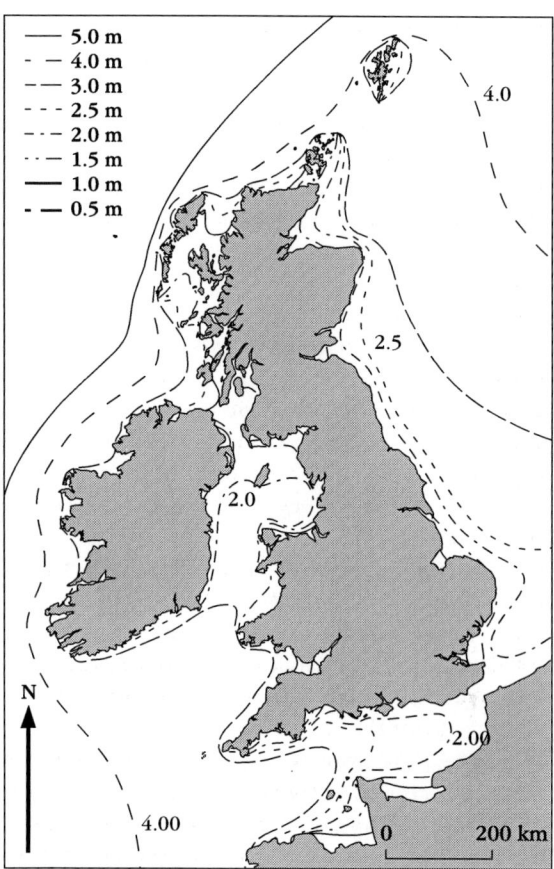

Figure 1.5 Significant wave height exceeded for 10% of the year (significant wave height is the mean value of the highest 1/3 of all waves). Wave height is one of the manifestations of the quantity of wave energy. (UKDMAP 1998 © NERC.)

of estuaries (or Firths in Scotland), and here salt-marshes also develop, though compared with areas of low tidal range they are generally steeper and show much stronger zonation of vegetation, such as in the Bristol Channel–Severn estuary. Seawards of the vegetated marsh, wide tidal flats of sand or mud occur. At such sites, not only are tidal streams more important in determining coastal form, but the relative shelter of the estuary also reduces the impact of waves and wave-induced currents in the shaping of marsh terraces and creek systems. At intermediate tidal ranges, features resulting from both extremes can be found, in places the pattern reflecting local exposure to waves or local protection, so changing the wave energy–tidal range balance from one place to another.

It seems unlikely that tidal range can affect cliff evolution, because coastal cliff form is mainly controlled by the basal removal of material during storms, especially those co-inciding with spring tides. However, the height range and perhaps also the slope of shore platforms fronting cliffs will vary with tidal range. Where wave action is spread over a greater vertical range it will be less effective at every level than where the tidal range is very small. Thus we would expect wide, and almost level, shore platforms where there is a very small tidal range, and more steeply sloping platforms where the tidal range

An introduction to British coastal geomorphology

is high (Trenhaile, 1997).

In partly enclosed seas, strong winds and accompanying changes in atmospheric pressure can produce unusually low or unusually high sea levels, known respectively as 'negative surges' and 'surges'. Negative surges can be dangerous for large ships that might run aground or lose steering control over sandbanks, but seems to have little morphological effect. Surges can cause serious coastal flooding, and also have the effect of allowing waves to break with greater force and at higher elevations at the beach or cliff, often producing the effect of several years of 'normal' cliff recession in a single tide. They can also lead to considerable changes along low coasts by cutting back dunes, removing sediment from beaches and producing washover fans where the outer coastal barrier is overtopped. Considerable changes followed the surge in the North Sea of 31 January–1 February, 1953, and severe tidal flooding was caused by an Irish Sea surge at Towyn on the North Wales coast in 1990. On coasts facing the Atlantic Ocean 'extreme waves' have been reported (e.g. those crossing Loe Bar early in the 20th century) and these may result from tsunamis generated far away by submarine earthquakes and by the interaction of large storm waves with local bathymetry (Hansom, 2001). High-level sand layers on the east coast of Scotland have been attributed to tsunamis triggered by the Storegga slides in the Norwegian Trough, some 7000 years ago (Long et al., 1989).

The waves reaching the coast are mainly generated by winds offshore. In the case of the semi-enclosed seas of the Irish Sea and the North Sea, most waves are generated by winds blowing across relatively restricted fetches, and so have a short wave-length, short period, and are relatively steep. Thus along the coasts of these seas, the varying pattern of length of fetch is an important control over wave energies from all directions offshore as well as the frequency with which winds blow from any one direction. Wave height is proportional to the square root of the length of fetch, so that in the North Sea waves from a northerly direction are generally the largest, with a secondary maximum in East Anglia for waves from the south-east. In the Irish Sea, a west-facing beach like Blackpool, Lancashire, gets its largest waves from a westerly direction, but these are always short-period waves and so put rather small volumes of water onto the beach as they break. As a result, the wide, sandy beach at Blackpool generally consists of a series of ridges and intervening runnels; the seaward slopes of the ridges may be in equilibrium with the short-period waves. Locally on the North Sea coast, ridge-and-runnel beaches are found in the shelter of a headland, limiting waves reaching the beach to those from the east or south-east; an example is in Bridlington Bay, which is protected from the larger northerly waves by Flamborough Head, Yorkshire.

On those parts of coast exposed to North Atlantic storm and swell waves, energies are much higher and the long-period waves put large volumes of water onto the beach as they break. Thus (as well as providing excellent surfing), such wave conditions often produce very wide beaches with a gentle slope, for example Rhossili in South Wales, the beaches of the Western Isles and in some of the more exposed Cornish bays. Where strong regional winds build large waves, energies are very high, but it is also possible for long-period swell generated far offshore, even in the South Atlantic, to reach the western beaches. Such swell loses height as it moves across the ocean, but it can be distinguished by its typically long period. Beaches exposed to the Atlantic Ocean tend to be dominated by the high energies associated with long-period waves, even where the exposure is indirect and the waves reach the coast after refraction. But they are also exposed to local storm waves with much shorter wave-lengths and period, though often steep and rather destructive. In such locations, considerable changes in the beaches can occur from one storm to another or from storm to calmer conditions dominated by swell.

High wave-energies can cause considerable erosion even of resistant rocks, and will exploit structural weaknesses such as faults, joints and bedding planes. Narrow inlets, caves, stacks and natural arches are found along our higher-energy coasts even in the most resistant rocks. Good examples of such forms, eroded into resistant lithologies, are found almost everywhere on the islands of the St Kilda group and in the Shetland Islands, such as Foula and Papa Stour. They can also be found in weaker rocks where wave energies are lower, for example, the Chalk cliffs of Thanet, Kent. Waves are also responsible for the longshore drift of sediment along the coast and this is described in the following section.

COASTAL SEDIMENT SUPPLY AND SEDIMENT CELLS

K.M. Clayton

Some comments on the geological sources of the coarser beach sediments have already been made above. In many places beach sediments are derived updrift from cliffs undergoing erosion (the so-called 'feeder bluffs'). In some cases the only possible source of beach sediment may lie offshore, whereas in upland areas a large part of the coastal sediment supply is eroded inland and delivered down the main rivers. Finally, some lengths of coarse-grained beach sediment (storm beaches) such as Chesil Beach in Dorset, Slapton Sands in Devon, and Blakeney Point in Norfolk must have been brought landwards during the Holocene marine transgression and now represent relict accumulations which are no longer being added to. However, finer-grained sediment can still move onshore, for example, from shoals within Carmarthen Bay in Wales.

As Bird (1985) has noted in a global review, sediment loss from beaches during the last few decades, resulting in decreased width and increased slope, has been far more common than cases where sediment is accumulating and progradation occurring. The reasons for this remain uncertain, although a role is played by slowly rising sea level, reduction in offshore sediment supply and also by interference with eroding cliffs and longshore transport by coastal structures. The same pattern of beach sediment loss is documented on the English coast; for example, careful measurement from 1:10 000 maps between Flamborough Head and the Thames Estuary showed not only that the length of coast being eroded greatly exceeded that at standstill or prograding, but that along almost all of this coast the low-water mark (LWM) has receded more than high-water mark (HWM), which in turn has receded more than the coastline itself, i.e. the cliff top or the solid line often close to HWST (high-water spring tides) mapped by the Ordnance Survey. Thus the beaches along this part of the North Sea coast are steeper and far less wide than they were a century ago. However, the reasons for this are far from clear, though again sea-level rise, reduction of both offshore and fluvial supply and coastal engineering structures must all play an important part. Indeed, in England particularly, the widespread

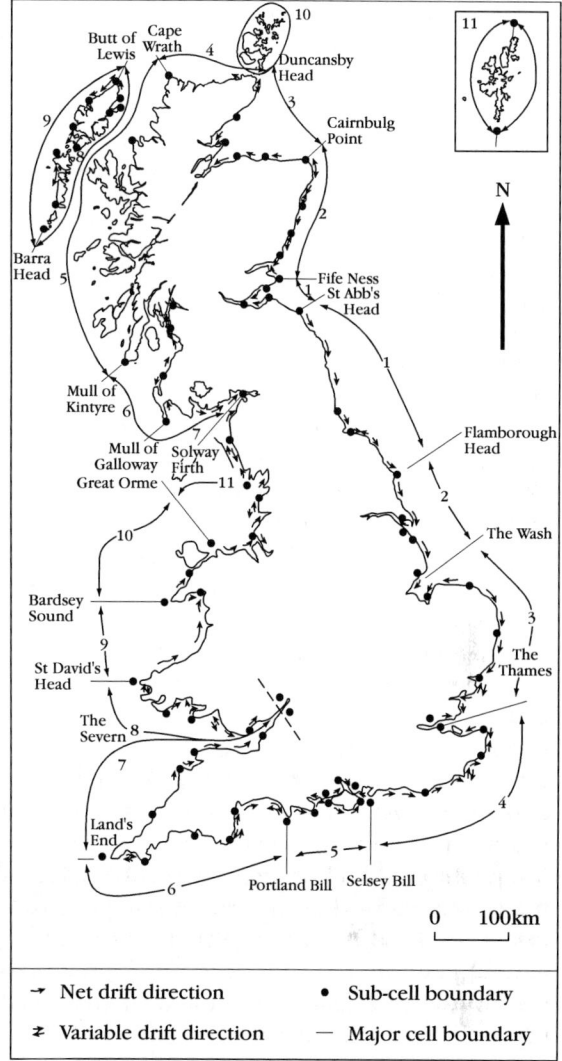

Figure 1.6 Littoral sediment cells and subcells and direction of littoral drift. After Motyka and Brampton, 1993 and HR Wallingford, 1997. Cells are numbered 1 to 7 anticlockwise from St. Abb's Head for Scotland and there are three subcells within the Orcadian cell and two within the Shetland cell (shown in the inset); clockwise from St Abb's Head, cells are numbered 1–11 for England and Wales.

construction of groynes, revetments and walls has interfered with coastal sediment movement and the coastal sediment balance.

The alongshore transport of sediment (littoral or longshore drift) is achieved by waves and the currents they induce within the breaker zone. The direction is determined largely by the angle of wave approach, i.e. it is related to the dominant fetch. Thus the general direction of transport is southwards on the eastern coast of England, and eastwards on the Channel coast.

In the northern North Sea, the pattern is westward movement along the Moray Firth and mainly southward movement along the Aberdeenshire and Angus coast. The pattern around the Irish Sea is a little more complicated since it is not open to the north as is the North Sea, and in general the same direction of transport is not maintained for such long distances as along the Channel and North Sea coasts (Figure 1.6). Locally where the coastal alignment changes, or shelter is provided by a major headland, the direction may be reversed. On some coasts the local direction of movement varies considerably from one month to another; this is generally where the fetch varies different directions (such as in the Irish Sea) and under these conditions the direction of longshore transport varies with the wind direction. But even if the pattern from one month to another is variable, such coasts generally show a consistent long-term pattern of movement, reflecting the dominant combination of frequency of wind direction and the length of fetch.

Beaches that are declining in sediment quantity, whether caused by sea-level rise, reduction of offshore and fluvial sources or anthropogenic interference, would normally revert to a dynamic equilibrium by allowing more rapid erosion of the coastal cliffs, thus improving sediment supply to the local beach system. That this is the natural pattern is suggested by the response to the southwards migration of ords (lengths of low beach volume) along the Holderness coast of England, which is accompanied by a rapid rise in the local rate of cliff recession. However, increasingly, such cliffs have been protected by structures such as revetments and sea walls, so the sediment supply remains restricted, even though beaches are losing sediment downdrift and offshore. This is of course just one aspect of the struggle between a natural coastline that will move in position as an adjustment to, for example, sea-level rise, and the desire of coastal managers to stabilize the position of the coast.

Despite this general understanding of longshore-transport patterns, the details of direction and rate of transport have only begun to be established over the last few decades. Part of the incentive has been the attempt to understand changing sediment volumes better, as the starting point for the improved management of the coast, whether through built structures or the feeding of sediment to beaches from offshore. Local studies of sediment transport include the use of radioactive tracers as at the southern tip of Orfordness and the adjacent Shingle Street beach, as well as various less successful attempts using dyed or fluorescent sand. More recently computer modelling has been used, involving the modelling of the offshore topography and the generation and refraction of waves based on offshore wind data iterated over several years. This has allowed estimates of potential longshore transport, though at many locations this is not reached due to the shortage of sediment.

In several cases, such work has enabled a sediment budget to be quantified, involving the calculation of sediment input from cliffs (and its partition by size into mud, sand and gravel), the modelling of the rate of transport downdrift, the measurement of the volumes of prograded sediment in zones of accretion and on spits, and thus, as a balancing element, offshore removal, which has been generally very difficult to measure directly (see Figure 5.5). This was achieved relatively successfully (Clayton *et al.*, 1983) for the Norfolk cliffs and coast southwards to Great Yarmouth, and fairly consistent estimates have also been made for the Holderness cliffs, where the total volume eroded is higher, though a much greater proportion is mud.

More recently the drive to understand coastal sediment budgets and their inter-relationship with coastal management schemes has led to the adoption of littoral sediment cells and subcells as the basic units of coastal zone management in Great Britain (Figure 1.6). The primary cells are large scale and the dividing points are the headlands, such as Cairnbulg Point or Flamborough Head, around which almost no sediment can pass, or embayments that act as sediment sinks (e.g. the Wash). Within these cells are secondary subcells; some are short lengths where the coastal alignment reverses the direction of drfit for a limited distance, others are where renewed cliff supply restores the dwindling longshore drift and produces larger beach volumes. In terms of shoreline management plans, the subcells are the most important units, but maintaining the natural integrity of the primary cells is a consideration, based on the principle that any interference with longshore drift can disadvantage beaches – and coastal stability – downdrift. A difficulty in adopting this approach is that some cells are already severely disrupted by engineering works, either by coastal defence structures that have prevented coastal cliffs providing a continuing supply of sediment to

SEA-LEVEL HISTORY

K.M. Clayton

We have a good general understanding of global sea-level history from the time of the last glacial maximum, some 18 000 years ago. The abstraction of water from the oceans to build the great land-based ice caps reduced global sea level to some 120–140 m below that of the present day. By the beginning of the Holocene Epoch, 10 000 years before present (BP), sea level was some 40 m below present, and as it continued to rise (the Holocene marine transgression) was within 10 m of its present stand at about 5000 years ago, and close to present level by 4000 years BP (Figures 1.7a,b). However, the precise changes at any one site will depart from this pattern for many reasons, including crustal stability (tectonic changes, the effects of loading or removal of load by ice sheets and the oceans themselves, local sedimentation) and tidal changes as the coastal configuration has changed, as well as many other lesser effects that may lead to local departures from the general pattern of sea-level rise. For example, Carter (1982) has shown that the post-1950 fall of sea level at Belfast of about 1 mm a^{-1} is probably the result of land-claim around the estuary, rather than a genuine fall in sea level.

Thus local evidence of sea-level history is of great importance, but the complications involved mean that our knowledge is still very incomplete and not always well linked to our understanding of the likely causes. Further, we require both good stratigraphical evidence (including reliable relationship to the sea level of the time as well as accurate levelling to determine altitude) and good age determination (generally based on radiocarbon, converted to sidereal years) if the local evidence is to be useful; so prograding coasts generally have a better record of sea-level history than those undergoing erosion. Errors in age determination can result from inaccurate ^{14}C dates (e.g. the 'hard-water effect') or from compaction of sediments causing misjudgement of the postulated height of former sea-level markers.

An independent source of evidence for the last century or so is the trend of long-term tide-gauge records. The variability of short-term sea level with changes in the weather and the longer-term periodicity of spring and neap tides means that records of less than a few decades are unreliable, or frustrating where tide gauges have been discontinued (such as at Felixstowe) or recently established to help in the surge warning system, resulting in data which cannot yet give a statistically reliable long-term trend. Where gauges have remained in place for many decades, reliable data indicate that parts of northern Britain are still showing relative rise in land level (and thus relative sea-level fall), a result of continued recovery from the ice load of the last glaciation, while southern Britain generally shows relative sea-level rise, in places faster than the average annual rise of sea level (*c.* 1.5 to 2 mma^{-1}) around the coast, indicating at least local land submergence. The result is the widely accepted pattern of relative sea-level fall in the north-west and rise in the south, with relative stability along a line just south of the Scottish border (Figures 1.8a,b). In Scotland, the pattern of relative sea-level rise is related to the distance from the isostatic uplift centre (see Figure 1.8b), with central Scotland undergoing uplift and subsidence characterizing the northern and Western Isles.

What is less clear is how far neotectonic movement, other than glacioisostatic recovery and the effects of water loading following sea-level rise, is of significance. The long-held view that Britain is tectonically stable has been challenged in recent years (Embleton, 1993) and several authors have reported evidence of relatively local neotectonic movements (e.g. Clayton and Shamoon, 1999; Ringrose *et al.*, 1991), but these have generally (though not universally) been inferred from evidence covering long periods since Mid- or Late Tertiary times. Thus the rates of coastal tectonic movement, while high enough to be detectable, are slow compared with both the known rates of postglacial isostatic recovery and submergence, and with the current rate of sea-level rise. Thus there is an 'inheritance effect' that has affected broad coastal (and near-coastal) form, but the direct effect of neotectonic movements on contemporary coastal change is likely to be small. Over time, further detail will emerge from comparison of local chronologies of Holocene sea-level change, adding to the significance of these studies.

An introduction to British coastal geomorphology

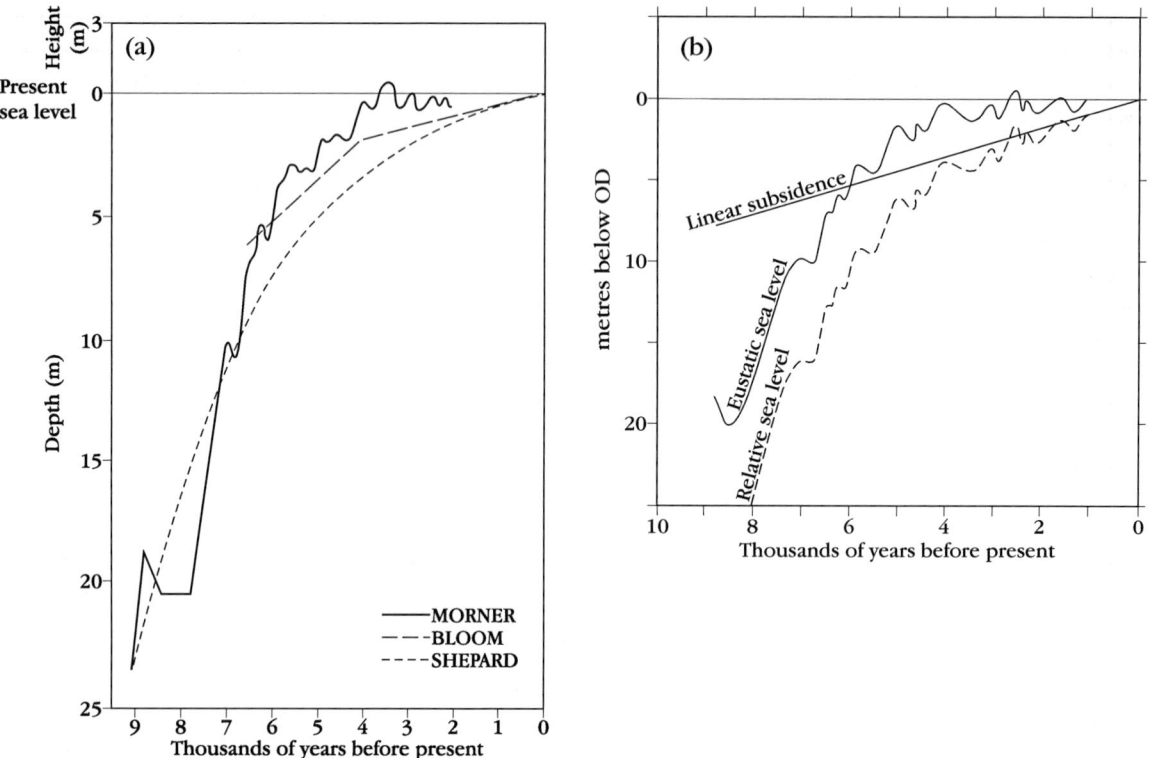

Figure 1.7 Holocene sea-level history: (a) global view; (b) sea-level history in the area around The Wash (Norfolk/Lincolnshire). Three views of the global change in sea level over the last 10 000 years are shown in (a); the smooth curves combine data from different areas, a reconstruction based on a smaller region will show an irregular pattern over time (Shepard, 1963; Bloom, 1978; Mörner, 1972). Because of the local effects of uplift and subsidence, it is increasingly recognized that such global sea-level curves have the potential to mislead and that local relative sea-level curves are generally more secure. The sea-level curve in (b) for the area around The Wash is based on an accreting sedimentary sequence preserved in an area that is subsiding at an average rate of 0.9 m ka^{-1}. If this subsidence has been at a steady rate, then the local relative sea-level curve (the pecked line) can be converted to a eustatic curve (the solid line) by subtracting the effect of subsidence. (Based on Chorley et al., 1984, and Tooley and Shennan, 1987.)

Detailed local histories of sea-level change have been established in several locations where the stratigraphical record is good, including around Morecambe Bay, the Fens and the north Norfolk coast, and research continues to add further data to the sea-level history of these sites. From these records it appears that the general rise of sea level has not always persisted throughout the mid-late Holocene period, with a sequence of relative transgression succeeded by relative regression repeated several times. An example of this is the Fenland sea-level curve, which, while showing a *general* rise over the last 7000 years that slowly reduced in rate towards the present day, nevertheless includes at least four intervals of sea-level fall (Figure 1.7b). Similar regressions are found in other records (e.g. the north Norfolk coast) though it is not always certain that they co-incide in time. This is not necessarily to be expected, for the Fenland area shows an average rate of subsidence, when the sea-level curve is compared with the average of several North Sea records, of 0.91 m ka^{-1}, and there are hints that this has varied over time. Further, there is good evidence of a time-lag between sea-level tendency and local sedimentation in Fenland, so allowance is needed when comparisons are made with other areas. Thus the breaks in the general record of transgression may well result from a local combination of sea-level and land-level changes. Attempts have been made to separate these two influences, and in time we may expect greater success as the quality of local records improves. For the moment, it would be rash to suggest we have a reliable and detailed record of sea-level changes

Sea-level history

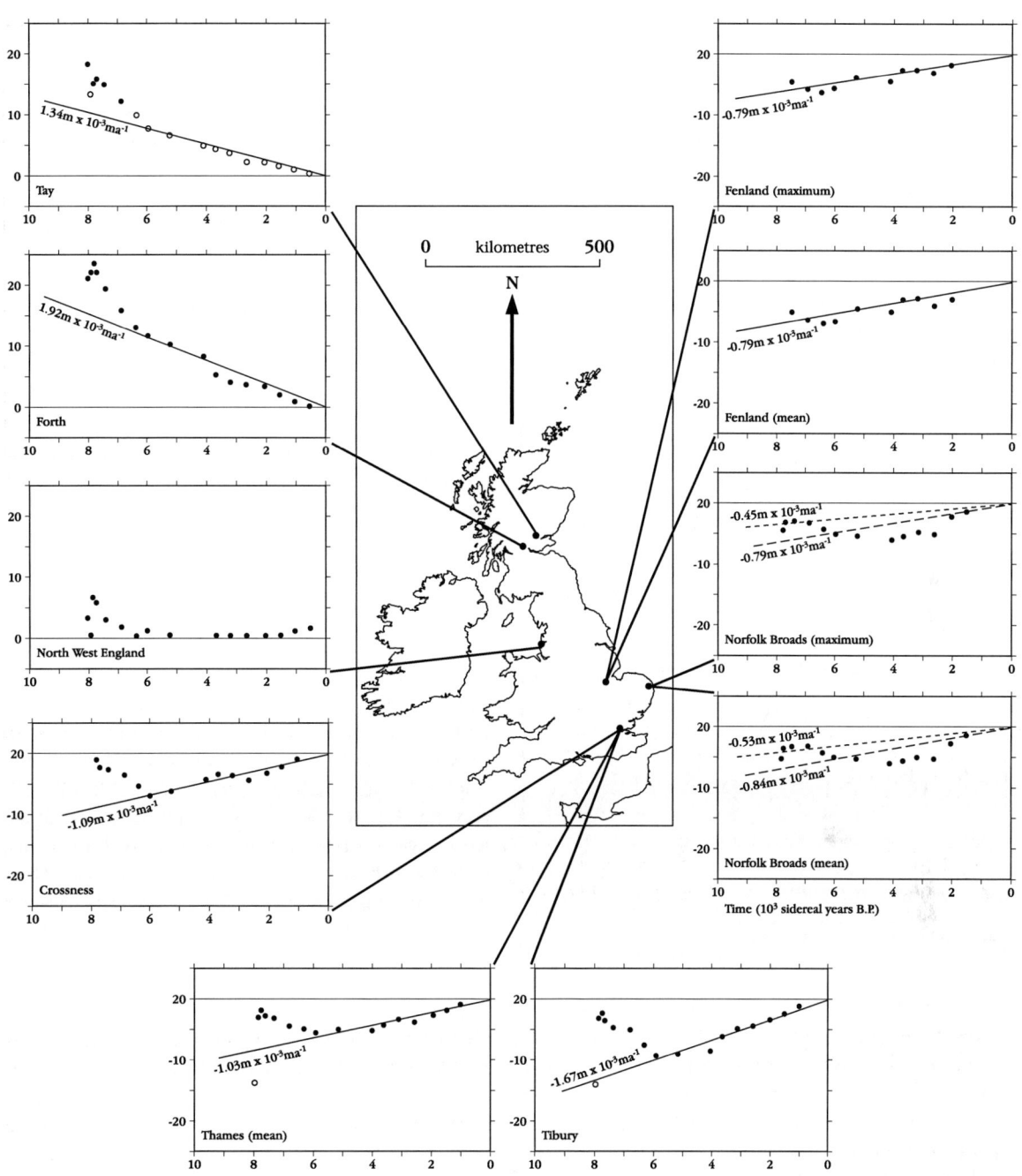

Figure 1.8a Uplift and subsidence, based on trends in sea level around the coast of Britain. These trends are established by comparing local sea-level curves with the eustatic trend shown in Figure 1.7b. Open circles indicate that the data were obtained by the author by extrapolation. The vertical scale on the graphs is in metres OD, the horizontal scale is in 10^3 sidereal years before present. (After Tooley and Shennan, 1987, p.136, fig. 4.9.)

over the last few thousand years.

Currently, a better understanding is being acquired of the role of past sea-level change on prograding coasts such as The Wash–Fenland, north Norfolk and the Thames estuary, and this is concomitant with improvements in our understanding of coastal evolution. The contrasting history of many Scottish prograding coasts undergoing relative sea-level fall over the Holocene Epoch (and in places the delivery of

An introduction to British coastal geomorphology

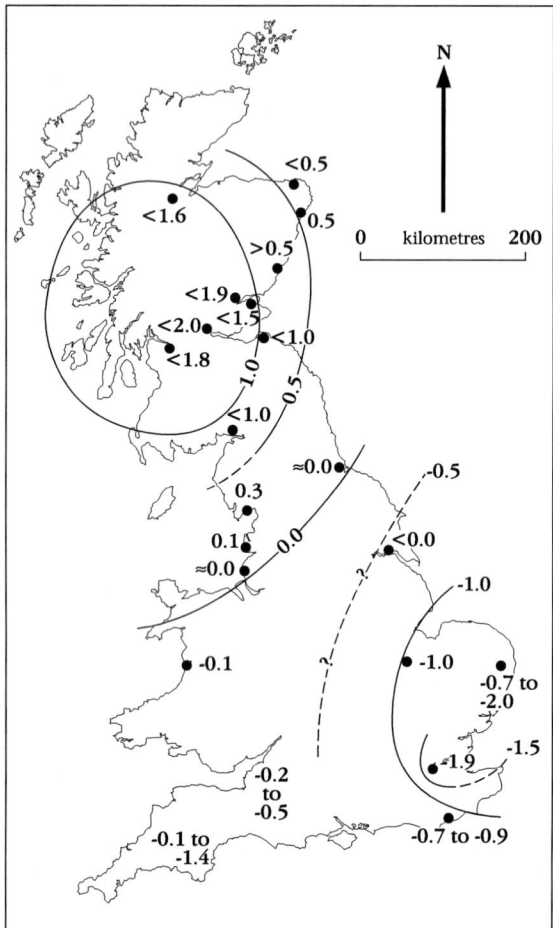

Figure 1.8b Map of isobases of uplift (positive values) and subsidence (negative values) following the Lateglacial and Holocene deglaciation of the British Isles. The rates shown (in mm a^{-1}) are of crustal movement in Britain. Isobases cannot be drawn for much of southern England; point estimates are shown for guidance. (After Shennan, 1989, fig. 9.)

relatively large volumes of sediment down rivers and from offshore) has yet to be compared to subsiding coasts farther south. Some Scottish coasts show progradation where large amounts of sediment arrived at a time co-incident with a relative sea-level fall, such as in the Dornoch and Moray firths, but some coasts in the northern and western islands have undergone almost continuous relative sea-level rise. Our knowledge of the variation from place to place of relative sea-level rise has yet to be applied to the form and dimensions of shore platforms on coasts undergoing erosion, even though this seems likely to be a factor in the evolution of cliffed coasts.

Our understanding of climate change, including the evidence for a rise in mean world temperature over the past 150 years, is linked to the evidence of contemporary sea-level rise across the world oceans (and matched by the North Sea data) of 1.5 to 2 mm a^{-1}. Climate change models suggest that this rate of rise may increase to at least 4 mm a^{-1} and perhaps higher, with half of this rise coming from the expansion of the warming ocean. Nevertheless, future rates of sea-level rise may reach or exceed the highest rates experienced during the Holocene Epoch, so greater understanding of past changes must aid our management of such coasts in the future.

One aspect of future sea-level rise is that as the coastline is driven landwards (which will usually be the case except where sediment supply can keep pace with the tendency to such adjustment), the process causes what is currently termed 'coastal squeeze'. That is, the natural coastal landform areas from intertidal mudflats to saltmarshes, beaches and sand dunes will be reduced in width between the low-water mark and whatever man-made or natural slopes mark the inland penetration of the highest tides. Unless conditions allow these landward limits to move (e.g. by the abandonment of flood banks), the zone within which natural coastal landforms can develop will be reduced in width, and those forms will be less well developed and no doubt liable to still greater damage in future. The engineering adjustment to threatened coastal squeeze by the removal of artificial structures is termed 'managed re-alignment' or 'managed retreat' (Hansom et al., 2001).

COASTAL MANAGEMENT AND COASTAL ENGINEERING

K.M. Clayton

Over the last two centuries, the land near the coast has seen many changes in land use. The recreational use of the coast first led to fashionable seaside resorts, and with the development of the railway network many of these grew into major coastal towns devoted to sea bathing and leisure pursuits, with promenades and piers. At the same time, the industrial revolution led to the growth of ports and of seaside industry, while the increasing skill and resources of the coastal engineer, allowed ever-larger schemes of land-claim. Saltmarshes, once reclaimed on a small scale for grazing, were seen as ideal sites

Coastal management and coastal engineering

for large-scale industry looking for level land. That local docks could be developed for the import of raw materials and the export of heavy finished products was an added advantage. With the increased mobility offered by the motor car, seaside villages have expanded into commuting dormitory towns or retirement areas epitomized by Peacehaven east of Brighton.

The result of this combination of coastal urbanization and the confidence of coastal engineers has led to the profound modification of much of our coast (Figure 1.9). Few of the sites described in this volume, especially those in England, are entirely free of coastal structures, though care has been taken to select those that are least disturbed. Despite the caution of the Royal Commission on Coastal Erosion and Afforestation, which reported from 1907 to 1911 and noted that a balance would need to be struck between defended coasts and the supply of sediment from unprotected cliffs, much more of the coast has been modified throughout the 20th century. Events such as the serious flooding and loss of life of the 1953 surge on the North Sea coast of England encouraged further expenditure on coastal defences.

Conditions changed slightly after 1985 when all supervision of coastal engineering schemes at a national level in England and Wales was transferred to the then Ministry of Agriculture, Fisheries and Food (formerly the Department of the Environment had borne responsibility for upland coasts) and the same cost/benefit tests were applied to defences protecting upland coasts as to coastal flood defences, though to date few of the existing structures have been removed. However, many are now deteriorating rapidly, and where they protect undeveloped agricultural land it seems unlikely they will be replaced. In the same way, the modern advice that planning law should be used to prevent developments of land liable to flood, or of coasts liable to undergo erosion, will take many years to take effect, though if applied consistently a reduction in developed coasts should emerge

Relatively little research has looked at the effects of engineering works on natural change at the coast. It seems likely that not only the production of sediment from cliffs undergoing erosion has been reduced (Clayton, 1989), but also the rate of transfer along the coast by longshore drift has been modified by the large number of groynes along our shores. Indeed, these groynes may not only have reduced longshore

Figure 1.9 Coastal engineering structures in Britain ('coastal protection' and 'sea defences'), shown as coastline with heavier line weight. Note the concentration on south-eastern England, Bristol channel, and north Wales and and north-west England. (Data from Halcrow Group Ltd for England and Wales, published with permission from Department for Environment, Food and Rural Affairs; Scottish data after Ramsey and Brampton, 2000a–f.)

transport, but in places seem likely to have allowed more beach sediment to be moved offshore and so be permanently lost to the coastal system. Sea walls increase wave reflection and often lead to sediment loss from the beaches fronting them. Land-claim of saltmarshes has reduced the volume of the tidal prism entering many estuaries and may have been a factor in the erosion of the remaining marshes in front of the sea banks (walls) protecting the areas that have been reclaimed. We lack full understanding of the reasons for the persistent erosion of many saltmarshes, although in places current schemes for the set back of the banks and re-establishment of saltmarsh may lead to a more general improvement in conditions.

The funding of coastal defence works in the UK has concentrated on the initial capital cost and not the cost of later repairs. This has made it difficult to finance beach nourishment as a method of coastal defence in the UK. Special arrangements were made to allow the early beach feeds, as at Portobello and Bournemouth, and the recycling of material as at Dungeness and on the eastern side of The Wash at Snettisham. It remains uncommon, although it is increasingly being utilized, for example along most of the Lincolnshire coast, and south of the artificial reefs on the Norfolk coast at Waxham and at Kingston-on-Spey in the Moray Firth.

The recognition that the British coast may be divided into a series of coastal sediment cells (some of which may have smaller subcells nested within them, see Figure 1.6) is now being applied to shoreline management. Under current arrangements, in England and Wales, a set of Shoreline Management Plans has been drawn up by Local Authorities, or groups of Local Authorities (under the auspices of the Department of Environment, Food and Rural Affairs), for each coastal sector, the sectors being based mainly on the recognized coastal sediment cells. In Scotland, some Local Authorities have undertaken similar work for the coast under their jurisdictions, although it is not a formal requirement. The philosophy behind this approach is that it is through the understanding and management of coastal sediment volumes that the greatest stability may be achieved, and that within a cell, local areas may be allowed to undergo erosion in the knowledge that this will improve sediment supply, and thus stability, elsewhere. Where the solution of feeding sand (and/or gravel from offshore) is adopted, this can be seen to benefit the whole cell if the feed is in an updrift location, whereas in a few cases (such as at Dungeness and on the eastern side of the Wash), sediment is collected downdrift and recycled updrift to move again through the system on a cyclical basis.

In England and Wales, the development of Shoreline Management Plans involved a complex consultative process led by the Environment Agency and utilizing engineering consultants. Maritime local authorities and bodies such as the Countryside Council for Wales, English Nature, and the Countryside Agency were involved. These plans are updated at regular intervals; the result is a more organized and longer-term strategic assessment of needs than before, though inevitably the resulting plans tend to require considerable compromises between the widely varying views of the bodies involved. How far compromises can be made to work on the coast remains to be seen, but it may be that in future a choice between determination to hold the line, or a willingness to allow natural change in the future, would be better choices than some of the compromise proposals currently being adopted.

Certainly something of a change in attitude can be recognized at a national level. Proposals subject to cost/benefit assessment must now include the 'non-intervention' option, and, facing the inevitability of future sea-level rise, the advantages of 'managed re-alignment' are being canvassed. Insofar as these changes lead to no further encroachments on the length of natural coast remaining in Britain they will be welcomed by the geomorphologist.

One aspect of the coast that has received greater attention (especially since the surge of 1953) is coastal hazard, both the local threat of rapid cliff recession, but more widely the threat of coastal surges, both in the North Sea (e.g. 1953) and the Irish Sea (e.g. the flooding at Towyn in 1990). It is recognized that water levels can be raised during a surge by as much as 2.5 m, and with current and future sea-level rise, and perhaps an increase in storm frequency with climate change, the matter is taken seriously. Well-designed warning systems based on meteorological forecasting systems are now in place, and tidal barrages have been constructed, for example, on the Thames at Woolwich and at Hull and Swansea. Education of the public at risk, however, is less developed.

FURTHER READING

For further information on coastal geomorphology processes and influences, the reader is directed to *Coastal Systems* (Haslett, 2000) *Coastal Geomorphology: An Introduction* (Bird, 2000); *An Introduction to Coastal Geomorphology* (Pethick, 1984); *Coastal Environments: an Introduction to the Physical, Ecological and Cultural Systems of Coastlines* (Carter, 1988), and *Coasts* (Hansom, 1988). These titles give a comprehensive treatment of coastal evolution and dynamics, providing background for the study of coastal landforms and how and why they are changing, with worldwide coverage of examples, numerous illustrations and extensive references to the scientific literature. French (1997) provides useful discussion of managed re-alignment and coastal zone management and Viles and Spencer (1995) discuss coastal problems.

Steers provides descriptive texts on the coastlines of England and Wales (1964a) and Scotland (1973).

GCR SITE SELECTION GUIDELINES

V.J. May and N.V. Ellis

The GCR site-selection exercise for coastal geomorphology followed four categories ('GCR Blocks'), one for each of England, Scotland and Wales and one for 'Saltmarsh Geomorphology'; although three of the 'Blocks' are country based, comparisons were made to ensure that certain types of site occurring in each were not over-represented in a Great Britain-wide context.

Before site assessment and selection began, the first stage in the project was to apply the ethos of the GCR – outlined below – in order to fine-tune GCR selection-criteria. The coastal geomorphological literature was reviewed to identify the most cited sites to assist in the compilation of lists of candidate GCR sites. The GCR site selection work also included a survey of the morphodynamics of the whole coastline, carried out in conjunction with the CORINE coastal erosion project (European Commission, 1998), which provided a means by which to judge the 'representativeness' (see below) of the short-listed sites.

Broad GCR site selection criteria

The general principles guiding GCR site selection are described in the introductory GCR volume (Ellis *et al.*, 1996), but can be encapsulated in three broad components:

- International geological or geomorphological importance (for example, internationally renowned 'type' sites, but other sites that have informal, but widely held, international recognition are also selected).
- Presence of 'classic' or exceptional features that are scientifically important (for example, 'textbook' examples of particular features or exceptionally unusual or rare types of features are included).
- Presence of representative Earth science features that are essential in comprehensively portraying Britain's Earth history. Thus, a site may be selected for showing the most complete regional representation of phenomena that are otherwise quite widespread.

It should be assumed that an 'internationally' rated site will also be representative of an event or process and may include exceptional features.

In order to ensure true national importance in the selected representative sites, site selection was underpinned by the premise that the *minimum number* of sites should be selected. By choosing only those sites absolutely necessary to represent the most important aspects of Britain's

Table 1.5 Morphosedimentological classification of the British coast (based on European Commission (1998 – the CORINE project érosion cotière).

Morpho-sedimentological type	Active (km)	Protected* (km)	Total (km)
Hard-rock cliffs	7990	7	7997
Soft-rock cliffs	1401	221	1622
Shingle beaches	818	225	1043
Sand beaches	1274	302	1576
Heterogeneous beaches	415	126	541
Beaches for which no data available	59	0	59
Muddy and estuarine coasts	999	484	1483
Totals	12956	1365	14321
Anthropogenic coasts (including harbours, land-claim)			2096
Total			16417

* i.e. modified by coastal defence/protection works.

coastal geomorphology unnecessary duplication was avoided.

Where several choices of representative sites exist, a series of weightings was applied to the general guidelines to help to distinguish the best (or most suitable) site for the GCR. For example, preference was given to sites with the most extensive or best-preserved record of a certain feature, the most detailed geochronological evidence or a particularly long history of study. Sites that have contributed to our understanding of coastal geomorphology or have significant potential for future research were also preferred.

In some cases, 'representative' sites were selected for the GCR as part of a group of related sites. Such a group of sites may show different aspects of one type of phenomenon, which shows significant regional variations in its characteristics, for example, sites with similar landforms have been selected from areas having different tidal ranges. Wave climates also vary greatly, the greatest contrast being between the North Sea and the Atlantic coastline. Within this, wave energies vary enormously, the highest being in the Outer Hebrides, west Orkney and Shetland. Thus suites of sites can be identified in the GCR that demonstrate the differences of wave climate and tidal range, as well as more local variations in wave and tidal conditions. In this case there may be 'core' sites, perhaps those showing the most extensive and best researched landform features, while other sites may demonstrate significant variations on the main theme. Nonetheless, it is the group of sites together that remains nationally important.

On an entirely practical level, all selected sites must be conservable, meaning in essence that development planning consents do not exist or that amendments can be negotiated.

Finally, extensive consultations were carried out with appropriate geomorphologists, and many sites were assessed carefully before the final listing was produced. The comments made by the specialists and the field observations were used to produce a modified site shortlist, and this was then slightly adjusted during preparation of the site descriptions for this volume.

GCR Networks

There are many problems inherent in producing a truly representative list of nationally important sites that merit conservation. In order to help provide a framework for selecting sites, the concept of GCR networks has been applied.

The geomorphology of the coastline is controlled by a complex interaction of factors – the dynamics of the coastal 'cell', geological controls (e.g. rock type and structures), the Pleistocene inheritance (isostatic and eustatic effects), sediment 'budget', tidal regimes as well as anthropogenic influence. It is the intention within the 'representativeness' rationale of the GCR to be able to demonstrate the interplay of these themes and their manifestations from the evidence present in the selected GCR sites. These themes can be thought of as providing a basis for individual *GCR Networks*, which link clusters of representative sites.

A broad categorization was devised for the European Commission CORINE project érosion cotière (European Commission, 1998), and the classification scheme is shown in Table 1.5. However, in order to establish a scheme of GCR networks for coastal geomorphology, a more detailed structure was required. Ultimately, some 26 networks (broadly following King, 1978) were identified for the GCR project. The themes were:

A. Cliffed coasts
1. Large-scale structural control: longitudinal and transverse coasts
2. Small-scale structural control: caves, arches, stacks, geos, zawns
3. Cliff forms and processes: plunging cliffs, slope-over-wall, hog's back, variety of rates of cliff retreat, differential erosion
4. Exhumed and emerged forms: cliffs, benches
5. Karstic development

B. Shore platforms (including both contemporary and emerged features)
6. Structurally controlled
7. Erosionally dominated

C. Beaches and intertidal sediments
8. Beach orientation: relation to wave direction, swell-dominated beaches
9. Beaches undergoing erosion
10. Prograding beaches
11. Beach phases
12. Pre-existing sediment sources, including pre-existing clasts
13. Emerged ('raised') beaches
14. Cliff-foot beaches

15. Dunes: rock-based, gravel-based, restricted sources, sand plains
16. Spits
17. Barrier beaches
18. Cuspate forelands and nesses
19. Tombolos and tied islands
20. Intertidal sediments
21. Mudflats, ridge and runnel forms
22. Saltmarsh morphology – creeks, saltpans, piping
23. Machair

D. Coastal valleys
24. Chines, truncated valleys, coastal waterfalls

E. Inlets and submerged coasts
25. Fjords, rias, estuaries

F. Semi-enclosed bays
26. Restricted sediment sources and transfers, submarine barriers, sediment sorting

In England and Wales alone, 186 sites were identified as candidates to represent these themes, each of which was visited and assessed before being reduced to a select 59 GCR sites. For Great Britain as a whole, 99 GCR sites were selected (see Figure 1.2).

Clearly, any one site may be helpful in elucidating several of these themes and therefore may contribute to more than one GCR Network (for example, Culbin, on the Moray Firth, provides information on sea-level change as well as gravel delivery data over the Holocene Epoch at a site characterized by well-developed dunes situated on top of spit structures; Carmarthen Bay demonstrates sea-level change and the geological controls in cliff development; see Table 1.6). Many sites demonstrate several themes, although some are dominated by a single key feature; this has been the basis for the arrangement of the chapters in this book. However, sites with particularly diverse assemblages of coastal features are grouped together in a final chapter.

The GCR sites described in this volume exclude major coastal landslides, such as Folkestone Warren; which are described in the *Mass Movements* volume of the GCR Series (Cooper, in prep.). Cross-reference to these sites, however, is made in Chapter 2 and the chapter on soft-rock cliffs (Chapter 4). Also, sites that are of interest for coastal features that are particularly important for elucidating the Pleistocene history of Britain are described in the Quaternary volumes of the GCR series (e.g. Campbell and Bowen, 1989, Gordon and Sutherland, 1993).

Similarly those sites on the coast that are important for their palaeontological, stratigraphical, petrological or mineralogical features are described elsewhere in the GCR series.

GCR site boundaries

Definition of the boundary of the sites was based primarily upon the extent of the landform suite, but on the more rapidly changing sites, the geographical area in which processes that produce and maintain the landform suite was included as far as possible. Therefore, on a dynamic coast with unconsolidated sediment, definition of boundaries was based initially upon the process-unit – the sediment-transport cell – where this can be recognized. However, technically this means that many coastal geomorphology GCR sites should extend seawards beyond the low-water mark (LWM), since the sediments and the processes are predominantly marine. However, a limitation is that areas below the LWM are not explicitly covered by the Site of Special Scientific Interest (SSSI) system by which GCR sites are conserved.

On hard-rock cliffs, the landward boundary is usually the level of spray action or of subaerial processes such as gullying, but on several sites it has been set back a few metres from the cliff-top edge. On cliffs undergoing rapid erosion, the boundary has been identified with a probable position no more than 10 years hence. In these cases, it will be necessary to adjust the boundary as cliff retreat continues. The inland boundary of dune systems is usually the edge of blown sand that has not been reclaimed for farming or forestry. Where dunes have been afforested, the zone over which coastline changes have been documented in the 20th century has been included.

ANTHROPOGENIC INFLUENCES AND THE GCR

V.J. May and N.V. Ellis

Prior to the GCR project, with the Coast Protection Act (1949) and the east coast floods of 1953, considerable attention was given to the

An introduction to British coastal geomorphology

Table 1.6 Main features of each GCR Site, broadly following the classification of King, 1978, to show where different features are represented.

	Chapter 3 Hard-rock cliffs	1. Large-scale structural control	2. Small-scale structural control	3. Cliff forms and processes	4. Exhumed forms: cliffs, benches	5. Karstic development	6. Shore platforms – structural control	7. Shore platforms – erosional control	8. Beach orientation	9. Beach undergoing erosion	10. Prograding beach	11. Beach phases	12. Pre-existing clasts	13. Emerged ('raised') beaches	14. Cliff-foot beaches	15. Dunes, including sandplains	16. Spits	17. Barrier beaches	18. Cuspate forelands and nesses	19. Tombolos and tied islands	20. Intertidal sediments	21. Mudflats, ridge and runnel forms	22. Saltmarsh morphology	23. Machair	24. Coastal valleys	25. Inlets and submerged coasts	26. Semi-enclosed bay
1	St Kilda Archipelago	×	×	×	×																						
2	Villians of Hamnavoe, Shetland	×	×	×	×																					×	
3	Papa Stour, Shetland	×	×	×	×			×																		×	
4	Foula, Shetland	×	×	×	×		×	×							×											×	
5	West Coast of Orkney	×	×	×	×		×	×						×	×											×	
6	Duncansby to Skirza Head, Caithness	×	×	×	×		×	×						×	×												
7	Tarbat Ness, Easter Ross	×	×	×				×							×												
8	Loch Maddy–Sound of Harris coastline			×		×																				×	
9	Northern Islay, Argyll and Bute	×			×			×						×	×												×
10	Bullers of Buchan, Aberdeenshire		×	×	×		×	×							×												
11	Dunbar, East Lothian	×	×	×	×		×	×						×	×												
12	St Abb's Head, Berwickshire	×	×	×										×	×												
13	Tintagel, Cornwall	×	×																								
14	South Pembroke Cliffs, Pembrokeshire		×	×	×		×	×																			
15	Hartland Quay, Devon	×	×	×																					×		
16	Solfach, Pembrokeshire			×				×													×					×	×	

Table of GCR site features

Chapter 4 Soft-rock cliffs

#	Site
17	Ladram Bay, Devon
18	Robin Hood's Bay, Yorkshire
19	Blue Anchor–Watchet–Lilstock, Somerset
20	Nash Point, Glamorgan
21	Lyme Regis to Golden Cap, Dorset
22	South-west Isle of Wight
23	Kingsdown to Dover, Kent
24	Beachy Head to Seaford Head, East Sussex
25	Ballard Down, Dorset
26	Flamborough Head, Yorkshire
27	Joss Bay (Foreness Point), Kent
28	Porth Neigwl, Gwynedd
29	Holderness, Yorkshire

Chapter 6 Gravel and 'shingle' beaches

#	Site
30	Westward Ho! Cobble Beach, Devon
31	Loe Bar, Cornwall
32	Slapton Sands, Devon
33	Hallsands, Devon
34	Budleigh Salterton Beach, Devon
35	Chesil Beach, Dorset
36	Porlock, Somerset
37	Hurst Castle Spit, Hampshire
38	Pagham Harbour, West Sussex
39	The Ayres of Swinister, Shetland
40	Whiteness Head, Moray
41	Spey Bay, Moray
42	The West Coast of Jura, Argyll and Bute
43	Benacre Ness, Suffolk
44	Orfordness and Shingle Street, Suffolk
45	Rye Harbour, East Sussex
46	Dungeness, Kent

25

An introduction to British coastal geomorphology

Feature	47	48	49	50	51	52	53	54	55	56	57	58	59	60	61	62	63	64	65	66	67	68	69	70	71	72	73
26. Semi-enclosed bay											×																
25. Inlets and submerged coasts								×			×		×														×
24. Coastal valleys																											
23. Machair										×	×		×														×
22. Saltmarsh morphology								×		×									×				×		×		
21. Mudflats, ridge and runnel forms							×	×		×				×					×				×		×		
20. Intertidal sediments					×			×	×	×	×		×	×	×	×		×	×	×	×		×		×	×	
19. Tombolos and tied islands																									×	×	
18. Cuspate forelands and nesses													×	×								×		×	×	
17. Barrier beaches														×													
16. Spits						×							×		×		×	×	×	×	×	×	×	×			
15. Dunes, including sandplains			×	×	×	×	×	×	×	×	×	×	×	×	×	×		×	×	×	×	×	×	×	×	×	
14. Cliff-foot beaches	×		×											×													
13. Emerged ('raised') beaches				×				×	×		×	×															
12. Pre-existing clasts												×															
11. Beach phases	×					×	×		×		×	×	×														
10. Prograding beach		×		×	×	×	×		×	×		×	×			×	×	×		×	×	×	×	×			
9. Beach undergoing erosion		×	×	×	×		×	×	×	×	×				×	×	×		×	×	×						
8. Beach orientation	×	×	×	×	×		×		×	×		×	×	×	×		×	×	×	×	×	×	×	×			
7. Shore platforms – erosional control	×		×																								
6. Shore platforms – structural control																											
5. Karstic development																											
4. Exhumed forms: cliffs, benches		×	×	×																							
3. Cliff forms and processes	×		×	×	×												×		×							×	
2. Small-scale structural control	×		×	×																							
1. Large-scale structural control																											

Chapter 7 Sandy beaches and coastal dunes
47. Marsden Bay, County Durham
48. South Haven Peninsula, Dorset
49. Upton and Gwithian Towans, Cornwall
50. Braunton Burrows, Devon
51. Oxwich Bay, Glamorgan
52. Tywyn Aberffraw, Anglesey
53. Ainsdale, Lancashire
54. Luce Sands, Dumfries and Galloway
55. Sandwood Bay, Sutherland
56. Torrisdale Bay and Invernaver, Sutherland
57. Dunnet Bay, Caithness
58. Balta Island, Shetland
59. Strathbeg, Aberdeenshire
60. Forvie, Aberdeenshire
61. Barry Links, Angus
62. Tentsmuir, Fife

Chapter 8 Sand spits and tombolos
63. Pwll-ddu, Glamorgan
64. Ynyslas, Ceredigion
65. East Head, West Sussex
66. Spurn Head, Yorkshire
67. Dawlish Warren, Devon
68. Gibraltar Point, Lincolnshire
69. Walney Island, Lancashire
70. Winterton Ness, Norfolk
71. Morfa Harlech, Merioneth, Gwynedd
72. Morfa Dyffryn, Merioneth, Gwynedd
73. St Ninian's Tombolo, Shetland

Table of GCR site features

74	Isles of Scilly		×	×				×		×	×		
75	Central Sanday, Orkney						×	×	×	×	×	×	×
Chapter 9 Machair													
76	Machir Bay, Islay, Argyll and Bute				×	×		×	×	×	×		
77	Eoligarry, Barra, Western Isles			×	×			×	×	×	×	×	×
78	Ardivacher to Stoneybridge, South Uist			×				×	×	×	×	×	×
79	Hornish and Lingay Strands (GCR name: Machairs Robach and Newton), North Uist			×	×	×		×	×	×	×	×	×
80	Pabbay, Harris, Western Isles			×				×	×	×		×	
81	Luskentyre and Corran Seilebost, Harris			×	×		×	×	×	×	×	×	×
82	Mangestra, Lewis, Western Isles			×					×	×		×	
83	Tràigh na Berie, Lewis, Western Isles			×	×			×	×	×	×	×	×
84	Balnakeil, Sutherland			×	×	×		×	×	×	×	×	×
Chapter 10 Saltmarshes													
85	Culbin, Moray			×	×	×		×	×	×	×		
86	Morrich More, Ross and Cromarty			×	×	×		×	×	×	×		
87	St Osyth Marsh, Essex			×	×	×		×		×	×		
88	Dengie Marsh, Essex			×	×	×		×		×	×		
89	Keyhaven Marsh, Hurst Castle, Hampshire					×		×		×	×		
90	Solway Firth (north shore), Dumfries and Galloway					×			×	×	×	×	
91	Solway Firth: Upper Solway flats and marshes (south shore), Cumbria					×			×	×	×	×	
92	Solway Firth: Cree Estuary (Outer Solway Firth), Dumfries and Galloway					×			×	×	×	×	
93	Loch Gruinart, Islay					×	×	×		×	×		
Chapter 11 Coastal assemblages													
Culbin, Moray – see site number 85													
Morrich More, Ross and Cromarty – see site number 86													
94	Carmarthen Bay (including GCR site Burry Inlet), Carmarthenshire	×	×		×	×		×	×	×	×	×	
95	Newborough Warren, Anglesey		×	×		×		×	×	×	×	×	
96	Morfa Dinlle, Gwynedd			×		×		×	×	×			
97	Holy Island (GCR name: Goswick–Holy Island–Budle Bay, Northumberland	×	×	×		×		×	×	×	×	×	
98	North Norfolk Coast			×		×	×	×	×	×	×		
99	The Dorset Coast: Peveril Point to Furzy Cliff	×	×		×		×						×

27

An introduction to British coastal geomorphology

design of coastal structures and the maintenance of sea defences. The former Nature Conservancy Council carried out a large number of surveys of beaches and dunes; sites such as Chesil Beach, Dawlish Warren and Orfordness became key field sites. In the 1950s, Gibraltar Point became established as a field site undergoing regular monitoring, as Scolt Head Island had since the 1920s.

Since then, management of the coastline has been an important focus for some coastal geomorphologists, often in association with coastal ecologists, engineers and planners. In particular, between 1969 and 1981, Ritchie and colleagues in the University of Aberdeen produced for the then Countryside Commission of Scotland a detailed inventory of all the beaches, sand dunes and machair of Scotland, their geomorphology and the pressures to which they were subjected. This work has formed a useful benchmark for many of the Scottish GCR sites described and interpreted in this volume.

It is difficult to ascribe the geomorphological conservation value to sites that have been significantly affected by human activities. Very many coastal sites have been affected directly and indirectly by coast protection, tidal defences and reclamation projects, by mineral extraction, or by intensive recreational use. Some remote or otherwise undeveloped sites are used for activities that interfere with access, such as nuclear power stations or military training areas. Some GCR coastal geomorphology sites certainly include minor artificial structures, and it is not always possible to select GCR sites that are 'complete', 'natural' and 'unspoilt'. But, in general, sites where interference has been widespread have not been included in the GCR. There are two substantial exceptions: Spurn Head and Dungeness, both heavily modified major landforms. Such sites have been included in the GCR because the classic importance and long-history of study of the site argue for inclusion; indeed ongoing study of such sites will be important if we are to improve our understanding of how coasts react to changes brought about by human activity.

LEGAL PROTECTION OF THE GCR SITES

V.J. May and N.V. Ellis

The list of GCR sites has been used as a basis for establishing Earth science Sites of Special Scientific Interest (SSSIs), protected under the Wildlife and Countryside Act 1981 (as amended) by the statutory nature conservation agencies (the Countryside Council for Wales, English Nature and Scottish Natural Heritage).

The SSSI designation is the main protection measure in the UK for sites of importance to conservation because of the wildlife they support, or because of the geological and geomorphological features that are found there. About 8% of the total land area of Britain is designated as SSSIs. Well over half of the SSSIs, by area, are internationally important for a particular conservation interest and are additionally protected through international designations and agreements.

About one third of the SSSIs have a geological/geomorphological component that constitutes at least part of the 'special interest'. Although some SSSIs are designated solely because of their importance to wildlife conservation, there are many others that have *both* such features *and* geological/geomorphological features of special interest. Furthermore, there are localities that, regardless of their importance to wildlife conservation, are conserved as SSSIs solely on account of their importance to geological or geomorphological studies.

Therefore, many SSSIs are composite, with site boundaries drawn from a 'mosaic' of one or more GCR sites and wildlife 'special interest' areas; such SSSIs may be heterogeneous in character, in that different constituent parts may be important for different features.

There are, therefore, coastal SSSIs not described in this volume that are important for saltmarsh, machair, dune and shingle features as habitat/wildlife sites, regardless of the underlying geomorphology. Although such habitat types are intrinsically linked to the geomorphology, these other sites were not deemed to achieve the GCR standard for their geomorphology alone, or they duplicated geomorphological features better seen at other sites (The 'minimum number' criterion of the GCR is an important factor here (see above)). Therefore, for example, although only 11 localities are described in this volume for 'saltmarsh morphology' there are many other SSSIs that have been designated because of the habitat/wildlfe value of their saltmarsh, which will also be of interest to the geomorphologist.

Conversely, many of the SSSIs that are

designated solely because of their Earth science features have interesting wildlfe and habitat features, underlining the inextricable links between 'the environment' and the underlying geology and geomorphology.

It is clear from the discussion in previous sections that the conservation interest of the geomorphological features is likely to be affected by shoreline management activities outside the site itself, especially where the GCR sites lie within larger sediment transport cells. However, since SSSI notification of GCR sites presently extends to mean low-water mark in England and Wales, and mean low-water of spring tides in Scotland, there is no statutory protection of the shallow water sediments that may be the main sediment source for beaches.

International measures

Presently, there is no formal international conservation convention or designation for geological/geomorphological sites below the level of the 'World Heritage Convention' (the 'Convention concerning the Protection of the World Cultural and Natural Heritage'). World Heritage Sites are declared by the United Nations Educational, Scientific and Cultural Organisation (UNESCO). The objective of the World Heritage Convention is the protection of natural and cultural sites of global significance. Many of the British World Heritage sites are 'cultural' in aspect, but the Giant's Causeway in Northern Ireland and the Dorset and East Devon Coast are inscribed because of their importance to the Earth sciences as part of the 'natural heritage' – the Dorset and East Devon site is of particular relevance here insofaras it was the outstanding geology and coastal geomorphology that led to its inscription. The St Kilda World Heritage site certainly has an important geological component contributing to its status.

In contrast to the Earth sciences, there are many other formal international conventions –

Table 1.7 Coastal Annex I habitats occurring in the UK (from McLeod *et al.*, 2002.)

EU code	Habitat name	Lay name	Priority habitat/ species	UK special responsibility
1130	Estuaries	Estuaries		×
1140	Mudflats and sandflats not covered by seawater at low tide	Intertidal mudflats and sandflats		
1150	Coastal lagoons	Lagoons	×	×
1160	Large shallow inlets and bays	Shallow inlets and bays		×
1170	Reefs	Reefs		×
1210	Annual vegetation of drift lines	Annual vegetation of drift lines		
1220	Perennial vegetation of stony banks	Coastal shingle vegetation outside the reach of waves		×
1230	Vegetated sea cliffs of the Atlantic and Baltic coasts	Vegetated sea cliffs		×
1310	*Salicornia* and other annuals colonizing mud and sand	Glasswort and other annuals colonizing mud and sand		
1320	*Spartina* swards (*Spartinion maritimae*)	Cord-grass swards		
1330	Atlantic salt meadows (*Glauco-Puccinellietalia maritimae*)	Atlantic salt meadows		
1420	Mediterranean and thermo-Atlantic halophilous scrubs (*Sarcocornetea fruticosi*)	Mediterranean saltmarsh scrub		
2110	Embryonic shifting dunes	Shifting dunes		
2120	Shifting dunes along the shoreline with *Ammophila arenaria* ('white dunes')	Shifting dunes with marram		
2130	Fixed dunes with herbaceous vegetation ('grey dunes')	Dune grassland	×	×
2140	Decalcified fixed dunes with *Empetrum nigrum*	Lime-deficient dune heathland with crowberry		
2150	Atlantic decalcified fixed dunes (*Calluno-Ulicetea*)	Coastal dune heathland	×	
2160	Dunes with *Hippophae rhamnoides*	Dunes with sea-buckthorn		
2170	Dunes with *Salix repens* ssp. *argentea* (*Salicion arenariae*)	Dunes with creeping willow		
2190	Humid dune slacks	Humid dune slacks		×
21A0	Machairs	Machair		×
2250	Coastal dunes with *Juniperus* spp.	Dunes with juniper thickets	×	
8330	Submerged or partially submerged sea caves	Sea caves		×

particularly at a European level – concerning the conservation of wildlife and habitat. Of course, many sites that are formally recognized internationally for their contribution to wildlife conservation are underpinned by the geological/ geomorphological character, but this fact is implicit in such designations. Nevertheless, some of the sites described in the present volume are not only geomorphological SSSIs, but also *habitat* sites recognized as being internationally important. These areas are thus afforded further protection by international designations above the provisions of the SSSI system. Of especial relevance to the present volume are those coastal habitat types that are dependent on coastal geomorphology and are conserved as Special Areas of Conservation.

Special Areas of Conservation (SACs)

In 1992 the European Community adopted Council Directive 92/43/EEC on the conservation of natural habitats and of wild fauna and flora, commonly known as the 'Habitats Directive'. This is an important piece of supranational legislation for wildlife conservation under which a European network of sites – Special Areas of Conservation (SACs) – is selected, designated and protected. The aim is to help conserve the 169 habitat types and 623 species identified in Annexes I and II of the Directive.

Of the Annex I habitat types, 76 are believed to occur in the UK. The habitat types are very variable in the range of ecological variation they encompass. Some are very narrowly defined, comprising a single vegetation type; others are large units defined on a physiographic basis, such as Estuaries and Machairs, encompassing complex mosaics of habitats, and correspond approximately to the 'Broad Habitats' and/or 'Priority Habitats' of the UK Biodiversity Action Plan (Jackson, 2000). Although the habitats are identified for the conservation importance of their biological features, individually or collectively, many also represent geomorphological features, and this relationship is particularly clear in the list of coastal Annex I habitats (Table 1.7). Taken together, these cover virtually the full range of intertidal sediments, saltmarshes, dunes, shingle structures and sea-cliff habitats which occur in the UK.

Where habitats have qualified for selection, a process has been applied to ensure that the examples selected are of high quality and adequately represent the feature across its geographical and ecological range in the UK (McLeod *et al.*, 2002).

GCR SITE SELECTION IN CONCLUSION

It is clear from the foregoing that many factors have been involved in selecting and protecting the sites proposed for conservation and described in this volume. Sites will rarely fall neatly into one category or another; normally they have assets and characteristics that satisfy a range of the guidelines and preferential weightings. A full appreciation of the reasons for the selection of individual sites cannot be gained from these few paragraphs. The full justification and arguments behind the selection of particular sites are only explained satisfactorily by the site accounts given in subsequent chapters of the present volume.

Chapter 2

The geomorphology of the coastal cliffs of Great Britain – an introduction

K.M. Clayton, E.C.F. Bird and J.D. Hansom

Introduction

INTRODUCTION

The published research literature on coastal cliffs in Britain is relatively limited, receiving surprisingly little attention in textbooks despite the high proportion, 80% (Emery and Kuhn, 1982), of the world's coastline that is cliffed. Carter (1988) is typical of this under-representation, with the subject of cliffs and shore platforms covered in 12 pages, or just 2% of the text. Specialized texts concerning cliffs are also few in number; the two most significant in English being by Trenhaile (1987) and Sunamura (1992). As well as the lesser need for coastal protection measures, the main reason for this lack of attention seems to be the slow geomorphological evolution of hard rock cliffs and thus the obvious limitations of a process-based approach in describing and interpreting them. More work has been done on the rapidly evolving cliffs cut into weak rocks, such as London Clay (Hutchinson, 1973) or Chalk (So, 1965; May, 1971a; May and Heeps, 1985), and the glacial deposits of East Anglia (Hutchinson, 1976) and Holderness, Yorkshire (see GCR site report, present volume). Indeed, much of Sunamura's text reports experiments carried out on cliffs, or models of cliffs, cut in *weak* rocks. Even so, field measurement and quantification are uncommon, despite the pioneering work of Robinson (1977a–c) and Mottershead (1989, 1998).

With the exception of 'plunging cliffs', which

Figure 2.1 (a) Clò Mór cliff (193m) to the east of Cape Wrath, Sutherland is a good example of a plunging cliff, with no shore platform development, which has been inherited from former sea levels. (b) Recession of the Chalk cliff at Sewerby, west of Flamborough Head, Yorkshire, has produced a steep lower cliff with a sloping shore platform whose upper junction is obscured by a gravel beach composed of chalk gravels together with glacial gravels derived from bevelling of the cliff-top till. (c) Rapid erosion of the soft and unconsolidated glacial till cliff at Atwick, Holderness, Yorkshire, progresses by undercutting and rotational failure that is accentuated when the cliff-foot beach is thin or absent. This view looking north shows a very thin upper beach veneer over an area of exposed till shore platform (locally called an 'ord') whose surface is strewn with till blocks eroded from the cliff. (Photos: J.D.Hansom.)

descend into deep water (Figures 2.1a and 2.2D), cliffs rise above a shore that may be irregular and rocky, or above a more or less well-developed shore platform created by the retreat of the parent cliff (Figure 2.1b). This platform may sometimes be obscured by boulders, or, in weak rocks especially, by a shallow veneer of beach material, such as at Holderness (Figure 2.1c). Whatever the form of the shore adjacent to the cliffs, more attention has been paid to the cliffs themselves than adjacent platforms or other shoreline elements, yet they have evolved together and may share common processes. Over time, platforms are lowered by a variety of erosional processes, and are widened as the cliff retreats (Figure 2.2A). Thus platform width may be related to cliff retreat at any one sea level. However, the relationship is complicated by erosion of the seaward edge of the platform (thus underestimating the amount of retreat), or by inheritance of the platform from a previous sea level (thus overestimating the rate of retreat over the Holocene Epoch (Trenhaile *et al.*, 1999).

More work has also been published on coastal landslides than on the geomorphology of coastal cliffs *per se*, partly because today active landslides in Britain are far more common on the coast than inland (Hutchinson, 1972, 1976; Brunsden and Jones, 1980; Hutchinson *et al.*, 1991; Chandler and Brunsden, 1995). Hutchinson (1973) shows that the rate of basal cliff erosion is a fundamental control on the different processes of cliff-slope erosion (and thus the varying slope forms) on lithologically uniform sediments like the London Clay. Because complex landslides and mudflows are best-developed and most easily examined at the coast, much of the work in such mass movements has tended to treat the cliff-slope processes for their intrinsic interest, neglecting or even ignoring marine processes at the cliff foot.

Coastal landslide GCR sites represent almost one-third of all sites selected for the *Mass Movements in Great Britain* (Cooper, in prep) and form an important supplement to the weak-

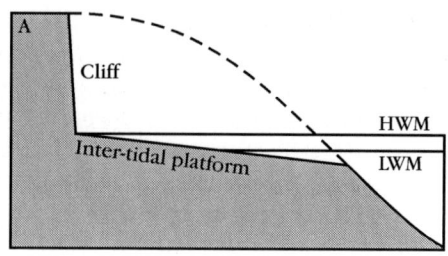

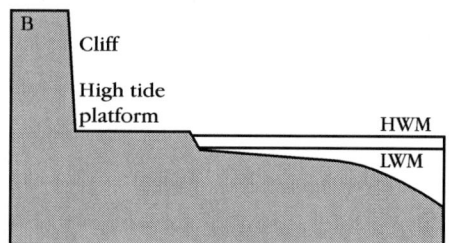

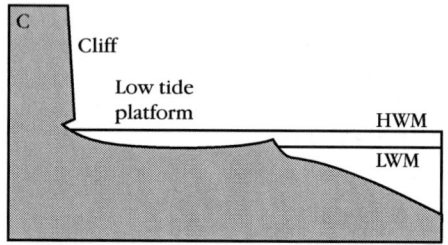

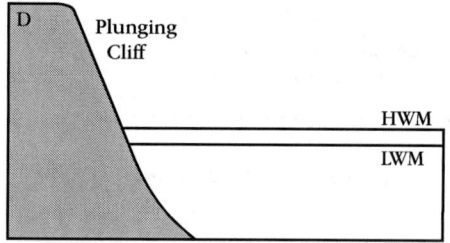

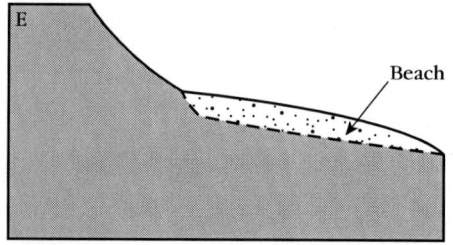

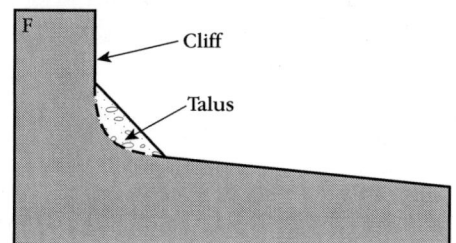

Figure 2.2 Coastal cliffs and their related shore platforms. A, cliff with intertidal ramp platform; B, cliff with shore platform at about high tide level; C, cliff coast with shore platform at about low tide level; D, plunging cliff with no shore platform; E, relict cliff with platform marked by emerged beach; F, typical inland cliff with talus at foot. (After Bird, 1984.)

rock sites selected for the GCR described in the present volume. For example, descriptions of the coastal cliffs subject to landslipping at Hallaig (Raasay), and along the Dorset coast (Axmouth to Lyme Regis, Black Ven, and Blacknor on the west side of the Isle of Portland), are included in the *Mass Movements* GCR volume. Only Lyme Regis to Golden Cap (see GCR site report) has been selected explicitly for the GCR both for its coastal geomorphology and mass movements features. Other sites covered in the *Mass Movements* volume include Trimingham, Norfolk, in Quaternary glacial deposits, Folkestone Warren, Kent, (Chalk over Gault Clay), Warden Point, Isle of Sheppey (London Clay) and the relict 'abandoned' cliffs, now fronted by saltmarsh at High Halstow, Kent.

One type of weak-rock cliff well represented in the GCR is cut in chalk, the commonest rock of south-eastern England and cropping out in coastal cliffs along a considerable length of the coast from the Isle of Thanet to Devon, as well as at Flamborough Head, Yorkshire, and in a short coastal cliff at Hunstanton, Norfolk, where the Red Chalk is well exposed. There is considerable variety of form within these representative sites, yet, despite a number of local studies, no integrated geomorphological study of our chalk cliffs has yet been attempted.

Because of the more rapid rate of retreat and the larger number of buildings along the cliffs on weaker rocks, they are often modified by drainage works and coastal engineering structures. If the value for research of as yet undisturbed cliffed coasts is to be maintained, and if their value in providing sections of geological importance is to continue, then engineering intervention must be minimized. For example, Hutchinson's work on the London Clay cliffs provides what is now a historical record of a series of coastal sectors that have been entirely modified by engineering works. Today that work would be impossible to carry out and it is therefore increasingly important that our remaining cliffed sites on weak rocks remain in their natural state.

INLAND AND COASTAL CLIFFS

It is perhaps obvious, but rarely stated, that coastal cliffs dominate the population of cliffed slopes, particularly in Britain. Inland cliffs are mainly of glacial origin, but there are also inland cliffs of tectonic origin in mountainous areas, and those that are the headwalls of active landslides. Other inland slopes occur as isolated features on the outer bluffs of active river meanders and as natural rock outcrops on steep slopes, as in the Weald. In all cases cliffs may form, and persist, where active basal removal continues – where material at the foot of the slope is removed by erosion, so maintaining the cliffed slope above. If this does not occur, the cliff will in time degenerate into a talus slope that progressively covers the cliff beneath (Figure 2.2F). This process is relatively slow on erosion-resistant rocks and much quicker in unconsolidated sediments. Yet glacially formed cliffs, on both trough sides and within corries, rise above talus slopes and only where streams erode their base do they stand as true cliffs from bottom to top of the slope.

That cliffs, even in the most resistant rocks, have been modified by talus accumulation within the postglacial period (at most 15 000 years, but commonly the 10 000 years of post-Loch Lomond Stade time) emphasizes the importance of active, ongoing basal removal of talus as an important factor in the maintenance of steep *coastal* cliffs. Even where part of the slope may survive from an earlier interglacial period of high sea level (commonly determined from the occurrence of dated interglacial deposits (emerged beaches or sediments accumulated in caves), if the cliffs descend to a shore platform without any large accumulation of talus obscuring their base, they probably owe part of their present form to Holocene basal erosion. The point is emphasized where, in areas covered by ice in the last glaciation, stacks rising from the shore platform now occur seawards of a talus-free cliff. Many such features (e.g. the Old Man of Hoy, Orkney) are Holocene in age, having been trimmed since sea level rose to its present position about 6000 years ago. However, there are also many cliffs and stacks that are till covered (and emerged stacks with till plugs, for example at Tarbat Ness, Ross and Cromarty) that clearly pre-date the last glaciation. Some of these have also been elevated as a result of local tectonic, and especially continuing glacio-isostatic, movements within the Holocene Epoch.

The fact that steep and active slopes are far more common on the coast than elsewhere has led many of those interested in slope processes and morphology to work on coastal cliffs. Often these researchers have had little interest in the process of marine erosion that maintains the slope as an active landform and it is common to find differences in process linked to slope angle without any reference to the rate of marine basal erosion. None of this helps to clarify the limited and often conflicting literature on coastal cliffs. In system terms, the cliff face processes are a cascade, by which the removal of sediment from the base by waves produces instability that is propagated upwards to the cliff top, (Figure 2.3), where rates of erosion are most readily observed and measured. The lack of understanding of this process of cascade is widespread. It has been stated more than once by those living at the top of 40 m-high cliffs in Norfolk, 'You can tell these falls have nothing to do with the sea – they are all at the top of the cliff'.

In a similar way, continued removal of sediment from the beach at the foot of cliffs undergoing erosion produces more instability (Figure 2.3) and persistent coastal retreat may only be maintained if the sediment is removed as rapidly as it is produced. Indeed, cliffs undergoing erosion may become protected by the formation of a wide and high beach, allowing their slopes to decline. Whereas the finer-grained sediment moves offshore, the coarser material (sand and gravel) forms a cliff-foot beach that may protect the cliff or be moved by alongshore drift. For example, estimates of material lost from the Norfolk cliffs suggest that about 25% of the sand and gravel moves directly offshore, while the remainder may travel as much as 60 km alongshore. In this particular case it is not the rate of removal of beach sediment that controls the rate of cliff recession, because the sections undergoing the most rapid erosion are also the highest (Cambers, 1973).

COASTAL SLOPE PROCESSES

As noted above, coastal slopes and cliffs can be defined simply as subaerial slopes that have been modified at their base by marine processes (Hansom, 1988), with the development of shore platforms at the foot of the cliffs representing the eroded remnant of the original coastal slope. Cliff and shore platform morphology is then dominated by the balance between marine processes and sub-aerial slope processes (Figure 2.3). However, this is an oversimplification since the efficiency of operation of both sets of processes derives in part from the relative resistance of rocks of different strengths (Table 2.1) and in part from the position of sea level on the profile. In hard-rock cliffs, it is often microstructures that control most of the differences in erosion rates. As a result, although cliff profiles vary greatly, there appears to be little correlation between rock hardness and cliff angle, steepness being as much a function of basal erosion rate as of rock resistance (Young, 1972). The position of sea level is also important since this controls not only the spatial location of process operation on the cliff-face, but also temporal shifts in such locations. Some cliffs plunge vertically into relatively deep water yet, in contrast, many other relict cliffs stand behind staircases of platforms that have emerged as a result of uplift or changes of sea level as occurs in Islay, Argyll and Bute.

Coastal slopes are affected by all of the main types of mass-movement processes that occur as slopes try to attain stable, equilibrium forms. However, they may possess shorter-term stability than their inland relatives because of undercutting, over-steepening, and the removal of basal debris by marine processes. Mass movements, ranging from the quasi-continuous fall of small debris to infrequent but extensive landsliding, play an important role in the development of rock cliffs. Fresh rock faces and the presence of talus at the base of cliffs attest to the importance of rockfalls on many coasts. Although they are more frequent than deep-seated slides, rockfalls tend to be smaller and more widespread and are probably the dominant form of mass movement on most rock coasts (Trenhaile, 1987). Rockfalls involve the detachment and fall of surficial material from steep rock faces and typically occur in well-fractured rock, especially where wave-cut notches have developed in lithologically or structurally weaker rocks at the cliff base. Rockfalls and coastal subsidence may occur where deep cave systems have penetrated into the cliff. Slab-falls occur in massive cohesive rocks with deep tension cracks parallel to the cliff face. Toppling or forward tilting is a common process in the Torridonian sandstone cliffs of Sutherland, and is characteristic of rock structures that consist of columns or are well-defined by joints, cleavage or bedding planes (Figure 2.4). All of these mass movements are essentially surficial failures induced by frost action and other types of subaerial weathering, basal erosion of the cliff, and unloading and hydrostatic pressures exerted by water in the rock clefts (Trenhaile, 1997). A high percentage of falls are consequent upon ongoing cliff erosion and

Coastal slope processes

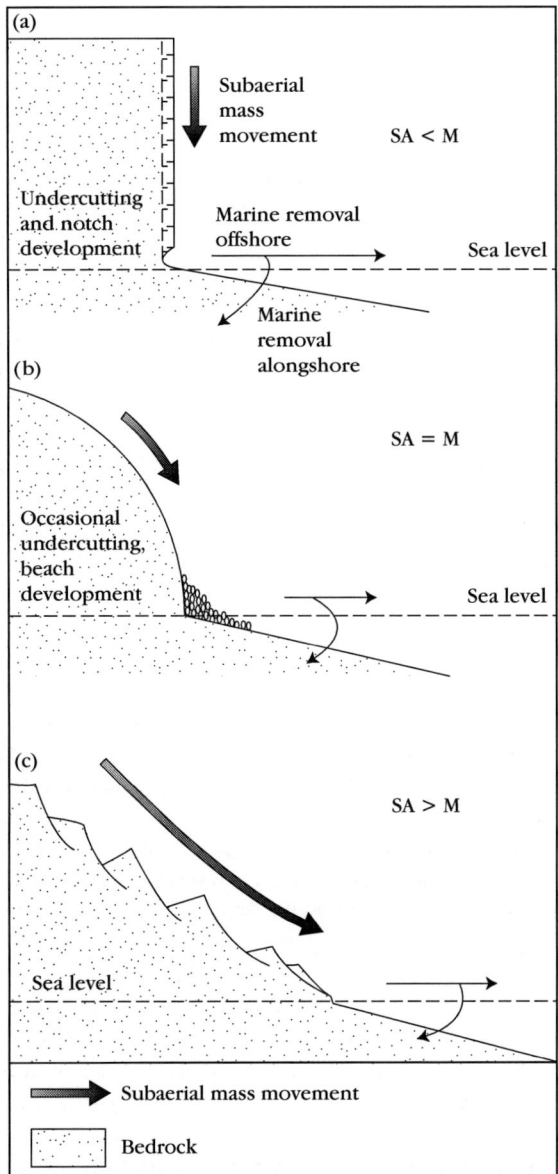

Figure 2.3 Processes of cliff retreat. SA = subaerial erosion of material, symbolized by the large arrow; M = marine erosion, symbolized by the fine arrows – eroded material is removed offshore and alongshore by marine process. (a) SA<M; here steep cliffs, undercut by marine processes, develop. (b) SA=M; here a balance between the two sets of processes allows small beaches to develop at the toes of sloping cliffs; (c) SA>M; here subaerial mass movements by sliding produce a low stepped profile and marine transport of plentiful debris. On most coastal slopes, the rate of erosion of material falls far short of the ability of waves and tides to remove it, so that the slope angles are maintained (a,b). However, on weaker rocks (c) material is delivered at a rate controlled in large part by the ability of the sea to maintain removal and thus the rate of basal erosion, in which case slope angle will decline until sediment input matches the rate of removal. (After Hansom, 1988.)

Table 2.1 Likely recession rates in different materials (compiled by Carter, 1988, from data in Sunamura, 1983).

Lithology	Recession rate (m a^{-1})
Granite	10^{-3}
Limestone	10^{-3} to 10^{-2}
Shales and flysch	10^{-2}
Chalk	10^{-1} to 1
Tertiary sedimentary	10^{-1} to 1
Quaternary sedimentary	1 to 10
Recent volcanic rocks	10 to 10^{2}

retreat: the formation of tension cracks parallel to the erosion surface arises from reductions in confining pressures as surface rock is removed. The 1999 failure of the chalk cliff at Beachy Head is a good example of this process.

Deep-seated mass movements are common in some coastal regions, where geological conditions are suitable, particularly where the compressive strength of the rock is exceeded by the load it must bear. Easily sheared rocks with low bearing strength are particularly susceptible to landslides and, as a result, they are more frequent in soft rocks and less common in resistant rocks (Trenhaile, 1987). However, a relatively common type of landslide in hard rock occurs where the cliff is characterized by seaward-dipping beds or alterations of permeable and impermeable strata, or where massive rocks overlie rocks with low load-bearing strength. In such situations, translational slides and 'dip-slip' slides, where failure occurs along a slope-parallel failure surface or bedding plane, produce large but often shallow features whose failure may often have been triggered by high pore-water pressures following prolonged rainfall. Spectacular examples of such landslides occur in the 30° westward-dipping beds of the Aberystwyth Grits near Aberystwyth, Wales.

Brunsden (1973) and Brunsden and Jones (1976, 1980) showed how complex coastal slopes may develop on coastal landslides. The cliffs of west Dorset are noted for the spectacular landslide systems that truncate NE–SW-trending ridges rising to between 140 and 170 m. The ridges are formed in chert and Upper Greensand overlying unconformably interbedded Lower Jurassic clays, marls, mudstone and thin argillaceous limestones. Large arcuate landslide scars form the upper part of the slope and are sepa-

Coastal cliff geomorphology

Figure 2.4 Toppling in the Torridonian sandstone cliffs south of Sandwood Bay, Sutherland. Dipping beds of well-jointed sandstones are subject to subaerial weathering and failure. Strong surf prohibits the debris from accumulating at the cliff foot. The stack in the distance is Am Buachaille (Gaelic for 'herdsman'). (Photo: J.D. Hansom.)

rated from nearly vertical sea cliffs by an undercliff marked by small rotational landslides, mudslides, large gullies and accumulations of debris. The landslide scars and the sea-cliffs retreat at similar rates despite the continually changing relationships and forms of the individual components of the landslide. Within these landslide areas, the role of the sea is crucial, the most active landslides occurring where marine action is most effective. In contrast, where landslides transport large boulders, the foot of the cliff may become protected by this natural rock armouring. In weaker rocks, Cambers (1976) demonstrated that on the Norfolk coast the highest rate of cliff retreat due to landslides is found where there is the greatest frequency of tides reaching the base of the cliff. Whether the slope maintains a particular angle or becomes gentler or steeper depends on the balance between the production of material by weathering and the rate at which it is transported out of the system (Carson and Kirkby, 1972). The role of the sea is critically important both for bedrock erosion and in removing sediments derived from slope processes (Figure 2.4).

PROCESSES OF MARINE EROSION OF CLIFFS AND SHORE PLATFORMS

Four main groups of marine erosion processes operate on cliffs and shore platforms: mechanical wave erosion; weathering; solution and bioerosion (Pethick, 1984; Hansom, 1988, Trenhaile, 1997). Although, these processes are often recognized and described, there remain relatively few quantitative estimates of their relative and absolute rates of operation with, for example, the studies of Trudgill (1986), Trenhaile (1987), Jerwood *et al.* (1990a,b), Sunamura (1992) and Stephenson and Kirk (2000a,b) as notable exceptions. Indeed, the relative importance of erosional processes has often been inferred from morphological evidence that is itself ambiguous (Trenhaile, 1980).

Processes of marine erosion

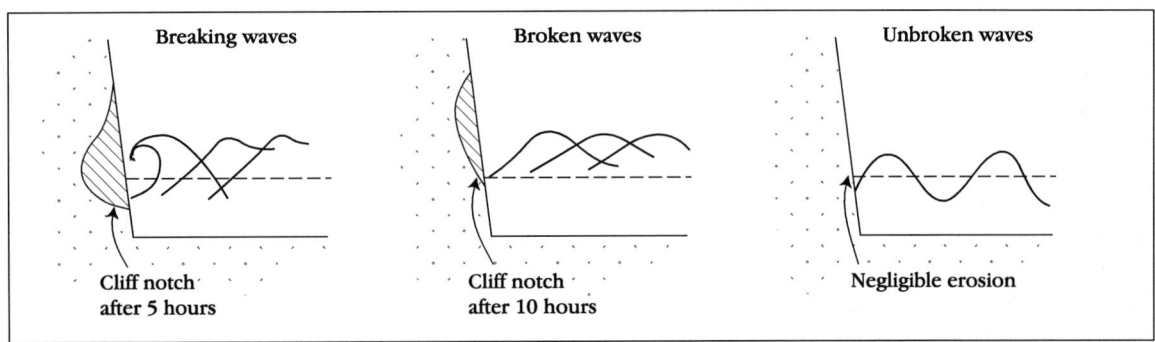

Figure 2.5 Cartoon depicting erosion of a vertical cliff under breaking-, broken- and unbroken-wave attack. Breaking waves cause the greatest amount of erosion. (Based on Sunamura, 1983, 1992 and Hansom, 1988.)

Mechanical wave erosion comprises two processes, wave quarrying and abrasion. Wave quarrying is the prising or pulling away of pieces of rock by the shock of impact of breaking waves. Secondary processes, such as pressure release and the pneumatic effects of air pressure in rock crevices, may also be involved in propagating rock failure along the crack-tip. Geological structure is important in this process since rocks with well-developed joints and fractures are more susceptible to quarrying. One of the few attempts to measure both the erosive process and the rates of cliff recession, so that a quantitative model could be prepared, is that of Sunamura (1975, 1977, 1981). Experiments on a model cliff in a wave tank and comparison with field data show that the pressure exerted by breaking waves causes maximum quarrying; broken waves are the next most effective; and unbroken waves reflected by the cliff in deep water cause negligible erosion (Figure 2.5).

Maximum wave quarrying is found on cliffs fronted by narrow, steep beaches, which induce wave breaking. However, over time, the production of a shore platform and beach composed of eroded debris, means that waves are progressively excluded from the cliff foot and the rate of erosion reduces (Sunamura, 1975). Freshly smoothed and scoured rock surfaces close to areas of loose sediment at the foot of

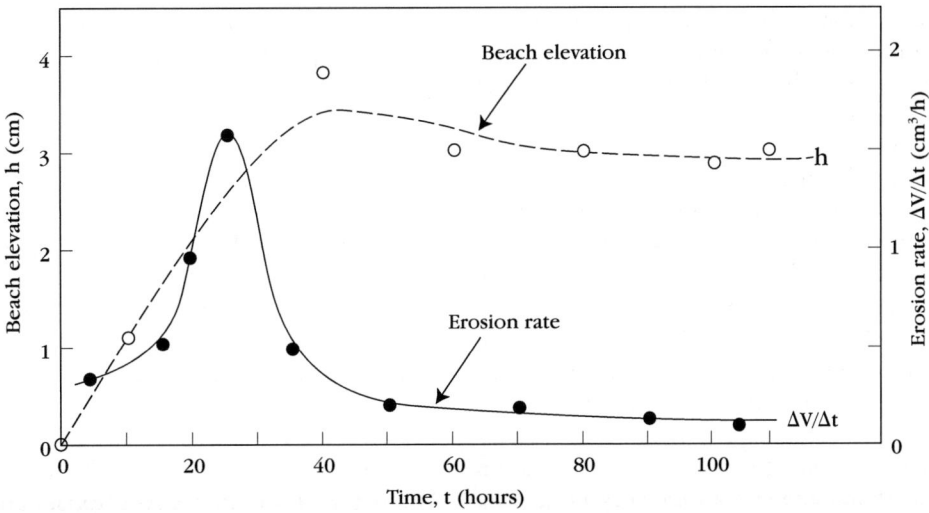

Figure 2.6 Temporal variations in abrasion rate ($\Delta V/\Delta t$), and beach elevation, h, expressed by the thickness of sand deposited at the cliff base, using data from wave-tank experiments. $\Delta V/\Delta t$ = volume of eroded material per unit time. (After Sunamura, 1976.)

cliffs are the most obvious demonstration of abrasional processes, particularly seen adjacent to the rough and angular appearance of sediment-free quarried surfaces (Tjia, 1985). Experiments reported in Sunamura (1992) show that abrasion rates are closely linked to beach elevation. Where production of sand from cliff erosion leads to the development of an embryo beach, abrasion erosion rates increase at first and then fall abruptly as enhanced beach volumes begin to protect the underlying rock (Figure 2.6). Clearly quarrying and abrasion work best together and in time produce notching and undercutting of the cliff base.

Cliffs and shore platforms are subject to two types of weathering process: water layer weathering and subaerial weathering. The first of these relates to the tidal wetting and drying of the cliff and platform by waves, spray and tides and may include chemical processes, such as hydration and oxidation, and mechanical rock breakdown caused by salt crystallization or the swelling of rock grains. Pitting or honeycombing of cliff faces within the spray zone is evidence of these processes and is particularly noticeable in sandstones or other sedimentary rocks where the cementing material becomes decomposed. Salt weathering in conjunction with frost action is a fairly potent force in the splash and spray zones and widespread cracking and spalling of a chalk shore platform in southern England was noted by Robinson and Jerwood (1987a–c). Jerwood *et al.* (1990a,b) used laboratory simulations to demonstrate how efficacious such process combinations could be in weathering softer rocks. Subaerial weathering relates to the normal weathering processes that loosen rock surfaces and deliver debris to the cliff foot. Subaerial weathering may affect rocks down to the level of permanent saturation (Bartrum, 1916); this weathered material is then easily removed by wave action. Russell (1971) notes that the water table marks a boundary between resistant rock below and weathered, easily erodible rock above.

Although they occur on all coasts to varying degrees, solutional and bio-erosional processes are most visible on tropical coasts. Since these processes often work in conjunction, the relative effects are difficult to separate. Solutional processes are most important in calcareous rock types where seawater, heavily charged with carbon dioxide in solution, aggressively dissolves the coastal rock. Work by Trudgill (1987) on the Irish coast has shown that, since seawater is undersaturated with respect to calcium carbonate, solutional activity is only prominent at night and in intertidal pools. As a result, on European coasts at least, chemical weathering is probably outweighed by the effects of a suite of biological processes. Such *bio*-erosion results from the activities of a huge range of organisms that either graze on algae on rock surfaces or bore into the rock in search of food or shelter. A range of overall erosion rates, ranging usually from about 0.5–1 mm a^{-1} on vertical and horizontal limestone surfaces (Trenhaile, 1997), has resulted from these studies but a key point is that bio-erosion serves to link the biological and geomorphological features, not only morphologically but also functionally. Organisms erode rock that then becomes involved in biogeochemical cycling, where abrupt changes in ecology or geomorphology will force the other components to react (Viles and Spencer, 1995). In some areas of the British Isles, such bio-erosion may be favoured by particular structural arrangements. Low-angled bedding planes in the limestones of County Clare, Ireland give rise to a series of stepped benches that control the vertical zonation of bio-eroders (Trudgill, 1987) and the same occurs in Welsh limestones at Hunts Bay and Rhossili Point. In turn, whereas the macro-scale controls on platform development are provided by structure, the meso- and micro-scale platform geomorphology is produced by bio-eroders. In spite of some very clear associations between geomorphology and biological processes, the effects of wave-related processes and their relationship with rock strength generally overshadow the impact of solutional and biological processes in Great Britain. The resultant cliff and platform morphology reflects this.

THE FORM OF COASTAL CLIFFS

The detailed form of a cliff coast is produced by a complex interaction of controls shown in Table 2.2, amplified below.

Geological controls on cliff form

The overall form of coastal cliffs depends strongly upon the nature of the materials forming them, but given the variety of contexts in which these occur, attempts to characterize cliff form

Form of coastal cliffs

Table 2.2 Primary, secondary and tertiary controls on cliff form (based on May, 1997a).

FIRST ORDER	SECOND ORDER	THIRD ORDER
Geological structure and lithology	Weathering and transport slope processes	Coastal land-use
Wave climate	Slope hydrology	Resource extraction
Subaerial climate	Vegetation	Coastal management
Water-level change (sea level and tide)	Cliff-foot erosion	
Geomorphology of the hinterland (landforms into which the cliffs are cut)	Cliff-foot sediment accumulation	
	Resistance of cliff-foot sediment to attrition and transport	

have not met with much success. The many combinations of process, lithology and structure and the variety of controls on cliff form in different climatic and sea-level situations make generalization of cliff form inherently difficult. Nevertheless, some types of cliff are more common in particular morphogenetic regions than in others. Steep cliffs are common in the wave-dominated environments of the north and west coast of Great Britain where the accumulation of cliff-foot sediment is restricted by wave-transport. Where wave activity is weaker and subaerial weathering stronger, coastal slopes tend to be gentler and more convex in form.

It is also possible to classify active sea-cliff profiles according to the interaction of process and geology. Emery and Kuhn (1982) propose a matrix of cliff-forms produced as a result of varying bedrock homogeneity and the relative importance of marine and sub-aerial processes (Figure 2.7). However, the shape and gradient of cliffs are also profoundly influenced by dip, strike, lithological variation, and structural weaknesses. Steep cliffs generally develop in rocks that are either vertically or horizontally bedded, whereas intermediate bed angles tend to produce more moderate slopes. Lithology also influences cliff morphology – high cliffs tend to be associated with more resistant rocks such as unbedded, impermeable, crystalline rocks that are highly resistant to wave erosion, whereas sedimentary rocks are more susceptible to wave quarrying, especially where dissolution of the rock cement or exploitation of weaker bedding planes aids disintegration. As a result of this complexity, the available models, although useful simplifications, take limited account of the infinite possibilities of structural variation. The topography of the cliff hinterland adds another dimension (see p. 44).

Characteristic medium-scale features of cliffs

The rate of mechanical wave erosion is particularly sensitive to variations in rock structure. Small bays, inlets and narrow gorges that develop along joint and fault planes and in the fractured and crushed rock produced by faulting are generally the result of accelerated erosion along these lines of structural weakness. Narrow inlets or geos, caves, arches and stacks are often found in close association with each other on coasts with well-defined and well-spaced planes of weakness. However, these features are less likely to develop in *very* weak rocks or in rocks with a *very* dense joint pattern, since the rock must also be strong enough to produce high, near-vertical slopes or to support the roofs of caves, tunnels and arches. If the joints or planes of weakness are very close together then long, narrow inlets develop such as the geos of northern Scotland. The angle of dip of the plane of weakness affects the occurrence and form of the erosional feature produced. For example, geo-like gorges with vertical sides are common in many horizontally bedded rocks with predominantly vertical joint planes such as occurs around the coast of Hoy in Orkney or at Skirza Head, Caithness (see GCR site report, Chapter 3). In steeply dipping rocks, where the planes of weakness are usually inclined, geos may either fail to develop or will be irregular in shape (Steers, 1962). Although

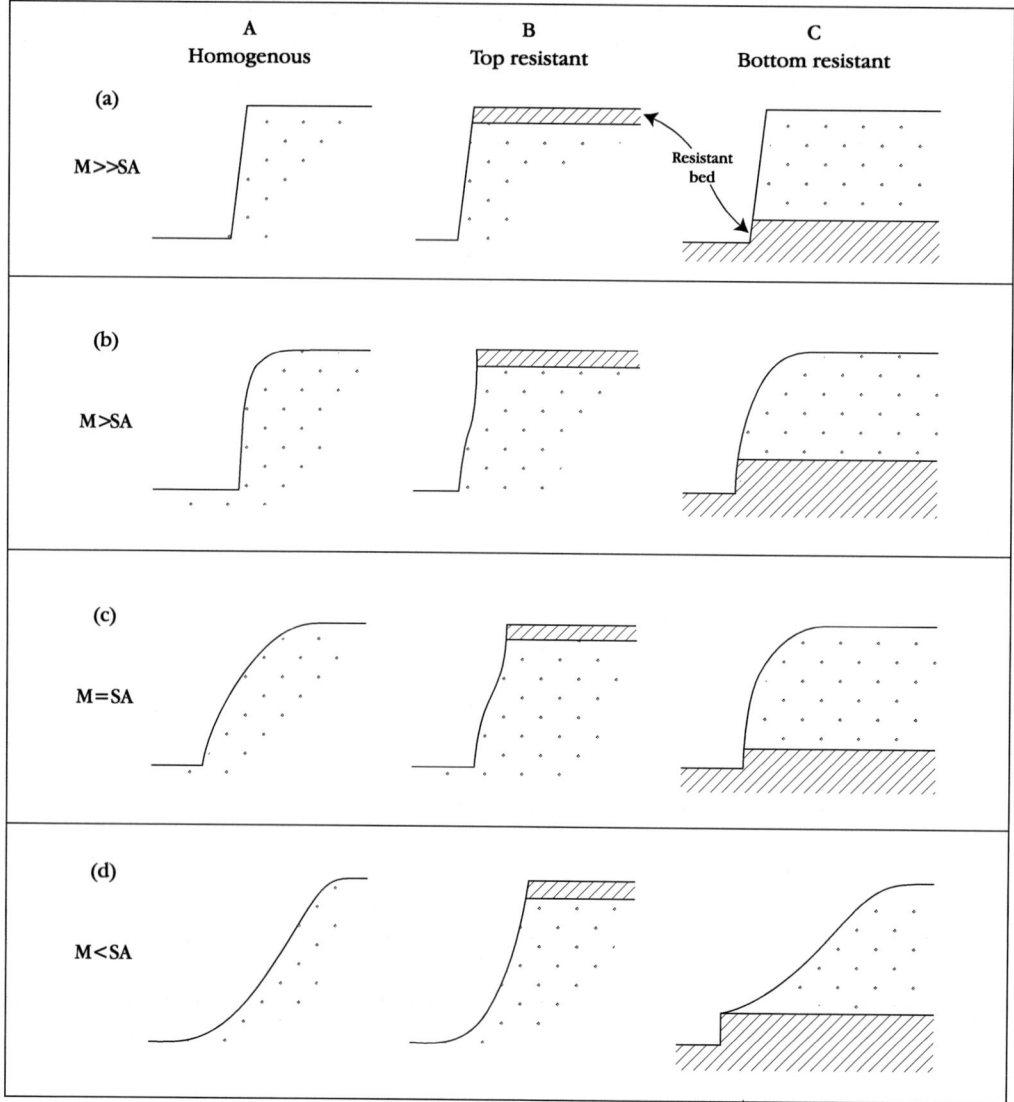

Figure 2.7 A classification of active sea-cliff forms according to comparative rates of subaerial erosion and marine erosion (SA = subaerial erosion; M = marine erosion). Type 'A' profiles are for cliffs of uniform resistance to erosion; type 'B', where a more resistant rock layer is present at the top; and type 'C', where there is a layer of more resistant rock at the base. (Based on Hansom, 1988, after Emery and Kuhn, 1982.)

joints and faults account for most narrow inlets, on igneous rock coasts geos may develop as a result of differential erosion of dykes and sills as occurs, for example, at Geo na h-Airde on St Kilda (see GCR site report in Chapter 3).

The occurrence of marine caves is again usually determined by structural weaknesses such as: joints; faults; breccias; planes; unconformities; irregular sedimentation; or the internal structure of lava flows. The form of caves, which can be tunnel or dome-shaped, reflects the number and inclination of these structures (Figure 2.8). Caves are particularly common in places where the rock is strongly jointed (Trenhaile, 1987) and vertical bedding or jointing tends to produce tall, narrow caves (Fleming, 1965). Lithology may account for the formation of some caves and they may develop as a result of the differential erosion of dykes that are weaker than the adjacent rock. Large caves in limestone regions may also have been formed by underground solution and later inherited and modi-

Form of coastal cliffs

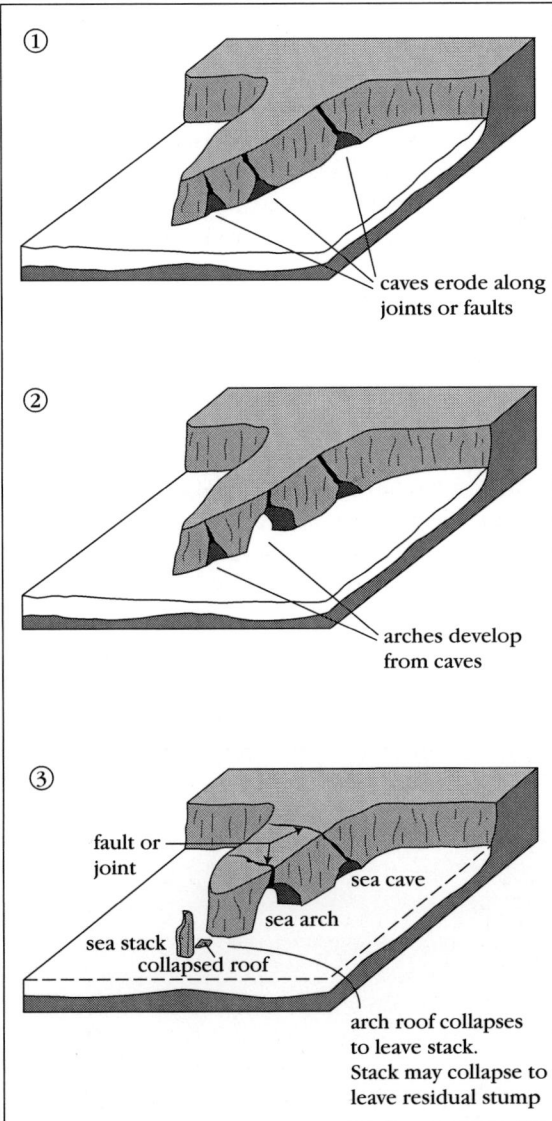

Figure 2.8 The development of caves, arches and stacks. Wave erosion is more effective along faults and joints where the rock is weaker, and so caves become excavated along these lines of weakness.

fied by wave action as a result of cliff retreat and changes in relative sea level. However, most sea caves are relatively small and have been excavated by wave quarrying, particularly during storm periods, in association with abrasive and hydraulic processes. If a cave becomes connected to the surface through a joint or fault-controlled shaft, a blowhole can develop, through which fountains of spray are blown out during storms and high tides, such as occurs at the Bullers of Buchan, Aberdeenshire. The presence of pebbles around the tops of blowholes in north-eastern Scotland (Steers, 1962) testifies to the enormous pressures generated during storms in narrow marine caves. In common with the cliffs into which they are cut, most caves, although modified by contemporary processes, may well pre-date present sea level and some (for example in the Gower peninsula, South Wales) are at least as old as the last interglacial period (Davies, 1983).

Tunnels and natural arches may develop from marine caves where wave erosion of either, or both, sides of a promontory may succeed in excavating a tunnel, usually along the line of a geological weakness, to produce a sea arch (Figure 2.8). The term 'sea tunnel' is appropriate when an arch is considerably longer than the width at the entrance. A typical example of a sea tunnel runs for about 100 m along a fault zone at Merlin's Cave, Tintagel, Cornwall (see GCR sitre report in Chapter 3; Wilson, 1952). Collapse of the roof of the arch or tunnel leads to the formation of a stack (Figure 2.8). Sea arches are ephemeral landforms, especially where they occur in weak rocks. For example, map evidence suggests that the Old Man of Hoy, in the Orkney Islands, developed from a promontory, since it did not exist prior to 1600. Yet by 1819 a prominent feature had developed with a stack and an arch with the twin legs that gave The Old Man its name (Hansom and Evans, 1995). The debris from an earlier roof collapse now litters the connecting platform between the base of the stack and the parent cliff. One of these legs was destroyed in a major storm in the early 19th century to produce a monolithic stack and the process has continued into the late 20th century with a large segment of its upper structure now in danger of collapse. Stacks often form quite quickly and sometimes can persist once they have become isolated from the mainland and the abrading effects of cliff-foot sediments. However, it is difficult to be precise about the age of stacks and in general it seems that most arches and stacks are short-lived features, judging from the rapid changes of morphology documented in many locations. Some arches last several hundred years (The Old Man), some only tens of years (e.g. Byrne, 1964) and others last barely over one year (Bird and Rosengren, 1986). The spectacular collapse of part of a famous double limestone arch at London Bridge, on the coast of Victoria,

Australia, occurred on 15 January 1990, leaving the other arch as an islet and stranding two tourists who were later rescued by helicopter. Stacks are generally produced from the collapse of sea arches, however some also survive as emerged or exhumed forms. When stacks eventually collapse, their bases will often survive for a time as reefs until these too merge into the developing coastal platform.

At a smaller scale, a wide range of features such as crevices, caves, clefts, and blowholes can form, together with even smaller-scale features such as tafoni and weathering forms.

Influence of inland topography on cliff form

It is also important to appreciate the role of hinterland landforms in determining cliff pattern and height. The obvious example of this is the site at Seven Sisters on the Channel coast of Sussex where an almost straight cliffline truncates the 'dip-slope' of the South Downs, crossing a series of dry valleys, the seven ridges between them forming the Seven Sisters (see Figure 2.9). Cliff height here is a function of the height of the land traversed by the cliff, not of the erosional energy of the waves at their foot. At Beachy Head, the eastern end of this line of cliffs, the maximum cliff height is $c.$ 160 m and major rockfalls and/or landslides are relatively common, yet the sea remains able to sweep clear the coastal platform at the foot, although for very large failures this clearance may take several years. Lower cliffs in Chalk are more stable and rockfalls occur less frequently (Hutchinson, 1972).

Whereas the role of landform in the varying height of the Seven Sisters is clear, this is more easily overlooked on a coast where varying geology has allowed development of a dissected landform where valleys grade to sea level. Commonly in southern England, lower and wider valleys are found in weaker mudrocks, and higher ground in the more resistant rocks. The Holocene drowning of this landscape has allowed the sea to cut high cliffs in Chalk and other more resistant rocks, whereas the clays and mudrocks that crop out within indented valleys and bays have much lower cliffs or none at all. Some accounts suggest that the sea has eroded these deep bays, rather than inheriting them in much their present form by the partial

Figure 2.9 Cliff height, and to some extent cliff form, is a function of the height of land cut by the cliffline. The photograph shows the cliff form of the Seven Sisters, Sussex, an almost straight cliffline truncates a series of dry valleys, the seven intervening ridges forming the Seven Sisters. (Photo: V.J. May)

submergence of a subaerially dissected landform.

A similar situation is the drowned and glacially sculptured landscape of western Scotland. Here cliff-like slopes on the sides of sea lochs, the deep fjord inlets, have been relatively little modified from the form left by the retreating ice, and similar forms occur on open coasts as in northern Skye. In other areas, the sea has partly drowned a glacially eroded landscape of low relief, as at the Loch Maddy GCR site in the Western Isles.

High-energy coasts – for example those exposed to the largest waves of the North Atlantic Ocean, as on St Kilda, or on the western coast of Orkney – have cliffs that extend to the full height of the eroded land and also plunge into deep water at their base. The highest cliff in Great Britain is Conachair, St Kilda, reaching 430 m in height. Such cliffs seem to be limited in height only by the elevation of the land behind the coast.

Elsewhere, cliffs only trim the lower part of the seaward slope, perhaps because the rock is particularly resistant to erosion, or because incident wave energy is restricted by limited fetch or shallow water, or because there is a lack of abrasive material. Plunging cliffs also retreat more slowly than cliffs that possess a shore platform, so these may not rise to the full height of the coastal slope. In some cases cliffs may be limited by the short time that the present-day coast has been exposed to wave attack. This is true of parts of the Scottish mainland coast, where Holocene emergence following crustal unloading shortened the period of sea-level standing close to its present position.

Clearly, cliffs are cut into a variety of pre-existing topographical situations (e.g. river or glacially scoured valleys and interfluves at all angles to the coast) and into a variety of geological structures (e.g. alternating layers of resistant and less-resistant, faulted and folded rock strata). Consequently, three-dimensional predictive models of cliff morphologies have not yet appeared.

SHORE PLATFORMS

Many cliffs undergoing erosion (other than plunging cliffs) stand behind shore platforms, sometimes wide, sometimes narrow, sometimes rather steeply inclined, but often rather gently sloping seawards (Figure 2.2). The platform is commonly intertidal, but a few plunging cliffs stand above a drowned shore platform that is not far below low tide level. Most weak-rock cliffs such as chalk have a platform in front of them (though it may be obscured by beach sediment). Plunging cliffs are found mainly on resistant rocks in the situation where the combination of wave energy and duration of the current sea level stand has been unable to develop a shore platform of any kind, and where any platform or rocky shore created in an earlier interglacial is not situated at present-day sea level. Once established, shore platforms may be readily extended, for they are exposed to quarrying, abrasion, weathering, solution and bioerosion.

Despite the significance of shore platforms and their relationship to the cliffs behind them, they have received less attention from researchers than the cliffs themselves. Accounts in textbooks such as those by Trenhaile and Sunamura are incomplete. Nevertheless, the description by Trenhaile (1987, pp. 206–39) of platform morphology, and of rates of erosion is a good summary to which little has been added in more recent literature. Thus the sites described in the present volume represent potential for future research. Most well-developed British shore platforms slope relatively steeply, and where the stratification or jointing of the rock imposes structural control, they are rather irregular.

Wide, sub-horizontal platforms seem underrepresented in Great Britain, probably because weathering processes are relatively less effective than wave processes and the forms are less likely to survive. Nevertheless, impressive sub-horizontal shore platforms occur within Robin Hood's Bay on the soft Jurassic rocks of North Yorkshire.

The processes involved in forming shore platforms have been much debated in the scientific literature, although it is now recognized that these are similar to those affecting marine cliffs, and, as a result, have largely been accounted for above. In common with cliff forms, a great range of shore platform morphologies created by local conditions occur. However, two fundamental types emerge: sub-horizontal platforms and sloping ramp platforms (Trenhaile, 1997; Figure 2.2).

Sub-horizontal platforms tend to have surfaces lying at either high or low-tide levels and terminate seawards at a low tide cliff whereas

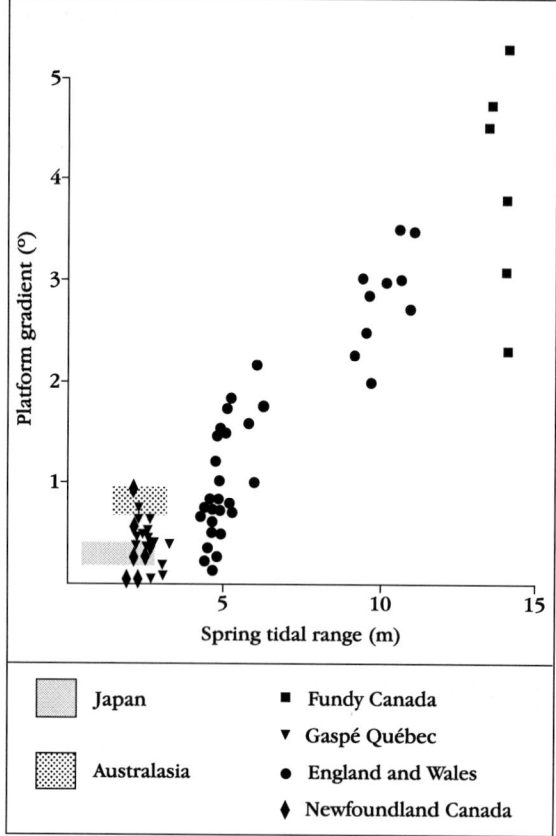

Figure 2.10 Plot of platform gradient against tidal range. Each point is a regional average of many surveyed profiles and suggests a direct relationship between gradient of platforms and spring tidal range. (After Trenhaile, 1987, p. 207.)

ramp platforms slope between the two tide levels with no major break in slope (Figure 2.2). In much of the literature, high-tide platforms (Figure 2.2B) are thought to be produced by water-layer weathering and develop best where rocks are permeable and horizontally bedded, where evaporation rates are high and where tidal characteristics allow a long drying period (Pethick, 1984). Since quarrying and abrasion processes rapidly destroy the features produced by slower weathering, high-tide platforms are thought to be best preserved where mechanical wave erosion is limited, and thus are rare in Britain. Low-tide platforms are thought to be produced by solutional and biological processes and are common in tropical areas, although they are also found on mid-latitude limestone and chalk coasts. Sloping intertidal platforms are very common in the storm-wave environments of the mid-latitudes (e.g. many parts of the British coast) and are produced by mechanical processes such as quarrying and abrasion. The slope or 'ramp' from the two tide levels is rarely regular, as variations in rock type and structure are exploited by wave erosion to produce an irregular surface.

Work by Trenhaile (1974a,b, 1978, 1987, 1997) has cast doubt on the traditional interpretations of shore platform types by highlighting the important, and perhaps controlling, role of tidal range in platform morphology. There is a strong positive relationship between platform gradient and tidal range, particularly in the storm-wave environments, but also in other climates and wave regimes (Figure 2.10): where tidal ranges are great, gradients are correspondingly steep. Such work suggests that, as a result, low tidal ranges can produce sub-horizontal platforms whether or not they occur in storm-wave environments (Trenhaile, 1997). Similarly, solutional, biological, water layer weathering, or even frost action in the intertidal zone may well produce platforms, but they are probably secondary to, and work via, the control of tidal range.

Trenhaile and Layzell (1981) suggested that the rate of intertidal erosion is determined by the time that still-water level occupies each elevation within the tidal range (the tidal duration factor), by an erodibility factor related to wave energy and rock hardness, by platform gradient and by the rate of submarine erosion. It was suggested that platforms develop best at mid-tide level because tidal duration, and thus erosion, is greatest in this zone (Figure 2.11). Carr and Graff (1982), however, demonstrated that maximum tidal duration is bimodal and associated with times of high and low water. Since tidal level varies most rapidly at the mid-tide position, wave action, and thus erosion, is least effective there. Trenhaile (1982) accepted this modification of the basic Trenhaile and Layzell model and added that this was consistent with the observations of Chastain (1976).

Relationships between morphology and factors other than tidal range are less clear and much of the field evidence is contradictory, particularly at the local scale. So (1965) described the varying slope of the coastal platforms of the Isle of Thanet in Kent, surveying them before the very extensive sea defences were completed. Platforms were found to decrease in height from

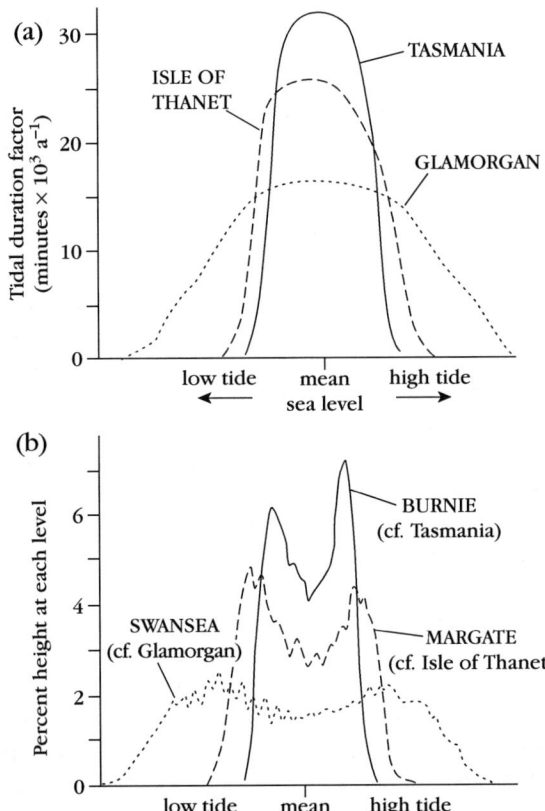

Figure 2.11 Tidal duration curves from three locations as plotted in varying detail by two sources. Tidal duration is the length of time that still-water level occupies each elevation within the tidal range. ((a) After Trenhaile and Layzell (1981); (b) after Carr and Graff (1982).)

headlands towards embayments and this was attributed to the greater energy of storm waves at headlands, allowing planation to a higher level. This was supported by the observation that platform height also varied with coast alignment, being highest where the coast faces north-east. Transverse gradient was greater in embayments than off headlands, and most platforms showed an upper concave and a lower convex form. Platform width, from their upper limit to the common low-tide cliff (locally a submerged cliff at the outer edge), was greatest on more resistant Chalk, and lowest on the weaker beds. So concluded that the very existence of a platform indicated that the recession of the cliff foot to form the platform was faster than the rate of landward recession of the low-tide cliff. It is remarkable that in the last 30 years no wider-scale work has been done on the English Chalk coastal platforms to confirm or modify the conclusions reached by So for the Isle of Thanet in Kent.

Geological effects, both structural and lithological, are clearly responsible for detailed variations in areas where the morphogenetic conditions are similar. Rock types that provide debris to the platform contribute to platform abrasion, yet large quantities of debris accumulating on the platform also serve to reduce abrasion and protect the surface (Sunamura, 1976; Figure 2.6). Robinson (1977b,c) found that platforms in north-east England became progressively narrower as sandy beaches, bare rock, boulder beaches and talus cones built up at the cliff foot. In addition, the compressive strength of the host rock needs to be matched against the stresses placed on it by mechanical wave processes. Other things being equal, igneous rocks such as granite or basalt have higher compressive strengths than sedimentary rocks such as limestones, chalk and sandstone and so are likely to be more resistant to wave erosion. Modelling suggests that platform morphology owes a great deal to structural control with fastest rates of erosion and thus of platform width associated with horizontally bedded rocks and where the dips are moderate and the strike is perpendicular to the rock face (Trenhaile, 1987). Slower rates are associated with vertical strata that strike obliquely or parallel to the cliff face. However, in some cases, mechanical strength may be secondary to chemical processes such as hydration. For example, although some mudstones are mechanically strong, they are susceptible to flaking due to hydration and dehydration (Suzuki *et al.*, 1970).

Some measurements have been made of the rate of lowering of coastal Chalk platforms using micro-erosion meters. However, since most platforms are stepped, erosion also occurs through the detachment of blocks from the front edges of steps and this also needs measurement if the real average rate of lowering is to be calculated. Observations near Brighton, Sussex (Robinson and Jerwood, 1987a,b) showed that frost contributes to the erosion of coastal platforms in Chalk, the combined effect of salt crystallization and frost-induced spalling declining towards low-water mark as the time for freezing within each tidal cycle is reduced.

On the cliffs in the Jurassic strata of North Yorkshire, Robinson (1977c) described four types of shore platform:

1. a sub-horizontal plane downwasting at 1–2 mm a^{-1};
2. a similar plane, with a ramp (slope >2.5°) at the foot of the cliff;
3. a ramp with beach and no plane to seaward;
4. complex forms dominated by geological structure, with no clear development of either ramp or plane, i.e. a rocky shore.

Platforms here ranged from 90 to 200 m in width; sandy beaches were only found landwards of wide platforms. He discovered older ramps above the present platform represent former cliff-foot positions, the ramp and platform both extending landwards over time by parallel retreat – i.e. preserving their angle of slope.

That platform width might represent the balance between the rates of basal recession and recession of the seaward margin below low tide level complicates the interpretation of measurements of platform width. Many authors assume that the current width represents the total retreat of the cliff since the completion of the Holocene marine transgression and others argue that platform inheritance from earlier interglacials means that the current rate of cliff retreat is close to zero. Comparison of the south and north sides of the Bristol Channel highlights the issue. The widest platforms occur on the southern side where energy levels are low. On the southern side, energy levels may be sufficient for basal erosion of the cliff-platform junction, but insufficient to erode the seaward edge of the platform at the same rate, so that the platforms have become very wide. In contrast, narrow platforms on the northern side (e.g. at Nash Point, Glamorgan) might be the result of effective erosion both at the outer edge of the platform as well as at the cliff base. Even over such short distances comparisons are problematic since it is possible that the Nash Point shore platform has evolved from a series of higher platforms and is thus partly inherited (Trenhaile, 1972, 1974a,b).

EVIDENCE OF INHERITANCE

Many cliff forms may also owe much to changes in sea level and climate in the geologically recent past and it is this diversity that creates some of the most spectacular and geomorphologically important cliffs in Britain. For example, some cliff profiles are 'compound', consisting of two or more major slope elements, whereas 'multi-storied' cliffs have two or more steep faces separated by more gentle slopes (Figure 2.12). Bevelled or 'slope-over-wall' cliffs are characterized by a convex upper slope above a steep lower face (Davies, 1980; Griggs and Trenhaile, 1994). Given the common occurrence of equifinality in geomorphology it is unlikely that a single explanation can account for all the British cases, let alone others elsewhere in the world. Nevertheless, many such cliffs have developed in resistant rocks over long periods and may be unrelated to the bevelling produced where a balance exists between marine and sub-aerial processes in the Emery and Kuhn (1982) model. Cotton (1951) suggested that, during the Quaternary Period, variations in sea level resulted in cliffs developed during high interglacial sea levels being abandoned during sea-level fall and then buried by ice or subject to paraglacial processes. This produced convex slopes above and an accumulation of talus below. When sea level rose again to interglacial highs, marine processes removed the talus to produce steep, lower cliff faces. As a result, there is a good match between the occurrence of bevelled cliffs and ice limits in the British Isles (Griggs and Trenhaile, 1994; Figure 2.13). Griggs and Trenhaile (1994) suggest that bevelled cliffs resulted where the talus reached the cliff top during the last glacial but that multi-storied cliffs resulted where the talus reached only partly up the cliff face (Figure 2.12).

However, our knowledge of past sea levels is incomplete and for British sites that were unaffected by isostatic recovery following the Last Glacial Maximum (c. 18 000 years BP) (or in Scotland, the Loch Lomond Stadial (11 000 years BP)), sea level during the last interglacial was probably between 3 and 6 m higher than present sea level. Earlier interglacial sea levels were closer to present sea level.

The assumption is that the normal situation is where sea level rose over the Holocene Epoch to reach the present coastline about 6500 years BP, with the shore platform of the last interglacial some 3–6 m above present mean sea level. However, there are very few cliffs where the sedimentary record (and completed research) allows detailed confirmation of this situation: sites with a well-established sea-level record tend to be low coasts. Present-day sea level

Evidence of inheritance

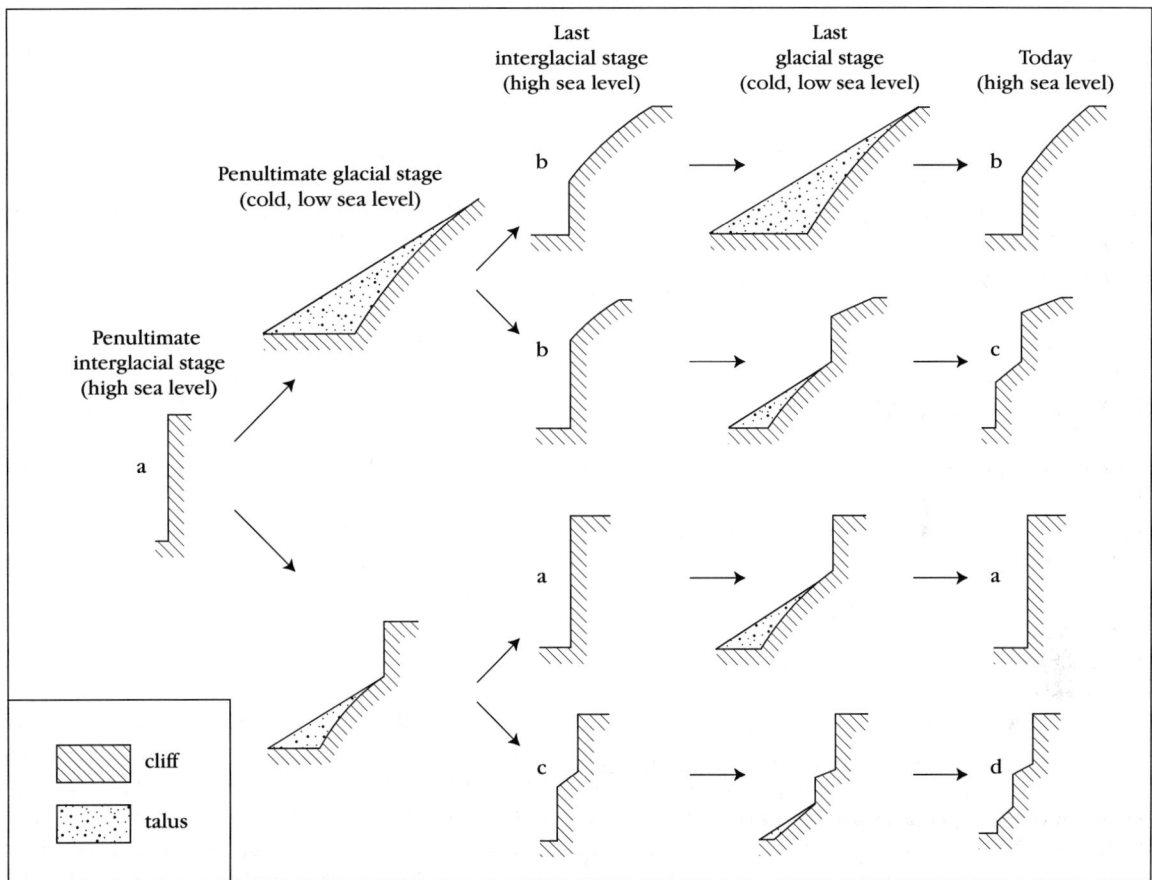

Figure 2.12 Flow-diagram model of coastal cliff development over two glacial/interglacial cycles, starting from a vertical, unbevelled cliff profile, (a). During low sea levels, periglacial activity results in talus accumulation at the bases of cliffs. During high sea levels, the talus is removed and the cliff trimmed and stepped, and bevelled profiles (b) develop where the talus reached the cliff top during the last glacial stage, whereas multi-storied profiles (c) develop where the talus extended only part of the way up the cliff face. Both (b) and (c) cliff forms can be affected by a subsequent interglacial-glacial cycle, leading to the numerous possible complex stepped profiles (d) that depend on the resultant level of talus development between cycles. (After Griggs and Trenhaile, 1994.)

around the British coast is rising at 1.5–2 mm a^{-1}, and although this has probably only been the case over the last 100 years or so, this has likely led to accelerated marine erosion of shore platforms and cliffs. Shennan (1989) draws isobases showing much of the Scottish mainland still rising slowl (see Figure 1.8b). However, over much of the coast this is currently less than the annual rate of sea level rise (Dawson *et al.*, 2001), so apart from the heads of Scottish inlets and estuaries, relative sea level in Britain is now either currently stable or rising everywhere.

Departures from the assumed normal pattern described above occur where land movements have been sufficient to affect the height and/or timing of Holocene sea levels. These movements are mainly glacio-isostatic and neotectonic movements of structural origin.

Glacio-isostasy delayed the arrival of the sea at its present elevation, though in many cases this has meant that the Holocene maximum was locally much higher than present-day level, achieved as a result of continuing slow uplift as the land has recovered from its ice load. For example, at Culbin on the Moray Firth coast of Scotland, the Holocene maximum reached 8 m OD and the cliff cut in glacial deposits is now elevated. Holocene emergence over much of the Scottish coast has limited the time available to create shore platforms adjusted to present-day sea level, especially in comparison with

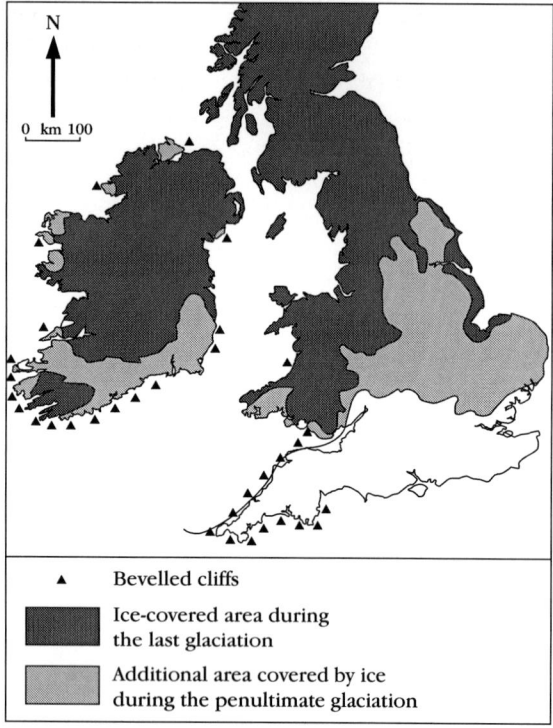

Figure 2.13 The relationship between the occurrence of bevelled cliffs and ice limits in the British Isles. (After Griggs and Trenhaile, 1994.)

more stable areas in England and Wales. On the other hand, the western and northern isles of Scotland, including St Kilda, did not suffer such late glacial unloading (or as large an ice load) and may have been assisted by relative sinking of the land (Flinn, 1969). As a result they have had considerably longer time periods available for shore platform formation. Yet in spite of this, shore platforms are relatively rare even in these locations and too little is known to say why.

Sea level is also influenced by neotectonic movements of the land, a relatively neglected area of study from the geomorphological perspective in Britain. Embleton (1993) concluded *inter alia* that Tertiary tectonism has continued into the Quaternary Period and is manifested today as seismic activity with maximum rates of present neotectonic movement being about 0.5 mm a^{-1}. Clayton (1998) came to very similar conclusions, though adding to tectonic causes the effects of uplift as a result of denudational unloading through erosion of the land.

Clayton and Shamoon (1999) propose that parts of the coast may have moved independently over the last 5 Ma or so. Whether they are still moving independently today, and whether the rate of movement has been enough to affect coastal landforms, remains to be established. The maximum rate of 0.5 mm a^{-1} postulated by Embleton (1993) implies a maximum movement since the end of the Holocene transgression of some 3 m, so the effect is small. Nevertheless, the influence of neotectonic movements on the form and elevation of individual cliffs and shore platforms, and on the hinterland processes such as fluvial incision, should be acknowledged even if it is difficult to quantify at present.

One emerging route to establishing the age and influence of neotectonism and coastal emergence is the dating of fluvial incision into hard rocks by means of cosmogenic isotope analysis. This may allow dating of erosional surfaces such as shore platforms, to establish whether they are old or young. However, old platforms that have been retrimmed will yield young ages.

From the above, it follows that there are two schools of thought relating to cliff and shore platform development and age. The first is that most of our hard-rock cliffs and platforms are contemporary and the result of erosion since Holocene sea levels stabilized some 6500 years ago. The counter-argument is that till-draped cliffs and stacks, as well as emerged caves and beaches that date to earlier interglacials, indicate that many cliffs and associated forms pre-date present sea level even though they are now trimmed by it.

Many cliffs and shore platforms in Britain, particularly those developed in softer and weaker rocks (such as the Holderness coast of Eastern England), are certainly Holocene in age and are well-adjusted to present day processes. Nevertheless, since several interglacial sea levels have occurred at similar altitudes to present, we must also acknowledge that many cliffs and shore platforms that now occur at present-day sea level may be inherited or partly inherited erosional surfaces that have been exhumed (Trenhaile, 1987). Certainly, many cliffs and shore platforms in the British Isles are demonstrably older than the last glaciation since they are draped with glacial till only partly removed by present-day wave activity (for example at Tarbat Ness and along the coast of Islay and Jura). Depending on the date and rate of elevation, emerged platforms may merge imperceptibly into the present-day shore platforms (e.g. on Islay, and at Dunbar).

Since Scottish coasts are affected by isostatic readjustments of various amounts away from the former ice depression centres, it is not surprising that inherited shore platforms and cliffs adorned with stacks and caves are variously elevated, submerged or consonant with the altitudes of the present platforms. Plunging cliffs are another aspect of the legacy of inheritance. Waves reaching such cliffs tend to be unbroken and so most wave energy is reflected, little erosion takes place and erosion rates may be correspondingly low. For example, the cliffs of Shetland and Orkney are thought to be largely inherited features (Flinn, 1974; Hansom and Evans, 1995). Although the cliff faces are composed of freshly exposed rocks owing to a combination of storm waves affecting the cliff-top (Hansom, 2001) and the current rise in sea level, the macro-profile and setting shows these cliffs to plunge into deep water. Thus there exists either no erosional step at present-day sea level or a very limited one, in spite of some 6500 years of erosion in often very exposed and high-energy positions.

The key point of disagreement may be identified using the apparent considerable retreat of the Orkney sandstone cliffs. Rapid retreat does and can occur; the Old Man of Hoy can be fairly convincingly dated as 400 years old (Hansom and Evans, 1995), yet the Orkney cliffs, like those of St Kilda and Shetland, are almost certainly inherited from previous sea levels and are now retrimmed to look new. Similarly, on the Gower limestones in Wales, steep cliffs are presently trimmed by waves at the headlands (e.g. Rhossili, Oxwich and Hunts Bay) yet the same cliffline can be traced into bays where they are relict and buried by either glacial till, periglacial deposits or Holocene beach deposits. In such situations whereas the cliffline is inherited, it includes some cliff forms that may be young

Nowhere does inheritance produce a more distinctive coastal geomorphology than in those locations where relative sea-level rise has submerged, but not necessarily modified, land surfaces initially shaped by glacial, fluvial or subaerial terrestrial processes. For example, Holocene submergence of the network of glacial troughs on the western seaboard of Scotland, has produced a spectacular fjordic coast that extends from Cape Wrath to the Firth of Clyde. The areally scoured glaciated erosion surfaces in north-west Sutherland and in the Loch Maddy area of North Uist, are now submerged to produce a bewildering, though distinctive, complex of aligned and moulded coastal forms and islets, skerries and streamlined ridges set within a range of sheltered and exposed coastal environments.

Given the conflicting evidence concerning the effect of interglacial inheritance versus Holocene development of cliff and platform in Great Britain, it seems clear that only through measuring the past and present rate of change on shore platforms and cliffs will clarification be achieved. Indeed only when this work has been done at a large number of locations, on different rocks, with varying wave energies and tidal ranges, can a satisfactory and convincing explanatory account of British cliffs and their shore platforms be assembled. For the present time, the site descriptions and interpretations assembled here provide some of the context for that work.

SEA CLIFFS AS BIOLOGICAL SSSIs AND SPECIAL AREAS OF CONSERVATION (SACs)

N.V. Ellis and C.R. McLeod

In Chapter 1, it was emphasized that the SSSI site series is constructed both from areas nationally important for wildlife, and GCR sites. An SSSI may be established solely for its geology/geomorphology, or its wildlife/habitat, or it may comprise a 'mosaic' of biological and GCR sites that may be adjacent, partly overlap, or be co-incident. There exist a number of sea cliffs that are of crucial importance to the natural heritage of Britain on account of their wildlife value but, implicitly, will have some features of regional importance for coastal geomorphology. Such sites are not included independently in the present geomorphologically focused volume because of the 'minimum number' and 'national importance' criteria of the GCR rationale (see Chapter 1).

Many sites that have been selected for the GCR for features other than coastal geomorphology are to be found on the coast of Britain. For example, where coastal cliffs provide sections through the successions of strata that are important for the geological research, the best, most representative sections have been chosen for stratigraphical GCR selection categories.

Coastal cliff geomorphology

Table 2.3 Candidate and possible Special Areas of Conservation in Great Britain supporting Habitats Directive Annex I habitat 'Vegetated sea cliffs of the Atlantic and Baltic coasts' and/or 'Submerged or partially submerged sea caves' as qualifying European features. Non-significant occurrences of these habitats on SACs selected for other features are not included. (Source: JNCC International Designations Database, November 2002.)

SAC name	Local authority	Cliff habitat extent (ha)
Ardmeanach	Argyll and Bute	125.9
Beast Cliff–Whitby (Robin Hood's Bay)	North Yorkshire	156.1
Berwickshire and North Northumberland Coast	Northumberland; Scottish Borders	†
Buchan Ness to Collieston	Aberdeenshire	62.2
Cape Wrath	Highland	299.6
Cardigan Bay/ Bae Ceredigion	Ceredigion; Penfro/ Pembrokeshire	†
Clogwyni Pen Llŷn/ Seacliffs of Lleyn	Gwynedd	65
Dee Estuary/ Aber Dyfrdwy*	Cheshire; Fflint/ Flintshire; Wirral	1
Durham Coast	Durham	120.4
East Caithness Cliffs	Highland	310
Exmoor Heaths	Devon; Somerset	85.6
Fair Isle	Shetland Islands	129
Flamborough Head	East Riding of Yorkshire; North Yorkshire	315.6
Glac na Criche	Argyll and Bute	50
Glannau Ynys Gybi/ Holy Island Coast	Ynys Môn/ Isle of Anglesey	111.1
Great Orme's Head/ Pen y Gogarth	Conwy	13.9
Hastings Cliffs	East Sussex	55.1
Hoy	Orkney Islands	94.9
Isle of Portland to Studland Cliffs	Dorset	579
Isle of Wight Downs	Isle of Wight	18.4
Limestone Coast of South West Wales/ Arfordir Calchfaen de Orllewin Cymru	Abertawe/ Swansea; Penfro/ Pembrokeshire	349.5
Lundy	Devon	†
Mousa	Shetland Islands	†
Mull of Galloway	Dumfries and Galloway	137.6
North Rona	Western Isles/ Na h-Eileanan an Iar	31.4
Overstrand Cliffs	Norfolk	28
Papa Stour	Shetland Islands	†
Pembrokeshire Marine/ Sir Benfro Forol	Penfro/ Pembrokeshire	†
Pen Llŷn a'r Sarnau/ Lleyn Peninsula and the Sarnau	Ceredigion; Gwynedd; Powys	†
Polruan to Polperro	Cornwall	192
Rigg–Bile	Highland	450.8
Rum	Highland	216.7
Sidmouth to West Bay	Devon; Dorset	807.5
South Devon Shore Dock	Devon	238.7
South Hams	Devon; Torbay	3.8
South Wight Maritime	Isle of Wight	198.6
St Abb's Head to Fast Castle	Scottish Borders	122.4
St Albans Head to Durlston Head	Dorset	28.7
St David's/ Tŷ Ddewi	Penfro/ Pembrokeshire	303.9
St Kilda	Western Isles / Na h-Eileanan an Iar	738.8
Strathy Point	Highland	169.3
Stromness Heaths and Coast	Orkney Islands	63.5
Thanet Coast	Kent	†
The Lizard	Cornwall	149.8
Tintagel–Marsland–Clovelly Coast	Cornwall; Devon	1457.9
Y Fenai a Bae Conwy/ Menai Strait and Conwy Bay	Conwy; Gwynedd; Ynys Môn/ Isle of Anglesey	†

* Possible SAC not yet submitted to EC.
† SAC proposed for sea caves; sea cliffs not a qualifying feature.
Bold type indicates a coastal geomorphological GCR interest within the site.

Cliffs in even the weakest rocks, such as at Alum Bay on the Isle of Wight and Barton Cliffs in Christchurch Bay where Tertiary rocks provide key exposures uncommon enough for these to be the type sites of the relevant Tertiary strata. The full list of selection categories can be found in Ellis *et al.* (1996). Ongoing coastal erosion at these sites is important in providing 'fresh' faces of rock to study. However, there are problems in balancing the need for maintaining a 'fresh' geological exposure with the fact that over time it may be eliminated by the very erosion that originally created the exposure in the cliff.

In addition to being protected through the SSSI system for their national importance, 'Vegetated sea cliffs of the Atlantic and Baltic coasts' are a Habitats Directive Annex I habitat, eligible for selection as Special Areas of Conservation (see Chapter 1). Two further Annex I habitats, 'Submerged or partially submerged sea caves' and 'Reefs', are important features of some cliff coasts. Furthermore, some sea cliffs are of international importance for breeding seabirds, and for this reason may be designated Special Protection Areas under the Birds Directive.

Sea cliff SAC site selection rationale

The UK supports a significant proportion of EU sea cliff vegetation (The National Vegetation Classification (Rodwell, 2000) describes 12 maritime cliff NVC types). In particular, the coast of England holds a major proportion of the European coastal Chalk exposures (113 km, compared with 85 km in France and shorter lengths in the Baltic area). All the selected SACs are extensive and have an exceptionally well-developed zonation of vegetation. They reflect the very wide ecological variation of the habitat type across the UK arising from the variability of cliff structure, geology and levels of maritime exposure. The SAC series includes rock types that range from unconsolidated drift or clay through soft shales, mudstones, limestones and Chalk, to acid igneous formations. In addition, both sheltered east-facing sites and exposed west-facing sites are included.

Chapter 3

Hard-rock cliffs – GCR site reports

Introduction

INTRODUCTION

J.D. Hansom

The selected hard-rock cliff GCR sites described in this chapter are formed in a wide range of rock types, from granites to sandstones, and, as described in Chapter 1, may be classified by overall rock resistance to denudation according to lithology and structure (see Figure 1.3). Strong lithological control and cliff development is seen where harder rocks form headlands and softer rocks form intervening bays, but rock structure and hinterland topography can be equally important. How far rocks at the coast depart from their overall mean position is a function of the importance of factors such as degree of rock jointing (which affects overall rock resistance to wave erosion) and/or the effectiveness of weathering processes, which may weaken the rock. Jointing and related structural controls are often involved in the development of headlands and bays, and over time lead to the isolation of headlands into islands, arches or stacks (see Figure 2.8). When stacks eventually collapse, their bases often survive for a time as reefs or skerries until these too merge into the developing coastal platform. In general, features such as arches and stacks are often found on actively eroding lengths of coast and on well-jointed rocks (see Chapter 2), but their absence is not necessarily an indication of slow coastal retreat. At a smaller scale, a wide range of features such as crevices, caves, clefts, and blowholes can form and even smaller-scale features such as tafoni and similar weathering forms also occur. Similarly, shore platforms have a range of features, from larger forms (ridges, scarps, runnels) related to structural controls to minor forms linked to abrasion or scouring (e.g. potholes and rock pools), those formed by weathering such as tafoni and solution basins on sandstones or limestones, and by bio-erosion, including the home scars and hollows of grazing molluscs.

The pattern of strata cropping out in cliffs affects their form (see Figure 2.7). In addition, the direction of the dip of rocks in relation to the coastline will affect cliff form. As occurs on inland slopes, steeper cliffs form where the strata are more of less horizontal; seaward dips result in a tendency to dip-slip rock failure and lower cliff angles. Unlike inland, these structural controls also affect the shore platform (the erosional stump left by a receding cliff), and study of the combined landforms over a wide range of rocks and structures should lead to a better understanding of the relationships involved. Intensely folded strata, as at Hartland Quay (north Devon) and on parts of the Pembrokeshire coast (see GCR site reports in the present chapter), can be eroded into complex forms that can be very different from the characteristic shore platforms and cliffs found nearby in the same strata without the folding. Some rock types develop characteristic cliff forms, such as the granite at the Bullers of Buchan or the sandstones of the Orkney Islands.

Just as local topography can fundamentally affect inland cliff form, the nature of the land adjacent to the coast affects coastal cliffs and rocky shores. Where coastal erosion has incised into former river valleys, a range of hanging valley features can occur, as at the Hartland Quay site in Devon. Submergence of previously glaciated landscapes creates a coastline of great diversity such as in the sea lochs of western Scotland or in the drowned, eroded glacial surfaces of Loch Maddy in the Western Isles.

The range of exposure to wave energy, the various geological strata involved, and the varying sea level history of these coasts, affected as they are by differential glacial unloading and different relative sea-level histories, form a good basis for future research work. Rates of cliff retreat and of platform lowering remain to be measured at almost all of these sites, though the existence of young features such as stacks imply that some coastal forms developed since the end of the last glaciation in spite of the host cliff coastline having been in existence for much longer and surviving several sea-level changes.

The conservation value of hard-rock cliff coasts

Unlike weaker-rock coasts, the pressure for coastal protection works is in general absent from hard-rock coasts and most sites will remain available for investigation without significant conservation activities. Nevertheless, it is also important to select a representative series of hard-rock cliff GCR sites to ensure that the Earth science conservation value, and geomorphological significance of such sites is recognized. Hard-cliff coasts are important to our understanding of the following processes:

Hard-rock cliffs

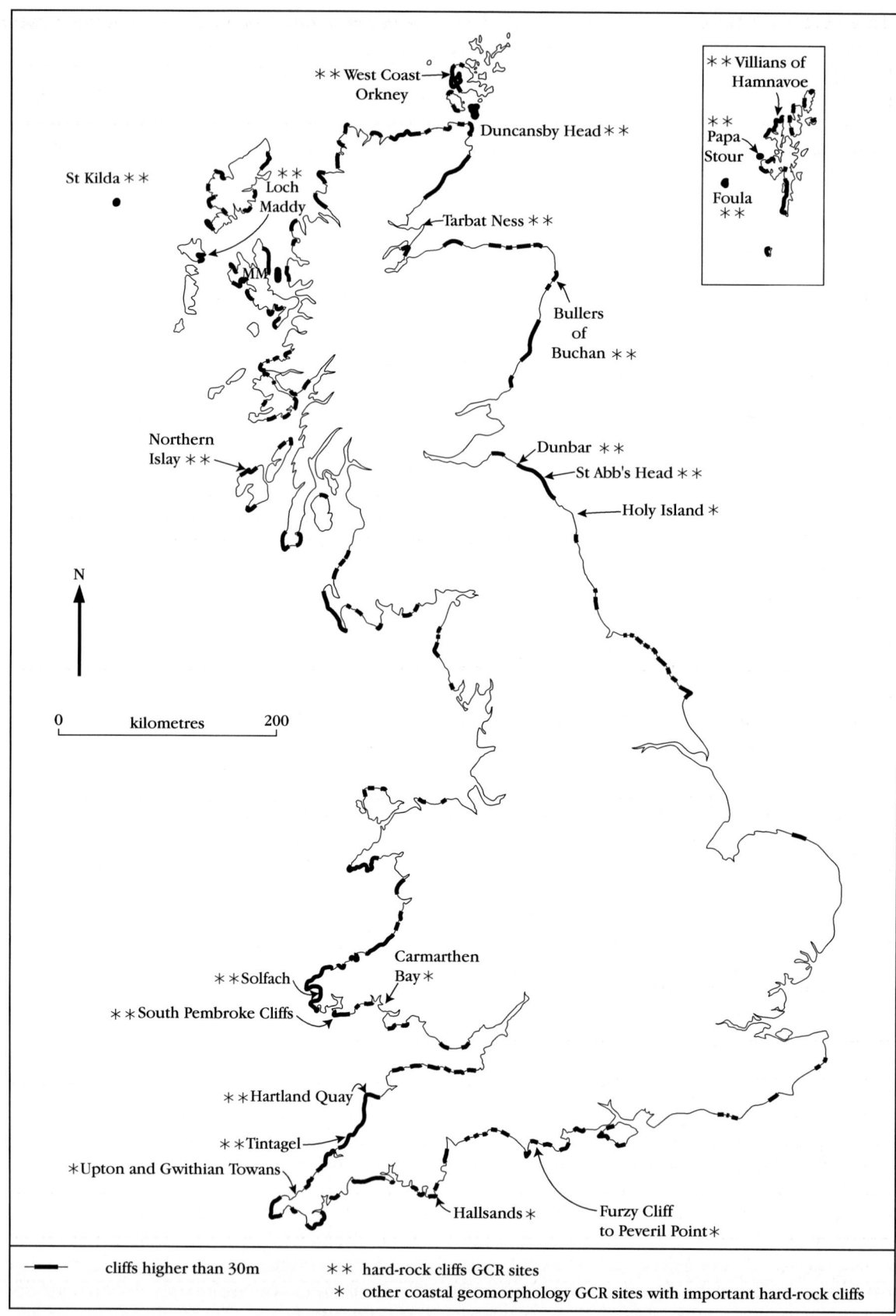

Figure 3.1 High-cliffed coast of Great Britain, showing the location of the sites selected for the GCR specifically for coastal geomorphology features of hard-rock cliffs. Other coastal geomorphology GCR sites that include hard-rock cliffs in the assemblage are also indicated.

Introduction

Table 3.1 Hard-rock cliff GCR sites, including those sites described in other chapters of the present volume that include hard-rock cliffs in the assemblage.

Site*	Main features	Main geological materials	Tidal range (m)
St Kilda Archipelago, Western Isles	Plunging cliffs, submerged caves and platforms; structural controls	Igneous complex of granophyres, basalts and dolerites	3.0
Villians of Hamnavoe, Shetland	Structural controls, wave stripping, cliff-top boulder beaches	Devonian extrusive andesites and ignimbrites	1.5
Papa Stour, Shetland	Diversity of cliff forms, caves, stacks, arches; inherited cliffs	Devonian extrusive rhyolite and ignimbrite	1.5
Foula, Shetland	Higher cliffs, shore platforms, geos; exhumed cliffs stacks and geos	Devonian sandstones and Dalradian metamorphic rocks	1.5
West Coast of Orkney	Structural control of steep overhanging cliffs; stacks arches; inherited cliffs; young individual features	Devonian Old Red Sandstone	3.0
Duncansby to Skirza Head, Caithness	Geos and stacks, shore platforms, blowhole	Devonian Old Red Sandstone	3.0
Tarbat Ness, Easter Ross	Weathering forms: tafoni and solution pits	Fault-controlled Devonian Old Red Sandstone	3.2
Loch Maddy–Sound of Harris coastline	Drowned surface of glacial erosion; rock basins, skerries and platform	Lewisian gneiss, faulted and crushed zones	3.5
Northern Islay, Argyll and Bute	Emerged shore platform and beach gravels	Precambrian quartzites and tillites; Dalradian Limestone	2.0
Bullers of Buchan, Aberdeenshire	Geos, caves, arches, stacks, platform, blowhole	Granite and dyke intrusions	3.5
Dunbar, East Lothian	Four shore platforms, some of which are glaciated	Devonian Old Red Sandstone, Carboniferous sandstone, igneous intrusions	4.5
St Abb's Head, Berwickshire	Steep cliffs, geos, fault-controlled inlets and headlands	Devonian extrusive felsites, tuffs, and grits; faulting	4.5
Tintagel, Cornwall	Longitudinal coast, structural control caves, arches, slope-over-wall cliff	Upper Devonian slates, siliceous sandstones, pillow lavas, tuffs and phyllites	6.5
South Pembroke cliffs	Structural controls, eroded karstic coast, stack, arch, cave, geo	Carboniferous limestones	6.0
Hartland Quay, Devon	Truncated valleys, waterfalls, slope-over-wall cliffs, shore platforms	Carboniferous interbedded fine-grained sandstones and shales	6.4
Solfach, Pembrokeshire	Ria, infilled ria	Cambrian and Ordovician flags and dolerites	5.9
Carmarthen Bay, Carmarthenshire	Ria, shore platforms	Old Red Sandstone and Carboniferous limestone	8.0
Furzy Cliff–Peveril Point, Dorset	Structural controls, longitudinal coast, slope-over-wall cliffs, truncated valleys	Portlandian and Purbeckian limestones and sandstones	1.9
Holy Island, Northumberland	Structural controls, shore platforms	Carboniferous sandstones and limestones	4.1
Upton and Gwithian Towans, Cornwall	Exhumed cliffs and stacks	Devonian slates	5.8
Hallsands, Devon	Emerged shore platform	Mica-schist and quartz-schist	4.4

*Sites described in the present chapter are in **bold** typeface

1. the processes of retreat in cliffs that are cut into rocks of varying resistance resulting from lithological and structural differences;
2. the effect of inheritance of cliffs and shore platforms from former sea levels
3. the controls on presence/absence of shore platforms and plunging cliffs
4. the detailed processes of wave quarrying, abrasion and weathering of rock surfaces at the coast
5. the processes of supply and transport of sediments from cliffs to beaches both below the cliffs and alongshore.

In the present chapter, the sites described represent a wide range of exposure to wave energy, a range of geological controls (structure and rock type) and varying sea-level histories. The order of the reports broadly reflects a reduction of wave energy and rock resistance, beginnign in the north and west, moving southwards into softer lithologies and lower wave energies

ST KILDA, WESTERN ISLES (NA 100 000)

J.D. Hansom

Introduction

The islands of the St Kilda archipelago rise out of the Atlantic Ocean 66 km WNW of Griminish Point on North Uist (see Figure 3.1 for general location). The four main islands of Hirta, Dùn, Soay and Boreray, together with their adjacent stacks of Stac an Armin, Stac Lee and Levenish have a coastline that is *c*. 35 km long in total. The coast is rugged and almost entirely cliff-bound, with Conachair (430 m) on the main island of Hirta forming the highest British sea cliff, and Stac an Armin (191 m) the highest stack. A variety of spectacular cliffs and cliff-related forms, including geos, arches, stacks, caves, blowholes and vertical cliffs characterize the coastline of the archipelago. Partly because of the dramatic sea cliffs, the entire archipelago was selected as Scotland's first natural World Heritage Site in 1987.

St Kilda is all that remains above sea level of a large ring volcano, thought to have been active about 60 million years BP (Steers, 1973; Miller, 1979; NCC, 1987). Although few fragments remain of the St Kilda volcanic complex, its geological history has been pieced together by comparing the precipitous remnants with the more complete and accessible igneous complexes of Mull and Ardnamurchan (Cockburn, 1935) and absolute dating based on the radioactive decay of minerals (Miller and Mohr, 1965). St Kilda is mainly composed of Tertiary intrusive igneous rocks, with coarse-grained gabbro in the southwest of Hirta and Dùn. Extensive exposures of igneous breccia intruded by dolerite occur on the north coast of Hirta, on Soay, Boreray and on the two large stacks (Figure 3.2). Granitic intrusions occur in the cliffs of Conachair and Oiseval. A late intrusive phase of sheets and dykes of dolerite and felsite has indirectly influenced the formation of arches, caves and ledges in the cliffs.

Unsheltered coasts in the western isles are exposed to high mean wind speeds and for 75% of the time, wind speeds exceed 4 m s^{-1} mainly from between west and south (BGS, 1977a) and so St Kilda experiences extreme weather conditions with high wave-energy levels that probably exceed any other site in the British Isles. The predicted 50-year wave height of 35 m for this area is significantly higher than for other parts of the UK (BGS, 1977a). Significant wave heights exceed 5 m and 1 m for 10% and 75% of the year respectively (UKDMAP, 1998) and although a maximum significant wave of 9 m was recorded 15 km west of South Uist (BGS, 1977a), individual storm waves regularly exceed this. For example, to the west of Shetland storm waves reached 15.13 m at 60°N, 4°W in January 1974 (Marex, 1975), and 24.4 m at 59°N, 19°W (Draper and Squire, 1967). Since the seabed around St Kilda lies at about 120 m depth and the islands rise from a plateau at a depth of about 40 m, the coastline is very exposed to large, high-energy Atlantic waves and this has resulted in some of the most dramatic and spectacular coastal geomorphology in the British Isles. In keeping with its oceanic location, spring tidal range at St Kilda is limited to 3 m and maximum tidal streams on spring tides are low at 0.25 m s^{-1} (BGS, 1977a).

The archipelago's present isolation dates from the end of the Devensian glaciation when sea levels rose to flood the intervening land surface. As a result of this isolation, little research has been carried out on forms and processes operating on St Kilda. To date, only descriptive works of the coastal geomorphology of St Kilda are available (e.g. Mathieson, 1928; Steers, 1973; Miller, 1979), although more exists concerning the Quaternary history of the islands (e.g.

St Kilda

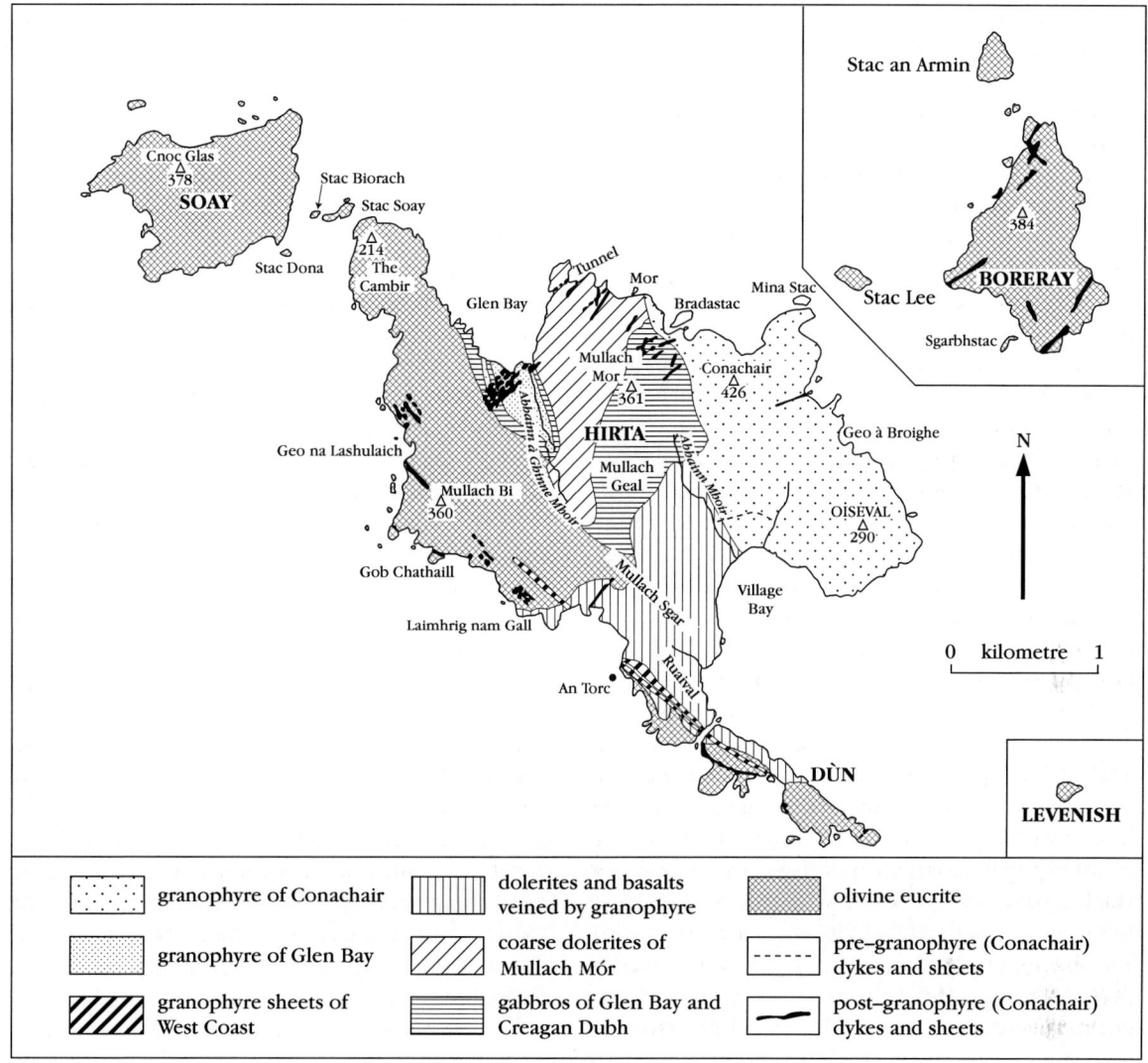

Figure 3.2 Geological sketch map of the St Kilda archipelago showing the dominantly volcanic nature of the bedrock geology and the controlling effect of the granophyre sheets of the west in producing an approximately linear coastline. For relative geographical positions of the component islands of the archipelago, see Figures 3.1 and 3.4. (After Nature Conservancy Council, 1987.)

Sutherland, 1984; Sutherland *et al.*, 1984; Peacock *et al.*, 1992).

Description

The complex and varied coastal outline of the St Kilda archipelago (Figure 3.2) is composed of a variety of cliffs, caves and geos, arches, tunnels, stacks and blowholes; since it would be over-exhaustive to describe the entire 35 km coastline in detail, the following account focuses on those features that are especially significant in the selection of this site for the GCR. Geos are deep inlets of the sea, often narrow with steep precipitous sides and often excavated along the line of a structural fault or less-resistant lithology. 'Geo' is derived from the Norse *gja*, meaning a crack or cleft, and so the name mainly occurs in those areas where Norse influence was most marked in the past, such as in the Northern and Western Isles and in the north of Scotland.

The southern coast of Hirta (from Village Bay to Caolas an Dun), together with the north coast of Dùn, comprises low cliffs that are characterized by a profile that is relatively unusual in the British Isles. The steeper upper section of the cliff is cut in unstratified Quaternary till, whereas the lower-angled slope beneath the drift is cut in solid rock. The angles of the lower slopes match the dip angles of the underlying sheets and shelves of dolerite and microgranite. These lower slopes have been stripped of their till cover by wave action, yet they continue to afford some protection to the upper slopes from waves. Nevertheless, the upper slopes remain unstable and produce high-angled failures and an unusual profile. The overall regularity of this stretch of coastline is broken by several small geos and caverns that are related to wave-quarrying of small-scale faults and intrusions.

The relative homogeneity of the cliff forms on the north coast of Dùn is in stark contrast to the south coast, where the vertical lower face continues below the water level and plunges directly into deep water. The characteristic profile of the cliffs is a smooth and vertical face at the base, often with an overhang. This is surmounted beyond the reach of wave impact by a 70° face that is very rugged and broken (Figure 3.3a). The detail of the coastline is extremely irregular and the only recognizable trend is related to wave-exploitation of the three major NE–SW fault lines that cross Dùn. The detail of the plunging cliff at sea level consists of a variety of stacks, caverns, arches, blowholes and narrow inlets at all stages of formation. Shore platforms are absent. Several deep geos pass into caverns that penetrate through the narrow island, an impressive example being the natural arch near Gob an Dùin, which is almost 50 m long and 24 m high. These caves and geos have generally been eroded along the line of dykes or thin, inclined, sheet intrusions where they crop out at sea level.

The south-west coast of Hirta (from Ruaival to The Cambir) displays local variants of the forms found on the south coast of Dùn, according to geological differentiation and differences in altitude. On the southern stretch of this coastline, the Mullach Sgar complex crops out and the many dykes and sheets of microdolerite and granite result in irregular notched and stepped cliff profiles, with remarkable colour contrasts. Farther north, towards The Cambir, the cliffs reach higher altitudes (up to 350 m at Mullach Bi) and are generally more uniform, although at water level the diversity of wave erosional forms resemble those of the south coast of Dùn. Free faces have developed in the high cliffs with massive scree deposits or block fields beneath, close to the angle of repose. Below these screes, the lower cliffs are vertical. Along most of this coast close to sea level, wave impact has excavated smoothed and shallow concavities into the gabbro and everywhere large columnar stacks, eroded stacks and arches are prominent. At the narrow neck of The Cambir, glacial till, consisting of large and angular boulders set within a sand and clay matrix, is being actively eroded by waves on both sides, together with wind-stripping of the surface soil and turf cover. In 1970, the neck measured 60 m from west scar to east scar but, by 1998, it had narrowed to 45 m, a rate of retreat of 0.26 m a^{-1} on each side.

The north-east coast of Hirta contains the highest and most impressive cliffs in the archipelago. The coarse dolerites cropping out between Gob na h-Airde and Mullach Mór, result in an irregular and stepped cliff profile, with many small protuberances. At the western end of this cliff, the natural arch at Gob na h-Airde is more than 91 m long and over 30 m high. The rectangular shape of the geo leading to this arch reflects the strong lithological control on this stretch of coastline. Farther east, the cliffs of Conachair and Oiseval are cut in granite and granophyre, intruded by dolerite sheets and dykes, creating a characteristic profile with short, near-vertical, free faces interspersed with longer, high-angled, grassed slopes and occasional screes. In places, horizontal sheets of dolerite have created pronounced ledges in the cliffs and preferential erosion of dykes has left several high altitude pinnacles. The continuous and very steep cliff of Conachair, the highest sea cliff in Britain, reaches a height of 430 m. This entire north-east coast is characterized by numerous stacks, caves, and overhangs with small horizontal cliff ledges at a variety of altitudes that appear to be related mainly to structure. Stripping of soil and turf by high winds occurs at a variety of altitudes east of Gob an h-Airde and reaches 400 m OD at Conachair.

The island of Soay to the north-west of Hirta is best defined as a complex and large stack (Steers, 1973) with no level ground. The coastline of Soay displays a similar sequence of features to the adjacent coast of Hirta, with the cliffs of north Soay reaching heights of 305 m. The arch in Soay Stac, together with the dog-tooth profile of Stac Biorach, in the narrow sound between Hirta and Soay, represent successive phases of coastal erosion. The island of Boreray, north-east of Hirta, has also been described by

St Kilda

Figure 3.3a The west coast of Hirta, the main island of St Kilda, looking towards Dùn, is characterized by stepped cliffs that steepen downwards to plunge steeply to well below sea level. (Photo J.D. Hansom.)

Figure 3.3b Stac an Armin, seen here with the vertical plunging cliffs of Boreray (the second-largest island of St Kilda) in the foreground. This is the highest sea stack in Great Britain. (Photo J.D. Hansom.)

St Kilda

Steers (1973) as a large stack, and the steeply cliffed coastline of the island is again honeycombed with caves and geos. The adjacent stacks of Stac an Armin and Stac Lee, tower 191 m and 166 m above sea level and are characterized by vertical faces plunging into deep water. The former is the highest stack in the British Isles (Figure 3.3b; Steers, 1973).

If the St Kildan landforms above sea level are impressive, then submerged beneath the waves is an equally spectacular suite of coastal landforms. Figure 3.4 reveals two relatively level seafloor platforms, the largest of which lies to the west as a virtually horizontal bedrock surface at –120 to –125 m depth with only a patchy sediment cover (Sutherland, 1984). This platform is

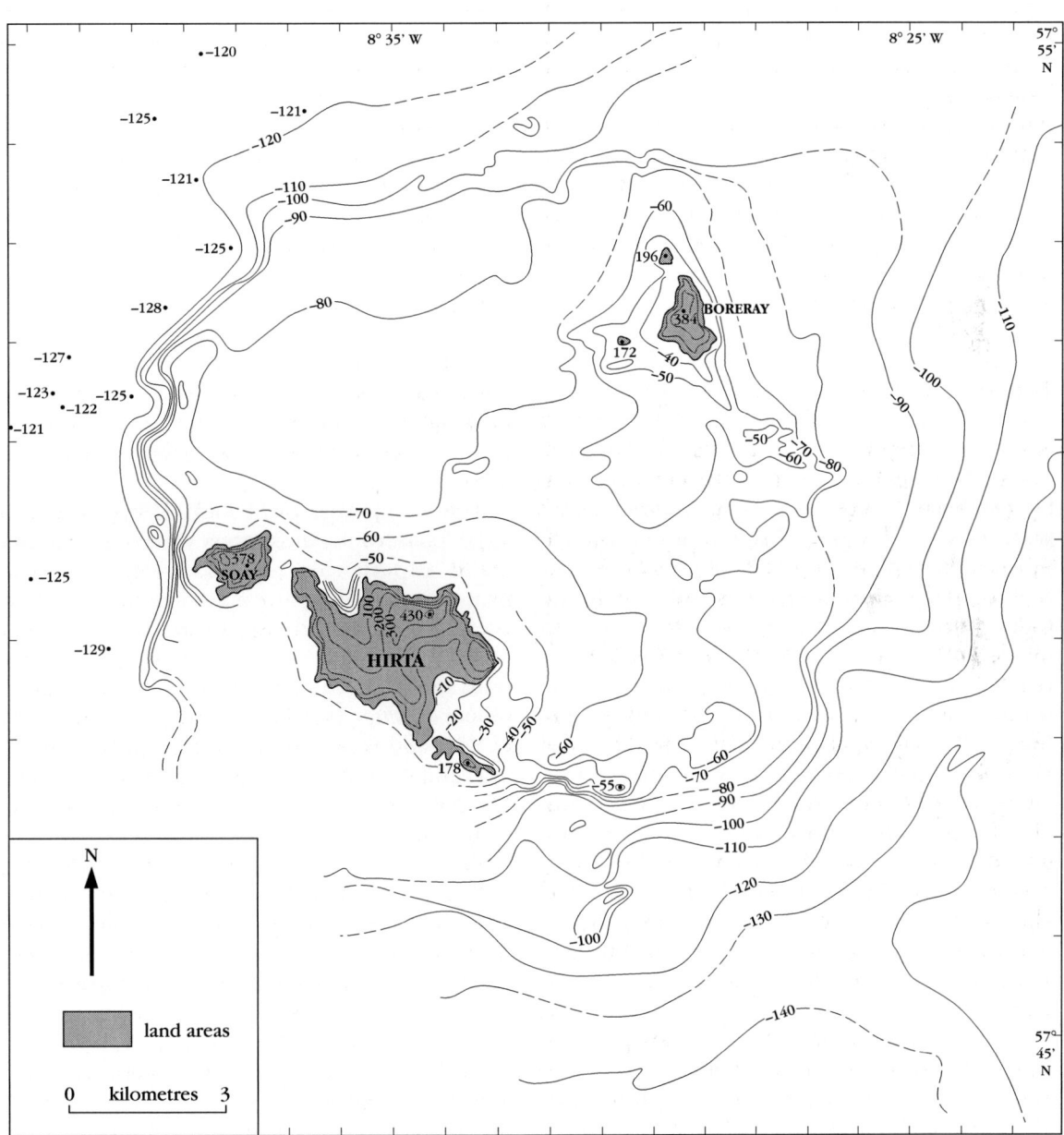

Figure 3.4 Bathymetric map of the St Kilda archipelago. Depths are given in metres. Note the prominent break of slope where the roughly circular igneous complex stands proud of the otherwise low-gradient seabed. The seabed lies at c. –120 m and abruptly gives way to steep submarine cliffs that rise to a c. –60 m surface. In turn this surface abruptly gives way to submarine cliffs that may rise above sea level. Bathymetry is in metres below OD. (After Sutherland, 1984.)

backed by a 40 m-high submerged cliff in the west and a series of 20–30 m-high benches in the south-east that separate the –120 m surface from a low-angled surface, which rises very gently from about –80 m in the north-west to –60 m in the south-east, although there is a fairly ubiquitous step in the profile at –70 m on most sides. The islands and sea stacks of the archipelago that rise steeply above sea level directly from this surface do so from a common depth of –40 m and the two valleys on Hirta can be traced down to its level. As a result there appear to be two platforms, one at –120 m and separated by a prominent submerged cliff from the second, which lies between –80 and –40 m. The second platform is itself backed by a prominent cliff that rises from –40 m to a maximum height of 430 m at Conachair. In spite of the extensive occurrence of submerged platforms and cliffs, at present sea level only a few low skerries exist, and shore platforms are largely absent from St Kilda.

Interpretation

Since the eruption of the ring volcano 60 million years BP, the igneous complex has been affected by numerous processes: submergence; glaciation; marine erosion; and subaerial erosion, all of which have played a part in its physical breakdown. Active erosion of the islands continues today with: wave quarrying of notches, caves, inlets, and geos along much of the cliff foot; rockfalls and scree formation especially on the south-west coasts of Hirta and Soay; and vegetation and soil-stripping at high altitudes, for example, at the Cambir and on Conachair. Wave quarrying and abrasion are reducing the island of Dùn to a series of stacks (Steers, 1973) and the stack of Levenish, to the south-east of Dùn, may be a former extension of the island. Miller (1979) states 'no-one knows how long these remote islands will resist the elements, but disappear they will in the fullness of geological time'.

It is the relationship between lithology, geological structure and the sheer range of erosional and mass movement processes that creates the distinctive and dramatic coastal scenery of present day St Kilda. The volcanic formation of the St Kilda igneous complex, the subsequent cauldron subsidence and the various intrusive stages that followed (see Miller (1979) for a fuller discussion) imparted a unique and complex geological structure. The gabbro that forms much of the islands has been injected with numerous sheets and dykes of basalt, dolerite and granophyres, many of which were further shattered, faulted and altered during each intrusive stage. Erosion of the gabbros, distinguished by their very coarse grain and dark colour, is responsible for the most jagged cliffs and stacks of St Kilda (Miller, 1979), while the numerous sheets and dykes provide weaknesses for preferential erosion, creating the complex of caves and geos at the level of wave attack. The geomorphology of the high cliffs, which often have an irregular notched and stepped profile, reflects a strong lithological control, but could also be regarded as an example of 'multi-storied' cliff profiles. Elsewhere, these are thought to reflect repeated cycles of sea-level rise and fall and intervening periods of paraglacial modification of the slopes (Trenhaile, 1997). However, much of the paraglacial veneer of material that has survived on lesser slopes elsewhere in Scotland has been stripped off in St Kilda, and only erosional evidence survives. There may well be remnants of past higher sea levels in the cliff profiles, but the co-incidence with lithology is too marked to ignore.

The archipelago of St Kilda is exposed to a large fetch from almost every direction and as a result experiences high wave-energy levels, probably exceeding any other site in the British Isles. Large, high-energy, relatively unimpeded Atlantic swell and storm waves impact directly onto the exposed cliffs. During storms a great plume of water and spray overtops the high cliffs of Dùn and falls into Village Bay (Miller, 1979), indicating the nature and scale of wave attack. The effect is so persistent that halophytic swards with species more characteristic of saltmarsh environments, such as sea plantain *Plantago maritima* and sea milkwort *Glaux maritima*, dominate the cliff vegetation communities, clothing not only the exposed south-west-facing coast of Mullach Bi, but also the lee-slopes of the Ruival cliffs facing Village Bay (Walker, 1984). Subaerial erosion and mass-movement processes, such as frost action and soil creep, are active in St Kilda, largely a result of the combination of steep slopes, complex geology, thin drift cover and extreme weather conditions. The operation of these highly active erosional processes upon structure and lithology creates much of the site's scientific interest and coastal geomorphology.

However, in spite of the apparently high-energy St Kildan wave climate, the presence of extensive platforms at −120 m and −80 to −40 m, both backed by precipitous cliffs the upper one represented by islands and stacks abruptly rising from −40 m, represents something of a paradox. A highly erosional system might be expected to produce a marked platform at, or close to, present sea level on the more gentle slopes, yet the final 40 m of Holocene sea-level rise seems to have accomplished relatively little by way of platform erosion (Sutherland, 1984), in spite of being close to its present level for about 6500 years. The suggestion clearly is that the planation of the two bedrock surfaces and of the cliffs that back them was achieved during two earlier periods of lower and stable sea-level conditions at −120 m and −80 to −40 m. Sutherland (1984) suggests the −120 m surface and its cliff, which rises to −80 m, to represent a Late Devensian sea level (i.e. about 18 000 years BP) and the −80 to −40 m surface and its cliff behind to represent the Loch Lomond Stadial sea level (i.e. about 11 000 to 10 000 years BP). There is a measure of agreement for these age allocations from glacio-eustatic and glacio-isostatic modelling studies of the Hebrides area (Lambeck, 1992). The predicted sea levels for the St Kilda area at 18 000 years BP lie at −120 m and, although not specifically modelled by Lambeck (1992), a figure of −80 to −40 m appears to be realistic for the St Kilda area at 10 000 years BP. However, whereas it is likely that the −120 m platform was indeed modified as recently as 18 000 years BP, it may date from glacial periods earlier in the Quaternary Period when sea levels were at similar lows. As a result, the shallower surface may also have been initiated much earlier than the 11 000–10 000 years BP date suggested above.

Since there are no shore platforms at present sea level in St Kilda, the interpretation that relates the development of the submerged St Kildan platforms to 18 000 years BP and 11 000 years BP carries with it the implication that erosional conditions were more severe during these times than during the final 40 m of sea-level rise of the Holocene Epoch (particularly for the Loch Lomond Stadial, because of the need to cut a substantial platform in only 1000 years). However, it may be that the severe erosional conditions experienced in these cold periods were more the result of ice-related processes than of wave-related processes. During both cold periods, low sea-temperatures, floating ice, and intertidal frost-shattering would have resulted in very efficient shore-platform development, similar to that experienced today on high-latitude shores (Hansom and Kirk, 1989) and also thought to have caused rapid shore-platform erosion in the Inner Hebrides during the Loch Lomond Stadial (Dawson, 1984). Research in high latitudes indicates that although such ice-affected shore platforms develop rapidly, the conditions for frost-shattering (which require standing water and low-gradient intertidal surfaces) are not favoured by high wave-energy environments (which produce intertidal erosional ramps). As a result, the platforms produced by ice-affected processes are progressively destroyed when the wave climate becomes more severe (Hansom, 1983). It is thus possible that a reduced St Kildan wave climate during the two cold periods was conducive to extensive platform development, but that the increasing wave energy of the Holocene Atlantic Ocean was not. The lack of St Kildan platforms at present sea level indicates only that present conditions are not well-suited to shore-platform development; it does not mean that no erosion occurs.

Conclusions

The geomorphological interest of the dramatic cliff coastline of the St Kilda archipelago lies in the height, scale and diversity of the cliffs and stacks, together with the wide range of erosional features that have formed in this high wave-energy environment, including numerous geos, caves, natural arches, blowholes and stacks at all stages of formation. A strong lithological and structural control of the coastal landforms adds further interest.

However, the present lack of knowledge concerning the detailed evolution of the St Kildan coast enhances the conservation value of the site, and more work needs to be done to unravel the story behind the existence of a spectacular suite of submerged cliffs and associated platforms, juxtaposed beside the present cliffs and the absence of any substantial shore platforms.

There is, of course, an additional historical, cultural and biological context to the cliffs of St Kilda that demands mention. The St Kildans maintained a 'hunting and gathering' economy on the islands up until the early 20th century, scaling the high cliffs to obtain seabirds for food. Cleits (store huts) still remain at various levels on the steep cliffs, such as Carn Mór and Aird

Uachdarachd. However since the 1930s, this inhospitable, but geomorphologically spectacular, environment has been uninhabited and is managed by the National Trust for Scotland. The cliffs are also of great ornithological significance and St Kilda is recognized as one of the most important seabird sanctuaries in the North Atlantic Ocean. This assemblage of cliff geomorphology, seabirds and cultural and historical contexts set within a land-ownership and management environment of high standard led to the designation of St Kilda as Scotland's first natural World Heritage site.

VILLIANS OF HAMNAVOE, SHETLAND (HU 240 810–HU 242 840)

J.D. Hansom

Introduction

The Villians of Hamnavoe, north-west Shetland, consist of over 3 km of almost vertical cliffs that rise from a height of 12 m OD at Whal Wick in the south to 45 m OD in the north. The coast is fully exposed to the west and north-west and receives the full violence and power of Atlantic storm waves. As a result, a range of cliff erosion features have developed along this stretch of coastline. The cliffs are cut in Devonian extrusive rocks, largely andesitic tuffs overlain by andesite lavas. In places the lavas, which are less resistant than the underlying tuffs, have been eroded to form a second cliffline fronted by a wide terrace up to 20 m above sea level, and yet remains so affected by storm waves that large blocks and slabs have accumulated to form a storm beach at the junction of the tuffs and lavas. In addition, differential erosion of the lava and tuff beds has led to a stepped cliff profile and marine exploitation of numerous cracks and fissures has resulted in the erosion of several geos and a large blowhole. Other parts of the site are characterized by large areas of wave-

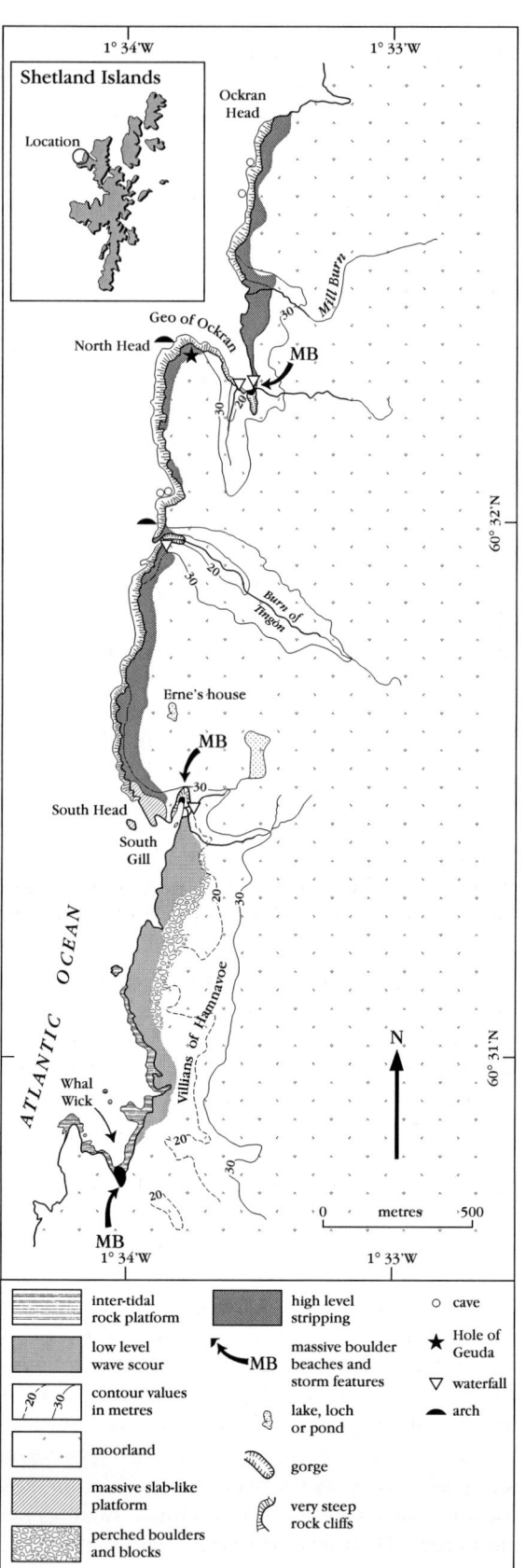

Figure 3.5 Geomorphological sketch map of the Villians of Hamnavoe showing extensive surfaces affected by both low-level wave-stripping in the south, and high-level wave-stripping in the north. For general location see Figure 3.1. (Modified from unpublished work by W. Ritchie.)

scoured bedrock that extend to altitude, together with individual wave-moved boulders, clusters of boulders and boulder beach ridges, all of which exist at altitudes well above those normally associated with wave processes (Figure 3.5).

The most characteristic feature of the Shetland climate is the frequency of strong winds. The mean wind speed is 6.5–7.5 m s^{-1} and gales occur on average for 58 days per year. Although not quite as exposed and windy as St Kilda, for 75% of the time the hourly mean wind speed exceeds 4 m s^{-1} with the most frequent strong winds from the south-west (BGS 1977a). Along the western coast of Shetland, a combination of exposure to prevailing winds and deep water close inshore produces a high-energy wave climate at the shore. For example, at the Villians of Hamnavoe, the sea floor falls steeply to depths of –50 m within 500–700 m of the shore.

The Holocene evolution of the Shetlands is dominated by submergence (Hoppe, 1965; Flinn, 1964, 1974; Mykura, 1976; Birnie, 1981), and numerous examples of intertidal and subtidal peats support this. As a result, the cliffs of Shetland are not characterized by features such as emerged ('raised') shore platforms or notches. The Villians of Hamnavoe was selected for the GCR because it vividly demonstrates the effects of high-energy storm-wave conditions on a low cliff coast. The high altitude abrasion and scour features, associated wave-shifted slabs and boulders and high-altitude contemporary storm beaches are of outstanding geomorphological significance. Even so, as with much of the hard-rock coast of Scotland, remarkably little geomorphological research has been carried out on these distinctive and outstanding examples of high-energy landforms.

Description

The near-vertical cliffs of the Villians of Hamnavoe, north-west Shetland, rise from a height of 12 m at Whal Wick in the south to 45 m in the north. South Gill (a boulder-filled geo)

Figure 3.6 The Villians of Hamnavoe looking north towards South Head. The scoured surface is littered with both eroded boulders and debris thrown up by waves. Since some of this debris is of modern human origin (plastic fishing floats etc.) the waves that sweep the surface and emplace the debris and boulders are likely to be recent. (Photo J.D. Hansom.)

marks the boundary between the higher (c. 20–45 m OD), steeper and sometimes overhanging cliffs, with a narrow basal intertidal platform to the north and the generally lower (c. 10–18 m OD) coastal rock platforms to the south that are adorned with wave-shifted slabs and boulders that comprise the contemporary storm beaches.

In the northern part, the Burn of Tingon cuts through the plateau in a ravine-like valley, which falls to sea level as a stepped waterfall, a dramatic illustration of the influence of geological structure on coastal erosion and fluvial forms. Additionally, along this section of cliff coast, there are numerous examples of caves, natural arches and a blowhole (the Hole of Geuda) whose roughly circular vent falls some 30 m from the cliff-top plateau. However, the most distinctive feature of these cliffs is the extent and height of contemporary cliff-top surface stripping. Stripping of turf and scouring of bedrock occurs up to 30 m inland of cliff edges that are themselves some 30 m OD. The exposed rock surfaces and the cliff edges are remarkably rough and irregular, reflecting variations in rock hardness and structure. In places on the cliff top, vesicles in the bedrock have been exploited by the high-level spray to produce micro-forms that seasonal frost action and solution have exploited.

At South Head (Figure 3.6), to the west of South Gill, an unusually wide, gently-sloping, smooth and slab-like platform, is backed by a vertical cliffline set back some 50 m from the coast. This 30 m-wide sloping terrace co-incides with the junction of lavas that overlie tuffs. Although the terrace is around 18 m OD, it is affected by storm waves, and large blocks and slabs of up to 2 m in diameter form a contemporary storm beach at the base of the cliffline. To the south of South Gill, the bare rock coastal edge is more ramp-like in appearance, reaching 20 m some 200 m inland, but is heavily scoured with surface vegetation having been stripped. However, the scoured rock surface is strewn with excellent examples of shifted rock fragments in the form of both individual boulders and imbricate clusters of boulders, as well as perched slabs at heights of up to 20 m OD and, where lower heights occur, up to 100 m inland. In all the above cases, the imbricate clusters demonstrate common orientations that are consistent with the general orientation of the host coastline, although minor local variations in boulder cluster orientation and dip reflect intricacies of the cliff top and cliff-edge gradient (Table 3.2). In several cases the boulder clusters incorporate modern human debris, such as fishing floats and timber spars wedged between, behind and underneath the boulders (Hansom *et al.*, in press).

To the south of the Villians of Hamnavoe, at Eshaness, Hansom (2001) has described the boulder deposits on the cliff tops. The most spectacular of these, at the Grind of the Navir, reach almost 20 m OD and are situated some 50 m inland at the rear of a 15 m OD subhorizontal rock platform, which is itself fronted by 15 m-high vertical cliffs. Three boulder ridges have been formed, the seawardmost of which reaches 3.5 m high and is composed of angular boulders of local ignimbrite that reach up to

Table 3.2. Altitude and orientation of some cliff-top boulder deposits in Shetland (after Hansom *et al.*, in press).

Location	Altitude (m)	Coastal orientation (degrees)	Mean orientation of boulder long axis (degrees)	Number of boulders	Mean long axis (m)
Virda Field, Papa Stour	35	5	300	15	0.7
South Head, Villians of Hamnavoe	25	0	315	25	1.1
Grind of the Navir 1 (beach ridge)	19	0	314	20	1.2
Grind of the Navir 2 (boulder clusters)	20	0	290	25	0.7
Esha Ness	35	20	275	15	1.0

2.1 m in length (Figure 3.7). Fresh scars of these dimensions occur in the cliff edge and on the sub-horizontal surface of the cliff top.

Interpretation

The Villians of Hamnavoe demonstrate the dramatic effects of high-energy storm waves on hard-rock cliffs, the relationship between geological structure and the high-energy process environment resulting in a distinctive and, in the British Isles, unique coastal geomorphology. The staircase cliff profiles, natural arches, caves and geos reflect a strong lithological and structural control, yet the resultant concentration of wave energy and power in geos and coastal valleys and the extensive wave run-up slopes on seawards-dipping rock platforms provide dramatic evidence of the extreme wave conditions experienced on these cliffs and shore platforms. The landforms that have developed in this high-energy wave environment are of great geomorphological significance and include: high altitude abrasion and scour features; contemporary storm beaches at the junction between the andesite tuffs and lavas; and the wave shifted slabs and boulders at high altitude.

In the north the stepped stairway inlet, together with the vertical blowhole of the Hole of Geuda, is a dramatic feature, particularly during storm conditions. The wave-quarried excavation of the inlet has proceeded inland along faults in the bedrock structure and intersected a fault or failure plane in the vertical dimension to result in the collapse of the cave roof to form the blowhole. To the south, the lower elevation of the cliffs, together with the low angle and exposure of the rock surface, facilitates wave run-up inland to remarkable distances and heights and this has resulted in wave-stripping of vegetation, together with wave-movement of boulders and gravels. Stripping by wind is also likely to play an important part in propagation of the stripping limit inland during storms. Turf edges at 40 m OD on South Head may mark the limit of wave wash of a major storm that occurred in

Figure 3.7 The largest of three wave-emplaced boulder ridges that occur on top of a 15 m-high cliff some 50 m inland of the cliff edge at the Grind of the Navir, to the south of the Villians of Hamnavoe. Note 1.8 m-high figure for scale. (Photo J.D. Hansom.)

1991/2, gravel thrown through the air during the same storm occurring up to 100 m inland of the upper limit of wave wash.

However, Hansom *et al.* (in press) show that the limits of scoured bedrock, and the clusters of imbricate boulders, closely follow the indentations of the cliff edge and are thus likely to be mainly related to extreme wave processes. Detailed wave-refraction modelling at The Grind of the Navir also indicates that deep-water offshore waves of between 15 and 20 m in height undergo enough nearshore refraction only to reach breaking at the cliff face itself and so they achieve maximum erosive power at this point. The implication is that 20 m-high, deep-water storm waves fairly regularly impact on the 15–20 m-high cliffs along this coast and are capable of constructing boulder beaches above 15 m OD and 50 m distance from the cliff edge (Hansom, 2001). Since the Holocene evolution of the Shetlands is dominated by submergence (Birnie, 1981), the boulder beaches are not emerged features. Other than the 7000 years BP tsunami produced by the Storegga slide (Smith, 1997), there is no convincing evidence of other tsunamis reaching this coast and so the most probable explanation for the high beaches relates to the effect of extreme waves during storm events suggested by Hansom (2001). The fresher blocks in the beaches are almost certainly excavated from fresh sockets in the fronting ramp and cliff top and the inclusion of modern fishing equipment wedged within the clusters of imbricate boulders strongly suggests a modern date. The distribution of larger, older blocks above and landward of the fresher blocks, suggests that storms of greater intensity have affected the coast of Shetland over the last 3000 years. Sands from underneath these boulders have yielded an Optically Stimulated Luminescence date of 1605 AD, and so may have been emplaced during the stormy conditions of the Little Ice Age (Sommerville *et al.*, in press).

Other than the very recent research noted above, there has been no other geomorphological work published for this outstanding environment, in spite of the geological and geomorphological importance of the site having been highlighted in the past (NCC, 1976). There remains great scope for research of international significance at the Villians of Hamnavoe, including: identification and measurement of the range and rates of processes operating on hard-rock coasts in extreme high-energy conditions; and assessment of the frequency of extreme storms and their effect on the rate of cliff retreat, denudation of the scoured rock surface, and transport of large fragments of rock.

Conclusions

One of the most exposed coastal sites in Mainland Shetland, the Villians of Hamnavoe provide some the best examples of the power of high-energy storm waves. Well-developed, high-altitude scour features occur at a range of heights above 15 m OD and up to 100 m inland, with associated wave-shifted slabs and boulders. Similar high energies occur during major storms on St Kilda, but the overall height of the cliffs reduces the amount of potential wave run-up and washover and so may reduce erosion since the waves reach the cliff unbroken. At the Villians of Hamnavoe, nearshore water depths are sufficient to allow breaking on the cliff and a higher net erosion rate. The products of erosion are evident in the accumulation of high-altitude contemporary beaches constructed out of large boulders at distance from the cliff edge. In addition, a unique staircase waterfall and associated system of arches, caves and overhung cliff ledges, together with the spectacular blowhole, the Hole of Geuda, demonstrate the strong lithological and geological control in this extremely high-energy wave environment.

PAPA STOUR, SHETLAND (HU 170 600)

J.D. Hansom

Introduction

The small island of Papa Stour (3.5 × 3 km), separated from the western Mainland of Shetland by the narrow Sound of Papa, contains a remarkable assemblage of hard-rock coastal forms of national importance (NCC, 1976). In many ways the coastal landforms of Papa Stour represents in microcosm the coastal landforms of the Shetland Isles and most of the distinctive features of the Shetland coastline are represented. Cliffs of various types, geos, stacks, skerries, subterranean passages, caverns, caves, natural arches and blowholes are all found within this relatively small, but highly scenic, area. Papa Stour and its surrounding seabed also displays many of the features of a submerged dissected plateau, which, on a rising relative sea level, has been actively and selectively eroded by a wide range of wave-energy environments.

Although Papa Stour is not quite as exposed and windy as St Kilda, it shares a wind and wave climate similar to the Villians of Hamnavoe and Foula. For 75% of the time the hourly mean wind speed exceeds 4 m s^{-1} with the most frequent strong winds from the south-west (BGS 1977a). At Papa Stour, a combination of exposure to prevailing winds and deep water close inshore produces a relatively high-energy wave climate at the shore and significant wave heights are about 3 m for 10% of the year and are less than 1.5 m for 75% respectively (Draper, 1991). However, storm waves at sea reach heights well in excess of this. The sea floor falls steeply away from the island to 50 m at about 1 km offshore but there are numerous skerries and stacks in the nearshore that serve to reduce wave energy. Wave energies are further reduced by the North Shoals and Ve Skerries that lie at 20–30 m depths to the west and south-west of Papa Stour.

Although the scientific interest of Papa Stour has been recognized for many years and the research potential is great, it has not attracted any detailed geomorphological research. Nevertheless, the cliffs and related features of Papa Stour warrant further investigation.

Description

The island of Papa Stour is composed almost entirely of Devonian extrusive igneous rocks, mostly rhyolites and ignimbrites, with smaller outcrops of basalts, tuffs and agglomerates. Locally, there are numerous small-scale changes in structural alignment and lithology. Small faults and fissures abound, facilitating differential erosion and differing levels of rock breakdown.

Although the entire coastline of Papa Stour is of geomorphological interest, an 8 km stretch of the western part of the island contains an impressive assemblage of hard-rock coastal landforms. The description that follows concentrates on this dramatic stretch of coast from Wilma Skerry (a low gradient wave-scoured promontory) on the south-west coast, clockwise round to Lamba Ness on the north coast (Figure 3.8).

At Wilma Skerry and around the offshore skerry of Swarta, the wave-scoured rock surfaces above the high tide line give way to sloping intertidal shore platforms, although why this should be the only development of sizeable shore platforms on the west coast of Papa Stour is unknown. North of Wilma Skerry, the cliffs are up to 20 m high and indented with caves, natural arches and geos. The three Galti Stacks, protrude prominently from the sea at the mouth of a narrow geo, all four features corresponding to exploitation of a distinctive fault line. Selective erosion of the many vertical joints in the rocks makes caves a common feature along this stretch of coast. North of the Galti Stacks, the relatively wide Brei Geo indents the coastline with steep, almost vertical, sides that increase in height to almost 35 m at the head of the inlet.

The coastline between Brei Geo and North Lunga Geo is highly indented, with alternating small geos and promontories. North Lunga Geo widens inland and has an isolated rock pillar (over 10 m high) in the centre of the inner geo. The backwall of the geo reaches c. 35 m high some 70 m from the outer coast and a gravel storm beach occurs at its base. This widening-inland characteristic is common to many geos on Papa Stour.

To the north, Christie's Hole provides a dramatic example of a geo inlet with a subterranean passage and collapsed cavern (Figure 3.8, inset). Complex relationships between marine erosional features and the structurally controlled pattern of shallow lochs on the plateau are clearly demonstrated at Christie's Hole. Marine erosion along a structural line of weakness has cut geos, caverns and subterranean passages that underlie depressions on the surface. The collapse of the roof of one of these subterranean passages resulted in the instantaneous drainage of one of the flooded depressions and the loss of an inland loch.

Farther north, two geos indent the coastline to the south of Aesha Head. Binnie Geo is a smaller version of North Lunga Geo and Hirdie Geo, with its extensive blocky scree slopes and basal boulder accumulations; it lies along a major fault-line. North-west of the fault the control of rock type on general coastal shape and evolution is perfectly demonstrated in the bay south of Aesha Head, where erosion of the basalt has produced a gently sloping and wide valley in the plateau and low-angled cliffs sloping inland. At low tide, a wide boulder beach extends

Hard-rock cliffs

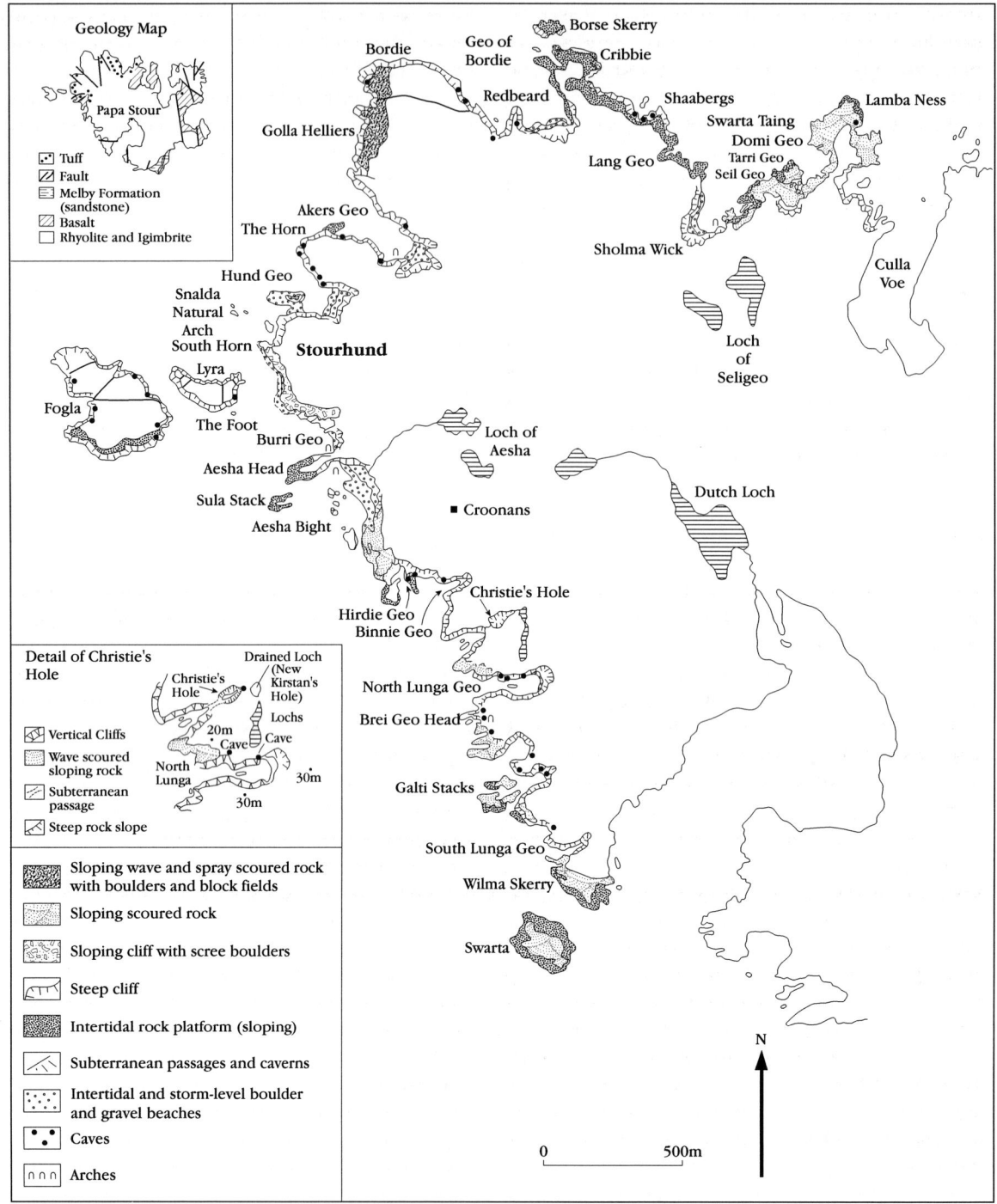

Figure 3.8 Geomorphological sketch map of Papa Stour showing extensive wave-scoured cliff-top surfaces, together with stacks, caves, arches and geos. For general location, see Figure 3.1. (Modified from unpublished work by W. Ritchie.)

almost as far as the low offshore rocky skerries where the two irregular-shaped stacks of Aesha and Sula protrude from the sea beyond. Aesha Head itself is a distinctive narrow promontory composed of rhyolite. A spectacular natural arch spanning the narrow neck of the promontory marks the rhyolite–basalt junction.

Northwards from Aesha Head the land rises

steeply to over 50 m OD at Stourhund, with its horn-like point. The view from here to the caves, arches and cliffs of Lyra Skerry and, the larger, Fogla Skerry to the west, provides striking and dramatic coastal scenery. These islands contain numerous subterranean passages. The north headland of Fogla Skerry contains a magnificent series of buttresses, arches and interlinked caves and many of the cliffs are much steeper than those on the adjacent mainland.

The steep, almost vertical, cliffs north and east of Stourhund are over 50 m high, with no shore platform or basal apron of boulders. Offshore there are spectacular narrow rock pinnacles and stacks, including an impressive natural arch at Snalda. The high basalt cliffs are indented with the spectacular geos of Hund Geo and Akers Geo, whose high, vertical rock walls extend far inland, again corresponding to erosion along faults. Access to the inner part of Akers Geo is restricted by a rock promontory lying transverse to the long axis of the geo and which is pierced by a natural arch. Even so, powerful waves rush into the geo, creating an extensive, wide, boulder beach at the base of the 40 m-high inner wall.

A 400 m-long subterranean passage extends through the outer cliffs close to the headland of Bordie that marks the north point of the island. East of this headland is the Geo of Bordie, not a geo in the true sense, but a compound north-facing bay. The distinctive and narrow promontory of Redbeard subdivides the bay of Bordie into two unequal parts. High, vertical, and often overhanging, cliffs, of uniform basalt lithology extend for over 500 m at the western side of the bay of Bordie, declining in elevation to the east. East of the precipitous cliffs of Redbeard, the bay is wider and lower and has a cobble beach. Here the free face of the rock cliff is masked with recent rock falls and screes and occurs up to 100 m from the beach. This represents the sub-aerial recession of the basalt cliff top by failure and rock fall rather than by marine undercutting, such that the debris delivered to the cliff foot is reworked into beach deposits.

The low peninsula of Cribbie, to the east of the Geo of Brodie, is wave-scoured, with highly dissected sloping rock surfaces. Shore platforms are virtually absent. Indeed, the outer peninsula is almost detached as marine erosion has carved a narrow, deep geo on the east side, which terminates in a distinctive blowhole (the Kiln) in the plateau surface. Near the Kiln, there is a block field 15–20 m above sea level on the south side of the narrow geo. There is also a spectacular natural arch at Cribbie.

The relatively low rocky coastline from Cribbie to Lamba Ness is highly indented and irregular, with numerous small inlets, geos, reefs, low stack-like pinnacles and wave-scoured rock surfaces with only small fragments of shore platform in evidence on the east side of Sholma Wick. A boulder beach is present on the west side of the inlet of Sholma Wick. The low, rough, bare rock headland of Lamba Ness is severely wave-scoured for almost 100 m inland and to over 12 m OD. Wave action from the north-west is clearly an extremely effective erosional process in this exposed location and numerous stacks have been created by erosion, commonly along fault lines (Figure 3.9).

Interpretation

Erosion of the mainly basaltic rocks of the island of Papa Stour has produced an impressive series of geos, stacks, blowholes, cliffs and sea caves (NCC, 1976). This site provides textbook examples of almost all of the main hard-rock coastal landforms and the juxtaposition of these and the wide range of landforms found on Papa Stour without doubt justify its inclusion in the GCR.

The coastal landforms of Papa Stour represent a microcosm of the Shetland archipelago and display many of the distinctive features of the Shetland coastline. The origin and evolution of the Shetland coast has been much debated (e.g. Flinn, 1964, 1969, 1974; Steers, 1973). Flinn (1964) highlights the absence of modern shore platforms at the foot of the cliffs of Shetland, echo-sounding showing that the cliffs descend below present sea level often to considerable depths. The sea floor around the archipelago is stepped or terraced with common occurrences of nearly horizontal surfaces commonly at depths of 24 m, 46 m and 82 m below present sea level (Flinn, 1964). These terraces are regarded as indicating erosional surfaces produced by earlier sea levels before submergence took place (Flinn, 1964). The lack of shore platforms at present sea level with the steep cliffs plunging directly into deep water to submerged platforms also implies that conditions are no longer suitable for the planation of platforms, either because sea level has been too mobile or that the processes of planation have changed as was suggested in the St Kilda GCR site report.

Hard-rock cliffs

Figure 3.9 Fault-controlled stacks at Lamba Ness, Papa Stour. (Photo J.D. Hansom.)

The surfaces at 46 m and 82 m below sea level have parallels with those at St Kilda but not the surface at 24 m and this may reflect a divergence of the relative sea-level histories of the two locations in late Quaternary times. Flinn (1974) regards Shetland as an erosional remnant standing above the North Sea floor with sea-level rise gradually drowning the valleys and re-activating the relict cliffs of former sea levels. Certainly, the cliffs are likely to be inherited features of earlier higher sea levels, an argument put forward in the GCR site report for St Kilda.

The erosional features of Papa Stour have great potential for research into the ways in which lithology and structure control the geomorphology of hard-rock coasts. Well-developed geos on Papa Stour often co-incide with the axes of major faults (e.g. Hund Geo and Akers Geo), while wave erosion differentially exploits the small-scale faults and fissures, local changes in lithology, alignment and bedding planes to produce geomorphological diversity. This, combined with the variations in exposure to wave energy of Papa Stour, produces a spectacular range of structural and erosional situations.

Conclusions

Papa Stour is of national geomorphological importance owing to the juxtaposition of a diverse range of excellent examples of almost all of the main hard-rock coastal landforms; ranging from low wave-washed skerries to impressive near-vertical cliffs. Within an 8 km stretch of coastline, the western part of Papa Stour contains a spectacular range of coastal forms, with: cliffs of a range of altitudes, some of which are wave-scoured, some with well-developed scree formations; geos and inlets of varying size and orientation, some with inlet head beaches and others with sheer cliffs; subterranean passages, some over 400 m long; caverns; caves; natural arches; blow-holes; offshore islands; skerries and stacks. This dramatic pattern of diversity has its roots in differences in wave exposure, the presence of major and minor faults and subtle changes in lithology, alignment and bedding planes that are unmatched elsewhere in Britain (NCC, 1976).

FOULA, SHETLAND (HT 940 400)

J.D. Hansom

Introduction

The island of Foula (13 km^2) lies 22.5 km west of the Shetland Mainland and is the most westerly of the Shetland Isles. The island's dramatic profile is dominated by the summit of The Kame (376 m), Britain's second-highest sea cliff. The entire coastline is exposed to the extremes of wind and wave energy as it shelves steeply into deep water (greater than 60 m). The high cliffs and fragmented cliff-foot shore platforms along the west and south-west coasts are exposed to the frequent and high-energy Atlantic storm and swell waves.

The island is primarily composed of Middle Devonian sandstones, which rest unconformably on, and are faulted against, a narrow strip of Dalradian metamorphic rocks along the east coast (Blackbourn and Russell, 1981; Blackbourn, 1985). The metamorphic rocks are cut by a series of microgranitic intrusions, the topographic expression of which adds protrusions to the coastal outline of the island. Erosion and faulting of the sandstone has led to the development of a series of approximately east–west-trending gentle south-facing dip slopes and steep north-facing escarpments (Flinn, 1978; Blackbourn, 1985) that form ridges and valleys that dominate the skyline and control the form of the dramatic cliffs along the north, south and west coasts of the island. Flinn (1978) concluded that Foula had been completely overridden by ice during the last glaciation, with ice flow from the south-east that had been deflected to the north and north-west by the higher land of Hamnafield and The Sneug. Localized ice deflection to the west around The Noup influenced the glacial erosion of the Daal valley.

Although Foula may not be quite as exposed and windy as St Kilda, it shares a wind and wave climate similar to Papa Stour. At Foula, a combination of exposure to prevailing winds and deep water close inshore produces a relatively high-energy wave climate at the shore and significant wave heights are about 3 m for 10% of the year and less than 1.5 m for 75% respectively (Draper, 1991). However, storm waves at sea reach heights well in excess of this. There are few skerries and shoals offshore and since the sea floor falls steeply away from the island to –80 m depth, onshore wave energies are high.

Description

The Foula coast is almost entirely cliffed with heights of 150 m at Wester Hœvdi (HT 937 388), 210 m at Soberlie (HT 951 410), 248 m at The Noup (HT 953 375) and 376 m at The Kame (HT 940 400). However, the crest of the cliffline lowers dramatically along the east coast where it ranges from 50 m to below 10 m in height. Local variations in cliff form are related to the slope of the inland topography, itself controlled by the structure of the bedrock geology and, to a lesser degree, past glacial erosion. Cliff-foot shore platforms are found along some sections of the coastline, although they vary greatly in extent and are partly a function of structural control. Classic examples of caves, arches, tunnels, stacks, reefs and skerries are also present around the coast. It is convenient to subdivide the coast into seven sections on the basis of the bedrock geology and/or surface topography (Figure 3.10, after Pirkis, 1963).

In section one the area of Dalradian metasediments and microgranite intrusions between Wurr Wick and Shoabill is geologically distinct from the sandstone bedrock that dominates the rest of the island. The two inlets of Wurr Wick and Shoabill have been cut along the line of the faulted contact between the sandstones and Dalradian metasediments and the cliffs are lower than elsewhere at no more than 50 m high. Inland limits of weathering and surface stripping associated with wave spray are variable and dependent upon the geology and inland topography. On the northern side of Strem Ness a tunnel has formed along the line of minor fault plane between Wurr Wick and Scarf geo. The nature of the cliffs around Strem Ness changes at the Head o'Ruscar, where a microgranite sill has been intruded into the host rock and is characterized by a gentle, seaward-sloping bedrock ramp, which facilitates rapid wave run-up during storms and erosion of the bedrock surface. Deep geos occur at Kubbi a'Skeld and Sloag of Ruscar, and to the south towards Ruyhedlar Head, a near-vertical cliff of granulite rises to 50 m. The coastline becomes indented by intersecting geos with many excellent examples of rock-coast landform development such as caves, arches stacks and stumps. South of Swaa Head, the altitude of the mica-schist cliff averages 20 m in height. The rocks are significantly less resistant to weathering and erosion than the psammites to the north and the coastal edge is a low, gently sloping platform, locally dissected by narrow geos cut along the boundaries of microgranite intrusions and scour and stripping of vegetation is evident up to 20 m above sea level and 50 m inland. Hedd o'da Baa, to the south of Ham Voe, comprises a low, flat, coastal edge up to 10 m high composed of peat resting upon glacial till, but to the south the altitude of the coast rises towards the large and complex geos of Ham Little, Selchie Puddle and Shoabill.

The second coastal section from Shoabill to Hellabrick's Wick (Figure 3.10) and inland has some of the lowest and flattest land on the island. The surface falls from 40 m at Heddicliff to 20 m at South Ness reflecting the gentle southerly dip of the underlying sandstone. The cliffs south from Shoabill to North Hœvdi are mainly composed of sandstones and shales capped by glacial deposits that feed debris to the cliff-foot boulder beach, known as 'South Wick', via extensive and unstable scree slopes. South of this, the coast is composed of harder sandstones capped by a thin layer of glacial till

Hard-rock cliffs

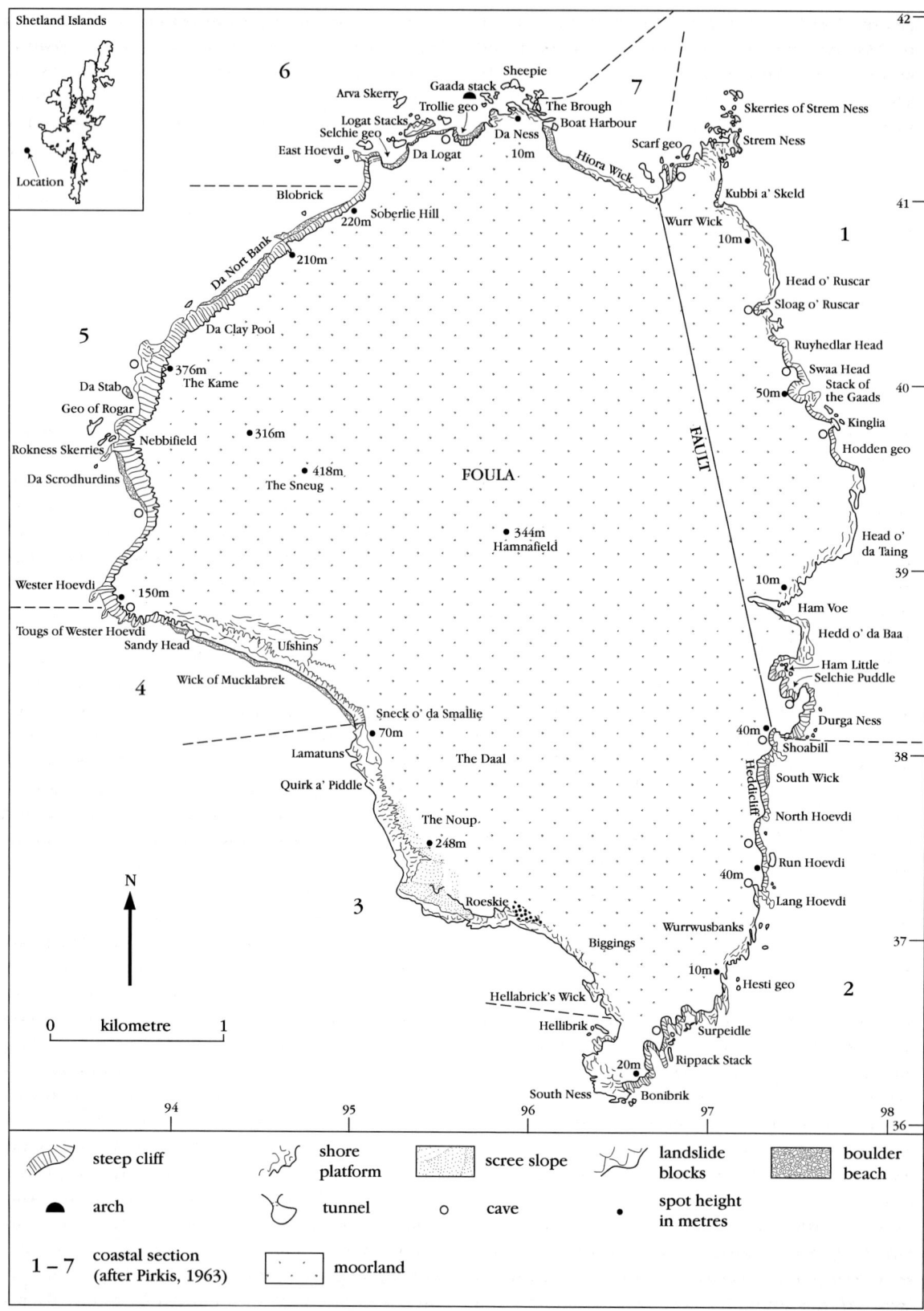

Figure 3.10 Coastal geomorphology of Foula. Sections 1–7 refer to descriptions in the text. The highest and most spectacular of these are Section 3 and Section 5 where the cliffs rise to 248 m at the Noup, and 376 m at the Kame respectively. See Figure 3.1 for general location. (After Pirkis, 1963.)

(Pirkis, 1963) and the cliffs are low, but sheer, and are deeply indented by geos with stacks offshore. The northern part of the third coastal section from Hellabrick's Wick to Smallie is mainly composed of the Noup Sandstone, the form and height of the cliff mirroring that of the topography inland and rises to The Noup at 248 m. The cliffs are stepped and formed of a series of landslided sandstone blocks and associated scree slopes. Steeply sloping (25°) cliff-foot platforms lead up to the base of the landslides. To the north the cliff edge drops to around 70 m where the Daal, a glaciated trough, intersects the coast. Large cliff failures are currently active at the western end and several large tilted slip blocks of sandstone appear to have failed along the upper junction of the intertidal ramp of the shore platform. One of the most impressive and distinctive is the Sneck o'da Smallie, a 60 m-deep and 1 m-wide cleft that extends for 50 m.

The fourth section of coast between Smallie and Wester Hœvdi is one of the most exposed parts of the Foula coastline and contains excellent examples of cliffs influenced by structural dip (Pirkis, 1963). The cliff profile shows a series of steps up to 250 m high, comprising at least one major rock slide. The southerly dip of the sandstone has produced a ramp-type shore platform along this section of coast that is narrow and structurally controlled. It is poorly developed and locally obscured by slabby boulder beaches and scree deposits. Just north of the Smallie, a large rock failure known as 'Ufshins' has slid at least 40 m down a section of cliff face defined by faults in both the east and west.

The fifth section of coast between the 150 m-high Wester Hœvdi and the 220 m Soberlie is composed of sandstone and has by far the most spectacular cliffs on Foula. The sheer face of The Kame at 376 m, is the second highest sea cliff in Britain (Figure 3.10). The cliffs at Wester Hœvdi are also sheer with numerous caves but no shore platforms. To the north, short lengths of sloping intertidal shore platform occur with skerries offshore. Beyond The Kame, the cliff crest drops rapidly to Soberlie Hill, its base indented with caves and headlands. Along the Da Nort Bank, localized accumulations of cliff-foot boulder beaches occur sourced from the joint controlled collapse of the upper cliff face. There is a general absence of shore platforms along this stretch of the coast.

The sixth part of the coast between Soberlie and Da Ness is composed of glacial till-covered sandstone into which have been eroded an abundance of caves, arches, tunnels, stacks and reefs. To the east of Da Logat, the Logat Stacks are well-developed examples of triangular stacks with rocks dipping steeply (around 45°) southwards. The stacks at Gaada, Sheepie and The Brough are similarly dip-controlled, Gaada stack being dissected by two arches the upper surfaces of which are capped by eroded and loosened boulders. The lower part of the uppermost bed remains unweathered towards the base of the landward face and protrudes as a low ridge, preventing downslope slippage of weathered material from above.

The final section of the coast to Wurr Wick is unique as it is both low-lying below 10 m and depositional being composed of a storm ridge composed of local sandstone, igneous and metamorphic *c.* 1 m-long boulders. During storms such boulders are thrown several tens of metres inland (Pirkis, 1963), for example at Boat Harbour.

Interpretation

Early work by Walton (1959) on cliff coasts along the east coast of Scotland suggested that many seemingly active cliff landforms are inherited features from earlier sea levels, buried by glacial till during the Quaternary Period, and subsequently exhumed and subject to further wave erosion over Lateglacial and Holocene times. Where conditions were favourable for preservation, even some fragile features such as stacks and arches were re-occupied and numerous examples of till-adorned stacks and till-choked emerged ('raised') stacks are known from the Scottish coast, for example at Tarbat Ness and Cullen in the Moray Firth. In Foula, cliffs with till caps and till plugs inside the heads of geos occur widely along the north and east coast and are a powerful argument in favour of the inheritance of cliffs from earlier times. This is supported by more recent analysis from a wide range of coastal settings including Antarctica, where fragile stacks have emerged as ice caps have retreated (Hansom, 1983), St Kilda (Sutherland, 1984) and Canada (Trenhaile, 1997). In spite of this, some of the more fragile features such as stacks and arches can also be shown to have developed, and in some cases were subsequently destroyed, entirely within the Holocene Epoch as marine quarrying and abrasion has progressed, for example the Old Man of Hoy in Orkney (Hansom and Evans, 1995). On Foula, The Brough stack supported an arch carrying a Bronze Age broch (fort) on its upper surface until the arch collapsed during a severe storm in 1965. This implies that whilst the overall form of a cliff coast may well be inherited from former conditions and reoccupied by present processes, the detailed form, development and change can be a relatively rapid process.

In Foula, the occurrence of inherited cliff features, especially along the north and east coasts, may be explained in terms of differential erosion rates leading to preservation. It is probably reasonable to assume the highest energies in Foula occur on the west and south coasts since these are fully exposed to waves from the dominant westerly and south-westerly directions. In comparison, wave energies may be relatively lower along the east and, possibly, north coasts. It follows that there is a greater likelihood of preferential preservation on the north and east coast and this broadly appears to hold. Foula would therefore provide an interesting site for a more detailed examination of the relative age of cliff coastlines in Scotland.

As with St Kilda, although shore platforms exist, the lack of extensive and well-developed shore platforms in such a high-energy environment as Foula, remains problematic and invokes similar arguments to those used for St Kilda. Pirkis (1963) argued that the absence of boulder beaches at the foot of the cliffs showed that erosion rates along the coast of Foula were low. However, beaches require a surface on which to develop and their absence here is probably related to the lack of an extensive and widespread cliff-foot shore platform, an absence itself related to unsuitable conditions for platform development. In Foula, where geological structures permit, limited shore platforms have developed, such as at Ufshins in the south-west. However, nowhere is there a shallow surface close to sea level from which these platforms, stacks and stumps commonly rise. Most of the cliffs of Foula plunge into deep water, and although minor erosional notching at present sea level exists, nowhere is it extensive or substantial.

In spite of the relative lack of shore platforms, several features of rock coast forms are well-represented and display good relationships with geological structure. For example, on the northern side of Strem Ness, a tunnel has been excavated headwards from two caves, one on either side of the Ness, to eventually coalesce along the line of minor fault plane between Wurr Wick and Scarf geo. At South Ness, some of the geos have deep and actively eroding caves at their head, suggesting that some geos form through cave formation and roof collapse along the lines of weakness in the sandstone. During south-easterly storms, wave action is intense within these geos, and cave formation is still active. The presence of large landslide-blocks along the west coast indicates that large-scale failure of the cliffs, which occurred in the past, probably continues as a result of failure along bedding planes in the sandstones and nowhere is this more dramatically displayed than at the Sneck o'da Smallie. Where shore platforms do exist in Foula, they tend to be structurally controlled. At the Noup, and to the north of here, the dip of the sandstone have been exploited by storm waves to produce a cliff-foot, ramp-type shore platform that is narrow and structurally controlled. This enhances wave run-up and has resulted in failure and slippage of the cliff-foot sandstone blocks.

With deep water offshore and adjacent steep cliffs, the boulders that comprise Hiora Wick beach are unlikely to be supplied with large quantities of fresh materials, and the boulders and gravels are likely be recycled. Pirkis (1963) used the distribution of clasts of different lithologies along the beach to infer that grey sandstone clasts eroded from Da Ness are fed southwards and alongshore to dominate most of Hiora Wick, whereas in the extreme south-east, igneous and metamorphic rocks sourced from outcrops in the south-east are found.

Conclusions

The island of Foula is outstanding for its assemblage of hard-rock coastal landforms, which include the second-highest sea cliff in Britain. With the exception of well-developed shore platforms, examples of most of the features and stages of coastal landform development in rock are found. The island experiences relatively high wave-energy levels and the west and south coasts are fully exposed to swell and storm waves generated in the Atlantic Ocean. These conditions have facilitated the development of a fine assemblage of sheer-faced and composite cliff forms, geos, sea caves, tunnels, arches, stacks and stumps, many of which show clear relationships with geological structure.

WEST COAST OF ORKNEY
(NY 229 188–NY 222 094 AND
NY 237 054–ND 173 991)

J.D. Hansom

Introduction

High-cliff coastlines are a feature of much of the Atlantic coast of the Orkney Islands. The 20 km stretch of cliffs between Rora Head, at the south-west tip of the island of Hoy, and the Hole o'Row, mid-way along the exposed west coast of the Mainland (see Figure 3.1 for general location and Figure 3.11), provide some of the best examples in Europe of Old Red Sandstone cliffs and associated features. Lithology and structure are major geomorphological controls and the cliffs and associated forms of west Orkney are good examples of the control exerted by geology on coastal landforms. The rich variety of cliff and cliff-related forms along this coast include steep and overhung profiles; sea-stacks; arches; caves; geos and shore platforms, all reflecting the dominant geological control of horizontally bedded, fractured and faulted, sandstone and flagstone.

The most characteristic feature of the Orkney climate is the frequency of strong winds. The prevailing winds are from between west and south-east for 60% of the year. Winds greater than 8 m s^{-1} occur for over 30% of the year and gales occur on average for 29 days per year. Along the south-west coast of Hoy, a combination of deep open water and exposure to prevailing winds produces a high-energy wave climate. Within Hoy Sound, conditions are more sheltered, especially from the north and south-west but on the outer coast of Mainland and Hoy during westerly and northerly storms, wave conditions are more severe. Because the sea floor falls steeply away from the west to 60 m, the coast is exposed to relatively high wave energies. Spring tidal range in the western Orkney Islands is 3 m (UKDMAP, 1998).

As with much of the hard rock upland coast of Britain, there have been few detailed geomorphological studies of the Orkney coastline. Nevertheless, the nature of the cliffs and associated features have been described in more general terms (Steers, 1973; NCC, 1978) and Hansom and Evans (1995) have examined the nature and development of the famous sea stack called the 'Old Man of Hoy'.

Description

The GCR site has three main sub-units (Figure 3.11):

1. The west coast of Hoy, from Rora Head to Kame of Hoy.
2. The north-west coast of Hoy, from Kame of Hoy to The Pow.
3. The west coast of Mainland, from Breck Ness to Hole o'Row.

The cliffs from Rora Head to Kame of Hoy are high, steep, and in places, vertical. The highest cliffs occur in the north at St John's Head, but no part is less than 50 m high. A narrow (40–70 m-wide) intertidal shore platform, with a cover of fallen boulders, is a common feature along much of this stretch of coast. The Old Man of Hoy, one of the tallest and most spectacular sea stacks in Britain (Figure 3.12) towers 137 m above the sea surface yet is only a few metres wide at the top. Its sides are composed of vertical or overhanging walls that fall sheer on all sides. The pinnacle itself is separated from the adjacent cliffs by a 60 m-wide chasm whose base is strewn with debris, which presumably has fallen from a collapsed arch (Hansom and Evans, 1995).

To the north of the Old Man, the cliffs of St John's Head rise vertically to 335 m. The rock is composed of alternating beds of relatively soft, sandy and pebbly sandstone with occasional beds of harder grey flagstone, bestowing upon the cliffs a slab-like, notched and often overhung profile. These variations in hardness, the near-horizontal bedding, combined with the multiple joints, cracks and faults common in sandstones and flagstones are important factors explaining the spectacular vertical cliffs, caves and stacks of this coastline. The cliff forms are dominated by blocky shapes and cut by deep steep-sided geos and ravines that often bisect headlands. Weathering has etched out vertical chasms in the cliffs along joint planes and near Kame of Hoy the coastline is extremely rugged with many minor sea stacks and inlets. Near the Old Man, a few near-horizontal platforms have been cut into the sandstone, but elsewhere the shore platforms are gently sloping at 10–20°.

The north-west coast of Hoy between Kame of Hoy and The Pow is a transitional coastline, between the high cliffs at Kame of Hoy and the lower, north-facing cliffs with platforms at the

Hard-rock cliffs

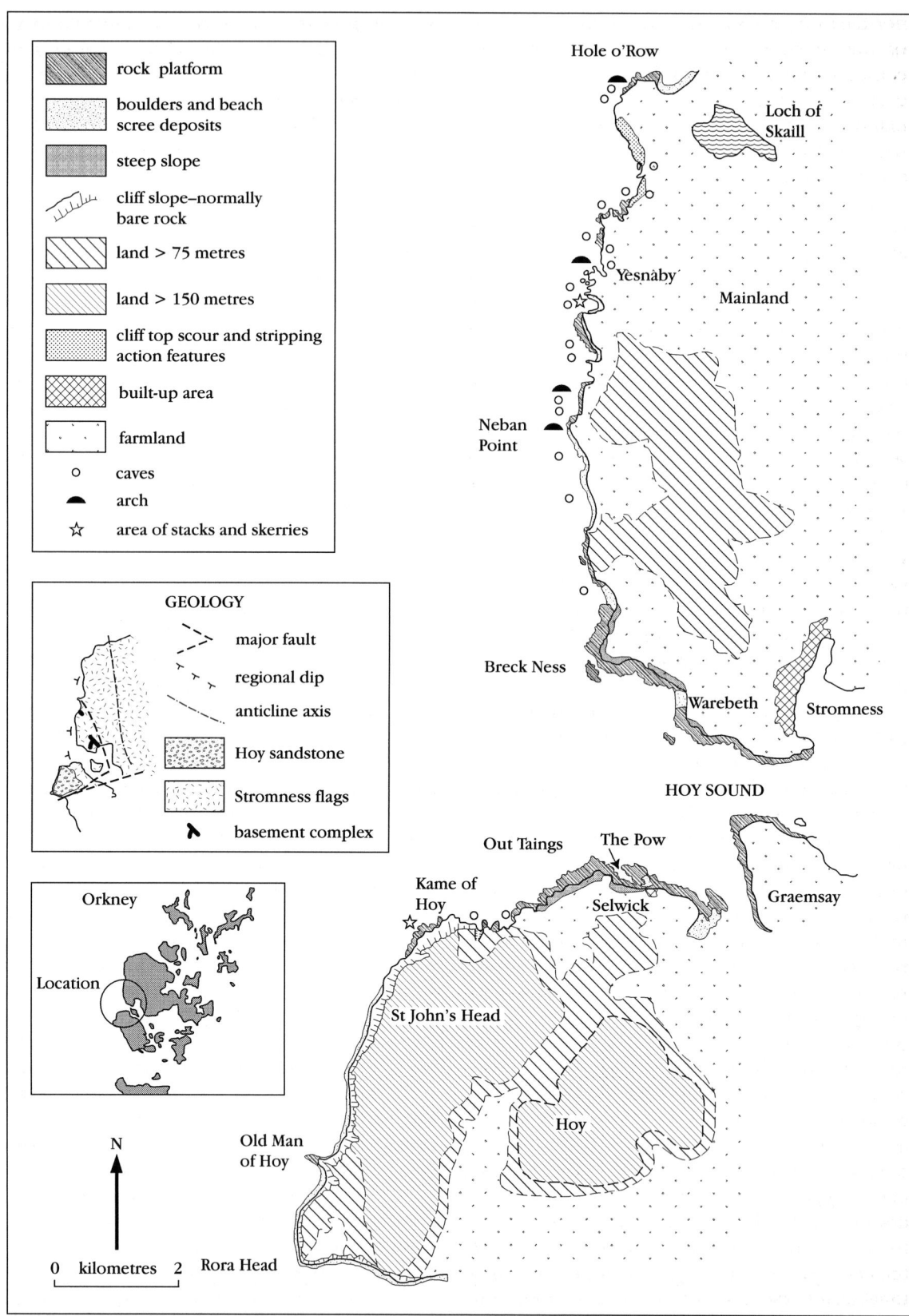

Figure 3.11 Coastal features of the West Coast of Orkney. Erosion of the Hoy Sandstone and Stromness flags (inset) has produced an impressive coast of steep cliffs, caves and stacks. (Modified from unpublished work by W. Ritchie.)

north-east of the island. The 1.1 km stretch east of the Kame contains some of the steepest coastal slopes in Orkney and there are excellent examples of high plunging cliffs with no shore platforms. This zone also marks the junction between the Old Red Sandstones to the south and the grey Caithness Flagstone Group to the north. These flagstones consist of rhythmic sequences of thinly bedded siltstones, shales and finely laminated sandstones and so, apart from the stretch of high cliffs east of the Kame, most of the remaining coastline towards The Pow consists of low cliffs, degraded terraces ranging from c. 5–10 m high, and some well-developed shore platforms, particularly at the Taing of Selwick. At Selwick, the rocks dip to the west, imparting to the 60 m-wide shore platform a well-defined ribbed appearance where the eroded fronts of the beds run north towards the water's edge. Numerous small-scale intertidal fissures and cracks have been excavated by the waves, some have been abraded by boulders and many of these have become lodged in fissures in the shore platform. The lower-lying areas of shore platform have become buried by accumulations of sand and gravel, for example at The Pow.

The west coast of Mainland between Breck Ness and Hole o'Row consists of a dramatic series of almost vertical cliffs reaching up to 60 m in height, within which the effects of erosion are well developed, with a great variety of geos, caves, arches, stacks and cliff forms. The underlying geology is the Caithness Flagstone Group but some basement granitic and gneissic rocks outcrop as inliers, the largest of which lies to the south of Yesnaby. This stretch of coast is typical of Orcadian cliffed coastline, in detail highly irregular but, in general, lacking substantial embayments. The cliff tops are lower than on Hoy at between 20–60 m, but occasionally reach 100 m above sea level, particularly in the north where the coastline intersects the generally rising relief of the landward plateau. The cliffs are mainly vertical or overhanging with a distinctive notch at wave level where the flagstones have been quarried by waves. The notch is best-developed where narrow sloping platforms front the cliffs and allow broken waves to impact on the cliff foot, such as at Alga Bar and Brough of Bigging. On the lower cliff tops, wave spray erosion is well developed and many cliff top areas have been stripped of their soil and drift cover up to 40 m inland. Gravel is rare along this stretch of coast, except occasionally within the heads of geos and at Billia Croo, where a small gravel beach occurs.

In general the shore platforms are steeply dipping and narrow, tending to occur as projections from the cliff foot rather than as a continuous fringe along the coast. In detail the platforms are controlled by intertidal outcrops of relatively more resistant flagstone beds. In the steeply dipping strata, these present coherent intertidal ramps (30°) over which waves surge to finally break on the cliff. The intertidal shore platforms have smooth seaward ramp surfaces whereas the landward surfaces are stepped and crenulated. Excavation of these landward steps is often well-advanced enough to have resulted in separation from the original platform to form offshore skerries. In some places deep water extends to the foot of vertical plunging cliffs, for example at Black Craig.

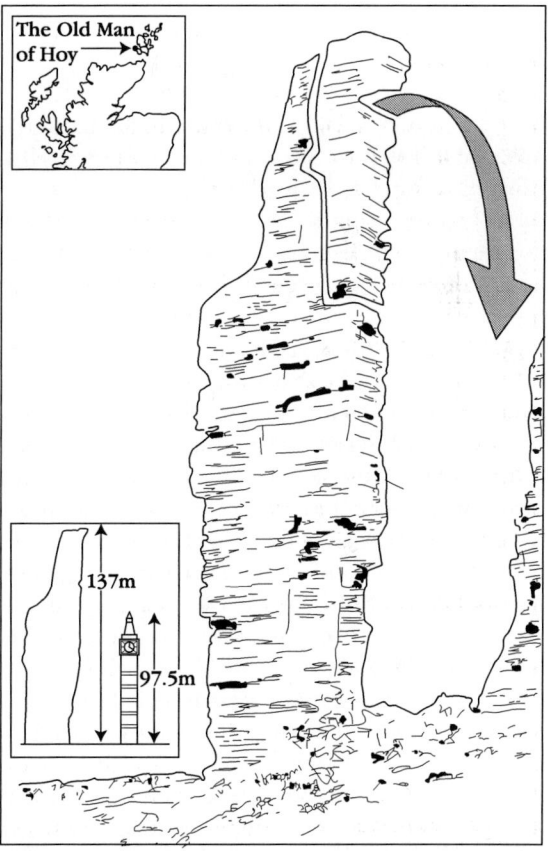

Figure 3.12 The Old Man of Hoy, West Coast of Orkney, showing incipient failure cracks. London's 'Big Ben' is shown for scale in the inset. (After Hansom and Evans, 1995.)

The variety of geos, caves, arches and stacks that occur along this coastline tend to be triangular or rectangular in form, rather than slot-shaped, reflecting the cross-cutting joint and faulting pattern of the flagstones. The stacks and arches at the Castles of Yesnaby and Qui Ayre are rectangular in plan and are known locally as 'castles' (Figure 3.13). Both narrow towards the base where arches have been eroded. At Qui Ayre, the arch is so large that the roof appears unstable. Variations in lithology and geological structure again are major controls on the coastal landforms and the resultant differential erosion and exploitation of weaker strata, joints, fissures or dykes are responsible for much of the meso- and micro-topography. To the south of the Hole o'Row wave and spray erosion is both active and effective (Steers, 1973) on the cliff face and the base of the cliffs and spray and wind erosion is important at the top of the cliffs. In summary, this is an actively evolving cliff coastline that encompasses, in a relatively small area, a consistent suite of cliff-type features developed in a uniform sequence of sedimentary rock.

Interpretation

In common with cliffs elsewhere in Scotland, the cliffs of Orkney are likely to be inherited features that have persisted over several changes in sea level (Sutherland, 1984; Hansom, 1988; Trenhaile, 1997). However, cliffs and stacks are also affected by present-day erosional wave processes, and features such as stacks are clearly ephemeral features (Hansom and Evans, 1995). There is no doubt that erosion is active on the west coast of Orkney (Steers, 1973; NCC, 1978; Hansom and Evans, 1995), however, the rate of erosion and thus the amount of cliff recession is unknown. The historical development of the Old Man of Hoy gives an insight into the rate and type of active erosional processes affecting these ancient cliffs. As late as the early part of the 19th century this famous stack had an arch on its landward side (Steers, 1973). The arch has since been lost and all that remains is a perceptibly thicker part of the column indicating where the arch was at one time attached. Hansom and Evans (1995) trace the historical development of the Old Man further. Maps dated c. 1600 and 1750 do not portray the Old Man as a stack but as a headland. However, by 1819, the headland had been eroded into a stack and arch with the twin legs that gave the 'Old Man' its name. Early in the 19th century, a severe storm washed away one of the legs (Miller, 1976) creating the free-standing stack. This pattern of ongoing erosion continues today and in 1992 a 40 m-long crack had opened up in the top of the south face leaving a large overhanging block that will eventually collapse. From the above it is apparent that the Old Man is a relatively young feature (less than 250 years old) and in geological terms a mere infant. It seems reasonable to assume that many stacks may have similar development patterns but variable life spans depending on exposure, structure and lithology. Certainly the dynamic nature of the processes that have shaped the Old Man will also lead to its eventual demise (Hansom and Evans, 1995). However, as erosion proceeds, other sea stacks will undoubtedly be eroded from the cliff face. Indeed, Steers (1973) noted a high pillar at Bre Brough, which had almost separated from the cliffs. In spite of this detail it remains difficult to estimate a rate of cliff recession from such intermittent activity.

It is recognized that active erosional processes, combined with variations in lithology and geological structure, have produced a unique variety of cliff and cliff-related landforms on the west coast of Orkney (Steers, 1973; NCC, 1978). For example, the coastline of Hoy is dominated by beds of red and yellow Hoy Old Red Sandstone that rise vertically above a pedestal of dark basalt lava (Kellock, 1969). Ritchie (1984) describes the lower Hoy beds as being soft and friable and the overlying beds to be harder forming prominent outcrops. Undercutting of such beds leads to sequential failure of beds above and the development of a vertical profile. The large stacks at the Castles of Yesnaby and Qui Ayre are rectangular in plan but both narrow downwards and have arches eroded through their base, an indication of marine erosion being more active than subaerial failure. Indeed the arch at Qui Ayre is now so large that it will soon fail and result in complete separation of the stack from the host cliff.

Variations in lithology and geological structure are major controls on the cliffs of Orkney and the resultant differential erosion and exploitation of weaker strata, joints, fissures or dykes are responsible for much of the meso- and micro-topography. Much of the detail depends directly upon marine erosion of joints, bedding planes, faults and dykes of igneous rock intruded into the sedimentary strata. The rock in con-

West Coast of Orkney

Figure 3.13 The spectacular arch at Qui Ayre, Yesnaby, West Coast of Orkney, is one of several arches and columnar stacks in the area in various stages of development. (Photo J.D. Hansom.)

tact with such dykes is often removed fairly easily, leaving impressive features such as the natural arch at the Hole o'Row. The diversity of caves, arches, stacks, geos and vertical cliffs along this short stretch of coast provide an excellent field site to study the development of erosional features and further our current understanding of coastal processes and forms on hard upland coasts. Although, to date, little research has been carried out on Orkney cliffs, the research potential is immense and the range, size and physical attributes of this spectacular coastline, together with its high wave-energy, justify its inclusion here.

Conclusions

The exposed Atlantic coastline of the west coast of Orkney is of national geomorphological importance. The 20 km stretch of coast includes some of the most spectacular and dramatic cliff forms in the British Isles, with numerous geos, inlets, caves, arches, stacks and excellent examples of the relatively rare phenomenon of cliff-top scouring. Although sea stacks are familiar features of many hard rock coastlines in Scotland, few are more spectacular or famous than the towering monolith of sandstone, the Old Man of Hoy, which reaches a height of 137 m. The scientific importance of this site lies in the range of spectacular cliff forms displayed over such a short stretch of coastline and the clear influence that geological structure plays upon their form. The evidence of contemporary coastal retreat is also very clear and there are opportunities in Orkney to establish the retreat rate of these sandstone cliffs.

DUNCANSBY TO SKIRZA HEAD, CAITHNESS (SD 398 710)

J.D. Hansom

Introduction

One of the finest stretches of cliff coastline in mainland Britain extends southwards for 6 km from Duncansby Head on the north-east extremity of the Scottish mainland (Steers, 1973). The spectacular cliffs and related forms provide excellent examples of the characteristic cliff forms of the Old Red Sandstones of north-east Scotland and show clear relationships between geological structure and coastal morphology. The cliffs, caves, geos, arches and stacks provide a dramatic coastline and the famous Stacks of Duncansby are cited frequently in the international literature (e.g. Trenhaile, 1987). In spite of this, there has been no detailed geomorphological research on these spectacular cliffs, although numerous descriptive accounts exist (e.g. Steers, 1973) and the geological memoir provides an account of the relationships between geological structure and coastal form (Crampton and Carruthers, 1914).

Similar in many ways to Orkney, the most characteristic feature of the Caithness climate is the frequency of strong winds. The prevailing winds are from between west and south-east for 60% of the year. Winds greater than 8 m s^{-1} occur for over 30% of the year and gales occur on average for 29 days per year. However, the exposure of this east-facing coast is less than that of the west and north and for much of the time the winds blow off the land and so help reduce wave energies. The sea floor falls away from the mainland to 60 m depth by about 5–10 km offshore. Along the eastern coast, shelter is afforded by the mainland and the Orkney Islands, and so the wave climate is not as severe as in the north or west and significant wave heights off Duncansby Head are 2 m for 10% of the year and about 0.5 m for 75% of the year (Draper, 1991).

Description

The coastal geomorphology is dominated by horizontal beds of Old Red Sandstone that are classified locally into a block of resistant Thurso flagstones cropping out between Skippie Geo and Fast Geo and the more variable, but generally weaker, John o'Groats Sandstone Series to the north and south. The coastline is best described using four geologically defined sections from south to north (Figure 3.14):

1. Skirza Head to Skippie Geo: a *c.* 1 km stretch of cliffs and deep geos cut in the John o'Groats Sandstones in the south.
2. Skippie Geo to Fast Geo: a *c.* 1 km stretch of Thurso Flagstones.
3. Fast Geo to Gibbs Craig: a *c.* 3 km stretch of cliffs in the John o'Groats Sandstones, with magnificent stacks.
4. Gibbs Craig to Duncansby Head: a *c.* 1 km stretch of cliffs and geos of the horizontally bedded Thurso Flagstones in the north.

Duncansby to Skirza Head

From Skirza Head to Skippie Geo, the compact, fissile, near-horizontal flagstones produce a coastal scenery dominated by 40 m-high cliffs, and an abrasion-ramp shore platform that extends more or less continuously for almost 1 km northwards from Skirza Head in the south to Sailor's Head. Four deep, near-vertical geos indent this stretch of coastline. Long Geo is the largest; farther north two shorter and rock-floored geos occur, one with a scree and boulder beach. Skippie Geo is an inlet part of which is raised *c*. 20 m above sea level. Flat skerries and an extensive intertidal shore platform are exposed below the caves, overhangs and slab-like walls of Skippie Geo.

Between Skippie Geo and Fast Geo the cliff height increases from 40 m to 55 m and the vertical cliffs are cut by Wife Geo, a 250 m inlet where an association of caves, arches, plunging vertical cliffs, rock pinnacles and buttresses have resulted in one of the finest compound geo features in Scotland.

Between Fast Geo and Gibbs Craig, the John o'Groats Sandstones form the famous high cliffs and stacks of Duncansby (Figure 3.15). The cliffs reach almost 80 m OD at Hill of Crogodale. In the south, between Girn and Hill of Crogodale, the steep cliffs rise from 55 m to 75 m and consist of vegetated slopes alternating with near-vertical rock buttresses. Low-gradient shore platforms form abrasion ramps of up to 100 m wide along much of this coastline, although north of Fast Geo and north of Crogodale, the shore platform is covered by extensive gravel and boulder beaches. Elsewhere the boulder beaches form a relatively narrow fringe at, or just above, the high-water mark. In the north, the impressive Stacks of Duncansby rise as pyramidal structures from the surrounding shore platform less than 100 m from the cliff base. The southernmost stack reaches in excess of 50 m and is higher than the adjacent cliff, on account of the landward slope of the mainland cliff edge. The stacks have distinctive outlines of almost square, castellated blocks of red sandstone.

The adjacent cliffs display considerable local variation in form and profile with steep buttress-type rock cliffs alternating with relatively low-angled vegetated slopes developed on a continuous cover of superficial materials. At the base of the extensive apron of slumped materials is a low rock cliff, succeeded by a low-angled shore platform with a variable cover of boulders and cobbles. Northwards, the cliffs gradually decline to *c*. 25 m OD at Gibbs Craig and the Duncansby fault boundary.

Between Gibbs Craig and Duncansby Head, the land rises to the north but the horizontally bedded flagstones display similar coastal forms to the flagstone area to the south (Section 2, Figure 3.14). The coastline is characterized by vertical, often overhanging, cliffs with irregular profiles formed as a result of differences in hardness and susceptibility to erosion of the flagstone beds. Pillar-like stacks occur close to the 35 m-high cliffs at Gibb's Craig and The Knee. The Geo of Sclaites, south of Duncansby Head Lighthouse, is a textbook example of this type of inlet with a natural arch at its entrance and a basal cave at its narrow head. Between the geo and Duncansby Head to the north, the vertical or overhanging cliffs display excellent examples of basal notches. Long Geo, north of the lighthouse on the north-facing coast of the headland, is long and steep sided with a distinctive overhanging profile. Farther west, at The Glupe, a large blowhole has developed, similar in form, although smaller, than The Pot at the Bullers of Buchan (see GCR site report).

Interpretation

The cliff coastline from Skirza Head to Duncansby Head is of high scientific and educational value for the following features:

1. The clear relationship displayed between coastal form and geological structure.
2. The spectacular and diverse range of plunging cliffs, stacks, arches, and caves.
3. The numerous deep, long, vertical-sided geos, some of which provide textbook examples (e.g. the Geo of Sclaites).
4. The erosional extension of the shore platform landwards into the cliff base.

Geological variations such as strike directions and dip angles at the coast, fault and joint patterns, differential hardness and resistance to erosion, and differential susceptibility to the processes of terrestrial and marine weathering and erosion all affect coastal form. At this site, the contrast between the fine-grained calcareous and argillaceous flagstones and the more-easily weathered, friable and varied sandstones appears to play a large part in determining

Hard-rock cliffs

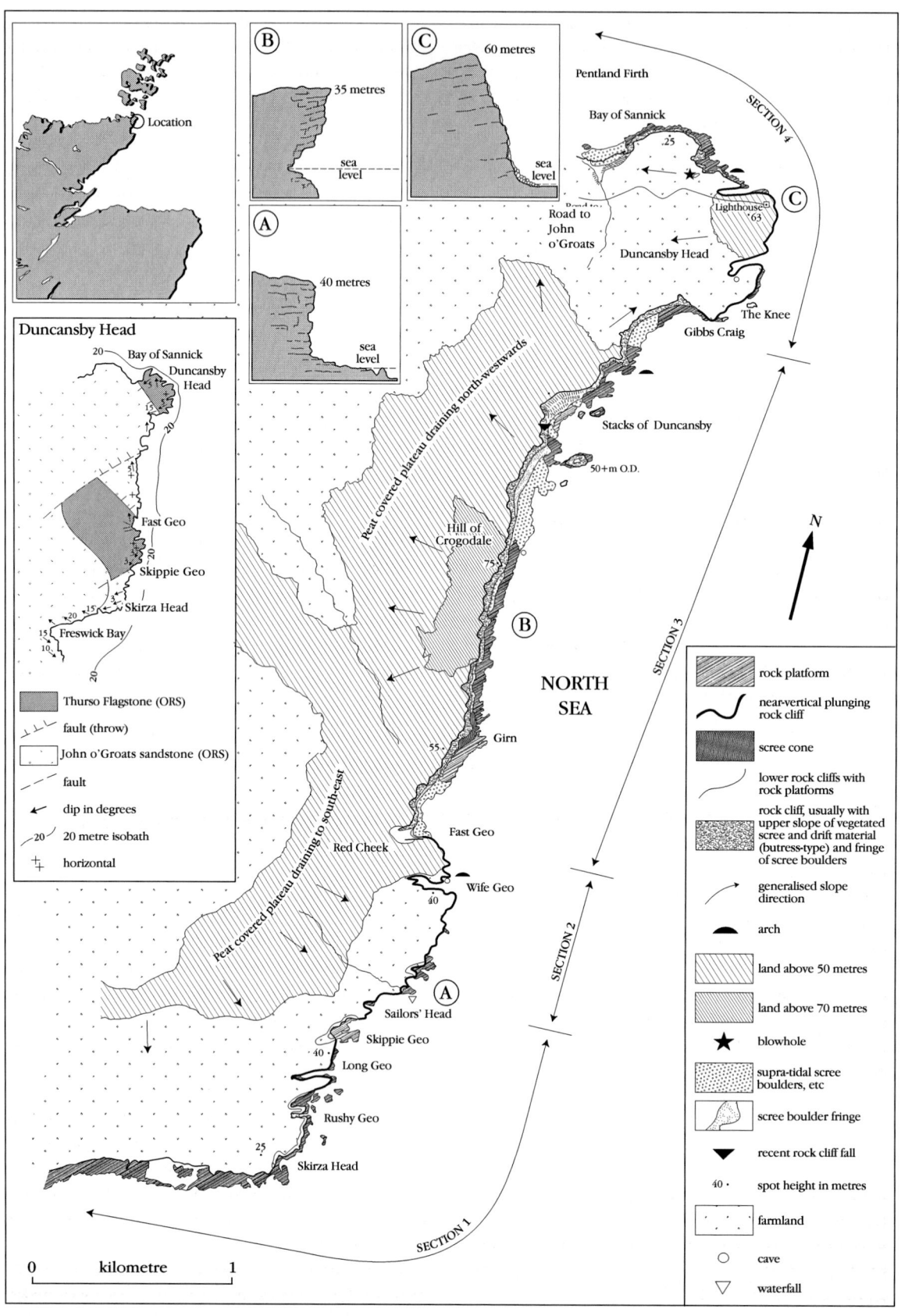

Duncansby to Skirza Head

Figure 3.15 The three large stacks of Duncansby stand in stark contrast to the otherwise bleak and smooth landscape of the north-east coast of Caithness. Looking north towards Duncansby Head and South Ronaldsay, Orkney, in the background. (Photo: courtesy of Ken Crossan.)

coastal form. For example, Crampton and Carruthers (1914) assert that the slight embayment that extends between the outcrops of Thurso flagstones at Gibbs Criag and Fast Geo, and includes the Stacks of Duncansby, is a result of the more rapid erosion of the intervening higher, but softer and more variable, John o'Groats sandstones. However, the rate of cliff retreat is unknown. The sandstone stretch (Section 3, Figure 3.14) appears to have retreated somewhat more than the flagstones to the north and south, probably due to lower resistance of the sandstone blocks allowing more effective wave erosion of the cliff base. Additionally, slope processes play an important part in the form of the sandstone cliffs, with frequent landslides and screes that alternate between rock buttresses, features that may have their origin in the pattern of master joints and other vertical lines of weakness. Erosion along bedding planes, in conjunction with the numerous vertical cracks and fissures, give rise to the often blocky and castellated appearance of the cliffs and stacks, best seen in the Stacks of Duncansby. The overall pyramidal shape of the stacks implies that subaerial weathering and the exploitation of joints and bedding planes higher up has been more effective than wave erosion of the base of the stack.

◄**Figure 3.14** Coastal geomorphology of north-east Caithness, Duncansby to Skirza Head GCR site. Descriptions of sections 1–4 and of representative profiles A–C are in the text. The geology of the area is predominantly composed of horizontally bedded Old Red Sandstones (ORS), which have been eroded into steep cliffs. (Modified from unpublished work by W. Ritchie.)

Large-scale structural weaknesses partly explain the formation of the numerous deep-set geos and rectangular stacks that are typical of the flagstone areas. The crush zones, rucks and strong vertical joints of the flagstones provide structural weaknesses that are exploited by marine action to form long geos, often with caves excavated in the backwall (e.g. at the Geo of Sclaites), although several geos have depositional beaches at their heads. Others have well-developed scree slopes indicating a local reduction in the efficiency of wave erosion and a relative dominance of subaerial slope processes. The irregular cliff profile characteristic of the flagstone cliffs again demonstrates the strong geological control on coastal geomorphology, the irregular stepped profile reflecting subtle differences in hardness and susceptibility to lateral weathering of the flagstone beds.

In both the sandstone and flagstone areas, shore platforms are relatively well-developed and in places are relatively shallow in angle, reaching 100 m wide. They appear to be best developed where the bedrock dips are low or horizontal in both lithologies and so their form is aided by structural control, in spite of extensive evidence of active abrasion of their surfaces. In the area between Fast Geo and Hill of Crogodale, the platforms appear to be actively extending the cliff base landwards, in spite of an intermittent covering of debris from above.

Conclusions

The cliffed coastline between Duncansby Head and Skirza Head is one of the finest and most spectacular stretches of cliff coast in mainland Britain. Lithological and structural control is important in determining cliff morphology. The flagstone cliffs cropping out in the north and central section are typically steep, near-vertical and sometimes overhanging, with irregular stepped profiles and numerous deep, long, vertical-sided geos. Although less steep, the cliffs of the John o'Groats sandstones are higher. The magnificent castle-like Stacks of Duncansby rise to above 50 m, but are less than 100 m from the base of the sandstone cliffs (Figure 3.15). Well-developed and actively evolving shore platforms extend along much of the cliff base.

This dramatic stretch of cliff coastline, with its complex of deep geos, caves, arches, stacks and shore platforms, provides textbook examples of many cliff forms and demonstrates the strong structural control on cliff morphology. It is also easily accessed via the road at Duncansby and so can be appreciated by most visitors. The scientific and geomorphological interest is very high, although, as with the majority of hard-rock coastal sites, it has not been thoroughly investigated. In this respect, the GCR site is of immense scientific importance, both in terms of its educational value and research potential.

TARBAT NESS, ROSS AND CROMARTY (NH 950 878)

J.D. Hansom

Introduction

Tarbat Ness forms the southern headland of the Dornoch Firth and juts out into the Moray Firth in a north-easterly direction (Figure 3.16). It is composed mainly of a peninsula of Upper Old Red Sandstone, separated from the underlying Middle Old Red Sandstone by a fault in the east and south. The cliff and shore platform features at Tarbat Ness are excellent examples of the operation of a range of pitting, saltspray and honeycomb weathering and tidally zoned biological processes that are relatively unusual in Scotland. The landward-dipping beds, occasional joints and minor faults in the Old Red Sandstone have been differentially eroded to provide a coastline of great variety, including some of the best examples of differential erosion processes on tilted sandstone strata in Scotland. In addition, the striking contrast between the coastal forms on the high-energy south-east and lower-energy north-west coast of the peninsula adds to the geomorphological interest. Emerged gravel beaches and platforms, together with a prominent emerged cliff are also found along the coastline of Tarbat Ness and emerged sea stacks are also well developed south of Wilkhaven pier on the south-east coast of the peninsula.

Hourly mean wind speeds in the inner Moray Firth at Tarbat Ness reach 3 m s^{-1} for 75% of the time and 18 m s^{-1} for 0.1% of the time but winds are mainly offshore. Onshore winds from the north-east (the longest fetch) account for only a small proportion of all winds, but winds from the south-east are almost as frequent as south-westerlies and have relatively long fetches of 25 km. Water depths off Tarbat Ness are relatively shallow, reaching 10 m depth at about 300 m offshore and 20 m at 5–10 km offshore (UKDMAP, 1998). The Ness is thus in a relatively sheltered location being subject to lengthy fetches only between north-east and south-east, with a maximum fetch to the north-west of 16 km. The north-west facing side is sheltered from the worst of the north-easterly storms that approach the ness obliquely and is mainly unaffected by easterly waves. Similarly, the south-east shore is also sheltered from the worst of the north-easterly storms but is more exposed to storm waves from the south-east.

The relict landforms of Tarbat Ness and the Dornoch Firth are of great significance for the interpretation of the glacial and sea-level history of the area. The Holocene development of the Dornoch Firth has been reconstructed based on these and other emerged marine features (Firth *et al.*, 1995; Hansom, 1991; Ogilvie, 1923; Smith, 1968; Smith and Mather, 1973). However, in spite of extensive research on the current processes of the coastline of the inner Dornoch Firth (Hansom and Leafe, 1990), there has been no geomorphological research on the active coastal forms and processes of Tarbat Ness itself.

Description

Tarbat Ness is composed mainly of a peninsula of Upper Old Red Sandstone, separated from the underlying Middle Old Red Sandstone that crop out in the east and south by a fault boundary to the south of the pier at Wilkhaven (P on Figure 3.16). There is a general decline in altitude of the headland from 17 m above sea level in the south and east to 10 m above sea level in the north and west. All of the features described below occur within the Upper Old Red Sandstone, which here is composed of a great variety of calcareous-rich layers, finer-grained red sandstone, grits and some conglomeratic beds. The beds individually range from a few centimetres to over one metre thick and some are fissile. The Ness itself is strike-aligned with a well-defined dip towards the WNW and mainland and this is reflected in the ridges and clefts in the shore platform that are angled obliquely to the coastline. Higher dips of 40° occur in the south and west, but these decline to 25° towards the north-east and the Ness itself. On the east coast south of the fault boundary at Wilkhaven Pier, there is greater variation in both strike and dip angles and a change in coastal morphology from the well-defined serrated shore platforms of Tarbat Ness to low and uneven platforms to the south.

The features of interest at Tarbat Ness centre on the development of distinctive shore platforms and on smaller-scale weathering features such as pitting and honeycombing. At the larger scale both the north-west and south-east sides of Tarbat Ness display excellent examples of serrated rock platforms that have been cut and weathered across dipping sandstone beds of varying resistances. On the north-west coast, differential and selective erosion of the steeply inclined sandstone beds of the shore platform has produced a staircase of steeply dipping parallel knife-edge ridges and narrow, linear clefts, typically no more than 2–3 m wide. At the upper levels of the platforms, the tops of the beds and protuberances are adorned with honeycomb weathering micro-forms and tafoni, while the clefts are occasionally partly filled with gravel deposits. The platforms on this coast reach 250 m wide at Camas Solais but taper to 50 m wide towards the north-east.

There is a greater variety of larger-scale rock forms on the more exposed south-east coast of the peninsula. To the north of the fault, the 12 m-high sandstone cliffs are in places overhanging due to active undercutting by waves. Farther north occur the remnants of higher shore platforms (e.g. Craig Ruadh) and several narrow inlets, one of which has a distinctive gravel beach at its head with steep rock cliffs on either side. Both contemporary and emerged shore platforms and the sides of the narrow inlets contain excellent examples of honeycomb, solution and abrasion micro-features. To the north, a series of seven sloping platforms, the upper surface of each representing the top of a more resistant bed of sandstone, have been cut into the receding cliffs producing a distinctive 'stepped' profile. These steps progressively lower and coalesce north-eastwards into a low angular broken shore platform that extends offshore as a series of dipping reefs towards Tarbat Ledge. To the south of the Wilkhaven fault a low, uneven and 300 m-wide rock platform is backed by an emerged platform covered by a grassy terrace of emerged gravels.

At a smaller scale there is a wide range of weathering micro-forms at Tarbat Ness, ranging from solution pits, saltspray tafoni and honeycomb features. The greatest variety of pitting features, ranging in size from millimetres to one or more centimetres in diameter, is found just above the high-water mark on both the north-west and south-east coast of the peninsula and

Hard-rock cliffs

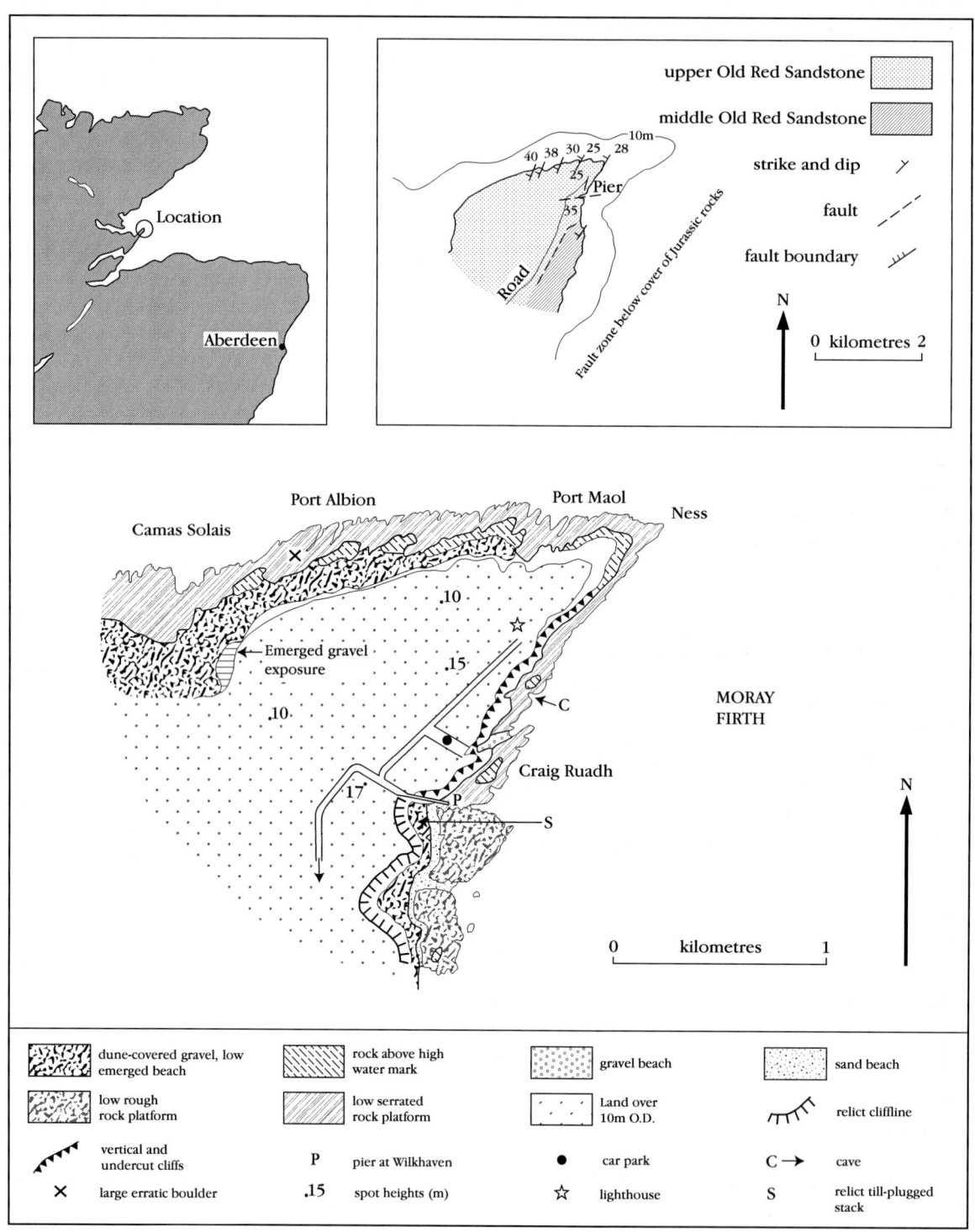

Figure 3.16 Geomorphological map and geological sketch map of Tarbat Ness, Ross and Cromarty, north-east Scotland. The eastern Moray Firth shore is fault-controlled and rocky with a prominent emerged cliffline. The northern Dornoch Firth shore has well-developed emerged gravel beach-ridges. At the Ness itself, the low rock shore platform is characterized by a range of well-developed weathering pits and tafoni that are rare on Scottish coasts. At 'S' an emerged till-plugged stack occurs in front of the relict cliff. (Modified from unpublished work by W. Ritchie.)

tend to develop selectively in certain strata. Larger pits and circular depressions are mainly found on the higher and flatter rock surfaces where they may become filled with stagnant water enriched by bird guano.

Interpretation

The rocky coastline of Tarbat Ness is dominated by varying amounts of the erosional processes of quarrying, abrasion, chemical weathering and biological weathering. However, none have been quantified or studied in any detail. As a result of the combination of exposure and dominant wave direction, there exists a fundamental contrast between the higher energy environment of the south-east facing coast and the more benign north-west coast. In addition, lithology and geological structure play an important role in determining coastal form. The landward-dipping beds, occasional joints and minor faults within the sandstones have been differentially exploited by erosion to produce a distinctive coastline, particularly on the exposed south-east coast. The variety of texture and composition of the tightly-bedded red and yellow sandstones, grits and conglomerate beds have responded differentially to coastal processes producing considerable local variation in form.

Differential and selective erosion of the calcareous sandstone beds has produced many of the larger-scale landforms such as the steeply dipping parallel platform ridges on the north-east coast and the 'stepped' platforms at the tip of the peninsula. Wave quarrying of small-scale faults or joints in the dipping beds has resulted in excavation along these joints to produce a series of characteristic knife-edge ridges and linear clefts. Geological structure, particularly the strike and dip of the strata, plays an important role in determining the morphology of the larger forms on each coast. A much more broken profile with large scarps and short dips characterizes the south-east coast, whereas the north-west coast has long, smooth-topped, dipping beds separated by small scarps. Except on the south-east coast, and where gravel is locally available, few platform surfaces show signs of fresh abrasion. The fresh scars that do exist have angular edges indicating that wave quarrying is important in places.

At Tarbat Ness, erosional micro-forms, such as small pitting features, are typically the product of karst-like solutional processes produced by spray action on the calcareous beds, although salt crystal growth is likely to account for the formation of the tafoni features. Wave-spray processes extend to greater altitudes on the south-east coast and at the Ness itself and, as a result, pits and tafoni are found at greater altitudes on these coasts. However, it is on the upper surfaces of the north-west coast platforms that several of the more delicate solutional features occur, principally because wave energy is less and the features have time to develop. Additionally, burrowing and boring by intertidal organisms plays an important role in the development of some micro-forms, particularly close to the water surface at the edges of rockpools where micro-notches have developed. Although unstudied, there is likely to be a biological zonation relationship between the types of organisms found, the morphology and altitude of the erosional forms produced, and the wave and spray processes operating. In addition, whereas some pits and circular depressions closely resemble karst-type solution hollows and cavities, some of the lower and larger pits and depressions now appear to be subject to mechanical abrasion from gravels within them. All of the lower altitude features are currently active although some, especially those at higher elevations, may be partially relict. There is great scope at Tarbat Ness for detailed research to determine the relationships between the factors responsible for the development of such microforms.

The relict cliff and emerged beaches that occur higher up on Tarbat Ness provide spectacular evidence of former relative sea-levels in Lateglacial and Holocene times. Although they are of great geomorphological interest in their own right (Hansom and Leafe, 1990; Hansom, 1991; Firth *et al.*, 1995), they are also of relevance to the sea-level context within which the platforms and micro-scale features of Tarbat Ness have developed. The southward extension of the active cliff at Craig Ruadh is represented by a relict cliff at Wilkhaven and comprises a rock cliff veneered with glacially derived tills and gravels that fill the gap between the cliff and a sea stack rising from the emerged shore platform. Thus it appears likely that the general morphology and association of cliffs and shore platforms at Tarbat Ness was largely in place before the last glaciation and as such is inherited. Tarbat Ness probably became ice free at about 14 000 years BP, and the glaciogenic sediments became trimmed by a high sea level at about 20 m OD. Gravel beaches were also constructed up to 20 m OD at this time (Hansom and Leafe, 1990). On the north-west side of Tarbat Ness, the high gravel beaches are cut by a prominent cliff whose base lies at 10 m OD. This cliff was probably first cut during the fall to a Lateglacial low sea level at about 10 500 years BP, but then reoccupied as the Holocene sea rose to 6 m OD at 6500 years BP. It is this last rise in relative sea level and the subsequent fall to present levels that has resulted in the erosional trimming of the present shore platform and the weathering of its surface. Present-day processes are thus likely to be engaged in the superficial trimming of an exhumed surface.

Conclusions

The principal geomorphological interest of Tarbat Ness GCR site lies in the range of active micro- and macro-cliff and platform forms. In addition, the juxtaposition of these actively evolving forms with the well-preserved emerged beaches and relict cliffs set back from the present coast adds to the scientific interest of this site.

The Upper Old Red Sandstone peninsula of Tarbat Ness displays a great variety of meso- and micro-scale forms on the cliff and shore platforms. The micro-forms in the rocks and platforms display excellent examples of pitting, salt-spray and honeycomb weathering and tidally zoned biological processes. Differential and selective erosion of dipping beds, occasional joints and minor faults in the sandstone have produced a coastline of great geomorphological variety. The 'stepped' profiles and parallel ridges of the shore platforms characteristic of this coast provide textbook examples of differential erosion processes on tilted sandstone strata. In addition, the contrast between the coastal forms on the high-energy south-east coast where abrasion is in evidence, and the low-energy north-west coast of the peninsula where weathering processes are more important adds to the geomorphological interest. The adjacent emerged gravel beaches and relict cliffs allow the development of both meso- and micro-forms at Tarbat Ness to be effectively placed into a temporal framework.

LOCH MADDY–SOUND OF HARRIS COASTLINE, NORTH UIST, WESTERN ISLES (NF 940 730)

J.D. Hansom

Introduction

At its maximum extent the GCR site of Loch Maddy is approximately 10 km wide and 10 km long, extending northwards from Loch Maddy (Loch nam Madadh) in North Uist to the south part of the Sound of Harris (see Figure 3.1 for general location). It covers slightly less than 100 km² in area of intricate and complex shoreline studded with innumerable islands, skerries and intertidal rock outcrops. The coastal geomorphology is both diverse and exceptional, with low cliffs, discontinuous shore platforms, sheltered sea-loch environments, intertidal sandflats, low rocky islands, reefs, skerries, isolated rock outcrops, rock pinnacles, intertidal rock and boulder pools (Figure 3.17). Almost all of these features are related to the submergence of rock surfaces close to sea level that have undergone intense glacial scouring in and around the main sea inlets of Loch Maddy, Loch Blashaval, Loch Aulasary and Loch Mhic Phàil. The coastal landscape produced is on a scale reminiscent of the Norwegian skjaergard or strandflat, and, with the possible exception of parts of western Ireland (Guilcher *et al.*, 1986), is not found elsewhere in the British Isles. The diversity of landforms within a general trend of Late Quaternary sea-level rise and land submergence is of particular geomorphological significance.

Unsheltered coasts in the Western Isles are exposed to high mean wind speeds, but the inner coast of the Minches is relatively more sheltered. The seabed offshore of Lochmaddy is shallow with numerous skerries and reefs. Spring tidal range at Lochmaddy is 3.5 m and maximum tidal streams are variable around 1 m s^{-1}, depending on location (UKDMAP, 1998). However, although the tides on the Atlantic side of the Sound of Harris are out of phase with those in the Minch, there are no strong currents flowing between the two. Along the east coast of the Western Isles, the irregular coastline produces a highly variable wave climate. Offshore of Lochmaddy, the outer coast of the Minch experiences moderate wave energies, particularly from the south and north-east, between Weaver's Point and Leac Na Hoe where the 20 m

Loch Maddy–Sound of Harris Coastline

depth contour comes within 300 m of the shore. The inner parts of the shoreline are very sheltered and are subject only to small locally produced waves.

Description

Several types of coastline occur in this area but in essence all are the product of submergence of a glaciated rock platform close to sea level. To the east of the site lies a line of glacially scoured hills that correspond to the line of the Outer Hebridean Thrust zone (Figure 3.18). Beyond this the outer Minch coast is high and rugged, reaching 281 m at South Lee but reducing in height northwards to 154 m at Leac Na Hoe. To the west of the site a line of hills rises to 190 m OD. The intervening inner Minch coastline (i.e. within the sea lochs of Maddy, Blashaval, Mhic Phàil and Aulasary) rarely rises more than a few metres above present sea level. The geology of North Uist comprises an ancient basement of metamorphic Lewisian gneiss that was intruded by basaltic sills and dykes during Tertiary times. The Outer Hebridean Thrust Plane occurs high on the west facing slopes of the hills of eastern North Uist and divides the island into two distinctive geological provinces. To the west, the rocks are relatively uncrushed gneisses, whereas in the east the rocks are crushed gneisses and mylonites.

Loch Maddy, the largest sea loch of North Uist, reaches over 20 m deep and is described as a 'fjard' (Earll and Pagett, 1984). Unusual in Scotland, fjards are similar in origin to fjords but occur in areas of low-lying land that have been subject to extensive erosional scour by glacier ice and have intensely serrated shorelines, interrupted by many peninsulas and inlets (Earll and Pagett, 1984). The low and fragmented shoreline of Loch Maddy has numerous inlets and small rocky islands, while the intertidal area consists predominantly of gravel and boulders, with limited coarse sand patches and outcrops of scoured bedrock. Many of the irregular sea loch inlets link with inland freshwater and brackish lochs, particularly in the vicinity of the small town of Lochmaddy, allowing limited tidal exchange of water (Figure 3.17).

The outer Minch coastline to the north and south of the entrance to Loch Maddy consists of cliffs rarely more than 30 m high and with gradients of 20–50° that continue underwater. The term 'pseudo-cliff' has been coined to describe such features. The cliff slopes are highly irregular both in plan and profile, however there are no caves, arches or stacks and only limited cliff-foot accumulations of gravel or boulders. Shore platforms are absent. There are few, if any, scree slopes. The cliffs are 'clean' bare rock cliffs and contrast markedly with most cliff areas elsewhere in Scotland.

North-westward of Loch Maddy lies the highly irregular coastline and numerous large and small islands of Loch Blashaval. Here, the seabed is made up of a series of ridges and deep, rock-floored, narrow basins (Admiralty chart 2825). The scale, alignment and relief of this submarine topography mirrors that of the patterns of adjacent subaerial lochans and ridges. The trend is mirrored in the islands of Keallasay More, Keallasay Beg, the Cliasay group, and Flodday, and several of the minor inlets and small islets. The outline of the Sound of Harris coast is very irregular, although the north-west to south-east geological trend is evident in an extensive series of low, intertidal, shore platforms and skerries, known as 'the Rangas', which stretch towards the island of Torogay and the Sound of Berneray. Extensive banks of submerged sand are associated with these reefs, indeed, much of the floor of the shallow Sound of Harris is sand-covered. The larger islands in the Sound (Torogay, Sursay, Tahay, Opsay, Vaccasay and Hermetray) have a patchy cover of glacial deposits. Tahay is a conical, rocky island rising to over 65 m OD, whereas Vaccasay is low and irregular with extensive intertidal rock platforms at sea level. A complex and extensive group of shore platforms, reefs, skerries and islands lie close to Opsay where a series of extensive intertidal boulder shoals combine to form a complex small-scale archipelago, with enclosed tidal pools and uneven rocky intertidal surfaces.

Two large, but shallow and wide, sea lochs, Loch Mhic Phàil and Loch Aulasary, penetrate into the north coastline either side of the low 'island' of Stromay. Stromay is joined to the mainland for most of the tidal cycle. The low irregular coastline of Loch Mhic Phàil is characterized by a multitude of narrow interdigitations of land, sea and low rocky islets rarely rising more than a few metres above sea level, each thinly veneered by till and peat. Tidal ponds are a characteristic feature of this area. The ponds are basins whose centres lie below low tide and which remain partly flooded when the tide

Hard-rock cliffs

Figure 3.17 The submerged landscape of North Uist looking north-west over Lochmaddy. Submergence of a low undulating rock surface has resulted in a landscape of low rock basins, platforms and skerries with a range of tidal and salinity conditions. (Photo: P. & A. Macdonald/SNH.)

recedes to reveal washed perimeters of boulders and rocks within the intertidal zone. The low, rocky and peat-veneered shoreline of Loch Aulasary is similarly irregular. At low tide Loch Aulasary is almost completely land-locked as a result of broken shore platforms in the north closing the gap between the island of Stromay and the long peninsula west of Leac na Hoe (Figure 3.18).

Interpretation

The essential character of this extensive and distinctive GCR site is a product of the submergence of a low-amplitude and intensely glaciated platform of ancient metamorphic rock. Accordingly, small-scale details of rock type and structure, patterns of previous glacial action and sea-level change, are all central to the explanation of the nature of any particular stretch of this intricate rocky coastal zone. To the west of the Outer Hebridean Thrust Plane the bedrock is composed of relatively uncrushed gniesses that are highly durable and resistant to erosion. As a result the relative durability of the underlying bedrock finds morphological expression in the orientation of the coastal rock skerries, headlands and reefs, all of which are strongly linked to the north-west to south-east regional foliation of the gneiss. For example, in the Lochs Blashaval and Siginish area, both subaerial and seabed topography is made up of a series of ridges and deep, rock-floored, narrow basins, the scale, alignment and relief of which mirrors the structural trend.

Several glaciations have moulded the rocky platform of North Uist into a complex of tightly packed linear depressions and ridges (Geikie, 1878; von Weymarn, 1974). The regional dispersion of ice during the last glaciation mirrored and enhanced the north-west to south-east geological trend (Gordon and Sutherland, 1993; Mactaggart, 1997a). The size, alignment and dimensions of the depression and ridges reflects both the direction of ice movement and the relative strength of the rocks. Since deglaciation of the Western Isles about 14 000–15 000 years BP, the dominant trend in the Outer Hebrides has been one of rising relative sea level, interrupted by temporary regressions (Sissons, 1967). Ritchie (1971) believes that the Holocene sea-level rise in the Uists was of the order of 80 m,

Loch Maddy–Sound of Harris Coastline

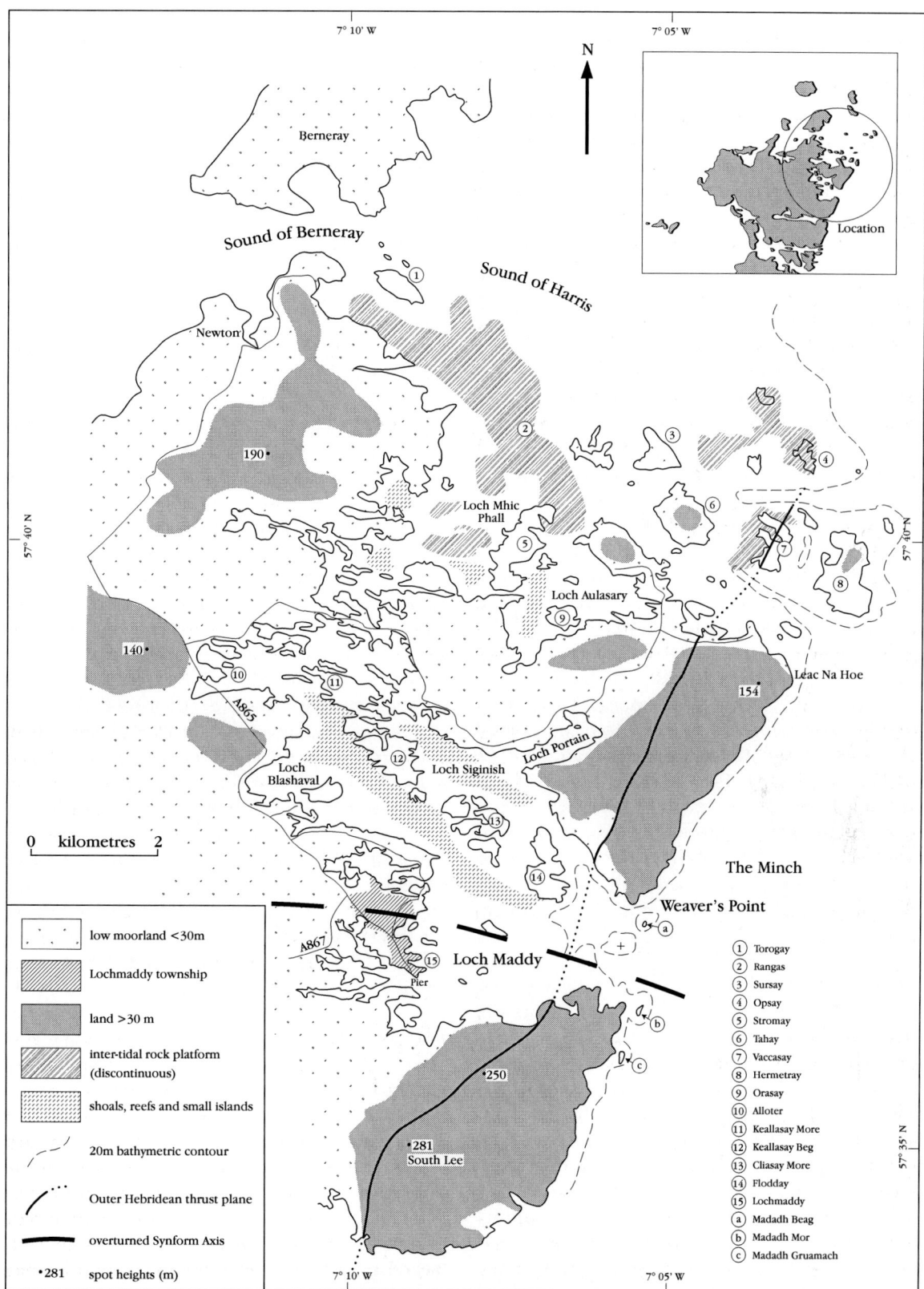

Figure 3.18 Coastal geomorphology of the Loch Maddy–Sound of Harris area, North Uist, showing the extensive areas of intertidal rock platform, small islets and skerries produced by submergence of a pre-existing low-lying rocky surface. The eastern coast is fault-controlled. (Modified from unpublished work by W. Ritchie.)

while Steers (1973) accepts a rise in sea level of between 61–73 m. However, both agree that most of this rise took place before 5700 years BP and led to the submergence of a surface assemblage of landforms near Loch Maddy whose morphogenetic affinities lie more with the Norwegian strandflat than with any landscape in the British Isles. At this time, the low glacially eroded terrain of Loch Maddy was transformed into a multitude of islands, skerries and convoluted inlets and straits.

The lack of erosional features such as caves, arches and stacks in the pseudo-cliffs of the outer Minch coastline is also a likely result of a fairly rapid Holocene submergence (Ritchie, 1968). In spite of the relatively exposed nature of the outer Minch coast and the occurrence of crushed gniesses and mylonites, erosional features are not well developed, even in the more exposed locations. In the absence of any characteristically marine cliff-foot features, it is most likely that the 'pseudo-cliffs' are drowned slopes that have been locally steepened by glacial and slope processes rather than by marine processes and basal undercutting.

As a result of Lateglacial and Holocene submergence, the Outer Hebrides do not show the well-developed suites of emerged ('raised') beaches and associated features so characteristic of much of the Inner Hebrides and Scottish mainland (Sissons, 1967; Ritchie, 1971; Steers, 1973). However, Godard (1965) recognized a number of shore platforms just above modern sea level (e.g. at 0.5 m above high-water mark on the south shore of Loch Maddy) which he suggested might indicate limited emergence. However, the platforms are undated and are more likely to be either interglacial in age, or glacial in origin and may be unrelated to marine processes. They may be simply rock surfaces that have been brought to their present altitude by subsequent submergence (Ritchie, 1968). Similarly, the rock platforms close to sea level in the northern part of the Loch Maddy area are also likely to be washed rock surfaces that now occur close to sea level, rather than shore platforms cut at this level by marine processes.

The inner bays of Loch Maddy are entirely sheltered from the storm and swell waves that sweep both the open Atlantic Ocean and the Minch. However, because the area experiences very strong winds, small but steep wind-generated waves are commonplace over short fetches and this results in very effective trimming of the overlying glaciogenic material and the development of boulder lags at high-water mark.

The scientific importance of this extensive coastal area does not centre on unique individual features such as the eroded remnants of Tertiary olivine rock pinnacles at 'the Maddies', the intertidal rock and boulder 'pools', or the scattered shore platforms and islands in the Sound of Harris. It is the *totality* of this diverse and low, irregular, rocky coastline that is of particular significance. The Loch Maddy–Sound of Harris coastline shows the response of various types of surfaces, essentially those shaped by glacial processes, to the submergence caused by Late Quaternary relative sea-level rise.

Conclusions

The Loch Maddy–Sound of Harris coastline displays an exceptional and distinctive range of submerged, glacially eroded, coastal landforms. With the possible exception of parts of western Ireland (Guilcher *et al.*, 1986), this assemblage of landforms is not found elsewhere in the British Isles. The low, irregular, rocky coastline with pseudo-cliffs, fjard inlets, sheltered sea loch environments, shore platforms, skerries, isolated rock outcrops and intertidal rock and boulder pools, displays many excellent examples of features produced by a marine transgression across a low glacially-scoured surface. It is this diversity of landforms, within a general trend of Late Quaternary submergence, that is of unique geomorphological and scientific significance in Britain.

NORTHERN ISLAY, ARGYLL AND BUTE (NR 363 766–NR 425 774)
POTENTIAL GCR SITE

J.D. Hansom

Introduction

The coastline of northern Islay in the Scottish Inner Hebrides is characterized by some of the finest examples of emerged shore platforms and emerged gravel beaches to be found anywhere in western Europe (see Figure 3.1 for general location). This coastline is the UK type locality for a feature that has come to be known in the literature as the 'High Rock Platform', a 650 m-wide shore platform formed during the Quaternary Period and now found at an elevation of *c.* 33 m OD. It is backed by a cliff of up to 70 m OD and fronted by two lower shore platforms, the Low and Main Rock platforms, respectively at, and slightly above, present sea level. The geomorphological interest of these emerged erosional landforms is enhanced by the presence of glacial and marine deposits resting on the platform surfaces. The emerged shore platforms and associated marine and glacial deposits of northern Islay are part of a network of sites from which the glacial history and pattern of isostatic uplift in Scotland was originally interpreted (e.g. Synge and Stephens, 1966; Sissons, 1974, 1976; Dawson, 1980a,b, 1991). The three distinctive platforms in northern Islay provide evidence of rates of isostatic uplift following deglaciation and this has implications for relative sea-level change in the region.

Unsheltered coasts in the Western Isles are exposed to high mean wind speeds, but the irregular form of the coastline causes great variation in wind climate. The northern coast of Islay is exposed between the north and west, but open water fetches are less from other directions. 68% of storm waves and 80% of swell waves come from the west (Ramsay and Brampton, 2000a). Much of the seabed around the west coast of Islay lies at about 50 m. In terms of exposure and open water fetches, the north-western coast is subject to high wave-energies whereas the eastern coast, south of Rubha a'Mhàil is much more sheltered.

Description

In northern Islay, a high shore-platform, eroded in quartzite rock, forms a continuous level feature between the headlands of Mala Bholsa and Rubha a'Mhàil (Dawson, 1993) (Figure 3.19). Along most of this coastline, the platform varies between 400 and 600 m in width and is backed by a cliff up to 70 m high. The height of the landward edge of the platform varies between 32 and 35 m OD, while the surface of the platform has a gentle seaward slope of *c.* 4° and is free of emerged stacks (Figure 3.20). In some areas, exposures reveal accumulations of till resting on the surface of the platform whereas in other areas the till is overlain by beach gravels. Elsewhere, accumulations of abandoned beach gravels rest directly on the high platform, but occur no higher than 27 m OD. Inland from Port a'Chotain, the platform is overlain by a prominent, arcuate end moraine at Coir Odhar (Figure 3.19).

The seaward edge of the High Rock Platform ends at a 20–35 m-high abandoned cliff, below which is a lower shore platform that extends almost continuously along the north Islay coast (Figure 3.20). The origin of this feature is complex, with many areas of the shore platform showing signs of moulding by ice (Dawson, 1991). Indeed, two distinct rock platforms exist in the intertidal zone of northern Islay (the Low and Main Rock platforms), both of which differ not only in width but also in morphology (Dawson, 1980a,b). The Low Rock Platform is the most conspicuous being regionally horizontal and generally about 100 m wide (but in places reaches 300 m wide; Dawson, 1980b). The smooth ice-moulded surface of the Low Rock Platform declines very gently seaward as a ramp. In places, for example near the lighthouse at Rhubha a'Mhàil, the landward edge of the low platform is marked by a 1–2 m-high cliff rising to a higher shore platform at *c.* 2 m OD (the Main Rock Platform) which here is 10–25 m wide. Unlike the Low Rock Platform, the surface of the Main Rock Platform is characterized by an absence of smooth rock surfaces (Dawson, 1980b). Instead its regionally tilted surface is characterized by protruding angular quartzite ridges with occasional stacks and caves and arches being cut into the backing cliff.

The abandoned quartzite cliffs that mark the landward edge of the Main Rock Platform and the seaward edge of the High Rock Platform, are typically crenulate and indented. The cliffs are adorned with emerged geos, stacks, natural arches and caves (e.g. Uamh Mhór near Port a'Chotain, and the complex cave network on the headland of Mala Bholsa). The floor of the emerged caves and much of the surface of the extensive Low and Main Rock platforms is mantled by gravel beach deposits. Between Mala Bholsa and Port a'Chotain gravel accumulations are locally banked against the cliff. Elsewhere the gravels occur as ridges, the most conspicuous ridge located in the Port a'Chotain embayment where it is succeeded landwards by the most extensive suite of Holocene beach sediments on the northern Islay coast (Figure 3.19).

Interpretation

Since the Low and Main Rock platforms occur at, and slightly above, present sea level, their ages are important in the debates surrounding the age of the present coastline and whether it has been cut during the Holocene Epoch, or whether it is an older feature simply re-occupied by modern sea level. The origin of the High Platform of northern Islay has long generated scientific debate (e.g. Johnson, 1919; Sissons, 1967). Early workers considered the High Rock Platform to have a warm interglacial origin (McCann, 1968; Dawson, 1979) whilst other opinion considered the platform to represent the product of periglacial shore erosion followed by glacio-isostatic uplift (Sissons, 1982). Similar debate surrounds the origins of the Main and Low Rock platforms (e.g. Wright, 1911; Dawson, 1980a,b, 1991). The regionally tilted Main Rock Platform can be traced throughout much of the Inner Hebrides and is now thought to have been produced largely during the severe cold conditions of the Loch Lomond Stadial some 11 000–10 000 years BP (Dawson, 1991), the tilt reflecting differential crustal recovery following deglaciation. The glaciated and regionally horizontal Low Rock Platform has been interpreted as interglacial in origin on account of its lack of isostatic recovery and tilt (Dawson, 1980a) and thus pre-dates the formation of the Main Rock Platform.

Debate surrounds the origins of the emerged shore platforms, beaches and end moraines. The High Rock Platform was described by Wright (1911) as representing 'a preglacial plain of marine denudation', and was interpreted as having formed prior to the only apparent general glaciation of the area. A later, more detailed discussion considered that the High Rock Platform was interglacial in origin and that the Coir Odhar end moraine was the product of a Lateglacial readvance (McCann, 1964). Synge and Stephens (1966) disagreed with this view, suggesting that the Coir Odhar Moraine was the product of ice-sheet decay, although they agree that the High Rock Platform formed prior to the last glaciation of the area. Sissons (1982) asserted that the High Rock Platform had been cut in periglacial conditions but Dawson (1993) calculated that to achieve this would take 28 000 years of relatively stable sea levels, a level of stability not shown by evidence elsewhere. Accumulations of till resting on the High Rock Platform also demonstrate that the platform must pre-date at least one period of glaciation (Dawson, 1991). Beach gravels, which overlie the till up to 27 m OD in places, indicate that a period of high sea level occurred after glaciation (Dawson, 1991). In northern Islay, these beaches are thought to have been formed during the Lateglacial period when sea levels reached 27 m OD (Dawson, 1991). As a result, only the seaward part of the High Rock Platform was exhumed from beneath the till cover and the inner edge of the High Rock Platform which lies at 32–35 m OD was left untouched. It is therefore most likely that the platform itself is the result of abrasion and quarrying during interglacial conditions and has been re-occupied several times since during periods of high relative sea level.

The Main Rock Platform identified in northern Islay is also well developed throughout much of the Inner Hebrides, cropping out along the coasts of Jura and Scarba (Dawson, 1980a) and can be traced intermittently to the Oban area, where it rises to maximum levels of 10–11 m OD (Dawson, 1991). Towards the west and south-west from Oban, the altitude of the platform decreases at a rate of between 0.13 and 0.16 m km^{-1}, until it passes below present sea level in northern Islay, Colonsay and western Mull (Dawson, 1991). The origin of this platform has also been hotly contested (e.g. McCann, 1964; Synge and Stephens, 1966; Gray, 1974, 1978; Dawson, 1980a,b, 1982, 1991). It is now generally considered that this feature, which exhibits no evidence of glaciation, is a relatively young feature that was formed by very rapid and efficient periglacial shore erosion during the Loch Lomond Stadial (Younger Dryas), some 11 000 to 10 000 years BP (Dawson, 1980a, 1991). This interpretation is supported by the rapid rates of erosion identified on similar platforms in modern polar environments (Hansom, 1983; Dawson et al., 1987) and from cosmogenic dating of the Main Rock Platform on Lismore, which indicates that it was cut during the Lateglacial period (Stone et al., 1996). A further suggestion in support of a periglacial origin is that the platforms are wide and well developed in sheltered locations. Such locations can be argued to disadvantage wave abrasion and quarrying and to favour periglacial conditions (Hansom, 1983).

It has been estimated that global sea level at this time was approximately 45–50 m below present, thus the present elevations of this tilted shoreline reflect a pattern of substantial, but differential, isostatic recovery in western Scotland over the 10 000 years since the platform was cut (Dawson, 1991). Isostatic uplift, following

Northern Islay

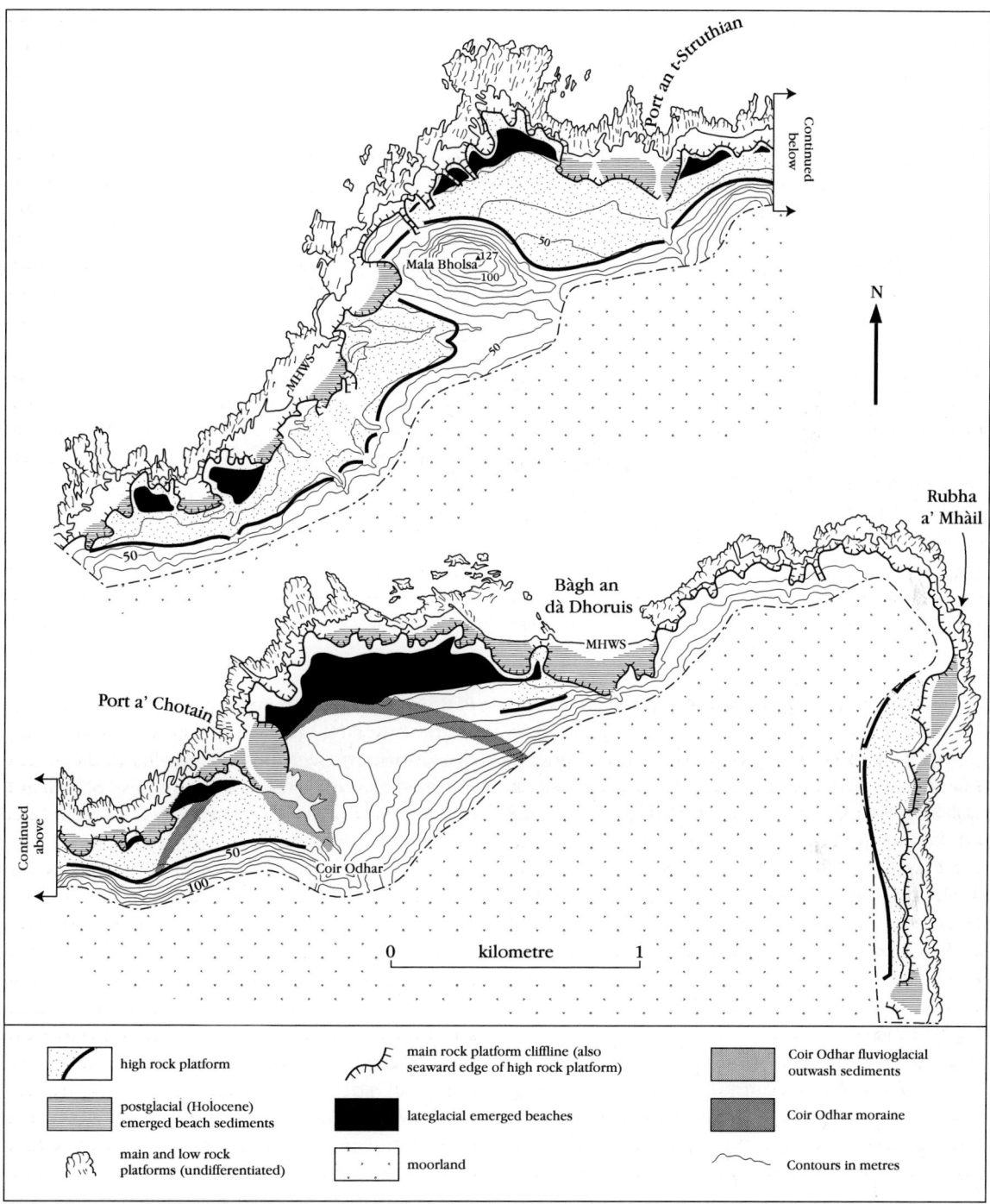

Figure 3.19 Geomorphological map of the coast of northern Islay between Mala Bholsa and Rubha a'Mhàil, northern Islay, showing a fine series of emerged rock platforms and beaches some of which have been capped by glacial moraines whose age informs the chronology for the platforms and beaches. MHWS = Mean High-Water Springs. For general location see Figure 3.1. (After Dawson, 1991.)

deglaciation, was greatest at the centre of the last (Late Devensian) ice sheet in Scotland and the tilt of the Main Lateglacial Shoreline (i.e. the Main Rock Platform and its emerged gravel beaches) reflects this pattern. Northern Islay is of particular geomorphological significance because it is the location where the regionally tilted Main Rock Platform crosses the regionally

Hard-rock cliffs

Figure 3.20 The coast of northern Islay, south of Rubha a'Mhàil, showing the High Rock Platform and its backing cliff. In the foreground the Main Rock Platform and its backing cliff is also well developed. Lateglacial and Postglacial emerged gravels also adorn parts of the coastline. (Photo: J.E.Gordon.)

horizontal and glaciated Low Rock Platform, forms that co-incide close to present sea level in places and so could be confused with 'modern' planation. To the west of the lighthouse at Rubha a'Mhàil, the Low Rock Platform and Main Rock Platform are essentially represented by the same surface. To the west of Mala Bholsa, the horizontal and glaciated Low Rock Platform is intertidal and the unglaciated, partly washed and regionally tilted Main Rock Platform lies below sea level.

Conclusions

Some of the finest examples of emerged shore platforms and gravel beaches in western Europe are well preserved on the northern Islay coast. The three distinct emerged ('raised') rock platforms of northern Islay, most strikingly developed between the headlands of Rubha a'Mhàil and Mala Bholsa, have provoked much scientific debate concerning their origin. This is the type locality in the UK for the High Rock Platform, here emerged to an elevation of c. 33 m OD. The High Rock Platform is mantled with till and marine deposits, demonstrating that the platform pre-dates at least one period of glaciation and a later period of high sea level (Dawson, 1991). The regional tilt of the Main Rock Platform, which can be traced throughout much of the Inner Hebrides, reflects the relative rates of isostatic recovery in western Scotland following deglaciation. The highly uneven platform is considered to have been produced by periglacial shore processes during the severe cold conditions of the Loch Lomond Stadial some 11 000–10 000 years BP (Dawson, 1991). The glaciated and regionally horizontal Low Rock Platform pre-dates the formation of the Main Rock Platform and has been interpreted as pre- or inter-glacial in origin (Dawson, 1980a). Since all of these platforms appear to be emerged features, the co-location of the lower two at present sea level strongly indicates that they are not Holocene or modern features but are older, inherited from former sea levels, and subject to Holocene and modern retrimming.

Notwithstanding the contrasting interpretations that surround the origin and age of these emerged marine features, the complete assemblage of well-preserved marine and glacial features in northern Islay is of outstanding geomorphological and scientific importance. The juxtaposition of the three platforms facilitates observation and study of cliff-slope evolution since abandonment, as well as rates of platform/cliff formation under varying process regimes.

BULLERS OF BUCHAN, ABERDEENSHIRE (NK 103 362–NK 116 388)

J.D. Hansom

Introduction

The granite cliffs at the Bullers of Buchan in north-east Scotland contain fine examples of many of the typical features of rocky coasts, such as the exploitation by erosion of joints, cracks and dykes in massive igneous rock. Selective erosion of lines of weakness in the otherwise uniform rock, such as intrusive dykes, and marine exploitation of minor differences in hardness and structure, has produced a wide variety of rock coastal landforms. The range of features is impressive at a variety of scales with numerous geos, inlets, caves, arches, stacks, platforms and cliffs. Unfortunately however, and in spite of a substantial body of regional knowledge of past sea level and climatic changes, there has been no detailed geomorphological research carried out in this area.

The coastline of the Bullers of Buchan faces east and so is exposed to North Sea gales from the north-east and east. The dominant wave approach directions on this coast are from the north-east to south-east (Buchan, 1976). Water depths offshore reach 60 m depth at about 5–10 km offshore. The indented nature of the coast results in a great degree of variability in the actual wave climate at any one location.

Description

This 3 km stretch of coastline is composed of pink granite and, although uncertainties exist concerning age, it is likely to be pre-Lower Old Red Sandstone. The rocks have rectangular jointing patterns with a dominant near-vertical and horizontal pattern and a secondary pattern that is inclined at *c.* 45° (Figure 3.21). Exploitation by marine and subaerial processes along these joints, fissures and cracks has resulted in angular, near-vertical and triangular cliff forms. In addition, later intrusion of igneous dykes has further weakened the host rock, leading to rapid erosion at such sites and a likely explanation for many of the geos and inlets (Steers, 1973). The coastal plateau is capped by a 1–3 m-thick cover of till and is subject to mass movement and failure at the coastal edge.

The cliffs of this dramatic coastline vary in height from between 20 m to 40 m OD. Bevelled cliff profiles occur over about 50% of the coastal length. The lower cliff is steep and cut in bedrock, while the upper part of the cliff is often composed of a more gently sloping till surface that has been subject to slumping and mass movement (Figure 3.21). The cliffs consist of two distinctive types.

The cliffs to the north of the island of Dunbuy are steep, although rarely vertical, with discontinuous intertidal or submerged shore platforms that are best-developed adjacent to the geos. The cliffs are capped by a relatively thin and sometimes absent till cover (e.g. on the exposed headland of Grey Mare the till cover has been stripped away for up to 100 m inland).

South of Dunbuy the cliffs are lower (*c.* 20 m OD) and have a more irregularly dissected plan and profile. These low cliffs often have a composite profile with a low gradient upper slope, a steeper (*c.* 45°) middle section and a much steeper basal element. The granite cliff top is severely weathered in exposed areas, but this may represent the exhumation of an ancient pre-glacial weathering surface that is widespread in north-east Scotland.

The otherwise continuous sweep of the cliffs is punctuated by the occurrence of several deeply incised geos of which three types are found:

1. Long, narrow and deep inlets with steep rocky sides and a rock headwall. These typically have little or no beach and may have caves or enlarged fissures at the head. Long Haven, in the south, provides a spectacular example of such a geo: the 300 m-long, narrow, steep-sided inlet has a small scree slope and boulder beach at its head and contains an extensive shore platform on its north side. Perhaps the best example of this is The Pot, where a deep rock-enclosed inlet is separated from the sea by a tunnel-like arch. The Pot resembles an enlarged blowhole and during storms this dramatic feature, which is *c.* 60 m deep and 15 m wide, is awash with a froth of white water as waves crash against the precipitous cliffs (Figure 3.22).

2. Wider more complex inlets (e.g. Robie's Haven, North Haven and Twa Havens) typically contain residual pinnacles, buttresses, skerries or stacks. Boulder beaches, till or scree slopes and slumped debris are often well

developed at the inlet heads. The boulder beach in North Haven extends to *c.* 6 m above present sea level. Dunbuy is a large till-capped stack with the same summit elevation as the adjacent mainland plateau surface 30 m away.

3 Numerous smaller irregular indentations in the cliffs also occur and usually have less steep walls and a more serrated and uneven surface (e.g. at Partans, south of Dunbuy).

The intertidal shore platforms are characterized by a jagged but gently sloping morphology. However, they are discontinuous and tend to occur in association with geos in the north and south of the area. The cliff coastline is characterized by so many geos, skerries, stacks, reefs, caves and arches that a detailed description of each is impossible. Perhaps the best example of a conical stack is the Temptin' at NK 110 384 on the north side of North Haven. Skerries at two distinct levels occur between here and the Grey Mare headland to the south. Jagged linear reefs are characteristic offshore from the narrow rocky headlands between Bowness Castle and Dunbuy. The caves of North and South Seals provide the best examples of caves, while spectacular natural arches can be found at Robie's Haven and Long Haven. However, the most dramatic example is the natural arch cut through the island of Dunbuy. Though now above high tide level, the Dunbuy arch may not be entirely an abandoned feature since the collapsed remains of part of the roof litter the base.

Where a supply of material has been available from above, boulder beaches have developed at the heads of the inlets and geos. Some of these are only accessed by waves in the severest of storms, such as the beach at North Haven, which lies at 6 m OD. Others are composed of well-sorted and rounded gravels such as at Twa Havens and Dunbuy.

Interpretation

The range of hard-rock coastal landforms found in the Bullers of Buchan area reflects an interplay between geological structure and exposure to wave activity. The rectangular joints of the mainly uniform granite, together with the presence of intruded dykes, have resulted in planes of weakness in the host rocks that are susceptible to differential marine quarrying and abrasion. The characteristic angularity of the cliff profiles appears to be the result of erosion along the main joint directions in the granite, a dominant near-vertical and horizontal pattern and a secondary pattern that is inclined at *c.* 45°. Steers (1973) states that some of the caves and inlets have been cut along dykes of dolerite, which are eroded more easily than the granite, specifically referring to the geo at Dunbuy as a good example. Buchan (1931) describes the exploitation of two porphyrite dykes at Robie's Haven and Lammylair. Although fully exposed to storm waves from the north-east to south-east (Buchan, 1976), the degree of development of particular erosional features is probably influenced more by detailed differences in structure than by differences in degree of exposure to waves. In addition, subaerial processes on the cliff faces, such as mass movements induced by cold and wet conditions or frost action, may also be of considerable importance.

A noticeable feature of this coast is the contrast between the higher cliffs in the north where a mantle of slumped till masks the upper section of the cliffs and the lower cliffs in the south where wave and spray action has removed most of this superficial layer. Cliff-top stripping occurs up to *c.* 30 m OD in Orkney and Shetland and where cliff heights are low and wave exposure is high at the Bullers of Buchan, it may not be unreasonable to expect stripping of cliff-edge till, at least on the cliffs in the south. Most of the tops of the narrow peninsulas and stack-tops are stripped of till cover but they are also occupied by nesting birds whose activities may accelerate the stripping process.

The presence of till on top of isolated islands, such as at Dunbuy, might be used to argue for an entirely Holocene age for the geos that now separate islands and stacks from the mainland. However, the glacial legacy in the north-east of Scotland is predominantly one of till deposition and of preservation of pre-glacial surfaces, rather than of glacial erosion. For example, Devensian glaciation failed to remove or substantially modify the Tertiary weathered bedrock surface and in places a substantial thickness of saprolite or superficial weathering material has been preserved (Hall, 1986). As a result, it is possible that pre-existing stacks, islands and geos may have been similarly preserved only to be re-occupied by the Holocene rise in sea level. Such an interpretation infers that the plugs of till that once filled and flanked such sites have subsequently been removed by Holocene processes and the

Bullers of Buchan

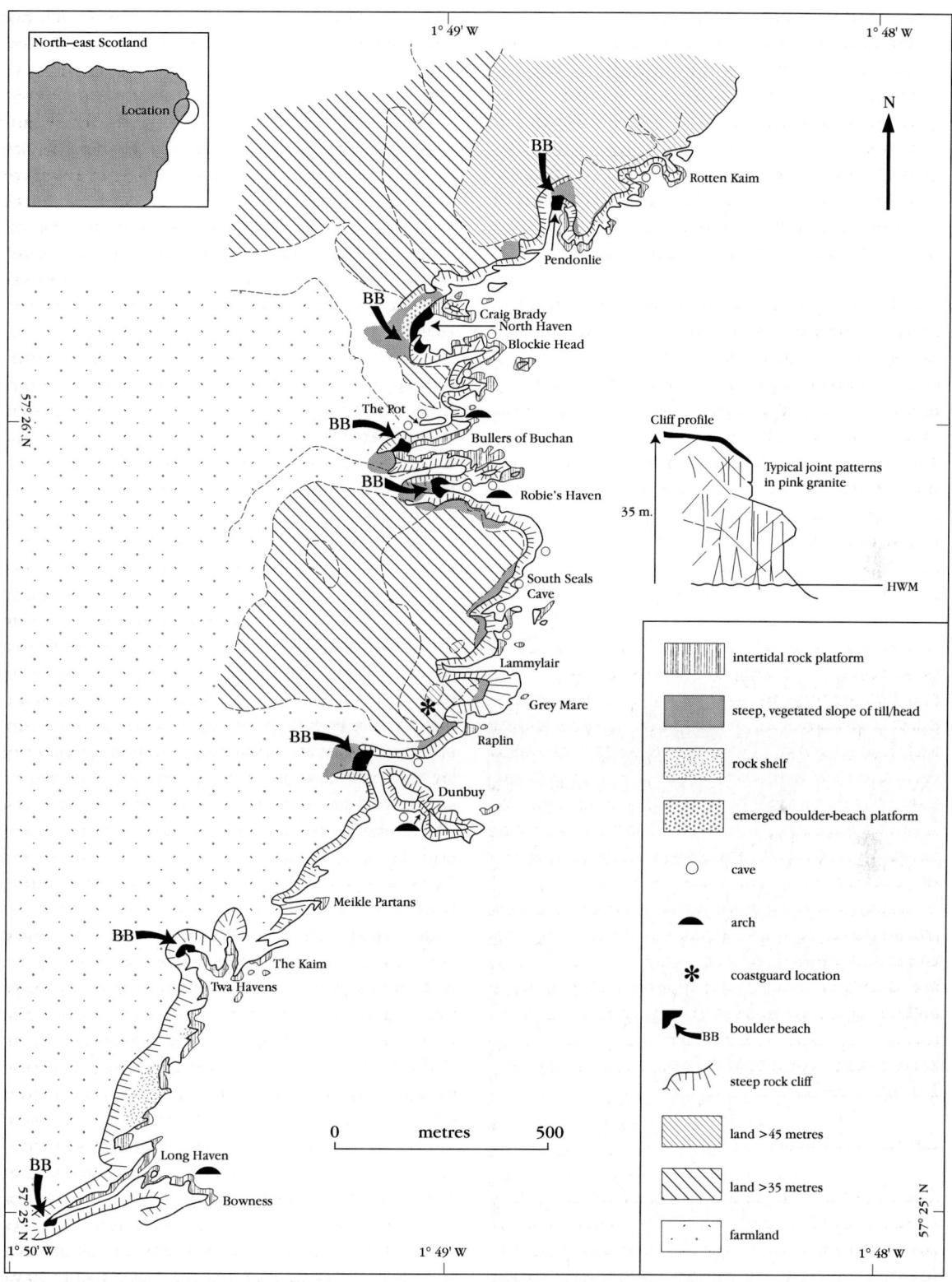

Figure 3.21 Geomorphological map of the Bullers of Buchan, north-east Scotland. The inset on the right shows the typical cliff profile relative to high-water mark (HWM). Much of the cliff tops are veneered by glacial till. (Modified from unpublished work by W. Ritchie.)

rock landforms exhumed.

The discontinuous nature of intertidal and submerged shore platforms along this coast also pose problems of interpretation since most occur at headlands in the north and south and few occur in the central section of coast, for example between The Kaim and Meikle Partans. The occurrence of such intertidal abrasion ramps might be argued to indicate efficient wave quarrying and abrasion, but if so why do such sites remain headlands? It is likely that the reason for discontinuous shore-platform development at the Bullers of Buchan is related to the occurrence of sites where structural weaknesses allow relatively rapid quarrying and the development of geos, inlets and caves. The development of these features promotes wave-breaking and enhances further platform cutting. Where the structure is less heavily jointed, the cliff morphology is more likely to be uniform and plunging, with more wave reflection and more limited platform development. It is also possible that these forms are inherited features from former sea levels.

There is also great geomorphological interest related to specific landforms, such as the enlarged blowhole of The Pot (Figure 3.22) and the isolated island of Dunbuy. These spectacular landforms provide dramatic field evidence of the strong structural control on sequential development of erosion of rocky shores. The axis of the natural arch, which currently separates The Pot from the open sea, corresponds to a distinctive vertical joint in the granite. It is likely that marine erosion has exploited this joint, eroding a deep and extensive cave into the cliffs. Over time, presumably as a result of wave quarrying and abrasion, the cave roof has become structurally weakened, leading to collapse along its length except at the entrance (the natural arch). The boulder-floored nature of The Pot is almost certainly the result of the collapse of this sea cave. It is also likely that the till-capped island of Dunbuy was formerly an exposed headland exposed to marine erosion from more than one direction. A succession of erosional forms, caves, arches, tunnels or multiple geos, may have developed, with eventual roof failure leading to the isolation of the headland from the mainland plateau surface some 30 m away, similar to the formation of many stacks.

Where boulders and gravels are locally available, such as at North Haven and at Twa Havens and Dunbuy, beaches have developed at the heads of the inlets and geos. Where the orientation of the geo allows the entry of high magnitude storm waves, particularly from the southeast, the boulder beaches remain active and are well developed. The inception of such boulder beaches is likely to have occurred during the Holocene transgression and although some are still accessed in the severest of storms, they are now mostly abandoned.

Conclusions

The Bullers of Buchan GCR site is a comparatively small area, which contains a fine range of rocky coastal forms that have developed in massive igneous rock. The rocks have rectangular jointing patterns with a dominant near-vertical and horizontal pattern, together with a secondary pattern inclined at *c.* 45°. Exploitation by marine and subaerial processes along these weaknesses has resulted in angular, near-vertical

Figure 3.22 The Pot, Bullers of Buchan is a 60 m-deep enlarged blowhole connected to the sea by a 15 m-wide tunnel-like arch. (Photo J.D. Hansom.)

and triangular cliff forms. The landforms reflect the complex relationships between the strong structural control of the granite and varying degrees of wave exposure. Marine erosion has selectively eroded igneous dykes and exploited minor differences in geological structure, producing a complex and spectacular coastline with numerous geos, caves, arches, stacks, shore platforms, skerries and isolated islands, including the dramatic, 60 m-deep, enclosed sea inlet of The Pot.

DUNBAR, EAST LOTHIAN (NT 661 778)

J.D. Hansom

Introduction

The GCR site of Dunbar contains an excellent range of rocky coastal landforms within a 2 km stretch of coastline. Of exceptional note is a series of emerged and submerged shore platforms, often backed by cliffs of varying heights, including features that pre-date the last (Late Devensian) glaciation. Four distinct shore platforms are preserved, ranging in altitude from 25 m above OD, to 11 m below OD (Rhind, 1965; Sissons, 1967, 1976; Hall, 1989; Gordon and Sutherland, 1993). These landforms are representative of erosional coastal features found along the east coast of Scotland and are important for the interpretation of former sea-level changes and processes of rock coast development. The site is important in terms of Quaternary reconstruction and is included in the Quaternary of Scotland GCR coverage (Gordon and Sutherland, 1993). In addition to intertidal shore platforms, there are stacks, cliffs, offshore skerries and coarse gravels.

Considerable local geological variation occurs and the eight main rock types can be generalized into three divisions: the Old Red Sandstone in the east and its igneous intrusions; the basaltic tuffs of the central headland; and the Lower Carboniferous sandstones, cementstones and cornstones in the west together with their numerous dykes and pipes (Francis, 1975) (Figure 3.23). The foreshore is characterized by numerous volcanic intrusions that run through the fractured sandstone, mudstone and cementstone, as well as well-defined shatter zones at the junction between the bedded tuffs and breccias and the adjacent sedimentary rocks. Complex variations in intertidal shore platform morphology occur and, while there is less variation in the geology of the cliffs, these also show changes in height, slope and indentation that are related to rock type.

The nearshore seabed slope offshore of North Berwick is relatively gentle at 1:60 out to −20 m depth. Tidal range is about 4.5 m at mean spring tides. Dunbar is relatively sheltered from the west and south but is dominated by waves from between 20°N and 60°N, the approach directions of 35% of the storm and 60% of swell waves (Ramsay and Brampton, 2000b). The only information on nearshore wave conditions is from the Torness sea-wall construction *c*. 10 km to the south, which shows the largest waves to approach from 45°N to 90°N.

Description

Although altitudinal overlap occurs, the landforms of the area are best described by treating the emerged shore platforms separately from the intertidal features. The highest emerged platform (A, Figure 3.23) is one of a number of fragments that occur at 16 to 25 m OD between North Berwick and Berwick-upon-Tweed (Rhind, 1965; Hall, 1989). This emerged shore platform lies at an elevation of *c*. 20 m OD at Dunbar and its surface shows evidence of ice-moulding (Hall, 1989) and a thin cover of glacial till and of the drift tail of a crag-and-tail is preserved on its surface (Sissons, 1967, Hall, 1989).

A second platform (B, Figure 3.23) co-incides with the present intertidal zone and reaches a maximum width of more than 300 m west of Long Craigs, where it truncates the underlying sediments and agglomerates (Clough *et al.*, 1910, Francis, 1975). For about 1 km west of the Harbour, the platform is backed by a 20 m-high cliff composed of volcanic tuffs and sandstones and into which are cut several shallow caves. Several stacks protrude above the platform surface. Present-day gravel and boulder beaches mainly occur at the heads of the embayments in the cliffline.

Farther west, the backing cliff of the intertidal platform is degraded and fronted by Holocene beach deposits resting on a third platform (C, Figure 3.23) at an intermediate level, separated from the lower intertidal platform (B, Figure 3.23) by a rock step 1–2 m high (Gordon and Sutherland, 1993). Fragments of this platform, whose front edge lies between 2 and 4 m OD are found intermittently along much of the coastline between Aberlady and Torness (Hall, 1989). The relationships between the three platforms are best seen in section at NT 6633 7899 (Gordon and Sutherland, 1993). The lowest platform (D, Figure 3.23) occurs offshore at depths of between −11 and −13 m OD.

The intertidal landforms of the area vary depending on location and geology: the offshore skerries; the Old Red Sandstone in the east; the tuffs and agglomerates in the central section; and the Carboniferous sandstones with dykes in the west.

The 12 low skerries that lie approximately shore-parallel offshore from the Yetts to Long Craigs are part of a single Late Carboniferous quartz-dolerite dyke. They provide a degree of protection from wave processes at low tide but serve to widen the zone of wave breaking at high tide and consequently result in wave energy dissipation on the shore at high tide.

Between Victoria Harbour and the peninsula to the west of the bathing pool, most of the coast consists of an intertidal abrasion platform developed on tilted Upper Old Red Sandstone, although steep rugged cliffs cut in basanite occur at Castle Rocks near the harbour entrance and low irregular stacks occur to the south of the harbour. The surface of the intertidal platform is very uneven and is crossed by bedding and minor joint systems that have been selectively eroded. A patchy scattering of subangular boulders occurs on the surface. In the west, the distinctive conical stack of Dove Rock is a volcanic plug composed of basanite girdled by rings of tuff and breccia.

The intertidal zone of the central section comprises a series of wide seaward-dipping platforms cut in mixed tuffs and agglomerates and crossed by quartz-dolerite dykes, minor faults and distinctive rectangular fracture patterns that have produced numerous clefts and slot-like inlets. Isolated residual rock pinnacles and stacks of up to 8 m high interrupt the surface, for example, at Bath Rock and Pincod. A great variety of micro-morphological features occurs on the platform surface. Close to the platform surface potholes and sand- and gravel-filled pools occur whereas higher up pitting and honeycomb features are found. The platforms are backed by low but bold, near-vertical cliffs, with the cliff–platform junctions often masked by accumulations of boulders and gravel.

In the west, between the shatter zone and the sandy beach of Belhaven Bay, a bold north-west facing conglomerate cliff gives way along the line of shatter to a rock projection, which extends into Belhaven Bay. A fringing sand and gravel beach extends from the end of the cliff westward at the upper edge of the shore platform to merge at its western extremity with a low emerged ('raised') beach and dune area. Traversing the main rock platform, in a south-west to north-east direction, is a narrow, but prominent, quartz-dolerite dyke that rises from the adjacent surface as a rocky ridge.

Interpretation

A great variety of rock coastal forms exists within a small area at Dunbar. The juxtaposition of structural and lithological settings, together with changes in wave exposure and abrasion, have produced a broad range of cliff forms that show striking variations in plan and profile. At the micro-morphological scale, there exists a wealth of forms on the shore platform surface that are related to weathering processes, such as solution and pitting, and also a range of micro-forms related to abrasional processes. For example, on the intertidal platform to the north and west of the headland, narrow linear pools separated by densely spaced 30 cm-high ridges are partly structurally and lithologically controlled and partly the result of differential abrasion. To the north and west of the headland, the intertidal zone consists of lower shore platforms with many local differences in form according to rock type and structure. The low angle of dip of the strata (5–15°) produces a corrugated platform with changes in form and colour according to lithology.

At the larger scale, the shore platforms at Dunbar demonstrate the relationship between marine erosion, sea level and glaciation. Shore platforms occur at four main levels in this area (Rhind, 1965; Sissons, 1967, 1976; Hall, 1989; Gordon and Sutherland, 1993) although differentiation between levels is often problematic. The highest platform (A) is ice-moulded and has a till cover indicating that it pre-dates the Late Devensian glaciation, although its age is unknown (Hall, 1989). Similarly, the ages of the next two lower platforms (B) and (C) are unknown, although evidence from elsewhere suggests that they too pre-date the last ice sheet (Hall, 1989), with the lower of the two (B)

Dunbar

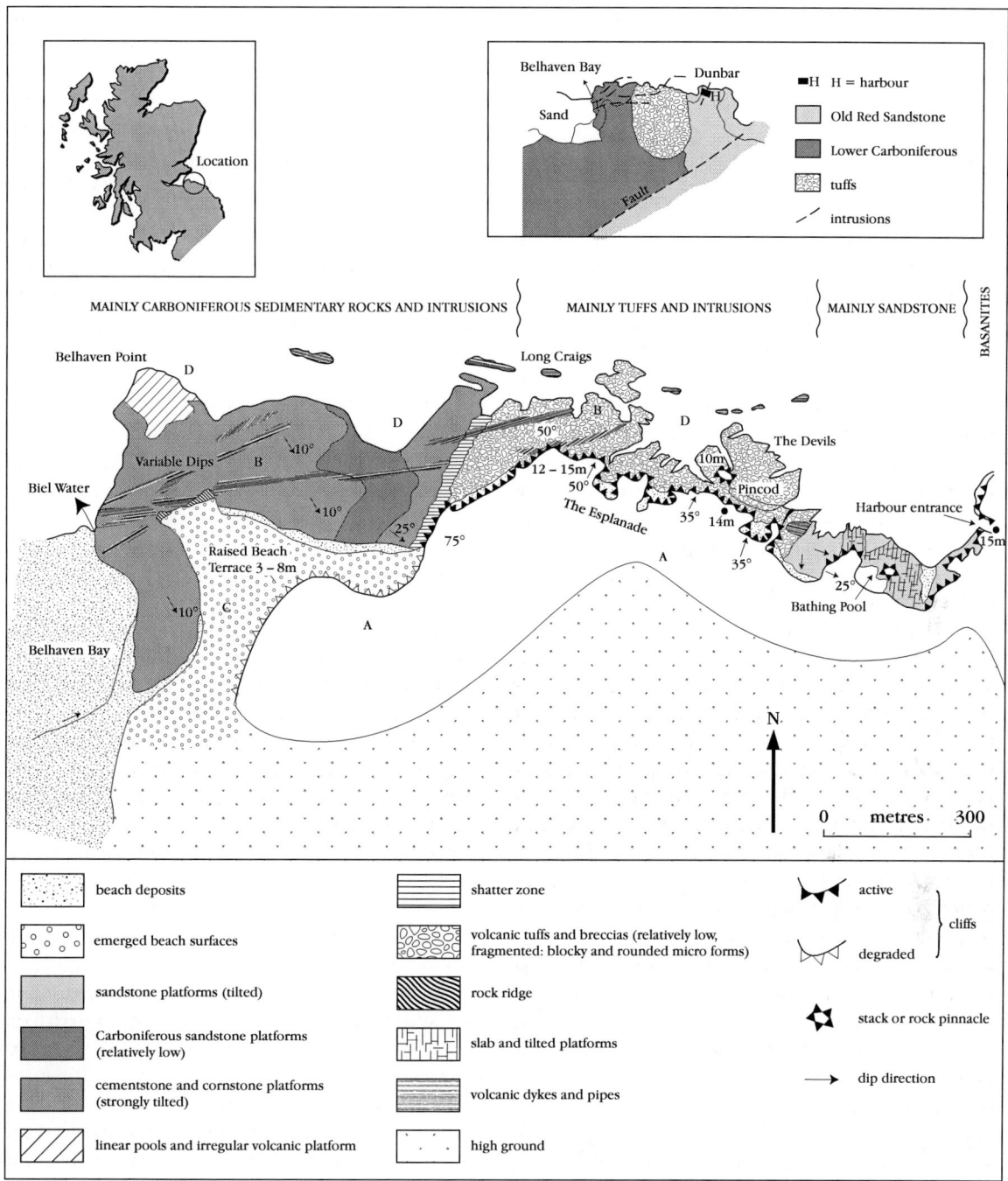

Figure 3.23 Geomorphological map and geological sketch map of the Dunbar GCR site showing a series of emerged rock shore platforms that have been eroded across a varied geology and are backed by cliffs of varying heights. Platform A lies at *c.* 20 m OD and is ice-moulded; Platform B is intertidal; Platform C underlies emerged Holocene beach deposits and Platform D is subtidal. (After Gordon and Sutherland, 1993.)

suffering partial stripping of its till cover and renewed marine erosion in the present intertidal zone during the Holocene Epoch. It is possible that both platforms form part of an intertidal platform equivalent to the Low Rock Platform of western Scotland (Dawson, 1980a). The offshore platform has been correlated with a buried gravel layer and platform farther west in the Firth of Forth and a submerged platform at Burnmouth farther south. As a result, it may be part of the Main Lateglacial Shoreline, submerged in south-east Scotland, and so may date from the Loch Lomond Stadial that occurred 11 000–10 000 years BP (Sissons, 1974).

Conclusions

The GCR site of Dunbar displays a wide range of rocky coastal landforms within a relatively small area. In addition to intertidal shore platforms, there are also coarse beaches, offshore skerries, stacks and cliffs together with a range of small-scale erosional forms that owe their genesis to variations in rock type and structure and the differential effects of wave processes on these. However, Dunbar is most notable for a series of shore platforms of different ages, including three examples that probably pre-date the Devensian glaciation. Two of these (B and C) lie close to the altitude of the present intertidal zone and consequently have been exhumed and eroded by the Holocene sea. The lowest platform (D) may have been eroded during the Loch Lomond Stadial some 11 000–10 000 years BP and is now isostatically submerged. The juxtaposition of these multiple shore platforms and their altitudinal overlap with the present intertidal area, makes Dunbar one of the best examples in eastern Scotland to illustrate the relationship of exhumed platforms to glaciation, sea level and the landforms of the contemporary coast. As a result, the site is also important in terms of Quaternary reconstruction and is included in the Quaternary of Scotland GCR coverage (Gordon and Sutherland, 1993). In addition, the coastal landform assemblage at Dunbar is representative of the erosional processes and forms characteristic of rock-coast development in south-east Scotland.

ST ABB'S HEAD, BERWICKSHIRE (NT 902 690–NT 917 677)

J.D. Hansom

Introduction

St Abb's Head, some 65 km east of Edinburgh, forms a coastline of magnificent rugged and precipitous cliffs cut by numerous clefts, gullies, geos, caves and coves with many offshore stacks, reefs and skerries. This serrated coastline demonstrates well the intricate relationship between marine processes and geological structure (Figure 3.24). Marine exploitation of geological weaknesses within the largely igneous rock mass has created a coastline containing a great variety of spectacular hard-rock coastal landforms, which display substantial local contrasts within a comparatively small area. In addition, the marked contrast in coastal form between the felsite headlands of Lower Devonian age at St Abb's Head and the sedimentary rocks of Silurian age to the north-west adds to the geomorphological interest of the site. Steers (1973) described this stretch of coastline, together with the steep cliffs of the Silurian sedimentary rocks, as 'one of the finest lines of cliff in these islands'. However, to date, almost all of the scientific literature concerns the geology and there has been little detailed geomorphological research.

St Abb's Head is relatively sheltered from the west and south but is very exposed to the north and north-east and so 35% of the storm and 60% of swell waves approach from between 20°N and 60°N (Ramsey and Brampton, 2000b). The only information on nearshore wave conditions is from the Torness sea-wall construction, some 18 km to the nort-wesuth, which shows the largest waves to approach from 45°N to 90°N. The nearshore seabed slope offshore of St Abb's Head is relatively steep at 1:45 out to –40 m depth and allows access of storm waves to the shore.

Description

Some 200 m inland, between White Heugh in the south and Pettico Wick in the north, the headland of St Abb's Head is bisected by a major geological fault that runs parallel to the outer coast in a north-west to south-east direction. The fault line is marked by a distinctive inland valley depression (occupied by Mire Loch) and the isolation on the seaward side of the valley of a series of volcanic ridges that reach over 75 m OD and meet the coast in a series of high cliffs. West of the fault boundary lie the Silurian sedimentary rocks, although in the south lies an area of Devonian conglomerates with inliers of Devonian–Carboniferous rocks and an area of Devonian intrusive rocks. East of the fault boundary are the Devonian extrusives of St Abb's Head, largely felsites with tuffs and grits. A series of minor faults run perpendicular to the major fault in a north-easterly direction.

St Abb's Head

The coastline is extremely serrated, complex and rugged (Figure 3.24), but can be subdivided into three sections depending on orientation. To the south-west of the fault boundary, the cliff altitude lies below 50 m OD and the cliffs are formed mainly of Silurian greywackes with a till cap. Below these lies a series of gravel and boulder pocket beaches masking an intertidal shore platform, outcrops of which appear landwards of a high elongate offshore stack (Craig Robin). At Hardencarrs Heugh, cliffs on the northern side of an inlet and deep, boulder-filled gully are matched by the high and narrow grass-covered peninsula of White Heugh on the south side. This marks the point where the St Abb's fault meets the coast. The complex and dissected headland of Wuddy Heugh lies to the north-west of the fault boundary. The cliff tops are till covered but the slope is generally steep, rocky and bare with no shore platform at the base. Small-scale structural features and minor lithological differences in the igneous rock have been etched out as numerous small indentations and irregularities at the cliff base.

The north-east coast, although only 2 km long, contains a great variety of spectacular rocky coastal landforms, displaying substantial local contrasts within a comparatively small area. The hinterland topography consists of three main ridges, each reaching over 75 m OD and sloping at various degrees towards the coast and showing distinctive benches and steep facets on the grass-covered hillsides. On the coast, local differences in rock type are translated into differences in resistance to erosion. Horsecastle Bay and Cauldron Cove are distinctive low-lying depressions, the former characterized by a well-formed gravel storm beach and boulder-strewn intertidal zone. Between these two depressions, the steep grass-covered cliffs and inlets form part of the slope of Kirk Hill (Figure 3.24). Distinctive finger-like rock peninsulas occur here (e.g. the long asymmetric ridge of Waimie Carr, vertical on its northern side and gently-sloping on its southern side), together with blocky scree slopes, finger-like shore platforms and high, vertical cliffs. North of Cauldron Cove, the 75 m OD 'lighthouse cliffs', consist of extensive and steep grass slopes punctuated by a series of grassy benches protruding as free faces. An intertidal shore platform occurs at the base of the cliffs, although in places this is replaced by near-vertical plunging cliffs. The offshore area is complex with numerous skerries and high, fractured stacks.

Between the lighthouse and Hope's Heugh the cliffs are generally higher, bolder and more deeply indented, with deep geos, inlets, prominent finger-like ridges and narrow peninsulas, some of which continue offshore as elongate ridges or stacks. Headland Cove is a good example of a long linear geo with near-vertical sides. At Hope's Heugh there is an spectacular example of a natural arch cut through an elongate ridge. The cliffs are commonly near-vertical at the base, below very steep (c. 20–25°) grass-covered inclines that slope to the main cliff tops. Between Hope's Heugh and the bay of Pettico Wick the upper parts of the high cliffs are steep and predominately grass-covered whereas the lower parts are bare rock above gravel and boulder beaches. In the south, close to the distinctive triangular-shaped stack of Staple Rock, the coast is very rugged and complex owing its morphology to the topography of the hinterland as well as marine erosion. West of the fault boundary at Pettico Wick to Broadhaven, the Silurian greywackes, siltstones and shales have steeply dipping and tightly folded sedimentary beds and the cliffs display excellent examples of bedding plane control, with high-angled, slab-like cliffs that reach heights of up to 152 m west of Broadhaven Bay. The upper cliffs are very steep and grassy although the lower parts are bare. The cliff base in Broadhaven Bay is characterized by a relatively extensive, and well-developed, intertidal shore platform, interrupted only by small embayments with gravel and boulder beaches.

Interpretation

St Abb's Head provides a good example of the effect of a major fault on the planimetry and geomorphological development of a rocky coastline. The fault marks a distinct change in lithology between the sedimentary province in the west and the St Abb's Head igneous province in the east. The hinterland topography of the headland consists of a north-west to south-east valley developed along the fault line and a series of high volcanic ridges composed of different lava flows running normal to the fault. This geological control has resulted in a serrated and intricate coast characterized by steep high cliffs that have grass-covered upper faces and bare lower faces. Marine exploitation of minor faults has produced a distinctive series of finger-like

Hard-rock cliffs

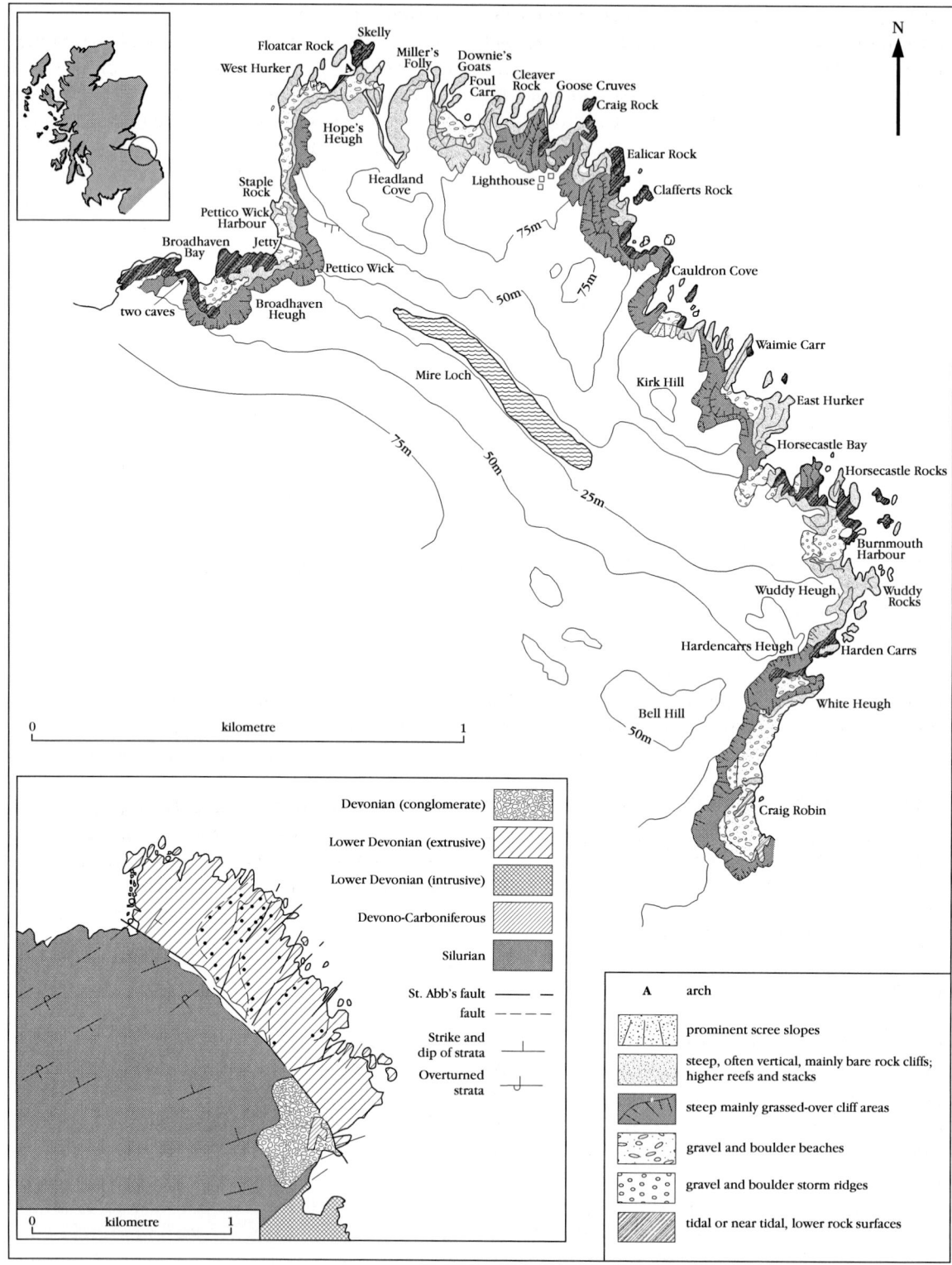

Figure 3.24 Geomorphological map and geological sketch map of St Abb's Head showing the heavily indented nature of the coast resulting from a strong structural control. (Modified from unpublished work by W. Ritchie.)

ridges, geos and inlets that trend normal to the axis of the main fault, producing a rugged coastline. In this respect, the coastline of St Abb's Head provides an exceptional example of the strong control of geological structure on coastal geomorphology.

There is a marked contrast of the cliff form on either side of the major fault. The abrupt geological transition from the felsite of the headland to the Silurian greywackes, siltstones and shales to the west of the fault is reflected dramatically in coastal form. The high-angled, slab-like cliffs cut in the sedimentary rocks strongly reflect the steeply dipping and tightly folded sedimentary beds, displaying excellent examples of bedding plane control. The unique contrast between these sedimentary cliffs and the rugged cliffs and associated forms of the igneous headland provide an excellent site for both research and educational purposes.

A complex interplay between geological structure, marine processes and subaerial processes is evident at St Abb's Head. Exposure to high-energy waves from the easterly quarter has resulted in very effective quarrying and abrasion processes that have exploited both major and minor lithological and structural differences in the igneous rock mass. Geos, caves, inlets, stacks, arches, rock peninsulas and ridges all reflect the differential resistance of the rock to marine processes. In addition, subaerial processes play a part in shaping this unique headland. The distinctive benched profile of the grassed upper coastal slopes at St Abb's Head, with numerous rock outcrops, reflects differential subaerial erosion of the sequence of lava flows that make up the volcanic headland. Free rock faces of lavas and grits often protrude from the grassy cliff slope, some of which are subject to active subaerial erosion, the rate of which is a function of minor differences in lithology and geological structure. The numerous scree slopes and boulder fields (e.g. the bay between Cleaver Rock and Foul Carr) are now largely inactive.

The origin of the shore platforms in this area is also worthy of note. Unlike at Dunbar, where four levels can be identified and approximately dated, only intermittent development of intertidal shore platforms occurs at St Abb's Head. The platforms appear to be best developed in both igneous and sedimentary rocks where the surface of a lava bed or bedding plane crops out in the intertidal zone. Irrespective of geology, all of the shore platforms are intertidal abrasion ramps. However, in spite of suitable structural conditions for the widespread development of platforms, the general distribution of platform remnants separated by embayments suggests that a once more extensive platform has undergone dissection. Some of these remain active under present conditions but, in common with many other east coast cliffs capped by till deposits, it is probable that both shore platforms and cliffs are exhumed features that have undergone modification by Holocene marine erosion.

In summary St Abb's Head is of high scientific importance for the following reasons:

1. The clear relationship between geological structure and coastal form.
2. The contrast between the coastal forms of sedimentary and igneous rocks indicating the strong relationship between lithology and coastal form.
3. The spectacular coastal forms produced in a large igneous rock mass.

Conclusions

The igneous mass of St Abb's head forms a spectacular rugged coastline with numerous clefts, gullies, geos, caves, stacks, reefs and skerries. Headland Cove provides a dramatic and textbook example of a near-vertical geo, the adjacent steeply benched coastal slopes rising to above 75 m. Of principal geomorphological importance is the clear relationship displayed between lithology, structure and coastal form. Marine erosion has exploited planes of weakness within the igneous rock (e.g. major and minor faults and local lithological differences) creating a complex, varied and highly indented coastline. The transition from the felsite of the headlands to the sedimentary rocks west of the fault boundary produces a dramatic and unique contrast in coastal form that enhances the geomorphological interest of the site.

TINTAGEL, CORNWALL (SX 043 858–SX 070 895)

V.J. May

Introduction

Much of the northern coastline of Devon and Cornwall is characterized by cliffs cut into rela-

Hard-rock cliffs

tively resistant rocks. Even on this resistant coastline, differences of rock strength are reflected in the development of a headland and bay topography. The rocky cliffed headland at Tintagel is one of the many locations where the coastal forms are strongly related to major structural features (see Figure 3.1 for general location). It is also one of the few locations where these relationships have been studied in an assemblage of coastal forms, including slope-over-wall cliffs, geos, caves, arches and stacks.

Description

The site extends from Start Point in Backways Cove (SX 043 858) in the south to Bossiney Haven (SX 070 895) in the north. There are sandy beaches at Trebarwith Strand and Bossiney Haven, but much of the coast is formed by cliffs that drop over 100 m directly to below sea level. Some lower cliffs form the lower element of slope-over-wall features: some of these latter forms are bevelled, whereas others form hogbacks. The cliffs and rock platforms are cut in Upper Devonian and Lower Carboniferous rocks, the former much affected by metamorphism. Although there are major thrust faults, the cliff form is most influenced by several roughly parallel normal faults. South of Tintagel Island, some short stretches of cliffline are true fault-line cliffs. Elsewhere erosion has cut back the cliffs from their original fault-controlled position. North of the Island, the coastline is more complex, with many inlets and headlands. Erosion along normal faults, less-resistant beds and joint-planes has produced an intricate set of bays, headlands, stacks, blowholes and caves. Local variations of structure and rock strength are the major control on the landforms. Dewey (1909, 1914) and Owen (1934) described the structures of the area, and Steers (1946a) outlined the main coastal features. Cotton (1951) saw much of this coastline as having two cycles of development, placing particular emphasis on the differences in the cliff profiles. The most important work was carried out by Wilson (1951, 1952) who described the relationship between the coastal features and structures (Steers, 1971a).

This very indented cliffed coastline (Figure 3.25) can be subdivided on the basis of its present cliff and beach morphology into six sub-units.

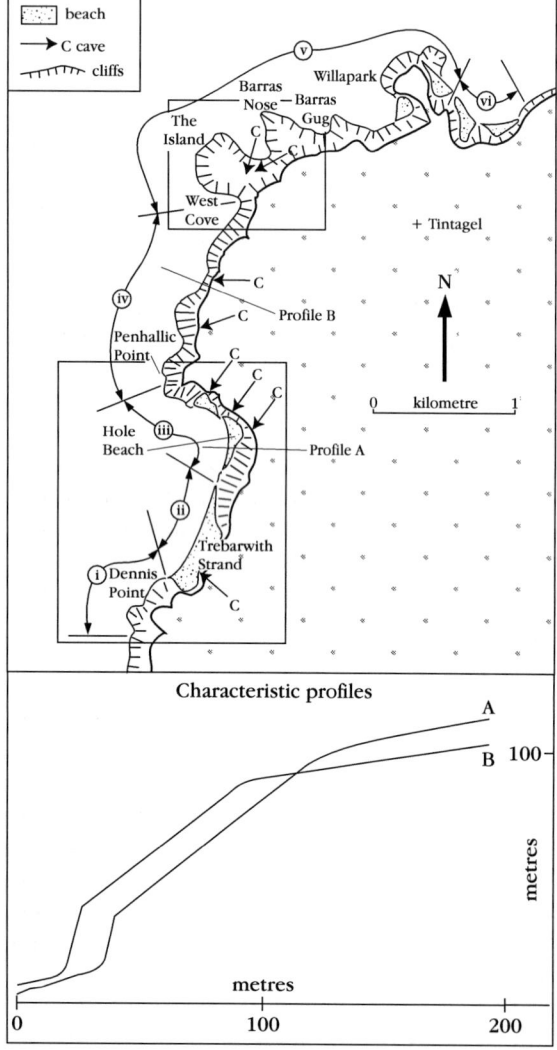

Figure 3.25 Main features of the Tintagel coast (i) Start Point to Dennis Point: vertical and slope-over-wall cliffs; (ii) Trebarwith Strand: sand beach backed by cliffs over 90 m high; (iii) Hole Beach: caves developed on line of faults and thrust planes; (iv) Penhallic Point to West Cove: slope-over-wall; (v) West Cove to Bossiney Haven: complex coast with peninsulas at different stages of separation from mainland; (vi) Bossiney Haven: geo and arch. The inset shows characteristic slope-over-wall forms between Trebarwith Strand and Tintagel Island.

(i) Between Start Point and Dennis Point (SX 043 858 to SX 045 863), the cliffs vary in both height and form. At Backways Cove near-vertical cliffs are only 15 m high at the mouth of a hanging valley, whereas at Dennis Point the overall height of a well-developed slope-over-wall form exceeds 80 m. Gull Rock is an isolated stack about 500 m offshore.

Tintagel

(ii) Trebarwith Strand (SX 045 863 to SX 049 868) is a sandy beach, backed by cliffs that rise to over 90 m. Access to the beach is via a hanging valley with a floor at about 14 m OD whose seaward end has been much degraded by paths and steps.

(iii) At its northern end the beach gives way to a complex of boulders, eroded volcanic rocks including elongated pillow lavas. The coastline has a right-angle bend at Hole Beach where it is cut by a normal fault, to the east of which is a major thrust plane. The lower near-vertical rock-faces usually give way to an upper slope but occasionally extend almost to a bevelled surface at about 80 m.

(iv) From Penhallic Point (SX 046 877) to West Cove (SX 050 889) slope-over-wall forms are dominant and the plan is characterized by several straight sections aligned from north-east to south-west *en echelon* and separated by shorter north–south sections.

(v) Between West Cove, Tintagel and Bossiney Haven (SX 065 896), the coastline is very complex. Three promontories, The Island, Barras Nose and Willapark, each with a narrow neck, are in different stages of separation from the mainland. The cliffs are mainly slope-over-wall forms bevelled at about 80 m OD, but at Willapark there is an excellent example of a hogback cliff.

(vi) Bossiney Haven has a strongly joint-controlled geo as well as the well-known Elephant Rock where a high vertical arch has formed almost separating a narrow 'trunk' of rock from the mainland (Figure 3.26).

The main features of the site were described by Wilson (1952) and this account is based largely upon Wilson's interpretation of the geological features of the area. The coast is cut into Upper Devonian slates, siliceous sandstones, pillow lavas and tuffs and phyllites, which have been overthrust towards the NNW (Wilson, 1951). The overthrust strata were affected by approximately parallel normal faulting. The beds dip generally to the west and the normal faulting throws the thrust-slices down to the west or north-west. The faulting at Tintagel (Figures 3.27 and 3.28) is dominated by two important fault zones: the Castle Fault between West Cove and Smith's Cliff, and the Caves Fault Zone, which cuts through The Island across

Figure 3.26 Elephant Rock, Bossiney, showing the relationship of cliff features to vertical jointing. (Photo: V.J. May.)

Tintagel Haven to Barras Gug. Similar fault zones affect the cliffs both north and south of Tintagel. The thrust planes lie at low angles, but the normal faults form sloping shear zones, which Wilson noted are easily worked on by marine erosion. Joints particularly with a general alignment towards 325–330° and north–south joints also play an important part in the coastal morphology of this site.

Much of the coastline is distinguished by narrow joint-controlled inlets (for example at Bossiney Haven), known locally as 'guts' or 'gugs', caves cut along fault zones, as well as landsliding on undercut seaward-dipping bedding planes and inclined fractures (Wilson, 1952). Marine erosion along such features has allowed the sea to reach relatively weaker materials, such as the slates, and to cut narrow inlets parallel to the coastline. Wilson compared the site with the coastline around Lulworth Cove. The development of caves is strongly associated

Figure 3.27 Major fault and thrust at Tintagel as the focus for marine erosion, cave and ultimately stack development. (Photo: V.J. May.)

with fault zones, but Wilson also considered that some inlets associated with faults had been the focus of more rapid erosion when they co-incided with former or present lines of drainage. South of The Island, the en-echelon form of the cliffline is strongly linked to faults and other weaknesses parallel to the general alignment of the coastline.

Interpretation

Wilson (1952) observed that active erosion took place along structurally controlled and preferred locations. Where structural weaknesses were flat or gently dipping, they only influenced the process of marine erosion if they occurred close to sea level. In contrast, steeply inclined lines or zones of weakness could control the direction of marine erosion over a large range of sea levels, for if the line of weakness continues through the cliffs both above and below sea level, any features associated with it can continue to develop whether sea level falls or rises. Normal faults appear to have been most important as they trend at an acute angle to the present-day coastline. Moreover, most of the faults on this coastline strike in a direction more or less parallel to the direction of maximum fetch. Once the sea had penetrated into these parallel fault-zones it began to cut back the cliffline by undercutting the harder rock bands between the inclined shatter zones (Figure 3.27). Since many of the faults dip seawards at about 45°, cliffs develop by removal of the material of the shatter zone material and the development of a structurally controlled sloping surface. The sea would subsequently cut a vertical wall in the lower part of the slope to produce the slope-over-wall form.

Wilson also considered the two-cycle model proposed by Cotton (1951) in which the structurally controlled slope was first eroded by the sea, but with a fall in sea level and the onset of periglacial conditions during the last glacial, the upper slope was affected by subaerial slope processes, a talus of debris accumulated to protect the former sea cliff. With a Holocene rise in sea level, the debris would be removed and the sea would exhume and retrim the former sea cliff. Wilson believed that Cotton's two-cycle origin for the cliffs explained many of the coastal features of the area. Unfortunately there is no evidence of relict talus or emerged ('raised') beach deposits within this site to corroborate it. Caves occur at or close to present sea level, but are absent at higher levels and there are no reports of submarine caves or the continuation of caves below sea level. Nevertheless, the alignment of 'guts' and coves with hanging valleys indicates that preferred lines of erosion were available to the sea. Along the foot of the cliffs, the debris of 'ancient rockfalls is still to a great extent protecting the base of the cliff from erosion' (Wilson, 1952, p. 39). Modern falls, however, have the same effect but, as elsewhere on the cliffed coasts of Britain, there has been no investigation of their longevity and their effectiveness in providing temporary armouring to the cliff foot. Between Penhallic Point and Tintagel the cliffs drop directly into the sea with only narrow steps forming the intertidal area. Elsewhere in the site, platforms occasionally reach 150 m in width and occur either at the foot of the cliffs or in the small bays where they underlie the sandy beaches. There has

South Pembroke Cliffs

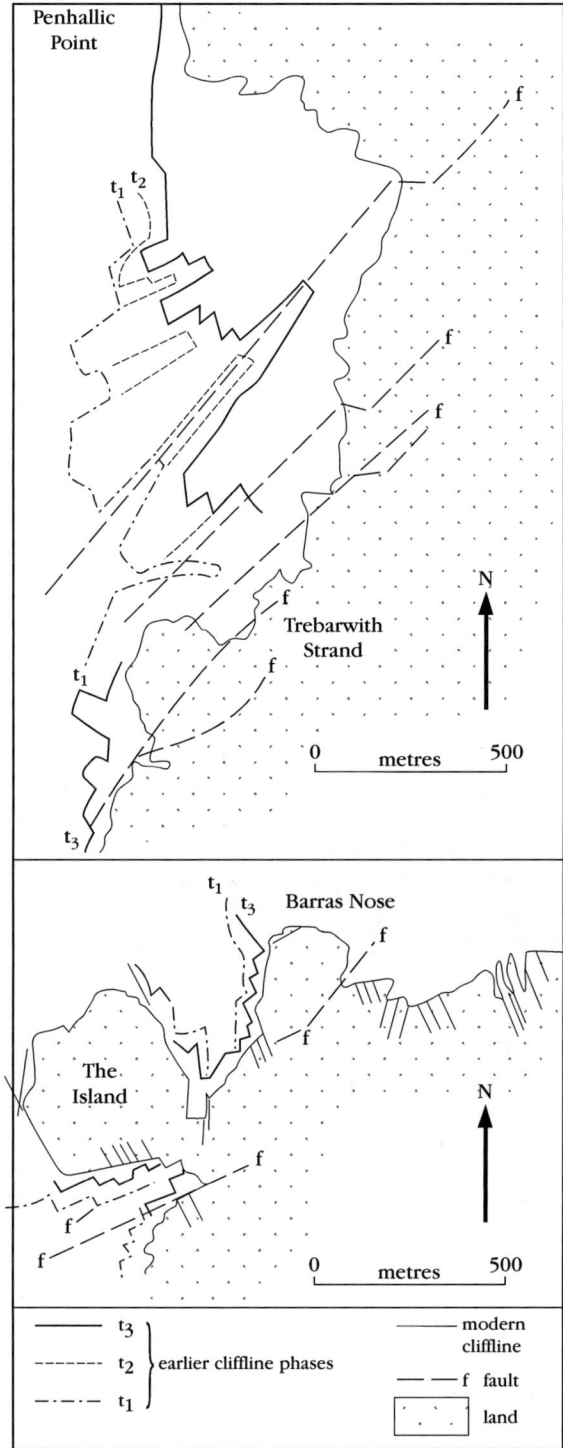

Figure 3.28 Examples of coastline development controlled by major faults, Penhallic Point and Barras Nose. See Figure 3.25 for general location. (After Wilson, 1952.)

been no detailed investigation of the shore platforms, but their form appears to reflect strongly the effects of rock strength and the detailed structures.

Wilson's (1952) paper remains unusual amongst the coastal literature in considering the detailed relationship between cliff development and rock structures on a hard-rock coast. There are comparable sites, for example, in south-west Wales and at Trearrdur Bay in south-west Anglesey, but no comparable work. Like Hartland Quay (see GCR site report) to the north, this site contains hanging valleys, waterfalls, hog's-back and bevelled cliffs. Unlike Hartland Quay, it also demonstrates very clearly the relationship of structure to cliff development. It is not generally regarded as a longitudinal coast, but in terms of the development outlined by Wilson (1952) it has similarities to the coastline around Lulworth Cove (see GCR site report for the Dorset coast in Chapter 11). Lulworth Cove is backed by relatively weak materials and so the effects of breaching of an outer resistant wall are followed much more strikingly by the development of bays than has occurred at Tintagel. Tintagel is thus important not only because of the links between structures and landforms, but also because it provides a contrasting example to Lulworth Cove.

Conclusions

Tintagel is one of the very few hard-rock sites where the relationships between major structural features and coastal development have been examined in detail. The landforms include geos, caves, stacks, arches and slope-over-wall cliffs. An excellent example of the way in which major structural features can control the development of coastal landforms, it is also a good example of a longitudinal coast, although not so strikingly obvious as the most commonly cited example at Lulworth Cove. Unlike the latter, it is predominantly in hard rocks and the rates of change are less.

SOUTH PEMBROKE CLIFFS, PEMBROKESHIRE (SR 958 932–SR 966 928; SR 922 944–SR 942 940)

V.J. May

Introduction

The coastline west of St Govan's Head contrasts with that of southern Gower because of its

Hard-rock cliffs

absence of emerged ('raised') shore platforms and the presence of steep active cliffs. The two sections of cliffed coastline that form this site (Figure 3.29) enclose some of the finest examples of coastal forms in England and Wales. Cut into massive limestones of Carboniferous age, the cliffs include exceptional examples of the development of geo, stack, cave and arch. Faults and other lines of weakness have been exploited by the sea to produce such well-known features as the Green Bridge of Wales, Elegug Stacks and the Huntsman's Leap. The importance of this site is greatly increased by the retreat of the coastline into an area of karstic landforms. Thus, the combined effects of solution, collapse and marine reworking of these landforms have produced an intricate and geomorphologically important assemblage of forms. Like a number of cliffed sites, the literature is limited (Steers, 1946a, 1969; Guilcher, 1958), and a single paper provides most information about the nature and origins of the site (John, 1978).

Description

There are two parts to this site (Figure 3.29). The first part (SR 958 932 to SR 966 928) includes the Huntsman's Leap (an excellent example of a geo). The second (SR 922 944 to SR 942 940) includes the Green Bridge of Wales, Elegug Stacks and the Devil's Cauldron. The cliffs rise to between 45 m and 50 m where they cut the Flimston 'coastal flats' – an erosion surface generally attributed (John, 1978) to marine erosion during Pliocene or early Pleistocene times. At the coastal edge, the structures of the Carboniferous Limestone are truncated not only by the cliffs, but also by this well-developed ero-

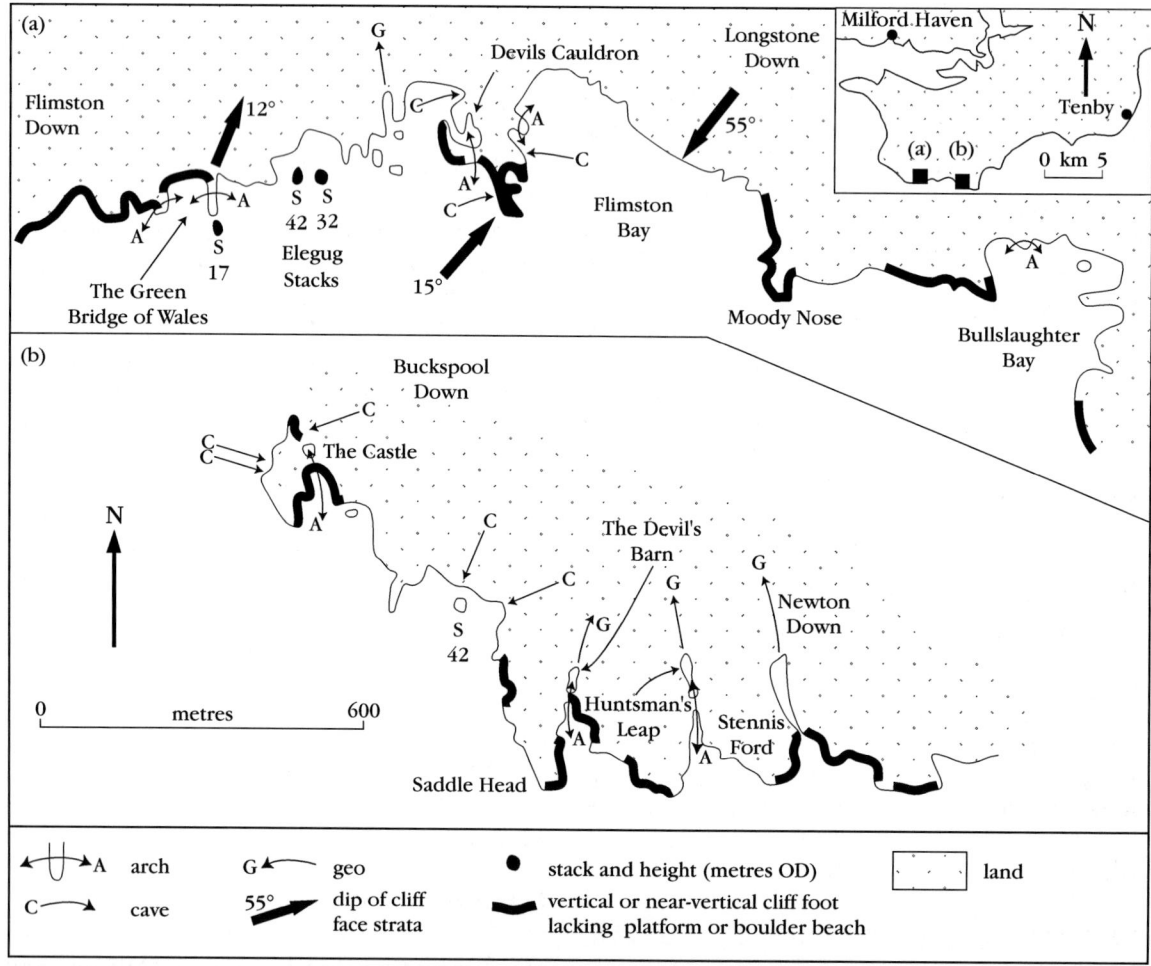

Figure 3.29 Erosional features of the south Pembrokeshire coast. (After John, 1978.)

South Pembroke Cliffs

sion surface. The dip of the Carboniferous Limestone exposed in the cliffs varies from landwards, west of the Devil's Cauldron, to seawards, east of Flimston Bay. Cliff forms that are steep, near-vertical and occasionally overhang, where the dip is to landward (Figure 3.30), are replaced by cliffs that are much gentler in profile and where the seaward-dipping beds largely control the cliff form. Much of the cliff foot is marked by a jumble of boulders from both recent and older rock falls.

The eastern part of the site is distinguished by the best-developed geos on the coast of England and Wales. In addition, the future development of similar features can be predicted as groups of aligned blowholes and caves provide the focus for marine erosion. The Huntsman's Leap and Stemmis Ford are two fine examples of geos, the latter extending about 180 m in from the coastline. The Devil's Barn includes two blowholes that are the cliff-top expression of an arch and marine erosion beneath them. At the Castle, there is a sequence of caves and arches as well as a blowhole. If their roofs collapsed much of this area would become separated from the mainland. They are probably solution forms that are being reworked by marine action.

The western part of the site includes some of the most unusual coastal forms of the coastline of England and Wales. The Green Bridge of Wales is an arch of about 24 m in height and it spans more than 20 m. Its upper surface slopes down from the cliff top. The outer limb of the arch rests on a broad pedestal-like base. Here the limestone dips inland. The Elegug Stacks, of which the higher reaches about 36 m, also rise from a broad, sloping pedestal. A fault runs through the eastern base of the larger of the two stacks. To the east of the Elegug Stacks, the sea has exploited a large number of faults and major shear planes, as well as deposits of gash breccia, to produce an intricate assemblage of caves, arches and geos (Figure 3.31). Of 52 faults and major shear planes recorded by John (1978), 22 co-incide with a cliff face, 11 form the axis of geos and 19 are associated with neither. Only four co-incide with a cave or arch. This pattern continues to the east in Flimston Bay. It is largely absent on Longstone Down where the Bullslaughter syncline produces seaward dips of up to 55°. From Moody Nose eastwards, the local control of erosion by faults and shear planes is also very evident.

At Flimston Castles, the coastline is extraordinarily complex and includes the Devil's Cauldron. Here a shaft, 45 m deep with a maximum diameter of 55 m, is open to the sea via an arch 18 m high and 21 m wide, a narrow fault-guided chasm connecting with the sea. John (1978) cites Thomas' estimate that some 113 000 m³ of rock was removed to produce the Devil's Cauldron. Its considerable interest arises from the fact that much of this coastline truncates solution features of the limestone landscape. The Devil's Cauldron, like several other features on this coast, is probably a karstic form that has been exploited by the sea as the coastline has retreated. Similar eroded features occur at Flimston Castles and at the Devil's Barn and the Castle.

Figure 3.30 Cliff profiles, South Pembroke Cliffs GCR site. Cliffs are steep, near-vertical and occasionally overhang where the dip is to landward. (Photo: S. Campbell.)

Interpretation

There is no comparable site in England and Wales, and, with a few exceptions, similar exam-

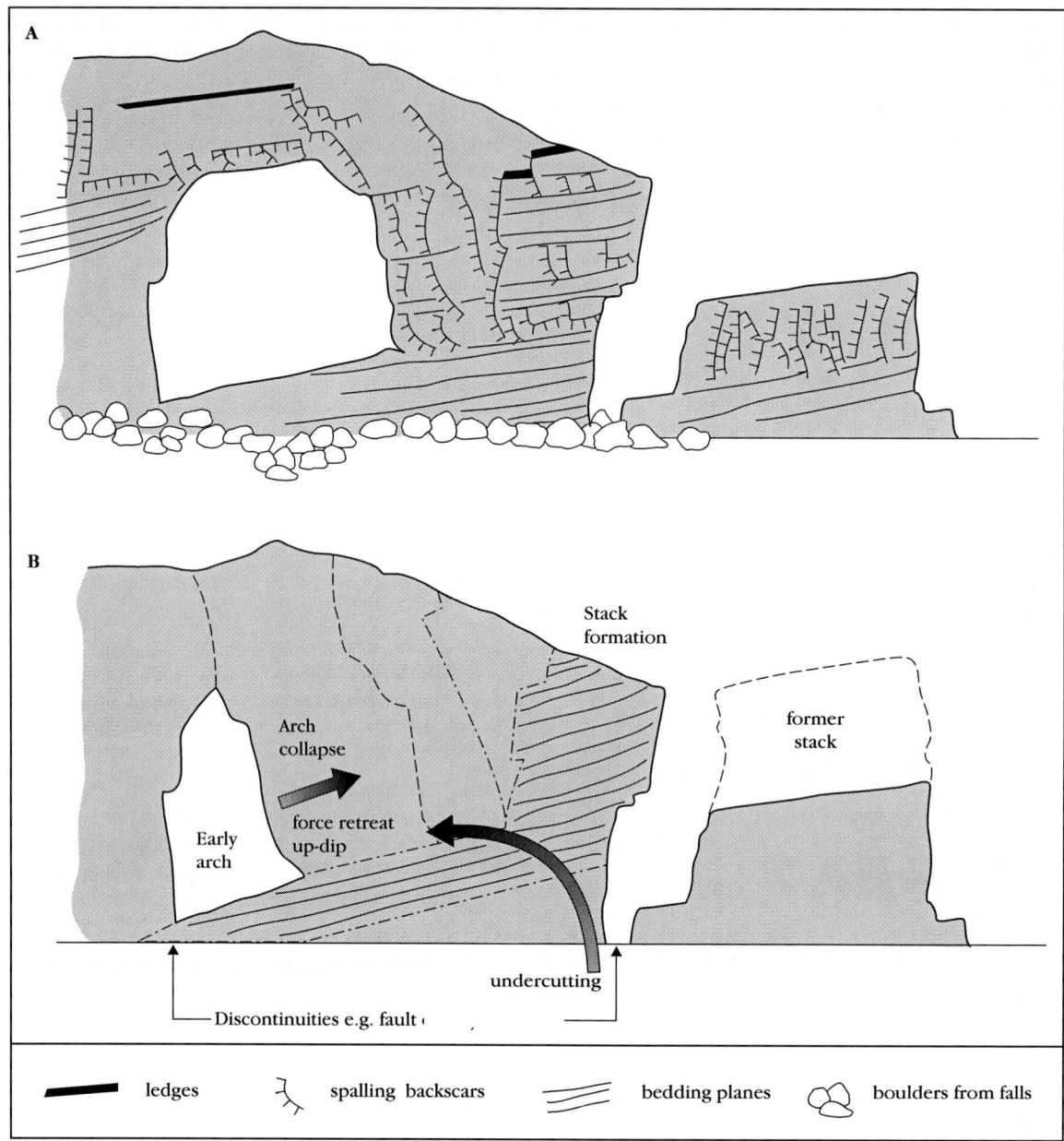

Figure 3.31 Arch and stack development. (A) Form of the arch and stack at The Green Bridge of Wales. (B) Interpretation of development of the feature. An initial arch develops on the line of a discontinuity, and extends up-dip by spalling and collapse of up-dip rock surfaces. The arch roof collapses and a new stack is isolated.

ples around the European coastline have rarely been described in detail. The only other similar site in Britain in which exhumation of erosional forms has been well described is the Bullers of Buchan in Scotland. There, however, the stacks and geos are often cloaked by, and infilled with, till and so their preglacial or interglacial origins can be accepted with little question. Here, the evidence is more circumstantial. A rock platform that may be reworking an earlier form is subject to erosion at present by both chemical and physical processes.

Marine erosion of karstic forms is common along the north-eastern Adriatic coast, and comparable features on the south coast of Gozo, Malta have been described briefly (May and Schwartz, 1981; Paskoff and Sanlaville, 1978). Although stacks and arches are well represented in Chalk, such cliffs generally lack the development of geos found here. Some features where karstic forms are truncated by marine erosion occur in Chalk north of Flamborough Head. The South Pembroke cliffs include a well-developed coastal landscape in Carboniferous Limestone and, unlike much of the Gower peninsula, is not characterized by emerged ('raised') beaches and platforms. Nevertheless, the coast both to the east and west provides evidence of considerable longevity. In West Angle Bay, pre-Devensian till may underlie Ipswichian emerged beach deposits and both lie above a rock platform, which could therefore be attributable to higher sea level during the last Ipswichian interglacial (Campbell and Bowen, 1989). Inland at Hoyles Mouth and Little Hoyle's Cave sediments have been interpreted as showing that the limit of the Late Devensian ice was close to this site. Furthermore these caves record the occupancy by humans in Upper Palaeolithic times (*c.* 18 000 years BP) and suggest that the area to the south must have been ice free. Similarly Marros Sands farther to the east preserves evidence of intense Devensian periglaciation. It thus appears that the South Pembroke coast was ice-free during Devensian time, was probably affected by intensive periglacial processes and that parts of the coast may have been reworked. However, although John (1978) has suggested that parts of the cliffs may be more than 5 million years old and that others may date from the last interglacial, there is no direct evidence that the south Pembroke cliffs preserve former features. For example, the present sea-level platform may not be entirely contemporary, but no evidence is available to support or reject the hypothesis that it is exhumed. There is similarly no evidence to date for the age of the karstic features into which the cliffs are currently being cut. Solution forms are well developed on some parts of the platform (Guilcher, 1958) and are generally regarded as being contemporary features. Any attempt to relate the levels of the platforms to past sea levels must take account of the rates at which solution takes place. There has been, however, little research here into the detailed evolution of these forms.

The present-day changes in features such as the Green Bridge of Wales as recorded by photographs show that most changes this century have occurred on the down-dip side of the arch wall. Undercutting and spalling have narrowed this part of the feature and provide an insight into both its future and more generally the development of stacks in this hard-rock context (Figure 3.31). The original break through the promontory from which the bridge formed probably occurred at the point where the pedestal rock occurs at the cliff foot. The seaward face of the cave and then the arch has retreated more rapidly. This face has retreated most rapidly at its base and in due course when the arch collapses the stack will stand on a pedestal several metres above sea level. The largest geos are characterized by a narrow neck and/or blowholes. They appear to have developed by widening and lengthening around these blowholes (which may co-incide with karstic features) rather than by progressive lengthening.

In summary, the regional evidence points towards a very long history for this coastline, but the local evidence supports a modern origin for the coastal features as the cliffline cut into the existing karstic landscape. There is little evidence in the cliff forms that the cliffs are anything other than modern surfaces resulting from the undercutting, toppling failures and rock falls of an older cliffline.

Conclusions

This is a rare assemblage of active coastal erosional features, whose origins are better documented than many other cliffed sites. Well-developed geos, stacks, arches and cliffs truncate a former karstic landscape. In addition, it forms part of a southern British suite of structurally controlled coastal landforms, which includes Tintagel as the least dynamic and Old Harry (see GCR site report for Ballard Down) as the most rapidly changing. The marine erosion of former karstic features to produce an intricate coastline of arches and stacks is not found on this scale elsewhere on the British coast.

HARTLAND QUAY, DEVON (SS 221 226–SS 230 278)

V.J. May

Introduction

The coastline of North Devon runs transversely across Devonian and Carboniferous strata, but at Hartland Point it changes direction abruptly towards the east (see Figure 3.1 for general location). Much of the coastline is cliffed, broken only by small valleys that have been eroded to present sea level (for example at Crackington Haven) or form hanging valleys (for example south of Hartland Quay). South of Hartland Point, the relationships between coastal valley systems and coastal retreat are of particular interest. This site contains fine examples of cliffs and shore platforms, and demonstrates clear relationships between cliff forms, platform development and lithological variations (Arber, 1911). Furthermore, it is also noted for a remarkable set of river valleys that have been truncated by the cliffline, so that their floors now lie well above present sea-level (Arber, 1911; Arber, 1949). Unlike similarly truncated streams in the south-west Isle of Wight (see GCR site report in Chapter 4), those in the Hartland Quay area have been unable to erode valleys to sea level and so many reach the shore via waterfalls (Arber, 1911). In some cases the streams have also cut gorges that include waterfalls. In common with other hard-rock coasts, Hartland Quay has been the attention of only limited research since the detailed monograph by Arber (1911). Keene (1986, 1996) and Goudie and Gardner (1985) have reviewed the development of the site in the light of more recent interpretations of Pleistocene geology in western Britain (Stephens and Synge, 1966; Kidson and Tooley, 1977).

Description

Described as 'perhaps the finest coastal scenery in the whole of England and Wales' (Steers, 1946a, p. 219), this site extends some 6 km from Longpeak Beach in the south (SS 221 226) to Hartland Point in the north (SS 230 278). Cliffs cut into Carboniferous interbedded fine-grained sandstones and shales vary in height between 25 m and 100 m. The structural features lie east–west, following the Variscan trend. Because the shales are eroded more easily than the sandstones, these structures are etched out both in the cliffs and the platforms. Caves have been cut in the weaker shales and mudstones, or along faults on the axial planes of the folds (Keene, 1996). Five valleys truncate the coastline and reach the sea via waterfalls. A platform, up to 300 m in width, dominates the intertidal zone. Its mean width is 160 m (based on 30 measurements at different localities), with headland platforms being on average 50 m wider than those in bays). Beach development is limited, being mainly confined to small bay-head accumulations of locally derived shingle and cobbles. There are also considerable areas of boulders resting both at the cliff foot and upon the platform. The cliffs are subject to much localized mass-movement, for example at Blagdon Cliff where there is a fault-controlled landslide. To the east of the site at Keivill's Wood (SS 352 237), the co-incidence of a large rotational slip scar and an almost flat boulder spit (The Gore) have been interpreted as the scar and lag deposit of a large landslide, which based on chart evidence pre-dates 1795 (Keene, 1996). The rate of retreat of the cliffs around Hartland Point has been estimated at between 20 and 40 mm a^{-1}. With sea level at or close to its present level for the past 6000 years, this suggests net retreat of up to 240 m, a distance close to the width of the platforms.

The cliffs at Hartland Point and Blagdon Cliff reach over 100 m, but decline to just over 30 m at the mouth of Tichberry Water (SS 228 267), which flows into the sea over the northernmost of the waterfalls that distinguish this stretch of the coastline. The narrow flat floor of its valley is continued southwards between a small hill known as 'Smoothlands' and the continuation of its southern valley side. A small stream fed from a spring flows along part of this hanging valley floor to enter the sea over a 22 m cliff on the northern side of Damehole Point (SS 226 265). Two streams, Blegberry Water and Abbey River, both flow onto Blegberry Beach via waterfalls, although of very different forms. At Blegberry Water the stream flows from about 35 m OD down a joint-controlled waterfall that Arber (1911) described as a 'primary sheer waterfall' (Figure 3.32). Abbey River in contrast has a flat floor that hangs some 12 m above beach level. The valley is underlain by solifluction debris, and the stream has cut (in Arber's terminology) a 'mature canyon'. Steep cliffs up to 100 m in height form the coastline to Hartland Quay.

Between Hartland Quay and Speke's Mill Mouth, the cliffs vary in height from below 30 m to over 70 m. Each of four small headlands (Hartland Quay, Screda Point, Screda Bay southside, and St Catherine's Tor) slope inland to a flat-floored valley that hangs at about 30 m OD above each of the intervening bays. Wargery Water flows along part of the hanging valley to drop to the sea at Childspit Beach. To the south, Milford Water flows over a flat-floor until it plunges into the sea over 'the most spectacular' waterfall at Speke's Mill Mouth (Arber, 1911). To

Hartland Quay

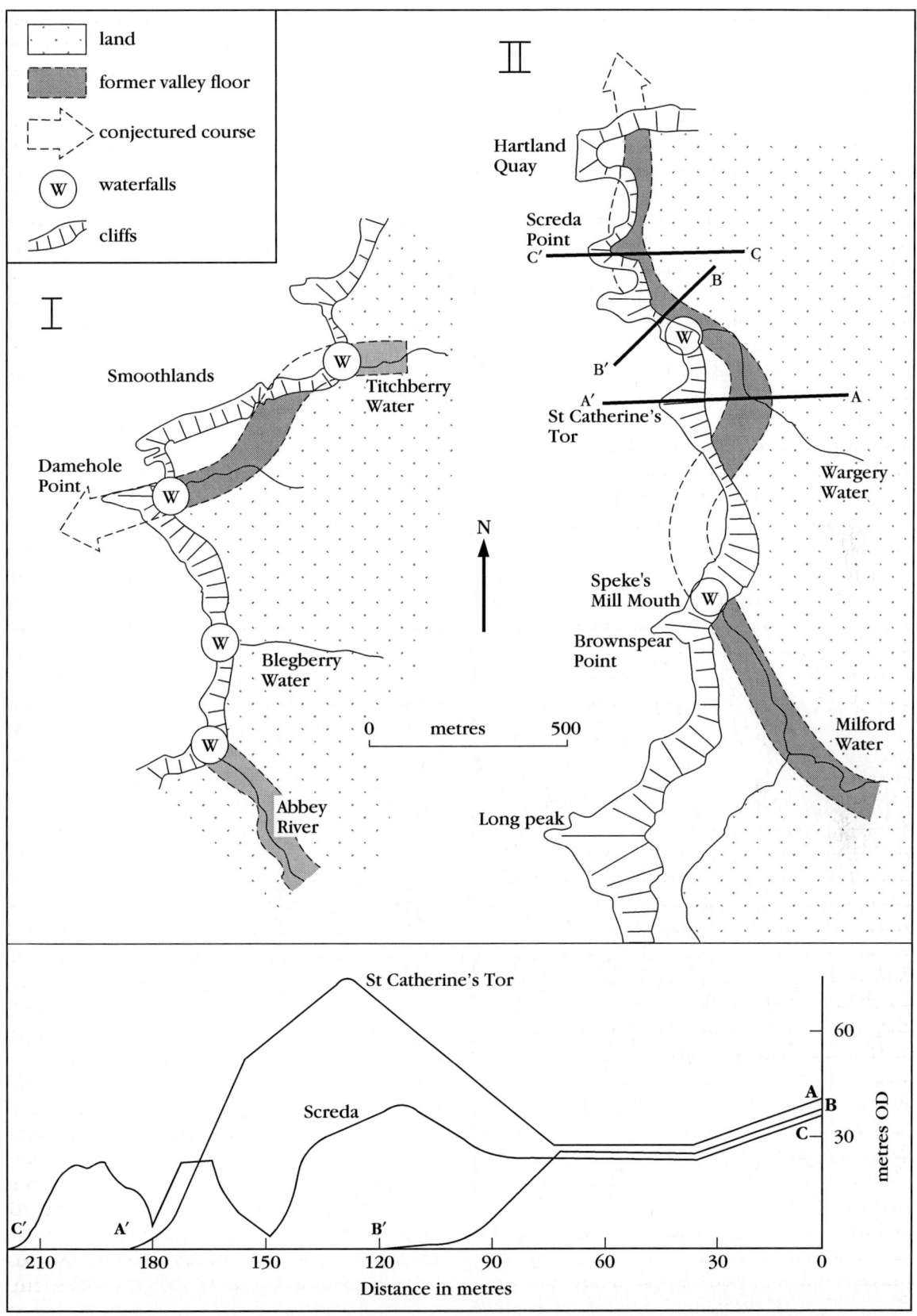

Figure 3.32 Hartland Quay GCR site – showing the pattern of truncated valleys. The profiles A–A', B–B', C–C' are shown at the bottom of the figure. Section I lies to the north of Section II. (After Arber, 1911.)

Hard-rock cliffs

the south the cliffs rise to over 100 m at Longpeak. Each of the small headlands is associated with reefs that run at right angles to the cliffline and extend across the platform. The platform is varies between about 250 m and 150 m in width although its elevation varies a great deal depending upon the arrangement of the beds across which it cuts. For example north of Hartland Quay it is a broad feature cutting across all the exposed beds, whereas to the south at Screda Point, buttress reefs are predominant (Figure 3.33). On the platform, these buttress reefs often appear as steeply dipping walls of rock.

The hanging valley floor between Hartland Quay and St Catherine's Tor is rock-floored with only a shallow depth of weathered material resting on it. There is neither solifluctued infill nor incised valley. The flat-floored hanging valleys become gradually lower in height towards Hartland Quay and are usually interpreted as representing the truncated remnants of the former floor of the Wargery and Milford Waters. The waterfalls vary from sheer falls across great slabs of rock to stepped features confined to very steep-sided narrow gorges or 'gutters' (Arber, 1911). The detailed form depends to a substantial extent upon the exact arrangement of the beds over which they flow as well as the nature of the material itself.

Interpretation

The well-developed platforms and cliffs offer ample evidence that this is an active coastline along which rockfalls, landslides and stream erosion all play a part. Arber (1911) interpreted the features here as resulting from the inability of the streams to erode sufficiently rapidly to compensate for the rapidly retreating cliffline. The truncated downstream courses often survive as dry hanging valleys. He described the hanging valleys as sea-truncated valleys, and reconstructed the former courses of both Titchberry Water and Milford Water. Streams that once flowed farther seawards were cut into by the retreating cliffline, and their water was diverted, usually resulting in the formation of waterfalls. Arber described the waterfalls as 'unique in Britain' and his investigation remains the only

Figure 3.33 Cliffs, platform, beach and truncated valleys south of Hartland Quay. (Photo: Lou Johnson, www.walkingbritain.co.uk.)

detailed examination of them. Although the detailed form of the waterfalls depends on local variations in rock strength and the dip of the strata, Arber divided them broadly between those where the sea was more active in eroding the cliffs than the stream was in downcutting. In contrast where the stream was the more effective agent, the waterfalls more commonly formed gutter or canyon falls. The differences in waterfall morphology may provide an indicator of the very variable rates of cliff retreat in comparatively hard coasts where cliff-top retreat is often recorded as minimal. Although coastal waterfalls occur elsewhere in Britain, they are uncommon and nowhere as common as here. The reasons for this remain speculative, but seem likely to relate to the high proportion of streams flowing towards or along the coast, the impermeability of the strata, and the relatively slow rate of downcutting compared to cliff retreat.

Steers (1981) argued that although storm waves reach to and above the junction of cliff and platform there was no reason to assume that the platform is of wholly modern origin. Since the emerged ('raised') beaches at Trebetherick and Fremington are only a little higher than the present platforms, Steers argued that there is no reason why the platforms should not be much older in origin than they appear. In contrast the erosional activity of the cliffs and the platforms and the site's exposure might suggest that this cliffline had retreated considerable distances. A consistent contemporary rate over the last 6000 years for example would, however, only place the cliffs between 250 m and 120 m farther out to sea. Farther north on the south coast of Wales, there are well-preserved emerged platforms and beaches. The shales and sandstones around Hartland present a significantly different surface for erosive processes. Whereas the Carboniferous Limestone of Gower is comparatively free of discontinuities, the Carboniferous shales and sandstones of Hartland are very thinly bedded, much folded and faulted and provide numerous opportunities for erosion by both marine and slope processes.

An ice-margin explanation is proposed by Goudie and Gardner (1985) as an alternative to the coastal retreat explanation. If the Fremington tills are Anglian in age, then ice entered the nearby Taw–Torridge valley about 450 000 years BP. Irish Sea ice probably extended far enough south during the penultimate glacial period to allow marginal drainage channels to develop between the ice and the coastal slope (Stephens and Synge, 1966; Kidson and Tooley, 1977; Keene, 1996). This coast was, however, ice-free during Devensian time (Keene, 1996). Goudie and Gardner (1985) outlined a possible alternative origin for the hanging valleys, for with the ice margin at or close to the coast, the usual outlets of the streams might become blocked. As a result a lake would build up until the lowest point of the valley side was overtopped. A new valley was then cut by the diverted stream. With greater discharge, higher impermeability and probably more and larger sediment loads, the streams would cut broad valleys. Once the ice retreated, the streams would revert to their former courses. The lack of infilling of the hanging valleys is seen as supporting this argument. Although this hypothesis, which was developed in order to explain the Valley of the Rocks west of Minehead (Mottershead, 1967), appears to offer a satisfactory explanation for that feature, its extension to the Hartland area appears less convincing.

The ice-margin hypothesis does not explain satisfactorily the dissection of the valleys in the Hartland area, where the former Milford Water has its left bank removed at four separate locations. The implication of the hypothesis is that the stream flowed over the ice at these points (since coastal retreat is not considered as a complementary process). The nature of the evidence and the origin of these unusual coastal landforms warrants further detailed investigation. Keene (1996) points out that the valley of Milford Water upstream of the truncated supposed meltwater section is also flat and steep sided. Valleys such as Abbey River are, in contrast, infilled by solifluED material, probably of Devensian age. Their rock floors lie much closer to present sea level. Unconsolidated angular material in a matrix of finer-grained materials is entirely local. Post-Devensian increases in stream activity account for the development of meander terraces in the soliflucted material (Keene, 1996). In both cases, subsequent retreat of the cliffs would have allowed truncation to have taken place leaving them hanging above the present beaches.

Farther south at Marsland Water and Welcombe Mouth, valleys are incised much nearer to present sea level and there is clear evidence that earlier valleys were filled by soliflucted debris (probably Devensian in age: Keene, 1996). This suggests that at least in that area pre-Devensian streams flowed to a similar local base-level to today, but does not necessarily confirm that the coast was near its present position. On balance, the ice-margin hypothesis is less likely unless either Devensian ice reached the area and was banked against the coast or the valleys preserve forms that derive from the Anglian glacial presence along this coast. The latter also seems unlikely given that there have been four major changes of sea level since the Anglian and that cliffs could retreat at least 200 m in each interglacial period. An intrinsic part of the debate arises from the anomalous relationship of the truncated valleys to the structures. The majority of valleys follow the strike of the rocks. The question to address therefore is whether the development of drainage in these patterns is anomalous. If not, then the simple explanation that the valleys result from truncation by coastal retreat would appear most likely.

Conclusions

This part of the north Devon coast displays excellent examples of cliffs, platforms and differential adjustment of stream systems to coastal retreat. The only site in Britain where the development of coastal waterfalls has been examined in detail, Hartland Quay is also important for the remarkable truncation of valleys running along, rather than towards, the cliffs. The shore platforms have been cut across the complex structures, but little research on them has been carried out.

This site has caused controversy in that the origins of one of its main features, Arber's 'sea-truncated valleys', remain open to discussion. It contains some of the best examples of coastal waterfalls in Britain, the cliffs are finely developed, and a series of hanging valleys give the site unusual characteristics. The platforms are also well developed, although, as Steers has pointed out, they may well owe their existence to more effective marine activity in the past. If the ice margin was sufficiently close to produce (or at least influence) the flat-floored valleys, there remains the possibility that sea-ice and later periglacial conditions may also have played a significant role in the development of this site. However, in the absence of clear evidence that Devensian ice was marginal to the coast, the ice-margin hypothesis for valley development is less convincing. These valleys differ from those along the south-west coast of the Isle of Wight where cliff retreat has cut across the upper courses of cliff-top valleys. On the western hard-rock coasts, similar beheaded valleys occur, for example at Dinas and Cemaes, north of Fishguard, but these have been explained as ice-marginal overflow channels (Steers, 1946a). The different forms of waterfall described by Arber (1911) add to the unusual nature of this site.

SOLFACH, PEMBROKESHIRE (SM 802 241)

V.J. May

Introduction

The small ria at Solfach (Solva) and its infilled counterpart, the Gwada valley, are the westernmost such features on the south coast of Wales, lying some 5 km east of St David's (see Figure 3.1 for general location). Rias (drowned river valleys) are common features of the coasts of the Bristol Channel, Devon and Cornwall. Many are large landforms such as Milford Haven, Pembrokeshire, and the Fowey River, Cornwall, but many more are small. Solfach is a good small-scale example of a ria. The site includes both the present ria of Solfach Harbour itself and the infilled ria called the 'Gwada Valley'. The proximity of the two features adds interest to the site, which has been little affected by human activity. Slope-over-wall cliff forms surround the present and former rias that have been cut into a near-horizontal surface at an altitude of about 60 m OD. There has been some infilling of the upper reaches of the present ria, whereas the Gwada Valley has been almost entirely infilled by alluvial sediments. Mentioned briefly by Steers (1946a), Solfach was described by Goudie and Gardner (1985).

Description

Solfach Harbour and the Gwada Valley are very good examples of two phases of the development of submerged rocky coastlines. Solfach is a good example of a ria: a former glacial melt-water channel that a subsequently became a river valley that was flooded by rising sea-level during Holocene times. The Gwada Valley is a comparable feature in origin, but sedimentation has filled almost all its length leaving a small bay with a sandy beach at its mouth. Solfach and the

Solfach

Gwada Valley are cut into a near-horizontal surface (commonly regarded as a former marine erosion surface) at about 60 m OD. From this surface, the land slopes at between 20° and 35° before dropping abruptly into the sea at the seaward end of the valleys. Both valleys are thought to be 'curved' segments of subglacial meltwater channel.

Solfach is flooded at high tide, but gravel, sand and mud are exposed at low tide. These sediments that have been dumped here in a narrowly confined delta will gradually fill in the whole ria just as has happened to the Gwada Valley. In the lower part of the ria, the intertidal forms are more related to marine action as waves penetrate the estuary. Shore platforms of rock occur, particularly on the eastern side.

The Gwada Valley has much the same general form but its valley floor infill is extended almost to its seaward end. Flat-floored with only a small stream across it, this valley has a much-reduced fluvial input compared to Solfach, and its beach is predominantly sandy.

Interpretation

The present-day features combine vertical cliffs in hard Cambrian and Ordovician rocks with sedimentation in a small marine delta at the head of Solfach Harbour. Steers (1946a) passes little comment on the site other than to note that it is a good example of a ria. The stream, though capable of carrying fine-grained sediments into the estuary, would require a long time to bring about the substantial erosion that was necessary to carve the steep-sided meandering valley. Similarly, present day marine action does not appear to be especially effective in eroding the shoreline. Several separate phases of development have to be invoked to explain the present assemblage of forms.

The first phase (Figure 3.34) appears to have been predominantly fluvial and produced the gentler upper slopes of the valleys. A second phase (Figure 3.34) then produced the steep sides of the lower valley, which in parts of the estuary extend well below sea level. If this phase was fluvial, it required much larger discharges than occur at present. The development of the forms to below present sea level suggests that they developed during a period of lower sea level, i.e. they derive from a glacial period. Much larger discharges at that time could help explain the incision. Alternatively, changed climatic conditions with higher rainfall and greater runoff during an interglacial could also generate the larger discharges needed. Goudie and Gardiner (1985) have suggested that an alternative explanation may be sought in the erosional vigour of sub-glacial streams that flowed beneath, or issued from, ice-sheets covering south-west Wales. If so, Solfach differs significantly in origin from the rias of southern Cornwall and Devon.

The differential infilling of the two valleys has not been explained. The stream carrying sediment into the Gwada valley appears to have been more active and been able to overcome the sorting and transporting action of the sea. The subsurface sediments and the depth of the

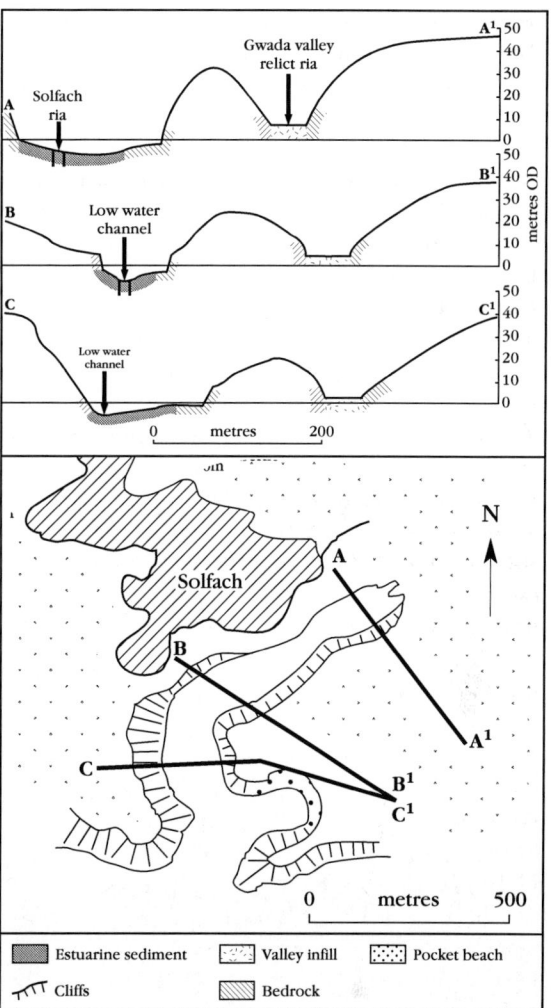

Figure 3.34 Cross profiles of Solfach and the Gwada Valley, showing the contrast between the ria of Solfach and the infilled former ria at Gwada.

underlying bedrock surface are not known. However, Goudie and Gardner (1985) state that it has been infilled more than the Solfach valley. They consider that the valleys were drowned about 6000 years BP, but may also have been drowned previously by earlier higher interglacial sea levels. Their discussion of the formation of the rias and their subsequent infilling may need to be re-thought in the light of more recent work on sea-level change in south-west Wales reported by Campbell and Bowen (1989)

Rias are represented in two other GCR sites, Carmarthen Bay (the estuaries of the Taf and Twyi) and Loe Bar (see GCR site reports in chapters 11 and 6 respectively). The former are larger meandering features, over 1 km wide near their mouths, whereas the Helston River flows into the Loe Pool, which is blocked by the bay–bar at its mouth. Together, these three sites exemplify different stages of ria formation and destruction. Solfach is a good example of ria development, but questions have been raised about its origins. The extent to which fluvial processes were associated with sub-glacial streams remains open to conjecture, nevertheless Solfach may represent a ria form that combines the effects of both glacial and fluvial environments. If so, it is a rare feature in Britain, and probably in Europe.

Conclusions

A small ria and its infilled neighbour form the site. Solfach and the Gwada Valley provide a distinct contrast in the development of coastal landforms, both being drowned water-worn valleys, but the Gwada Valley has been infilled in contrast to the tidal Solfach. Solfach is a distinctive example of a ria, not least because of the combined effects of glacial and fluvial processes in its formation.

Chapter 4

Soft-rock cliffs – GCR site reports

Introduction

INTRODUCTION

V.J. May and K.M. Clayton

Cliffed coastlines undergoing rapid erosion characterize much of the south-eastern British coast where they are cut into relatively 'soft' geological materials such as sandstones, clay, shale and chalk, as well as many weak superficial deposits, notably the extensive glacial till of the east coast of England (see Figure 1.2 and Table 4.1). There are also many locations around the British coast where short lengths of rapidly changing cliffs occur wherever softer materials crop out, for example part of the coastline between Weymouth and Lulworth (see GCR site report for the Dorset Coast, Chapter 11). It is also common for bayhead cliffs in upland coasts to be formed in weak materials, often of glacial or fluvio-glacial origin, which produce pocket beaches. In such locations, pocket beaches are often entirely dependent on the erosion of truncated valley deposits as local sources of sediment, for example at Crackington Haven, Cornwall. In contrast, some pocket beaches may depend entirely on adjacent headlands as their sole source of sediment. Additionally, small cliffs, of up to a metre or so in height, often develop at the margins of saltmarshes (see Chapter 10), exhibiting on a micro-scale many of the features of the larger soft cliffs described in this chapter.

Many cliffs are associated with shore platforms that may extend seawards for several hundred metres from the cliff foot. Most actively developing platforms are associated with cliffs that are undergoing active retreat, and so many of the sites described in this chapter are also important for their platforms. A discussion of platform development is provided in Chapter 2 of the present volume.

The selection of soft-cliffs for the GCR used the following classification, based on the cliff lithology, form, and recession rates. It is comparable to the later Jones and Lee (1994) classification of cliffs on the basis of coastal recession.

1. Cliffs that are steep in profile and are retreating rapidly. Their steep profile is largely a function of the rapidity of retreat, which commonly exceeds 0.04 m a^{-1} and may attain 1.82 m a^{-1}. Formed mainly in weak or unconsolidated sands, clays, and gravels, the sediments are often of glacial origin. They are significant sources of beach sediment, for example along the Yorkshire and East Anglian coasts. Their mass-movement features are well represented at Trimingham (described in the *Mass Movements* GCR volume; Cooper, in prep.).

2. Cliffs cut into stiff mudrocks, such as Kimmeridgian clay, the Gault Clay, and Tertiary clays such as the Barton Clay. Often affected by shallow slides and mudslides, these cliffs provide very little sediment for beaches because most of the fine-grained material is carried into deeper water as the toe of slides is removed. They typically retreat at rates between 0.25 m a^{-1} and 1.50 m a^{-1}. As coastal features they are represented best by parts of the cliffed coasts of the south-west Isle of Wight and Dorset, although the best example of the mass-movement features is at Warden Point on the Isle of Sheppey (Dixon and Bromhead, 1991; Cooper, in prep.).

3. Cliffs cut into sandstones, shales and chalk that retain steep profiles despite a variety of retreat rates, which range from very slow to 1.20 m a^{-1}. Parallel retreat (over a timescale of 5 to 10 years) is their most common behaviour. Although they provide potential beach sediments, these are often quickly reduced by attrition to sand or smaller-sized particles. Chert, flint and other hard materials within these rocks may provide important components for beaches. In contrast, when the cliff failures produce boulders, they may form substantial elements of the intertidal landscape and may persist for very long periods of time. These cliffs are often associated with wide shore platforms which have attracted a substantial literature. The effects of small-scale structural features such as faults and joints often contribute to the formation of buttresses, caves, arches and stacks in these coasts. Stacks are often the most distinctive persistent elements of otherwise rapidly retreating coasts.

4. Major landslides in clay that carry extensive volumes of more-resistant overlying material such as chalk, greensand, flint and chert to the shoreline and provide major inputs of potential gravel beach sediment. Typically they retreat at rates in excess of 0.35 m a^{-1}. The harder components of the debris delivered to the beach provide very important beach material, whereas the fine-grained materials are usually quickly dispersed off-

Soft-rock cliffs

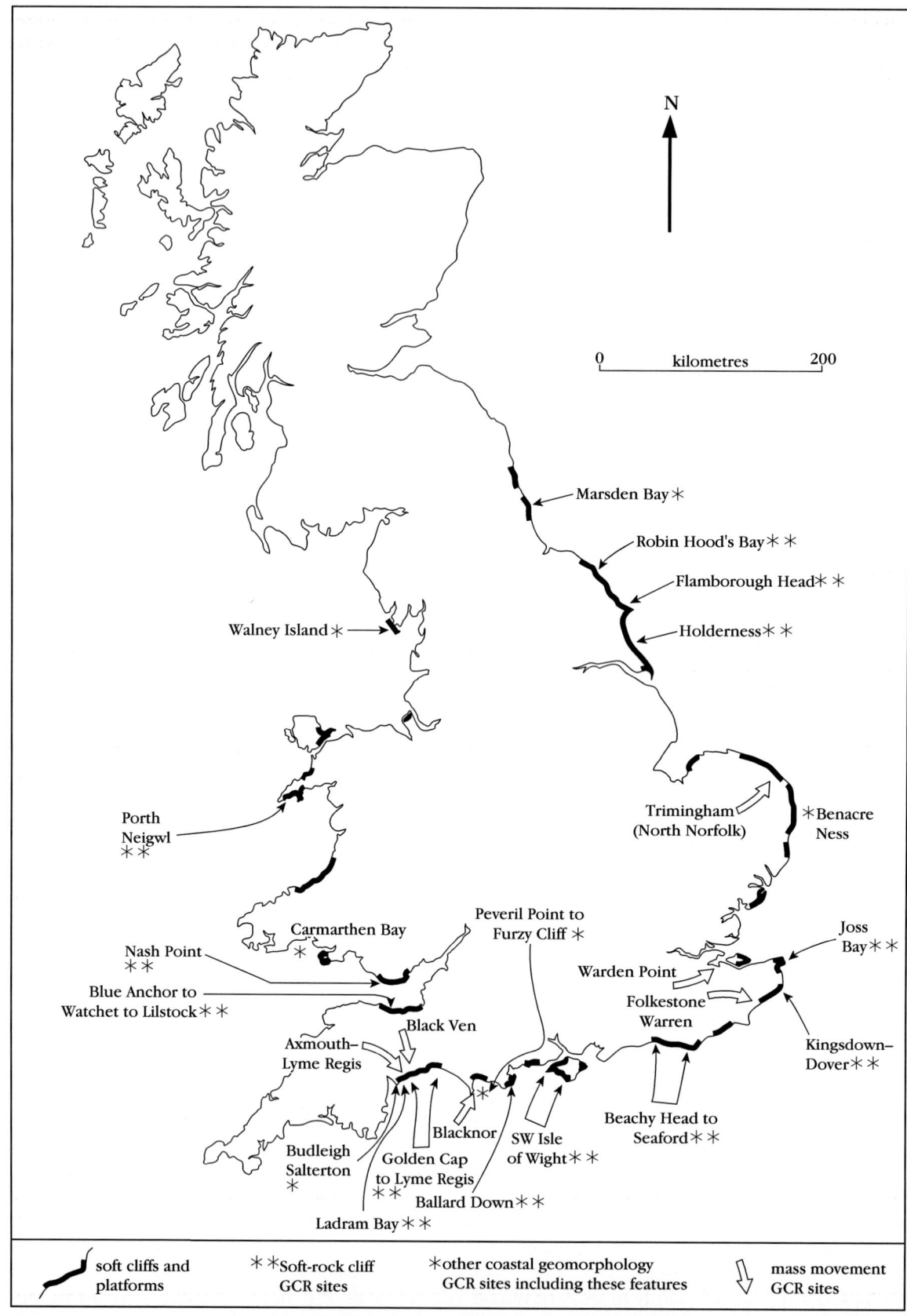

Figure 4.1 Location of significant soft-cliffed coasts and platforms in Great Britain, indicating the sites selected for the GCR specifically for soft-rock cliff geomorphology. Other coastal geomorphology sites that include soft-rock cliffs and sites selected for the Mass Movements GCR 'Block' that occur on the coast are also shown.

Introduction

Table 4.1 The main features of soft-rock cliff coastal geomorphology GCR sites, including coastal geomorphology GCR sites described in other chapters of the present volume that contain soft-rock cliffs in the assemblage. Sites described in the present chapter are in **bold** typeface.

Site	Main features	Other features	Mean rate of cliff-top retreat (m a^{-1})	Tidal range (m)
Budleigh Salterton	Cliff erosion feeding Budleigh Salterton Pebble Beds into local and regional beaches	Shingle beach (see Chapter 6)	0.30	4.0
Ladram Bay	Cliff–stack–platform development in Triassic sandstone and mudstone		0.20	3.7
Robin Hood's Bay	Cliffs in till resting on Liassic shales. Till/platform junction	Platform across Liassic shales	0.03	4.8
Blue Anchor–Watchet–Lilstock	Rapid retreat in Liassic shales with very unusual 'washboard' topography in macro-tidal environment	Platform development	Up to 1.20	9.4
Nash Point	Rapid cliff retreat in Liassic shales. Cave development	Platform development	0.2–0.10	6.0
Lyme Regis to Golden Cap	Intensively researched landslide and related beach coast	Major mass-movements	0.60–0.96	3.5
Peveril Point to Furzy Cliff	Rapidly eroding cliffs in range of materials from Chalk to Oxford Clay. Longitudinal coast	Semi-enclosed beaches. Submarine rock reefs. Landslides)	0.00–0.41	1.7 (east)–2.0 (west)
South-west Isle of Wight	Differential erosion in materials from Chalk to Wealden. Contrasts between relict and modern beaches. Stacks. Chines	Major mass-movements	0.20–2.10	3.3 (east)–2.2 (west)
Kingsdown to Dover	Cliff and beach development in high (over 30 m) cliffs. Recent beach depletion	Flow failures	0.20–0.60	5.9
Beachy Head to Seaford	Cliffs of variable height in Upper Chalk. Narrow platforms. Locally limited sediment supply. Recent beach depletion		0.40–1.26	5.3
Ballard Down	Classic cave–arch–stack site in Upper Chalk. Transverse coast	Pocket beach formation	0.01–0.60	1.7
Marsden Bay	Cliffs and stacks	Beach phases		4.2
Flamborough Head	Highly complex chalk cliffs overlain by Devensian till. Caves and stacks	Extensive platforms	0.30–0.90	4.0
Joss Bay	Cliff and platform development in Upper Chalk		0.30	4.0

Table 4.1 – *contd*

Site	Main features	Other features	Mean rate of cliff-top retreat (m a^{-1})	Tidal range (m)
Carmarthen Bay	Both hard-rock cliffs and easily eroded cliffs	Major dunes, sand-spits and barrier beaches, rias, emerged beaches, intertidal sandflats, saltmarsh		8.0
North Norfolk Coast	Rapidly eroding cliffs in chalk and till, latter feeding regional sediment budget	Major spits, beaches and saltmarsh (see Chapter 11)	0.30–0.42	4.7 (E)–6.4 (W)
Benacre Ness	Rapidly eroding till cliffs resulting from longshore movements of ness and subsequent reduction of natural protection	Shingle ness (see Chapter 6)	0.42–0.96	2.1
Porth Neigwl	Rapidly retreating glacial drift cliffs, chines, beach cusps	Contemporary beach cementation (see Campbell and Bowen, 1989)	Up to 1.00	3.9
Walney Island	Till cliffs, rapid erosion	Barrier islands, recurved spits		9.0
Holderness	Rapidly eroding cliffs, mainly in till	Till shore platform, ords, thin beach	Up to 2.22	4.0

shore. The coastline between Golden Cap and Lyme Regis is the best representative of this coastal system, but the mass movements are especially well-represented by Folkestone Warren and the coastline between Axmouth and Lyme Regis (Cooper, in press).

Each of these types is represented by a GCR site, or GCR sites described in the present chapter (Figure 4.1, Table 4.1).

Retreat in soft-rock cliffs

Cliffs in weaker materials retreat at rates that range from 0.01 m a^{-1} to over 3 m a^{-1}. Although average values for cliff retreat have been used to compare the magnitude of retreat in weak cliffs, it is essential to recognize that the rate of change in such cliffs, or indeed in any cliffs, is rarely regular (see Figure 4.2). Competing types of geomorphological processes affecting soft cliff sites operate at different rates, or are episodic, so the local form of cliffs can change quite considerably over time; it is common to observe morphological change seasonally. Many of these cliffs are affected by large mass-movements, which produce temporarily protective areas of debris at the cliff foot, or enhance beach volumes sufficiently to provide protection against wave attack for a time. Table 4.1 identifies both the sites that represent soft-cliff coasts specifically and those which are described in other chapters or in the *Mass Movements* GCR volume (Cooper, in prep.).

Two examples demonstrate the irregularity in the long-term mean and short-term variations of cliff recession. At Birling Gap, six-monthly surveys of the cliff top over a decade (from 1952 to 1961) showed that there had been considerable temporal and spatial variation in the amounts lost, although over the ten-year timescale there is a high degree of consistency in the average retreat rate overall (see Table 4.2; May, 1971a). However, at Hengistbury Head, rates of retreat – as well as cliff-face changes – were recorded by the author at both the cliff top and foot, and these measurements demonstrate that although there is also a close similarity between cliff-top and cliff-foot retreat rates, there are considerable variations in the magnitude and frequency of the retreat event. These two examples show that the mixture of materials, structures, wave climates, beach characteristics and platform development is such that rapid retreat cannot be ascribed to any single rock type or location. Cliffs cut in rocks that retreat at the highest rates in one loca-

Introduction

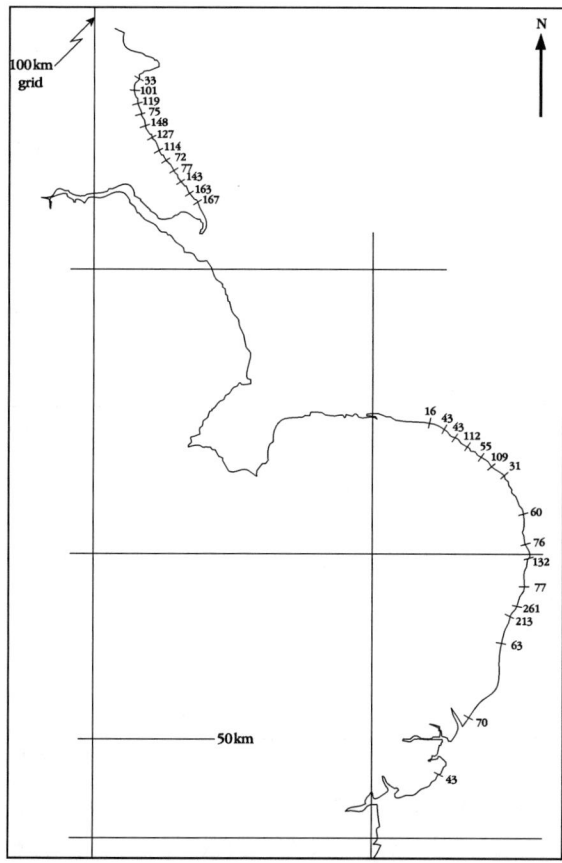

Figure 4.2 Rates of retreat along the North Sea Coast of England from Bridlington to Clacton-on-Sea. Rates are shown as averages for each length of cliff; where the length of cliff exceeds 5 km, values are every 5 km along the coast. Values are totals (metres) for 100 years to 1980. See also Table 2.1 and Table 4.2 (Compiled by K.M. Clayton)

tion may show minimal rates of change elsewhere.

It is also easier to reconstruct the development of the rapidly changing cliffs of Holderness and East Anglia, largely cut into glacial deposits, than the hard-rock cliffs of western Britain. Rates of erosion vary from 0.25 m a^{-1} to 3.5 m a^{-1} in Holderness (Figure 4.2), and an average of 6 m a^{-1} since the 1930s at Covehithe in Suffolk. Such cliffs undergoing rapid erosion suffer cliff failure in large part by rotational landslides, and their significance for study of these processes has formed a major reason for their inclusion as GCR sites. Thus many are described in the GCR volume on mass movement sites (Cooper, in press.).

From the marine-process viewpoint, two features are particularly noteworthy. First, the pattern of erosion over time and space is complex (Cambers, 1973). Despite such spatial and temporal variations, overall data for the whole of the Norfolk cliffs imply a long-term average rate of retreat close to 1 m a^{-1}. Certainly cliff positions are difficult to establish prior to the first Ordnance Survey maps, but evidence of vanished villages near Cromer in Norfolk described in the Domesday Survey (1086 AD), or maps showing the steady erosion of the streets of the medieval town of Dunwich since 1589 AD (Robinson, 1980a; Figure 4.3) strongly suggests long-term persistence of an erosion rate comparable with that found today. This implies both the long-continued effectiveness of the longshore and offshore removal of sediment, and the continuation of wave-energy levels at the coastline similar to those today.

Yet, second, when the extent of coastal retreat since the slowing of the Holocene rise in sea level at about 6000 years BP is considered, it is clear that a third factor has been at work – the gradual and persistent deepening of the offshore zone. Along the North Sea coast of England (e.g. Holderness and north-east Norfolk), some of this change has been contributed by relative sea-level rise, but part may also be attributed to sea-floor erosion, probably by abrasion and bio-erosion. Insofar as the rate of cliff retreat has been sustained, the gradual deepening of this submerged offshore zone (from both erosion and sea level rise) and so the maintenance of offshore gradients may well have been the basic control on wave energy and so on the rates of coastal erosion. Along these coasts a shore platform also underlies the beach, but it is often seen only after severe storms, since erosion contributes enough sediment to maintain a thin covering beach (Figure 2.1c).

An intermediate position is held by the Chalk cliffs of England. Chalk is the commonest rock of south-eastern England and crops out in coastal cliffs along a considerable length of the coast from the Isle of Thanet to Devon, as well as at Flamborough Head and at Hunstanton, Norfolk, where the Red Chalk is well exposed. Several lengths of Chalk cliffs are included in the GCR sites described in this chapter, including the steeply dipping (and rather resistant) Chalk of the Isle of Wight and Dorset. The rate of retreat tends to be ≤1 m a^{-1} with the more sheltered sites undergoing erosion at about 0.2 m a^{-1}. Chalk cliffs differ from weaker rocks (where the platforms are usually buried by a beach) in commonly displaying shore platforms at their foot.

Soft-rock cliffs

Sand can usually only accumulate in bays, although considerable lengths (as for example the Seven Sisters, Sussex) can be fronted by a rather patchy beach of flint pebbles or cobbles. In addition, the greater coherence of Chalk means that cliff failure is generally by falls (toppling) rather than by rotational slides, although where mudrocks underlie the cliff section, as at Folkestone Warren (Hutchinson *et al.*, 1980), or on the southern coast of the Isle of Wight (Hutchinson *et al.*, 1991), huge rotational slides have occurred, extending from below sea level to the cliff top at 200 m.

There is considerable variety of form within the examples described here (Flamborough Head, Thanet, the Seven Sisters and the folded Chalk of the Dorset coast), yet, despite a number of local studies, no integrated study of our chalk

Table 4.2 Rates of cliff-top retreat of soft-cliffed coasts (from various sources).

Cliff-top retreat (m a^{-1})	Rock type	Location	Period (years)
0.01	Upper Chalk	North Ballard Down	100
0.01	Upper Chalk	East Ballard Down	100
0.03	Bracklesham Beds	Highcliffe Castle	92
0.07	Upper Chalk	Kingsdown–St Margaret's Bay	84
0.07	Upper Chalk	Thanet	85
0.09	Middle/Lower Chalk	Dover to Folkestone	90
0.16	Upper Chalk	Cuckmere to Seaford	120
0.18	Chalk	Hambury Tout to White Nothe	98
0.19	Upper/Middle Chalk	St Margaret's Bay	84
0.27	Hamstead Beds	North-west Isle of Wight	95
0.28	Glacial drift	North Yorkshire	72
0.29	Glacial drift	Holderness	100
0.37	Jurassic clays	Furzy Cliff–Shortlake	98
0.39	Kimmeridge clays and shales	Kimmeridge	100
0.41	Upper Chalk	Newhaven–Rottingdean	89
0.41	Wealden	South-west Isle of Wight	125
0.41	Kimmeridge clays	Ringstead	99
0.42	Glacial drift	Weybourne–Cromer	100
0.57	Glacial drift	Gorleston–Corton	100
0.57	Glacial drift	Holderness	100
0.58	Barton Clay	Barton	62
0.68	London Clay	Reculver	79
0.83	Glacial drift	Gratby–Caister	100
0.85	Glacial drift	Holderness	100
0.88	London Clay, crag and glacial drift	The Naze	100
0.96	London Clay	Northern Isle of Sheppey	79
0.96	Glacial drift	Cromer–Mundesley	100
1.05	Glacial drift	Pakefield–Kessingland	100
1.06	Chalk	Beachy Head	90
1.08	Sandstone	Cliffend	75
1.11	Glacial drift	Holderness	100
1.19	Hastings Beds sandstones	Ecclesbourne Glen	75
1.20	Glacial drift	Holderness	100
1.22	Chalk	Birling Gap	120
1.26	Chalk	Seaford Head	120
1.43	Hastings Beds clays	Fairlight Glen	75
1.75	Glacial drift	Holderness	100
1.96	Glacial drift	Holderness	100
2.22	Glacial drift	Holderness	100
3.00	Glacial drift	Covehithe	100

Introduction

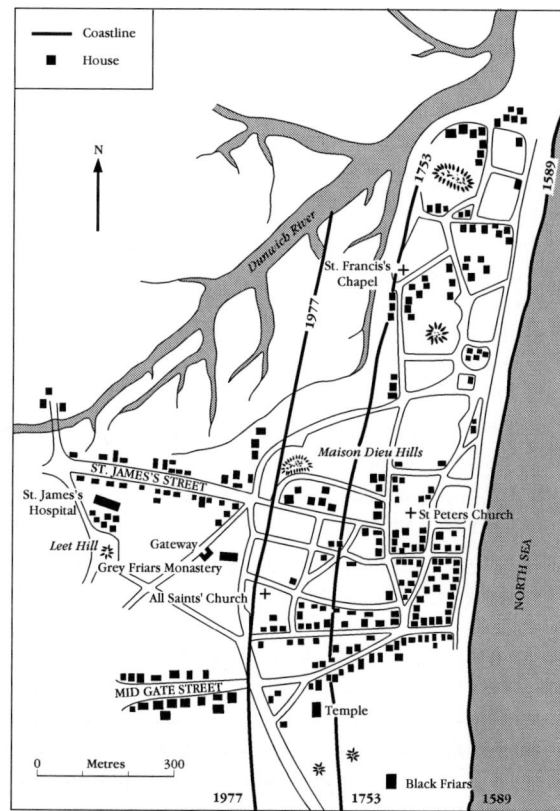

Figure 4.3 Retreat of the coastal cliff at Dunwich, Suffolk, plotted on the 1589 map of Agas; the 1977 cliff top as surveyed by A.H.W Robinson. (After Robinson, 1980a, p.141)

cliffs from a geomorphological viewpoint has yet been attempted (the stratigraphy of the Chalk is described in the GCR volume by Mortimore *et al.*, 2001). Again, the present GCR volume may stimulate such work.

As mentioned above, several of the weak-rock cliff sites are described within the *Mass Movements* GCR volume. Further soft-rock cliff sites in the GCR are those important for the sections that they provide in deposits reviewed in the Quaternary GCR volumes (Campbell and Bowen, 1989; Gordon and Sutherland, 1993; Campbell *et al.*, 1998).

Anthropogenic influences

Because of the co-incidence of soft-rock cliffs and human occupation of the south and east coasts of England, these areas are commonly modified by drainage works and coastal engineering structures aimed at arresting erosion. Current rules for funding these works are making coastal protection works more difficult to justify than has been the case over recent decades. Nevertheless, it remains important for undisturbed cliffed coasts to be protected from anthropogenic intervention if their value for geomorphological research is to be maintained, and indeed if their value in providing sections of importance to geological research is to continue. Hutchinson's work on the London Clay cliffs provides what is now a historical record of a series of coastal sectors that have been entirely modified by basal engineering works. Today that work would be impossible to carry out and it is therefore increasingly important that our remaining cliffed sites on weak rocks remain in their naturally changing state. To some extent their designation as Sites of Special Scientific Interest (SSSIs) can help to facilitate debate on options available to avoid intervention or manage the land in a way sympathetic to the conservation of the scientific features of interest.

There are now few locations along the coast between the Exe estuary and the mouth of the Tees where rapidly retreating cliffs remain unaffected by human intervention. Even in areas where they have not been affected by the construction of sea-walls, their dynamics have been altered by the obstruction of longshore sediment transport. Thus erosion of the chalk coasts of the South and North Downs has been reinvigorated by a reduction in cliff-foot beaches following the construction of major harbour walls and coast protection works at Newhaven, Seaford, Folkestone and Dover. The south-west Isle of Wight is one of the very few coastlines where there has been minimal modification both to the cliffs and the sediment transport system.

In contrast to many soft cliffs that have been investigated in detail before coast protection works were emplaced (e.g. Clements, 1994; Barton, 1991), some of the remaining unprotected cliffs have been less well investigated, despite their critical role as feeder-bluffs.

Although much interesting work has been published, there is still more research needed before we achieve an integrated understanding of the links between cliff-foot erosion, rock type, slope processes and slope form on cliffs in weak rocks. At least on the steeper cliffs (and these are usually those undergoing the most rapid erosion) within each rock type, landslides are the major process delivering material down the cliff slope. This reflects magnitude rather than frequency, though they are spatially common along the coast concerned. As a result, casual inspection of the cliffs, especially in winter when the cliffs are wet, will suggest that small streams and mudflows contribute proportionately more to slope transport than is actually the case, for though they are common, they are individually far smaller in size than the landslides (e.g. Cambers, 1973).

The conservation value of soft-rock cliff coasts

The geomorphological significance, and hence the Earth science conservation value, of soft-cliff coasts arises from their importance to our understanding of three linked processes:

1. the processes of retreat in cliffs that are cut into rocks of varying resistance;
2. the processes of platform development;
3. the processes of supply and transport of sediments from cliffs to beaches both below the cliffs and alongshore.

The rates at which cliffs and platforms produce sediment and the rate at which it is reduced and/or transported provides a strong feedback mechanism on cliff recession and platform lowering. The three processes are linked first by the sediment pathway from cliff to cliff foot to beach to down-drift beaches, second by the role of the sediment pathway from platform to beaches, and third by the inter-relationship between beach sediments and platform morphology and development. It is not usual to regard the erosional slope extending across the intertidal zone in poorly consolidated materials as a platform, but it is predominantly a surface of active erosion and a source of sediments. It exerts considerable effects on wave-energy dissipation, runoff and sediment transport. On many soft coasts, the erosion of cliffs provides the major source of beach sediment. Without erosion of cliffs, many beaches will cease to exist. Thus the continuing conservation of many sand and gravel beaches depends upon the continuation of cliff erosion.

In the present chapter, sites are arranged so that the soft-rock cliffs cut into the oldest rocks are described first, followed by others in decreasing stratigraphical age; in this way the important Chalk cliffs sites are grouped together and the Chapter ends with the cliffs cut into the Quaternary sediments of the Holderness coast.

LADRAM BAY, DEVON (SY 096 847–SY 104 858)

V.J. May

Introduction

Ladram Bay comprises a series of well-developed stacks, cliffs and platforms cut into the red sandstones of the Triassic succession, one of very few assemblages of such features in southern Britain. To the east, similar forms are well developed in the Chalk at Ballard Down in the Isle of Purbeck, at the Needles at the western tip of the Isle of Wight and near Joss Bay in the Isle of Thanet (see GCR site reports in the present volume). Ladram Bay is unique in Britain in having been cut in the comparatively easily eroded sandstone, but the forms have been preserved largely because this is a relatively low-energy site. On the west coast of Britain, most examples of stacks and associated forms are cut into more resistant sedimentary rocks of the Carboniferous, Devonian and Torridonian successions. In recognition of the importance of the site for geology and coastal geomorphology, it is part of the Dorset and East Devon Coast World Heritage Site.

The cliffs vary in height from about 25 m to over 40 m. The shore platform is rarely wider than 150 m, even where it extends below low tide levels. A beach of large shingle and cobbles masks the cliff–platform junction at some points, there being a tendency for the beaches to be better developed where the platform is absent or narrow. The platforms are structurally controlled to the extent that some surfaces co-incide with near-horizontal joint planes. Erosion along near-vertical joints has played a major role in the isolation of the stacks from the mainland.

Description

Ladram Bay is a small site comprising cliffs, stacks and platforms between Smallstones Point (SY 096 847) and High Peak (SY 104 858; see Figure 4.1 for general location). The southern part of the site (just over 1 km in length) has cliffs that are, for the most part, about 25 m in height, whereas they rise to over 120 m at High Peak. The lower cliffs here, as elsewhere in the site, are steep (with angles of inclination generally in excess of 80°). At High Peak, however, the upper part of the cliff is more complex, with

Ladram Bay

Figure 4.4 (a–c) Undercutting of the cliffs at Ladram Bay. (a) General view looking north showing the stacks associated with headlands; small pocket beaches occupy the bays (b) Two natural arches as they appeared at the beginning of the 20th century in a picture postcard, and (c) the present-day equivalent, view looking SSW. The strata are dipping seawards. (Photos (a,c): V.J. May.)

multiple mass-movements producing a stepped profile (Figure 4.4a). Stacks occur at Ladram Bay itself, and off High Peak, where they are known as 'Little and Big Picket Rocks'.

A series of platforms occur at Smallstones Point, at Ladram, and below High Peak in association with more resistant layers of the sandstone. Their slope reflects the dip of the beds forming them (about 4°). Erosion along near-vertical joints appears to have been important for the separation of the stacks from each other. Therefore, this is a coastline that is strongly controlled by structure. Between the headlands, there are small pocket beaches composed mainly of large mostly locally derived shingle and cobbles, but very little flint. Any sand that occurs is derived locally from rock outcrops.

Interpretation

The development of stacks and associated forms in the more easily eroded materials such as chalk and sandstone depends upon the ability of the sea to exploit weakness in the rocks and the

Soft-rock cliffs

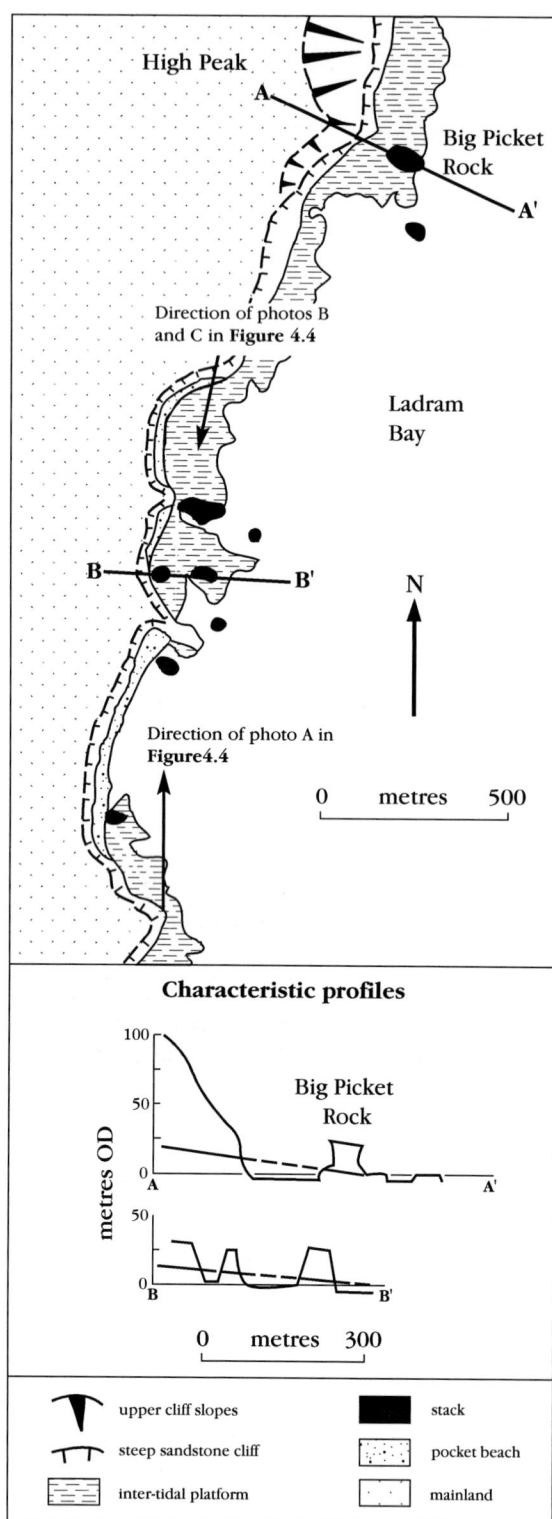

Figure 4.5 The cliffline, platforms and stacks at Ladram Bay. Characteristic profiles are shown (A–A' and B–B'). Of particular note are the absence of stacks below the high cliffs, the presence of strata with fewer discontinuities in the lower stacks, and the tendency for stacks to be associated with headlands.

resistance of the rocks to undercutting at the points where the stacks occur. At Ladram Bay, the stacks appear to result from a combination of:

1. local structural weaknesses,
2. wave energy sufficient to exploit rock weaknesses,
3. the occurrence of resistant strata at the base of the stacks. Owing to the dip of the strata, stacks only occur where the harder strata crop out at sea level (Figures 4.4 and 4.5).

The combination of mass-movements and a platform is unusual, according to Wright (1969), yet at High Peak both occur. It could be argued that the existence of the platforms here owes more to structural effects than to marine processes. This site is scientifically important as:

1. a representative of the coastal landforms in the Triassic and in sandstone, both of which crop out only to limited extent on the coastline of Britain;
2. a good example of stack development in relatively weak material in a sheltered location, the nearest comparable site being in the rather less well-sheltered chalk at Ballard Down;
3. an excellent example of the relationships between structural features and the development of stacks and platforms;
4. part of a suite of erosional landforms in contrasting energy and rock settings, namely the Magnesian limestone at Marsden Bay, the Chalk at Flamborough Head, Joss Bay, the Needles, and Ballard Down and more resistant materials at Gwithian Towans and Tintagel.

As a result, this is an important site in the national GCR network of coastal landforms which, although not as well documented as coastal sites within the Chalk, demonstrates how erosional features such as stacks and arches may develop in low-energy environments when a resistant gently-dipping basal stratum, with intersecting vertical joints or faults, and less resistant upper strata occur together. The presence of the slightly harder pedestal-forming bed appears to be particularly important to stack formation in relatively weaker strata no matter what the wave climate may be.

Conclusion

Ladram Bay is a small but important locality for geomorphology, because of the development of stacks and associated features in lithologies different from many better-known examples. The site demonstrates the role of slightly harder pedestal-forming beds in the formation of stacks in generally more easily eroded coastlines such as those comprising sandstone and chalk. The site is unusual in that mass movements and platforms are found together.

ROBIN HOOD'S BAY, YORKSHIRE (NZ 965 030)

V.J. May

Introduction

The coastline of the North York Moors is one of the most scenically dramatic in England and Wales. It transects a large part of the Jurassic succession from the Lower Lias (Lower Jurassic) strata at Saltburn-by-the-Sea in the north to the Corallian (mid-Upper Jurassic) deposits at Filey in the south. Few parts of this coast have been examined geomorphologically in detail, except for the cliffs and shore platforms around Robin Hood's Bay, where well-developed platforms cut across outcrops of Liassic shales. The cliffs mainly comprise till resting on the Lias and are subject locally to considerable mass-movement and rapid cliff-retreat. Much of the geomorphological interest in the site arises from the platforms and their relationship to the cliffs. Robin Hood's Bay contrasts with other 'active platform' sites: first, it is affected exclusively by the North Sea wave climate; second, it has been subject to glacial and postglacial processes prior to sea level reaching its present position, and third, it is close to the point along the east coast where isostatic stability rather than uplift or subsidence is predominant.

Studies of shore platforms have been a major focus of geomorphological research since the 1960s (e.g. Agar, 1960; Trenhaile, 1972, 1974a,b, 1983; Trenhaile and Layzell, 1981; Robinson, 1977a–c; Sunamura, 1983) and this GCR site has been among the most frequently examined in such studies. It forms an essential member of the network of 'active platform' GCR sites because, although Liassic shales are represented elsewhere at other GCR platform sites, no other North Sea site includes such a marked relationship between upper till cliffs and an underlying pre-glacial bedrock platform that is currently being reworked. In addition Robin Hood's Bay is by far the most studied site, particularly because of the contrast between this site and others in the 'active platform' network in England and Wales (Trenhaile, 1974b). It has been identified as the only location in 225 km of platformed coast with a continuous platform extending around a well-defined headland–embayment sequence (Trenhaile, 1974b). Unusually, the rate and nature of erosion in a cliffed coast with extensive platforms has been the focus of attention (Agar, 1960; Robinson, 1977a–c).

Description

Robin Hood's Bay lies between NZ 956 051 and NZ 908 037 on the north-east coast of Yorkshire (see Figure 4.1 for general location). Its main feature of interest is its well-developed platform cut across the geological structure known as the 'Robin Hood's Bay Dome' comprising Lower Lias shales (Figure 4.6). The bay is backed by low cliffs in the shales overlain by extensive deposits of till. The southern boundary of the site lies at South Cheek where the Peak Fault is crossed by the platform and exposes a small area of ferruginous shales. From calculations for four points around the bay, Agar (1960) estimated that the coastal retreat rate varied from a maximum of 0.305 m a^{-1} to a minimum of 0.046 m a^{-1}. Here, as elsewhere along the north Yorkshire coast, bays were retreating more rapidly than headlands (Table 4.3). In addition, there was a considerable difference between the rates of retreat of the till and the Lower Liassic strata at the cliff foot.

The cliffs are about 50 m in height in the northern part of the bay where they are cut by two steep-sided valleys, Mill Beck and Stoupe Beck (see Figure 4.6). These are cut mainly in till. Although the Lias forms the lower part of the cliff, it is commonly masked by debris from the landsliding clays above it, and by a storm beach of shingle. South of Stoupe Beck, the cliffs rise steadily to reach a maximum of 107 m at the southern end of the bay. Here the Lias forms most of the slope, with near-vertical lower cliffs comprised entirely of Lower Lias rocks.

Soft-rock cliffs

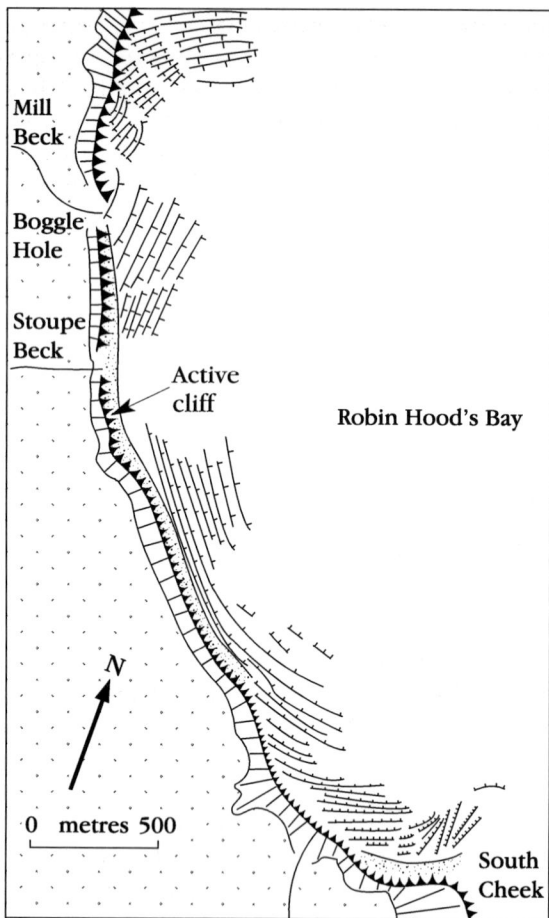

Figure 4.6 Pattern of seaward-facing micro-cliffs on the landward-dipping strata (the strike of the strata is indicated) on the low-gradient intertidal platform in Robin Hood's Bay.

The Peak fault runs through South Cheek, with the result that the southern part of the cliff is dominated by Toarcian rather than Pliensbachian shales. Agar (1960) did not measure change here although his paper suggests that erosion was substantially less than elsewhere in the bay.

The platform is distinguished by a series of curving ridges and troughs that reflect the differential erosion of the shales and the structural form of the Robin Hood's Bay Dome (Figures 4.6 and 4.7). Trenhaile (1974a) suggests that the platform on the northern side of the bay is concave in form, a feature that he attributes to ramp development. However, rather than having a truly concave form, the platform is made up of two elements. The ramp is related to harder material forming the upper part of the platform. Elsewhere in the bay, linear forms up to 500 m in width are more characteristic of the platform (Trenhaile, 1974b) and extend for considerable distances offshore with a gradient of about 1 in 100 to depths of about 40 m (Agar, 1960).

Interpretation

The broad geological structure of the Robin Hood's Bay Dome has not affected the macro-morphology of the platform cut across it, whereas the micro-morphology of the platform is strongly dependent upon the structure across the Dome (Figure 4.7). In contrast to the platforms between Watchet and Lilstock (see GCR site report for Blue Anchor–Watchet–Lilstock in the present chapter), this site does not have a well-developed 'washboard' surface but demonstrates well how spatial arrangements of the micro-morphology are controlled by the varying dip and strike of the beds of the outcrops. In detail this may affect the refraction of waves, especially at lower stages of the tide. In turn this affects the transport of sandy sediment within the intertidal zone. The development across a very complex structure of platforms that display similar characteristics to those cut across simpler ones gives this site an important place in the debate about shore-platform development.

Most writers have concentrated on the development of the cliffs and the platform, rather than the links between the platform and geomorphological processes. The single exception is Robinson (1977c, see below). Agar (1960) used his measurements of cliff erosion in combination with an assessment of the degree to which the upper cliff had changed in postglacial (Holocene) times to judge the development of the coastline over the past 10 000 years. He regarded present-day conditions as 'optimal', i.e. the sea breaking on the gradually sloping foreshore and attacking the vertical face of the cliffs at high tide to develop a cliff-foot notch. Erosion rates could thus be interpreted as being maxima. Agar argued that a slightly lower sea level would result in the action of the waves being concentrated on the platform and having a much less important role in coastline retreat. A higher sea level would similarly have only a limited effect because waves would be reflected from the vertical cliffs. Taking account of the contemporary interpretations of the curve of sea-level change, he argued that apart from a short period around 7000 BP, the past few centuries are the only postglacial period 'in which favourable conditions for formation of the present foreshore have existed' (Agar, 1960, p. 422). Extrapolating from the measured rates of retreat, he argued that most of the local erosion has occurred only during the last six centuries. As a result, many profiles, including those of South Cheek, would have been affected by only limited postglacial erosion. Their upper slopes were not regarded by Agar as contemporary forms, but as probably of last interglacial age. Both the discussion following Agar's paper and later comments cast doubts on his interpretation of the coastal features.

Straw and Clayton (1979) consider that if Agar is correct then the present coastline must approximate in location to that of the Ipswichian (Eemian) interglacial. They cite the resistance of the rocks to marine erosion and recognize the difficulty of ascribing the platform solely to late Holocene marine erosion. They thought it inevitable that the platforms must have been prepared during preceding interglacial periods. However, Robinson (1977c) was not convinced by the view that many of the platforms have been reworked and that notches revealed beneath the till show that the platforms are at least Weichselian in age. Robinson counters by arguing that many of the features are recent, some less than 200 years old, and that much of the alleged pre-Weichselian glacial form has been buried by postglacial landslipped material that has then been removed, exhuming the pre-talus surface.

In a wider discussion of shore platforms, Trenhaile (1974b) describes this site as the only location in 225 km of platformed coastline with 'a continuous shore platform extending around a well-defined headland–embayment sequence'. He also records that the platform gradient increases towards the headlands, especially in the north, typically from about 35' to 2.5°. The headland site is more rugged than the Lower Lias shales of the embayment and is also more exposed to greater wave activity. From such evidence here and on other shore platforms around the coastline of England and Wales, Trenhaile (1974b) concludes that the platform gradient is being maintained in dynamic equilibrium. This appears to cast doubt on the claim that many of these platforms, including the platform at this site, have been inherited from previous forms (Trenhaile and Layzell, 1981). However, they argue that the evidence suggesting that shore platforms are partially inherited features is not incompatible with the evidence indicating that they are at or close to a state of dynamic equilibrium with a morphology finely tuned to their present environments. Despite some debate (Carr and Graff, 1982; Trenhaile, 1982), this argument appears to hold good for Robin Hood's Bay – that the platforms are likely to be be reworked earlier platforms, retrimmed by Holocene seas. Unfortunately neither Agar, nor Trenhaile and Layzell, take sufficient note of the role of debris on the platforms either in its erosional, or its protecting and roughening, role.

Robinson (1977a–c) argued that the morphology of the platforms resulted from the presence of sand debris rather than the nature of the rocks forming the platform. The width of the platform is controlled primarily by the protection afforded to the cliffs by the deposits at their foot (Figure 4.7). Where debris is absent, the platform has a low angle of inclination, characteristically about 1°. Robinson calls this the 'plane'. In contrast, where there is a beach, the slope is greater, usually up to 15°. This is the 'ramp'. Trenhaile (1974b) believed that the steeper ramp was produced by harder materials. Robinson identified five erosion processes here:

1. micro-quarrying;
2. the expansion and contraction of clay mineral lattices by hydration and desiccation. He estimated that processes 1 and 2 together low-

Table 4.3 North Yorkshire coast cliff retreat rates in m a^{-1} (based on Agar, 1960).

		Cliff top	**Cliff foot**
Whole coast		0.02	0.05
Headlands only		0.01	0.04
Bays only		0.04	0.07
Robin Hood's Bay	Lower Lias	0.02	between 0.07 and 0.16
	Glacial drift	0.31	between 0.05 and 0.31

Soft-rock cliffs

Figure 4.7 Shore platform at Robin Hood's Bay looking east from Mill Beck (see Figure 4.6 for location). (Photo: J.D. Hansom.)

ered the platform by 0.144 cm a^{-1};
3. wave-quarrying, by which removal of small blocks from the cliff foot lowered the platform surface by 2.3 cm a^{-1};
4. corrasion: direct abrasion of the in-situ rock by wave-transported sediment lowered surfaces by 5.79 × 10^{-3} cm tide^{-1};
5. wedging, in which small sediment particles forced into cracks in bedrock gradually force it apart. This lowered surfaces by 11.05 × 10^{-3} cm tide^{-1}.

Robinson (1974) showed that erosion was more rapid when a thin beach was present, but seasonal variations in wave action also affect the efficiency of erosion of the ramp. In contrast the plane is affected by desiccation and contraction at low water – especially in summer – and expansion at high water. The annual rates of lowering estimated by Robinson's use of a micro-erosion meter are about 1.5 times faster than those obtained by longer-term comparisons of platform levels on the Chalk around the Isle of Thanet (see GCR site report for Joss Bay in the present chapter).

Conclusions

Robin Hood's Bay is a very important site for study of platformed coastline development, because the platform cuts across a complex structural geological structure, the 'Robin Hood's Bay Dome'. Despite the complexity of the underlying structure, the platform displays many of the features observed elsewhere in much simpler structures. Unlike many other platforms, it has been the focus of detailed investigation of the erosion processes, in particular the varying role of beach sediments in either erosion or protection of the platform and cliff foot. It demonstrates well a relationship between headland and embayment in which preglacial erosion of the platform may have produced an equilibrium form that is being reworked today. This remains, however, the paradox of this site, for it is not possible to determine the extent to which the platform is being exhumed or reworked. The cliffs are cut both into the Lower Lias bedrock and glacial materials and provide an excellent example of a cliffed coastline where the comparatively recent weaker

BLUE ANCHOR–WATCHET–LILSTOCK, SOMERSET (ST 034 436–ST 070 438 AND ST 116435–ST 169 455)

V.J. May

Introduction

The southern cliffed coastline of the Bristol Channel between Blue Anchor and Hinckley Point contrasts with the higher hog's-back cliffs of the Exmoor coast and the low, estuarine, wide mudflats and fringing dunes around the mouth of the River Parrett to the east. This site, which comprises two areas east and west of Watchet, is characterized by cliffs rising to a maximum of 84 m and fronted by a particularly well-developed series of intertidal platforms varying in width from 120 m to over 500 m. At St Audrie's Bay, the cliff has been proposed as the type locality and section for the base of the Jurassic System (Warrington *et al.*, 1994). The base of the Hettangian Stage at the base of the *Planorbis* chronozone is placed at the horizon in which ammonites of the genus *Psiloceras* appear.

The platforms are veneered in part by shingle, sand and mud, and reflect in detail the variable resistance to erosion of the Turassiched Marls, Penarth Beds and Lower Lias bedrock. A key feature of the platforms is their development in a macrotidal environment and their different exposure from narrower platforms in similar rocks on the northern side of the Bristol Channel at Nash Point. Whereas there has been considerable research into the nature of the platforms on the northern side of the Bristol Channel, it has been singularly lacking on the southern coast. Ussher (1908) was the first to describe the main features and Steers commented that the cliffs and shore features were of 'considerable interest' (Steers, 1946a, p. 211).

Description

The western part of the site extends from the eastern end of the sea-wall at Blue Anchor (ST 034 436) to just west of Watchet (ST 070 438). Near-vertical cliffs rise eastwards to Blue Anchor Point (ST 040 437) where they give way to higher cliffs that are much affected by many small landslips. From their highest elevation of 84 m, they fall steadily towards sea level at Watchet. At Warren Bay (ST 057 434), they truncate a valley that is left hanging about 25 m above the base of the cliffs.

The alignment of the coastline of the western part of the site has little relationship to the direction of wave attack from the Atlantic Ocean. The coastal plan is primarily a function of the varying strengths and structures truncated by the cliffs and platforms. Differential erosion is a dominant force both in the general form and the detail of the coastal features. The platform varies between 300 m and 500 m in width. The general slope of the platform reflects the process of marine planation in cutting across the outcrop, but the varying strength, dip and strike of the beds give rise to a varied micro-relief. Parts of the platform warrant the description 'washboard-like relief', a form that has been described elsewhere (for example, Suzuki *et al.*, 1970), but rarely reported in Britain (Figure 4.8).

The eastern part of the site extends from the eastern side of St Audrie's Bay (ST 116 435) to Lilstock (ST 169 455). Much of the cliff is near-vertical, rising from about 25 m to over 50 m in height at Quantocks Head. The cliffs decline in height to near sea level at Kilve Pill (ST 143 444) and then maintain a height of about 30 m to the eastern boundary of the site. The platform is between 120 m and 300 m in width, and demonstrates similar relationships between rock type, structure and micro-relief to the area west of Watchet. This part of the site, however, shows a stronger relationship to the prevailing direction of wave attack from the WNW, both in the alignment of a shingle beach on the eastern side of St Audrie's Bay and in the alignment of the cliffs to the north-east of Kilve Pill.

Interpretation

Although the coastline is one on which erosion dominates, there have been few measurements of change. Mackintosh (1868) estimated the rate of cliff retreat on the Lias cliffs as 1.2 m a^{-1}. Retreat is far from uniform, with very little change at some points whereas others attain current rates of change comparable to those noted by Mackintosh. The platforms here are of considerable interest and they warrant further inves-

Soft-rock cliffs

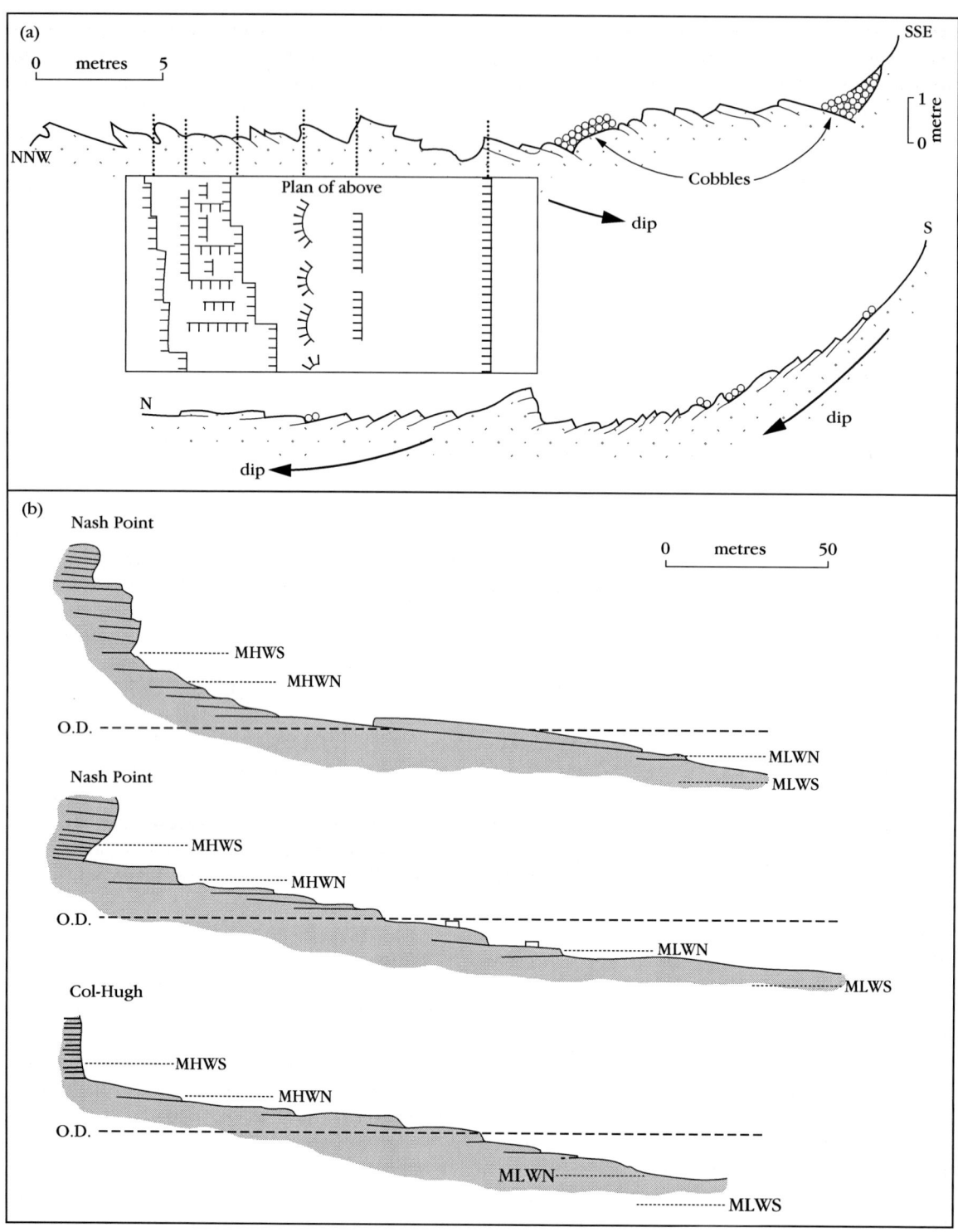

Figure 4.8 (a) Cross-sections, showing characteristic forms of the platform east of Watchet, where the dip of strata to landward or seaward strongly affects the pattern of micro-cliffs, (b) three characteristic platform profiles at Nash Point, Vale of Glamorgan (see GCR site report in the present chapter) where dip of strata is more uniform than at Watchet. Mean high- and low-water spring tide levels (MHWS and MLWS) and mean high- and low-water neap tide levels (MHWN and MLWN) are shown. (Part (b) is after Trenhaile, 1972.)

Blue Anchor–Watchet–Lilstock

Figure 4.9 Cliffs and shore platform at Kilve, Somerset (Photo: V.J. May)

tigation. Both the platforms and the cliffs in the Lias on the northern side of the Bristol Channel have been investigated in some detail (see GCR site report for Nash Point below). Although the Watchet sites lie in a similar tidal regime to that of Nash Point, they are much less exposed to the high wave-energy levels reported by Williams and Davies (1987). The maximum fetch of this site is just over 300 km to the WNW, whereas Nash Point has a maximum fetch of 5000 km to the south-west. Atlantic waves approaching the Watchet sites undergo considerable refraction and approach the shore at an angle, whereas Nash Point receives the full undiminished energy of Atlantic storms. In the less vigorous environment of Watchet, processes of intertidal weathering are more important and there is less movement of particles broken from the bedrock.

The detailed nature of the platform reflects the minor structures of the rocks forming it, as well as the dip of the strata (Figure 4.8a). For example, parts of the platform are distinguished by a blocky structure in which large numbers of small vertical joints about 0.25 m to 0.35 m apart create a series of irregular three- to six-sided polygons. The platform cuts across folded strata that dip variously seawards, landwards (Figure 4.9) and alongshore, and there are important variations in both the platform morphology and its effects upon wave action and shingle and cobble movements. Where the strata dip landwards, for example east of Kilve, the intertidal area is characterized by a series of micro-cuestas, up to 1 m in height (Figure 4.8). As the edges of these up-tilted strata have been broken up by marine erosion and intertidal weathering, they form a cobble field between the minor cuestas. The size of material in the cobble fields varies from the almost unaltered newly-quarried blocks, through sub-rounded and rounded blocks and pebbles to sand. Because the beds are rarely horizontal and often lie at an angle across the beach, many of them also show signs of the action of flowing water along the base of the micro-cuesta dip-slope. The blocky nature of the beds also imposes a maximum height on the micro-cuestas. As soon as a block is partly undermined it begins to slide along the joint surfaces and frequently topples forward. Wave action is insufficient to remove most of the blocks, which appear to remain close to their original site, until they have been worn down to a threshold size and shape which allows movement.

Where the strata dip seawards, for example, west of Watchet and between Quantocks Head and Kilve, the lower cliff sometimes forms a sloping rampart formed by unbroken strata. On the platform, micro-cuestas are formed with the scarp facing landwards, and large accumulations of shingle and cobbles are retained on the landward side of the micro-cuestas. Most erosion of the scarps is achieved by wetting and drying processes; the erosional product is readily transported along the sloping micro-vales. The scarps are very effective in preventing shingle and cobbles moving down the slope of the platform towards the sea. Even where the scarp is no higher than 0.1 m, its alignment is clearly marked by the line of cobbles resting against it.

Where the strata dip alongshore, the platform is also marked by micro-cuestas, but this pattern allows waves to reach well up the shore along the micro-vales between ridges. The most active parts of the cliffs characteristically co-incide with these more exposed locations, for example to the west of Lilstock. Although they differ between seaward-and landward-dipping topographies, the large inter-cuesta sediment fields in this site contrast very strongly with comparatively bare platforms on the northern side of the Bristol Channel. Although this probably reflects the lower wave-energy environment of the Watchet sites, the morphology and slope of the platforms is also important. Most of the clasts are subangular. The only smoothed surfaces occur on the more resistant strata at the foot of the cliff, which are commonly cloaked by large beaches of more rounded clasts.

The platforms in this site show a morphological pattern in which the harder and wider, jointed and bedded strata form the upstanding forms. Suzuki *et al.* (1970) found that, in contrast to the generally expected relationship of rock hardness or strength to the extent to which strata protruded above the platform, the micro-cuestas at Arasaki, southern Japan, were formed in an apparently weaker tuff than the surrounding mudstones. The ability of the mudstone to absorb greater quantities of water, and the greater stresses that occurred as a result, caused them to be eroded more efficiently than the tuffs, which became the micro-cuestas. On the Watchet sites, the micro-relief is strongly controlled by the thickness of the in-situ strata, so that only those beds that are thicker than about 0.25 m form micro-cuestas. The variation (between 0.5 m and 0.02 m) in bed thickness, which characterizes the Lias in this site, thus appears to be the critical factor in the development of the washboard-like relief of these platforms. They warrant detailed study to develop a fuller understanding of the complex relationship between platform relief and the role of sediment as an abrasional agent.

Conclusions

This is a fine example of shore platforms developed in a macrotidal environment. The site includes one of the best examples in Britain of 'washboard' platform relief. The development of the platforms depends on the way in which the cliffs retreat. Although the site has received very little attention in the literature, the rapidity of retreat along much of this coastline is important for geomorphological study. Similar features have not been described elsewhere on the British coast, partly because platform studies have concentrated upon the gross morphology of platforms rather than their micro-forms, but also because such forms are comparatively rare. The platforms at Robin Hood's Bay are comparable, but are affected by a tidal range almost half that of this site. The other shore platforms, especially those in the Chalk, lack the resistant strata to produce micro-cuesta topography, and the platforms in the Carboniferous Limestone and Portland Stone are more commonly affected by weathering by dissolution. This site is thus important to the network of active platform sites both in its macrotidal location and its varied morphology.

NASH POINT, GLAMORGAN (SS 934 677–SS 905 699)

V.J. May

Introduction

Much of the coastline of the Vale of Glamorgan is formed of cliffs cut in the Lias limestones and mudstones and fronted by platforms that attain widths in excess of 500 m. The line of cliffs is broken by a number of small steep-sided valleys. This site (see Figure 4.1 for general location) comprises the cliffed coastline east and west of Nash Point and is cut mainly in limestone and

Nash Point

mudrocks of the Blue Lias. The cliffs vary in height from 62 m to less than 30 m, and are commonly near-vertical, even overhanging in places. Intertidal platforms are generally between 200 m and 250 m in width (Figure 4.10). Although they slope seawards, their micro-relief is largely controlled by the relative strength of the limestones and the argillaceous beds across which they are cut. Variations in cliff-form are not always directly associated with variations in rock type. Similarly, the coastal plan does not always accord with the terrestrial landforms that it transects. Because of its exposure to the Atlantic Ocean, this is a high-energy environment; both the cliffs and the platforms have been the foci of much recent investigation (Trenhaile, 1969, 1971, 1972, 1974a,b, 1983; Trenhaile and Layzell, 1981; Carr and Graff, 1982; Sunamura, 1983; Williams and Davies, 1984, 1987; Davies and Williams, 1986; Davies *et al.*, 1991; Williams *et al.*, 1993).

Nash Point has been described as an example of a site with structurally controlled platforms (Davies, 1972). Trenhaile (1969, 1971, 1972, 1974a,b, 1983) and Sunamura (1983) debated shore-platform development by reference to this and other British sites, and cliff and beach features were described by Williams and Davies (1987). A review of the literature suggests that the processes operating on this site have probably received more direct and regular attention than any other vertical cliff site in the UK (Mackintosh, 1868; Keatch, 1965; Trenhaile, 1969, 1971, 1972; Williams and Davies, 1984, 1987; Davies and Williams, 1986; Williams and Caldwell, 1988).

Description

The site extends from the western side of St Donat's Bay (SS 934 677) to Cwm Nash (SS 905 699). To the east of Nash Point, the continuous line of cliffs is broken by the valley of the Marcross Brook. The western cliffs are aligned

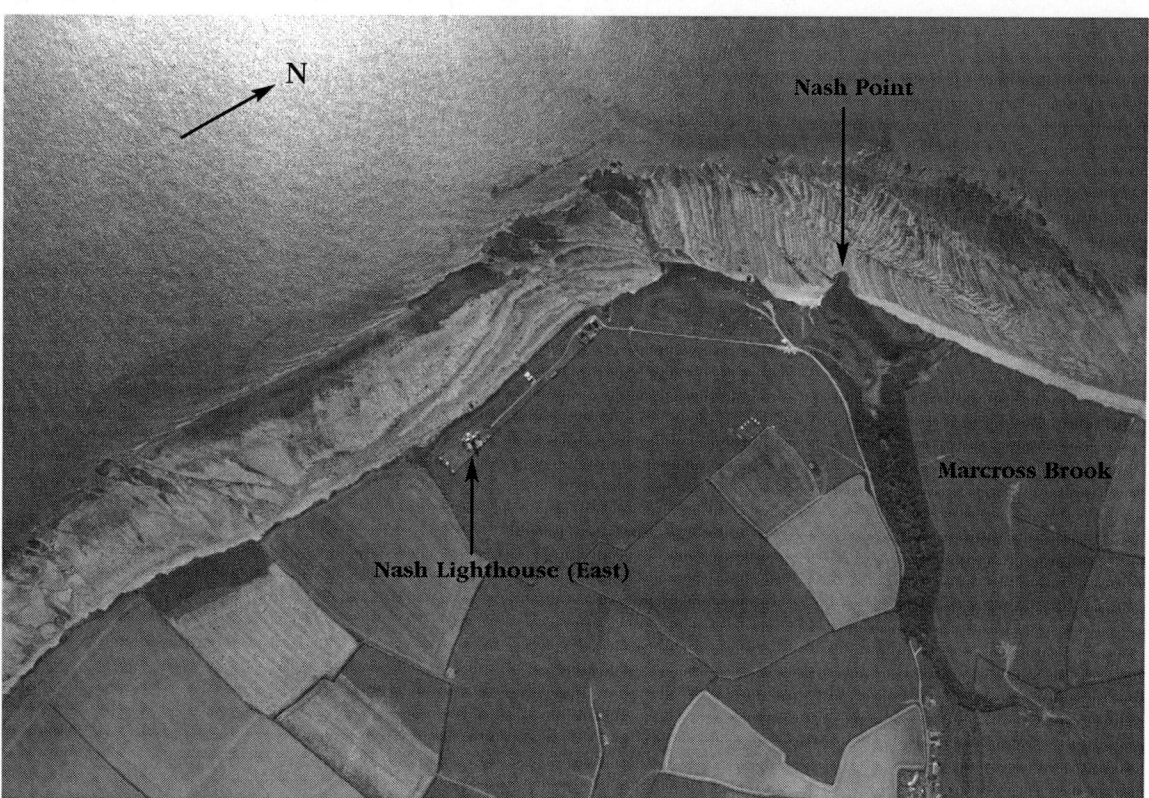

Figure 4.10 Nash Point, this view from directly above the site demonstrates the near-vertical nature of the cliffs and the width of platforms at low water. The micro-relief of the shore platforms is controlled largely by the the relative strengths of alternating beds of limestone and argillaceous rocks and jointing patterns, on this photograph particularly noticeable in the vicinity of Nash Point itself (see also Figure 4.8b). (Photo: CCUCAP, © the Countryside Council for Wales.)

towards the south-west, facing the dominant and prevailing wind (over 40% of all winds) across a fetch of 5000 km. The eastern cliffs trend east–west. The dominant and prevailing winds blow alongshore but the southerly onshore winds have a fetch of only 24 km. According to Trenhaile (1972), this part of the coastline is characterized by longshore drift.

The tidal range is 6 m. Surges associated with low atmospheric pressure have been recorded, increasing water levels by up to 1.5 m (Williams and Davies, 1987). This is a relatively high-energy environment, having recorded cumulative wave energy densities of 68×10^5 joules m^{-1} crest width^{-1} over one day, and wave power of 85,000 joules m^{-1} s^{-1} (Williams and Davies, 1987). Assuming a still-water level at mean high-water neap tides, the total breaking wave force recorded in one storm was of the order of 7×10^5 Pa (7 bars). Cliff retreat has been estimated by several writers (summarized by Williams and Davies, 1987) and varies between 0.1 m a^{-1} and 0.02 m a^{-1}.

The Lias around Nash was divided by Trueman (1922, 1930) into:

1. The *Arietites bucklandi* biozone – thick concretionary limestones alternating with thin mudstones, and
2. the *Sclotheimia angulata* biozone – mainly thick mudstones alternating with thin limestones.

The limestone beds reach almost 1.0 m thick in places, whereas the mudstones rarely exceed 0.5 m. There are three main groups of near-vertical joints trending NNE–SSW, NE–SW, and SE–NW. Joints resulting from pressure-release rebound lie most commonly parallel to the cliff face (the direction of greatest stress release). The platforms grade seawards at low angles (i.e. around 2°), and are broken only by small scarps and shallow solution features (Trenhaile, 1972). The platforms owe much of their uniformity to the exposure of single beds of limestone and so contrast dramatically with the platforms in the Lias on the southern side of the Bristol Channel. Characteristic profiles and the location of major breaks of slope are shown in Figure 4.8b. On the platforms, small scarps (generally less than 0.25 m in height) are associated with erosion of thin shale horizons and undercutting of the thicker beds.

Interpretation

A substantial literature concerning this site has focussed on two separate morphological units of the coast, the platform and the cliff. Trenhaile (1974a,b) discussed the development of shore platforms here and elsewhere in Britain (for example Robin Hood's Bay, see GCR site report in the present chapter). At Nash Point, the shore platform has probably evolved from the destruction of a series of higher platforms and is partly an inherited feature (Trenhaile, 1972). However, the rate of cliff recession and the present erosion of the platform indicate that more extension of the platforms has occurred than in the southern part of Gower. The development of platforms around Nash Point contrasts also with that on the southern side of the Bristol Channel. The platforms at Nash Point are strongly related to the dip of the strata, their surfaces being structural, whereas on the Somerset coast they are cut across steeply dipping strata. At Nash Point, lowering of the platform is mainly brought about by retreat of low steps, and Trenhaile (1972) concluded that contemporary scarp retreat is of the order of magnitude necessary to bring about parallel slope retreat. The platforms could be regarded as an example of the principle that the timescale of observation affects the significance of time in landform development (Schumm and Lichty, 1965): Johnson's (1919) model of platform development is not necessarily supported by the evidence from this site when long periods of time are considered. Over shorter timescales (i.e. under 100 years), there is a trend towards dynamic equilibrium in which there are 'fairly high correlations' between platform gradient and elements of platform morphology.

Williams and Davies (1984, 1987) demonstrated that retreat of the cliffs and thus extension of the platforms at high-water level resulted from several processes. Large-scale cliff failures usually occurred as a result of toppling and translation failures. The latter are usually very complex, low frequency and high magnitude events. The detachment of joint blocks also affects the cliffs. Although small in extent, their product is removed quickly by wave action, and so it is difficult to estimate the volumes of individual movements. Williams *et al.* (1993), having analysed rockfalls along 22 km of adjacent coastline, developed numerical models of cliff failure. Translation failure was predicted in cliffs where the *angulata* series formed a high proportion of the cliff mass and where cliffs are buttressed by limestone of the *bucklandi* biozone. Toppling failures were predicted for vertical and overhanging cliffs that were undercut at the base.

Caves cut into the cliffs are restricted to the low-energy environment of the W–E-trending cliff sections. Davies and Williams (1986) showed that interaction between the presence of particular limestone beds at the cliff base and within the cliff, and protection from the most direct wave attack is crucial. The most suitable basal limestone strata are the most massive (Trueman Bands, 28, 39, 47, 48 (Trueman, 1922/1930), but other limestone strata also form cave lintels. Caves do not develop where there is a high proportion of mudstone, requiring an average of over 66% limestone strata in exposed cave walls. Caves do not usually develop where there is a wide or low angle shore platform. 76% of the caves are associated with joints perpendicular to the cliff face. Cave retreat appears to operate at a similar rate to cliff recession. This examination of cave development is especially important to coastal geomorphological studies, because it goes beyond the usual suggestion that caves are associated with lines of weakness. Furthermore, the role of the basal beds can be shown to be consistent with the characteristics of arch and stack development described elsewhere in the present chapter (see GCR site reports for Flamborough Head, Ballard Down and Ladram Bay in the present chapter). The rate of cliff retreat may be the controlling factor, since any general retreat of the cliff will increase the wave energy available for cave excavation, and lowering of the platform in front of the cave will enhance the available wave energy. The three morphological units, the platform, the cliff, and the cave therefore appear to be functionally linked, but further investigation is required.

In terms of exposure to wave energy, this is the most-exposed, and best-documented Lias cliff site in southern Britain. It contrasts with other Lias sites at Robin Hood's Bay and Watchet (see GCR site reports in the present chapter) in its exposure and relatively high-energy environment. Although platforms around the Chalk (for example, Joss Bay) have also been described in some detail (So, 1965), they rarely display the marked alternation between hard and weaker beds that characterize the Lias and produce the sloping stepped form of the platforms at Nash Point. Although caves are a common feature of many cliffed coasts, few are described in the geomorphological literature, and Nash Point is an exception.

Conclusions

Nash Point is situated in a relatively high wave-energy environment and is dominated by platforms over 500 m in width and vertical cliffs over 60 m in height. The cliffs have received much less attention in the geomorphological literature than the platforms, but have many features in common with other cliffs undergoing active erosion. This is one of few sites where cave development has been investigated in detail. Like the platforms of the Isle of Thanet, the platforms at Nash Point are simple in form when compared to others elsewhere on the British coast. The initial surveys by Trenhaile (1969, 1971, 1972, 1974a,b) provide a basis upon which later studies have been able to build. As a result the site is important within British coastal geomorphological studies because of both the repeated surveys and an increasing understanding of the processes that occur here. Few vertical-cliff sites have been as examined and the processes so elucidated as at Nash Point, and, as a result, it is internationally important for its coastal geomorphology.

LYME REGIS TO GOLDEN CAP, DORSET (SY 380 927–SY 428 913)

V.J. May

Introduction

Between Ridge Cliff, to the east of Seatown, to Lyme Regis, there are four main cliffed areas, Ridge Cliff, Golden Cap, Cain's Folly and Black Ven, separated by valleys at Seatown, St Gabriel's Water and Charmouth (see Figure 4.1 for general location, and Figure 4.11). There are two pocket-type beaches, the larger between Golden Cap and Lyme Regis, the smaller to the east of Golden Cap at Seatown. This group of cliffs and beaches is geomorphologically important because:

1. The cliff changes (especially the landslides at Black Ven) are probably the most fully investigated of any in the world. The international

contribution to geomorphology is outstanding.
2. There are excellent examples of arcuate beach ramparts, formed by the boulder content of landslides.
3. The beaches are fed by chert and flint from the cliffs, so that it is possible to monitor the links between landslides, cliff erosion and beach-sediment budgets.

The landslides of this coast are well documented (Arber, 1941, 1973; Lang, 1914, 1928, 1942, 1944, 1955; Wilson *et al.*, 1958; Brunsden, 1973, 1974, 1996; Brunsden and Jones, 1972, 1976, 1980; Conway, 1974; Denness *et al.*, 1975; Brunsden and Goudie, 1981; Allison, 1990, 1992; Koh, 1992; Lee, 1992; Brunsden and Chandler, 1996; Brunsden *et al.*, 1996; Pile, 1996). Many coastal texts refer to the landslides here (e.g. Bird, 1984; Steers, 1964a, 1981), but there has been much less attention to the beaches (Lang, 1914; Bird, 1989; Bray, 1986, 1990a,b, 1996) and the offshore zone (Darton *et al.*, 1981).

As well as its geomorphological significance, the site is famous stratigraphically and palaeontologically and it is one of the GCR sites that form the Dorset and East Devon Coast World Heritage site, established in 2001.

Description

In general, the coastline truncates a series of NE–SW-orientated ridges that rise to between 140 m and 170 m OD. The area is composed of interbedded, firm, fissured clays, mudstones, marls and thin bands of hard argillaceous limestone of Lower Jurassic age. These are overlain unconformably by silty clays, fine-grained silty sands and chert beds of the Cretaceous Gault and Upper Greensand. Whereas the Lias dips ESE at 2–3°, the plane of the unconformity and the Cretaceous beds above it dip to the south-west at 2–2.5° (Brunsden and Jones, 1976). The sides of the ridges have a thick cover of solifluction and landslide debris. The ridge-tops are covered by a superficial layer of flint and chert head.

The eastern limit of the site is Ridge Cliff (SY 428 913) where the cliffs attain a height of 100 m in sands and clays of the Upper and Middle Lias. They decline westwards towards Seatown where a small stream, the River Winniford, enters the sea. They then rise to the highest point on the Dorset coast at Golden Cap (188 m OD; Figures 4.11 and 4.12). Here the Upper Greensand forms a steep upper section to the cliff profile, but the main part of the cliff is formed by the Eype Clays. It has been greatly affected by landsliding, but less dramatically than the cliffs to the west. The intertidal area is characterized by rock ramparts that represent the remnants of landslides that have carried Upper Greensand blocks to the foot of the cliff. Whereas the clays that form the bulk of the slide debris have since been eroded, the curved boulder aprons remain. Darton *et al.* (1981) indicated that similar features occur offshore, and Bray (1996) has confirmed this. At both St Gabriel's and Ridge Water small streams flow in hanging valleys at about 65 m before finding their way across the slipped cliff face. To the east of Charmouth, the cliffs at Cain's Folly rise to 145 m, whereas to the west their highest point is 177 m.

Seatown Beach, to the east of Golden Cap, is formed mainly in flint and chert shingle, but it also contains pebbles of Lias shales and limestones. It lies between two headlands that inhibit longshore transport of sediment both into the beach from the west and out of it towards the east (Figure 4.12). The beach gravels are sparse and poorly sorted at the western end and there are patches of sand and exposures of the underlying strata (Bird, 1989). Towards the centre of the beach at Seatown itself, the predominantly flint and chert pebbles are found in zones of contrasting size parallel to the beach face. Cusps are often well developed. At the eastern end of the beach, beneath Ridge Cliff, the beach is higher and wider with coarser and better-sorted shingle, but following periods of easterly wind can be denuded to reveal a deeply incised platform (D. Brunsden, pers. comm.).

Over the period 1901 to 1987, this beach had an input of 190 000 m^3 of shingle, mostly from intermittent transport around Golden Cap, particularly between 1932 and 1962 when the annual input was up to 8600 m^3 a^{-1} (Bray, 1996). In the past shingle was mined from Seatown Beach (Bird, 1980). Bray (1996) estimated that between 125 000 and 175 000 m^3 were extracted during World War II and a further 34 000 m^3 between 1956 and 1986. The extraction permit expired in 1987. Extraction of shingle, together with modest entrapment and attrition losses, has produced a complex series of volumetric changes with an overall deficit. Bray (1996) esti-

Lyme Regis to Golden Cap

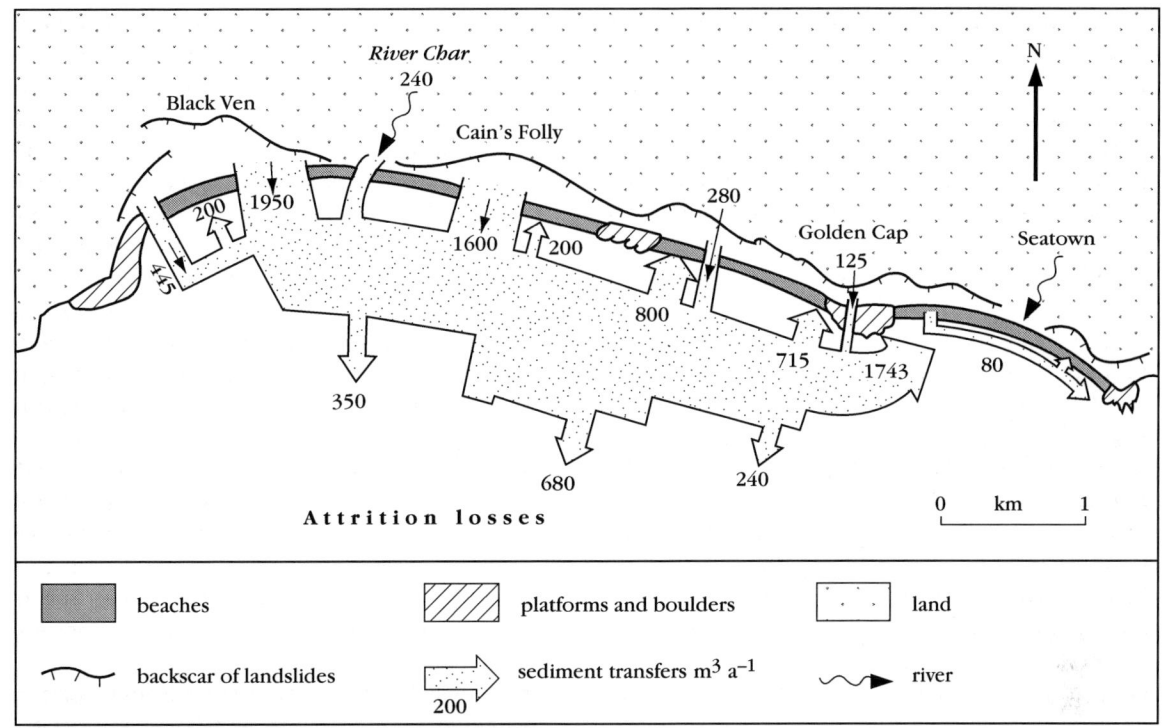

Figure 4.11 The sediment budget of beaches between Lyme Regis (to the westmost part of the map) and Seatown. (After Bray, 1990a.)

Figure 4.12 View looking south-east from Golden Cap, showing the depleted shingle beach at Seatown, platforms that are cut across folded strata, and the residual boulders at the west end of the beach (foreground). (Photo: V.J. May.)

mated that, with the cessation of extraction and no input from the west, the net deficit had fallen to 39 $m^3 a^{-1}$ by 1989. An unaccounted factor in beach and cliff erosion is the erosional lowering of the platform when it is intermittently exposed.

To the west of Golden Cap, the beaches are interrupted beneath the landslides by large lobes of debris from the slides and mudflows, but they give way westwards to an intertidal area that is dominated by structurally controlled shore platforms, especially just to the east of Lyme Regis. Most of the beach material is derived from the cliffs. Bray (1986) estimated that between 1901 and 1987, about 420 000 m^3 of mainly chert gravel with a B-axis greater than 10 mm was eroded from the cliff back-scar between Lyme Regis and West Bay. The beach between Golden Cap and Charmouth is well-sorted laterally, changing from a mixed sand and shingle beach at Charmouth to a predominantly cobble beach below Golden Cap (Bird, 1989). Supply of gravel is concentrated at the western end of this site, with transport towards the east. The volume of the beach increases towards

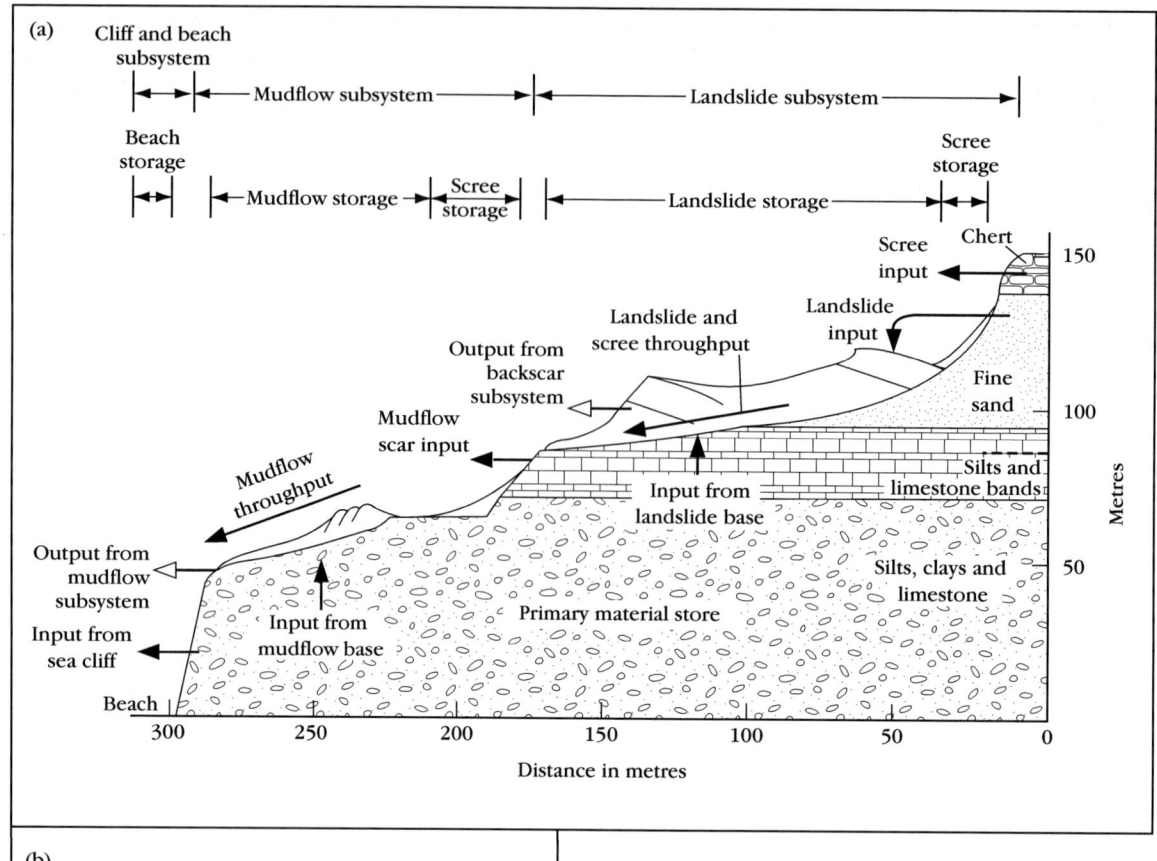

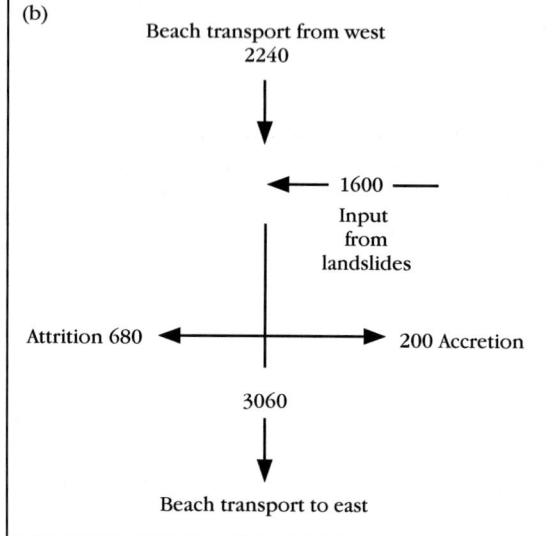

Figure 4.13 (a) Cross-section of the Black Ven system (Lyme Regis to Golden Cap GCR site) and sediment supply to its beach. See also Figure 4.11.
In (b) the volumes of sediment (in $m^3\ a^{-1}$) moving through the Black Ven beach are given. (Based on Brunsden, 1973 and Bray, 1990a.)

Golden Cap, but some material may leak around the headland when landslide snouts have been eroded (Bray, 1990b, 1996). However, this has not occurred since the mudslide of 1962.

The dominant features of this coast are the rapidly changing cliffs east and west of Charmouth. At Fairy Dell, Brunsden and Jones (1976) described both the historical and more recent history of the landslides. They identified three sub-parallel clifflines separated by benches covered by debris. These benches were attributed to lithological variations, the upper bench being associated with the sub-Gault unconformity and the lower bench with the outcrop of the Green Ammonite Beds. Three morphodynamic zones were identified.

1. Zone 1, an upper zone with an arcuate scar up to 45 m in height. The lower part of the scar was covered by a partially vegetated scree slope. The upper bench consisted of several large rotational landslide blocks separated by screes of chert.
2. Most of the central Zone 2 is extremely complex with deep V-shaped gullies, rotated

landslide blocks, areas of gentle relief in which debris in accumulated and mudslide basins.
3. Zone 3 is a lower zone of near-vertical sea cliffs with a variable amount of accumulated debris at their foot.

The upper arcuate scar and the sea cliffs retreat at similar rates (between 0.29 and 0.71 m a^{-1}).

At Black Ven, the cliffs are marked by very active mudslides (Brunsden, 1968; Figure 4.13). The upper part of the cliff is cut into Greensand sands and cherts that lie unconformably on Lower Lias clays. Limestone layers within these clays act as local base levels over which the debris moves. The impermeability of the clays allows groundwater seepage at the base of the Greensand, which produces high levels of instability in the cliffs (Denness et al., 1975). Retreat of both the upper cliff and the sea cliff has been rapid, with annual rates in excess of 1 m averaged over periods longer than ten years. Over shorter periods of about two years the rate of retreat has been greater than 5 m a^{-1}. The present rapid activity is a relatively recent phenomenon, for large parts of the present upper cliff had been relatively stable prior to the major failures of 1957 and 1958 (Bray, 1996).

Interpretation

The cliffed coastline and landslides of West Dorset are important internationally because they have been the focus of research that has influenced understanding of both geomorphology in general and the coastal system in particular. Brunsden, and other workers, have demonstrated the applicability of 'systems methodology' to rapidly changing complex landforms and this led to a better appreciation of timescales in geomorphology. Denness (1972) discussed the reservoir principle of mass-movements in relationship to the cliffs at Black Ven, and later work identified the importance of secondary reservoirs to an understanding of complex mass-movements (Denness et al., 1975). Bray (1986, 1990a,b, 1996) has described the sediment budget of the beaches and shown how they both depend upon and affect the landslides.

At Fairy Dell, Brunsden (1973) demonstrated how the application of system theory could be used both to elucidate the inter-related forms and processes in a complex mass-movement and to fashion the design of field experimentation. Fairy Dell, it was argued, could be explained by the following evolutionary model (Brunsden and Jones, 1976). Marine erosion both removes debris derived from mass-movements higher up the slope and maintains a more or less steady rate of cliff retreat. This increases the rate at which debris is carried across the lower bench in morphodynamic Zone 2 and brings about failure of undercliffs so that arenaceous materials move from the upper zone. Retreat of the undercliffs and removal of material from the slipped blocks of morphodynamic Zone 1 causes them to move seawards. The upper arcuate scar may also fail as a result. Thus marine erosion initiates a 'zone of aggression' which gradually affects the whole slope. The erosional events of the sea cliff were shown to be high frequency and low magnitude, whereas the events in the upper slopes were large, but infrequent (Brunsden and Jones, 1980). The effect of such events was dramatically revealed in 1994 when a rotational slip in the upper cliff at Black Ven loaded debris below and triggered a high-velocity sand avalanche across the beach west of Charmouth. The dry sand-fall fluidized and flowed seawards over a distance of 800 m (Brunsden and Chandler, 1996).

Cambers (1976) demonstrated that large landslides often provide large sediment stores at the foot of the landslide slope. Until sufficient sediment has been removed for unloading of the slope to occur, landsliding will be reduced. Thus, sediment storage can have a critical role in regulating the transmission of the 'zone of aggression' through the landslide system. Brunsden and Jones (1980) developed this point further to illustrate the concept of the 'formative event', which shapes the landform most effectively. At Fairy Dell, the formative events are the large movements that both produce large features in the slope and are recognizable over long periods of time, except in the sea cliffs where the formative events are small and frequent. As a result the sea cliff lacks sediment storage and has a relatively smooth form, but the mass-movement slopes are distinguished by considerable storage of sediment and great irregularity. The landslides at The Spittles (Figure 4.14) and Black Ven have been described by Brunsden and Chandler (1996) as re-activated features from the last interglacial period.

Denness (1972) and Conway (1974) examined the relationship between groundwater

Soft-rock cliffs

Figure 4.14 The Spittles, east of Lyme Regis. (A) Main landslide scar – sand and chert cliff; (B) landslide storage and throughput system; (C) sea cliff and mud flows; (D) beach; (E) dissected shore platform. (Photo: V.J. May.)

flows and the extensive instability of the cliffs at Black Ven. The reservoir principle of mass-movement argues that where a permeable rock capable of holding and discharging ground water rests on an impermeable layer, a supply of water, independent of rainfall is introduced into areas of instability. Not only is landsliding more rapid than if surface water alone is involved, but there may be accelerated weakening of the rock fabric. At Black Ven, this 'primary reservoir' is the Upper Greensand resting on the Gault Clay. Debris accumulations within both active and relict mass movements can also act as more localized 'secondary reservoirs'. At Charmouth, the Higher Sea Lane landslip involves re-activation of relict mudflows that overlie the Lias clay (Denness *et al.*, 1975). The mudflows, acting as a secondary reservoir, supply water to the Lias and movements take place perpendicular in direction to the original flows. Thus, as the sea cuts the cliffs back into valley-side slopes, which are characterized by older inactive mass movements, new formative events have been triggered.

Analysis of 12 000 beach pebbles from beaches in this site shows that west Dorset beaches have similar pebble lithology and size distributions. Littoral drift from the west is suggested by an increase in roundness and sphericity towards the east (Bray, 1990b, 1996). Taken together these characteristics suggest that the beaches, including Chesil Beach, were probably interconnected in the past at a lower sea level, about 5000 years ago.

The beach most variable in volume over time is at Charmouth, which, in spite of being closest to the input from the landslides, has rapid throughputs of sediment because of the dominant drift towards the east. Accretion of shingle reduces erosion of the foot of the cliff and the landslides, and retreat rates diminish. Reduced retreat rates lessen the input of shingle to the beach and so accretion is also reduced. As a result the zones of active landsliding may migrate in the direction of longshore drift, as exemplified by an eastward shift in the area of intensive activity from the Spittles to the central part of Black Ven during the first half of the 20th century (Bray, 1990b). However, the lack of sediment at the western extremity of this beach at Lyme Regis and the progressive installation of groynes has aided the extension of landsliding towards Lyme Regis. Although not part of this site, Lyme Regis itself is underlain by numerous slides and shows substantial activity (Lee, 1992; Pile, 1996).

Within the Lyme Regis to Golden Cap GCR site, the application of systems modelling can be further developed, first by examining the sensitivity of the cliffs to high magnitude, low-frequency oceanographic events, and second, by investigating the entrapment processes within the boulder arcs left by erosion of landslide lobes. At Golden Cap, the Greensand has a much more restricted outcrop than on the cliffs between Lyme Regis and Charmouth. Mass-movements are more infrequent. The foot of the cliff is protected by a substantial accumulation of boulders, many of them in arcuate ramparts (Bird, 1985; Bray, 1986, 1990a,b, 1996), which also occur on the sea floor (Darton *et al.*, 1981; Bray, 1990a, Brunsden *et al.*, 1996). Thus this part of the coastline becomes more irregular as its formative events occur less frequently. Moreover, the modification of waves by refraction over such boulder zones concentrates wave energy around the flanks of the headland. Some of the sand- and shingle-sized product from the landslides travels alongshore and may accumulate against the updrift side of debris fans, thus offering some additional protection to the foot of the cliff. Lateral sorting of the beach sediments has been described (Bird, 1989), and the effects of different local sediment sources outlined, but this requires further examination. Future work should not only continue to elucidate the development of the cliffs, but also develop Bray's (1990b) integrated sediment-budget model of the whole site so that the effects of debris inputs at the western end upon the behaviour of the cliffs farther east can be predicted better. Similarly, the effects of surges in the English Channel and large waves such as those reported at Chesil Beach warrant further investigation, for until now most geomorphological investigation at this site has considered the terrestrial processes rather than integrating them with the marine processes. Recently, these cliffs have been used for the development of a model that aims to estimate the future erosion of soft-rock cliffs with accelerating rates of sea-level rise (Bray and Hooke, 1997).

This is a very actively changing site that has a long record of geomorphological investigation upon which future research and education can build. It offers opportunities not only for fine-tuning and evaluation of existing models of cliff behaviour but also the development of more complete models of the whole coastal system from cliff top to offshore. Although other cliffed coasts which are affected by mass-movements occur around the British coast, some have been greatly modified by coast protection works (for example at Folkestone Warren) and in some the longshore sediment-transport system has been modified significantly (for example on parts of the East Anglian coast). This site has been little affected by human modifications although the coast protection works at Charmouth have introduced a salient that acts as a barrier to sediment eastwards, thus further subdividing the system. The landslides between Lyme Regis and Charmouth have been investigated in more detail than any other site worldwide and as a result the complex inter-relationships between active mass-movements, marine erosion and beach development are better understood here than elsewhere. The comparatively well-understood modern processes at this site are also important in throwing light upon the development of the other coastal features of Start Bay, especially Chesil Beach, Slapton Beach and Hallsands. All these beaches contain flint and chert clasts, yet only in this site can they be shown to be contemporary in origin, and the rate of supply estimated. As a result it becomes possible to judge the extent to which beaches in this area were formerly interconnected (e.g. Bray, 1996).

The significance of this GCR site to the study of coastal geomorphology is considerable, since the research here has:

1. focused attention on the appropriate timescale for coastal studies in this site, i.e. 100 years;
2. demonstrated the critical role of the processes at the cliff foot in activating the larger, formative events of the landslide systems;
3. examined the role of the beaches in affecting

the process alongshore;
4. developed models and concepts that have much wider application.

Conclusions

Major landslides at Golden Cap, Fairy Dell and Black Ven dominate this cliffed coast, but the site is also of particular interest because, unusually, the sediment system from cliff top to beach has been investigated more thoroughly than any other site in Britain, and probably worldwide. This is a very important site for coastal geomorphological studies, especially as it is possible to interpret both terrestrial and submarine landforms. Estimates of the sediment budget at the site have integrated the cliff system with the associated beaches, making this a site of considerable importance for monitoring the effects of medium-term change.

SOUTH-WEST ISLE OF WIGHT (SZ 493 755–SZ 306 852)

V.J. May

Introduction

The south-west coast of the Isle of Wight (see Figure 4.1 for general location) is geomorphologically rich in features of interest. It contains examples of most of the cliff types undergoing active erosion described from other GCR sites. The site demonstrates well how coastal processes have produced different cliff forms related not only to variations of lithology and geological structure, but also variations in the intensity of coastal processes and the timescales over which coastal evolution occurs.

Much of the coast of the Isle of Wight is affected by rapid retreat and landslides. Damage to property remains a constant hazard and stabilization has been a priority on the urbanized parts of the coast (Clark *et al.*, 1993). However, the GCR site area is mainly unprotected, except at Freshwater Bay.

The site extends from Chale in the east (SZ 493 755) around the Needles (its westernmost point at SZ 289 849) to Alum Bay (SZ 306 852; see Figure 4.15) and crosses outcrops of Chalk, Upper Greensand, Gault, Lower Greensand and the Wealden exposed on both sides and in the core of the Brighstone anticline.

The general plan of the coastline is controlled by the relative resistance of the Chalk in the west and the Upper Greensand in the east, and the effectiveness of the prevailing and dominant south-westerly wave systems in maintaining the alignment of the shoreline. There are many, small irregularities associated with locally resistant outcrops, and the beaches comprise both locally derived materials and some residual flints. Erosion has been rapid, so that small streams have been unable to keep pace with continually steepening gradients. 'Chines' (small coastal gorges) and waterfalls are common.

The cliffs vary in height from about 15 m in Brook Bay to over 145 m at Tennyson Down. At Freshwater Bay, the former Yar Valley is truncated and coast protection works have been constructed, but elsewhere interference with the beach system and its feeder bluffs has been negligible. The structural impact of variations in dip and rock strength is well exemplified in the stacks at Freshwater Bay and the Needles, as well as in the differential erosion of the Chalk itself along Tennyson Down.

Platforms are poorly developed on the Chalk coast, the cliff foot being masked by extensive boulder accumulations. However, parts of the coast intersecting the Greensand and Wealden strata have large platforms, for example at Hanover Point, whose seaward extensions affect wave refraction locally. This is one of six major south-west facing beach systems in the English Channel. It is distinctive by reason of its rapid retreat and the differential feeding of sediment to it, as well as a limited flint content. Flint is important only on the beach between Atherfield Point and Blackgang where major landslides feed the beach. The Chalk cliffs in the north are very slow to change; they feed very small quantities of flint into the beach, except at Scratchell's Bay and on the northern side of Tennyson Down.

Research has focused on the cliff processes, with increasing emphasis on the landsliding (Lyell, 1835; Steers, 1946a; May, 1964; Hutchinson, 1965, 1987, 1991; Hutchinson *et al.*, 1981; Barton, 1990, 1991; Clark *et al.*, 1993; Bromhead *et al.*, 1991; Chandler, 1991; Hutchinson, 1991), and the development of chines (Englefield, 1816; Lyell, 1867; Bristow, 1889; Bury, 1920; Cotton, 1941; Steers, 1953a; Flint, 1980, 1982).

South-west Isle of Wight

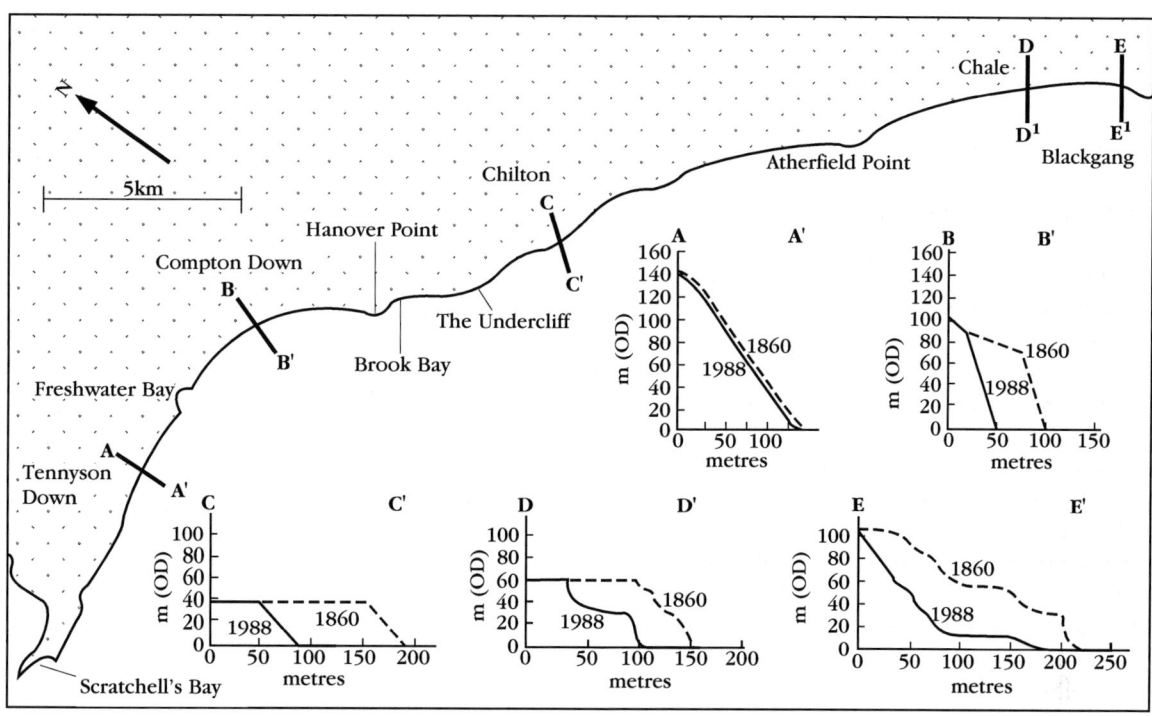

Figure 4.15 Variations in the rates of cliff retreat from Blackgang to the Needles (to the west of Scratchell's Bay), Isle of Wight. Cliff profiles for sections A to E are shown. (After Hutchinson, 1984.)

Description

The site can be divided into eight sections (Figure 4.15), as follows.

1. St Catherine's Point to Blackgang Chine, forming the western part of the Undercliff, which rises to over 180 m in height. The steep upper cliffs comprise mainly Upper Greensand, underlain by Gault Clay and Lower Greensand, all dipping gently to the south-east. There have been many large landslides (Hutchinson, 1965).
2. Blackgang Chine to Atherfield Point. Cliffs that are undergoing very active erosion at the eastern end give way westwards to steep fairly stable cliffs in the Ferruginous Sands. Two chines break the cliffline. This is the only substantial fringing flint-shingle beach within the site.
3. Atherfield Point to The Undercliff (Brighstone Bay) is dominated by active cliffs in the Atherfield Clay and the Wealden beds. The rate of cliff retreat has been estimated at greater than 1 m a^{-1} (May, 1964). Chines are distinctive features both here and in the next two segments of the coast.
4. Brook Bay, which is cut mainly into the Wealden Marls (Wessex Formation). Its eastern side is marked by an undercliff of slipped blocks up to 7 m thick known as 'Roughland'. On its western side, the cliffs are about 16 m in height. The dip of the strata is to the south.
5. Hanover Point to Freshwater Bay. The cliffs vary in height from about 15 m at Hanover Point to over 80 m at Compton Down. The dip is towards the north and so there is a gradual transition from the Wealden Marls and Shales (Wessex and Vectis formations) through the Lower Greensand to the Chalk at Compton Down. This is a very active coastline and the coast road at Compton Down is so seriously threatened that complete realignment has been considered (Barton, 1990). Chalk from the eroding cliffs is transported south-eastwards towards Hanover Point, but is virtually absent within 1 km of the Chalk cliffs.
6. Freshwater Bay forms a small semi-circular bay between relatively resistant Chalk headlands. Stacks and caves have formed on either

side of the bay, but erosion of considerably altered chalk along the former valley side as well as the risks of flooding of the Yar Gap have led to the construction of a sea-wall within the bay. There is a small beach of flint shingle, much of which is angular or subangular.

7. Tennyson Down (South) is formed entirely of high, steep, Chalk cliffs rising to over 140 m. There is a very narrow beach mainly of boulders and cobbles, and falls of rock appear to occur only occasionally. However, much of the cliff top is deeply broken by large tension cracks and extensive falls could occur at any time.

8. Scratchell's Bay to Alum Bay includes the world-famous stacks of 'The Needles' (Figure 4.16). As well as their aesthetic appeal, they have been widely reported in the literature as being of interest geomorphologically. For example, Lyell (1835) reported falls of some of the pinnacles in 1764 and 1772, and Huxley (1884) included a sketch of them. Scratchell's Bay to the south has a narrow beach of flint and chalk shingle, but there is limited beach development along the northern side of Tennyson Down. This is, however, a very active Chalk cliff, up to 100 m in height, which continues inland at the southern corner of Alum Bay. This part of the site demonstrates well the contrast between the purely sub-aerial products of erosion of the non-marine cliffs and the combined marine and sub-aerial product of the sea cliffs to the west.

Interpretation

The following aspects of the coastline have attracted specific attention in the geomorphological literature.

1. the rapidity but differential nature of cliff erosion;
2. the landslides;
3. the development of the chines;
4. the sources and transport of sediment (Figure 4.17).

The rapidity and diversity of erosional processes around the coastline of the Isle of Wight is especially well demonstrated along its

Figure 4.16 The Needles and Scratchell's Bay, Isle of Wight, with narrow flint and chalk beach fed by contemporary rockfalls. (Photo: J.E. Gordon.)

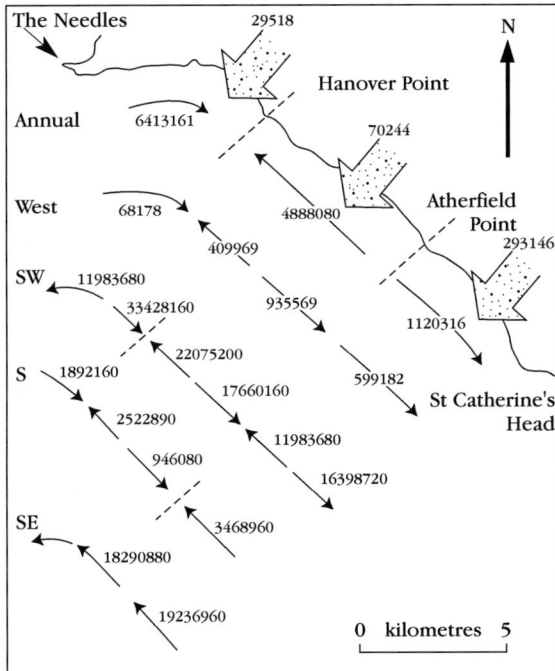

Figure 4.17 Sediment inputs from cliff retreat (m³ a⁻¹) annual longshore potential sediment transport and variations with wind direction. See text for explanation. Total sediment input = 392 908 m³ a⁻¹ (After Davies, 1997.)

south-western coast. May (1964) summarized the main features and estimated that rates of cliff retreat during the previous 100 years ranged from 0 to more than 2 m a⁻¹. The resistance of the intertidal area and the cliff foot to erosion are particularly significant, for it can be shown that the more resistant layers (such as the Perna Bed hard band and the 'Pine Raft' at Hanover Point) not only form platforms but also reduce rates of cliff retreat. Thus headlands have formed at Hanover Point and Atherfield Point where the cliffs are formed in materials that are intrinsically weak. Their existence as headlands owes much to the greater relative resistance of the cliff foot and intertidal platforms. Much of the cliff consists of easily eroded materials, and so variations in the resistance of the cliff foot or platforms associated with these slightly harder bands play a very important role in the overall outline of the coast.

Hutchinson (1965) reconnoitred the documented landslides of the Isle of Wight indicating that this site contained 13.5 km affected by rockfalls, 0.6 km affected by seepage erosion, about 4.25 km affected by base failures, and 5.5 km affected by slope failures and mud-flows (Figure 4.18). Only 5.1 km comprised relatively stable, soft rocks. The main concentration of base failures occurs at the eastern end of the site around Blackgang (Figures 4.18 and 4.19). The largest landslide recorded in the Isle of Wight took place in February 1799 when about 40 ha subsided by as much as 10–12 m (Anon., 1887; Cooke, 1808; Webster, 1816). Hutchinson (1965) regarded the movement, known as the 'Gore Cliff landslip', as a renewal of movement in the ancient failed mass forming the Undercliff. The slide has been affected by several subsequent movements. Colenutt (1928) described a large toppling failure at Gore Cliff that blocked the road. Hutchinson (1987) suggested that the continuing severe toe erosion, the narrowness of the Undercliff and the opening of joints behind Gore Cliff, indicate that a period of renewed regression was approaching. Bromhead *et al.* (1991) showed that the 1978 landslide was the most recent in a series that had affected the coastline between Rocken End and Blackgang Chine during the last 200 years. It had a basal slip surface controlled by bedding about 18 m above the base of the Gault Clay and re-activated earlier slides. The large landslides that form the Undercliff result from the interaction between marine erosion initiated during the Holocene transgression and the seaward-dipping synclinal structure and detailed lithology of the Cretaceous rocks (Hutchinson, 1991).

To the north-west, seepage erosion appears to have been most important in giving rise to benches in the cliff profile. Fitton (1847) suggested that less permeable beds within generally more permeable strata would encourage water from these beds to undermine the cliff. As a result, the cliff is characterized by an upper active cliff, a bench and a lower sea-eroded cliff. The upper cliff and the sea cliff retreat at different rates, and the rates of recession trebled from the 19th to the 20th centuries. Hutchinson *et al.* (1981) and Hutchinson (1987) describe the processes in detail, as follows. The upper scarp of the Ferruginous Sands cliff at Walpen collapses as a result of seepage erosion in fine layers in the Foliated Clay and Sand. The resulting debris moves across the undercliff bench, usually obliquely, by compound slides and by rotational slips in its seaward edge. Mudslides and stream action also carry debris across the undercliff. As the bench surface dips below sea level towards the south-east, the bench is broken by deep-seat-

Soft-rock cliffs

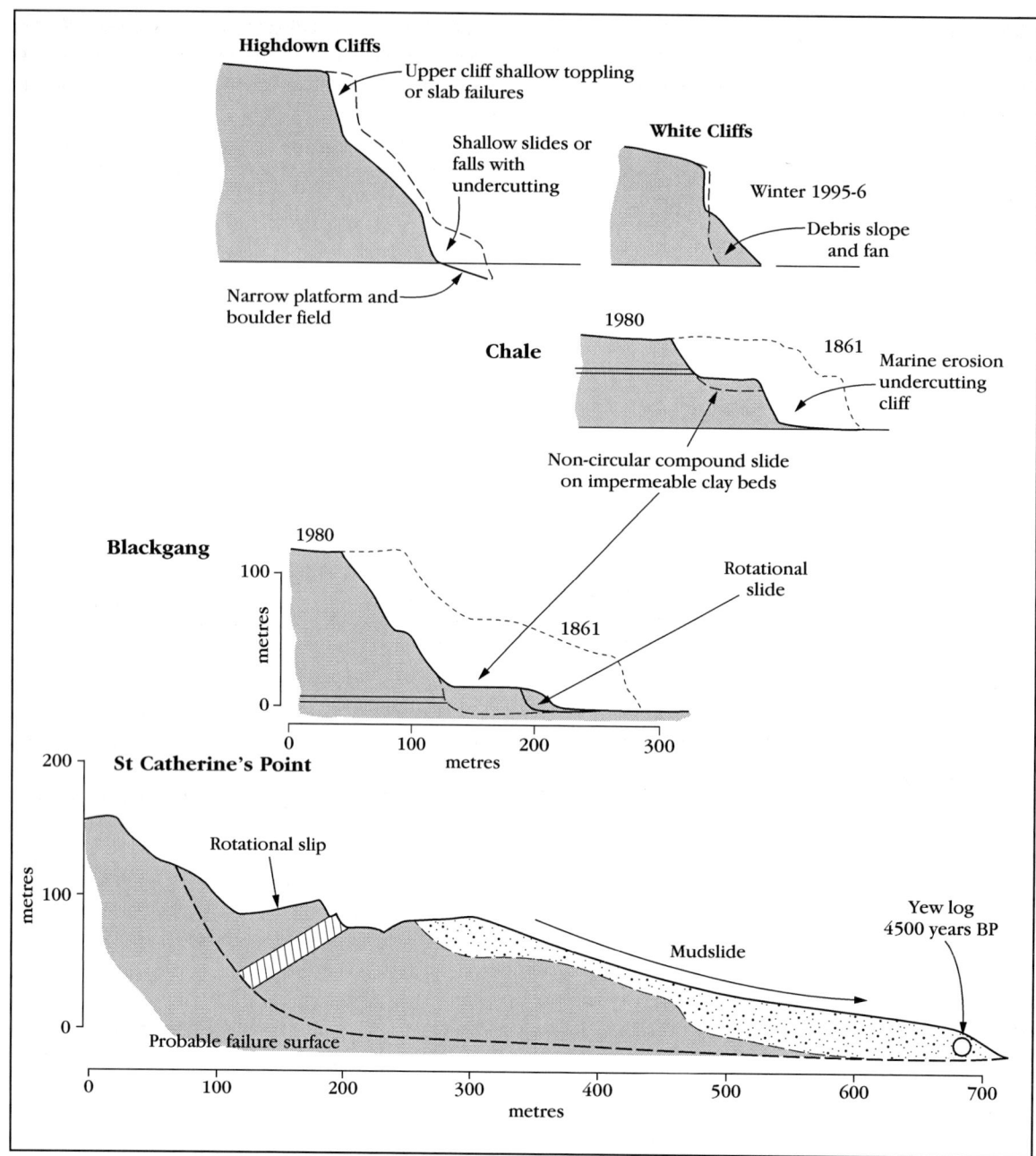

Figure 4.18 Differences of failure in the cliffs of south-west Isle of Wight, ranging from large rotational slides to shallow failures. (After Hutchinson, 1984.)

ed base failures.

North-westwards from Atherfield Point, the cliffs are marked by many slides, some shallow, others more deep-seated and giving rise to 'staircases' of slipped blocks, but they have received rather less attention in the geomorphological literature than the cliffs around Blackgang. The cliffs towards Alum Bay also warrant much fuller examination, not only because of the threats to roads at several points, but also because there is considerable evidence in the cliff-top tension fissures, particularly on Tennyson Down, that major cliff failures could occur (Figure 4.21). At Compton Down, the chalk cliffs are undergoing more active erosion than any other part of the chalk section of this coastline. Cliff failure

South-west Isle of Wight

Figure 4.19 Characteristic slope failures at Compton Down, looking west, showing shallow slides in chalk rock. (Photo: V.J. May.)

Figure 4.20 View looking east from Compton Down where chalk pebbles typically survive for little more than 1 km owing to their erosion during longshore drift. Well-developed cusps commonly characterize this beach. (Photo: V.J. May.)

occurs in massive units that give rise to open fissures behind the cliff edge (Figure 4.20; Barton, 1990).

Explanations for the origins of the chines range from landsliding (Lyell, 1867), enlargement after rainfall (Gardner, 1879), wind erosion (Englefield, 1816), and spring-sapping (Bristow, 1889). Bury (1920) concluded that they were formed recently, but did not suggest their probable age. In his view, where the rate of headward erosion was faster than that of cliff retreat the chines were lengthening. The valley-within-a-valley form he attributed to the role of much larger volumes of water forming the upper open valleys whereas the chines were the result of present-day conditions, the size of the chines being a function of their present-day small streams, which misfit the larger upper valleys. He argued that the different levels were a result of changes of sea level, a view supported by Steers (1953a). Flint (1980, 1982) concluded, in contrast, that they are a function of basin characteristics (stream power and geology) and cliff retreat-rate, not the result of cliff retreat alone. The critical drainage basin area for the maintenance of a base-levelled chine varies with rock type. In sands, this area is about 2.5 km^2, in marls 0.64 km^2 and in shales greater than 0.73 km^2. Below these values, stream power decreases, waterfall height increases and the chine is removed by cliff retreat (Flint, 1980). Landsliding is not a primary agent in initiating chines. Flint (1980, 1982) refuted Bury's (1920) view that the different levels of the chines were due to sea-level changes alone.

Because of the variety of beds, the sources of beach material can often be identified comparatively easily. Apart from Scratchell's Bay, there is no part of the Chalk coastline that has a beach composed solely of flint and chalk shingle.

Soft-rock cliffs

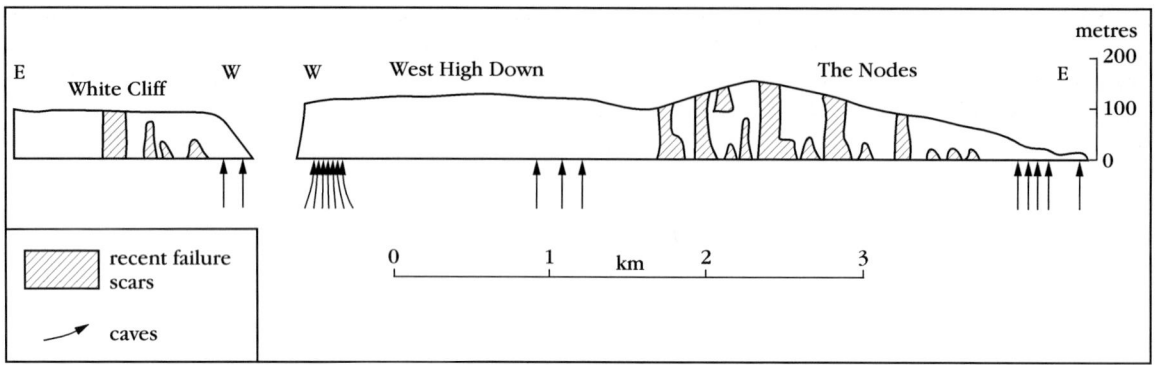

Figure 4.21 Cliff-face failures west of Freshwater Bay. (Based on British Gas aerial survey, February 1996.)

Elsewhere, the beach is minimal or is mixed with shingle and larger material derived from the Lower Greensand. From Compton Down, chalk pebbles indicate a transport direction towards the south-east, but are virtually absent 1 km downdrift from the cliff. Most beaches along the Wealden coastline are sandy, but contain varying amounts of shingle and clay depending on their relationship to the chines and landslides. Most beach material appears to be locally derived and recently produced. East of Atherfield Point, however, the beach is formed of flint shingle. There has been no detailed investigation of the sedimentological characteristics of this anomalous beach. Possible sediment sources to the south-east, but longshore transport has normally been described as being from north-west to south-east. However, Davies (1997) has shown that under south-westerly waves there are three sediment-transport cells with little exchange between them. In contrast, under easterly waves, there is net movement alongshore towards the west (Figure 4.17). The bulk of the material entering the beaches is fine-grained sand or clay, and much is transported away from the beach in suspension. There is a marked lack of shingle to the west of Atherfield Point, which acts as a groyne to westerly movement. Alternatively, this beach may represent part of a former larger beach system that extended throughout Brighstone Bay at an unkown distance offshore. Hutchinson (1987) showed that the foot of the Undercliff at the eastern end of the site includes a former sea cliff and platform probably related to a period of relative sea-level still-stand about 7500 to 8000 BP. A period of active landsliding occurring about 4500 BP protected this feature. Of particular interest is the suggestion that the shoreline position at the eastern end of the site was not different from that of today, but with sea level perhaps as much as 7 m lower. The western Yar was already open to the sea, having been cut to a depth of at least −13.4 OD (Nicholls, 1987) during the Pleistocene Epoch.

The south-western coastline of the Isle of Wight stands apart from the other south-west-facing beaches of the English Channel in the rapidity of its retreat (Figure 4.17). Although within broadly the same wave conditions as the Seven Sisters in East Sussex, both the mechanisms and the coastal plan of the south-west Isle of Wight differ as a result of the variety rock types and their different responses to sub-aerial and marine processes. Another key difference is that whereas the Seven Sisters display a more-or-less uniform rate of retreat over the 100 year timescale, the south-west Isle of Wight cliffline has retreated at varying rates that have accentuated coastal crenulations rather than reducing them. Finally, the Isle of Wight site can be regarded as the type area for chines in sands and clays, for they are well-developed common features about which a considerable amount is known.

This site is the best example in Britain of a coastline that cuts across an anticline in relatively weak geological materials. Erosion follows the core of the Brighstone anticline, but the variations in rock strength exposed in the cliff and foreshore have produced considerable differential erosion. The importance of the site lies mainly in the range of responses to erosion in different lithologies within the larger scenario of a shoreline that is strongly controlled in its general outline by the dominant south-west wave regimes. In contrast to the coastline of east Dorset, which is either clearly longitudinal (around Lulworth) or has a strong headland–bay pattern (the transverse coast around Swanage), this coastline is one where headland–bay topography is slight, but retains similar amplitudes as it retreats at comparable rates throughout its length.

The relationship between the old landslides at the Undercliff and former shorelines gives this site another distinctive feature. The resistance to erosion of the exposed southern side of Tennyson Down contrasts strongly with the much more sheltered northern side of the headland and demonstrates very well the importance of both structure and lithology. This also deserves more detailed investigation, because of its relationship to sediment supply and the potential for catastrophic change to the southern cliff.

The lack of large-scale anthropogenic interference with the coastal processes makes this site particularly important for investigation of the links between cliff retreat and beach development, especially since the different lithologies of the retreating cliffs provide natural markers for examination of longshore transport processes.

Conclusions

The south-west coastline of the Isle of Wight is dominated by cliff and beach features related to the south-westerly wave climate of the English Channel, but also demonstrates well the effects of differential erosion both where rock-types change and where coastal retreat outstrips the erosional ability of coastal streams. The site includes the Needles and extends from the Chalk of Tennyson Down in the west to the Lower Greensand cliffs of Blackgang Chine in the east. There are rapidly retreating cliffs, well-developed cliff-foot beaches and steep-sided valleys (the chines). This is the type area for chine formation. Differential erosion in relatively weak rocks is affected by more resistant bands, except at the extremities of the site where the Undercliff overlies an older shoreline in the east and the stacks of the Needles and the high cliffs of Tennyson Down resist erosion. Although not as well-recognized internationally as the very active landslide coast east of Lyme Regis, this coast contains examples of all the soft rapidly retreating coastal types that occur in Britain and parts such as the Undercliff are renowned internationally. It is one of the best examples of a coastline that cuts across a major anticline in generally weak materials.

KINGSDOWN TO DOVER, KENT (TR 382 472–TR 374 450 AND TR 368 443–TR 340 422)

V.J. May

Introduction

The cliffs between Kingsdown and Dover (see Figure 4.1 for general location) show an excellent example of structural controls on coastal cliff-form. These cliffs, broken only by the deep valley and bay at St Margaret's, rise to between 30 m and 110 m OD. Retreat of the cliffs has been about 0.2 m a^{-1}, but this takes place mainly as large slides affecting the whole cliff face. A well-developed platform extends to below low-tide level. Beaches are formed mainly of clasts of rounded chalk and a mixture of rounded and angular/unrounded flint. Little evidence now remains of a former fringing beach that extended from Dover to Kingsdown. The present beaches depend upon the contemporary erosion of cliffs and platforms. When major cliff-falls occur, boulder-sized debris usually forms a protecting rampart with the smaller chalk cobbles and flints being added to the beaches (Figure 4.22).

Research on the Kingsdown to Dover site has focused on the nature and processes of cliff retreat (May, 1964; May and Heeps, 1985; Hutchinson, 1980, 1983; Birch, 1990; Leddra and Jones, 1990) and the relationship between cliff-form and major structures in the Chalk (Middlemiss, 1983). Like many such cliff sites, it is referred to by Steers (1946a) and Bird (1984). Comparable surveys have been made of other Chalk cliff sites in England and Wales (see GCR site reports for Flamborough Head, Joss Bay, Ballard Down and Peveril Point to Furzy Cliff) and on the Normandy coast (Precheur, 1960).

Description

The vertical cliffs between Dover and Kingsdown are currently undergoing the most active change

Soft-rock cliffs

in England and Wales. They are cut through the Upper Chalk of the *Micraster cortestudinarium* and *Micraster coranguinum* biozones. Unlike the high Chalk cliffs at Beachy Head, for example, they are generally a simple near-vertical face up to 110 m in height. Erosion rates have increased in recent decades. Cut across the eastern end of the North Downs, their height reflects mainly the gradual slope of the Chalk landscape towards the north-east. Thus, between Dover and St Margaret's Bay the cliffs reach heights of 110 m, although at the Dover end of the site they also have lower faces where the cliffs are cut into coombs such as that at Langdon Bay. North of St Margaret's Bay, the alignment of the cliffs is closer to that of the strike and so they reflect more closely the dip of the Chalk surface, falling gradually to a height of about 30 m at Kingsdown.

Middlemiss (1983) describes the relationship between the structure of the Chalk and the failures of the cliffs. Between Kingsdown and the windmill at South Foreland, 300 measured joints fall into four main groups.

1. Striking at about 300°: major strike joints.
2. Striking at about 205° to 210°, dominantly vertical or very steeply dipping: tensional dip joints complementary to the first group.
3. Striking at 205° to 210° with dip angles between 50° to 70°, and
4. Striking at about 280°.

Middlemiss demonstrated that cliff-form corresponds both in plan and profile with the attitude of the jointing. Group (1) and (4) joints that are roughly perpendicular to the coastline in plan view are often important as the sites of caves and minor changes of the cliffline (e.g. Ness Point and White Fall). Almost every length of the cliffline is related to at least one group of joints, with groups (2) and (3) dominant (Figure 4.23).

To the south of South Foreland, there has also been a number of large falls in recent years, where the cliffs turn increasingly towards the west. The Chalk is greatly jointed but the cliff patterns have not been described in the same detail as those in Middlemiss's paper. Where

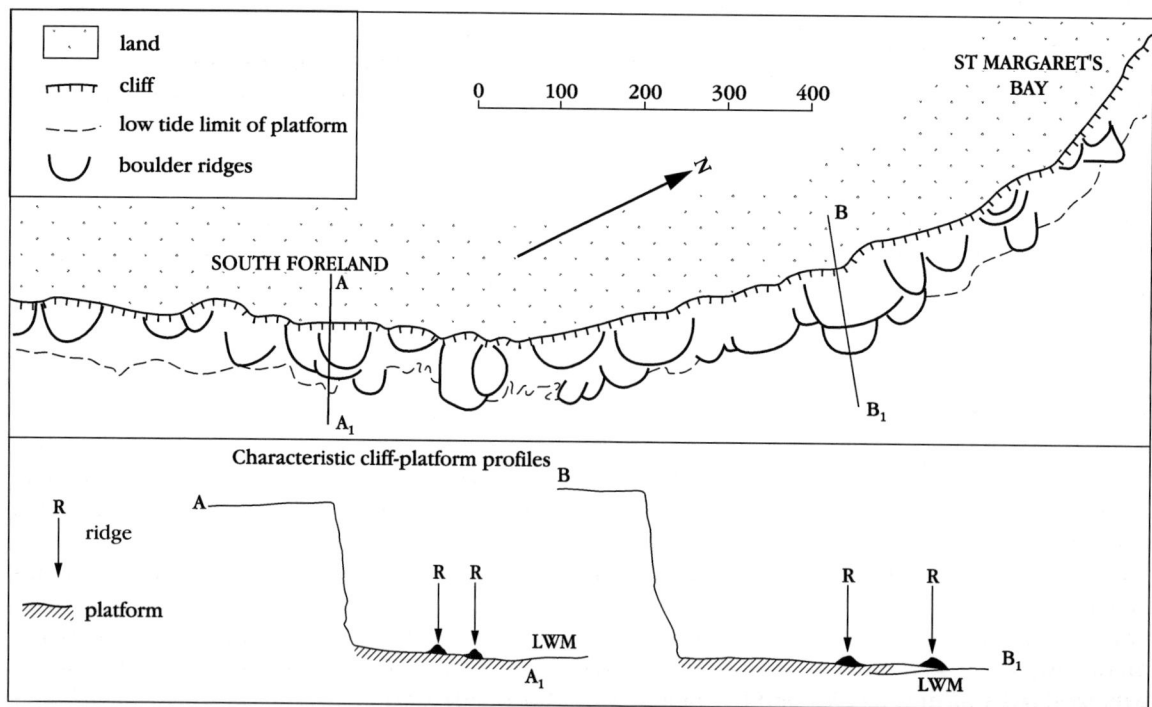

Figure 4.22 Sketch map of boulder ridges, South Foreland to St Margaret's Bay within the Kingsdown to Dover GCR site. Characteristic cliff-platform profiles through A–A₁ and B–B₁ are shown in the lower part of the diagram.

Kingsdown to Dover

Figure 4.23 View looking north of St Margaret's Bay, Kingsdown to Dover GCR site. (A) Small, fringing beach of flint, mostly derived from recent cliff falls; movement alongshore is restricted by fall debris; (B) large toe of a slide extending beyond low-water mark; (C) cliff being eroded where previous rock fall has been completely removed; (D) vegetated slope that developed behind a former slide toe and debris; these features then protected cliff-foot bedrock from erosion; (E) typical upper cliff profile above debis slopes. (Photo: V.J. May.)

recent falls are absent, the cliff has a basal notch and a very steep profile (Figure 4.24). The platform in these areas is free of small chalk debris, but has small pockets of flint, both rounded and angular, the latter derived directly from the platform. Platform surfaces are generally of two types. The first of these is co-incident with tabular flint horizons in the Chalk and follows their dip so that flint forms the surface of the platform. The second type is steeper, lacks a flint surface and cuts directly into the Chalk. Both types of platform are dissected by sub-parallel channels up to 0.4 m in depth. There is considerable biological activity in the erosion of the platforms, with borers such as piddock *Pholas* spp. riddling the surface layer to depths of several centimetres. Substantial lengths of the site, in contrast, have platforms that are cloaked by debris. This is often boulder-sized and suffi-ciently stable to carry a cover of algae. Much of the large debris is derived from frequent rock-falls that affect this coastline.

Middlemiss (1983) identified four main areas north of South Foreland where falls have been particularly frequent:

1. South Foreland (TR 363 434) and the cliff immediately to the north-east. The structure of the cliff is determined by three sets of joints striking at 303°, 233° and 202°.
2. At Leathercote Point (TR 374 451) and for about 300 m northwards to The Cut (TR 375 454), where master joints striking at about 301° and 232° are the major factors.
3. At White Fall (TR 378 457), where two small faults, striking at about 301° with a throw to the north of about 1.5 m, are intersected by a major joint striking at 232° and dipping at 70°

to the south-east. Middlemiss suggested that the presence of about a 1.5 m thickness of calcareous downwash at the top of the cliff may be a contributory factor here.

4. East of Hope Farm (TR 376 462) and for 400 m northwards, where major joints determine the structure of the cliff. An additional set of joints striking at about 342° affect the outline of an embayment (TR 379 465). The valley bottom deposits of Hope Farm Valley, some 2–3 m of residual gravel and calcareous downwash, which not only cap the cliff but also fill several large solution pipes, may play a contributory role.

At the northern end of the site, joint-controlled channels cut into the platform are filled by angular flint and chalk mixed with rounded, oxidized, flint shingle together with shelly fragments in a chalky matrix. This is characteristic of the deposits occurring beneath chalk debris fans that preserve, albeit temporarily, the existing beach sediments beneath. The chalky matrix comprises the fine-grained materials at the base of the fall and may include fines that have washed through the debris above or have been derived by erosion of the overlying debris. These deposits are spatially associated with the remnants of a large talus fan at the cliff foot that has mostly been destroyed. Although the sub-debris sediments are normally ephemeral, their presence emphasizes the variability of sources for the beaches.

Interpretation

The rate of retreat of these cliffs is moderate, but in recent years retreat has accelerated. May (1966) reported the average cliff-top retreat over a period of 84 years from 1873 to 1957 as 6 m between St Margaret's and Kingsdown, and 16 m between St Margaret's and Dover. May (1971a) expressed cliff-top retreat at South Foreland as about 0.19 m a^{-1} between 1817 and 1962. Hutchinson (1972) compared chalk falls on the Kent coast in general between 1810 and 1970 with the number of days with air frost and effective rainfall. Falls tended to co-incide with, and to follow, periods of maxima of air frost and to follow periods of maximum water surplus. Bird and May (1976) stated that between Dover and Walmer there had been intermittent recession of the cliffs by rockfalls, which occur most frequently during the winter and are associated with periods of high average numbers of days with air frost. May (1964) described a state in which large falls produced extensive debris slopes that were sufficiently long-lived to develop a stable vegetation of chalkland grasses and coastal plants. The debris survived for several decades, sometimes in excess of 50 years. However, such longevity of debris no longer appears to be the case. Almost without exception the cliffs fall abruptly to a distinct junction with a platform, which is about 200 m in width. Where present, the beach is narrow and formed of rounded chalk and both rounded and angular flint. There is no evidence, apart from in the northernmost part of the site, of the once-continuous fringing rounded flint shingle beach described by Austen (1851). The construction of the harbours at Dover and Folkestone during the second half of the 19th century prevented any continuing supply from the south-west. Hutchinson et al. (1980) recorded the links between this interruption to the littoral drift and major changes at Folkestone Warren, but there has been no comparable record for the coastline east of Dover except at the northern end of the site. The effect has been less dramatic but no less important. Furthermore, the construction of groynes at St Margaret's and the former Royal Marines firing range at Kingsdown virtually prevent any supply of flint from the erosion of the present-day cliffs to reach the beaches at Walmer and Deal. The site is thus characterized by reduction in a protective lag deposit of flint shingle and depends for its basal protection entirely on the contemporary products of cliff-falls. These are now insufficient in volume to provide other than localized protection since much of the Chalk is dissipated by attrition. Residual boulder arcs reduce wave energy but do not provide the cliff foot protection that the 19th century beaches previously offered.

Two main processes promote the instability in these cliffs, namely wave action along the master joints, and the effects of percolating water within the joints. The latter effect is accentuated, according to Middlemiss, where residual gravel and calcareous downwash form the cliff top, as reservoirs for percolating water. The tendency for falls to follow periods of hard frost suggests a combination of mechanisms for the falls. Water freezing in these zones exerts pressure through its expansion. Alternatively, pressure is exerted by the ponding of ground-

Kingsdown to Dover

Figure 4.24 Langdon Bay. (a) Boulder rampart residue from earlier debris tongue; (b) in the foreground, talus from a clif failure is seen; in the background, residual boulder fields from flow-type failures are present; (c) parallel ridges bounding a large flow-failure that left the platform comparatively clear of large debris. (Photos: V.J. May.)

water in a joint behind a plug of ice.

Hutchinson (1980, 1983) showed that as cliff height and slide volume increase, a 'degree of flow' appears in the debris. In falls from the higher cliffs (70–150 m), a 'chalk flow' can occur, which may carry debris for distances in excess of four times the height of the cliff across the near-horizontal shore platforms. Hutchinson's working hypothesis was that these flow slides occur because high pore-water pressures were generated through the crushing impact of relatively weak blocks of high-porosity, near-saturated, chalk. Leddra and Jones (1990) show that steady-state flows can result from rapid or undrained loading of the Chalk, with the result that debris from high chalk cliffs can flow for considerable distances. The debris produced by cliff-falls, such as those described by Hutchinson, varies in size from fines to boulders over 1 m across. The fines are quickly dispersed by wave action and longshore currents. Shingle-sized debris is commonly rounded within a few days and is worn down to sand-sized fragments over a period of several months. It is often too mobile to attract algal growth. The larger material, however, remains *in situ* for very long periods of time and becomes colonized by algae and molluscs. While these boulders gradually diminish in size, the platform retains a substantial cover, commonly for many years. One fall in 1982 produced a fan of debris across the platform from which much fine debris was quickly removed; the main boulder pattern remains today. On one 1500 m length of the cliffs south of St Margaret's Bay, some 30 residual arcuate forms have been identified (Figure 4.24).

Kingsdown to Dover is an important coastal geomorphology GCR site because it demonstrates:

1. the role of structure in controlling cliff development in a site undergoing comparatively rapid erosion;
2. the significance of relict beaches in retarding recent cliff development;
3. the limited present-day sediment supply from the flint-rich Chalk;
4. the role of cliff-fall debris in modifying the form of the platform and controlling the wave refraction;
5. the role of structural features in affecting platform morphology and the role of flint layers;

Conclusions

At Kingsdown to Dover, vertical cliffs in the Chalk undergoing active erosion rise above well-developed intertidal platforms. Beaches are mainly contemporary in origin and depend upon frequent rockfalls. Boulder ridges occur on many platforms and mark the former extent of cliff-falls. The site is distinguished by the rapidity and nature of the changes within it, and by the relatively simple morphology of the cliffs and platforms. This is an important cliff site that demonstrates well the role of structures in controlling cliff development.

This geomorphologically active site is important for understanding the nature of cliff failure and its effects upon platform development, and unlike many similar cliffs it can be shown to have become much more active during the past 100 years as the supply of shingle from the west and the residual protection from the older beaches has diminished. Unlike other Chalk cliff–platform sites, it has a large and active sediment supply but much of the platform is cloaked by boulders that affect wave-energy distributions. Internationally, this is one of a very small number of near-vertical cliffs that have been investigated in some detail. It is also potentially very important for understanding the links between cliff erosion and the supply of sediment to adjacent beaches.

BEACHY HEAD TO SEAFORD HEAD, EAST SUSSEX
(TV 490 980–TV 600 968)

V.J. May

Introduction

The Chalk of the South Downs reaches the sea between Brighton and Eastbourne in a series of cliffs that become generally higher towards the east and culminate in the 150 m-high cliffs of Beachy Head (see Figure 4.1 for general location). This site comprises a cliff–beach–platform system developed on the Chalk, includes the classic world-renowned cliffs of Beachy Head and the Seven Sisters, and cuts across the mouth of the Cuckmere Valley (Figure 4.25). The cliffs reach a maximum height of 156 m at Beachy Head (TV 588 956) and vary between 14 m and 79 m in height along the Seven Sisters. Retreat

Beachy Head to Seaford Head

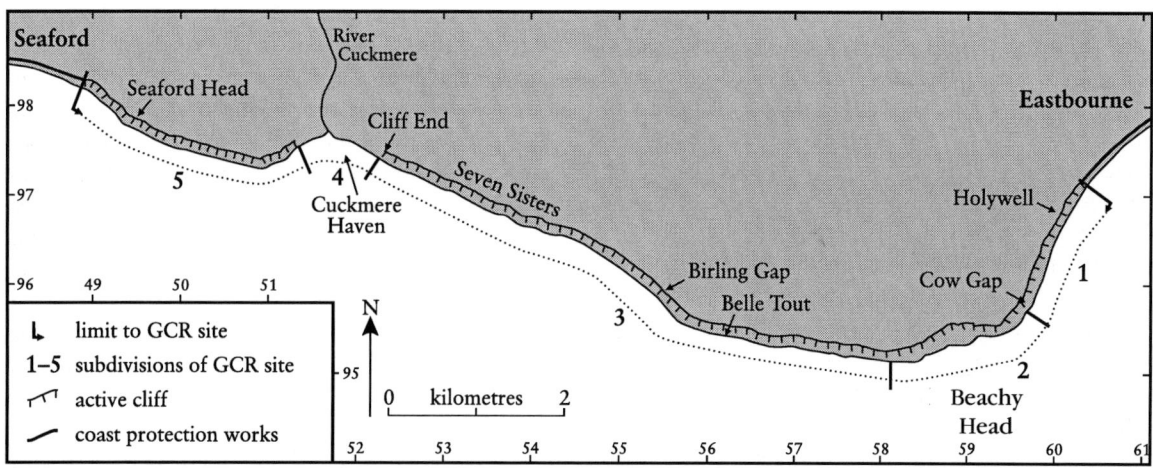

Figure 4.25 Sketch map of the Beachy Head to Seaford Head GCR site, showing the five subdivisions of the site as described in the text.

of the cliffs has been estimated at 0.42 m a^{-1}, reaching a maximum of 0.91 m a^{-1} at Birling Gap (May, 1971a). A narrow fringing beach of chalk and flint rests upon the cliff–platform junction, except where major falls, especially at Beachy Head, extend below low-tide level (Figure 4.26). Debris ramparts from cliff-falls are common at Beachy Head. The coastline plan is controlled primarily by dominant and prevailing wave energy from the south-west, with Seaford Head and Beachy Head acting as more resistant points between which a predominantly wave-energy controlled shoreline has developed. Structural variations seem to have had little effect upon the overall coastal plan, the cliff materials being sufficiently rapidly eroded to ensure control by wave action. However, in detail the cliffs are controlled by a rectilinear jointing pattern (see Figure 4.27). The beach is one of six major SW-facing beaches in southern England, all of which differ significantly in their geological characteristics. Most of the site faces south-west and is exposed to Atlantic waves. The fetch from the south-west is similar to that which affects all the major beaches in the English Channel. The cliff and beach of Seven Sisters in particular is unusual in having a very strong similarity in its alignment to such beaches as Chesil Beach and Dungeness. Of the beaches, Seven Sisters beach is the most rapidly and consistently fed by flint from cliff-falls. The most directly comparable cliffs also comprise Upper Chalk along the Normandy coast, but they lack the same aspect and degree of exposure.

Research here has been mainly restricted to understanding the nature and rates of cliff retreat (May, 1971a) and the relationship between the cliffs, platforms and the processes operating on them (Robinson and Jerwood, 1987a,b). They are nonetheless cited by a number of texts including Bird (1984), Holmes (1965), Small (1978), and Precheur (1960), the latter considering the contrasts with the similar Chalk coast of Normandy.

Description

The site comprises five sub-units (Figure 4.25):

1. The Chalk cliffs and Greensand and Gault Clay platforms below Cow Gap and Whitebread Hole (TV 600 968 to TV 595 955).
2. The high complex cliffs of Beachy Head (TV 595 955 to TV 579 953).
3. The truncated dry valley mouths, cliffs and platforms of the Seven Sisters, including Birling Gap (TV 579 953 to TV 521 976).
4. The shingle beach and marine delta at Cuckmere Haven (TV 521 976 to TV 514 976).
5. The Chalk cliffs and platforms at Seaford Head and Hope Gap between TV 514 976 and TV 490 980.

Although most of the site is cut into the Upper Chalk, at Beachy Head the orientation of the coastline trends across the escarpment and virtually the whole of the Chalk from Lower to Upper Chalk is exposed (Figure 4.26). In addition, the Upper Greensand and the Gault Clay beneath crop out on the shore platform east of Beachy Head.

The easternmost part of the site is aligned south-west–north-east and has a cliff height of 24 m. The upper part of the cliff is formed by a staircase of narrow slipped blocks. At the foot of the cliffs, both the Upper Greensand and the Gault Clay crop out, failures in the latter giving rise to the upper cliff features. The platform is distinguished by several repeated outcrops of these rocks, which appear to be related to mass-movements that have penetrated below sea level. At the northern end of the site several springs flow from the top of the Plenus Marl and may affect the erosion of the cliffs. The beach is formed of flint and chalk rubble, and extends northwards towards Eastbourne where it is dominated by flint, the Chalk typically being worn down over a distance of about 1 km.

On the eastern side of Beachy Head, the coastline is aligned WSW–ENE and cuts across the line of the South Downs escarpment. The cliffs reach a height of 160 m and have three, and sometimes four, distinct slope segments.

1. An upper vertical or near-vertical cliff, much broken by tensional fissures. In the past, it was distinguished by a series of pinnacles known as the 'Seven Charleses', which stood about two-thirds up the vertical face. Although the pinnacles are said to have disappeared before the end of the 19th century (Castleden, 1982), similar forms continue to occur on this slope. Failures of the cliff carry

Figure 4.26 (a) Beachy Head, cliff top view looking east, the cliffs are characterized by slab failures in the lower cliff that gradually undermine the upper cliff. (b) Cliff collapse at Beachy Head, early 1999; the failure affected the whole cliff face and produced a very large debris area at the cliff foot. (Photos: V.J. May.)

Beachy Head to Seaford Head

Figure 4.27 Relationships between joints, cliff morphology and retreat near Birling Gap. (Photo: V.J. May.)

Figure 4.28 Detail of the chalk and flint platforms east of Birling Gap. (Photo: V.J. May.)

debris to its foot.
2. A middle vegetated segment, which has angles around 60°. The length of the slope varies considerably, sometimes forming the majority of the cliff. At some points, it is almost totally bare of vegetation and appears to be a shear plane along which failure has occurred.
3. A lower steep bare rock cliff, which is currently attacked by wave action, but may have been exhumed from beneath a debris fan. Wave energy at the foot of the cliff is much-modified by the alternation of debris with cliffed embayments.
4. A debris slope formed of material that has fallen from the upper cliffs.

Generally having angles of about 30°, these debris slopes often extend from below low-water mark, sometimes joining the middle slope segment. They provide considerable protection to the foot of the cliff. Older debris has often left a residual cover of boulders on the platform. At the western side of Beachy Head, these act as a natural groyne reducing the transport of shingle eastwards from the Seven Sisters. Pocket beaches of flint and chalk rubble occur between the debris fans.

The Chalk shore platform extends beyond low-water mark, providing the foundations for the Beachy Head Lighthouse. The water around this part of the site is characterized even during gentle seas by greatly increased turbidity because chalk fines are commonly in suspension.

Between the western side of Beachy Head (TV 579 953) and Birling Gap, the cliffs are similar to those of the Seven Sisters. In contrast to the Seven Sisters, however, these cliffs produce large debris fans and many residual boulders form curved low ridges on the platform, a pattern noted along the coastline east of Dover. Beachy Head itself is distinguished by many small high-frequency rockfalls, but there are also infrequent very large events, the most recent occurring on 11 January 1999. A mass-movement event affected the whole height of the cliff over 200 m in length. The debris from the slide extended seawards to the base of the lighthouse and was marked by a seaward boundary of very large boulders (in excess of 3 m in diameter). Similar events are recorded on maps dating from the earlier part of the 20th century (Figure 4.26b).

The Seven Sisters have been cut across the dip of the Chalk and the cliffline truncates seven dry valleys (Figure 2.9). The cliffs rise to over 72 m in height, but their lowest points occur at the valley mouths (heights of 18 m at Crowlink and 12 m at Birling Gap). The dry valleys are underlain by Coombe Rock, and the Chalk has also been affected by periglacial heaving and shattering. At both Crowlink and Birling Gap, this sub-valley weaker material passes below sea level and the platform is cut lower here than elsewhere as a result. At Birling Gap, the rate of retreat has been higher than on the other cliffs (Table 4.2). The cliffs are vertical, with a narrow platform upon which rests an intermittent beach of chalk and flint shingle. Much of this sediment is derived from the erosion of the cliffs.

May (1971a) showed that between 1950 and 1962 cliff-top retreat at Birling Gap varied substantially both from year to year and between winter and summer. Of cliff-top land-loss, 87% occurred during winter. Most retreat took the form of long narrow strips or small lens-shaped areas. Over the timescale of this survey the cliff-top retreat was close to parallel, whereas over the seasonal scale it was much more spasmodic in time and location. Very large falls, such as at Baily's Brow in 1925, estimated at half a million tonnes (Castleden, 1982), are less common. Along this cliff, marine erosion is very effective in removing the debris from cliff-falls of all sizes. The cliff is undercut and frequently collapses along the line of joints, which are at a slight angle to the cliff-face. The failures appear to have a toppling nature, responding to the slight seawards dip of the Chalk. The strong winter frequency of cliff retreat may also be attributed to the effects of frost, and increased pore pressures. Weathering of the cliffs and platform produce substantial quantities of smaller debris, which together with the debris from falls, provide the main input to the beaches. In contrast to the generally narrow and discontinuous beaches below the Seven Sisters, large beaches occur at Birling Gap (Figure 4.27) and below the western cliffs of Beachy Head. Like the others, they are predominantly composed of rounded chalk pebbles and angular flint. The beach at Birling Gap fills the slight embayment formed by the more rapid retreat of the cliffs and lowering of the platform. Its alignment is strongly related to the dominant and prevailing waves. As a result it tends to smooth the plan of the shoreline between the cliffs east and west of the beach. The platform is lower at Birling Gap partly as a result of the deep weathering of the Chalk associated with the dry valley and also the absence of flint layers within the platform (Figure 4.28). To both east and west, the beds dip slightly eastwards and flint bands form sloping surfaces that are truncated.

Although the platforms along this coast have been commonly described as erosional features, little account has been taken of the role of the flint bands in controlling the micro-relief. Thus although there is an accordance of heights along the platform, the platform is a series of gently sloping micro-cuestas (Figure 4.28), with scarps on their western side at a point where the undercutting of the underlying chalk brings about collapse of the flint cap.

At Cuckmere Haven, the lower River Cuckmere is now channelled artificially between two groynes. In the past however, the river was deflected to the east by a shingle spit to enter the sea below Cliff End. Both in the past and at the present, the river forms a series of frequently changing distributaries that produce a small delta. This is an unusual feature, both on the coast of south-east England and on a coast undergoing rapid erosion. It suggests an effective fluvial sediment supply to the beach as well as considerable reworking of beach sediments by the river distributaries. This is the only beach within the site that contains rolled oxidized flint shingle as the predominant constituent. Elsewhere, high proportions of the beach comprise rolled chalk and much angular and sub-angular flint recently derived from the cliffs and platform.

The westernmost part of the site is formed by Upper Chalk cliffs that rise to 85 m at Seaford Head. The cliffs east of Seaford Head itself are affected by a series of vertical joints and have a generally vertical form. At Short Cliff, the cliffs are less steep and include gentler facets. At Hope Gap, the cliffs truncate a dry valley exposing the weakened chalk beneath the valley floor to erosion both in the cliffs and in the platform. Parts of the Chalk are overlain by deposits of orange-brown sands, silts and clays that are well-exposed at Short Cliff. They have been described as Palaeogene beds modified during the early and mid-Pleistocene (Castleden, 1982) and appear to have played an important part in the development of pipes in the Chalk. The pipes can be seen in the cliff face, sometimes descending, as at Short Cliff, to high-water mark. On the platform, Castleden has noted up to a dozen circular holes up to 1 m in diameter which show that piping extended well below present sea level. Their rims sometimes stand several centimetres above the platform (see also GCR site report for South Haven Peninsula in Chapter 7). The platform is well-developed, and its micro-relief is strongly affected by differential erosion along joints that run at right angles to the shoreline. As much as 1 m in depth, these eroded joints act as drainage channels for the platform.

Interpretation

The cliffline of the Seven Sisters shows strong adjustment to the primary direction of wave approach from the south-west, the headlands at Seaford and Beachy Head acting as strongpoints between which it is aligned (see Figure 2.9). The mouth of the Cuckmere River also affects the coastal alignment. Nevertheless, this coast retreats more-or-less parallel and at a similar rate to shingle beaches that share the same wave climate, such as the southern side of Dungeness. The modern production of beach sediment is small and does not offer much protection to the cliffs. At Birling Gap, the beach should provide greater protection, but this is not the case. The role of the debris overlying the cliff foot and the platform may be considerable, both in supplying tools for abrasion and erosion and also in affecting water flow over the platform. There is little understanding of the sub-beach weathering of chalk platforms (but see GCR site report for The Dorset Coast: Peveril Point to Furzy Cliff in Chapter 11), in contrast to Robinson's (1977a–c) work on Lias platforms around Robin Hood's Bay. Robinson and Jerwood (1987a,b) have demonstrated the importance of subaerial weathering of the platforms during severe winters, when freezing and thawing produce considerable breakdown of the platform surface. This process releases both chalk and flint clasts, but as elsewhere on the Chalk coasts, only flint makes any long-term contribution to the beach sediment budget. The flint bands play a critical role in controlling the micro-relief of the platforms and so allow higher wave energy inputs to the cliff foot at lower points.

Erosion along this coast has been very active throughout its recent history, but despite the considerable activity at Beachy Head the Head itself has remained salient. The large debris fans provide a substantial degree of protection, with wave energy reduced slightly as the coastal alignment changes. Even so, the equilibrium between effective removal of debris and its retention is finely balanced. It appears to be strongly related to the size of debris produced by the rockfalls, for debris of small size is rapidly removed. Where a rockfall contains a large boulder element, it appears that the rate of boulder reduction (which is a function of boulder size) brings about greater roughness of the intertidal platform and thus more effective wave attenuation.

The longshore transport of beach sediment, particularly rounded flint shingle, from the west is largely prevented by long jetties at Newhaven and a long groyne at the eastern end of Seaford Beach. The older rounded and oxidized flint shingle, which forms beaches in West Sussex and to the east of this site at Pevensey, has been replaced almost exclusively in this site by recent subrounded and subangular flint and chalk clasts. The effects of longshore transport are such that only limited quantities of shingle derived from the falls at Beachy Head can travel to the west. As a result, this beach system now relies upon contemporary inputs of sediment. Both Seaford Head and Beachy Head act as headlands between which the less-protected more-active cliffs and beach of the Seven Sisters have adapted to the dominant and prevailing south-westerly waves. In this respect, these Chalk cliffs contrast strongly with all other chalk cliffs in England. The majority are more strongly controlled in their alignment by structural features (e.g. Kingsdown to Dover), the nature of the cuesta into which they are cut (e.g. south-west Isle of Wight) or have very complex shorelines resulting from structural weaknesses at high angles to the shoreline (e.g. Flamborough Head). The rapidity of erosion and the relatively sparse beach mean that much of the Seven Sisters coast acts in a similar way to major beaches in adjusting to the alignment of the dominant wave direction. Among British cliffed coasts, other than those in weak sands and clays, this is unusual. This site is a member of a suite of some six beach or beach–cliff English Channel sites that show a similar alignment towards the south-west and comparable adjustment to the wave energy input, even though they occur in different rocks or have variable supplies of beach sediment.

The geomorphological interest of this site is very high, even though, like most cliffed sites, it has not been thoroughly investigated. The potential for increasing our understanding the ways in which cliffs and platforms develop through studying this site is substantial.

The Beachy Head to Seaford Down GCR site is of particular importance to coastal geomorphology because of the following features:

1. the well-developed cliff–beach–platform sequence, in which platform development is directly related to cliff retreat and both processes depend upon, but also control, beach volume;
2. well-developed platforms, notably at The Mares and Birling Gap;
3. solution piping in the cliffs and platforms;
4. a marine delta at Cuckmere Haven;
5. truncated dry valleys in the Seven Sisters;
6. rapid cliff erosion associated with efficient removal of debris from falls and a limited supply of flint to the beaches;
7. very active high cliffs at Beachy Head supplying substantial quantities of sediment to the local beaches;
8. the development of a platform across slipped blocks of Gault Clay and Upper Greensand, an especially rare feature because it preserves several major mass-movement events.

Conclusion

The cliffed coast between Beachy Head and Seaford Head is a GCR site of worldwide landscape importance, and international importance to research into coastal geomorphology because of the links between cliff, beach and platform development, the alignment of the shoreline to the dominant wave conditions, the contemporary sediment supply to its beaches, the almost total lack of relict sediment ources, and its contrasts with other coastlines formed in the Chalk.

BALLARD DOWN, DORSET (SZ 041 825)

V.J. May

Introduction

Ballard Down (see Figure 4.1 for general location) forms a distinct promontory at the eastern end of the Isle of Purbeck where differential erosion of sands and clays to the north and south of the Chalk cuesta has produced Studland Bay and Swanage Bay respectively. Strahan (1898), Davies (1935) and Arkell (1947) described the stacks and cliffs around Old Harry Rocks, and Steers (1946a) described its main features. Precheur (1960) related the formation of the stacks to jointing in the Chalk. May and Heeps (1985) describe some of the changes that have taken place. Despite the comparative lack of research at this site, it is widely used as a textbook example of the cave–arch–stack–stump sequence (Figure 4.29).

Ballard Down is a key site for coastal geomorphology, and one of many GCR sites that collectively form the Dorset and East Devon Coast World Heritage site, which was declared on account of its Earth science features of interest. Ballard Down includes a series of predominantly Chalk cliffs, platforms and associated beaches, best known for the classic assemblage of stacks, arches and caves at Handfast Point (May and Heeps, in press). The site is also important for revealing not only the relationships between local tectonic structures and coastal form, but also the effects of different wave dynamics on the north and south sides of the peninsula. In terms of wave energy, Ballard Down is the most sheltered of the major Chalk cliff systems and forms a key element in the network of such sites.

Description

The Ballard Down GCR site lies to the north of Swanage and is cliffed throughout its length. The cliffs rise from about 30 m on the Wealden clays in Swanage Bay to 117 m at Ballard Point (SZ 040 813). They then fall to little more than 20 m at Handfast Point where they turn west and continue at a similar height to the northern end of the site towards Studland on the Reading Beds and London Clay in Studland Bay. These cliffs can be divided into six different sections.

First, the clay and sand cliffs that cut across

Ballard Down

the northward-dipping Wealden and Lower Greensand rocks at the northern end of Swanage Bay, and are marked by many small mass-movements as well as some larger slides that have brought about a scalloped outline to the plan form of the cliff top. These unstable cliffs collapse onto the predominantly shingle beach, and small lobes and fans of debris occasionally spread onto its upper surface. There are several springs within these cliffs that give rise to local gullying, as well as the larger surface movements.

Second, from SZ 040 810 to SZ 048 813 (Ballard Point), the cliffline cuts at a very acute angle across the south-facing Chalk scarp of Ballard Down. The Chalk dips northwards here at angles ranging from 60° to 90°. The slopes have vertical sections both near to the top and particularly towards Ballard Point, but much of the cliff is characterized by shallow slides, scree slopes and a cover of scrubby vegetation (Figure 4.30a,b). Marine action is only effective at the eastern end where these subaerial features become less dominant and the cliff-foot protection afforded by boulders and a shingle beach diminishes.

Third, from Ballard Point (SZ 048 813) to the southernmost of the Pinnacles (SZ 053 820) the cliffs are close to vertical and cut across the Chalk either side of the Ballard Down Fault. Despite the considerable changes in dip either side of the fault, there is no significant change in cliff-form. Throughout this section the cliffs fall directly into the sea. There is no intertidal platform, although there is occasionally a narrow beach of Chalk and flint shingle exposed at low tide. A submerged platform with a veneer of boulders extends seawards from the foot of the cliffs. The cliff foot is undercut in parts and there are several small caves, including one where the Ballard Down Fault reaches sea level.

Fourth, from the Pinnacles to Handfast Point, the coastal forms become increasingly more complex, with five small bays, several stacks, and many small caves and arches (Figure 4.31). The largest cave, Parson's Barn, is 12 m in height at its mouth. May and Heeps (1985) mapped the changes between 1887 and 1982, showing how the stacks at Handfast Point have developed (Figure 4.31). Precheur (1960) considered that the stacks developed where the sea has eroded a series of major vertical joints. These can be seen in the blocks that have not yet separated from the mainland. Small caves usually develop at sea level; arches cannot form, according to Precheur, because the Chalk forming the upper part of the cliffs is too weak to form permanent roofs. The stacks, in contrast, are relatively

Figure 4.29 The cave–arch–stack sequence at Handfast Point, looking north-east, with Old Harry Rocks to the right. (Photo: V.J. May.)

Soft-rock cliffs

Figure 4.30 Ballard Down. Views looking east from SZ 038 810(a) taken on 12 January 2001 and (b) on 16 January 2001, showing the development of the landslip over four days. In (a) note the chalk scar formed by the failure of the slope. In (b), note the rectangular scar of the shallow rockslide that followed removal of bedrock and weathered slope materials at the back of the earlier failure. (Photo: V.J. May.)

Ballard Down

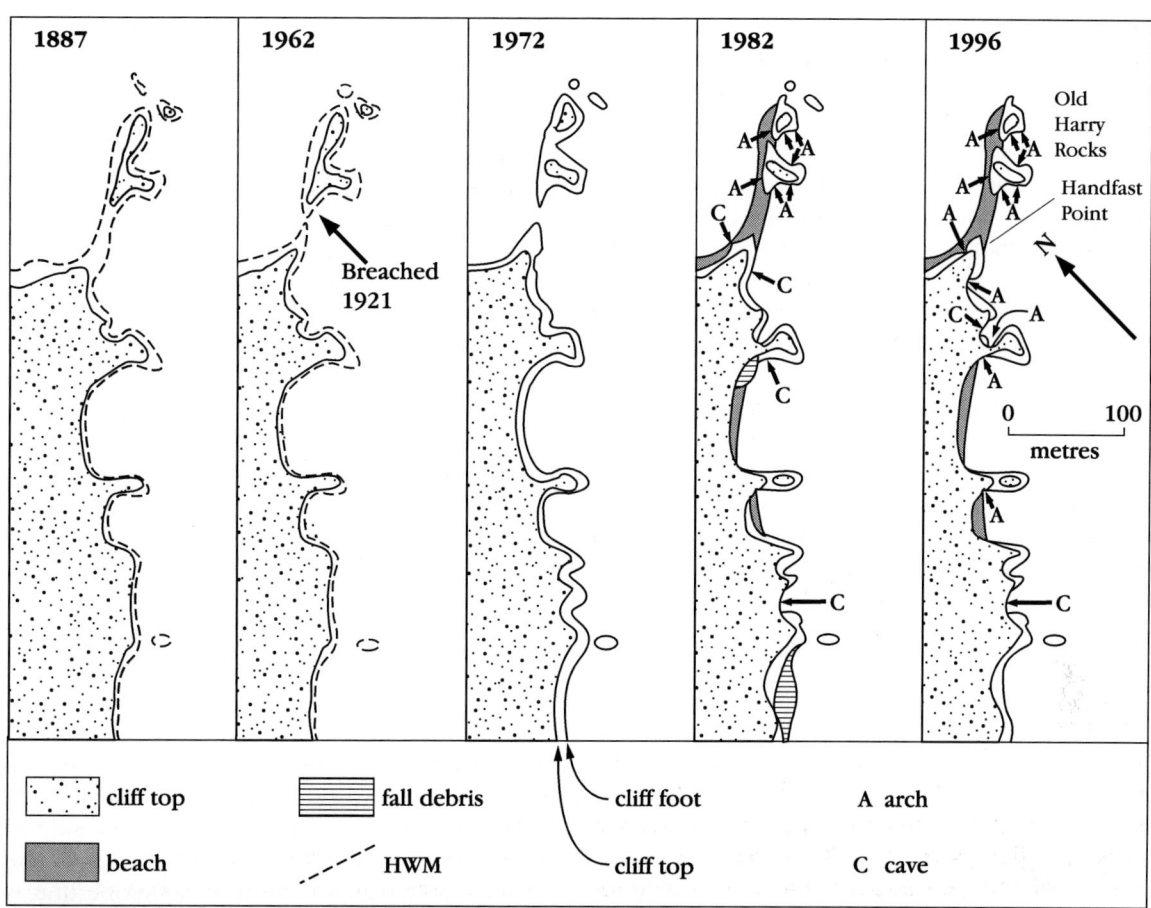

Figure 4.31 Cave–arch–stack development at Handfast Point 1887–1996. (Sources: 1887 Ordnance Survey and May and Heeps, 1985)

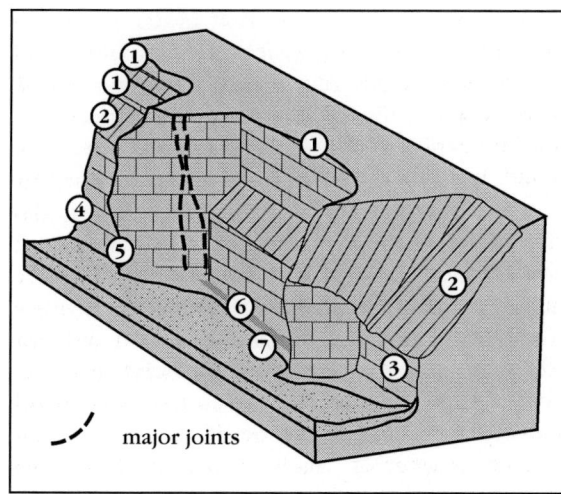

Figure 4.32 Multi-faceted northern cliffline west of Handfast Point towards Studland. 1. Vertical upper cliff; 2. vegetated debris slope; 3. lower vertical cliff; 4. smooth cliff-platform junction; 5. notch; 6. flint and chalk pocket beach; 7. chalk platform.

resistant to erosion as their foot is formed of harder Chalk. May (1971b) outlined the relationship of the erosional forms to the jointing pattern.

Fifth, the cliffs forming the northern sheltered side of Ballard Down, to the west of Handfast Point, appear to have a simple vertical form, but detailed surveys show that their form is made up of several facets (Figure 4.32). May and Heeps (1985) described the cliffs as affected only by rainwash, frost-action and gentle wave action. Notching and undercutting is rare. According to May and Heeps, many of the cliff profiles have a central truncated debris slope upon which further research is required.

Finally, at the western end of the Chalk cliffs towards Studland, the cliffs are cut in the Reading Beds and the London Clay and are affected by small slides and gullying, some of the latter associated with paths between the beach and the cliff top. Erosion of the lower cliff at the junction of the Chalk and Reading Beds provides large broken flints that litter the upper beach.

The beaches of the site also vary considerably. South of Ballard Point, they are formed mostly of flint, but chalk pebbles and cobbles are predominant close to the cliffs at Ballard Point. Their proportion decreases southwards until they form less than 10% at the southern end of the site. This beach includes a great mixture of other materials. Calkin (1968) reported that 'rounded pebbles of Purbeck Marbles and others of limestone from the reefs at Peveril Point can be picked up at the north-east end of Swanage Bay'. Sandstones and quartz pebbles from the Wealden are common. There are also granite pebbles, which are generally regarded as derived from ballast carried into Swanage Bay by vessels arriving to pick up quarried stone from Swanage. North of Ballard Point, the beaches are very limited in size and formed mainly of chalk and flint recently eroded from the cliffs. North of the Pinnacles, the beaches are almost entirely formed of chalk and flint. Much of the flint is angular, retaining the form of flints derived directly from erosion of the cliffs. The chalk pebbles tend to be more rounded, but are quickly reduced in size. There is also a thin veneer of sand on parts of the intertidal platform, which increases in width towards Handfast Point. West of Handfast Point, the beach and platform are narrow, but increasingly cloaked westwards by cobbles of chalk and flint. Below the cliffs of Reading Beds and London Clay, the beach includes more sand, although there are considerable spreads of chalk and flint cobbles as well.

Interpretation

Research at this site has focused on three issues:

1. the question of the development of the stack–arch–cave complex at Handfast Point, and its relationship to the strength and structures of the Chalk,
2. the nature of the cliff profiles along the northern side of Ballard Down,
3. the relationship between cliff erosion, beach development and protection of the cliff foot by beaches and debris.

Like several other Chalk sites, Ballard Down has a set of well-developed stacks, arches and caves associated with a headland. The penetration of headlands by cross-joints and faults aids the development of these erosional features at Handfast Point. May (1971b), Precheur (1960) and May and Heeps (1985) have described changes in these cliffs and discussed their location. Precheur, in particular, considered that a hard pedestal band was important for the longevity of caves and arches as well as such features as the Pinnacles. As the Chalk dips northwards, harder bands dip below sea level and become less effective. The pedestal band is important both because it provides greater strength at the base of rock columns and because it channels water flow through the joints. Well-developed, near-vertical joints are opened up at their base by the sea, and very close jointing in the upper part of the cliffs encourages small blocks to fail. The joints are harder than the surrounding bedrock, but individual clasts within the joints are vertical and have a greater tendency to drop out than the surrounding horizontal and better supported blocks. The stacks are a result of the narrowing of the headland, jointing patterns and the relative resistance of a pedestal band to erosion.

The cliff profiles along the northern side of Ballard Down are, like many other cliffs in the Chalk, marked by a central grassy slope that is cut across the bedrock and is veneered with chalk debris. The size of this debris slope tends to increase towards the western end of the cliffs. Marine undercutting of the lower cliff has been too slow to destroy this segment. The upper vertical segment can only retreat under the influence of subaerial processes. It is possible that the debris slope represents a former slope formed when these cliffs were less exposed to marine action. These forms have not been dated, but two hypotheses were put forward by May and Heeps (1985). First, with a lower, glacial, sea level the dip slope of the Chalk would be affected by frost-shattering and other periglacial processes. As a result, a debris slope could form, but it is surprising that it did not affect the whole slope given the relative weakness of the Chalk. A rise in sea level would reactivate these cliffs in the bevelled cliff hypothesis (see Chapter 2). Such an hypothesis ignores the effects of the rather late opening of Poole Bay. A second alternative is that these cliffs have been active during the period since sea level reached its present position, but were protected temporarily. This is a possibility since there have been very large accumulations of sand in

Studland Bay and features such as the cliffs at Redend Point (see GCR site report for South Haven Peninsula below) have been protected from the sea in the recent past.

The third focus for research at the site concerns the rate of supply of flint and chalk from Chalk cliffs to the beaches and the longevity of sediment fed to them. May and Heeps (1985) described the sediment budget of a rockfall and its debris just south of Handfast Point. Unlike the cliff profile described at Joss Bay by Hutchinson (1972) (see GCR site report), this fall produced a significant amount of flint that entered the pocket beach. The flint supply was insufficient to build up a beach that could protect the foot of the cliff, and chalk pebbles were quickly reduced by attrition and disappeared within a matter of months. Even if longshore transport were possible, the supply to other beaches would not be significant. Along the northern cliffline in contrast, there is a constant supply of small platy chalk clasts, which rapidly become rounded into small shingle-sized fragments. The minimal wave activity limits the extent to which they are then moved further. Attrition appears to be the most important process at present, although at the time of writing, further investigations were in progress.

In summary, these cliffs offer an interesting contrast to two other east-facing cliffed sites in the Upper Chalk, Joss Bay and Flamborough Head. At Joss Bay, there is less variation in dip and there are very wide platforms. At Flamborough Head, the cliffs are much more intricate in plan and are more exposed. Ballard Down is a textbook example of the development of coastal forms such as stacks. The contrasts in cliff morphology within the site give it further geomorphological importance.

Conclusions

Chalk cliffs and platforms in the northern part of Ballard Down are comparatively simple when contrasted with very complex stack, cave and arch forms around Old Harry Rocks. With the eastern part of the Furzy Cliff to Peveril Point GCR site, this forms a fine, internationally renowned, example of a transverse coast. The development of caves, arches, and stacks is a major feature of this site, but it is also an excellent example of the ways in which steep cliffs change as a result of both marine and subaerial processes. Unlike many such sites it is well-documented. It is one of a network of contrasting cliff sites in which stacks are a key feature, but this is one of the few locations where the dynamics of erosional forms continue to be monitored.

Its international significance is recognized with its inclusion in the Dorset and East Devon World Heritage Site.

FLAMBOROUGH HEAD, YORKSHIRE (TA 182 746–TA 202 686)

V.J. May

Introduction

Flamborough Head is the largest promontory on the North Sea coast of north-east England, projecting some 10 km eastwards from the Holderness coastline, which is undergoing rapid erosion, to its south (see Figure 4.1 for general location). Flamborough Head is treated as a separate site from Holderness because of its importance as one of the suite of Chalk cliff sites. Holderness is discussed separately owing to its importance as a cliff undergoing rapid retreat.

Flamborough Head is the northernmost coastal outcrop of the Chalk and the most extensively affected by glacial conditions. Like the Chalk cliffs of the North Downs in Kent, Ballard Down in Dorset and Tennyson Down in the Isle of Wight, the cliffline at Flamborough Head cuts across the Chalk cuesta and many different parts of the Chalk succession are exposed. This situation, combined with the effects of different levels of exposure to wave action, has brought about considerable variety of coastal forms within the site.

Flamborough Head forms part of the GCR network of Chalk coastlines and it lies within the zone of North Sea wave climate, unlike the majority of other GCR sites, which are partly or wholly affected by Atlantic swell and English Channel wave climates. Winds are generally offshore, but important secondary wind and wave directions are from the south, east and northeast, the latter being important in winter. The fetch for many waves generated in the southern North Sea is generally less than 700 km, whereas waves generated from a northerly sector may have a fetch extending into the Arctic area. As a result, much of the site is affected by long-refracted swell (Figure 4.33; see also GCR site reports for Holy Island, Chapter 11, and

Soft-rock cliffs

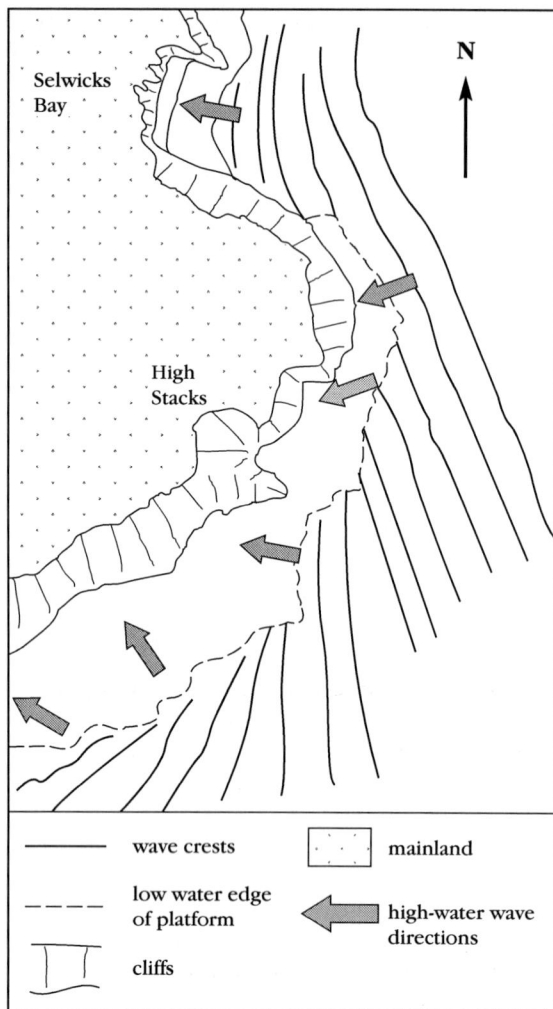

Figure 4.33 Wave refraction at Flamborough Head, showing variations in wave direction crossing the platform owing to wave refraction. See Figure 4.34 for location. (Based on aerial photographs in Pethick, 1984.)

Marsden Bay, Chapter 7). It is also the only GCR coastal geomorphology Chalk locality that is extensively overlain by glacial deposits. The northern cliffs are relatively simple, both in plan and profile (Figure 4.34); they feed small amounts of flint to their fringing beaches. The Chalk is extensively faulted, with some 1340 faults within one 6 km length (Peacock and Sanderson, 1993, 1994). Many excellent examples of caves, arches and stacks are associated with this faulting, and a number of blowholes have developed where the overlying till has collapsed into caves that intersected the Chalk–till junction. One contributing factor to the large numbers of caves has been the hardness of the Chalk. Secondary diagenetic deposition of calcium carbonate in the chalk pore spaces has produced chalk cliffs that are much more resistant to erosion than the Chalk of similar age in southern England. Shore platforms are well-developed both in this area and along the southern shoreline, where the beach is mainly sandy, and lacks flints. Marine processes vary from north to south: the southern cliffs are less active than those to the north.

As with many cliffed coastlines, there are more passing references to Flamborough Head in the literature than detailed studies of it. Nevertheless, the nature of both its cliffs and platforms has been commented upon in more general descriptions of the coast (Steers, 1946a; Straw and Clayton, 1979; Pethick, 1984) and discussions of platform morphology (for example, Trenhaile, 1974b).

Description

This site has three main subdivisions (Figure 4.34):

1. The northern cliffs between Bempton (TA 182 746) and Long Ness (TA 228 725), where the dip of the Chalk is to landward at about 20° (Figure 4.35b);
2. the complex coastline around Flamborough Head itself between Long Ness (TA 228 725) and Cattlemere Hole (TA 256 703; Figure 4.35a); and
3. the southern cliffs from TA 256 703 to the western boundary of the site at Sewerby, where the dip of the strata in the cliffs is to seaward (Figure 2.1b).

The northern cliffs, known as 'Bempton Cliffs', fall southwards from about 110 m at the northern edge of the site to about 65 m at Long Ness (TA 228 725). A narrow platform, Chalk and flint shingle beach and debris from rockfalls extend about 75 m from the foot of the cliff seawards to the low tide mark (Figure 4.35b). Both their plan and profile are simple. Straw and Clayton (1979) have suggested that wave erosion has been particularly severe on this coastline 'where, opposite a long northern fetch, the Bempton cliffs rise a sheer 130 m'. Steers (1946a) described the cliffs as being in greatly contorted flinty Chalk, but the contorted nature of the Chalk has not affected the cliff-form to any significant extent. At Staple Nook, there is a slight indentation of the coast associated with

Flamborough Head

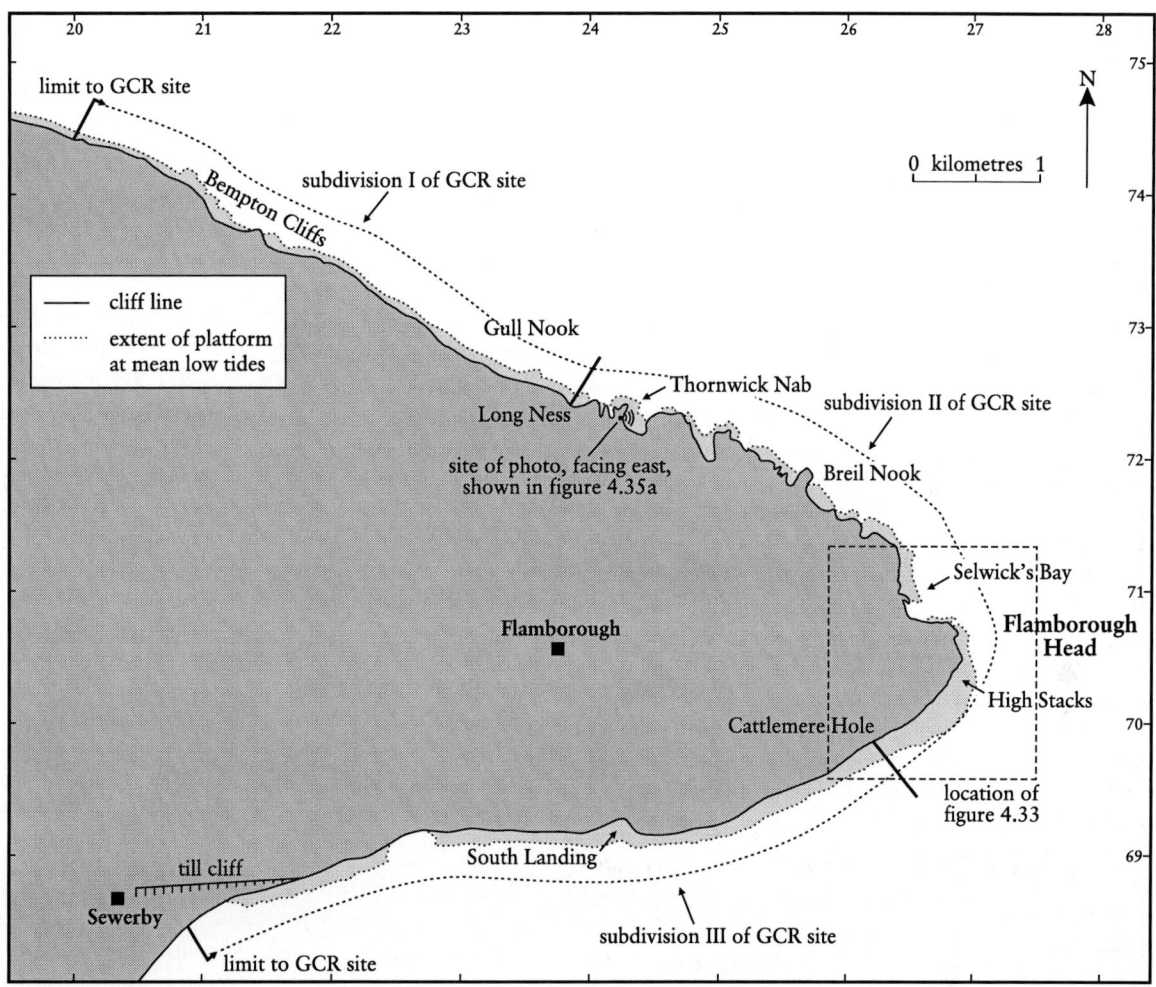

Figure 4.34 Sketch map of the Flamborough Head coastal geomorphology GCR site, showing the three main divisions of the locality.

weaker contorted Chalk.

The central cliffs are cut in the Middle and Upper Chalk (*Terebratulina lata-Micraster coranguinum* biozones) and are characterized by numerous caves, arches and stacks. There are well-developed, structurally controlled, platforms with many vertical joints and small faults exposed both in the surfaces of the platforms and in the cliffs (Figure 4.35a). Marine erosion along the joints and faults has been especially effective in developing a very large number of erosional forms. There are some 50 distinct inlets along this section of the coast. They vary in nature from caves and narrow steep-sided geo-like forms to small bays, over 55% of which are aligned towards the NNE and north-east and about 15% towards WNW. Devensian till cappings give the upper cliffs a complex profile that has been much affected by landslips. The lower part of the profile is steep, often with a tendency to overhang at its base. The plan of the cliffs is affected both by the erosion of the many structural weaknesses and by the form of the subglacial surface. Where former valleys have been truncated by the sea, and also where the dip of the Chalk brings the subglacial boundary closer to sea level, the sea has been able to erode inlets and bays more effectively. The dolomitized Chalk has also been affected by periglacial processes and much of it is deeply shattered, thus reducing its general resistance to present-day weathering and marine erosion.

There are several blowholes within large hollows in the till. These have developed where caves in the Chalk have grown upwards to the boundary between the Chalk and the till. Whereas the blowhole outlet would remain small if it were in the Chalk, the lower slope sta-

Soft-rock cliffs

Figure 4.35 Flamborough Head, (a) looking east from Thornwick Nab. The upper cliff is in Devensian tills, the lower cliff in chalk with numerous caves, arches and platforms. (b) Looking WNW at Bempton Cliffs; steep cliffs with a short upper vegetated facet in tills. Pipe-like forms extend down the whole height of chalk cliff; the cliffs have a narrow platform with a cobble and boulder beach. (Photos: V.J. May.)

bility of the clay material in the till overlying the Chalk has produced more open hollows. On the north side of Selwicks Bay several blowholes appear to have merged and the intervening Chalk has collapsed to produce a complex inlet (Figure 4.36). Some of the caves are associated with gullies across the platform on the line of the joint or fault controlling the development of the cave. Other caves lack this relationship, their floors being formed by slightly dipping beds of more resistant Chalk standing above the general level of the platform.

The cliffs on the southern side of Flamborough Head are simple in plan and profile, but are less active than those to the north at Bempton (Figure 2.1b). Sheppard (1912) placed the coastline of Roman times 2000 years BP over 1.6 km offshore from Sewerby and close to the present shoreline south of Flamborough village. There is, however, no direct evidence to corroborate this, although farther south documentary evidence emphasizes the large land-losses since Roman times. Steers (1946a) noted that the rate of retreat was small because of the hardness of the Chalk forming the cliffs and platform and because the old pre-glacial cliff was still being exhumed. Valentin (1954) estimated that the cliffs at Sewerby retreated by 18 m between 1852 and 1952. The upper part of many of these southern cliffs is well-cloaked with vegetation, with a lower angle of slope reflecting the presence of the till above the very steep Chalk cliffs. There is a preponderance of small falls of Chalk, rather than the more substantial rockfalls of the northern cliffs. The platform is a true abrasion platform: it is not structurally controlled since it cuts uniformly across the strata (Figure 2.1b). Where backed by Chalk cliffs, the platform is cut into chalk, but is replaced by a till platform of the same gradient and width at Sewerby where the cliffs are in till. The beach has two elements, an upper narrow beach mainly of chalk pebbles, and patchy, but few, flints, and a thin, sandy veneer resting on the platform. Trenhaile (1974b) has shown that the gradients of the platforms around Flamborough Head are higher than at any other part of the north-east Yorkshire coastline and other Chalk coastlines in England. The dominance of waves from northerly directions means that most waves are refracted around the headland and approach these southern cliffs from the east (Figure 4.33). As a result, this part of the site tends to be swept clear of much of its surface sediment, which is transported southwards. Erosion rates are relatively slow (less than 0.3 m a^{-1}) and so there is only a small input to the Holderness sediment budget.

Interpretation

This site contains the largest assemblage of active coastal erosional forms anywhere in the English Chalk, coast protection works having removed most of the very complex features on the north coast of the Isle of Thanet at Birchington. The situation at Flamborough Head, like others on the English coast, presents a puzzling question concerning the apparent resistance of a promontory that is otherwise riddled with structural weaknesses. The many faults have given rise only to the large number of inlets because the Chalk is sufficiently hard to prevent collapse. It behaves more like a hard limestone coast than a weaker chalk coast. Thus despite the deep incision by caves and other inlets into the cliffs, it is the most prominent feature of the eastern coastline of England north of the Wash. In part its form is accentuated by the rapid erosion of the weak materials forming the coastline of Holderness to the south. The southern side of the promontory has certainly undergone some erosion because it cuts across the Ipswichian shoreline at Sewerby (Catt, 1977). Nevertheless, the central section described above does not appear to be undergoing rapid erosion. Comparison of the photograph of Selwicks Bay in Steers (1946a) with the present cliffs suggests that although there have been small changes, there have been no major changes. The dip of the Chalk here varies between 10° and 15° but the coastline is so complex that there is no simple relationship between the cliff-forms and the local dip. The platforms are complex with considerable variation in relief both towards the sea and along the platforms. The development of the cliffs cannot be considered without discussion of the platforms because they affect the distribution of wave energy over each tidal cycle, most particularly in reducing the energy available for marine erosion of the foot of the cliff and the removal of talus from its foot.

Trenhaile (1974b) demonstrated that the platform gradient here was higher than might be expected from consideration of both the geology and the morphogenic environment. Analysis of covariance shows that tidal range correlates

Soft-rock cliffs

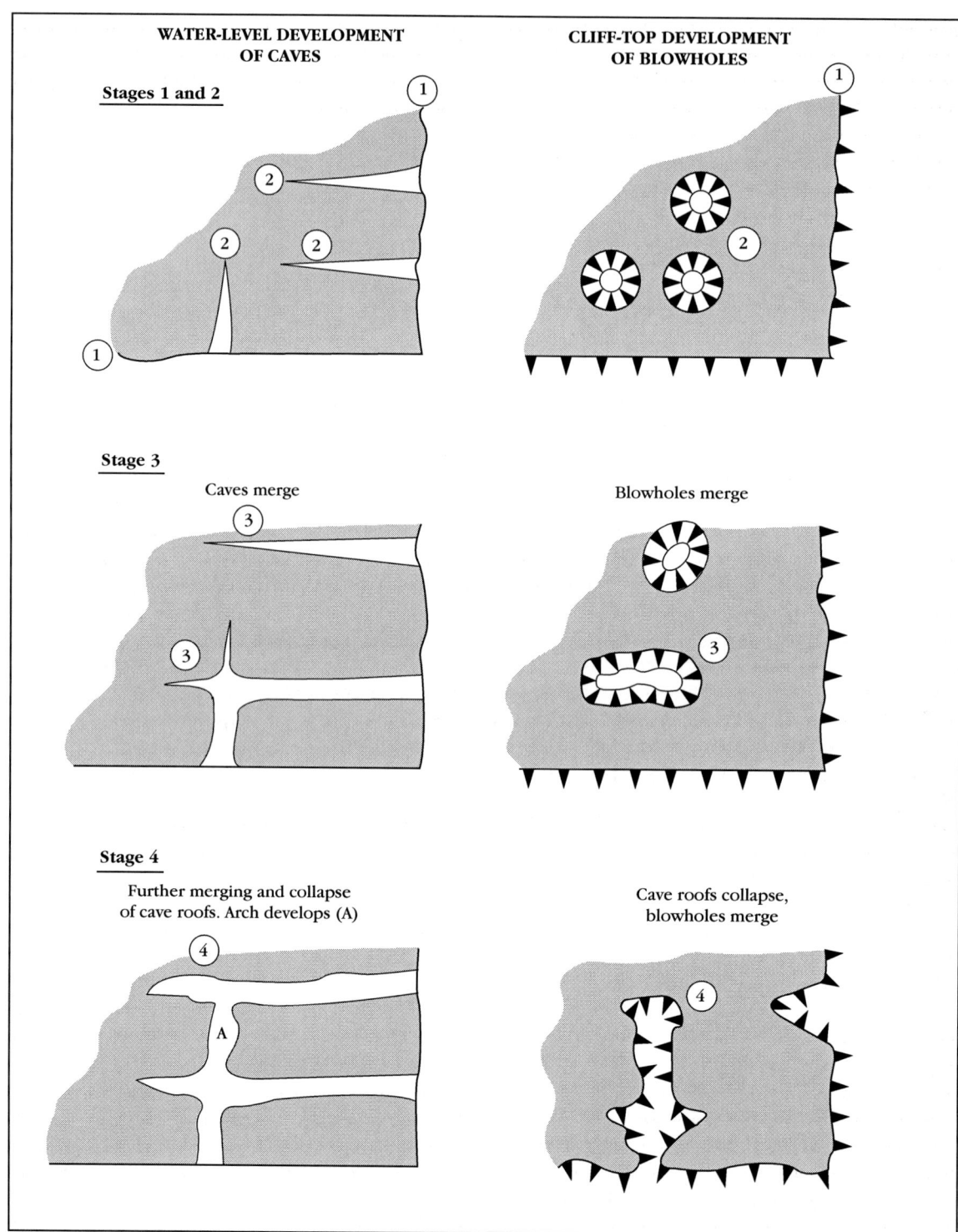

Figure 4.36 Cave and blowhole development at Flamborough Head, shown schematically in plan view. There are several stages in the development of blowholes here. Stage 1: caves develop along major joints or faults. Stage 2: caves extend upwards into the overlying till, which begins to collapse allowing hollows to appear in the till. Stage 3: caves merge and blowholes coalesce. Stage 4: Further merging of caves, cave rooves collapse, arches and/or geos develop. Subsequently, isolated blocks or stacks may develop.

strongly with platform gradient, but this correlation is not dependent upon rock type. For the same *Micraster coranguinum* Chalk biozone and a similar fetch and tidal range, the platforms at Flamborough Head have much greater gradients than those around the Isle of Thanet. This variation could be attributed in part, however, to differences in the lithology between geographically separate parts of the same biozone, an issue not discussed by Trenhaile. He has suggested that waves that approach with the least energy, owing to refraction, are most significant for platform development. This must, however, be modified by the roughness of the platform and surrounding intertidal areas. Much of the intertidal area in the central section of this site is distinguished by rocky outcrops that owe their features to the differential action of marine and sub-aerial processes upon them. There have been no detailed field surveys of these features, but the very rough nature of the surfaces has been observed to have the effect both of channelling water flow, particularly during backwash of waves and drainage on falling tides, and of dissipating much of the energy contained within waves crossing the intertidal area. The channelled flow of water along joints, into and out of inlets and caves, has a very localized effect.

When the sides of such channels are undercut sufficiently they may collapse, but the length of many of the channels and caves suggests that the penetration along them is carried out much more efficiently than widening of them. Most waves approaching Flamborough Head are strongly refracted. Their behaviour in crossing the intertidal area is very complex and, except during periods of storms, inefficient in attacking the innermost parts of the bays. As a result it could be argued that it is the very complexity of the coastline here that contributes to its relative resistance to recession.

In contrast, the platform along the southern shoreline of Flamborough Head westwards to Sewerby shows many of the features that have usually been associated with shore platforms. Its slope is not complicated by strong micro-relief or debris accumulations. Most waves that affect it are strongly refracted around the headland and tend to approach from the south-east and east thus travelling at an angle across the platform rather than at a normal orientation to the cliffs. This area lacks flints in the Chalk and most of the sediment available to be used by the natural system as erosional tools is sand or chalk pebbles. This area thus raises important questions for the debate concerning platform development, which further modelling and observation should consider.

Conclusions

Flamborough Head is a very important site for the following reasons. First, it is the most complex cliffed Chalk coastline in England, with numerous caves and arches. Second, it is the largest such site that has been affected by glacial processes, which have not necessarily contributed directly to the coastal forms but may have affected the nature of the Chalk itself. The Chalk is overlain with Devensian tills, a combination that gives this site further interest because of the effects this has on the nature of the cliff-forms. Third, it exemplifies well the effects of different wave climate upon coastal forms. Fourth, it provides an excellent site for the study of coastal erosional processes and the linkage of cliff–beach–platforms processes that Pethick (1984) suggests is needed if platforms are to be placed in context and better understood. Finally, it is the only Chalk cliff GCR site that is affected solely by North Sea wave systems.

JOSS BAY, KENT (TR 383 716–TR 402 696)

V.J. May

Introduction

This site comprises the most extensive Chalk intertidal platform in England and is backed by near-vertical cliffs of Upper Chalk. It is one of the least-modified parts of the coastline on the Isle of Thanet between Margate and Broadstairs (see Figure 4.1 for general location). The cliffs have retreated at about 0.3 m a^{-1} in historical times, and the platform close to the cliffs has been lowered by about 0.03 m a^{-1}. Coastal retreat takes place by a combination of small rockfalls, shallow rock-slides and marine undercutting. The cliffs have two main orientations, which conform with joint patterns, the Chalk itself being well jointed and closely bedded with a northerly dip of about 1°. Beaches formed of small quantities of flint and chalk occasionally mask the cliff foot, whilst there is an accumulation of predominantly biogenic, shelly sand in

Soft-rock cliffs

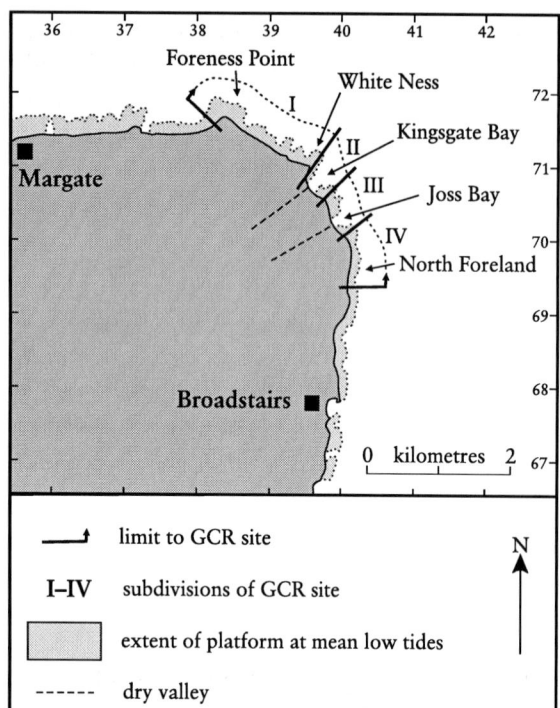

Figure 4.37 Sketch map of the Joss Bay coastal geomorphology GCR site.

Description

The site extends from Foreness Point (TR 383 716) to Hope Point (TR 402 696) south of North Foreland (Figure 4.37). The characteristic form is a vertical or near-vertical cliff about 20 m in height fronted by a sandy beach of varying width up to a maximum of about 180 m, and a well-developed Chalk intertidal platform up to 500 m wide. The plan form of the coast is marked by sharp changes in cliff orientation, which allow the site to be subdivided into four sections:

1. Foreness Point and Botany Bay, which includes several stacks,
2. Kingsgate Bay,
3. Joss Bay, and
4. the cliffs of North Foreland.

There has been some coast protection work, notably in Kingsgate Bay, to protect a cliff-top road and Kingsgate Castle, and in Joss Bay also to protect the cliff-top road. South of Joss Bay, very small segments of cliff have been bricked-up or walled-in in the past. Cave entrances have also been bricked over to prevent undermining of the cliff top and the further development of blowholes. At Foreness Point, sewage outfalls at the western boundary of the site cross the platform and appear to reduce sand transport westwards across the platform. As a result, it is possible to examine the contrasts in platform development with or without a significant sand veneer.

Although the whole of this site is cut into the Upper Chalk, there are differences between its northern and southern parts. Between Foreness Point and White Ness, the cliffs are cut mainly in the *Marsupites testudinarius* biozone above the *Uintacrinus socialis* biozone. Harder bands of the latter are associated with ledges on the platforms, thresholds at the foot of stacks and ramp-like cliff bases (So, 1965). South of White Ness, the coastal alignment is at right angles to the dip and the cliffs are formed mainly in the *Micraster coranguinum* biozone of the Chalk overlain by the *U. socialis* biozone. Within both Kingsgate Bay and Joss Bay, parts of the cliffs are cut across dry valleys that are underlain by periglacially modified chalk and slope wash. Hutchinson (1972) noted that the Chalk in the upper 3 m of the cliffs at Joss Bay appeared to have been frost-shattered, more or less *in situ*, as occurs

the bays. In Botany Bay, a duniform ridge formed at the cliff foot in the early 1980s became sufficiently stable to be colonized by vegetation, but this is not characteristic of the site as a whole.

Although the geomorphology of the Chalk coast has attracted general attention (e.g. Steers, 1946a; Bird, 1984), there have been two groups of detailed studies, one focusing upon the platforms and their development (So, 1965; Wood, 1968; Trenhaile, 1974a,b), the other upon cliff-retreat processes (May, 1964; Hutchinson, 1972). There has been considerable engineering interest in this coastline in general because of the coast protection issues related to cliff retreat and in particular the penetration of caves beneath urban areas, such as at Broadstairs. In reality, cave development within this site is very restricted. The best examples were to the west of the site at Birchington (May, 1964; So, 1965) but they were destroyed as part of a coast protection scheme. Nevertheless, this site still contains good examples of cave–arch–stack development that can be traced in historical records for about 140 years.

Joss Bay

Figure 4.38 One of two stacks in Botany Bay. This stack was joined to the mainland in 1842 and became separated during the 19th century. (Photo: V.J. May.)

elsewhere in the site. Peake (1961) recognized two sets of major joints that are sub-vertical and generally lie within 10° of the main joint directions of 10° and 290°. The Chalk is closely but irregularly jointed between these major joints. Except in the *Micraster* Chalk, flints are uncommon in the cliffs. Their presence south of North Foreland affects the form of the platform, since they appear to provide a degree of armouring to its surface.

May (1964) described the coastal changes around the Isle of Thanet, and So (1965) examined the platforms in detail. They agree that the cliffs are generally retreating at about 0.3 m a^{-1} and the platform is being lowered at about 0.03 m a^{-1}. In detail, the changes are much more varied with many small cliff-falls taking place as well as localized lowering of the platform or accretion of sand. At Kingsgate, May (1964) noted that there had been little change since about 1870 when a brick facing- and buttress-wall was constructed below the castle. In the bay, however, the cliff-top road had been so undermined that it was supported on a concrete bed buttressed above the beach. Between White Ness and Foreness Point, the cliffs run parallel to the main direction of jointing, and rockfalls from the *U. socialis* biozone are frequent. South of White Ness, falls appear to be less frequent, but Hutchinson (1972) described one such fall in detail. This fall, in early 1966, cut the cliff top back by about 2.3 m along about a 20 m length. Hutchinson's analysis demonstrated that a notch of 0.5 m could produce such a failure given the shear strength of the Chalk. By February 1971, a notch 1 m deep was measured at the site of the fall. Thus not only had about 500 m³ of rock been removed from the cliff, but the sea had also cleared this away and undercut the cliff once more. The average 0.3 m a^{-1} quoted above thus disguises some larger but less frequent events.

Despite their very exposed location, two stacks have formed at the eastern end of Botany Bay. Cartographic evidence shows that the larger of these separated from the mainland after the tithe survey of 1842 when it was joined by a neck of land about 22 m wide. The date of formation of the smaller stack is not known, but it is younger (Figure 4.38).

So (1965) showed that the platforms are gradually extended landwards by cliff recession. They show variable micro-relief often as a result of pitting, block-loosening and more resistant flint patches. Their height decreases from headlands to embayments, particularly at White Ness and Hackemdown Point, even (according to So) when the dip and jointing do not favour this pattern. At Foreness Point, the platform is also lower on the western side of the headland than on the eastern side. Irrespective of the initial height of the platforms, the gradient is greater in the embayments than at headlands. Profiles are characterized by a concave upper and a convex lower section, but are also steeper at Foreness on its eastern side. Wood (1968) showed that this platform steepened markedly seawards. The 120 m closest to the cliffs was very gently-sloping, with an average gradient of about 1 in 400, but below the +0.3 m OD contour it steepened to 1 in 70. At White Ness and Hackemdown Point, low cliffs also occur at the outer edge of the platform. Recession of these low-tide cliffs is also less rapid than the high-water cliffs (So, 1965).

There is often a sandy veneer on the platform, much of it composed of shelly fragments, some from reworking of the Chalk and some from present-day molluscs. In the southern part of the site, flints also form small patches often close to the foot of the cliff. Wave direction tends to keep the platform south of Hackemdown Point swept clear of sand, apart from small patches infilling hollows in the platform. In Botany Bay, however, considerable quantities of sand can accumulate and during the early 1980s an unusual cliff-foot dune built up and became temporarily stabilized by vegetation. This protected the cliff foot from erosion, although sub-aerial processes of rainwash and frost-shattering continued. As a result a series of small debris-slopes grew on top of the dunes, which themselves rested on the cliff–platform junction and beach.

Interpretation

This site represents very well the relationship between cliffs and platforms on the English coast, especially because the rates of change in both have been documented for a longer period of time than elsewhere. Cliffline crenulation is present but was much better represented in the past by the cliffs at Birchington in north-west Thanet, where a comprehensive set of caves, geos and blowholes had developed. These were mostly destroyed during coast protection works in the late 1960s. As a result some upper parts of the platform described by So (1965) have been destroyed. Because So showed that the platforms on Thanet had a generally upper concavity, and a lower convexity, and other platforms in England and Wales appear to lack this (Trenhaile, 1974a,b), the remaining platforms that are represented by this site are all the more important members of this suite of active shore platforms.

So (1965) argued that the coastal platforms of Thanet were the result of storm waves, both past and present. As water-levels change with the tides, so the zone of effective wave attack and of marine planation also shift vertically. The height at which the water-level associated with storm-wave action remains longest is within the mean neap tidal range. It was argued by So that wave attack would be most frequently focused at mean tide-level and storm-wave platforms would rise up to 1 m above this height; platforms within the site range in altitude between OD and +0.91 m OD, corresponding to a mean tide level of +0.17 m OD. Platforms around the Thanet coast, however, extend below low tide levels, decreasing in height comparatively rapidly and lacking signs of planation. Furthermore, the low-tide cliffs have not receded as rapidly as the rate of cliff recession. These platforms, So concluded, must relate to a lower sea level, but So did not explain their co-incidence with low tide levels. Debate elsewhere (Trenhaile and Layzell, 1981; Carr and Graaf, 1982) concerning the duration of wave attack related to the tidal duration curve may throw light on this issue in general, but has not been considered in relation to this site. Similarly, since waves lose energy in crossing the platform, a rising sea level is required to ensure that past platform-widening continues. Wood (1968) considered that the notch at the foot of the cliffs marked the 'true level of present day erosion' and that the platform close to cliffs had been cut with the sea near its present level. The greater retreat of the bays was attributed to the occurrence of sand and the potential for abrasion, whereas the headlands, lacking such aids to abrasion, would erode less rapidly. The width and steepening of the platforms may have resulted from a rise in sea level of as much as 6 m since c. 2700 years BP, according to Wood (1968).

Neither Wood (1968) nor So (1965) considered the origins of the bays. The eastern end of Botany Bay, Kingsgate Bay and Joss Bay are each associated with the truncated mouths of dry valleys that are underlain by frost-shattered chalk and slope wash. It is possible that not only were the bays already related to the drainage pattern of eastern Thanet, but also that the more weakened Chalk would aid retreat in the bays. South from White Ness the platforms have an extensive cover predominantly of flint, much of which is *in situ*. The extent to which this provides an armouring to the chalk surface has not been investigated. Between 50 m and 75 m from the main cliff foot there are, however, some small landward-facing micro-cliffs (up to 0.3 m in height) that are capped by this flint layer, which in places co-incides in its slope with flint layers in the main cliffs. These flinty layers are uncommon in the bays. South of North Foreland the micro-cliff is close to the position of the main cliff of about 100 years ago. The presence of these higher sections of platform poses a question about the way in which the platforms have developed, since it is evident from the above that the platform is a simple result of cliff retreat (see also GCR site report for Beachy Head to Seaford Head in the present chapter). Other processes appear to be important in the greater lowering close to the cliff foot. One possibility is that waves reaching this area from the north-east, despite refraction across the platform, travel strongly along the foot of the cliff rather than approaching it from seaward. Additional scouring in this position could accelerate platform lowering. The site thus provides continuing opportunities to investigate further the mechanisms by which platforms develop. Most models treat platforms two-dimensionally, concentrating upon the profile of the platform, and thus ignore the three-dimensional form and the behaviour of waves crossing it. In addition, the roughness of the surface itself affects the erosional efficacy of waves crossing it (see GCR site reports for Flamborough Head, and Kingsdown to Dover in the present chapter). The lack of boulders at Thanet means, for example, that platforms are given little protection from wave action, unlike some of these other platforms.

Like other active cliff and platform sites, Joss Bay has to be viewed as part of a network of such features in different tidal and wave environments. It is one of the few sites where there have been both detailed surveys of the platforms and geotechnical investigation of the cliffs. In common with the other sites, its importance also lies in its contribution to understanding of the complex relationships between cliffs, platforms and beaches, especially since this site lacks the considerable shingle cover found on other sites along the English Channel coast.

Conclusions

As the most extensive Chalk intertidal platform in England and Wales, Joss Bay provides several insights into the links between platform width, wave energy and platform extension. First, it is very wide and, according to some early literature, at the maximum limit of platform extension. Second, there have been a number of investigations of both the detailed morphology of the platform and of the geotechnical qualities of the cliffs. The rates of retreat of the cliffs and of lowering of the platforms may be causally linked. Third, it has a very restricted sediment veneer. Taken together, the investigations of form and process of the platform and cliff demonstrate the influence of the detailed lithological variations across the platform and the role of platform morphology in affecting the direction and nature of wave attack upon cliffs. Because of the detailed studies, this site provides a reference site against which other platforms can be compared.

PORTH NEIGWL, GWYNEDD (SH 270 274)

V.J. May

Introduction

The coastline of Llŷn (the Lleyn Peninsula, north-west Wales) is characterized by both hard rock and weaker till cliffs, and a number of distinctive beaches. Between the mouth of the Afon Glaslyn and the Menai Strait, there are some 18 sand, shingle and cobble beaches that are bounded by rocky headlands. Those along the south-eastern coast are the best-developed set of zeta-curve beaches associated with strong wave refraction on the coastline of England and Wales, but almost all are affected by coastal protection works. Aberdaron Bay (about 6 km to the west) and Porth Neigwl (Hell's Mouth) are more symmetrical in form than the others, Porth Neigwl in particular facing almost directly into the dominant south-westerly Atlantic swell (see Figure 4.1 for general location and Figure 4.39). It is a rare example of a cliff–beach system on the coast of Great Britain confined by long headlands where waves and swell are little affected by refraction (Guilcher, 1958). As a result, the predominantly till cliffs have developed a plan-form

Soft-rock cliffs

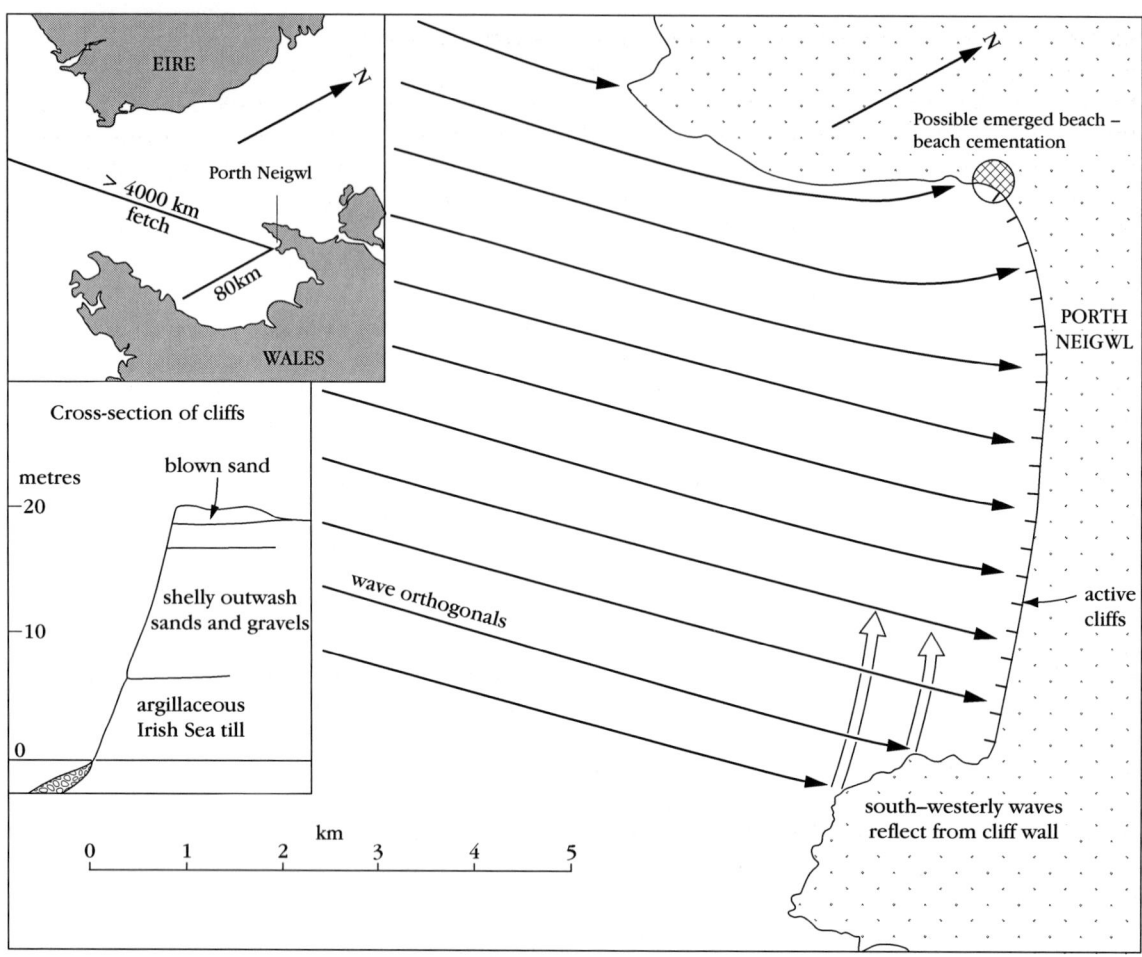

Figure 4.39 Wave refraction and reflection in Porth Neigwl. Wave orthogonals show the direction of travel of waves and are drawn at right angles to the wave crest. Open arrows are also orthogonals for reflected waves.

that is controlled strongly by the dynamic relationship between south-westerly swell and waves, and the strength of the till (Figure 4.39). The narrow beach is usually subdivided into a lower beach, formed mainly of sand, and an upper beach dominated by cobbles and boulders. A common feature of the beach is a series of bars aligned at an acute angle to the beach itself. Beach cusps are also a characteristic feature. The beach is unusual amongst British west-coast beaches in having no associated dunes (Steers, 1946a). Some controversy surrounds a possible emerged ('raised') beach at the western end of the locality (Whittow, 1957, 1960, 1965; West, 1972; Campbell and Bowen, 1989), where there may be present-day cementation of the beach (West, 1972).

Description

The Porth Neigwl (or 'Hell's Mouth') coastline is about 11 km in length and comprises three main elements; (a) a narrow beach, below (b) cliffs of glacial sediments, which lie between (c) cliffs of Cambrian and Ordovician bedrock.

The western side of the bay is formed by cliffs up to 60 m in height, cut partially into sandstones and partially into glacial sediments that rest upon the bedrock slope. The cliff runs SW–NE; this area is the most sheltered part of the bay. There is a narrow cobble and boulder beach. The main beach faces southwards at its western end but gradually curves to face southwest at its eastern end. The cliffs are over 30 m in height at its western end, but are more usually about 18 m high. Between SH 276 268 and

SH 283 263 they are only 10 m in height. The cliffs are cut mainly into thick blue-grey and brown Irish Sea till, but east of SH 283 263, there is a higher proportion of gravels in the cliff. Holocene peat and sands are also exposed in the cliffs (Campbell and Bowen, 1989).

Along the eastern side of the bay, the cliffs are cut mainly in Cambrian Hell's Mouth Grits, and attain a height of over 110 m. The lower parts of the cliffs are almost vertical, but, owing to the strata dipping at between 30° and 45° into the cliffs, only limited development of very narrow shore platforms has occurred. There is little evidence of active erosion in the cliffs of this part of the bay. Rockfalls are infrequent and small in magnitude. The till cliffs that form the central part of the bay are, by comparison, easily eroded and have retreated rapidly, undermining cliff-top tracks and fields.

The beach itself is formed by an upper berm composed mainly of cobbles derived from the erosion of the cliffs, and a lower finer-grained beach, which has a maximum width of about 100 m. Two regular features of the beach are well-developed cusps and small bars on the lower beach, which are aligned sub-parallel to the beach itself. They normally merge with the beach at their western end, and they disappear and reform over time depending upon the wave conditions.

Waves usually approach the beach from the south-west because of the effect of the restricting headlands, but the fetch varies between over 4000 km to the south-west to much shorter distances to the south and south-east (80 and 50 km respectively; Figure 4.39). Waves approaching from these directions are less strongly refracted than those of the long Atlantic swell, but there is some sheltering of the eastern corner of the bay under south-easterly conditions. The western corner of the bay, in contrast, is very exposed to the south–east waves and by refracted south-westerly waves. Despite these modifications of wave approach, wave energy appears to be often spread evenly along the whole beach and the similar plan of both the cliffs and the beach reflects this.

Interpretation

Guilcher (1958) described Llŷn, and Porth Neigwl in particular, as one of several examples where the coastline has become irregular as a result of the exhumation of the underlying rock surface from beneath a cover of drift. The broad outline of the bay results from the rapid retreat of the glacial infill between the two headlands to east and west, but there is no direct relationship between cliff height and cliff retreat along the glacial cliffs. The rapidity of erosion is such that cliff-top streams have been unable to keep pace with the rate of retreat and so hanging, truncated, valleys into which streams have incised their lower courses have developed (for example at SH 269 274). Similar features have been described elsewhere in the present volume (see South-west Isle of Wight and the Dorset Coast GCR site reports), and there is considerable debate about their origins (Flint, 1982). Unlike the truncated valleys in the area around Hartland Quay (see GCR site report), the valleys at Porth Neigwl are most akin to those of the south-west Isle of Wight and tesitify to the local rapidity of cliff retreat. This contrasts with the evidence discussed below concerning cementation of the beach.

The detail of the bay results mainly from the longshore movements of sediments within a *swash*-aligned system (see Chapter 5). Furthermore, water movements are strongly constrained by the confining headlands. Waves are little affected by refraction within the bay except along the foot of the two headlands, but reflection from the headland cliffs, especially to the east, may produce complex wave patterns.

The beach is notable for the very common occurrence of large beach cusps along its length. Cusp development has been attributed to edge waves, in which cusp spacing is related to interactions between the edge-waves and the incident waves (Darbyshire, 1982). The regularity of beach cusps has been described by many coastal scientists (see for example Komar, 1976; Pethick, 1984). Bowen and Inman (1969) suggested that the rhythmic beat of the incoming waves on the water of the near-shore zone creates a secondary set of waves at right angles to the incoming waves. The combination of incoming waves and edge waves produces a regular series of undulations in wave height along the beach. The resultant differences in wave-energy distribution produce the regularly-spaced cusps.

In Porth Neigwl, the lack of variation of the direction of wave approach means that cusps are likely to be broken down or change their wavelength mainly as a consequence of variations in the period of incoming waves rather than any directional change. They are, as a result, a common characteristic form on this high-energy beach. However, reflection from the eastern wall of the bay also produces waves that travel obliquely across the bay at regular intervals. Waves travel into Porth Neigwl from comparatively deep water, in contrast to many other similar beaches which are related to south-westerly swell (see, for example, GCR reports for Carmarthen Bay and the English Channel sites such as South-west Isle of Wight and Loe Bar). The site provides an excellent field-study site for future research into the effects of interference of reflected waves with incoming and edge waves. Because of the limited refraction and deep water close inshore, it also provides a good site for investigation of wave behaviour and beach and cliff responses to rapid sea-level rise.

Porth Neigwl contrasts with the other beaches of Cardigan Bay in lacking any significant development of cliff-top dunes. Some swash-aligned beach–cliff systems develop small cliff-top dunes that are maintained by wind transport from both the beach and the cliff-face. There is no evidence of this process here. The rate of cliff erosion and the narrowness of the beach inhibit intertidal drying and so wind action is insignificant. There is also no evidence of an offshore source of beach sediment. It is one of the best examples in England and Wales of a high-energy beach with local sediment feed. Although it has a similar wave regime to the south-west facing flint shingle beaches of the English Channel, it differs from them in having deep water close to the shore and in depending entirely on erosion of the cliffs for its sediment supply.

In one other aspect, this site is unusual. It is the site where cementation of beach materials was thought first to have occurred in an emerged beach situation (Whittow, 1960), but was later shown to be part of the present-day beach (West, 1972). Whittow (1957, 1960, 1965) described a shelly conglomerate as a postglacial (Holocene) emerged beach, although he recognized that a wave-cut notch could not be seen because of the masking effect of landslips in the till that formed the cliff above the site. He also noted that coastal shelly drifts terminated inland at a height of about 3 m against a steep rock cliff, which he suggested might represent the old sea cliff of the Great Interglacial (Hoxnian) emerged beach. West (1972) demonstrated that inorganic calcite had been deposited in the western part of the beach, but gave no evidence of it being a Holocene emerged beach. Campbell and Bowen (1989) accepted this interpretation. It is unusual to find present-day beach sediments cemented in this way, but there is evidence from elsewhere (for example, east of Dover) that it can take place beneath debris from cliff falls.

Whittow (1965) also suggested that the presence of the emerged beach indicated that the till cliffs could not have retreated more than about 800 m since the end of the transgression about 6000 years BP. There are, as yet, no ^{14}C or amino-acid dates for the cemented material. The presence of hanging and incised valleys indicates that the rate of retreat of the cliffs has been faster than the rate of down-cutting, but this does not provide evidence of either the rate of retreat or the magnitude and frequency of retreat events. West's (1972) re-interpretation of the cemented material as contemporary suggests that retreat has taken place at marked intervals, for sufficient time must have passed without disturbance of the beach to allow cementation to take place. In this respect, the site poses interesting and as yet unanswered questions about the nature, magnitude and frequency of retreat of the till cliffs in this very high-energy environment. There is no other site on the coastline of England and Wales where contemporary cementation has been reported in a comparable location below cliffs. For this reason alone, the site is of considerable scientific importance.

Conclusions

Porth Neigwl is a rare example of a cliff–beach system, confined by long headlands, in which waves and swell approach from deep water and are little affected by refraction. Porth Neigwl is one of the few beaches on the coastline of England and Wales where waves travel in deep water sufficiently close to the shoreline to be little affected by refraction.

The till cliffs are retreating rapidly, but despite ample supplies of coarse sediment, the beach remains very narrow. This locality is an excellent example of a very high-energy environment that lacks intertidal platforms (contrast with Nash Point, for example – see GCR site report in the present chapter). It is also of considerable interest because it is probably the only recorded site of possible contemporary cementation of a cliff-foot beach in England and Wales.

Holderness

HOLDERNESS, EAST YORKSHIRE (TA 182 660–TA 142 190)
POTENTIAL GCR SITE

K.M. Clayton

Introduction

Cliffs cut into weak Quaternary rocks undergoing rapid erosion occur along the North Sea coast of Britain and locally around the Irish Sea. The Holderness Cliffs (see Figure 4.1 for general location) stretch from Bridlington in the north some 60 km to Kilnsea in the south, where the coast continues southwards as a spit to Spurn Head (Figure 4.40). Most of this line of cliffs remains undefended, though walls and groynes have been built along relatively short frontages at Bridlington, Skipsea, Hornsea and Withernsea, and more recently at Mappleton (south of Hornsea) and in front of the gas terminal site at Easington close to Kilnsea. The contemporary rate of erosion increases from north to south; from less than 0.5 m a^{-1} just south of Bridlington to as much as 3 m a^{-1} at Easington. A feature of this coast is the sectors with an unusually low beach profile; these are locally known as 'ords' and over time they migrate southwards down the coast. As the ords pass by, waves are able to erode the cliffs more effectively and the rate of erosion speeds up, to slow down again when a higher and wider beach replaces the ord (Figure 2.1c).

Description

The Holderness cliffs front an undulating till plain deposited during the last (Devensian) glaciation. The cliffs themselves cut through various till facies and related fluvio-glacial gravels.

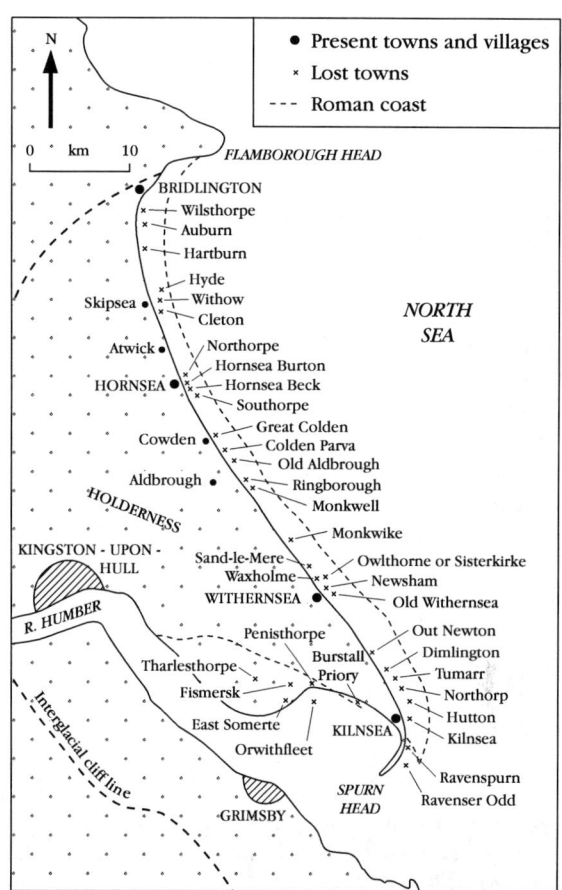

Figure 4.40 Lost villages of the Holderness coast. As the till has been easily eroded for hundreds of years at rates of 2 m a^{-1}, there has been substantial loss of agricultural land and villages. (After Hansom, 1988)

They begin in the north at the exposed Sewerby cliff of Ipswichian age, where the Chalk cliffs of Flamborough Head end, and continue southward for 61.5 km to Kilnsea at the northern end of the Spurn Head spit. The average height is

Table 4.4 Land-loss by natural sections of the Holderness coast, 1852–1952 (Valentin, 1954, 1971).

Section	Annual cliff recession (m)	Shore length (m)	Annual land-loss (m^2)	Average cliff height (m)	Annual loss in volume (m^3)
A. Sewerby to Earl's Dike	0.29	8100	2357	11.0	25 927
B. Earl's Dike to Hornsea	1.10	13 650	15 015	11.8	177 177
C. Hornsea to Withernsea	1.12	24 250	27 160	16.2	439 992
D. Withernsea to Kilnsea Warren	1.75	15 525	27 200	13.2	359 040
Entire coast (approx.)	1.20	61 500	72 000	14.0	1 000 000

about 13.5 m, and the maximum cliff height 35 m. The highest cliffs are between Atwick and Roos, with a secondary peak at Holmpton, south of Withernsea. From Barmston, some 8 km south of Bridlington to the southern limit of the cliffs at Kilnsea, the −10 m submarine contour runs parallel to the coastline and some 600 m offshore.

Given the rapid rate of erosion along these cliffs and the consequent loss of land and buildings, accounts of the coastal erosion can be traced back over several centuries, although only over the last 150 years can these be regarded as reliable. Reid (1885) and Sheppard (1906) summarized knowledge at the time they wrote, while the Royal Commission on Coastal Erosion and Afforestation (1907–1911) early in the 20th century referred to the relentless erosion of this cliffline, regarding it as 'the most serious around the coast of the British Isles'. The first account that deals comprehensively with the full length of the cliffs is by Valentin (1954). An independent account that was read to a Yorkshire Geological Society meeting in York a year earlier is Dossor (1955), but Valentin's paper is more detailed and far more widely quoted.

Valentin's account summarizes land-loss by rate of cliff recession, by area and volume (the last two on a parish basis). He divided the coast into four sections: Sewerby to Earl's Dike; Earl's Dike to Hornsea; Hornsea to Withernsea; and Withernsea to Kilnsea Warren. As Valentin's 1952 research post-dates the first edition of the six-inch map (1:10 560) by 100 years, he was able to summarize the pattern of erosion over a century (Table 4.4, Figure 4.41)

Valentin went on to discuss the reasons for the persistent cliff recession, concluding that wave attack was the dominant factor. Increasing exposure southwards (the northern sector being protected from the north by Flamborough Head) was thought to account for the steady increase in the rate of cliff recession from north to south.

Phillips (1962, 1964) described the ords and their relationship to coastal recession. The occurrence of these low beach sectors is of considerable importance in determining the local pattern of cliff recession, and their episodic migration down the coast eventually leads to their progression down the narrow spit at Spurn, threatening breaching as they pass southwards. More detailed studies of ord development and migration are to be found in Scott (1976).

The type of cliff failure varies with the lithology of the cliff materials, cliff height, the local water table, and rate of recession. These issues are addressed by Richards and Lorriman (1987), and from a soil mechanics standpoint in Robertson (1990), while the pattern in time and space is analysed by Pethick (1996). Most, though not all, authors link the passage of ords with increased rates of cliff recession.

Winkelmolen (1978) concluded from a study of lithological variations in samples collected from the cliffs, beach and offshore zone that the postulated north–south longshore drift of beach sediment could not be established, and concluded as a result that most erosion and sediment sorting was associated with easterly storms.

Short lengths of the cliffs in front of built-up areas have been protected for some 80 years, with groynes and sea-walls at Bridlington (4 km), Hornsea (2 km) and Withernsea (1.8 km) and as recently as 1992 a short length at Mappleton south of Hornsea was added. Many proposals have been put forward to increase the length of protection along this coast, including the construction of an offshore barrier utilizing colliery waste and a proposal for defended headlands separating eroding bays. To date the defences have remained restricted to the small coastal towns as a result of national policies on the funding of coast-protection structures that result in the limitation of protection to built-up areas through cost–benefit tests (see also Ramsay et al., 1977).

A general account, setting the Holderness coast in its wider setting, is provided by Pethick and Leggett (1993).

Interpretation

There is widespread (though not universal) agreement on the major controls operating along this coast; the importance of the protection provided from the north by Flamborough Head (and the Smithic sandbank offshore), and the role of the beach (including the progression of ords) in controlling the rate of cliff recession. Recent studies have tended largely to confirm these features, while increasing the detail in which we understand the controlling coastal processes. However, the contribution from Winkelmolen (1978) shows that the field data are capable of alternative interpretations. His evidence does not preclude southerly littoral transport as surely as he claims – the effect of sporadic (in time and place) inputs of ill-sorted sediment derived from the rapid erosion associated with ords makes it unlikely that a north-south pattern of sediment size could develop. Pethick (1996) has provided a more detailed examination of the pattern in time and space of cliff recession, which is discussed below.

All authors agree that the recession along this coast has been occurring for a long period of time, no doubt moving landward rapidly during the earlier part of the Holocene transgression, and perhaps continuing at a relatively constant rate over the last 6000 years or so that sea level has stood close to its present elevation. Valentin utilized the recession rate for 1852–1952 to estimate that, 3000 years BP, the coast was some 7 km east of its present position at Dimlington, and perhaps half that distance at Skipsea. He also noted that the morainic ridge that forms the high ground at Dimlington declines westwards, and that while it reached 42 m on the 1852 map, the ridge top was at only 38 m at the cliff edge in 1952 and will be at only 30 m by 2052. Thus at the very least the present rate of erosion here is likely to continue and a much smaller volume of sediment will be contributed to the beach and offshore system in the future. Valentin notes that Sheppard (1912) regarded the Roman coastline (c. 2000 years BP) as lying 2.5 to 3.5 km east of the present-day cliffs, a view repeated by other authors since then.

The recognition of ords by Phillips (1962, 1964) has undoubtedly helped to explain variations in both the style and rate of cliff recession over time. Research on the Spurn Head spit showed the effects of an ord to be particularly noticeable. Similar features occurred in front of the cliffs farther north. Phillips reported their length as 45–55 m and that they were moved southwards by severe storms at an average rate of movement of about 1.6 km a^{-1}. She also noted that ords are unknown in Bridlington Bay (where a ridge and runnel beach occurs), but become common towards Hornsea, some 15 km south, explaining this by the southward increase in exposure to northerly waves. Phillips (1964) describes ords as a departure from a perceived normal beach form with a high upper berm and a lower beach ridge; where ords occur, the upper beach is missing.

It would seem equally legitimate, given the persistent erosion, to describe the low beach as the norm and the sectors with slower cliff recession where the beach is high as anomalous. However, Phillips' viewpoint was supported by Scott (1976) who studied ords in 1974–5 and noted that they had first been described (as 'hords') by Thompson (1824). In contrast to Phillips' 45–55 m, Scott found their average length to be 1–2 km, though they varied considerably. Approaching an ord from the north, the upper beach becomes lower and narrower, ultimately merging into the cliff foot beach. The lowest point of the ord has till exposed across the width of the beach. He noted that ords first appear between Skipsea and Hornsea, and are then found at intervals all the way south to Spurn Head. From aerial photograph analysis he identified ten ords along the coast in 1961, eight in 1966 and nine in 1972. The detailed field study of individual ords in 1974–5 established that though they might remain stationary for long periods, they never move to the north and in the long-term all of them moved southwards. Motion was by filling-in by southward-moving beach sediment at the northern end and the moving away of the higher beach at the southern limit of the ord by storm-wave induced longshore transport. Pringle (1981, 1985) reviewed the movement of an ord, first located in 1969, 250 m south of Withernsea, monitored between 1973 and 1976 (Pringle, 1985), and surveyed every six months from April 1977 to April 1983; it moved between 68 and 668 m in six months (with an annual average of 496 m) to the south. Where the ord occurred, the beach level was on average 3.9 m lower than elsewhere, allowing even neap tides to reach the cliff at high water. Cliff-top erosion of 10 m a^{-1} was associated with the passage of the ord (with a maximum of 15 m a^{-1}) and the site of maximum cliff recession moved southward as the ord passed by. The total till volume eroded from within the lengths affected by the migrating ord averaged 254 000 m^3 a^{-1} and from the inter-ord areas, 55 600 m^3 a^{-1}, despite the fact that the total length of inter-ord cliffs was nearly three times the length affected by the ord.

Description of the style of cliff failure and its development into a model of recession with active basal erosion is provided by Richards and Lorriman (1987). This includes a section based on cliff recession in Holderness: the process involves (a) toe erosion by marine action which both undercuts and steepens intact clay and removes failure deposits; and (b) a range of mass movement mechanisms including relative-

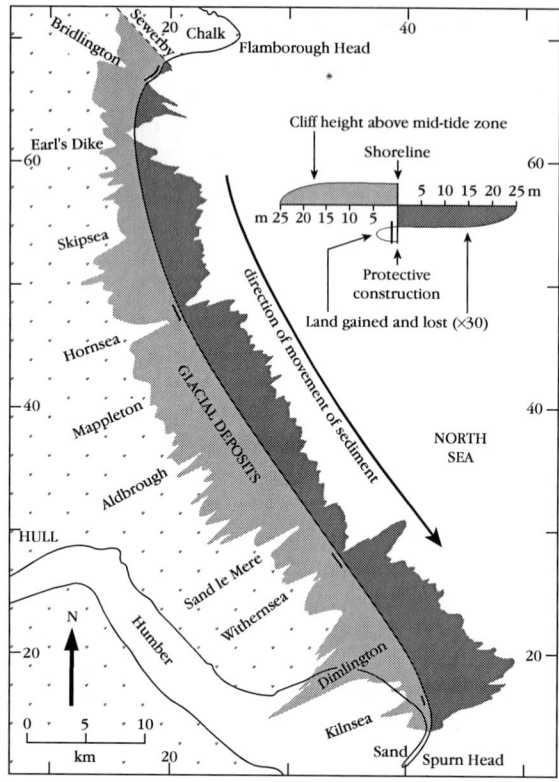

Figure 4.41 The relationship between cliff height and erosion along the Holderness coast. (After Valentin, 1971, in Steers, 1971a). For the cliff height profile, the vertical exaggeration is × 30.

ly deep-seated wedge and rotational failures, slumps, spalling and superficial mudflows. As in the case of the Norfolk cliffs (Cambers, 1976), there is a strong relationship between cliff recession rates and the frequency with which high tides reach the cliff base. The rate of recession is increased by the current rise in relative sea level, estimated as 2–3 mm a^{-1} (Suthons, 1963).

The changing nature of cliff recession as ords move along the coast was investigated by Robertson (1990). Although the cliffs of Holderness are appreciably lower than those of Norfolk, his description of the processes at work is very similar to the styles of recession observed in the Norfolk cliffs (Cambers, 1976). He found cliff height and beach volume to be the major controls on cliff failure, a relationship also identified by Valentin (1971; Figure 4.41). Cliff slope was generally 40–50°, because of the generally low pore-water pressures. The main method of recession is by deep-seated failure caused by weakening of the till by stress relief as a result of unloading and the removal of basal sediment by wave attack. Where the beach is high (between ords), recession is by mudslides and shallow slips. When a large landslip occurs, time is needed for it to be removed completely by wave action before further recession will occur at that point.

The changing nature of the beach fronting the Holderness cliffs was investigated by Mason and Hansom (1989). Using time-series surveys of a small stretch of coast at Atwick the occurrence and disappearance of areas of beach stripped of sediment (ords) was used to predict beach behaviour over different wave conditions and seasons. Using the Holderness beaches, they demonstrated that a Markov model was capable of describing and predicting transitions between beach types and different time periods.

Valentin's approximation of the sediment eroded from the Holderness cliffs as 1 000 000 m^3 a^{-1} has been refined and developed by later studies. Pringle (1985) noted that the average proportion of coarser sediment (sand and gravel) for the three tills present near Withernsea was 31.3%. A detailed field study of a 4 km length between Skipsea and Atwick by Mason and Hansom (1988) identified beach areas stripped of sediment and quantified changes in beach volume as well as inputs from cliff erosion over one year. They estimated an annual output (1850–1968) of 1 340 000 m^3 a^{-1}, and an input to the beach of 462 000 m^3 a^{-1} from the cliffs on the assumption that 33% of the eroded till was sand and gravel. Southward sediment movement by longshore drift in their sectors averaged 28 000 m^3 a^{-1} to the south. Offshore transfer from the two central beach sectors, each 1875 m long, averaged over 50 000 m^3 a^{-1}, an offshore proportion of two-thirds.

One outcome of Pethick's (1996) study was his estimate of 340 000 m^3 a^{-1} for the net annual potential longshore transport to the south. He estimated the non-cohesive sediment input at 280 000 m^3 a^{-1}, giving an excess potential longshore rate at the southern end of the system. He concluded that the orientation of the Holderness coast provides the maximum possible potential longshore sediment transport, and an excess of potential over actual that increases steadily to the north. He estimated the total sediment store in the Holderness beaches as about 2 000 000 m^3, the equivalent of only eight years of input from cliff recession.

Holderness

An informative analysis of the variation of coastal cliff recession in time and space was provided by Cambers (1976) based on her work on the Norfolk cliffs. Pethick reached similar conclusions in his 1996 study, using the recession data collected by the local councils from the 'erosion posts' located every 500 m along this coast. Though installed in 1951, annual measurements run from 1957 and the database available to Pethick (1996) was thus 45 years. Given the limitations of this database he initiated the collection of more detailed measurements from 1993. Pethick concludes that apparent variations in cliff recession rates are better explained by the spacing of the measurements in time and space than by the progression of ords. While accepting that ords are linked to more rapid recession, he does not believe they explain the 5–8 year periodicity he found and that relatively small landslips produce a pattern of migrating embayments separated by inclined promontories, and that the southward migration of these embayments leads to the observed periodicity. At any one site, 2.7 cliff failures each leading to average cliff-top recession of 0.68 m yields the overall average recession of 1.82 m a^{-1}.

Conclusions

Holderness is the longest and least-defended length of rapidly eroding cliffs cut into weak sediments in Britain. It has been studied intermittently over the past 150 years or more and although in general the relationships between wave energy, beach volume and recession are understood, and the style of cliff failure has been related to the strength and pore-water pressures of the Quaternary sediments, there is no doubt much still to be learnt from such a natural and extensive site. In addition, little is so far known of the fate of the eroded sediment, or the processes which, by deepening the adjacent sea floor, have allowed recession to continue over a distance of as much as 10 km over the last 5000 years.

The short defended lengths of this coast produce sectors where little or no erosion has occurred since construction of the defences. However, immediately to the south of each defended section, accelerated erosion occurs where the beach is depleted because of the retention of sediment by groynes to the north and the lack of cliff input behind sea-walls. These limited interruptions to an almost continuous length of eroding cliffs totalling almost 60 km do not detract from the value of this site; the comparable cliffs of Norfolk are defended for all but a few kilometres. The largely rural nature of Holderness has made it difficult to justify the expenditure of public money on coastal defences and this is likely to remain the case for the foreseeable future. Further, the erosion of the cliffs provides the sediment that maintains the Spurn Head GCR site, and in addition are the likely source for sand arriving on the prograding North Norfolk Coast GCR site.

Chapter 5

Beaches, spits, barriers and dunes – an introduction

E.C.F. Bird, K.M. Clayton and J.D. Hansom

Introduction

INTRODUCTION

The coasts of Britain support a wide variety of beach types and materials (Table 5.1). Some are sandy, but many include coarser material (granules, gravels, cobbles and occasionally boulders), often forming a steep upper beach behind a flat, sandy foreshore exposed at low tide, such as on the eastern shores of Dungeness, Kent (see GCR site report in Chapter 6) and at Ynyslas in Wales (Chapter 8). Others, such as Chesil Beach, Dorset and Spey Bay in Moray (Chapter 6), consist entirely of well-rounded beach gravel, in Britain often called 'shingle'. Shelly deposits, with the shells either intact or broken, or comminuted to shell grit and calcareous sand, also occur on British beaches, particularly on the Atlantic coasts, where extensive dune systems and the Scottish machair have formed from sediment derived from calcareous beaches. Chapter 7 of the present volume is devoted to sandy beach and dune sites, and Chapter 9 to machair beaches and dunes.

There have been beaches ever since the oceans first formed and coasts began to take shape along the margins of the land. Beach deposits are found in sedimentary formations of various geological ages, as in the Tertiary sediments of the London and Hampshire Basins. During the Quaternary Period, the 'emerged' or 'raised' beaches around Britain formed when the sea level was higher relative to the land, and have emerged as the result of land uplift, or a lowering of sea level, or some combination of the two processes. Beaches that formed on coastlines during low sea-level stages are now submerged on the sea floor, but few traces of these persist because of wave reworking or concealment by later sea-floor sediments. The present-day existing beaches began to form on the British coast about 6500 years BP, when the Holocene marine transgression (also known as the 'Flandrian' or 'Late Quaternary' marine transgression) brought the sea to a level where wave action shaped the present coastline, cutting back some parts and depositing sediment on others.

Beaches are also found along cliffed coasts – except where the cliffs plunge into deep water, or where the shore is too rocky and rugged to have retained a beach – but most occur on low-lying coasts, except where wave energy is weak and the shore has become marshy and muddy. Beaches can be regarded as occupying coastal compartments or sediment cells delimited by headlands that prevent longshore sediment movement where they extend into deep water (e.g. Portland Bill), and further restricted and subdivided by lesser promontories (e.g. Hengistbury Head) past which sand and gravel drift, particularly during stormy periods (Bray et al., 1995).

Beaches are characterized by accumulation of sediments that extend from the point at which wave-accumulated sediments first apear (the lower limit of wave activity) to the upper limit of wave activity. Operationally, the intertidal zone co-incides with the visible beach, but depositional beach forms extend below this level. The intertidal zone is often characterized by a series of ridges and troughs of sediment culminating in a beach face affected by the uprush/swash and backwash of waves (see Figures 5.1 and 5.2).

Beaches may protect the land that lies behind

Table 5.1 Classification of beach structures based on their plan form (after Pethick, 1984); outline definitions are provided in the glossary of the present volume.

Rhythmic beach morphology	Cusps
	Crescentic bars
	Cell circulation topography
Shoreline beaches	Pocket beaches – swash-aligned (Davies, 1980)
	Open beaches – drift-aligned (Davies, 1980)
	Zeta-form or fish-hook beaches (Silvester, 1960; Swift, 1976)
	Combined swash and drift alignment
Detached beaches	Spits
	Cuspate forelands, nesses and tombolos
	Barrier beaches and islands

Beaches, spits, barriers and dunes

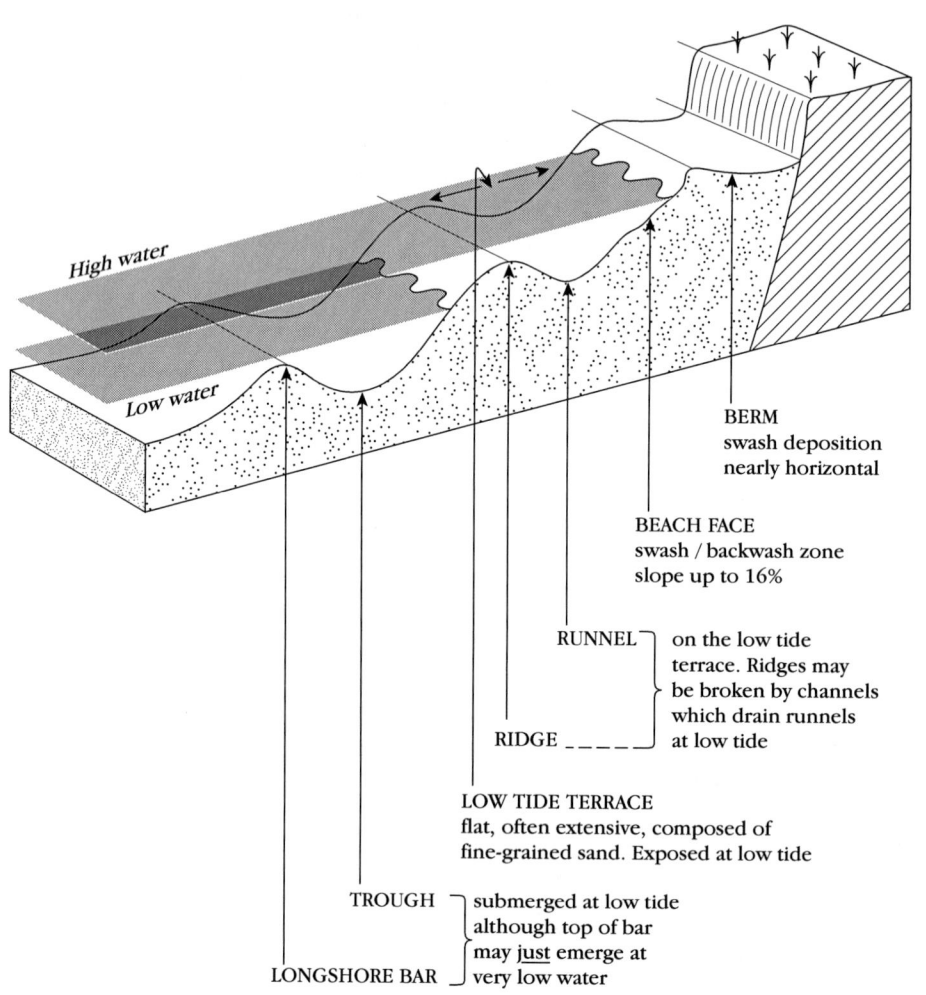

Figure 5.1 Beach morphology. Synonyms: The term 'ridge-and-runnel' is sometimes used for 'bar and trough'; 'ball and low' is the old name for 'bar and trough'; 'bar', 'offshore bar' etc., are old names for barrier islands, not to be confused with *longshore bar*; 'swash bar' is the old name for 'berm'; 'high-tide beach' is used for 'beach face'; 'low-tide beach' is used for the seaward edge of low-tide terrace. See also Figure 5.2. (After Pethick, 1984, p. 93.)

them from erosion by waves. Gravel (shingle) beaches are permeable, and absorb or reflect much wave energy, but where the sand supply has been sufficient to form wide sandy beaches with a very low transverse gradient, wave energy is dissipated (Figures 5.3 and 5.4). Many gravel beaches are on parts of the coast that receive high wave energy from occasional storm waves, but wave energy is also high on sandy beaches exposed to Atlantic swell and storm waves. The distribution of gravel (shingle) beaches and sandy beaches depends on the nature and sources of available beach material and the patterns of waves and currents that have delivered it to the coast (Figure 5.4).

PROVENANCE OF BEACH SEDIMENTS

The main sources of beach material are cliffs and rocky shores that are undergoing erosion, rivers that carry sediment down to the coast, particularly during floods, and the sea floor, although the relative importance of each source varies geographically (Figure 5.5). Some beach sediments are similar to those in rock formations exposed in the nearby cliffs and shore outcrops from which they have been derived. Beaches occupying coves on the south-west coast of England have a mineralogical composition indicating that they have been derived from nearby cliffs and coastal slopes (Stuart and Simpson,

Provenance of beach sediments

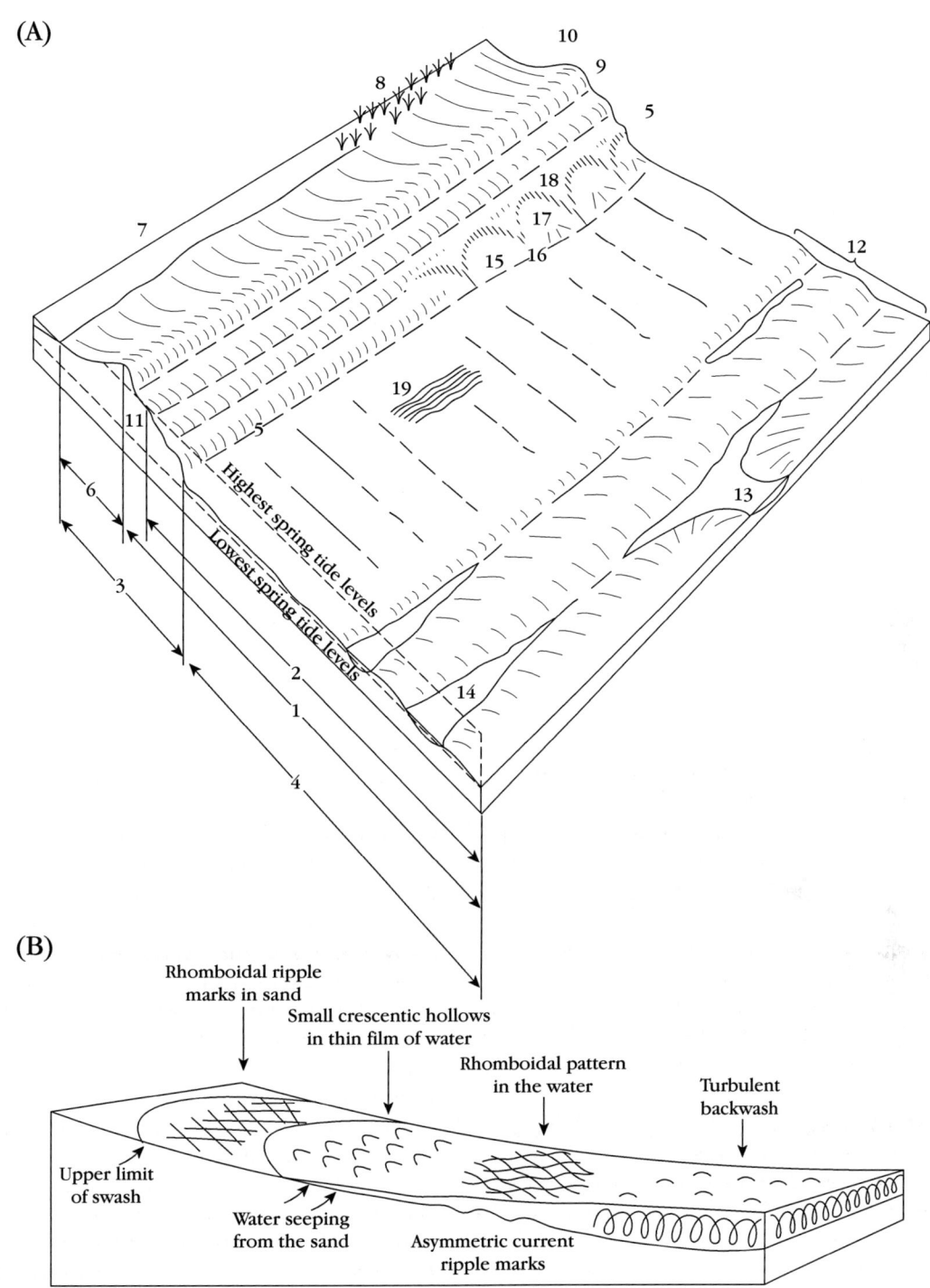

Figure 5.2 (A) Beach terminology: (1) beach, (2) shore, (3) upper beach (cordon littoral), (4) foreshore, (5) break of slope between upper beach and foreshore, (6) inner side of beach ridge, (7) lagoon, (8) marsh, (9) berms, (10) storm beach, (11) coastline, (12) ridges and runnels on the foreshore, (13) channel on foreshore, (14) pool in runnel of foreshore, (15) beach cusp, (16) apex of cusp, (17) bay of cusp, (18) horn of cusp, (19) ripple marks. (B) Formation of rhomboidal ripple marks. (After Fairbridge, 1968, p. 67.)

Beaches, spits, barriers and dunes

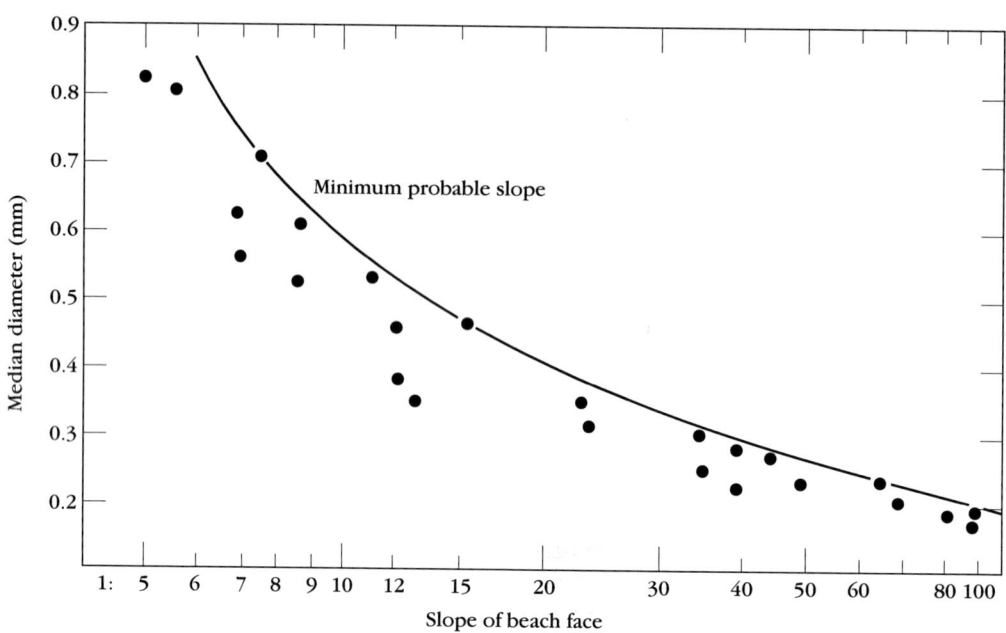

Figure 5.3 The relationship between mean grain size of sand and beach slope, (beach slope is given as a ratio, from 1:5 to 1:100). (After King, 1972a, p. 325.)

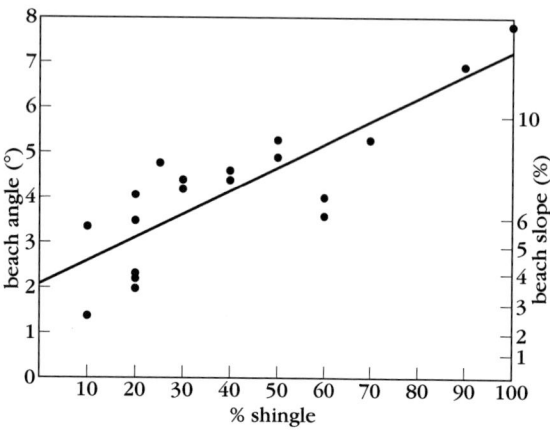

Figure 5.4 Beach steepness in East Anglia as a function of the proportion of shingle. The scatter of points is largely a function of variations in exposure to higher wave energies. (After Clayton, 1992, p. 64.)

1937) as have the pocket beaches at the foot of cliffs in Foula, Shetland. By contrast, many beaches on the south and east coasts of England contain quartz sand and flint gravel that have been either carried many kilometres along the coast by longshore drift, or brought onshore as sea levels rose. On the Moray coastline of Scotland, the present and emerged beaches of the Spey and Culbin (see GCR site reports in Chapters 6 and 11, respectively) beach system have been fed by longshore movement of glaciogenic sands and gravels during the Holocene Epoch.

Measurements have been made of the rates of sediment supply to beaches from cliff erosion and the volumes of beach material moving along the coast (Clayton, 1980, 1989b). In Lyme Bay, gravel from cliff-top deposits feed shingle beach compartments within which there is a predominant eastward drifting, such as between Lyme Regis and Golden Cap (see GCR site report in Chapter 4) (Bray, 1992). For the Norfolk cliffs, with an average height of about 20 m, a length of 40 km, and an average retreat rate just below 1 m a^{-1}, the natural input of sediment is estimated to be about 750 000 m^3 a^{-1} about a century ago of which some two-thirds is sand and gravel, the remainder mud (Cambers, 1973). However, with the gradual extension of coastal defences during the second half of the 20th century, this volume has fallen by 50%. Because more than half of the sand and gravel leaves the cliff system via the beach and is transported by longshore drift to Great Yarmouth, some 40 km farther down the coast, the reduction in output has resulted in reduced beach volumes downdrift, leading to erosion there. The persistence of cliffs along the south and east coasts of Britain where the rocks are weaker indicates that in general a combination of offshore removal and longshore transport downdrift is able to remove all of the sediment produced by cliff retreat.

Much attention has been given in the British coastal literature to beaches of gravel or shingle (sometimes called 'coarse clastic beaches') (e.g. Carter and Orford, 1984). Some shingle beaches are found where coastal rock outcrops have yielded debris of suitable size, such as fragments broken from thin, resistant rock layers. Others, such as at Whiteness Head and on the island of Jura in Scotland (see GCR site reports in Chapter 6) have been produced by the delivery of large amounts of glaciogenic gravels to the coast, as well as by erosion of cliffs and rocky shores cut into intricately fissured igneous or metamorphic rocks. At Nash Point in South Wales and east of Watchet in Somerset (see GCR site reports in Chapter 4) beaches of grey limestone gravel piled up at the base of the cliff have been derived from the weathering and dissection of Liassic limestone shore ledges and cliff outcrops. Flints released from layers or nodules in Chalk outcrops dominate the shingle beaches of southeast England between Beachy Head and Seaford, Kingsdown and Dover and on Thanet (see GCR site reports in Chapter 4). Flint cobbles and pebbles are black or blue (sometimes retaining white rinds) where they have been recently released from the Chalk (such as on the shores below Ballard Down in Dorset), but the brown flints that dominate shingle beaches in south-east England, notably at Dungeness and Chesil Beach (see GCR site reports in Chapter 6), have had a longer and more complicated history. They have been weathered (with oxidation of ferrous to ferric compounds, and some leaching of silica) during residence in various Tertiary and Quaternary gravel deposits on land and on the sea floor during low sea-level phases before they arrived on the present coastline. Flint cobbles are gradually reduced by attrition to pebbles and sand, comminuted flint sand being a major constituent of the brown beaches on the south-west coast of the Isle of Wight (see GCR site report in Chapter 4) (Bird, 1997).

In much of Great Britain, sand and gravel eroded from offshore deposits or cliffs cut in Pleistocene glacial drift, for example on the Holderness coast, glaciogenic gravels and sand have been supplied to local beaches and drifted south to the spit at Spurn Head (see GCR site report in Chapter 8). Similarly, at Porth Neigwl in north Wales, a beach of sand backed by pebbles and cobbles has been derived from erosion of cliffs cut in Pleistocene till and much of the sand and gravel in beaches in Robin Hood's Bay, Yorkshire (see GCR site reports in Chapter 4) has come from glacial drift deposits rather than from the underlying Liassic limestones and shales.

In south-west England beaches also include sand and gravel derived from periglacial deposits, the frost-shattered earthy rubble that mantles coastal slopes, such as at Tintagel (see GCR site report in Chapter 3), and quartzite pebbles from disintegrating outcrops of vein quartz in the Devonian rocks. Many beaches in Devon and Cornwall incorporate sand and gravel from Pleistocene beach deposits that now stand a few metres above high tide level: beaches that were buried by periglacial deposits, then exposed and dissected by marine erosion during and since the Holocene marine transgression.

Sand and gravel are also supplied to beaches by rivers on many coasts (see Figure 5.5), particularly where swift streams flow from inland mountain ranges, such as in Wales and Scotland. Many beaches have received sand or gravel from rivers, either directly, or where wave action has sorted and carried shoreward fluvial sediment first deposited off river mouths during floods. The Tyne and the Tees are among the rivers whose sediment has nourished beaches on the north-east coast of England, but the sandy structures in the estuaries of some rivers, such as the Tay, Eden and Forth also include shelly material that originated on the sea floor, moved in by waves and inflowing tides. In fact, much of the sandy sediment in the rias of south-west England has been swept in from the sea floor by wave action rather than deposited by rivers or eroded from cliffs. In Scotland, the Spey and Findhorn rivers (see GCR site report for Spey Bay in Chapter 6) continue to deliver gravel (largely derived from glaciogenic deposits) to beaches on the north-east coast of Scotland as they have done over the Holocene Epoch. However, those beaches on the coast between Nairn and Burghead also include large amounts of sediment carried shorewards from once extensive glaciogenic deposits on the sea floor.

On some coasts, beaches have received wind-blown sand from the backshore dunes, such as at Cheswick Sands, north of Holy Island in Northumberland. On the north coast of Cornwall dunes have spilled from Constantine Bay across Trevose Head and supplied sand to the beach in Harlyn Bay (Bird, 1998). A similar

Beaches, spits, barriers and dunes

process occurs at Balnakeil, in Sutherland (see GCR site report in Chapter 9), where winds drive sand eastwards from Balnakeil Bay over the peninsula of An Fharaid to cascade over a cliff and onto the shore at Flirum.

A great many beaches consist partly or wholly of sediment moved by waves from the sea floor during the Holocene marine transgression, and some still receive sea floor sediment. Sand and gravel that had been deposited by rivers, periglacial solifluction and melting glaciers on the emerged sea floor around Britain in Pleistocene times during low sea-level phases, together with weathered material from sea-floor rock outcrops, were reworked by waves and currents, and carried shoreward by wave action during the Holocene marine transgression to form beaches. This is considered to be the predominant source of sand and gravel on most Scottish beaches. This shoreward drifting is also the likely cause of a proto-gravel barrier at Chesil Beach, and contributed to the flint-dominated shingle beaches at Slapton Ley and Loe Bar near Porthleven in Cornwall (see GCR site reports in Chapter 6), which are now far from any shore sources of flint. The shingle at Slapton was carried onshore from a river terrace or beach gravel deposit on what is now the floor of Start Bay, and Reid and Flett (1907) suggested that the Loe Bar shingle came from Tertiary or Pleistocene deposits of flint gravel on the floor of Mount's Bay. Like many of Britain's shingle beaches these are now relict, no longer receiving gravel from the sea floor.

On some coasts wave action still moves sand and gravel from shallow sea-floor areas onto beaches. Johnson (1919) quoted John Murray's observation that shingle and chalk ballast gravel dumped by ships in water about 20 m deep 11 to 16 km off the north-east coast of England drifted onto the shore between Sunderland and Hartlepool. Various experiments have indicated that where the sea floor is gently sloping, sand can be moved shorewards by long ocean swell from a depth of up to 10 m (King, 1972a), and van Straaten (1959) found that waves moved sand onto beaches from a depth of 9 m. Shoreward drifting of sand and gravel is also indicated where shelly material (or other marine biogenic sediment, such as algae or foraminifera) are constituents of calcareous beaches, such as on the Atlantic coasts of Britain. In the Isles of Scilly (see GCR site report in Chapter 8), where cliffing is limited and sediment inflow from rivers is negligible, Barrow and Flett (1906) realized that the sand and gravel beach deposits had been carried in from the surrounding sea floor by wave action; they are calcareous beach sands with an admixture of quartzose sand from weathered granites on the sea floor. The same is true of many beaches on the Atlantic coasts of Britain, particularly the long curving white sandy beaches shaped by ocean swell on Sanday in Orkney and on the west coast of the Hebrides (see GCR site reports in chapters 8 and 9 respectively).

An extensive area of active coastal sand deposition occurs on the Northumbrian coast in the vicinity of Holy Island (see GCR site report in Chapter 11). Sand has been derived from shoals consisting of glaciofluvial drift deposits (including eskers), reworked and carried shoreward by wave action to form wide sandy beaches, generally backed by dunes that spread seawards as progradation continued. The prograding beach at Tentsmuir (see GCR site report in Chapter 7) in Fife has also been supplied with sand swept in from sea-floor shoals of glacial drift, but (at least in the northern part) also includes fluvial sand deposited off the mouth of the River Tay: the proportions of sand received from recent fluvial deposits and relict glacial deposits have not been determined. In recent decades the sandy shore at Holkham Bay in Norfolk has prograded as the result of inflow of sediment from the sea floor. Tràigh Mhór on the Hebridean island of Barra (see GCR site report for Eoligarry in Chapter 9) is a wide intertidal sandflat, which is a habitat for cockles *Cerastoderma edule*, and the adjacent beaches consist largely of in-washed cockle shells. Shelly beaches are also found on the Essex coast, such as at St Osyth Marsh (see GCR site report in Chapter 10), where there are no other sources of sand or gravel.

It is being recognized increasingly that the sediment sources of many British beaches are no longer as plentiful as they were earlier in the Holocene Epoch. In some cases this is because of recent coastal protection schemes that have reduced the sources of sediment from erosion, but in many others the reasons lie in an overall reduction of sediments sourced both from rivers and the seabed. Over Holocene times, the spread of vegetation resulted in river banks becoming more stabilized and fluvial sediment fluxes fell, a process that has recently been reinforced by artifical bank protection, so that rivers now contribute much less sediment to beaches than in earlier times. Similarly, glaciogenic seabed sediments were plentiful as the sea-level rise slowed at the end of the Holocene transgression about 6500 years ago. Since then a stable or only slowly rising sea level has resulted in the progressive reduction of offshore sediment volumes so that the seabed now supplies to the sediment budget only a small percentage of the previous amounts (Figure 5.5). Because of these reductions in the two major beach sediment sources, sediment fluxes to beaches have also reduced and consequently many beaches are now erosional (Hansom and Angus, 2001; Hansom, 2001). The implications of these reductions are described in the introduction to Chapter 9 since the machair dunelands of the Western Isles have been greatly affected by these changes in sediment economy.

Some beaches have received sediment from waste generated by coastal or hinterland mining and quarrying. In Cornwall the beaches at Par and Pentewan prograded during the past 200 years as the result of deposition of sand and gravel derived from mining waste brought down by rivers draining areas of tin and copper mining, and later china clay quarrying (Everard, 1962), and the beaches at Porthallow and Porthoustock have received in-washed gravel from quarry waste spilling over nearby cliffs (Bird, 1987). Beaches on the Durham coast have been augmented by the dumping of colliery waste (Carter, 1988). Near Workington in Cumbria there is a beach dominated by basic slag deposits from an old steelworks (Empsall, 1989), and beaches near ports may include pebbles from ships' ballast, such as at Charlestown in Cornwall and at Tentsmuir in Fife.

COASTAL SEDIMENT MOVEMENTS

Apart from the western coasts, where beaches are exposed to Atlantic Ocean swell and storm waves, much of the British coastline faces narrow or enclosed seas, and variations in fetch (the extent of water across which waves may be generated by winds) and in wind, wave and tide regimes result in coastal sediment fluxes alongshore. Beach sediment moves alongshore when waves arrive at an angle to the beach, producing oblique swash followed by orthogonal backwash. Drifting of sand and gravel along the shore under such conditions can be readily observed from patterns of beach accumulation

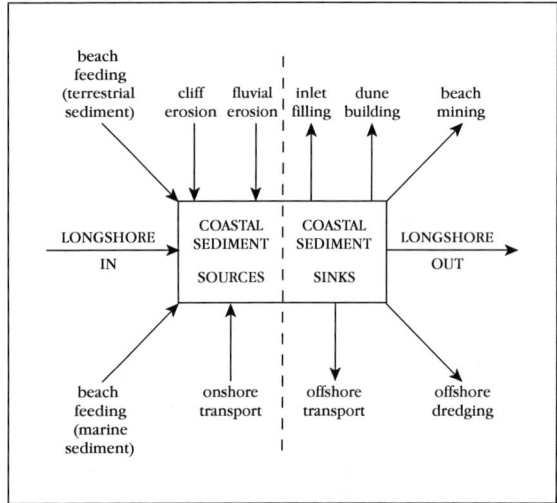

Figure 5.5 Sources and sinks of coastal sediment can be quantified to produce a sediment budget. Note the human element in the coastal sediment budget. (After Davies, 1980.)

against groynes, or followed with the use of tracers, when identifiable materials placed on the beach move with the drifting beach material (Jolliffe, 1961). Oblique waves also generate longshore currents strong enough to transport sand, and sometimes gravel, along the coast in the nearshore zone (Figure 5.6). Coastal features that indicate the net direction of longshore drift include deflection of river mouths, such as at the mouth of the River Spey in Moray, Scotland, accretion at breakwaters, such as at Newhaven in Sussex and the growth of spits, for example, at Culbin in Moray and at Calshot Castle in Southampton Water.

In recent years it has been recognized that longshore sediment movement is also affected by currents produced by forced resonance within the nearshore. Such 'edge-wave' activity interacts with the incoming waves to produce longshore currents within circulation cells that result in the formation of beach cusps and other rhythmic forms. The currents produced by edge-wave activity can co-exist with longshore currents produced by oblique wave approach and may control the net transport of sediment. Together, the longshore currents that result from both processes usually increase from the shore to reach a maximum just beyond the mid-surf position (where the contribution from oblique waves reaches a maximum), before declining rapidly to zero outside the breaker zone (where

Beaches, spits, barriers and dunes

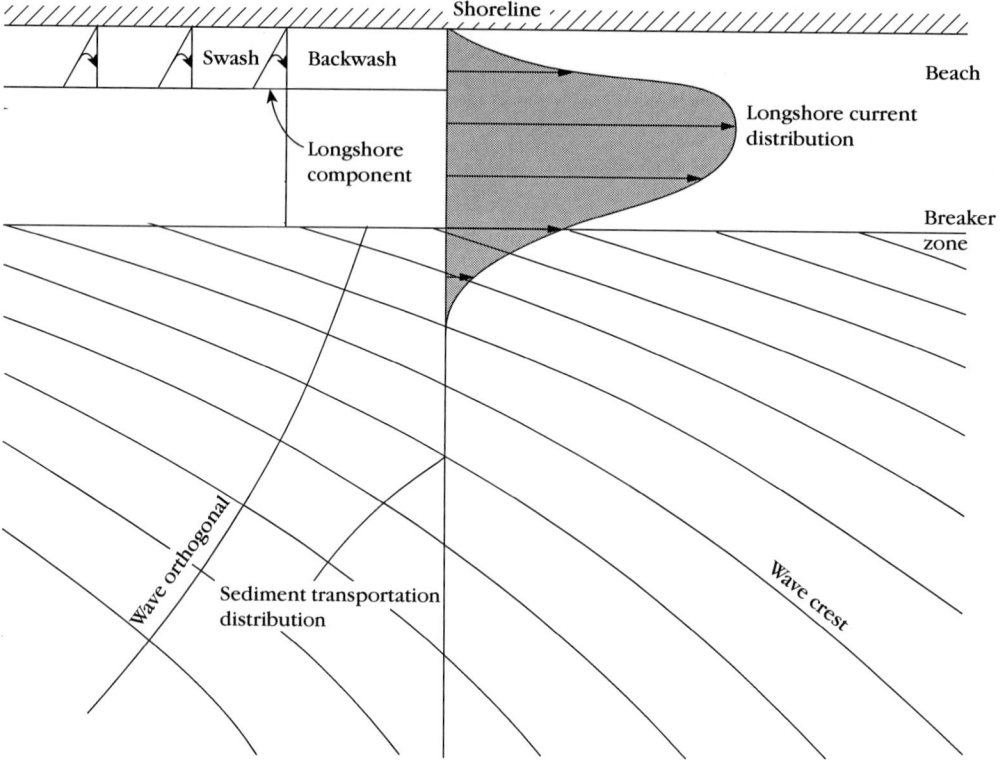

Figure 5.6 The run-up (oblique swash) and longshore current contributions to longshore or littoral drift. The amount of sediment moved alongshore depends on the wave energy component oblique to the shore. (After Fairbridge, 1968 and Komar, 1976.)

edge wave activity is negligible) (Trenhaile, 1997).

On the south coast of England the dominant waves are from the south-west, and are responsible for moving gravel eastward along the coast, such as between Bognor Regis and Rye in Sussex. The lobate foreland at Langney Point, near Eastbourne, and the cuspate foreland at Dungeness, east of Rye (see GCR site report in Chapter 6), have both been nourished with gravel that has travelled in this way. In Christchurch Bay sand and gravel beaches, supplied with sediment eroded from cliffs cut in gravel-capped Tertiary formations, have drifted eastward along the coast to accumulate in Hurst Castle Spit (see GCR site report in Chapter 6). On the Suffolk coast dominant north-easterly wave action has generated southward longshore drifting of shingle along Orfordness, and in Lincolnshire there is similar drifting of sand southwards to Gibraltar Point (see GCR site report in Chapter 8); (Figure 5.7).

Longshore drift usually alternates, moving sand or gravel first one way, then the other, as waves come in at different angles to the shore; this process can separate sand from gravel along the coast. On the south coast of England, beaches show alternations of eastward drift (by the dominant south-westerly waves) interrupted by westward drift (by weaker and less frequent south-easterly waves), resulting in a net eastward drift, such as on the north coast of Lyme Bay (Bird, 1989).

In the inner Moray Firth in Scotland, waves from the east are responsible for longshore drift to the west that has deflected river exits west and forced the migration of spits such as at Culbin

Figure 5.7▶ Some examples of English spits: (A) Spurn Head; (B) Orfordness; (C) Hurst Castle; and (D) Dawlish Warren. While the plan form of spits varies greatly, they all require an updrift sediment feed to form. In most cases, especially shingle spits, the sediment supply has now greatly decreased. (After Pethick, 1984, p. 108.)

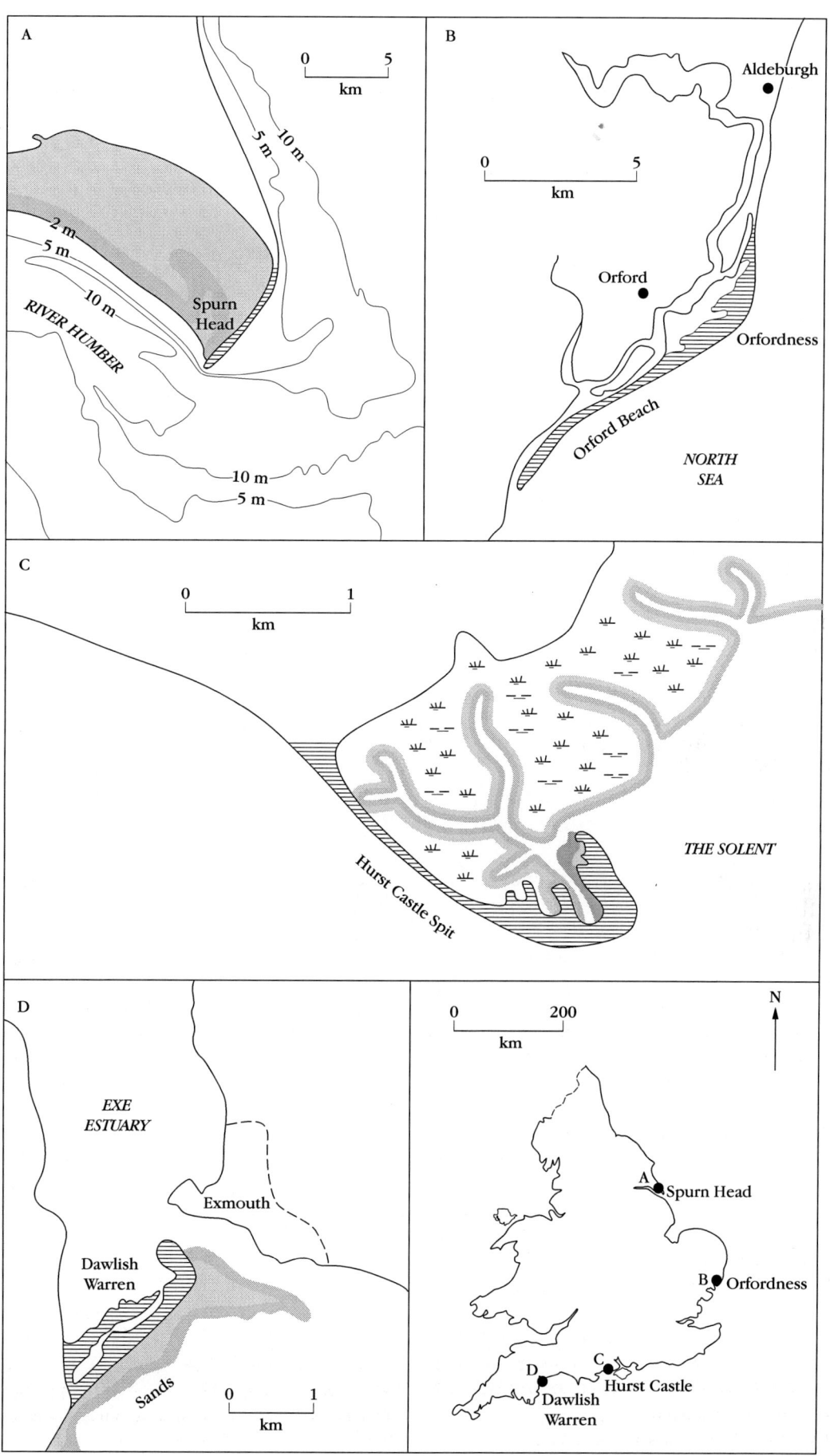

and Whiteness Head (see GCR site reports in chapters 11 and 6).

Offshore and onshore movements of beach sediment occur when waves form plunging or surging breakers, with a strong backwash that withdraws sand and gravel to the sea floor, whereas gentler spilling wave swash moves sediment shorewards and onto the beach. This sequence is known as 'cut-and-fill', and can be observed on most beaches, erosion during phases of storm wave action being followed by accretion in calmer weather. In Marsden Bay (see GCR site report in Chapter 7), King (1953) showed that currents generated by low, gentle waves when winds blew offshore moved sand shorewards on to beaches, whereas seaward movement occurred when steeper waves were generated by winds blowing onshore.

BEACH PLAN

Many beaches have gently curved outlines, shaped by incident waves refracted in such a way that they anticipate, and on arriving fit, the beach plan (Davies, 1958): these are termed 'swash-aligned beaches'. Loe Bar in Cornwall (see GCR site report in Chapter 6) is an example, and in the Outer Hebrides a sand and gravel beach shaped by Atlantic Ocean swell extends for more than 20 km along the west coast of South Uist, between Ardivachar and Stoneybridge (see GCR site report in Chapter 9). Breaking almost uniformly along such a beach, waves produce swash and backwash that generates onshore and offshore movements of beach material, the beach outline being maintained through sequences of cut-and-fill, with only minimal short-term alongshore movement.

Beaches between long promontories, for example, in Oxwich Bay in southern Wales (see GCR site report in Chapter 7), have been shaped by refracted waves that develop a curved outline as they move into the embayment. Beaches in the lee of headlands are often asymmetrical (zeta-curved), such as on the north Wales coast east and west of Pwllheli, where they have been shaped by south-westerly waves refracted around headlands. Within archipelagoes, such

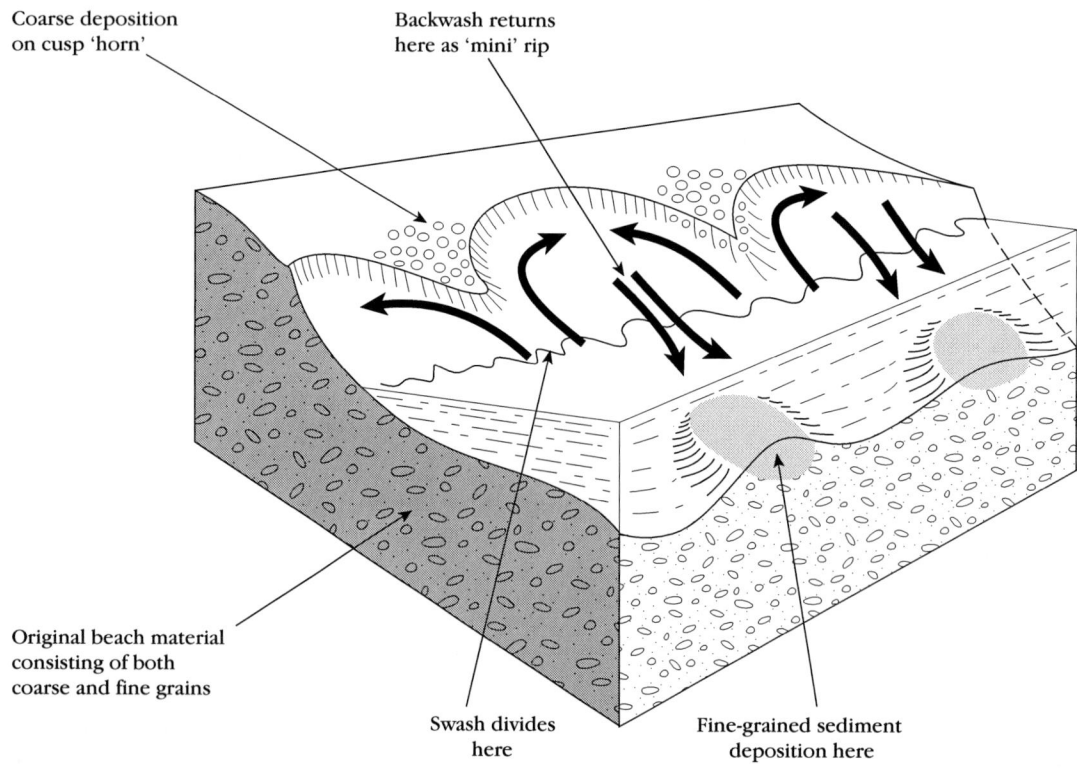

Figure 5.8 The formation of beach cusps. Cusps vary in size, but typical separation is in the range 2–10 m. (After Pethick, 1984, p. 112.)

Beach profiles

as the Isles of Scilly, beaches face various directions and have outlines that become orientated at right angles to the maximum fetch. Examples of this are seen in the inlets and voes of Shetland, for example at the Ayres of Swinister (see GCR site report in Chapter 6), and in the Orkney Islands at Sanday (see GCR site report in Chapter 8).

Many other beaches are drift-aligned, being shaped by waves that arrive at an angle to the coastline. The beach at Rye Harbour in Sussex (see site report in Chapter 6) is an example of a drift-aligned beach, on which the dominant south-westerly waves move sand and shingle alongshore. Drift-aligned beaches have less stable planforms than swash-aligned beaches because they gain and lose sediment alongshore, as well as onshore and offshore. Some bay-head beaches (as in Lulworth Cove, Dorset) are entirely swash-aligned, but most beaches in Great Britain are subject to alternations of swash and drift domination as the direction of incident waves varies. Even on Orfordness (see GCR site report in Chapter 6), which has been dominated by longshore drift, there are sectors with parallel ridges built by swash (Carr, 1969a).

Beach cusps (Figure 5.8) are minor and ephemeral features on beaches. Many shingle beaches, or beaches with mixtures of sand and shingle, develop beach cusps under certain wave regimes, particularly when edge waves interact with incoming waves, the nature and effects of which were described by Carter (1988). The cusps increase in size with larger incident waves; at Loe Bar in Cornwall (see site report in Chapter 6), cusps spaced at intervals of up to 20 m formed by strong Atlantic swell have been seen on the beach.

BEACH PROFILES

Beach profiles reflect both the nature of beach sediment and wave conditions. Studies of the response of gravel beaches to changing wave conditions have been made on several beaches around the coasts of Britain, particularly on

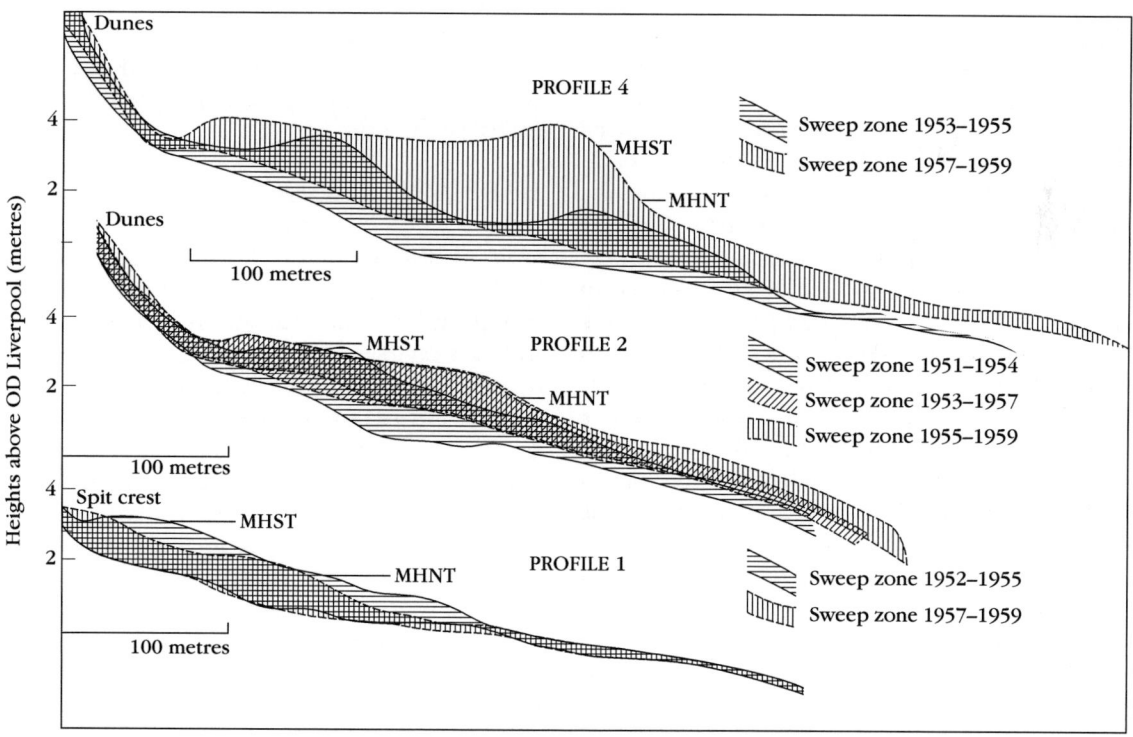

Figure 5.9 Sweep zones are a means of establishing long-term trends in beach form and sediment volume. These examples show the recovery of the south Lincolnshire beaches after the storm surge of early 1953. The sweep zones mark the vertical range of successive profiles (usually surveyed every few months), and by establishing two or more time periods, longer-term trends can be separated from short-term changes linked to changing wave climate. MHST: mean high spring tide; MHNT: mean high neap tide. (After King, 1972a, p. 359.)

Beaches, spits, barriers and dunes

Chesil Beach and Dungeness, but the dynamics of sandy beaches have received less attention in Britain than overseas, where sandy beaches are much more extensive.

There are contrasts in the response of gravel and sand beaches to alternations of cut-and-fill. In stormy episodes, waves reduce the gradient of the lower part of a gravel beach as strong backwash withdraws pebbles from the beach face, but the upper beach may be steepened and raised as gravel is carried shorewards by storm swash, forming a ridge or terrace known as a 'berm'. During calmer weather waves move finer nearshore gravel back onto the lower beach, restoring the swash-built lower slope, but leaving the steep upper beach unaffected.

The response of sandy beaches to these alternations of wave activity is a little different. On the sandy beaches at Ainsdale in Lancashire or Braunton Burrows in north Devon (see GCR site reports in Chapter 7), storm waves are almost entirely destructive, lowering and cutting back the beach, whereas berms are built by constructive wave action in calmer weather. Often swash-built sand bars are prominent at high and low tide levels, where wave action is more prolonged. Similar sequences have been observed on east coast sandy beaches, such as at Marsden Bay and Gibraltar Point (see GCR site reports in Chapter 8). On Scolt Head Island and Blakeney Point, a 1953 storm surge scoured and lowered sandy beach sectors (producing cliffs in backshore dunes) at the same time as driving the crest of the gravel beach landwards, forming washover fans on the backing saltmarshes (Steers, 1964b). The contrast may be related to greater percolation on gravel beaches, making wave swash less effective, and less able to build a berm than on sandy beaches. The outcome is that the outline in planform of a gravel beach is determined by storm waves, whereas sandy beaches are more influenced by the constructive waves and swell that arrive in calmer periods (Figure 5.9).

BEACH STATES

Analysis of incident wave regimes has shown that steep (> 3°) beaches (especially gravel beaches) tend to reflect waves, whereas gently sloping (generally sandy) shores dissipate their energy, the waves breaking and spilling across a wide surf zone (Wright and Short, 1984). Beaches may thus be described as exhibiting reflective or dissipative states (an intermediate category has also been recognized), and these can be defined using parameters related to wave power (Masselink and Short, 1993), and related to particular shore morphologies. These relationships are most clearly seen in low tidal range beaches, and on swash-dominated beaches, and are complicated on large tidal range and drift-aligned beaches by laterally migrating features such as lobes and bars.

Developed in Australia and the USA, this classification has been applied to beaches on various coastlines, but has so far been little used in Britain, perhaps because here gravel beaches are normally reflective, storm waves are more common than long swells, and tide ranges are relatively large. Nevertheless, swash-aligned beaches around Britain include some that are normally dissipative, such as the broad sandy beaches in the vicinity of Holy Island, and Lingay Strand in North Uist (see GCR site reports in chapters 11 and 9 respectively) and others (as on the Lancashire coast) that pass frequently from reflective to dissipative states in the course of cut-and-fill sequences. Some, such as the gravel-backed sandy beach at Porth Neigwl (see GCR site report in Chapter 4) are usually reflective at high tide and dissipative at low tide, whereas the beaches of Spey Bay, Moray (see GCR site report in Chapter 6), are usually reflective at most stages of the tide.

BARS AND TROUGHS

One of the common features of dissipative beaches is the presence of one or more bars and troughs exposed at low tide, such as Luce Sands in Galloway (see GCR site report in Chapter 7). A 'bar' is defined as a ridge or bank of sand (sometimes gravel) built up by wave action offshore and parallel to the coastline to a level where it is exposed at low tide but submerged at high tide. Where waves arrive obliquely, bars and troughs are aligned at an acute angle to the beach, such as on the shore at Porth Neigwl (see GCR site report in Chapter 4).

It should be noted that some coastal features called 'bar' as part of the place name (such as Loe Bar in Cornwall) are actually not bars *sensu stricto*, but *barriers* (see below).

The term 'ridge-and-runnel' has been used to describe multiple broad intertidal bars and swales running parallel to the coastline, as seen on the shores of South Lancashire (Gresswell,

Lateral grading

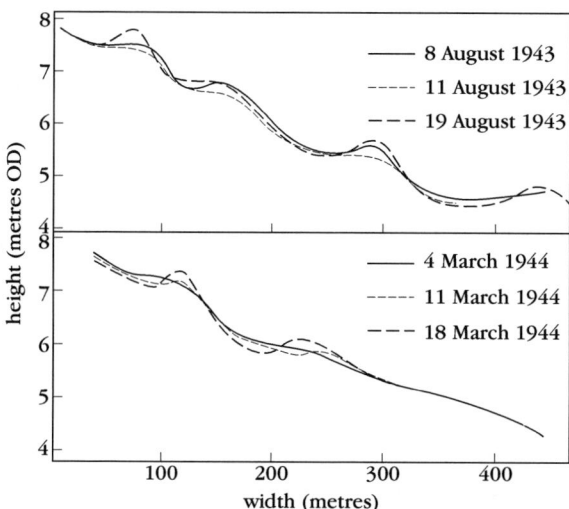

Figure 5.10 Profile of ridge-and-runnel beach (Blackpool). Ridge-and-runnel beaches occur where wide sandy beaches are dominated by local waves and swell is excluded, as here within the limited fetch of the Irish Sea. The short-period waves require a steep beach slope for equilibrium, and this is achieved through the formation of a series of ridges separated by runnels. (After King, 1972a, p. 342.)

1953b; Figure 5.10). The ridge crests are typically spaced at intervals of about 100 m, with an amplitude of about 1 m, the intervening swales becoming lagoons that drain out through transverse channels as the tide falls. They are essentially swash-built bars, formed by dissipating wave action, which represent an adjustment between the shore profile and the oscillatory turbulence produced by spilling waves that break over the bars, re-form over swales and break again over the next bar. The number, spacing and amplitude of multiple bars varies with the height and period of breaking waves, but bar topography, in turn, influences where and how incident waves break. The morphology of sand-bars and troughs changes in relation to variations in wave incidence and energy as the tide rises and falls. Similar sub-parallel bars and troughs off Gibraltar Point in Lincolnshire (see GCR site report in Chapter 8) differ from ridge-and-runnel in that they have been built by north-easterly waves and associated currents, the bars having grown southwards as intertidal spits (King and Barnes, 1964).

LATERAL GRADING

Some beaches show lateral grading of sediment, with finer-grained sediment towards one end and coarser towards the other. Chesil Beach (see GCR site report in Chapter 7) famously displays excellent longshore sorting, with small pebbles at the western end increasing in size to large cobbles at the south-eastern end (Carr and Blackley, 1974a; Figure 5.11). Beaches in Lyme Bay (Charmouth, Seatown) show similar lateral grading, while those on the Dorset coast east of Weymouth are graded in the opposite direction, possibly because the general dominance of south-westerly wave action on the south coast of England is replaced by south-easterly wave action in the lee of Portland Bill. There has been much discussion of the causes of lateral grading on beaches around Britain, but it should be noted that most beaches display only poor gradation and some no grading at all.

It is possible that lateral grading results from longshore sorting by oblique waves and associated currents, the coarser sediment being retained updrift as the more readily mobilized finer-grained sediment is carried downdrift. Alternatively, progressive downdrift reduction of initially coarse beach sediment by breakage and attrition may occur as it moves alongshore. This may explain the diminishing grain size downdrift of pebbles derived from Lias limestone on the beach east of Stolford in Bridgwater Bay, Somerset (Kidson, 1960) and the reduction of cobbles to pebbles northwards along the spit at Westward Ho! in north Devon (see GCR site report in Chapter 6), but such grading could also be achieved by longshore sorting. Some sandy beaches show coarsening of sediment downdrift, such as in Norfolk where the movement of sand and gravel along the coast is accompanied by preferential removal of sand to offshore bars (McCave, 1978a). Lateral grading can also result from alternations of longshore drift, the sorting of pebbles on Chesil Beach possibly as a result of the more frequent and stronger south-westerly waves carrying beach sediment eastwards and the less frequent and weaker south-easterly waves returning the finer-grained fractions westwards (Jolliffe, 1964). It has been suggested that beach sediment coarsens towards sectors of higher wave-energy, but this is not an explanation of lateral grading, unless it leads to sorting as the result of drifting from higher to lower wave-energy sectors. It is also possible that beaches like Chesil Beach may have originally developed at a lower sea level and have since been driven onshore retaining

Beaches, spits, barriers and dunes

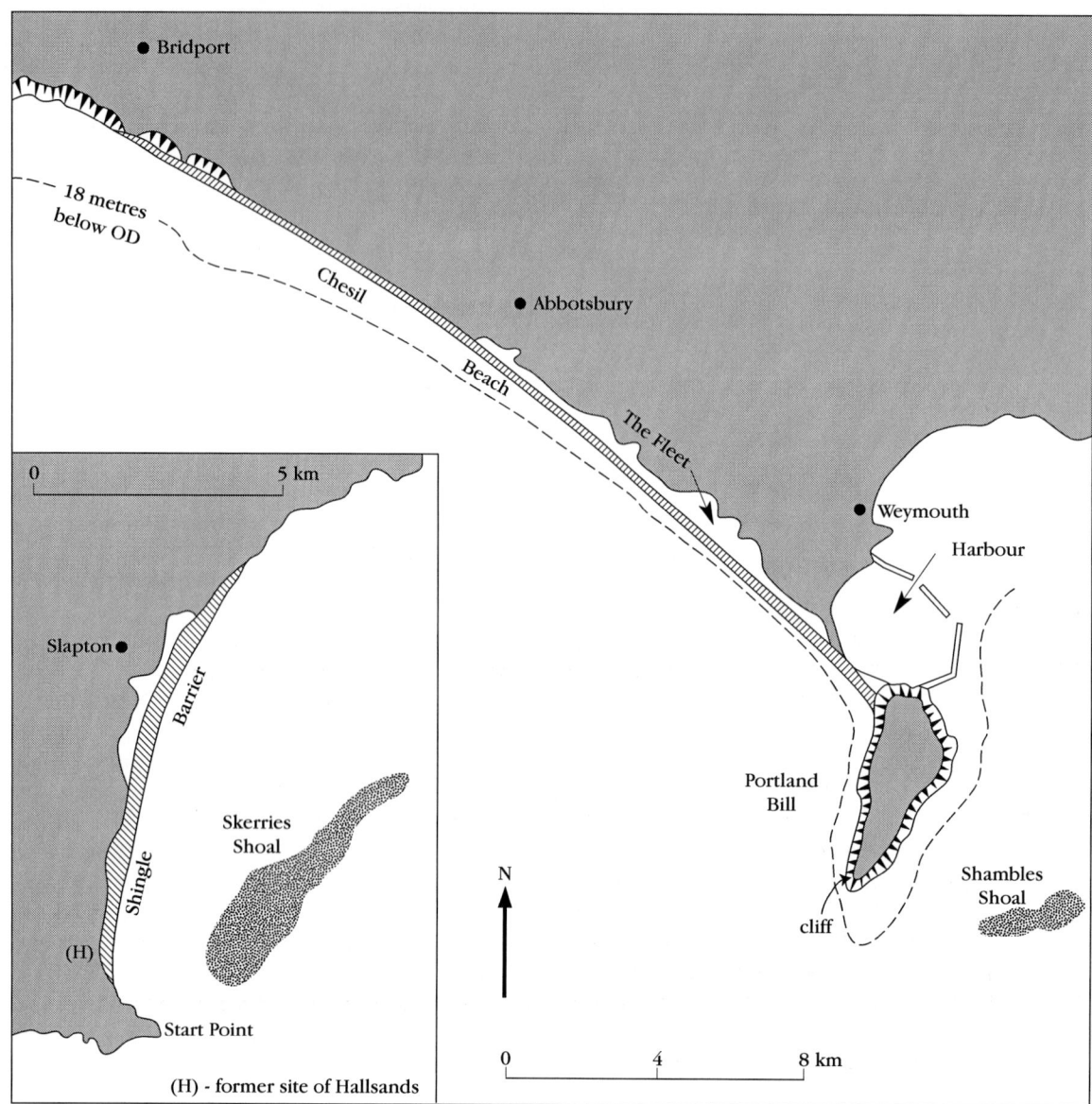

Figure 5.11 Coastal barriers enclosing coastal lagoons, Slapton, Devon and Chesil Beach, Dorset. Each of these barriers show gradation in pebble size along the barrier; coarse material is at the southern end of the Slapton barrier at Hallsands, and at the eastern end of Chesil Beach. (After Bird, 1984, p. 144.)

longshore gradation inherited from a more unidirectional wave regime than that of the present-day. The grading remains because of the lack of any new sediment that may 'dilute' the grading. Most beaches show poor lateral grading, however, and beaches such as Chesil Beach are the exception.

PROGRADING BEACHES

Progradation of beaches occurs where there is a continuing supply of sand and gravel from long-shore or offshore sources, and a predominance of constructive wave action. Beaches also prograde on actively emerging coasts, stimulating shoreward drifting of sediment, such as in northern Britain where uplift due to postglacial isostatic rebound is continuing, for example, Morrich More (see GCR site report in Chapter 11). This has probably contributed to the extensive progradation on the sandy coasts bordering the Moray Firth at Culbin, Spey Bay and Whiteness Head, and in the Dornoch Firth (see GCR site reports in the present volume). A well-documented prograding beach occurs at Luce Sands, in Dumfries and Galloway, where a uni-directional wave approach sweeps sediment into a re-entrant trap (see GCR site report in Chapter 7)

Few beaches are naturally prograding in southern Britain, although local progradation occurs in estuaries and river mouths and on the ends of spits. The northern part of the sandy beach on South Haven Peninsula in Dorset (see GCR site report in Chapter 7) has advanced seawards, but this is largely due to accretion alongside a training wall built at the entrance to Poole Harbour in 1924. Similar local progradation has occurred updrift of harbour breakwaters at Lyme Regis, Newhaven and Rye on the south coast of England. There are prograding sandy beaches on the shores of Carmarthen Bay in southern Wales (see GCR site report in Chapter 11), where sand is being swept onshore from extensive shoals to form a beach that is over 1.6 km wide at low tide. Most sandy beaches in southern Britain are mainly undergoing erosion, although some are stable.

BEACH RIDGES

Sandy beach ridges originate as berms built by constructive wave action, whereas gravel beach ridges have been piled up by storm waves. On prograding coastlines multiple beach ridges (with intervening swales) may form, their height and spacing being determined by the rate of progradation (depending on available sources and patterns of sediment supply) and the upper swash limit of the waves that built them.

There are numerous roughly parallel shingle beach ridges on Dungeness (see GCR site report in Chapter 6), each marking a former coastline. It has been suggested that variations in the crest levels of these beach ridges may be indications of former sea levels. Lewis and Balchin (1940) found that the crests of some of the older shingle ridges on Dungeness were 2 to 3 m lower than those formed more recently along the eastern shore, possibly indicating that sea level had since risen, relative to the land. If a coast is emerging (as the result of land uplift or a falling sea level) the crest heights of successive beach ridges are likely to decline seawards, but there will also be variations related to differing swash limits in each constructional phase. On parts of the coast of Scotland that have been rising because of isostatic rebound following deglaciation, parallel shingle beach ridges have crests that decline seawards, such as in Spey Bay east of Lossiemouth (see GCR site report in Chapter 11; Comber, 1995), and on the west coast of Jura (see GCR site report in Chapter 6), (McCann, 1964).

Most beach ridge systems in Britain are swash-aligned, having been built parallel to the incoming waves, but sub-parallel beach ridges can also be formed by the successive addition of longshore spits, built by waves arriving at an angle to the shore. On the South Haven Peninsula in Dorset (see GCR site report in Chapter 7) during the past three centuries, the formation in stages of three broad sandy beach ridges (surmounted by dunes), traced from historical maps by Diver (1933), has included a component of northward longshore spit growth. The growth of Morrich More in Ross and Cromarty is related to the same process, the later stages of which can be charted using historical maps dating from 1730 to show a series of beach ridges capped by dunes and now separated by saltmarsh (see GCR site report in Chapter 11).

A particular kind of beach ridge is found on saltmarshes, particularly on the Essex coast near St Osyth (see GCR site report in Chapter 10), where storm surges have swept sand and shells up across the marshland, and left them as a ridge emplaced at the swash limit. Such ridges also occur in France near Dinard and are similar to the sandy ridges known as 'cheniers', deposited on marshes and deltaic plains in Louisiana and elsewhere. Their rarity in Britain could be due to the lack of river deltas and deltaic coastal plains, on which cheniers are typically found.

SPITS, TOMBOLOS AND CUSPATE FORELANDS

Spits are found where beaches diverge from the coastline. Their recent evolution can be traced from historical maps and aerial photographs, and various studies have related their shaping to incident wave regimes and the effects of occasional storm surges. Some spits are almost straight, like the southern part of Orfordness (see GCR site report in Chapter 6), where the mouth of the River Alde has been deflected nearly 17 km southwards, but most end in one or more recurves, representing earlier terminations, such as at Hurst Castle spit (Figure 5.12). Blakeney Point and Scolt Head Island are shingle spits with the remains of several former recurved

Beaches, spits, barriers and dunes

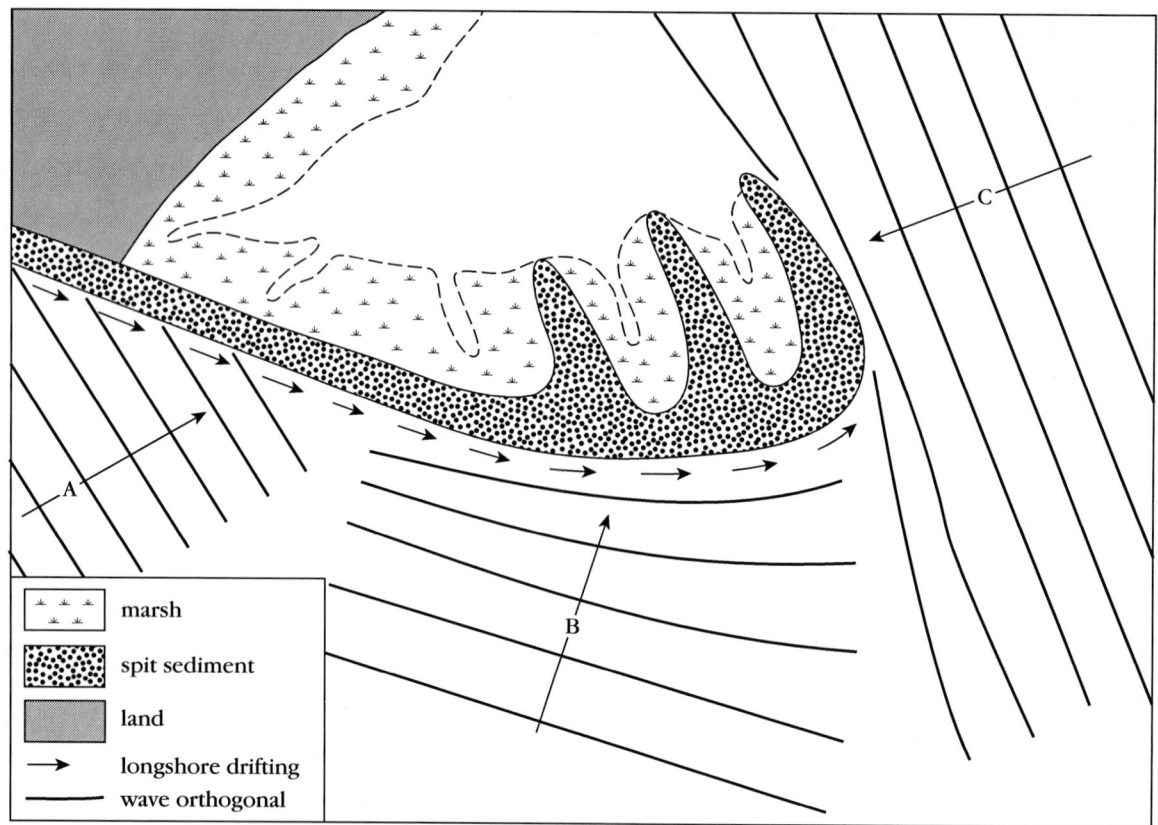

Figure 5.12 The shaping of a recurved spit, based on the outline of Hurst Castle Spit (see GCR site report in Chapter 6). Waves from A, arriving at an angle to the shore, set up longshore drifting which supplies sediment to the spit; waves from B and C determine the orientation of its seaward margin and recurved laterals respectively. (After King and McCullagh, 1971 and Bird, 1984, p. 148.)

terminations, which show that they have grown intermittently westwards, shaped by alternations of north-easterly and north-westerly wave action (see GCR site report for North Norfolk Coast in Chapter 11). These spits have been derived from morainic deposits at the margin of the Last Glacial ice-sheet (which crossed the Norfolk coastline in this area), the glacial drift deposits having been sorted and rearranged in the course of the Holocene marine transgression by wave- and tidal current-action.

Gravel beaches and beach ridges on the coast of Spey Bay and Burghead Bay in north-east Scotland have been supplied with gravel by the rivers Spey and Findhorn, carried westwards along the shore to form spits with recurves at their western ends, as the result of waves arriving from the north and north-east. Since the ongoing development of spits is linked to ongoing sediment supply to fuel distal accretion and spit extension, many are characterized by updrift erosion that may truncate earlier ridges or breach through the spit at the proximal end. The Bar, at Culbin in Moray, is a fine example of this process (see GCR site report in Chapter 11; Hansom, 1999). To the west, Whiteness Head (see site report in Chapter 6), on the southern shores of the Moray Firth, is a recurved spit of well-rounded gravel, derived from glacial drift deposits and similar to Blakeney Point in Norfolk. It has been built by westward longshore drifting and driven landward by storm surges so that the older recurves (projecting from the inner side) have been partly overrun (Steers, 1973).

The shaping of Hurst Castle spit (see GCR site report in Chapter 6) in relation to the direction of approach of dominant waves was demonstrated by Lewis (1931): it is exposed to south-westerly and southerly waves from the English Channel and easterly waves along the Solent, but is protected from south-easterly waves by the Isle of Wight. A computer simulation (SPIT-SIM) of the growth of this spit indicated the

Coastal barriers

importance of constraints on the landward growth of the recurves (King and McCullagh, 1971). The main shingle bank has been driven landwards by storm surges, so that saltmarsh peat now crops out on the beach face (Nicholls and Webber, 1987a,c) (Figure 5.12).

On Rattray Head in Aberdeenshire, Scotland (see GCR site report for Strathbeg in Chapter 7) there is evidence of spit growth, first to the south-east, and later to the north-west, implying a reversal of longshore drifting, but it is possible that the sediment came in from the sea floor, and that the spits were largely swash-built. Paired spits of the kind seen at Poole, Christchurch and Pagham Harbours on the south coast of England (Robinson, 1955), and at Braunton Burrows (see GCR site report in Chapter 7), may result from a convergence of longshore drift produced by such alternations, but the entrance to Pagham Harbour was formed by the breaching of a shingle barrier in a 1910 storm. Paired spits can be shaped by waves refracted into the mouths of bays or estuaries (Kidson, 1963).

Tombolos are wave-built ridges of sand or gravel that link islands, or attach an island to the mainland. Some have formed as the result of the growth of a spit in the lee of an island until it reaches the mainland, others from the fusing of paired spits, and others as barriers (or augmented bars) built up across a strait. Stages in the evolution of tombolos can be seen in the Isles of Scilly (see GCR site report in Chapter 8), several of which are linked by depositional features of sand or shingle, some being partially submerged banks known as 'swashways'. On Samson, a sandy barrier links two former islands, and a shingle barrier ties Gugh to St Agnes. The curved shores of the sandy St Ninian's tombolo in the Shetland Islands (see GCR site report in Chapter 8) i have probably been shaped by a combination of refraction and diffraction of waves around St Ninian's Isle off the south-west coast of the Shetland mainland (Flinn, 1997).

Cuspate forelands (sometimes known as 'nesses') have formed by the deposition of beach sediment in protruding, more-or-less symmetrical, structures shaped by bi-modal wave-approach directions. Usually there is erosion on one side and longshore drifting round the point to an accreting shore on the other, such as on Dungeness, where the ridges have been truncated along the southern coastline (see Figure 6.43), and new ridges have formed on the pro-graded eastern shore. Dungeness probably originated as a spit on the coast off Rye about 5000 years BP, and has been built up and consolidated by waves arriving from the south-west and from the east, through the Strait of Dover (Lewis, 1932).

Morfa Dyffryn and Morfa Harlech (see GCR site reports in Chapter 8) are large cuspate lowlands on the north Wales coast, but they differ from Dungeness in that they originated from lobes of glacial drift, the margins of which were re-shaped by wave action, which built fringing beaches that are backed by dunes. Their points have grown northward as the result of erosion of sand and shingle from their southern shores and longshore drifting to the north.

The direction in which cuspate forelands migrate depends on wave patterns and local conditions. Winterton Ness on the north-east coast of Norfolk (see GCR site report in Chapter 8) is a lobate sandy foreland that has been migrating southwards as the result of erosion of its northern shore and accretion on its southern shore, supplied by longshore drift generated by the dominant north-easterly waves. In contrast, Benacre Ness (see GCR site report in Chapter 6), in a similar situation on the Suffolk coast, has been migrating northwards as the result of accretion on its northern side of sediment supplied by the predominant southward longshore drifting, and erosion on its southern side (Steers, 1964a). Migration of offshore shoals may also have influenced incident wave patterns and the evolution of these two nesses (Robinson, 1966). In Scotland, Buddon Ness at the mouth of the Tay has been built as a result of the seaward movement of sediment moved by the River Tay and the southwards movement of sediment on the outer coast. Falling sediment supply to the seaward coast had initiated chronic erosion, now arrested by artificial structures (see GCR site report for Barry Links in Chapter 7).

COASTAL BARRIERS

A coastal barrier is a prominent ridge or bank of sand or gravel built up by wave action to above high-tide level, backed by a lagoon or marsh. Some barriers (swash-aligned) have been built by waves arriving parallel to the coastline, with beach material supplied mainly by shoreward drifting from the sea floor; others (drift-aligned) have grown as longshore spits shaped by waves

Beaches, spits, barriers and dunes

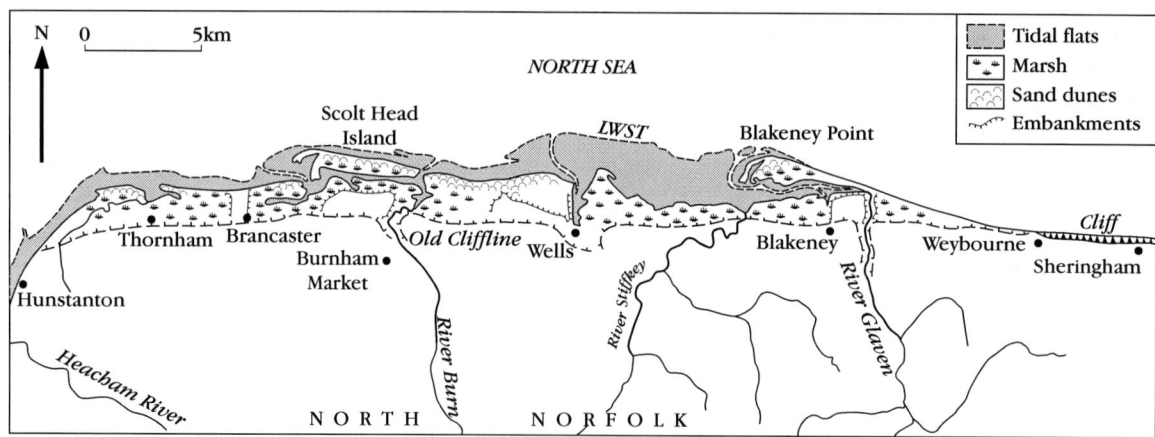

Figure 5.13 Coastal barriers backed by saltmarsh, North Norfolk Coast GCR site (see GCR site report in Chapter 11). The barriers and recurves carry sand dunes; behind are sheltered tidal inlets and extensive areas of saltmarsh, part of which has been reclaimed for grazing. (After Bird, 1984, p. 149.)

arriving at an angle to the coastline; and some are combinations of these two principal types. 'Barrier islands' are separated by transverse channels that are flooded at least at high tide.

The Loe Bar (see GCR site report in Chapter 6), near Helston in Cornwall, is an example of a swash-aligned barrier. It is 180 m wide, and consists largely of flint gravel and shingle moved from the sea floor, with only a small proportion derived from adjacent cliffs of slate and greenstone. The barrier is part of a beach that extends northwards from Gunwalloe to Porthleven, and runs across the mouth of a former ria, enclosing the lagoon known as the 'Loe Pool'. Loe Bar is sometimes overwashed by large waves from the south-west during storms, which have produced fans of gravel projecting into the lagoon.

Chesil Beach is another coastal barrier, consisting largely of pebbles that migrated shorewards from the sea floor during the Holocene marine transgression. It is essentially swash-aligned, but subject to alternations of longshore drift. It encloses a lagoon, The Fleet, the inner shore of which has low promontories and bays that have never been exposed to the open sea. Chesil Beach is migrating landwards, partly by occasional washover in storm surges, and partly because water driven through the permeable barrier by strong wave action forms gravelly fans spilling into The Fleet. Landward movement is confirmed by the presence of outcrops of lagoonal peat on the seaward slope of Chesil Beach.

In the Orkney and Shetland Islands there are many small gravel barriers, known as 'ayres', that have been deposited across the mouths of embayments to enclose, or partly enclose lagoons, known as 'oyces', which rise and fall with the tide. These are well represented in the Ayres of Swinister, Shetland, and in Central Sanday, Orkney (see GCR site reports in Chapter 6), where a complex series of proto-barriers have evolved to form the present barriers enclosing tidal lagoons at Little Sea and Cata Sand.

Most British barriers are of gravel, which may be topped with sand dunes, but in Sandwood Bay (see GCR site report in Chapter 7) on the north-west coast of Scotland there is a wide swash-aligned, dune-capped, barrier beach built of sand that has moved in from the sea floor, enclosing a freshwater loch at the mouth of Strath Shinary, which is a glacial trough.

Barriers that originated as longshore spits are usually distinguished by the presence of recurves that mark former terminations, such as on the barrier island known as 'The Bar' on the Culbin coast (see GCR site report in Chapter 11), which formed during the 18th century, and has grown eastwards and westwards (Comber, 1995). Orfordness may be of composite origin: it has grown southwards as a longshore barrier backed by the deflected River Alde, but it includes parallel beach ridges formed during phases of swash-dominated progradation.

Scolt Head Island (see GCR site report for North Norfolk Coast in Chapter 11) is a good example of a barrier island consisting of shingle ridges, sandy intertidal areas and dunes, backed by saltmarshes (Figure 5.13). Its changing outlines have been much studied, particularly the evolution of Far Point, a western outgrowth in the form of a recurved spit (Steers, 1960). Although Walney Island (see GCR site report in Chapter 8) has the appearance of a barrier island it is actually a deposit of glaciofluvial gravel and till, cliffed on the seaward side, with dune-covered spits of sand and gravel that extend from either end.

COASTAL DUNES

British coastal dunes occupy an area of about 70 000 ha (Dargie, 2000) and occur on 7.4% of the coastline of Britain (Doody, 1989). Of this, the dune area of Scotland (*c.* 50 000 ha) far exceeds that of England (*c.* 11 900 ha) and Wales (*c.* 8100 ha). Scotland holds 71% of the British resource and has the largest and highest British dune systems (Dargie, 2000).

The dunes have formed where sand winnowed from a wide beach has been blown landwards and deposited on the coast above high-tide level, such as on the Ainsdale Dunes (see GCR site report in Chapter 7) in south-west Lancashire. The most extensive dune systems in Great Britain occur behind the sandy beaches of the Atlantic coasts of the Outer Hebrides, where ocean swell has formed wide, gently shelving beaches from which the prevailing westerly winds have blown sand onshore. High sand dunes have formed behind Tràigh Luskentyre and Tràigh Seilebost (see GCR site report for Luskentyre and Corran Seilebost in Chapter 9), Hebridean bays where broad sands are exposed at low tide on the south-western shore of the island of Harris. Other extensive and high dunes occur where plentiful sediment has been available in the past, such as between Aberdeen and Fraserburgh at the Forvie and Strathbeg GCR sites (see reports in Chapter 7). Exensive dune systems also occur at estuary mouths such as Barry Links and Tentsmuir (Forth of Tay); Culbin (Moray Firth); Morrich More and Dornoch (Dornoch Firth) (see GCR site reports in chapters 7 and 11)

Large dunes have also formed behind sandy beaches on the North Sea coast in the vicinity of Holy Island and in Lincolnshire, East Anglia and Wales. Some new dunes are forming behind prograding sandy beaches, but, more typically, where beach erosion is occurring, the seaward margins of coastal dunes are cliffed and receding. Since only a very few beaches in Britain are accreting, then most of the dune systems are erosional, as a result of both sea-level rise and a reduction in sediment supply.

There are several kinds of coastal dune topography (Figure 5.14). In an accreting sequence, the first colonizers of the small mounds of sand that accumulate around the flotsam of the high-water mark are salt-tolerant species such as sand couch *Agropyron junceiform* and sea rocket *Cakile maritima*. The presence of these pioneers serves to enhance sand deposition and the embryo dune grows to a foredune that may coalesce laterally to form a foredune ridge. The foredune ridge is built up immediately behind a sandy shore and held in place by vegetation, typically marram *Ammophila arenaria* or sea lyme grass *Leymus arenaria*. There has been much discussion of whether foredunes are initiated when vegetation colonizes wave-built berms of sand or gravel, or whether their original alignment depends on the growth of sand-trapping vegetation along a seed-bearing strandline of plant litter on the beach. As the dune continues to receive sand, and marram and other dune grasses are well equipped to cope with sand inundation, the foredune grows in height to produce a first dune ridge that is well covered with sand-trapping vegetation. Depending on the continuity and rate of sand accumulation, there may be several dune ridges formed in this way, although the farther from the sand source the more its sand supply is intercepted by the growth of dunes to seaward. Eventually, the older dunes become virtually stable with a dense ground vegetation cover, little new sand arriving and moribund stands of marram being progressively replaced by vegetation more suited to stable environments, such as mosses, lichens, a variety of grasses, and eventually shrubs and trees (Figure 5.14). At this stage the older dunes are susceptible to erosion if their vegetation cover is disrupted and many dunes show signs of such point-erosion, which leads to the development of blow-outs and the formation of parabolic dunes (Figure 5.15) However, a more serious type of erosion is the frontal erosion of the dune faces above the beach, since it leads to removal of younger parts of the dune system. Figure 5.14E shows the result of the latter type of erosion, and probably results because a combination of sea-level rise and reduction in sediment supply to the fronting beach forces the landward translation of the backshore into the

Beaches, spits, barriers and dunes

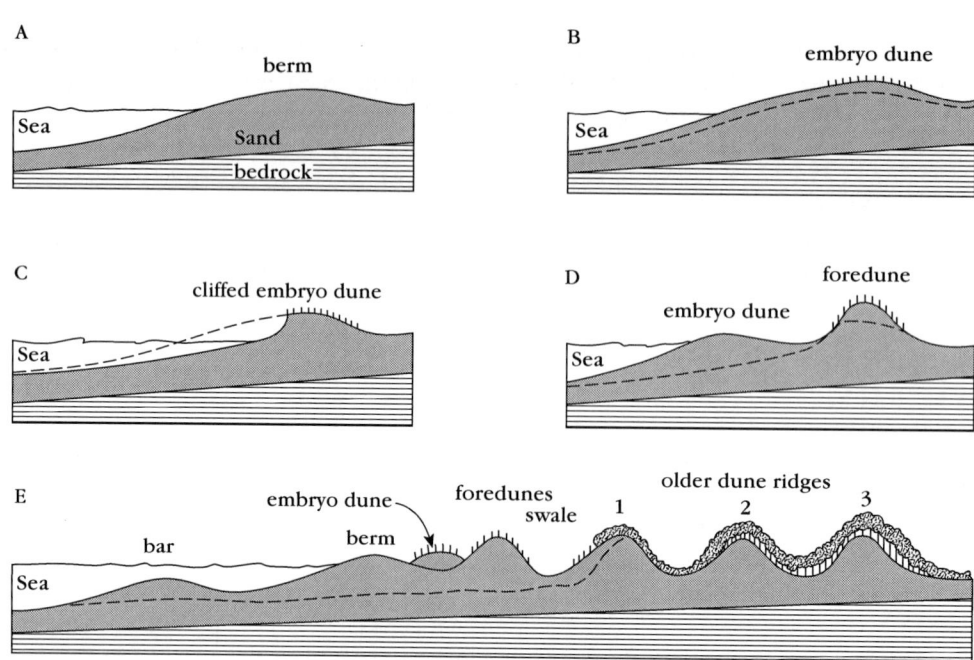

Figure 5.14 The formation of coastal sand dunes on a prograding coast (A–D). The older dunes farthest inland develop relatively mature soils and vegetation, and often the sand differs in colour from the younger dunes nearest the beach. The pecked line on E shows the effect that rising sea level and reduction in sediment supply has on sand dunes. Most of the dunes of Britain show such frontal erosion to a greater or lesser degree. Most dunes on the Scottish western and northern coastas are erosional. (Based on Hansom, 2001; Hansom and Angus, 2001, after Bird, 1984, p. 180.)

dune system. In Scotland, this latter situation is common and many mature dune systems have erosional edges characterized by steep sand faces (such as occurs at Dunnet in Caithness and at Barry Links in Tayside (see GCR site reports in Chapter 7). Dargie (2000) reports the extent of bare sand and mobile dune vegetation to total a mere 3.6% of the windblown sand extent of Scotland. Embryo and foredunes are thus rare.

Ranwell (1972) suggested a classification of dune systems that has been followed here for the sites in England and Wales (see Table 7.1). It is based on the combined effects of sand source, restrictions to longshore sand-movement and the geomorphological feature. Under this scheme, dunes are divided into two main groups: foreshore dunes, which are found most commonly on spits, nesses and offshore islands, and hindshore dunes, where most of the dune system lies on land behind the beach. In northwest Scotland, the dunes are dominated by 'machair', a flat sandy plain behind a narrow cordon of undulating dunes (Ritchie and Mather, 1984; see Chapter 9). Wind direction and speed affect the extent to which dunes (given an adequate and continuing supply of sand) can build in height. However, there are important differences in the dune patterns, well exemplified at Newborough Warren (see GCR site report in Chapter 11). Newborough Warren has a large reliable supply of sand and the coastal linear dunes have migrated inland. In contrast, Tywyn Aberffraw, a few kilometres to the north-west (see GCR site report in Chapter 7), has a distinctly different pattern of linear dunes that, despite their presence immediately behind the beach, have an alignment that is broadly perpendicular to the shore and which is associated with the development of parabolic dunes and blowthroughs. The time taken for a full cycle of linear dune construction and migration was estimated by Ranwell at about 80 years, and whilst this explains the particular features of Newborough Warren, it may also provide an insight into some of the separate zones found in other dunes such as Tywyn Aberffraw and South Haven Peninsula, Dorset (see GCR site reports in Chapter 7).

Many dunes are associated with the development of large accretional structures especially where there are extensive intertidal sandflats. Shingle or gravel spits, nesses and offshore bar-

Coastal dunes

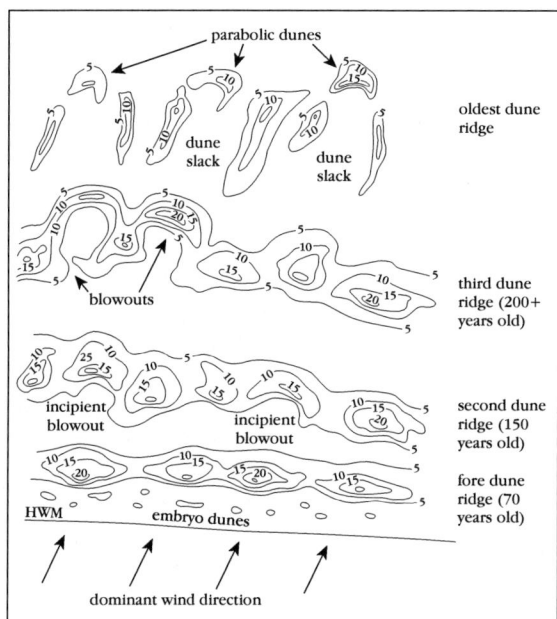

Figure 5.15 Sequential development of the dune ridges in a dune system. The blowthroughs of the second and third dune ridges eventually form into parabolic dunes in the older ridge. (After Pethick, 1984.)

rier structures have provided a base upon which sand beaches and dunes have developed. Particular examples occur on the distal ends of cobble spits at Westward Ho! and Ynyslas, on large estuary-mouth sand and shingle spits such as Spurn Head, Morfa Harlech and Morfa Dyffryn, on offshore islands for example around Lindisfarne, and on barrier beaches such as at Blakeney Point (North Norfolk Coast GCR site) and Pembrey (Carmarthen Bay GCR site). Extensive hindshore systems such as at Ainsdale, Lancashire, appear to have grown atop sandbanks that developed offshore. The dunes capping the emerged and modern sand ridges of Morrich More, Ross and Cromarty, may have developed in a similar way (see GCR site reports in the present volume).

Where the foundations of these dune systems rely primarily on longshore sediment transport, many are affected by erosion at their proximal end. This may result from up-drift coast protection structures (e.g. groyne-fields) or from reductions in the natural rate of longshore sediment supply brought about by changes in wave direction or changes in the nature of the sediments in the source region. For example, along parts of the west coast, the erosion of till and periglacial deposits has exposed hard-rock coasts and as a result the sediment supply is now much reduced. Similarly, estuarine sediment supply may be significantly reduced or enhanced as a result of changes in catchment management. Many dune systems rely on both continued positive sediment budgets of their foundation materials and of the sand supply. Offshore sand supplies have also reduced in volume as the Holocene supplies have become exhausted and longshore sediment supplies have been cut off or reduced.

Successive dune ridges may form on a prograding coast as parallel foredunes separated by dune swales. The evolution of parallel dunes can be traced at such sites as Tentsmuir in Fife or Winterton Ness in Norfolk (see GCR site reports in chapters 7 and 8). As they are of increasing age landwards they can be used to study the evolution of soil profiles, the yellow or brown sand of recently formed dunes becoming grey or white on the older ridges as podzol profiles form and deepen. The rate of soil evolution depends on the nature of the sand and the type of colonizing vegetation: it is relatively slow on calcareous sands, but more rapid on quartzose sands as on South Haven Peninsula in Dorset (see site report in Chapter 7), where dunes formed within the past three centuries and colonized by heath vegetation already have white podzolic A horizons.

Coastal dunes are often interrupted by blowthroughs, which are unvegetated or sparsely vegetated hollows excavated by the wind, where sand has been driven landwards to form a looped ridge (Figure 5.15). There are good examples at Dunnet Bay (see GCR site report in Chapter 7). Blowthroughs can form naturally, during stormy periods when the seaward margin of the dunes is cut back by wave scour and the dune locally breached or overwashed. Some blowthroughs have originated where the dune vegetation has been depleted by grazing animals, by cutting, burning or trampling, or damaged by vehicles. Blowthroughs can grow into larger parabolic dunes with noses of sand advancing landwards and vegetated trailing arms on either side of a corridor formed by deflation. Parabolic dunes have been studied in the Sands of Forvie in Aberdeenshire, Scotland (see GCR site report in Chapter 7), where their development is accompanied by an adjustment between the dune morphology and the wind-flow patterns (Robertson-Rintoul, 1990). The orientation of parabolic dunes in relation to regional wind regimes has been demonstrated, for example on Barry Links in Angus (see GCR site report in Chapter 7), where they have SW–NE- trending axes (Landsberg, 1956).

Where onshore winds are strong and the vegetation cover sparse there is more general movement of sand landwards in the form of transgressive dunes, with sand spilling down a generally steeper leeward slope. These are well developed in Newborough Warren (see GCR site report in Chapter 11), where three successive sand ridges have formed and migrated inland at rates of up to 16.7 m a^{-1} (Ranwell, 1958), and in the Sands of Forvie (see site report in Chapter 7), where several roughly parallel E–W-trending sand ridges have been migrating northeastwards within a corridor bounded by high N–S-trending dune ridges. It is not clear how these migrating sand ridges, separated by low-lying slacks, were initiated.

On Braunton Burrows (see GCR site report in Chapter 7) various dune forms, including foredunes, parallel ridges, swales, blowthroughs and parabolic dunes persist despite devegetation and damage during a phase of intensive military use. However, like Culbin, dune management has tended to take greater care of the vegetation than a geomorphologist might wish, and geomorphological diversity is lost if processes are arrested. There is similar diversity and dune fixing behind Luce Sands (see GCR site report in Chapter 7) on the southern side of the isthmus that links the Rhinns of Galloway to the mainland, where low parallel foredunes are backed by high transverse older dunes and many swales. Detailed surveys of such dunes, such as those by Single and Hansom (1994) at Luce Sands are necessary to determine rates and patterns of change. Kidson *et al.* (1989) made a photogrammetric survey of the Braunton Burrows dunes (scale 1:2500) in 1983 and used it to calculate volumes of gain and loss since an earlier survey in 1958.

Dune vegetation and habitats

Historically, many dunes were used as breeding grounds for rabbits, an activity recognized in the common usage of the terms 'burrows' or 'warren' in dune names. Grazing both by rabbits and other domestic animals has affected the vegetation of many dunes. The removal or absence of grazing often leads dunes to become dominated by scrub, thus losing much of the species diversity associated with grazing and the attendant geomorphological instability and interest. Shrubs, especially sea buckthorn *Hippophae rhamnoides*, *Rhododendron*, willow *Salix* spp., and tree lupin *Lupinus arboreus*, and trees especially pine *Pinus* and birch *Betula* have invaded dunes or been introduced to increase dune stability. Many coastal dunes are now stabilized by a vegetation cover, much of it marram *Ammophila arenaria* that has colonized the sandy surface.

In the past few centuries attempts have been made to arrest drifting sand by planting marram or laying brushwood and planting pine trees, as at Culbin (see GCR site report in Chapter 11) (Steers, 1973). The Royal Commission on Coast Erosion and Afforestation (1907–1911) had encouraged the planting of conifers on many dunes, both in order to extend the forest resource in Britain and to 'fix' dunes both as sea defences and to prevent them from invading agricultural land. Stamp (1947) commenting on the use of the British landscape described dunes as 'rarely providing land of high value' (p. 231), argued that afforestation was a safer use of many dunes than grazing and regarded the management of dunes as golf courses as obviating the 'risk of erosion' (p. 162).

The vegetation of dunes is significantly affected by the proportion of calcium carbonate to silica in the deposited sand (see Table 7.3), the hydrology of the dunes, climate (especially wind) and the ways in which dunes are managed.

The development of dunes in the Holocene Epoch

There is historical and archaeological evidence that coastal dunes have at times been more geomorphologically active, either because larger quantities of sand were arriving or the vegetation cover was sparser, so that bare and mobile dunes drifted more than they do now. At Skara Brae, in the Orkney Islands, dunes overran a Neolithic settlement about 4700 years BP. We know of its location only because it has since eroded out of a retreating dune face suffering frontal erosion. In Cornwall, drifting sand buried farms, villages and churches behind St Ives Bay and Perran Sands in medieval times, possibly as a result of destruction of vegetation that had previously fixed the dunes, or because

Coastal dunes

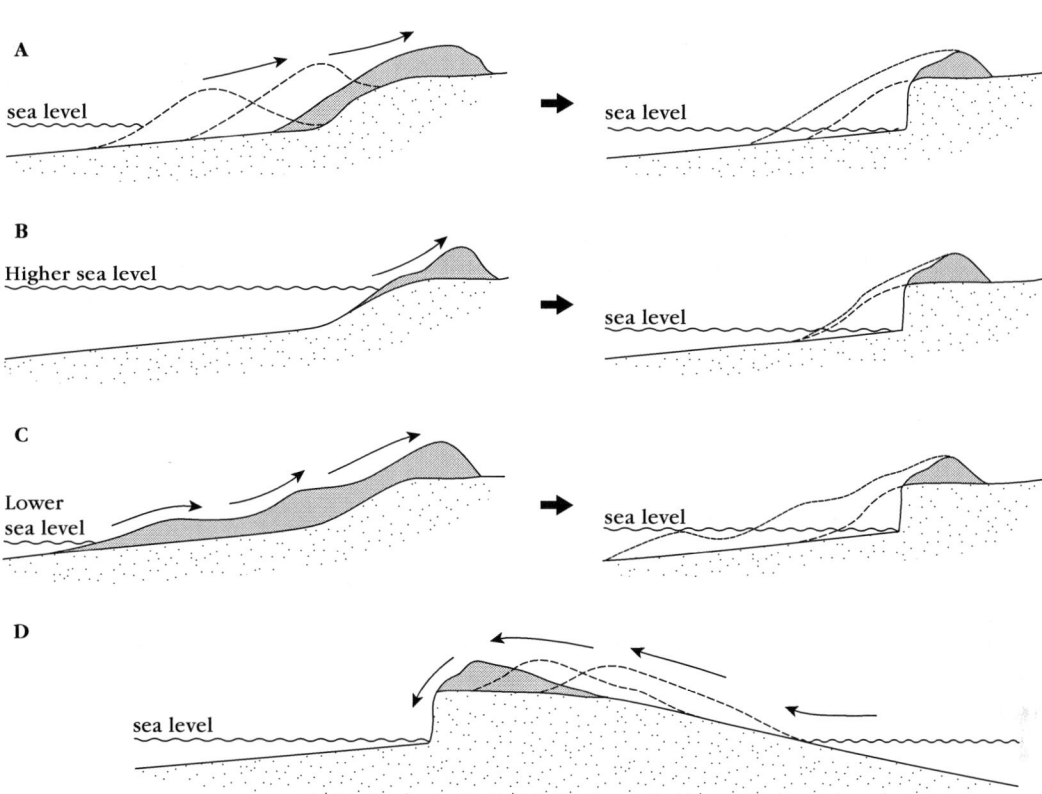

Figure 5.16 Ways in which cliff-top dunes may develop. (A) A transgressive dune truncated by cliff recession; (B) a dune formed at higher sea level stranded by cliff recession after a fall in relative sea level; (C) a dune formed during a lower sea level truncated by cliff recession as sea level rises; (D) a dune that has advanced from a neighbouring beach across a headland. (After Bird, 1984, p. 190.)

of a phase of stormier climate. Stormier conditions during the 17th and 18th centuries were responsible for the inundation of farmland by sand at Culbin in Moray, the subsequent instability resulting in dunes that exceed 45 m in height in Aberdeenshire and 30 m at Lady Culbin in Moray.

Dune fields where dunes have been blown up underlying rock surfaces, can reach altitudes in excess of 100 m on parts of the Highlands and Western Isles coast of Scotland. On some coasts dunes have migrated up and over cliffs. In the past few decades a climbing dune has been heaped against the Thanet cliffs near Foreness Point (see GCR site report for Joss Bay in Chapter 4). Some cliff-top dunes were built from sand blown up from the beach in this way, but the link has been removed by marine erosion (Figure 5.16). An example of this is seen at Upton and Gwithian Towans (see site report in Chapter 7) in Cornwall. The grassy Northam dunes at the northern end of the Westward Ho! cobble spit are also relict, having formed when there was a contiguous sandy beach from which westerly winds supplied sand. The dunes at Balnakeil in Sutherland formed in this way and now cascade over a headland onto the beach beyond (see GCR site report in Chapter 9).

Machair

On the coasts of Scotland there are areas of almost level, calcareous sandy plain, known as 'machair' (Ritchie and Mather, 1984). Typically these have developed behind a cordon of coastal dunes, such as at Eoligarry (see GCR site report in Chapter 9) on the northern part of the Hebridean island of Barra. The geomorphology of machair has been recently reviewed by Hansom and Angus (2001) and the machair GCR sites are described in Chapter 9. The machair sites have also been recognized for the contribution to wildlife conservation as a habitat 'type' protected by European law. This further significance of the machair sites is also discussed in Chapter 9.

On the Atlantic coasts of Britain dunes derived from beaches of calcareous sand may contain cemented (calcarenite) horizons. This dune calcarenite is a form of lithified dune sand found mainly on many warm temperate and tropical coasts. In Britain, and has been reported on Balta Island (see GCR site report in Chapter 8), off the east coast of Unst, Shetland Islands, and forms large outcrops at Evie, Orkney, and at Hough Bay, as well as occurring at several machair locations in western Scotland.

CAUSES OF BEACH AND DUNE EROSION

During the past few centuries, beach erosion has been prevalent on the coasts of Britain (as on much of the world's coastline), with only a few sectors of continuing progradation. Beaches show alternations of cut-and-fill, with erosion during stormy periods followed by accretion in calmer weather, but on many beaches erosion now chronically exceeds accretion. The reduction of beaches as buffer zones has led to erosion of formerly protected coastal land, and the introduction and extension of 'coastal defence' structures that often have further adverse effects on beaches, such as at Dawlish Warren in Devon (see GCR site report in Chapter 8).

Several causes of beach erosion have been identified, but further research is necessary to understand which of these have operated on particular sectors of coastline. Beaches lose sediment when it is carried along the coast in longshore drift or withdrawn to the sea floor during stormy periods. A beach is also depleted when sediment is swept landwards by wave action as washovers into estuaries and lagoons or onto coastal plains, or when winds blow sand from beaches to backshore dunes that are moving inland. Some beaches that prograded when sand or gravel moved in from the sea floor during the Holocene marine transgression (and for a period after the ensuing stillstand was established) are now being eroded because this source of sediment has diminished. This may be the explanation for beach erosion on the north coast of Cornwall, such as at Upton and Gwithian Towans, and at Braunton Burrows in Devon (see GCR site reports in Chapter 7).

Slowly rising sea levels also lead to landward translation of the shoreface and frontal dune erosion. It can be argued that beach erosion is now a widespread natural phenomenon, and that stable or prograding beaches are anomalous on coasts where a major marine transgression has been followed by a stillstand. Certainly many beaches in Britain are now erosional in character, probably as a result of the combined effect of sediment reduction and sea-level rise and, in some areas, possibly an increase in storminess (Dawson *et al.*, 2001). Unfortunately, many of our efforts to reduce erosion have, in fact, exacerbated the problem.

Anthropogenic factors

Present-day inputs of sediment are increasingly restricted by coast protection works, but the present natural supply from offshore and cliffs undergoing erosion is also insufficient to produce and maintain the volume of gravel beaches at the levels of the late 19th century. Major exceptions are the cliff-beach system east of Lyme Regis, fed by cliff-top chert and transported by major landslides to the beach, and beaches fed by coarse, clastic, glacial sediments along parts of the upland coasts of Scotland. Many beaches have transgressed as Holocene sea level rose and have become compartmentalized by headlands (Hansom, 2001) or coast protection works.

At Dungeness, there is virtually no modern material feeding into the beaches. Longshore transport along its southern shore is removing and redistributing shingle from older parts of the structure and artificial beach feeding has helped maintain this transport stream since at least the 1950s. Even where there are chalk cliffs undergoing erosion that produce flints the present supply is insufficient to produce large fringing beaches. For example along the coast east of Dover, longshore transport has carried most of the 19th century fringing beach to a sink north of Deal. With further supplies from the west cut off by the harbour arms at Folkestone and Dover, the cliffs have become more erosionally active, but the supply of flint remains small.

The economic value of sand beaches as a major attraction for coastal recreation and tourism and the susceptibility of soft cliffs to erosion has led to the construction of extensive coast protection works to combat erosion at many coastal resorts.

Causes of beach erosion

The response to erosion has generally been to build structures such as sea-walls and groynes, which often exacerbate the problem, notably where groynes and breakwaters have reduced or cut off longshore drift and thereby sediment supply. The pebble beach between Dover and Kingsdown (see GCR site report in Chapter 4) that used to be maintained by eastward drifting of shingle along the shore beneath the cliffs has almost disappeared as the result of the interception of longshore drift by the breakwaters at Dover Harbour. The proximal end of Orfordness is now only prevented from breaching by a sea-wall and groyne-field, and the main ridge at Blakeney has been breached in recent years, possibly as a combined result of reduced longshore sediment supply and its natural transgressional tendency.

Beach erosion on the Bournemouth coast became severe after the building of sea-walls halted recession of the sandstone cliffs, and cut off the supply of sand and gravel to these beaches; sandy spoil from the dredging of the main channel into Poole Harbour (May, 1990) was used to feed the beach. Farther to the east, the sand spit at the mouth of Christchurch Harbour, which was formed by longshore sand transport from the erosion of Bournemouth's cliffs, now requires both walls and groynes along most of its length owing to reduced sediment supply.

At Spey Bay, Moray, the supply of fluvial gravels to the beach has further diminished as a result of bank protection in the River Spey and this has contributed to erosion and thinning of the beach, which protects the village of Kingston (Gemmell *et al.*, 2001a,b).

Sand and gravel extraction

Extraction of gravel has taken place from offshore banks, beaches and large landward deposits, such as Dungeness and Rye Harbour, to supply the considerable demand for aggregates. Elsewhere, deepening of nearshore water by the dredging of sand or gravel has been followed by beach erosion, such as at Hallsands (see GCR site report in Chapter 6) in Devon (Robinson, 1961).

In Cornwall and in the Hebrides, beaches have been depleted where calcareous sand has been removed from beaches for use as lime on farmland, and similar erosion occurred after the extraction of pebbles from shingle beaches at Gunwalloe in Cornwall and Seatown in Dorset.

The volume of sediment removed from these coastal sediment stores and transport pathways has been considerable and it is not surprising that some beaches are now seriously in deficit. For example, Bray (1986) estimated that between 100 000 and 200 000 tonnes were extracted from Seatown Beach during World War II and a further 50 000 tonnes between 1956 and 1986, an annual loss of 2095 m^3 a^{-1}. Material has been removed from Chesil Beach at West Bay for over 700 years, with about 1 million tonnes of gravel removed between the mid-1930s and 1977 (Hydraulics Research Station, 1979).

In many areas of Scotland, particularly the western and northern Isles, sand has been traditionally removed for agricultural purposes either to 'lighten' a heavy glacial clay soil or to provide a ready source of lime using shelly sand in an otherwise acidic environment. Although now much reduced, it is known that such activity continues in the more remote places, in spite of chronic beach and dune erosion in those areas.

Beach replenishment

In recent decades, more attention has been given to beach replenishment as a means to mitigate the effects of coastal erosion, using sand and gravel obtained from the sea floor or inland quarries, such as at Bournemouth, Weymouth, Sidmouth and Minehead in south-west England, or recycled from downdrift accumulations, such as at Rye in Sussex (Bird, 1996). The beach replenishment scheme at Spey Bay in Moray has successfully used the seawardmost gravel ridges of the Spey mouth gravel complex to recharge a depleted downdrift section (Gemmell *et al.*, 2001a,b). Although not on the same scale as replenishment programmes in North America or the Netherlands, the British examples suggest that such an approach represents an environment-friendly and sustainable way to manage beaches, particularly since traditional engineering approaches may cause negative side effects on adjacent coasts.

Chapter 6

Gravel and 'shingle' beaches – GCR site reports

Introduction

INTRODUCTION

V.J. May

The gravel structures and beaches of the British coast are among its best-known and longest-studied geomorphological features. In England, deposits of well-rounded beach gravel are known as 'shingle', and this less geomorphologically precise word is retained for the English and Welsh sites described in the present chapter, where the usage is more common in the literature.

'Shingle' is characterized by grain sizes between 4 and 64 mm (–2 to –6 phi). Many shingle structures are formed predominantly of clasts within this size range, but even the most distinctively sorted such as Dungeness and Chesil Beach contain clasts of many different sizes. Some beaches are characterized by clasts whose long axis exceeds –6 phi and are described as 'cobbles'. The eastern part of Chesil Beach, and the majority of the materials at Budleigh Salterton and Westward Ho! exceed shingle size. Many shingle beaches are, in reality, of mixed clast sizes, with varying quantities of both finer- and coarser-grained materials.

Although most shingle beaches in England and Wales are formed of flint or chert, many include clasts comprising relatively weak materials such as sandstone or chalk or other harder materials; a wide range of clasts formed from harder materials are characteristic of western and northern beaches and in Scotland.

Scolt Head Island, Chesil Beach and Dungeness are three of the most scientifically well-known coastal shingle structures of international renown, but there are many other geomorphologically important types of shingle and gravel features in Britain including small cliff-foot beaches, bay-bars, small recurved spits and beach plains. Some coastal gravel/shingle structures in Britain are at least 5500 years old. Others are relatively young, such as recently formed cliff-foot beaches.

About 1040 km of the British coast is formed by gravel structures (excluding any cliff-foot beaches). But gravel structures also form the base of many sand spit and dune features. If the sand structures with a gravel *base* are added, then the British coast is fringed by about 2900 km of gravel-dominated beaches.

The extent to which gravel beaches are well-sorted affects permeability and the extent to which they behave as reflective or dissipative structures.

Gravel/shingle features have been classified according to their plan-form (e.g. Pethick, 1984; see Figure 6.1), and their profile (e.g. Wright and Short, 1983). Pethick (1984), following many earlier writers, summarized the plan-form of beaches at different scales, ranging from the small rhythmic features such as *cusps* that occur on many beaches, to the very large detached beaches such as spits. Wright and Short (1983), in contrast, focused on the relationship between beach profiles and wave conditions, with a specific emphasis on the differences between a dissipative domain in which beaches display a flat shoaling slope and wide surf zone, with multiple parallel nearshore bars and a reflective domain, characterized by steep beaches (> 6°), with no nearshore bars. Between these extremes there are several intermediate domains that are dominated by longshore bar-troughs, rhythmic bars, transverse (welded) bars or low-tide terraces.

The GCR sites descibed in the present chapter (Figure 6.2 and Table 6.1) contain many short-timescale features, but as major landforms owe their origins to processes acting over considerable periods of time. They include cliff-foot fringing beaches, pocket beaches, bay-bars, cheniers, spits (some with complex recurves), barrier beaches and cuspate forelands.

The simplest gravel and shingle structures are cliff-foot fringing beaches. They fall broadly into two groups:

1. beaches at the foot of cliffs that are the present-day source of most clasts in the beach.
2. cliff-foot beaches where the gravel/shingle is derived from longshore transport. In southern Britain, many shingle beaches are formed from flint that was eroded from the chalk under periglacial conditions and, in northern Britain, under glacial conditions from a range of lithologies, and then transported landwards by the postglacial transgressing seas.

Gravel and shingle beach ridges

The process of migration of a ridge across the sea floor during a marine transgression has often been used to explain the establishment of large linear features such as Slapton bar, Chesil Beach, and Blakeney Point. They may be parts of barrier beaches that have assumed their present form

Gravel and 'shingle' beaches

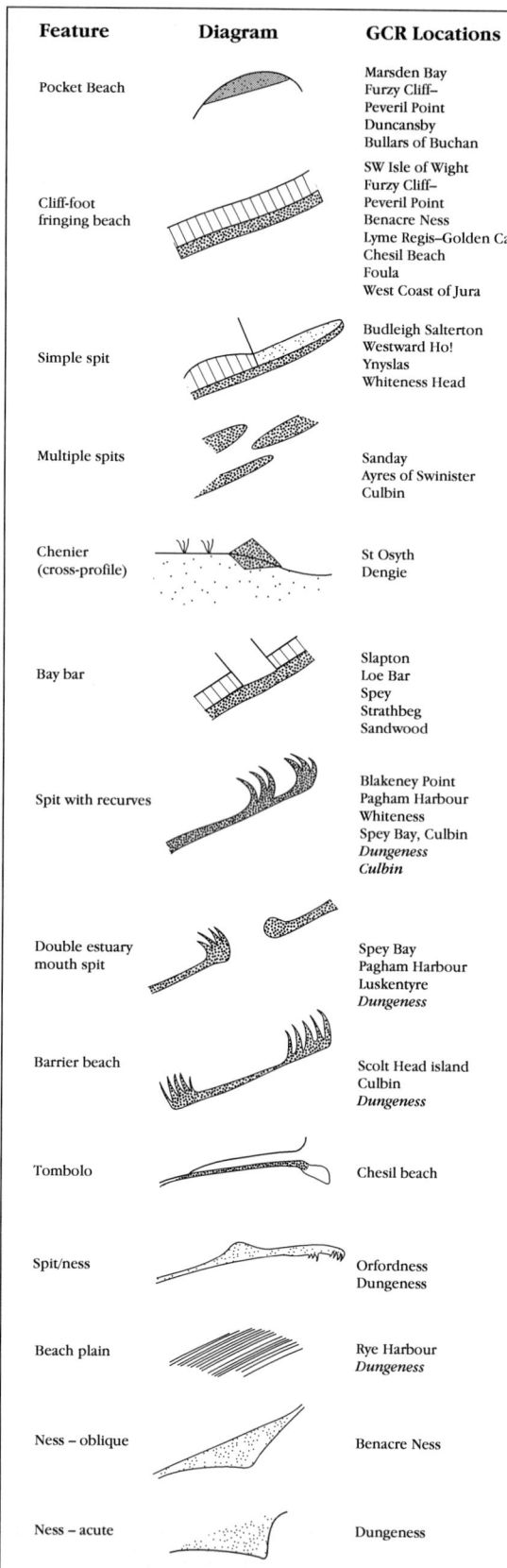

because they have been built up against the coast. As they transgress the pre-existing landscape, these fringing beaches can become compartmentalized by headlands into smaller embayment beaches, as for example between Osmington and Kimmeridge and in Spey Bay. Chesil Beach is undergoing a similar change between West Bay and Abbotsbury. Slapton bar is attached to headlands at both ends and is in continuity with a cliff-foot fringing beach, whereas Chesil Beach fringes cliffs for much of its western half and then stands separately as a tombolo joining the Isle of Portland to the mainland. In contrast, Blakeney Point is only attached at its eastern end where it continues the line of fringing beach below Sheringham cliffs. It then becomes aligned towards the dominant waves from the north-east. However at its distal end it lies in deeper water, and waves from the north and north-west have been instrumental in developing a series of modern and relict recurves. Gradual elongation of a barrier partially blocking an embayment or the extension of a spit may give rise to a bay-bar.

Larger shingle structures are characteristically built up of series of beach ridges, several hundred in the case of Dungeness, which preserve earlier episodes of beach-construction. Beaches that are initially oriented alongshore as drift-aligned features show a tendency to swash-alignment through time (Davies, 1972). This process involves erosion updrift and the truncation of former ridges, often at a significant angle to the present-day ridge orientation. As spits extend into deeper water, they develop recurves under the influence of different wave directions. Recurves are often grouped, as for example at Scolt Head Island and Blakeney Point, and may be related to pulses of greater sediment flux. Such pulses have been identified at Spey Bay by Gemmell *et al.* (2001a,b) along the main gravel structures but the extent to which these might contribute to recurving at the western end of Spey Bay is less clear.

On some coasts, large cuspate forms develop.

Figure 6.1 Forms and typology of gravel and shingle structures and the GCR sites that represent them. The schematic diagrams show the plan form of the structure concerned. Italic type indicates presence of relict features at a site. In some cases gravel forms the core of the feature, and is now covered in sand.

Introduction

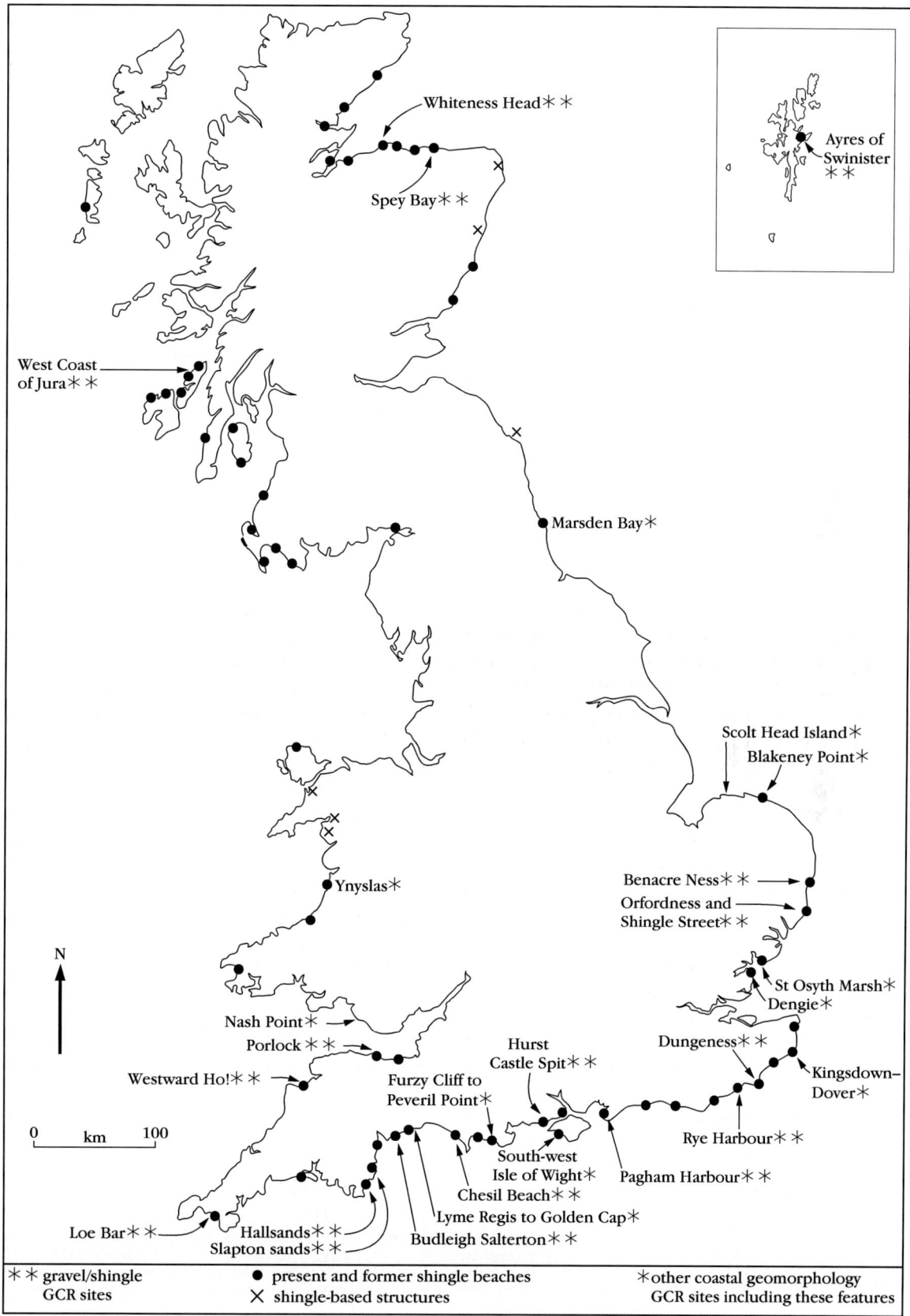

Figure 6.2 Coastal shingle and gravel structures around Britain, showing the location of the sites selected for the GCR specifically for gravel/shingle coast features, and some of the other larger gravel structures.

Gravel and 'shingle' beaches

Table 6.1 Main features and sediment sources of gravel/shingle beach and ness GCR sites, including coastal geomorphology GCR sites described in other chapters of the present volume that contain shingle beach/ness structures in the assemblage.

Site*	Main features	Other geomorphological features	Present day natural sources of sediment	Tidal range (m)
Marsden Bay	Beach phases	Cliff, stack	Local cliff erosion – small	4.2
Furzy Cliff to Peveril Point (Dorset Coast)	Shingle pocket beaches	Cliffs/platforms Mass movements	Cliff erosion – small, restricted	1.7 (E)– 2.0 (W)
Nash Point	Cobble and shingle pocket beaches	Platforms, caves	Local cliff/platform erosion – small	6.0
Kingsdown to Dover	Cliff-foot beach	Cliffs and platforms	Cliff erosion – small	5.9
Seven Sisters, (Beachy Head to Seaford Head)	Cliff-foot fringing beaches	Cliffs and platforms	Cliff/platform erosion – small	6.0
South-west Isle of Wight	Cliff-foot beach and feeder cliffs	Cliffs	Chalk and sandstones – small	3.3 (E)– 2.2 (W)
Lyme Regis to Golden Cap	Shingle beach sediment supply and budget	Feeder cliffs	Significant inputs of flint/chert	3.5
Ynyslas	Sand and shingle spit	Dunes	Reworking till – restricted	4.0
Westward Ho!	Cobble beach and spit	Dunes	Reworking of emerged beach – restricted	7.9
Loe Bar	Shingle bay-bar	Cliffs, ria	Local cliff erosion – small	4.7
Slapton Sands and Hallsands	Shingle bay-bar Beach destruction	Emerged beach, relict cliff and platform	Minimal	4.4
Budleigh Salterton	Shingle beach and spit Major former feeder to south coast beaches	Soft cliffs	Cliff erosion – maintains budget	4.0
Chesil Beach	Barrier beach Tombolo		Minimal – local	2.0
Porlock	Retreating shingle barrier with both swash-aligned and drift-aligned longshore sections	Recent breached tidal inlet allowing active back-barrier saltmarsh development	Minor source of gravel from updrift coastal slides. Main solifluction source of sediment now exhausted until future sea-level rise creates new supply	9.3
Hurst Castle Spit	Shingle spit and recurves	Saltmarsh	Possible from offshore	2.2
St Osyth Marsh	Cheniers	Saltmarsh	Localized reworking of gravels and chenier root	3.8
Dengie Marsh	Cheniers	Saltmarsh	Localized reworking of gravels and chenier root	3.8
Blakeney Point (North Norfolk Coast)	Major shingle spit	North Norfolk coast assemblage	Cliff erosion – restricted Longshore transport – large	6.4 (W)– 4.7 (E)
Scolt Head Island (North Norfolk Coast)	Barrier beach and spits	North Norfolk coast assemblage	Longshore transport – large	6.5

* Sites described in the present chapter are in **bold** typeface

Introduction

Table 6.1 – *contd*

Site*	Main features	Other geomorphological features	Present day natural sources of sediment	Tidal range (m)
Pagham Harbour	Double spit development		Local cliffs – restricted Kelp rafting	3.4
Ayres of Swinister	Complex of bay bars and spits		Local tills – small	1.5
Rye Bay	Spit developments Shingle beach plain		Reworking proximal end Longshore – minimal	5.8
Benacre Ness	Shingle ness	Rapidly retreating cliffs	Cliff erosion – maintains input	2.1
Whiteness Head	Spit		Longshore transport – large	3.5
Spey Bay	Spits, bay bars, emerged gravel ridges		Longshore – now partially restricted – fluvial input	3.5
West Coast of Jura	Over 11 000 year sequence of emerged gravel ridges	Emerged shore platforms	Local, between headlands	2.5
Orfordness and Shingle Street	Major shingle ness and spit		Longshore – restricted by groyne fields	1.9 (N)– 3.4 (S)
Dungeness	Major cuspate foreland Relict barrier beach Over 5000 year sequence of beach ridges		Re-distribution within site	6.2

* Sites described in the present chapter are in **bold** typeface

Often known in Britain as 'nesses', these cuspate forelands may result from the convergence of opposing movements of sediment alongshore, as at Buddon Ness (Barry Links GCR site), Angus, or may be a horizontal wave-form that migrates alongshore progressively transferring sediment from the windward face to the opposite side. Some nesses are fringing features, for example Benacre Ness, which lies at the foot of cliffs cut in Quaternary tills and gravels. In contrast, the longshore-parallel spit at Orfordness has developed a distinct cuspate feature or ness at a point where there is a change in shore alignment combined with the effects of wave refraction by offshore banks. Most nesses are strongly associated with substantial offshore banks, although Dungeness is unusual amongst such features in lacking an associated offshore bank. These banks affect wave refraction, but it is not possible to state unequivocally whether the shoals develop as a result of offshore transport of sediment from the foreland as it aligns itself at an angle to the shore, or if their presence is a contributory factor in the development of the foreland. The forms of gravel and shingle beaches are predominantly the result of wave action, with the small-scale features responding to each individual wave. Over longer timescales, however, gravel beaches are strongly controlled by the dominance of particular wave directions and the effects on longshore transport of clasts.

Where isostatic uplift has been substantial, emerged gravel ridges occur where supply is, or has been, plentiful, such as on the west coast of Jura, in the Inner Hebrides (see GCR site report).

On many British upland coasts, gravel and shingle structures form the base upon which sand spits and dune fields have accreted, but with a few exceptions such as at Culbin, Moray (Comber, 1993) and at Central Sanday (Rennie and Hansom, 2001) (see GCR site reports in chapters 11 and 8) many of these buried gravel structures have not been interpreted. Gravel extraction from these locations may put the sand structures at risk of accelerated erosion, as happened at Spurn Point prior to the 1849 breach (IECS, 1992).

Past management also influences the ability of gravel structures to adjust to sea-level change and storminess, Porlock being a good example of a free-standing gravel structure undergoing post-management adjustment.

The conservation value of gravel and shingle beaches

In spite of reductions in sediment supply, many gravel/shingle beaches remain scientifically important, and worthy of conservation-protection measures so that they can continue to evolve and provide information about the development of coastal gravel systems, coastal form development and the effects of coastal management. It is important that such sites are managed wisely so that the systems can be allowed to develop as naturally as possible. The sites are of high conservation value because

1. internationally, they are among the most well-known coastal features of Britain, especially Chesil Beach, Dungeness and Culbin,
2. they have a distinct flora and support several endemic species of invertebrates,
3. they continue to act as sources of sediment for adjacent beaches,
4. they preserve several millennia of recent coastal deposition and changes in their form reflect variations in wave and wind climates.

In the present chapter the GCR sites (Figure 6.2 and Table 6.1) follow a sequence from shoreline or fringing beaches to the more complex forms of detached beaches.

Gravel and shingle structures as biological SSSIs and Special Areas of Conservation (SACs)

In Chapter 1, it was emphasized that the SSSI site series is constructed both from areas nationally important for wildlife, and GCR sites. An SSSI may be established solely for its geology/geomorphology, or its wildlife/habitat, or it may comprise a 'mosaic' of biological and GCR sites that may be adjacent, partly overlap, or be co-incident. There are a number of coastal SSSIs that are crucially important to the natural heritage of Britain for their wildlife value, but which implicitly contain interesting geomorphological features – such as gravel/shingle structures – that are not included independently in the GCR because of the 'minimum number' criterion of the GCR rationale (see Chapter 1). These sites are not described in the present geomorphologically focused volume.

In addition to being protected through the SSSI system for their national importance, certain types of gravel/shingle habitat are eligible for selection as Special Areas of Conservation (SACs; see Chapter 1) under the 'Habitats Directive'. The principal Annex I SAC coastal gravel/shingle habitat present in the UK is 'Perennial vegetation of stony banks', but on gravel/shingle beaches commonly fringing this habitat, the more transient 'Annual vegetation of drift lines' also occurs.

Coastal gravel/shingle SAC site selection rationale

The Habitats Directive Annex I habitat type most relevant to the present chapter is 'Perennial vegetation of stony banks'. Ecological variation in this habitat type depends on stability, the amount of fine material accumulating between clasts, climatic conditions, width of the foreshore, and past management of the site. The ridges and lows formed in gravel/shingle structures also influence the vegetation patterns, resulting in characteristic zonations of vegetated and bare gravel/shingle. The presence of the yellow horned-poppy *Glaucium flavum* and the rare sea-kalc *Crambe maritima* and sea pea *Lathyrus japonicus*, all species that can tolerate periodic movement, is significant. In more stable areas above this zone, where sea spray is blown over the gravel/shingle, plant communities with a high frequency of salt-tolerant species such as thrift *Armeria maritima* and sea campion *Silene uniflora* occur. These may exist in a matrix with abundant lichens.

On the largest and most stable structures the sequence of vegetation includes scrub, notably broom *Cytisus scoparius* and blackthorn *Prunus spinosa*. Heath vegetation with heather *Calluna vulgaris* and/or crowberry *Empetrum nigrum* occurs on stable structures, particularly in the

Introduction

Table 6.2 Candidate and possible Special Areas of Conservation in Great Britain supporting Habitats Directive Annex I habitat 'Perennial vegetation of stony banks' and/or 'Annual vegetation of drift lines' as qualifying European features. Non-significant occurrences of these habitats on SACs selected for other features are not included. (Source: JNCC International Designations Database, July 2002.)

SAC name	Local authority	Gravel/shingle habitat extent (ha)
Bae Cemlyn/ Cemlyn Bay	Ynys Môn/ Isle of Anglesey	1.3
Chesil Beach and the Fleet	Dorset	96.2
Culbin Bar	Highland; Moray	122.5
Dee Estuary/ Aber Dyfrdwy*	Cheshire; Fflint/ Flintshire; Wirral	1
Dungeness	East Sussex; Kent	2266.1
Isle of Portland to Studland Cliffs	Dorset	1.4
Lower River Spey–Spey Bay	Moray	65.2
Minsmere to Walberswick Heaths and Marshes	Suffolk	8.8
Morecambe Bay	Cumbria; Lancashire	57.5
North Norfolk Coast	Norfolk	98.4
North Uist Machair	Western Isles / Na h-Eileanan an Iar	3
Orfordness–Shingle Street	Suffolk	553.3
Sidmouth to West Bay	Devon; Dorset	4.4
Solent Maritime	City of Portsmouth; City of Southampton; Hampshire; Isle of Wight; West Sussex	226.5
Solway Firth	Cumbria; Dumfries and Galloway	8
South Uist Machair	Western Isles / Na h-Eileanan an Iar	†

* Possible SAC not yet submitted to EC
† Feature is minor component of SAC
Bold type indicates a coastal geomorphology GCR interest within the site

north. This sequence of plant communities is also influenced by natural cycles of degeneration and regeneration of the shrub vegetation that occurs on some of the oldest ridges.

Vegetated stony banks are scarce. There are only a few large sites in Europe, and the UK hosts a significant part of the European resource of this habitat. Although there are only some 4000 ha of stable or semi-stable vegetated gravel/shingle around the whole of the coast of the UK, the habitat is widely distributed and also exhibits a wide range of variation. The selection of sites reflects the UK's special responsibility for conservation of this habitat type and aims to cover the geographical range and variation of the habitat type. All the largest examples with good conservation of structure and function have been selected, together with additional smaller sites to complete the coverage of range. Site selection has also favoured gravel/shingle structures that support vegetation sequences ranging from pioneer communities to heath and scrub. The selected sites represent a substantial proportion of the European resource.

The vegetation that colonizes drift lines of gravel/shingle at or above mean high-water spring tides is dominated by annual plants. The types of deposits involved are generally at the lower end of the clast-size range (2–200 mm diameter), with varying amounts of sand interspersed in the gravel/shingle matrix. These deposits occur as fringing beaches that are subject to periodic displacement or overtopping by high tides and storms. The distinctive vegetation, which may form only sparse cover, is therefore ephemeral and composed of annual or short-lived perennial species. At most sites where it occurs, the habitat is naturally species-poor, and there is a limited range of ecological variation. Many gravel/shingle beaches are too dynamic to sustain drift-line vegetation. Many of the fringing beaches supporting drift-line vegetation are small, and annual vegetation may exist in one location in one year but not another.

Therefore, although widespread around the UK, sites where this Annex I type is persistent are rare, and even the largest sites probably support less than 10 ha of this habitat. Sites have been selected to reflect the more constant occurrences of drift-line vegetation, normally found in association with larger, more stable areas of gravel/shingle structures. The selected sites represent the majority of the more persistent examples of this habitat type in the UK. They all exhibit good conservation of structure and function (i.e. they are relatively unmodified and are less prone to human disturbance) and represent the range of variation in substrate type and physical structure.

Table 6.2 lists coastal shingle SACs, and indicates which of these sites are also (at least in part) important as part of the GCR and are described in the present chapter.

WESTWARD HO! COBBLE BEACH, DEVON (SS 440 310)

V.J. May

Introduction

The cobble ridge at Westward Ho! extends northwards across the mouth of the Taw–Torridge estuary (see Figure 6.2 for general location). The major sand dunes of Braunton Burrows lie to the north. Westward Ho! ridge forms a major, classic, coastal landform, more by reason of its sediment size than of its form. Few spits are formed by large cobbles at the back of an extensive sandy intertidal zone. The cobble ridge has retreated steadily during the last 100 years (Figure 6.3) and is now much modified at its landward end by artificial gabions. There has also been some artificial beach feeding since the 1980s. Prior to the engineering works, some of the cobble material reached the beach from natural sources to the west where erosion of rock cliffs, platforms and Pleistocene pebble deposits provided a source for much of the material. Dunes form much of the northern part of the spit, known as 'Northam Burrows', but sand gravel and cobbles move towards the distal end, forming a spatulate feature within the estuary of the rivers Taw and Torridge. Both the unusual beach and the nature of its retreat have been the focus of scientific study (Spearing, 1884; Rogers, 1908; Kenyon-Bell, 1948; Stuart and Hookway, 1954; Slade, 1962; Keene, 1986, 1989, 1992, 1996). The processes and forms of the Taw–Torridge estuary have been reviewed by Comber *et al.*, (1993).

Description

The ridge described by the early writers extended from the cliffs at Westward Ho! itself to its distal end, but retreat of the spit and the threat that this posed to the village led to coast protection works (Keene, 1986, 1996) which have transformed the southern proximal part of the cobble ridge into an artificial structure. This section has been excluded from the GCR site boundary. Nevertheless, the four main elements of the original feature are still represented within the site (see Figure 7.10 for local geomorphology):

1. A single ridge of cobbles about 25 m wide, WNW-facing, which is transgressing onto blown sand overlying blue clay. The transgressional beach rests on this base at about high-water springs, but cobbles form a frontal slope to just above high-water neaps (Figure 6.3). Historically, this ridge has been associated with shoreline erosion at its landward end.
2. An area of former cobble ridges that lie at an acute angle to the present-day shoreline and represent earlier phases in its growth. They underlie much of the blown sand of Northam Burrows.
3. A spatulate distal end, known as 'Grey Sand Hill', which extends into the Taw–Torridge estuary.
4. A wide intertidal area that includes areas of sand and cobbles, particularly towards the estuary.

The tidal range at spring tides is 7.9 m. The maximum is from the Atlantic Ocean, but most waves approach the beach from slightly north of west, towards which the main ridge is aligned.

Interpretation

The origin of the ridge is not known, although Keene (1996) regards it as a comparatively recent feature. Hall (1879) described a peat and blue clay deposit about 400 m seawards of the cobble ridge and Rogers (1908) recorded remnants of former forest, a kitchen midden and a submerged pebble ridge in the intertidal area.

Westward Ho!

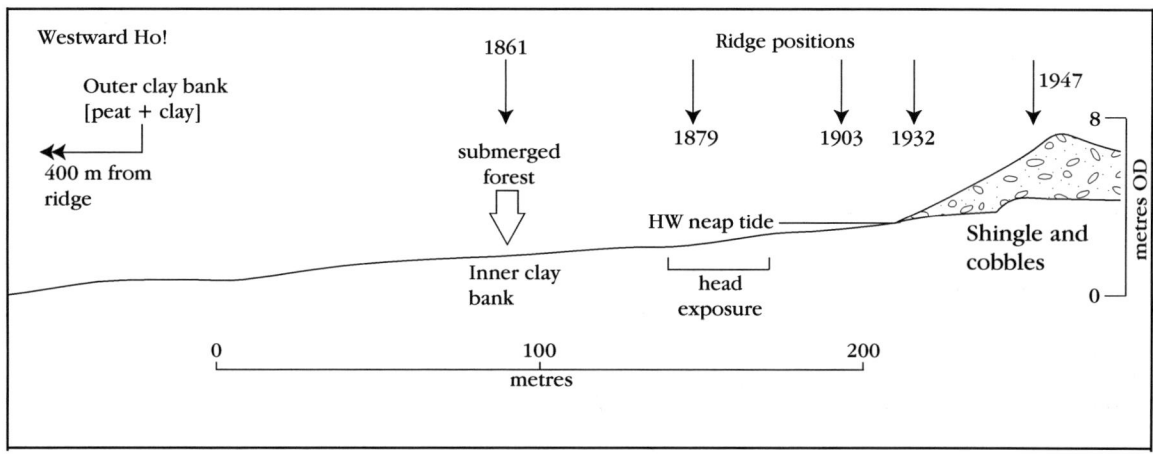

Figure 6.3 Historical recession of position of beach crests at Westward Ho! (Based on Campbell and Bowen, 1989 and Keene, 1996.)

The peat contained leaves, seeds and fruits of iris, oak *Quercus*, hazel *Corylus*, alder *Alnus*, elder *Sambucus*, sea aster *Aster tripolium*, common orache *Atriplex patula*, blackberry *Rubus*, dogwood *Cornus* and lesser spearwort *Ranunculus flammula*. Shells of oyster *Ostrea* and limpets *Patella* were also recorded. The blue clays were reported to include flint flakes (thought to be Neolithic in age), small pebbles and angular fragments of Carboniferous rocks. The mud-snail *Hydrobia ulvae* was common and bones were said to include red deer *Cervus elephas*, and Celtic shorthorn cattle *Bos longifrons*. Since most of the sediment disappeared before modern interpretation and dating techniques were available, the records have to be taken as a possible indication that the cobble ridge lay seawards of the position of the submerged forest and that the land behind it was well colonized by vegetation. The similarity with the development of the cobble ridges at Clarach and Ynyslas on the Welsh coast (see Figure 8.15) may indicate a much earlier age (about 4000 years BP) for Westward Ho!. In contrast to summer depths of sand on the western beach of up to 1.2 m, erosion during the winter of 1983–1984 exposed a thick band of head, suggesting that a wide apron of periglacial debris may have extended some distance seawards of the present-day shoreline (Keene, 1996) and this may provide an alternative explanation for the location of the submerged forest. Erosion of such deposits would provide the source for a transgressing postglacial beach. Samples from the top of the peat bed were radiocarbon dated to 6585 ± 120 year BP (Q-672; Churchill and Wymer, 1985) and 4995 ± 105 years BP (Kidson, 1977). The Holocene sequence is described in detail by Campbell (1998) following description of inner and outer peats by Balaam et al. (1987) which indicates that the outer peat was inundated by marine/estuarine conditions about 5200 years BP.

Steers (1946a) noted that the burrows at Northam were first recorded by Ridon in 1630. Though limited, this evidence indicates a predecessor to the present feature farther seaward than the 19th century ridge. Campbell (1998) emphasizes that Westward Ho! shows clear evidence for a transition from a lower Late Devensian sea level, followed by an intially rapid rise of the Holocene sea that swamped a coastal forest dating from about 6000 years BP. With a slowing of sea level rise, the present coastal configuration was established.

Stuart and Hookway (1954) cited Risdon's 17th century survey of Devon as the earliest reference to a ridge at Westward Ho!. The first useful cartographic evidence is dated 1861 when the ridge was 1.8 km long: in the following 100 years the cobble ridge migrated landwards some 152 m (Figure 6.3). The northern part of the ridge grew from four sub-ridges in 1884 to twelve in 1952. Accretion at the northern end was accompanied not only by retreat at the southern end but also a reduction in overall volume. Up to 1884, it is reported (Spearing, 1884) to have stood just over 2 m above the Burrows and about 6.5 m above the level of the sand beach. It was about 48 m wide. By 1954, it was about 25 m wide and stood a similar height above the Burrows. Steers (1964a) suggested, without quantification, that the average grain size of the sediment might also have diminished. Between 1886 and 1947 the shoreline moved seawards about 25 m at its northern bend facing the Taw–Torridge estuary, but between 1959 and 1996 the ridge crest retreated 30 m (Keene, 1996). This is thought to be the result of increasingly focused wave energy at this point. Short period waves, especially in storms from the north-west, are refracted by sand and shingle banks in the estuary. As erosion has continued unabated and coast protection measures have been undertaken, most reports concentrate on the rate of erosion and the protection measures to be adopted (Slade, 1962; Kenyon-Bell, 1948; Stuart and Hookway, 1954; Halcrow, 1980; Comber *et al.*, 1993). The beach and intertidal changes and sediment budget estimates have been reviewed by Comber *et al.*, (1993), who report a net loss of shingle and cobbles on the eroding part of the ridge of 1500–5000 m^3 a^{-1}. The net annual loss between 1886 (when beach volume was estimated at 300 000 m^3) and 1974 has been estimated at 1200 m^3 a^{-1}, and between 1981 and 1986 the local authority transported about 15 000 m^3 a^{-1} of material from the distal end of the spit to re-inforce its proximal end. This has since reduced to 5000 m^3 a^{-1} to offset erosional loss. Excepting Keene (1986), little attention is paid in the literature to the sources and origin of the cobble ridge, but there may be some similarities to the history of the Ynyslas ridge (see GCR site report in Chapter 8).

There appear to have been three main natural sources for the cobbles:

1. The emerged ('raised') beach and solifluction apron to the west of Westward Ho!, from which pebbles and cobbles have been and still are eroded and are subsequently moved by longshore transport to the modern beach.
2. offshore cobble ridges such as that described by Rogers (1908) and comparable intertidal cobble fields, from which waves would transport material landwards.
3. Materials derived directly from the erosion of the cliffs to the west. Much of the material forming the cobbles in both the emerged and modern beaches is derived from the Culm Measures.

The intermittent growth of the ridge is attributed by Keene to fluctuations in the supply of pebbles from the west, depending upon both the magnitude and frequency of cliff-falls and on the rate of movement of pebbles along the shore from the west.

There are some similarities between Westward Ho! and the pebble beach at Budleigh Salterton (see GCR site report in the present chapter) in that both owe their main constituents to the reworking of pebble or cobble beds. However, Budleigh Salterton has both ends and most of its length resting against the cliffs. It can only move as the cliffs retreat, a process that provides a fresh supply of pebbles to the beach. At Westward Ho!, the supply of cobbles from the emerged beach is limited. Furthermore, as the beach has migrated inland, both the distance and the area of foreshore over which the material can travel have changed. Cobbles could be moved across the intertidal area just as effectively as alongshore in the storm beach. Those within the storm beach are likely to move inland during periods of washover or to migrate alongshore towards the distal end. There is little evidence to suggest that any return towards Westward Ho!

In northern Britain, many sand spits and beaches rest upon shingle and cobble ridges. At Westward Ho!, however, insufficient sand has been transported from the intertidal zone to build up the dunes on the proximal part of the spit, which rests on clay and head above a rock platform. Beaches formed of large shingle and cobbles are found at several locations along the western coasts of England and Wales (for example, the beaches at Newgale and Llanrhystyd – see Orford, 1977). There are none that form a narrow spit such as that at Westward Ho! The presence of the spit affects the alignment of the Taw–Torridge estuary and this affects the distal development of the sand beach at Braunton Burrows (see GCR site report in Chapter 7). Carter (1988) suggests that such beaches originate as a result of either marine reworking of glacial deposits by the rising Holocene sea or primary sedimentation from adjacent eroding cliffs or seabed. At Westward Ho! the latter appears to be most important, but since the source includes clasts from an emerged beach, that deposit may be attributable to either of the origins suggested by Carter.

Conclusions

Westward Ho! is a rare and excellent example of a narrow cobble spit. Although cobble beaches are not uncommon in Britain, they rarely occur as spits. Second, much of the material forming the spit includes clasts that were already substantially modified as a result of erosion of an emerged ('raised') beach. Third, there is a well-documented Pleistocene–Holocene history for the intertidal sediments.

LOE BAR, CORNWALL (SW 643 241)

V.J. May

Introduction

Loe Bar lies about 4 km SSW of Helston, Cornwall (see Figure 6.2 for general location). It encloses a lagoon occupying part of a former ria and forms an integral part of a beach system extending from Porthleven (SW 627 254) to Gunwalloe (SW 653 223; Figure 6.4). The site is important to coastal geomorphology on two counts. First, Loe Bar is a rare example (in England and Wales) of a bay–bar, and second, it is a key member of a suite of major beaches formed and maintained by predominantly south-westerly wave regimes. The beach is formed of rounded, fine shingle and coarse sand predominantly comprising flint or chert for which there is no local source on land. Present-day inputs of sediment from the adjacent cliffs are small, and overall the beach is in deficit. The bar itself is washed over during periods of high wave-energy and a series of washover fans occurs behind the bar. The periodic breaching of the bar has been described in the literature (Ward, 1922; Toy, 1934; Steers, 1946a; Goudie and Gardner, 1985), but the origin of the bar has attracted less attention (Toy, 1934; Goudie and Gardner, 1985).

Description

The site extends for 4.3 km south-eastwards from Porthleven, but the Loe Bar itself forms only 400 m of this beach. The site is divided into three parts: the cliffs and beach known as 'Porthleven Sands', the bar itself, and the cliffs and beach known as 'Gunwalloe Fishing Cove Beach'.

The beach throughout the site forms a single sediment cell that can neither gain nor lose sediment from alongshore because of confinement by the harbour arm and rocky headland at Porthleven in the north, and the headlands to the south-east at Gunwalloe (Figure 6.5). Most of the sediment (over 90%) that forms the beach is flint, which falls into two size classes: medium to very coarse sand (between +0.5 and –1.0 phi) and small pebbles (above –4.0 phi). Towards the eastern end of the beach, the proportion of quartz and serpentine increases slightly. The cliffs are cut mainly in Devonian Mylor Beds (mainly grey slates) west of the Bar, and Lower Gramscatho Beds (which weather more readily than the Mylor Beds) to its east, but supply only small quantities of material to the beach. Both clifflines have a 'slope-over-wall' or bevelled form, the lower 'wall' being cut both into the very resistant schist and quartz that crop out at the foot of the cliff, and the much less resistant solifluction material that forms much of the cliff either side of the Bar, but especially towards Porthleven. There the local erosion has been retarded by the construction of a wall and gabions. Gabions have also been used to protect the foot of the cliff below Bar Lodge just to the west of the bar.

The Bar was described by Leland during the 16th century as having a tendency to be regularly breached by the River Cober. Borlase described the bar in 1758 as forced up against the mouth of the valley by south-west winds. From time to time, the people of Helston excavated a channel through the bar in order to reduce the risk of flooding from the lake. The cost for this activity, according to Borlase, was three halfpence! The last known artificial breaching of the bar took place in the winter of 1867–1868 after which an overflow adit was cut at the north-western end of the bar. The bar has moved inland since the mid-19th century and may have increased in height (according to Goudie and Gardner, 1985, although no details are given). Washover features can be seen on its landward side, the result of overtopping. However, because of its steep front and gentle upper and back slopes, such overtopping occurs at the extreme of the swash phase and so only limited amounts of sediment are transferred over the crest. The crest itself is only a few centimetres higher than the Loe Pool behind it. The level of water in the Loe Pool does not appear to be affected by tides and so may be perched in a similar way to that at Slapton (see GCR site report, this chapter). The orientation of the beach, and the limited amount of sediment in the beach and the bar are such that sediment

Gravel and 'shingle' beaches

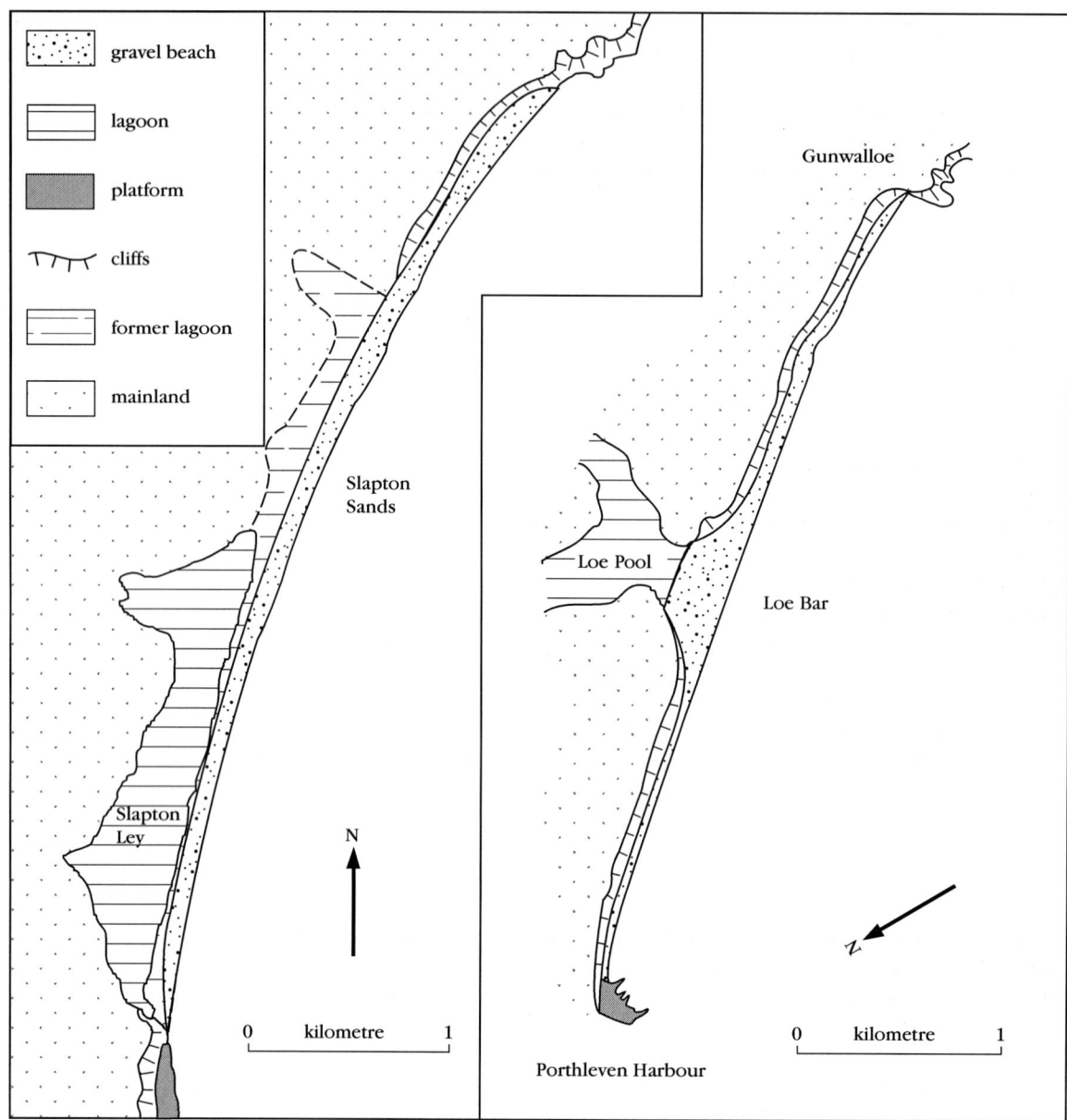

Figure 6.4 Comparison of geomorphological form between Slapton Sands and Loe Bar. Slapton Sands encloses a large lagoon, part of which has been infilled by sediment and become a brackish wetland. At Loe Bar, a cliff-foot beach confined between headlands has blocked off a narrow estuary.

tends to move towards the centre of the bay at present. This process depletes the north-western and south-eastern parts of the bay, but maintains the sand and shingle supply to the bar. The accumulation of shingle blocking the Pool appears to be considerable, for Rogers (1859) describes a boring made in 1834 that reached 22 m (just over 9 m below low-water mark) without reaching bedrock.

Interpretation

Although the Loe Bar has frequently been cited as a textbook example of the comparatively rare British case of a bar completely blocking a lake (e.g. Wooldridge and Morgan, 1937; Steers, 1946a; Monkhouse, 1965; Barnes, 1977), its formation has never been satisfactorily explained (Steers, 1981). Curiously, it is quoted more often than Slapton Ley, which has similar

Loe Bar

Figure 6.5 Loe Bar, looking approximately south-east, showing the bar and its washover features. (Photo: V.J. May.)

characteristics (Figure 6.4). King (1972a) regarded it as exemplifying the closure of an estuary that only occurs where discharge is very low or there is a very permeable barrier. The latter suggests general seepage through the bar, but this does not appear to be the case. Toy (1934) suggested that two spits had developed from east and west across the valley. In this model, tidal action would move sand and shingle eastwards from Porthleven towards the bar, but strong south-easterly winds would bring about a reverse movement from the east. Eddies set up by the flow of the river and tidal scour would keep the channel clear, but would encourage the growth of spits on opposite sides of the valley mouth. The bar would be completed when the gap was closed by either gradual processes, or severe waves, perhaps during storms. The so-called 'tidal wave' reported by Toy appears to have been an example of particularly large swell which pushed large quantities of sand and shingle up and over the bar. Although the occurrence of tsunamis in the North Atlantic Ocean has been considered as a possible mechanism for the formation of some beaches, there is no evidence for the influence of tsunamis here. The height of the bar is such that it would only be possible for very large run-up to carry material to the top of the bar. Steers (1946a) was convinced that the development of the bar was more likely to have resulted from wave action than from tidal action. He noted later (1981) that its position relative to wave action is comparable with Chesil Beach, i.e. it is only the relatively infrequent, largest waves that can carry material on to the top of the bar to build it up or even to move its crest inland.

Explaining the presence of a high proportion of flint in the beach is a problem as indeed it is at Slapton Sands. Bird and Schwartz (1985, p. 364) described the Loe Bar as a beach of flint shingle 'washed up from the sea floor (there being no flint sources in the adjacent cliffs or hinterland areas)'. King (1972a, p. 519) described the Loe Bar as made up of 'unusual material', intermediate between sand and shingle in size. Other flint-dominated shingle beaches along the southern coast of England are also found isolated from present-day sources of flint. The most likely origin for such beaches is their development as barrier beaches that were gradually moved onshore as Holocene sea level rose, the flint coming from drowned terraces of the former river flowing down the English Channel. During its movement across the sea floor in front of the cliffs the beach could have been breached and re-formed, and might even have been more spit-like. The outline of the bay is such that longshore movements could carry shingle towards the bar, and maintaining its form. Thus the beach tends to maintain an equilibrium form with the dominant and prevailing waves that means a more or less smooth curve throughout its length. The bar itself is thus a sediment sink as far as the overall beach budget is concerned. It remains unclear why the beach has such a strongly bimodal size distribution.

Conclusions

Loe Bar is a rare and excellent British example of a shingle bay–bar, and it is one that has international renown as a type example of this form. The site has additional interest because the bar blocks a ria, and the sediments in the adjacent cliffs suggest that the beach occupies an interglacial embayment. The beach is in overall deficit, but the bar is likely to survive as long as shingle remains within the main beach to maintain it. Breaching may occur from time to time, although that has not been recorded since the 1830s. The continued existence of the bar is dependent upon the maintenance of sediment supply from the wider area. As sediment is transported and, in effect, trapped in the bar, it can be replaced by sediment moving in from the main beach. As a result, the stability of the bar depends on natural and unimpeded sediment supply and transport from the wider beach system. In turn the ecological interest of the Loe Pool depends upon the unimpeded natural evolution of the geomorphological processes operating at the site.

Comparable bay–bars with a lagoon occur at Slapton and at the western end of Chesil Beach. At one time the beach at Pagham Harbour also formed a bay–bar with a lagoon, but this was breached and insufficient shingle has been supplied to it for the bar to redevelop. The differences between Loe Bar and Pagham Harbour are, first, their oceanographic context, Loe Bar being directly exposed to south-westerly Atlantic swell, whereas Pagham Harbour is affected by refracted swell, and second the tendency for the Loe Bar to rebuild after breaching, from both natural and anthropogenic causes. The evolution of Loe Bar most likely relates to migration of barrier beaches across the present-day seabed to more-or-less their present position with changes in the present-day beach mainly associated with overtopping. It is also evident that the outline of much of this coastline pre-dates present-day sea level and the Loe Bar is probably reoccupying an interglacial embayment.

SLAPTON SANDS AND HALLSANDS, DEVON

V.J. May

Both Slapton Sands and Hallsands lie on the east-facing coastline of Start Bay, south of the Dart estuary and Torbay (see Figure 6.2 for general location). Though they are distinct sites both in location and in their geomorphological features (Figure 6.6), their relationship to the past and present-day processes is intrinsically linked. This common introduction describes their setting and the debate about their origins. Site-specific details are contained in the individual site descriptions that follow.

Start Bay was described by Hails (1975a–c) as an asymmetrical embayment, about 60 km^2 in area, similar in form to many bays elsewhere for which the term 'zeta-curved bay' is frequently used. The coastline transects Lower Devonian rocks, which strike east–west and form an intermittent series of cliffs. Between Hallsands and Pilchard Cove a sequence of barrier beaches forms the main feature. These beaches are composed predominantly of granules and small pebbles (i.e. between −1 and −4 phi, with flint a dominant grain type (Gleason *et al.*, 1975). There are four main beaches, Hallsands, Beesands, Slapton Sands and Blackpool Sands, all formed of shingle coarsening to cobbles in the south. Beesands and Slapton include barrier beaches, between 100 m and 140 m wide at high tide, which impound lagoons. The crest of the barriers is generally at about 6.0 m OD ± 0.5 m. The ridges become narrower and lower towards the south, in the same direction as

Slapton Sands and Hallsands

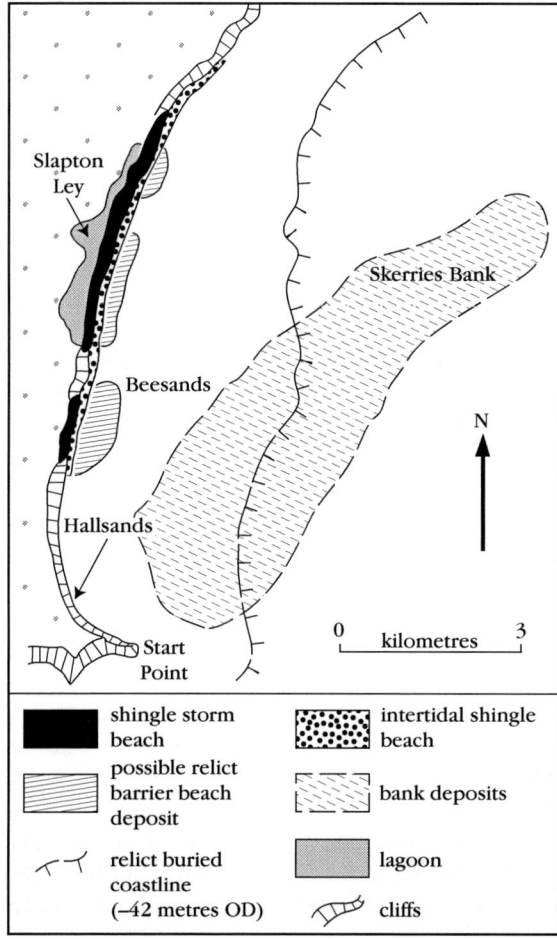

Figure 6.6 Hallsands and Slapton Sands represent parts of a once-continuous gravel beach. Offshore, there is evidence of buried shorelines and a possible former barrier beach. The present-day shingle beach is separated by rock headlands. (After Hails, 1975a.)

offshore, there is a pronounced break of slope in the bedrock slope that forms the seabed at an average depth of −42 m OD. This has been interpreted as an ancient coastline that may have been exposed during periods of lower sea level during the Pleistocene Epoch. Opposite the central part of Slapton Ley, there is a second break in the bedrock slope at about 28 m depth. An offshore bank, known as 'Skerries Bank', extends north-eastwards for over 6.5 km and appears to be linked to Start Point. Its maximum height is −4.8 m OD, but for most of its length the crest lies between −7.5 m to −9.0 m OD. Waves from all directions, except north-east, are liable to break on the bank, and it is responsible for significant refraction of waves inshore (Figure 6.7).

Kelland and Hails (1972) identified in Start Bay three lithological sub-environments as a result of grain-size analysis: 'barrier', 'bay' and 'bank' deposits. The barrier deposits consist mainly of gravels that occupy a narrow zone from the front of the barrier beaches to about 200 m offshore, except near Torcross Point and Limpet Rocks where they extend 500 m seawards. Flint and chert (40%) and quartz (46%) comprise most of these sediments, other materials including mica-schist, slate and shale, as well as rhyolite, felsite, granite, and quartz porphyry. The bay deposits are mainly medium- to fine-grained sands, with varying proportions of silt, and whole and comminuted shells.

Hails suggested that the Skerries Bank occupied its present-day position during the later part of the Holocene transgression. During the past 5000 years or so ephemeral barriers were probably constructed, destroyed and submerged and only rather limited amounts of gravel were transported landwards across the floor of Start Bay. Today, no new material is entering the bay, either alongshore or from offshore. As a result, the Bay was considered by Hails (1975a) to be a closed system under present-day conditions. Hails' analysis showed that both Hallsands and Beesands are located at points in the bay where wave energy is focused during north-easterly storms (Figure 6.7). This, combined with high spring-tides and five years of large-scale extraction of gravel, can probably be blamed for the 1917 Hallsands disaster (see GCR site report, below).

Both Hallsands and Slapton Sands thus have origins that depend to a substantial degree on the wave regimes within the bay, the effects of

sediments become coarser. Worth (1904) and Hails (1975a–c) also recognized a sharp decline in sediment size offshore immediately away from the coast. Although the offshore slope varies near the beaches, depths of −14.5 m OD are encountered by 600 m offshore. There are short lengths of coast protection works at Beesands, Torcross and Blackpool Sands, but these may have little effect on the overall sediment budget of the beaches. In contrast, the commercial removal of about 640 000 tonnes of gravel at Hallsands between April 1897 and December 1902 has had a much greater impact (Worth, 1907).

Kelland and Hails (1972) have described in some detail the submerged form of Start Bay and its offshore deposits. Between 0.5 km and 3 km

Gravel and 'shingle' beaches

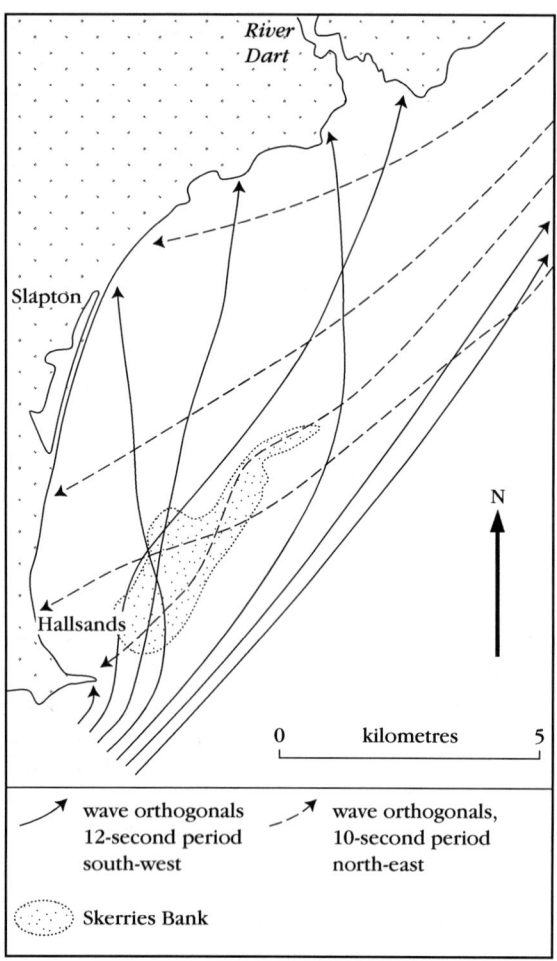

Figure 6.7 Wave refraction along the coast between Slapton Ley and Hallsands. The Skerries Bank affects waves entering Start Bay from the south-west. Wave energy is concentrated in locations such as Hallsands and Beesands during north-easterly winds. (After Hails, 1975a– c)

SLAPTON SANDS, DEVON (SX 823 417)

V.J. May

Introduction

A shingle barrier beach enclosing a lagoon, the beach ridge at Slapton Sands comprises mainly flint, chert and quartz shingle that extends some 5.6 km from Limpet Rocks, just south of Torcross, to Shiphill Rock at Strete. The beach at Torcross has been artificially strengthened by a wall to protect the hamlet against wave attack notably during north-easterly gales, but otherwise the beach remains little affected by human intervention although the A379 road runs along its crest. The southern 2.2 km separate the lagoon, Slapton Ley, from the sea, whereas to the north the ridge is backed first by an infilled former arm of the lagoon (see Figure 6.4) and then by cliffs of Lower Devonian slates and grits. Very little locally derived material is found in the beach sediments. In the English Channel, Slapton Sands is unusual in combining shingle material with an easterly aspect. It has been the focus of considerable research effort (Steers, 1946a; Hails and Carr, 1975; Morey, 1976, 1980, 1983) and is a major site for educational studies.

Description

The barrier beach is predominantly shingle despite the name 'Slapton Sands', and encloses a freshwater lagoon, Slapton Ley. Divided into two parts, the Higher Ley to the north, and the Lower Ley to the south, the lagoon occupies a former marine embayment bounded by degraded cliffs. The Higher Ley is mainly covered by reeds *Phragmites* whereas the Lower Ley is open water. It is up to 500 m wide and has a maximum depth of –4.0 m OD. The beach varies from 100 m to 140 m in width at high tide with a crest that is generally at about +6.0 m OD ± 0.5 m. The gradient of the intertidal beach persists below low-tide level to about –7.5 m OD. It then slopes gently to reach –14.5 m OD at about 600 m offshore. There are barrier beach deposits for about 500 m offshore, according to Kelland (1975), which exceed 5 m in thickness.

Within the lagoon, there are extensive sheets of washover gravels derived from the barrier (Morey, 1983) except in the central part known

the Skerries Bank and the transgression of a flint–quartz barrier beach which now lacks external sources. The interest of the two sites arises from their 'closed' nature, the effects of gravel extraction on the coastal sediment budget and coastal erosion, and the development of barrier beaches.

The beaches form a single beach system, but owing to the modification of the coast at Beesands, itself separating a small lagoon from the sea, by the dumping of coastal protection material, this area was excluded from the GCR site boundary. Nonetheless, the continuity of the system is evinced by the consistent reduction in the mean sediment size from south to north.

Slapton Sands

Figure 6.8 View, looking north-west, of the shingle barrier beach of Slapton Sands, enclosing the freshwater lagoon, Slapton Ley. Artifical sea-walls protect Torcross in the foreground. (Photo: V. J. May.)

as 'Ireland Bay'. This is the widest part of the lagoon, and is the mouth of the Start Stream. Morey described the following sedimentary sequence in the lagoon:

1. Light-grey, silty, estuarine muds that may be an extension of the lower bay deposits (Hails, 1975a). The fauna suggests a salinity gradient with restricted water circulation behind a growing barrier or spit. The tidal entrance was not located by Morey (1983), but he suggested that it was probably in the southern part of the Lower Ley.
2. A thin brown organic silt with a sharp lower boundary, but passing upwards into fen peats. The presence of pollen of *Chenopodium* and reeds *Phragmites* has been interpreted by Morey as suggesting a local transition from vestigial saltmarsh to reed swamp.
3. Fen peats about 1.3 m thick. An early reed swamp stage gradually changes to a sedge (*Carex*)-dominated fen community. The top of the peat has been dated at 1813 ± 40 years BP and the base at 2889 ± 50 years BP.
4. A layer of muddy sand, thickening seawards.
5. The upper layer is formed by lacustrine muds of terriginous detrital origin.

The present-day lake is perched presumably on its own sediments (Morey, 1976), despite the permeability of the shingle barrier.

Interpretation

Although Ward (1922) provided an early description of the site, Steers (1946a) was emphatic that there was no very conclusive hypothesis for the origin of the beach. The sedimentary characteristics of Start Bay have been described above, but the development of Slapton Sands and Slapton Ley is based mainly upon Morey's papers (1976, 1980, 1983). During the early Holocene, a transgressional shoreline of saltmarshes, estuaries and ephemeral lagoons developed in a macrotidal environment within the shelter of the Skerries Bank. It is thought unlikely that major barriers developed until the shoreline was close to its present-day position about 5000 years BP.

The rate of transgression declined partly as a result of a reduction in the rate of eustatic sea-level rise and the proximity of the coast to a relict pre-Holocene cliffline. Major accumulations of gravel can only occur where bedrock is below modern sea level and where overwashing can spread gravels across submerged infilled Holocene valleys. Without a substantial eustatic rise, there would have been insufficient space for gravels to accumulate except at Beesands and Slapton Sands.

The site was also the location for some detailed observations of the relationships between nearshore sediment dynamics and nearshore motions of the water itself (Huntley and Bowen, 1975a). They observed edge waves with a longshore wavelength of 32 m and with a period twice that of the incident waves. Swash interaction in the narrow surf zone on this steep beach was proposed as the process generating these waves. Nearshore circulation cells were also observed here. On steep beaches (such as this one and other GCR sites, for example, Hurst Castle Spit, Loe Bar, Porth Neigwl), the short-period edge waves that are observed may be responsible for small-scale topographical features such as beach cusps (Huntley and Bowen, 1973, 1975b).

Conclusions

Slapton Sands is one of the few examples of a bay–bar in Britain. The barrier beach, which encloses a freshwater lagoon, Slapton Ley, is predominantly shingle despite the name 'Slapton Sands'. The locality has been investigated not only as geomorphologically interesting in its own right, but also as part of larger-scale studies of Start Bay as a whole. As a result of the degree of study of the site, a great deal more is known about its dynamics than in many other sites. It contrasts strongly in location with the other major example of this type of landform, Loe Bar, in that it is sheltered from the main Atlantic wave systems. It demonstrates better than many other localities around the British coast the links between seabed features and the shoreline landforms, both in their Holocene history and their effects upon modern-day wave behaviour. Together with Loe Bar, the two sites demonstrate clearly how similar coastal landforms may develop in different wave conditions.

HALLSANDS, DEVON (SX 819 382)

V.J. May

Introduction

Hallsands lies at the south-western end of Start Bay (Figure 6.6); the scientific interest in this site arises from:

1. its location at a point where wave energy is focused at the shoreline by offshore banks,
2. the catastrophic destruction of Hallsands village, and
3. the formerly buried cliff forms that were exhumed by removal of gravel and shingle during storms in January 1917.

The cliffs are cut in mica-schist and quartz-schist, an emerged ('raised') platform of which provided the foundation for part of the former hamlet of Hallsands, other parts having been built on the shingle beach itself. The site is regarded by many coastal scientists as being a classic and vivid exemplar of the dangers of beach sediment extraction, as well as having intrinsic geomorphological interest in its landforms. Work by Hails and Carr (1975) has shown, however, that the concentration of wave energy on this part of the coastline by the offshore Skerries Bank during north-easterly gales was of primary importance in bringing about the rapid localized erosion at Hallsands (as well as at Beesands and Torcross to the north). The destruction of Hallsands village has provided a stimulus for research that has led in turn to a substantially better understanding of both this site and the wider geomorphological history and hydrodynamics of Start Bay and its coastline.

Hallsands is unusual among British coastal erosion sites in being very well documented and internationally renowned (e.g. Komar, 1976; Bird, 1984, 1985; Holmes, 1965). Its renown results from the much-reported destruction of the cliff-foot hamlet of Hallsands in January 1917. Detailed surveys were carried out between 1903 and 1923 by Worth (1904, 1909, 1923), during 1956–1957 by Robinson (1961) and in the 1970s by Hails and Carr (1975). There are few erosional sites in the UK that have been studied and reported in such detail. The wider physical links with Slapton Sands and the question of the anthropogenic origins of the Hallsands disaster have ensured that it has

Hallsands

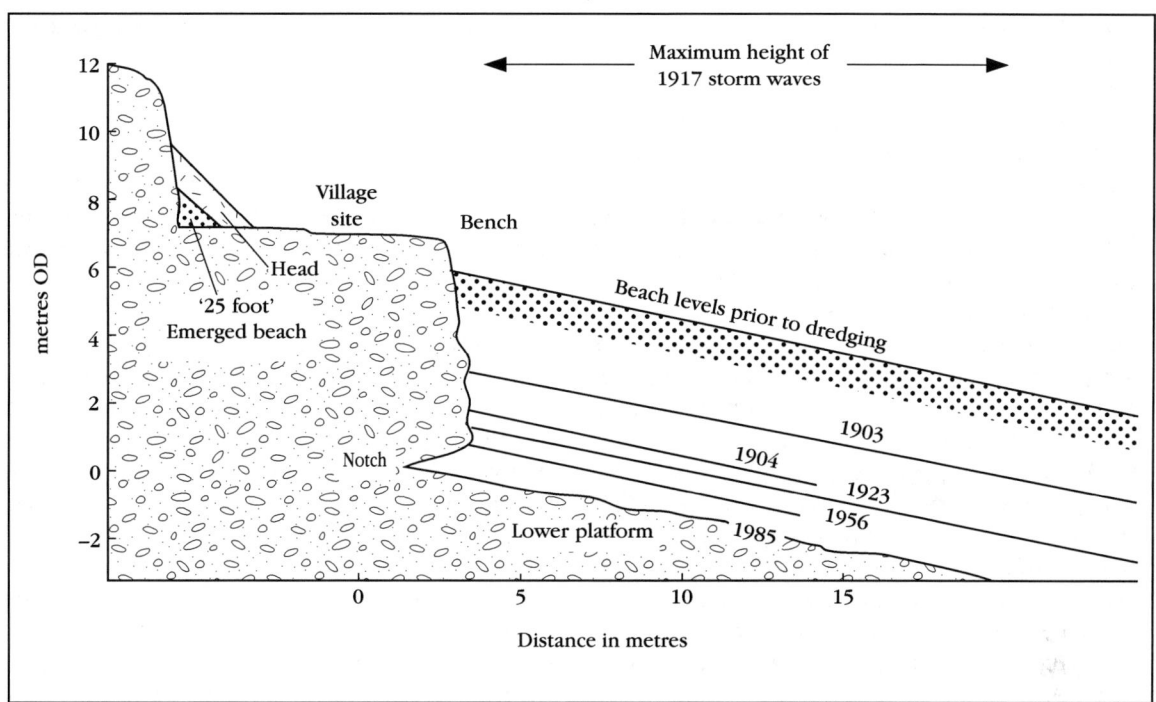

Figure 6.9 Cross-section of beach at Hallsands, showing the historic beach levels prior to dredging. (After Mottershead, 1986.)

retained its interest as a site of international importance.

Description

The site is about 500 m in length, and is part of a cliffed shoreline between Start Point and Hallsands. Although the cliffs have a slope-over-wall form, the lower wall is distinguished at locations such as Hallsands by a bench about 7 m in height above sea level (Figure 6.9). Photographs in Mottershead (1986) show that the village stood on the rock bench, about 1–2 m above the shingle beach, contrary to the impression given by the beach profiles in Worth (1904), Robinson (1961) and Goudie and Gardiner (1985) that the beach and platform were a continuous unbroken form. Robinson (1961) described the rock bench as a wave-cut bench overlain by a considerable thickness of head. The bench is discontinuous, with promontories separated by deep ravines. At two locations, emerged beach deposits with bands of rounded pebbles occur beneath the head. The steeply sloping cliffs behind the bench have been modified by solifluction since their formation.

Mottershead (1986) described a notch at the top of the lower platform that may be as much as 2 m deep and certainly pre-dates the extraction of shingle, which was removed at the end of the 19th century. The notch is at about Ordnance Datum on the promontories, but rises to +1 m OD within the ravines. These ravines can be traced in places to the base of deep gashes containing rotten rock in the upper high cliffs. Freshly fallen debris, large boulders, small stones and clay often fill the upper end of the ravines. Mottershead suggested that the ravines represent the former location of deeply weathered rock now eroded by wave action.

Worth (1904, 1909, 1923) surveyed the beach at Hallsands immediately after the cessation of commercial gravel extraction and subsequently after the disaster of 1917. In 1897, a local contractor, Sir John Jackson, was licensed by the Board of Trade to remove sand and other materials from the beach of Start Bay at Hallsands and Beesands. Up to 1600 tonnes was removed daily for the extension of the Royal Dockyard at Devonport. Worth estimated that 395×10^3 m^3 was removed before the licences were withdrawn in 1902, when the beach level had fallen by at least 3 m. During the winter of 1900–1901 storms undermined sea-walls and removed sand and shingle from the rock ravines behind. Buildings situated at these points collapsed and

Gravel and 'shingle' beaches

Figure 6.10 The ruins of the landward row of the houses of the former village of Hallsands. The seaward row of houses has completely disappeared. Compare with Figure 6.9. (Photo: V.J. May.)

as a result, dredging was stopped in 1902. By 1904, Worth estimated that the beach had fallen by as much as 6 m, and that 97% of the former beach volume had been removed. Continuing damage occurred with each major storm and a sea-wall was built during 1904. This appears to have been effective until January 1917, when, during a north-easterly gale, waves over 12 m in height destroyed much of the remainder of the village.

Interpretation

At the time, local opinion attributed the disaster to the effects of gravel extraction. Hails (1975a, p. 3) commented that this 'view of the reckless exploitation of shingle...has never been scientifically substantiated'. Worth (1904, 1909, 1923) had never accepted the official view that the beach would be naturally replenished and so set out to monitor the post-excavation changes. He was able to estimate the former extent of the beach by using photographs taken before and after the dredging occurred. In particular, a small stack or stump known as 'Wilson's Rock', which was covered by shingle before the start of dredging, stood afterwards over 3.5 m above the lower rock platform. Up to 1907, there was some gradual rebuilding of the shingle barrier beach to the north of the village, but elsewhere there was little change in the beach. The 1917 storm lowered much of the beach by almost 2 m. Robinson (1961) repeated Worth's surveys and found that the beach was lower in parts than it had ever been previously. Today, the beach has become very limited in volume and a rock platform is usually exposed below the bench on which the shells of the houses stand (Figure 6.10). Robinson noted that the most depleted conditions occurred after a period of easterlies, but did not discuss the reasons for the 1917 disaster. The very limited supply of shingle to the beach could not result from longshore transport since there are no sources to the south and any shingle moving southwards was probably retained by Tinsey Head about 800 m to the north.

Hails (1975a) also questioned whether the village had been constructed on bedrock that was sufficiently resistant to withstand storm wave attack. Robinson (1961) reported that at the Coastguard Cottages the cliff top retreated almost 7 m between 1907 and 1961, commenting that there had been a surprising amount of cliff recession. The schists possess many structural features that weaken the cliffs, and the sea has exploited these weaknesses. To the south of Wilson's Rock, however, there has been substantially less general cliff retreat. Mottershead (1986) described the tendency of houses that had been built on the sediment-filled clefts to collapse during storms, and demonstrated the variable strength of different parts of the village site. Nevertheless, the bedrock exhumed from beneath the shingle shows considerable resistance to erosion. The importance of this site is based on the following:

1. The detailed survey record, which is rare among coastal sites. Indeed it appears to be the longest time-series of beach profiles recorded in the British literature. (There are of course much longer series of repeated *plan* surveys.)

Budleigh Salterton Beach

2. The debate about the causes of the Hallsands disaster, and the explanation by Hails (1975b) that it resulted from a combination of gravel extraction, focused wave energy and high wave and tide conditions.
3. The evidence of the timescales at which coastal systems adjust to change. As Mottershead (1986) pointed out, large-scale changes did not occur until a combination of high wave-energy from a particular direction took place. Once the beach was eroded, a further 15 years elapsed before the final disaster occurred.

Conclusions

This is one of the world's best-known coastal erosion sites, mainly because of the catastrophic destruction of Hallsands village following gravel extraction. Erosion of the beach has exhumed buried cliffs and platforms. It is a unique site, combining exhumation of earlier coastal landforms with a long record of surveys that show how this exhumation took place. It is especially important because it is a rare location in which the effects of beach erosion related to both wave conditions and gravel extraction can be demonstrated convincingly. In the context of coastal management worldwide, Hallsands is especially important because it shows the environmental impact of gravel extraction at a site where the coastal processes are not fully understood.

BUDLEIGH SALTERTON BEACH, DEVON (SY 040 801)

V.J. May

Introduction

The coastline of east Devon is characterized by several river valleys that are infilled by alluvium, and their mouths partially blocked by shingle or cobble beaches. Although the beaches at Seaton and Sidmouth have some similarities in form with the beach at Budleigh Salterton, they have suffered considerable erosion in recent years and sea defences and coast protection works have been erected to protect the low-lying towns behind them. However, Budleigh Salterton Beach (see Figure 6.2 for general location) remains largely undisturbed. The beach is formed primarily of shingle- and cobble-sized material derived from the erosion of cliffs cut into Triassic sandstones and pebble beds at the western part of the site. The beach is unusual among English beaches because it is fed with material derived entirely from erosion of cliffs, which cut into Triassic strata. There is a noticeable lack of flint and chert in the beach clasts. The plan form of the beach shows a strong relationship to the refracted Atlantic waves and more direct, but less frequent, waves from the eastern English Channel. The cliff–beach–estuary assemblage was once a very common feature of the coastline of south and south-east England, but most examples have been modified by coast protection works. At Budleigh Salterton, the relative stability of the beach and the lack of natural evolution of the estuary mouth have not required artificial protection.

Research on the site has been limited, although distinctive clasts derived from the Budleigh Salterton Pebble Beds occur in many of the beaches within Lyme Bay and are reported in the literature (Ward, 1922; Steers, 1946a; Bird, 1989; Carr, 1974; Carr and Blackley, 1974a).

In recognition of the site's importance for

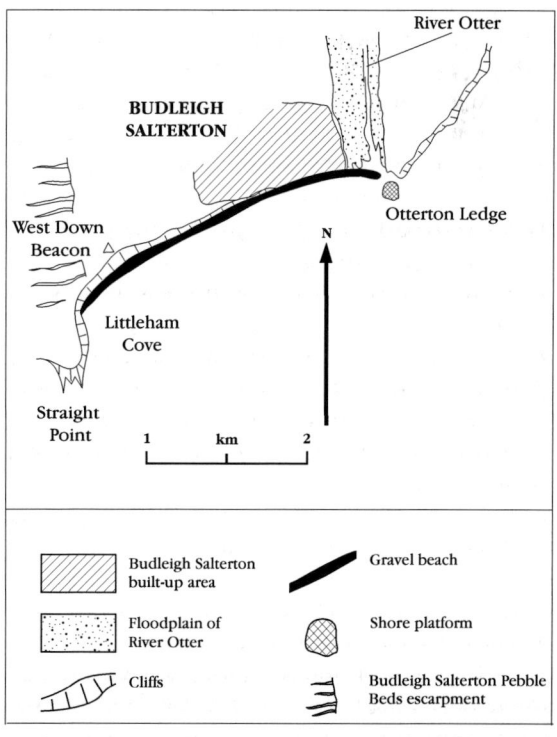

Figure 6.11 Sketch map of the Budleigh Salterton Beach GCR site.

Gravel and 'shingle' beaches

Figure 6.12 Budleigh Salterton beach, looking west, also showing the cliffs that provide the source material for the gravel beach. (Photo: V.J. May.)

coastal geomorphology and Triassic stratigraphy, it is one of the GCR sites that form the Dorset and East Devon Coast World Heritage site.

Description

The site comprises five morphological units.

1. The western cliffs around Littleham Cove, where a narrow platform formed of more resistant sandstones provides a headland against which a zeta-shaped beach is beginning to form.
2. The high cliffs at West Down Beacon, which reach over 125 m in height and are affected by considerable landsliding, a process that feeds sand, clay and pebbles to the beach below.
3. The eastern cliffs that are more affected by rock falls and are increasingly protected by the beach itself.
4. The barrier beach, which rests on the terraces of the Otter Valley and extends across the floodplain as a spit.
5. The cliffs and platform at Otterton Point.

As erosion of the cliffs has taken place, particularly by landslipping around West Down Beacon (Kalaugher *et al.*, 1995), clasts have been added to the beach and distributed throughout its length. The beach is lower, narrower and sandier at its western end in Littleham Cove and becomes higher, wider and less sandy eastwards, especially to the east of West Down Beacon. Although the cliffs decline steadily towards the Otter valley so that the pebble beds do not crop out there, there has been sufficient input of eroded material from the western cliffs to maintain a supply, albeit spasmodically. At the eastern end of the site, the pebbles not only form a barrier across the valley, but also spread seawards around the river mouth lodging against Otterton Ledge, a rock shore-platform. The shingle bank blocked the Otter estuary by the mid-16th century (Ward, 1922). There is no record of substantial erosion or flooding at the town. It appears that, although there has been some retreat of the coastline to the west and that the barrier beach has moved slightly landwards over the valley floor, the clasts are retained within the bay and provide an effective form of coast protection.

Interpretation

The distinctive form and petrology of the clasts, which distinguish them from the predominantly flint or chert content of many English south coast beaches, make them a useful 'tracer' for studies of longshore movements of sediment, especially because the clast source is very restricted in outcrop, both at Budleigh Salterton on the coast and offshore on the seabed.

The beach is unusual in that it shows a very high degree of stability, which is probably a result of its position between two relatively stable headlands, and the fact that the beach rests for much of its length against the cliff foot.

The Triassic Budleigh Salterton Pebble Beds were described by Henson (1970) as a poorly sorted, braided river deposit consisting mainly of ellipsoidal quartzite pebbles with subordinate pebbles composed of vein quartz, 'schorl', sandstone and porphyry. They have a maximum dimension between 19 mm and 100 mm (Carr, 1974) and all show a high degree of rounding. The formation dips to the east at about 5° and forms a marked escarpment running northwards from the coast at West Down Beacon. Carr and Blackley (1974a) described more fully what they identified as metaquarzite clasts from this beach. Pebbles from this distinctive formation have been identified within many other beaches on the southern coastline of Britain, notably at Chesil Beach, Langney Point and Dungeness (Ward, 1922; Steers, 1946a; Lewis and Balchin, 1940). Steers (1946a) suggested that the sites most distant from Budleigh Salterton contain pebbles probably transported as ballast, but there is no reason to reject natural processes as a source for the Chesil Beach examples (Carr and Blackley, 1974a).

The plan form of the beach is largely controlled by the wave-energy distribution between Littleham Cove in the west and Otterton Point in the east where the beach diverts the river Otter eastwards. The alignment of the beach is also affected by the shore platform at Otterton Ledge.

The small estuary of the Otter may allow some sediment to be stored in this embayment. More important is the very high permeability of the pebble ridge at its eastern end so that the ridge is little disturbed except by the largest waves. Unlike the part of Chesil Beach formed of large clasts, Budleigh Salterton beach is not exposed to the full strength of the south-westerly waves. It is aligned towards the SSE. Periods of storm-wave attack from the more easterly directions move sediment along the beach and on occasions denude parts of it. At West Down Beacon, Kalaugher et al. (1995) have shown that intermittent movements in a mudslide at the base of the cliffs can be linked to large-scale collapses of the conglomerate that forms the upper cliff. In stormy conditions the mudslide is triggered at high tide. The landslides interrupt longshore transport and add new material to the beach. Bird (1989) identified this beach as the only one within Lyme Bay that is not laterally graded. Carr (1974) considered that when significant grading was observed, it occurred along the whole beach, with the smallest grade material at the eastern end. He noted that in June 1972 mean clast size increased from the centre of the beach. This observation is consistent with the patterns of sorting described by Heeps (1986) in similar confined beaches in south-east Dorset. Carr also demonstrated that not only was sediment graded in size from the centre of the bay, but it also changed in shape from the end of the beach.

This site is one of the few beaches of the English Channel coast that has avoided any significant coast protection works; furthermore, the size of its clasts appears to have made it less attractive for commercial extraction. These circumstances make it an important locality for further geomorphological investigation of a 'natural' pebble and cobble beach system.

Moreover, it is unusual to find a beach where the sediment source can be so readily identified and the inputs to the beach monitored. It is all the more unusual for the sediment to enter the beach system already well rounded. Unlike other small bayhead beaches on the southern English coast, Budleigh Salterton has a single main source of clasts. It is dissimilar in that it is sheltered from the main wave-energy inputs from the Atlantic Ocean. Whereas those of south-east Dorset (Heeps, 1986) have a reduced energy input as a result of submarine barriers, it is the effects of refraction that have been most important in reducing the energy inputs to Budleigh Salterton Beach. Large beaches of cobbles are comparatively rare on the English coast, but they are commonly found in association with large south-west fetches. This site thus provides a substantial contrast with cobble beaches at Westward Ho! and Porth Neigwl, as well as with the cobbled part of Chesil Beach (see GCR site reports in the present volume).

As a site for the investigation of sediment budgets, beach adjustment to wave conditions and the effects of beach permeability on both beach and cliff stability, Budleigh Salterton offers considerable opportunities for field investigation and coastal modelling. Owing to the distinctive clast source, the site provides a rare opportunity to observe the ways in which clasts survive or change in shape following their introduction into the marine environment. The clasts can be readily identified amidst large volumes of flint gravel, confirming their longevity. Furthermore, it is the only site that has large well-rounded clasts dominated by a single rock type other than flint and chert. The unusual quantity, hardness and shape of clasts from the Budleigh Salterton Pebble Beds have given rise to a unique beach.

Conclusions

This pebble and cobble beach is uniquely fed by pebbles and cobbles derived entirely from Triassic sediments. This is the only point where these pebbles enter the coastal system at present, although they are found in beaches along the length of the south coast of England. The lack of anthropogenic influence greatly increases the geomorphological importance of this site as one in which an intact and virtually unmodified natural system can be studied. Owing to its important geology and geomorphology, it is part of the Dorset and East Devon Coast World Heritage site.

CHESIL BEACH, DORSET (SY 462 903)

V.J. May

Introduction

Chesil Beach has been described as 'unique' – and is of considerable international renown and scientific significance. Its sheer size (over 18 km in length and exceeding 14 m in height), the systematic longshore size-grading of beach material, the evidence for the south-westerly provenance of its pebbles, the availability of historical records, and the sedimentary record in the adjacent lagoon ('The Fleet'), each contribute to the geomorphological importance of the site and help to explain why there is a vast scientific literature about it. In recognition of the importance of the site for coastal geomorphology, it is one of the GCR sites that form the Dorset and East Devon Coast World Heritage site.

Chesil Beach is one of five major gravel/shingle features along the British coast, together with Dungeness, Orfordness, Spey Bay and

Figure 6.13 Chesil Beach. View looking north-west, from Portland, with Chesilton in the foreground. The beach reaches 14 m OD and over 150m wide at its eastern end, where limited washover still occurs in spite of artifical modifications. (Photo: J.D. Hansom.)

Figure 6.14▸ Map and sections of Chesil Beach. For general location see Figure 6.2. (Based on borehole information in Carr and Blackley, 1969, 1973; and Carr and Seaward, 1990.)

254

Chesil Beach

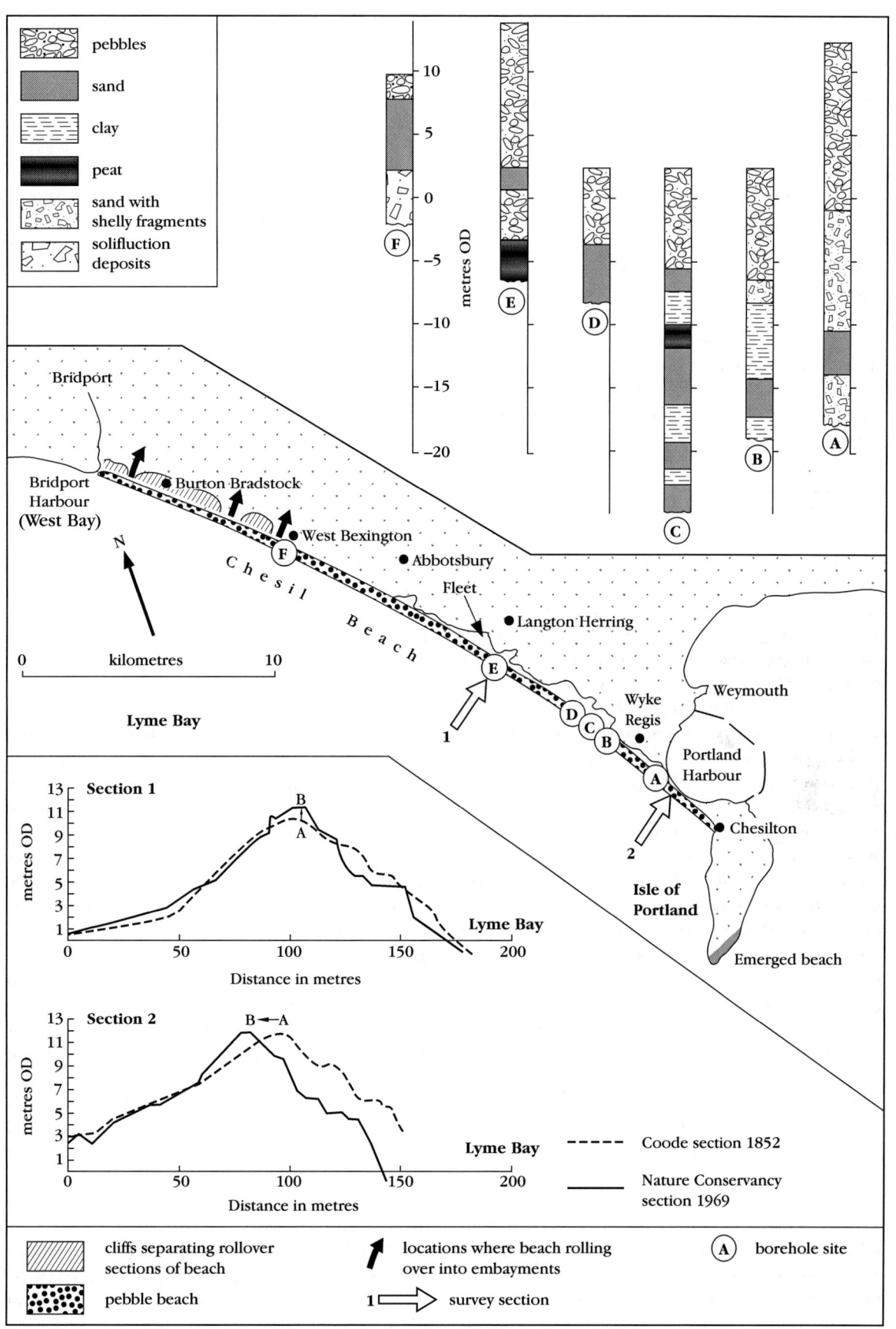

Culbin (see GCR site reports). It has been described as a 'prodigious accumulation of gravel', 'probably the most extensive and extraordinary accumulation of shingle in the world', 'an heroic piece of natural engineering' and 'unique' (quoted by Carr, 1983a). Only one coastal feature in the British Isles – Scolt Head Island – has been written about more than Chesil Beach. There have been a number of reviews of the literature, which currently totals over 75 published accounts, e.g. by Coode (1853), Strahan (1898), Arkell (1947) and Carr and Blackley (1974b). Many other writers have concentrated on specific aspects of the site including de Luc, 1811; de la Beche, 1830; Austen, 1851; Coode, 1853; Rennie, 1853; Bristow and Walker, 1869; Codrington, 1870; Fisher, 1873; Groves, 1875; Prestwich, 1875; Black, 1879; Cornish, 1898a,b; 1912; Reid, 1898; Strahan, 1898; Richardson, 1902; Johnson, 1919; Prior, 1919; Ward, 1922; Baden-Powell, 1930; Lewis, 1931, 1938; Steers, 1946a, 1962, 1964a; Bond, 1951; Rimmer, 1953; Arkell, 1954; Adlam, 1961; Jolliffe, 1964, 1979, 1983; Neate, 1967; Carr, 1969a, 1971a, 1974, 1981, 1983a, 1999; Carr and Blackley, 1969, 1973, 1974b; House, 1969; Carr *et al.*, 1970; Bird, 1972, 1989; Carr and Gleason, 1972; Hardcastle and King, 1972; Hydraulics Research, 1979, 1985, 1991a,b; Brunsden and Goudie, 1981, 1997a; Ladle, 1981; Carr *et al.*, 1982; Draper and Bownass, 1983; Goudie and Gardner, 1985; Bray, 1986, 1990a,b, 1992, 1996, 1999; Hannah, 1986; Carr and Seaward, 1990, 1991; Heijne and West, 1991; Hook and Kemble, 1991; Brunsden, 1999). In addition, The Fleet's origins have been investigated by Ladle (1981), Whittaker (1980), Robinson *et al.* (1983) and The Fleet Study Group (including Coombe, 1996; Goudie pers. comm., Whittaker, pers. comm.).

Description

Chesil Beach is a simple, linear, pebble and cobble storm beach, which, because it links the so-called 'Isle of Portland' with land farther west at Abbotsbury, is frequently quoted as an example of a tombolo (e.g. Holmes, 1944; Monkhouse, 1965; Twidale, 1968). The mean spring tidal range at Bridport is 3.5 m with mean high-water spring tides (MHWS) at +1.8 m OD and mean low-water spring tides (MLWS) at –1.7 m OD (Nunny, 1995). Waves from the south-west have a fetch in excess of 4000 km and surges have produced wave heights in excess of 6.5 m (return period = 1 in 5 years) and 9 m (return period = 1 in 50 years). The beach extends at least 18 km from Chesilton in the east, where it ends against a sea-wall and the cliffs of the Isle of Portland, to an arbitrary boundary in the west. This limit depends upon the criteria used to define it and may, as Prior (1919) first suggested, have changed over time. Bird (1989) discussed the possibility that the beach extended farther west to a cliffed boundary at Eype, and Brunsden and Goudie (1997a) speculate that it may have reached Golden Cap. This part of the shoreline has been separated from the modern Chesil Beach since at least 1742 when the first harbour breakwaters were built at West Bay (Hannah, 1986). The western limit of the GCR site has been taken as West Bay to ensure that it includes the full range of features and sediment grading characteristics of the Chesil Beach.

The pebble and cobble feature is joined to the mainland at Abbotsbury and Chesilton, and is backed over the intervening 13 km by the shallow tidal Fleet lagoon (Ladle, 1981). Opposite The Fleet, Chesil Beach is between 150 m and 200 m wide, but it is narrower both adjacent to the cliffs in the west (e.g. 35–60 m at Burton Bradstock, SY 485 890) and between 40 and 54 m at its extreme eastern end. The beach crest is intermittent at the western end, but becomes continuous from midway between West Bexington (SY 530 865) and Abbotsbury (SY 570 837). Ridge height increases progressively from about 7 m at Abbotsbury to a maximum some 14 m above mean sea level at Chesilton (SY 680 735). Offshore the beach drops at a broadly similar gradient to that of the seaward face above low-water mark before shelving gradually to about –18 m OD some 270 m offshore at Wyke Regis and –11 m at a similar distance off West Bexington. The offshore slope between –25 and –50 m is almost linear and steepens to about 1 in 20 off the eastern end of Chesil Beach between –25 and 0 m OD (Nunny, 1995). Boreholes (Carr and Blackley, 1969) show that only in the vicinity of Wyke Regis is bedrock anywhere near the surface, contrary to suggestions earlier in the literature (Figure 6.15).

Although it has been suggested by a number of writers that little gravel-sized material now appears to be available to nourish the beach from offshore and maintain the present-day

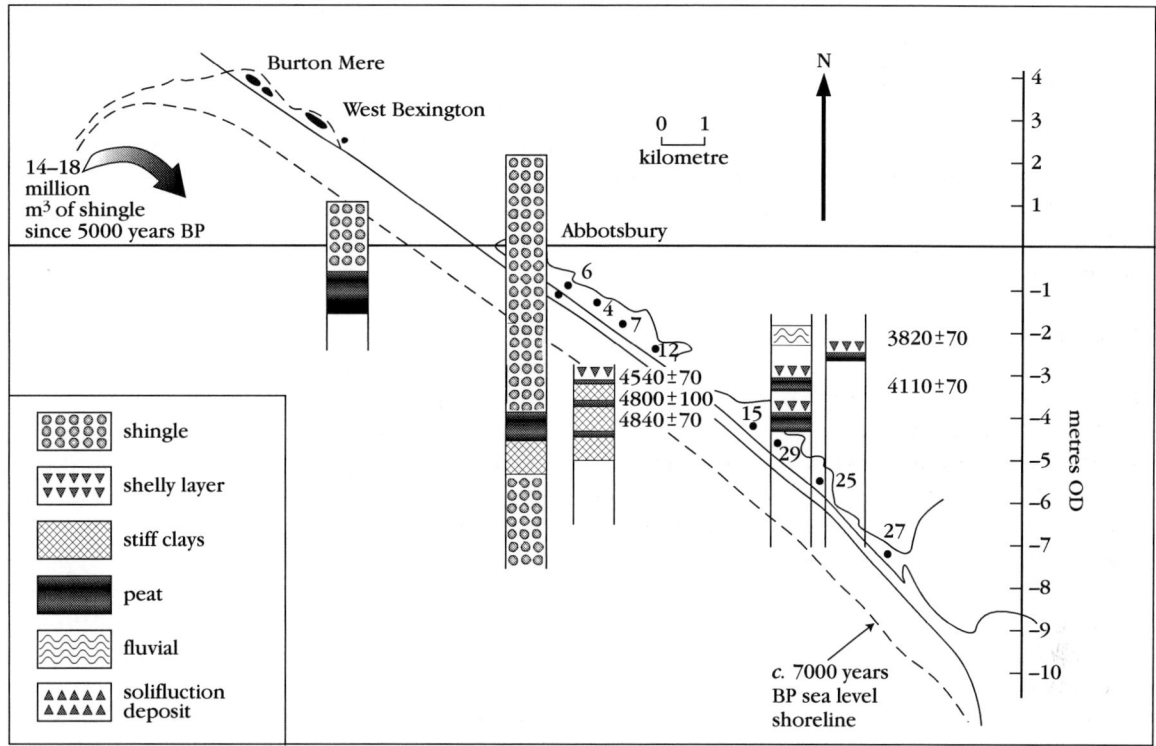

Figure 6.15 Sediment profiles of Chesil Beach and The Fleet. Sample cores are shown in sequence along the beach and The Fleet. Some peat layers have been dated in cores from the bed of The Fleet (dates are given in years BP). (Based on Carr and Blackley, 1973; Coombe, 1996 and Whittaker, pers. comm.)

characteristics of its clasts, more than 50% of the sediments in depths of less than 25 m between Abbotsbury and Burton Bradstock are gravel formed into wave-oriented mega-ripples. There are also extensive areas of bare rock (Nunny, 1995). Opposite The Fleet there are a number of rock outcrops broadly at right angles to the beach. At lower sea levels these outcrops may have affected wave refraction and thus sediment transport in a different way from the present-day shore-parallel wave approach. Borehole samples suggest that flint and chert pebbles become more angular with depth (Carr and Blackley, 1969). However, at these lower horizons, samples are largely derived from more local, less resistant, Jurassic strata. This implies that attrition is of some importance as a cause of loss of volume of the beach, at any rate in the long term. The boreholes also indicate that the massive pebble and cobble deposits are concentrated in the exposed, i.e. subaerial, part of Chesil Beach. Although shingle is present below low-water mark, it occurs as limited, discontinuous horizons. Estimates for shingle volume range between 15 and 60 million m^3, mainly because the volume of deposits below sea level is not adequately known from borehole evidence.

Like Chesil Beach itself, much has been written concerning the origin of The Fleet tidal lagoon. Carr and Blackley (1973) provide information on the form of the bedrock underlying Chesil Beach and The Fleet, and the material with which The Fleet is infilled, which mainly comprises silt, sand, pebbles and peat (some of which has been dated, Figure 6.15). They clarify some of the earlier ideas about its origin. The slope of the former coastal platform, largely planed-off bedrock, continues underneath the Chesil barrier to meet the hills inland. The break of slope, where the two join, and an associated ancient pebble and cobble storm beach (Carr and Blackley, 1973) are found at a depth of about −15 m OD opposite East Fleet, and at comparable depths elsewhere, as far west as Abbotsbury.

The Fleet is recognized as a marine Special Area of Conservation (SAC).

Gravel and 'shingle' beaches

Interpretation

Although Chesil Beach and the associated Fleet lagoon have considerable botanical and zoological interest and importance, it is for their physical features that they are best known. There is a continuing debate about the origins and development of the features. The key issues under debate are:

1. the sources of the material forming Chesil Beach
2. the cause and extent of the distinct gradation in the size of clasts
3. the role of longshore sediment transport
4. the role of extreme events in the formation and present-day development of the feature
5. the origins of The Fleet, and
6. the origins and development of Chesil Beach itself.

The presence of beach pebbles of similar lithology to the cliffs to the west in Devon suggested to many authors that transport was eastward from the source areas by littoral drift along former shorelines (de la Beche, 1830; Fitzroy, 1853; Rennie, 1853; Pengelly, 1870; Baden-Powell, 1930; Arkell, 1947) or crossed Lyme Bay (Strahan, 1898). Prestwich (1875), however, suggested that sediments were transported north-west from a precursor of the Portland emerged ('raised') beach (SY 675 684), situated south of the Isle of Portland. Prior (1919) suggested three possible sources: an earlier Chesil-like beach from Start Point in Devon to Portland, erosion of the east Devon coast or river gravels deposited in Lyme Bay by a river of which The Fleet was once part. Bond (1951) argued for an ancient Exe–Teign river, flowing up to 10 km offshore from the present-day coast to a mouth south-east of Portland. Arkell (1947) thought shingle rafting by ice could account for some of the more exotic pebbles. The geological evidence indicates that there have been various potential sources of the pebbles and cobbles, including fluvial as well as marine deposits, and that the relative significance of these sources is likely to have varied over the long term. About 98% of the material is flint and chert that could have been derived from a number of primary (and secondary) local sources, but the remaining 2% (including Triassic quartzites) probably was derived from the south-west. 95% of the quartzite material is derived from the Budliegh Salterton Pebble Beds (Carr and Blackley (1969). Opposite The Fleet, there was a higher percentage of pebbles other than flint or chert, but Carr and Blackley (1969) could not explain this; they concluded (1969, 1974b) that all lithologies represented in the Chesil Beach could be derived from either local sources or older sources in south-west England.

The way in which the mean size of the pebbles forming Chesil Beach broadly increases towards the eastern, Chesilton, end has attracted considerable attention in the literature, but few quantitative measurements have been published. De Luc (1811) described the range from that of a 'hen's egg' at Chesilton, through 'horse beans' near Abbotsbury, to coarse sand at Burton Cliff. The mean long-axis of pebbles at Chesilton is of the order of 50 mm, falling to 35 mm opposite Portland Harbour and rather under 25 mm seawards of Herbury Point (Carr, 1969a). Thereafter, the exponential fall continues slowly as far as West Bexington. West of West Bexington, longshore grading is less systematic and varies locally, as does beach crest height. These variations may reflect minor changes in beach orientation, commercial exploitation of beach material, and/or local sources of supply from the cliff to landward. Reid (1898, 1907) explained the higher proportion of Budleigh Salterton quartzites at Chesilton by their greater resistance to abrasion. Arkell (1947) thought that their greater durability or later replenishment could account for their increased frequency here relative to the Portland emerged beach. Carr and Blackley (1969) stated that the percentage of quartzites did increase towards Chesilton but was less than appeared to be the case in the field. The quartzite pebbles tend to be flatter with a larger surface area that may make them more conspicuous (Carr and Blackley, 1969).

Carr (1965) took samples at 27 locations, 1.6 km apart (0.8 km apart at the eastern end), with between 3 and 11 samples per section at crest, high and low water. During 1965–1966 pebble samples were taken along a series of transects across Chesil Beach, as part of a wider research programme (Carr, 1969a). The results showed that on surface profiles between Smallmouth and the Bridging Camp, there were areas near The Fleet where pebble size was smaller, and degree of sorting greater, than nearer the beach crest. It was suggested that these samples might represent the legacy of a different beach relationship from the present-day one (a similar point is made by Brunsden, 1999). Borehole samples along the beach, covering the area between approximately Smallmouth and Herbury showed angular, local limestone pebbles at depth, reaching as much as 47% just west of the Bridging Camp, and 33% farther east. Both these maximum values occurred at about −15 OD. In the case of Carr's core (see Figure 6.15), this limestone-dominated material was separated from the present-day beach by sand, suggesting a different regime at a different depth level, and hence time (even if equal height does not always imply contemporaneity). Brunsden (1999) comments that graphical data compiled by Babtie from a survey in 1996 in which sediment was sampled at 1 km intervals reveals important facts about the sediment wave hypothesis. The printed Babtie data link only as a line graph those points that show the trend and similarities of size. Points that deviate from the overall trend are left as major outliers, revealing large variation from the otherwise westward decreasing trend. Thus there are areas where sediments are distinctively larger than would be expected from the normal trend.

Studies over very short periods of time using brickbats (Richardson, 1902) and painted pebbles (Adlam, 1961) showed net eastward wave-induced transport and preferential movement of larger materials. Carr's (1971a) experiments using clasts of a foreign or exotic lithology confirmed that larger material moved more rapidly eastwards, showing particularly strong eastward drift at Wyke Regis (SY 660 760), becoming more variable and random nearer Portland. Carr concluded that there was no consistent drift near Portland, thus explaining the absence of a large accumulation at Chesilton. Both pebble measurements and tracer experiments (Carr, 1971a) have shown that thickness (B axis) appears to be the most significant dimension affecting pebble transport alongshore by waves. Although movement seems predominantly to the east, negative correlations have also been recorded. Near Portland, travel eastwards is reduced by the more random nature of longshore movement as compared with sites farther west, where waves usually approach the shoreline more obliquely. The length of burial time of particles also varies between the two areas. These factors, coupled with the absence of sizeable amounts of new material, probably account for the lack of any permanent drift and thus of the absence of large quantities of pebbles being deposited at the eastern end. Rejection onto the surface of clasts larger than the general population is most marked under conditions of long-period swell. Thus material that is unduly coarse for a particular part of the beach would be given more opportunity for lateral transport than the remainder. Small pebbles would work their way down into the beach matrix. Together, these processes and effects, according to Carr, produce the longshore grading pattern. However, the lateral sorting could simply be related to the ability of the most powerful waves to produce eastwards drift offset only by westward movement as far as the smaller cobbles and shingle are concerned.

Experiments by Gleason and Hardcastle (1973) using the indigenous material show vertical sorting to be dependent on the wave frequency and square root of the significant wave height (highest third of all waves). Longshore sorting was dependent upon the angle of swell approach. Carr and Blackley (1974b) argued that fresh sediment inputs would tend to diminish size grading, although Bray (1990b) considered that new inputs of larger pebble sizes may be necessary to counter the continuous reduction by attrition. Carr and Blackley (1969) showed how limestone clasts derived from quarry waste from the western Isle of Portland were virtually unrepresented in the natural beach population at a distance of some 3–4 km northwest of Chesilton (a reflection of both attrition and net longshore transport).

Most explanations for the sorting concentrate on the commonly held view that:

- there is a continuous size change along the beach,
- sorting occurs by size and shape,
- rates of pebble movement depend on pebble thickness (B-axis dimension),
- different wave energy of storms from the south-west and the south-east cause the sorting, and
- different depths of water offshore and therefore differences in available energy are a fundamental cause of the clast distribution and sorting.

However, the above model may be based on simplistic notions of sediment transport with pebbles moving singly by longshore drift that assumes an open system with a continuous sup-

Gravel and 'shingle' beaches

Figure 6.16 West Bay, Chesil Beach, showing the retreat of the shoreline and lack of sediment at the western end of the modern Chesil Beach. (Photo: V.J. May.)

ply of material for drift (Brunsden, 1999). He argues first that the present-day beach cannot be regarded as an open system, for it is closed at the western end by the harbour walls at West Bay. Even before they were constructed, restriction of longshore movement by headlands to the west existed. Brunsden suggests that on Chesil Beach, groups of pebbles move both west and east as the wave approach varies but with a net westward movement. Groups dominated by small sizes move over groups of bigger ones and along different storm ridges, forming and reforming as conditions dictate. Such a process has also been observed on the shingle beaches at Ringstead (May, 1999) and at Spey Bay, where 'slugs' of gravel move alongshore (Gemmell et al., 2001b). As pebbles are eroded from each beach ridge, remnants of pebble groups are left at different beach levels according to the severity and sequence of storms. When major storms washover the beach, surface beach material is moved over the crest regardless of the size and composition of pebbles that occur on the beach at that place at that time. As the beach moves landwards this sediment may re-emerge on the beach face many years later. As a result, Brunsden argues that any sorting model must be a complex, episodic spiral of individual movement and movement of gravel groups.

On the cliffed coast to the west, gravel inputs of approximately 5000 m^3 a^{-1} have occurred over the past 4000 to 5000 years (Bray, 1990a,b). Assuming that losses by attrition and entrapment have remained constant through time at approximately 1500 m^3 a^{-1}, Bray estimates that between 14 million and 18 million m^3 of sediment has been supplied to the coast by landslides since 5000 years BP. The beaches between Lyme Regis and West Bay store slightly less than 1 million m^3 of shingle at present. Since the long-term mean rate is similar to present-day figures, Bray argues that current shingle budgets support the hypothesis that erosion of the west Dorset coast provided a major sediment source for the creation and replenishment of Chesil Beach. Today, however, a continuous supply of material no longer exists. Shoreline transport is regulated by landslide activity occurring at the main headlands of Golden Cap and Doghouse Hill. Bray's (1990b) model envisaged intermittent pulses of gravel bypassing these headlands at intervals of 30–50 years, most recently at Golden Cap between 1949 and 1962. Relict landslide deposits (boulder aprons) identified some 2–3 km offshore from Golden Cap (Brunsden and Chandler, 1996) may indicate the extent of past cliff erosion (Figure 6.17b). Bray (1990a) estimated that about 32 million m³ of gravel would be supplied to the retreating shore. Some materials would also have been contributed from the East Devon coast, giving a combined total of 58 million m³. Chesil Beach is the only significant gravel accumulation within this part of the Lyme Bay cell.

According to Bray, the hypothesis is supported by a variety of evidence. Surveys in Lyme Bay have failed to reveal alternative offshore shingle sinks. Analysis of the size and lithology of clifftop gravels in the Charmouth area indicate that the material is comparable with Chesil Beach shingle, and the size, shape and lithology of pebbles on Charmouth, Seatown, Eype and Chesil beaches support the hypothesis that the beaches were formerly contiguous (Bray, 1990b). The contemporary pattern of eastward littoral drift has probably existed over the last 5000 years. There is no evidence that the coastal orientation was significantly different in the past or that the frequency of west and south-west storms was less (Bray, 1996).

The present-day volume of Chesil Beach is estimated at 15 to 60 million m³ (Carr, 1980). The estimated surplus by coastal landsliding is 14 to 18 million m³ over the past 4000 to 5000 years. Although potential shingle supply from landslides is significant by comparison with the present-day volume of Chesil Beach, it is unlikely to have been the main supply (Carr, 1980). Chesil Beach had already formed by 7000 years BP (Carr and Blackley, 1973). It is therefore suggested (Bray, 1990b) that sediment supply from terrestrial sources, such as coastal cliffs, updrift or feeder bluffs/cliffs) was a mechanism by which Chesil Beach has been nourished and enlarged. The original gravel source was probably fluvial and periglacial deposits on the floor of Lyme Bay, which gradually decreased in importance as the rate of sea-level rise slowed down. Subsequent erosion of the cliffs provided a supply of flint and chert that helped to maintain Chesil Beach. Shingle supply from the west by littoral drift was possible until as recently as the mid-1860s when longshore transport was halted by the construction of the piers at West Bay. The supply process up to 1860 may have offset attrition losses and assisted in the maintenance of the unique Chesil Beach size grading (Carr, 1969a) because it ensured that a wide range of clast sizes was always available. However, mineral extraction has been important in the past, especially at Seatown and West Bay. Material has been removed from the beach at West Bay for over 700 years, with about 1 million tonnes of gravel removed between the mid-1930s and 1977 (Hydraulics Research Station, 1979). Of this, more than 470 000 tonnes were taken from East Beach, Bridport Harbour, and 370 000 tonnes from Cogden Beach. At one time, pebbles were removed from other locations, e.g. during 1905–1907 from the back-slopes of Chesil Beach. Perhaps the most significant of these activities was the selective pebble picking carried out nearby, not because of the absolute quantities, varying between approximately 100 and 350 tonnes per year from 1944 to 1972, with a recorded total of some 9400 tonnes, but rather because of the removal of particular sizes and shapes. This may well have produced a disproportionate weakness in the beach as well as affecting locally the longshore grading pattern and distorting the geomorphogical processes.

Changes in crest height of Chesil Beach over the last 300 to 400 years and that at one time the crest may have been lower over most of the length between Abbotsbury and Portland. Although the total volume of beach material appeared to change very little between 1852 and 1968–1969 (Carr and Gleason, 1972), the crest height between Abbotsbury and Wyke Regis showed a substantial increase. This was of the order of 2 m at Langton Herring (SY 605 810). Between Langton Herring and east of Wyke Regis (SY 650 770) there was a rise, typically, of 1.5 m. However, near Chesilton, a drop of 0.5 m, reaching an extreme fall of 3.5 m at one point, was recorded. Carr and Gleason (1972) found difficulty in explaining this phenomenon although it gave credence to early 19th century reports that the beach used to be overtopped more frequently. A comparison of the 1968–1969 profile and associated data with that of March 1979 shows that the single winter of 1978–1979 was capable of producing the same order of change at the south-east end of Chesil Beach as that indicated between 1852 and 1968–1969. Thus at one location there was a maximum fall of 2.7 m in crest height between

September 1978 and March 1979 surveys. Coupled with the known stability of the crest between 1955 and September 1978, it suggests that one event could be enough to produce the scale of change observed over the period from 1852 to 1968–1969. Such an event appears to have occurred in 1904 under similar long-period swell conditions to those recorded on 13 February 1979. A possible mechanism to account for these height changes is that where atypically large swell waves arrive parallel to the beach, the crest is overtopped, lowered, and rolled inshore (i.e. towards Portland Harbour). Farther west, towards Abbotsbury, the same swell would arrive more obliquely so that instead of clasts moving from low-water mark, over the crest, and down the backslope, the material would simply be transferred from the face to the crest, by which time the wave energy would be expended. During this process, the crest would become higher than before and there would be some net longshore transport of clasts towards the east.

The western end of Chesil Beach shows considerable variation, which is related not only to the construction of the piers and mineral extraction, but also the reaction of the beach to prolonged periods of different wind direction. Between 1901 and 1984, based on comparisons between maps and field survey, accretion on the eastern side of the West Bay piers was not continuous. There were brief periods of erosion as for example between 1961 and 1964 (Hydraulics Research, 1979, 1985; Jolliffe, 1979). Littoral drift between West Bay (SY 462 904) and Cogden Beach (SY 504 880) was investigated using a mathematical beach transport model, and a hindcast wave climate model based on Portland wave data covering the period from 1974 to 1984 indicated mean net eastward transport at 8000 m³ a⁻¹, similar to the documented trend for accretion against the east pier at West Bay (Hydraulics Research, 1985). The analysis is open to question, however, because it ignored swell waves and waves under 1 m and used shingle transport equations that had been calibrated on other beaches. Analysis of beach profiles and aerial photographs covering the period 1977 to 1990 showed that there was a marked switch from previously recognized patterns of accretion immediately east of the pier at West Bay in about 1982, to erosion, which resulted in retreat at mean high-water level by 40 m by March 1990 (Hydraulics Research, 1991a). The wave climate changed after 1982, with fewer south-east storm waves and an increase in westerly waves, i.e. a return to the generally accepted historical prevailing pattern. Littoral drift calculations confirmed net westward drift before 1982 and net eastward drift of up to 14 000 m³ a⁻¹ after 1982 (Hydraulics Research, 1991a). It appears from these studies that net littoral drift is very delicately balanced at both ends of Chesil Beach, especially at the western end where slight wave climate and storm frequency variations may produce major reversals of drift. Beach morphometry may react relatively slowly to changes in drift regime because of the large volume of Chesil Beach, and so trends may be identifiable only at the ends of the beach. They contribute therefore, little to an understanding of the mechanisms along the whole beach.

The environmental history of The Fleet is critical to an understanding of the origins and subsequent development of Chesil Beach. Within The Fleet, sedimentation of clays, silts and sands is evident above the −15 m level to about −3 m OD where commonly thick common-reed *Phragmites* peat layers (dated at *c*. 5000 years BP) occurred (Carr and Blackley, 1973). As peat becomes exposed it is eroded and thrown up on the beach between Abbotsbury and West Bexington. The pollen from a sample at −13.4 m OD in Carr and Blackley's Langton Herring (E) borehole (Figure 6.14) at the boundary between sand and bedrock was tentatively dated as early Pollen Zone VI. They interpreted this as showing that both the peat formation and the sedimentary infill above bedrock must have been very rapid. At the eastern end of the Narrows (SY 650 772), peat deposits, with a high pine-pollen content, were found underlying the landward side of Chesil Beach at a depth of *c*. −5.3 m and dated at *c*. 6200 years BP. Bedrock at this locality, occurred at −13.1 m OD, although it was as shallow as −7.8 m OD at a neighbouring location. It is unclear how these data fit into the evolutionary picture – does it imply, for example, that the relatively deep, narrow channel between the army Bridging Camp and Chesil Beach was cut as some sort of overflow feature? (Carr, 1999).

The detailed investigation of the sediments within The Fleet shows that its evolution has been a complex one (Goudie, pers. comm.; Whitaker, pers. comm.). Coombe (1996) provides details of 26 boreholes that include lagoonal and pre-lagoonal phases. A series of radiocarbon dates in peats at depths of −3.00 m OD, −3.60 m OD and −4.32 m OD have been dated between 4540 ± 70 and 4840 ± 70 years BP (Figure 6.15), indicating rapid accumulation such as Carr and Blackley (1973) had inferred. Two samples in the East Fleet (cores 25 and 29) at −3.00 m OD and −3.15 m OD were dated at 3820 ± 70 and 4110 ± 60 years BP respectively. Mean grain size in Coombe's cores decreased westwards, indicating that energy lessened along The Fleet, supporting the hypothesis that the infill was derived mainly from the south-east in Weymouth Bay. The pollen in a sand sample beneath the lagoonal sequence included a high frequency of pine *Pinus*, together with oak *Quercus*, hazel *Corylus* and birch *Betula*. These are similar to the Carr and Blackley (1973) findings that placed these basal sediments in Pollen Zone VI. This suggests that these sands found below the lagoonal sequence may be older than 6000–6500 years BP. In all the cores, there is evidence of saltmarsh above the sands. This predates the brackish-marine lagoonal phase, which is not older than 5000 years, and is the dominant feature today. Throughout The Fleet narrow shell beds rest on top of each peat bed. Goudie *et al.* (pers. comm.) consider that the peats either represent stillstands or slight falls in the rising Holocene sea level or indicate that The Fleet was closed by a barrier and became a freshwater lake or a reed swamp cut off from the sea. West Fleet was behaving more like an estuary than a lagoon. At West Bexington, foramifera and ostracods show that a tidal, near-marine, water body existed well to the west of the modern Fleet around 4000–5000 years BP (Whitaker, pers. comm.). However, in order to allow for such a body to exist, it is necessary to have a barrier well seaward of the solifluction slope between Abbotsbury and West Bexington.

Although there is at present insufficient evidence for this to be more than conjecture, it would help explain the nature of the West Bexington and Burton Mere environments, both of which appear to be extensions of The Fleet. A further complication in identifying the origins of The Fleet and Chesil Beach comes from the solifluction slope itself, which extends the alignment of Chesil Beach towards West Bexington. If, as Brunsden and Goudie (1997a) have suggested, the solifluction materials overlie evidence of the Portland emerged beach, the western part of an extended Fleet could be a very old feature. To the west of West Bexington, present-day cliffs probably extended farther seawards and would have acted as an early headland from which sandy materials could supply a low sandy barrier beach, it also provides a basis for some re-interpretation of the classic transgression model for the development of Chesil Beach.

The longer term evolution of Chesil Beach is still not completely understood, but the chronological sequence compiled by Carr and Blackley (1973), modified by Bray (1990b) makes it possible to put forward the following scenario for the initiation and development of Chesil Beach.

1. Around 210 000 years BP, the Portland emerged beach is formed and the slopes between Abbotsbury and the Narrows are trimmed. The beach is re-occupied about 125 000 years BP. Westwards from Abbotsbury, there is also an emerged beach and so a coastline very close to the present-day extended at least between Smallmouth and West Bexington. A forerunner of Chesil Beach may have existed as a bank well offshore of the present beach some 120 000 years BP (Carr and Blackley, 1973), contemporaneous with the development of the Portland emerged beach.

2. From the emerged beach level of between +7 and +15 m OD, sea level fell to about −120 m. The seabed is weathered by periglacial processes and a series of gravel-rich deposits, probably comprising material from the Portland emerged beach, solifluction deposits, river gravels and fluvioglacial deposits were deposited on the floor of Lyme Bay. Degraded landslides and solifluction deposits extend 2 to 3 km seawards of the modern shoreline position and mantle the former coastal cliffs.

3. From about 20 000 years BP, sea level rose by 1 mm a^{-1}, and by about 10 000 years BP sea level was at approximately −45 m. The proto-Chesil Beach approaches the former shore of Lyme Bay. The sea level then rose rapidly increasing by an average of 1.5 m every 100 years.

4. About 7000–6500 years BP, the model suggests that closer to the modern coastline, the transgressing Chesil Beach overrode existing sediments as sea level rose. A shallow lagoon that became The Fleet was rapidly filled with silt, sand, pebbles and peat from 7000 to 5000 years BP. About 7000 to 6500 years BP, with

Gravel and 'shingle' beaches

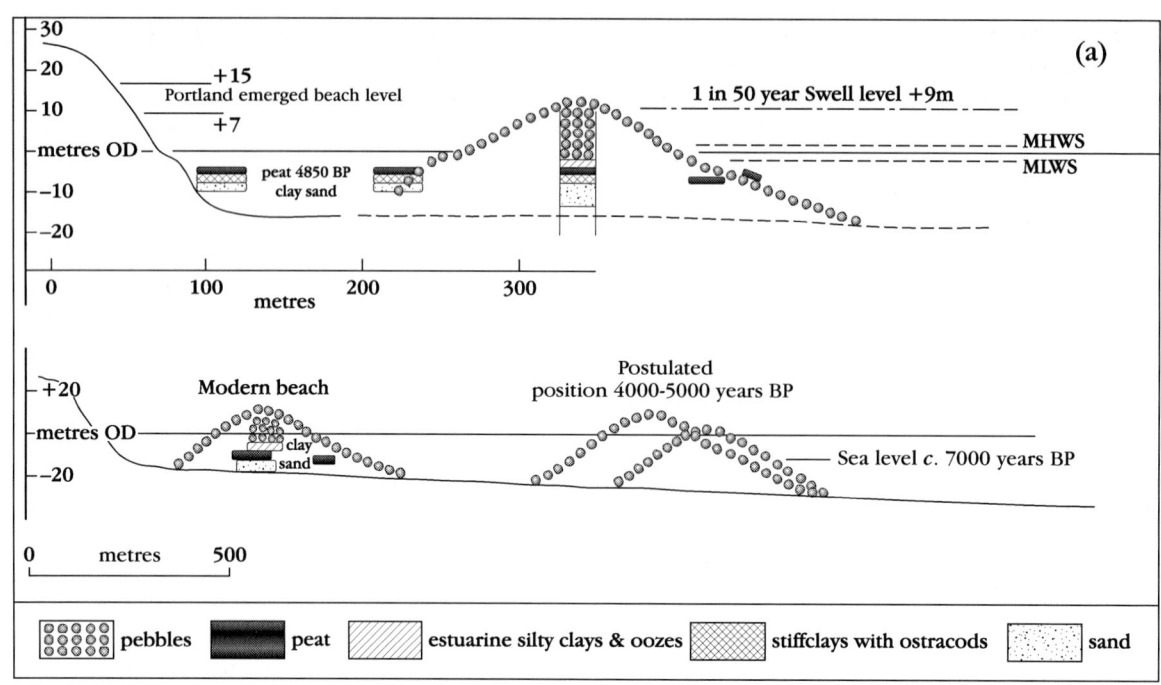

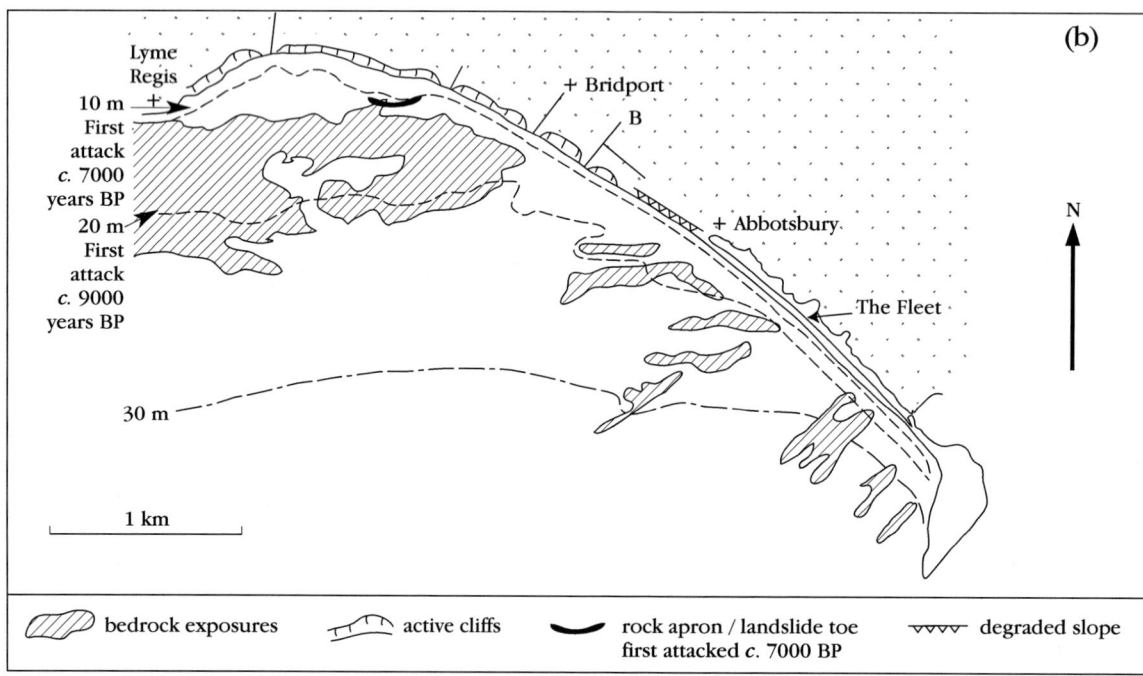

Figure 6.17 Chesil Beach (a) relationships between modern beach and dated peats and water levels (mean high-water springs and mean low-water springs, MHWS and MLWS, are shown). By *c.* 5000 years BP, the supply of flint was able to create a barrier beach atop an earlier sand ridge and estuarine peats. (b) Seabed features of eastern Lyme Bay and their relationship to Chesil Beach. Note the relation of bedrock exposures and seabed contours to the present shore, which probably affected the development of the earlier beach form. 'First attack' indicates the bathymetric contour representing the shoreline first attacked by the sea at the date shown.

sea level between −12 and −4 m OD, a low sand and gravel beach developed at about 1 km offshore (Figure 6.17a). Relict cliffs farther to the west, abandoned by falling early Devensian sea levels, were re-activated by marine erosion (Figure 6.17b) and large quantities of gravel were transported eastwards, to feed and enlarge the new Chesil Beach (Bray, 1990b). The Fleet forms behind this barrier beach. West of Abbotsbury, wave-energy distributions are modified by refraction and higher energies occur much closer to the eastern end. Solifluction deposits and landslide toes begin to be trimmed by the rising sea and gravels begin to travel eastwards.

5. Between 6500 and 4000 years BP, the lagoon is closed and a freshwater lake forms in the East and West Fleets. There was probably dry land between Weymouth and Portland, and The Fleet extended farther west to Burton Mere. The implicit shelter needed for peat deposition is taken as an indication that by 4000 to 5000 years BP, Chesil Beach had formed close to, or a short distance seaward of, its present position (Bray, 1990b)

6. As sea level rose, between 4000 and 2000 years BP, tidal conditions developed in The Fleet. There was now a continuous beach to Portland from Abbotsbury and also a separate beach facing eastwards into Weymouth Bay.

7. From about 2000 years BP to 1850 AD, marine erosion to the west supplies increasingly large volumes of clasts to the beach, which therefore grows in volume and height to establish the current form. The beach became a shingle ridge overlying a finer-grained, impermeable core. This continued until the mid-19th century.

8. From the mid-19th century, the building of the Cobb and West Bay harbours cut off longshore sediment transport at the same time as the beaches became more fragmented by the gradually emerging headlands at locations such as Golden Cap. Combined with the effects of beach mining and attrition, this leaves Chesil Beach as a relict feature suffering slow decline in volume.

This scenario implies that the classic model of a transgresing gravel beach could be replaced by a two-phase model in which the early Chesil Beach is a low sand and gravel barrier that provided a base upon which the more massive gravel and cobble structure was constructed as large supplies of these materials became available. Better understanding of the wave climate of Lyme Bay still leaves questions about the effect over many centuries of major events, such as those of 13 December 1978 (a 1 in 50 year event with 9 m swell) and 13 February 1979 (when 18-second period waves arrived without warning out of a moderate sea).

Despite its magnitude, it is likely that in common with many other coastal features Chesil Beach at a comparatively late stage in its evolution. Carr and Blackley (1973) suggest that during the last interglacial, any proto-Chesil Beach must have been entirely reworked by marine action. Such may well be the ultimate fate of the present-day structure.

Screening of coarse indigenous shingle for gabion and mattress-fill for the trial length completed in November 1981 could also have had an effect. It is difficult to determine the proportion of beach volume lost by extraction, but Carr (1981) estimated that between the mid-1930s and 1977, something of the order of 2% was likely to have been removed overall. Although more research needs to be carried out to determine how losses through attrition compare with this, it is very unlikely that the latter are as great on this essentially flint and chert beach.

Most anthropogenic pressures on the beach have been concentrated at the extremities (Carr, 1983a). Those at the Portland end are most critical because it is there that the beach is subject to maximum fetch from the Atlantic Ocean, and, of scientific importance, the beach crest is at its highest and the rate of change in grading is greatest along the most easterly 2 km.

There can be little doubt that Chesil Beach is in a fragile state and is finite in amount. No more material is being supplied and loss - continues through attrition and removal offshore. The logical prediction must be that Chesil Beach will now steadily move onshore and break up into separate beaches and bays. It may rotate at Weymouth, breach at several places, allowing The Fleet to become saline and disappear or develop into lakes like those behind Cogden and West Bexington. The main new headland may be at the Narrows. Severe erosion will take place at East Cliff and Burton Bradstock, where the next bays will develop. The processes that happen when a barrier beach comes onshore, already seen between Lyme Regis and West Bay, will be the model for the rest of the beach.

Gravel and 'shingle' beaches

Conclusions

Chesil Beach is a massive, linear pebble and cobble coastal barrier/tombolo backed by cliffs at its western end and a lagoon, The Fleet, in its central and eastern parts. The beach ties the Isle of Portland to the mainland of Dorset. It is renowned worldwide especially for the size grading of its constituent clasts. There are four British gravel structures that rival Chesil Beach in scale, Culbin, Spey Bay, Dungeness and Orfordness, but none displays the simplicity of form or the simple barrier shape evident at Chesil Beach. Chesil Beach is very unusual in lacking any development of recurves, even within the lagoon between Abbotsbury and Wyke Regis. The fact that it is simple in form offers enormous scope as a baseline exemplar for studying many other more complex structures.

Despite the number of cores taken through the beach and The Fleet, the internal structures remain speculative. Furthermore, the three-dimensional form of the beach, the surfaces on which it rests, the nearshore seabed and the processes that produced and maintain the beach also warrant more detailed investigation.

The origins of the beach are open to debate, but can be summarized as follows (Bray, 1990b). With sea level at the end of the Devensian about 100 m below its present-day position, a barrier beach formed as a result of the erosion of river gravels and other offshore sediments. About 7000 years BP, infilling of the Fleet began and was virtually complete by 5000 years BP (Carr, 1974). According to Bray (1990b), Chesil Beach was thus formed before there was significant erosion of the west Dorset coastline. It was then maintained by longshore sediment transport from west Dorset. An alternative view suggests a two-phase development with a sand-dominated barrier offshore upon which the cobble ridge was then established.

Even so its origins are still a matter of debate. Despite the local modifications to the beach, Chesil Beach remains a remarkable coastal landform that is regarded worldwide as having extremely high scientific value because of its form, size, composition and documentary record. Chesil Beach is included in its entirety in the Dorset and East Devon Coast World Heritage Site, declared in December 2001. Few other sites are more cited or visited by coastal scientists.

PORLOCK, SOMERSET (SS 858 484–SS 899 492)

J. Orford

Introduction

The coastal gravel barrier and beach at Porlock is the longest continuous coastal gravel barrier system on the western coast of Britain (see Figure 6.18). It is 5 km long and fronts low-lying farmland, which is being flooded daily by the tide as a result of a major breach that occurred during a severe storm in 1996. The breach has not 'healed' and has resulted in saltmarsh development and clay deposition in the back-barrier area associated with the upper elevations of a prevailing macro-tidal regime (9 m at mean high-water springs). The existence of the barrier relies upon a complex and dynamic interaction of geomorphological processes operating in an environment that is sediment-constrained and storm-wave dominated. Sporadic breaching and 'healing' events had been the natural cycle of evolution of the barrier before the onset of anthropogenic activities that attempted to stabilize the barrier and prevent it from breaching.

The barrier shows evidence of longshore segmentation, and of sediment erosion and reworking caused by the long-term failure of longshore sediment supply, which lead to well-developed swash-aligned and drift-aligned sections that are rarely found adjacent to each other on UK gravel barriers.

The geomorphological importance of, and wider interest in, the Porlock coastal gravel barrier has grown significantly over the last two decades, owing to debate about the effects of earlier coastal management strategies. There can be little doubt that the catastrophic breaching event of 1996 occurred as a consequence of barrier weakening through cumulative management activities (over several decades) intended to protect the coast. Although the barrier is returning to a natural state it is unlikely that the breach will heal, and part of its present-day interest is the development of the geomorphology following the breach.

Description

The Porlock barrier fringes the seaward edge of a coastal embayment at the eastern end of the Exmoor plateau (Figure 6.18). The crest height

Porlock

west of Gore Point is typically 5–6m OD, and 12 m OD to the east, towards Hurlstone Point. The local lithologies west of Porlock are red-purplish or grey fine-grained sandstones, and thick flaggy grit beds interspersed with shale and pebble beds. All of these lithologies are represented in the sediment of the barrier. The pattern of sediment facies (shape and size) differentiation across the barrier (cf. Bluck, 1967) is well developed, but has been disturbed locally by recent management activity.

The coastal edge of the Exmoor plateau (c. 40 km long) shows a distinctive 'hog's-back' form with a 350 m fall from the plateau top to sea level, with a steep and unstable convex coastal slope (Arber, 1911) c. 1 km wide. Arber commented on the thick wooded cover of the coastal slope, but during the 20th century anthropogenic disturbance of this cover resulted in an increase in the frequency of coastal landslides in the unconsolidated sediment. The sediment size arriving at the hog's-back shoreline from these slides has a considerable range from blocks (>20 m) to mud. The Porlock foreshore shows a well-developed outer boulder frame in the intertidal zone for all but the last kilometre of its proximal (eastern) end, suggesting the landward retreat of sections of the barrier. The recent increase in sediment supply from landslides has not alleviated the sediment deficit characteristic of the barrier, so it is likely that the apparent surplus of sediment west of the Porlock barrier is a recent phenomenon and not the typical situation throughout late Holocene times (i.e. over the last 2000 years), although landslides will have supplied sediment episodically in the past history of the feature.

The Exmoor coastal slope was covered by Devensian solifluction sediment, forming fans that coalesced at the foot of the plateau, which have been reworked by Holocene relative sea-level rise. The present barrier is transgressing across such a fan at Porlockford, and the western end of the barrier is developing on top of the intertidal scar of an eroded fan (Gore Point).

The rising ground of the relict solifluction fan at Porlockford cliffs controls the present-day embayed position of the barrier. Fan resistance to wave action can lead to barrier segmentation around the flanks of the fan. This western foreshore shows a mixture of boulder scars as the rising sea level has eroded its way through the old fans, and a lag of large boulders (outer boulder frame) that resist swash action and incorporation into the barrier. The frame dissipates wave energy but its low position (<–1 m OD) means that it does not protect the barrier during high tides and surges. The exposed position led to attempts to build up the barrier crest east of Porlockford and New Works in the early 1990s by reshaping the profile (making it higher to c. 8 m OD) but thinner (Figure 6.18C; profile P4); replacing washover sediment back onto the ridge top; and renourishing the ridge sediment volume with dredge material (gravel, sand and mud) from the entrance of Porlock Weir harbour. Attempts to export down-drift gravel back to this swash-aligned section were successfully resisted by the National Trust, the landowners of the easterly drift-aligned section. The weakness of this swash-aligned bay was the core management problem in the 1990s, as potential barrier breakdown and associated back-barrier flooding were perceived to be central issues for local economic well-being and aesthetics.

Between Porlock Weir and the tidal sluice gate (New Works; Figure 6.18B), the continuous easterly longshore depletion of sediment has reduced the volume of the barrier and accelerated its retreat onshore by rollover (where sediment is carried over the ridge top by storm washover), forcing the embayed barrier into a more swash-aligned structure. The Porlock barrier has retreated most at this section. Here, the barrier is low (c. 6–7m OD) and narrow (<40 m wide), commensurate with the loss of volume as the barrier is 'stretched' between Porlock Weir and New Works without new material being added.

Although New Works appears to have been built at the transition between the up-drift swash-aligned and down-drift drift-aligned sections, it is unlikely that the works *per se* were of a sufficient age to have played a part in the development of this transition point. New Works was probably sited here because of its low elevation with respect to back-barrier drainage.

After 1950, the ridge east of the New Works area was in constant need of remedial attention due to washover forcing the remnant ridge into a swash-aligned posture commensurate with the westward barrier. The barrier was rebuilt on several occasions: by regrading washover fans; adding new gravel (from the old recurves) to the upper beach; and installing groynes to reduce westward movement of sediment. All this activity reflected the continuing movement of gravel eastwards as the swash-aligned section was

Gravel and 'shingle' beaches

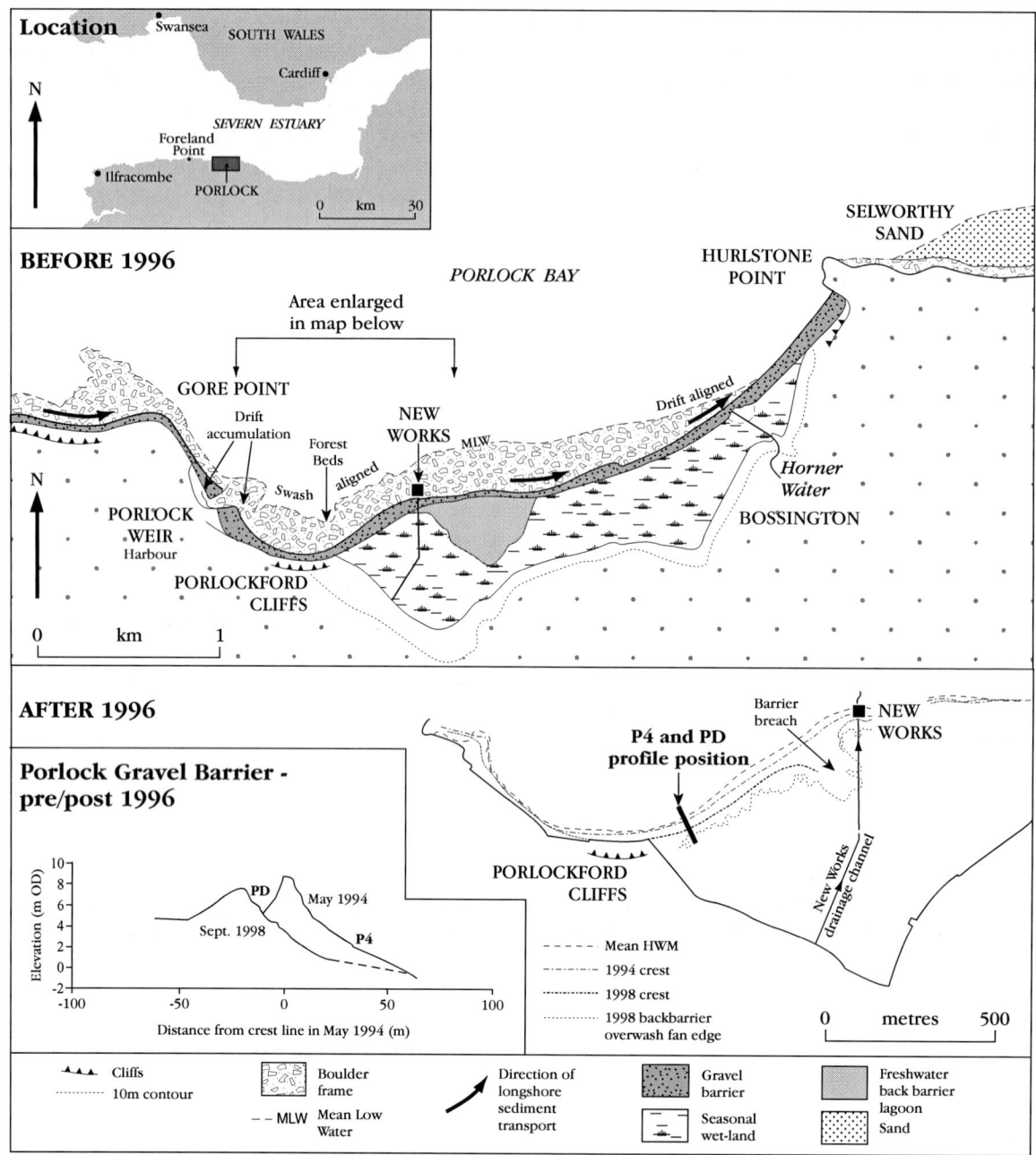

Figure 6.18 (A) Porlock barrier and back-barrier; (B) barrier crest and back-barrier changes before and after the 1996 barrier breaching; (C) barrier profile changes due to the 1996 storm.

evolving further eastwards. The barrier has retreated marginally at this section since the 1980s and is currently held by old shore-parallel stone walls now being consumed by barrier rollover. This anthropogenic intervention (stopped in 1990s) makes it difficult to determine the natural position of the barrier in the decade before 1996.

During late October 1996 Hurricane Lillie moved across the North Atlantic Ocean and degenerated into a deep depression before moving across southern Ireland and the UK. Storm surge levels superimposed on the high tide exceeded the height of the managed barrier

Figure 6.19 Overview of Porlock barrier (October, 1997) looking east. The 1996 barrier breach is identifiable, as are the storm generated washover fans at point NW (New Works sluice gate); HP is Hurlstone Point. (Photo: W. B. Whalley.)

between Porlockford Cliffs and New Works on October 26th. Massive overwashing demolished the barrier crest and moved gravel onto the back-barrier area. The volume of overwashing waters was sufficient to fill the back-barrier area and during the falling tide, forced a breach west of New Works. The extent to which the gravel ridge was also pushed back has been partially identified by differences in crest surveys undertaken by Orford *et al.* (2001) in July 1994 and September 1998, and in a series of detailed measurements post-1999 by Bray and Duane (2001). In outline the original managed crest of the barrier was demolished and pushed back in the form of classic washover fans (Figure 6.18B). During February 1997, a series of major westerly cyclonic storms helped to reconsolidate the beach ridge at a new position 15–25m farther landwards.

Since 1999, ridge changes west of New Works have been slight, the lack of movement and restructuring of the ridge into a broader and lower (*c.* 6–7 m OD) feature suggests a period of stability. The barrier will continue to retreat as a function of the elevation reached by extreme run-up of breaking waves, which will also increase with future storminess and relative sea-level rise. Past lowland management and land-use in the area has meant that there is space for barrier retreat without any overwhelming demand for coastal protection. In order to understand the development of gravel barriers that are not impeded by protection measures, Porlock is a key 'open air laboratory' site that will enable geomorphologists to better predict the outcome of management activities on gravel barriers. The freedom to evolve now will mean the site is once again moving towards a more 'natural' pathway and therefore a key site to see how a barrier naturally responds to past anthropogenic forcing.

Interpretation

The barrier probably initially formed as a drift-aligned spit building eastwards; remnants of the recurves from this sediment-surplus phase can be identified at positions between New Works and the old limekilns near Horner Water, but there is no available dating evidence for this barrier growth phase. As sediment supply from the hog's-back area diminished owing to a reducing rate of the rise of relative sea level in mid-late Holocene times, then barrier reworking would have taken place (Orford *et al.*, 1996). Barrier sediment from updrift positions (west) was transported to continue the spit growth at its eastern end, until reaching Hurlstone Point (which acted as a natural groyne) and sediment was trapped in major beach ridges. These latter forms have been steadily drift-aligned (20 m accretion between 1880 and 1980) into high (12 m OD) beach ridges. Their seaward growth means that there is no shore platform exposed at low water and thus bigger waves can approach closer inshore without dissipation, allowing storm waves at high tide to generate swash sufficient to move gravel to the crest top.

The west–east Exmoor coastal slope is the main source of gravel for the Porlock barrier. There is a net easterly beach drift system powered by Atlantic swell waves and depression-generated storm waves running from the north-west to west into the Severn Estuary. The slight shift in coastal alignment east of Foreland Point might indicate that the contemporary source area for Porlock is somewhat less than the whole Exmoor length. There is little fine-grained sediment in the Porlock barrier system because much of the sand-sized load is moved offshore, or deposited to the east of Hurlstone Point. It is feasible that Holocene relative sea-level rise has now eroded the low-elevation fans at the foot of the coastal slope and it is the reduction of this source that has now pushed Porlock's sediment budget into deficit. The 20th century rise in sediment supply (due to up-drift changing coastal slope management practice) seems only to be reaching as far as Porlock Weir. The cannibalization of the Porlock barrier and development of the major swash-aligned unit deep into the western end of the Porlock embayment means that sediment transport from the old western source area has now virtually ceased. This has not been the case in the past when the barrier was probably developed as a spit extension of the coastal plateau edge into Porlock embayment.

Jennings *et al.* (1998) explored the relative sea-level rise identified by a palaeo-ecological reconstruction of organic deposition found exposed in the foreshore, and in the pre-1996 back-barrier marsh at Porlock. Their results fit into the broader relative sea-level rise envelope identified by Heyworth and Kidson (1982) for the Severn Estuary. Tree roots bedded into a fen-type freshwater environment are exposed on the foreshore of the western swash-aligned barrier, and cores taken from the pre-1996 seasonally wet back-barrier area (now tidally flooded) identify a mixture of environments related to whether the back-barrier area was open to intermittent tidally induced flooding and mud deposition, or was closed such that a freshwater fen environment was generated. The thesis of Jennings *et al.* (1998) was that barrier coherence (hence barrier strength to resist breaching) and potential for tidal incursions into the back-barrier area was related to the rate of relative sea-level rise. They suggested three domains of barrier activity related to relative sea-level rise, noting that as the rate of rise decreased there would be an increasing tendency for spatial stability of the barrier, a decreasing longshore sediment supply rate, and as a consequence a greater potential for the barrier to be cannibalized and disturbed sufficiently to allow storm breaching sufficient for tidal incursion. Prior to 7000 years BP, relative sea-level rise rates of $c. >6$ mm a^{-1} were likely to generate high longshore sediment supply but reduce the ability of the barrier to maintain any longshore coherency sufficient to act as a barrier to tidal influence in the back-barrier area – hence the evidence of fine-grained, marine, back-barrier deposition. Between 7000 and 5000 years BP, relative sea-level rise rates of 6–2mm a^{-1} allowed sufficient sediment to seal any barrier breaching, while the spatial translation of the ridge was so reduced that the barrier maintained its coherency to act as a buffer to saline influences, thus letting a back-barrier freshwater regime operate in which fen-deposition predominated. When storm breaches were intermittently open, the back-barrier area was exposed to saline waters, thus allowing thin marine mud-sequences to be inter-digitated with organic fen materials. By late Holocene times (<2000 years BP) decelerating rates of relative sea-level rise (<1 mm a^{-1}) reduced the longshore sediment supply allowing barrier breaches to remain open and the fen to be replaced by tidally dominated back-barrier sedimentation.

The 1996 breach is unlikely to be sealed by the existing low longshore transport rates. It has been widened and deepened with headwater erosion by tidal flows into the consolidated

Holocene clays of the old back-barrier. Bray and Duane (2001) have mapped the retreat of the breach and the extension of the inlet sidebars. They have also monitored tidal elevations and associated sedimentation within the now active upper tidal frame behind the barrier. The high turbidity of the water column in the Severn Estuary and its macro-tidal regime ensure that fine-grained sediment enters behind the barrier on almost all tides. Annual deposition rates of 10 mm a^{-1} measured during 1999–2000, suggest that this is a site of great potential for saltmarsh development. Whether this sedimentation rate is the initial result of a forced change in the system and will decline as the back barrier adjusts to the new tidal regime is uncertain. However the lack of coastal squeeze at this site suggests that this will be an important test site for evaluating saltmarsh growth and adjustment to accelerating relative sea-level rise in future decades. Bray and Duane (2001) also underline the potential for barrier change immediately east of New Works. The implications for further breaching in this area are intriguing, though current inlet efficiency may be reduced if more breaches occur, thus limiting more persistent breaches.

The historical state of the barrier is unknown, though the 'stabilized' barrier during the 20th century produced a local view of back-barrier stability that has not been the norm for most of its Holocene existence. The tidal sluice-gate system (New Works) indicated the measures taken to ensure the drainage of freshwater from the back barrier to allow pasture development. Even then, the pre-1996 intermittent wetlands identified a problem in fully draining the back-barrier area and the resultant small mere and fen that did develop provided the interest for the original Porlock SSSI biological designation. This past anthropogenic intervention has flavoured perspectives as to how Porlock barrier should be managed, although the latest storm-forced changes to the managed barrier section show that a natural mode of barrier evolution is now appropriate, and that this should be of prime concern in future management strategy.

Conclusion

The Porlock barrier is central to geomorphological studies into how a freestanding gravel barrier responds to relative sea-level rise and storminess changes. The barrier is likely to remain a centre of coastal interest for its combination of evolving swash-aligned and drift-aligned longshore sections; its post-management adjustment to a stable cross-barrier profile in relation to relative sea-level rise and storm-wave climate; its rollover dynamics and washover response; its breaching behaviour; its developing tidal inlet control on barrier segmentation and longshore transport impedance; its back-barrier fine-sediment deposition and saltmarsh development. It represents one of the best UK examples of how managed 'stabilized' barriers are non-sustainable at the decade-timescale. It also exemplifies the likely mode of barrier failure, if coastal gravel-dominated barriers are not allowed to adjust freely to changing relative sea level.

HURST CASTLE SPIT, HAMPSHIRE (SZ 310 900)

V.J. May

Introduction

Hurst Castle Spit extends the shingle fringing beaches at the eastern end of Christchurch Bay across the western arm of the Solent (see Figure 6.2 for general location). Its seaward end is marked by Hurst Castle, constructed in the mid-16th century, and threatened from time to time by erosion since the mid-19th century. The spit protects Keyhaven Marshes (see separate GCR site report in Chapter 10 of the present volume).

Following the seminal paper of Lewis (1931) in which he argued that beaches align themselves at right angles to the direction of approach of dominant waves, Hurst Castle Spit is often used as an example of a multi-recurved spit (Johnson, 1919; Wooldridge and Morgan, 1937; Steers, 1946a; Sparks, 1960; Bird, 1968; King, 1972b; Komar, 1976; Bird and Schwartz, 1985). King and McCullagh (1971) used Hurst Castle Spit as the basis for an early computer model – 'Spitsim'). More recently, Clark and Small (1967), Clark (1974), Nicholls (1984, 1985) and Nicholls and Webber (1987a–c, 1989) re-examined its features. Nicholls and Webber suggest that this is not a complex recurved spit, but owes its detailed form to variations in local sediment supply and to changes in sea level that had not been considered in previous work.

During the winter of 1989–1990 the whole ridge was overtopped and moved inland by up to 80 m. The risk to the recently completed coast protection works at Keyhaven and Lymington, the ecologically important saltmarsh behind the ridge and low-lying residential areas from flooding was judged to be so severe that major coast protection works were put in place during the late summer and autumn of 1996. 120 000 tonnes of imported Norwegian rock (with boulders up to 1.5 m across) were placed along a 550 m-long section of the proximal end of the spit. 500 000 tonnes of shingle dredged from the Shingles Bank were placed along the remaining length of the barrier beach to double its width and raise it to a height of 7 m (see Figure 6.21). A rock revetment 100 m in length was constructed to protect the western wing of Hurst Castle and regrettably shingle was excavated from the most recent distal recurves to be placed along the remainder of the frontage of the castle. As a result, despite the recognition of the importance of this GCR site, much of its natural interest has been removed. Although much weakened at its proximal end by erosion combined with a reduction in the littoral supply of shingle as a result of coast protection works, the spit nevertheless retains its characteristic form.

Description

The natural spit had two main parts (Figure 6.20):

1. Hurst Beach, a single transgressive shingle ridge orientated towards the dominant south-westerly waves of Christchurch Bay. Along much of its inner length, the shingle rests upon saltmarsh and earlier gravels which are occasionally exposed on the foreshore. Its maximum height varies between 3 m and 5 m OD. The direction and energy distribution of the waves approaching the beach is affected by both the Isle of Wight and a shallow offshore shingle shoal, known as the 'Shingles Bank'.
2. An active recurve, behind which are three groups of preserved recurves, which may pre-date the construction of Hurst Castle (AD 1541–1544). The recurves are aligned towards the dominant north-easterly waves of the western Solent.

The beach is formed mainly of subangular to subrounded flint pebbles with subsidiary fine- to medium-grained sand, derived from the erosion of Pleistocene sandy gravels farther to the west. Net littoral drift is towards the east, although much of the sand is lost offshore. Nicholls and Webber (1987a) reported a littoral sediment transport sub-cell boundary at Hordle Cliff to the west. This, they believed, means that the spit is much more dependent upon local sources of sediment than earlier writers had supposed. At Hurst Point, much of the shingle is lost offshore into Christchurch Bay or the western Solent. The nearshore slope is steep and tidal energy in the western Solent is high, the tidal streams between Hurst Point and the Isle of Wight attaining 2.3 m s^{-1} at spring tides and so capable of moving small shingle. Dyer (1970) reported that much of the material forming the Shingles Bank is similar in composition to that forming Hurst Beach and most writers agree that it is likely that much of the sediment lost from Hurst Point reaches the Shingles Bank.

The shingle ridges forming the recurves are generally about 1 m lower in altitude than the transgressive Hurst Beach and the present active recurve, except over the last 100 m at its distal end. Within the main area of shingle north of Hurst Castle it is still possible to identify two strongly recurved distal ridges, which suggest that this area includes at least three ridge groups. The innermost of these must pre-date the construction of the castle, and the second may pre-date the extension of the castle between 1861 and 1873. The second group of ridges are not continuous, having been removed in their central section by erosion. At present the active recurve appears to be marked by zones of accretion at Hurst Point and its distal end with a zone of erosion between them (Figures 6.20 and 6.21).

The landward end of the spit was most recently armoured by boulders in 1996 and both beach replenishment and reshaping of the ridge have been adopted as measures to prevent breaching of the ridge. The beach fronting the castle has been subject to considerable erosion and the foundations of the outer works of the castle have been undermined. A series of groynes had been in place for several decades, but these had decayed badly and were renewed by English Heritage during the late 1980s together with a programme of beach replenishment using shingle from the zone of accretion at Hurst

Hurst Castle Spit

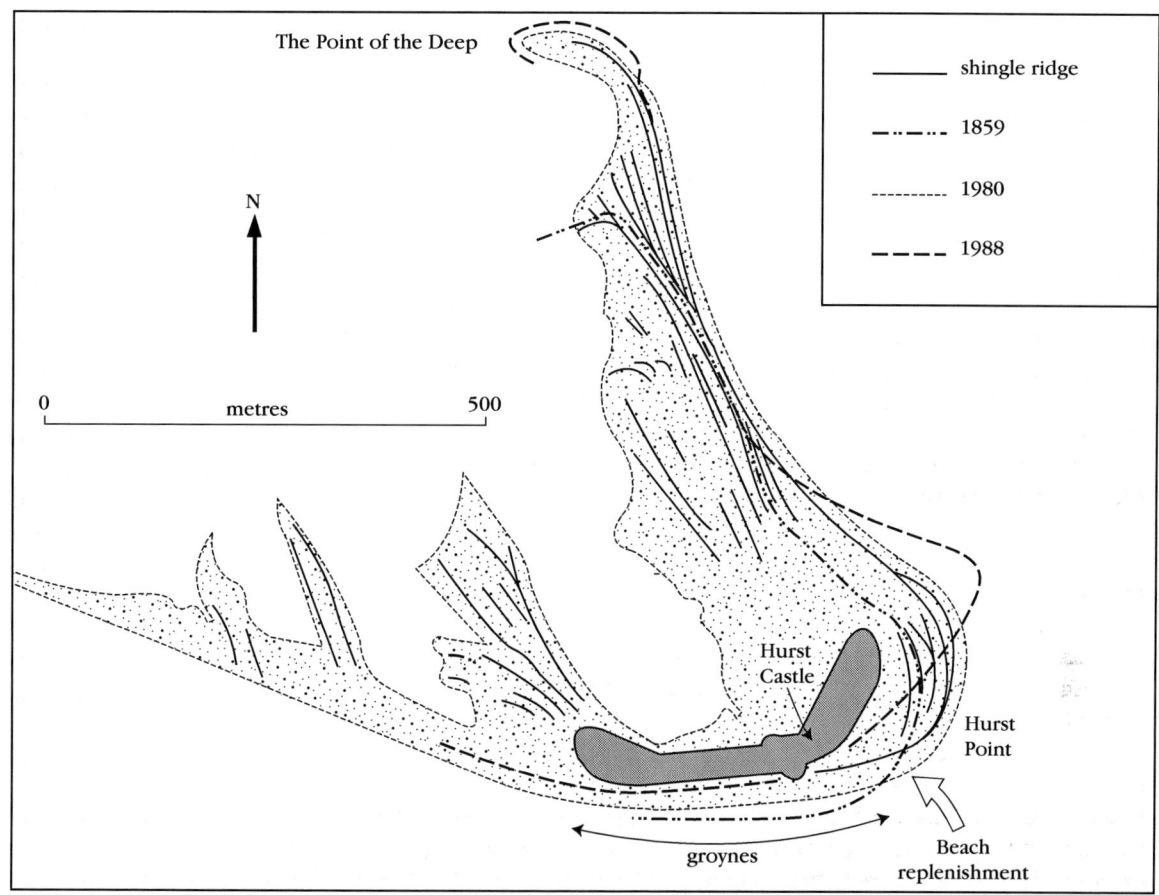

Figure 6.20 Distal recurves at Hurst Castle Spit – the history of geomorphological development. (After Nicholls and Webber, 1987a.)

Point. Further replenishment was needed by 1996.

The natural spit was affected by several processes. Hurst Beach is transgressive (Figure 6.21), moving over the saltmarsh and Pleistocene gravels, which were reworked, providing a local source of shingle. The transgressive ridge is affected by overwashing and seepages, both of which can lower the ridge crest. Some shingle is moved along the spit with littoral drift increasing towards Hurst Point where it attains a maximum of 15 000 m^3 a^{-1} for the shingle fraction alone (Bray *et al.*, 1992). Shingle is moved seawards from Hurst Point onto the Shingles Bank, which, as it changes shape and height, affects the wave energy distribution along Hurst Beach (Nicholls and Webber, 1989). North of Hurst Point, the active recurve continues to grow towards the Point of the Deep, but this appears to be at the expense of the central part of the active recurve.

The spit protects a large area of saltmarshes, known as 'Keyhaven Marshes' (see separate GCR site report in Chapter 10 of the present volume) which are drained by an intricate pattern of creeks dominated by three major creeks: Mount Lake, alongside the spit, Keyhaven Lake and Hawkers Lake. The first two merge and drain into the Solent after being diverted by the modern recurves of the spit. Active marsh-edge beaches (cheniers) are formed mainly of shells and shingle with a low sand content. Much of the saltmarsh edge is being eroded rapidly, resulting in patches of unvegetated mud. The surface of the marshes is characterized by a high proportion of eroded marsh, pans, and broad channels. There are only small areas of higher-level, species-rich saltmarsh, located mainly

Gravel and 'shingle' beaches

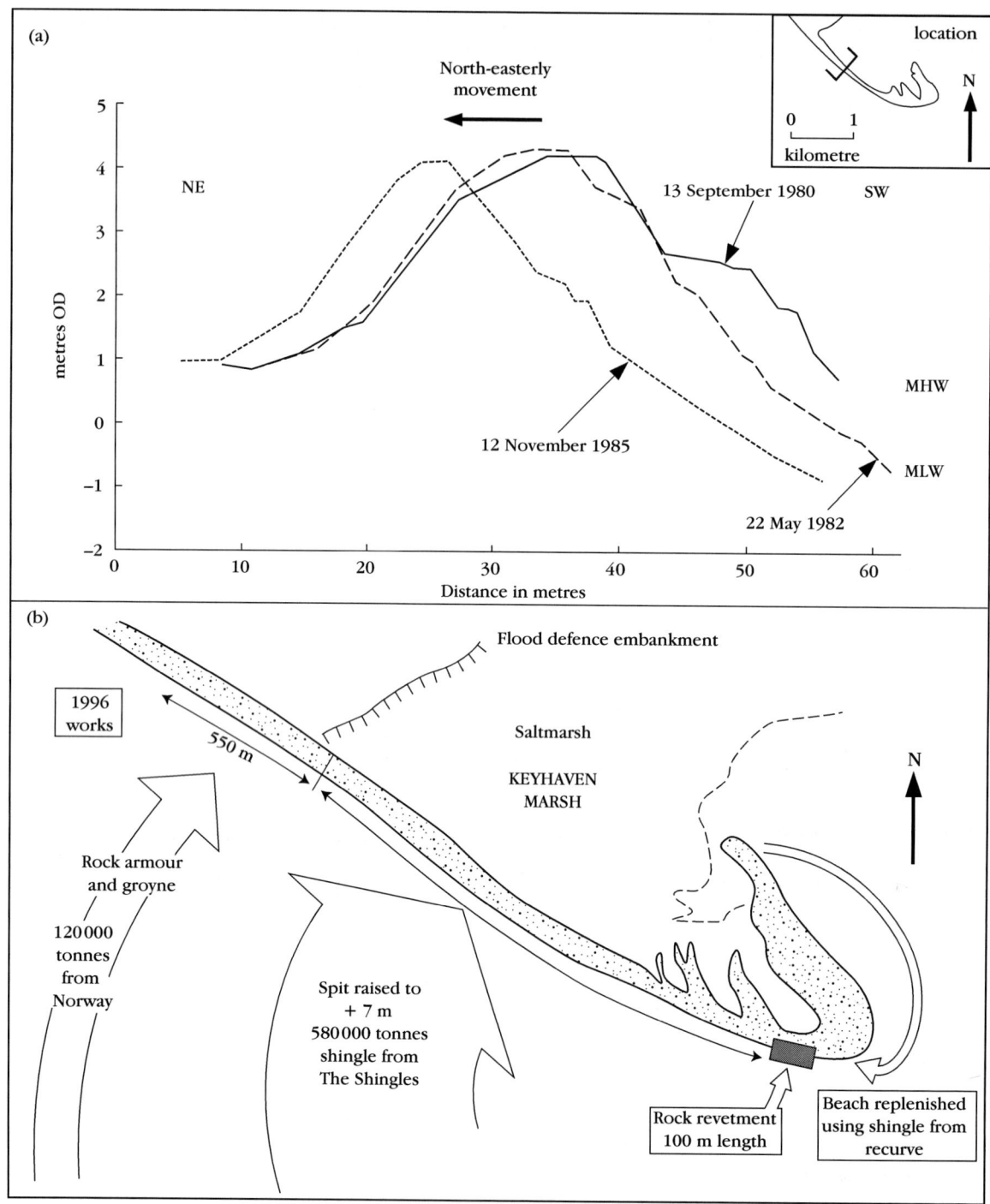

Figure 6.21 (a) Changes in the profile of Hurst Castle Spit. (After Nicholls and Webber, 1987a.) (b) 1996 coast protection works at Hurst Castle Spit. The pecked line in (b) delimits the saltmarsh edge.

close to the spit and on its older recurves. Sea purslane *Atriplex portulacoides*, common sea-lavender *Limonium vulgare*, sea plantain *Plantago maritima*, sea meadowgrass *Puccinellia maritima*, annual sea-blite *Suaeda maritima*, samphire *Salicornia* spp., and sea aster *Aster tripolium* are common throughout these higher marshes. In contrast, the lower more extensive marshes are species-poor and dominated by common cord-grass *Spartina*

Hurst Castle Spit

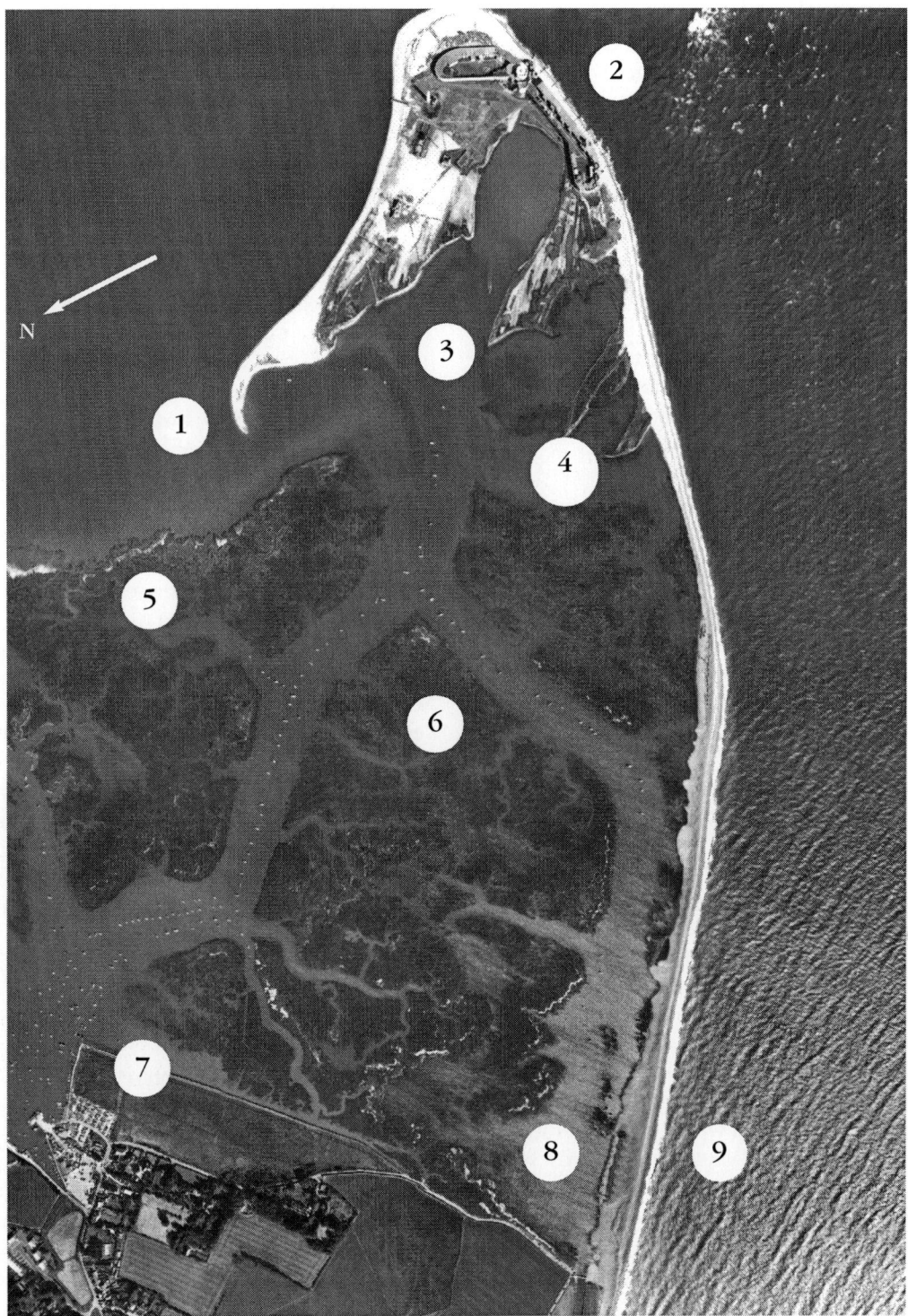

Figure 6.22 Aerial photo of Hurst Castle spit. 1. Distal end of modern beach; 2. Groynes protecting Henrician (16th century) castle; 3. and 4. earlier recurves; 5. saltmarsh – the seaward edge of saltmarsh is undergoing retreat; 6. *Spartina anglica*-dominated saltmarsh, declining in area; 7. coastal defences at Keyhaven; 8. most commonly overtopped and artificially rebuilt section of beach ridge; 9. waves approaching from south-west. For discussion of the saltmarsh features, see GCR site report in Chapter 10 for Keyhaven Marsh. (Photo: courtesy Cambridge University Collection of Aerial Photographs, Crown Copyright, Great Scotland Yard.)

anglica. The intertidal area close to the spit is often a stony mud (see GCR site report for Keyhaven Marsh in Chapter 10).

Before the late 19th century, much of this marsh lay as much as 1 m lower and was dominated by eel grass *Zostera*. Colonization by *Spartina anglica* following its hybridization from the native *Spartina maritima* and the introduced *Spartina alterniflora* in Southampton Water led to a rapid build-up of the saltmarsh surface. The area of *Spartina*-dominated saltmarsh reached a maximum about 1930, after which the area declined. As the recurves of the modern spit have extended into the westernmost creek, they have increased local accretion of mudflats.

Interpretation

Ward commented (1922, p. 114) that 'for many centuries, the spit has in size and position suffered no substantial permanent change, so that accretions of fresh shingle must, now that the period of growth is over, be balanced by wastage from the spit...'. Both the history and the future development of Hurst Castle Spit depend upon the balance of sediment supply and loss. Lewis (1931, 1938) showed that the main ridge was aligned at right angles to the direction from which storm waves approach along the English Channel with a fetch extending across the Atlantic Ocean from where the strongest winds and waves also come. The recurves face the waves from the north-east along the Solent. Lewis suggested that the spit grew towards the south-east as shingle was moved by oblique waves alongshore from the eroding coastline of Christchurch Bay. Large waves from the direction of maximum fetch would build up the main part of the ridge. Constructive waves would build the beach up to high tide level and storm waves would deposit shingle onto the crest of the ridge, building it above the reach of normal waves. Storm waves would also comb shingle down onto the lower beach. Constructive waves following the storm would push shingle up the beach. The result of this combined wave action would be a gradually consolidated spit probably moving slowly towards the north-east. Lewis considered that, as the spit curved more towards the east and the angle of waves to the beach changed, the rate of drift would increase. Shingle would move rapidly around Hurst Point, but travel more slowly along the recurves. As water depth increased, more shingle would have been required to extend the spit. Refracted waves would move some shingle along the recurves, while north-east waves in the Solent with their longer fetch could build up the ridges. The distinctive sharp angle at Hurst Point (Figure 6.22) is the result of the main ridge-building waves being restricted by the shelter of the Isle of Wight to two main directions, the south-west and the north-east.

The role of tidal streams was thought by Lewis 1931 to be very small, though affecting wave height and the angle of wave approach to the beach. He concluded that the tidal role in the growth of the spit was negligible. This was contradicted by Williams (1960) who claimed that the effect of east–west tidal streams in Christchurch Bay could explain the eastward growth of the spit. Clark and Small (1967) concluded from a drifter study in Christchurch Bay that seabed movements of shingle could occur in depths as great as 20 m both directly onto the spit and across the floor of the bay towards Mudeford, from where they could move alongshore back towards Hurst Castle. King (1968b) and King and McCullagh (1971) followed Lewis' (1931, 1938) hypothesis for the growth and shape of the spit in their development of a computer model to simulate the effects of different wave directions, refraction and increases in the depth of water. Clark (1974) questioned whether the spit was either a complex recurved spit, as had been commonly accepted, or merely a storm beach resting on the edge of one of the partially submerged former terraces of the Solent River (Everard, 1954). Nicholls (1984, 1985) and Nicholls and Webber (1987a–c, 1989) reviewed the history and present development of Hurst Castle Spit arguing that previous models of its evolution emphasized longshore growth at the expense of other factors, especially changes of sea level. The submergence of a Pleistocene valley system along the line of the present-day western Solent brought about a major transformation of Christchurch Bay from a low to a high tidal energy environment. The Shingles Bank was particularly important because, in refracting waves crossing it, it influenced wave energy along Hurst Beach. However, the date of its formation remains unclear. If it acts as a sink for shingle from the spit then it may be relatively recent. Substantial local supplies of sediment were also available from Pleistocene sandy gravels, which lie both in the cliffs at Milford and beneath the root of the spit.

Nicholls and Webber (1987b) suggested that the second youngest recurve formed during a possible period of sea-level still-stand between about 4500 years BP to 3000 years BP (West, 1980; Devoy, 1982; Nicholls and Clark, 1986). Nicholls and Webber (1989) acknowledge that sub-shingle sediment compaction may be important in affecting ridge height and follow Lewis (1931, 1938), Lewis and Balchin (1940: at Dungeness) and Carr (1970: at Orfordness) in their interpretation of increasing height with decreasing age of ridges as indications of a rising sea level.

On Hurst Beach, there are two types of overwashing: crest-maintaining overwashing and throat-confined overwashing (Nicholls and Webber, 1989). The backslope of the beach ridge is affected by collapse features associated with seepage, but this does not appear to affect the development of washover throats as has been reported elsewhere (Eddison, 1983b; Carter et al., 1984). Nevertheless, the washover throats are significant forms. The landward movement of shingle during a single storm was estimated by Nicholls and Webber (1989) as twenty-five times greater in throat-overwashing than in crest-maintaining overwashing. Washover throats have been observed to have a preferred longshore spacing, and wave run-up has exhibited regularly-spaced maxima along the shore, implying edge waves or some similar effect.

The present-day processes described by Nicholls and Webber at Hurst Castle Spit suggest that much of the spit should be regarded as a barrier beach rather than as a spit in the traditional sense, because it has not arisen simply as a result of longshore transport. Nevertheless, the alignment of the transgressive beach and the recurves depend upon the two different sets of dominant waves from the south-west and the north-east, as described by Lewis. Sediment was derived from erosion of the shore to the west. The present-day reduction in littoral drift is partly a result of the protection of the coastline at Milford and the construction of groynes. Longshore input of sediment has now been reduced and some beach replenishment has been undertaken. Beach nourishment will have to be repeated in the future, as the decision to protect Hurst Castle itself requires stabilization of that part of the beach. Much of the surface of the shingle has been disturbed and the recurves have been damaged by vehicles. Parts have been affected by the various phases of castle and lighthouse construction. The interpretation of the history of the spit during the Holocene Epoch will depend upon investigation of the recurves, and considerable care will need to be taken to avoid further damage. The main ridge is now almost entirely managed, but provides an interesting situation for case-studies of the effects of diminished littoral transport, and barrier management by beach replenishment and rock armouring.

Conclusions

Although Hurst Castle Spit is commonly regarded as a classic example of a complex recurved spit, recent work suggests that it was initiated in response to a combination of processes, among which a rise in sea level appears to have been most important. It may thus be better interpreted as a form of barrier beach. It is nevertheless a very important site because of its seminal role in coastal studies and its simplicity of form when compared to other shingle beach sites such as Orfordness and Dungeness. Its place in coastal studies rests upon the classic studies of Lewis (1931, 1938). The revised model of Nicholls and Webber is a development of earlier work and offers a basis for understanding of other shingle beaches. Studies of the detailed forms and processes on Hurst Beach have brought about a better understanding of shingle beach development. Hurst Castle Spit, Orfordness and Dungeness each display different aspects of shingle spit and barrier beach development, but the similarities of ridge patterns and their relationships to sea-level variations make them a unique suite of sites in Europe. The spit provides shelter for the intricate system of creeks in the very important Keyhaven saltmarshes where *Zostera*-dominated marshes have been colonized by *Spartina anglica*. The site forms part of a Ramsar site and both an SAC and an SPA.

The spit's present-day largely artificial form provides an opportunity to analyse changes from a natural feature to a mainly anthropogenically influenced landform, and research into its re-establishment of equilibrium conditions will be of significant interest, particularly when compared with the formerly managed Porlock GCR site.

PAGHAM HARBOUR, WEST SUSSEX (SZ 880 960)

V.J. May

Introduction

Pagham Harbour (see Figure 6.2 for general location) is the easternmost of a series of drowned river valleys and shallow estuaries that characterize the coastline of southern central England. They include Poole Harbour in the west, the Solent, and Langstone and Chichester harbours. With the exception of the Lymington River and the Medina estuary at Cowes, they all have sand or shingle spits at their mouths, and many are distinguished by double spits extending from both sides of the estuary. The shingle spit across the mouth of Pagham Harbour comprises a series of sub-parallel shingle ridges and recurves, which mark different phases of extension and accretion. Shingle reaches the beach via the intertidal zone. The behaviour of the spit and the so-called 'Pagham Delta' (an area of deposition associated with the mouth of the estuary) are intimately linked with water and sediment circulation around the Selsey peninsula. The area also provides an excellent example of the role of weed-rafting of shingle in coastal sediment budgets.

Ward (1922) and Steers (1946a) described the main features of the development of the estuary and the spit; Robinson (1955) compared the development of the Pagham site with the spits at Christchurch and Poole. Kidson (1963) challenged the hypothesis for double spit development. May (1964) showed how the spit had developed over a period of several years. The development of double spits is not uncommon at the mouths of shallow estuaries, but Pagham Harbour is distinguished by having changed from spit to bay-bar to spit again. Its double form is probably a result of breaching rather than a result of opposing directions of longshore sediment transport. The supply of shingle to the spit has been and continues to be dominated by transport from the direction of Selsey Bill (Harlow, 1979; Hooke *et al.*, 1996), supplemented by kelp-rafted pebbles (Jolliffe and Wallace, 1973).

Description

The site comprises a shingle beach that extends from the east side of Selsey Bill (SZ 870 941) to the Pagham Beach Estate (SZ 895 975), part of the Pagham Harbour estuary, and its extensive intertidal gravels. The intertidal gravels occur as irregular extensions of the beach at Inner Owers and a bank known as 'The Spit', but form a distinctive delta-like form at the mouth of Pagham Harbour. At Church Norton, the beach is formed by a number of ridges or 'fulls' (as they are called locally). Shingle is characteristically larger locally on the ridges. The shingle spit extends across the mouth of the estuary, with a series of short recurves marking periods of advance. The estuary forms the easternmost part of a valley system which extends to the west of Selsey Bill where it has been truncated by the retreat of the coastline and is now only prevented from periodic inundation by a shingle ridge. In 1910, this western ridge was breached and the sea flowed through the low-lying land into Pagham Harbour, breaching the shingle ridge, which until then completely closed the harbour entrance.

The present-day spit across the mouth of Pagham Harbour is a much smaller version of a longer feature that grew north-eastwards from at least the mid-17th century and gradually forced the outflow from the estuary farther towards Pagham, until it was breached in 1910. By 1934, two separate ridges extended from each side of the estuary. The intertidal gravels diverted the outflow towards the north-east, but ten years later the outflow was located at the south-western end of the southern ridge (Robinson, 1955). This spit gradually decayed and was replaced by a newer structure that grew from the south-west (Figure 6.23). This not only diverted the outlet north-eastwards but also changed the wave patterns within the entrance to the harbour and thus the alignment of the decaying former spit.

The intertidal area at the harbour mouth extends over 800 m seawards and is largely composed of gravel locally derived. The outflow from the harbour appears able to maintain a channel through these deposits. The intertidal area has had much the same width throughout its history, but both the outlet and the form of the delta have altered their positions. Thus, as the spit has changed its shape, so it has affected the form of the intertidal area. As a result, wave refraction has also altered.

Severe erosion along the eastern side of Selsey Bill has meant that the present area of the harbour mouth has been exposed increasingly

Pagham Harbour

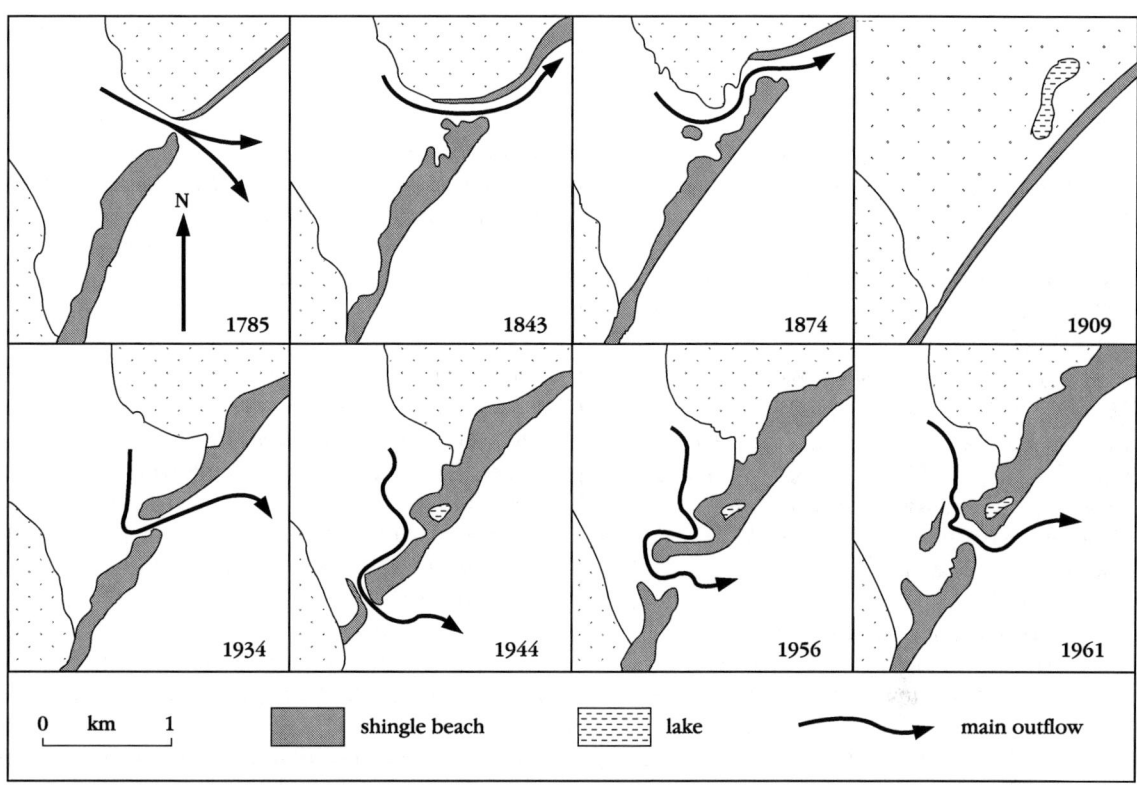

Figure 6.23 Historical changes at Pagham Harbour 1785–1961. (After Robinson, 1955.)

to waves approaching from the south. The limited documentary evidence suggests that Selsey was connected to the mainland only on its south-western side at Medmerry. Gradual silting of the harbour and its land-claim was associated with the growth of the beach across the mouth of the harbour by 1909.

Interpretation

Steers (1946a) noted the different sized spits, the larger extending from the south-west, the much smaller from the north-east. He suggested that they appeared similar to the opposing spits at the mouth of Poole Harbour and were probably of similar origin. It was not clear, however, if a local counter-movement of shingle from the north-east was responsible for the smaller spit. Robinson (1955) considered the double spits not only at Poole and Pagham, but also at Christchurch. From a detailed consideration of cartographic and field evidence, he argued that the spits resulted from unidirectional drifting followed by breaching. This model depends upon longshore transport that forms, maintains and usually extends a spit across an estuary and also diverts the river outflow. At Christchurch and, Robinson believed, elsewhere, wave and river conditions would bring about breaching, such that the distal end of the spit would eventually be attached to the mainland and its proximal end modified so that it assumed the form of a spit. At Pagham Harbour (Figure 6.24), it can be shown that the spit has normally grown from the south-west fed by the very large quantities of material eroded from the cliffs at Selsey. Between 1956 and 1961, such a phase of growth was accompanied by gradual decay and transgression of the former north-east spit on to the saltmarsh within the estuary (May, 1964). Although Kidson (1963) challenged the general applicability of Robinson's unidirectional view, he acknowledged that in some areas growth was predominantly from one direction and that, since breaching could be shown to have taken place, attached beaches could co-exist on both sides of an estuary. Robinson's discussion focused on Poole Harbour as well as Pagham Harbour, and the debate is covered more fully in the description of South Haven Peninsula GCR site in Chapter 7 of the present volume.

Davies (1972, p. 140) suggested that 'discus-

Gravel and 'shingle' beaches

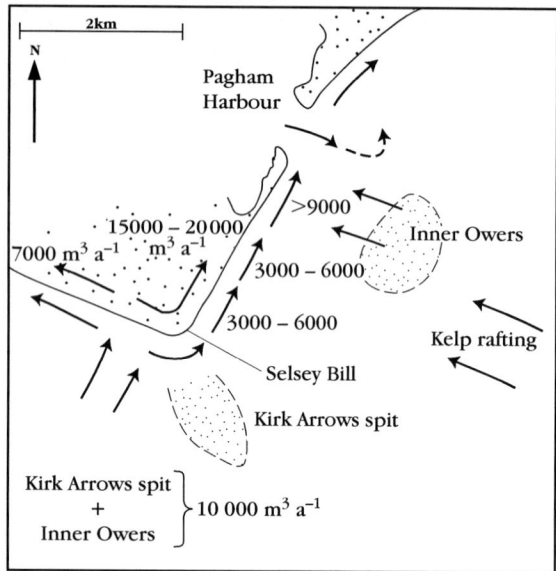

Figure 6.24 Sediment pathways at Pagham Harbour. Arrows show sediment pathways with estimated annual volumes. (Based on Lewis and Duvivier, 1976; Hooke *et al.*, 1996; and Harlow, 1979.)

sion of apparently anomalous inlet locations on the British coasts for instance (Robinson, 1955; Kidson, 1963) may possibly have been clouded by lack of consideration of the swash deflection process'. Bascom (1954) argued that, on beaches where drift is minimal, the position of inlets is determined by berm height. This in turn is determined by the distribution of wave energy along the beach. Where wave refraction is greatest, the berm is lowest and therefore a breach is most likely. At Pagham, there is considerable refraction of long waves crossing the delta with the result that at high tide they approach the beach from a more south-easterly direction than at low tide. Nevertheless, drift is considerable. The mechanism proposed by Bascom does not appear to apply here because drift is substantial, and berm height has tended to be lowest towards the distal end of the spit.

Until the construction of coast protection works at Selsey Bill in the late 1950s, erosion of the emerged ('raised') beach deposits had contributed about 4000–5000 m³ sand and shingle annually to the westwards drift (Harlow, 1979). Larger amounts probably travelled towards Pagham. With annual cliff retreat in excess of 6 m a^{-1} between 1932 and 1951, the eastern cliffs at Selsey Bill were probably supplying about 9000 m³ a^{-1} to the beach leading to Pagham.

Jolliffe and Wallace (1973) described a process in which kelp-rafted shingle was trapped by the seabed off Selsey Bill. Two small denuded anticlines are the focus of shingle accumulation. Clasts travel up the gentle dip slopes, and are prevented from escaping by the ratchet-like scarp slopes of the individual beds. Shingle is then moved by wave action along the strike of the beds, to arrive ultimately on the beach. Harlow (1979) estimated that about 1000 m³ a^{-1} is added to the westward drift from Selsey Bill by this mechanism. The division of longshore transport at Selsey Bill probably means that shingle also travels towards Pagham from this source. Hooke *et al.* (1996) show that between 3000 m³ a^{-1} is added by wave-driven onshore transfers to a longshore component of a similar magnitude. There is some kelp-rafting as well as some transport seawards from within the estuary. The rate of longshore transport east of Pagham was not quantified by them. It is likely to have diminished considerably as a result of the construction of groynes between Selsey Bill and East Beach (SZ 874 948). Storm waves, for example during early January 1998, have overtopped the shingle ridge and moved the main crest landwards.

The historical evidence for the Pagham site demonstrates that a double spit form can result from unidirectional sediment movement. Breaching of the spit, nevertheless, produced a feature upon which the smaller-scale structures, for example small recurves, are a result of local longshore movement contrary to the general regional pattern. Thus the larger feature is the result of one set of processes, but its detailed form is a result of the modification of the smaller-scale processes. The importance of different time and spatial scales is well exemplified by the site. Pagham spit is the best-documented member of a considerable number of small paired spits in southern England, which together enhance our understanding of estuary-mouth sediment dynamics. It is the only such site where the sequence from unidirectional shingle spit to breaching and the resultant formation of a double spit have been documented definitively. In other cases, one spit has been described but the other ignored (e.g. at the mouth of Chichester Harbour), or there has been no investigation at all (e.g. the spits at the mouth of Newtown Harbour on the Isle of Wight), or coast protection works have radically altered one or both of the spits (e.g. Christchurch Harbour). As a bay-bar, it was a comparable form to the Loe Bar (see GCR site report), but unlike the latter was dominated by strong longshore sediment transport. In contrast to other double spits in England and Wales where sand is the main sediment, Pagham spit is formed predominantly of shingle. The development of shingle ridges has allowed the extension, breaching and repositioning of the detached ridges to be traced with greater certainty than is possible with sandy structures. Pagham Harbour thus adds considerably to the understanding of spit development.

Conclusions

Well known for the double shingle spit, Pagham Harbour is an excellent example of spit growth and breaching associated with both longshore and offshore sources of sediment. Today the natural sediment supply has largely ceased as a result of anthropogenic influence, but the ridge patterns preserve the earlier history well.

THE AYRES OF SWINISTER, SHETLAND (HU448 723)

J.D. Hansom

Introduction

The gravel beaches of the Ayres of Swinister ('ayre' is a local Shetland name for a spit or barrier) together form an exceptional example of a barrier complex, connecting the north-east mainland of Shetland to the small offshore island of Fora Ness (Figure 6.25; see Figure 6.2 for general location). Of the three gravel barriers, only the South Ayre forms a complete connecting tombolo. The other two extend out from the mainland but do not reach Fora Ness. However, they are classic examples of bay-head and mid-bay barriers (Shepard, 1952). A tidal basin called 'The Houb', lies below mean low-water springs between the South Ayre and North Ayre and contains submerged peat deposits, the dates of which have provided important information concerning the Holocene sea-level history of the area (Birnie, 1981). This site is a classic example of a coastline undergoing submergence.

Tombolos, bay-head and mid-bay barriers are relatively common features of the inner coastline of the Shetland archipelago (Flinn, 1964). However the complexity of the gravel landform assemblage at the Ayres of Swinister, with three substantial features occurring within a small area containing submerged peat deposits and salt-marsh remnants is unique both nationally and internationally. In spite of these credentials and numerous descriptive accounts, the site has failed to attract detailed geomorphological research. Nevertheless, Birnie (1981) provides stratigraphical details of the submerged peat at the site as part of a wider study to determine past environmental changes in the Shetlands, and Bentley (1996a) speculated on the evolution of the ayres.

Description

In contrast to the west coast of Shetland, the east coast is less exposed to storm-wave activity from the west yet remains open to waves from the east and north-east. Since the most frequent strong winds and storm wave activity come mainly from the south-west (BGS and Scott Wilson Consultants, 1997a), the eastern voes of Shetland and the inner reaches of Dales Voe, where Fora Ness is located, are relatively sheltered and subject mainly to locally generated waves. On account of the shelter provided by Fora Ness, sea conditions adjacent to the North Ayre are benign enough for the safe mooring of

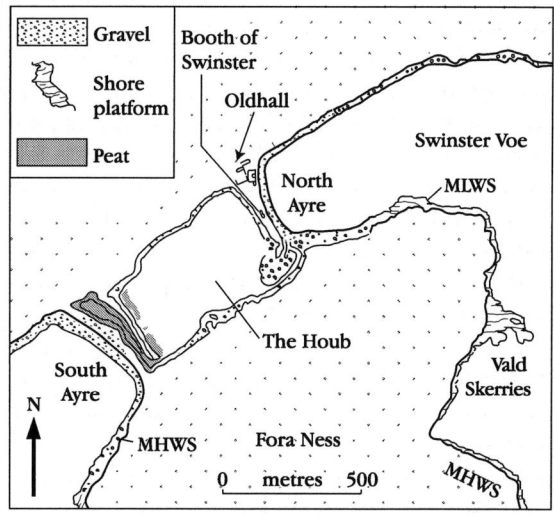

Figure 6.25 The Ayres of Swinister: a triple gravel barrier. Only the southern barrier is a true tombolo, the others are spits that enclose The Houb, a tidal basin. For general location, see Figure 6.2.

Gravel and 'shingle' beaches

Figure 6.26 South Ayre and North Ayre at high tide looking north-east towards Swinister Voe, showing the very sheltered nature of the site. Fish farms can be seen at the North Ayre (upper left of the photograph) (Photo: J.D. Hansom.)

floating pontoons associated with a fish farm (Figure 6.26). In spite of this, severe storm events from the north and east can affect the east coast of Shetland, especially the outer reaches of the eastern voes. Spring tidal range on the east coast of the Shetland Islands is limited to 1.5 m and maximum tidal streams on spring tides are generally between 0.25–0.5 m s^{-1} (BGS and Scott Wilson Consultants, 1997a) although these speeds may be exceeded where the voes narrow.

The relative sea-level history of Shetland is incompletely known (Firth and Smith, 1993), but the work of Birnie (1981) provides more local information than is available at other similar sites in Shetland. The general lack of emerged ('raised') marine sediment and landforms combined with classic drowned river valleys (the voes) and a local tradition of marine submergence in historical times, have long been accepted as evidence for continuously rising sea-level since the decay of the late Devensian ice-sheet (Mykura, 1976). This view is supported by observations of now submerged freshwater peats in many of the sheltered voes (Hoppe, 1965; Birnie, 1981).

The Ayres of Swinister consists of a tombolo (South Ayre) which connects the mainland to Fora Ness, a bay-head barrier (unnamed) and a mid-bay barrier to the north-east (North Ayre); all of which are mainly composed of angular and subangular gravel (Figures 6.26 and 6.27). The bay-head barrier and mid-bay barrier are both hinged on the mainland side of the voe and enclose the shallow lagoonal area (The Houb), which lies below MLWS and does not dry out. A smaller lagoon is enclosed between South Ayre and the bay-head barrier, although this dries out at low tide.

The gravel tombolo of South Ayre, which is 300 m long and c. 50 m wide connects the island of Fora Ness to the north-east mainland of Shetland (Figure 6.25). The outer (south-west) side of the tombolo is exposed to waves from the south-west and Dales Voe, and the gravel beach maintains a smoothly curving swash-aligned arc, whereas the inner (north-east) side is characterized by a grassy turf, which is affected by spray and waves from the south-west

The Ayres of Swinister

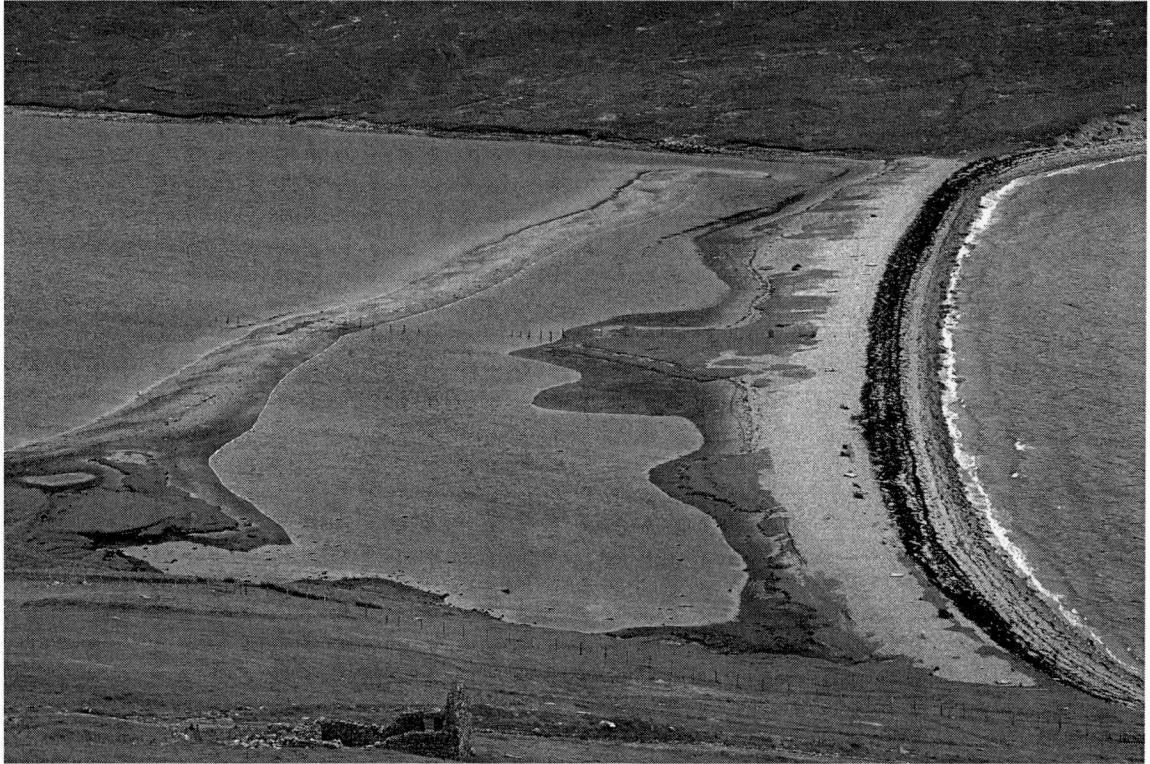

Figure 6.27 Washover lobes of gravel on the tombolo of South Ayre (to the right), with Fora Ness in the distance. Intertidal peats, which extend subtidally, are exposed at low tide in the lagoon between South Ayre and the unnamed barrier to the north-east (on the left). (Photo: Lorne Gill/SNH.)

(Bentley, 1996a). The inner side is characterized by a series of well-defined washover lobes and, in places, the colonizing vegetation has been subsequently partly or wholly buried by gravel. It is not known whether the peat exposed in The Houb to the north-east of the South Ayre extends to any extent under the feature itself, since it is absent from the exposed intertidal foreshore on the south-west.

A narrow, *c.* 20 m-wide lagoon separates South Ayre from the unnamed bay-head barrier immediately to the north-east (Figure 6.27). Indeed, as a result of the proximity of these two gravel features, South Ayre has often been described as a double ayre (Smith, 1993). However, the bay-head barrier is breached by a narrow tidal channel at its recurved southern tip connecting the lagoon to The Houb. The northeast side of the barrier describes a series of small arcs in response to the low-energy waves and shallow waters within The Houb. This narrow lagoon has been described as an intertidal peat flat (Smith, 1993) on which drying cracks readily develop at low tide. The peat substrate, exposed beneath the intertidal muds of the lagoon, extends between the South Ayre and the bay-head barrier.

The North Ayre lies 500 m to the north-east of South Ayre and extends from the mainland towards the island of Fora Ness (Figure 6.26). A *c.* 40 m-wide tidal channel with a substantial flood-tide delta separates the gravel barrier from the island and connects the intertidal basin (The Houb) to Swinister Voe. North Ayre is recurved at its south-east end and a smaller, *c.* 30 m-long barrier extends towards it from Fora Ness. North Ayre appears to be relatively stable; a derelict crofthouse is located on the ayre and a grassy turf mat covers much of the ayre's surface. In places the turf is covered by lobes of gravel, which have been deposited by waves from both north-east and south-west directions.

On the eastern shore of The Houb, hill peat overlying till and weathered bedrock extends below the high-water mark and floors the intertidal basin. The submerged peat contains in-situ tree stumps complete with stems and roots (Smith, 1993). The dissected peat apron (parts

283

of which have been subject to peat cutting operations in the past) has in places been transformed into pseudo-saltmarsh. Coring of one of the uncut peat areas in the lagoon showed approximately 3 m of organic material overlying a grey, gritty clay, and yielded a radiocarbon date on *Betula* (birch) fragments 1.6 m below the surface of 4586 ± 40 years BP (Birnie, 1981). The intertidal zone on the western shore of The Houb is characterized by a low-gradient slope of sands and gravels that extend south-eastwards from the flood-tidal delta at the North Ayre.

Interpretation

Individual tombolos, bay-head and mid-bay barriers are characteristic features of the submerging inner coastline of the Shetland archipelago (Flinn, 1964), however the Ayres of Swinister provide a unique assemblage of all three of these landforms in close association. It seems likely that given its angularity and match with local glacial tills, the sediment that comprises the North and South Ayres has been eroded from the flanks of Swinister Voe by waves refracting from the north-east and south-west. Waves approaching the North Ayre comprise both swell and storm waves generated in the larger Voe in the east and thus energy levels on the northern shore are likely to be relatively higher than on the south shore. As a result the Ayre is wider and higher on this side, and it tapers to the south where the tidal prism of The Houb has given rise to flows strong enough to keep open the tidal narrows. On the other hand, waves approaching the South Ayre are unidirectional having been generated wholly within Dales Voe and travelling north-east. As a result, energy levels are evenly distributed across beach and a uniformly wide barrier has been constructed, which joins the mainland and Fora Ness to form a tombolo.

Over a longer period of time the development of The Houb must have been linked to changes in sea level. The Holocene period in Shetland is generally considered to have been characterized by a rising relative sea level (Firth and Smith, 1993). Modelling of sea-level change seems to support this view of an early and rapid rise in sea level at Shetland sites, slowing towards about 6500 years BP, but at no time undergoing relative sea-level fall (Figure 6.28; Lambeck, 1993). During the early phase of rapid Holocene relative sea-level rise, shoreline migration in

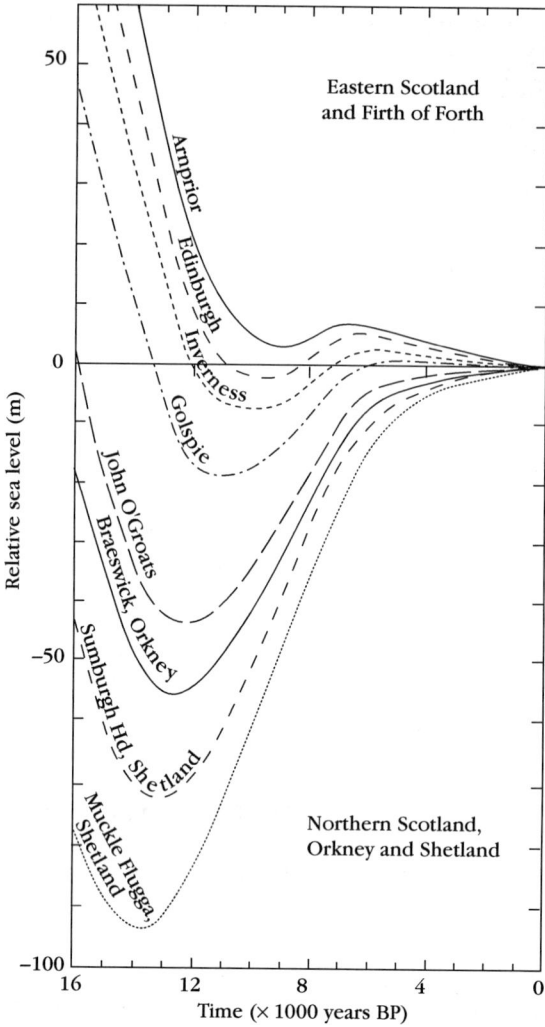

Figure 6.28 Graphs of modelled relative sea level against time over the last 16 000 years, along a south–north transect from Shetland to the Firth of Forth. (After Lambeck, 1993; Hansom, 2001.)

Scotland was probably too rapid for significant shoreface modification (Hansom, 1999, 2001) and so it seems reasonable to suggest that these constructional features were not emplaced until after the rate of sea-level rise began to slow down at some time around 6500 years BP. A radiocarbon date of 4586 ± 40 years BP (Birnie, 1981) from the underlying peat in The Houb suggests that although subsequent sea-level rise flooded The Houb and arrested further peat development, it is possible that earlier rising sea levels drove pre-existing gravel barriers onshore over any pre-peat backshore wetland. Barrier growth before about 4600 years BP may have resulted in the water table alterations that led to peat growth.

Unfortunately, the outcrops of intertidal peat that might be expected to be eroding out of the foreshores of both the North and South Ayres, are nowhere to be seen. However, the existence of peat between the South Ayre and the bay-head barrier (Smith, 1993) strongly suggests that it is continuous below the latter and that the bay-head barrier was constructed after the peat developed and thus it may post-date the development of both the South Ayre and North Ayre, although its relationship with the latter is less clear. This would suggest that the bay-head barrier may be the product of sediment eroded from the underlying glacial till and reworked alongshore. Further, the low-energy undulating planform of the feature suggests that it has never been subject to strong swash-aligned wave action from the north-east, although, if the Houb were open, this would be the direction of approach of storm waves. It may, however, be subject to waves of sufficient power to move sediment alongshore within The Houb to construct the bay-head barrier.

An alternative explanation is provided by Bentley (1996a) who has proposed the construction of the South Ayre and bay-head barrier at a high relative sea level, and the North Ayre at some time later as a result of sea-level fall. Sea-level rise then consolidated the North Ayre, flooded The Houb at some time after 4600 years BP and reworked all three features. A slowly rising sea level is suggested to have resulted initially in the development of a tombolo (South Ayre) either by normal tombolo constuction or by 'roll-over' as the gravel barrier mounted a peat-covered gap connecting Fora Ness to the mainland. A bay-head barrier then formed on the north-east side of the tombolo as a result of gravel moved from the north-east into the newly formed bay. Prior to c. 4600 years BP a suggested fall in sea level resulted in shoreline migration to the north-east and a new bay-head barrier (North Ayre) formed allowing peat to develop in the dry area (The Houb) between the two barriers. A later rise in sea level resulted in the breaching of the North Ayre, the flooding of The Houb, the submerging of the peat and gravel washover of the structures.

The above two evolutionary models of the Ayres of Swinister are speculative and raise several unanswered questions. For example, are the levels of longshore wave power and sediment supply of Model 1 sufficient to result in construction of the inner barrier? Model 2 requires a sea-level fall before 4580 years BP followed by a rise, for which there appears to be no other evidence in Shetland. Much scope remains for further research at this site and the triple assemblage of gravel landforms together with the associated peat beds rich in plant remains provide an excellent site for detailed research in evolutionary chronology and in stages of submergence of the Shetland coastline. Establishing these details would enhance the scientific interest of the site. Nevertheless, this triple assemblage of a gravel tombolo, bay-head barrier and mid-bay barrier, unique in Britain, forms a classic example of gravel construction on a submerging coastline.

Conclusions

Although individual barriers and tombolos occur around much of the Shetland Isles, the Ayres of Swinister is a rare example in Britain of a triple gravel tombolo-barrier complex where a tombolo, bay-head barrier and mid-bay barrier exist in close association. In addition, the site includes important peat deposits that record the sea-level history of the area and have already provided an important constraining date on sea-level rise in this part of Shetland. For these reasons, the site is justly regarded as a classic example of a submerging coastline and is of international geomorphological importance. The site is also important for its biological interest, including lagoonal flora and fauna (Thorpe, 1998).

WHITENESS HEAD, MORAY (NH 802 587–NH 842 568)

J.D. Hansom

Introduction

Whiteness Head on the south shore of the Moray Firth, north-east Scotland (for general location see Figure 6.2), is a classic example of an actively prograding spit of c. 3.5 km long. It is composed entirely of gravel ridges whose ends recurve southwards into an area of saltmarsh and intertidal sandflat (Figure 6.29). The gravel ridges are capped by small amounts of wind-blown sand. Westward sediment transport along this shoreline has resulted in the progressive westerly growth of Whiteness Head (Stapleton and Pethick, 1996; Ramsay and Brampton,

Gravel and 'shingle' beaches

2000c), the historical evolution of which has been well documented (Ogilvie, 1923; Steers, 1973; Smith, 1974; Halliwell, 1975; Stapleton and Pethick, 1996; Bentley, 1995). From a short gravel bar in the 1880s, the end of the spit at Whiteness Head has migrated in a westerly direction at rates of between 10 and 30 m a^{-1} over the last 150 years. Such downdrift growth of the distal end of the spit is currently fuelled by updrift erosion of its proximal end (Ritchie, 1983; Bentley, 1995). The Whiteness Head SSSI comprises one area on the eastern side of the Ardersier Platform Construction Yard (an oil rig construction facility), and another area on the west side of the yard. The GCR site lies entirely within the eastern part of the SSSI.

Description

The gravel spit at Whiteness Head protrudes out from the southern shore of the inner Moray Firth, in a north-westerly direction. At its broadest the 3.5 km-long gravel spit is *c*. 200 m wide, although it narrows to less than 15 m close to the updrift end, and encloses *c*. 1 km² of saltmarsh and intertidal sandflat on its southern side (Stapleton and Pethick, 1996). To the west, extensive mudflats extend to the glacial till headland of Ardersier, whereas to landward, the spit and saltmarsh are backed by a complex of emerged Holocene sand and gravel deposits backed by a prominent, 10–12 m-high, emerged cliff (Ritchie *et al.*, 1978).

Whiteness Head spit consists of at least seven recurved gravel-spit complexes, each one partially truncated at its proximal end by the succeeding one (Ritchie *et al.*, 1978) (Figure 6.30). They lie at altitudes of between 3 m and 5 m OD and are separated by saltmarsh and dry gravel slacks. In general, the highest ridges form the active beach and lie on the seaward (northern) margin of the spit. The upper beach face consists of a gravel storm-ridge with a steep seaward slope of over 10°, increasing to 18° in the centre of the spit (Bentley, 1995). The lower beach face is characterized by a low-gradient sand slope with intertidal bars. The beach face is also punctuated by well-developed, but ephemeral, suites of rhythmic cuspate forms. The offshore hydrography indicates a gentle offshore gradient passing into deep water beyond a narrow sloping shelf with no evidence of a submarine bar system in the nearshore zone.

Extending along the entire length of the spit

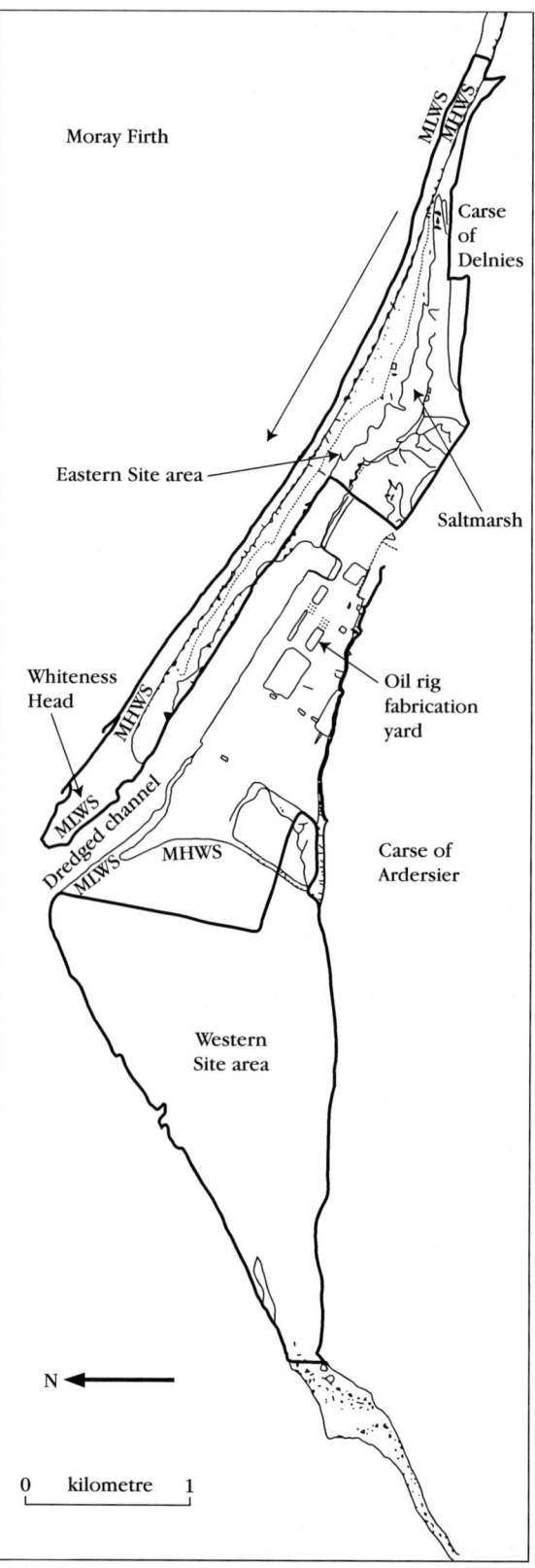

Figure 6.29 Sketch map of the Whiteness Head area. The GCR site lies entirely within the eastern site area; the arrow indicates direction of net longshore sediment movement.

Whiteness Head

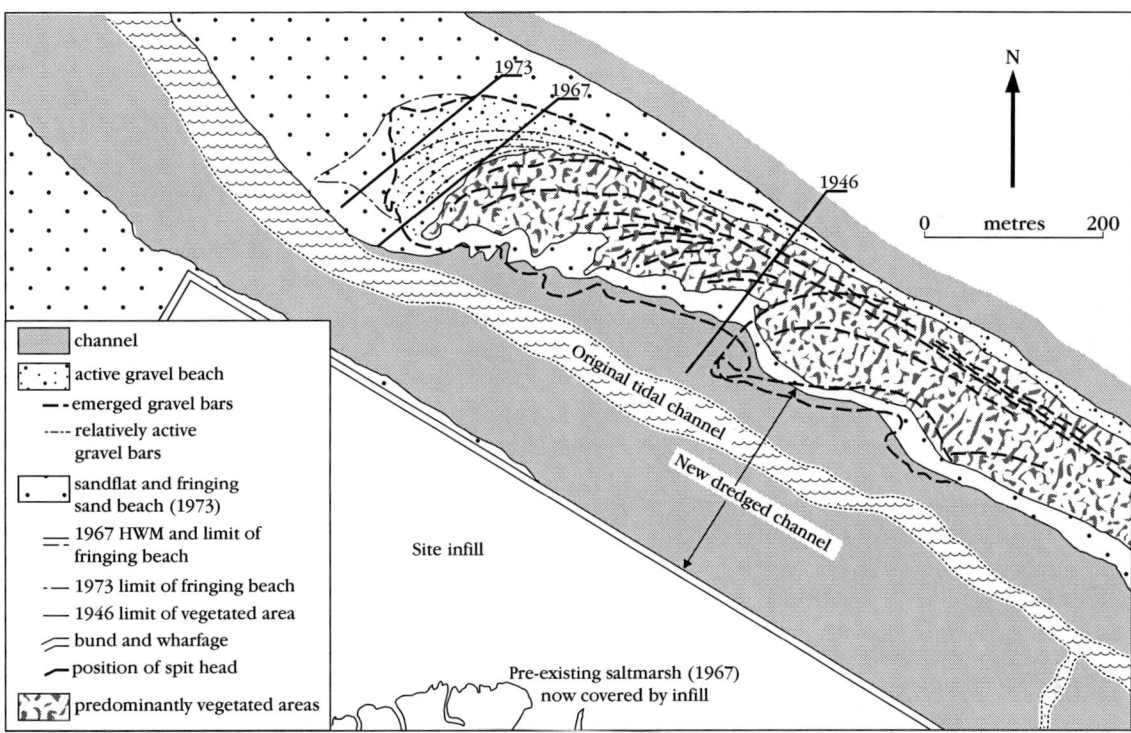

Figure 6.30 Map of historical changes in the Whiteness Head Spit between 1946 and 1973. (After Smith, 1974.)

parallel to the outer coast, a prominent gravel ridge marks the limit of storm-wave action. This laterally mobile storm ridge is almost 2 m higher than the older recurved ridges and is overtopped by waves during storm-wave activity (Bentley, 1995). It declines in height eastwards towards the proximal end of the spit where a low plateau of gravel extends landwards into the saltmarsh as a series of lobate washover fans. This has imparted a crenulate inner edge to the spit at its narrowest point. Known locally as 'the Gut', the gravel storm ridge forms the only barrier preventing waves from entering the saltmarsh behind, although wave over-topping and erosion is common during storm activity (Bentley, 1995). This eastern part of the spit is also characterized by erosion of the older gravel ridges, which trend at an angle to the present-day coast. This results in truncation of the seaward ends of ridges and slacks to form an undulating gravel cliff whose height is usually less than 0.7 m (Ritchie *et al*., 1978; Bentley, 1995). Accretion and growth occurs at the distal (western) end of the spit where the storm ridges of the outer beach form a series of landward-trending gravel recurves, which are rapidly colonized by pioneer plants. Ongoing westward extension is now curtailed by maintenance of the tidal channel at the distal end by dredging.

Although the spit is composed almost entirely of gravel, the distal (west) and inland portions are locally overlain by a thin cover of blown sand, seldom exceeding 2 m in thickness (Bentley, 1995). The dune forms in this location are subdued as a result of episodic deflation and wave overtopping and the underlying gravel base is exposed in several places. In the place of a true foredune system there is a rather irregular sand mantle of amorphous mounds capped by marram grass *Ammophila arenaria*, although small areas of narrow linear dunes capped by blue lyme-grass *Leymus arenaria* have formed atop the gravel ridges, particularly at their distal ends. In the east, a dense cover of *Ulex–Calluna* (gorse–heather) communities occurs but this changes westwards through *Ulex–Ammophila* communities to *Ammophila* and eventually to a dominance of *Leymus* on the most recent embryonic dunes at the extremity of the spit. For a fuller description of the vegetation of Whiteness Head see Currie (1974).

The inner (southern) beach of the spit is characterized along its length by small bays that have formed between the recurved ends of the

Gravel and 'shingle' beaches

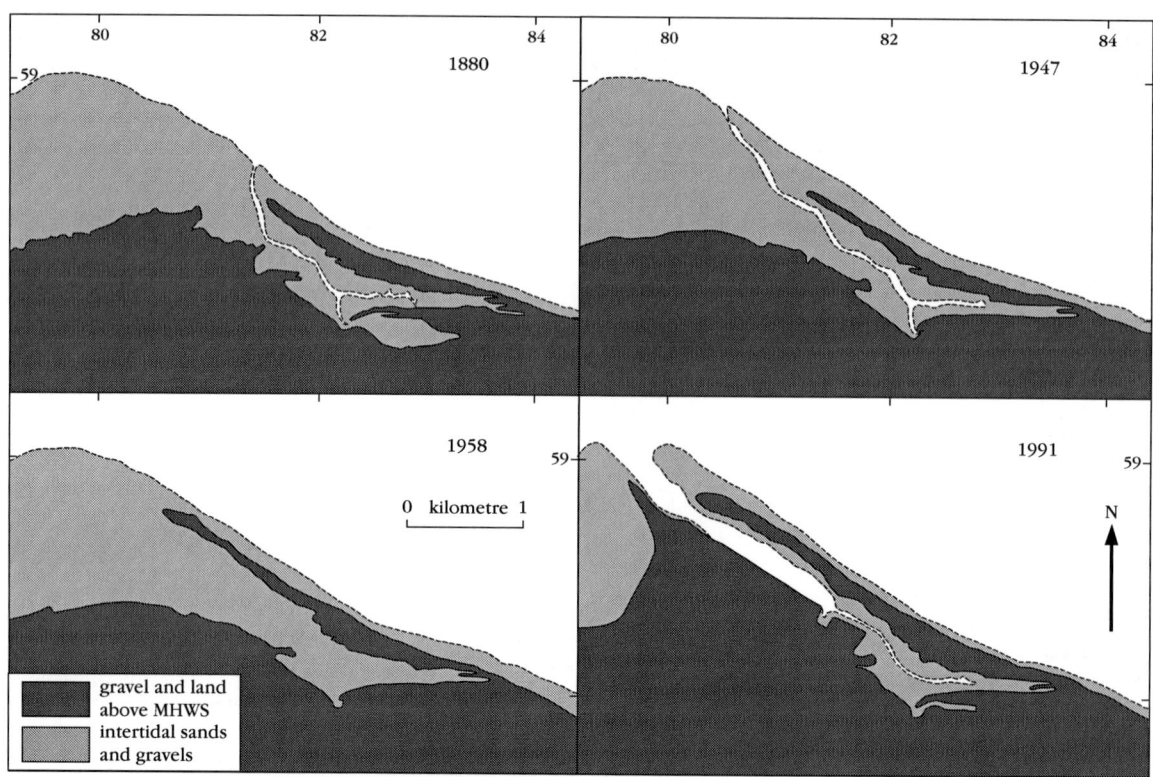

Figure 6.31 Historical evolution of the Whiteness Head Spit between 1880 and 1991. In 111 years the spit has lengthened considerably and the creek morphology has changed. Note the pronounced change between 1958 and 1991 when the McDermott construction yard was built and a prominent channel was dredged on the south side of the spit. (After Stapleton and Pethick, 1996.)

relict gravel ridges marking former positions of the distal end of the spit (Figure 6.30). Along this inner edge a wide intertidal sand-flat is exposed at low tide although this merges landwards into gravels that underlie saltmarsh. The saltmarsh is drained by a system of creeks and there is a marked absence of saltpans. Dominant species on this marsh include samphire *Salicornia*, common saltmarsh-grass *Puccinellia maritima*, sea aster *Aster tripolium*, sea plantain *Plantago maritima*, greater sea-spurrey *Spergularia media*, red fescue *Festuca rubra*, sea-milkwort *Glaux maritima*, thrift *Armeria maritima* and cord-grass *Spartina anglica* (Stapleton and Pethick, 1996).

Interpretation

The Moray Firth is landlocked except in the sector 000°–090° and so north-easterly swell and wind waves dominate. Although smaller wind waves generated in other sectors within the firth can be significant, the dominant direction of wave energy is from the north to north-east sector and incident waves meet the shoreline obliquely. This has strongly influenced the alignment of coastal features, the net westward drift of sediment being manifest in the orientation of spits along the middle and inner Firth including at Speymouth, Lossiemouth, Findhorn Bay, Buckie Loch, The Bar at Culbin and Whiteness Head. Erosion of the coast is thus at its most severe under north and north-easterly storms. Tides in the Moray Firth lie in the range 3.5 m at springs, but they generate relatively weak currents because, rather than being directed into the firth, the tidal wave crosses its entrance (BGS, 1996). However, the co-incidence of a north-easterly gale and high spring tides can elevate water levels considerably along the coast, producing locally significant erosion along the southern Moray Firth.

The sea-level history of the Moray Firth is relatively well known and newly deglaciated areas were being flooded by the sea at *c.* 13 000 years BP (Firth, 1989; see Figure 6.28, Inverness curve). However, rapid isostatic recovery during this period outstripped the rate of sea-level rise, producing a fall in sea level, which was already low by the onset of the Loch Lomond Stadial. The onset of the Holocene transgression is dated at 9610 ± 130 years BP from the inner Moray Firth (Haggart, 1986, 1987), and at 8800 years BP in the Beauly Firth near Inverness (Figure 6.28). The culmination of this rise at c. 6400 years BP was marked in the Beauly and inner Moray Firths by the formation of the Main Postglacial Shoreline (MPS) and a series of emerged gravel ridges at up to c. 9 m OD at Spey Bay (Comber et al., 1993). Since the peak of the Holocene Transgression, sea level has displayed a falling trend to the present-day level, and is marked by a series of emerged shoreline features around the inner Moray Firth at successively lower altitudes (Firth and Haggart, 1989).

The prograding gravel spit of Whiteness Head is of outstanding geomorphological significance, and Whiteness Head, together with Spey Bay (see GCR site report below) and Culbin (see GCR site report, Chapter 11), form an important suite of GCR sites on the southern shore of the Moray Firth. Wave-induced westerly longshore drift predominates along this coastline (Ramsay and Brampton, 2000c) and is reflected in the coastal landforms of all three GCR sites with westerly deflected spits at Speymouth, Findhorn Bay, Buckie Loch, The Bar at Culbin and Whiteness Head. Although extensive coastal protection works at Nairn, 6 km east of Whiteness Head, has undoubtedly reduced the natural supply of longshore sediment to Whiteness Head, it has been suggested that the spit represents the early stages in the development of a 'flying barrier' similar to The Bar at Culbin (Ritchie et al., 1978; Ritchie, 1983; Bentley, 1995).

The historical evolution of Whiteness Head spit is relatively well documented (Figure 6.31). The spit has been developing since before 1853 (Ogilvie, 1923), although no information is available concerning the time of its initiation. There are reports that in 1823 no gravel was present at Whiteness Head, and although Steers (1973) considered this statement to be highly unlikely, his reasons are unclear. There is great variation in the published figures of the rate of westerly growth at Whiteness Head, which probably results from either differences in the methods used to calculate changes, or in the actual growth of the spit, which may have been pulsed, with some years experiencing more rapid growth than others (Bentley, 1995). For example, Ogilvie (1923) states that the spit had grown by 1070 m in 70 years, a rate of 15.2 m a^{-1}, while Steers (1973) reported the spit to have extended 823 m between 1868 and 1937, a rate of 11.9 m a^{-1}. In the 10 years between 1955–1965 the spit is reported to have extended 304 m, an annual rate of 30.4 m a^{-1}, whereas in the following 10 years (1965–1975) this rate of westerly growth had slowed to 18.2 m a^{-1} (Halliwell, 1975). Based on a comparison of the earliest (1880) and most recent (1977) Ordnance Survey maps, the overall westerly extension of the spit is 1.48 km, giving an average annual growth rate of 15.2 m a^{-1} (Stapleton and Pethick, 1996). This long-term average growth rate corresponds exactly with the rate quoted by Ogilvie (1923). The range of westerly growth rates of Whiteness Head are remarkably similar to those measured at The Bar, Culbin, which has been shown to extend westwards at a mean rate of 14.6 m a^{-1} between 1976 and 1989 (Comber, 1993). Both spit features can be seen to be predominantly moving to the west rather than landwards, although proximal erosion is a feature of both.

As occurs elsewhere in the Moray Firth, westerly longshore drift of sediment has been the driving force behind the creation and continued migration of the gravel spit at Whiteness Head, its westward growth resulting from updrift erosion fuelling gravel accretion at its distal (western) end where newly deposited recurved gravel barriers develop. Most of the sediments responsible for the construction of the features of this coast were originally derived from rivers and offshore sources accumulated in the Moray Firth basin during Lateglacial and Holocene times, and plentiful before about 6500 years BP (Hansom, 1999). However, in common with coasts elsewhere (Carter, 1988, 1992), the amount of sediment has since diminished and on the Moray Firth coastline there has been a progressive reduction in both gravel and sand supplies to the coast. In a situation of sediment starvation coastal development proceeds via internal re-organization of existing sediments and so current spit extension is probably entirely fuelled by the truncation of older ridges at the proximal end and recycling of sediment downdrift (Hansom, 2001). Continued growth of the spit is thus at the expense of updrift erosion.

The impact of this process on coastal alignment can be identified in the erosional truncation of the older gravel ridges, which trend at an angle to the present-day ridge orientation. There has also been a north-eastwards movement of the whole system as the spit builds out seawards and an increase in intertidal width at the distal end of the spit (Stapleton and Pethick, 1996). As a result of updrift erosion and downdrift accretion, the spit has 'rotated' clockwise over time from an east–west alignment to a south-east–north-west alignment (Bentley, 1995). Where sediment supply is restricted, this rotation is consistent with the concept of a wave-driven progression in the evolution of beach alignments from an original state of drift-alignment towards an equilibrium state of swash-alignment (Davies, 1972). As there is no reason to assume that the gravel shortage is a temporary phenomenon, it has been suggested that a continuation of the present mode of spit development (i.e. proximal erosion fuelling distal accretion) will eventually result in the severance of the spit near its hinge-point and the creation of a detached barrier similar to that at Culbin (Smith, 1974; Ritchie *et al.*, 1978; Ritchie, 1983; Bentley, 1995).

A significant proportion of the previous extent of the saltmarsh was reclaimed during the building of the Ardersier Platform Construction Yard in 1973. The yard lies to the landwards of the GCR site and occupies a 120 ha site of former saltmarsh and emerged mudflat (carse) (Smith, 1974). Land-claim was achieved by suction-dredging of sand from the intertidal zone landwards of the spit and deposition into settling ponds behind sand bunds, which were infilled to a level of *c*. 4 m OD (Smith, 1974). A dredged channel is now maintained to a depth of up to 12 m and the yard edge adjacent to the channel is protected with a steel wall (Smith, 1974) (Figure 6.29). Erosion is currently a problem at the eastern end of the wall and parts of the claimed land are suffering erosion (Bentley, 1995). Although the geomorphological interest of the intertidal flats and saltmarsh behind Whiteness Head has been diminished by the dredging and land-claim operations, the outer beach remains a classic example of a rapidly prograding gravel spit complex. It is an important site for the study of the processes of active spit development and migration, in a context of sea-level fall, relative sediment starvation and active longshore drifting. As a result, the site is of importance to the understanding of the long-term evolution of dynamic gravel spits.

Conclusions

Whiteness Head on the south shore of the Moray Firth, north-east Scotland, is a classic example of an active, rapidly prograding gravel spit. The *c*. 3.5 km-long dynamic spit complex, which has been migrating westwards at rates of 10–30 m a^{-1} over the last 150 years is of outstanding national geomorphological importance. The relatively well-documented historical evolution of Whiteness Head adds to the scientific interest and provides a key to understanding the long-term evolution of gravel features in areas of active longshore drift. Today, in a period of relative sediment starvation and sea-level fall, updrift erosion of the spit is currently fuelling downdrift accretion and driving the continued westerly migration of the spit complex. It is possible that a continuation of the present-day processes will result in the severance of the spit from the mainland creating a detached barrier similar to that at Culbin (see GCR site report in Chapter 11).

SPEY BAY, MORAY (NJ 264 688–NJ 388 642)

J.D. Hansom

Introduction

Spey Bay (see Figure 6.2 for general location) is one of the most important gravel coastal geomorphology sites in Great Britain for several reasons. The extensive and well-developed gravel ridge complex is recognized as the finest in Scotland, providing examples of dynamic coastal processes and active fluvial supply of gravels that are unparalleled in the British context (Comber *et al.*, 1994; Gemmell, 2000; Gemmell *et al.*, 2001a,b). In addition, the active coastal margin is backed by a magnificent emerged strandplain of Holocene gravel ridges that record the progressive history of coastal development and sea-level fall in this part of Scotland over the last 10 000 years (Ogilvie, 1923; Steers, 1973). The Speymouth delta and related forms have a complex and well-documented history of dramatic change (Grove, 1955) and provide an excellent example of the fluvial-coastal interaction and sediment interchange of an actively braided gravel-bed river entering a high-energy coastal environment (Gemmell *et al.*, 2001a,b). Indeed,

Spey Bay

the lower River Spey has also been selected as a GCR site in its own right on account of its unique fluvial geomorphology (see GCR site report in the GCR volume *Fluvial Geomorphology of Great Britain* (Gregory, 1997)). Nowhere else in the UK is there such a dynamic example of an actively braiding gravel-bed river delivering sand and gravel to a wide coastal gravel beach and backed by a suite of emerged gravel shorelines (Figure 6.32). The wealth and scale of juxtaposed features provide a unique insight into the Holocene development of this part of the Scottish coastline.

The GCR boundary follows the Spey Bay SSSI boundary along mean low-water springs (Figure 6.33) for 13 km from Porttannachy (Portgordon) in the east to 2.5 km east of Lossiemouth in the west, and is paralleled by a landward boundary along tracks and fencelines up to 250 m inland from MLWS. The site is selected for its gravel features and although the western limit is located some 2.5 km east of Lossiemouth, the ongoing migration of gravels beyond this boundary is set to continue. As a result, the obvious geomorphological boundaries of this stretch of coast lie at the rocky headlands at Porttannachy (Portgordon) and Lossiemouth.

Description

As already noted in the description of the Whiteness Head GCR site above, the Moray Firth is landlocked except between north and east and so north-easterly waves dominate and their oblique incidence produces a westerly movement of sediments. Erosion of the coast is at its most severe under northerly and north-easterly storms. For further details on waves, tides and

Figure 6.32 The extensive gravel ridges and emerged coastal and fluvial terraces of the Spey mouth in 1963. At this time, the river was diverted west by over 1 km, threatening the village of Kingston in the right centre of the view. (Photo from Gemmell *et al.*, 2001b)

Gravel and 'shingle' beaches

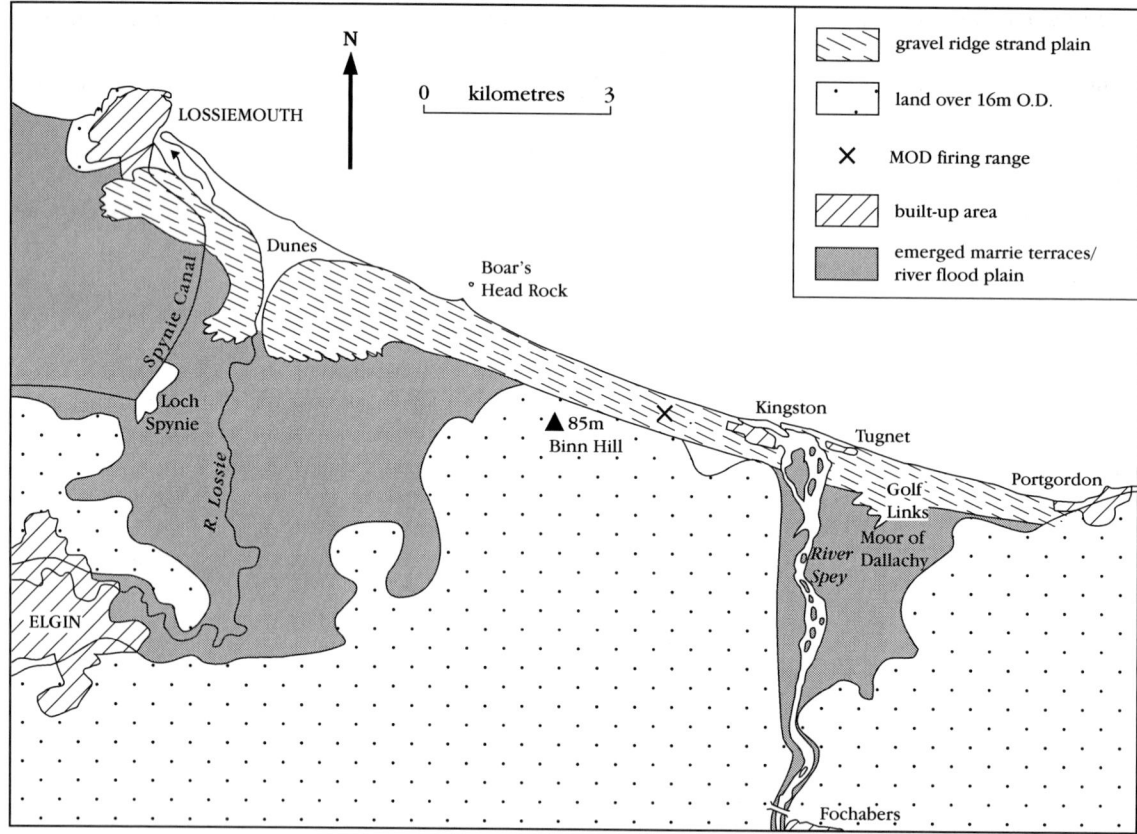

Figure 6.33 Spey Bay showing coastal gravel strandplain backed by emerged marine (and fluvial) terraces. Land over 16 m is mainly glaciofluvial sands and gravels. MoD is a Ministry of Defence weapons testing range. (After Ritchie, 1983.)

general sea-level changes, see site GCR report for Whiteness Head (present chapter).

Spey Bay lies on the southern shore of the Moray Firth and extends for a distance of c. 16 km between rock headlands at Lossiemouth in the west and Porttannachy/Portgordon in the east (Figure 6.33). The present-day coastline trends WNW–ESE and is juxtaposed against a magnificent emerged strandplain of gravel ridges (Figure 6.34) which have isolated a series of glaciogenic deposits, glacio-fluvial and recent river terraces and residual pockets of low marshy ground (Ritchie, 1983). The contemporary gravel beach extends for c. 13 km from Porttannachy/Portgordon in the east to about 2.5 km east of Lossiemouth, where an abrupt transition occurs to a low-angled sand beach backed by sand dunes. The gravel beach is punctuated between Tugnet and Kingston by a complex of gravel ridges that mark the delta of the River Spey (Figure 6.32). Along much of Spey Bay, the average altitude of the main gravel ridge crest is c. 6 m OD, although this varies considerably.

Crestal overtopping of the gravel storm ridge is common along much of the coastline, particularly west of Porttannachy/Portgordon and west of Kingston, and coastal recession is a constant concern. The width of the gravel beach varies considerably, from c. 10–15 m immediately west of Kingston, to up to 70 m wide, west of Boar's Head Rock.

West of the Speymouth delta, close to the Ministry of Defence firing range, several low altitude vegetated ridges are truncated by the present gravel beach, and curve gently landwards. These low altitude recurves occur continuously from this point to the westward extent of the gravel beach beyond Boar's Head Rock and become higher and more prominent westwards. West of Boar's Head Rock up to five well-defined gravel ridges curve gently landwards from the rear of the present-day active ridge. In the west, the ridges reach c. 6 m and stand up to 2 m above the adjacent intervening troughs. To landward, the ridges are sparsely vegetated by mosses and grasses and are eventually buried by

Spey Bay

Table 6.3 Westerly extension of the active gravel beach (West Spey Bay). (From Gemmell et al., 2001b.)

Time period	Westerly growth (m)	Growth per annum (m a^{-1})
1870–1903	1360	41
1903–1967	2090	33
1967–1994	720	27
July 1994–December 1995	30	20
1870–1995	4200	34

high sand dunes and forest. The full landward extent of ridges is known to be substantial, continuing into the Spynie area to the south of Lossiemouth and as far west as Burghead Bay (Gemmell et al., 2001a,b). Within Spey Bay, the junction between these younger, more recently deposited ridges and the ridges of the emerged gravel strandplain, is marked by a 1–2 m rise in altitude and a distinct break in slope, best seen 2–3 km east of Kingston.

On the seaward face of the gravel beaches, cusp forms of different wavelengths are well developed, the size and spacing of these ephemeral features altering in response to short-term processes that vary with wave and tidal conditions. Beach-face slope angles, the degree of sediment sorting and crest elevations also alter in response to wave and tidal conditions. Sediment sorting is well developed down the beach face, with finer-grained, well-sorted gravel lying in the intertidal zone whereas larger calibre, but more poorly sorted, gravel, occurs at or above high-water mark or in the horns of the cusps. However, there is no obvious alongshore trend in beach sediment size, until the abrupt transition from gravel to sand close to Lossiemouth. The median grain size of the gravel varies from 30 mm to 50 mm along the beach, whereas the sand has a median grain size of 0.22 mm (Gemmell, 2000; Gemmell et al., 2001b).

Along much of its length, the gravel ridge is subject to washover during storms and at several places washover throats occur in the main gravel ridge that allow coarse gravel lobes to accumulate landwards of the main ridge. This roll-over effect is widespread along the coast. Gravel is also being moved westwards under the influence of westerly waves. According to Grove (1955) 'the most recent gravel bank on the west side [of Spey Bay] appears to have grown steadily along the beach towards Lossiemouth over a distance of one and half miles (2.4 km) since 1870' at an average rate of westerly extension comparable with that of the gravel spits that grow across Spey mouth (Grove, 1955). Using map and field evidence the total westerly extension of the gravel beach was 4.2 km between 1870 and 1995, an average annual extension rate of 33.6 m a^{-1} (Gemmell et al., 2001a,b; Table 6.3). Where there was no gravel present in 1903, today there is a 60 m-wide gravel beach, consisting of an active beach ridge of c. 4 m OD behind which lie several landward-recurving ridges at about the same altitude.

Changes in the position of mean high-water springs (MHWS) and MLWS at Spey Bay between the first (1870) and the latest current Ordnance Survey (1970) reveal that the eastern side of Spey Bay has been eroded over the intervening 100-year period. This erosional trend declines to the west beyond the Spey delta until it gives way to accretion c. 4 km west of the delta (Gemmell et al., 2001a,b). Recession rates since 1975 of 1–1.5 m a^{-1} have been recorded both east and west of the river exit (Riddell and Fuller, 1995). The replacement of sand by gravel, discussed above, is reflected by accretion over the 1870–1970 period in the west of Spey Bay along a 4 km stretch in the vicinity of Boar's Head Rock. Farther west, the sandy beach and dunes at Lossiemouth are wholly erosional over the map period, a trend which continues today (Gemmell et al., 2001a).

At the mouth of the Spey, complex fluvial and coastal processes interact to create a dynamic and highly active system (Figure 6.35). Historical records (Grove, 1955) suggest that:
'the mouth of the Spey alters more rapidly from year to year than almost any other section of the coastline of Britain and ... the position of the mouth of the Spey has fluctuated violently throughout the last two centuries'.

Changes in the position of the river mouth

Gravel and 'shingle' beaches

Figure 6.34 The emerged gravel ridges of Spey Bay descend in a 'staircase' from 9–10 m OD to the present-day beach. The greatest extent of the unvegetated gravel occurs to the west of Kingston (see Figure 6.35), where this picture was taken. (Photo: J.D. Hansom.)

between 1726 and 1995, documented in Gemmell *et al.* (2001a,b) show a natural tendency for the river mouth to shift westwards towards Kingston, driven by the wave-driven westward migration of the spit across the mouth. If this natural process is uninterrupted, the mouth migrates by up to 1.2 km west of its 'central' (*c.* 200 m west of Tugnet) position. According to local tradition the Speymouth spit was *c.* 5 km long in 1798 (Hamilton, 1965). The river has always returned to a central location through natural breaches of the gravel spit, as recorded in 1829 and 1981 (Riddell and Fuller, 1995), but recent breaches have been artificially engineered to a 'central' location in order to reduce the threat of flooding and erosion at Kingston. In spite of this, there are also several documented examples (e.g. 1870, 1989 and 1995) of temporary easterly drift at Speymouth, which, although generally short-lived, demonstrate the sensitivity of the longshore drift system to local variations in the wind and wave climate. A complex suite of gravel ridges is present both to the east and west of the Spey outlet, enclosing tidal lagoons. The orientations and recurves of these ridges relate to the interaction of coastal and fluvial processes during the varying positions of the river over time, allowing former positions of the Spey mouth to be identified.

As a result of the erosional nature of much of Spey Bay protection measures have been implemented, mainly in the east. A vertical sea-wall fronts the village of Porttannachy/Portgordon in the east, a 400 m-long section of rip-rap backs the beach immediately to the east of the river mouth at Tugnet and a programme of beach replenishment using gravel excavated from the ridges at the delta mouth has recently been implemented along a 2 km stretch at Kingston (Gemmell *et al.*, 2001a,b).

Interpretation

The contemporary coastal development of Spey Bay is juxtaposed against a fine suite of older emerged Holocene landforms and provides a unique site for the study of former sea levels and how these relate to both past and contemporary sediment budgets. In the UK context, Spey Bay

Spey Bay

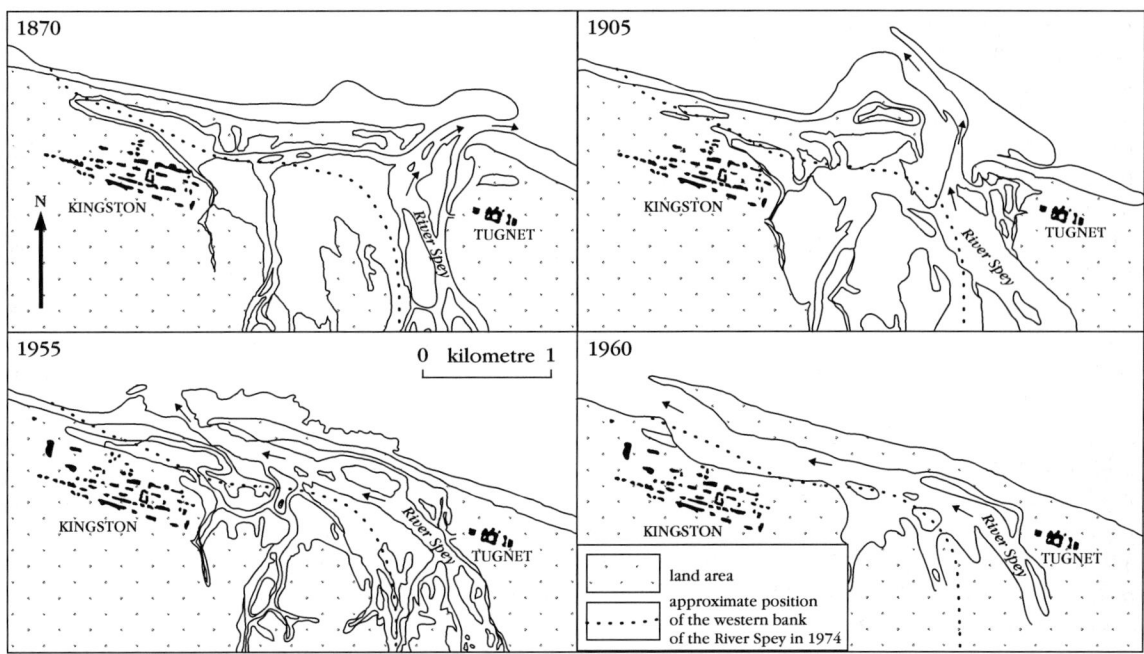

Figure 6.35 Movement of the River Spey mouth between 1870 and 1960. (After Grove, 1955 and Gemmell *et al.*, 2001a.)

also provides a unique insight into the under-researched area of deltaic processes that occur as a large gravel-bed river exits into a highly dynamic open coast situation.

Ogilvie (1923) provided the first interpretation of the gravel strandplain of Spey Bay in its wider regional and Holocene context, although the most recent and detailed work is that of Riddell and Fuller (1995), Gemmell (2000), and Gemmell *et al.* (2001a,b). Initial emplacement of the Spey Bay ridges against the foot of the Holocene cliff probably began before the peak of the Holocene Transgression (*c.* 6500 years BP), when rising sea levels delivered large quantities of offshore material to the coast. Together with River Spey gravels, this material infilled the low and flooded areas south of the present-day coast in the area of the Moor of Dallachy, almost as far as Fochabers upriver and to the south of Kingston along the foot of Binn Hill (Gemmell *et al.*, 2001a,b; Figure 6.33). Along its entire length, the Holocene coastline was marked by a very prominent cliff, last occupied *c.* 6500 years BP, and now fronted by a seawards-falling staircase of younger emerged beaches (Figure 6.34). Progressively, the bay of the lower Spey was closed off by gravel storm ridges across the mouth between Porttannachy/Portgordon and Kingston and infilled by fluvial accretion behind.

Extending westwards from Kingston, gravel ridge accretion extended beyond the lower Lossie–Spynie area, which was an inlet of the sea, as far as Burghead Bay and the Findhorn area. However gravel accretion and sea-level fall progressively cut off the inlet of the Spynie area and River Lossie from the open sea. Eventually an 800 m-wide swathe of ridges developed to separate Binn Hill from the sea. Altitudes of the ridges suggest that the majority of the sequence was deposited after the peak of the Holocene transgression, with levelled transects displaying a stable and then rapidly falling trend in altitude (Comber, 1993; Gemmell *et al.*, 2001a,b).

There remains great potential for further research at this site and elsewhere on the Moray Firth to produce a quantified Holocene sediment budget for the Firth. For example, research at Culbin *c.* 40 km west of the Spey (see GCR site report for Culbin, Chapter 10) recognized the Holocene contribution of sediment from the River Spey and Findhorn to the development of the large gravel ridge strandplain at Culbin (Comber, 1993).

Contemporary coastal processes and landforms are dominated by a strong westerly movement of gravels and truncation of the gravel recurves in west Spey Bay by the present-day beach suggests that the preceding generations of gravel ridges were subject to similar driving forces. However, since the older ridges were probably deposited along a coastline that trended along a west–east axis, rather than the present-day WNW–ESE, there is a strong suggestion of long-term erosion and planimetric re-adjustment of this part of Spey Bay (Hansom and Black, 1996). These recurves were also noted by Ritchie (1983) who posed the question of '…whether or not the present beach ridge is another gravel beach ridge that continues the pattern of progradational ridges or, as is more likely, it is largely a product of the reworking of the front of one of the ridges of the emerged gravel foreland'. The suggestion from both Ritchie (1983) and Hansom and Black (1996) is that while the eastern part of Spey Bay is eroded and rotates landwards, it fuels a seawards rotation of the west Spey Bay gravel ridges.

Gemmell (2000) and Gemmell *et al.* (2001a,b) produced a preliminary contemporary sediment budget for Spey Bay (Figure 6.36), using a combination of map and field evidence and results from published computer modelling. Sediment input into the contemporary system derives from updrift erosion of the present-day beach and from the River Spey itself, which from modelling studies contributes an estimated 8000 m^3 a^{-1} to the coast (Riddell and Fuller, 1995). These input sediments contain approximately 80% gravel and 20% sand, and so the annual input of *gravel* into the system from the river is *c.* 6400 m^3 (Figure 6.36). When added to the amount contributed by recession of the adjacent shore (9496 m^3), some 15 900 m^3 of gravel is input to the Spey system annually. However, the annual accretion of gravel in the ridges at Boar's Head Rock and beyond only accounts for approximately 5600 m^3 and so there is an apparent annual net loss of 10 316 m^3 of gravel from the Spey system. This apparent loss may be a function of error in the amount and periodicity of material delivered or in the estimate of gravels held at Boar's Head Rock. If losses are occurring then the final destination of the gravels is unknown. There is also an apparent annual net loss of *sand* from the Spey Bay system that amounts to 8864 m^3. The final destination of the sand is probably to sand dune accretion behind the beach and infill of the Lossie saltings, however an unknown amount is almost certainly lost to the offshore zone and may bypass the local sediment cell boundary (Ramsay and Brampton, 2000c) at Lossiemouth.

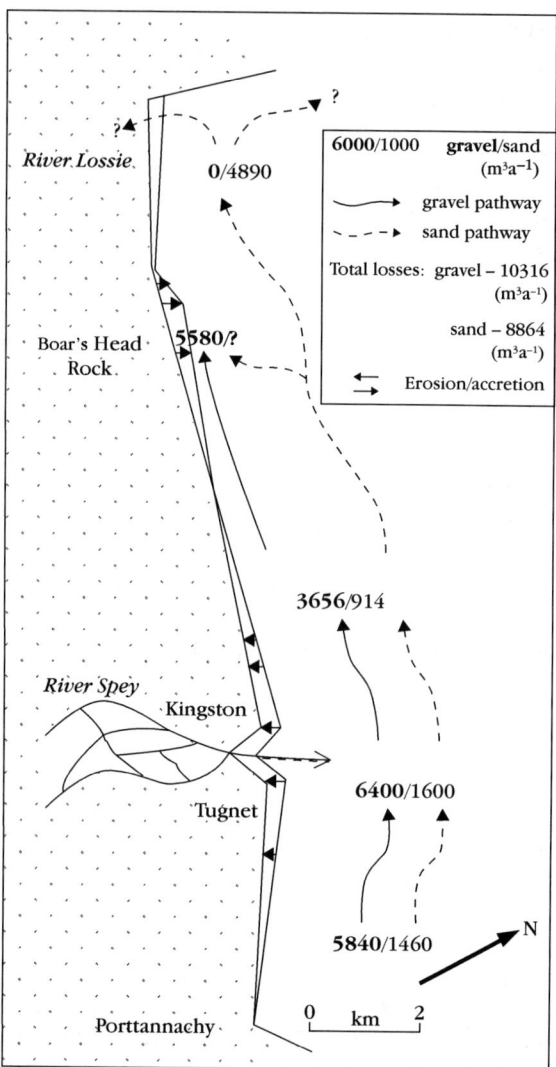

Figure 6.36 Diagrammatic representation of the Spey Bay sediment budget. Scale approximate. (After Gemmell *et al.*, 2001a.)

The above sediment budget, although preliminary and under revision, is instructive because it supports the geomorphological evidence and indicates that the entire length of Spey Bay functions as a discrete sedimentary unit, with erosion of one section influencing accretion at another. However, the supply of gravel alongshore is not constant and is subject to pulsing depending upon fluvial supply and storm events (Gemmell *et al.*, 2001a). Further, since sediment supply from the offshore to Scottish coasts is now much diminished, the supply to the Spey system is no longer added areally to the shoreface as before. Instead, reduced volumes of sediment are now added mainly as point sources at the Spey delta and at erosional sites (Hansom, 2001). The result is that periodic alongshore re-distribution of discrete plugs of gravel occurs, and this may give rise to local areas of gravel surplus and deficit.

It follows that interference in the natural transit of gravels will inevitably affect the geomorphological evolution of the Spey system. For example, the hard coastal defences at Porttannachy/Portgordon already interrupt sediment transport and effectively starve the downdrift beach of feeder sediment, thus contributing to accelerated erosion. As a result, proposals to erect coastal defence structures in mid-Spey Bay and protect Kingston are likely to impact negatively on the downdrift beaches of west Spey Bay. Work by Gemmell *et al.* (2001a) has also shown that increasing fluvial protection of the banks of the River Spey over the last century has probably reduced the amount of sediment entering Spey Bay and may now be beginning to affect the natural geomorphological evolution of the coastal system.

Conclusions

Spey Bay is an important site for coastal geomorphology and of particular interest because the large input of fluvial gravels is unusual in a UK context. The active gravel storm ridges of the present-day coast are some of the finest in Scotland and their constant adjustment to waves in Spey Bay demonstrates both short- and medium-term dynamic coastal processes. The active coastal margin is juxtaposed against a magnificent emerged strandplain, with a suite of gravel shorelines relating to the progressive history of coastal development within the Moray Firth as adjustments took place in Holocene sea levels. Additionally, the site is important on account of the unique fluvial–coastal interaction displayed at Speymouth, where the coarse sediments of the dynamic and actively braided River Spey enter a high-energy, open-coast gravel beach.

At Spey Bay, there is great scope to provide more accurate contemporary and Holocene sediment budgets for the river and coast. In addition, the contemporary development of Spey Bay has three unique features, all of which have great potential for future study: the loss of the sand beach at Lossiemouth and the progressive replacement by the westerly accretion of gravel; the gradual change in coastal orientation of Spey Bay, as updrift erosion in east Spey Bay fuels downdrift accretion; and the dynamics of fluvial-coastal interaction and periodic release of gravels at Speymouth.

THE WEST COAST OF JURA, ARGYLL AND BUTE
(NR 659 985–NR 442 724)

J.D. Hansom

Introduction

The west coast of Jura (see Figure 6.2 for general location) contains a remarkable assemblage of emerged coastal landforms including shore platforms and some of the most extensive areas of well-developed Lateglacial gravel ridges in Britain. The area is noted for small areas of machair-like dune surfaces and for the finest example of a medial moraine in Great Britain at Sgriob na Caillich (NR 475 765). The emerged coastal landforms provide valuable information for understanding changes in Late Devensian and Holocene relative sea level.

Description

The 37 km-long stretch of the west coast of Jura, between Glengarrisdale Bay (NR 659 985) in the north, and Rubha Aoineadh an Reithe (NR 448 751) in the south, together with a small area (0.2 km²) at Inver (NR 442 724), is one of the classic localities in Great Britain for emerged coastal landforms. Spectacular unvegetated spreads of Late Devensian and Holocene emerged beach gravel are juxtaposed with excellent examples of three emerged shore platforms, the High Rock Platform, Main Rock Platform and Low Rock Platform. The emerged gravel beaches were described by Ting (1936, 1937), however, the first major study of the emerged beaches was by McCann (1961, 1964), who sought to describe and explain the origin of the western Jura gravels by relating them to Lateglacial relative sea-level change. More recently, the emerged shorelines of western Jura have been investigated in detail by Dawson in several papers, most of which are reviewed by Dawson (1993) in the *Quaternary of Scotland* GCR volume.

The High Rock Platform and associated cliff that extend continuously between Shian Bay

Gravel and 'shingle' beaches

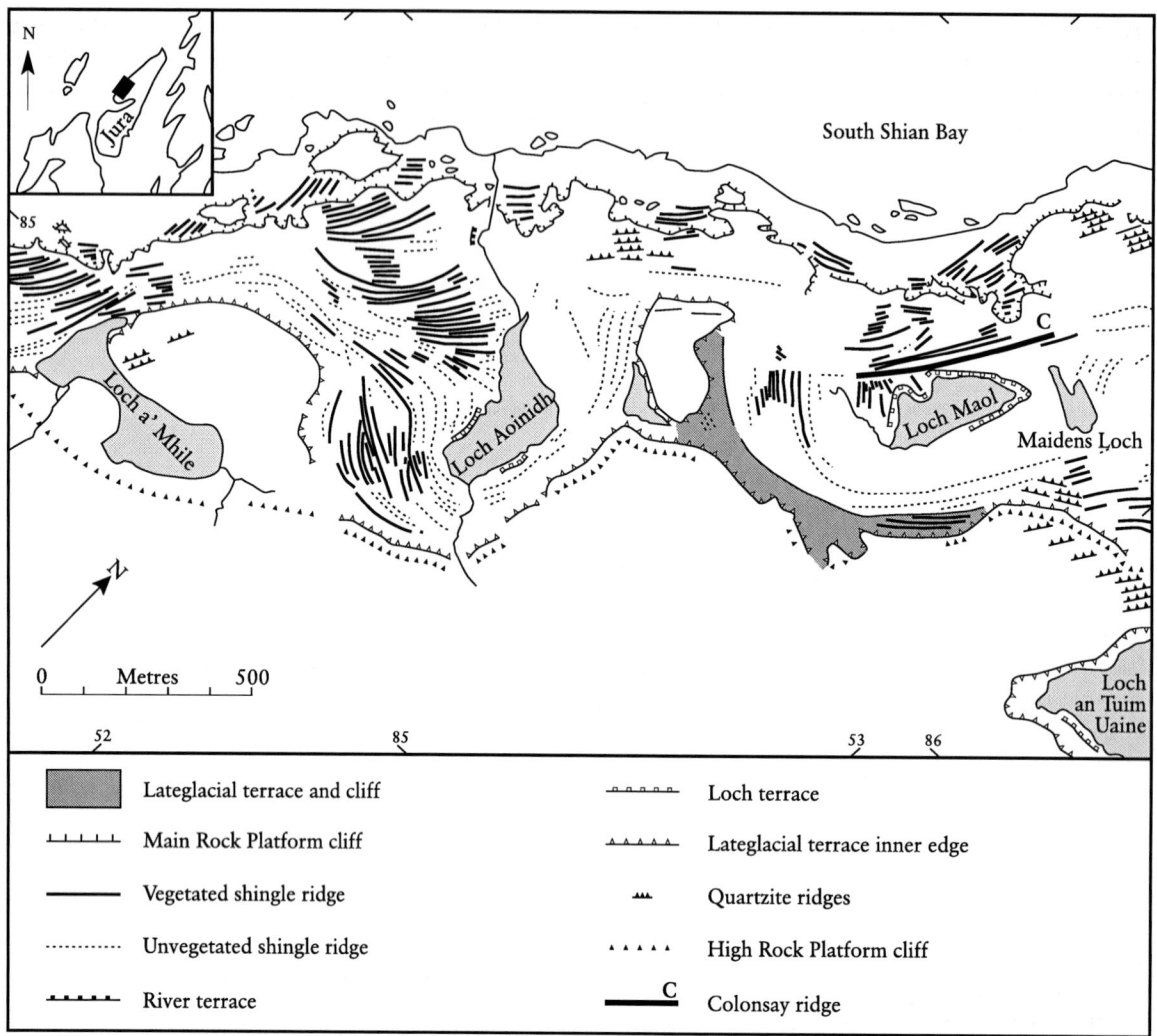

Figure 6.37 Geomorphology of western Jura in the area of South Shian Bay, showing the 'staircase' of emerged gravel ridges. (After Dawson, 1993.)

(NR 530 875) and Ruantallain (NR 505 833) are are discussed extensively by Dawson (1993) and so only a brief account is included here. The platform has an average width of 350 m but reaches 600 m and the backing cliffs are usually 5–15 m high. The inner edge of the platform lies at an altitude of 34.1–32.1 OD and its seaward slope is about 4°. The surface is covered in places by gravel spreads, which overlie not only rock platform but also patches of lodgement till. Elsewhere the surface is ice-moulded and a striated bedrock sea stack occurs immediately north of Loch a' Mhile (NR 514 850).

Between Shian Bay and Ruantallain the seaward edge of the High Rock Platform forms the cliff of a lower platform, the Main Rock Platform. The lower platform is 50–150 m wide and the inner edge occurs at 3–5 m OD. It is locally overlain by Holocene emerged beach sediments, and the crenulate cliffs of 10–15 m high are indented by numerous emerged sea caves. The platform has a serrated and uneven surface with no signs of glacial moulding. The platform is also continuous between Shian Bay and Glendebadel Bay.

The intertidal rock platform fragments of the Low Rock Platform are conspicuous along long stretches of the Jura coast. Typically 100 m wide, they are best developed on the foreshore between Rubh' Aird na Sgitheich and Allt Bun an Eas. The platform surfaces are locally ice-moulded and in most places pass inland beneath till. Between Rubh'Aird na Sgitheich and Glenbatrick the platform is overlain by up to 15 m of Late

The West Coast of Jura

Figure 6.38 An unbroken 'staircase' of unvegetated emerged gravel beaches falls from c. 30 m OD to sea level on the West Coast of Jura. Looking eastwards towards Glenbatrick. (Photo: J.D. Hansom.)

Devensian emerged beach gravels.

The coast of western Jura is dominated by conspicuous emerged beach terraces and 'staircases' of unvegetated beach ridges. Although discussed by Ting (1936, 1937), the most detailed studies of these emerged coastal features are by McCann (1964, 1968) and Dawson (1979, 1982). There are two main sets of gravels: a higher suite and a lower suite. The higher suite of emerged coastal terraces can be traced almost continuously southward from Shian Bay (Figure 6.37) to Inver but other areas occur at Corpach Bay, Glendebadel and in the Glenbatrick area (Figure 6.38). In most cases the emerged marine deposits are ridges of unvegetated quartzite gravels that decline in altitude from north-east to south-west, from 40 m OD at Corpach Bay to 24.5 m OD at Inver. The altitudes of the higher suite of beaches suggests the existence of two shorelines, the higher of which declines in altitude from 40 m OD at Corpach in the north to 34 m OD near Ruantallain in the south-west, a regional gradient of 0.56 m km^{-1}. A separate and lower set of ridges occurs in the south-west, from 31 m OD at Glenbatrick to 24 m OD at Inver, a regional gradient of 0.53 m km^{-1}. Along most of the west coast of Jura the higher suite of emerged ridges terminate at the cliff of the Main Rock Platform and so it is unusual to find them below 20 m OD. However, at South Shian Bay (Figure 6.37), emerged beach gravels descend to 11 m south-west of Loch Maol, probably on account of the unusually low altitude of the rock platform on which they sit (Dawson, 1993). The detailed pattern of ridge crests on both sets of gravel beaches bears a close relationship to the intricacies of the rock platforms on which they sit and of the rock headlands that separate individual beach units.

The lower suite of beach gravels is widespread throughout the west coast of Jura and falls in altitude from between 10 and 12.3 m OD to merge with the gravel ridges of the modern beach. Only a few distinct coastal terraces exist, but emerged gravel banks and beach accumulations everywhere mantle the rock surfaces of the Main and Low Rock Platforms. Spectacular staircases of emerged gravel ridges commonly occur, the best example of which is present north of

Inver where 31 individual unvegetated ridges descend from 12.3 m OD to the present-day beach.

Small areas of windblown sand occur in several of the small pocket beaches. Of these, Corpach is of most geomorphological interest in that it contains a small area of rare cliff-foot dunes that resemble a machair surface. These dunes mantle the Main Rock Platform and are locally banked in great ramps of bare sand against the emerged cliff landwards. Although showing active deflation in places, the dunes are mostly vegetated and contain buried palaeosols together with several features of archaeological interest including grave burials in the emerged gravels. The dunes at Corpach, together with dune areas at Shian, Bagh Gleann nam Muc and Glengarrisdale show some of the geomorphological attributes of machair, with a close sward of grasses and an absence of the normal dune grasses such as marram *Ammophila*.

Interpretation

The High Rock Platform of the southern Inner Hebrides, best seen in western Jura and northern Islay, was considered by Wright (1911) to be 'pre-glacial' in age on account of the emerged beach gravels that rest between its surface and a superficial cover of till. McCann (1968) suggested instead that it was 'interglacial'. Dawson (1979) accepted an interglacial origin but considered that the shoreline had been warped by neotectonic activity. Sissons (1982) proposed that the platforms that comprise the High Rock Platform were produced by cold-climate shore erosional processes and that the various platform fragments are part of a series of glacio-isostatic tilted shorelines. However, the altitudes from the West Coast of Jura (and northern Islay) do not demonstrate any platform tilt and so it is probably a single feature. At present there exists no general agreement on platform origin or age. Dawson (1983) argues that formation of the western Jura platform by cold-climate shore erosion would have taken a minimum of 8000 years and that such a lengthy period of relative sea-level stability during a single period of cold climate was unlikely. It would therefore appear that the western Jura High Rock Platform may represent the product of several periods of Pleistocene coastal erosion (Dawson, 1993).

The inner edge of the platform, which abuts the lower cliff of the High Rock Platform, occurs at 3–5 m OD and constitutes part of a glacio-isostatically tilted shoreline that declines in altitude to the south-west, from 6 m OD in northern Jura to sea level in northern Islay at a regional gradient of 0.13 m km^{-1} (Dawson, 1993). Together with its altitude, jagged nature and freshness of form, this gradient suggests that the platform conforms to the Main Rock Platform identified elsewhere in the Inner Hebrides (Gray, 1978). Dawson (1980a) reports the platform to be unglaciated although where the regionally tilted Main Rock Platform merges with and crosses the regionally horizontal intertidal Low Rock Platform the distinction becomes blurred (Dawson, 1979, 1980a; see below). The intertidal ice-moulded rock platforms that occur in north-west Jura and northern Islay are referred to by Dawson (1980a) as the 'Low Rock Platform', its regional horizontality explained as having been produced by marine processes during interglacial periods.

Since the lower set of platform fragments has been ice-moulded, they were produced prior to the last glaciation. This platform, first noted by Wright (1911) as a '… preglacial plain of marine denudation…', termed the 'Low Rock Platform' by Dawson (1979). Dawson noted that its presence as an ice-moulded intertidal feature, unaffected by glacio-isostatic tilting along many parts of the Scottish coastline, implied interglacial origins. Sissons (1981) argued that the glaciated intertidal features represented a set of platform fragments of different ages that had been subject to glacio-isostatic deformation and then exhumed in the intertidal zone as a result of present-day marine activity. According to this hypothesis, the rock platform features were initially produced by cold-climate, shore-erosion processes. The higher platform fragments were considered part of the glacio-isostatically tilted Main Rock Platform that was regarded as having been produced during the cold climate of the Loch Lomond Stadial (Dawson, 1993). This shoreline is generally considered to pass below sea level on Islay, owing to its glacio-isostatic deformation (Dawson, 1993).

The regional variations in altitude of the highest emerged beach terraces on the west coast of Jura suggest the existence of two shorelines. The older emerged shoreline at 34–40 m OD is also thought to occur in northern Islay and has a regional gradient of 0.56 m km^{-1} (Dawson, 1993). A separate and slightly younger shoreline is present in south-west Jura declining in altitude to the south-west from 31 m to 24 m OD at a regional gradient of 0.53 m km^{-1}. Dawson (1982) inferred from these that both south-west and north-west Jura remained ice-covered while the higher shoreline formed between Corpach Bay and Shian Bay. Deglaciation of south-west Jura took place at a slightly later date and was accompanied by the formation of the lower shoreline and its gravel ridges. The regularly declining ridge-crest altitudes of the western Jura gravel 'staircases' indicates that, although stillstands may have occurred during the fall in the sea level from 35 m to 20 m OD, no major sea-level oscillations occurred. Most of the west Jura gravel spreads below this altitude terminate at the cliff of the Main Rock Platform and so patterns of sea-level change below 20 m OD cannot be established except at South Shian Bay where McCann (1964) proposed that a prominent gravel spit at 19 m OD called the 'Colonsay Ridge' (Figure 6.37) represented a pause in the overall fall in sea level.

The presence on the west coast of Jura of extensive spreads of emerged gravel is primarily due to the glacio-isostatic uplift of the higher shoreline and its altitudinal relationship with the till-covered High Rock Platform. These relationships indicate that both high and low gravels are of Late Devensian age. On deglaciation, the maximum sea level along this coast (34–40 m OD) stood several metres higher than the inner edge of the High Rock Platform. Wave erosion of the till cover resulted in extensive gravel deposition, a process enhanced by the gentle sloping nature of the underlying platform surface and its open exposure to westerly waves.

The trend of falling sea levels in the Lateglacial was reversed later in the Holocene with the highest ridges of the extensive suites of lower beach gravels probably representing the culmination of this rise. The culmination of the sea-level rise at about 6500 years BP in the west coast of Jura produced ridges that now lie at 12 m OD. Subsequently a fall of sea level to its present-day level, the result of ongoing isostatic uplift, deposited a staircase of 31 individual gravel ridges and indicates that about 12 m of uplift has occurred in the area over the last 6500 years (MacTaggart, 1998a).

The cliff-foot dunes at Corpach are of interest in that they, along with similar features in other bays along the west coast of Jura, have been interpreted as machair (Ritchie and Crofts, 1974). The dunes mantling the Main Rock Platform are mostly stable and vegetated but bare sand has been blown into highly dynamic climbing dunes up the emerged cliff behind. The dunes at Corpach, together with dune areas at Shian, Bagh Gleann nam Muc and Glengarrisdale, all show some of the geomorphological attributes of machair, with a close sward of grasses and an absence of dune grasses (e.g. marram *Ammophila*). However, their status as machair has been questioned from a botanical perspective because few of the dune areas have even moderate amounts of carbonate sand and do not support classic machair vegetation communities (Dargie, 2000; Angus, 2001). Without this supporting evidence, it seems more appropriate to regard the blown sand deposits of the west coast of Jura as dune systems rather than machair.

Conclusions

The west coast of Jura is outstanding for its assemblage of emerged coastal landforms and gravel beach deposits. Both the range of features and their extent and degree of development are exceptional and include not only fine examples of the three major rock platforms recognized in western Scotland, the High, Main and Low Rock Platforms, but also extensive spreads of unvegetated Lateglacial and Holocene gravel beach ridges unparalleled elsewhere in Britain for the length of their morphological record of sea-level changes.

BENACRE NESS, SUFFOLK (TM 532 824–TM 535 831)

V.J. May

Introduction

Benacre Ness (see Figure 6.2 for general location) is a good example of a ness formed in shingle and associated with rapid coastal retreat of part of the beach and of nearby cliffs. Williams (1956) outlined the history of the ness (which had been referred to by Ward (1922) under the name 'Covehithe Ness'), showing how it had progressively moved from south of Covehithe to its present-day position (see Figure 6.39); almost 6 km in 200 years. The site comprises three landform units: namely cliffs cut mainly in fluvioglacial sand with a fringing beach of sand and shingle, a beach ridge fronting Benacre Broad and The Denes, and Benacre Ness itself, formed of sand and shingle ridges. Although the evidence from longshore sediment transport is that material moves towards the south, the ness form itself has moved northwards. As well as being a classic landform, therefore, Benacre Ness is of considerable importance for studies of coastal form-process dynamics (Steers, 1946b; Russell, 1956; Williams, 1956, 1960; Hardy, 1966; Cambers, 1975; McCave, 1978b; Carr, 1981; Onyett and Simmons, 1983).

Description

The southern part of this site is formed by the cliffs at Covehithe, which are undergoing very rapid erosion. Cambers (1973) estimated the historical (100-year) rate of retreat at up to 4.25 m a^{-1} and it has exceeded 6 m a^{-1} since 1929. The estimated loss of beach volume during a single 24-hour period at Covehithe was 300 000 m^3 (Williams, 1956). The rapidity of cliff-top retreat is shown vividly by the truncation of the lane leading from Covehithe itself and the loss of autumn-sown crops by the next spring. The cliffs decline northwards to Benacre Broad. Here the beach is a single fringing ridge that blocks the mainly infilled Benacre Broad (between TM 532 824 and TM 535 831). This ridge has similar rates of retreat to the cliffs to the south, but the rate of change declines along The Denes until Benacre Ness where accretion at rates up to +2.46 m a^{-1} has occurred since 1880. The northern part of the ness is marked by erosion, but at significantly lower rates (up to 0.36 m a^{-1}) than the southern part. Steers (1981) has drawn attention to the very high rates of retreat that have occurred during storm surges. Between 19 March 1977 and 11 March 1978, at three separate points, 9 m, 8 m and 14 m were lost. During 1979 and 1980, the same points lost 2.7 m, 4.6 m and 8.8 m. Even larger values of 12 m and 27 m were recorded by Williams (1956) during the storm surge of early 1953. A surge in 1990 caused overnight recession of 35 m (K. Clayton, pers. comm.).

Interpretation

Much of the early description of the ness and its changes was based upon cartographic evidence (Ward, 1922; Steers, 1946b; Williams, 1956). Both Williams (1956) and Steers (1964a) found inconsistencies in the cartographic record, and Steers inclined towards the view that the ness has only existed in a form similar to present since about 1826. Steers (1946a) suggested that the feature had originated as a spit across the Kessingland River but offered no evidence for this or for the direction of the spit's growth. Williams (1956) suggested that a northward transport of sediment would occur immediately following surges. The large quantities of material eroded from the cliffs would be transported 'under the action of an abnormal north-going pull as the level of the sea falls'. Sediment would accumulate seawards of the ness and would then be gradually pushed up the beach by wave action. Steers accepted that this might be the case today, but could not see how this process could occur at earlier stages of ness formation when the ness was protecting the cliffs. The beach throughout the site is composed mainly of flint shingle but there are varying amounts of sand, except at the ness, which is formed almost entirely of shingle. Whereas the beach fronting Benacre Broad and The Denes is tending to move landwards, the ness itself has moved progressively northwards. Onyett and Simmons (1983) described Benacre as moving rapidly to the north, but were unable to provide evidence as to the net change in its volume.

In his discussion of Ward's paper, Russell (1956) suggested a simple mechanism for the apparent conflict between southward movement of sediment and northwards movement of the ness. The alignment of the northern face of the ness would mean that drift was zero, whereas the alignment of the southern face would increase longshore transport. Material would be accreted at the northern face, but the southern face would be eroded (Figure 6.39). As a result the ness would move northwards. The same mechanism was proposed by Cambers (1975) at Winterton Ness. Williams (1960) accepted Russell's suggestion, adding that the rate of movement of the ness is probably slowing as the amount of material available for transport reduces. However, Russell's concept does not explain how the ness forms in the first instance. At some point along the coast, wave direction must have been affected by refraction and the alignment of the coastline in such a way that the beach aligned itself to face the dominant waves. If, as Steers (1946a) suggested, there was a spit along this coast, the slight change in alignment

Benacre Ness

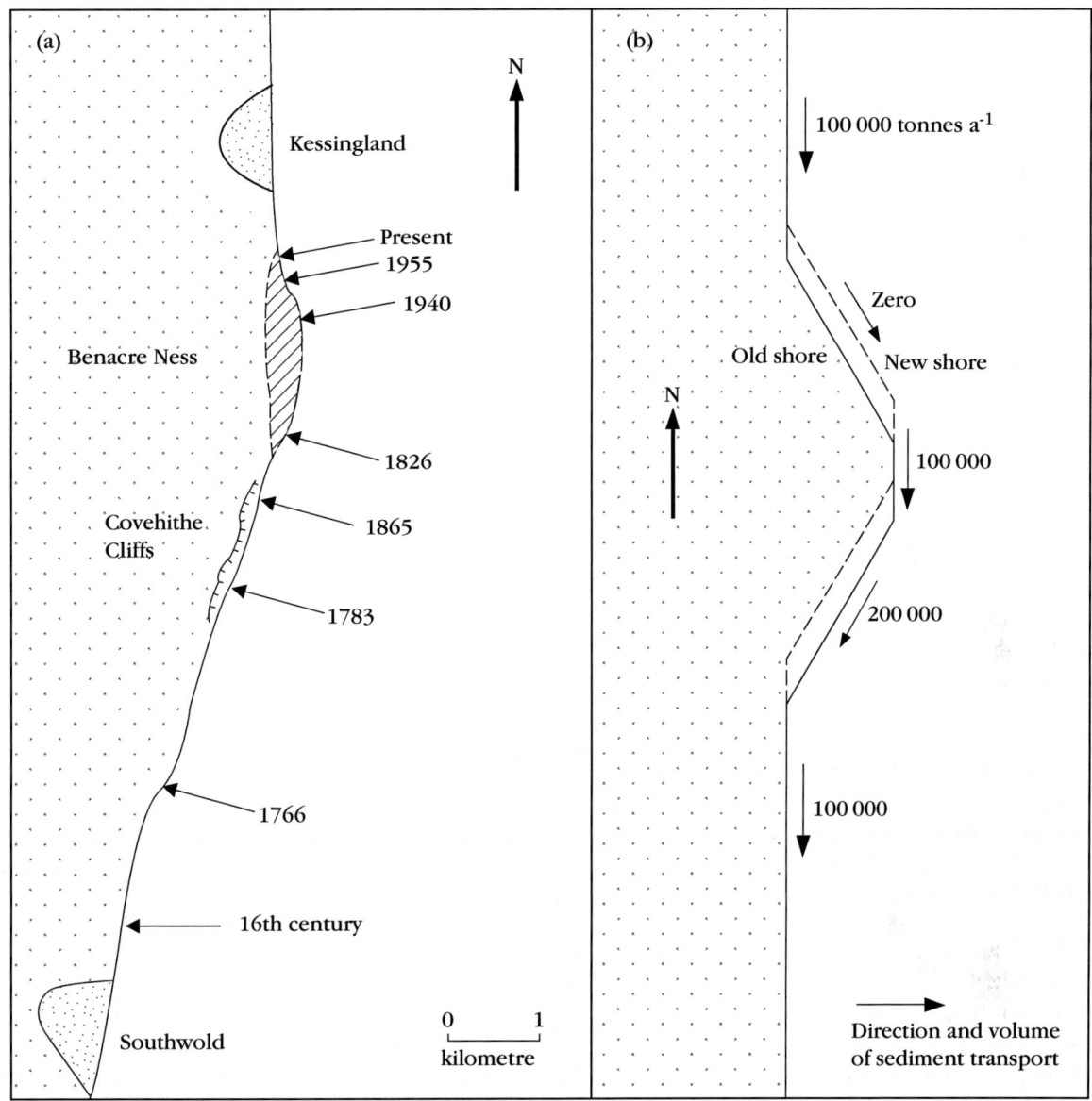

Figure 6.39 Cliff erosion and ness migration at Benacre Ness. The ness moves at 25 m a^{-1} to the north. The early accounts interpret the movement of the ness northwards as a result of accretion on the updrift side of the ness. The alternative view is that transport is towards the north (see Figure 6.40) and that accretion occurs on the lee (northern) side of the ness. Hardy (1966) suggests a reversal of movement of both the spit and direction of transport. (After Williams, 1956.)

of the coastline south of Benacre could have been sufficient to have caused the beach to change its alignment slightly to face the waves. Once this had occurred, the mechanism proposed by Russell would ensure the maintenance of the ness and its movement northwards.

However, Hardy (1966) demonstrated that the ness had migrated southwards since the beginning of the 20th century, in association with a complex pattern of erosion and deposition in neighbouring areas. A northward-trending offshore bank had developed since 1945 that was aligned away from the coast in the immediate vicinity of the ness. It was likely that a flood channel had developed between the ness and this bank that would allow sediment to come ashore in the area of the ness. An ebb channel carried most material offshore on the eastern side of the north–south-trending bank. Changes in the foreshore appeared to have occurred before the formation of the bank. Robinson (1966) argued that a flood channel carrying material southwards would feed sediment on to the northern flank of the ness, whereas the ebb channel carried material northwards on to its southern flank. Changes in offshore relief show that between 1824 and 1956, the ebb channel was established off the 19th century position of the ness. A gradual northerly shift of the ebb channel was followed by northward migration of the ness. McCave (1978b) identified the nesses as a possible location at which fine-grained material was removed from the longshore sediment drift, with coarsening in the direction of sediment movement. Carr (1981), however, regarded McCave's argument as unproven.

Benacre Ness differs from other nesses in England and Wales in two important aspects. First, it is clearly a migratory feature, unlike many others where the ness marking the change in direction in the beach appears to be more stable. Second, it moves in the opposite direction to the long-term direction of sediment transport. Like other similar features of the East Anglian coast, it is important for research into the links between longshore sediment transport and sediment transport offshore, but because of its different characteristics provides additional information for the overall understanding of the East Anglian coast.

Conclusions

This is an important example of a small cuspate foreland moving counter to the direction of sediment transport. The movement of the ness has an accompanying effect on the degree of protection afforded to the cliffs. To the south, Covehithe is the most rapidly eroding area on the English coast.

ORFORDNESS AND SHINGLE STREET, SUFFOLK (TM 358 400)

V.J. May

Introduction

The shingle ridges that form Orfordness (see Figure 6.2 for general location) extend about 15 km south from Aldeburgh on the Suffolk coast and divert the River Ore for a similar distance (Figure 6.40). South of the mouth of the river, the shingle ridges at Shingle Street continue southwards towards Bawdsey. Orfordness

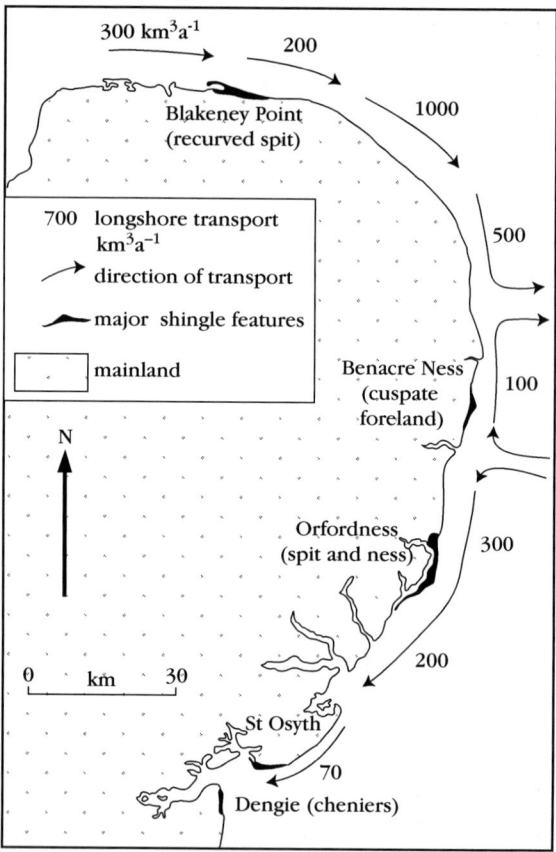

Figure 6.40 Longshore transport data for the East Anglian coast, showing estimated volumes and transport directions related to major shingle features. (After Cambers, 1975).

comprises three elements: storm beach, undergoing erosion, to the north; cuspate foreland; and shingle spit to the south, terminating at North Weir Point. On the opposite side of the estuary of the River Ore is the complementary, but gradually disappearing, Shingle Street ridge and lagoon complex. The Orfordness ridges provide evidence for oscillations in sea level, and research work on the spit and in the estuary has helped clarify many of the processes that are relevant in spit development worldwide.

The site has been well documented (Redman, 1864; Redstone, 1908; Steers, 1926a; Grove, 1953; Cobb, 1957; Kidson *et al.*, 1958; Kidson and Carr, 1959; Kidson, 1961, 1963; Carr, 1962, 1965, 1967, 1969b, 1970, 1971c, 1972, 1973, 1986; Carr and Baker, 1968; Randall, 1977; Green and McGregor, 1986) but with the exception of Steers (1926a) and Carr (1965, 1967, 1969b, 1970, 1971c, 1972, 1973, 1986), most have considered either only part of the shingle structure, or have concerned themselves with it in a wider context (Redman, 1864; Kidson, 1963). Generally, writers agree that this is one of the largest and most important shingle structures on the British coast. They include Redman (1864) who referred to 'This extraordinary mole of shingle', Steers (1964a) who described it as '...the largest of the east coast shingle spreads', and Carr (1969b) who commented that it is 'one of the most important shingle formations on the coast of the British Isles'. In the HMSO report (1947) Orford Beach and Shingle Street was described as being 'of the very greatest importance and interest physiographically'.

The area had been used for military purposes from 1914 and as a bombing range during World War II (1939–1945), but this seems to have not greatly harmed its scientific interest. However, post-war military pressures were particularly damaging to the northern and ness areas of the site. At Orfordness, though it was still possible to map the sequence of ridge heights in the late 1960s, the evidence near Stonyditch was damaged by the extraction of shingle there and its transference by light railway to Slaughden, south of Aldeburgh, for a beach-nourishment scheme. Further damage resulted from the construction of an abortive early-warning system on Lantern Marshes in 1971. This affected the neighbouring shingle spreads, which were practically destroyed. More recently, Green and McGregor (1986) have assessed the geomorphological quality of features within the site, where significant geomorphological interest still remains. The lagoons noted by Cobb (1957) at Shingle Street have largely disappeared. Mostly the legacy of a previous phase of development of the River Ore, the lagoons have been the victims of the natural changes that take place at Shingle Street as the distal point of the spit changes its position. The southern part of the spit is outstanding, for it is virtually undisturbed.

Description

Orfordness comprises three elements: a storm beach undergoing erosion in the north, together with some intermittent shingle spreads; an extensive spit in the south where the shoreline is either accreting or being slowly eroded; and linking these, the ness proper. This is a cuspate foreland situated where there is a change in the orientation of the beach from approximately north–south to north-east–south-west. It consists of a complex series of ridges piled one against another. Such ridges extend from opposite Lantern Marshes (TM 458 525) as far as North Weir Point (TM 376 436) and across the mouth of River Ore to Shingle Street (TM 374 434) and Bawdsey (TM 359 401). Over much of this length, there is a systematic overall series of sub-parallel ridges, except at Shingle Street where ridges are less continuous and more ephemeral.

South of the lighthouse, with its complex pattern of shingle ridges, the spit narrows to reach its minimum width of under 50 m at high-water mark about 1.5 km north of the present-day river mouth. Over the whole of this length, as far as the distal end of the spit, the structure consists of a series of sub-parallel ridges, on the river side of which recurves may be present. Very rapid growth of the distal point and its linking with estuarial banks, or its breaching, may result in an extreme form of these with the recurves separated by tidal pools. The spit terminates at North Weir Point. Between North Weir Point and Shingle Street lies the River Ore, with one or more channels and extensive shoals. These shallow banks are areas of considerable size, which are exposed for at least part of the tidal cycle. They are subject to considerable change, which reflects both the previous environmental conditions and the stage in the development of the spit. Between 1955 and 1970, annual growth of the distal point varied between zero and 88 m. Over the long-term, variability is far greater.

In the northernmost part of the site (Figure 6.41), the ridge crests all fall below 1.5 m OD, that is, they are generally below present-day high water on spring tides. The next shingle structure to the south is below 2.0 m OD and most of the remaining area north of Stonyditch (TM 455 497) is less than 2.5 m OD. The broad increase in height continues immediately south of Stonyditch where most of the ridges fall into the ranges 3.0 m to 3.5 m and 3.5 to 4.0 m OD. There is an extensive area towards Stonyditch Point, where the ridge crests are below 3.0 m OD. Most of these ridges, which fall in the range 2.5 to 3.0 m OD are either cut by ridges on the seaward side or are truncated near The Crouch, but one or two extend along the spit as far as the present-day Havergate Island where the relationship between the ridges again becomes more complex.

Apart from the fairly extensive area of ridges in the range 3.0 to 3.5 m OD on the ness, where

Gravel and 'shingle' beaches

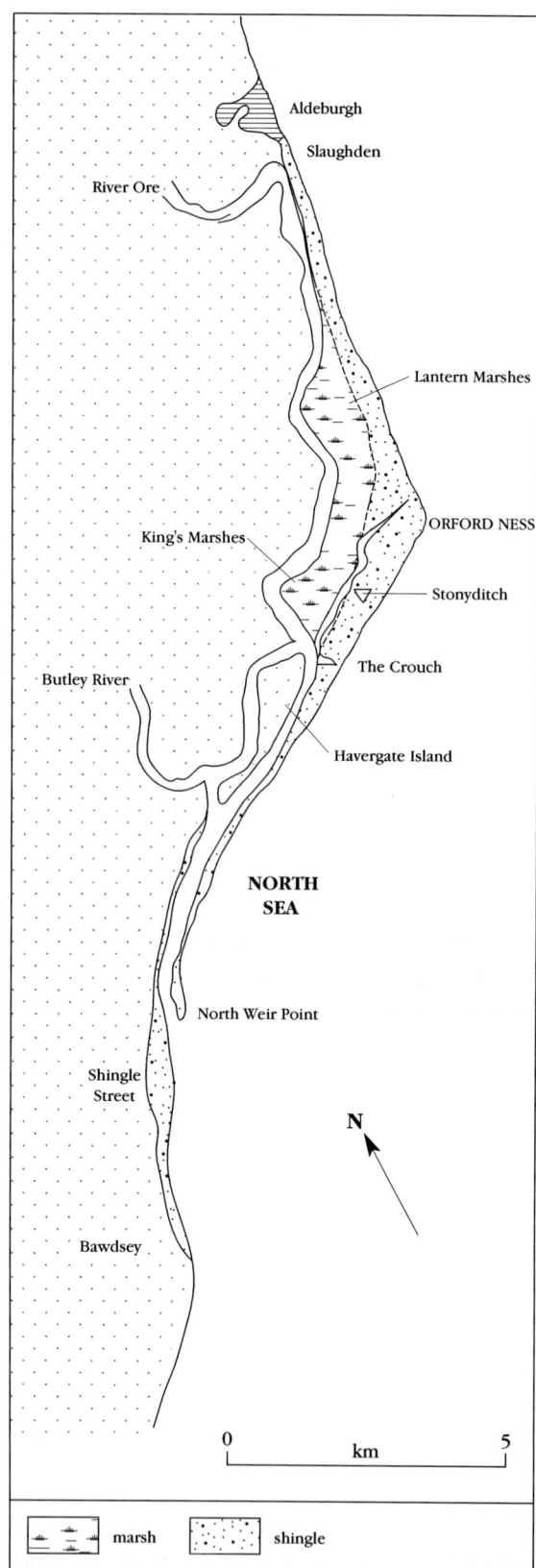

Figure 6.41 Sketch map of the Orfordness–Shingle Street area.

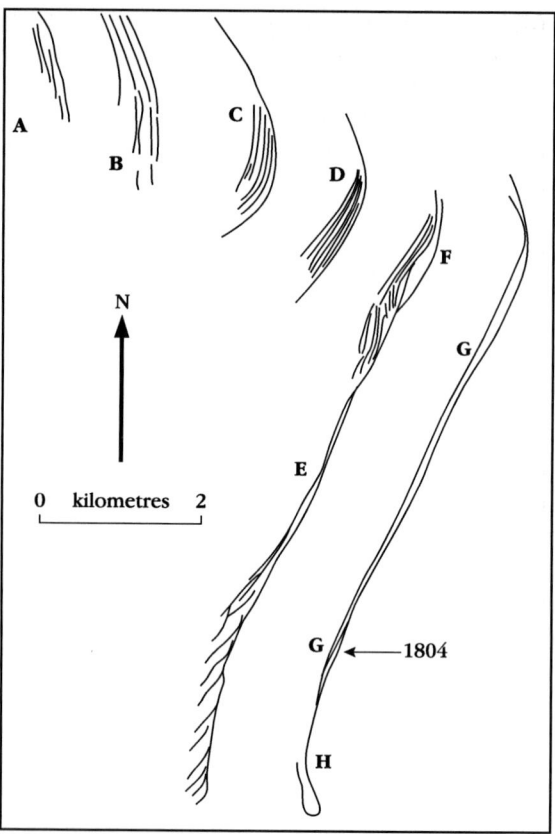

Figure 6.42 Schematic diagrams of shingle ridge groups representing developmental phases of Orfordness (A–B) spit development; (C–D) ness development; (E) extended spit with distal recurves; (F) additions to ness; (G–H) storm beach additions to spit. Each diagram portrays the north–south position accurately; the east–west position is arbitrary. (Based on Carr 1969b, 1972, 1973.)

they fall in height to seaward, only small areas of similar height occur elsewhere along the length of the spit. At the ness, both 2.5 to 3.0 m and 3.0 to 3.5 m groups are truncated on the seaward side by ridges between 3.5 and 4.0 m OD (Figure 6.43). These, together with the highest category (greater than 4.0 m OD with a maximum of approximately 5.0 m OD), extend throughout the spit to the distal point.

The height between the shingle ridges ('fulls') and the intervening hollows ('swales' or 'lows') varies. Generally there is about 0.3 m difference between adjacent ridges and hollows but occasionally this reaches a maximum of 1.5 m. The base of the shingle under the ness, and much of the nearby area, lies between –0.2 and –10.7 m OD. It appears to rest on marine planation surfaces of various ages. Shingle forms the river

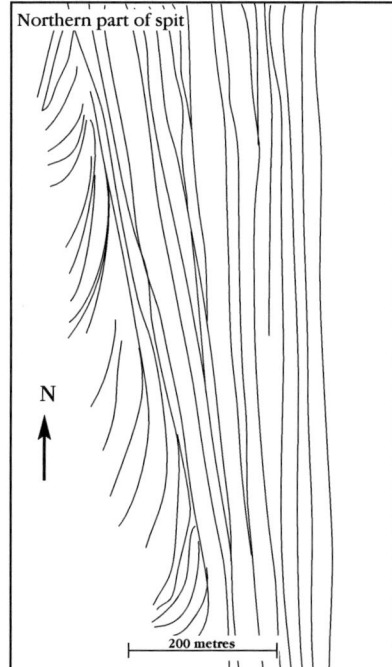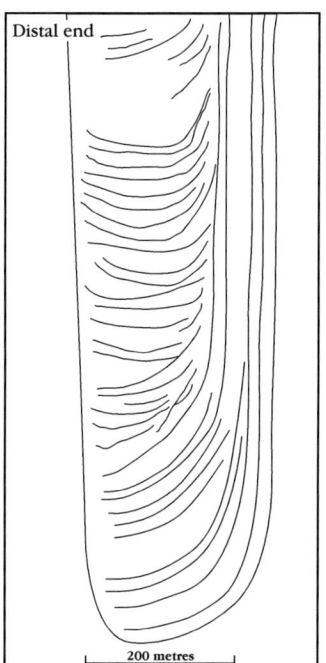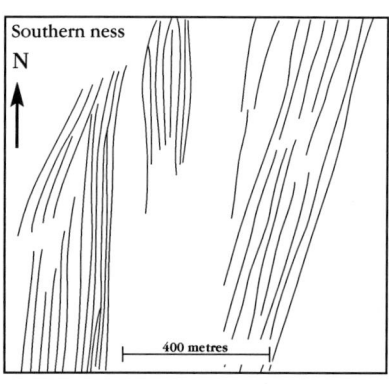

Figure 6.43 Variations in ridge patterns of Orfordness, in the southern part of the ness, the northern part of the spit with an earlier recurved spit fronted by individual shingle ridges, now largely destroyed, and also at the distal end of the spit, showing recurves. (Based on Carr, 1973; Green and McGregor, 1988.)

bed at The Narrows (TM 420 475), and one borehole records 'stone' under the estuarine clays on the northern part of Havergate Island from −10.1 to −14.0 m OD.

The well-rounded clasts of the ness and the spit range between 4 mm and 75 mm in length. Over 99% are flint. Elsewhere in the site, if pebbles are present, they are found in small quantities only: they are frequently rectangular rather than rounded, and are slightly more varied in composition. Most occur in a primarily sandy matrix.

The River Ore runs roughly parallel to the spit from Slaughden to North Weir Point (Figure 6.41). Its bed ranges in depth from −5.0 m to −12.5 m OD. Except near the present-day river mouth, the maximum depth at any given place has remained almost constant for the last 160 years and probably longer. The mouth of the Ore appears to have been displaced towards the south as the shingle structure lengthened in that direction. This impression is correct only in part. Displacement is likely both immediately south of Aldeburgh and south of The Crouch, although at the latter site it was complicated by the precursors of Havergate Island and the Butley River. Elsewhere, existing creeks, which ran approximately parallel to the coastline, were joined together as their exits became blocked.

Marshes that have been the subject of land-claim are present on the landward side of the River Ore throughout its whole extent. They vary in width from 0.4 to 2.4 km. Marshes also occur at Havergate Island (the interest in which is now almost entirely ornithological), and from Slaughden to The Crouch on the southern side of the river. On the spit farther south, there are only small areas of saltings. The reclaimed marshes are at −0.3 m to +0.6 m OD and the saltmarshes at 1.2 m to 1.5 m OD. King's Marshes are separated from the shingle of the ness by a tidal creek (Stonyditch), which runs approximately north-east–south-west. The truncated head of this creek rises in an area of saltmarsh between two series of ridges, so that northward of this point the shingle ridges and marshes are adjacent. This is the only instance where the major shingle structure abuts the marshes. Borehole logs suggest that at least some of the shingle and estuarine clay were deposited contemporaneously. The differing depths at which bands of shingle occur both in nearby boreholes and within the same borehole indicate how complex the sequence of events must have been. Although the ridges south of The Crouch are about 1000 years old or less,

Gravel and 'shingle' beaches

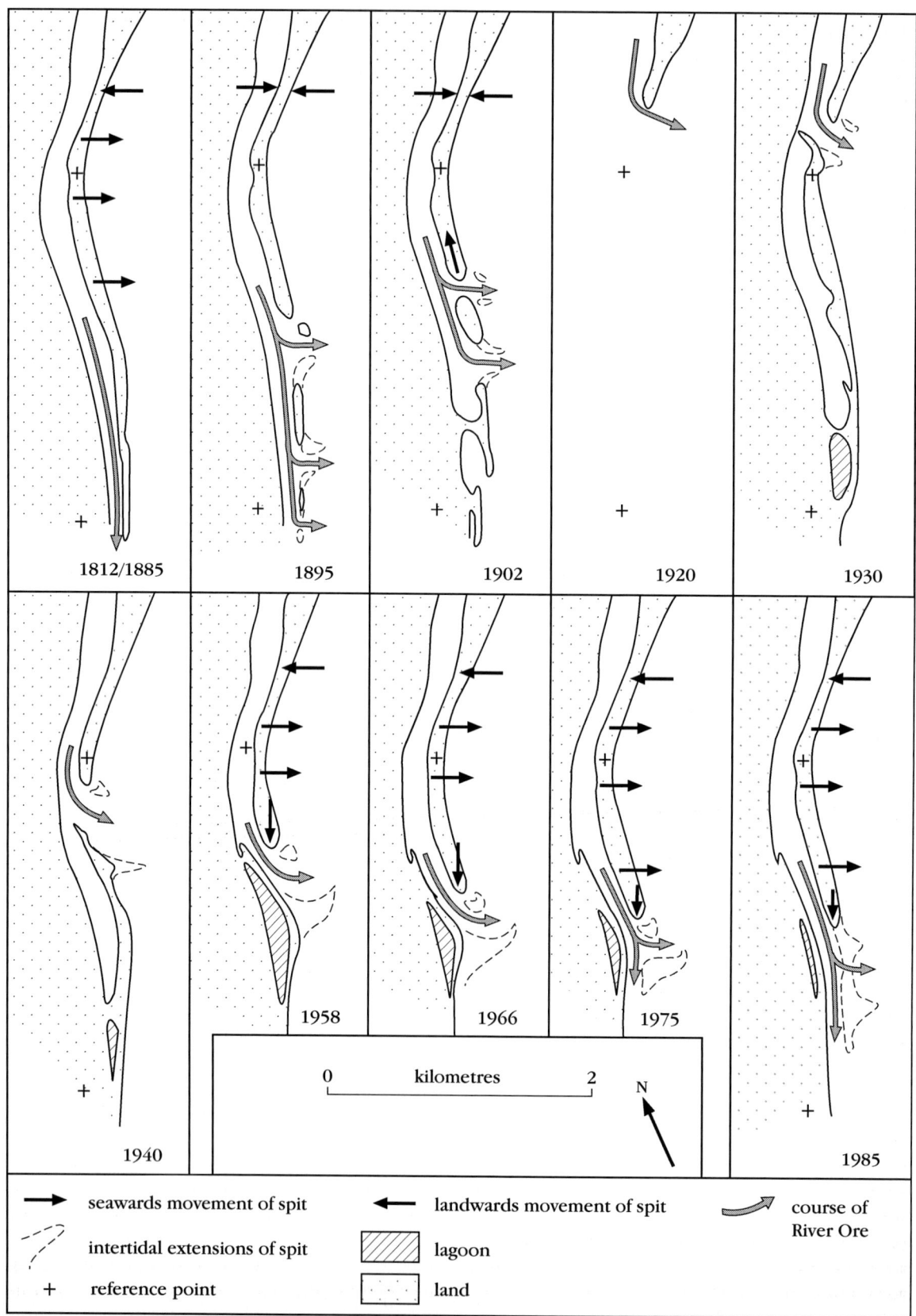

Figure 6.44 Historical changes in the position of distal features at Orfordness. (After various authors, mainly Carr, 1965; and Green and McGregor, 1988.)

radiocarbon dates from the Aldeburgh Marshes suggest some form of barrier may have been in existence to seaward as early as 6500 years BP. There is evidence for such a feature existing by about 3500 years BP.

Historical evidence suggests rapid growth of the spit towards the south-west in the later 16th century. Until 1800 AD, cartographic evidence is rather inconclusive, but since the 19th century, a widely fluctuating distal point (Figure 6.44) can be seen in maps. Over the period 1812 to 1921, fluctuations within the range of approximately 2900 m in total length took place, the maximum recorded southerly growth being at the beginning of that period, although a comparable length was also attained about 1892. It seems likely that similar fluctuation took place in the period immediately before accurate maps and charts were available, a view supported by Hodgkinson's map of 1783. The position of the distal point in 1980 was comparable to that of 1804 and 1902.

Interpretation

The stretch of coast that comprises the Orford shingle spit and the estuary of the River Ore has been the subject of extensive geomorphological research, especially during the period 1955–1970, and several new concepts or modifications of previous ideas about such environments have resulted (e.g. the 'counter-drift' concept of Kidson, 1963). In Britain, Carr has argued that it is the only remaining natural, dynamic and sustainable cuspate foreland, as well as being an outstanding example of a shingle spit.

Carr (1962, 1965, 1967, 1969b, 1970, 1971c, 1972, 1973, 1986) demonstrated the path of the shingle across the estuary, the absence of supply of material from offshore, the way in which new ridges may be melded onto earlier ones without obvious trace, and the inter-relation of spit, bar, banks and the Shingle Street features. The southward progression of the distal point results in the landward recession of the shoreline at Shingle Street. Nevertheless, each sequence of spit development has occurred in the same lateral position, probably due to the artificially constrained channels of the River Ore, which prevent landward migration of the position of the spit. In this respect, the spit differs from many other sites, such as Spurn Point, which have a history of spit breaching and lateral displacement.

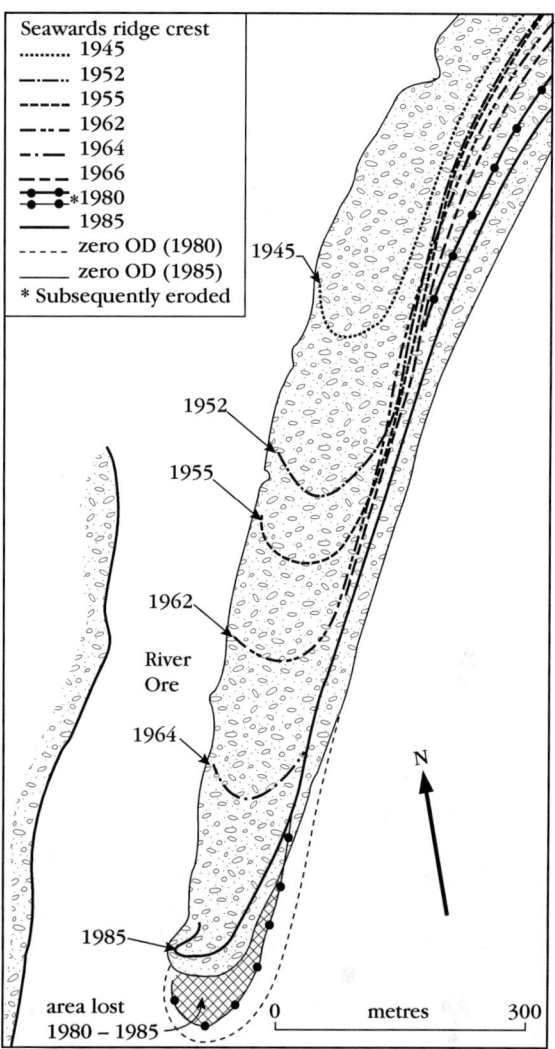

Figure 6.45 Historical distal changes at Orfordness. showing development of major ridge crests.

There appears to be no direct relationship between the extension of Orford spit and wave incidence, for instance, there is no correlation between annual southward growth and the prevalence of winds from a north-easterly quarter, a relationship that might have explained the longshore movement of shingle along the spit (Steers, 1926a). Carr (1986) suggested that the changing position of the distal point followed a cyclic pattern of development (Figure 6.44). It is not possible to confirm this model because of a lack of early records and a gap in the recent record. Carr suggests that the last breach occurred in 1920, but there is no mention of this in Steers' account. Carr's model would predict a rapid southward extension of the spit in the near future followed by breaching and an equally rapid reduction in the spit.

Carr (1972) also explains the periodic recession of the spit. As the spit grows southwards, both its shoreline and that of Shingle Street become straighter, thus allowing material to leave the system more quickly, especially as the offshore banks are eliminated. Counter-drift (Kidson, 1963) would be unlikely and the spit would become thinner and more susceptible to breaching. The protection of the offshore Whiting Bank would also be reduced as the spit extends southwards, the offshore zone would become steeper and waves would affect the spit from a greater range of directions. There would be a greater likelihood that the river would be blocked at its mouth and also that accretion in the river would become more rapid. The lengthening of the spit reduces the time for river discharge during periods of higher runoff, and increases both the hydraulic gradient between the two sides of the spit and the time-lag between high or low tide in the river and on the seaward side of the spit. During surges, there could be greater susceptibility to seepage and overtopping, and thus a greater likelihood of breaching of the spit.

Carr (1970) suggested development stages of the shingle, and Green and McGregor (1986) proposed a stratigraphical classification of the shingle. Each shingle ridge or group of ridges represents a stage in the development of the shingle complex, but not all stages are represented at Orfordness because of natural erosion and human activity. Green and McGregor (1986) argue that, in any area of Orfordness, an individual ridge or group of ridges can be classified in terms of the extent to which the stage that it represents is found elsewhere in the complex. The continuity and relationships of groups of shingle ridges can be traced using aerial photographs, and the shingle system can be divided into development stages and sub-stages. These are either individual ridges partly or entirely isolated in marshland, or groups of ridges separated from one another by a major erosional contact. In several cases, sub-stages consist of several groups of ridges (usually 5–10 ridges per group) with similar alignments, but separated from each other by erosional contacts. Green and McGregor (1986) believe that their interpretation confirms, and in places refines, the pattern of development stages proposed by Carr (1970).

Where successive stages are in direct contact with each other, the sequence and relative age of the ridges can be ascertained. Continuity of development is poorly represented in contrast to Dungeness, reflecting the limited extent of the shingle and its greater susceptibility to natural change. In particular, the limited seaward progradation of the ness itself over less than 1 km has allowed only 50 to 60 ridges to develop in contrast to more than 500 at Dungeness. Ridge continuity has been interrupted by natural erosion, usually as a consequence of the narrowness of the spit and its extreme elongation. The ness is also migrating towards the south. Green and McGregor (1986) have regarded overlapping groups of ridges in erosional contact as being separate development stages whenever their age relationships are uncertain. The small, frequently isolated, areas of shingle on the landward side of the estuary are also regarded as separate development stages.

Conclusions

Orfordness comprises three elements: a storm beach in the north, the ness itself, and a spit in the south. One of the largest and most important shingle structures on the British coast, it is an outstanding example of a shingle spit and shingle-spit cuspate foreland complex. Orford Beach and Shingle Street have the very greatest physiographical importance and interest. The international reputation of Orfordness rests primarily upon the history of scientific investigation and the existence of detailed records of its early growth and present-day dynamics.

DUNGENESS AND RYE HARBOUR, KENT AND EAST SUSSEX

V.J. May

Introduction

The large shingle cuspate foreland of Dungeness and the associated beaches at Rye (see Figure 6.2 for general location and Figure 6.46) provide a record of the development, partial destruction and reconstruction of a very large shingle barrier beach and spit system. Former shorelines, which are thought to have formed during approximately the last 5500 years, include both exposed shingle ridges and a large number of

Dungeness and Rye Harbour

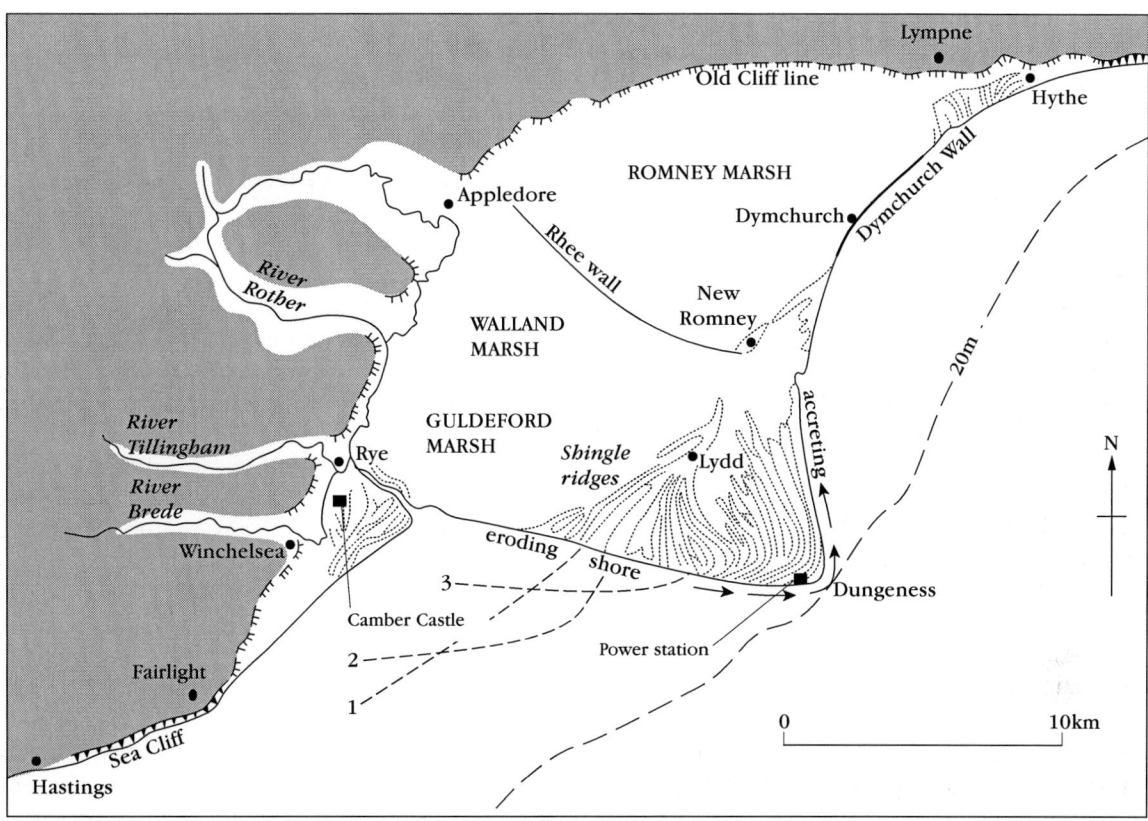

Figure 6.46 The cuspate foreland, Dungeness, Kent. The pecked lines 1 to 3 indicate former positions of the original spit over time, showing the downdrift extension of the spit across the bay. Saltmarsh has formed behind the outer shingle barrier. Over time, updrift erosion and downdrift deposition led to rotation of the feature from position 1 to 3. Land-claim of the marsh occurred in two phases – in the north it was drained in the Roman period, and in the 13th century diversion of the River Rother from its course north of Lydd to its new exit at Camber Castle led to the draining of the southern marshes. (After Bird, 1984, p. 159.)

buried ridges, Dungeness itself being especially noteworthy for its sequence of some 500 progressively younger, but increasingly cuspate eastwards, ridges. Bare shingle occurs at Rye Harbour and Camber, Dungeness and Hythe. At Rye Harbour lies the westernmost group of shingle beaches, which extend across the former Romney Marsh embayment between the Fairlight Hills east of Hastings and the former sea cliffs at Hythe. Unlike the beaches of Dungeness and Hythe, the development of the beach ridges at Rye Harbour has taken place mainly since the 16th century. At Camber Castle, the exposed shingle ridges generally post-date the mid-16th century (Lovegrove, 1953), whereas at the western end of the Lydd Ranges, they may be over 3500 years old (Eddison, 1983a,b). Between New Romney and Hythe (see Figure 6.46), a large number of ridges at high angles to the present-day shoreline were described by Elliott (1847) and have been detected in the distribution of the Beach Bank soil series and in the exposed shingle ridges west of Hythe. Both dunes and beach ridges are found at the foot of former sea-cliffs at west of Hythe, where there is considerable archaeological interest in the relationship between the Roman and Saxon forts at Lympne and the nature of a navigable inlet behind the shingle ridges (Shackley, 1981). Much of the human history of the marshlands has involved land-claim and drainage. The development of the marshes, land-claim and drainage within Romney Marsh (*sensu lato*) are discussed in Eddison (1995), Eddison and Green (1988) and Eddison *et al.* (1998).

Although the development of the features at Rye, Camber, Dungeness, Romney and Hythe are interrelated, they form separate physiographic units. Eddison (1983a,b) summarized the main phases of the evolution of the barrier beaches, expanding upon the detailed surveys carried out during the 1930s by Lewis (1932, 1937), by

Gravel and 'shingle' beaches

Lewis and Balchin (1940), and by Lovegrove (1953) for Camber, and the Soil Survey of England and Wales (Green, 1968) for Romney Marsh during the 1950s and 1960s. This wider view is necessary for an understanding of the development of Dungeness and Rye Harbour.

The idea that there had been a former continuous beach from Fairlight to Hythe was suggested during the early 19th century by Elliott (1847), but later Gulliver (1897) and then Lewis (1932, 1937) attempted to explain the gradual development of the barrier beach into the cuspate foreland of today. Longshore transport processes brought flint from the west where very large volumes may have been deposited during the Pleistocene Epoch on the floor of the English Channel to be carried towards the present-day shoreline by the waves of the rising Holocene sea (Eddison, 1983a). Some of this shingle was trapped in the Pevensey embayment, but much of it accumulated to the east of Hastings. As the western part of the barrier beach weakened, shingle was re-distributed to more eastern beaches. The beach is thought to have gradually changed its alignment towards the predominantly south-westerly up-Channel waves (Lewis, 1931).

Eddison (1983a,b) suggested that the earliest features, a submerged forest at Cliff End and the low-lying shingle deposits at Broomhill and Sandylands, represent an early barrier beach, probably dating from between 5500 and 4000 years BP. The barrier progressively extended towards Hythe by a series of recurves, identified by Elliott (1847). Thus with the shingle emplaced close to the present-day sea level, longshore transport gradually moved much of the shingle into the Romney embayment. The eastern end of the barrier beach extended towards, but never joined, the former cliffs around Lympne, so providing shelter for both Roman and Saxon vessels approaching the fort at Lympne, a site now about 3 km from the coastline.

Between Jury's Gap (TQ 993 180) and Dungeness itself, there are about 500 individual ridges in four main groups: their alignment changes by about 10° between each group. Eastwards from Galloway's Lookout (TR 045 172), the ridges are characterized by increasing curvature and the preservation of a ness form. Some of the western exposed shingle (for example, Jury's Beach, The Forelands and Holmstone) probably represents recurved sections of the early ness form. Near the power stations (Figure 6.46), gravels and sands were deposited for about 1900 years beginning at least 3270 years BP (Greensmith and Gutmanis, 1990). These deposits suggest that there was already a spit or ness feature here by the end of the British Roman period.

During a series of 13th century storms, the shingle ridges at Camber were destroyed. At this stage, the developing ness was isolated from the beaches to its west. Subsequently, shingle ridges grew north-eastwards from the area to the south of Winchelsea and westwards from Broomhill to enclose the wide estuary of the River Rother. The Camber Castle group of ridges fan out, changing their alignment through about 50°, whereas at Camber Sands, on the northern side of the River Rother, a series of narrow ridges, including short distal recurves, extended westwards from the shoreline at Broomhill (see Figure 6.47).

By the mid-17th century, it is probable that the present-day pattern of longshore sediment movement was established. Shingle moved towards the mouth of the Rother from Fairlight to its west and Broomhill to its east, from Broomhill along the southern shoreline of Dungeness to the ness itself and thence northwards to Greatstone, and from St Mary's Bay southwards towards Littlestone. At Hythe, shingle moved eastwards towards the Lympne inlet, as appears always to have been the case. Much of the barrier beach at Romney has been built upon and the ridges at Hythe retain little of their original form. The modern shoreline is reinforced by sea-walls at Winchelsea Beach, Broomhill, Littlestone and Dymchurch, with artificial beach feeding at Cliff End, Jury's Gap, the Dungeness power station and St Mary's Bay, with over 110 000 m^3 of material being added in 1979 (Eddison, 1983a), mostly by re-distribution from within the site. Nevertheless the southern shoreline continues to be eroded and shingle added to the eastern shoreline, the landward movement of the southern shoreline often taking place in stormy periods when it may move several tens of metres inland.

The present-day features of Dungeness and the associated beaches at Rye, Romney and Hythe thus result from changing sediment transport rates and deposition over some 5000 years, the different alignments of the shingle ridges and the buried beach ridges demonstrating gradual development from a barrier beach to the

Rye Harbour

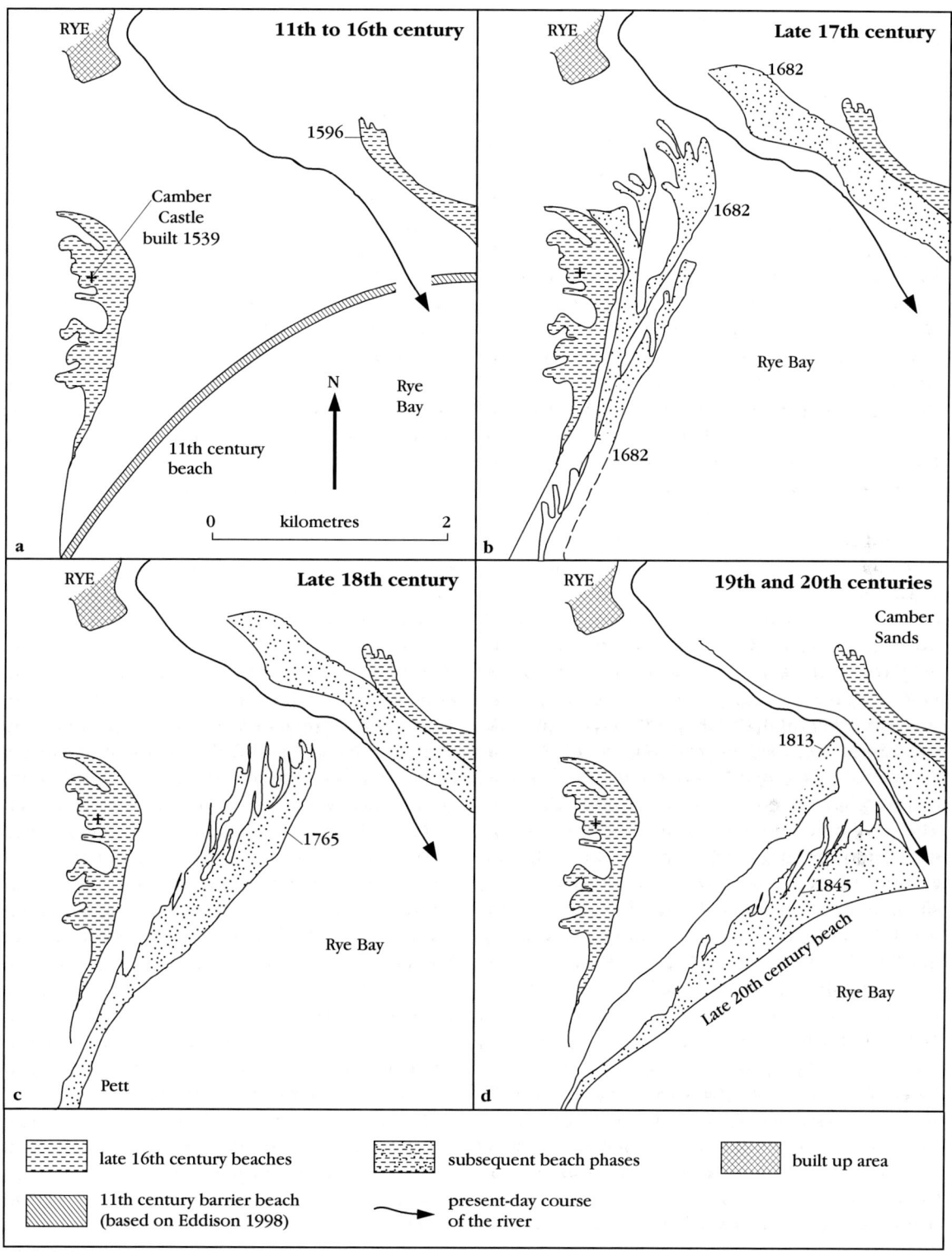

Figure 6.47 The historical evolution of Rye Bay. Dates indicate shoreline and beach area from contemporary maps and charts. (After Lovegrove, 1953; and Eddison, 1998.)

present artificially strengthened and nourished cuspate foreland.

RYE HARBOUR, EAST SUSSEX (TQ 935 180)

V.J. May

Introduction

Rye Harbour is the westernmost of the shingle beaches that extend across the former Romney Marsh embayment between the Fairlight Hills east of Hastings and the former sea cliffs at Hythe. The development of the beach ridges at Rye Harbour has taken place mainly since the 16th century. The shingle ridges at Camber are found in four main groups (Figure 6.47), the oldest at the site of Camber Castle (built in 1539), the most recent associated with the modification of the river mouth by training walls. In focusing upon the development of Dungeness, coastal geomorphologists have tended to overlook this site. Ward (1922) and Steers (1946a) provide general accounts, and its historical development was elucidated by Lovegrove (1953). It is a good example of the development of double spits at harbour mouths (see also GCR site reports for Pagham Harbour, South Haven Peninsula and Dawlish Warren in the present volume) but has never been recognized as such. This may be because its development has been overshadowed by greater emphasis on the larger features of the Dungeness complex. Parts of the site have been disturbed by gravel extraction and the longshore movement of shingle has been modified by coast protection works as well as management of the river mouth.

Description

In 1287, the former town of Winchelsea, which stood on a low shingle bank four or five kilometres south-east of the present-day town, was destroyed by storms. These storms also appear to have deflected the River Rother to its present-day course past Rye (Ward, 1922). Winchelsea was rebuilt on its present-day site, which was then accessible by sea. Shingle ridges grew north-eastwards from the area to the south of Winchelsea and by the mid-16th century provided the site for Camber Castle (Figure 6.47).

The Camber Castle group of ridges are mainly recurving distal features of a spit that gradually extended to the NNE. Most of the laterals end in narrow ridges, which trend from WSW to WNW and are truncated by the later storm ridges that form the seaward side of the group. The second group of ridges lie behind a seaward ridge, which trends to the north-east. Many of the laterals here run predominantly south–north before curving inland to the west. The third group, known as 'Nook Beach', are aligned more towards north-east. The final group of ridges is progressively aligned more and more with the present-day beach, which faces SSE. The ridges as a whole fan out from the area of Winchelsea Beach, with a change in orientation through about 50°. On the northern side of the River Rother, a series of narrow ridges, including short distal recurves extended westwards from the shoreline at Broomhill. These are now cloaked by dunes fed by a wide, intertidal, sand beach at Camber Sands.

The natural supply of shingle to the westward end of the beaches was always restricted and a narrow single ridge at the proximal end was frequently breached. There are many embankments in the area of Winchelsea Beach and seawards of Camber Castle, which indicate apparently successful attempts to prevent permanent breaching of this narrow neck. Most of the shoreline between Cliff End at the eastern foot of the Fairlight Hills and the western end of this site is protected by coast protection works. Artificial beach replenishment material taken from borrow pits within the relict shingle plains and ridges has been used at least since the 1950s on the beach at Pett (Thorn, 1960). Some of this shingle feeds into the modern beach at Rye Harbour.

Interpretation

The general interpretative context for this site is described in the previous section (p. 310).

This is a site in which the historical record is particularly important in demonstrating the evolution of a system of spits on opposite sides of a river estuary following breaching of a barrier beach. Because the historical record at the mouth of the River Rother is comparatively good, it is possible to recognize the growth of two beaches, terminating in recurved spits, into the estuary (Figure 6.47). By the end of the 16th century, according to the cartographic evidence examined by Lovegrove (1953), the Rother estuary was about 2 km across and was bounded by two well-developed spits. By the end of the 17th century, they had grown farther into the estuary narrowing it to about 400 m. Such long-term evidence is rare, and so this site is of considerable importance for helping to understand the process of coastal change over a timescale of several centuries. Much of this site has developed as a series of beach ridges that gradually alter their alignment towards the dominant waves as the beach builds into more exposed waters. However, as sediment has been transported alongshore with a diminishing supply from the west, the proximal end of these beaches has narrowed and from time to time breached. This is a recurrent feature of shingle spits, well exemplified here. Its development into a long spit parallel to the coast such as at Orfordness has been prevented by the location of the Rother, and it has not yet prograded sufficiently to form a new cuspate foreland such as Pevensey. It thus represents an intermediate stage in the development of shingle features. Equally importantly, it records very well the development of part of the Dungeness complex of barrier beaches during a period when the cuspate foreland at Dungeness itself was developing its most distinctive form.

Conclusions

Rye Harbour is important because:

1. It demonstrates the behaviour of part of the Romney Marsh barrier beach system as spits, in contrast to the development of the cuspate foreland at Dungeness. This is important because it helps improve our understanding of the longer-term development of cuspate forelands and barrier beaches. Both features show how beaches align themselves towards the dominant waves (Lewis, 1931). At Rye there was a restriction on the lengthening of the spit by the estuarine flow of the River Rother.
2. It complements the evidence of double spit formation, which has been examined elsewhere on the southern English coast (Robinson, 1955; Kidson, 1963).
3. It provides a good example of beach plain development where the dominant forms are large numbers of laterals, unlike eastern Dungeness where successive storm beaches form the beach plain.
4. It exemplifies the problem of the sediment supply to spits, especially gravel spits, where the longshore supply is restricted and breaching of the narrow neck at the landward end is a potentially frequent event.

DUNGENESS, KENT (TR 050 180)

V.J. May

Introduction

Dungeness is the largest cuspate foreland in Britain, and globally very unusual because it is formed predominantly of flint shingle. Beaches ridges date from about 5500 years BP and the best-preserved sequence can be traced from the 8th century AD. In addition to exposed shingle covering about 2158 ha, there are also buried shingle banks, which underlie a further 1150 ha. Other large shingle structures such as Chesil Beach, Spey Bay and Orfordness are comparable in terms of the length of coastline that they occupy, but they do not contain the enormous volume of shingle stored in the shingle ridges at Dungeness. The feature is often regarded as an integral part of a system of former barrier beaches that extend about 40 km from Fairlight in the west to Hythe in the east. Other well-known cuspate forelands, such as the Darss peninsula on the German Baltic coast, Cape Kennedy in Florida, Cabo Santa Maria on the Portuguese Algarve coast and Cabo Rojo on the Mexican coast, rival and exceed Dungeness for size, but Dungeness is unique globally because it has a number of features that are absent or less well developed elsewhere.

Dungeness is formed almost entirely of flint shingle and is a relatively advanced form of cuspate foreland, much of the shingle having been re-distributed from barrier beaches to form a ness with a particularly acute angle between its two main shorelines. It has long been recognized internationally as a major example of its type. For instance, as early as 1913, de Martonne described it as 'le type le plus connu: la pointe de Dungeness'. Standard texts from all parts of the world refer to Dungeness as the best-known example of a cuspate foreland (e.g. Holmes, 1944, 1965; Zenkovich, 1967; Bird, 1968, 1984; Paskoff, 1985).

No area inland of beaches to have been occupied and land-claimed over so long a period of

Gravel and 'shingle' beaches

time (about 1200 years) has been documented so intensively as Dungeness, and the documentary record extends over a far longer period than for any comparable site.

Finally, in contrast to many similar features, it lacks an offshore shoal that might extend its form seawards.

The Soil Survey of England and Wales (Green, 1968) has shown that shingle ridges often extend many hundreds of metres beyond the area of exposed shingle, the Beach Bank soil series representing the distal parts of successive beach ridges. Parts of the Lydd soil series also lie above shingle, while the Lydd series itself and parts of the Greatstone series are dominated by sand and loamy sand, which may be derived from sandy beaches associated with the shingle beaches in much the same way as sandy beaches are found today on the eastern shoreline of Dungeness. Recent archaeological and geomorphological studies have built on the work of the Soil Survey.

Large areas have been damaged by gravel extraction, vehicle tracks, military training areas and the construction of the Dungeness group of nuclear power stations. Detailed assessments of the damage have been made by Fuller (1985) and Green and McGregor (1986), both reports being drawn upon extensively in the assessment of this GCR site.

Description

The present-day shoreline at Dungeness is formed by a southern beach that faces SSW and is gradually moving north and inland over older relict beach ridges, the acute bend of the ness itself, which is migrating SSE, and an eastern beach, which has gradually migrated eastwards as the ness has grown. Much of this eastern beach is fronted by a wide intertidal sand beach. Landward of the present-day beach, there is a sequence of buried and exposed shingle ridges, which become both younger and more curved towards the east. Waller (1993, 1994) suggests that peat dating from 1100 to 2000 years BP helps date the development of both Romney Marsh and Dungeness. The oldest beaches, buried at Broomhill and Sandylands, have been tentatively dated between 5500 and 4000 years BP (Eddison, 1983a). Between Jury's Gap (TQ 993 180) and Dungeness itself, about 500 indi-

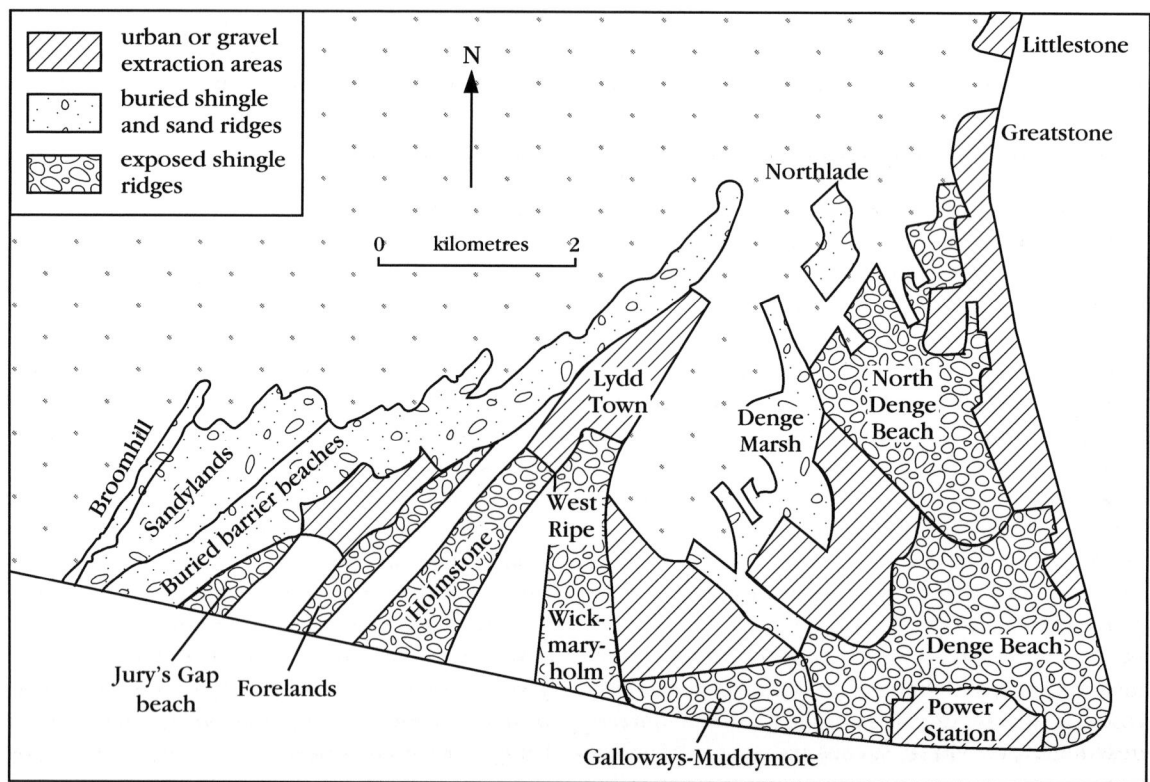

Figure 6.48 Major zones of shingle at Dungeness.

Dungeness

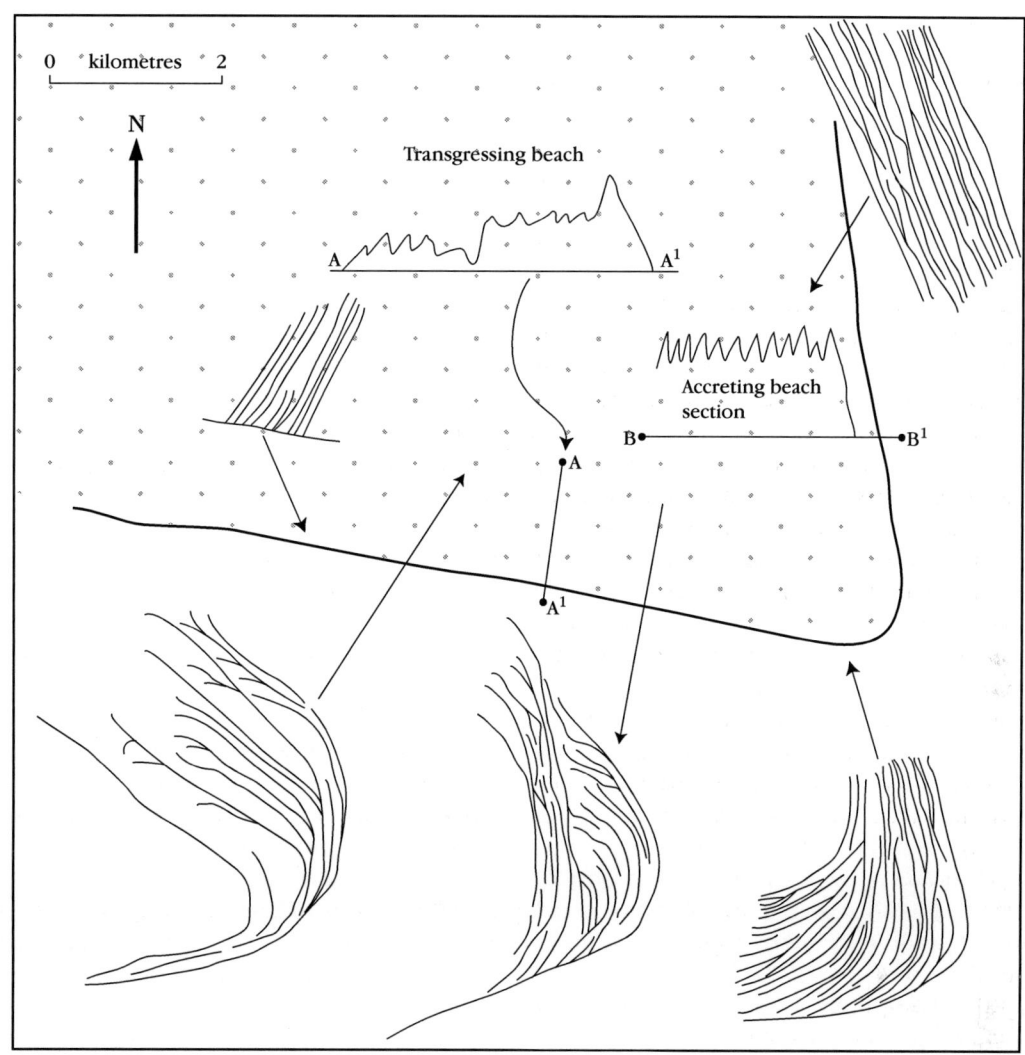

Figure 6.49 Schematic representation of the characteristic shingle ridge patterns and profiles at Dungeness. The vertical variation in ridge altitude is typically about 3m.

vidual ridges form four main groups, Jury's Gap beach, the Forelands, Holmstone and West Ripe–Wickmaryholm (Figure 6.48) with a change in ridge alignment of about 10° towards the north-west between each group. Eastwards of Galloway's Lookout (TR 045 172), the ridges become more curved and preserve earlier ness forms. At their seaward (southern) edge, the ridges are truncated by a single continuous ridge that forms the present-day shoreline. Inland, many of these curved ridges are buried, but represented by the Beach Bank soil series, which is described as associated with 'old inland pebble banks and beaches' (Green, 1968). The mapped distribution of these old beaches portrays them as having relatively straight seaward boundaries (Figure 6.49), whereas their landward edges are marked by lateral features at angles of between 30° and 40° to the main ridges, probably representing recurves in the distal part of the beaches. Only in the area around Open Pits (TR 073 180 and TR 074 185) are distal recurves completely preserved in the bare shingle. To the north of the Open Pits and the present-day ness, the shingle is found in over 100 sub-parallel south–north-trending ridges that extend northwards to very short, buried, distal forms. Although parts of this area have been disturbed by gravel extraction, the undisturbed parts remain the finest example of a gravel strandplain on the coastline of Britain, and have few rivals elsewhere in Europe.

The shingle is almost entirely (over 98%) rounded flint shingle (Steers, 1946a), with pebbles of cherty sandstones derived from the Upper Greensand, fine-grained sandstones from the Hastings Beds, and red, grey and liver-coloured quartzites, including examples of the Triassic Budleigh Salterton Pebble Beds. Some of the latter may have been brought here as ballast. The Soil Survey (Green, 1968) recorded, unusually, subangular gravels as well as rounded pebbles at Northlade where a large distal complex is buried north of Lydd Airport. The material forming the ridges (known locally as 'fulls') is often smaller in size (down to 8 mm) than in the troughs between them ('lows' or 'swales' (Lewis and Balchin, 1940). Green and McGregor (1986) noted concentrations of coarser material (up to 150 mm maximum diameter). There is no published explanation of this phenomenon.

Lewis and Balchin (1940) showed that there are considerable differences in the heights of the ridges, which vary between 3.7 m and 6.3 m OD, and that there are differences between groups of ridges (Figure 6.49). In contrast the buried ridges are at much lower altitudes, for example, ridges west of Hythe reach between +0.6 m and −1.0 m OD, at St Mary's Bay +1.0 m OD and at Broomhill +1.5 m OD.

Individual ridges are normally defined by a ridge–swale relief of between 0.5 m and 2.0 m. A ridge frequency of 60 to 100 per mile (32–62 per km) is characteristic (Lewis and Balchin, 1940; Eddison, 1983a). Typical ridge widths are estimated at between 16 m and 28 m (Green and McGregor, 1986). Relief frequently deepens towards the edge of individual ridge systems, particularly towards the distal ends of ridges, which feather out into alluvial areas and where ridge-swale relief may reach 3 m (Green and McGregor, 1986). In some parts of the area, individual ridges are rarely continuous over long distances (Lewis and Balchin, 1940). Parasitic ridges are common, inter-ridge recurves also occur in many sectors, and natural pits have often formed where shingle-branching or recurve patterns give rise to enclosed hollows. Irregular, anastomosing ridge patterns also occur (Green and McGregor, 1986). These features increase in frequency towards the distal end of individual ridge systems. Natural pits whose formation is encouraged by the deepening of relief often observed in such situations contain an infilling of fine-grained alluvium varying in thickness from 0.5 m in the smaller pits to 3.5 m in the larger ones. The basal alluvium (0.3 m) contains evidence of marine conditions, but there is a transition upwards into freshwater deposits (Waters, 1985). Green and McGregor (1986) examined the ridges in more detail in a sample area of 0.3 km, concluding that ridge length may be bimodal. Long continuous ridges are separated by one or more subsidiary ridges of not more than a few hundred metres in length. Ridge–swale relief varied between 0.5 m and 1.75 m, but ridge crest elevation, even on the longer ridges, rarely varied by more than 0.5 m.

The most detailed accounts of the sedimentology of the Dungeness shingle are by Hey (1967) and Greensmith and Gutmanis (1990). Borehole data, the lack of mention of sub-shingle conditions in the literature and the trenched exposures observed by Hey all indicate that the shingle in the vicinity of the power station commonly varies in thickness between 3 m and 7 m, has a locally irregular base, and generally overlies sand. The sub-shingle surface was almost a plane surface with some irregularities caused by shallow channels and it fell by about 2 m from north to south in the excavations (Hey, 1967). The shingle was in beds from 7.5 cm to 75 cm in thickness, with beds sharply defined by changes of average particle size. Average bed thickness was about 15 cm and the beds dipped uniformly to the south-east at an average angle of 8° to 10°. The strike of the bedding planes was exactly parallel to the alignment of the shingle ridges in the immediate vicinity. Average particle diameters for individual beds varied between 8 mm and 40 mm (Hey, 1967). Much of the shingle includes impermeable beds of sand which give rise to a locally important freshwater aquifer.

The Soil Survey of England and Wales (Green, 1968) has shown that shingle ridges often extend many hundreds of metres beyond the exposed shingle, the Beach Bank soil series representing the distal parts of successive beach ridges. Parts of the Lydd soil series also lie above shingle, while the Lydd series itself and parts of the Greatstone series are dominated by sand and loamy sand, which may be derived from sandy beaches associated with the shingle beaches in much the same way as sandy beaches are found today on the eastern shoreline of Dungeness. The ridges often display outlines that indicate clearly patterns of primary depositional morphology (Green and McGregor, 1986). There is a broader tract of buried shingle on the western side of the GCR area, where both aerial photographs and Green's (1968) soil mapping indicate patterns of primary depositional morphology, similar to the exposed shingle. To the northeast of Broomhill, the shingle is replaced by the Midley Sand, but an outlying buried ridge close to the western boundary of the GCR area forms the most westerly identifiable element of the Dungeness shingle system. The buried shingle at Scotney Court and north of Lydd varies in thickness from 2 m to over 5 m increasing in thickness northwards. The shingle increases in thickness towards the south and attains depths in excess of 15 m in the area of the power station. Some boreholes (Green, 1968) reported 'sand with gravel', which continued for another 12 m. A series of mounds in Green's Newer Marsh between Lydd and New Romney have been identified by Vollans (1995) as accumulated remains of the spent sediments cleaned out from filter pits or troughs used in 11th century salt-making. At Belgar, one such ridge extends for almost 2 km in front of the distal end of the Lydd spit.

Dating of the shingle relies on cartographic sources and organic deposits; very few dates have been measured for the areas outside the exposed shingle. Tooley and Switsur (1988) date a marsh infilling of a shingle low at 3410±60 years BP. Peat overlying gravel at Broomhill has been dated at about 3600 years BP and in Scotney Marsh at around about 4000 years BP. Within the shingle-bank complex near Scotney Court (TR 023 202), Callow et al. (1966) dated in-situ woody roots lying above shingle at 2740 ± 400 years BP: this occurred beneath silty clay loam overlain by peat. This places the Early Barrier Beach at a date earlier than between 3100 and 2300 years BP.

The development of the foreland has been described (Lewis and Balchin, 1940; Eddison, 1983a,b) by using the trend patterns of the shingle ridges, assuming that each ridge is a former storm beach and so represents a former position of the shoreline. The rate of progradation of Dungeness has been estimated at between 4.1 m a^{-1} and 5.5 m a^{-1}, both Redman (1852) and Hey (1967) estimating the higher value for the period from the early 17th century to the early 19th century. The lower value was estimated for the period 1878 to 1938 by Swallow (Lewis and Balchin, 1940). The morphological patterns of the shingle indicate different modes of deposition associated with different positions in the coastal system at the time of deposition and/or variations in the rate or direction of progradation (Eddison, 1983a,b). Within the area known as 'Denge Beach' (Figure 6.48), the full sequence of ness forms, from their early development to the present-day, is preserved. Few areas anywhere preserve such a complete sequence of beach ridges known to have been formed over at least 2000 years. Many of the ridges can be traced almost without break from this area northwards to their distal ends around Lydd Airport. North of the Dungeness power station, Hey (1967) reported thicknesses of about 10 m, whereas the Soil Survey noted that as much as 17 m depth of shingle had been found in the vicinity of the power station. The clean shingle forms only the upper 1–1.5 m (Hey, 1967), being composed beneath this depth of 'closely-packed pebbles with interstices filled with sand'. Sandy deposits beneath the shingle contained a few scattered pebbles and some marine shells. The sandy gravel is described by Hey as being in beds of between 0.1 m and 1 m in thickness. Each bed was composed of similar mixed material, but each bedding-plane was marked by a distinct change in pebble diameter. The bedding planes had a constant dip of 8°–10° towards SSE, the strike being almost the same as the alignment of the surface ridges. Greensmith and Gutmanis (1990) show, following analysis of 80 boreholes in the vicinity of the power stations, that the 40 m-thick marine Holocene succession can be divided into basal gravels, middle sands and upper gravels that rest directly on a pre-Holocene erosion surface cut across the Lower Cretaceous Hastings Beds between −32 and −35 m OD.

Dix et al. (1998) show that high resolution seismic (Chirp) surveys in Rye Bay indicate a dominant seaward-prograding shelf sand body (SSB) with only minor amounts of gravel. The presence of buried gravel beaches at Broomhill dating from the mid-Holocene (Tooley and Switsur, 1988; Long and Innes, 1995b) and studies of drowned Holocene barriers elsewhere (e.g. Forbes and Boyd, 1987; Oldale, 1985; Browne, 1994; Forbes et al., 1995) pointed to the possible preservation of early Holocene barrier structures in Rye Bay. The bedrock surface undulates between −25 m and −35 m OD (Lake and Shepherd-Thorn, 1987; Greensmith and Gutmanis, 1990; Long *et al.*, 1996). NW–SE-trending channels with maximum depths of *c.* –45 m OD may be offshore extensions of the former valleys of the Rother, Tillingham, Brede and Pannel (Dix *et al.*, 1998).

At Dungeness point, Greensmith and Gutmanis (1990) and Basa *et al.* (1997) describe a basal gravel (0.5 to 1.0 m thick) overlain by very fine- to fine-grained, moderately well-sorted sands (20–30 m). Their upper surface is channelled. These 'Middle Sands' are capped by gravel up to 5 m in thickness. At the Open Pits (TR 073 180 and TR 074 185), the ridges are very short and are truncated by a single south–north ridge suggesting that spit extension was more important here than the formation of individual storm beach ridges. There is no other part of the Dungeness beach-complex where distal features occur other than at the landward end of the very long linear ridges (Fuller, 1985; Green and McGregor, 1986).

The area of North Denge Beach broadly occupies the area between the former Southern Railway line from Lydd to New Romney, Lydd Airport and the residential buildings along the coast. It is a fine example of a shingle beach-plain, comprising over 100 sub-parallel shingle ridges, which run northwards to end in very short buried distal features. Some parts of this landform have been excavated for gravel, but a complete set of the ridges straddles the track from the Water Tower (TR 068 202) to Lade (TR 083 208), with only the most recent ridges being obscured by housing and the coastal road. Most of the ridges post-date the mid-8th century shoreline postulated by Lewis and Balchin (1940), the distal features south of Greatstone having marked historically the south side of the gradually silted and reclaimed area known as 'Romney Sands'.

Interpretation

The general interpretative context for this site is described above in the previous section (p. 310 ff).

There are three major issues to be addressed at Dungeness: the description and interpretation of the pattern of shingle ridges, the age of the features and their relationship to the development of the beaches, and the relationship between marsh sediments and the shingle structures.

A summary of the phases of development of Dungeness is presented in Table 6.4.

The earliest discussion of the formation of Dungeness (Elliott, 1847; Gulliver, 1897; and Lewis, 1932, 1937) regarded the ness as having evolved from barrier beaches crossing Rye Bay. These beaches were regarded as having aligned towards the dominant south-west waves and grown by redistribution of sediment from proximal and seaward areas to the recent locations straddling the mouth of the River Rother. Lewis's (1931) conceptual model has provided the basis for the early evolution of the beaches. However, Dix *et al.* (1998) argued that there is little evidence offshore to support Lewis' view.

The western shingle structures described above appear to represent the barrier spit extending towards Lydd and Hythe. The change in growth direction of the ness towards the south-east has not been explained adequately, and some of the western exposed shingle (for example, Jury's Beach, The Forelands and Holmstone) probably represents recurved sections of the early ness form. Green and McGregor (1986) show that these shingle areas are separated by alluvium, which often attains depths of more than 2 m within 10 m of the shingle margin. They consider that these areas of alluvium imply rapid eastwards growth of the ness, which would not have allowed sufficient time for closely spaced recurves to develop. The former southern shore of the ness is first identifiable where it is intersected by the modern southern shoreline about 1 km east of the Galloways Lookout. Northern parts of the sharply curved shingle ridges represent proximal areas of recurves, with, in several places, deep natural pits separating the curved ridges at the point of greatest inflection. This probably indicates shorter periods of rapid eastward growth of the ness (Green and McGregor, 1986), but it may also reflect reduced supplies of shingle from the west and reduced wave energy inputs to the southern shore. Most of the northern extremities of the recurves forming Denge Beach appear to have ended in deep water. The distal parts of the recurves gradually changed alignment towards the north from an early orientation of about 310° to 320° to a modern beach alignment of about 340° to 350° (Figure 6.50). The Holocene sediments in the vicinity of the power stations are consistent with a prograding, upwards-coarsening, barred shoreline, laid down under mixed wave-tidal conditions and predominantly rising sea levels (Greensmith

Dungeness

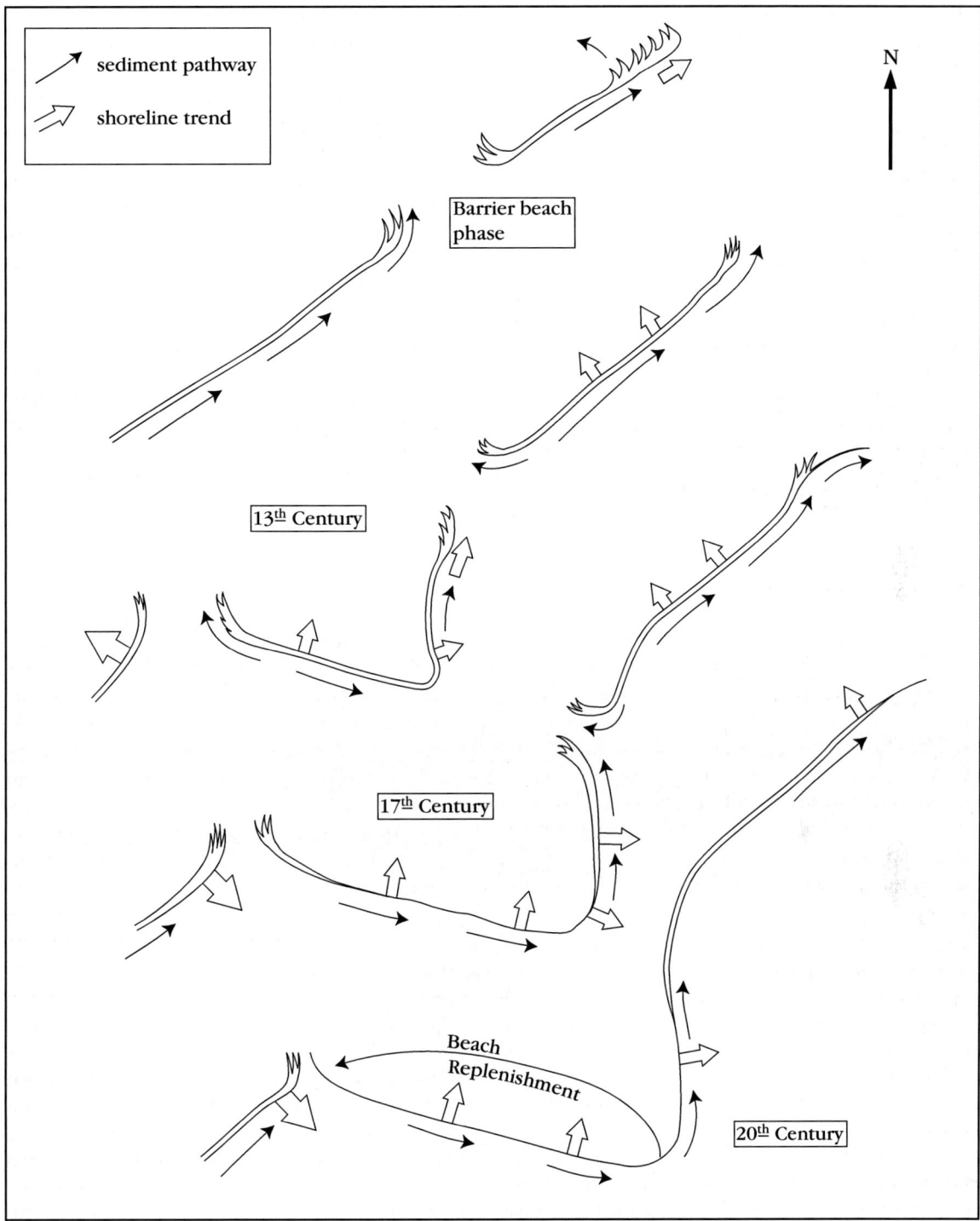

Figure 6.50 Historical sediment pathways and development at Dungeness. Each schematic map shows the probable sediment movements associated with the erosional and accretional trends in the shoreline.

and Gutmanis, 1990). The basal gravels and middle sands were deposited over a period at least 1900 years, a process that began at least 3270 years BP. Radiocarbon dates (1370–3270 years BP) from levels between –32 and –34 m OD are regarded as anomalously young and interpreted as arising from intertidal and high subtidal shells being swept, probably during prolonged stormy period, into depths greater than 20 m.

Gravel and 'shingle' beaches

Table 6.4 Development phases at Dungeness. Ridge height data are mainly from Lewis and Balchin (1940).

Phase		Preserved as	Shingle ridge height (m OD)
1	Low barrier beach associated with Midley Sands, stretching from Fairlight to St Mary's Bay and thence to Hythe. Dating uncertain but placed between 5500 and 4000 years BP by Eddison (1983a)	(i) Broomhill and Sandylands (ii) Recurves at St Mary's Bay (iii) Low-level shingle at West Hythe	Max = +1.5 Max = +1.0 +0.6 to –1.0
2	Higher level barrier system, dated c. 3000 years BP Overlain in parts by peat dated c. 2700 years BP	(i) Shingle ridges at Jury's Gap and the Wicks, and Beach Bank Soil Series west and north of Lydd (ii) Shingle recurves at Hythe	Average = +4.11 Max = +5.00 +2.8 to +3.5
3	Slightly higher beaches, younger than peat. Dated c. 2000 years BP	(i) Holmstone Beach and its extensions as Beach Bank Soil Series west of Lydd (ii) Recurves at Hythe	Average = +4.31 No published data
4	(a) Ness development with eastern shore trending south-east–north-west to Lydd	(i) Wickmaryholm eastwards to Muddymore Pit	Average (west of Galloways) = +4.69 Average (east of Galloways) = +3.81
	(b) Barrier beach with spit and recurve development to north and south	(i) New Romney (ii) Recurves at Hythe	No published data No published data
5	(a) Ness development with long NW-trending ridges. Eastern limit dated at about 750 AD.	(i) Areas south and west of Open Pits (ii) Beach Bank series in Denge Marsh	Max = +6.28
	(b) Land-claim	(i) Areas mainly around Lydd within embankments (ii) Open Pits	
6	Spit extension and recurves	(i) Open Pits	
7	(a) Ness and beach plain to distal recurves	(i) Denge Beach to Northlade (by c. 1250 AD) (ii) Greatstone Point (by c. 1800 AD)	+4.5 to +6.0 Average = +5.33
	(b) Dune development	(i) Romney Warren (ii) Camber	
	(c) Spit development	(i) Littlestone Point (ii) Broomhill Farm, Hythe	
	(d) Land claim	(i) West of Lydd (ii) Caldecot–Belgar area (iii) Romney Hoy	
	(e) Beach ridges associated with longshore drift	(i) Camber and Rye Harbour (ii) Romney Hoy: Littlestone and Greatstone Points (iii) Hythe Ranges	
8	(a) Modern sea-wall construction	(i) Dymchurch Wall is earliest example	
	(b) Beach-feeding	(i) Broomhill, (ii) Pett (iii) Power Station (iv) St Mary's Bay	

The processes of longshore transport at Dungeness have been modified, first by a system of beach replenishment and second by coastal protection structures defending the power station site. The replenishment programme, where shingle near the ness is returned to the western end of the beach near Brommhill, has been operating since the 1950s (Thorn, 1960) and is one of the longest running schemes anywhere. Figure 6.50 offers an interpretation of the probable sediment pathways both now and at earlier stages of beach development. They warrant further investigation to evaluate the effects of wave climate, storm events, different sea levels, and changes in sediment supply.

The preservation of so many beach ridges has tempted a number of writers (Gilbert, 1930; Lewis, 1932; Lewis and Balchin, 1940) to invoke changes in sea level as the cause of their varying height. Surveys of ridge altitudes by Plater and Long (1995) largely confirm the variations in altitude reported by Lewis and Balchin (1940). Plater and Long do not, however, agree with the earlier interpretation. They recognized that the altitude of the ness could be as much as 1.2 m below both the adjacent west–east (proximal) ridges and the south-east–north-west (distal) ridges. The latter also fell towards their north-western ends. Plater and Long (1995) observed an overall rise in ridge-swale altitude of about

Dungeness

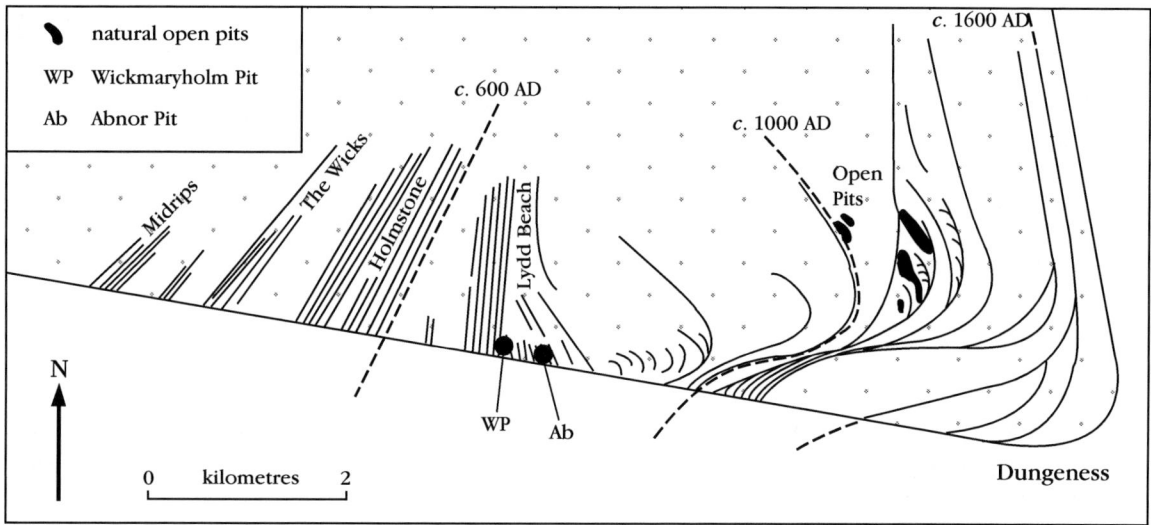

Figure 6.51 Eastward development of Dungeness. The orientation of the beach ridges change and the ness forms is preserved in ridges dated between 600 AD and 1000 AD. The natural 'Open Pits' are areas of naturally lower and enclosed land that is seasonally or in some cases permanently freshwater. (Based on Steers, 1946a and Eddison, 1983b.)

1.5 m, between Galloway's Lookout and Denge Marsh Sewer, which they explain as a function of sampling location rather than real altitudinal change. Having taken measurements from a consistent point beyond the ness of the mapped ridges, they found an overall rise in ridge height of about 1 m (from c. 4.0 m on the Roman shoreline to 5.0 m OD on the AD 750 shoreline of Lewis and Balchin (1940)). In the central part of their transect there is evidence for a fall from c. 4.5 m to 3.9 m OD followed by a rise to c. 5.1 m OD. Plater and Long (1995) emphasized that because shingle ridge morphology and sedimentation are controlled by a number of interdependent variables (Carter *et al.*, 1989; Jennings and Smyth, 1990), temporal variation in any single parameter is unlikely to explain the altitudinal trends.

Although the roughly 1 m increase from west to east in ridge altitude in Denge Marsh may be interpreted as related to sea-level rise and storm event magnitude between the Roman period and the mid-8th century, along-profile morphology accounts for much of the variability in ridge altitude (Plater and Long, 1995). Their stratigraphical, magnetic and diatom evidence indicates relatively uniform and widespread phase of marsh sedimentation. They propose that sedimentation took place on a surface extending from lower marsh to intertidal mudflat. Coarser laminations resulted from increased wave energy or velocities of tidal flow. At Galloway's Lookout–Greenwall and Brickwall Farm, marsh sedimentation was preceded by shingle emplacement, but later phases of ridge construction took place towards the end of the sedimentation phase. A high sediment supply from the Romney Marsh catchment during the mid-to late-Holocene provided much of the marsh sediments (Plater and Long, 1995). The intertidal flat then provided a surface upon which subsequent ness development could occur. The shingle-marsh interface in Denge Marsh appears to have moved eastward with the prograding shingle foreland as a series of advancing depositional environments (Plater and Long, 1995). Comparison of the altitude of the marsh surface and the base of a mottled facies with present-day mean high-water springs (MHWS) indicates that these sedimentary markers were close to MHWS about the times of the AD 774 charter and the great storms of AD 1287–1288. The uppermost mottled facies may have been deposited by the 13th century storms onto 8th century marsh surfaces (Plater and Long, 1995). This, according to them, contradicts the north-easterly younging trends at Brickwall Farm and Denge Marsh.

Thus long-term sea-level rise may have driven progressive tidal sedimentation (Plater and Long, 1995), taking advantage of storm-induced recurve emplacement and consequent back-barrier deposition. The intertidal flat seaward of the ness provided a base for this recurve formation. Others (Hey, 1967; Greensmith and Gutmanis, 1990; Plater, 1992; Long and Hughes, 1995) investigated the frequency and patterns of the ridges in terms of sedimentation rates and storminess. Shingle deposition is influenced by sediment composition and supply, prevailing wave climate and tidal dynamics, basement controls and inheritance controls such as influence of headlands on wave refraction and the need for back-barrier lagoon drainage (Lewis and Balchin, 1940; Carr and Blackley, 1973; Carter and Orford, 1981, 1993; Carter *et al.*, 1987, 1989; Jennings and Smyth, 1990; Orford *et al.*, 1991). In contrast, ridge morphology is controlled mainly by storm event magnitude and frequency (King, 1973; Orford *et al.*, 1991). Ridge orientation is largely affected by sediment budget and transfers alongshore and wave climate (Lewis, 1931, 1933).

Marsh accretion results from incremental deposition of tidal lag sediments (settling at high tide) (Pethick, 1981; Allen, 1990a). Development of the marsh-shingle complex thus depends on processes at different ends of magnitude–frequency scales. Long and Hughes (1995), for example, argue that alternating gravel and marsh sediments result from changes in storm incidence and rates of gravel supply. Plater (1992) suggests that argillaceous and arenaceous sediments above buried shingle ridges in Denge Marsh result from storm breaching of the shingle complex. Most ridges (in their final form) are the result of major storms and so may indicate changing patterns of storminess, but changes in sea level have undoubtedly also been involved. Wass (1995) considers, on the basis of an investigation of sediments and microfauna, that the channel mapped by Green (1968) was a sheltered arm of a tidal inlet in which low-energy conditions prevailed. He concludes that this is inconsistent with the Rother (or any major distributary) crossing the northern part of Romney Marsh since the peat formed there about 3000 years BP. Plater and Long (1995) coupled stratigraphical investigation of Denge Marsh with diatom, mineral magnetic and radionuclide analyses to attempt to establish a chronology of marsh development, the nature of the palaeoenvironments and the primary sediment sources. Spencer *et al.* (1998a) utilized 3400 boreholes and pollen, diatom and radiocarbon dating to interpret the sedimentary record of Walland Marsh. Gravel lies beneath much of Scotney Marsh, and peat directly above the gravel accumulated between *c.* 3900 and 2400 years BP.

Boreholes near Rye show a pronounced coarsening-upwards sequence between –12 and –4 m OD, which pre-dates the main marsh peat in Walland and Romney Marsh, which formed after 6000 years BP (Long *et al.*, 1996, 1998). Long *et al.* (1996) propose three hypotheses for this coarsening-upwards sequence:

1. a rapid rise in relative sea level;
2. landward migration of a coastal barrier or dune;
3. initiation of large-scale sand movement from the west after the opening of the Strait of Dover.

Dix *et al.* (1998) argue that the Rye Bay sand body has many similar characteristics to shelf sand bodies (SSBs) of south-east Australia (Roy *et al.*, 1994). Rye Bay lies in a similar high-energy environment, has a steeper (more than 1°) shoreface and was affected by stable relative sea-level rise and may have had large-scale sand transport: all features of SSBs. Dix *et al.* (1998) argue that there is no evidence from their Chirp survey to suggest landward-migrating barriers in the early Holocene Rye Bay. Rather the evidence points to seaward progradation, with gravel largely absent. Thus Lewis' (1932) former extrapolated positions of the shoreline may not have existed. Dix *et al.* (1998) argue that any barriers probably existed much closer to the present-day shoreline. They identify a need for further work on the processes that allowed SSB progradation during a marine transgression and in-situ examination of the buried intertidal and subtidal stratigraphy of Rye Bay. They consider that an early Holocene complex could have been reworked by relative sea-level rise during the mid- and late Holocene and that this stopped close to the upper depositional surface of the underlying sandy body. A second possibility is that Rye Bay was too deep to allow early Holocene inter- and supertidal gravel deposits to accumulate. Such barriers would only form much closer to shore and rapid SSB progradation occurred with the slowing of relative sea-level rise after about 6000 years BP. A third hypothesis suggests that the early Holocene barrier was sand not gravel and that this is represented by the Midley Sand. Long and Innes (1993) have shown, however, that the Midley Sand is one of the youngest elements of the stratigraphical sequence of the marsh. Any such sand ridge would also have to be rapidly reworked to account for the absence of landward-inclined reflections in the Chirp profiles.

Evidence of a linkage between barrier formation and early marsh deposition is provided by the infilling of a shingle low at Broomhill (dated 3410 ± 60 ^{14}C years BP (Tooley and Switsur, 1988) and by recent stratigraphical evidence from Midley (Long and Innes, 1993; Plater and Long, 1995).

Litho-, bio- and chrono-stratigraphical investigation of the Midley Church bank (Innes and Long, 1992; Long and Innes, 1995a) show that the near-surface and surface outcrop of sand (Green's (1968) 'Midley Sand') must have accumulated after deposition of the lower sand and the younger marsh sediments. Peat began to accumulate beneath the Midley Church bank as marine influence declined from about 3700 years BP until about 2700 years BP, after which there was a gradual return to marine conditions; peat accumulation had ceased by *c.* 2200 years BP. Cereal-type pollen, other herbs and ruderal pollen types within the peat may indicate local Bronze Age farming. Long and Innes (1995a) suggest that the Midley Church bank was either an aeolian deposit or a water-lain sandbank (possibly within a former course of the Rother).

According to Eddison's (1983a,b) model, the first of the high-level shingle ridges were emplaced by about 3000 years BP with approximately 500 ridges over 10 km deposited owing to the co-incidence of storms and high tides. Discrete populations of ridges with similar orientation can be identified within the Dungeness complex. These might be the result of extended periods of high storm-frequency rather than individual events. Greensmith and Gutmanis (1990) also note this phase of shingle deposition on the seaward flanks of the ephemeral Midley Sound barrier complex from *c.* 3400 years BP. This could be linked to a more regional scale control on shingle deposition via changes in wave climate between 5000–300 years BP (Jennings and Smyth, 1990). Eddison (1983b) implies progressive or pulsed development, other authors have proposed much shorter periods of time for upper shoreface and storm beach deposits near the ness (i.e. 750 years – Greensmith and Gutmanis, 1990, and *c.* 350 years – Hey, 1967).

The more recent (eastern) ridges can be dated from cartographic evidence, but the accuracy and precision of the chronology of the western (older) ridges is more problematical (Plater and Long, 1995). Although documentary and archaeological evidence provides reasonable indications of age, linking this to particular ridges or groups is also problematical, the broad similarity between sediments of Denge Marsh (Long and Fox, 1988; Plater, 1992) and the 'post-peat' deposits of Romney and Walland marshes (Green, 1968; Burrin, 1988; Waller *et al.*, 1988) suggests that marsh sedimentation may have taken place in the lee of the shingle foreland following the phase of peat deposition in much of Romney Marsh which culminated about 2000 years BP. Brooks (1988) suggests that Denge Marsh (which lies entirely with Green's (1968) 'New Marshland') was emplaced by Saxon times. However, Bronze Age axes in shingle north of Lydd (Needham, 1988) and evidence of Roman occupation of shingle west of Lydd (Cunliffe, 1988: Green, 1988) imply earlier marsh deposition. An alternative view (Cunliffe, 1980, Lamb in Eddison, 1983b) is that the most recent sediments of Denge Marsh were deposited during a series of storms in the 13th century. This largely confirms the view of Lewis and Balchin (1940) and Steers (1946a) that the marsh at Denge was largely deposited around AD 744 (Table 6.4).

Understanding of the development of the Dungeness foreland over time has a very practical application today. At the Public Inquiry held in 1958 into the proposed siting of a nuclear power station, it was pointed out that the new construction would be on an eroding shore. In spite of this the construction proceeded and, together with subsequent development, now requires to be protected from frontal erosion of the beach by the annual addition of up to 30 000 m^3 of gravel (Summers, 1985). The gravel is sourced from the accreting east side of Dungeness and transported artifically to nourish the south side where the reactors are sited (Figure 6.46).

Conclusions

Dungeness is a large, complex and geomorphologically important site, first because of the shingle ridges, and second for the shingle foreland. Beach ridges such as those found at Dungeness are not confined only to cuspate forelands, shingle ridges with recurved distal ends being found at many scales around the British Isles (for example, Blakeney Point, Orfordness, Hurst Castle

Spit, and Pagham). The complex overlapping – and associated truncation – of sets of ridges that can be dated is extremely well-developed at Dungeness, where it occurs on a large scale over a known timescale.

Shingle structures of such complexity are unusual globally. Dungeness is a cuspate foreland of intermediate size in global terms, but features the size of Dungeness are rare on the coasts of Britain.

Although none of the individual geomorphological features of Dungeness is unique, their association together gives the site its special interest. The considerable damage to much of the original feature (Fuller, 1985) has not obliterated the most important features and every part of the sequence of ridges is still preserved at some point. The as yet little-analysed archival and archaeological evidence provides a potentially rich field for further interpretation of the development of this large and complex feature.

Chapter 7
Sandy beaches and dunes – GCR site reports

Introduction

INTRODUCTION

V.J. May

Sandy beaches and their backing dunes are a common feature of the British coast. Although the European Commission's CORINE project recorded 9.6% of the British coast to be sandy beach (European Commission, 1998), this statistic did not include any cliff-foot beaches. Sand beaches and dunes occur throughout the British coast, but are concentrated mainly on the northern and western coasts. For example, 75% of coastal dunes, by area, occur north of the Tees and Solway Firth and sand beaches occur in association with dunes and other sandy structures. Sand beaches also commonly form the lower parts of beaches where shingle ridges occur close to high-water mark. They also occur below many cliffs, for example, the chalk around the Isle of Thanet, Kent, the cliffs of eastern England from Flamborough Head to Essex, and along much of the Cornish and Welsh coasts, as well as in association with sand cliffs and other strata that yield sand as a major fraction of weathered debris (Figure 7.1).

Since sand supplies from the upper beach are usually required to build sand dunes, the fact that sandy beaches and dunes commonly co-exist is unsurprising. However, some sandy beaches are not backed by dunes, mainly owing to limited throughput of sand, an unfavourable wind regime or lack of availability of a site suitable for deposition.

The relationship between sandy cliffs and beaches has typically been described in the context of beach sediment budgets, beach management and coast protection (e.g. Clayton, 1989b; Psuty and Moreira, 1990; Bird, 1996). Extensive sand cliffs (for example at Bournemouth, Dorset, and Culbin, Moray, and, more widely, along the coast of the Algarve in Portugal and much of the coast of California) often have substantial sand beaches at their foot. Much of the supply of sand from sand cliffs in southern and eastern Britain has been reduced in recent years by coastal protection schemes. For example, before they were progressively protected by sea walls during the 20th century, the Bournemouth cliffs (some 11km in length) produced about 115 000 m^3 of sediment annually (of which 80% was coarse enough to stay on the beaches). By the mid 1980s, this supply had fallen to 4000 m^3 a^{-1}, coming mostly from the unprotected cliffs to the east (Halcrow Maritime, 1999).

A number of writers have argued for a 'systems approach' to sandy beach study, a point also made strongly in respect of cliffed coasts by Brunsden (1973). Such an approach allows each of the influencing factors affecting beach and dune form to be examined in isolation in order to determine its effect. This methodology allows the links between process and form to be better identified. Therefore, much of the investigation of sandy beaches has focused on changes in beach profiles in response to variations in weather conditions, especially wind, and on beach sediment budgets. Long-term trends in beach morphology and the relationship between beach and dune morphology and ecology typify many other studies. However, it is also apparent from the evidence of GCR sites described in this chapter that change in many sand and dune systems is associated with high magnitude/low frequency events superimposed on the more routine processes. Similarly the relationship between many subsystems that make up these features function over different timescales and with different intensities. For example, the sandy beach and dunes at Gibraltar Point, Lincolnshire, comprise many different subsystems, which include nearshore tidal ridges, a ridge-and-runnel foreshore, and a backshore with arcuate foreshore dune ridges and dune slacks. The spit protects the upper beach ridge sheltering an area of mature (Old Marsh) from New Marsh by a storm beach, which resulted from an occasional extreme event in the evolution of the area. More change occurred in a few hours in 1922 than during years of normal sedimentation; isolation of the storm effects helps in gaining an understanding of the relative importance of both frequent and infrequent events and evolution. This is a theme that is common to many other sand coast GCR sites: Spurn Head on the Holderness coast responded dramatically to a surge in 1849 and both Spurn Head and Gibraltar Point showed different reactions to the 1953 surge. In terms of the development of sub-parallel dune ridges, Gibraltar Point offers considerable contrasts to the GCR site at South Haven Peninsula, Dorset, mostly because of different tidal and wave conditions and differences in sediment supply. In particular, Gibraltar Point lies in a macrotidal and South Haven in a microtidal environment. In both of these sites the processes operating in one subsystem have important repercussions in

Sandy beaches and dunes

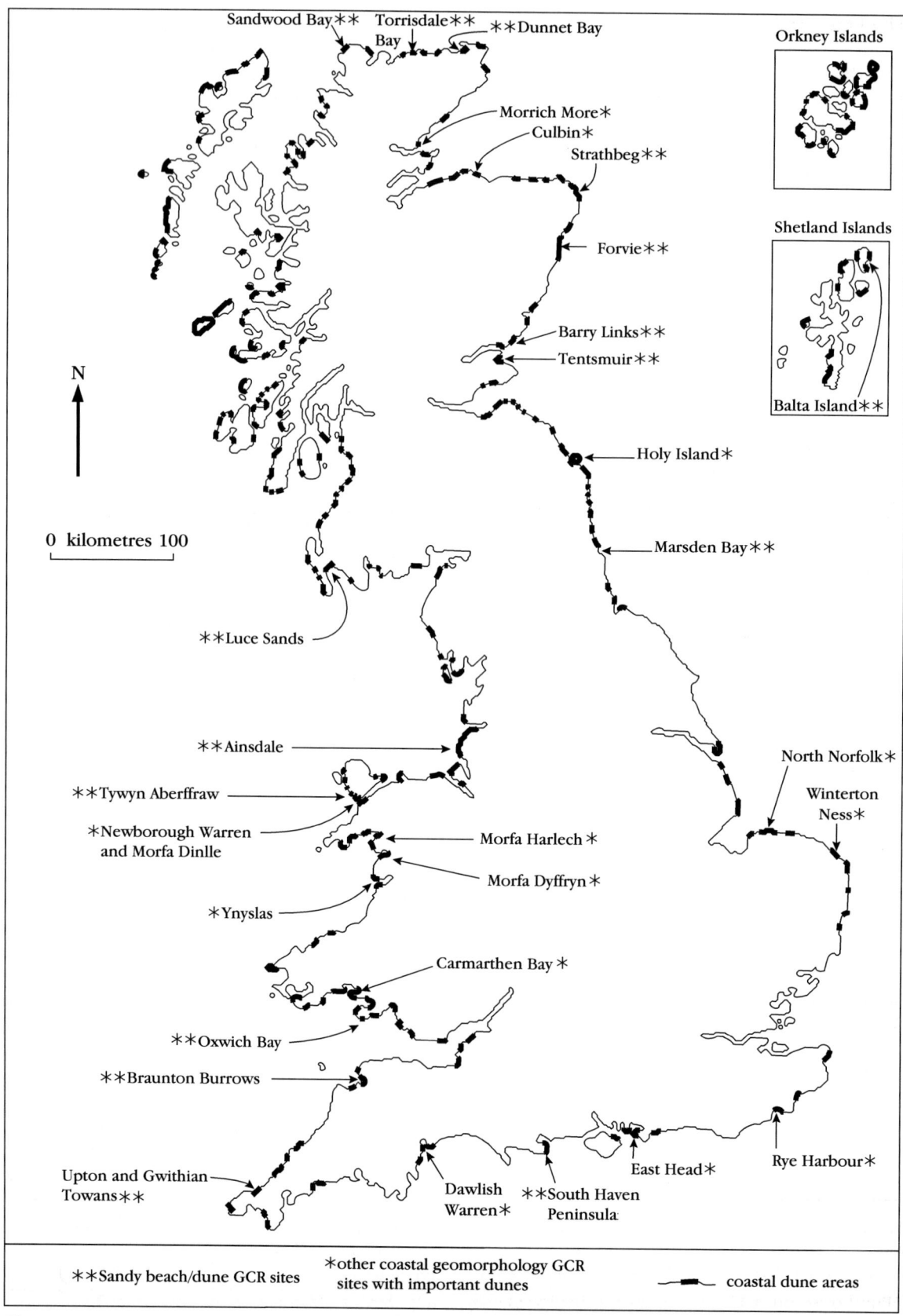

Figure 7.1 Great Britain sandy beaches and coastal dunes, also indicating the location of GCR machair-dune sites (see chapter 9) and other coastal geomorphology GCR sites that contain dunes in the assemblage.

Introduction

all of the others.

It is evident from the GCR sites described in this chapter (Figure 7.1 and Tables 7.1 and 7.2) that both beach and dune features co-exist and depend upon the availability of sand that may come from the seabed, from fluvial sources and from cliff erosion, depending upon their geomorphological setting. Small sand beaches can develop with very limited sediment supplies. For example, small sand beaches form localized pockets within embayments of the Thanet chalk coast and the indented rocky coasts of southwestern Britain and northern and western Scotland. Sand commonly forms a veneer on some shore platforms and displays a range of minor current- and wave-related forms.

In the Chalk, sand derives from attrition of flint and from the release of fossil shell fragments from the chalk itself. Elsewhere, sandstone and soft sediment cliffs provide large quantities of sand to their beaches, which may then be transported alongshore. Erosion of the till coast and shallow seabed off Holderness provides very large volumes of sand and gravel annually that are transported both alongshore to form a large sand spit at Spurn Head and into the North Sea. Along the coast of East Anglia, very large volumes of sand and gravel are derived from erosion of till cliffs, but there are also large volumes in offshore banks that result

Table 7.1 Main features and present-day sediment sources of dune types. Exemplar sites described in the present chapter are in **bold** typeface. See also Table 7.2. (Based on Ranwell, 1972.)

Type	Sediment sources	Geomorphological setting	Wind directions	Exemplar GCR sites
Foreshore dunes				
Spit dunes	Intertidal banks and longshore	On promontories at estuary mouths with near-parallel or radiating ridges and slacks	More common with onshore prevailing and dominant, but not restricted to this	**Forvie, Strathbeg, South Haven Peninsula**, Morfa Harlech, Holy Island (Goswick and the Snook), Culbin, Morrich More
Prograding ness dunes	Accretion at ness, possibly with longshore sediment supply from opposite directions alongshore	On open coast	Prevailing and dominant winds from opposite directions (offshore/onshore)	Winterton Ness, **Barry Links, Tentsmuir**
Offshore island dunes	Offshore, longshore and intertidal drying banks	Offshore or barrier islands narrow, subject to washover, often display time-series development in main direction of longshore transport	Can occur with both onshore and offshore prevailing winds	Scolt Head Island, Blakeney Point recurves (North Norfolk Coast), Pembrey (Carmarthen Bay), Culbin, Morrich More
Hindshore dunes				
Bay dunes	Restricted in longshore direction	Usually at bay head on indented coasts	Prevailing onshore	**Dunnet Bay, Luce Sands, Upton and Gwithian Towans, Tywyn Aberffraw, Oxwich Bay Sandwood, Balta Island, Torrisdale Bay and Invernaver**
Hindshore dune system	Offshore and intertidal	Extensive sandy coasts	Prevailing and dominant winds from the same direction	**Braunton Burrows, Newborough Warren, Ainsdale**, Holy Island (Ross Links)
Hindshore sand plains	Offshore, intertidal and beach	Bay-head and low-lying rocky coasts	High wind-speeds that restrict vertical development	**Tywyn Aberffraw**

331

from the offshore transport of longshore sediment. The sand beaches here largely result from the continued throughput of sand. On the more indented coast of the western and northern British Isles, sand beaches are commonly found in embayments where sand cannot escape, and in estuaries and firths, where sand from landward and seaward sources is locally plentiful. Many beaches that are dominated by gravel at the shoreline are also characterized by extensive, sandy, lower beaches. Similarly many beaches formed in heterogeneous materials are sorted locally into sand and gravel for short periods of time and the sand may be blown into sand dunes to the rear of the beach.

Although this chapter covers sandy beaches as well as dunes, there is little further introduction to beaches that has not been covered in Chapter 5. However, is Marsden Bay, County Durham, is exceptional, where a sandy (and locally mixed sandy gravel) beach lies at the foot of Magnesian limestone cliffs. This was the site of pioneering work on beach mobility in response to variations in wind and waves over 50 years ago (King, 1953), and for that reason is the first site covered in this chapter. For the rest, wide, sandy beaches are usually associated with – and indeed allow the formation of – dunes, but given their varied location, their varying exposure to waves and their range of tidal conditions, they show considerable differences from place to place.

Sand dunes are most likely to be associated with stable and accreting beaches, with a wide upper beach that allows drying and sediment movement by strong winds. A typical example is the west-facing beach of Dunnet Bay in Caithness, a sand trap with onshore winds. Other wide beaches, especially where they are not fully open to the ocean (as around the Irish Sea) and so have waves with more limited fetch, are frequently barred, with ridges and runnels, as at Ainsdale, Lancashire (see GCR site report). Other barred beaches are found at Holkham Bay, North Norfolk, which is a prograding beach, and Braunton Burrows, Devon, as the aerial photograph (Figure 7.9) demonstrates.

Most beaches are more likely to be suffering erosion than progradation (Bird, 1985), and this is certainly true of the UK. The exceptions are in northernmost England (e.g. Holy Island) or parts of Scotland, where postglacial isostatic rebound has offset present-day sea-level rise. As a result, these wide, prograding beaches are backed by some of the largest dune fields in Britain, particularly where sediment was moved onshore during the later part of the Holocene sea-level rise, such as in much of Scotland. It is no surprise that 71% of the dune area of Britain is in Scotland. With the virtual stabilization of sea level, many beaches have lost volume and dune cliffing has become more common throughout Britain. In places, climatic and/or sea-level changes have led to an oscillation between dune cliffing and dune growth on varying timescales, such periodic cliffing maintaining some dynamic stability via contributions of sand to the fronting beach. In general, present sea-level rise and lack of new sediment means that cliffed dunes are more common than active foredune growth in Britain.

The sandy beaches described in this chapter are only a small sample of the important beach sites included in the coastal geomorphology 'Block' of the GCR, since the great majority of the GCR sites have sandy beaches of one type or another. Chapters 9 and 11 also include descriptions of sandy beach and dune sites where such features are an important part of the coastal geomorphological assemblage. The great depositional sites of Morrich More in the Dornoch Firth and Culbin in the Moray Firth, the Northumbrian coast around Holy Island, the North Norfolk coast and Rhossili Bay (Carmarthen Bay GCR site) all provide unmodified, dynamic examples of some of the finest sandy beaches to be found in the UK.

Coastal dunes

There are over 295 separate coastal dune sites around Great Britain (shown on the small-scale map in Figure 7.1), the largest of which attain over 8000 ha in area. Their total area is about 70 000 ha of which 71% by area are in Scotland (Dargie, 2000).

Most British dune systems originated when substantial seabed deposits were moved onshore during the early and middle part of the Holocene Epoch and began to be deposited close to their present locations from about 6500 years BP. In some areas where the sea-level history is more complex, such as in the Western Isles of Scotland, the arrival of dune sands first began about 8700 years BP and may have been non-synchronous between sites (see Chapter 9; Hansom and Angus, 2001). Dune systems such as those at Ainsdale and Braunton Burrows can be shown to have developed over the past six

Introduction

Table 7.2 Main features, sediment sources, tidal ranges of sandy beach and dune GCR sites, including coastal geomorphology GCR sites described in other chapters of the present volume that contain dune features in the assemblage. It should be noted that all of the machair sites in Chapter 9 have dune features (see Table 9.1). Sites described in the present chapter are in **bold** typeface.

Site	Main features	Other features	Present-day sediment sources	Tidal range (m)
Marsden Bay	Beach phases	Cliffs and stacks	Local cliff erosion – small	4.2
South Haven Peninsula	Shore-parallel dune ridges, originating from the 16th century, slacks, sand-spit	Relict and active cliffs, caves, rock platform	Longshore – restricted Offshore – significant	1.5
Upton and Gwithian Towans	Climbing dunes, exhumed bedrock base	Stacks	Offshore – restricted	5.8
Braunton Burrows	Large dune field, parabolic dunes, slacks	Ridge and runnel	Intertidal and estuarine	7.3
Oxwich Bay	Bay-head beach and dunes	Cliffs and emerged platform	Offshore – limited	8.2
Tywyn Aberffraw	Sand plain, isolated parabolic dunes shore-parallel linear dunes		Offshore, probably in deficit	4.7
Ainsdale	Large dune field, slacks, ridge and runnel, long dated history		Offshore – limited – in deficit	8.3
Luce Sands	Bay-head dunes	Holocene emerged gravel ridges	Onshore and longshore – significant	5.6
Sandwood Bay	Dynamic beach–dune complex, climbing dunes	Gravel-cored bar, blowouts	Offshore and recycled – limited	4.2
Dunnet Bay	Bay-head dunes and sand plain	Blowouts	Offshore – limited	4.0
Balta Island	Climbing dunes	Beach–dune–grassland continuum	Local – limited	1.9
Strathbeg	Shore-parallel dune ridges, large blowouts	Holocene emerged gravel ridges	Longshore – restricted, loch outlet source	3.3
Forvie	Shore-parallel dune ridges, originally moved as waves northwards		Longshore – cycled from estuary	3.1
Barry Links	Foreland sand plain, linear parabolic dunes		Estuarine, longshore – limited	4.4
Tentsmuir	Shore-parallel dune ridges–intertidal sands		Estuarine and longshore – significant	4.4
Torrisdale and Invernaver	Beach–dune, hill-top dunes, glaciofluvial terraces	Archaeological context	Offshore and fluvial recycled – now limited	4.0
Morrich More	Shore-parallel beaches and dunes: sandplain	Holocene beaches and cliffs	Offshore – restricted	4.3
Culbin	Shore-parallel dunes, large dune field now stabilized by forest	Holocene emerged gravel ridges and spits	Longshore –restricted, offshore – limited	3.6
East Head	Small spit-based dunes		Intertidal	3.4
Holy Island	Dune field, spits, barrier beach	Cliffs, Holocene saltmarsh, intertidal mudflats	Longshore, offshore – significant	4.1
Dawlish Warren	Parallel spit-based linear dunes	Recurved spit	Intertidal and possibly estuarine In deficit	4.1
North Norfolk Coast	Major mainly linear dunes	Spits, barrier beach	Longshore and offshore	6.4–4.7
Morfa Harlech	Linear shore-parallel dunes		Longshore – restricted, estuarine	4.5
Morfa Dyffryn	Linear shore-parallel dunes, blowouts, dunes invading slacks		Longshore – restricted, offshore	4.3
Winterton Ness	Linear dunes on cuspate foreland		Longshore	2.6
Ynyslas	Spit-based dunes		Longshore – restricted, estuarine	4.3
Pendine	Shore-parallel linear dunes		Offshore, estuarine to distal end	8.0
Pembrey	Large dune field, spit-based linear dunes		Offshore and estuarine	8.0
Whitford spit	Estuary-mouth spit		Offshore, drying intertidal	8.0
Laugharne Burrows	Cliff-top dunes		Local redistribution, drying intertidal	8.0
Newborough Warren and Morfa Dinlle	Major dune field, parabolic and linear dunes, spit, tied island and slacks	Saltmarsh	Offshore and estuarine	4.7

333

Sandy beaches and dunes

Table 7.3 Calcium carbonate content of upper beach/foredune in selected coastal geomorphology GCR sites. Sites described in the present chapter are in **bold** typeface. (Based in part on Goudie, 1990, and various sources cited by Ritchie and Mather, 1984.)

Dune location	CaCO₃ (%)	Median grain size (phi)
Culbin	0.0	2.0
South Haven Peninsula	0.015	?
Lossiemouth	0.26	2.0
Tentsmuir	0.4	2.5
Luce Sands	0.5	2.4
Forvie	0.55	1.9
Buddon Ness (**Barry Links**)	1.0	2.0
Walney Island	1.51	2.21
Morfa Dyffryn	3.34	2.31
Ainsdale	3.57	2.13
Invernaver	3.8	1.9
Morfa Harlech	3.96	2.13
Newborough Warren	4.56	2.50
Ynyslas	4.98	2.29
Strathbeg	7.86	2.0
Rattray (**Strathbeg site**)	9.10	1.9
Laugharne (Pendine)*	11.15	2.40
Morrich More	12.0	2.4
Pembrey*	12.04	2.33
Oxwich Bay	12.45	1.93
Tywyn Aberffraw	13.20	2.47
Llangennith*	15.65	1.63
Braunton Burrows	19.59	2.13
Dunnet Bay	20.4	1.7
Dunbar	20.4	1.5
Westward Ho!	21.79	2.45
Machir, Islay	33.6	2.2
Mangersta, Lewis	38	1.4
Luskentyre, Harris	44	2.0
Traigh na Berie, Lewis	47	2.4
St. Ninian's Tombolo, Shetland	47.5	2.0
Balnakiel	52.0	1.8
Hayle (**Upton and Gwithian Towans**)	56.80	1.56
Loch Gruinart, Islay	59.0	2.1
Eoligarry, Barra	80.0	2.0
Ardivachar, South Uist	84.0	1.7
Balta Island, Shetland	95.5	1.8

* Carmarthen Bay

millennia, especially from the evidence of preserved peat associated with dune slacks and larger wetlands that developed shorewards of the coastal beaches. In contrast, other dunes are more recent, for example at South Haven Peninsula the dunes have formed since the 16th century. Some dunes, for example at Culbin, Moray, Newborough Warren on the Isle of Angelsey, and Hayle and Upton and Gwithian Towans, Cornwall, have migrated inland covering buildings and farmland. British dunes tend to be located:

1. in areas of high tidal range,
2. where prevailing winds provide the main means of landward aeolian transport, and
3. in association with estuary mouths dominated by large sandy sediment loads or at the heads of inlets and bays,
4. on north-eastern coasts, where strong winds from the north and east provide the means for landward aeolian transport e.g. the coasts between Aberdeen and Fraserburgh and Northumberland.

Narrow, linear-dune systems occur along eastern coasts that are associated with sandy estuaries or high tidal ranges, but the size of the dunes is generally much less than those of the exposed and windy western coasts, even though the intertidal sandy area may be very extensive.

There are few significant dunes on the eastern coast of England, apart from the dunes around Holy Island, Northumberland, and along the Lincolnshire and north Norfolk coasts. Between the Tees and the Tamar there are 24 dune sites (*c.* 8%) and between the Tamar and the Mull of Galloway 67 dune sites (*c.* 23%). The remaining 204 (*c.* 69%) sites lie along the coast of Scotland and the English coast north of the Tees. The largest area of dunes is in north-west Scotland, particularly in the Outer Hebrides where machair predominates (Ritchie and Mather, 1984; Dargie, 2000; see Chapter 9). Of 43

Table 7.4 Variations in calcium carbonate content and pH in foredunes and main dunes. (Based on Salisbury, 1952; and Willis, 1985)

Location	Calcium carbonate content of dunes		pH	
	Foredunes	Main dunes	Foredunes	Main dunes
South Haven Peninsula	0.015	0.01	7.0	3.6
Southport (near Ainsdale)	6.0	0.2	8.2	5.5
Braunton Burrows	20.0	8.5	9.05	8.2
Blakeney Point, North Norfolk Coast	0.6	0.02	7.3	4.2

Introduction

nationally important sand dune sites, only six lie on the south or east coast (Doody, 1985).

The foredunes around the coast of England and Wales are notable for their generally low calcium carbonate content (Table 7.3). Goudie (1990) shows that of 42 foredune areas in England and Wales, 29 had less than 20% CaCO$_3$. The highest values occur between Land's End, Cornwall, and Woolacombe, Devon, and along the south coast of Pembroke, with many greater than 50%. The highest CaCO$_3$ content in England and Wales occurs in Constantine Bay, Cornwall (87.5%). Studland Bay, Dorset, in contrast, has almost no CaCO$_3$ (only 0.015%). There is also a tendency for the main dunes to have lower CaCO$_3$ and pH than the foredunes (Table 7.4). The very high CaCO$_3$ content of the foredunes of the south-west coast is probably a result of the high concentrations of shell debris. The more carbonate-rich sands also tend to be coarser with mean D$_{50}$ (median grain size value) of 1.75 phi (Goudie, 1990). This, with their comparatively low density and often platy form, may make them more readily transported by wind (Goudie, 1990). Where the main source is estuarine, the grain size is usually smaller. Scottish dunes and beaches, and especially machair, tend to follow a pattern of very high CaCO$_3$ content where biogeneic sources predominate often reaching extremely high values (Mather and Ritchie, 1984; e.g. Balta Island has 95.5% shell sand, see GCR site report in the present chapter).

On much of the southern coast of Britain, sand was in plentiful supply for dune building at the end of the main Holocene rise when sea level attained present levels about 6500 years BP. In recent centuries, however, the supply of sand has diminished significantly and erosional conditions generally prevail.

In England, few southern or eastern dunes are accreting, the most important exceptions being at Holy Island and South Haven Peninsula, and even the latter is affected by erosion of its older southern beach and dunes. In contrast, on western and northern coasts, dunes are common features, reflecting the combination of plentiful sand supplies mainly from the seabed in the past, but also from upland river catchments, and the effects of prevailing onshore winds. However, many are now affected by wave erosion of their fronts either by occasional storms or by long-term changes in sea-level and storminess, together with reduced sediment supply. Prior to 6500 years BP sand supply for dune building was plentiful, but it is now much reduced, and, as a result, frontal dune erosion is commonplace (Hansom, 1988; Hansom and Angus, 2001).

The conservation value of sandy beaches and dunes

Dunes are geomorphologically important because of:

1. their natural dynamism and the relationship with their ecology
2. their role in preserving and then exhuming Holocene sedimentary sequences and
3. their role in coast protection.

The selected GCR sites (Table 7.2) include the beach and dune sites that best exemplify the different ways in which the physical coast responds to the effects of climate, waves and currents when there is a substantial and continuing provision of sand-sized sediments. They are areas of both progradation and erosion which provide a highly dynamic foundation for some of Britain's most important sites for fauna and flora. Internationally, they have been recognized by geomorphologists as exemplifying especially well the ways in which coastal dunes form, change and are modified.

Most dune systems around the British coast are complex, and very few have individual isolated stable dunes within them. English east coast dunes are generally narrow, have only limited periods of onshore winds, and lack large and constant sand supplies. Many of those on the west coast lie upon bedrock surfaces of varying height and so lack the level foundations of sand plains. They also have usually had ample supplies of sand in the past that have produced a complex dune topography in which dunes are at many stages of development and sand is transferred from erosional phases to depositional ones (for example at Newborough Warren and Morfa Dyffryn). Tywyn Aberffraw is an important member of the network of dune systems because of its relatively limited sediment supply and restricted development of dunes. In this respect it contrasts especially strongly with its neighbour at Newborough Warren.

In this chapter the site reports are ordered in a clockwise fashion starting with the Marsden Bay GCR site.

Dunes and sandy beaches as biological SSSIs and Special Areas of Conservation (SACs)

In Chapter 1, it was emphasized that the SSSI site series is constructed both from areas nationally important for wildlife, and GCR sites. An SSSI may be established solely for its geology/ geomorphology, or its wildlife/habitat, or it may comprise a 'mosaic' of biological and GCR sites that may be adjacent, partly overlap, or be co-incident. There are a number of sand dune and beach sites that are crucially important to the natural heritage of Britain that are notified as SSSIs primarily for their wildlife value, but implicitly will contain interesting coastal

Table 7.5 Candidate and possible Special Areas of Conservation in Great Britain supporting Habitats Directive Annex I coastal dune habitat(s) (other than machair) as qualifying European features. Non-significant occurrences of these habitats on SACs selected for other features are not included. (Source: JNCC International Designations Database, July 2002.)

SAC name	Local authority	Dune habitat extent (ha)
Barry Links	Angus	447.6
Braunton Burrows	Devon	767.5
Carmarthen Bay Dunes/Twyni Bae Caerfyrddin	Abertawe/ Swansea; Caerfyrddin/ Carmarthenshire	780.2
Coll Machair	Argyll and Bute	409.0
Culbin Bar	Highland; Moray	612.9
Dawlish Warren	Devon	28.2
Dee Estuary/ Aber Dyfrdwy*	Cheshire; Fflint/ Flintshire; Wirral	4.0
Dornoch Firth and Morrich More	Highland	974.4
Dorset Heaths (Purbeck and Wareham) and Studland Dunes	Dorset	95.9
Drigg Coast	Cumbria	519.8
Durness	Highland	386.7
Humber Estuary*	City of Kingston upon Hull; East Riding of Yorkshire; Lincolnshire; North East Lincolnshire; North Lincolnshire	529.0
Invernaver	Highland	54.2
Kenfig/ Cynffig	Pen-y-bont ar Ogwr/ Bridgend	673.8
Limestone Coast of South West Wales/ Arfordir Calchfaen de Orllewin Cymru	Abertawe/ Swansea; Penfro/ Pembrokeshire	397.1
Monach Islands	Western Isles / Na h-Eileanan an Iar	215.1
Morecambe Bay	Cumbria; Lancashire	220.5
Morfa Harlech a Morfa Dyffryn	Gwynedd	228.6
North Norfolk Coast	Norfolk	387.3
North Northumberland Dunes	Northumberland	1078.6
North Uist Machair	Western Isles / Na h-Eileanan an Iar	963.3
Oldshoremore and Sandwood	Highland	165.3
Penhale Dunes	Cornwall	422.4
Saltfleetby–Theddlethorpe Dunes and Gibraltar Point	Lincolnshire	265.6
Sands of Forvie	Aberdeenshire	469.7
Sandwich Bay	Kent	258.3
Sefton Coast	Sefton	1072.7
Solent Maritime	City of Portsmouth; City of Southampton; Hampshire; Isle of Wight; West Sussex	113.2
Solway Firth	Cumbria; Dumfries and Galloway	32.6
South Uist Machair	Western Isles / Na h-Eileanan an Iar	545.7
Tiree Machair	Argyll and Bute	237.4
Torrs Warren–Luce Sands	Dumfries and Galloway	819.5
Winterton–Horsey Dunes	Norfolk	44.7
Y Twyni o Abermenai i Aberffraw/ Abermenai to Aberffraw Dunes	Gwynedd; Ynys Môn/ Isle of Anglesey	672.3

* Possible SAC not yet submitted to EC.

Bold type indicates a coastal GCR interest within the site.

geomorphology features that are not included independently in the GCR because of the 'minimum number' criterion of the GCR rationale (see Chapter 1). These sites are not described in the present geomorphologically focused volume.

The importance of dunes as areas of national ecological significance was recognized and described by Tansley (1939, 1945) and Steers (1946a, 1953a). Soon after the Nature Conservancy was established in 1949, it designated a number of major dunes as National Nature Reserves, including Braunton Burrows, Newborough Warren, Ainsdale and Holy Island. The *Nature Conservation Review* (Ratcliffe, 1977) confirmed the great importance of dunes as part of the network of nationally significant sites.

In addition to being protected through the SSSI system for their national importance, certain types of dune are Habitats Directive Annex I habitats, eligible for selection as Special Areas of Conservation (see Chapter 1). The Directive identifies a suite of dune vegetation types (see below), representing the succession from dune initiation to mature, stable dune habitat. Collectively, these types encompass almost the full range of coastal dune habitats present in the UK.

Dune SAC site selection rationale

The sites are, for the most part, the most extensive examples in the UK and have the best conserved structure and function, demonstrating transitions between Annex I types, while also representing the range of geographic and ecological variation of each habitat type.

- **Embryonic shifting dune** vegetation exists in a highly dynamic state and is dependent on the continued operation of physical processes at the dune/beach interface. It is the first type of vegetation to colonize areas of incipient dune formation at the top of a beach.
- **Shifting dunes along the shoreline with** *Ammophila arenaria* ('white dunes') encompass most of the vegetation of unstable dunes where there is active sand movement. Under these conditions sand-binding marram *A. arenaria* is always a prominent feature of the vegetation and is usually dominant.
- **Fixed dune vegetation** occurs mainly on the largest dune systems, being those that have the width to allow it to develop. It typically occurs inland of the zone dominated by marram *Ammophila arenaria* on coastal dunes, and represents the vegetation that replaces marram as the dune stabilizes and the organic content of the sand increases.
- **Decalcified fixed dunes with crowberry** *Empetrum nigrum* represent the later, more mature, stages of the successional sequence characteristic of sand dunes. Exposure to rainfall over long periods means that there is leaching of the surface layers, causing a loss of calcium carbonate and increased soil acidity.
- **Atlantic decalcified fixed dunes (***Calluno-Ulicetea***)** occur on mature, stable dunes where the initial calcium carbonate content of the dune sand is low. The surface soil layers rapidly lose their remaining calcium carbonate through leaching, and become acidified.
- **Dunes with** *Hippophae rhamnoides* comprise scrub vegetation on more-or-less stable sand dunes in which sea-buckthorn is abundant.
- **Dunes with** *Salix repens* **ssp.** *argentea*, where creeping willow is dominant, forming prominent, low scrubby growth.
- **Humid dune slacks** are low-lying areas within dune systems that are seasonally flooded and where nutrient levels are low. Dune slacks are often rich in plant species.
- **Coastal dunes with juniper** *Juniperus* spp. comprises common juniper scrub in a variety of dune situations.
- **Machair** – see Chapter 9 of the present volume.

Table 7.5 lists coastal sand dune SACs, and indicates which of these sites are also (at least in part) important as part of the GCR and are described in the present chapter.

MARSDEN BAY, COUNTY DURHAM (NZ 400 650)

V.J. May

Introduction

Marsden Bay (see Figure 7.1 for general location) includes beach, rock and cliff features and is a classic locality for beach process studies, based on the work of C.A.M. King over 50 years ago (King, 1953). Until very recently, it was the only site where the behaviour of beaches resting against relatively resistant cliffs had been studied intensively. King's analysis of the relationship between beach profiles and wave conditions influenced much subsequent work on beaches, and as a result, the location is frequently cited in the literature. It is also notable for a suite of cliffs and shore platforms cut into the Permian Magnesian Limestone, which crops out only on this part of the British coast between the River Tyne and the River Tees, and it contains the best examples of stack and cliff development in this rock type. The Permian concretionary limestone is most common in the headlands, stacks and arches, whereas the bays are cut into a weaker dolomite. An intricate assemblage of forms has developed as local small-scale joints have been exploited by marine erosion. Although Marsden Beach itself is dominated by sand with some shingle, there are also cobble and boulder accumulations associated with both the cliff-foot and former stacks. This site remains better documented than any of the other small cliff-foot beaches that occur along the north-east coast of England.

Description

Marsden Bay is a small bay some 1200 m in length in the limestone cliffs between the rivers Tyne and Wear (see Figure 7.1 for general location). The site extends southwards beyond the bay itself to include parts of the cliffs towards Lizard Point (Figure 7.2). At the northern and southern extremities of the site the cliffs are about 15 m in height, but rise behind the bay to between 25 and 30 m. High spring tides reach the foot of the cliff throughout the site. There are no other sand or shingle beaches close to the site, and King (1953) suggests that it is unlikely that longshore transport carries any sediment into the bay. As a result, the only sources are the cliffs or the offshore zone; the cliffs themselves appear incapable of supplying the sand volume although they do supply limestone clasts. The beach faces north-east with a maximum fetch to the north of at least 1900 km. King chose the beach to compare the effect of the prevailing offshore south-westerly wind with that of the dominant onshore northerly wind.

The northern part of the site is dominated by cliffs much broken by caves, arches and a large stack. A platform cut in the limestone extends offshore for about 200 m and strongly refracts all waves approaching the cliffs. The foot of the cliff

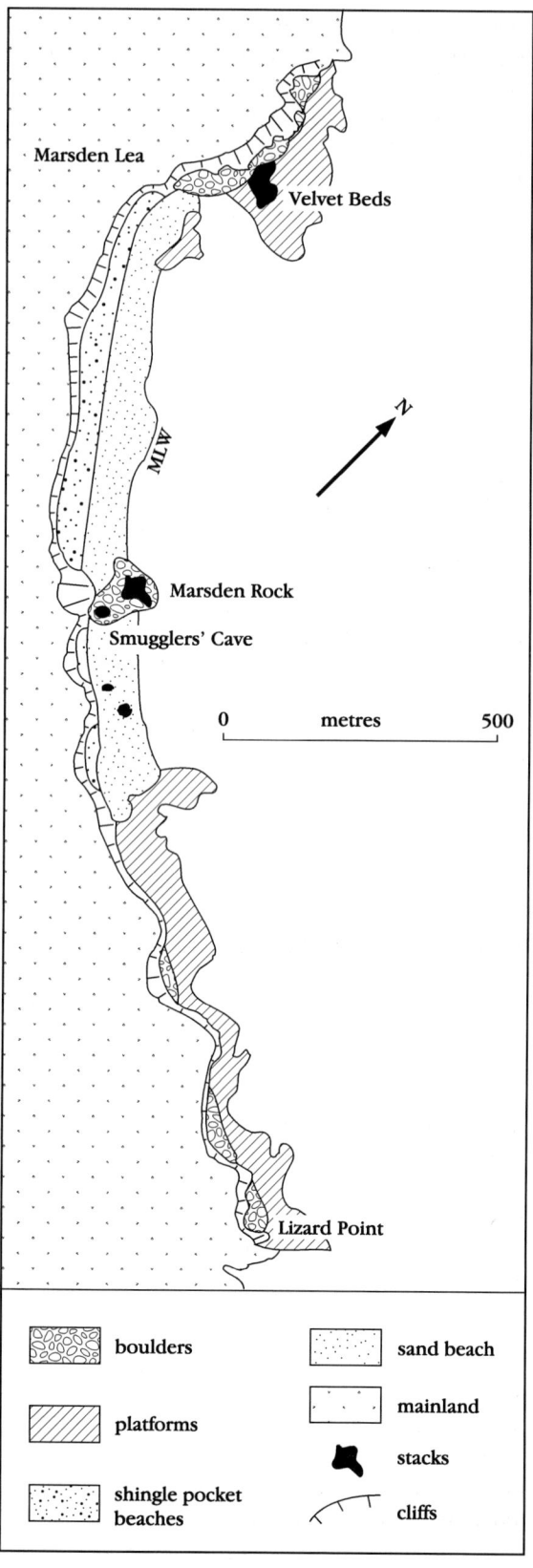

Figure 7.2 Key geomorphological features of Marsden Bay, Marsden Lea to Lizard Point.

Marsden Bay

Figure 7.3 Marsden Bay – view looking towards the north-west showing the Magnesian Limestone cliffs and stacks and stumps. (Photo: V.J. May.)

is littered by cobbles and boulders of limestone, their size and shape being strongly controlled by the blocky nature of the local rock. Marine quarrying and abrasion exploit the numerous small joints in the limestone. There are several stacks and stumps towards the centre of Marsden Bay (Figure 7.3), which show evidence of being more strongly controlled by larger discontinuities. The two largest stacks have upper surfaces at 27 m and 24 m, i.e. about the same as the adjacent cliff top. In February 1996, the roof of the largest arch collapsed and Marsden Arch became a stack. The cliffs in the southern part of the site are more strongly controlled by the jointing pattern with short straight sections separated by sharp almost right-angle joint-controlled bends. Cobbles and boulders have accumulated within the resulting small bays.

King's 1953 paper describes the changes that occurred in the sandy beach profiles under certain well-defined conditions as follows.

1. Swell with an onshore wind produced long high waves. These were steep high-energy waves that proved very destructive, carrying sand seawards. Because these waves have a long period and are steep, the resultant beach was low and relatively wide.
2. Swell with an offshore wind produced long low waves. These flat waves are constructive, though with only moderate energy. Sand is moved landwards from the lower to the upper beach. The beach gradient remained low.
3. Local stormy sea with an onshore wind produced short high waves. These waves are very erosive, cutting into the upper beach and moving sand to the lower beach. Beach gradient is flat.
4. Locally generated waves with offshore winds produced short low waves. These waves are very flat and have little energy. Although they are constructive, the action is limited to a narrow zone, in which the beach gradient is steep as a result of the combination of both short period and flat waves.

King concluded that on Marsden Beach the prevailing offshore wind is associated with constructive wave action, and the dominant onshore wind with beach erosion.

Interpretation

King's 1953 paper was a significant step in the understanding of beach processes. Previously, most studies had concentrated upon spits and barrier beaches, such as those at Hurst Castle, Hampshire, and Scolt Head, Norfolk, and the behaviour of beaches that were free to transgress the land behind them. At Marsden Bay, the beach (composed of fairly coarse sand and gravel) rests against a cliff and so its long-term movements are controlled by the rate of retreat of the cliffs and the associated supply of sediment from them. Within those constraints short-term changes in beach profile are related to wave type and wind direction. Marsden Bay offers an opportunity for assessment of the effects that future sea-level change may have on cliff-foot beaches. As sea level rises, the tidal duration curve on a cliff-foot beach should move up the beach slope. Since the beaches at Marsden Bay appear to have retained a similar volume since King's surveys, they provide an opportunity for use as a baseline for process-response studies as the rate of sea-level rise changes.

The second reason for the importance of this site is the development of the stacks and caves in limestone dominated by small, closely spaced joints. Although the southern part of the site is controlled by larger, more widely spaced, joints, there are few stacks there. Other sites with comparable forms (e.g. the GCR sites South Pembroke Cliffs, Flamborough Head, Kingsdown to Dover, Ballard Down) have major joints that control either or both cliff form and cave–arch–stack development. In relatively weaker materials such as the Chalk and the Keuper Sandstone (e.g. the GCR site Ladram Bay) the presence of a more resistant layer at the foot of the cliff appears to be especially important in increasing the likelihood that stacks will develop (Precheur, 1960). Since these conditions appear to be unimportant in Marsden Bay, other factors such as the local structure of the Magnesian Limestone, its strength, the role of boulders in modifying the behaviour of waves locally, and the nature of the intertidal platform may all play a part. Much of the platform, for example, is covered by small boulders derived from the Magnesian Limestone. Their dimensions are strongly related to the blockiness of the limestone. Their presence makes the platform surface particularly rough when compared with many of the Chalk platforms farther south (for example Flamborough Head, Joss Bay, Kingsdown to Dover), and so wave quarrying at most states of the tide is likely to be less effective than where blocks are absent. However, abrasion might be enhanced where blocks occur. Variations in jointing along the cliffs may give rise to differences in rock strength that in turn affect the development of bays and headlands in the site, and these deserve further investigation.

Conclusions

There are two key features of this site:

1. It provides a benchmark for beach studies both because of its suitability for surveys continuing King's work in the 1950s and because that work provided a frequently cited demonstration that wind and wave conditions affect beach behaviour on much shorter timescales than the seasonal beach models elsewhere.
2. It represents cliff–beach–platform development in a rock type that is little represented along the British (or European) coastline, and offers an additional example of forms such as stacks and arches.

SOUTH HAVEN PENINSULA, DORSET (SZ 033 848)

V.J. May

Introduction

South Haven Peninsula on the southern side of the entrance to Poole Harbour (see Figure 7.1 for general location and Figure 10.2) is an excellent example of a prograding sandy beach that has been well documented in both the historical record and in more recent field surveys (Diver, 1933; Steers, 1946a; Arkell, 1947; Robinson, 1955; Carr, 1971b; May and Schwartz, 1981; Bray et al., 1995; May, 1997b). Three main former ridges occur, each with dunes fronted by a seaward slope extending beneath alluvial deposits. Much of the seaward dune-system is prograding, and accretional forms characterize much of the strandline. However, the northern part of the site, known as 'Shell Bay', has been affected by erosion and the alignment of the beach as a result of the construction of a training bank alongside the main navigable channel to Poole Harbour. The southern part of the site extends to low cliffs at Studland that form the southern limit of the sediment cell that feeds the beach (May and Schwartz, 1981).

Description

This is one of the few prograding beaches in southern Britain; despite some erosion at its northern and southern extremities and intensive recreational usage, it retains many of the key features which were described by Diver (1933). South Haven Peninsula contains five sub-units.

1. A sandstone cliff (SZ 041 825–SZ 038 828) to the south of Redend Point cut in Bracklesham Group rocks of Eocene age that stands behind a wooded shingle and sand ridge and a modern beach of angular flint and some chalk and sandstone pebbles.
2. Redend Point (SZ 038 828–SZ 036 829). The most resistant part of the site, the headland comprises the Redend Member (formerly Redend Sandstone – Arkell, 1947) and has well-developed platforms, low cliffs and a series of small caves. The cliffs decline to the north and are affected by small landslides.
3. A narrow, fronting sand beach (SZ 036 829–SZ 034 837), which links the cliffs to the main beach at South Haven. Erosion along this section was sufficiently rapid during the 1980s and early 1990s to have stimulated some attempts to retard it using a vertical wooden revetment and gabions. Marram *Ammophila arenaria* was planted to help build up some dunes in front of holiday chalets. These occupy an area that once formed the southern end of the South Haven dunes; it is now severely degraded by trampling and so has been excluded from the site. During early 1996 the wooden structures were outflanked and destroyed and the front edge of the dunes cut back in excess of 4 m; a pattern repeated in early 1997.
4. The main area of the dunes (SZ 034 837–SZ 042 860) faces south-east and is prograding, apart from some erosion at the southern end in the vicinity of the main recreational facilities and the National Trust car park. This part of the site includes the former dune ridges described by Diver (1933), Little Sea, large wetland areas between the dunes, and the former sea cliffs which pre-date the formation of the dunes (Figure 7.4).
5. Shell Bay (SZ 042 860–SZ 036 867). This is the northern part of the sand dunes where they have been limited in their northwards growth by the deep-water entrance to Poole Harbour.

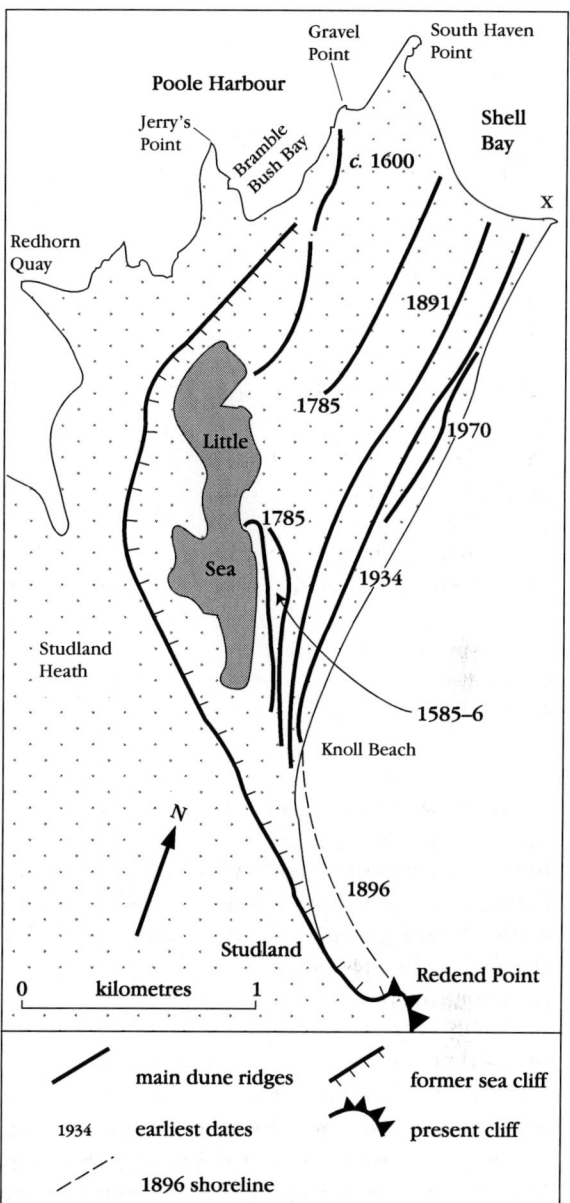

Figure 7.4 Historical dune development at South Haven. The 'Training Bank' extends south-eastwards from point X. (After Diver, 1933.)

Interpretation

Diver (1933) used the many maps of the entrance to Poole Harbour, from Saxton (1575) onwards, to interpret the history of the ridges of blown sand that form South Haven Peninsula.

Sandy beaches and dunes

Figure 7.5 View looking north-eastwards towards Sandbanks and Poole Harbour, with Shell Bay (SB) to the right foreground, to the east of South Haven Point (SHP). Gravel Point (GP) lies in the forground to the left (see Figure 7.4 for sketch map).

Brownsea Island, in the centre of Poole Harbour (see Figure 10.2, Chapter 10, for map), lies to the WNW of Sandbanks, just out of view. (Photo:
© ukaerialphotography.co.uk.)

The 17th century shoreline was represented by a low bluff or cliffs of sands and clays within the Bracklesham Group. By the time of Avery's 1721 survey, the first of the sand ridges had begun to form enclosing a tidal inlet that became a lagoon (Little Sea) by the end of the 18th century, and is now totally enclosed. Its western shore is the mid-17th century cliffline. However, a detailed estate map of Studland parish drawn by Ralph Treswell in 1585 and 1586 shows Little Sea as a narrow arm of the sea along whose banks there are no fewer than fourteen fields held as tenancies. The seaward coast of Little Sea was formed by a narrow northward-trending ridge named 'Burnet poynte'. The western shoreline extended northwards to form a large recurve that, according to Treswell, had its distal end at 128 perches (641 m) from Brownsea Castle. Today, although the present tip of South Haven Peninsula lies twice as far (1200 m) from Brownsea Castle, a series of low intertidal gravel and sand ridges are exposed between South Haven Point and Brownsea Island. The northernmost tip of these ridges, known as 'Stone Island', lies about 500 m from Brownsea Castle. On the Poole Harbour side of the ridge there are four small headlands known in order from south to north as 'Rede orde', 'Coke orde', 'Geries orde' and 'Rickmans orde'; these co-incide in position with the modern Redhorn Quay, an unnamed ridge, Jerry's Point and Gravel Point. The southern three are formed in Bagshot Beds (Poole Formation) and gravels similar to others associated with former terraces of the Frome and its tributaries. There is no field sedimentary evidence to suggest that they represent former distal features of the South Haven spit, although it is possible that at one time the beach at Jerry's Point was linked to the spit. In contrast, the last, Rickmans orde (i.e. Gravel Point), lies in the approximate position where Diver located the distal end of the spit portrayed on Camden's 1607 map. In light of Treswell's survey it is now possible to regard the dunes at South Haven as having an earlier origin than Diver (1933) or later writers have suggested. The presence of salterns (sites of salt production) at the head of Little Sea (recorded as early as the Domesday survey (1086 AD; Thom and Thom, 1983) and tenanted enclosures on the ridge to seaward of the 16th century Little Sea suggest at least some stability to the feature.

The early dune ridges were fronted by a wide intertidal area of drying sand and the low tide limit was close to its present-day position. Indeed, even allowing for the limitations of comparison of maps and charts, the low-water mark has remained in more or less the same position

South Haven Peninsula

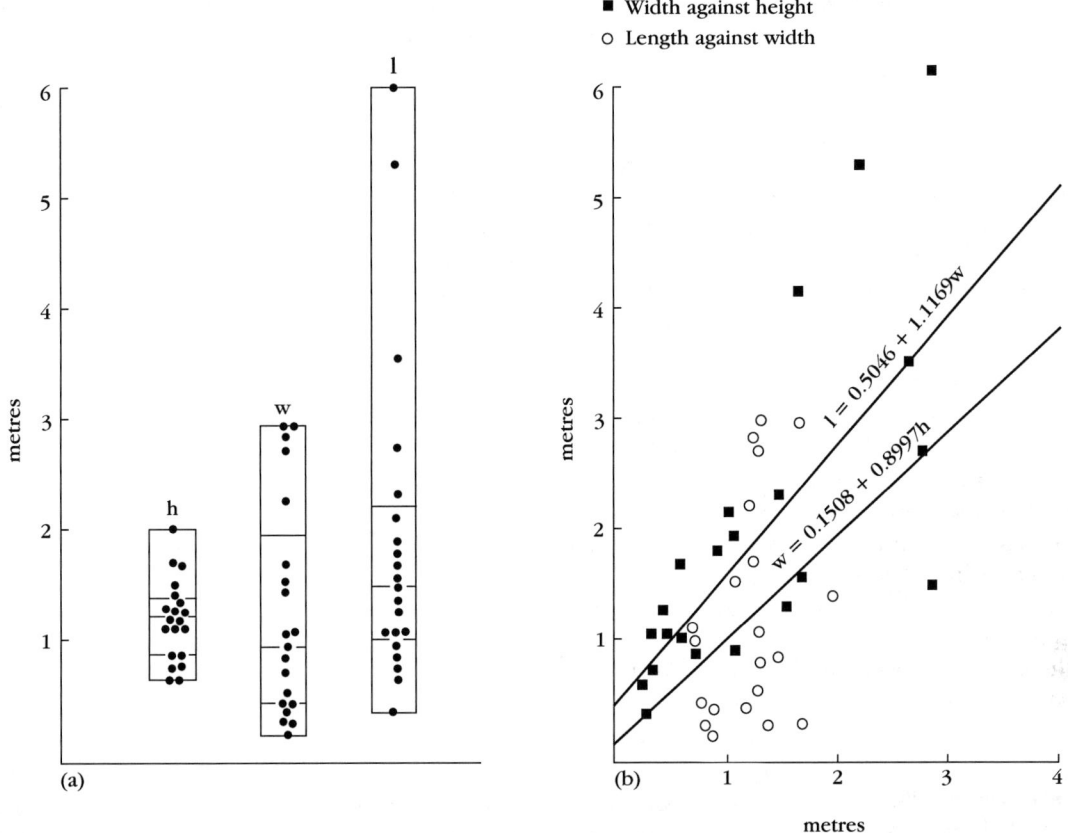

Figure 7.6 Cave relationships at Redend Point, South Haven Peninsula GCR site (see Figure 7.4 for location). (a) Cave height, h; width, w; length, l. (b) Relationships between cave height (h), and w and l.

for at least 200 years, a fact apparently overlooked by all previous writers. Robinson (1955) noted that the one fathom (c. 2 m depth) line had moved shorewards between 200 m and 300 m in the six decades following Mackenzie's survey of 1785. Ward (1922) considered that the northern spit at Sandbanks was supplied by sand moving from the east; Steers (1946a) supported the idea of a counter-drift of sand, i.e. from both south and north, towards the mouth of Poole Harbour. The entrance to Poole Harbour is deep, with a maximum ebb-tide current of 2.5 m s^{-1}. There is no evidence to suggest that its position has changed during the last few centuries. Robinson (1955) concluded that there was no evidence of such counter-drifting and proposed that the development of South Haven Peninsula resulted from a process of frontal accretion in which a series of beach ridges were built up parallel to the dominant waves, providing the foundation for the dunes. He went on to suggest that the apparent double spits at the mouth of Poole Harbour could have resulted from a single embankment across the harbour mouth that resembled a partial bay-bar, even though there was no documentary evidence to support this. Kidson (1963) took issue with the frontal accretion hypothesis and supported the view of Steers (1946a) that counter-drifting was a possible explanation for the double spits.

Using aerial photographs taken between 1936 and 1970, Carr (1971b) showed that erosion had occurred at the southern end of the dunes (a maximum of 0.8 m a^{-1}) and in Shell Bay (maximum 0.67 m a^{-1}), but accretion had been more typical of the main part of the dunes, gaining an average of 2.15 m a^{-1} and a maximum of 4.3 m a^{-1}. Since 1970, these trends have continued, although phases of accretion alternate with erosion, with some large erosion events affecting even the normally accreting shoreline. During periods of prolonged east winds (e.g. during early 1996), the dunes between Redend Point and the Knoll car park were cut back by over 4 m. Eyewitness accounts suggest that the beach was lowered in January 1996 to levels which had last been exposed in the 1940s during preparations for the Normandy D-Day landings.

May and Schwartz (1981) indicated that a maximum of 10% of the total volume of sand added to the beach between 1933 (Diver's survey) and 1971 (Carr's 1971b survey) could be accounted for by the presumed pattern of erosion and longshore transport. Although Bray *et al.* (1995) and the Shoreline Management Plan (Halcrow Maritime, 1999) use the general model of longshore transport associated with shoreward transport on the shallow sea floor of Studland Bay, they did not quantify in detail the rates or volumes of sediment movement.

Surveys by BP (British Petroleum) over 15 months in 1990 and 1991 carried out in connection with a proposal to build an offshore island for drilling indicated that, for a zone extending seawards 450 m from the dunes between the Sandbanks Ferry and the chalk cliffs at Redend Point, there was a net gain of volume between July 1990 and May 1991 of *c.* 90 000 m³ and between May 1991 and October 1991 a net loss of volume of *c.* 91 000 m³. The foredune–beach–shallow water area system was therefore in balance. The largest gains occurred within Shell Bay (more than 40 000 m³ gained over a length of about 370 m) and the greatest losses in Shell Bay (over 32 000 m³) and in the southern parts of the dunes (over 40 000 m³). Over the 3 km from the Training Bank (the northern beach) to Knoll Beach, however, there was a gain in the first period of 23 890 m³ and a loss in the second period of 19 436 m³, a net gain of 4454 m³. The net vertical gain over the whole of this area was 3 mm. There were considerable movements of sand within this zone, with erosion on individual profiles sometimes balanced by accretion within the same profile. The differences in quantities in individual profiles suggest that there are important local movements of sand within the shallow water zone that are not a simple longshore process of transport. These are not yet fully understood.

Since the 1990–1991 surveys, the general pattern of accumulation in Shell Bay and southwards from the Training Bank has continued. The northern beach has suffered several phases of erosion, as has the whole frontage southwards to Middle Beach. As a result, Middle and Knoll beaches are very narrow at high water. In contrast, the beach north of Knoll Beach is much wider and is able to absorb most waves without significant erosion of the dunes themselves. However, the continuing retreat at Knoll Beach is changing the alignment of the shoreline. This puts the southern end of the dunes north of Knoll Beach at progressively greater risk.

Farther south around Redend Point, the late 19th century Geological Survey maps and the First Edition of the OS 1:2500 plans of 1886 show that the dunes had earlier extended over 80 m seawards of the present cliffline. Although there is no direct evidence to describe the earliest relationship between this cliff and the development of the dunes, Treswell's 1585–6 survey indicates a cliff with 'furzy ground' where the oldest dunes survive today. However, to the south of Redend Point, the cliffs have not been eroded by the sea during this century and they have developed a small talus slope of sand that is pitted with small holes and casts (between 6 and 10 mm in diameter) of sand-wasp burrows. To seaward of the cliff, trees several decades old stand on a sand and shingle ridge. A stack of Redend Sandstone that rose from the present-day beach was partly demolished by a fall from the cliffs above it in 1995. There is at present no evidence for the last date at which the sea cut the cliffs.

The development of caves and a shore platform in the Redend Sandstone demonstrates a clear link with near-vertical joints in the sandstone. Most of the caves are narrow clefts in the rock, but some are semi-circular in cross section with considerable evidence of the effects of abrasion in the morphology of the cave floors. The floors are usually continuations of the platform across which it is possible to trace some of the joints. Figure 7.6 shows that the longer caves tend to be both higher and wider than shorter caves, probably caused by structural controls. Potholing occurs both on the platform and within the caves, sometimes in association with the truncated parts of pipes within the sandstone. This is a very rare feature, the only other site where planed-off pipes have been reported being to the west of Cuckmere Haven, Sussex (Castleden, 1982). Marine and subaerial erosion is taking place at the cliffs, but it is slow compared to the changes that have taken place in the dunes to the north. The contribution to the sediment budget of the South Haven beaches by sediment derived from cliff erosion is relatively small. The occurrence of chalk pebbles on the southern part of the beach seaward of the

Upton and Gwithian Towans

National Trust car park is a good indication that sediments do travel from the southern part of the site. Nevertheless, the main source of sand for the beach appears to be the seabed where there have been extensive sandbanks around the mouth of Poole Harbour throughout the time for which there is a cartographic record.

It remains unclear why there was a sudden onset of dune-building in the 17th century as envisaged by Diver (1933), nor is there any explanation of the wide intertidal area that formed a base for them. The Treswell map shows an earlier sand spit enclosing an inlet that now forms the southern part of Little Sea. The early history of the site is therefore far from clear, despite the detailed historical record. Further investigation is required of both the sub-dune surface and the sedimentation processes in the South Haven area.

One of the rare prograding beaches of southern England, largely nourished by seabed sources of sand, this site has been well documented and its history since the 16th century has been described. A series of sub-parallel dune ridges are the major features. The site also includes low cliffs which were protected by the growth of the dunes at Studland and have since been re-exposed to marine action. The dunes are often quoted as an example of ecological succession, the sequence of dune ridges being characterized by a change inland from dune plants such as lyme-grass *Leymus arenarius* and marram *Ammophila arenaria* to heath species and finally oak scrub. Much of the site is a National Nature Reserve because of its ecological features, but the geomorphological characteristics of the site are a key component. In particular, it is one of the very few east-facing dunes on the English Channel coast.

Conclusions

This site is important because, unusually in southern Britain, it is prograding, and the sequence in which it developed has been well documented. Furthermore, the sediment budget for the beach suggests that much of the sand is derived from offshore, also a relatively unusual feature of beaches in southern Britain. It contrasts with many other British dune systems in lying in an area dominated by offshore winds.

UPTON AND GWITHIAN TOWANS, CORNWALL (SW 575 406)

V.J. May

Introduction

Much of the shoreline of St Ives Bay is formed by dunes, known as the 'Hayle, Upton and Gwithian Towans', banked against and covering bedrock to heights of over 60 m (see Figure 7.1 for general location). Blown sand also covers parts of the western side of Godrevy Point at the northern end of St Ives Bay. Documentary evidence indicates that the dunes spread inland covering small houses from the 12th century onwards (Steers, 1946a). Dunes in the southern part of the site are gradually replaced northwards by rock cliffs, caves, stacks and arches overlain by blown sand and dunes. These features have been exposed as the covering dunes have been eroded and the shoreline has retreated. Remnants of former dunes are still preserved on the stacks, but are gradually being removed by subaerial processes. There has been only limited research on this site (Steers, 1946a; Balchin, 1954; Hosking and Ong, 1963); nevertheless the site is important as an example of a relict cliff coastline. It also allows examination of the interface between the dunes and the sub-dune surface.

Description

The site lies at the northern end of St Ives Bay. It is formed at its southern end (SW 572 043) by active climbing dunes which reach over 25 m in height and at its northern end (SW 580 416) by a series of cliffs, stacks, caves and rocky platforms known as 'Strap Rocks'. Between these two contrasting forms, the dunes undergoing erosion are gradually replaced at the shoreline by a small rock cliff upon which they rest. This cliff reaches about 20 m in height south of Peter's Point before declining towards 15 m around Strap Rocks. Between Peter's Point and the northern boundary of the site (SW 530 417), the cliff is broken by small coves, stacks and caves associated with lines of weakness in the Lower Devonian rocks. The stacks appear to have developed as marine action has attacked joints and other weaknesses (Figures 7.7 and 7.8).

Steers (1946a) described the area as a 'mass

Sandy beaches and dunes

Figure 7.7 Upton and Gwithian Towans GCR site. Both on the mainland and on the stack the sequence a–d is as follows: (a) dune grasses on blown sand; (b) thin sandy soil on weathered clay and angular intermittent gravel-sized clasts; (c) weathered bedrock; (d) bedrock. (Photo: V.J. May)

of high and well-developed dunes', and makes no reference to the erosional forms that now characterize the northern part of the site. Indeed, the dunes have suffered considerable erosion not only since the 1940s but also over a much longer period, for there appear to have been dunes well to seaward of the present shoreline in the early 19th century. Even allowing for the inadequacies of topographical maps as evidence for coastal change (Carr, 1962) both the Ordnance Survey and the Geological Survey maps point to considerable retreat of the shoreline.

Erosion of the rocky coast is slow when compared with the inferred rate of retreat of the dune shoreline. No dating of the interface between the dunes and the surface beneath them has been attempted. Many west coast dunes lie upon a rocky base and often owe their height to their rocky foundations; similar sub-dune rocky cliffs have been observed in Brittany, though without the intricate forms of Strap Rocks.

The dune sands are carbonate-rich (Table 7.3). The beach sands contain both tin and other heavy metals which have presumably been carried to the beach down the streams both to north and south (Hosking and Ong, 1963). The potential fluvial sediment supply to the beach has not been quantified, but may have been significant in the past. De la Beche (1830) indicates that up to 100 000 tonnes a^{-1} was removed from the Camel estuary in the 1820s, and extraction from this area may have removed comparable volumes.

Interpretation

This site is unusual in that it contains both active dunes and intricate erosional forms, the latter exhumed from beneath a retreating sandy shoreline. The relatively rapid changes in this site's cliffed coastline today contrast with the much slower changes farther north at Tintagel. It is the exhumed cliffline that makes this site particularly important for coastal geomorphology. There are several other locations where there is clear evidence of exhumation (Hallsands and Redend Point (South Haven Peninsula) in England, and Tarbat Ness and the Bullers of Buchan in Scotland – see GCR site reports). The dune–rock interface is poorly preserved at Redend Point, whereas in this site it is well exposed. At the Bullers of Buchan (Walton, 1959), stacks and geos are cloaked and infilled by till, indicating that they are at least older than

Upton and Gwithian Towans

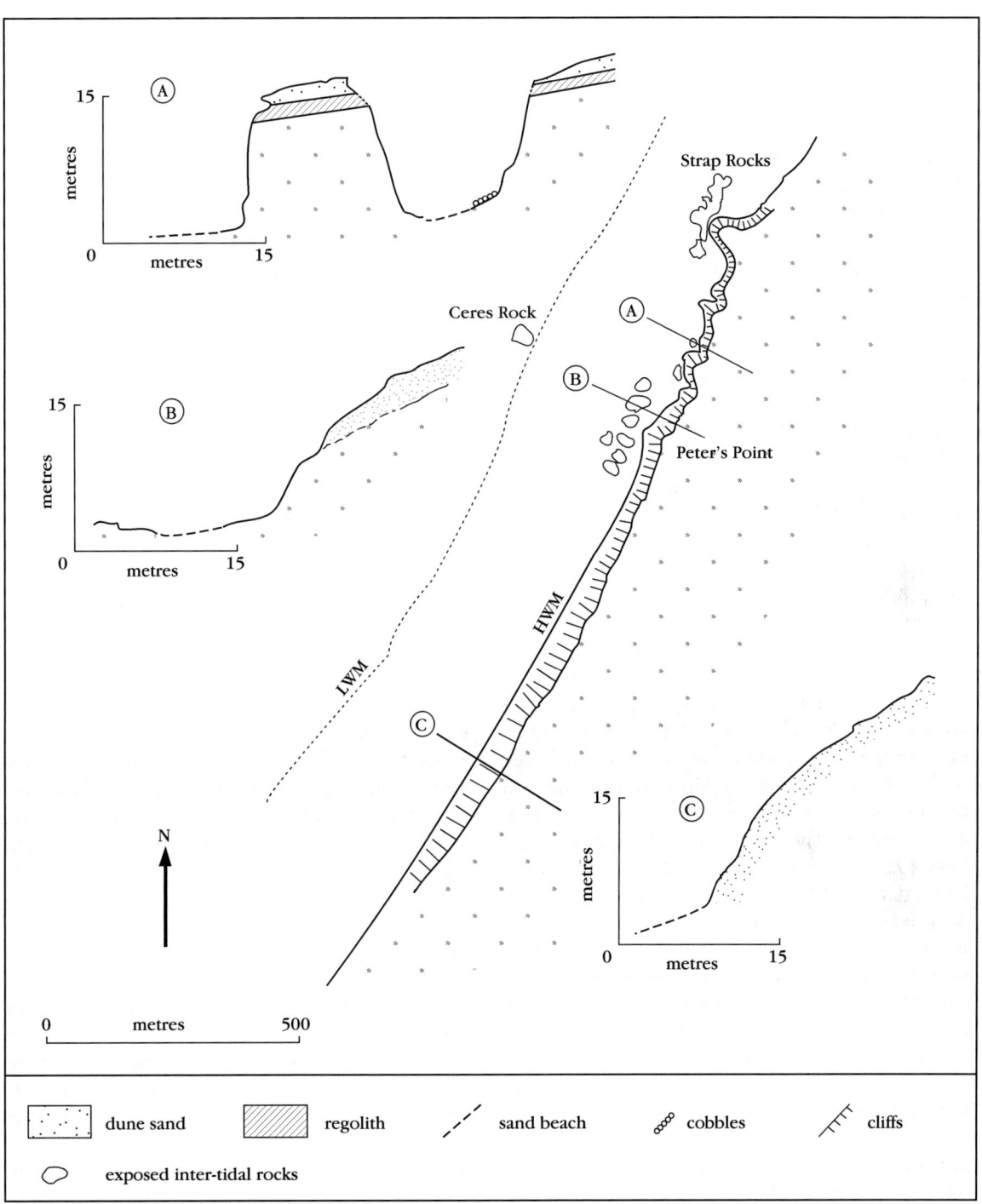

Figure 7.8 Relationships between dunes and cliffs at Peter's Point. Profiles through section A, B and C are shown.

the last glaciation.

Taken as a whole the site demonstrates a sequence from sandy dune shoreline through progressively dominant erosional rock forms to stacks with residual dune deposits atop them. As such, it is a good example of the cyclic nature of the processes affecting the British coastline.

Erosional processes were replaced by depositional marine or periglacial processes, the resultant forms then being eroded and the earlier erosional forms exhumed. Even on a shorter timescale of less than 100 years, much of this coastline is marked by oscillation between erosion and deposition.

Sandy beaches and dunes

The age of the exhumed cliffs is not known. It is possible that they merely pre-date the dune growth and migration recorded along much of the Cornish coast during medieval times. Steers (1946a) notes for example the spread of the dunes in Perran Bay that engulfed St Piran's Chapel. Balchin (1954) suggests that St Piran's Chapel was buried before the 12th century. Leland (1535–1543) described St Ives in the 16th century as 'sore oppressed or over covered with sandes…'. Steers (1946a) also refers to 'the east side of St Ives Bay where the dunes have buried St Gothian's Chapel and by 1907 had banked themselves around the walls of Millook churchyard'.

In the absence of a firm date, the contemporary origin of the cliffs must be considered. Assuming that the erosion of the dunes was sufficiently rapid during the late 1940s to bring the sea to the foot of the sub-dune rocky topography, all of the features now in existence at Strap Rocks could have formed since the 1950s. However, although erosion is undoubtedly taking place there is no evidence that the cliffs are eroding at a sufficiently rapid rate to produce the forms during this time. The presence of a layer of regolith and small angular clasts (possibly head) beneath the dunes that is continued on to the top of the stacks suggests that this was laid down before they were isolated. This suggests a substantial period of cold conditions followed by warmer conditions after the cliffs were formed, and so the cliffs could be pre-glacial and have been re-occupied in the Holocene Epoch. Subsequent erosion has removed the surface between the stack and the mainland. On balance it is likely that the forms substantially pre-date the dunes, have been exhumed and are being reworked at present.

Conclusions

The Upton and Gwithian Towans GCR site contains an unusual set of forms that warrant further investigation. Cliff-forms and erosional features are being exhumed at this site from beneath a formerly more extensive dune system. They are unusual within Great Britain because there are few sites where erosional forms are being exposed by the removal of dunes at present. As dunes to the south have been eroded they have exposed former cliffs, caves and stacks.

BRAUNTON BURROWS, DEVON (SS 440 350)

V.J. May

Introduction

Braunton Burrows (see Figure 7.1 for general location) is one of the three largest dune systems on the west coast of Britain. The dunes extend about 6.5 km southwards from Saunton Down across the lower valley of the rivers Taw and Torridge. The dune belt is about 1.3 km wide throughout this length and is fronted by a sand beach, which, in places, exceeds 1 km in width at low tides. At the north end, the structure of the dune system is influenced by the hills of Down End, at its southern end by the Taw–Torridge estuary. Individual dune ridges, which sometimes exceed 30 m OD, are best developed in the central part of the system. Although wartime use did extensive damage, this proved to be short-lived, and the dune complex remains quite outstanding (Figure 7.9).

Although considerable research has been carried out in adjacent areas (e.g. Prestwich, 1892) into the 'raised beach' erratics, Mitchell (1960), Kidson (1977) into the interglacial and periglacial deposits of the Taw estuary, and McFarlane (1955) and Davies (1983) into the 'buried channel' of the Taw–Torridge, little geological or geomorphological data specifically concerning the dune system at Braunton Burrows has been reported in the literature. The presence of former dunes and glacial erratics along the northern cliffed coast of the site (Campbell *et al*., 1998) raises important questions about the long-term history of the site. There is, however, an important ecological literature, which led Ratcliffe (1977) to record Braunton Burrows as one of the best-described dune systems in Europe. This includes some geomorphological information (see for example Willis *et al*., 1959a,b; Willis, 1963, 1965, 1967, 1985, 1989; Willis and Jefferies, 1963; Hope-Simpson and Jefferies, 1966; Hewett, 1970, 1971; Hope-Simpson and Yemm, 1979; Hope-Simpson, 1985, 1997; Boorman, 1993). Kidson and Carr (1960) provided a brief summary of the structure of the system in the context of the dune stabilization programme carried out mainly in the central area during the 1950s, and Greenwood (1969, 1978) examined the textural contrasts between the beach and dune deposits.

Braunton Burrows

Figure 7.9 Aerial photograph of dunes and Crow Point. 1, Westward Ho! cobble beach; 2, Taw–Torridge estuary; 3, Crow Point; 4, Airy Point; 5, Braunton Burrows showing main dune ridges and blowthroughs; 6, ridge-and-runnel beach. (Photo: courtesy Cambridge University Collection of Aerial Photographs, Crown Copyright, Great Scotland Yard.)

A number of unpublished reports have been written concerning the aggregate extraction around Crow Point at the mouth of the Taw–Torridge estuary (Figure 7.9). Probably the most relevant in the present context is that by Blackley *et al.* (1972) on the movement of sediment labelled with tracers within the Taw–Torridge estuary area. A review of the geomorphology and management of the Taw–Torridge estuary including Westward Ho! and Braunton Burrows includes details of erosion and sedimentation (Comber *et al.*, 1993)

An extensive programme of topographical surveying of the dunes was undertaken by Kidson and co-workers between 1957 and 1960, during which time the whole dune system was mapped on a scale of 1:2500. Revisions using aerial photography were carried out subsequently. A series of profiles was surveyed across the central and southern dunes annually from 1956 to 1962, primarily to monitor the effects of replanting. Thereafter surveys were less extensive and less frequent because over most of their section-length the dunes had been stabilized. Measurements of blowthrough orientation, and an investigation into the nature and periodicity of the changes in the underlying water-table, were also carried out by Kidson and Hewett, respectively, as part of the work of the Physiography Section of the former Nature Conservancy. Similarly, Kidson and Carr examined the form of the underlying rock surface by means of earth resistivity and refraction seismography, supplemented by boreholes. Very little information concerning these studies has been published. Some maps were produced for the 1964 International Geographical Congress (Kidson, 1964a), but most information remains in the form of partially processed field data. This was remedied to some extent by Kidson *et al.* (1989) who contrasted the earlier surveys with further surveys carried out in 1983, although the description of the dunes that follows is based in part on unpublished English Nature data. Sarre (1989) concentrated specifically on foredune processes and the ways in which they are affected by variations in relief and vegetation.

Description

The dunes at Braunton Burrows are formed mainly of sand with grain size generally between 0.2 mm and 0.3 mm diameter (Willis 1985). They extend from the cliffs below Saunton

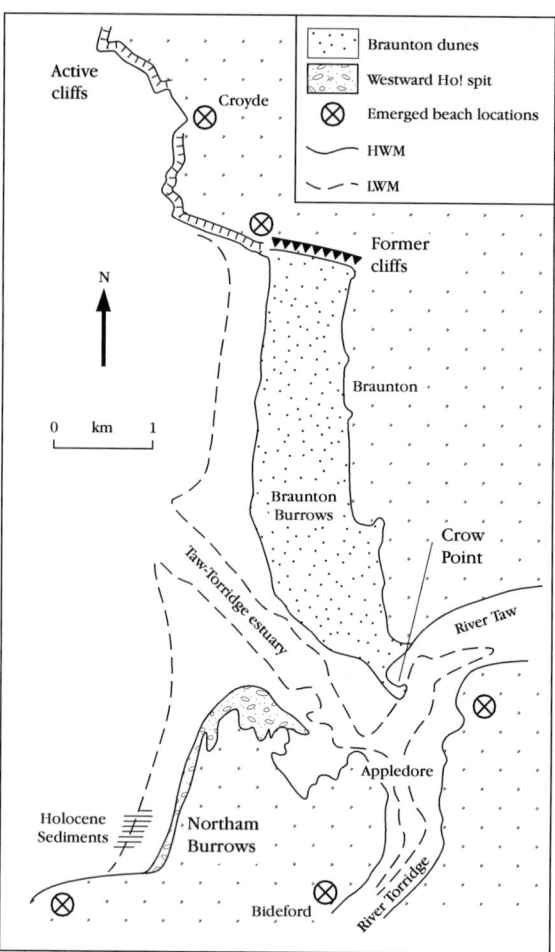

Figure 7.10 Braunton Burrows and Westward Ho! GCR sites, showing locations of emerged beaches and generalized geomorphology. See also Figure 7.11 for photograph of the area around Crow Point.

Down, south for some 5 km to Airy Point (SS 448 330), and thence south-eastwards for a further 1.5 km or so to the narrow strip of land that culminates in Crow Point (SS 466 317; Figures 7.9 and 7.10). In the central area, where the dune structure is best developed, the Burrows consist of three ridges, separated by slacks. The ridges lie parallel to the shore and to each other, within an overall width of about 1.3 km. It is in this area that the highest parts of the dunes occur. In both the north where the dunes abut on to the high land of Saunton Down and in the south, where the shoreline trend changes in response to the Taw–Torridge estuary, the system of three ridges is replaced by a less well-defined double ridge system. Throughout the dunes, but especially in the south, there are a number of sub-ridges perpendicular to the main dune alignment and the coast. They appear to be the legacy of major blowthroughs in the main dune alignments, and may form the northern and southern boundaries of the slacks. There are a small number of parabolic dunes, particularly in the southern part of the dunes (Figure 7.9).

At the rear of the system, there is an extensive area of low dunes and slacks. This is best developed in the central zone; to the south, it becomes narrower, while to the north, it has been modified into a golf course. Both this low area and the slacks elsewhere may be extensively flooded during winter.

For much of its length the seaward dune ridge, usually rising to about 15 m OD, is fronted by a more or less continuous line of foredunes rising to some 4.5 m above OD. The elevation of the slacks is highest in the middle of the central zone of the dunes at about 9 m OD. Towards the north, the surface falls slightly (by the order of 2–3 m) and by rather more at the southern end. There, the lowest slack areas are at approximately 4 m OD, so that if the seaward dune ridges were breached, the slacks would be inundated at high water on spring tides (tidal range being 7.3 m). In the central zone, in particular, slacks nearest the shore are somewhat lower. Willis *et al.* (1959a) observed that the water-table underlying the system was dome-shaped, being some 6 m higher in the centre than at its margins (Figure 7.11). They believed wind deflation took place down to this level. Unpublished work (by D.G. Hewett) shows that the neap-spring tidal cycle can be reflected in the water levels under the dunes for a considerable distance inland.

The central zone of the dunes underwent the heaviest pressure through military usage during World War II, and thereafter. Comparison between maps made by the Ordnance Survey in 1885 and those of the Nature Conservancy produced between 1957 and 1960 indicate major topographical changes over the intervening

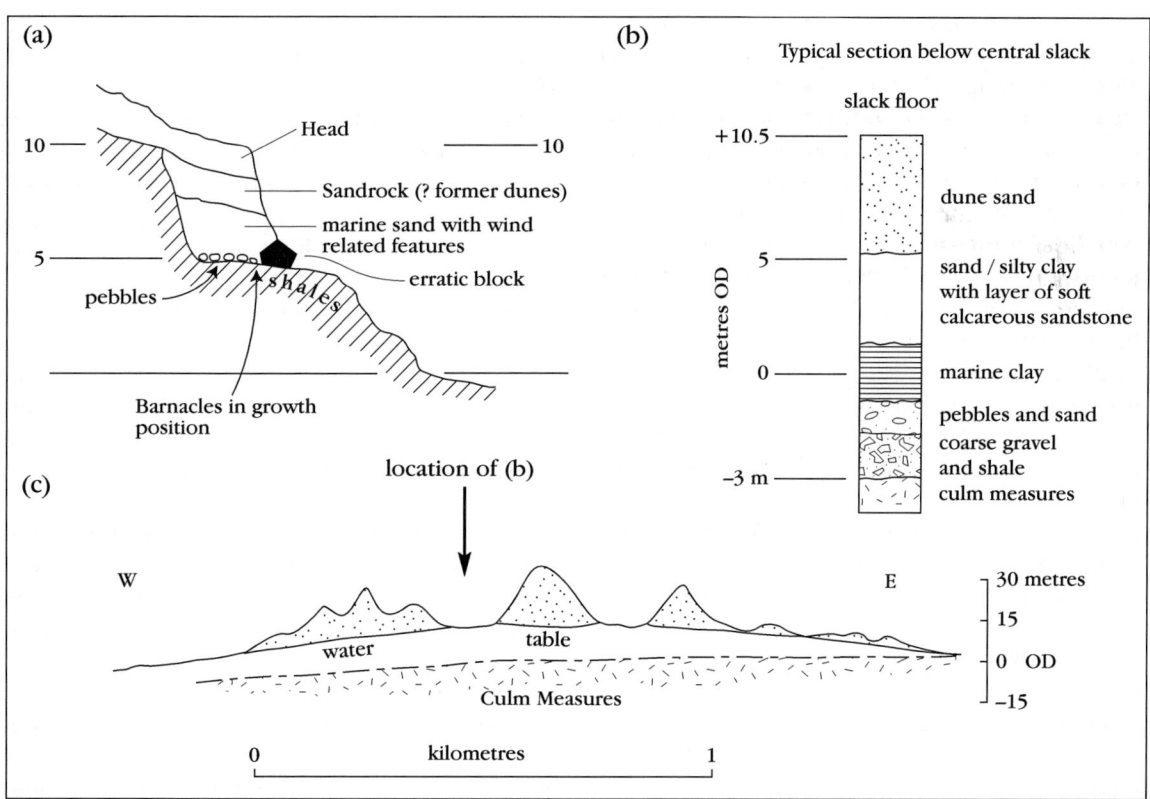

Figure 7.11 Emerged beach profile and dune features at Braunton Burrows GCR site. (a) Section through emerged beach and possible former dunes at Saunton Down; (b) section through the central slack within the main dunes, showing that the dunes lie on both marine clay and gravels and sand resting on the underlying Culm Measures bedrock.; (c) cross-section of the dunes showing the relationship of the slacks to the water table. (Based on Keene 1996; Willis 1985; and Willis *et al.*, 1959a.)

period. These were particularly acute in this central zone where some parts showed an actual inversion of topography with slack areas in 1885 becoming the intermediate ridge of 1957. Mobile dunes continue this process and are encroaching on the slacks at present. Kidson *et al.* (1989) noted that the main dune crest moved eastwards by up to 60 m between 1885 and 1958. Between 1958 and 1983 there was a net gain of sand along the central foreshore, although the source is uncertain.

During the early 1940s, the widespread use of the dunes for military training prior to 'D-Day' (6 June, 1944) caused extensive damage, leaving a 'semi-desert of shifting sand' (Breeds and Rogers, 1998). Furthermore, mine clearance in 1946–1947 was carried out using high-pressure hoses. This destroyed foredunes and rebuilding was encouraged by emplacing brushwood or hessian fencing in the early 1950s. Similar techniques, followed by planting of marram *Ammophila arenaria*, were used to redevelop the middle and landward (main) dune ridge to a more uniform crest height (Hewett, 1970). Some 5–6% of the dune system was replanted between 1952 and 1963, but the overall impression is of a much greater proportion having been so treated. In some respects, the dunes have been over-stabilized so that the characteristic contrast between dynamic growth of both foredunes and major ridges, compared with slacks and older dunes to landward, has been reduced, greatly diminishing their geomorphological interest. This over-stabilization is also linked to the reduction of the rabbit population through myxamatosis in 1954–1955. During the inter-war years, the Burrows had been heavily disturbed by rabbit warrens. Reports in the local papers describe the need to move the (then) lifeboat house in 1857, 1862, 1882, and 1892 due to burial or erosion in the foredunes area and suggest that instability already existed prior to military use in the 1940s. All this supports Willis *et al.* (1959a) who argued that the dynamic character of the system was not solely a result of the recent military activity.

Probably the most striking change between the 1885 and 1957–1960 surveys was the increase in the area occupied by high dunes. It has been calculated that there was a 35-fold increase in the area of land in the central Burrows exceeding almost 30 m OD over that period (Kidson and Carr, 1960), while in the northern third of the system isolated areas topped the 30 m contour for the first time. The more recent survey by Kidson *et al.* (1989) suggests that this trend had continued up to 1983, with the system as a whole having a positive sediment budget.

Willis *et al.* (1959a) describe the contrast in the soils of Braunton Burrows between the highly calcareous (pH 9) high dunes with their low moisture content, and the gley soils of the often far wetter slacks. While the slacks have a somewhat lower pH, they are still alkaline. Plant cover, which depends upon root depth, water availability and nutrient status, strongly affects the likelihood of aeolian erosion. Water availability to plants depends largely on the depth of sand above the water table. During dry periods, sand more than 0.5 m above the water table has a water content less than 5%, but sand within 0.3 cm of the water table is maintained close to saturation and between 0.3 and 0.5 m above the water-table, capillary action maintains moderately high water-content (Willis, 1985). Fluctuations in the level of the water table are strongly correlated with rainfall (Willis *et al.*, 1959a). The combination of a period of lower-than-average rainfall between 1983 and 1992, deepening of the western Boundary Drain at the eastern edge of the Burrows in 1983, and drainage on the golf course, may have led to a general lowering of the water table (Burden, 1997; Packham and Willis, 2001).

It appears that underneath much of the dune system there is a relatively flat bedrock surface at approximately −3 m to −4 m OD, i.e. typically about 10 m below slack level (Figure 7.11). There is some evidence to suggest lower bedrock surfaces not only in the beach–foredune area, but also in parts of the estuary and inland. Keene (1996) suggests that surfaces lie at about −10 m OD below the Burrows either side of the estuary. However, the extreme depths (about −30 m OD) for the nearby estuary bedrock channel suggested by McFarlane (1955) have been disputed by Davies (1983) who suggested −20 m to −21 m OD as more realistic.

A borehole (Figure 7.11) sited in a slack with its floor at about 10 m OD and almost in the centre of the dunes passed through approximately 5 m of sand. Below this level, there were some 4 m mainly of sand, silt and/or clay, but including a 1 m-thick band of soft calcareous sandstone. Underlying these deposits were 2.5 m of marine clay. This, in turn, was underlain by pebbles and sand and then coarse gravel and shale;

less than 2 m in total. Bedrock (Culm Measures) was ultimately reached at −3 m OD. Another borehole located at the extreme northern end of the dune passed through head before reaching a landward extension of the emerged ('raised') beach occurring in the cliff sections of Down End, Saunton. Such cliff sections may show remains of an earlier lithified analogue of the present dune system in juxtaposition to the emerged beach. A coastal platform underlies this feature, which is overlain by head. Kidson (1977) regarded the head as Devensian, the emerged beach as Ipswichian.

Interpretation

The history of Braunton Burrows, both recent and long term, raises a number questions about its origins, its long-term stability, its dynamics and resilience during recent decades of disturbance and the relationships between its ecology, geomorphology and hydrology. Packham and Willis (2001) believe Braunton Burrows to be over 2000 years old, but the Quaternary sediments along its northern fringe indicate that a dune system was here in Early Devensian times (Campbell and Gilbert, 1998), i.e. about 70 000 years ago. During the second half of the 20th century, the natural changes in the dunes have been modified by the effects of mine clearance, movements of heavy military vehicles, grazing and alterations to the hydrology.

The evidence of emerged beaches and other sediments along the southern coast of Saunton Down is important for the interpretation of the development of Braunton Burrows. The emerged beach has been the focus of attention since the late 19th century (Sedgwick and Murchison, 1840; Hall, 1870; Hughes, 1887; Stephens, 1966, 1974; Gilbert, 1996). Campbell and Gilbert (1998) have described the shore platform that extends from Saunton to Croyde as one of the finest examples anywhere in Britain. However, its three main surfaces are, as yet, of undetermined age. Erratics, which rest on the platform, indicate an origin earlier than the last glacial period. The presence of calcarenite up to 3 m thick that includes preserved flute features similar to modern wind-sculpted features in the dunes indicates the presence of earlier dunes banked against the rock slopes (Evans, 1912; Keene 1996). The fact that these possible old dunes are overlain by head (Figure 7.11) supports the suggestion of preservation of interglacial dunes but points to the area having had a similar coastal landscape to that of today. The emerged beach sequence at Saunton has been attributed an Early Devensian age (*c.* 70 000 years BP; Campbell and Gilbert, 1998). In contrast to the southern part of the bay at Westward Ho! (see GCR site report in Chapter 6), there has been only limited description of intertidal submerged forest remnants. Rogers (1946) records that an oak trunk 9 m in length was identified at Braunton Burrows in 1630 (Steers, 1946a). However, there is no record of peat beds being exposed in the seaward edge of the dunes and boreholes through the dunes have not yet proved peat. Dating of the dunes has not yet been possible as a result.

Interpretation of the more recent evolution of the site is mainly dependent upon cartographic evidence from the beginning of the 19th century. These show that over the northern two-thirds of the site the position of the foredunes is substantially the same as it was 150 years ago in spite of post-1945 destruction. However, erosion has occurred near the estuary. Sand and gravel extraction in the estuary appears to date from the 18th century or earlier, although cartographic evidence suggests that Crow Point continued to grow until at least the mid-1850s. By 1874 it was necessary to erect the first groynes on Braunton Burrows in the vicinity of the lighthouse. At a public inquiry held in 1905 evidence was given that erosion was such that high tides could virtually reach the lighthouse. Nevertheless, extraction continued to be permitted. During the decade 1960–1970, some 600 000 tonnes of sand and gravel were removed from the Braunton side of the estuary, yet the apparent cause and effect of dune erosion and estuarine extraction were denied. A tracer experiment in 1971 showed that, with waves from the north-west, sediment from Airy Point reached the extraction site at Crow Point. Since that time the Crow peninsula has been breached.

Braunton Burrows, apart from the area near the estuary, thus demonstrates a surprising degree of resilience to disturbance. Many other coastal dune systems were also disturbed by wartime military activity and post-war mine clearance operations and have also recovered, but in general they were less exposed to onshore winds than Braunton Burrows. Since World War II, the general location and breadth of the beach has changed little suggesting that despite sand migration inland there continues to be some replacement from the nearshore zone (Kidson *et al.*, 1989). The progressive expansion of more woody plants over the dunes has also added to their stabilization (Packham and Willis, 2001). It has been suggested (Packham and Willis, 2001) that the ecological significance of the site depends upon management that encourages an increase in biodiversity and this may be achieved by increased levels of sand movement. Similarly, the geomorphological interest will be enhanced by allowing more dynamism to occur. Even with the extraction at the distal point, and the associated erosion, the dunes as a whole appear to have remained in a positive sediment budget unlike some smaller systems such as Dawlish Warren.

Conclusions

Braunton Burrows is one of the three largest dune systems in western Britain, over 6 km in length, up to 1.5 km wide and attaining heights over 30 m. It is not only one of the largest and most complete dune systems on the coastline of England and Wales, but also unusual in its resilience to serious disturbance and its maintained growth. It is also one of the few such areas where changes in the dune topography have been surveyed regularly. Its ecological importance was recognized in 1964 in its designation as a National Nature Reserve, but lack of grazing decreased its nature conservation interest, and it ceased to be an NNR in 1996. Nonetheless, an agreed experimental programme of grazing was established in 1998 and the site retains its SSSI status; it is a candidate Special Area of Conservation. In terms of geomorphological interest, a conservation policy that allows greater sand mobility within the dunes will only serve to enhance its value as a GCR site.

OXWICH BAY, GLAMORGAN (SS 510 870)

V.J. May

Introduction

Oxwich Bay, on the south coast of the Gower Peninsula (see Figure 7.1 for general location) supports a well-developed system of dunes. As at Pwll-ddu, to the east (see GCR site report in

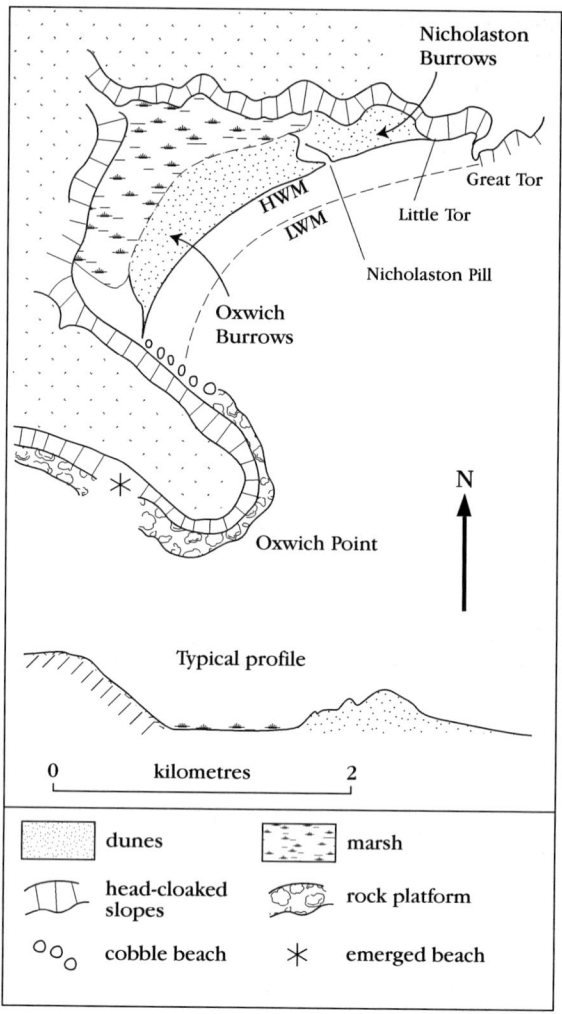

Figure 7.12 Key geomorphological features of Oxwich Bay, together with a typical profile.

Chapter 8), the beach and dunes have diverted a small stream towards the east. Oxwich Bay includes part of the Oxwich Burrows National Nature Reserve (NNR), a wetland site behind the dunes, but its main geomorphological interest lies in the association of the dunes with bounding cliffs which restrict the transport of sediment into and out of the bay and modify the behaviour of waves within the bay. Research has concentrated mainly on the Quaternary deposits that rest at the foot of the coastal slopes and cliffs (Campbell and Bowen, 1989), and demonstrates that this cliff is mainly a Pleistocene relict feature variously reworked along its foot during the Holocene Epoch.

Oxwich Bay

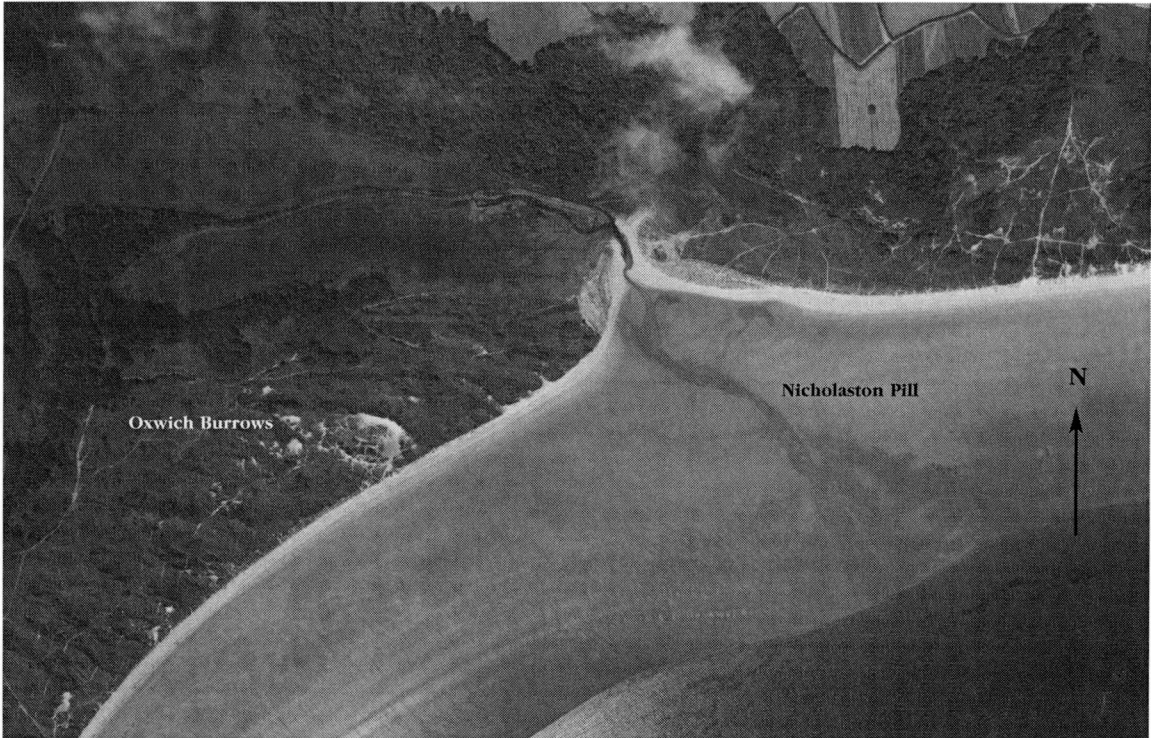

Figure 7.13 Oxwich Burrows. Linear dune ridges are evident, with alignment close to right angles to the shore at the western end of the dunes. Towards the east, the ridges retain this orientation close to the shore, but have a more east–west alignment inland. Similar ridges are absent from the dunes east of the stream mouth. (Photo: courtesy Cambridge University Collection of Aerial Photographs © Countryside Council for Wales.)

Description

The site lies between Great Tor (SS 530 876) in the east and Oxwich Point (SS 513 850) in the west. The re-entrant form of the bay is controlled by a syncline in Namurian strata aligned north-west–south-east. There are three main landforms (Figure 7.12).

1. An area of dunes that rests against slope-over-wall cliffs in Nicholaston Burrows and between Little Tor and Great Tor. At the extreme eastern end of the site the cliffs, over 60 m in height, fall directly to an active beach.
2. A single major dune system, Oxwich Burrows, and an intertidal sandy beach, backed by scrub and saltmarsh.
3. Cliffs, mainly of slope-over-wall type, between Oxwich Point and Oxwich Castle. Only the lower part of the cliff, the 'wall' and shore platforms have been included in the site.

Largely sheltered from the main energy of waves approaching the Gower Peninsula from the south-west, the beach in Oxwich Bay swings from a east-south-easterly aspect at its southern end to a southerly aspect at its more exposed eastern end. The beach extends eastwards beyond Great Tor at low tide into Three Cliffs Bay. The dunes are crossed by a small stream, Nicholaston Pill, which drains marshland behind the dunes. It is likely that sand moves out of the bay towards the east at low tide, especially when wave energy is high. Since the dunes remain, this implies that there is a balancing supply from offshore. The shallowness of the bay (mainly less than 10 m in depth), and the unidirectional wave energy at the beach also serves to enhance the sand supply to Oxwich Bay.

Interpretation

The generally accepted history of the site (Ratcliffe, 1977) suggests that the main beach ridge developed about 2500 years BP enclosing a brackish lagoon. This was land-claimed during the 16th century and, subsequently, fish-ponds were dug in the wetlands behind the dunes. Since the formation of the ridge, the dunes appear to have been stable, even growing in extent. Sufficient sand has accreted to fill the area of Nicholaston Burrows on the eastern side of Oxwich Bay, most of which was probably a rocky coast when the ridge was first formed. To the west of this site, the 60 m coastal platform is truncated seawards by a relict cliff, which is partially buried by superficial deposits (Campbell and Bowen, 1989). A shore platform cut in Carboniferous Limestone at about 10 m OD forms the lower part of the slope. At Nicholaston Burrows, the slopes have a similar upper form, but the lower slopes are masked by the dunes banked against them. This is the only substantial length of former cliffed coast on the southern side of the Gower Peninsula that is protected in this way by dunes.

This is undoubtedly a beach that has had a positive sediment budget until recently. The increased recreational use of the dunes during the last three decades has brought about much localized erosion, associated with trampling and blowthroughs. The erosion of the frontal dunes might suggest that the whole site is in deficit or at least is trending towards such a state. However, a brief consideration of the sediment budget of the site suggests that this is not likely, when the whole bay is considered. First, there is only a limited possibility for longshore transport of sediment out of the beach. Second, sand is transferred within the site, either by wind or by waves so that erosion in part of the beach and dune area appears to be balanced by deposition elsewhere. Third, this beach could be expected to act like other bay-head beaches that lie in locations where waves are strongly refracted during their approach towards them.

Unfortunately, the most vulnerable part of the shoreline is the main point of access to the beach. Most trampling occurs at the most likely point for erosion. Sand is transferred mainly from beach to dunes and *vice versa*, but because the total volume of sand is limited the dunes have been constrained both in height and in their growth seawards. Equally there has been no tendency for the dunes to migrate inland since they are largely stable.

Conclusions

Oxwich Bay is a structurally controlled bay within which nationally important dunes have developed despite restricted longshore sediment sources. It had, until very recently, a positive sediment budget, unlike many beaches worldwide. Trampling of the dunes by visitors appears to have led to instability, hence its current tendency for erosion at its western end. Its sheltered, low-energy, environment has allowed the growth of low – but nationally important – dunes, despite a restricted sediment input. The combination of a low-energy beach in a macrotidal setting, relict cliffs against which it is banked, its bay-head location and the biologically important but restricted dunes make this a nationally important site. It is an important member of the national GCR network of bay-head beaches.

TYWYN ABERFFRAW, ANGLESEY (SH 362 685)

V.J. May

Introduction

Many of the structurally aligned valleys that cross Anglesey (Ynys Môn) form shallow estuaries on the south coast, including the largest, the Afon Cefni, which lies within the Newborough Warren GCR site (see site report in Chapter 11). Tywyn Aberffraw comprises a small beach and area of dunes, which lie to the west of the much larger dunes of Newborough Warren, and occupy a confined valley site (see Figuer 7.1 for general location). Because of the physical constraints of the locality, there is little possibility of sand entering or leaving the bay alongshore, and the bounding cliffs supply very little sediment to the beach (Robinson, 1980b). Tywyn Aberffraw provides an excellent opportunity for the study of beach and dune relationships within an area of restricted sediment supply. The site is also distinguished by the relative isolation of individual fixed parabolic dunes upon a sand plain (Steers, 1946a; Ranwell, 1955; Robinson and Milward, 1983). This landform assemblage has few comparable equivalents in Britain.

Description

Tywyn Aberffraw is a small site, with a beach only 700 m wide between hard-rock headlands, but the sand plain and dunes extend about 2.5 km inland (Figure 7.14). The site fills the western end of Traeth Mawr, one of several low-lying basins that cross Anglesey from south-west to

Tywyn Aberffraw

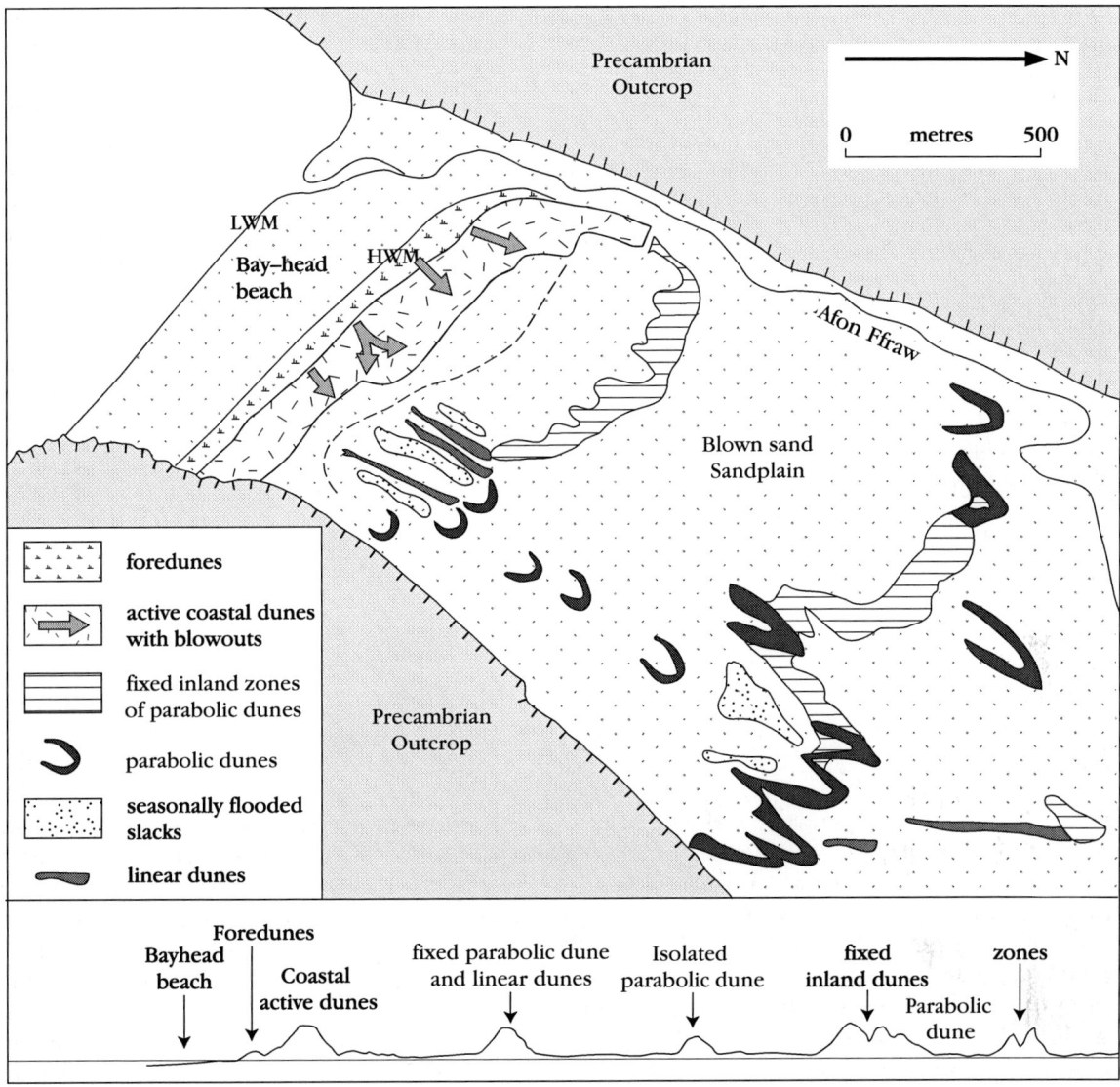

Figure 7.14 Key geomorphological features and profile of the Tywyn Aberffraw GCR site. (After Robinson and Milward, 1983.)

north-east and cut here into the Precambrian grits, shales and lavas of the Gwna and Fydlyn groups. The valley is filled at its south-western end by sand, almost all lying within the rocky confines of the valley, but there has been very little migration of sand on to the higher rocky surfaces on either side. Broadly, the site falls into four morphological units: the cliffs, the beach (swell-aligned towards the south-west) the active dunes, which form a triangular area less than 250 m wide at the western end of the beach but widening to about 1 km along the eastern boundary of the site, and the sand plain and mainly fixed dunes that form most of the site.

The cliffs form low (rarely more than 7 m) vertical features cut into the Precambrian rocks, above which there are more extensive slopes rising to about 40 m OD. These slopes bound the site on both sides, with a small stream, the Afon Ffraw, flowing at the base of the western slope. Although the slopes are not included within the site, they have confined sand to the valley floor and provide some shelter from winds from the west and east. The beach is about 300 m wide at low water and is composed almost entirely of sand. Waves can only approach the beach from a very narrow range of

Sandy beaches and dunes

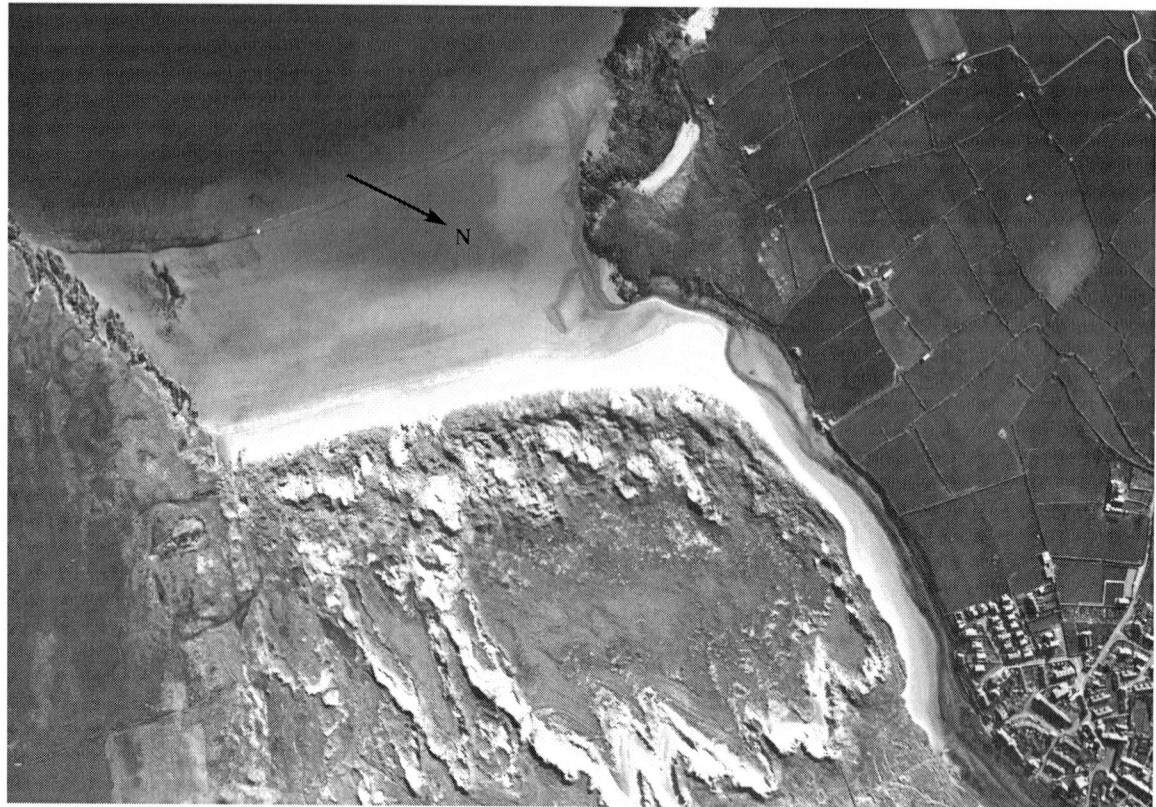

Figure 7.15 Aerial photograph of Aberffraw, Anglesey, for comparison with Figure 7.14. (Photo: Cambridge University Collection of Aerial Photographs © Countryside Council for Wales.)

directions (south to WSW). Waves approaching from the south-west undergo no refraction and fetch to the south-west exceeds 4000 km. The beach appears to have undergone only limited retreat in recent decades. The active dunes are affected by some recreational trampling and are eroded periodically by wave action. Wind action carries sand along the eastern side of the site, while on the western side the Afon Ffraw it prevents accumulation along the foot of the bounding slope. The western slope is, however, also sheltered from the strongest winds. As a result, along the eastern edge of the site the active zone extends almost 1 km inland. The main part of the site is dominated by a sand plain that slopes from over 14 m OD in the east to about 8 m OD in the west. Several isolated vegetated parabolic dunes rise above the sand plain, one of the best examples being at SH 367 694.

Interpretation

This site occupies a similar position on the coast to that which Ranwell (1955) identified at Newborough Warren, a few kilometres to the east, as being an ideal location for maximum sand movement by wind and therefore a prime site for the study of dune development and migration. Unlike Newborough Warren, the beach at Tywyn Aberffraw is extremely restricted laterally with limited sand removal by longshore drift. The beach retains an almost constant orientation because the adjacent headlands restrict the direction of wave approach. As a result, this can be described as 'an almost perfect bay-head beach' (Robinson, 1980b, p. 42) for which the offshore zone is the most likely source of sediment for the accumulating beach. In turn the dunes rely entirely on the supply of sand from the narrow beach and some recycling by the stream. The restriction of the site between higher ground has also affected the patterns of wind transport because the western side is more sheltered from westerly winds and the valley widens away from the beach. The sand available for maintenance of dunes, which are

Ainsdale

migrating inland, is therefore limited both in quantity and in its spatial distribution. However, the site is grazed by sheep, cattle and rabbits (Ashall *et al.*, 1995; Potter and Hosie, 2001), which may affect sand mobility locally. Unlike Newborough Warren, where there is a tendency for linear dunes to migrate inland (Ranwell, 1955), dunes at Tywyn Aberffraw tend to be isolated and lower. This is almost certainly a result of a limited sediment input and a minimal longshore supply. Parabolic dunes stand in isolation above the plain in a landform assemblage that is uncommon in England and Wales.

The internal structure of one of the largest parabolic dunes consists mainly of landward-dipping accretion surface, both on the windward (foreslope) and the leeward (rearslope) of the dune (Bristow and Bailey, 2001). A large area of trough cut-and-fill identified by ground-penetrating radar on both the windward slope and the dune crest indicates that sediment is being moved from the windward side and transported to the leeward side. As the dunes migrate landwards, they preserve some of the older landward-dipping surfaces of earlier phases of accretion.

Apart from Newborough Warren, probably the best comparable coastal sites are the Sands of Forvie and Barry Links on the north-east coast of Scotland, where a number of parabolic dunes stand above a sand plain (see GCR site reports, this chapter).

Tywyn Aberffraw is an important member of the GCR network of dune systems because of its relatively limited sediment supply and restricted development of dunes. In this respect it contrasts especially strongly with its neighbour at Newborough Warren.

Conclusions

Tywyn Aberffraw is an almost perfect geomorphological exemplar of a bay-head beach that is backed by dunes that include probably the best example of a sand plain with isolated parabolic dunes in England and Wales. The site combines the unusual attributes of a near-perfect sandy bay-head beach and a sand plain with isolated parabolic dunes. Furthermore, its continued development can be shown to depend almost exclusively upon offshore sand sources and fluvial recycling rather than any longshore sediment transport.

AINSDALE, LANCASHIRE (SD 285 105)

V.J. May

Introduction

Much of the coastline of Merseyside and Lancashire is dominated by dunes and very wide intertidal sand and mudflats, but it has also been developed as residential areas and/or resorts, for example Formby, Southport, Blackpool and Morecambe. Serious problems of erosion and coastal flooding have led to parts of the shoreline being strengthened by sea-walls and embankments, and many of the original dune areas have also been damaged or destroyed by afforestation and urbanization. Ainsdale is a National Nature Reserve primarily because of its important dune flora and fauna, but it also includes features of considerable geomorphological importance; predominantly its dunes and the multi-barred ridge and runnel foreshore (see Figure 7.1 for general location).

Much of the shoreline is affected by erosion, but there are relatively stable bar forms in the intertidal zone. Transport in this zone is predominantly alongshore but has had little influence on the erosion of the shoreline. This contrasts with the effects of changes in the intertidal zone both at Spurn Head and on parts of the Belgian coast (de Moor, 1979). There are many different bedforms displayed upon the foreshore. This is not unusual in itself, but the importance of the site lies in the considerable research that has been carried out. In this respect, it offers an excellent opportunity for comparisons with the sandy shoreline at Gibraltar Point (see GCR site report in Chapter 8) which also lies in a macrotidal environment, but which has a different wave climate.

This coastline was the focus of one of the first regional coastal monographs (Ashton, 1920). Its importance in studies of the evolution of the coastline has been continued to the present time (e.g. Gresswell, 1937, 1953a,b, 1957; Tooley, 1974, 1976, 1978, 1982; Parker, 1975; Kidson and Tooley, 1977; Bird, 1985; Bird and Schwartz, 1985; Innes and Tooley, 1993). In addition, considerable attention has been given to the forms and processes of the intertidal areas and the dunes (e.g. Sly, 1966; Parker, 1971, 1975; Wright, 1976, 1984; Pye and Smith, 1988; Pye, 1990, 1991; Pye and Neal, 1993, 1994; Pye *et al.*,

1995). Unlike many other coastal geomorphological sites, there is also a substantial history of detailed oceanographic investigation in the offshore area (Darbyshire, 1958; Bowden, 1960, Murthy and Cooke, 1962; Lennon, 1963; Lennon, *et al.* 1963; Halliwell and O'Connor, 1966; Belderson and Stride, 1969; Draper and Blakey, 1969; Ramster and Hill, 1969). Its ecology is summarized in Atkinson and Houston (1993) and Smith (1999). Hansom *et al.* (1993) reviewed the dune morphology in the context of general erosion and sedimentation patterns in the Ribble estuary area.

Description

The site falls into eight zones (Figure 7.17a) described below as a seaward transect (Parker, 1975):

1. Inland the site is dominated by extensive largely stable dunes that rise to over 23 m OD. Many of the ridges are aligned east–west and generally reach about 16 m OD. Although there are hummocky dunes, true parabolic dunes are poorly developed.
2. To seaward, the dunes are characteristically aligned with the shoreline in a belt up to 200 m wide. They are separated by narrow slacks.
3. An active eroding dune zone up to 80 m in width. Blowthroughs (Parker, 1975), some of which form deep gullies, affect the local movement of sand inland. Damage by trampling also occurs.
4. A narrow, upper foreshore plane area described by Parker (1975) as 'a planar seaward-sloping zone lying between the most landward runnel and the sand dunes at high water mark'.
5. A zone of ridges and runnels, including as many as four ridges, 0.5 m to 1.2 m in height.
6. A lower foreshore formed mainly by intertidal sandflats, which is terminated by a low-water berm.
7. A subtidal slope.
8. Sublittoral sand ridges, 0.5 m to 1.0 m in height with a wavelength of 300 m to 500 m.

Pye (1990) recognized three phases of dune development:

1. Before 1800 – irregular hummocky dunes with incipient blowthroughs and parabolic dunes. These are fed by a positive beach sand budget, and with incomplete vegetation cover, prograde gradually seawards.
2. 1880–1906 – a series of dune ridges parallel to the coast were produced by a positive beach-sand budget and sand-trapping vegetation provided by brushwood fencing and marram *Ammophila arenaria* planting. The dunes vary from mobile to semi-fixed, and embryo dunes are still developing where beach accretion occurs.
3. Post-1906 – erosion around Formby Point and disruption of the vegetation cover produced large transgressive sand sheets. These result (Pye, 1990) from: (a) little resistance by vegetation to blowthrough development or to sand encroachment on a broad front; (b) the large directional variability of wind, and (c) the limited development and maintenance of high dunes at Formby because of heavy pedestrian-pressures.

This is a macrotidal environment, with a range at high spring tides of 8.2 m. Occasional surges raise the level of high water (Lennon, 1963). Maximum local waves occur with strong south-westerly to north-westerly winds. The most common waves have a significant wave height of 0.6 m to 1.0 m and a period of 4.0 s to 4.5 s (Parker, 1975). The highest waves may exceed 9.0 m (Murthy and Cook, 1962; Draper, 1966). Pye (1990), however, suggests that severe storm waves with period 8 to 7 s do not exceed 5.7 m in height. The intertidal slope is about 1:244. The dunes include both active and stable areas, though they are generally more stable inland. This stability is reflected in a soil chronosequence that culminates in podzols under sandy heath at Freshfield (Kear, 1985). Slacks are affected by high water-tables. Ranwell (1972) noted that a slack dominated by a mosaic of semi-aquatic plant communities described by Blanchard in 1952 was being threatened by erosion of the dunes on its seaward side. As a result of such changes in the dune morphology, former slack deposits appear from time to time on the foreshore. Fossil dune slacks have been identified on Formby foreshore (Tooley, 1976). Holocene silts and clays underlie the foreshore and affect the drying and wetting of the sand ridges. They are commonly exposed in the runnels. At Downholland Moss marine transgressive and regressive overlaps were dated to 6890

Ainsdale

Figure 7.16 Ainsdale National Nature Reserve, view looking towards the west, North of Fisherman's Path. The site important for geomorphology (it is one of the three largest dune systems of the west coast of England and Wales) as well as for wildlife. In the middle distance a 'toadscrape' has been created to encourage natterjack toads. (Photo: copyright English Nature.)

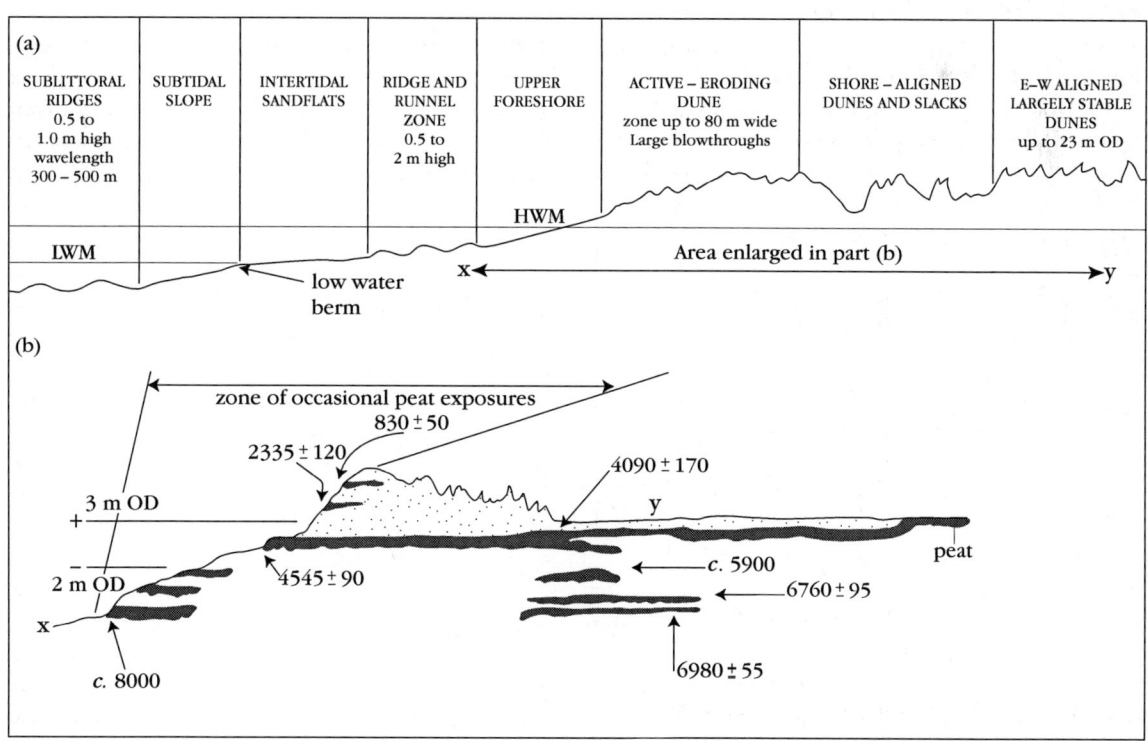

Figure 7.17 (a) Modern cross-section and zonation (eight zones) of active dune shore and nearshore zone. (After Parker, 1975.) (b) Historical schematic summary of dated peats. (After Tooley, 1978.)

± 55 years BP and 6790 ± 95 years BP at –0.87 m and –0.36 m OD (Tooley, 1976). Tooley (1978) refined this interpretation, recognizing five periods of marine transgression designated as Downholland I–V, with radiocarbon dating of key horizons as index points for sea level (Tooley, 1978).

Parker (1975) described the main processes working on both the active dunes and the foreshore. Only limited sand from the foreshore is fed directly to the dunes because even though the ridges dry, blowing sand cannot reach the dunes where the runnel between the most landward ridge and the plane area remains wet at all times. Landward sand movement from the ridge and runnel zone to the dunes is blocked. Sand is blown from the plane area along the face of the dune cliff to travel inland along the gaps in the ridge. Erosion of the sand cliff is strongly associated with wave undercutting (Figure 7.18). Most rapid retreat occurs when tides exceed +5.2 m OD. Lennon (1963) showed that tides in excess of this level are rarely produced by undisturbed astronomical tides, and so most erosion of the sand cliffs appears to be associated with storm surges at high water. In the foreshore area, the underlying Holocene sands and silts are often exposed in the runnels. They are eroded as the beach ridges move across the foreshore. The coastal profile between high and low-water mark is retreating under wave conditions that commonly approach the beach at a large angle. Retreat of the shoreline is thus associated with the processes that affect longshore sand movement and lead to lowering of the foreshore.

On the multi-barred foreshore, waves and tidal streams are the most important sources of energy for the movement of sediment, together with the abundance of sand and the influence of the short fetch on wave length. Breakers are dominant on the ridges, whereas currents (both wave-induced and tidal) predominate within the runnels. Sediment transport is predominantly alongshore within the runnels, but there is only limited movement onshore. Parker (1975) commented on the lack of understanding of the role of mud in processes of sub- and intertidal sedimentation. He found little evidence to support the suggestion (Robinson, 1964) that some channels are dominated by ebb flows in contrast to others that are dominated by flood-tide flows.

Interpretation

The general form of Ainsdale has come about as a result of progradation associated with sea-level rise. Sea floor deposits of sand were gradually transported landwards to broadly their present-day positions between 5000 and 7000 years BP (Figure 7.17). In this it is similar to many of the features of the coastline of England and Wales. It attains its status as a member of the network of coastal dune GCR sites in the dynamism of present-day processes that affect changes in the shoreline, the intertidal area and ultimately the stability of the dune system. The role of sea-level

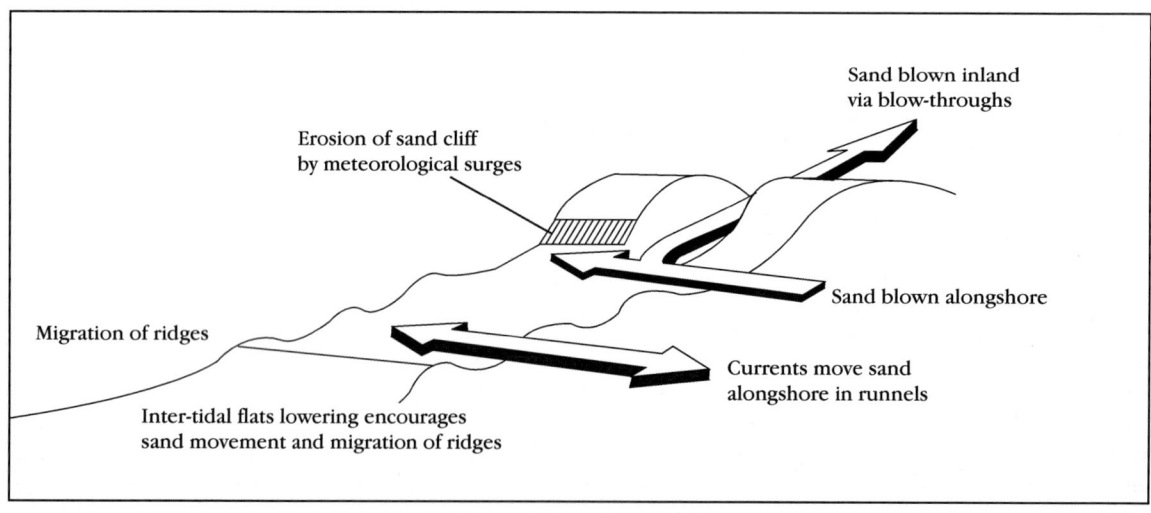

Figure 7.18 Dune-front processes at Ainsdale.

Ainsdale

change, the development and breaching of coastal barriers and progressive sedimentation have been the subject of local reports since the 17th century (Binney and Talbot, 1843; de Rance, 1869, 1872, 1877, 1878; Reade, 1872, 1881, 1902, 1908). During the early 20th century there were many studies of the biogenic sediments (Travis, 1908, 1922, 1926, 1929; Erdtman, 1926; Blackburn in Cope, 1939) and the stratigraphical record (Cope, 1939; Wray and Cope, 1948; Hall, 1954–1955). Gresswell (1937, 1953a,b, 1957, 1964) developed a model of coastal evolution based on an initially low sea level about 18 000 years BP followed by rapid sea-level rise to about 5000 years BP. Subsequent glacio-isostatic uplift caused the shoreline to retreat westwards and the sandy coast developed as a regressive wedge (Gresswell, 1953a). Gresswell (1953b) identified the former coastline of southern Lancashire at about +5.2 m OD, his 'Hillhouse Coast', but this was rejected by Tooley (1978) who argued that evidence for this former coastline was seriously flawed. Tooley (1976) showed that in the Martin Mere basin, the 'Hillhouse' coastline is not related to a marine event, but to a period of elevated lake levels.

Since Gresswell's work, the palaeogeography has been comprehensively reconstructed (Tooley, 1969, 1970, 1971, 1973, 1974, 1976, 1977a,b, 1978, 1982, 1985a,b; Tooley and Kear, 1977; Huddart and Carter, 1977; Huddart, 1992; Innes and Tooley, 1993; Pye and Neal, 1993). Tooley (1974) for example suggests that most of the constructional landforms of the Lancashire coast, such as the shingle spits, sandbars and sand-dunes (of which Ainsdale is the outstanding remaining example), were associated with extensive transgressions (probably four or five according to Tooley, 1978) from 9200 years BP to 5000 years BP. Estuarine and saltmarsh environments resulting from the transgressions before 4500 BC are preserved as the Downholland silt as much as 2.2 km inland. Although Huddart (1992) argues that there was an early barrier, a view supported by Pye and Neal (1993), Tooley (1978) and Innes and Tooley (1993) consider that although early sedimentation occurred, the main development of the sand dune barrier took place about 5000 years BP. A slightly lower sea level allowed large-scale transport of sand from the exposed intertidal and nearshore areas, probably a large offshore sandbank that was in place by 6800 years BP (Pye and Neal, 1993), to form the dunes. The seaward edge of Downholland Moss (now about 4 km from the coast) was covered by sand about 4090 ± 170 years BP (Tooley, 1978) and peat deposits at Sniggery Wood were probably buried about 4510 ± 50 years BP (Figure 7.17). 'Fossil' dune slacks or peat exposed within the beach have been dated (for example the former at Formby dated at 2335 ± 120 years BP and 830 ± 50 years BP: Tooley, 1978; Innes and Tooley, 1993 and the latter at Alt Mouth at 4545 ± 90 years BP and below low tide level dated at about 8000 years BP: Tooley, 1978). Innes and Tooley (1993) summarize the pattern as follows:

1. An initial period of sand migration and dune-building between 4600 and 4000 years BP.
2. After several centuries of sand migration, a coastal dune in place by 4000 years BP.
3. Continuing sand accumulation interrupted by marine transgressions about 3500, 2335, 1795–1370 and 800 radiocarbon years BP.
4. A recent erosional phase.

The present erosional phase is generally identified as commencing at about the beginning of the 20th century. It is reworking a substantial store of sand, but the predominant movement is offshore. Sly (1966) suggested that the area off Formby Point was marked by a divergence of bed load transport, and Ramster and Hill (1969) identified it as a zone of divergence of near-bed residual water drift. Hansom et al. (1993) studied positions of LWST and HWST at Formby Point to demonstrate erosion at LWST and accretion at HWST over the period 1841–1946, and erosion at both LWST and HWST between 1946 and 1989. Since 1906, 400 m has been lost at Formby Point with attendant foreshore steepening. This erosion has fuelled accretion in the Ribble to the north. Ainsdale has been more severely affected by erosion than other dune systems of the west coast. This appears to result in part from storm surges in the northern Irish Sea, and especially in the Mersey estuary, which affect the patterns of shoreline erosion. Between 1842 and 1906, accretion dominated. Between the 18th century and the early 20th century, the climate was relatively quiescent in terms of storm events (Lamb, 1982). Although Binney and Talbot (1843) suggested that storm events were probably the most important formative events for the evolution of this coast, sea-level rise has until recently been the more favoured explanation (Pye, 1991, 1992). Plater *et al.* (1993) however, demonstrate that storm surges have played a critical role in the erosion of the dune frontage and also in the sediment transport dynamics of the Formby–Ainsdale coast (Figure 7.18). Since about 1900 the whole frontage of about 5 km has been eroded by up to 3 m a^{-1} (Pye and Neal, 1994). However, single storm events can cause the dunes to retreat between 6 and 14 m (Pye, 1991). The high frequency of strong westerly winds has been a factor, but the construction of training walls and the dumping of spoil offshore has also played a role by focusing wave energy onto the north-central part of Formby Point (Pye and Neal, 1994).

The site is well known for its many smaller-scale features in the intertidal area, and has been the key site in Britain for the description and interpretation of ridge and runnel forms. The coastline at Blackpool and Ainsdale to its south were the location of extensive studies into the interpretation of aerial photographs prior to the Normandy landings in 1944 (Williams, 1947; King and Williams, 1949). At this time, King developed her swash-bar interpretation of ridge and runnel on equilibrium-seeking beaches, which was further elucidated in 1972 and 1982. Wright (1976) and Orford and Wright (1978) showed that, on the basis of detailed studies at Ainsdale, ridges and runnels as quasi-stationary features resulting from swash processes could be distinguished from break-point bars and troughs associated with breaking waves. More recently, Orme and Orme (1988) have argued that three models for ridge and runnel formation can be identified:

1. Swash-bar deposition of ridges. This follows King's (1959) model, but is most common in macrotidal areas.
2. Ridges and runnels result from the onshore migration of longshore bars and troughs.
3. Runnel erosion rather than ridge accretion can also occur. This appears unlikely at Ainsdale because of the role of the muddy subsurface in sand movements.

The contrast between the shore-parallel dune ridges and the west–east alignment of dune ridges farther inland suggests that the role of blowthroughs is important. As these semi-mobile dunes become stabilized by vegetation, they become fixed features of this landscape. A specific issue that warrants further research is the extent to which this fixed linearity is established with the development of each new seaward ridge and its associated slack.

Conclusions

Ainsdale is a nationally significant site in that the development of this extensive dune system depends upon not only long-term changes during the Holocene Epoch, but also on the detailed effects of surges, sand movements and especially the ridge and runnel of its multi-barred foreshore; it could be regarded as the type area for such forms in Britain. Ainsdale greatly increases our understanding of coastal processes and their relative roles at many different time and space scales. The detailed interpretation of the Holocene history means that this site is of international significance for understanding of the effects of changing sea levels during the Holocene Epoch. In addition, the detailed monitoring of the site provides a nationally important location for the development of strategies for coastal management in the face of global climate change.

One of the three largest dune systems of the west coast of England and Wales, Ainsdale is a National Nature Reserve because of its dune flora and fauna, but its coastal geomorphological interest is considerable. The place of Ainsdale in British coastal geomorphology is very significant, for research has focused on both the present-day processes and the changes during the Holocene Epoch and it provides a key site for interpretation of coastal change in north-western England.

LUCE SANDS, DUMFRIES AND GALLOWAY (NX 150 555)

J.D. Hansom

Introduction

Luce Sands represents an exceptional assemblage of dynamic coastal landforms and contains examples of both contemporary and Holocene marine features. In this respect, it is similar to the GCR sites of Culbin, Morrich More and Spey Bay in the north-east of Scotland (see GCR site reports in Chapters 11 and 6).

Luce Sands

Luce Sands is the largest and most complex system of beach and dunes in the south of Scotland and the juxtaposition of landforms are unique to this site. The geomorphological interest of Luce Sands and the reasons for its selection for the GCR include a series of Holocene emerged gravel ridges, the extensive, complex and dynamic dune system of Torrs Warren overlying the gravel ridges and the diversity of the contemporary coastal features. The relatively large size of the 'soft' coastal landforms would be of interest purely as static landforms, but the highly dynamic nature of the beach system imparts especial interest. Additionally, the ongoing accretionary processes at Luce Sands (Mather, 1979; Single and Hansom, 1994) identify this site as one of few in Britain (and a minority on an international scale) that displays long-term progradation.

Description

Luce Bay is situated to the south of Stranraer, south-west Scotland, between the Mull of Galloway and Burrow Head (see Figure 7.1 for general location). The beach of Luce Sands and the extensive Torrs Warren sand dunes occupy almost the whole of the head of Luce Bay, extending 11 km from Sandmill on the edge of Sandhead in the west to the coastline east of the Water of Luce. The site covers over 2409 hectares of land and intertidal sandflat (Figures 7.19 and 7.20).

The width of the intertidal beach varies and is greatly dependent on the state of the tide. At mean low-water springs, the beach is on average 750 m wide, but in the east the intertidal flats widen to nearly 2000 m at the exit of the Piltanton Burn. At mean high-water springs, the beach narrows to 0–10 m along much of the foreshore with the drift line skirting the toe of the backing dunes. The beach sediment is well-sorted, fine–medium-grained mineral sand (D_{50} = 0.2 mm) with a very low shell content. A series of well-developed sand-bars with intervening channels are present across the wide, very gently sloping intertidal beach. These bars and channels run along the shore in a generally shore-parallel fashion, although they are not regularly shaped features. Up to six sets of bars, which appeared to be migrating onshore, were noted in February 1993 (Single and Hansom, 1994).

The coastal edge along most of the beach has a low, subdued and accretional form. A gently sloping apron, clad in wheatgrass *Agropyron* and marram *Ammophila*, grades almost continuously from the upper beach and backshore. Although there are clear traces of short episodes of erosion during storms and unusually high tides, accretion has been the dominant process in the recent development of this part of the system, and both the scale and extent of accretion are unusual. However wave-induced toe erosion is evident on either side of the Sandmill Burn embayment in the west of Luce Sands (Figure 7.20) (Single and Hansom, 1994).

The Torrs Warren–Luce Sands dune system is the largest and most complex system of acidic dunes in the south of Scotland. The dune system at first sight appears to be chaotic, with no order to the main direction of the dune crests, blowthroughs, or dune slacks. However, closer examination shows that a general sequence of dune forms begins along the seaward edge where low accretional foredunes exist, especially at the rear of the central and eastern beach (Figure 7.20). These foredunes are arranged in a series of parallel ridges, each individually discontinuous and variable in height, rarely rising above 5–6 m. The younger and more seaward of these dunes are orientated and organized in distinct coast-parallel lines, demonstrating that the winds responsible for their construction blow onshore from the south and south-west (Single and Hansom, 1994). The more landward dunes are securely fixed under marram *Ammophila arenaria* and heather *Calluna vulgaris* (Figure 7.21). Behind much of the length of the seaward dune ridges lie extensive areas of dune slack, which separate the low foredunes from higher dunes to landward. These poorly drained slacks extend over 6 km and are often over 500 m in width. The floors of the slacks are characterized by several low, semi-continuous ridges, composed largely of sand around which freshwater marsh has developed. Since they run sub-parallel to the main beach, they may well represent older, flooded, foredune ridges. Peat is known to exist to unspecified depths in the dune slacks, which now support dense scrub, bushes and thickets.

Towards the east of the beach complex, the damp slack grades into an extensive and well-developed dune area to landward, showing high dune relief and large blowthroughs. The scalloped residual faces of these blowthroughs rise to c. 15 m but few of these run shore normal as

Sandy beaches and dunes

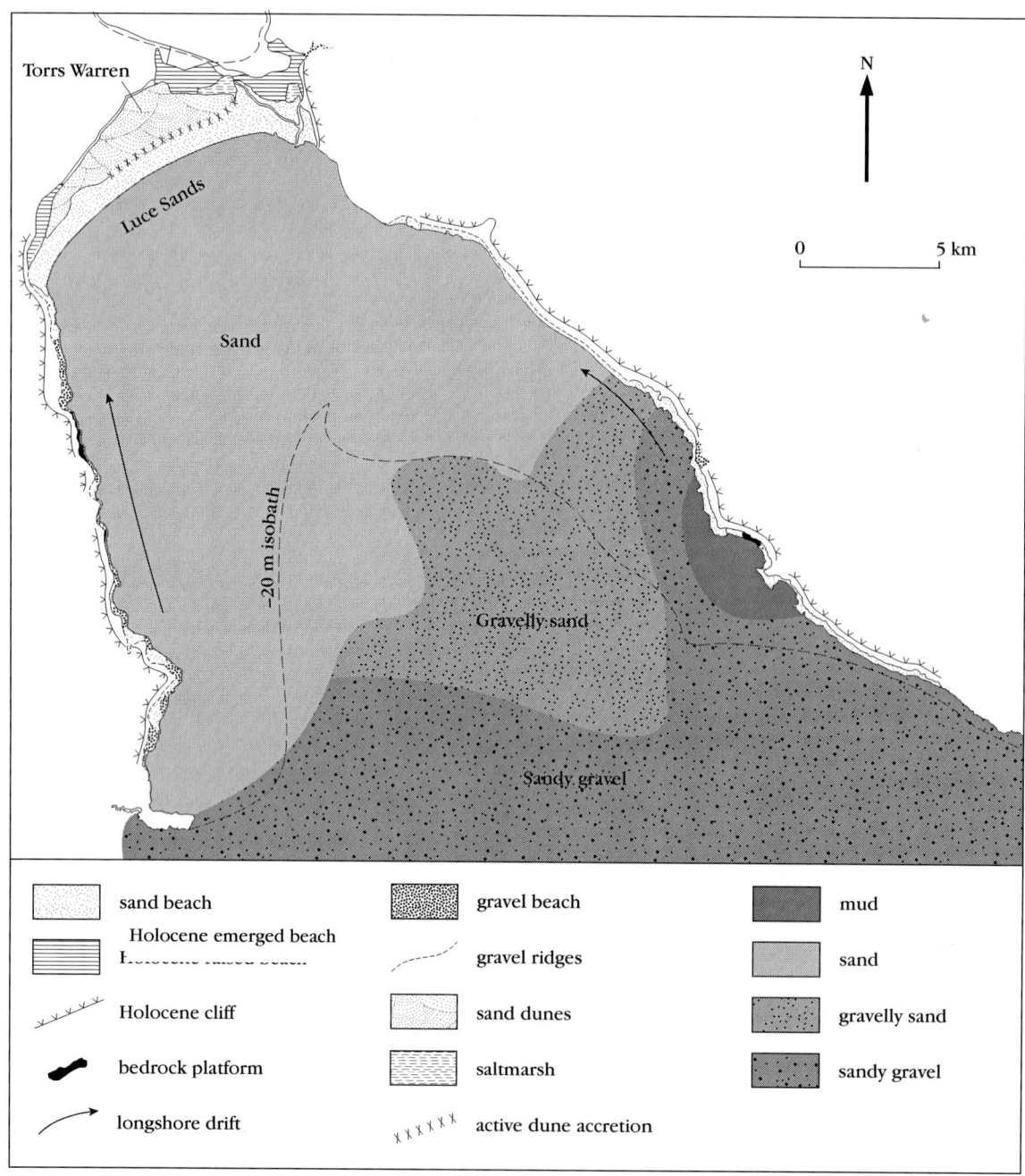

Figure 7.19 Luce Sands is located at the head of a long linear embayment that is floored by extensive areas of sands and gravels. The result of unidirectional wave activity is that sediment is transported northwards on to the beach at Luce Sands. (After Single and Hansom, 1994.)

would be expected if they were controlled by an onshore wind system. The main blowthrough direction is from WNW and relates to winds blowing from this sector. Severe wind erosion has occurred in this area in the past leading to blowthrough excavation down to the water table and the development of substantial erosional slacks. Both dunes and slacks are now largely stabilized, but several areas remain active and active blowthroughs are not uncommon. Nearer the coast, the dunes have a tendency to run beach-parallel and wind-blow activity increases here.

This area of high dunes grades almost imperceptibly landwards into an inner sand plain and old sandhills. This landward part of the dune system consists of a series of rolling sand surfaces interspersed with stabilized sand plains.

Luce Sands

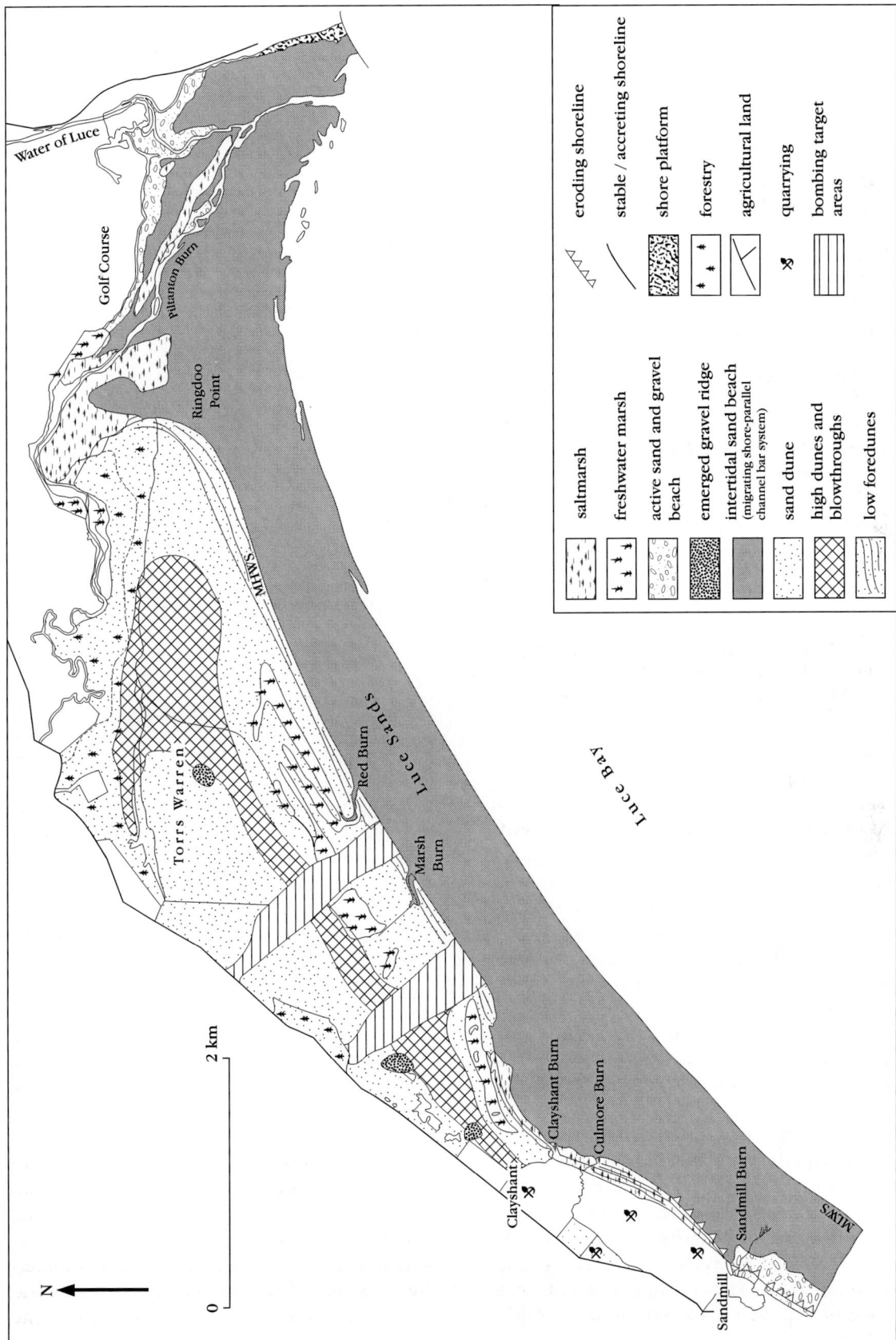

Figure 7.20 The generalized coastal geomorphology of Luce Sands and Torrs Warren showing the wide intertidal area backed by extensive, largely stabilized sand dune. In the central section of the bay, two large areas of dune have been levelled for military use, and access to these areas and to the adjacent intertidal area is restricted. (After Single and Hansom, 1994.)

Sandy beaches and dunes

Figure 7.21 The extensive and well-vegetated dune system of Torrs Warren has developed atop a series of emerged gravel ridges. Sections of these ridges are found in swales within the dune system and on the floors of healed blowthroughs. (Photo: J.D. Hansom.)

This area of varied and chaotic relief has suffered severe erosion in the past, but is now mainly quiescent. Erosion appears to have proceeded from all directions, although weak alignment to the WNW can be seen that may represent an older eroded dune system now blown into a sand plain (Single and Hansom, 1994). Erosional dune forms, such as squat cones, ridges, healed blowthroughs and slacks, are now stabilized by dune heath and scrub. The resultant chaotic topography rises in places to 15 m OD and rests on top of emerged gravels at up to 11 m OD.

To the rear of the beach well-developed and striking emerged gravel ridges extend from west to east and underlie the dunes of Torrs Warren in a broad but discontinuous arc of exposure that connects an eroded Holocene cliff in the west with its eastern counterpart (Single and Hansom, 1994). The extensive windblown deposits of the Torrs Warren dune complex obscure most of the gravel ridges, but from several exposures the general features, orientations and altitudes of the ridges can be determined (Single and Hansom, 1994). Levelling of the ridges indicates a high and well-defined suite of at least 13 gravel ridges at 9–11 m OD, with the heights of the ridge crests declining to seaward. Borehole evidence and other scattered exposures of gravels indicate that a lower set of gravel ridges may exist at altitudes of between 5 and 7 m OD (Figure 7.21).

Interpretation

Torrs Warren–Luce Sands has significant potential to further the understanding of contemporary coastal processes and to establish the patterns of Holocene coastal change for this region of Scotland. To date, however, there is a dearth of detailed geomorphological process studies for the Luce Bay area. Mather (1979) describes the geomorphology of the Torrs Warren–Luce Sand beach complex. Single and Hansom (1994) go further to interpret the process regime and Holocene development of the beach complex and place the site in its wider regional context. However, there remains much scope for innovative research on this complex coastal system.

The Holocene coastal deposits of Torrs Warren–Luce Sands provide an impressive group of emerged features related to a higher sea level. Superimposed on these deposits has developed the largest Scottish dune system south of the River Clyde. Together with a beach complex that is the largest in Scotland, Luce Sands represents an exceptional landform assemblage that records continuous and vigorous coastal deposition during the Holocene Epoch. Plentiful sediment supply has been a characteristic of the Holocene development of Luce Sands. The response of gravel systems to plentiful sediment conditions is generally to add extra ridges onto the seaward face rather than to rework the gravels into even higher ridges (Carter *et al.*, 1987). With a falling sea level, this is manifest at Luce Sands by a multi-ridged strandplain that decreases in altitude to seaward. These conditions of plentiful beach sediments have continued unchecked until the present time, although the earlier gravel sedimentation regime has been supplanted by a sand sedimentation regime producing a wide sandy beach and dunes (Single and Hansom, 1994).

The wide intertidal beach at Luce Bay is an example of a bay-head beach (Single and Hansom, 1994), produced where sediment is driven by unidirectional wave approach and accumulates at the swash limit. In Luce Bay, this process has gradually infilled the bay-head, initially by the deposition of arcuate cross-bay ridges of gravel and subsequently by the deposition of sands (Mather, 1979; Single and Hansom, 1994). The well-nourished nature of Luce Sands is enhanced by three factors. Firstly, much of Luce Bay is no deeper than 20 m (the 9 m isobath lies some 3 km offshore), and so the contributing area for onshore movement is great (Figure 7.19). Secondly, the floor of the bay is thickly veneered with unconsolidated material, such as glaciogenic sands and gravels that are relatively easily transported by shoaling wave activity (Figure 7.19). In addition, Luce Bay appears to function as a major trap for sediment transported along the Rhinns of Galloway coast (Single and Hansom, 1994). This sediment is moved northwards along a major tidal flood channel on the west side of the bay (Mather, 1979), where it accumulates at the bay-head in wide, well-nourished sandy beaches characterized by a positive beach sediment budget (Single and Hansom, 1994).

The Torrs Warren–Luce Sands dune system represents the latest stage in the progressive build-up and redistribution of unconsolidated sediments in Luce Bay. The most landward dunes overlying the emerged gravels are the oldest (Mather, 1979; Single and Hansom, 1994) and their formation immediately post-dates the mid-Holocene sea level fall. Since then, foredune ridges have been added progressively. The chronology of development of these foredunes, and their relationships with the dune slacks and high dune field farther landwards, are as yet imperfectly understood, but it is clear that the development has not been continuous or steady (Mather, 1979). Further research at Luce Sands may help assess the mode of dune development in conditions of plentiful, but pulsed, sediment supply.

Much of the dune system of Torrs Warren, with the exception of the low accreting foredunes near the coastal edge, has been subject to phases of severe wind erosion. It is highly likely that the processes of dune blowthrough activity characterizing much of Torrs Warren are directly related to land use changes and the widespread prehistoric use of the area. Archaeological evidence suggests that removal of the original woodland cover triggered early phases of sand-blow (McInnes, 1964) and several phases of sand-blow in the old dune areas have resulted in the burial of a number of former soil surfaces (Smith, 1903; Callander, 1911). In medieval times, human settlement together with a variety of land-uses took place on Torrs Warren (Jope and Jope, 1959) and the related grazing pressures probably maintained elements of dune instability within the system. Informal grazing and sheep rearing continued until the mid-1930s and dune heath management practises of rotational burning created further instability at this time (Idle and Martin, 1975). Since the present use of Torrs Warren–Luce Sands as a Ministry of Defence (MOD) weapons range, grazing has been curtailed and the dunes are now exceptionally stable over much of the area. Forestry in the northern part of the dunes has aided this stabilization process.

Anthropogenic influence is probably also responsible for the erosion at Sandmill Burn. Here the dunes seaward of the Sandhead caravan park were levelled and a raised flat platform was built out seaward of the coastal edge. Rubble and stacked concrete blocks were used to protect this artificial promontory. The presence of this protected section of coast has led to flank erosion of the adjacent coastline. In an attempt to alleviate erosion, *c.* 1 km of coast has been substantially altered. For example, sand-filled plastic barrels were placed along the eroding dunes in front of both the Sands of Luce and Sandhead caravan parks in 1991 (Single and Hansom, 1994). Not only was this an inappropriate method of coastal protection but it exacerbated erosion to the east as the sand which was used to fill the barrels was dug from a now-eroding remnant dune island (Single and Hansom, 1994).

In common with several other large coastal GCR sites the majority of Luce Bay is owned and managed by the MOD and public access is restricted. With the exception of the two bombing ranges, this land use has conserved much of the site in its natural state, due to access restrictions and the limited recreational use of the beach and dunes. However, small-scale interference along the eroding stretch of coast in the west of Luce Sands may affect the long-term natural evolution of this dynamic system. Any further artificial protection of this coast may result in the reduced transfer of sediment that maintains downdrift accretion.

Conclusions

The principal scientific importance of the Luce Sands GCR site lies in the large and complex system of beach and dunes. The rich variety of contrasting dune morphology includes: low parallel foredunes, dune slacks, high transverse dunes with well-developed blowthroughs, and a complex area of older dunes overlying emerged beach gravels. The dynamic relationships between these components lead to the distinctiveness and importance of the site. The emerged gravel strandplain beneath the dunes, deposited under a higher sea level, adds further interest to the site giving insights to the Holocene development of the complex. Additionally, the ongoing accretionary processes at Luce Sands impart a wider interest as this site is one of few in Britain that displays long-term progradation.

SANDWOOD BAY, SUTHERLAND (NC 220 650)

J.D. Hansom

Introduction

The beach–dune complex of Sandwood Bay, north-western Sutherland (see Figure 7.1 for general location), is among the most dynamic in Britain. The high level of activity of the beach and dune landforms, in a situation where human interference is limited, is of great geomorphological interest. Sandwood Bay is enclosed by cliffed headlands to the north and south and contains a diverse assemblage of spectacular soft coastal landforms. To the landward side of the wide sandy beach, a gravel-cored bar capped with highly dynamic sand dunes impounds the freshwater of Sandwood Loch. Other features of interest include extremely mobile sand dunes with large blowthroughs and climbing dunes that reach altitudes of over 100 m OD on adjacent hilltops (Ritchie and Mather, 1969).

Description

Sandwood Bay GCR site, western Sutherland, encompasses the seaward end of the glacially modified valley of Strath Shinary, and lies seawards of the north-western limit of Sandwood Loch (Figure 7.22). The beach and dune machair complex of Sandwood Bay separates the flooded lower part of this depression (Sandwood Loch) from the sea. The GCR site represents only the western part of the much larger Southern Parphe SSSI. The southeast to north-west orientation of Sandwood Bay corresponds to a structural depression along the junction of the near-vertical cliffs of Torridonian sandstones to the south and the bold convex cliffs cut in Lewisian gneisses to the north (Figure 7.23). The steep sandstone cliffs, which form the south-west limit of the bay, rise to over 90 m and are variously subject to block failure and granular disintegration, giving rise to talus cones in places. As a result, a wide textural range of sediments is contributed to the inshore zone from the crumbling cliffs (Ritchie and Mather, 1969). The gneiss rocks of the north are more massive and resistant and provide little detrital material. Exposures of bedrock crop out in several places within the beach and dune

Sandwood Bay

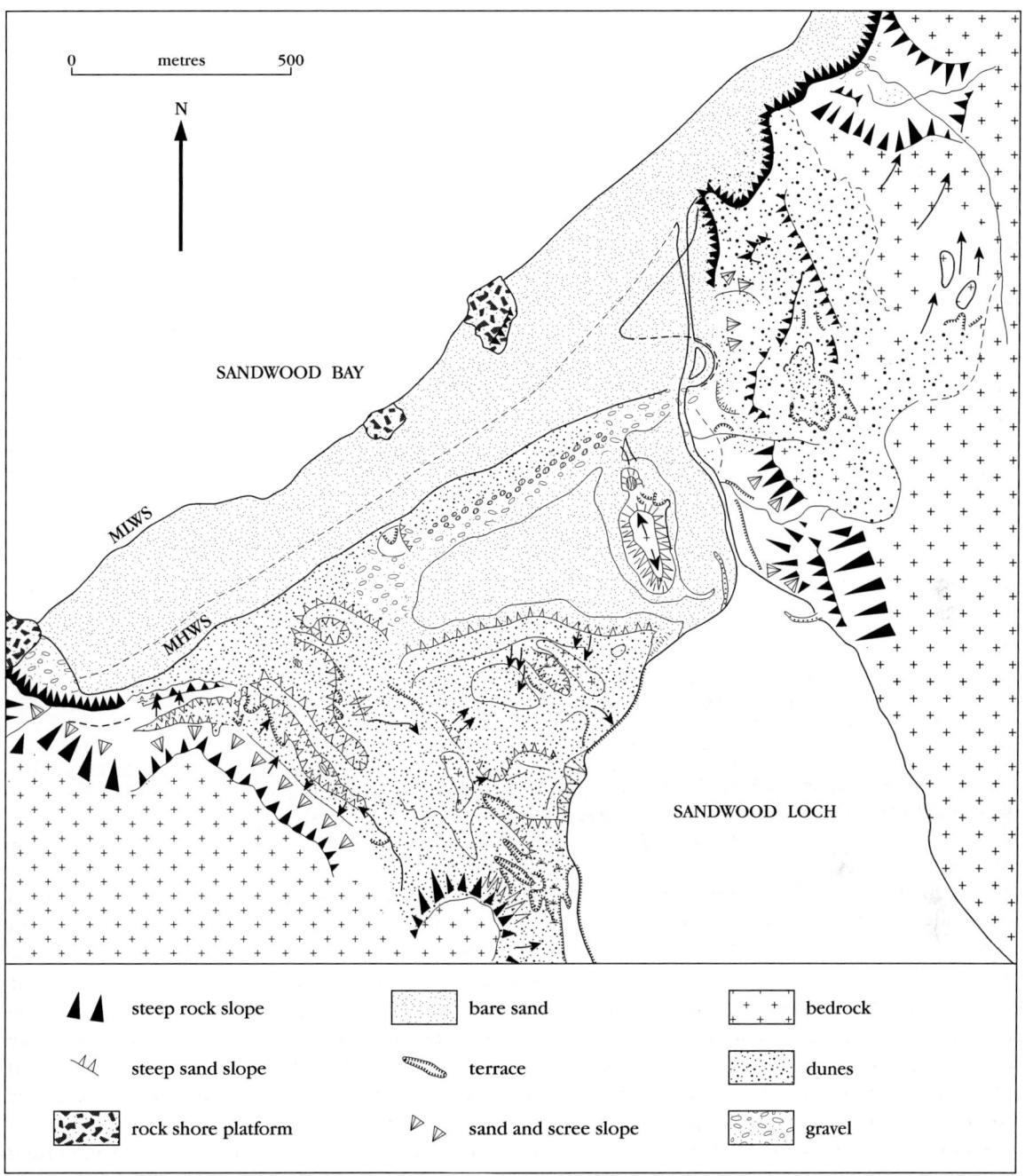

Figure 7.22 Sandwood Bay, Sutherland, is dominated by a large and highly dynamic area of blown sand and machair that lies between the sea and the freshwater Sandwood Loch. Arrows show slope direction. (After Ritchie and Mather, 1969.)

complex (e.g. the low rock skerries at mean low-water springs). In the hinterland the bedrock supports a thin and discontinuous cover of gravelly till. Erosion scars in stream sections within the gneiss reveal the presence of a thin veneer of gritty red till and Torridonian erratics on top of the gneiss testify to ice movement towards the north and north-west.

A wide sandy beach, with an average intertidal width of c. 250 m, has developed in this natural structural embayment (Figure 7.23). This exposed Atlantic beach is among the most dynamic in Britain and is characterized by ephemeral bars that develop and erode depending on prevailing wind and wave conditions (MacTaggart, 1996). There is some development

Sandy beaches and dunes

Figure 7.23 This view of the broad sweep of Sandwood Bay from the south shows the large areas of bare sand that indicate a high degree of dynamism at the beach-dune edge and within the dune-complex. Note the development of low tombolos linking the skerries to the beach crest (arrowed). Depending on the state of the tide these can be quite prominent features. (Photo: J.D. Hansom.)

of rip currents in the nearshore zone, possibly controlled by the partially submerged rock skerries. The lower beach has a relatively steep slope and the upper beach has a convex-up profile with a well-developed summer beach berm (MacTaggart, 1996). The reddish-coloured, medium-grained sand has median diameter of 0.46 mm within the dunes, but although the shell content is unknown it is likely to be about 40–50%, similar to most other west coast beaches of Sutherland.

Landwards of the beach a distinctive dune-capped gravel bar impounds the freshwater Sandwood Loch (Figures 7.22 and 7.23). In the centre of the bay, the dunes have an interesting and peculiar form, consisting of a series of upstanding vegetated dune pillars standing on a sand and gravel base. The dune pillars are likely to represent the erosional remnants of a more extensive dune system that previously covered the gravel bar. In July 1996, low embryo dunes were accreting seawards of the dune pillars (MacTaggart, 1996) (Figure 7.24). Gravel is periodically exposed in several other locations in Sandwood Bay, suggesting the gravel bar is laterally extensive and connects the cliffs in the south of the bay to the cliffs in the north (MacTaggart, 1996). The gravels of the bar are largely composed of Torridonian sandstones (Ritchie and Mather, 1969) seen best where the stream outlet of Loch Sandwood traverses the beach at its northern extremity. Between 1969 and 1996, this channel moved frequently and is now constrained by outcrops of gneiss in the north. During the high loch levels of late winter and early spring, several ephemeral streams develop and drain through the dune-capped gravel bar (Ritchie and Mather, 1969; MacTaggart, 1996).

In the lee of the dune-capped gravel bar there is an ephemerally flooded flat surface of bare sand (Figure 7.22). The flooding of this area is mainly due to the freshwater outlet from Sandwood Loch being impounded by the tide, but marine incursions may also flood this area (Ritchie and Mather, 1969). The low sand-bar

Sandwood Bay

Figure 7.24 Looking south from the dune-capped gravel bar of Sandwood Bay towards the stack of Am Buachaille ('the Herdsman') in the distance. The low embryo dunes in the foreground lie adjacent to dune pillars, he eroded remnants of a more extensive dune cordon. (Photo: Lorne Gill/SNH.)

that separates the ephemerally flooded tidal sand from the main body of Sandwood Loch, is best seen in summer when drainage of this area occurs. The accumulation of aeolian landforms at Sandwood Bay has been favoured by a location exposed to an open part of the Minch and Atlantic Ocean. An extensive sand-dune system has developed in the south-west of the bay as windblown sand is piled up against the Torridonian cliff escarpments and screes. Almost continuous sand recycling has produced exceptionally dynamic sand dunes in this part of the bay and several large elongate blowthroughs trending north-west–south-east are extremely active. Large areas of bare sand characterize this area; in the summer of 1996 there was evidence of considerable sand accretion and sand recycling within the system. The gravel or bedrock deflation bases of the main blowthroughs were covered with drifting sand and large sand aprons extended seawards onto the beach face (MacTaggart, 1996). The immaturity of the dune vegetation gives an indication of the dynamism of the dune system. The main dune area is characterized by abundant and vigorous marram *Ammophila arenaria* that stretches inland for 500–900 m (Dargie, 1994), and only locally do patches of more mature dune vegetation survive. In the south, marram-dominated dunes are piled up against the Torridonian cliff and blown sand colonized by marram extends inland and south-east onto the blocky scree and talus slopes.

Strong winds from the west and north-west have resulted in the extension of dune and aeolian activity inland, not only to develop dune surfaces high onto the Lewisian ridge to the north of the bay, but also to infill the northern part of Sandwood Loch. The lower parts of the northern Lewisian gneiss ridge consist of an assemblage of screes, glacially abraded and smoothed rock surfaces and climbing dunes. An active blowthrough has developed in the climbing dunes. At over 100 m OD, the upper slopes and ridge crest are covered by a well-developed climbing dune that supports a heath-type vegetation. Numerous erosion scars and terracettes characterize the surface as a result of sheep and rabbit grazing and scraping. Subsequent re-deposition of exposed sand has created localized accretion and embryonic dune forms within the dissected dune topography (MacTaggart, 1996). On the northern and western shores of the loch the dunes are cliffed and eroded as a result of wave action within the loch (Ritchie and Mather, 1969).

Interpretation

Sandwood Bay is perhaps the best example on the mainland of Britain of a naturally unstable and dynamic beach–dune system. Its relative remoteness has resulted in a system that is now largely unmodified by direct interference by humans and offers the rare opportunity to study natural rates of change in this high-energy and dynamic coastal system. Steers (1973) highlighted the fact that the site requires further investigation; however, perhaps as a result of the relative inaccessibility of the site, no detailed geomorphological work has been carried out to date other than the descriptions by Ritchie and Mather (1969) and MacTaggart (1996). In spite of this it is possible to interpret the landforms of Sandwood Bay in a systematic context.

Sandwood Bay has been glaciated several times in the past, the most recent Devensian ice passing northwards from Torridonian to Lewisian rocks leaving behind a legacy of polished and plucked valley sides and floor, a discontinuous till cover dominated by sandstone material and widespread occurrence of perched sandstone erratics on both the high and low ground (Ritchie and Mather, 1969). The exposures of bedrock on the foreshore and at the base of the dune complex suggest the existence of a discontinuous sill of rock, running transverse to the main structural corridor of Strath Shinary, and forming the foundation of the coastal and aeolian landforms that separate Sandwood Loch from the sea (Ritchie and Mather, 1969).

In common with beaches elsewhere in the Highlands and Islands of Scotland, Sandwood Bay was probably first closed by the development of a gravel barrier beach whose sediments were derived from adjacent rocky coasts and from glaciogenic deposits on the seabed (Ritchie and Mather, 1969). At Sandwood Bay, since the passage of ice was south to north, the local provenance of this glaciogenic material was Torridonian sandstone. The gravel was deposited on and between the various outcrops of bedrock that now underlie the beach and dune system. Since sea-level rise began to slow down in mid-Holocene times, it is likely that the gravel ridges date from this time and were overwhelmed by large amounts of sand that began to arrive from offshore to develop a wide beach and large dune system behind (Hansom and Angus, 2001).

Open to the north-west, and in a wind and wave environment dominated by westerly and north-westerly activity, Sandwood Bay is effectively a sediment trap for both onshore-moving sediments within the bay and for longshore-moving sediment from the cliffs to the south. However, frequent storm wave activity from the north-west is likely to result in a foreshore characterized by periodic reversals in onshore–offshore sediment exchange. Ritchie and Mather (1969) suggest that the Torridonian sandstone has been and remains a continual source of sediment for the coastal landforms of Sandwood Bay. The underlying gravel bar is composed predominantly of Torridonian clasts and the relatively coarse, reddish-coloured, quartzose fractions of the dune sand are derived from the subaerially weathered Torridonian cliffs to the south. The process continues today and in July 1996, angular and freshly weathered granules of Torridonian sandstone, blown and fallen from the cliffs behind, covered much of the adjacent beach surface (MacTaggart, 1996). Sand is probably still delivered to Sandwood from offshore, but since shell content is unknown it is difficult to estimate the offshore contribution, other than to suggest that it is now likely to be declining. Other sources of contemporary beach and dune sand come from the cliffs to the south and from sand recycled through the dune system by streams.

The exposed Atlantic location of the bay has also favoured the accumulation of aeolian landforms, and the natural structural embayment to landward has channelled windblown sand inland and uphill to cover the scree and high rock slopes to the north and south of the bay. The highly dynamic nature of the sand dunes and blowthroughs also suggest that there is continual sediment recycling within the system (MacTaggart, 1996). Where the dissection is greatest within the dune system, distinctive upstanding dune pillars (Figure 7.24) are likely to be the result of erosion either by ephemeral streams that drain over the bar during high loch levels or by wave action during storms and extreme high spring tides (MacTaggart, 1996).

Several changes have occurred since Sandwood was mapped by Ritchie and Mather in 1969, particularly in the south. The elongate multiple blowthrough system of 1969 has been modified into three large coalesced blowthroughs that have an amphitheatre-like form. Vertical accretion is widespread and, in 1996, 4–5 m thick aprons of sand had accumulated seawards and to the north of the main faces. Since the main axis of the blowthroughs in 1969 was north-west–south-east, the main direction of advance of the sand removal is assumed to be landwards to feed the dunes behind. However, it is also clear that substantial amounts of sand are also returned to the beach during winds from the south and south-east, and this is also the direction of advance of a high-altitude blowthrough on the northern side of the bay (MacTaggart, 1996).

Conclusions

Sandwood Bay, western Sutherland, contains a spectacular assemblage of soft coastal landforms that have accumulated at the head of Strath Shinary impounding the freshwater Sandwood Loch. The principal geomorphological interest of the site rests in the very high levels of geomorphological activity in the beach and dune landforms, in a situation where human interference is limited, and thus offers a rare opportunity to study natural rates of coastal change. Individual features of interest include the dune-capped gravel bar, highly dynamic and mobile sand dunes, large blowthroughs and climbing dunes that reach hilltop altitudes of over 100 m OD (Ritchie and Mather, 1969). The cliffs that enclose Sandwood Bay are integrally linked to the past and current evolution of the geomorphological system, the sandstone cliffs that are undergoing eosion to the south providing an important sediment source.

TORRISDALE BAY AND INVERNAVER, SUTHERLAND (NC 690 620)

J.D. Hansom

Introduction

The diverse assemblage of beach and dune landforms at Torrisdale Bay, Sutherland, northern Scotland (see Figure 7.1 for general location), is of national geomorphological importance. The dune landforms, which demonstrate various stages of development and dynamism, lie landwards of a wide intertidal sand beach and sit on top of the high, central, glacially scoured rock ridge and the terraces of the River Naver that drains into the east part of the bay and the River Borgie that drains into the west part. Dunes have formed on the hilltop at altitudes of up to 110 m OD on the central rock ridge and are of geomorphological and botanical importance. The site is also of importance from an archaeological perspective because the river terraces contain numerous cairns, hut circles and cist burials that may allow minimum dating of the landform surfaces. Despite the enormous research potential, which is enhanced by ecological and archaeological interests, the site has failed to attract detailed geomorphological research although several descriptive accounts highlight the site's significance (Ritchie and Mather, 1969; Steers, 1973; Bentley, 1996b).

Description

The Torrisdale Bay and Invernaver GCR site (Figure 7.25) encompasses two bays, at the mouths of the rivers Naver and Borgie, which drain into the east and west of the bay respectively. The two rock headlands that enclose Torrisdale Bay (Creag Ruadh on the east and Aird Torrisdale on the west) are formed of highly resistant metamorphic rocks of the Moine series (Ritchie and Mather, 1969). In the centre of the bay, a glacially scoured bedrock ridge is formed of strongly foliated Moine Schists (Ritchie and Mather, 1969). This 110 m OD bedrock ridge is cut by a series of parallel east–west-trending fractures that have been exploited by glacial action to produce a series of depressions or gullies along its flanks. These gullies form important access channels for blown sand to climb to the top of the ridge. The ridge has been extensively glacially scoured in a south–north direction and contains excellent examples of roches moutonnées, with smooth abraded surfaces on the south side (up-glacier) and rough plucked surfaces on the north (down-glacier). Smoothed bedrock surfaces abound and perched blocks, some of which are erratics, are common. Both the River Naver and River Borgie have well-developed glaciofluvial sand and gravel terraces. Ritchie and Mather (1969)

Sandy beaches and dunes

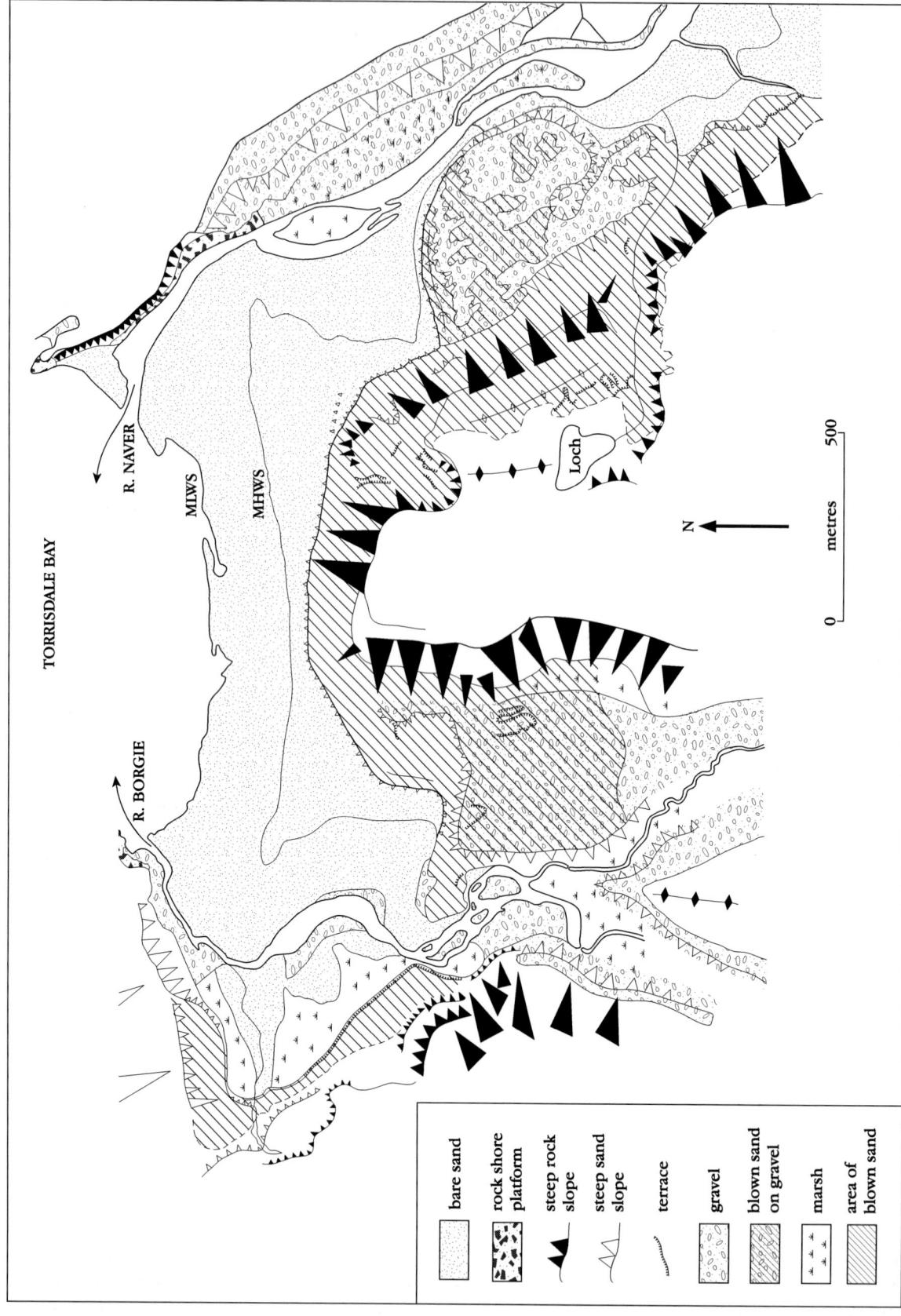

Figure 7.25 The geomorphology of Torrisdale and Invernaver is bisected by a glacially scoured rock ridge that is flanked on either side by glaciofluvial terraces that are capped by windblown sands. The unvegetated upper beach is wide and backed by low dunes. Areas of saltmarsh occur along the exit of the River Borgie. (After Ritchie and Mather, 1969.)

Torrisdale Bay and Invernaver

Figure 7.26 The large glaciofluvial terrace at Invernaver viewed from the east is flanked and capped by blown-sand deposits that also climb the ridge behind. The surface of the terrace also supports a wealth of archaeological remains including hut circles and cist burials. (Photo: J.D. Hansom.)

suggest the terraces on the east side of the lower Naver valley correspond to different early sea levels. The most prominent is an extensive flat-topped gravel terrace on the west side of the river at *c.* 15–20 m OD (Figure 7.26). The terrace surface is pitted by a number of kettle holes up to 10 m in diameter (Bentley, 1996b). The south and east sides of the terrace slope steeply down to the valley floor and the steep seaward cliff is fringed with vegetated sand dunes above the wide intertidal beach. A flat-topped gravel terrace at a similar altitude to the extensive Naver terrace forms the eastern side of the Borgie valley.

The wide and flat sandy beach of Torrisdale Bay extends over 1 km in length from the mouth of the River Naver in the east to the mouth of the River Borgie in the west. Extensive areas of intertidal sands characterize both bays and several areas of saltmarsh have developed in the inner reaches (Figure 7.27). At low tide the beach at Torrisdale Bay can be as wide as 950 m, with about 40% of the total beach area lying above high-water mark. The sand has a median diameter of 0.2–0.3 mm and has only 3–4% shell-derived calcium carbonate. Prominent sand-bars extend seawards across the mouths of both rivers. On the bar at the mouth of the River Naver, an area of low (3–4 m high) marram-clad sand dunes has experienced vigorous growth since the 1970s (Bentley, 1996b). Landwards and south of the beach at Torrisdale Bay an extensive dune system has developed, the detailed morphology of which is controlled by the eroded form of the bedrock ridge and the flat-topped glaciofluvial terraces that flank it. The Naver terrace contains a wealth of important archaeological artefacts including cairns, hut circles and cist burials. In contrast, the Borgie terrace has no known archaeological interest and is capped by an extensive (0.5 × 0.5 km) flat dune grassland surface used for grazing.

At the rear of the beach, the coastal edge is characterized by a frontal apron of young vigorous dunes that are mostly relatively stable with vigorous marram *Ammophila arenaria* growth although there is some localized evidence of erosion where the vegetation cover has been stripped. The main dune ridge extends along the back of the beach and drapes the seaward

Sandy beaches and dunes

Figure 7.27 The intertidal saltmarsh and sandflats of the River Borgie exit looking north-west over the low dune area and beach of Torrisdale Bay in the middle distance. (Photo: J.D. Hansom.)

edge of the Naver terrace before curving northwards round the flanks of the central rock ridge to continue into the Borgie estuary. In places the fringing dunes extend southwards onto the Naver terrace surface itself, while farther landwards some low isolated dune features rest on the top of the gravel terrace. An extensive area of dunes has also developed on the west side of Torrisdale Bay, in the triangular-shaped area lying between the Borgie terrace and the central rock ridge (Figure 7.25). The marram-clad dunes of this area are characterized by an irregular and hummocky surface topography and have no preferred alignment. However, the dune slacks occur at a common low altitude and contain standing water or damp surfaces and probably mark the position of the water table (Bentley, 1996b). Gravel is exposed at the base of some slacks (Ritchie and Mather, 1969). This dune system drapes the northern part of the Borgie terrace for up to 50 m southwards before giving way to a much smoother and vegetated dune grassland that extends over the terrace top and thins to the south.

The steep (up to 20°) rock slopes of the central ridge are host to a diverse array of landforms including dunes, bare screes of sand and rock debris and patches of heath. On the northern part of the ridge at the rear of the beach, 9–12 m-high coastal dunes merge into the dunes that are being blown uphill. These climbing dunes are most extensively developed on the north-east and east flank of the ridge, especially where rock depressions permit deeper sand accumulation. For example, the deep gully occupied by the stream flowing east from the loch to the lower River Naver, acts as a funnel for windblown sand. A small dune area has developed near the crest of the ridge at 110 m OD. This unusually high hilltop dune is characterized by numerous erosion scars and terracettes, the result of a combination of wind erosion and sheep scraping. The dune heathland of the ridge are also of considerable ecological interest with unusual associations of mountain avens *Dryas*, heather *Calluna*, crowberry *Empetrum*, sedge *Carex* and juniper *Juniperus* (Ritchie and Mather, 1969) providing a classic example of the 'altitudinal descent' of montane vegetation (Bentley, 1996b).

Interpretation

Ritchie and Mather (1969) describe the Torrisdale Bay area as a 'bewildering melange of landform and landscape elements'. The site comprises a fine assemblage of landforms that relate not only to the Quaternary evolution of the area but also the shorter-term dynamic beach–dune processes. Although no detailed geomorphological research has been carried out, it is possible to interpret the general evolution of this magnificent site from the morphological accounts of Ritchie and Mather (1969) and Bentley (1996b).

The site lies outwith the accepted limits of the Loch Lomond Stadial and so the last ice to override the site is most likely to be of Devensian age, flowing over the central bedrock ridge in a south–north direction and producing scouring, striations, and roches moutonnées (Ritchie and Mather, 1969). During the latter stages of the Devensian glaciation, the valleys of the Naver and Borgie acted as conduits for large volumes of meltwater and sediment discharged from the northern margin of the Scottish ice-sheet. As a result both the Borgie and Naver valleys contain large flat-topped terraces, that are the remnants of larger outwash terraces grading to former sea levels. The lower parts of these terraces were probably trimmed during the higher relative sea-levels of the Lateglacial period and were subsequently isolated as sea levels fell from 15–20 m OD (Ritchie and Mather, 1969). Ritchie and Mather (1969) suggest that fluvial reworking of these outwash terraces provided the sand and gravel for the large intertidal expanse of beach. However, the relative absence of gravel on the beach compared to its great abundance in the terraces may imply that offshore sources of sand were equally important in the initial development of the beach and dunes of Torrisdale Bay. This argument has been rehearsed elsewhere (Hansom, 1999, 2001) but broadly involves onshore delivery of large amounts of glaciogenic sand and gravel, at a time when the sea-level rise slowed during mid-Holocene times. The early arrival of gravel initiated development of gravel ridges that were subsequently inundated by large quantities of sand, which was then distributed into extensive dune and machair systems. However, in contrast to most Scottish beaches and dunes, Torrisdale Bay appears to be relatively stable or accreting. It is likely that the general decline in offshore sediment sources late in the Holocene Epoch has been offset by a ready source of sand recycled from the Naver and Borgie glaciofluvial terraces. The low percentage of shell-derived sand also suggests that onshore rather than offshore sources now comprise the main sand supply to the beaches and dunes.

On the main beach at Torrisdale Bay, a tendency for the beach axis to rotate clockwise over time was observed by Ritchie and Mather (1969), where sand from the east side of the beach on the Naver exit moves west towards the Borgie exit. This may be a function of westerly waves impinging on the east of the beach undergoing less refraction than the waves impinging on the west and so the energy gradient causes longshore transport of sand to the west, an area of lower wave energy. If this is indeed the case, then it may also provide a supply-driven explanation for the striking contrast in the extent and development of windblown depositional landforms on the west and east sides of the bay.

The extensive climbing dunes on the central bedrock ridge probably formed soon after the sandy beach was established, as a result of dune development at the base of the ridge being forced uphill by strong winds from the north, north-west and north-east, assisted by bedrock gullies and depressions that subsequently channelled the windblown sand to altitudes of 110 m OD (Ritchie and Mather, 1969). Dune accretion appears to be continuing today (Bentley, 1996b). The frontal edge of the coastal dunes is relatively stable with low embryo dunes to the seaward side. The wide sand beach and the area of new dunes close to the mouth of the River Naver suggests that sand is still available for dune formation (Bentley, 1996b), in stark contrast to most Scottish beach–dune systems where erosion is the dominant process. The Borgie terrace has a stable dune grassland on its surface, whereas the Naver terrace consists of bare gravel with discontinuous low dune hillocks. This is probably due to the greater exposure to onshore winds at the Naver terrace, but the frequency of cairns, cist burials, grave mounds and other archaeological features may offer another partial explanation in terms of antiquity of anthropogenic influence (Ritchie and Mather, 1969). Nevertheless, similar undiscovered features may lie beneath the dunes of the Borgie terrace.

In summary, the Torrisdale Bay site is of great geomorphological importance on account of the diversity of the landform assemblage and the juxtaposition of glacial, glaciofluvial, and coastal landforms. The combination of dunes that have been blown onshore onto glaciofluvial terraces and, blown to considerable altitude on the central bedrock ridge where dune grasslands have formed, is of considerable interest. These interests are further enhanced by the ecological and archaeological importance of the site. There is considerable scope for further geomorphological research at Torrisdale Bay.

Conclusions

Torrisdale Bay, Sutherland, northern Scotland, contains a diverse assemblage of dune landforms draped over a complex subsurface morphology comprising a glacially scoured bedrock ridge and the glaciofluvial terraces of the River Naver and River Borgie. A wide variety of dune landforms are well developed, demonstrating various stages of evolution and stability. Individual features of particular interest are the dune forms that have developed on the terraces, and the climbing dunes and high-level hilltop dune grassland. It is the juxtaposition of impressive glacial, glaciofluvial, and coastal landforms at Torrisdale Bay that is of outstanding geomorphological significance.

DUNNET BAY, CAITHNESS (ND 215 710–ND 201 682)

J.D. Hansom

Introduction

The wide sand beach of Dunnet Bay, Caithness, (see Figure 7.1 for general location) is backed by a massive, sharp-crested, coastal dune ridge with a gently sloping links plain on its landward side. The general morphology and scale of this extensive beach–dune–links system is unique in Britain. The coastal dune ridge is dissected by numerous spectacular, wide, deep blowthroughs at various stages of development. As blowthrough stability ranges from stable to extremely active, Dunnet Bay provides a key site for studies of blowthrough initiation, growth and natural or artificial stabilization. The dune and links also support important species-rich vegetation and invertebrate communities. Despite the enormous research potential at Dunnet Bay it has failed to attract any detailed geomorphological research, although two mainly descriptive accounts of the site exist (Ritchie and Mather, 1970a; Bentley, 1996c), and Hansom and Rennie (2003) have recently quantified coastal changes.

Description

The GCR site of Dunnet Bay lies at the head of a 4 km-wide, 6 km-long embayment on the north coast of Caithness. Inland, the Dunnet Bay structural depression extends south-eastwards across the country to Sinclair's Bay on the east coast of Caithness. As a result of the enclosed nature of Dunnet Bay, between the high sandstone cliffs of Dunnet Head to the north and the low flagstone platform to the south, it can been described as a sediment trap (Ritchie and Mather, 1970a). In addition, the deep penetration of the embayment means that incoming waves are almost completely refracted, arriving at the beach with their crests parallel to the arcuate beach.

The beach at Dunnet Bay is one of the largest in northern Scotland (Ritchie and Mather, 1970a), extending for over 4 km in a broad symmetrical curve (Figure 7.28). The intertidal zone is wide, with an average width of 180 m, and has a uniformly low gradient of 1–2°. Offshore the seaward gradient is also gentle, at about 1:124 (Ritchie and Mather, 1970a). The beach is composed predominantly of relatively fine-grained sand (D_{50} = 0.31 mm) of which 20% is $CaCO_3$ and is flanked by low rocky shore platforms at either end. Immediately adjacent to the rock platforms the upper beach is composed of coarse gravel that then grades to sand.

The wide beach is backed by a massive, steep (>30°) and high dune ridge (Figure 7.29). The backslope of the dune is less steep (12–16°) and has a slightly concave-up profile. The dune ridge reaches a maximum height (up to c. 20 m OD) and width (c. 350 m) towards the middle of the bay. Both the height and width decline towards the southern and northern ends. Frontal erosion of the dune ridge is evident along almost its entire length (Figure 7.29). In October 1996 the dune face was partially revegetated, and low embryo dunes had formed in places at the back of the beach, a

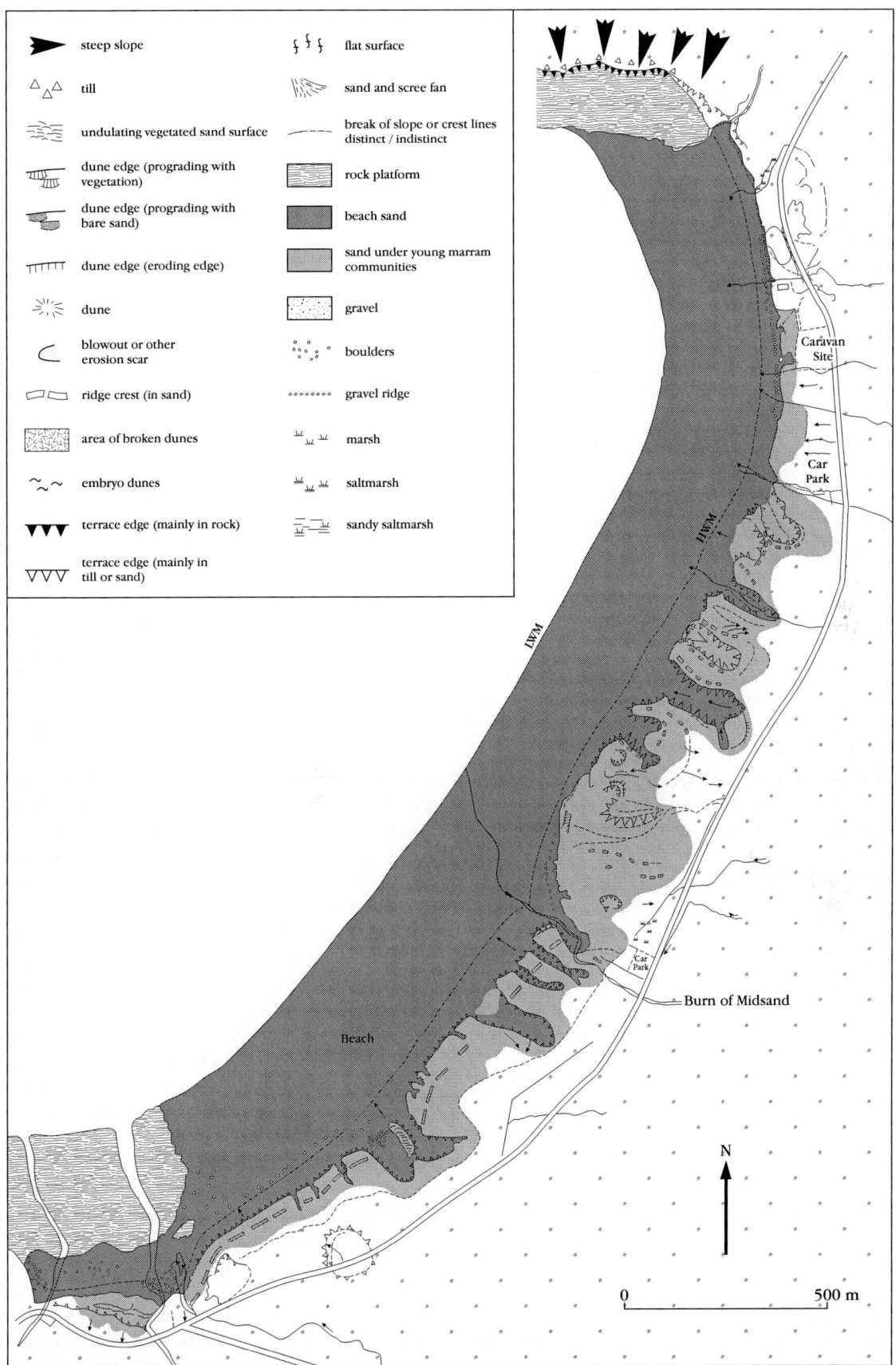

Figure 7.28 The coastal landforms of Dunnet Bay and dunes showing a coastal dune edge that is both undercut by frontal erosion and punctuated in several places by large, linear, blowthrough corridors. (Based on Ritchie and Mather, 1970a and Hansom and Rennie, 2003.)

Sandy beaches and dunes

Figure 7.29 The wide expanse of Dunnet Bay looking west over the indented exit of the Burn of Midsand. Much of the coastal edge comprises mature dunes whose edge is now steep and undercut and whose surfaces now support re-invigorated marram growth. (Photo: J.D. Hansom.)

result of calm summer conditions. Frontal dune erosion is predominantly due to wave attack at the dune base particularly during winter storms. Erosion is not a new phenomenon at Dunnet Bay and in 1970 the greater part of the dune front was over-steepened and in places undercut during storms (Ritchie and Mather, 1970a).

The dune ridge is dissected by nine streams flowing from the links plain to the sea. The largest stream, the Burn of Midsand, flows into the centre of the bay through a prominent break in the dune ridge, while the others have cut narrow V-shaped valleys through the dunes (Bentley, 1996c). Low embryo dunes have developed at the mouth of the Burn of Midsand as a result of the increased local sediment supply.

Morphological diversity of the main dune ridge is created not only by streams but also by numerous spectacular blowthroughs that dissect the dunes. The blowthroughs form several flat-floored, steep-sided erosion 'corridors' through the dune ridge and several saucer-shaped depressions on the windward slope of the wide dune ridge. The blowthroughs, which are often 10–12 m deep, up to 30–40 m wide and often devoid of vegetation, are some of the largest in Scotland (Bentley, 1996c). At least seven large blowthroughs at various stages of activity dissect the dune ridge. Several of the blowthroughs are compound, where two or more have joined laterally leaving only residual pinnacles of the former dune ridge between the areas of bare sand. A large blowthrough at the northern end of the bay forms a narrow corridor through the dunes. As this blowthrough is visible from the main road it is utilized as an access track to the beach for both pedestrians and vehicular traffic (i.e. quad bikes, motorbikes etc.), exacerbating the natural erosion processes. The blowthrough is extremely active, with evidence of wind being channelled through the corridor, scouring the sand and re-depositing it over the ends of the blowthrough as lobes of unconsolidated material. Attempts have been made to stabilize this blowthrough, along with several others, using sand fences and marram planting. This has not been entirely successful; in October 1996 the fences were full of sand and thus were no longer effectively trapping new sediment. The steep slopes flanking the active blowthrough corridors are generally unvegetated and extremely unstable, with evidence of loose sand slumping downslope. Gravel is exposed at the base of several of the larger blowthroughs. Between the large blowthroughs a number of smaller V-shaped blowthroughs penetrate the dune ridge.

Two blowthroughs to the north of the central stream can be described as relict blowthroughs, as although they are stable and fully vegetated they have retained their original form. The stabilization of these blowthroughs was undertaken by the Forestry Commission who used brushwood and the planting of coniferous trees (Ritchie and Mather, 1970a). Although this stabilization has been successful in that the landforms have effectively been frozen *in situ* at a previous stage of high instability, the remaining presence of over-steep slopes and topographical depressions that still channel onshore winds at high velocities may lead to instability at a later date.

Landwards of the wide coastal dune ridge a gently undulating links surface extends for up to 5 km inland. The main Castleton to Dunnet road, which lies landwards of the dune ridge, separates the dune environment from the more stable links area to the east. The long erosional blowthrough corridors and their associated re-depositional sandhills have been known to reach the main road and during winter storms sand is often blown across the road to the links area (Ritchie and Mather, 1970a). The GCR site includes a small representative area of the links to the east of the road in the northern part of the site. The links is formed entirely of blown sand that has been deposited over peat, till and bedrock (Bentley, 1996c). The beach–dune–links system of Dunnet Bay is unusual as the relatively steep dune backslope grades directly into the low-lying area of dune pasture, with an absence of secondary or older dune forms farther inland.

Interpretation

Dunnet Bay forms part of the GCR network of coastal sites on account of its unique dune morphology. The single, massive, sharp-crested coastal dune ridge is dissected by numerous spectacular large blowthroughs at various stages of activity and is backed landwards by an extensive dune pasture and links topography. The scale and range of activity in the various forms of dune blowthroughs and the relatively frequent occurrence of direct wave attack at the base of the dune ridge add geomorphological diversity and enhance the scientific interest. There remains much scope for research at Dunnet Bay particularly concerning the initiation, growth and stabilization of both wind- and wave-induced erosional forms.

Steers (1973) describes Dunnet Bay as 'a feature of primary importance in the coast of Scotland' on account of its enclosed nature between the high sandstone cliffs of Dunnet Head to the north and the low flagstone platform to the south. Since it is so enclosed it acts as an effective sediment trap: the only escape for sand from the bay is landwards (Steers, 1973). It is thus not surprising that a wide, high dune system and extensive links plain has accumulated landwards of the enclosed bay (Ritchie and Mather, 1970a; Steers, 1973). It has been suggested that the gentle offshore gradient (1: 124) of the sand-covered seabed implies there is a continuing reserve of sediment in Dunnet Bay (Ritchie and Mather, 1970a). However, the presence of the large active blowthroughs and frontal erosion of much of the dune face suggests that there may now be a diminution in the offshore sediment supply.

The large blowthroughs at Dunnet Bay are naturally induced erosional forms, although human activity may have exacerbated the natural process by utilizing blowthrough corridors as access tracks to the beach. Ritchie and Mather (1970a) found no positive relationship between blowthrough location and drainage conditions, offshore sediment supply or local wind patterns and conclude that, in the absence of any known trigger mechanism, the blowthroughs have a random stochastic distribution and the dune barrier as a whole is migrating landwards. Frontal erosion together with blowthrough advance appears to be moving the total volume of the dune landwards and will continue until a new stable equilibrium position is reached (Ritchie and Mather, 1970a). This erosion may be due to an increase in wave and wind energy from the north-west or a decrease in the offshore sediment supply (Ritchie and Mather, 1970a). As with the majority of coastal dune systems in Scotland, the dominant process at Dunnet Bay appears to be one of erosion and coastal retreat, although recent artificial stabilization methods have affected the natural evolution of this system. Stabilization of the blowthroughs by planting has been relatively successful: two of the larger blowthroughs are no longer active. The use of sand fencing in several of the larger active blowthroughs, although not entirely effective and in need of maintenance, has limited landward sand transfer. Hansom and Rennie (2003) have recently quantified the rate of retreat of the coastal edge at Dunnet Bay: between 1968 and 1998, 3.6×10^5 m^2 was lost at rates of 20 m a^{-1} mainly in the centre and north of the beach.

The relatively steep dune backslope of the main dune ridge and the absence of secondary ridges or extensive dune forms inland has been attributed to the frequency of strong winds from the south-east channelled by the structural depression between Sinclair's Bay and Dunnet Bay (Ritchie and Mather, 1970a). The to-and-fro nature of winds through this structural corridor may account for the vigour and rate of development of the blowthrough erosion corridors, which once initiated may be attacked by winds from both directions. Strong onshore storm winds from the north-west appear to be more dominant. Further research on the initiation, growth and development of these blowthroughs is required.

Gravel is exposed at the base of several blowthroughs. It has been suggested that the dune system at Dunnet Bay may rest on a gravel basement of emerged ridges and bars (Ritchie and Mather, 1970a; Steers, 1973) as is the case in many Scottish coastal dune systems (e.g. Culbin, Luce Sands and Strathbeg, see GCR site reports). This remains to be fully investigated and as no detailed height measurements are available it is unknown if these gravel forms are related to the present or to a former higher sea level. More research is warranted.

Conclusions

The unique general morphology of Dunnet Bay, which consists of a single, massive, sharp-crested dune ridge leading inland to a gently-sloping extensive links plain, is of immense geomorphological importance. The 4 km-wide arcuate sand beach is backed by a massive, steep (>30°) sharp-crested dune ridge reaching a maximum height of c. 20 m OD and width of c. 350 m. The low links plain extends up to 5 km landwards of the dune ridge. The morphology of this massive ridge is extremely diverse. It is cut by several small streams draining into the bay and, perhaps more significantly, at least seven large blowthroughs dissect the ridge. The blowthroughs, which are often 10–12 m deep and up to 30–40 m wide, are some of the largest in Scotland (Bentley, 1996c) and are at various stages of activity, ranging from stable to extremely active. The effects of artificial stabilization of the landforms is evident – at least two large blowthroughs are now essentially relict landforms stabilized by dune planting, while others have been partially stabilized by the use of sand fencing. The scale, dynamism, range of activity and diversity of the blowthroughs in this massive coastal dune ridge is of great geomorphological interest.

BALTA ISLAND, SHETLAND (HP 660 075)

J.D. Hansom

Introduction

The small uninhabited island of Balta is 5 km long and lies in a north–south orientation at the mouth of the Balta Sound, off the west coast of Unst, the northernmost island of the Shetland archipelago (see Figure 7.1 for general location and Figure 7.30). The island contains a continuous veneer of vegetated sand extending across Balta Island from a north-west facing bay at South Links through two low cols almost to the 45 m-high eastern sea-cliffs. The sand beach grades into a gravel storm ridge on the upper beach, with the coastal edge being marked by an erosional dune scarp. An extensive dune grassland plain lies behind and, although frontal dunes are absent, it is the most complete dune grassland system in Shetland (Mather and Smith, 1974; NCC, 1976). However, the complex is in an advanced stage of dissection due to a combination of rill dissection and severe wind deflation, probably due to overgrazing by the large rabbit population. In places the dune grassland has been deflated down to a base-level of aeolian calcarenite (Mather and Smith, 1974). This, together with the high rates of deflation, make this site important as a dynamic example of a beach–dune grassland–cliff continuum (MacTaggart, 1999).

Description

The morphology of Balta Island is markedly asymmetrical with the east coast characterized by 45 m-high cliffs, deeply indented with geos, contrasting with the low, more sheltered west-facing coast. The island is composed mainly of

Balta Island

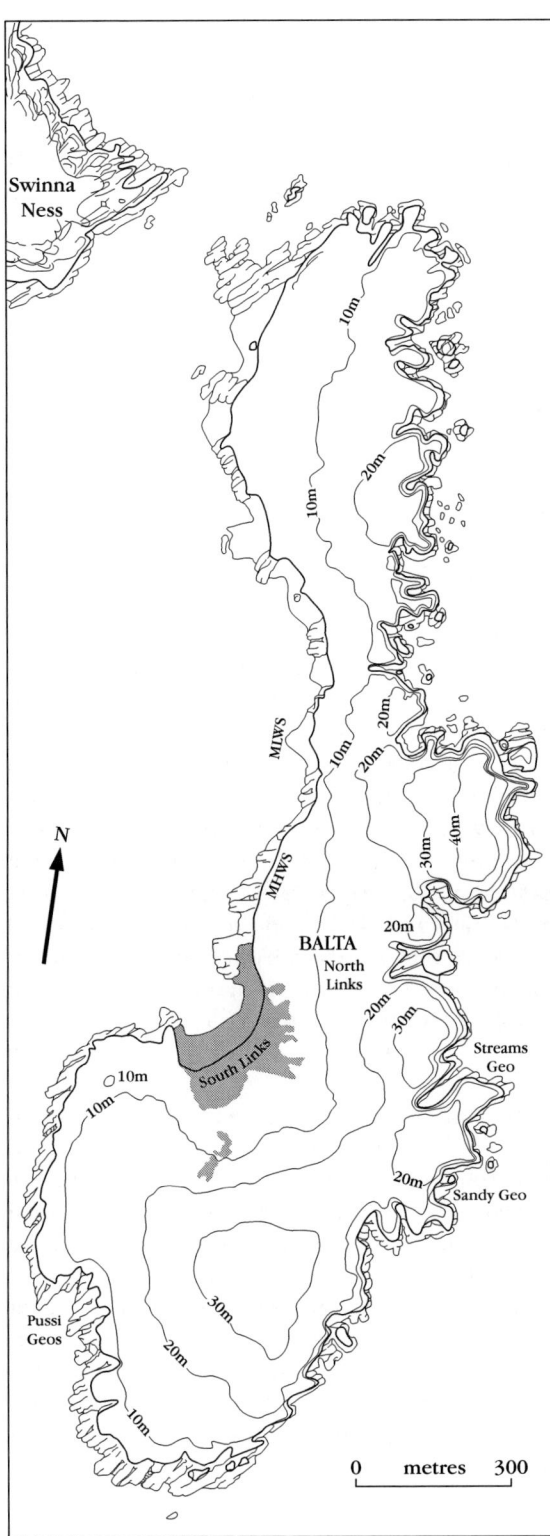

metagabbros (NCC, 1976) and the terrain generally consists of ice-scoured bedrock with a patchy till cover and many perched blocks. The only beach on Balta is a 200 m arc of sand at South Links on the west coast. This is backed by extensive windblown sand deposits of variable thickness that sweep across the island via two cols almost to the eastern sea cliffs. It is this extensive beach–dune grassland–cliff continuum that forms the GCR site of Balta Island (Figure 7.31).

Two intertidal rock platforms form the northern and southern limits of the beach at South Links. Fine-grained shell-sand dominates the lower beach and a gravel storm ridge, partly concealed by an apron of blown sand, forms the upper beach. Dunes are absent and the coastal edge backing the beach consists of eroded dune remnants separated by bare sand areas. These remnants, although indicative of severe erosion

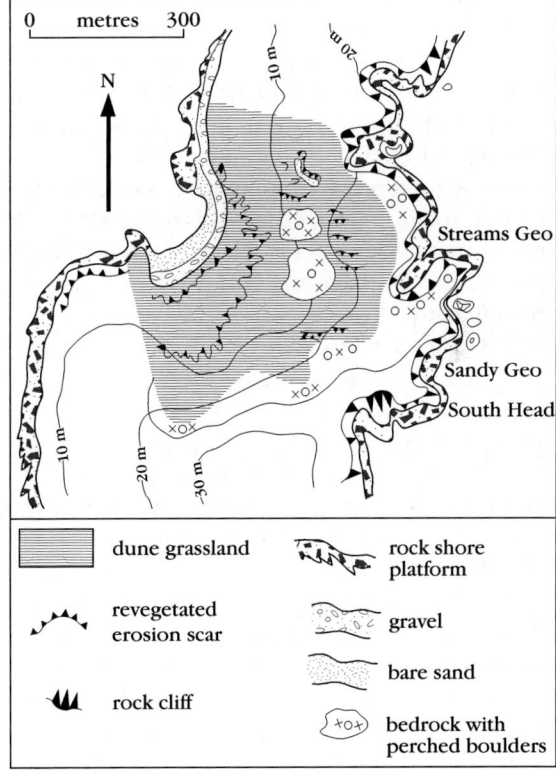

Figure 7.30 Balta Island, Unst, Shetland, is low in the west and high in the east. It is mainly rocky except where sand is blown up-slope from the beach at South Links. (After MacTaggart, 1999.)

Figure 7.31 The geomorphology of Balta, Unst. There are no dunes but instead the site supports a wide expanse of climbing dune grassland some of which has been eroded into low escarpments. In places the dune surface has been eroded down to a base level of calcarenite by both wind deflation and rill erosion. (After MacTaggart, 1999.)

in the past, appear to be relatively stable and well-vegetated (Mather and Smith, 1974). The underlying gravel storm ridge affords partial protection to the dune grassland toe during storm wave action but the presence of erosional remnants suggests that this may not always be so.

Landwards, some 8 ha of dune grassland veneers a large shallow amphitheatre in the ice-scoured bedrock. Several bedrock knolls carrying perched blocks protrude through the plain from the irregular bedrock surface beneath and this is also reflected in the varying gradient and thickness of the blown sand deposits. The sand thickness decreases rapidly eastwards from depths of over 2 m on the lower parts of the system, thinning to 0.5 m towards the cols on the eastern side (Mather and Smith 1974). These slopes of climbing dune grassland are often in excess of 30° (Mather and Smith, 1974). In places, where the lower levels of the dune sands are close to the water table, they have undergone cementation processes into aeolian calcarenite that rests directly on top of the bedrock.

The grassland surface is heavily dissected by finger-like erosion scars and a rill drainage network. For example, approximately 100 m from the coastal edge, a 1–1.5 m-high active erosion scarp gradually extends inland along a crenulated front. The scarp is characterized by numerous erosion scars whose distal extension inundates the adjacent turf with re-deposited blown sand. The erosion of the surface is also initiated by the numerous small rills that form a radial pattern converging on the lower centre of the South Links amphitheatre. These small rills are ephemeral features, and appear as overland flow becomes concentrated on the lower surfaces after heavy rain. Higher up, subsurface drainage is facilitated by the numerous rabbit burrows that pit the dune surface. Inland of the crenulated erosion scarp, erosion scars and deflation are less common and tend to be localized on the floors of depressions where subsurface drainage is concentrated (Mather and Smith, 1974). It is likely that heavy rabbit and sheep grazing and scraping has exacerbated the processes of wind erosion at South Links.

Interpretation

Balta Island contains the most complete dune grassland system in the Shetland Isles (Mather and Smith, 1974; NCC, 1976). The complex is at an advanced stage of natural dissection, as a combined result of wind deflation, rill activity and heavy grazing. The severity of the Shetland climate means that rates of change are likely to be more rapid than on equivalent mainland dune systems. As Balta Island is uninhabited, the beach–dune system has evolved naturally, with extremely limited direct human impact from ploughing and ditching. Indirect human influence via sheep grazing and scraping, together with a large rabbit population, has probably accelerated wind erosion processes. In spite of this, the predominantly natural and dynamic dune-erosion system at South Links is of geomorphological importance. In addition, the absence of dunes at the coastal edge is unusual and Balta represents an excellent example of a beach–dune grassland continuum. Aeolian calcarenite, the presence of which is unusual, provides a depth control on deflation and rill dissection (Mather and Smith, 1974).

Balta Island is an important site for the study of the natural process of wind deflation and as such has great research potential. Some degree of management of the rabbit and sheep population could be implemented as part of a research strategy designed to achieve a better understanding of the processes involved in natural deflation. However, as yet Balta has failed to attract any detailed scientific research, although the outstanding geomorphological importance of the site has been recognized (Mather and Smith, 1974; NCC, 1976).

Conclusions

A continuous dune grassland veneer extends across the small uninhabited island of Balta from the north-west-facing sand beach through two low cols to the 45 m-high eastern sea cliffs. A combination of rill dissection, severe wind deflation and a high rabbit population has resulted in dissection of the dune surface. A crenulate erosional scarp with linear 'finger' erosion scars is gradually extending landwards into the dune grassland. In places the surface has been deflated to a base level where an unusual outcrop of aeolian calcarenite occurs (Mather and Smith, 1974). The high rates of natural deflation and dissection at a site where there has been limited human interference is of geomorphological importance and Balta Island provides an excellent research area to study the end results of such erosion.

Strathbeg

STRATHBEG, ABERDEENSHIRE (NK 075 595)

J.D. Hansom

Introduction

The dune forms of Strathbeg, north-east Scotland (contain some of the most impressive parallel linear dunes in Scotland. The aeolian processes that created this suite of linear dunes remain active in parts of the beach today. Former coastal progradation has resulted in the isolation of the Loch of Strathbeg, one of the largest freshwater lochs in Britain (Bourne *et al.*, 1973), which now lies *c.* 1 km inland and is separated from the open coast by the spectacular dune field. The landward dunes lie on top of a series of Holocene emerged gravel beaches, the initial deposition of which resulted in the enclosure of an inlet now occupied by the Loch of Strathbeg (Walton, 1956; Ritchie *et al.*, 1978). In addition, Strathbeg contains spectacular examples of wind erosional processes in large-scale coastal dune ridges. Large blowthroughs and deflation plains have been excavated naturally down to the underlying gravel ridges in the southern part of the dune system.

Although the general evolution of Strathbeg has been described (Walton, 1956; Ritchie *et al.*,

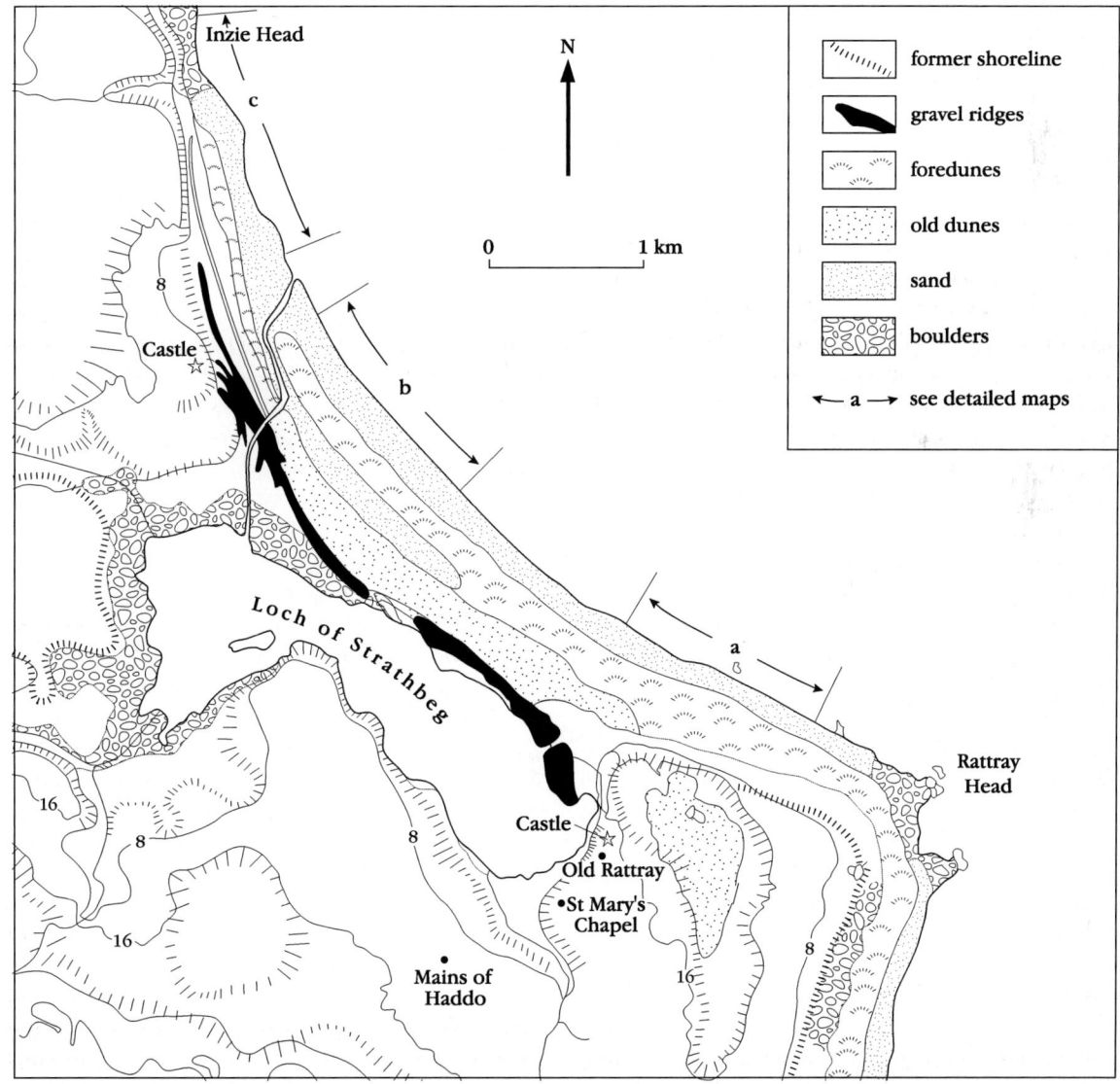

Figure 7.32 Generalized coastal features of Strathbeg, showing enclosure of the Loch by gravel ridges and a series of old dune ridges fronted by lower foredunes. Heights are in metres OD. The detailed sections a–c are shown in Figure 7.33a–c. (After Walton, 1956.)

1978; Ritchie, 1983), no detailed recent research has been undertaken concerning either the processes operating or the chronology of its geomorphological evolution.

Description

The GCR site of Strathbeg covers a total coastal length of *c*. 7.1 km between Rattray Head in the south and Inzie Head in the north, this north-eastern facing coastline marking the transition between the open North Sea and Moray Firth coasts (see Figure 7.1 for general location and 7.32 for detailed geomorphology). In the north the beach is fronted by a low intertidal rock platform and in the south, at Rattray Head, by offshore rock and banks of boulders. In both the north and south, the beach and dune complex is backed by a relict cliff cut in glacial deposits, although in the central section, the cliff lies inland and the beach and dunes are backed by the loch (Figures 7.33a–c). The sandy intertidal beach is relatively narrow (on average *c*. 90 m wide) with a slightly steeper backshore zone. The beach widens at the mouth of the stream that drains the inland Loch of Strathbeg and an extensive intertidal lagoon with a small saltmarsh has developed. Elsewhere the beach has a regular profile but is characterized, especially in the south, by a series of beach ridges and subparallel beach depressions. Strathbeg beach is highly dynamic experiencing many short-term changes in both longshore and offshore–onshore sediment transport (Ritchie, 1983; Rendel Geotechnics, 1995). Ritchie (1983) notes an offshore complex of sand-bars lying just below low-water mark and that the beach shows evidence of accretion, particularly to the south of the stream outlet. This outlet appears to form an important transition zone. To the north, the coastal edge consists of a steep sand-cliff (4–7 m high) which is undergoing active erosion (Ritchie, 1983). For several hundred metres on either side of the tidal sand-floored lagoon and meandering stream channels of the outlet, the coastline is prograding with fine examples of embryo and young foredune ridges, probably as a result of local sand feed from the lagoon. South of the outlet there is a large asymmetric dune ridge rising to over 8 m OD (Ritchie, 1983), which encloses the long narrow tidal lagoon and small, sandy saltmarsh associated with the loch outlet (Figure 7.33b). There is evidence of active sand accumulation on the dune crest and slopes (Ritchie, 1983).

Landwards of the beach there is an extensive and complex series of dune ridges backed by low dune grassland. At least seven parallel linear dune ridges, with a relatively regular summit altitude of 6–9 m above beach level and separated by shallow depressions occur in the north and central part of Strathbeg (Figure 7.33a). This general pattern of parallel dune ridges is broken in places by irregular dune topography as represented by areas of hillocky and transverse dune ridges. The dune morphology close to the loch outlet is particularly complex and 18 separate dune crests occur between the outer beach and the loch margin some 1.2 km inland (Figure 7.33b) (Ritchie *et al.*, 1978). A marked change in dune morphology occurs farther south where a series of large blowthroughs cut deeply into the dune system (Ritchie *et al.*, 1978). High residual dune ridges create a spectacular, active coastal landscape and large deflation features have formed where major blowthroughs have coalesced (see Figure 7.36). Deflation processes affect the entire dune system (Ritchie *et al.*, 1978) and the surface morphology has extensive low-altitude flat sand plains flanked by a wide zone of dune hillocks associated with re-depositional activity. There are excellent examples of re-depositional processes with sand spilling from dunes onto the flatter adjacent areas. These flat areas are also subject to extensive winter flooding and are locally described as 'winter lochs'.

Prominent emerged gravel ridges can be traced intermittently throughout the area and underlie the landward part of the dune system. The gravel ridges appear to be hinged to the higher ground at St Combs and in the past extended southwards to progressively enclose a former inlet whose position is now occupied by the Loch of Strathbeg (Walton, 1956; Ritchie *et al.*, 1978). At the northern end, a number of parallel gravel ridges terminate, sometimes with recurved ends, a short distance to the south-east of the present outlet of the loch. South of this point the gravel bars coalesce to form one main ridge. Farther south the gravel forms the south-east margin of the loch and recurves at the southern distal end terminate in the loch itself (Walton, 1956).

The freshwater Loch of Strathbeg is shallow (1–2 m deep) approximately 3 km long and 1 km wide (Bourne *et al.*, 1973). The loch is

Strathbeg

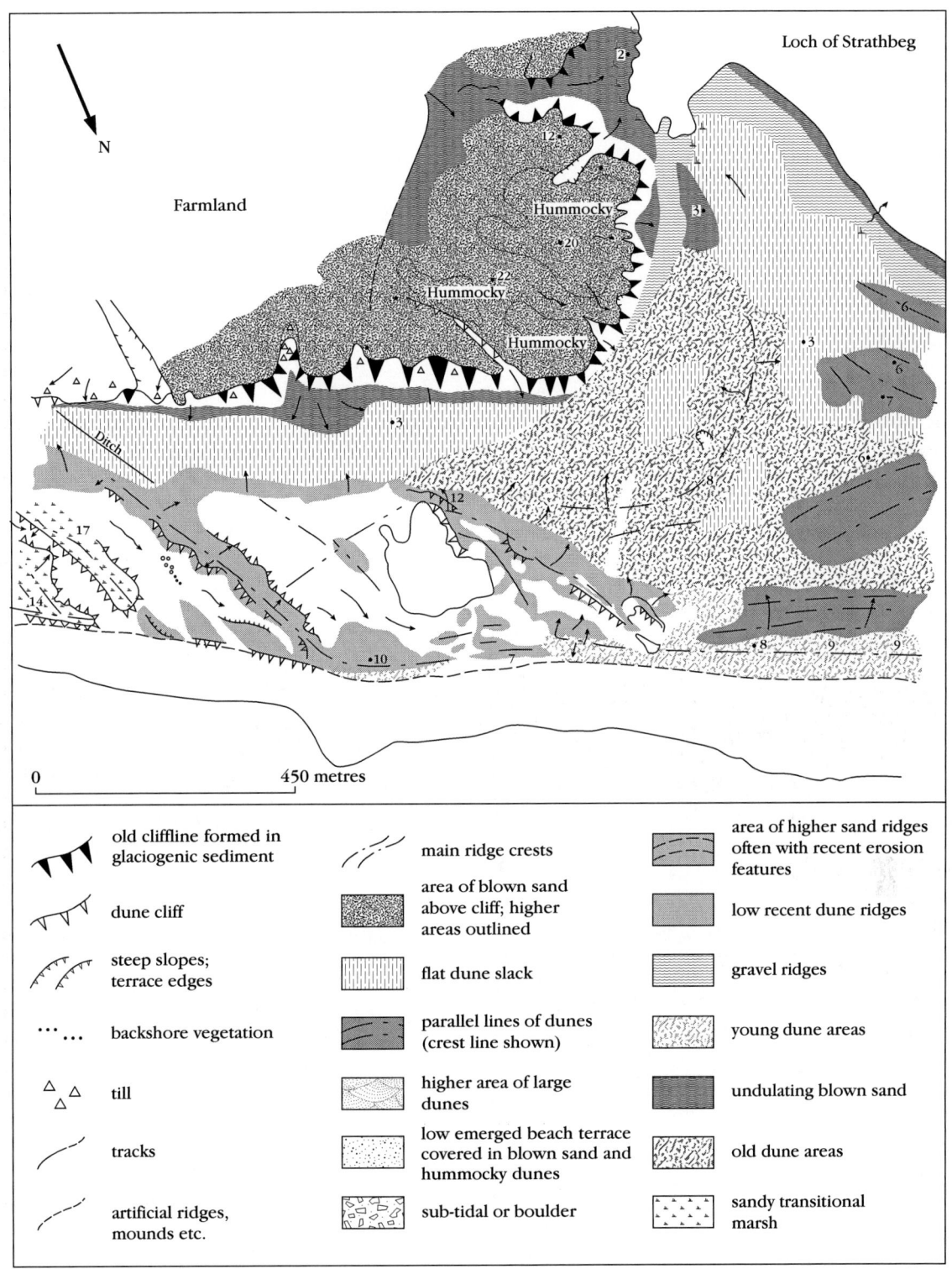

Figure 7.33a–c Detailed coastal geomorphology of the (a) south, (b) central and (c) north sections of Strathbeg, showing the extensive series of shore-parallel dune ridges punctuated by the outlet from the loch. Representative sections through ×–× are also shown. Arrows indicate direction of slope. The figure is continued overleaf. (After Ritchie *et al.*, 1978.)

Sandy beaches and dunes

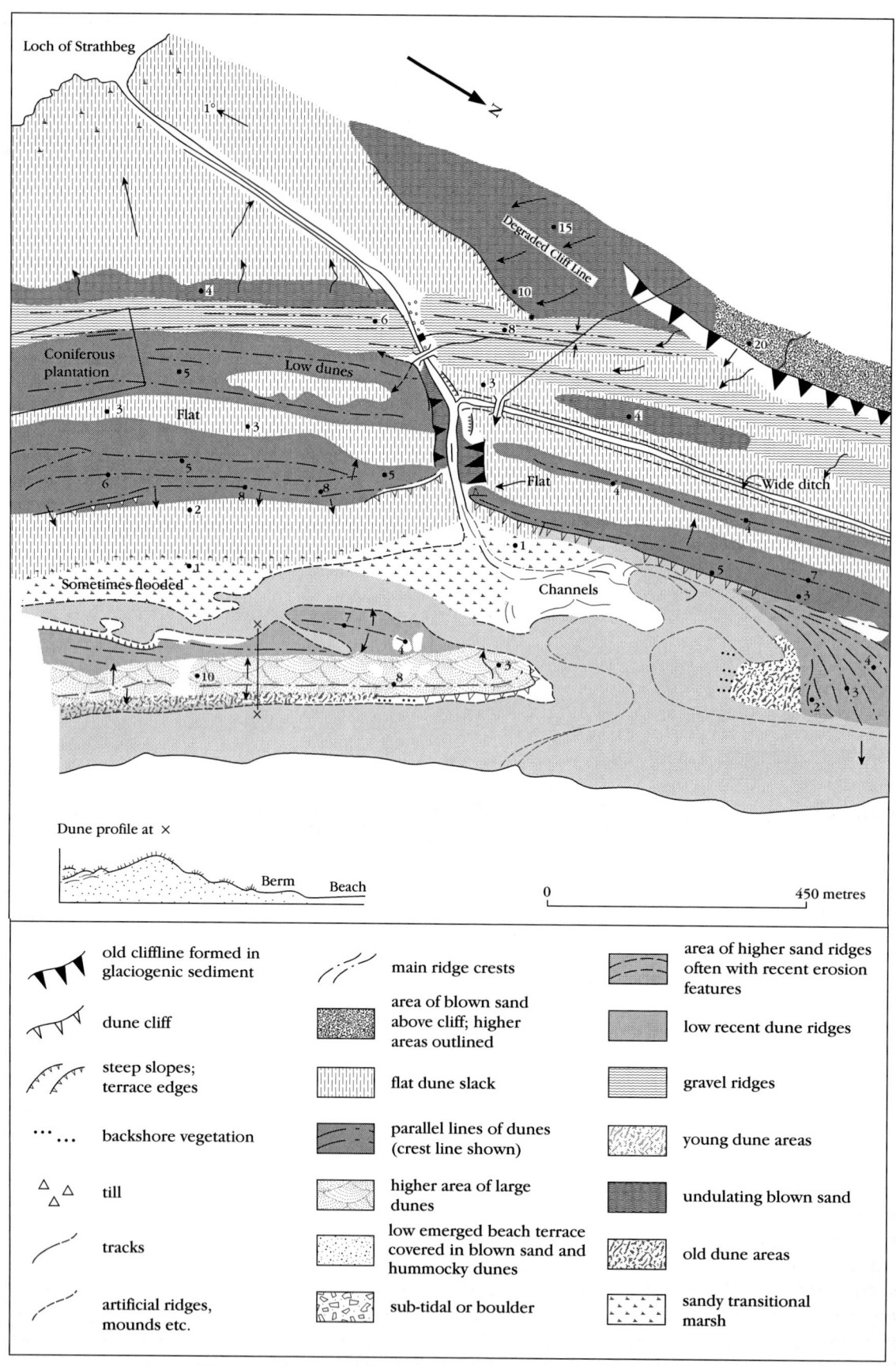

Figure 7.33b – *contd.* Coastal geomorphology of Strathbeg.

Strathbeg

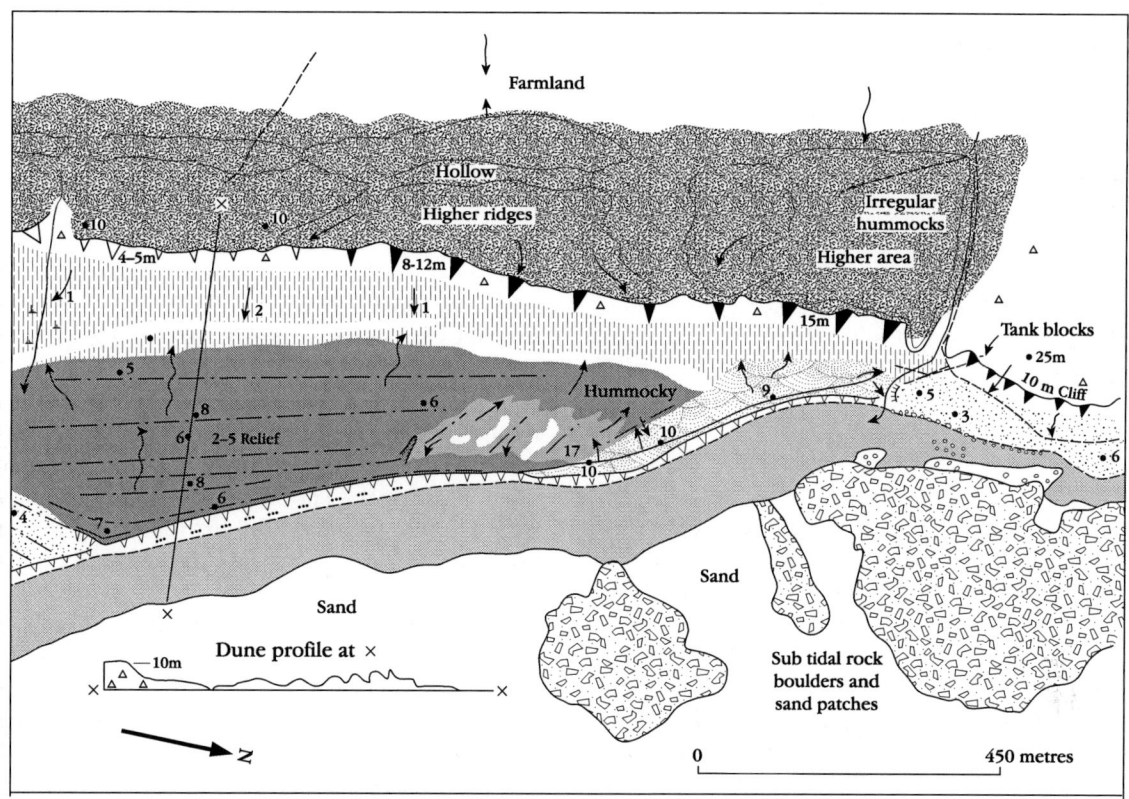

Figure 7.33c – *contd.* Coastal geomorphology of Strathbeg.

bounded by gravel bars, dune plain and dunes in the east and by a 5–9 m-high relict cliff cut in till in the west (Figure 7.32). The loch has a high nutrient level (Forteath, 1977) and is an important staging post for thousands of migratory wildfowl, particularly geese (Ritchie *et al.*, 1978) and is of international ornithological and ecological importance.

Interpretation

Walton (1956) first interpreted the complex evolution of Strathbeg, suggesting that the present coastline was the 'result of the gradual enclosure of a deep indentation of the coast in Late-glacial times, culminating in an smooth dune-fringed littoral behind which is now impounded the freshwater Loch of Strathbeg'. Walton (1956) identified two higher relative sea levels in the area by using remnants of degraded relict cliffs cut in till and now draped in vegetated blown sand deposits. Higher sea levels following deglaciation resulted in inundation possibly to a height of *c.* 16 m OD (Walton, 1956) and at this time (15 000–14 000 years BP), the Strathbeg

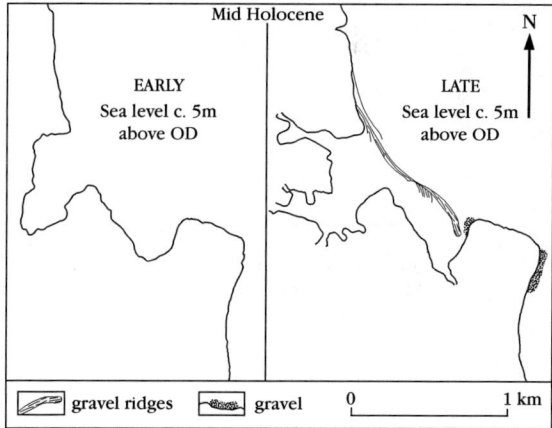

Figure 7.34 Possible evolution of the Strathbeg area during the Holocene Epoch showing the southward extension of gravel ridges and progressive closure of the former embayment. (After Walton, 1956.)

area was a large inlet, possibily with several offshore islands (Figure 7.34). Relative sea level then fell to below present before rising again sometime possibly around 7000 years BP. A dis-

Sandy beaches and dunes

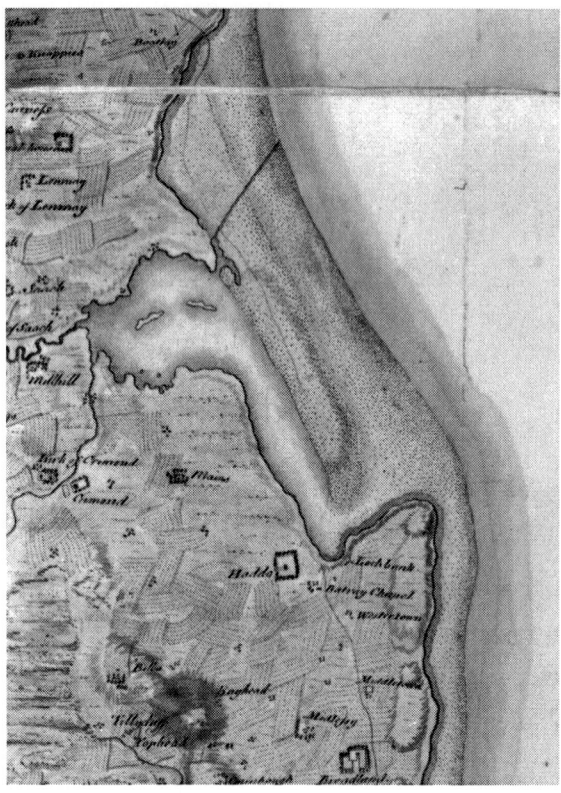

Figure 7.35 The Loch of Strathbeg, as mapped in 1755 by the military surveys of William Roy. Note the loch exit is located in the south, but also that an artificial channel across the north end of the beach was in existence to allow the loch to drain northwards. (Photograph © The British Library from the British Library Special Collections, Maps C9b 31.)

tinctive lower relict cliff forms the western margin of the present Loch of Strathbeg and Walton (1956) interprets this cliff as representing the margin of a later, possibly mid-Holocene, sea level at 5 m. The ridges of a gravel spit began to extend from the north to partially enclose a tidal lagoon with a narrow inlet (c. 45 m wide) between the southernmost limit of the gravel bar and the relict cliff near Old Rattray (Figure 7.32) (Walton, 1956; Ritchie et al., 1978). The later arrival of sand encouraged the development of the sand-dune complex.

Archaeological and historical evidence documents the final closure of the bay of Strathbeg. The tidal inlet of Strathbeg, sheltered from the open coast by the extensive gravel spit and overlying dunes, provided an obvious natural harbour and it is not surprising that a small coastal fishing village developed (Figure 7.35). The earliest evidence of settlement at Rattray dates to around the end of the 13th century (Walton, 1956) and during the 16th and early 17th centuries the harbour at Strathbeg flourished (Walton, 1956). However, the inlet was almost certainly shallowing over time and as early as 1654 there is evidence that the exit was threatened by sand deposition (Gordon, 1843). The final closure of the bay is thought to have occurred by the deposition of windblown sand during a storm in the 1720s (Walton, 1956). The final sealing of the inlet was apparently so sud-

Figure 7.36 A large coalesced blowthrough in the southern part of Strathbeg. The loch is visible in the top right. The figure provides the scale. (Photo: J.D. Hansom.)

den that a vessel is reputed to have been trapped within the harbour and its impounded cargo of slates then used to roof a nearby house at Mains of Haddo (Walton, 1956). Roy's map of 1747–1755 shows the enclosed Loch of Strathbeg with a new outlet at the north end of the lagoon (Figure 7.35). The decline of the settlement of Rattray followed and it had ceased to exist by the middle of the 18th century (Walton, 1956). At the end of the 18th century the high water-level of the Loch of Strathbeg threatened inundation of agricultural land and led to an artificial outlet channel being cut through the gravel bar (Walton, 1956). Since then this artificial cut has been maintained.

Since the final closure of the Loch of Stathbeg there has been over 1 km of sand accretion to seaward, involving the development of an extensive dune system consisting of numerous parallel linear dune ridges (Walton, 1956; Ritchie *et al.*, 1978; Ritchie, 1983). Dune morphology suggests that the dune ridges developed at the rear of a wide beach, which then abutted the emerged gravel bar and spit features (Ritchie *et al.*, 1978). Such processes have their modern counterparts, particularly at the outlet of the loch in the north–centre part of the bay, where young embryo and foredune ridges are developing seawards of the main dune ridge, although net sand drift is now considered to be low (Ramsay and Brampton, 2000d). Dune face erosion is largely confined to the north end of the beach although there is also a zone of active wave undercutting in the extreme south where erosion has produced high sand cliffs (Ritchie *et al.*, 1978). Elsewhere, the coastal edge tends to undergo cyclic seasonal effects of dune undercutting in winter and accretion in summer (Ramsay and Brampton, 2000d). Away from the major blowthrough corridors and deflation areas in the south (Figure 7.36), there is either minor aggradation or stablility, as occurs on the inland dune surfaces (Ritchie *et al.*, 1978).

Strathbeg provides one of the best examples in Scotland of a suite of parallel linear dune ridges with intervening depressions produced by progradation. These progradational processes also contributed to the final isolation of the Loch of Strathbeg and remain active today, particularly around the loch outlet. It is apparent from the above discussion that although the general evolution of this area has been interpreted (Walton, 1956) there remain substantial gaps in our knowledge and the exact sequence and chronology of the geomorphological evolution of Strathbeg remains uncertain. In addition, the spectacular erosional forms in the southern part of the dune system present a valuable opportunity to assess the processes and forms of erosion in a large-scale dune system.

Conclusions

The extensive and varied dune morphology of Strathbeg is of outstanding geomorphological interest. The site contains excellent examples of parallel linear dune ridges, with up to 18 separate dune crests and intervening depressions between the outer beach and the freshwater Loch of Strathbeg some 1.2 km inland. The progradational processes that created this suite of linear dunes remain active today. The dune system overlies a base of emerged gravel ridges, which were originally responsible for the partial closure of the inland loch. The southern part of Strathbeg contains spectacular examples of wind erosional processes in large-scale coastal dunes, and the relatively undisturbed nature of the Strathbeg dunes enhances the scientific interest of the site.

FORVIE, ABERDEENSHIRE (NK 020 270)

J.D. Hansom

Introduction

The Sands of Forvie, north-east Scotland (see Figure 7.1 for general location), form the fifth-largest and least-disturbed sand-dune system in Britain (Dargie, 2000). This vast site covers 810 ha and contains a remarkable assemblage of blown-sand landforms, some of which are unique in Britain, for example, the classic parabolic dunes at north Forvie and the unvegetated sand dunes of south Forvie. Others are representative of much of the dune coastline of north-east Scotland. The mode of evolution of the Sands of Forvie has sparked much scientific debate (e.g. Landsberg, 1955; Kirk, 1955; Steers, 1973; Walton and Ritchie, 1972; Ritchie, 1992). Early work suggested a series of large sand waves migrating successively from the south end of the peninsula to the north end

Sandy beaches and dunes

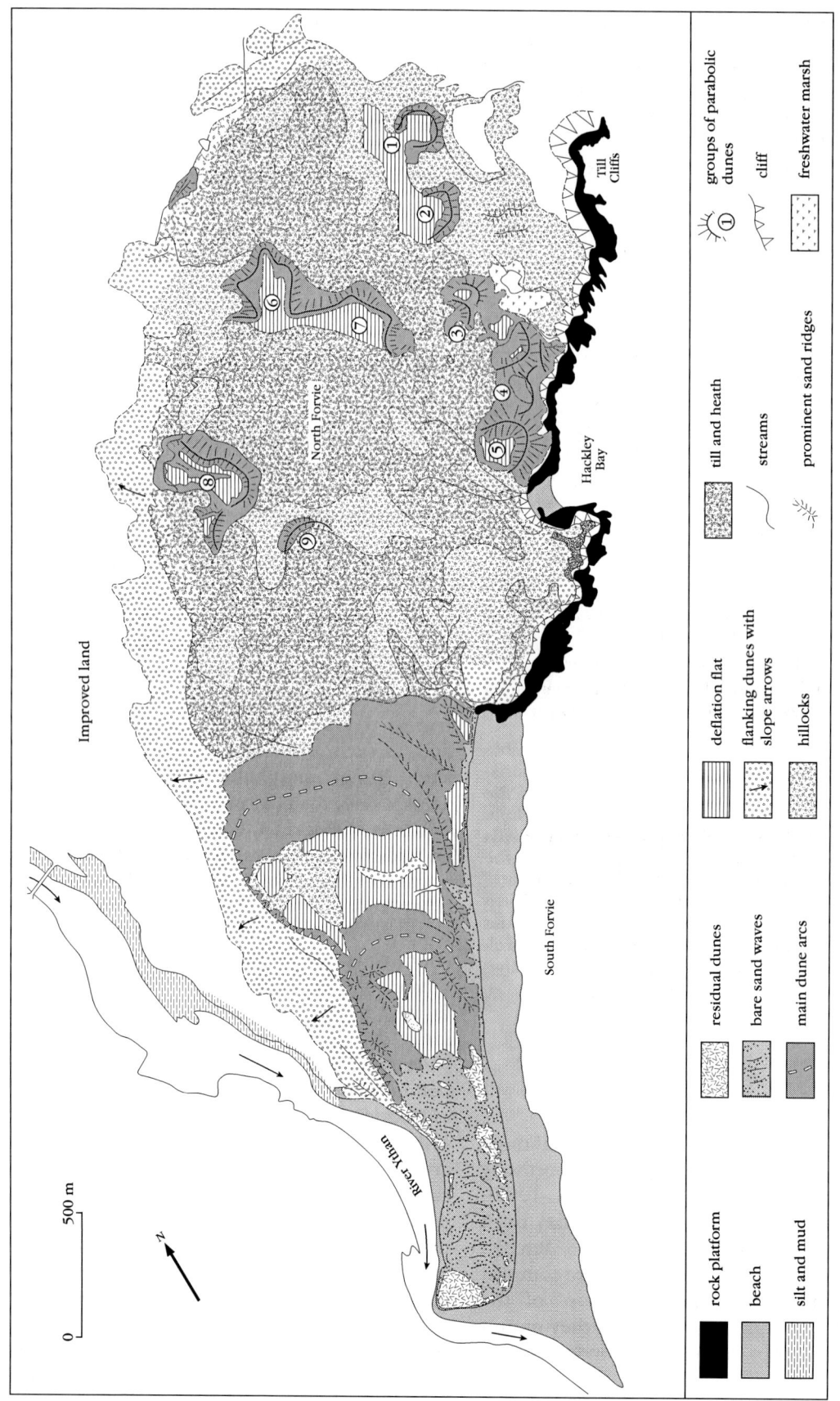

Figure 7.39 The coastal landforms of the Sands of Forvie showing the bare sand and dune-arc dominated southern part, and the largely stabilized and vegetated northern part, which also hosts the nine groups of parabolic dunes. (After Ritchie *et al.*, 1978.)

396

Forvie

general south–north migration direction of the great sand arcs of south Forvie (Ritchie, 1992). The extensive dune surfaces of north Forvie are mostly vegetated by acidic heaths growing over a thin sand veneer that is often less than a metre thick. The flanking dunes that form the western edge of south Forvie continue into north Forvie and have actively eroding and dissected east-facing slopes that in places merge with the parabolic dune complexes described above.

South of Rockend (Figure 7.37) the massive sand peninsula of south Forvie is underlain by low ridges of glacial deposits and emerged beaches, fronted in the east by an active beach and limited in the west by the tidal estuary of the River Ythan (Figures 7.37–7.39). Marking the southern end of south Forvie is a series of dynamic spits and bars where the River Ythan enters the North Sea. South Forvie is a large sand-dune system characterized by an outer zone of active coastal dune ridges and an inner zone of great sand arcs. The active coastal dunes of south Forvie range in height from 2 to 15 m and are all dissected to varying extents. Several large V-shaped blowthroughs occur and some of the dune ridges have been completely removed by deflation. Detailed study reveals that most of the foredune face at south Forvie shows a high degree of instability characterized by general retreat and erosion (Esler, 1976, 1983), although it is unclear whether this represents a long- or short-term trend.

The exceptional, and perhaps the most distinctive, feature of south Forvie is the great dome of bare sand that covers the south end of the peninsula (Figure 7.40). This extensive ridge of sand is more than 1 km long, 200 m wide and over 25 m high; it dominates the south Forvie peninsula. Surface instability is indicated by an absence of vegetation and the occurrence of sand-wave and ripple forms that trend in a south to north direction. However, active sand transport occurs not only northwards onto the adjacent deflation plain, but also westwards towards the Ythan estuary where it cascades down steep unvegetated slopes into the estuary and enters a semi-closed sediment circulation cell described by Wetherill (1980).

The central and northern parts of south Forvie consist of three very large arcuate sand ridges that extend across the entire width of the peninsula. The southern arc is 500 m wide and 20 m high, while the northern arc is c. 1500 m wide and exceeds 35 m high. In detail, the surfaces are very complex features with a series of vegetated dune ridges and bare deflation

Figure 7.40 The great dome of bare sand that dominates south Forvie is subject to active aeolian activity and sand movement. (Photo: J.D. Hansom.)

surfaces superimposed onto the main form. Parts of the south-facing slope are often severely deflated, particularly on the western side where the underlying till basement is exposed. Erosional forms (such as deep linear blowthroughs and V-shaped hollows) at various stages of activity are common. Functionally linked to the processes and forms of wind erosion are a series of depositional forms, the best example of which is a mass of bare sand that spills northwards from the north-west side of the northern arc, as a steep sand slope. These massive sand arcs are subject to rapid change and a detailed study of the northern arc showed a complex series of rapid alterations from vegetated to unvegetated status within a decade (Wright and Harris, 1988).

On the eastern side of south Forvie, the sand arcs meet the coastal dunes causing higher dune elevations that are termed 'nodes' by Ritchie *et al.* (1978). The western margin of south Forvie consists of flanking dunes with steep, eroding and actively dissected east-facing slopes, and stable, concave western slopes that grade steeply down to an emerged beach terrace above the tidal flats of the estuary.

South of the Ythan lies Foveran, a further area of beach and dunes is included within the GCR because of the intrinsic geomorphological interest and since it feeds sediment northwards into the south Forvie system and beyond (Figures 7.37 and 7.38). The area consists of a series of sub-parallel lines of massive, 10–12 m-high dunes with a well-developed wet slack between the broad coastal dune ridge and the sand-covered Holocene cliffline, which lies a few hundred metres inland. Low cliffs cut into glaciogenic deposits and emerged ('raised') beach deposits are conspicuous features underlying much of Foveran. The beach comprises a series of shore-parallel intertidal bars with intervening runnels whose migration has deflected small streams northwards. At the Ythan exit the northward drift has resulted in accretion so that the beach is now 250 m wide and backed by actively accreting embryo dunes (MacTaggart, 1998b).

In contrast to this essentially accreting area close to the Ythan exit, the area to the south is characterized by a discontinuous and severely undercut foredune. High eroded dune faces are produced along the seaward edge and extensive unvegetated sand aprons have accumulated between and behind the eroding dunes, indicating significant movement of sand landwards. At the north end of the beach, concrete anti-tank blocks dating from the 1940s are partially exposed by erosion of the coastal edge. Since these traps have been buried by sand accumulation prior to exhumation, then at least one cycle of accretion followed by erosion is suggested. Progradation in the north-east is suggested by the occurrence of a zone of stable dunes up to 25 m high fronted by foredunes that reach 21 m in width. Probably the most distinctive landform at Foveran is an extensive area of bare sand that extends 0.5 km inland from the northernmost fragmented foredune ridge. Ritchie *et al.* (1978) consider this area to be comparable to the more extensive bare sand area at south Forvie, its scale and height being partly determined by undulations in the underlying glaciogenic landforms and emerged, marine, gravel ridges. However, sand passing through breaches on the foredune ridge continues to migrate upslope in a northerly and north-westerly direction.

Interpretation

It is uncertain when the dune system at Forvie first began to develop (Stapleton and Pethick, 1996), although archaeological evidence indicates that blown sand accumulations existed near the south end of south Forvie about 5000 years BP (Ralston, 1983). The distribution and pattern of the Forvie dunes have been the subject of intermittent research since the pioneering vegetation study of Landsberg in 1955. Landsberg (1955) postulated an evolution whereby the great arcs of dunes, including the northern group of parabolic ridges, spread northwards from the beach, bar and spit sand sources of the mouth of the River Ythan. Sand also fed into the system from the extensive North Sea beaches on the east side of the south Forvie peninsula. Landsberg (1955) identified seven arcs of sand accumulation in the region and postulated that each wave of sand formed in the south and moved northwards at a migration rate which decreased progressively to the north as a result of vegetation colonization.

Although there is evidence for a chronology of sand drifting northwards (Landsberg, 1955; Kirk, 1955; Ralston, 1983), the hypothesis of a series of northward migrating dunes has since been rejected (Steers, 1973; Walton and Ritchie, 1972; Ritchie, 1992). The formation of new sand waves to the windward of a pre-existing wave would deprive the latter of aeolian sand, leading to vegetation colonization and dune stabilization. In addition, there is only patchy morphological, historical and archaeological evidence of recognizable sand waves or sand arcs in north Forvie and so the assumption of a consistent south to north aeolian transport mechanism is likely to be over-simplified and takes little account of the dune morphology at south Forvie.

Walton and Ritchie (1972) and Ritchie (1992) suggest that whereas south Forvie may have developed in a similar manner to that proposed by Landsberg (1955), north Forvie developed by a process of 'scatter and break-up'. This envisages sand moving from south Forvie into north Forvie as a series of events with a strong northerly component to form discrete and separate dune complexes (Figure 7.39). The series of dunes and dune ridges became increasingly isolated as the sand was forced onto the higher altitudes and more open topography of the northern plateau (Ritchie, 1992). The dunes that form the groups of parabolic dunes (Figure 7.39) are thought to have been fed by the periodic migration of sand from the north end of the south Forvie peninsula. A major influx seems to have occurred in 1413 when the Old Kirk of Forvie near Rockend and the surrounding cultivation rigs are known to have been abandoned (Ritchie *et al.*, 1978). Following this major influx, the dunes spread rapidly northwards, progressively infilling two lochs in the 18th and 19th centuries (Figure 7.41) and, as chronicled by Landsberg (1955), encroached on to farmland at Collieston around the end of the 18th century. However, the orientations of the parabolic dunes of north Forvie suggest a swing towards the east as the dunes migrated and stabilized, possibly due to vegetation colonization as the water table was exposed.

Today sand continues to spill northwards on the north-west side of the south Forvie peninsula (Wright and Harris, 1988), providing present-day analogues for past processes. The active sand movement in the north part of south Forvie suggest that similar areas were locally active in the past, providing pulses of sand that drifted onto parts of the north Forvie plateau. Separate pulses, in time and space, would facilitate the development of discrete masses of sand to then evolve into detached parabolic systems and other sand dune complexes (Ritchie, 1992).

Detailed studies on the dynamics of the Ythan

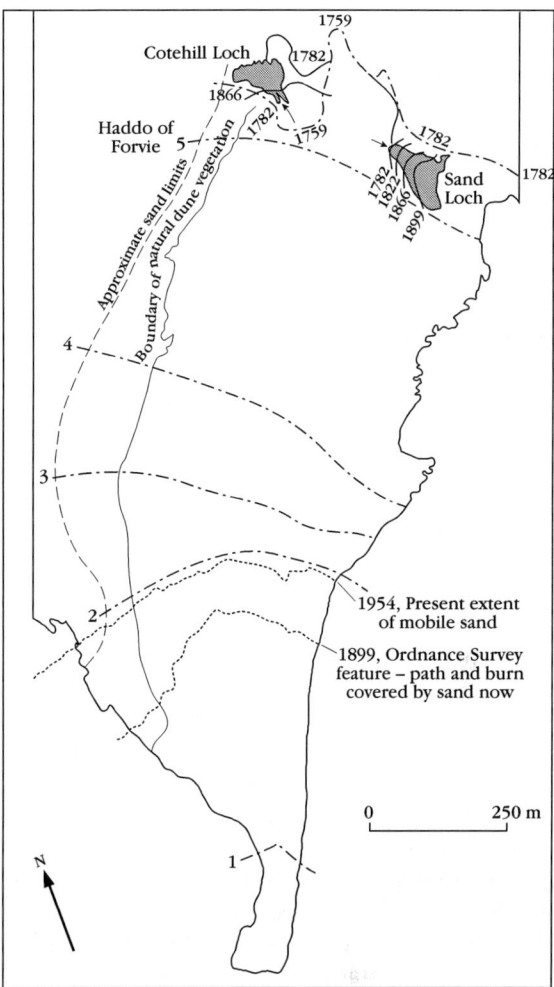

Figure 7.41 The postulated phases of sand movement over Forvie. The lines relate to sand limits as follows: 1 = the northern limit of sand before about 0 BC; 2 = the sand limit a few hundred years after 0 BC (there was little further northward encroachment until at least the 8th century AD); 3 = the limit of the area inundated by sand early in the 15th century; 4 = position of the sand front by the end of the 15th century; 5 = line reached by 1688. Further small advances are shown by dated boundaries. (After Ritchie, 1992.)

estuary (Stove, 1978; Wetherill, 1980) demonstrate that the sand peninsula and dome of bare sand of south Forvie forms part of a semi-closed sediment circulation cell. This involves east to west aeolian transport of sand over the dune and into the estuary. River flow and ebb tides carry the sand onto the estuary mouth spits and bars for transport back onto the beach. Based largely on cartographic evidence, Wetherill

(1980) suggests that the sedimentary regime of the Ythan estuary has been relatively constant for at least 150 years and that this semi-closed sand-transport cell has ensured relative stability in the position and form of the south end of the south Forvie peninsula. Although the Forvie dune system probably originates from a period when substantially more sand was available in the Ythan outlet and nearshore sedimentary environment, the river mouth dynamics and general wave climate throughout the last 5000 years may have been essentially similar to the present day. Nevertheless the longevity of the bare dome of sand that dominates south Forvie is remarkable. The lack of pioneer plant species suggests that the unvegetated sand dome represents a landform in dynamic equilibrium with its current sand budget, the rapid throughput of sand over the surface preventing colonization. Another hypothesis is that the dynamism and efficiency of aeolian erosional processes are such that the sand dome would have been removed long ago unless it were not underpinned by glaciogenic sediments or possibly bedrock (Ritchie, 1997). A programme of coring or ground-penetrating radar may resolve such questions and provide insights into the subsurface stratigraphy of the most spectacular remaining area of bare sand in Britain.

Recent experimental wind-flow measurements on the dune complexes of north Forvie provide an insight into the geomorphological processes of parabolic dune development (Robertson-Rintoul, 1985, 1990). The development of wind jets at crestal locations probably limits dune height, with eddies important on steep leeside and windward slopes affecting forward sand movements. Spiral vortex flows along the windward arms are likely to be responsible for lateral expansion of the parabolic dune.

The Sands of Forvie form the fifth-largest and least-disturbed sand-dune system in Britain (Ratcliffe, 1977). The blown-sand morphology of north Forvie is unique in Britain, with massive sand hills and dune complexes on the high rock plateau that appear to have migrated from the south. It is this unique evolution, which is not fully understood, that is of outstanding geomorphological interest and warrants the inclusion in the GCR. In addition, the parabolic dune forms of north Forvie are spectacular landforms (Ritchie, 1992). They provide an interesting contrast to the textbook formation displayed by the parabolic dunes at Barry Links and Morrich More. The dynamic interchange of sediment between the dunes of south Forvie and the extensive sand beach and spit complex at the mouth of the River Ythan add to the geomorphological interest of the site.

Conclusions

The Sands of Forvie represent a classic site for coastal geomorphology. The remarkable assemblage of windblown landforms, some of which are unique while others are representative of much of the dune coastline of north-east Scotland, are of outstanding scientific interest both individually and as an assemblage, and provide an excellent field site for innovative research at a variety of scales. The mode of evolution of this vast system, with huge volumes of sand migrating northwards from the 'normal' sand-dune complex of south Forvie onto the high rock-plateau of north Forvie, is unique. The parabolic dune forms of north Forvie are classic landforms that have developed in a different way to the classic textbook descriptions and this enhances their scientific interest.

BARRY LINKS, ANGUS (NO 550 320)

J.D. Hansom

Introduction

The Barry Links dune system has developed on an extensive broad triangular foreland (c. 11 km^2) on the northern side of the Firth of Tay, eastern Scotland (Figure 7.42). Although Barry Links contains representative examples of many beach, dune and links landforms, it is the exceptional series of well-developed parabolic dunes that is of outstanding geomorphological significance. Parabolic dunes are relatively rare in the Scottish coastal dune environment and are extensively developed in only three areas: Barry Links, Sands of Forvie, and Morrich More. The parabolic dunes of Barry Links are unique in that they have a pronounced V-shaped form with a mean length-to-width ratio of 3.3, compared to the more U-shaped forms of the dunes of Forvie and Morrich More (see GCR site reports), with ratios of 1.2 and 2.0 respectively (Ritchie, in MacTaggart, 1997b). The Barry Links parabolic dune systems have spectacular, elongated,

Barry Links

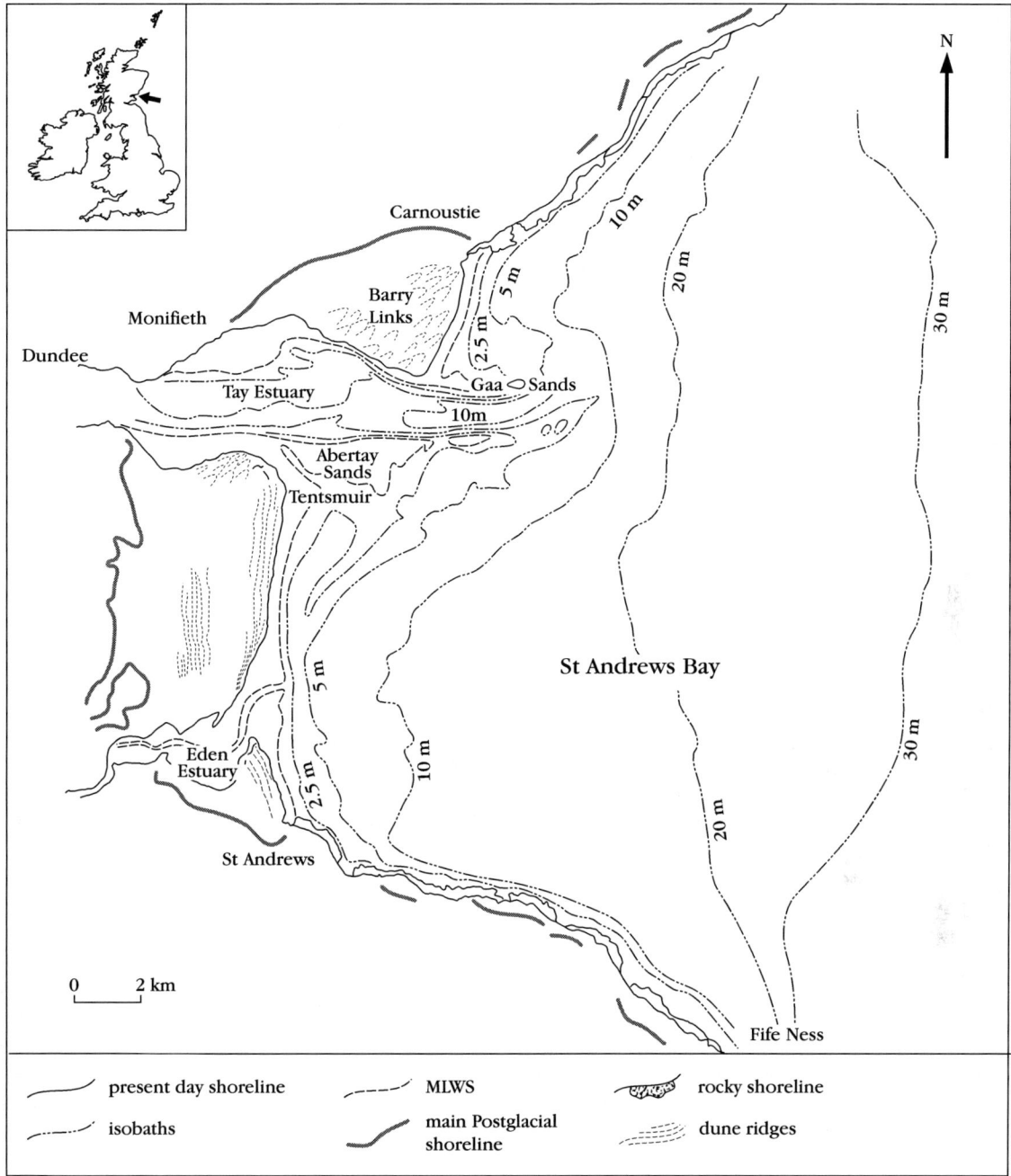

Figure 7.42 Location of Tentsmuir and Barry Links in St Andrews Bay. Tentsmuir and Barry Links have built out eastward of the main Postglacial (Holocene) shoreline at the mouth of the Tay estuary. Extensive intertidal and subtidal sand banks have also accreted at Abertay and Gaa Sands in the zone where river discharge interacts with open coast tides and waves. (After Ferentinos and McManus, 1981.)

hairpin shapes that are unique in Britain. The geomorphological features of Barry Links complement those of the Tentsmuir dune system to the south of the estuary, which is also of outstanding scientific merit (see GCR site report).

Description

The extensive sand-covered triangular foreland of Barry Links juts out on the northern side of the Firth of Tay, on the east coast of Scotland

Sandy beaches and dunes

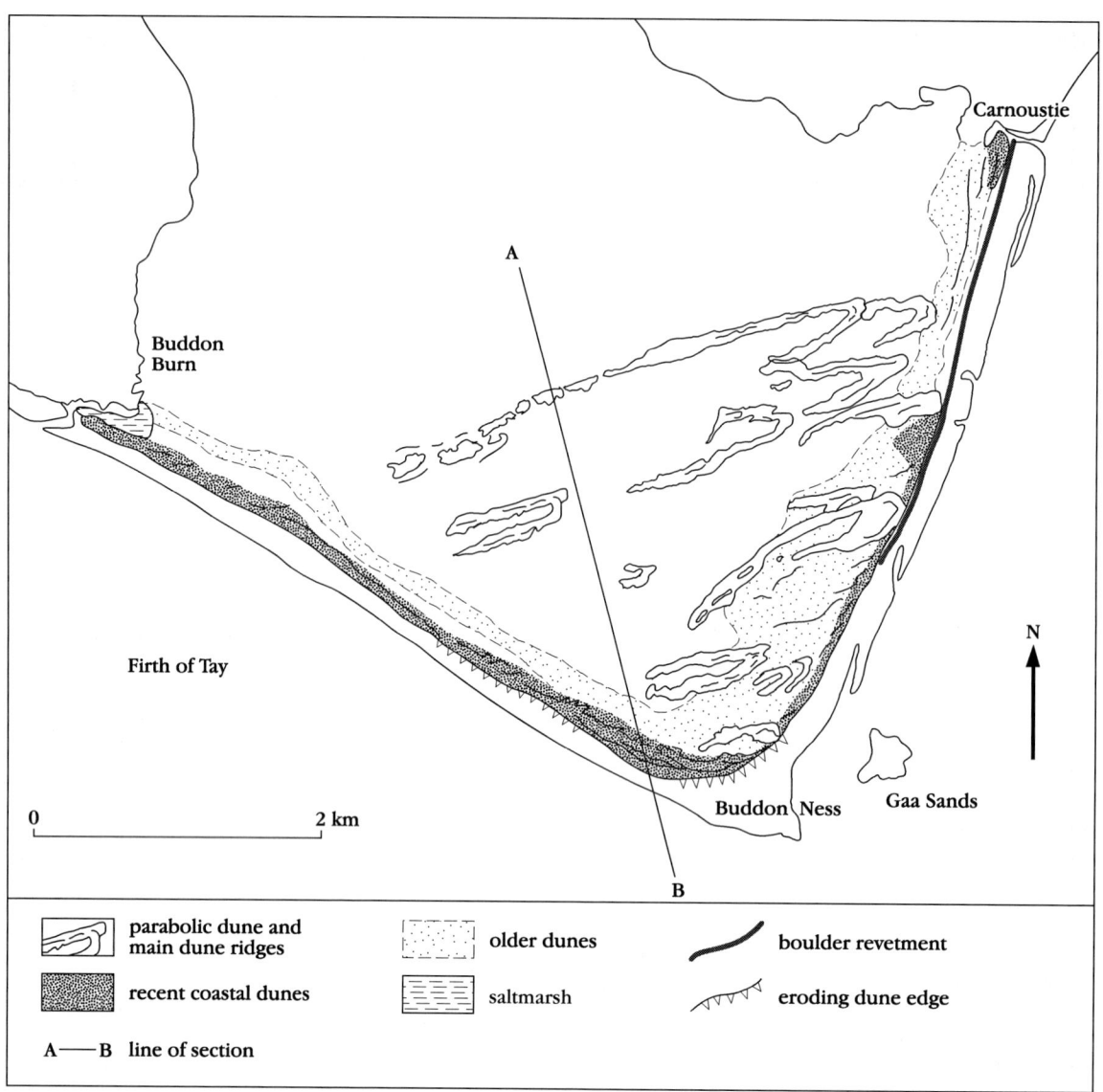

Figure 7.43 Generalized coastal geomorphology of Barry Links in 1981 showing the erosion of the narrow cordon of recent dunes and the linear nature of the series of older dune ridges, some of which are associated with parabolic forms downwind. As a result of concerns over erosion, a boulder revetment was built in 1992/1993 from the town of Carnoustie to extend along c. 3.5 km of the eastern shore. The section through A–B is shown in Figure 7.45. (After Wright, 1981.)

(Figures 7.43 and 7.44). Two 4.5 km-long sand beaches converge at Buddon Ness at the southern tip of the foreland with the extensive intertidal sandbanks of Gaa Sands, submerged during most of the tidal cycle, lying to the east of Buddon Ness (Figure 7.43). Both the east-facing and estuarine (south-facing) beaches are composed of medium-grade, non-calcareous sand (D_{50} = 0.24 mm) with occasional patches of gravel. The foreshore of the east-facing beach is c. 300 m wide, and flat with several intertidal shore-parallel bars. The extensive northern coast has a history of severe erosion, the dune face recorded to have retreated up to 10 m in one year (Wright, 1981). Early attempts in 1978 to combat erosion using gabions were rapidly overridden and the later 'solution' of 1992–1993 was to place rock armour on the beach and dune face along a c. 3.5 km length of coast (Hansom, 1999). Although aesthetically unattractive the

Barry Links

Figure 7.44 Barry Links looking north showing the high dune edge at Buddon Ness itself and dune ridges of the south (estuarine) side. Clearly visible are the long linear dune ridges, some with parabolic forms, that have been truncated by erosion on the eastern (North Sea) shore. (Photo: P. and A. Macdonald/SNH.)

rock armour appears to be serving its purpose of preventing further erosion along the protected stretch of coast and by 1994 there were no signs of slumping of the rock surface (ASH Consulting Group, 1994). However increased scouring at the toe of the armour combined with the loss of sand supply to the backshore from the now inactive dune face may have resulted in lowering of the backshore (ASH Consulting Group, 1994).

At the southern end of the rip-rap, an erosional bight has resulted in 50 m of recession landward of the rip-rap alignment and erosion extends to within 100 m of the Ness itself (Hansom and Rennie, 2003).

To the north of the point at Buddon Ness there are areas of local coastal accretion (Wright, 1981) where pioneer vegetation is colonizing the blown sand on the backshore. At Buddon Ness itself, where erosion appears to dominate, the beach is markedly steeper (7–8°) and narrower. Historically, the coastline around Buddon Ness has undergone considerable change (Wright, 1981), which is not surprising on account of its sensitive location at the point where the south-facing estuarine coastline changes to an open North Sea orientation. Local evidence shows considerable fluctuations in the position of the coastline. For example, by the early 19th century the site of the original Buddon Ness lighthouse, which was located on the southern extremity of the point during the early 16th century, was 6 m under water and 2 km to the south-east of the current one (Wright, 1981). West of Buddon Ness the beach is narrower (200 m) and lacks some of the morphological variety of the North Sea coast. The backshore is steeply sloping and fronts a foredune that has both actively accreting and eroding sections.

A series of long, narrow, well-vegetated coast-parallel dune ridges back the estuarine (south-facing) beach. The coastal dunes are 5–11 m high at Buddon Ness and decrease in height westwards, lowering to 1–2 m near the Buddon Burn. The topography of this coastal dune system is complex. Towards the western end of the shoreline for a distance of *c.* 2 km there are three clearly defined sub-parallel dune ridges. Farther east the dune ridges are characterized by old blowthroughs and associated re-depositional sandhills. Close to the point of Buddon Ness this mature dune complex is fronted by a relatively narrow line of actively accreting dunes (Figure 7.43). The single coastal dune ridge along the east-facing North Sea coast has a more varied morphology. Severe wave erosion has caused relatively rapid retreat along most of this length of coastline, although recent protection works along the northern 3.5 km stretch of coastline have effectively stopped activity in this part of the eroding dune face (see above). The vegetated coastal dune ridge is discontinuous with signs of intermittent marine breaching. A complex, high relief dune morphology has developed where the parabolic dune arcs of the interior coalesce with the coastal dune ridge. The relative proximity of several of the inland parabolic dunes to the eastern coastal edge (Figures 7.43 and 7.44) poses an interesting question concerning the relative importance of coastal retreat and the eastwards migration of the parabolic arcs in producing the truncated high dune cliffs characteristic of this coastline.

Inland from the coastal dunes is an extensive area of low undulating vegetated links (generally under 6 m OD) covering most of the triangular foreland of Barry Links. A well-developed system of parabolic dunes (Figures 7.43 and 7.44) has developed on this undulating links topography. The parabolic dunes of Barry Links are unique in Britain with a well-developed and pronounced V-shape. These dunes are long and narrow with a fairly regular outline in plan view and the extent to which secondary blowthrough development has occurred is minimal. The pattern of dune forms on Barry Links displays an unusual degree of regularity. Two distinctive morphological attributes contribute to this relatively ordered appearance. Firstly, the measured length-to-width ratios are closely and evenly distributed about the mean value of 3.3 for the dune system and, secondly, the dune orientations as represented by the directions of their long axes are remarkably uniform (Figure 7.44). The 243° orientation of the Barry parabolics suggest that they have migrated in the past from the south-west to north-east, towards the eastern coastline.

The Barry parabolic dunes are now almost completely vegetated and stabilized, with the exception of the large dune that is utilized by the Ministry of Defence (MOD) as a firing range. In the northern part of the foreland the parabolic dunes stand as discrete units. Towards the south, in the vicinity of Buddon Ness, some adjacent dunes overlap although this does not appear to disrupt the general parabolic shape. Some breaks in the orderly dune pattern occur at the southern and eastern margins of the foreland, where some of the parabolic dunes have intersected the present-day coastline. The convergence of the parabolic dunes with the coastal dune ridges has resulted in the production of a high relief and complex dune topography. Isolated SW–NE-trending elongated dune ridges suggest coastal erosion has truncated a former, more extensive, system of parabolic dunes.

Interpretation

The evolution of the large foreland of Barry Links remains speculative. Its general triangular shape, comparable to accumulation features in southern England such as Dungeness, Kent (see GCR site report in Chapter 6), suggests that the area has developed as a result of extensive deposition of beach materials in the past. There

Barry Links

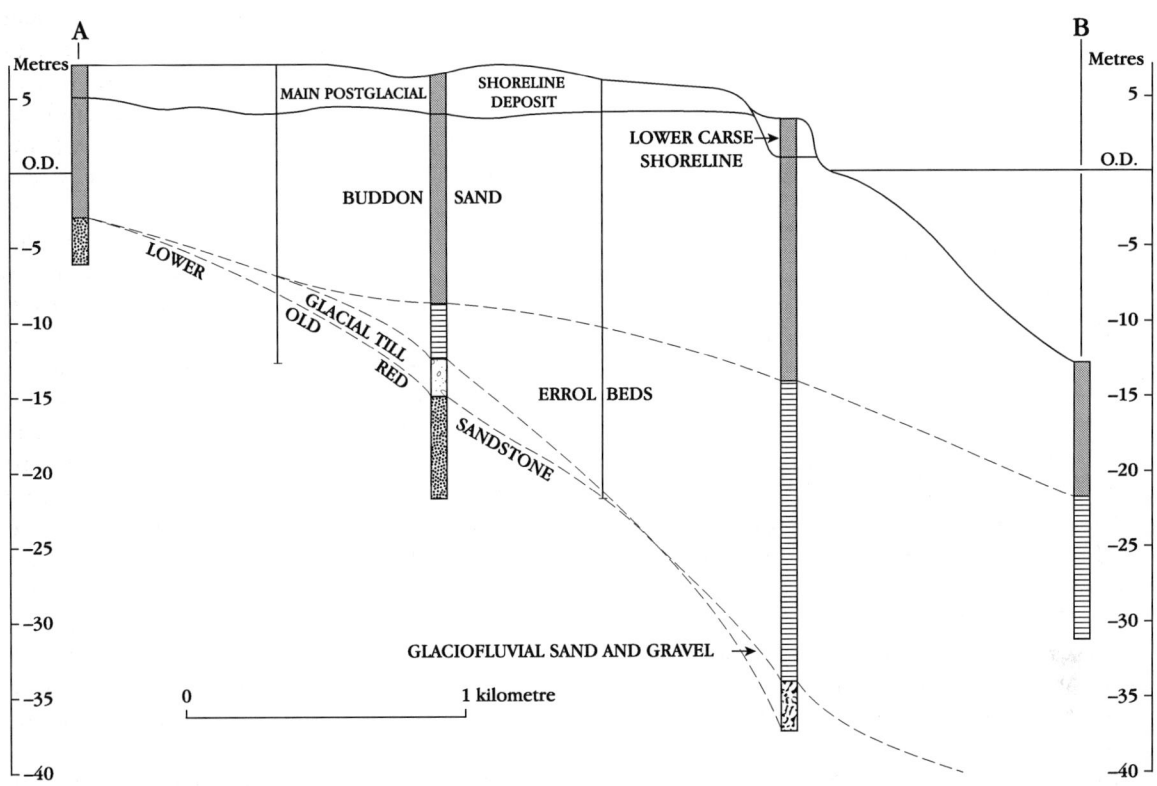

Figure 7.45 Stylized 3.5 km cross-section (along the line A–B on Figure 7.43) of Barry Links and Buddon Ness as reconstructed from borehole data. Barry Links sits atop substantial thicknesses of marine and shoreface deposits and suggests that this estuary-mouth site has undergone continued deposition over much of the Holocene Epoch. (After Paterson, 1981.)

appears to be little doubt that, as with much of the North Sea coast of Scotland, large parts of the foreland consist of emerged beaches. Steers (1973) posed the question of the stability of this distinctive triangular foreland. Analysis of old maps of the area covering the last 200 years show change only at the margins, suggesting that the main body of the feature is based on a more stable foundation, possibly an ancient beach or rock platform (Steers, 1973), although there is no surface indication of underlying rock. Borehole evidence (Figure 7.45) suggests that beneath the dunes lies a series of emerged shorelines cut into a thick sequence of marine sands (Buddon Sand) which themselves overlie marine clay (Errol Beds) (Paterson, 1981). The physiographical evolution of Barry Links clearly requires further investigation, particularly in the context of changing relative sea levels.

The parabolic dune system of Barry Links is one of the finest and well-developed in Britain, but there has been surprisingly little research carried out on these spectacular forms. Early work by Landsberg (1956) shows that wind regime is a major factor in the orientation of the Barry dunes (Figure 7.46). Each wind direction in proportion to its sand-moving power was plotted and the resultant vector (shown by the arrow on Figure 7.46) completes the wind direction polygon and indicates the direction of the dominant wind effect. This resultant vector conforms almost exactly with the mean dune orientation of the Barry dunes (Landsberg, 1956). The open exposure of the Barry Links foreland and the lack of topographical interference with formative winds from the south-west may explain the regular form, orientation and pattern of the Barry Links parabolic dunes (see Figure 7.44). The undulating links surface is remarkably even; the parabolic dunes that have developed on this sandy plain represent the primary relief features in the area so, in the Scottish

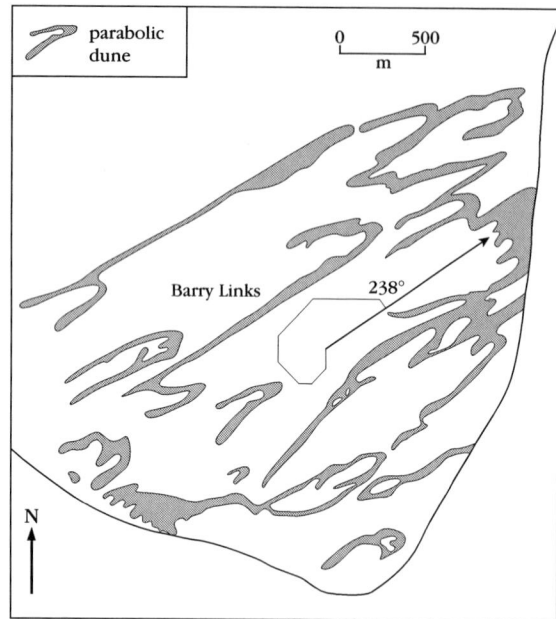

Figure 7.46 The relationship between mean parabolic dune orientation and the resultant vector of the wind polygon using dune orientations and locations in 1956. Note the eastern limits of the parabolic dunes in 1956 in comparison with their positions in 1981 as plotted in Figure 7.43. (After Landsberg, 1956, from Hansom, 1988.)

by released gabion fillings on the beach face (Sarrikostis and McManus, 1987). Tidal currents also transport sediment towards the Ness from the North Sea coast on the ebb as well as the flood (Figure 7.47). The extensive coastal protection works along the northern part of this coast will clearly affect the natural balance of the coastal system. Potential sediment supply from the previously eroding dunes to the downdrift beaches (i.e. Buddon Ness) has effectively stopped through protection works and this may have long-term implications for the entire system (Hansom and Rennie, 2003).

Barry Links is owned by the MOD with restricted public access. As a result, the majority of Barry Links, with notable exceptions near the firing ranges on the east coast, is relatively undisturbed and the existing land-use has produced a unique conservational environment. The site has been selected for the GCR on account of the well-preserved parabolic dune system. The pronounced V-shaped parabolic dunes are unique in Britain and demonstrate a close relationship between wind regime and dune orientation (Landsberg, 1956). The site also provides representative examples of beach, dune and links landforms and offers a valuable complement to the study of Tentsmuir on the south of the Tay estuary, which is also of outstanding geomorphological interest (see GCR site report, below).

context, the wind regime is unusually free from topographic effects. In addition, the Barry parabolic dunes lack the complications of subsurface control, a common feature of many coastal dune systems (e.g. Machir Bay). This may also help explain the orderly form of the Barry parabolic dunes. Map and field evidence shows an apparent migration of the dune forms towards the eastern coastline. However the present stability of the dunes suggests that this, certainly in the recent past, may be more a result of rapid marine erosion in the east, rather than the recent downwind migration of the dunes themselves (compare Figures 7.43, 7.44 and 7.46).

Potential patterns of longshore sediment transport on the coasts north and south of the Tay estuary have been calculated using wave refraction modeling (Sarrikostis and McManus, 1987). The model predicts a south-westerly drift down the exposed North Sea coast of Barry Links to Buddon Ness, where deposition occurs. Field evidence appears to support this model; the southward transport of material from Carnoustie to Gaa Sands has been demonstrated

Conclusions

The extensive sand-covered foreland of Barry Links contains an exceptional series of well-developed and preserved parabolic dunes which are of outstanding geomorphological significance. The dune orientations show a close relationship with local wind regime (Landsberg, 1956) and the pronounced V-shaped form of the Barry parabolic dunes are unique in Britain. The parabolic dunes, which have a mean length-to-width ratio of 3.3:1, have spectacular, elongated, hairpin shapes with an exceptional regular and orderly pattern. These unique characteristics may reflect the open exposure of the foreland, the lack of topographic interference with formative winds and the lack of subsurface control. In addition, Barry Links provides a representative assemblage of many beach, dune and links landforms offering valuable opportunities for studies of coastal evolution.

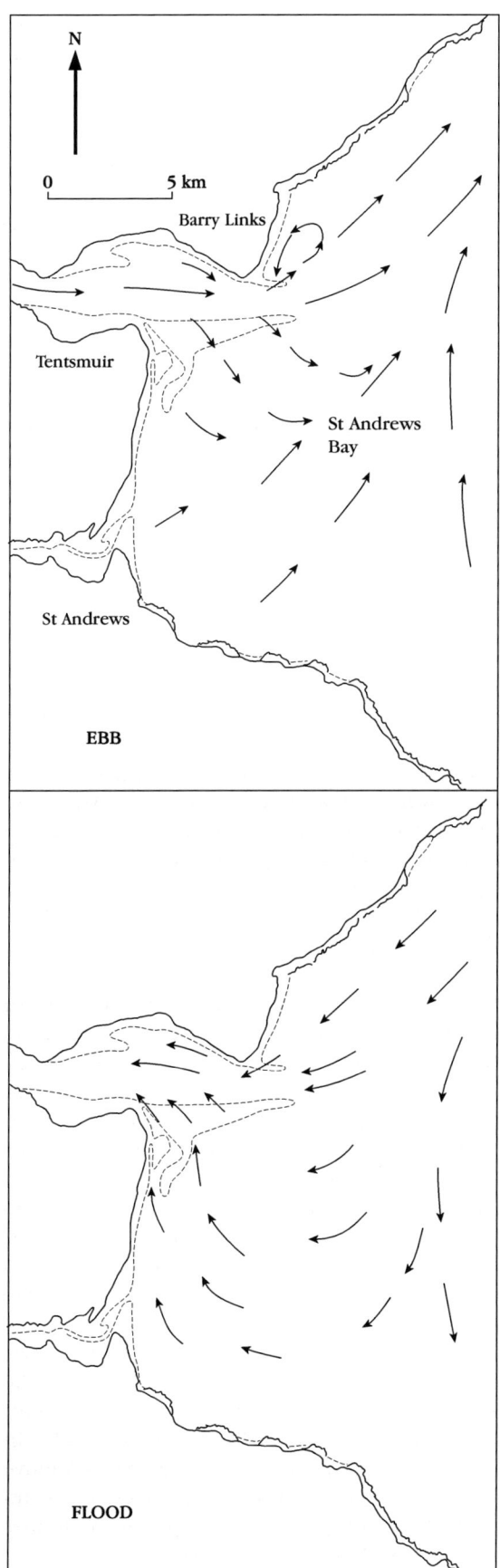

Figure 7.47 Mid-flood and mid-ebb tidal stream patterns in St Andrew's Bay based on a combination of direct measurement and hydraulic modelling. The open coast at Tentsmuir is affected by northward movement on the flood and south-eastward movement on the ebb, whereas the open coast at Barry Links is affected by southward movement on both the flood and the ebb. (After Ferentinos and McManus, 1981.)

TENTSMUIR, FIFE (NO 500 275)

J.D. Hansom

Introduction

The extensive lowland surface of Tentsmuir lies between the Tay estuary in the north and the sandstone headlands of St Andrews in the south (see Figure 7.1 for general location) and is one of the largest areas of blown sand in Scotland (Dargie, 2000). Tentsmuir is a site of long-term accretion with over 3.5 km of shoreline advance in 5000 years (Ferentinos and McManus, 1981). The Tentsmuir GCR site (see Figure 7.42), which includes the vast intertidal sand spits, banks and bars of Abertay Sands, forms the point where the coastline turns from the open sea into the Tay estuary and is of outstanding geomorphological interest. The rate and amount of coastal progradation at Tentsmuir is unique in Britain (Crawford and Wishart, 1966; Ritchie, 1979b) and has been documented by several workers, notably Grove (1950), Deshmukh (1974), Wal (1992) and Whittington (1996). The area is known to have been accreting in a north-eastward direction since 1812 at an average rate of 4.8 m a^{-1} (McManus and Wal, 1996). Long-term net accretion at Tentsmuir is the result of the integrated impacts of several natural processes acting in concert, both wind and wave activity resulting in the accumulation of sediment at the Point (McManus and Wal, 1996). Close relationships between the geomorphological and ecological evolution of Tentsmuir (Crawford and Wishart, 1966; Garcia-Novo, 1976) enhance the scientific interest of this highly dynamic and outstanding site.

Description

The Tentsmuir Point GCR site forms a relatively

small proportion of the 3300 ha of sand dune in the greater Tentsmuir area (Dargie, 2000) (Figure 7.48). Low, emerged beach sands and silts form the substrate materials for much of the Tentsmuir links and dune system, however the boundaries between the emerged ('raised') beach sand and blown sand are imprecise (Ritchie, 1979b). Most of the Tentsmuir area was stabilized by afforestation in the 1920s and as a result much of the morphological detail has been obscured. Nevertheless, there is evidence that beneath the forest cover there are sets of sand ridges running parallel to the coast (Ritchie, 1979b), and to the south and west, intervening lochs that have since been drained (Hutcheson, 1914). Structurally, Tentsmuir is composed mainly of two sequences of dune ridges arranged approximately parallel to the nearby coast with intervening slacks. In the north, bordering the Tay estuary, they trend east–west and along the open east-facing coast of St Andrews Bay they trend north–south. At Tentsmuir Point the often rather poorly developed 2–4 m-high dunes are weakly aligned north-west–south-east (McManus and Wal, 1996).

The morphology of Tentsmuir Point is intimately linked to the intertidal sand spits, banks and bars of Abertay Sands that stretch eastwards for 6–7 km beyond the Point (Figure 7.48). Abertay Sands are more than 1 km in width and incorporate a substantial island area at the southern entrance. These sand formations are highly dynamic and respond to the complex interplay of the three main variables, estuarine discharge, tidal streams and wave climate (Ritchie, 1979b). The complex development of the Abertay sand spits and bars, including an analysis of the main ebb and flood channels, is investigated in detail by Green (1973).

Tentsmuir Point provides a complex topography where it is possible to identify fragments of earlier phases of development, including former coastlines that evolved as a result of processes that are essentially similar to those operating today. The low-gradient sand beach at Tentsmuir Point reaches widths of up to 400 m and the lower foreshore typically shows ridge and runnel structures trending north–south (McManus and Wal, 1996). Tentsmuir Point is largely composed of medium-grained sands (D_{50} = 0.28 mm) that are highly susceptible to wind action. Sediment transport by wind is very important in the upper foreshore and backshore zones where the development of dune systems has led to an increase of vegetated land surfaces elevated above high spring tide levels. Active 2–4 m-high dune accumulations in the lee of the beach are found extensively at Tentsmuir Point. These low, hummocky marram *Ammophila*-clad dunes grade landwards into a *c.* 50 m-wide zone of low dunes where four separate ridges can be identified (Ritchie, 1979b). To landward, there is a distinctive flat dune-slack zone at *c.* 1–2 m OD that is of considerable ecological interest. Landwards of the dune slack there is a line of broader mature dune ridges that correspond approximately to the 1941 line of concrete anti-tank blocks (Ritchie, 1979b). The mature dune systems are heath-covered whereas the younger forms have characteristic *Ammophila*-dominated vegetation. The average surface elevation of the dunes and slacks at Tentsmuir Point is around 3–4 m OD (Ritchie, 1979b).

The south part of Tentsmuir Point (i.e. the east-facing coastline) is affected by the rhythmic changes of erosional coastal sections alternating with progradational sections, which is typical of the coast southwards as far as the Eden estuary (Ritchie, 1979b). Coastal erosion has been documented along different stretches of the Tentsmuir coast since 1964, notably north of the Eden estuary, but also at the southern end of the GCR site (McManus and Wal, 1996). The 3–4 m-high sand cliff cut in the coast-parallel dune ridges just north of the entry of the Powie Burn suggests this section of the coast was undergoing a period of recession in the 1970s (Ritchie, 1979b). The north coast of Tentsmuir is subject to substantial changes in response to the pattern of ebb discharge and flood tide channel migration associated with the dynamics of the south side of the Tay estuary (Figure 7.49).

The Tentsmuir area is a site of long-term net coastal accretion with over 3.5 km of shoreline advance in 5000 years (Ferentinos and McManus, 1981). Based on analysis of historical and recent data sources Tentsmuir Point is known to have been accreting in a north-eastward direction since 1812 (Grove, 1950; Deshmukh, 1974; Wal, 1992; Wal and McManus, 1993). The rates and form of coastal progradation at Tentsmuir have been reconstructed in Figure 7.50). In the earliest documented growth phase, 1854–1912, the high-water mark advanced north-eastwards by about 40 m on average, although in the south the shoreline receded. The next documentary evidence of the

Tentsmuir

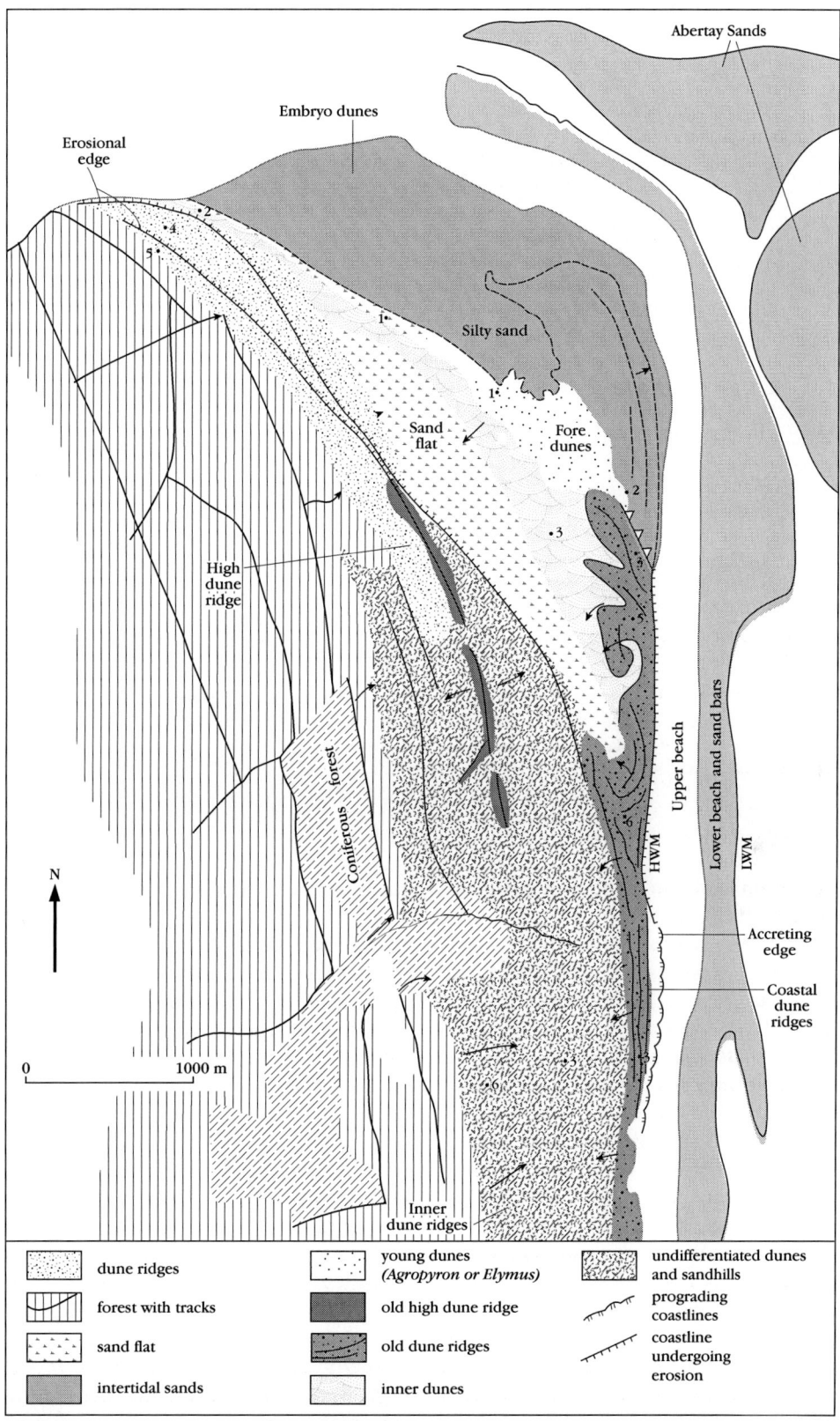

Figure 7.48 The coastal landforms of Tentsmuir showing the extensive areas of sandflat, foredunes and intertidal sandbanks that extend out to Abertay Sands. Erosional edges are found in the south of Tentsmuir and along parts of the Tay estuary coast. (Based on Ritchie, 1979b, and McManus and Wal, 1996.)

Sandy beaches and dunes

Figure 7.49 A spectacular oblique aerial photograph looking east towards the exit of the Tay at low tide with Tentsmuir and Abertay Sands extending into the distance on the south side and on the north Barry Links with Gaa Sands extending beyond. The recent sand accretions of Tentsmuir Point can be seen in the foreground. (Photo: P. and A. Macdonald/ SNH.)

shoreline position is in 1941, the year when the line of anti-tank traps and the low ridge (the Defence Dune), which lies 80–160 m seawards of the traps, were constructed at or above the high-water mark. Aerial photographs show a c. 500 m-wide flat beach surface seawards of the dunes in 1948. By 1962 hummocky aeolian sand accumulations supporting pioneer vegetation were present along the backshore, separated by narrow channels occasionally occupied during high tide. This dune growth had extended 40 m seawards of the Defence Dune. By 1972, the isolated hummocks had largely amalgamated so that a continuous vegetated area had been created extending a further 25–30 m seawards. Again a series of sand mounds supporting pioneer plants were present on the backshore. By 1978 the dune margin had advanced a further 60 m seawards as the mounds became incorporated within the vegetated dune area. New actively accreting mounds lay to the seaward. To the east a 400 m-long dune-covered spit extended northwards from Tentsmuir Point, providing shelter from waves. By 1985 the spit had broadened from 25 m to 80 m and the northern extremity had separated to create a recurved 'islet' over 300 m long. By 1990 the southern part of the spit had linked with the accreting sand mounds and the northern 'islet' had extended in all directions, although a narrow tidal channel remained between it and the vegetated land area.

Overall, in the 178 years between 1812 and 1990 the vegetated land area at Tentsmuir Point advanced 870 m in a north-eastward direction perpendicular to the coastline (McManus and Wal, 1996), and eroded about the same distance inland in the south. The average long-term (178-year) accretion rate is 4.8 m a^{-1}, although by plotting forward growth against time McManus and Wal (1996) show that accretion rates have increased greatly through time. The very high rates of accretion have been achieved by the retention of sand upon an already high beach surface. The vegetation has responded to rapid coastal accretion at Tentsmuir, with an outward movement of vegetation zones over time (Crawford and Wishart, 1966; Ritchie, 1979b). The pattern of floristic development is matched closely to the distribution of slacks in relation to coastal accretion (Crawford and Wishart, 1966).

Interpretation

Tentsmuir is one of the most rapidly accreting parts of the British coastline (Crawford and Wishart, 1966). Continuing coastal progradation is relatively rare in Britain and most dune systems are currently undergoing a period of retreat. Thus, the natural dynamism of the north-eastwards accretion observed at Tentsmuir Point has attracted considerable scientific interest and research. Grove (1950) first mapped the coastal progradation at Tentsmuir and suggested three main possible sand sources: sediment entering the coastal system from the Rivers Tay and Eden; offshore sediments; or sediment derived from coastal erosion. Later research established the detail of coastal changes between 1854 and 1990 (Figure 7.50) and the recent evolution of Tentsmuir Point is now relatively well documented (Deshmukh, 1974; Wal, 1992; McManus and Wal, 1996) and is summarized above. More recently research has focused on the processes and mechanisms fuelling the observed coastal accretion at Tentsmuir (e.g. Ferentinos and McManus, 1981; Sarrikostis and McManus, 1987; Wal, 1992; Wal and McManus, 1993; McManus and Wal, 1996).

The vast Tentsmuir dune and links system is the result of massive Holocene progradation, comparable to the formation of the Morrich More and Culbin systems (Ritchie, 1979b). Ferentinos and McManus (1981) note that the Tentsmuir shoreline has advanced over 3.5 km in the last *c.* 5000 years. The exact mechanism of coastal progradation is unknown, but Ritchie (1979b) envisages three possibilities. Firstly, as the sea level fell from a Holocene high of *c.* 10–15 m OD, successive beach zones built seawards and continue to do so. Secondly, the falling sea level left a wide beach zone upon which dune systems developed, or thirdly, some form of spit–bar complex curved southwards from the Tay enclosing a broad lagoonal area that was subsequently infilled from the east by blown sand. A detailed stratigraphical and geomorphological investigation is required to elucidate the Holocene evolution of the Tentsmuir system.

Recent process studies at Tentsmuir may provide a key to understanding the past. Tide, wave and wind activity are the major constructive processes contributing to the current growth of Tentsmuir (McManus and Wal, 1996). The floor of St Andrews Bay, an area within which sediments have been deposited during Late-glacial and subsequent times (Browne and Jarvis, 1983), may provide the immediate source of sediment for the Tentsmuir area (McManus and Wal, 1996). Sediments are also swept northwards onto Tentsmuir by a gyre in the flood tide (see Figure 7.47), whereas ebb tides may sweep sediments south, across Abertay Sands (Ferentinos and McManus, 1981). Based on wave refraction analysis, Sarrikostis and McManus (1987) demonstrated that wave fronts approaching the Fife coast from most directions become deformed in such a way that they sweep towards the Tentsmuir area. Consequently, as the result of sediment movement by waves approaching the shore at an angle, Tentsmuir Point experiences accretion due to the transport of bed material not only shorewards from the bed of the embayment but also northwards along the shore (Sarrikostis and McManus, 1987). Longshore drift also transports sediment from the eroding sections of the south Tentsmuir coast northwards to the Point. Based on a comparison of aerial photographs and maps, McManus and Wal (1996) estimated that the volume of sediment eroded from the dune margins along the Tentsmuir coast between 1978 and 1990 was 46×10^4 m^3. The volume of sediment that had accumulated at Tentsmuir Point over the same period was estimated to be 33×10^3 m^3. Thus the coastal erosion on the Tentsmuir beaches to the south could have readily supplied the material accreted at the Point. Long-term natural progression of sediment northwards along the beach face to Tentsmuir Point has led to the creation of a wide beach-surface that has extended north-eastwards and is protected behind the offshore Abertay Sands (McManus and Wal, 1996).

Wind activity is also a major constructive process contributing to coastal accretion at Tentsmuir (Wal and McManus, 1993; McManus and Wal, 1996). By 1990, over 300 m of the 500 m-wide beach surface noted in 1948 was covered with low vegetated dunes. The present area of accretion is now well above the level of spring high-water mark, indicating that the sediment covering the beach surface has been trans-

Sandy beaches and dunes

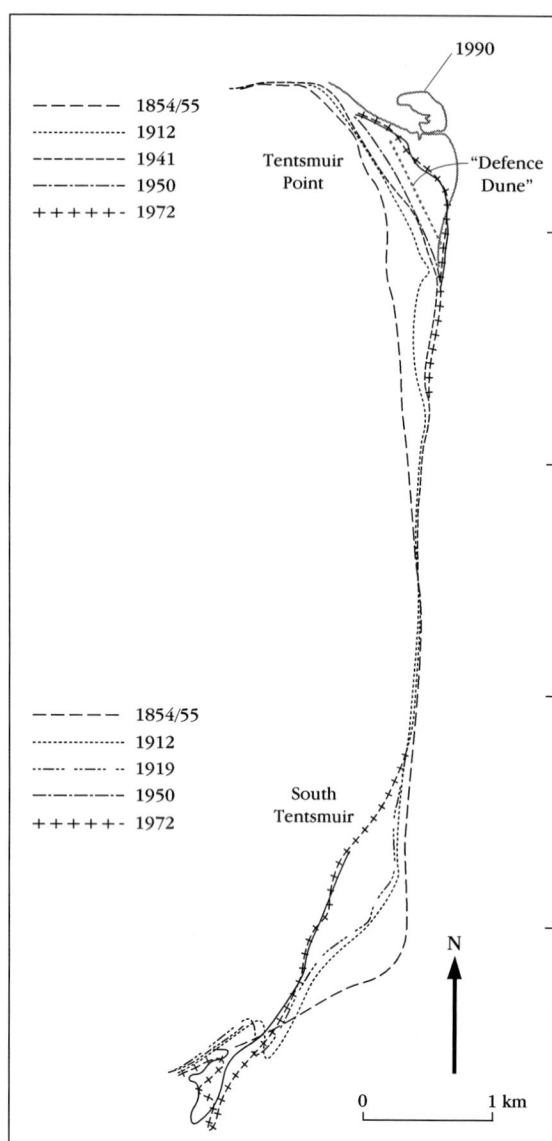

Figure 7.50 Long-term changes in the position of south and north Tentsmuir showing a general trend of erosion in the south and accretion in the north. (Compiled from McManus and Wal, 1996.)

form a distinctive group of landforms. The commonest offshore winter winds may carry sand from the dunes to create wind-shadow foredunes at the back of the upper beach. Offshore winds also carry sand onto the lower beach and into nearshore tidal waters where wave and tidal activity recycles it to the beach. Therefore, the principal geomorphological impact of offshore winds is the production of shadow foredunes (McManus and Wal, 1996). Onshore winds, common in spring and autumn, transport sand up the beach face, enhancing foredune growth and carrying sand landwards into the dune systems. Longshore winds from the south can transport large volumes along the coast towards Tentsmuir Point (McManus and Wal, 1996). For example, strong winds in November 1968 carried continuous sheets of sand northwards along the Tentsmuir coast for at least three hours forming a swarm of barchan dunes up to 1 m high on the beach at the Point. It is calculated that as a result of this storm, more than 40 000 tonnes of sand were transported to Tentsmuir Point (McManus and Wal, 1996) although much of the sediment was swept north into the channel of the Tay estuary, perhaps to be recycled into the system at a later date.

The close relationship between the geomorphological and ecological evolution of Tentsmuir Point has also attracted considerable research (e.g. Crawford and Wishart, 1966; Desmukh, 1974; Garcia-Novo, 1976; Whittington, 1996). From the wide intertidal beach to the margins of the forest (which is encroaching naturally onto the older dunes) there are excellent examples of the interaction of vegetation and landform, with particular emphasis on accretionary forms and processes. Vegetation zones have encroached seawards gradually stabilizing the accreting coastline. For further details of the biological interests and the complex vegetation successions at Tentsmuir Point see Crawford and Wishart (1966), Steers (1973) or Garcia-Novo (1976).

Conclusions

Tentsmuir Point, Fife, marks the southern limit of the Tay estuary and is one of the most rapidly accreting parts of the British coastline. In contrast to the majority of dune systems in Britain, which are generally undergoing retreat, Tentsmuir Point is actively accreting. Since 1812

ported to the site by wind action (McManus and Wal, 1996). A detailed study of the wind regime at Tentsmuir identified high-energy seasonal 'unimodal' (offshore or onshore) and 'bimodal' (both offshore and onshore) wind regime patterns (Wal and McManus, 1993). In addition, there can also be a 'unimodal' longshore wind and a 'bimodal' one that possesses a longshore component. Each of the major wind directions initiate sand transport at certain velocities and

Tentsmuir

Tentsmuir Point has advanced 870 m in a north-easterly direction at an average rate of 4.8 m per year in part, fuelled by erosion of South Tentsmuir. The very high rates of accretion have been achieved by the retention of sand upon an already high beach surface. Net accretion at Tentsmuir Point is the outcome of the integrated impacts of several natural processes acting in concert. Wave activity in St Andrews Bay results in the transport of sediment to the head of the embayment and northerly longshore drift along the Tentsmuir coast encourages progression of sediment along the beach face towards Tentsmuir Point. Wind action acts simultaneously, resulting in the accumulation of low dune forms on top of a wide beach surface. Onshore, offshore and longshore winds are all important in dune formation at Tentsmuir. The outwards movement of dune vegetation zones is related to long-term coastal accretion at Tentsmuir, and this close association between the geomorphological and ecological evolution enhances the scientific interest.

The site is also important as a National Nature Reserve and is part of a Special Area of Conservation.

Chapter 8

Sand spits and tombolos – GCR site reports

Introduction

INTRODUCTION

V.J. May

Sand spits and tombolos in Britain are associated with

1. areas of wide intertidal sandflats
2. estuary mouths
3. intensive erosion of cliffs that provide copious longshore sand supplies, and
4. comparatively sheltered locations.

Typically, they are low in height: even when dunes occur on them, the main structures are only a few metres in height. They are dynamic features of the British coast, for although sandy structures have been in their present sites for many centuries, they have changed in detail, undergoing erosion, breaching and accretion. Although some are still extending, they are also often marked by narrowing and breaching of their proximal (landward) ends. The term 'sand spit' is typically used for any low ridge of sand extending from the shore across an embayment, estuary or indentation in the coast and they have a number of forms (Figure 8.1), ranging from those that cross the mouths of estuaries or bays, such as at Forvie, Aberdeenshire, or to those that form barrier islands. Some sand structures, in contrast, link hard-rock features to the mainland or to islands (for example St Ninian's Tombolo, Shetland, and parts of the Isles of Scilly). Small bay-head beaches often form as ridges deflecting small streams alongshore. Although these have been described as spits, they are often the result of shore-parallel beach ridge construction (e.g. Pwll-ddu) rather than longshore transport that

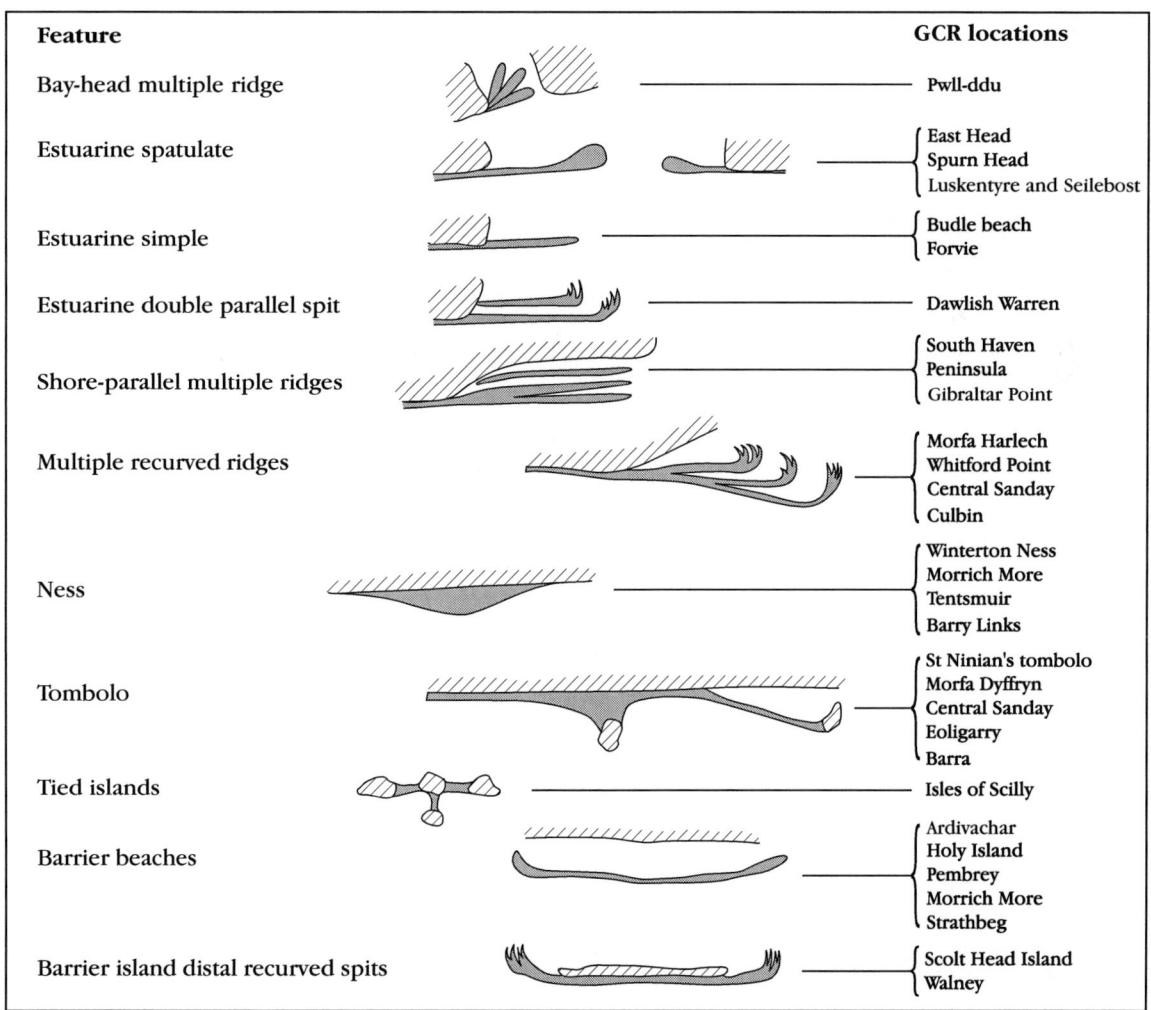

Figure 8.1 Sand spits and their associated structures, indicating some key representative GCR sites.

Sand spits and tombolos

has extended a spit across an estuary. Some sand spits have a distal 'spatulate' form that does not display individual recurve ridges. Typically, these occur where there is a base on which the sand transported to the distal end can accumulate over a wide area. This base may be salt-marsh or mud-flats. Many spits have been built upon gravel ridges, or, in Scotland, emerged beaches as their foundation, and in some cases the presence of morainic gravels provides a basis for the distal parts of these spits (for example Spurn Head, Yorkshire and Whitford Spit,

Table 8.1 The main features of sediment sources and tidal ranges of sand spit GCR sites, including coastal geomorphology GCR sites described in other chapters of the present volume that contain important sand spit structures in the assemblage of features. Many machair sites have small sandspits – see Chapter 9. (Sites described in the present chapter are in **bold** typeface)

Site	Main features	Other features	Present-day natural sources of sediment	Tidal range (m)
Pwll-ddu	Sand spits		Local fluvial and shallow nearshore	8.2
Ynyslas	Sand spit	Dunes	Estuarine, longshore (reduced)	4.1
East Head	Sand spit, distal dunes		Restricted alongshore: mainly from offshore banks	3.4
Spurn Head	Major spit in macro-tidal environment	Dunes	Longshore and offshore	6.4
Dawlish Warren	Sub-parallel double spit	Dunes	Intertidal banks	4.1
Gibraltar Point	Series of spits, effects of extreme events	Dunes	Longshore and offshore banks	7.0
Walney Island	Barrier islands recurved spits	Till cliffs	Cliff erosion	9.0
Winterton Ness	Linear dunes on cuspate foreland		Longshore	2.6
Morfa Harlech	Spits and recurves, ridge and runnel	Dunes	Longshore limited, intertidal estuarine banks	4.5
Morfa Dyffryn	Tombolo and dunes, sarn	Dunes	Longshore limited, offshore possible but unconfirmed	4.3
St Ninian's Tombolo	Tombolo	Dunes, climbing dunes	Nearshore and some local reworking	1.1
Isles of Scilly	Tied islands, spits	Emerged beach	Local feeder cliffs and platforms	5.5
Central Sanday	Tombolos, spits, sandflats, dunes	Gravel ridges, machair, dunes	Local reworking and nearshore machair	3.0
Eoligarry	Emerged tombolo	sand dunes and machair, bowthroughs	Local and offshore, biogenic sources from the east	4.0
Culbin	Bluckie Lock spit	Emerged gravel strand-plain, dunes, saltmarsh	Nearshore and erosional recycling	3.6
Morrich More	Innis Mhor sand spit	Emerged strandplain, dunes, saltmarsh	Fluvial, glaciogenic and offshore	4.3
Tentsmuir	Shore-parallel dune ridges, ness	Sand dunes, intertidal sands	Estuarine and longshore, significant	4.4
Luskentyre–Corran Seilebost	Sand spit	Sand dunes and machair	Nearshore, intertidal to the east	3.8
Forvie	Shore-parallel dune ridges, spit	Unvegetated and parabolic dunes	Longshore and recycled from estuary	3.1
Torrisdale Bay	Dune landforms, climbing dunes	Sandspits, intertidal sandflats, saltmarsh	Fluvial and offshore, limited	4.0
Holy Island	Barrier beaches, spits	Emerged beach, dunes	Longshore and offshore	4.1
Scolt Head Island, North Norfolk	Barrier beach, recurved spits	Dunes	Longshore and offshore	5.6
Newborough Warren	Spits, modern and relict	Dunes	Intertidal estuarine banks offshore, local reworking	4.7
Carmarthen Bay	Spits	Dunes, cliffs	Fluvial/estuarine, offshore and intertidal banks, local reworking	8.0
Braunton Burrows	Distal estuarine shore-parallel spit	Dunes	Fluvial/estuarine, offshore and intertidal banks, local reworking	7.3

Introduction

Carmarthen Bay). In some cases, the spits also form the base upon which important dune systems have developed, such as in Central Sanday, Orkney or at Luskentyre, Seilebost and Gualan in the Western Isles (see GCR site reports).

According to Pethick (1984), British coastlines with a tidal range of less than 3 m are noted for their spit development. However, sand spits are not restricted to areas with low tidal ranges: spits both in sand and gravel are a common feature of the high tidal ranges of the eastern English Channel and also occur on the Scottish coast. Pethick (1984), Goudie (1990) and Goudie and Brunsden (1994) provided incomplete maps of British major spits, defining a 'major' spit as being longer than 1.5 km (Pagham Harbour being the smallest mapped). Of 34 sand spits on the British coast south of a line between the Solway Firth and Fraserburgh, (thus omitting all the machair sites with spits and the numerous small spits in sealochs and voes of the Scottish Highlands and islands) 12 lie on coasts with a tidal range less than 3 m, and 22 on coasts with a tidal range greater than 3 m (Table 8.1 and Figure 8.2).

Where sand spits are supplied primarily by longshore sediment transport, many are affected by erosion at their proximal end. This may result from up-drift coast protection structures (e.g. groyne fields) or from reductions in the natural rate of longshore sediment supply brought about by changes in wave direction or changes in the amount and nature of the sediments in the source region. For example, along parts of the English west coast, erosion of till and head deposits (the former sources of sand) has exposed hard-rock coasts from which the sediment supply is much reduced. Reductions in sediment supply have also forced adjustments in coastal orientation with updrift erosion of many spits, as occurs at Culbin and Whiteness Head (see GCR site reports in Chapters 11 and 6). Similarly, estuarine sediment supply may be significantly reduced or increased as a result of changes in catchment management. East Head (Chichester Harbour, Sussex) contrasts strongly with many other spits in continuing to grow in volume even when longshore transport to it has been substantially reduced.

Many sand spits are associated with extensive intertidal areas of sand banks and submerged bars at the mouths of estuaries. It is evident from the studies of some sand spits that they depend to a significant extent upon the transport of sediment from these areas. Sand nesses (e.g. Winterton Ness), although less common than the gravel forms, are associated with offshore shoals, but the directions and quantities of sand moving between them are uncertain.

The problem of breaching of the proximal end of spits, and the potential demise of the feature, besets the management of many other sites (e.g. Spurn Head, Hurst Castle Spit, Dawlish Warren). Although it has been argued by de Boer (1964) in relation to Spurn Head that this can be shown to be part of a natural cycle of events, at other spits it is attributable to the reduction of longshore transport resulting from cliff or beach protection works. Kidson (1963), however, suggested that many spits were dominated by erosion and were well into a final stage of development leading to their extinction. The geomorphological interest of spits thus lies partly in their potential for self-destruction.

One recurrent feature of many sand structures is the development of separate and distinct ridges, seen for example at a small scale at Pwll-ddu, in parallel double spits at Dawlish and in multiple recurved ridges at Morfa Harlech. Similarly, many of the features are marked by recurrent breaching of the spit. Whereas de Boer's cyclic breaching hypothesis for Spurn Head has now been re-evaluated (see below), other sand spits such as East Head show periodic breaching of the main features often at their proximal ends. Many of the spits have not grown simply as a result of longshore transport extending a spit gradually across an embayment. Most show a characteristic of sudden rapid extension possibly resulting from rapid shoreward movement of sand ridges, followed by localized reworking and a period of comparative quiescence. Breaching or the construction of another ridge often then takes place. However, there are documented instances of the inlet becoming permanently sealed by longshore extension, as occurred at Strathbeg, Aberdeenshire in the 18th century, see GCR site report in Chapter 7.

Evolution of sand spits and structures

Although the GCR sites described in the present chapter show – in their alignment facing the dominant waves – a similar tendency to the beaches described by Lewis (1932, 1938), sand

Sand spits and tombolos

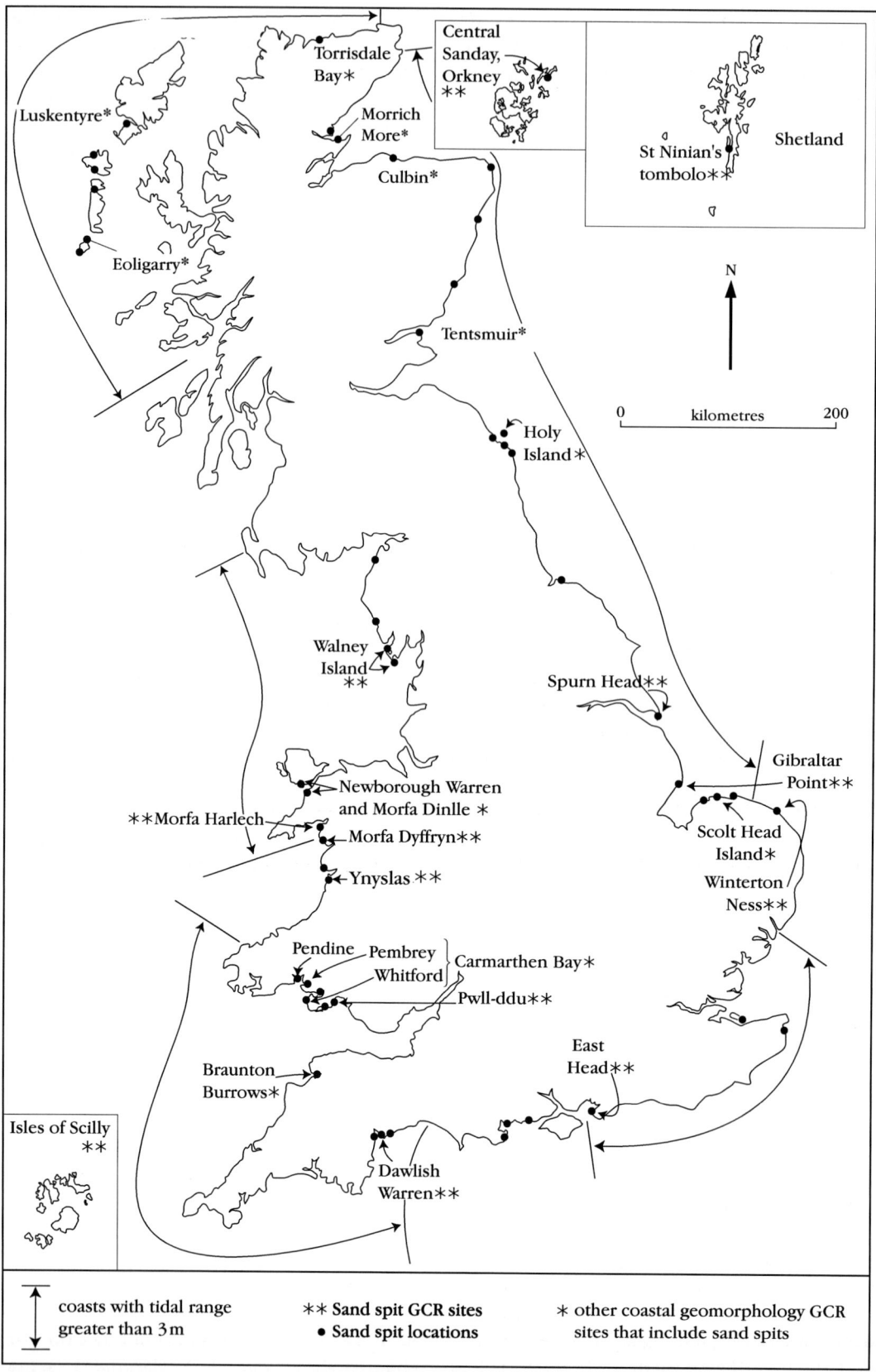

Figure 8.2 The location of sand spits in Great Britain, also indicating other coastal geomorphology GCR sites that contain sand spits in the assemblage. (Modified after Pethick, 1984).

Introduction

spits are much more dependent upon the foundations provided by lag deposits from erosion of rocky coasts and in particular glacial deposits. For example, Spurn Head, Morfa Dyffryn and Whitford Burrows depend in part upon the presence of remnant Devensian moraines. Similarly, the development of transgressive gravel ridges and the erosion of coastal platforms on low-lying coasts have provided the foundations for both transgressive sand ridges and for the extension of sandy beaches across estuaries. Roy *et al.* (1994) distinguish between the flux of sediment on wave-dominated sandy coasts and the movements of barrier sand masses during sea-level changes (which include phases of transgression, stillstand and regression). Many of the sand structures described here may result from a combination of these processes. Some features, which have been regarded in the past as the result of longshore transport, appear in fact to result from the transgression of barriers. These became restricted in further onshore movement by the pre-existing topography and have been re-shaped by subsequent wave conditions. The debate in Robinson's 1955 paper about the formation of double spits at the mouths of estuaries in southern England focused on two separate processes: longshore spit development and subsequent breaching as opposed to 'frontal accretion' (as Robinson called it, which can be regarded as a form of barrier development). Sites such as East Head and Spurn Head, although of different scale, have developed as the result both of longshore sediment transport and transgression, and such origins need to be considered for other similar structures.

Carter (1988) considered the concept of the coastal cell as a framework for the long-term development of spits. In his view, spits, like all beaches, depend upon the balance between the flux of wave energy (total shoreline wave energy per unit wave crest per unit time ECn) towards the shore (shore-normal P_N) and along the shore (shore-parallel P_L). The angle made with the shore by the breaking wave (α_b) affects the magnitude of P_L whose spatial distribution alongshore can be mapped. Where waves break parallel to the shore, $\alpha_b = 0$ and sediment is simply transported up and down the beach. However, on a spit, the breaking angle increases rapidly around the recurve, although at the same time wave height decreases due to wave refraction. As a result, P_L remains constant except at the farthest distal curve. This implies that the spit will only survive as a long-term feature as long as there is a sufficient longshore supply of sand to maintain the longshore component of sediment transport. Few British sand spits fit this model exactly. For many, the longshore supply of sediment is interrupted either by periods of weaker, or longer-term reduction in, longshore supply and transport or by direct interruptions as a result of the construction of structures such as groynes. For example, in parts Scotland, sediment reduction and sea-level rise has led to smaller coastal cells than before, and has forced internal re-organisation of sediments, manifested by updrift erosion and downdrift accretion of spits at Culbin and Spey (Hansom, 2001). Furthermore, where spits enclose large open bodies of water, waves also affect the behaviour of the spit on its landward side and can bring about significant changes in the overall development of the spit (e.g. Spurn Head). This model also largely ignores the role of the intertidal and offshore routes by which sand is transported often with different values of P_L. For example, both Dawlish Warren and Morfa Harlech (see GCR site reports in this chapter) display different patterns of wave breaking at low water from those affecting the upper beach and the main spit form. The effects of very long-period swell, high-energy events, surges and short periods of waves from opposite directions from the prevailing waves may each provide explanations for some of the sand structures around the British coast. Once developed, many of these features are very persistent forms. Although some features have developed in their present locations during the last 1000 years, many others are built upon a foundation that is considerably older: the spits at Culbin have ancestors that span most of mid–late Holocene times.

The conservation value of sand spits and associated structures

The conservation value of sand spits and structures arises from:

1. their historical role as areas of accumulation of sediment, so providing the basis for pioneer plant species to colonize the area,
2. their links with intertidal and offshore banks and bars as part of the sediment transport system,
3. their association with dunes and ecology,

Sand spits and tombolos

4. their role in narrowing the entrances to estuaries and providing protection for the development of extensive mud flats and saltmarshes in the resulting shelter,
5. their place in coastal education and research. Three of the longest-running continuous coastal university research programmes are based on major sandy structures at Spurn Head (University of Hull), Gibraltar Point (University of Nottingham) and Scolt Head Island (University of Cambridge) since the 1920s, and
6. the fact that many remain largely undisturbed by artificial structures and development.

Sand spits are also very important:

1. in providing sheltering structures at the mouths of navigable estuaries,
2. in their role as natural coast protection structures providing low-cost protection to low-lying coastal land.

Despite their dynamism, many of these structures are of considerable age, with documentary and archaeological evidence for their existence and growth reaching back over many centuries. Some of the structures have built up during the historical period and so it is possible to assess the ways in which these features have resulted from environmental changes within their coastal and river settings. Some of the sand structures are important assemblages of many geomorphological forms that demonstrate the different ways in which sandy coasts adjust to differences in sediment availability and changes in climate and oceanic conditions. Their conservation value derives from their present-day features and in demonstrating the ways in which the coastal environment adjusts to change can be observed and their persistence understood.

In this chapter the site reports are ordered so that the more simple forms precede the more-complex ones: Pwll-ddu, Ynyslas and Spurn Head are simple spits with spatulate form, lacking recurves; Dawlish Warren, Gibraltar Point and Walney Island are double/subparallel spits with some recurves; Winterton Ness has a cuspate form; Morfa Harlech is a large cuspate form with extensive recurves, Morfa Dyffryn a simpler cuspate-like spit, more accurately described as a tombolo; St Ninan's Isle displays classic tombolo forms, and the Isles of Scilly tied islands; and Central Sanday has an assemblage of features.

PWLL-DDU, GLAMORGAN (SS 580 970–SS 570 963)
V.J. May

Introduction

The coastline of south Wales is characterized by dramatic cliff scenery and some nineteen beach and dune systems between Merthyr Mawr Warren at the mouth of the Ogmore River in the east, and Broomhill Burrows near the mouth of Milford Haven in the west. They fall into three broad categories:

1. large spit and beach systems extending across the mouths of estuaries (for example Laugharne Burrows, Carmarthen Bay)
2. extensive low hindshore dunes that rest upon bedrock (e.g. Broughton Burrows, Carmarthen Bay)
3. small bay-head beaches and dunes, usually with an easterly aspect (e.g. Oxwich Bay, Glamorgan).

Pwll-ddu is the smallest and least well-developed representative of the final category. Its importance arises from its place within this group of sites. It contains a wide variety of coastal forms within a very small area: shore platforms, slope-over-wall cliffs, other cliff forms, and sand and shingle beaches. In addition to this important assemblage of features, the site includes a series of small shingle and sand ridges on the west side of Pwll-ddu Bay that have diverted a small stream to the east (Strahan, 1907; Ward, 1922; George, 1933; Steers, 1946a; Guilcher, 1958; Potts, 1968).

Description

Pwll-ddu (see Figure 8.2 for general location) is among the smallest of the coastal sand accumulations along the Welsh coastline, and, unusually, has not developed any significant dunes (Potts, 1968). There are three main morpho-sedimentological units within the site:

1. former and present-day sea cliffs on the eastern side of the bay,
2. sand and shingle beach and ridges, and
3. slope-over-wall cliffs and shore platforms at several levels on the western side of Pwll-ddu Bay.

The cliffs within this site appear to be receding

Pwll-ddu

Figure 8.3 Pwll-ddu Bay. See Figure 8.4 for explanation. (Photo: Cambridge University Collection of Aerial Photographs © Countryside Council for Wales.)

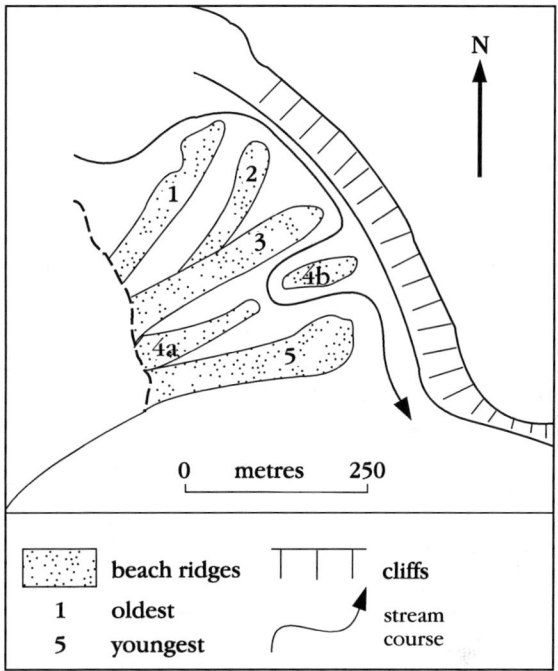

Figure 8.4 The succession of beaches at Pwll-ddu Bay: (1) oldest, (5) youngest (as yet undated). 1, 3 and 5 are the higher ridges that dominate the site.

very slowly, as there are well-preserved remnants of former platforms and slopes that are cloaked with periglacial scree material. On the eastern side of the bay, the foot of the slope has been trimmed, probably by a combination of marine and fluvial erosion. The small stream that flows along the eastern side of the valley appears to have occupied this position throughout the period of progradation that led to the growth of the beach ridges. The most landward (oldest) ridge is aligned SW–NE and successive ridges swing progressively to an alignment closer to the present-day position of the shoreline. The fourth ridge (4a,b on Figure 8.4) is formed of two separate features, the eastern one diverting the stream across the valley. Ridges 1,3 and 5 are the highest and longest ridges (Figures 8.3 and 8.4). They have changed little in appearance since the early part of the 20th century (Strahan, 1907).

Interpretation

Guilcher (1958) suggested that when a beach is prograding, several successive ridges may be left, and he described Pwll-ddu as a very fine example of this. Many dune sites rest upon sand and shingle ridges (e.g. Oxwich Bay and Morfa Harlech GCR sites), but the very limited development of dunes at Pwll-ddu means that this early stage of growth in bay-head beach–dune systems has not been buried by dunes. The site is sheltered from storm waves approaching from the south-west, and is aligned towards the SSE. Apart from waves generated from the south-east or south within the restricted fetch of the Bristol Channel, all other waves are refracted around Pwll-ddu Head. Sediment sources are limited. The stream entering the bay generally carries only fine-grained materials, longshore transport of sand is restricted by both Pwll-ddu Head and the headland to the east, and little is known about possible seabed sources.

The Pwll-ddu valley is possibly an advanced stage of a ria in which the rock floor, which lies below sea level, has been cloaked by infilling assisted by the blocking action of the growing beach ridges. As sand and shingle was deposited in the bay, the ridges appear to have migrated into the valley and aligned more towards the north-east as a result of refraction (Figure 8.4). Later ridges have protected the older ridges, overriding them in some cases. There is no dating of the features, though the order of formation is clear, but they offer an excellent location in which to demonstrate the sequence of non-marshland progradation within the macrotidal coastline of southern Wales (tidal range is about 8.2 m).

Pwll-ddu is one of the few British sites where a series of beach ridges has not been obscured by dune development. At Pwll-ddu the ridges remain clearly visible, probably as the result of a rather restricted sand supply. The small scale of the site will aid studies of the relationship between the shingle ridges and the associated sedimentation that both preceded and followed the growth of each ridge. The sheltered position minimises longshore sediment transport and wave direction is relatively constant. As a result, the site provides an excellent location in which both the past and present-day sedimentation processes and the sediment budgets can be described and, in the latter case, monitored.

Conclusions

Pwll-ddu is important because of its generally unmodified series of small beach ridges. Seen in the context of bay-head beach and dune development on western coasts, it is valuable for comparative purposes, because it represents an early phase in the development of these features. The site, although small, contains a wide range of coastal forms: shore platforms, slope-over-wall cliffs, sand and shingle spits.

YNYSLAS, CEREDIGION (SN 605 919)

V.J. May

Introduction

The spit at Ynyslas, north of Borth (see Figure 8.2 for general location), forms part of the Dyfi National Nature Reserve. (Watkin, 1976). It is a good example of a sand spit built upon a gravel base, but it is also important because it is possible to show that a similar feature has been in existence here since about 6500 BP (Wilks, 1977, 1979). The southern part of the spit is dominated by a shingle ridge upon which there has been some accumulation of sand. The central part of the spit is dominated by vegetated dunes, whereas the northern, dista, end forms a low sandy flat upon which there are some small vegetated dunes. The behaviour of the spit seems to be related not only to the general tendency in Cardigan Bay for sediment to move northwards, but also to the patterns of water movement within the lower Dyfi estuary (Dobson, 1967; Chesnutt and Galvin, 1974; Williams *et al.*, 1981).

Description

The spit extends about 3 km from the southern side of the Dyfi estuary. The main line of the spit is formed by gravels that are exposed at high-water level along the southern part of the spit. They are veneered with sand on the northern part of the spit, but re-appear north of the distal end of the spit at Cerrigypenrhyn (SN 611 953). The dunes form a narrow fringing ridge about 100 m in width that extends northwards (from SN 606 927 to SN 605 938) whence it swings more and more south-eastwards towards a former distal end at SN 615 936. Gravel and shingle are exposed both as a fringing high-water deposit and as a large ridge extending into the estuary (Figure 8.5). Former recurve and swale topography is exposed in this area. The northern part of the spit also extends further into the estuary as an area of dunes up to 9 m OD. Parts of the dunes have been eroded by recreational trampling. Extensive sandflats east of the dunes, parts of which are used for car parking, provide a reservoir of sand for the dunes and the estuarine sandbanks. The Afon Leri flows into the estuary today through a canalized course. However, in the past, before drainage works were carried out in this area, the Afon Leri entered the sea at Ynyslas Turn. This may indicate that there was insufficient sediment transport alongshore to divert the stream mouth farther north.

Offshore from the distal end of the spit, there are intertidal banks, including the South Banks, which cause reflection and refraction of waves approaching the spit from most directions. Tidal range at springs is about 4 m and tidal streams can reach 0.5 m s^{-1} near the shore (Williams *et al.*, 1981). Under storm-wave conditions from the south-west, wave energy is focused at the distal end of the spit, whereas with north-westerlies the distal end is less affected and wave energy is concentrated on the shingle ridge farther south (Williams *et al.*, 1981).

Submerged forest beds (best observed

Ynyslas

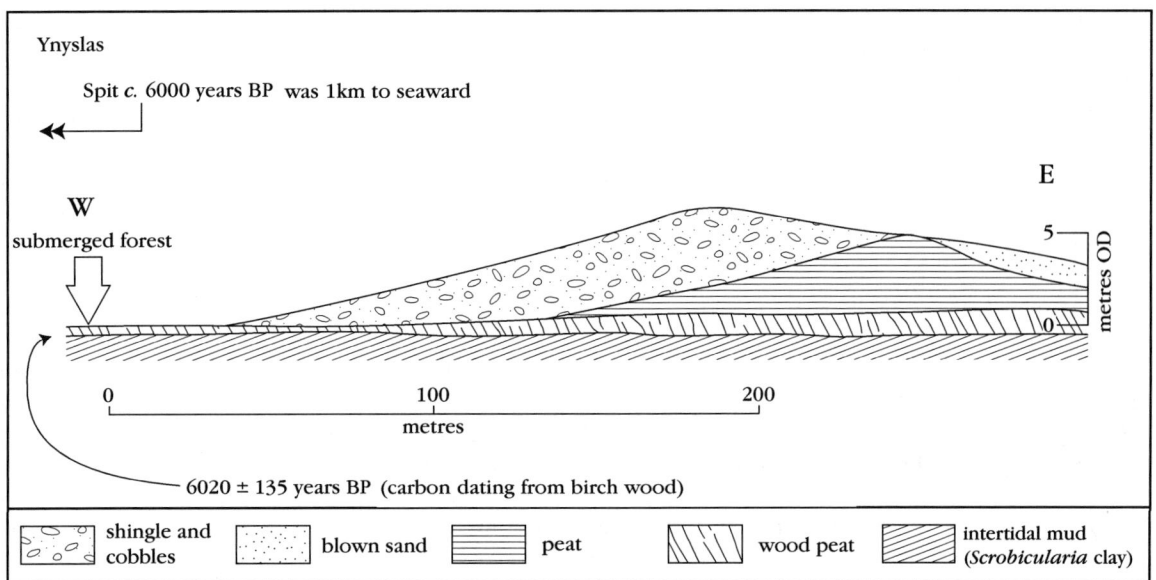

Figure 8.5 A east–west beach section at Ynyslas. The large arrow indicates the position of the submerged forest beds. (After Campbell and Bowen, 1989.)

between about SN 604 924 and SN 604 933) have been exposed on the foreshore as the spit gradually moved about 150 m landwards during the 19th century (Campbell and Bowen, 1989). The basal peat was dated at 5898 ± 135 BP (Godwin and Willis, 1961) and birch *Betula* wood *in situ* near the base of the forest bed was dated at 6026 ± 135 BP (Godwin and Willis, 1961). Borth Bog, a very important Quaternary site, lies to the east of the spit and owes its development largely to the protection afforded by the spit and its predecessors.

Interpretation

The earliest investigations of the Dyfi estuary focused on the estuary itself and its saltmarshes (Yapp et al., 1917; Richards, 1934; Burd, 1989). The sedimentary history of the estuary, including the behaviour of the area around the spit have demonstrated the longevity of the sedimentation within a sheltered microtidal estuary (Shih, 1991, 1992).

The shingle ridge was established during the Holocene transgression (Williams *et al.*, 1981), when coastal conditions stabilized sufficiently about 6500 BP to allow the creation of a sand and shingle spit that extended northwards from the cliffs at Borth (Wilks, 1979). This earliest position of the spit is thought to have been about 1 km seawards of its present position (Campbell and Bowen, 1989). Interpretation of the stratigraphical, radiocarbon and pollen data from the submerged forest beds and Borth Bog suggests that since about 4000 years BP the spit has been maintained by shingle supplied by the eroded material from the cliffs at Borth and has moved landwards across the submerged forest and peat at about 0.25 m a^{-1}. Churchill (1965), however, suggested that this coast had been elevated by about 3 m since 6500 years BP by isostatic uplift. This could have the effect of increasing the positional stability of the spit by reducing the effects of wave energy inputs.

Detailed analysis of wave and tidal conditions and the associated changes in its form have shown the feature to be relatively stable in recent years (Williams *et al.*, 1981). Although storm conditions are destructive, the spit recovers quickly, with the dunes acting as a sand reservoir. On this site, as elsewhere (for example Hurst Castle Spit, Hampshire and Ainsdale, Lancashire), surges at high springs play a particularly important role in modifying the form of the coast. The storm of 11 November 1977, for example, saw a surge of 1.1 m that co-incided with high spring tides. Unlike Ainsdale (see GCR site report), Ynyslas rests on a relatively permeable base and so there is a greater chance of the intertidal sands drying and being blown onto the dunes. In contrast, sand blows around the northern end of the dunes over the wide, distal sandflats, where current action across the intertidal and shallow submerged banks may re-distribute sand into the offshore banks. From here, it can return to the beaches. The longshore provision of sand and shingle from the south is now limited, especially as a large groyne-field now extends northwards from Borth. Just as the distal position of the Ynyslas spit affects the channel at the mouth of the estuary, so also the movement of sand into the estuary from the northern beaches affects the channel. The Shoreline Management Plan suggests that sedimentation causing reduction in estuary capacity is probably more concentrated on the northern side of the estuary than at earlier periods when Ynyslas was accreting more rapidly. The relative stability of the spit appears to be a result of its gravel base, as well as the reworking of sand between dunes and beach, the movements of sand between dunes, flats, banks and beaches and the continued but limited sediment supply from the cliffs to the south, although this is now restricted by groynes at Borth. The role of the estuarine water movements and their interaction with waves need to be investigated more fully.

There are similarities with the cobble ridge at Westward Ho! (see GCR site report in Chapter 6) where the ridge has also moved landwards and exposed older sediments in the intertidal area, but Westward Ho! lacks the critical evidence of the age of the feature, which is provided by dates from Borth Bog and the submerged forest. Ynyslas is especially important because of the links between local coastal processes and other aspects of Quaternary geomorphology.

Conclusions

This shingle and gravel ridge – or a similar feature – may have existed here or slightly offshore since about 6500 BP and has become the base for dunes at the mouth of the Afon Dyfi. It is of particular interest because of its age and effects on other features of the local landscape. It is not always possible to demonstrate that a beach has maintained much the same position at the mouth of an estuary while migrating landwards. Although dates for the origin of the spit have been suggested they allow only a limited estimate of the average rate of migration (between 0.15 and 0.25 m a^{-1}). The probable age of the Ynyslas spit, and its similarities with the feature at Westward Ho!, make it important for our understanding of the timescale over which many features of the British coast have developed.

EAST HEAD, WEST SUSSEX (SZ 761 985)

V.J. May

Introduction

East Head is a low ridge of sand-dunes and beaches that lies on the eastern shoreline of Chichester Harbour whose mouth it restricts (see Figure 8.2 for general location and Figure 8.6). Despite extensive interference with the process of longshore drift by groynes to the east, this site has been dominated by accretion during its recent history. In common with several other such features, it has a broad distal end, partly formed by recurves, and a narrow neck at its landward (proximal) end. Unlike many such features, it has prograded steadily even though there has been breaching of its neck. Today its volume is probably greater than at any previous period. Its development appears to be associated with changes in the intertidal 'delta' at the mouth of Chichester Harbour. Shingle ridges, which successively change orientation from NNW–SSE towards NE–SW, form the base of the dunes, which attain about 3.5 m in height. The area of the spit, which declined steadily during the 19th and early 20th centuries, has increased since 1972. Cartographic evidence suggests that a comparable period of progradation after progressive lateral migration of the spit also occurred during the 18th century. Like many of the shingle spits of the central south coast of England, East Head is paired with a spit on the opposite side of the estuary, but it has received less attention until recently. Searle (1975) outlined its key features and particularly the effects of occasional recent breaches in its form. Examination of the cartographic evidence during the last 200 years coupled with more recent aerial photography suggests some unusual features about the development of the spit, not least because what has appeared to be an erosional phase is in reality a depositional one.

Description

East Head is narrow (under 100 m) at its landward end but widens at its distal end to over 400 m. It is backed by saltmarsh onto which it has encroached. To seaward the beaches extend into intertidal areas known as 'The Winner' and 'East and West Pole Sands'. The coastline of

East Head

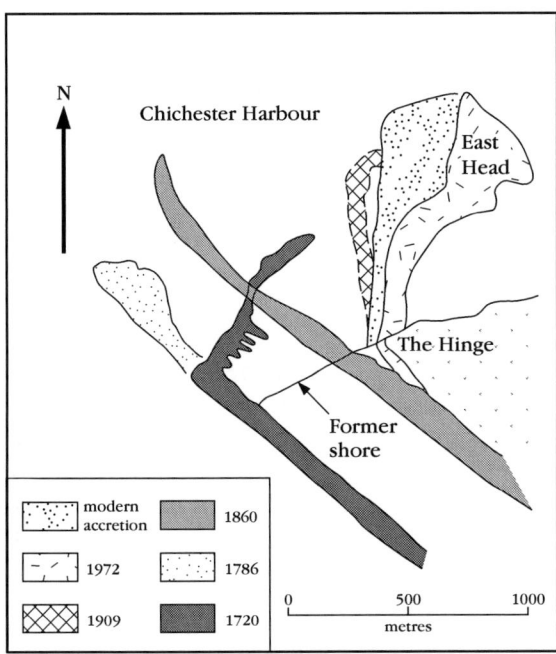

Figure 8.6 Historical changes at East Head. (After May, 1975.)

which East Head in part is low-lying with occasional low cliffs cut into drift deposits, including substantial quantities of gravel and sand. The shoreline is characterized by a flat and wide sand beach resting on a clay platform. Narrow banks of shingle mark the boundary between land and sea. The relatively weak Tertiary and Quaternary sediments offer little resistance to marine erosion, and consequently much of the eroding shoreline has been protected by walls. There is almost no part of the coastline of this site from which groynes are absent. Selsey can claim to have had one of the most rapidly eroding coastlines in the British Isles: over 9 m a^{-1} between 1932 and 1951 (Duvivier, 1961). At all points between East Head and Pagham, the shoreline during the past century was dominated by erosion. This erosion provided a major source of shingle feed to the beaches to east and west, until it was gradually cut off by coast protection works (Duvivier, 1961; May, 1964). In contrast to this pattern of continuous erosion to the east of East Head, the shoreline of Hayling Island on the western side of the mouth of Chichester harbour was much more stable and between 1875 and 1960 its beaches on Hayling Island grew seawards in direct contrast to those on East Head. The rate of change was not as great between 1933 and 1960 (May, 1966). Thus East Head lies in an area where retreat of the shoreline is characteristic, where large amounts of sand and gravel have been supplied to the beaches by cliff erosion, and where protection of the cliffs undergoing erosion has been a priority for coastal management during recent decades.

As a result of its location at the mouth of a navigable entrance to a once small but busy port, East Head has been regularly recorded on various maps, plans and charts produced since the 16th century. These form the basis for the present account of its development, described in more detail by May (1975). The earliest map that identifies East Head as a harbour mouth spit is the so-called 'Armada map' of 1587. It shows a stony ridge following the general alignment of the coastline before bending towards the north into Chichester Harbour. It was large enough to form the base for a small defensive battery. By the time of Avery's 1721 chart, the spit had a complex distal end and several recurves. The probable position of the spit in about 1720 is shown in Figure 8.6. The northward-trending part of the spit was breached sometime between 1720 and 1759 when the recurve is shown as broken. Between 1759 and 1846, the spit retained a similar shape with a wide distal end connected to the mainland by a narrower neck from which the remnants of the main recurve project northwards. Comparison of the surveys of the 18th century with the Tithe Map of 1846 suggests erosion of the whole shoreline except for the distal end of the spit. By 1875, all evidence of the earlier spits had disappeared except for a low ridge that projected from the northern side of the shingle ridge below highwater mark. This probably represents the remnants of the most northerly part of the 1721 recurve. From 1875 onwards, the spit swung towards the position now occupied by vegetated dunes. Vegetation was first mapped on the plans of 1911. The present alignment was first recorded in the OS 1:2500 survey of 1933, and has altered little since then.

Breaches have been frequent; the latest occurred in 1963. The 18th century maps reveal a breach in a north-trending ridge, and Ramsey (1934) reported several breaches during the early part of the 20th century. In 1963 a breach separated the two distinctive parts of the spit.

1. The proximal ridge, which has a similar alignment and dimensions to all the spits mapped since 1887. Although the ridge has moved

Sand spits and tombolos

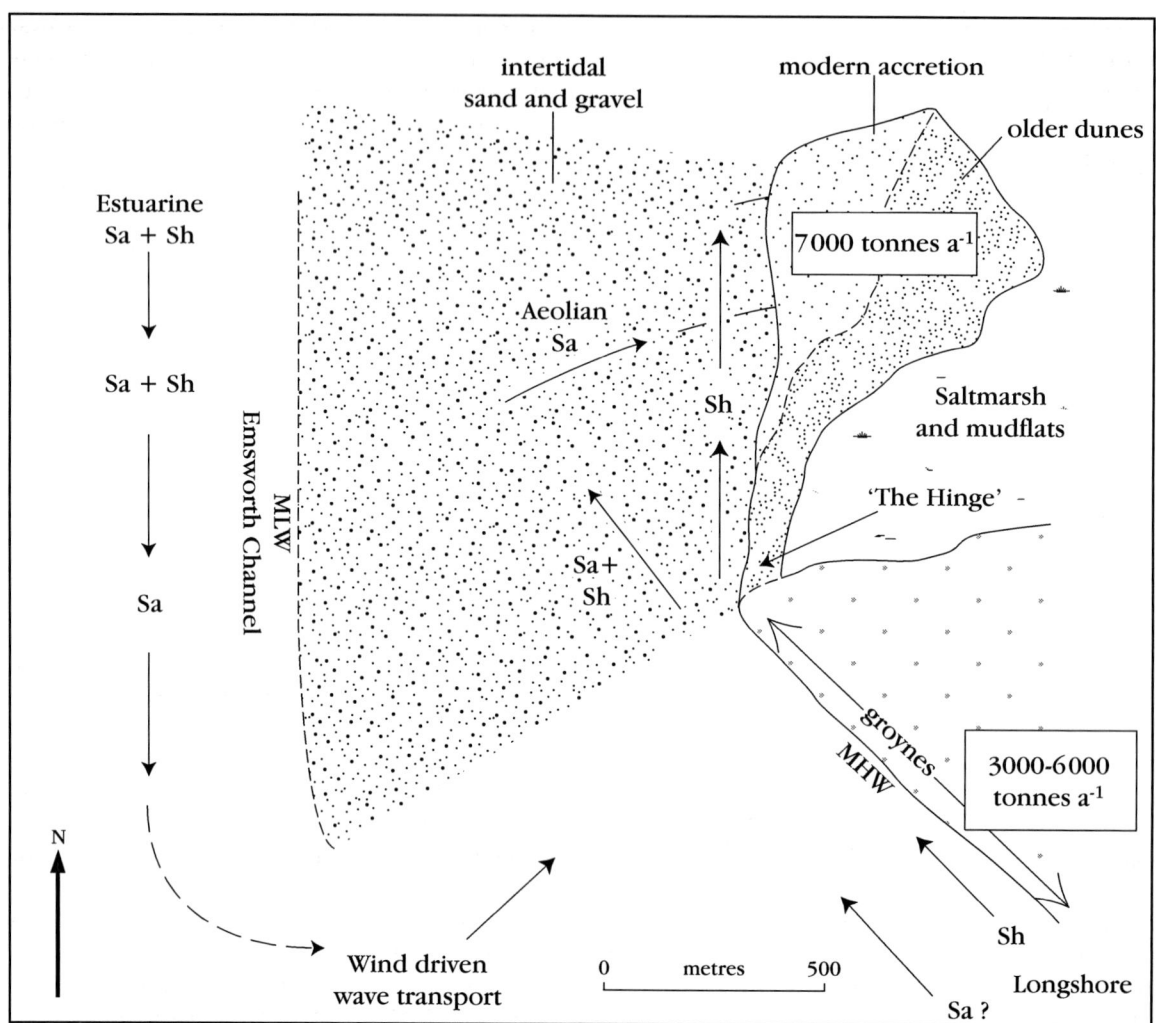

Figure 8.8 Schematic diagram showing the key features including sediment transfers at East Head. (Sa = sand; Sh = shingle). Sand and shingle transported out of Chichester Harbour may be added by wave and aeolian action to longshore transport from the south-east. This may account for the excess of sediment reaching the spit over longshore transport. (After Harlow, 1982.)

landwards, it has retained much the same plan and has rotated only slightly towards the north.
2. The distal end, which is roughly triangular in shape. In contrast to the proximal ridge, the outer spit swung rapidly towards the north-east during the early 20th century.

The differential movements of the two main parts of the spit have produced a 'dog-legged' feature in which the distal end is set back from the shoreline of the landward end. There appear to be two 'hinge points' on the spit, one at its junction with the mainland and known locally as 'the Hinge', the other about halfway along the spit. Interpretation of aerial photographs taken since 1963 shows that there has been a steady growth of sand ridges in front of the older vegetated ridge. The spit has a larger area now than at any previous time since the mid-18th century (Table 8.2). The Hinge has moved landwards with erosion of the cliffs to its east, but it has not moved alongshore.

The form of the spit appears to have been been of two types that appear to alternate. The first (Type A) is a long narrow ridge, the second (Type B) a narrow ridge turning landwards leading to a broadly triangular distal end. Thus Type B appeared first and was replaced by Type A by the mid-18th century. This was subsequently breached. From about 1760 to about 1850, Type B predominated, whereas Type A was the dominant form until 1933 and was breached several times (Ramsay, 1934). Type B then

East Head

Table 8.2 Area of East Head – historical data from 1846 to 1996

Date	Area (ha)	Data source
1846	8.9	Tithe map: property 541
1875	5.3	OS Area 83
1898	6.5	OS Area 310
1911	2.3	OS Areas 310 and 310a
1933	17.5	OS Areas 309a, 310 and 310a
1975	30.7	Searle (1975)
1996	c. 40	May (1997b)

became established once more. Steady erosion reduced the spit to its narrow form, which was breached in 1963. The establishment of groynes to the east may have been an important factor in speeding the onset of breaching. Since 1963, the more stable Type B has been the characteristic form (see Figure 8.6).

Interpretation

East Head owes some of its present-day form to the coast protection activities that have taken place both within and beyond the boundaries of the site. The National Trust has taken steps to ensure that the dune system is stabilized and that the vegetation, in particular, is not seriously disturbed. After the 1963 breach, brushwood and small scrub windbreaks were constructed to aid sand deposition (Searle, 1975). Harlow (1982) outlined the more substantial coast protection works that have been undertaken to the east of the site, at Medmerry and Bracklesham, and their effects upon sedimentary processes. Both activities have been intended to retain sediment within specific parts of the coastline. Whereas the National Trust action has been concerned with the retention of sand and shingle within the site, the action of coast protection authorities to the east has been designed to prevent (or at least retard) movement of sediment away from other sites towards East Head (Hooke et al., 1996). Shingle has tended to move northwards along the spit at East Head, but has not been replaced by shingle arriving from the east as happened in the past. Harlow (1982) estimated that about 7000 m^3 a^{-1} of shingle was added naturally to the front of East Head between 1975 and 1978.

After 1965, there was a marked increase in sand seaward of the old vegetated ridge of the dunes (Figure 8.6) and there was also considerable intertidal accretion that provided a source of windblown sand for the dunes. Harlow's (1982) sediment budget analysis for this coastline confirmed that this sand supply and the changes in intertidal areas of East Pole, West Pole and The Winner are related.

This site is unusual among small estuary-mouth spits in that it continues to grow even though the main longshore sediment source has been curtailed by extensive groyne-fields. It has been viewed as an erosional site by its managers, although the overall volume of sand and shingle has increased. The reason for this is that much of the sediment added to it has so far accumulated on the intertidal banks at the mouth of Chichester Harbour. Recent changes in the position of the spit have tended to assist progradation by making both wave and wind transport from the intertidal area perpendicular to the shore and dunes.

This site is also important because of the juxtaposition of shingle beach, spit, dunes and saltmarsh. A similar assemblage is found at Gibraltar Point, Lincolnshire, but East Head is smaller and in a different tidal and current environment. It has undergone considerable anthropogenic modification, as measures have been sought to manage and preserve the dune system. The shingle beach processes seem to have been independent of the management activities. The site is an excellent example of beach dynamics in circumstances where, despite interference with longshore transport, the planform of the sediment cell is adjusting towards a new dynamic equilibrium with changes in sediment availability and alterations in wave direction. As a result, the shoreline has swung back towards its earlier north-west–south-east alignment. The intertidal area is a mobile sediment store and forms part of a transport system by which sediment crosses the mouth of an estuary (Harlow, 1982). Progradation of the intertidal area provides a source of sediment for the beach and thence the dunes.

At a regional level, the contrast between East Head and the shingle spit at the mouth of Pagham Harbour, West Sussex, is an important one. The Pagham site lacks sand and dunes and has well-developed shingle ridges and fulls. The lack of sand is mainly a result of the limited volume in the intertidal area seaward of the spit. East Head has a shingle base, but its dunes owe their development very largely to the presence of intertidal sources of sand. In addition, East Head is on a windward shore in contrast to the lee position of the Pagham site. Together, the two features are important sites that help to elucidate the way in which spits have developed at the entrance to shallow-water estuaries.

East Head contrasts strongly with many other spits in continuing to grow in volume even when longshore transport to it has been substantially reduced. The problem of breaching of the proximal end of spits, and the potential demise of the feature, besets the management of many other sites (e.g. Spurn Head, Hurst Castle Spit, Dawlish Warren, see GCR site reports). Although breaching can be shown at some spits to be part of a natural cycle of events, in many other cases it is attributable to the recent reduction of longshore transport resulting from cliff or beach protection works. East Head is a rare example of continued progradation and a total increase in volume in an area of rising sea level and an increasing frequency of storm surges. This appears to arise from two related effects: the re-alignment of the spit to the dominant waves, that is, to face more towards the south, and a substantial local supply of intertidal sediment.

Conclusions

East Head is a small, growing, mixed sand and shingle spit upon which low dunes have developed. Unlike many such features, it has increased in size over recent decades, despite the reduction of longshore movements of beach material towards it from the cliffs undergoing erosion in Bracklesham Bay. Maps made during the past five centuries show considerable variation in its growth and decay with an important exchange of sediment between the spit and intertidal areas and the spit itself. Nourishment from the intertidal banks at the mouth of Chichester Harbour has been very important in the continued growth of East Head. It demonstrates well the need to conserve both the shoreline and the intertidal zones that are linked to it.

SPURN HEAD, YORKSHIRE (TA 420 130)

V.J. May

Introduction

The sand and shingle spit of Spurn Point or Head lies on the north side of the mouth of the River Humber (see Figure 8.2 for general location), together with an area of till and alluvium to the north. The northern part of the site is formed by low till cliffs that are being eroded at rates in excess of 2.5 m a^{-1} and which feed sediment to the spit. The spit extends for about 5.5 km south-westwards across the Humber estuary, mainly as a narrow feature about 150 m in width but widening at its distal end to over 350 m. Its maximum altitude reaches about 9 m OD, but for much of its length it rarely exceeds 6 m.

De Boer (1963, 1964, 1967, 1968, 1981) argued that the spit has been characterized by recurring 250-year cycles of partial washing away and re-growth (Figure 8.8a). The Institute of Estuarine and Coastal Studies (IECS, 1992) challenged this, suggesting that the present-day morphology of the spit results from 19th century construction work that followed a number of breaches of the spit in the 1840s (Figure 8.8b). The comprehensive documented history of the site arises both from the recorded losses of land and villages and the regular need to relocate lighthouses marking the entrance to the Humber (e.g. Smeaton, 1791). There is no spit of comparable form and length to Spurn Head in a macrotidal environment in the British Isles or probably in Europe. The tidal range in the lower Humber estuary reaches 6.4 m.

Description

The GCR site includes both the spit itself and a cliffed area of till and alluvium to the north-east of Kilnsea. Along the seaward side, the very low cliffs expose sections of Devensian till and alluvium and occasional patches of peat and tree remains. Kilnsea Warren, which forms the northern end of the spit proper and extends a little over 1 km southwards to the narrow neck, is a flattish area between 5 m and 6 m OD, mostly covered by marram *Ammophila arenaria* and sea buckthorn *Hippophae rhamnoides*. The coast changes its alignment from NNE–SSE off How Hill to north–south at the southern end of Kilnsea Warren. At the southern end of High Bents at the curving neck of the spit, a further 0.8 km farther south, the sea coast faces almost south-east. At this point, the narrow neck, only about 30 m wide between high-water marks yet rising to 9 m OD, has been greatly modified by management activities. The remainder of the peninsula is very nearly straight for over 3 km,

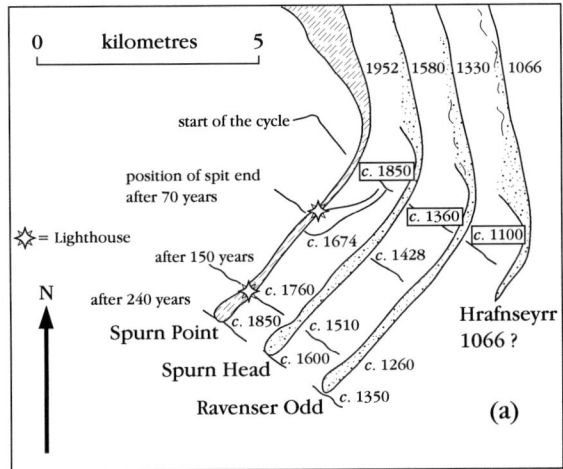

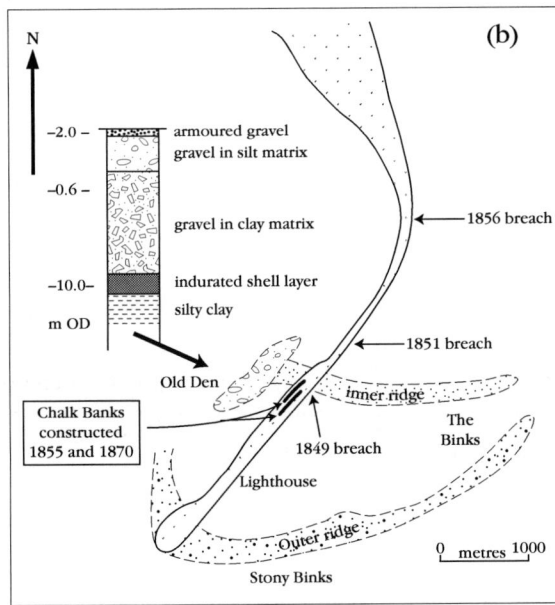

Figure 8.8 (a) The cyclical evolution of Spurn Head as envisaged by de Boer (1964). Over a period of about 240 years, the spit extends, beginning to develop a spatulate distal end after about 150 years. The neck of the spit is breached and a new cycle of spit growth begins. (b) The key features and 19th century development of Spurn Head. The log shows the sediments underlying Old Den. (After IECS, 1992.)

into a ridge and runnel beach, with the ridges curving across the beach. Four hundred metres offshore in the Humber, beyond a shallow muddy channel ('Greedy Gut'), lies 'Old Den', now a shoal of muddy shingle submerged by every tide. In the 17th century it was an island with dunes and vegetation. The area of most vigorous dune growth lies near a shorter jetty, where the prevalent south-westerly wind carries sand from the wide beaches on the Humber side. These dunes increased in height by about 6 m between 1960 and 1974.

Near the tip of the peninsula is an arcuate system of shoals, the Stony Binks, which branch off seawards at a tangent. This complex appeared to de Boer (1964, 1968) to be a shoal system related to the ebb tidal stream running past the point where it has scoured a deep hollow (depth at least 24 m) immediately off the tip. IECS (1992) show that the Binks lie along the line of an arcuate ridge of glacial till (about 15 m below present HWMOST), which extends seawards of Spurn Head north-eastwards along the line of the Binks (Figure 8.8b). A second till ridge about 2.5 km to the north follows a similar curved path running beneath Spurn Head (at about 6 m below HWMOST) near the Chalk Banks and co-incides with the position of the Old Den. During the breach of 1849 this ridge was exposed as a basal sill to the breach. The glacial ridge at the Point is overlain by 15 m of sand, gravel and cobbles. The gravels are exposed along the line of the Binks and the Old Den.

Three cores in the area of Kilnsea Warren to the north of the spit passed through sand and gravel, or through silt and clay with silt bands: all ended in clay at depths of 18 m below the surface (IECS, 1992). Whereas as far south as the Chalk Banks, cores passed from sand and gravel into clay at depths of 10 m to 12 m from the surface, i.e. −4 m to −6 m OD, cores to depths of 18 m in Spurn Warren ended in sand, gravel and boulders. A core taken in 1971 on the foreshore near the lifeboat station near the spatulate tip (TA 398 110) passed from sand and gravel into a firm sandy, silty clay, possibly the Skipsea Till, at between 10.5 m and 12 m from the surface (−12.5 m to −14 m OD) and into Chalk at about 28.5 m from the surface (−30.5 m OD) (IECS, 1992).

and is aligned approximately north-east–south-west. It consists of irregular ridges of sand dunes, usually highest on the North Sea coast, and intervening hollows. The foreshore on the Humber side of the peninsula consists of mudflats with patchy cord-grass *Spartina anglica* as far south as the Chalk Banks (Figure 8.8b). Farther south, the muddy shingle of Old Den contrasts with a wide beach of fine-grained sand that extends almost to the tip, commonly built

During the 17th, 18th and 19th centuries, the low ground between Kilnsea and How Hill at the narrow northern neck of Spurn Head was a part-

Sand spits and tombolos

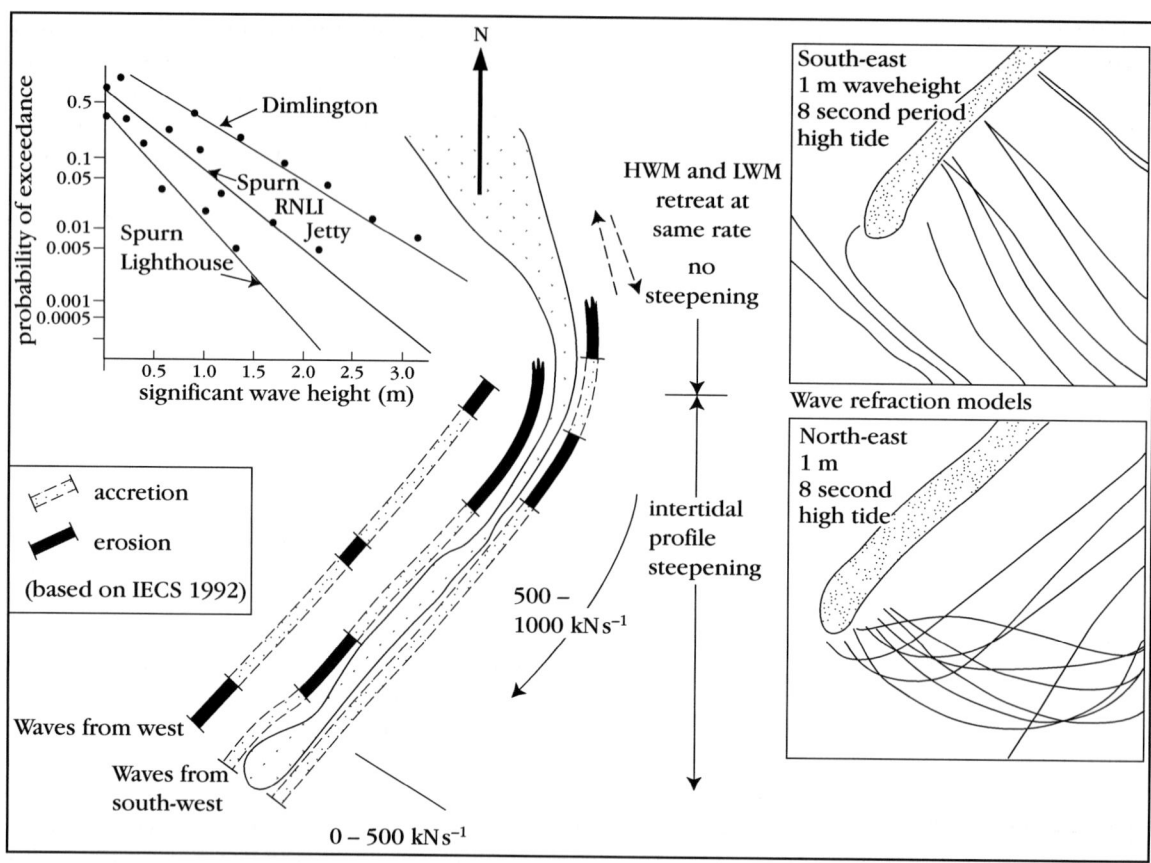

Figure 8.9 Active processes at Spurn Head. Wave-refraction models indicate areas upon which wave energy concentrates. With waves from different directions (shown by the wave orthogonals), the zones of erosion and accretion change, giving rise to possible breaching from both North Sea and Humber sides of the spit. The wave-refraction models show wave convergence and divergence for waves from south-east and north-east, in both cases for waves 1 m-high, of period 8 seconds and at high tide. (After Halcrow, 1988; and IECS, 1992.)

ly vegetated wave-swept bank of sand and shingle. During the 18th century ballast-diggers removed gravel and cobbles mostly from this area of the neck. Up to 50 000 tons (c. 51 000 tonnes) per year were removed, exceeding the natural rate of removal by up to seven times (IECS, 1992). The selective removal of the larger material greatly reduced the natural gravel foundation of the neck of the spit, weakening it substantially.

The Chalk Bank lies across the site of the great breach that opened in December 1849, which attained a width of 460 m and a depth of 5 m at high water before it was closed in 1855. The dunes in this section remain lower than elsewhere. A series of groynes at 250 m intervals were constructed between 1864 and 1926 and revetments were added about 1884: stakes and wattling were placed along the Humber (west) side, and blocks of chalk dumped. Dune growth was stimulated by the sowing of seeds of marram *Ammophila arenaria* and by thatching, and the surface may have been raised even higher and graded when the Spurn–Kilnsea railway was built around 1915. The neck widened by almost 40 m between 1846 and 1878 (IECS, 1992). The irregular river margin consists of the remains of the bank by which the breach was first closed in 1855: the straighter Chalk Bank was built in 1870. An area of saltmarsh between these banks is connected to the Humber by a tidal creek at the north-eastern end. The seaward side was strengthened by concrete revetments from 1942 onwards: these are now collapsing in places.

Spurn Head, from the lighthouse south-westwards, has been much affected by the establishment of the lifeboat station, by lighthouse construction, and by military fortification. The effects of fortifications are most concentrated within the spatulate tip of Spurn Head. A high and vertical sea-wall separates the beach on the Humber side of the peninsula from the interior of the spit where there are the remains of both civilian and military buildings and gun pits.

Interpretation

When the Devensian ice retreated probably about 14 000 years BP, leaving the Skipsea Till (Catt and Penny, 1966; Penny *et al.*, 1969; Madgett, 1975; Madgett and Catt, 1978), which forms the cliffs at the north of the site, sea level was low and much of the area was land. Sea level rise during the Holocene Epoch led to flooding of the North Sea and the Humber and the process regime that resulted in the formation of a spit ancestral to the present one (de Boer, 1981).

As the early Holocene sea level rose, the transgression of the gravels would have been prevented by the presence of pre-exisiting glacial ridges and two islands probably formed along the line of the Binks and Old Den. With sea level close to present-day levels by about 6000 years BP (IECS, 1992), longshore transport from the cliffs undergoing erosion at Holderness carried sand southwards to Spurn Head. From a sandy beach veneering the gravel, aeolian processes carried sand inland to form small dunes. A proto-Spurn Head would have comprised an intertidal gravel ridge from the mainland to an island in the present-day position of the Chalk Banks, with the island of Old Den to the south and west.

The earliest historical record of a peninsula occurs in the 7th century AD and four predecessors of the present-day feature may have succeeded each other at the mouth of the Humber (de Boer, 1964). Hydrographical conditions at the mouth of the Humber in the 16th century were probably similar to those of the present time (de Boer and Carr, 1969; de Boer, 1973; de Boer and Skelton, 1969). A series of breaches reduced Spurn Head in the 1850s to a string of small islands according to the Admiralty Chart of 1852. Thereafter, the peninsula was maintained by artificial defences against erosion and thus differs from its predecessors (de Boer, 1981).

The earlier development of Spurn Head, as described by de Boer (1964), combined extension of the peninsula by growth at the tip and a westwards retreat of the spit as a whole. The breaches on which de Boer based his 250-year cycle occurred in 1360, 1600 and 1849. The first 'breach' occurred when Ravenser Odd (in the vicinity of the present-day Chalk Banks) was abandoned after 60 years of continued erosion. IECS (1992) noted the co-incidence of the erosion of Spurn Head with the flooding of the Broads, attributing both to the 12th and 13th century sea-level rise, and speculated that it is this rather than the over-lengthening hypothesized by de Boer that accounts for the abandoning of Ravenser Odd. According to de Boer, the present-day peninsula developed after its predecessor had been breached about 1608 (de Boer, 1963, 1964, 1967; de Boer and Carr, 1969). The evidence for the 'breach' of 1600 is disputed by IECS (1992), which contradicts the view that the island, Old Den, came into existence when the predecessor of the present-day spit was breached around 1600 (de Boer, 1963, 1964, 1968, 1981; de Boer and Carr, 1969). The Old Den is clearly portrayed on maps of 1540 and so the view that it originated as a result of the breach must be questioned. Subsequently, the spit grew rapidly southwards to reach the position of the present-day lighthouse by about 1750. Since at least 1290, it had ended at the Chalk Banks. IECS considers that this rapid extension may co-incide with fall in sea level from the mid-17th to mid-19th centuries, which may have resulted from the 'Little Ice Age'. The intertidal gravel beaches between the two till ridges would have been exposed, allowing sand accumulation and dune building. As a result the sand supply to the Old Den dunes was reduced. They gradually eroded during the 17th and early 18th centuries and, by 1750, Old Den had been reduced to an intertidal gravel mound.

Remarkably, the 1953 storm surge failed to breach the peninsula, though causing some damage, and a comparison of the peninsula's condition in 1849 and 1953 is instructive. Before 1849, little if anything had been done to keep erosion in check. The neck of the peninsula was bare of dunes and wave-swept at high spring tides. This condition had become more pronounced since the beginning of the 19th century, and it appears that waves broke across the neck of the peninsula and carried with them beach material that would otherwise have been moved south. The spatulate tip of the peninsula was attacked by waves from both the sea and the river side, and reduced to a strip little wider than the rest of the peninsula. The broader tip at present is the result of accretion on the Humber (west) side of beach material carried round the tip, and when this supply is cut off the tip becomes attenuated.

A combination of several factors appears to have caused the 'breach' of 1849, including the selective removal of gravel and cobbles from the neck, and the loss of intertidal and aeolian sand supplied on the western side of the spit. Before 1808, when the North Channel was finally closed, it carried sand and fine-grained gravels during the flood tide into the channel and beyond. Aeolian transport carried sand from the banks not only to the western side of the spit but also along the entire estuarine shoreline of the Bight. Aeolian transfer is of the utmost importance to the western shoreline sediment budget, especially in the vicinity of the Old Den (IECS, 1992). The 1849 breach (de Boer, 1964) occurred sufficiently far south for the effects of wave erosion on the seaward side to be added to wave erosion of the river side, and far enough along the spit to be exposed to wave erosion in the lower Humber produced by north-westerly gales. The residual effects of riverside accretion farther south prevented the whole of the distal end of the spit from being swept away in such conditions: the results of the storm surge were thus limited to the opening of the breach across the neck.

The 1849 breach, once formed, developed into a tidal channel with streams of possibly 3–3.5 m s^{-1} running through it. It rapidly enlarged to a width of about 60 m and a depth of 5 m at high-water spring tides. Small ships began to use it instead of going round the point. The Old Den was enlarged at that time by the deposition of material swept through and scoured from the breach by the flood tide (de Boer, 1964). The development of the spit was very considerably changed by the closing of the breaches in the 1850s by the erection and maintenance of groynes and revetments and by the encouragement of dune-colonizing plants, especially marram *Ammophila arenaria* (de Boer, 1981).

Up to the present time, the distal part of the spit has been held in its mid-19th century position. However, over the past 6000 years the spit has migrated westwards at an average rate of about 0.2 m a^{-1}. Comparison of the shoreline on 19th century maps showing Smeaton's lighthouse and the present-day shoreline shows that Spurn Head has been retreating more recently at about 0.5 m a^{-1}. In contrast, the mainland Holderness coast has been retreating at about 2.0 m a^{-1} (IECS, 1992). As a result, Spurn Head now has three parts, a proximal section responding to the high retreat rates of the Holderness coast and a distal spatulate area largely fixed by coast protection activities, separated by a central section which is increasingly exposed to destructive waves coming from northerly and even west of northerly directions. These waves, generated by the strong to gale force winds that accompany the barometric pressure conditions associated with storm surges, are especially destructive. They occurred during both the 1953 storm surge, and the 1849 breach.

The relationship of Spurn Head to the Holderness coast is therefore different now from that in 1849. Erosion has continued unchecked north of Kilnsea: the destruction by the sea of the defences at Kilnsea has also caused rapid retreat of the coast. Relative to the Holderness coast, therefore, Spurn Head as a whole stands in a position more to seaward than it did in 1849, and this may be the primary cause of the present-day rapid erosion, from the northern boundary towards the Narrow Neck.

The wave and tidal processes operating along the shores at Spurn Head have been discussed by Phillips (1962, 1963, 1964), Robinson (1964, 1968) and Hardisty (1982). Along the seaward side, particularly after periods of constructive wave activity, the beach is built into a high, relatively steeply sloping upper beach zone of sand and shingle, and a lower flatter zone of finer-grained sand, sometimes with an intervening runnel. The lower edge of the upper beach is often a sharply demarcated line of seepage, locally called the 'grope'. The height of the beach varies along its length and lower sections, known locally as 'ords' or 'hords' (Phillips, 1964; Scott, 1976; Pringle, 1981), migrate slowly towards the tip of the spit. Erosion is often particularly severe where they occur.

The waves around the spit are responsible for

1. washover processes that transport sand from the eastern shore across the spit to its western shore,
2. longshore transport in the nearshore zone by wave-driven currents southwards to the southern extremity of the spit, and
3. transport by waves in combination with tidal currents onto the western shore (IECS, 1992).

As a result, most sand accumulates along the

southern part of the south-western shore.

The predicted 1-year wave height return periods of 4.75 m at Dimlington (15 km to the north), 2.25 m at Spurn Lighthouse and 3.25 at the RNLI jetty show that the Binks have a crucial role in reducing wave energy inputs to Spurn Point (IECS, 1992; Figure 8.9). Modelling wave refraction patterns IECS show that between Kilnsea and the Warden's cottage, with the greatest exposure to the north, wave energy and, as a result, sediment transport increases. Erosion of both the tills and beach sediments follows. As the shoreline swings towards the south-west, however, wave-energy potential decreases and sediment transport is reduced. Deposition increases towards the lighthouse and decreases towards the Point where it is zero (IECS, 1992). This is contrary to the view of de Boer (1964) who described deposition as occurring at the Point. IECS further suggest that there is sufficient sediment reaching the spit to maintain a positive sediment budget. The transport processes do not, however, carry the sediment to the areas with the greatest sediment deficits and that are therefore most likely to undergo erosion.

Conclusions

Spurn Head is an outstanding example of a dynamic spit system and is very unusual, if not unique, in that it extends well across the mouth of a macrotidal estuary. Unusually also, there is an exceptionally long historical record extending back to the 7th century AD. Though there are many spits of sand or shingle along the British coast, there are some features of Spurn Head that are exceptional and give it international importance. These features are as follows.

1. It derives its character as an outstanding example of a dynamic spit system from the coastline undergoing rapid erosion of Holderness whence it grows, where the mean annual rate of retreat over the century 1852–1952 rises in places to 2.75 m, an extreme figure that is among the highest over a comparable period of time anywhere in the world (Valentin, 1954, 1961). Its length and volume reflect the massive longshore transport from the Holderness coast.
2. It is exceptional, nationally and internationally, in that it extends across the mouth of a macrotidal estuary. Few spits are able to maintain comparable size and length in a setting with such a large tidal range.
3. It has an unusually long recorded history of more than 1000 years, exceptional for Great Britain and probably internationally. The comparable length of record for Dungeness offers considerable scope for unrivalled comparative studies.
4. The cyclic pattern of development proposed by de Boer is disputed by a view that the spit has a much more stable position due to Holocene gravel ridges and wave energy distribution.
5. The breaches in the spit are explained by different circumstances: in 1360 by sea-level rise, in 1650 by sea-level fall and in 1849 by gravel extraction and changes in longshore sediment transport.

DAWLISH WARREN, DEVON (SX 985 795)

V.J. May

Introduction

The coastal spit at Dawlish Warren is a classic landform that extends from the western side of the Exe estuary (see Figure 8.2 for general location) and diverts the main channel towards Exmouth. This complex sand spit at the mouth of the estuary is dominated by two parallel ridges, the more seaward of which has a broad distal end. Extensive sandbanks to seaward affect the low-tide and intertidal wave-energy distribution, but the beach form is largely the product of a combination of wave patterns at high tide levels and the discharge of the estuary, so that currents may control the sediment distribution more than waves (Figure 8.10). The site is now partly modified by gabions buried beneath the shoreline dunes and by a wall at its proximal end. Erosion has become acute here in recent years following protection of the cliffs to the south-west that had formerly provided at least part of the former sediment supply.

Description

The spit (Figure 8.10) extends for about 2 km north-eastwards across the Exe estuary from cliffs originally cut into Permian breccia and conglomerate, but now entirely formed by a artificial shoreline of boulders, timber structures

Sand spits and tombolos

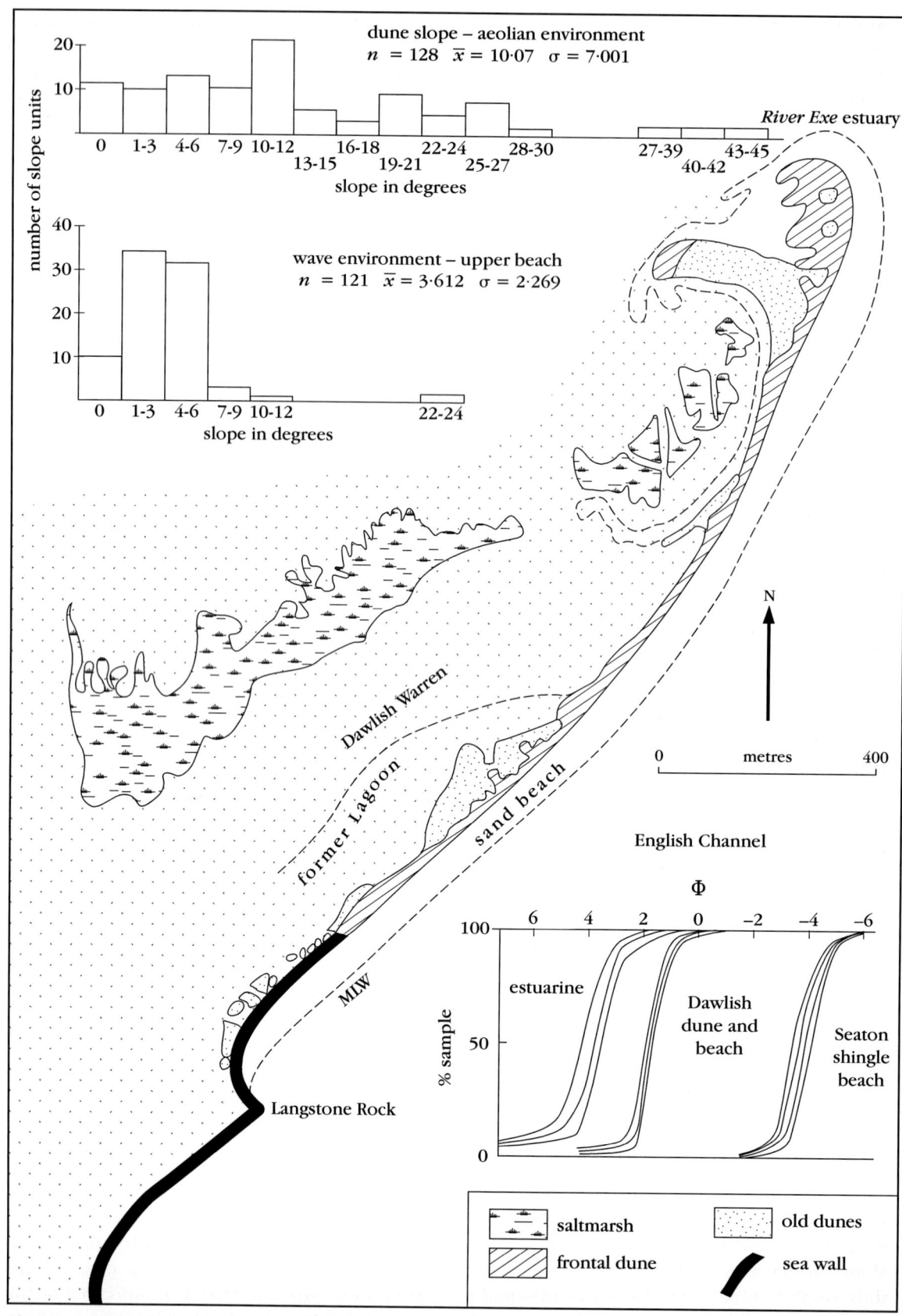

Figure 8.10 Key geomorphological features of Dawlish Warren, showing differences in slope on dunes and the upper beach, and differences in sediment sizes. n = number of observations of slope angle; $\bar{x}$ = mean slope angle; σ = standard deviation. Φ = $-\log_2$ (grain diameter in mm); the grain-size profile for estuarine material and for Seaton, Devon, are shown for comparison.

and a concrete sea-wall. This area is mostly excluded from the GCR site. The spit is about 500 m wide throughout its length, but is made up of several distinct units. The landward side of the spit supports an area of saltmarsh that has developed in its shelter. To seaward, the Inner Warren, is a low hummocky area of former sand hillocks resting upon clay, probably of estuarine origin, and 0.6 m to 0.8 m-thick shingle layers (Kidson, 1964b). The Outer Warren comprises a line of semi-fixed dunes of varying width behind a discontinuous line of sand hills between 25 m and 50 m in width and rising to a maximum of about 6 m in height. The distal part of this ridge widens into a triangular area that preserves several former shoreline ridges (Figure 8.11). There is a wide intertidal beach, which is connected at low tide to a large sandbank, the Pole Sand. Within the estuary, another large sandbank, Bull Hill Sands, is separated from the distal end of the spit by the channel of the Exe. Both sandbanks include substantial quantities of gravel at depths of −1.3 m to −1.6 m OD. The sand of the spit itself is as much as 20 m in thickness and rests upon Devensian gravels (Durrance, 1969), which in turn rest upon and fill deep channels cut into Triassic breccia. The bedrock slopes from about 0 m OD at the inner end of the spit to about −20 m OD beneath the distal end. A series of NW–SE-trending palaeochannels that have been cut by fluvial processes to below −40 m OD meander across it. Kidson (1964b) reported that the bedrock of Checkstone Reef, which underlies part of the Pole Sand, is occasionally exposed.

The spit appears to have existed in its present-day position on the western side of the Exe estuary since at least the 16th century. Martin (1893) was unable to confirm local reports that at one time it extended across the estuary from the eastern shore at Exmouth. Throughout its documented history (i.e. since 1869) Dawlish Warren has been undergoing erosion. The Outer Warren has been breached frequently and the shape of Warren Point changed considerably. When high spring tides are accompanied by south-easterly gales driving high onshore waves, the lowest part of the ridge is overtopped and breached. The breach is subsequently rebuilt as waves from the south and south-west move sand along the spit and extend it. At the same time the face of the beach and dunes is eroded and so the spit retreats. Sand is also blown from the dune ridge towards the distal end. Sand eroded from Warren Point may be transported into the estuary or may travel towards the Pole Sands. Kidson (1964b) pointed out that erosional phases at Warren Point were not associated with any increase in the volume of sediment in the Pole Sands. The interplay of different wave directions has produced a highly dynamic form.

Interpretation

The nature and behaviour of the sand spit at the mouth of the Exe has been a focus of geomorphological attention since the 1860s (Peacock, 1869; Martin, 1872, 1876, 1893; Kidson, 1950, 1964a,b; Mottershead, 1986). Steers (1946a) commented that an unusual feature was the formation of two spits – an inner and an outer – which was then unexplained. Pethick (1984) described it as a detached beach and Bird (1984) identified it as an example of a spit that has been artificially armoured. Most attention has been given, however, to the erosion and expected demise of the spit. Kidson (1964b, p. 178) described it as the 'outstanding example of a depositional feature which has passed through this period of stability and is now well advanced in the final stage leading to ultimate extinction'. Peacock (1869) had already forecast its ultimate extinction.

Some authors in discussing the dominance of erosion (Martin, 1893; Clayden, 1906; Steers, 1946a) have suggested that the construction of the railway towards Teignmouth in 1849 reduced cliff erosion and cut off the supply of sediment to the spit. Kidson (1950) has shown, however, that the rate of retreat of the face of the spit over the 100 years before and after the construction of the railway was comparable, and concluded that its impact was negligible. Comparison of the shorelines mapped by Kidson (1950) suggests that the spit has migrated landwards as a unit rather than suffering greater erosion at its proximal end, as might be expected if the reduction in longshore transport were the key factor in its retreat. Interestingly, however, Kidson suggested that much of the sand forming the spit was derived from erosion of the cliffs to the west. Rapid erosion of the cliffs at the end of the Holocene rise in sea level would probably have supplied most of the sand for the spit. Even if erosion was already active on the spit by the mid-19th century, the reduction of sand supply from the west cannot be ruled out as one factor in the continuing decline

Sand spits and tombolos

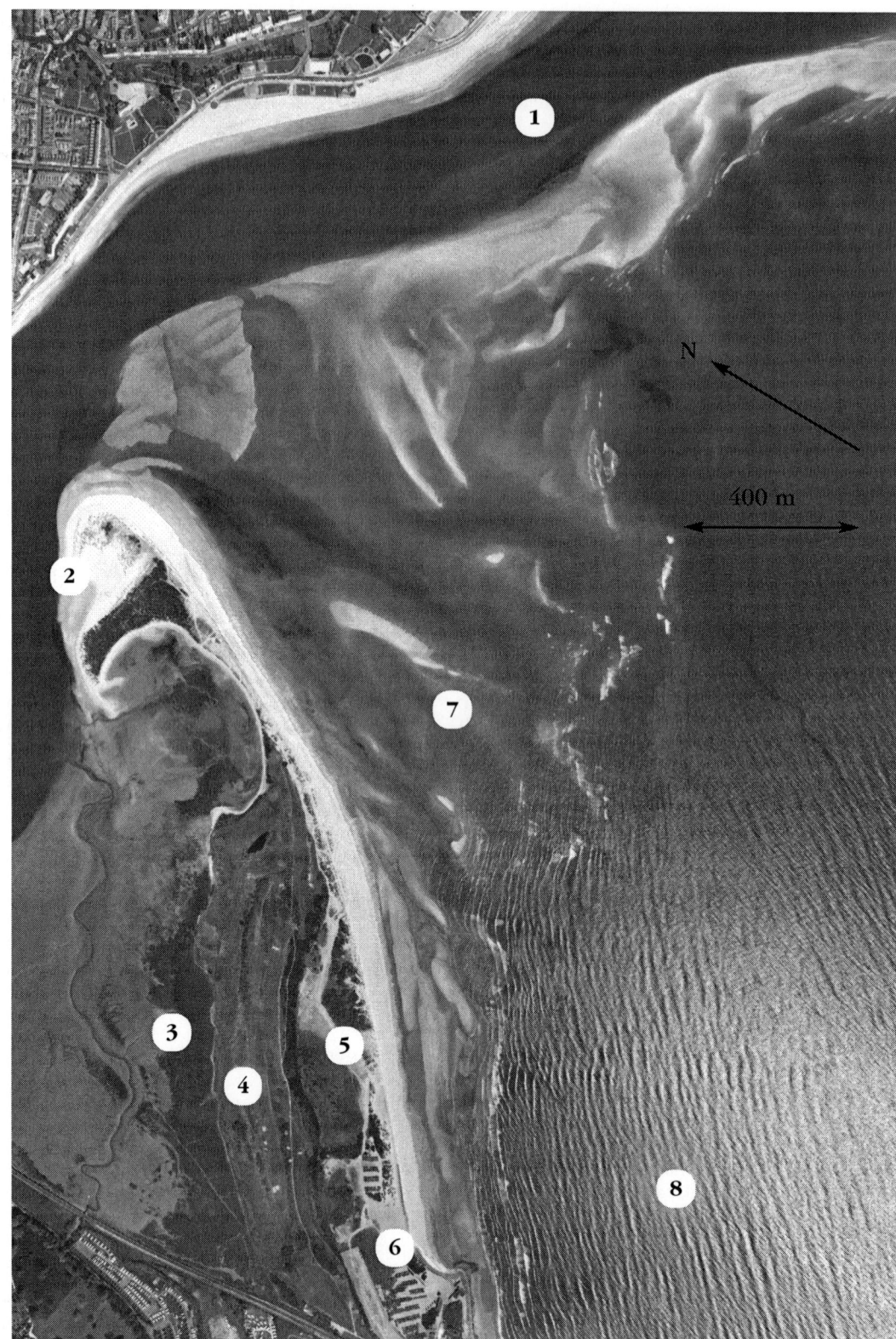

Figure 8.11 Aerial photograph of Dawlish Warren with the main geomorphological features numbered. 1 = Exe estuary, main channel; 2 = active recurved distal end (Warren Point); 3 = saltmarsh; 4 = inner spit (largely modified); 5 = outer spit; 6 = proximal end coastal protection works; 7 = intertidal sandbanks; 8 = prevailing and dominant wave direction (from the south-east). (Photo: courtesy Cambridge University Collection of Aerial Photographs, Crown Copyright, Great Scotland Yard.)

of the spit.

The Warren is sandy, unlike other beaches in the region, which are shingle (Mottershead, 1986). Mottershead noted that the River Exe has a drainage basin an order of magnitude larger than any other rivers flowing to this coastline. As a result it would have been able to deliver a much larger volume of sediment. As sea level rose during the Holocene Epoch these sediments could have been driven landwards to form the spit. This was, therefore, a 'once-only' mechanism according to Mottershead, who saw this as a realistic hypothesis in the absence of any published analysis of the mineralogy of the sands forming the Warren. This would suggest that sediment is no longer being supplied to the spit.

A second issue, which was noted by Steers (1946a), concerns the double (parallel) nature of the spit. Mottershead (1986) suggested that Kidson (1963) believed that the Inner Warren was probably an accumulation of windblown sand derived from the Outer Warren. Kidson, however, saw the area around Warren Point as the receiving area for sand blown from the Outer Warren. He suggested that the Inner Warren was a normal spit that built across the estuary with both wave-borne and windblown sand built into dunes on a shingle base. Neither author examined the reasons for the development of two separate spits.

The cartographic and documentary evidence points to the Warren as being developed from two spits. The Inner Warren could have developed as a spit across the estuary, but set well back from its mouth. During a period of reduced sediment supply, it might have migrated into the estuary. The intercalation of clay and sand layers suggest that this spit, like most others, migrated on to the marshland behind it, but that more rapid marshland sedimentation transgressed the landward sandy beaches before the present-day pattern developed. With the older spit pushed back into the estuary and sand supply to it cut off because of the presence of Langstone Rock, it would have been possible for a new spit or beach ridge to develop. The effect upon wave energy distributions of the underlying gravel and bedrock could produce sedimentation and longshore transport, which would initiate a new spit in much the same original position as the earlier one. The new one would then migrate up the estuary, overlapping the older feature.

The spit has now been armoured to reduce the chances of breaching and to ensure its survival as part of the sea defences in the lower Exe. As Kidson pointed out this may well be a futile exercise, as the long-term pattern here has been an erosional one. There is a need to understand much better the sedimentary pathways around the spit especially in the intertidal area.

Conclusions

Dawlish Warren is an important site for four reasons. First, it is a good example of a spit in the later stages of development. In this, it complements such features as Hurst Castle Spit, Hampshire, and Orfordness, Suffolk. If sea-level rise becomes more rapid, it can be expected that many more coastal features will show the changes that have occurred at Dawlish Warren. The efficacy of coastal engineering works in similar situations needs to be carefully evaluated and so this site offers a natural test-bed for measures to adjust to the rise of sea level over the next century. Second, it is unusual in having a double parallel form, for there are few such features in Britain. The nearest analogous site is at South Haven Peninsula, Dorset, and Gibraltar Point, Lincolnshire, has some similarities. Neither, however, has developed across a major estuary. Third, its predominantly sandy sediments make it unusual on the south coast of England where most beaches are of shingle and sandy spits are rare. Finally, the intertidal banks both within the estuary and to seaward form an integral part of this beach system, with the result that, unusually for the coastline of southern England, fluvial as well as marine sediments are a sediment source.

GIBRALTAR POINT, LINCOLNSHIRE (TF 568 562)

C.A.M King and V.J. May

Introduction

The Gibraltar Point area covers a wide range of types of accretion on a coast of low topography. It is one of the few stretches of relatively natural coastline on the east coast of England between the Humber and the Wash (see Figure 8.2 for general location). It has been studied in detail over several decades, the initial surveys being carried out over 50 years ago (Barnes and King, shallow nesses of accumulation, with a generally convex plan curvature to the sea, to form. The nesses gradually migrate south causing the variation of accretion at different profiles over time, as recorded in the repeated surveys. This longshore drift is responsible for the formation of the spit at Gibraltar Point.

The development of the spit has been studied by annual surveys, which provide evidence of its change in length, height, volume, and vegetation cover (Barnes and King, 1957; King, 1970). The new saltmarsh that developed in its shelter has been studied in detail (Harper, 1976, 1979; Hartnall, 1982), using stakes to measure accretion rates, as well as measuring currents, sediment concentration and details of vegetation (to assess their roles in the growth of the marsh and its creeks). The spit prolongs the upper foreshore where the coast turns abruptly into the Wash. It is a small spit of sand with a little shingle, derived from the glacial deposits and is the latest of a series of spits, each of which has been built farther to seaward than its predecessor. Armstrong's map of 1779 shows a spit prolonging the mature western dune line, which then formed the frontal system. During the 19th century this spit was preserved as the newer dunes developed east of the old system, a new spit prolonging these eastern dunes at the end of the 19th century. The end of this second spit was destroyed by the 1922 storm surge, which created the storm beach and truncated the end of the eastern dunes. The present-day spit has developed since this date.

Until about 1965, the spit continued to grow in length, height and volume, and it moved landwards by about 70 m. Since the spit became stabilized, vegetation has become established on its crest. During the 1980s this vegetation became denser, thus helping to preserve the feature despite a reduction in volume caused by the growth of a ness to the north. It will probably be starved of material as another spit develops seawards of it. This will continue the type of development recorded for former spits, and helps explain the small size of this spit compared with others on the east coast, such as Spurn Head and Orfordness. The spit at Gibraltar Point has formed by stabilization of a ridge on the upper foreshore, in a similar position to the sand dunes, which form one of the distinctive elements of this coastal system.

Sand dunes can be seen in a wide range of stages of development. The foredunes form from the uppermost beach ridges as they become stabilized by the growth of ridges to seaward. They become arcuate in form as their southern end is driven landwards once the northern end is stabilized. Once they are stable, windblown sand adds to their height, and this is trapped by vegetation. Grasses are the earliest colonizers: Lyme-grass *Elymus juncea* is followed by *Leymus arenarius* and marram *Ammophila arenaria* when the height is further increased and the frequency of tidal inundation reduced. As the dunes become more mature, sea buckthorn *Hippophae rhamnoides* becomes dominant, covering the stabilized eastern dunes. *Hippophae rhamnoides* also covers the dune slacks when they are blocked from the sea by the formation of the storm beach. The western dunes, which are about 100 years older, show a much more mature and mixed vegetation, including elder *Sambucus nigra*.

Saltmarshes of various types and in various stages of development occur at Gibraltar Point. Marsh slacks are elongated strips of marsh between the foredunes. They represent the runnels that have accreted owing to the deposition of silt in the shelter of the stabilized ridge to seaward. Vegetation again plays an important part in the accumulation of silt in the runnels. All stages of development are visible at Gibraltar Point, including the changes from initiation of a slack to the stage when only the highest tides can inundate it. Wider areas of marsh occur in the area. One lies between the main western and eastern lines of dunes. This is now mature, the inner part being a freshwater marsh, whereas the outer part between the old and the new spits is still covered by the high spring tides that flow up and spread out from the creek systems. The marsh exhibits a typical east coast mature marsh vegetation at the salting stage (see Figure 10.4). Sea purslane *Atriplex portulacoides* dominates the lower interfluves while the slightly higher creek levees are covered by sea couch *Elymus atherica*, sea lavender *Limonium vulgare*, sea aster *Aster tripolium*; other mature salting plants are present locally. The most northerly British presence of shrubby seablite *Suaeda vera* occurs in the marsh. This shingle plant indicates former shingle and sand ridges in the marsh, outlining their curved form. The marsh also provides examples of various types of pans. The drainage of the outer edge of the marsh was reversed by the building of the storm beach across its seaward end during the storm surge of 1922, which blocked one of the main creeks.

The New Marsh has developed since the 1922 storm surge when the new spit started to develop and provide the shelter necessary for deposition of fine-grained sediment. An aerial photograph taken in 1946 (the earliest evidence of new marsh development) shows mudflats devoid of vegetation apart from a narrow strip near the western end of the storm beach. Up to about 1960 the marsh was dominated by cordgrass *Spartina* spp. and creek development was also taking place (Barnes and King, 1961). By 1970, vegetation covered most of the marsh, including large stands of glasswort *Salicornia* spp., annual sea-blite *Suaeda maritima* and common saltmarsh-grass *Puccinellia maritima*, and the creek pattern was well developed. The marsh sediments include medium silt to sand, the latter being washed through the low proximal end of the spit in stormy high tide conditions, or blown from the spit by easterly winds. Since 1922, 1.4 m to 1.6 m of marsh sediments have accumulated. The mean annual rate of accretion, as measured in detail in the 1970s, is in excess of 40 mm a^{-1} in the centre of the marsh, falling to less than 20 mm a^{-1} in the upper part of the marsh. The mean rate over the marsh as a whole was 17.8 mm a^{-1}. The winter rate of accretion was found to be three times the summer rate, despite the die-back of some annual species of vegetation. The increase can probably be explained by the greater silt content of the incoming tidal waters during the winter owing to increased storminess. A strong correlation was found between data for wind speed, wave height and monthly suspended sediment. Some parts of the marsh are changing more than others, the central area showing the greatest changes and the strongest seasonality of deposition. Changes are most erratic in the lowest part of the marsh; there the River Steeping is gradually changing course, causing rapid erosion in places and equally rapid deposition elsewhere as its meanders shift. The New Marsh thus provides a dynamic environment in which saltmarsh processes and vegetation development are well displayed.

Conclusions

Gibraltar Point is one of a small number of sites around the British coast that have been the focus of more than 50 years of continuous geomorphological research. It includes intertidal sandbanks offshore, well-developed ridge-and-runnel forms on the foreshore, a spit, sand dunes and saltmarshes. The interaction between tidal and other coastal processes has been a key focus of research. One of the main reasons for its importance from the geomorphological point of view is the dynamism of the environment. Changes can be measured over short time spans. For example, seasonal variation of accretion on the New Marsh, annual movements of the beach ridges, and changes of the spit can be recorded relatively simply. The vegetation of the foredunes, marsh slacks and New Marsh also repays close study, since variations and development are rapid.

Gibraltar Point consists of many different subsystems, which include the nearshore tidal ridges, the ridge and runnel foreshore, the backshore with its arcuate foreshore dune ridges and dune slacks. The spit prolongs the upper beach ridge and shelters the New Marsh. The mature marsh (Old Marsh) is separated from New Marsh by the storm beach, a feature that illustrates the importance of the occasional extreme event in the area's development, when more change may occur in a few hours than years of normal sedimentation.

WALNEY ISLAND, LANCASHIRE (SD 194 646–SD 236 624 AND SD 167 715–SD 175 727)

V.J. May

Introduction

Walney Island (see Figure 8.2 for general location) is one of the largest islands around the coastline of England and Wales, exceeded in size only by the Isle of Sheppey, the Isle of Wight and Anglesey. The GCR site at Walney Island has two parts, which represent the main features of the island, in particular the distal features of an island erosion-deposition system. Walney Island itself is the product of erosion and reworking of glacial sediments rather than of coastal deposition (Steers, 1981), but the spits at North End Haws and South End Haws result from transport and deposition of eroded sediments. The spits are important in several respects:

1. They represent the distal features of the island and occur in a macrotidal environment.

2. They differ both in form and sediments. North End Haws is fed by sandy sediments in the intertidal zone and has small dunes on its surface, whereas South End Haws comprises mainly shingle with only limited dune development.
3. They are associated with 'scars' (boulder- and cobble-dominated areas of the intertidal zone) that are a characteristic form on this coast.

Research has concentrated mainly upon the changes in the distal features (Kendall, 1907; Whalley, 1977; Phillips, 1969; Phillips and Rollinson, 1971; Steers, 1981).

Description

The southern part of this GCR site extends from Hillock Whins (SD 194 646) in the north to Haws Hole (SD 236 624) in the south. The beach is set back about 150 m from the low cliffs to the north of Hillock Whins and runs in an almost straight line for 3.5 km (to SD 209 620), swings through 40° and runs for another 1000 m (Figure 8.13), beyond which it changes direction and forms a broadening beach for about 1.3 km. At South East Point, it turns abruptly northwards. Between Hillock Whins and South End, a beach of shingle and boulders rests against remnants of a series of low hills composed of till. The beach is about 50 m wide, fronts low bouldery till cliffs and separates a wide intertidal area up to 750 m wide from low-lying pasture, marshland and dunes. Much of the area behind the present-day distal area has been commercially exploited for its gravel and is now characterized by long shallow lakes. Steers (1981) comments that between 1895 and 1905 about 1.2 million tonnes of gravel and cobbles were removed from the Haws Point area.

Changes in the position of the Haws Point Spit were investigated by Whalley (1977, reported in Steers, 1981) and show that between 1907 and 1976 the spit grew by about 565 m. There were annual growth rates of 3.7 m a^{-1} between 1907 and 1919 and also between 1964 and 1976. These rates were far exceeded by almost three times between 1919 and 1946 (9.5 m a^{-1}) and 1946 and 1964 (10.0 m a^{-1}). Kendall (1907) estimated that between 1737 and 1889 the rate of growth had been about 4.4 m a^{-1}. The drift of material is mainly southwards from the till cliffs that are undergoing erosion, which form most of

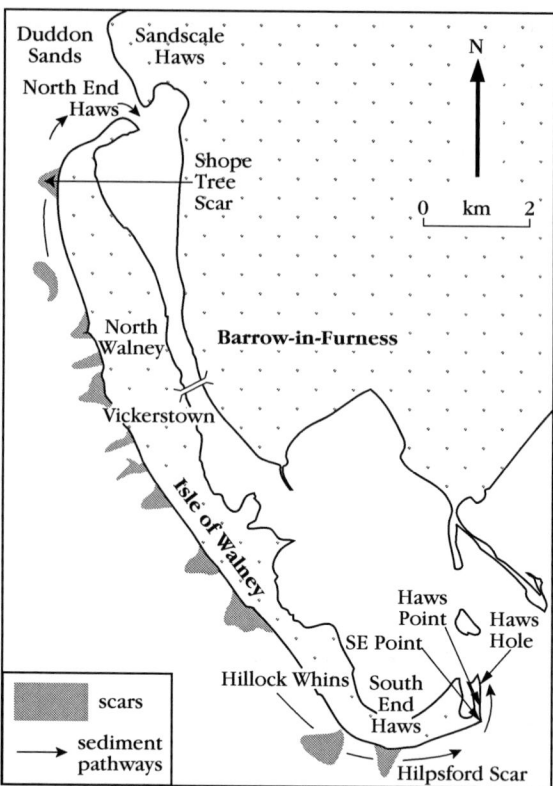

Figure 8.13 Location of scars along the coast of Walney Island, Lancashire.

the western side of Walney Island. With a tidal range of 3 m at neaps and 9 m at springs, sufficiently rapid currents develop in the channel beyond the distal end of the spit to move finer grades of sediment in suspension.

The northern part of the site (from SD 167 715 to SD 175 727), in contrast, is much sandier with a low fringing dune ridge resting on a shingle base. The dunes form a broad distal feature about 250 m in width fringed by a narrow shingle beach. The intertidal area is very wide, forming the southern part of Duddon Sands. The northern end of the site is separated from a further area of dunes, Sandscale Haws, by a 400 m-wide tidal channel.

Interpretation

Steers (1946a) described Walney Island as having several features characteristic of an offshore bar. However, the western shoreline of the island appears to have formed as a series of spits and tombolos linking several islands formed of till and related glacial sediments, for example, between Hillock Whins and South End. Erosion in the recent past has been rapid, up to

0.3 m a^{-1}, and variable. Steers suggested that the pattern of tidal streams gave rise to a predominantly southward drift of beach material. He was uncertain about a possible counter-movement to feed the northern spit. This is not unusual on comparable features with similar wave refraction in the mouths of estuaries (e.g. the North Norfolk Coast, East Head, South Haven Peninsula, and Dawlish Warren).

The presence of extensive deposits of gravel and boulders, sometimes known locally as 'scars', within the intertidal zone may reflect earlier positions of the retreating shoreline, probably related to high points on the eroded glacial sediments. The 'scars' also influence the distribution of wave energy by causing local refraction and offering more resistance to intertidal erosion. Each of the sharp changes of direction in the southern beach is associated with an extensive boulder covered area. The right-angle turn at South East Point may be partly a result of the effects of the deep water channel, but also of the change in wave direction to which this part of the beach is exposed.

Experiments using seabed drifters (Phillips, 1969) suggest that accretion on the southern spit results from transport from the seabed. This is important because it means that management of this feature and changes in its form are likely to be affected by offshore conditions, including the effects of gravel or sand extraction. According to Phillips, tidal streams assisted by the stronger waves that accompany the prevailing and dominant westerly winds bring about transport from the seabed. Drifters released on the ebb moved to the outer part of Morecambe Bay, whence they could reach Walney Island. On the flood, in contrast, they moved farther into the bay. Within the bay, movement was mainly to the north and north-east, but at the mouth of the bay, movement is in an anti-clockwise direction towards the north-west corner of the bay. In both cases, sediment would be transported into intertidal areas, whence it could be supplied to the spit at the southern end of Walney Island.

Walney Island has been the focus of several studies, mainly of coastal changes, but as Steers (1981) pointed out, they say little about the changes to North End Hawes or about its relationship with Sandscale Haws across the estuary. Future investigations should consider the evolution of this point and its relation to Sandscale Haws. There remains a need to consider the whole system, although in this volume much of the western cliffed coastline has been excluded from the GCR site because of the coastal defences and its urbanized nature. The spits at either end of the island exemplify well the unique nature of their sources of sediment and their development at opposite ends of this substantial barrier island.

The origins of Walney Island are the subject of some debate. Tooley (1978a) regarded the island as having been separated from the mainland during Lytham II (8390 to 7800 years BP) or possibly even earlier during Lytham I (9270 to 8575 years BP). Steers (1981, p. 132) was unconvinced that a single long island was separated from the mainland as suggested by Tooley, arguing that the island is a 'series of hillocks joined by beach drifting'. Walney Island is not a barrier island in the traditional sense of an entirely depositional feature, because much of its length is not composed of recent beach sediments, but rather of older glacial materials that are being reworked by marine erosion. This is consistent, however, with barrier development in higher latitudes where sediment is largely derived from the erosion of adjacent cliffs cut in glacial materials (Bird, 1984). Similar features were described in New England by Johnson (1925) and Sakhalin by Vladimirov (1961). Walney Island thus provides an unusual contrast with the other barrier sites along the coastline of England and Wales.

Conclusions

The Walney Island GCR site has two parts, both containing the distal features of a barrier island. Walney Island, however, differs from the usual characteristics of barrier beaches in being mainly the result of erosion and the reworking of glacial sediments rather than the result of coastal deposition. The spits at the northern and southern extremities of Walney Island form the distal features of an unusual barrier island. They are of considerable interest because of their sediment sources and the changes that have taken place within the spits themselves. Walney Island is a unique feature of the English coast, in that small eroded hillocks have been joined by a series of sand, gravel and cobble beaches to produce a single island. It warrants more research, for better understanding of its development will facilitate a better interpretation of the recent evolution of the coastline of north-west England.

Sand spits and tombolos

WINTERTON NESS, NORFOLK (TG 489 216–TG 506 181)

V.J. May

Introduction

The term 'ness' (an Old Norse word) is commonly used, particularly in south-east England, to describe either a headland, for example White Ness (Thanet), or a low-lying foreland or promontory, for example Dungeness. Derived local terms include 'nothe' (Dorset) and 'naze' (Essex). Technically it has been applied most usually where a narrow cuspate foreland occurs with a high obtuse angle between its two shorelines. Such features occur on the East Anglian coast at Winterton Ness, Benacre Ness and Orfordness. Winterton Ness is unusual because of its modern dynamism, its predominantly sandy beach and its migration, often in an opposite direction to that of the longshore drift.

Description

Winterton Ness (see Figure 8.2 for general location) is significant both for the well-formed dunes, which are its most characteristic landform, and for the processes that affect its continuing development. At Winterton Ness, there appears to be a slight sediment budget surplus and some growth in the volume of sediment retained in the ness. There is both erosion and deposition within the site and an important aspect of the interest of the site is its dynamism, a feature that has been the focus of much of the research here (Cambers, 1975; Craig-Smith, 1971a,b, 1973; Green *et al.*, 1953; McCave, 1978b; Onyett and Simmons, 1983; Robinson, 1966; 1980a; Steers, 1927, 1939a, 1964a; Steers and Jensen, 1953; Ward, 1922; Williams, 1956). It is one of a small number of such features cited in the wider coastal literature (Bird, 1984, 1985; Bird and Schwartz, 1985).

The site extends from TG 489 216 in the north to TG 506 181 in the south. From a narrow dune ridge at its northern end, the ness widens to over 500 m in its central section around the ness itself before narrowing again southwards. North-east of the village of Winterton-on-Sea, the site is formed of linear dunes and slacks. Much of the dune landscape was greatly altered during the 1953 floods, and many of the blowthroughs and other features that existed before 1953 were

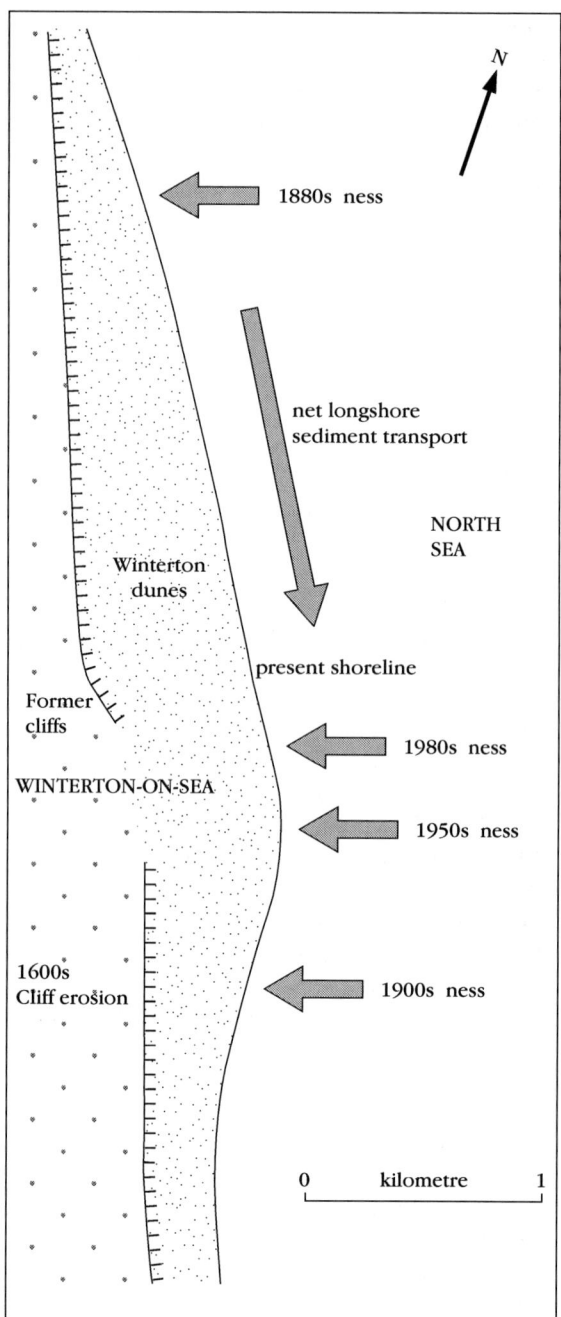

Figure 8.14 Former positions of the ness at Winterton, indicating a rapid southwards change in position between the 1880s and 1900s, but a subsequent movement northwards to the 1980s, and then a return southwards in the 1990s.

modified or eliminated (Steers, 1964a). Ridges of shingle also occur, albeit occasionally. Towards the ness, low dunes rest upon parallel sand and shingle ridges about 1.0 to 1.3 m in height. South of the village, the old cliffline (which is up to 15 m in height) is separated from a line of dunes by a valley, the origins of which remain obscure.

Cambers (1975) reported that there was some slight accretion around the Ness. Onyett and Simmons (1983) indicated that accretion at Winterton had passed its peak with the area of positive growth moving north to the Horsey area. Halcrow (1988) described the coast as retreating at rates up to 2 m a^{-1} on both sides of the ness, whereas at the ness the coast had advanced at about 0.5 m a^{-1} (Figure 8.15). The foreshore was steepening. They confirmed that the ness was migrating towards the north. The Shoreline Management Plan (North Norfolk District Council et al., 1996) reports beach retreat on the northern side of Winterton Ness of up to 1 m a^{-1}, in contrast to general accretion on its southern side. Sediment transport is towards the south. This suggests a return to the patterns described by the earliest descriptions of the Ness. It also accords with Cambers' view that the Ness results from a change in the rate of littoral drift resulting from the change in beach alignment. The northern edge of the site lies just south of a point where there have been breaches of the dune line (Steers, 1964b), as in 1938 and 1953, but the main dune area remains unscathed. There are several ebb-flood channels offshore from the Ness.

Interpretation

The development of nesses along the East Anglian coast has been the subject of some debate, but it is difficult to develop a general hypothesis for their formation as a group because their individual sedimentary characteristics are not similar. The beach at Winterton Ness, for example, is mixed sand and shingle, whereas Benacre Ness to the south is usually veneered by shingle. Both the movements of sand and shingle and the forms associated with them differ. Benacre Ness shows a strong tendency to migrate northwards, but the movements and history of Winterton Ness are less certain. Robinson (1966) demonstrated that ness features are associated with offshore ebb-flood channel systems and suggested that material reaches the nesses via these systems. The process thus involves a complex interaction of offshore tidal streams and wave action. Cambers (1975), however, suggested an alternative hypothesis. His sediment transport calculations show that in the vicinity of Happisburgh 1 000 000 m^3 of sediment could move annually towards Winterton. To the south of Winterton at Hemsby 390 000 m^3 could move southwards. Cambers interpreted this to mean that Winterton Ness could be an area where part of the balance is added to the ness, whereas the remainder moves offshore via the ebb-flood channel system. This is completely the reverse of Robinson's hypothesis. Despite some conflicting and inconclusive evidence from the analysis of drifter releases between Winterton Ness and Benacre Ness (Craig-Smith, 1973), Cambers proposed a simple hypothetical model for the sediment budget of Winterton Ness. Allowing for an accretion rate of 1 m^3 per metre length of coastline per year, 2000 m^3 would accumulate annually at the Ness. In practice such rates are unlikely to be achieved today with much of the cliffed coast now artificially protected to the north.

Ward (1922) linked the changes in the offshore banks to the intermittent pattern of erosion along the East Anglian coast. This view had been expressed earlier by the Royal Commission on Coastal Erosion and Afforestation (1907–1911), which attributed periods of increased erosion to the lowering of offshore banks. The relationship between the offshore banks and ness maintenance may therefore be a very complex one. Steers (1964b) put forward the possibility that the contrast between the northern and southern parts of Winterton Ness might result from the presence of a sandy spit or bank, or even a shingle bank, on which the dunes south of Winterton could form. If this were the case, the ness would form at a point where following the trend of the coastline would cause the beach to extend into deeper water and for refraction around it to assist the rate of transport of sediment southwards. What is of particular interest at Winterton is that, unlike other nesses, it does not appear to have been a gradually growing feature, but may be more akin to features such as Orfordness, Suffolk, where a ness is accompanied by a spit at its down-drift end.

The significance of this site lies in the contrast between its northern (erosional) and southern (aggrading) parts and the processes that affect its continuing growth. It has been suggested that Winterton Ness, like other similar forms, marks a location where there is net offshore transport of sediment. As far as the longshore

Sand spits and tombolos

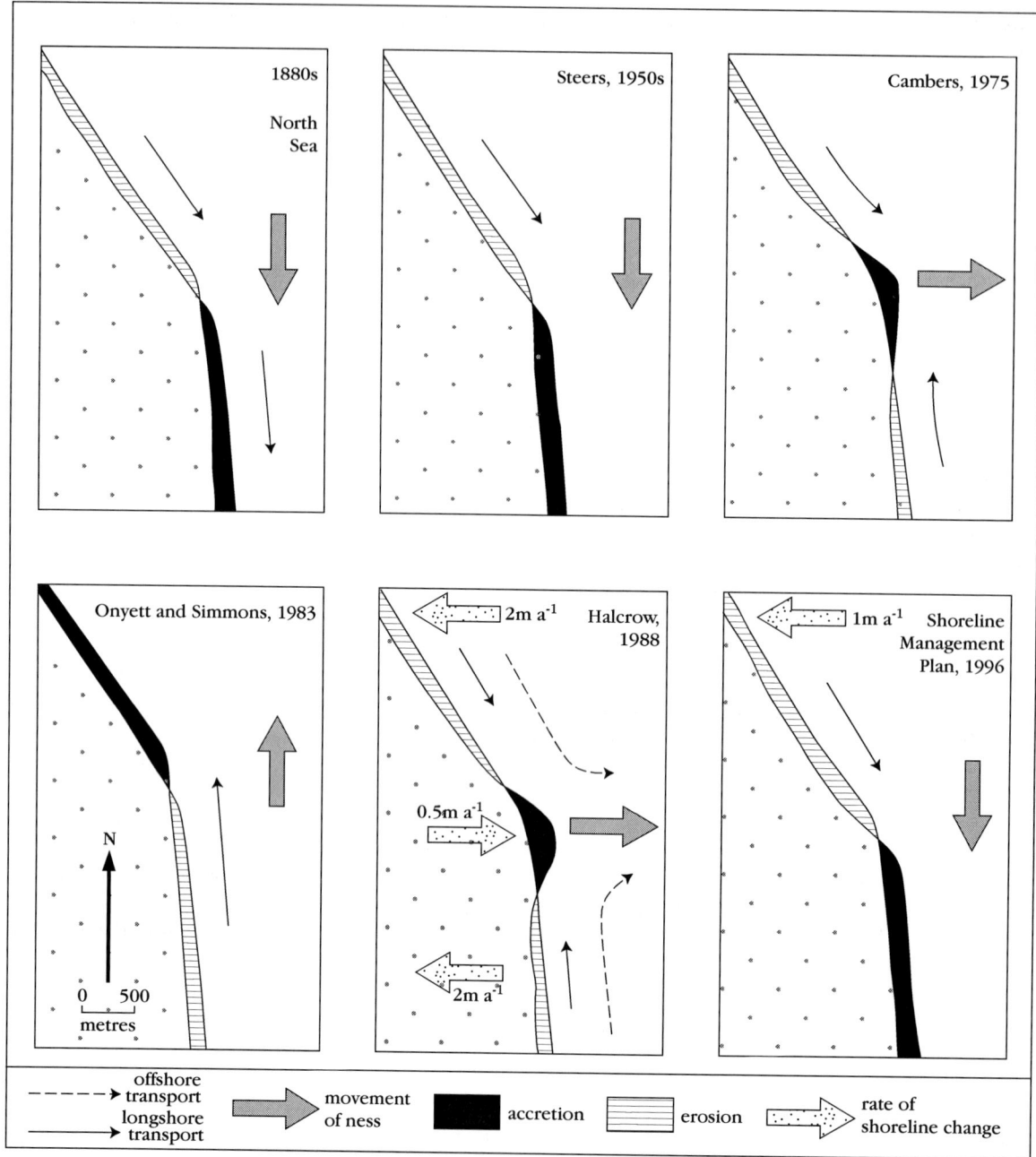

Figure 8.15 Different interpretations of the sediment transfers at Winterton Ness. In the 1880s, according to Steers (1964a) and the Shoreline Management Plan (North Norfolk District Council *et al.*, 1996), net sediment transport was southwards and the ness moved in the same direction. Others have suggested that transport is from the south, and Cambers (1975) and Halcrow (1988) agree on transport from both south and north with a transfer offshore and the ness extending seawards.

sediment budget is concerned, it is a sediment sink. However, the role played by offshore banks may be such that sediment returns elsewhere to the shore via ebb-flood channels.

The dynamism of the feature and its place within a continuum of longshore sediment transport makes the definition of its northern and southern extremities difficult. They have, however, been set, for GCR purposes, so as to include the processes maintaining the ness as well as the form itself. The offshore limit of the site should be related to the processes in the ebb-flood channels and on the offshore banks, but since the evidence of their precise role is conflicting they have not been included in the site. If sediment reaches the ness from offshore, as in Robinson's (1966) hypothesis, then it is essential that the offshore is offered the same protection as the ness itself for without the former the latter will remain at risk. If, however, Cambers' (1975) hypothesis is correct, designation of the offshore zone is not critical for the Ness. In the early 1980s the evidence seemed to favour the latter position, but Halcrow (1988) suggest that the ness is the result of littoral sediment supply from the north-west associated with converging tidal residual currents. There is a need for further investigation to determine not only the relationship of the Ness and the offshore area, but also the relationship of the probable offshore transfer of sediment to the sediment budget farther south.

Conclusions

Winterton Ness is a narrow cuspate foreland dominated by well-developed dunes and a sandy beach. It has been identified as an area with a sediment budget surplus and of considerable sediment transfer offshore. Winterton Ness differs from the other similar features of the East Anglian coast in being predominantly sandy and in having a slight sediment budget surplus. It also differs from other nesses because it has not had a consistent pattern of growth. Although there has been an historical pattern of movement towards the south, this has not been maintained in recent decades. The shoreline dunes are very geomorphologically active at the site, since they migrate inland on shorelines undergoing erosion or build seawards where accretion takes place. Winterton Ness has been cited by a number of writers (e.g. Ranwell, 1972; Goudie, 1990) as a key example of a prograding ness dune system. Behind the linear coastal dunes, the dunes of the central part of the ness are of considerable ecological importance because of their relative stability and alkalinity, and much of the GCR site co-incides with the Winterton Dunes National Nature Reserve. It is an important member of a group of narrow cuspate forelands that play an important role in the longshore sediment transport of the East Anglian coast.

MORFA HARLECH, GWYNEDD (SH 574 303–SH 550348)

V.J. May

Introduction

Morfa Harlech forms a large triangular area of sandflats, beaches and dunes, and claimed land between an abandoned cliff north of Harlech and the estuary of the Afon Glaslyn and Afon Dwyryd (see Figure 8.2 for general location and Figure 8.16). The present-day beach and dunes form a narrow fringing system in the south of the site, but widen northwards into several sub-parallel ridges. The alignment of a sand beach and dunes at an acute angle to the former cliffs has encouraged extensive sedimentation. Inland there are several recurved zones of former shoreline and dunes. Morfa Harlech is significant for the relationship of the ridges to sediment inputs from local rivers and the seabed. Though progradation is prevalent, there is also some localized erosion, both at the proximal end, near Harlech, and at the distal end of the spit. Morfa Harlech is little-affected by anthropogenic intervention into littoral sediment transport, and is part of a suite of beaches that are aligned to Atlantic swell in the Irish Sea. The first description of the site was by Steers (1939b). This was developed further by Steers (1946a) and King (1972b), but much of the description that follows is based on more recent examination of the site both in the field and on aerial photography taken at various dates since the late 1940s.

Description

The sandy beach at Morfa Harlech extends about 7 km NNW from the coastline at Llanfair towards the estuary of the Afon Glaslyn. The landward edge of Morfa Harlech is formed by a line of former cliffs upon which the 13th century castle at Harlech was built. Between the old cliffline and the beach there is a triangular area of reclaimed marshland, saltmarsh and both geomorphologically active and relict dunes. Within this area there are several rocky outcrops, such as Ynys Llanfihangel-y-traethau, former islands enclosed within the marsh. The main geomorphological interest lies in the beach and dunes that form the seaward part of Morfa Harlech (Figures 8.17 and 8.18).

Sand spits and tombolos

Steers (1939b, 1946a) described the development of the Morfa from its earliest days as a small spit of shingle and sand providing some shelter for vessels arriving at the castle watergate. As the spit grew northwards, sedimentation and the development of saltmarsh was accompanied by land-claim. This was not well documented until the early 19th century when embankments were constructed between the north-east corner of the high ground at Llanfihangel and the coast road at Glyn Cywarch. An Act passed in 1806 allowed embankment, and Steers (1939b) recorded that banks were constructed at both Morfa Harlech and farther up the estuary at Talsarnau soon afterwards. The 1808 embankment from Llanfihangel to Glyn Cywarch finally closed the creeks. Steers considered the relationship of the castle and its port to the growth of the spit and the marshland behind it. He concluded that the spit grew northwards and that small boats were able to reach the castle for several centuries after it was built.

Steers used a number of maps to interpret the historical development of the spit up to 1939. Since the 1950s, maps and aerial photographs have augmented his description of the development of the spit, but there has been little other work on its geomorphology. The beach and dunes fall broadly into three main parts: a southern section, about 3.25 km in length, formed mainly by sub-parallel dune ridges; a central section, formed mainly of recurved vegetated and generally low-lying dunes, and a northern area of mobile sand, also characterized by recurved sandy ridges. There are frequently several curving ridge and runnel forms in the intertidal area at the northern end of the spit. The north-eastern part of the site is formed by saltmarsh (Figure 8.17). These zones are well depicted on the aerial photo-mosaic (Figure 8.18). Although movements of sand along the spit towards its distal end have contributed in part to its extension across the Afon Glaslyn, changes in the position of the river channel have contributed also to the growth and erosion of the spit.

Bird and Schwartz (1985) included Morfa Harlech as one of several important British depositional structures in their review of the world's coastlines, and it is one of several beaches in Cardigan Bay that were reported by King (1972a) as being swash-aligned. It was also noted by Guilcher (1958) as one of the many beaches along this coastline where the larger

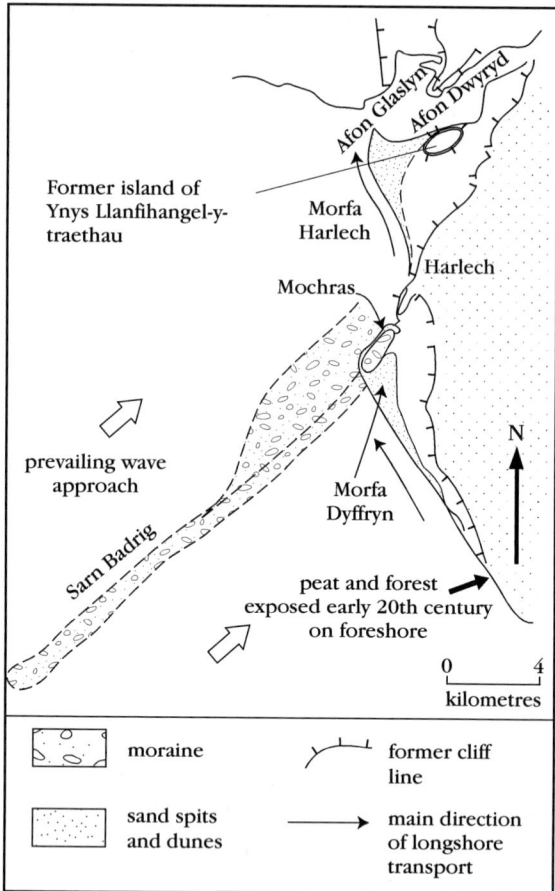

Figure 8.16 Context of Morfa Harlech and Morfa Dyffryn – key geomorphological features.

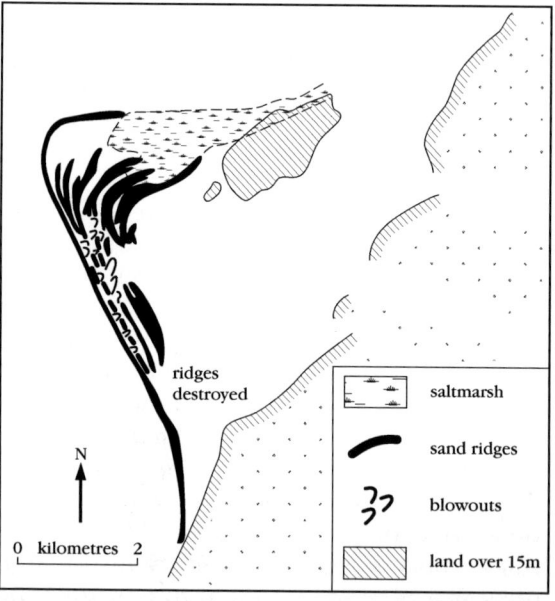

Figure 8.17 Key features of Morfa Harlech. (After Steers, 1946a.)

Morfa Harlech

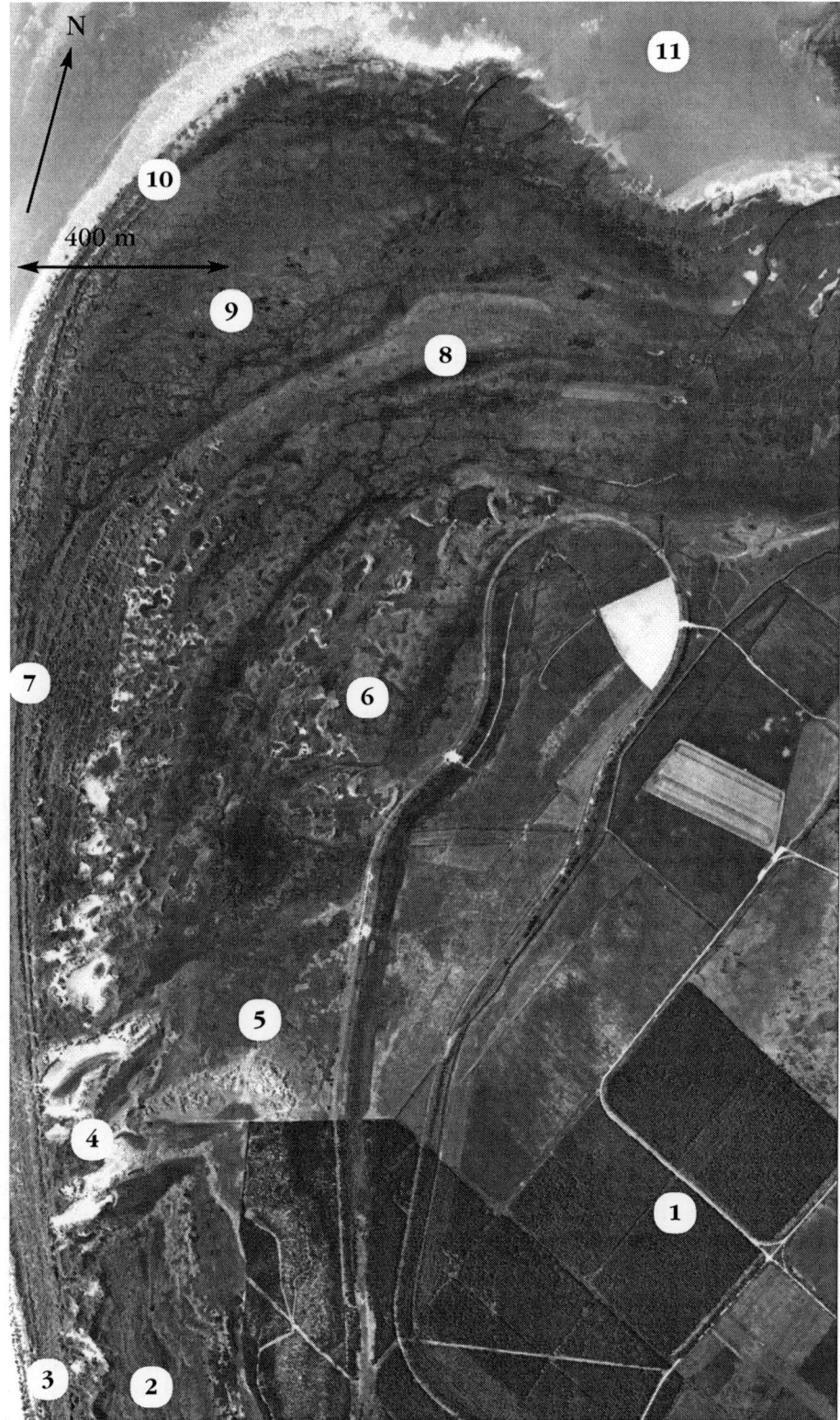

Figure 8.18 Aerial photograph of Morfa Harlech with the main geomorphological features numbered. 1 = former mainland; 2 = linear stable dunes ('grey dunes'); 3 = active 'yellow' dunes; 4 = zone of active blowthroughs; 5 = relict blowthroughs with SW–NE-aligned linear dunes; 6 = dune and slack topography; 7 = recurved linear dunes; 8 = former distal spits; 9 = 19th century distal features; 10 = modern distal dunes; 11 = intertidal sandflats. (Photo: courtesy Cambridge University Collection of Aerial Photographs, Crown Copyright, Great Scotland Yard.)

clasts are mainly slate and shale. As a result its sediments are characteristically platy in form rather than rounded, and this is reflected in the detailed structures of the beach ridges. Steers (1946a) described the southern part of the spit as a dune area bordered by a belt of coarse cobbles. In the 1940s, this was the only shingle visible in the whole area, and Steers estimated that it had moved north about 450 m since 1901. He considered that, prior to the construction of the railway embankment, the erosion of the till cliffs to the south would have provided a source for these cobbles. He also discussed the possibility that the dunes were underlain by shingle or in-situ till, but acknowledged that there was little evidence to resolve the issue. The way in which the sand dunes recurve in the northern part of the spit, decrease in height and fan out, led Steers to believe that they were not underlain by shingle. The northern end of the spit had grown an estimated 200 m in the previous 100 years, and Steers did not expect it to grow much further. In fact, since then the spit has grown much farther to the north-west and then retreated to about its present position. There is now a large area forming the distal end that is made up of a series of curved low ridges. They are overwashed at spring tides or during periods of high wave energy and have not become vegetated. The outer part of the distal area is formed by several ridges and runnels, features that have been discussed elsewhere by Orme and Orme (1988) and may provide a mechanism by which sediment is transferred from the intertidal zone to the beach (Figure 8.19). The changing position of the north-western part of the spit appears to be related to movements in the position of the ebb and flood channels of the Glaslyn estuary. This area has extensive sandy intertidal flats that receive sediment from the rivers of the Vale of Ffestiniog, but sources from Snowdonia via the Afon Glaslyn, north of Porthmadog, were restricted greatly during the 19th century when much of Glaslyn Valley was reclaimed. There are nevertheless extensive intertidal areas that could serve as sources for the spit.

The vegetated dunes are affected by shoreline erosion at their southern end, but this has not been a major problem within the site. There is some damage to the dunes as a result of recreational trampling, and blowthroughs occur along the dunes south of the main public access and at several separate locations farther north, both on the foredunes and on the older inner

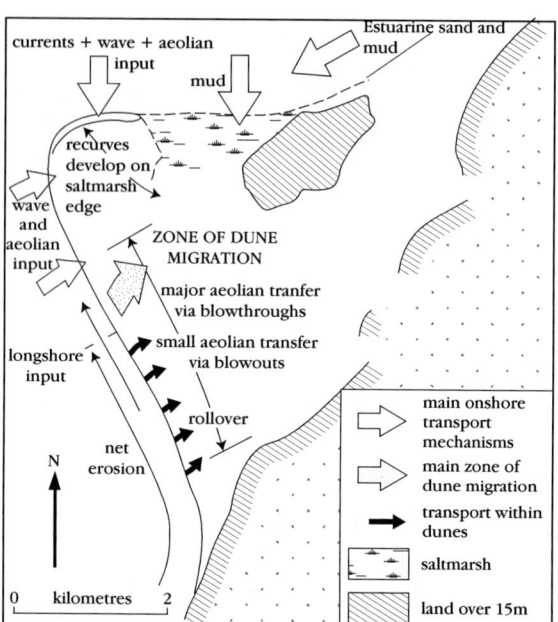

Figure 8.19 The main processes and sediment transfers at Morfa Harlech.

ridges. Steers (1946a) noted that parts of the inner dunes north of the Cefn mine were affected by blowthroughs, and described the outer seaward line of dunes as wind-eroded. The inner edge of the dunes was migrating at about 4 m a^{-1} on to the pasture behind the dunes. Following afforestation of part of the inner dunes, these dunes appear to have stabilized.

Interpretation

The changes in the area of the spit both during the last 150 years and during the last four decades indicate that Morfa Harlech as a whole is in a state of progradation. The amount of erosion at its southern end is small. The spit appears to be very close to equilibrium with the dominant and prevailing south-west waves, with very little net movement of sediment alongshore. Despite the substantial growth of the distal features, the general line of the southern beach and its position have remained similar for several decades. Its alignment depends to a considerable degree on its relationship to the Glaslyn estuary and the rocky shoreline on the northern side of Cardigan Bay, both of which affect the direction of waves approaching the spit. The source of sand for the spit is probably mainly from the substantial submarine glacial deposits that floor the bay, with some exchange also taking place between the estuarine sands and the distal end of the spit.

Morfa Harlech is a fine example of a sand spit developing across an infilling estuary. Most of its growth appears to have occurred during the last 700 years, but, unlike many other such forms, it does not appear to have been seriously affected by the worldwide tendency for such features to be affected by erosion (Bird, 1985). This is attributed to a large probable source of seabed sediment in Cardigan Bay and the large quantities of sandy sediment in the Glaslyn estuary that may have increased due to mining inland. The spit has been little affected by coast-protection works, although there is some confined damage resulting from recreational trampling. The processes that are geomorphologically active on Morfa Harlech have not been investigated in detail, but its largely pristine character makes it particularly important as a site for coastal geomorphological studies. There has not been a detailed investigation of the stratigraphy landward of the beach, especially in the Harlech area, which would allow the early history of the beaches to be described. Nevertheless, the historical evidence suggests that most of the beach is a much later development than Ynyslas to the south. It contrasts also with rock-based dune systems at Newborough Warren.

Morfa Harlech is the result of several phases of as yet undated spit growth, and the progressive sedimentation and land-claim of the area between the beach and the former rocky sea cliff upon which Harlech Castle stands. Unlike the other major depositional features of the coastline of Cardigan Bay, it appears to depend upon sediment supplies from the sandy estuary to its north. Morfa Harlech displays several phases of growth, a similar characteristic to several other beach systems (for example Pwll-Ddu, South Haven Peninsula and Gibraltar Point).

Conclusions

The Morfa Harlech GCR site comprises a well-developed spit across a major estuary whose sediment load may contribute significantly to the coastal sediment budget; it has several distinctive recurved zones that relate to its development during the last 150 years. It is a fine example of a multi-phase, gravel-based, sand spit that has gradually built across a major infilling estuary. Much of its growth has taken place during the last 700 years, and continues to show a positive sediment budget, largely as result of the large quantities of sand available on the shallow sea floor and in the Glaslyn estuary. Its almost totally unspoilt character makes it especially important for coastal studies.

MORFA DYFFRYN, GWYNEDD (SH 557 271–SH 579 213)

V.J. May

Introduction

Wave conditions in Cardigan Bay are dominated by Atlantic swell from the south-west, but locally generated waves may approach from the west or north-west. Cardigan Bay is bounded by the Lleyn Peninsula in the north and St David's Peninsula in the south; the bed of the bay is marked by three major SW-trending cobble and boulder banks, known as the 'Sarns', which are believed to be of glacial origin (Foster, 1970; Bowen, 1974) and thought to confirm an extensive westward flow of Late Devensian Welsh ice from the uplands (Campbell and Bowen, 1989). These sarns affect both wave behaviour and sediment movement in the bay. Morfa Dyffryn is linked geomorphologically to Sarn Badrig (see Figures 8.16).

The beach and dunes at Morfa Dyffryn front a cuspate foreland, which is about 3 km wide at Llanbedr. The beaches developed as a spit extending across the mouth of the Afon Artro, but today they link the morainic hill of Mochras to the mainland, following diversion of the river by an embankment in 1819. Near its southern end, Morfa Dyffryn comprises a narrow fringing beach of shingle, cobbles and sand upon which there are low dunes. Northwards, the dunes are wider and higher enclosing large slacks. At Mochras, the shoreline is formed of low cliffs of glacial material and the beach is dominated by cobbles and boulders. To seaward, Sarn Badrig extends from Mochras as a shallow, submerged ridge for about 17 km. Like many of the spits and cuspate forelands of England and Wales, Morfa Dyffryn was first described by Lewis (1938) and Steers (1939b; 1946a), but has subsequently received little detailed attention. Guilcher (1958) regarded it as a good example of a cuspate foreland, although this ignores the position and role of Sarn Badrig and the historical development of the feature.

Sand spits and tombolos

Description

Morfa Dyffryn (see Figure 8.2 for general location) is a broadly triangular area extending from Llanaber in the south to Llandanwg in the north. It is widest at Llanbedr where it extends westwards about 3 km from a probable former cliffline to its apex at Mochras. At its northern end, it encloses the much-modified estuary of the Afon Artro. Much of Morfa Dyffryn is excluded from the GCR site because it is agricultural land or forms part of RAE Llanbedr, whose construction in the 1940s destroyed much of the area of inland dunes.

The beach is virtually straight, faces south-west and extends for about 5 km from its southern boundary (at SH 579 214) to about 0.5 km south of Mochras (SH 552 255). At this point, it is aligned towards WSW before a sharp change of direction (at SH 550 262) so that the northern side of Mochras faces north-west. The southernmost part of the site is formed by a spit of sand and shingle that diverts the mouth of the Afon Ysgethin northwards. The spit is progressively extending northwards (Figure 8.20). North of the Ysgethin's outlet, the dunes gradually widen from a narrow fringing ridge about 120 m to over 1.2 km wide in the north. The dunes were described by Steers (1939b) as gradually extending inland.

The geomorphologically active dunes attain heights in excess of 20 m, with semi-parabolic ridges enclosing large slacks. Once the dunes reach a critical height (Ranwell, 1972), they tend to migrate inland and blowthroughs become dominant (Figure 8.21). The lowered areas are gradually replaced by new dunes (Ranwell, 1972). The northern area of the dunes first began to accumulate after the mouth of the Afon Artro was diverted in 1819. The 1838 first edition of the Ordnance Survey One Inch map shows Mochras as an island. With the opening of the present-day river mouth, the sand beach extended northwards to the low cliffs at Mochras. Apart from some changes in direction around the mouth of the Artro, the low-water line followed a very similar alignment to the present-day shoreline. The dunes have continued to migrate inland, but the beach is now stable in position. Sufficient sand is reaching the beaches to maintain their position and to continue to supply the landward-moving dunes.

Interpretation

Large cuspate features are rare on the British coast. Standard texts on coastal geomorphology from Johnson (1919) to Pethick (1984) refer to three such features: Benacre Ness, Dungeness and Morfa Dyffryn. Lewis (1931) regarded this beach as a good example of orientation towards the dominant waves. Others, such as Morfa Harlech to the north, show less well-developed orientation towards the dominant waves, as these other sites are more affected by the refracted waves and currents at the mouths of the estuaries. Other examples of similarly orientated beaches occur in Carmarthen Bay. King (1964) considered that this beach was controlled by dominant waves related to the coastal outline, but that Sarn Badrig also played a part in affecting wave alignment and energy. Morfa Dyffryn is one of several beaches in Cardigan Bay in which the coarse sediments are dominantly made up of slate and shale derived from local sources (Guilcher, 1965). The beach at Morfa Dyffryn is, nevertheless, predominantly sandy, particularly towards its northern end. Moore (1968) described the patterns of sedimentation in Cardigan Bay between Aberystwyth and Mochras. Sand was transported northwards along Morfa Dyffryn, but also entered the sand-floored area south of Sarn Badrig at both its shoaling and seaward ends. Moore regarded tidal streams as the most important agents of sediment dispersal in Cardigan Bay. Geochemical and mineralogical analyses suggested that the estuaries of both the Afon Mawddach and the Afon Dyfi were being filled by sediments from the sea rather than from the rivers.

The relative stability of the shoreline, despite a strong tendency for the dunes to migrate inland and a limited supply from littoral drift, poses questions about the sediment supply to Morfa Dyffryn. There is no longshore source of any volume to the south and the southern part of the beach shows signs of being generally in deficit. There is little evidence to support the hypothesis that sand may be transported from the north into this site. Moore (1968) supported the possibility of offshore sources. Sediment movements across Sarn Badrig from the northern part of Cardigan Bay would provide one mechanism for maintenance of the sediment supply to Morfa Dyffryn. Sarn Badrig and its landward expression at Mochras have

Morfa Dyffryn

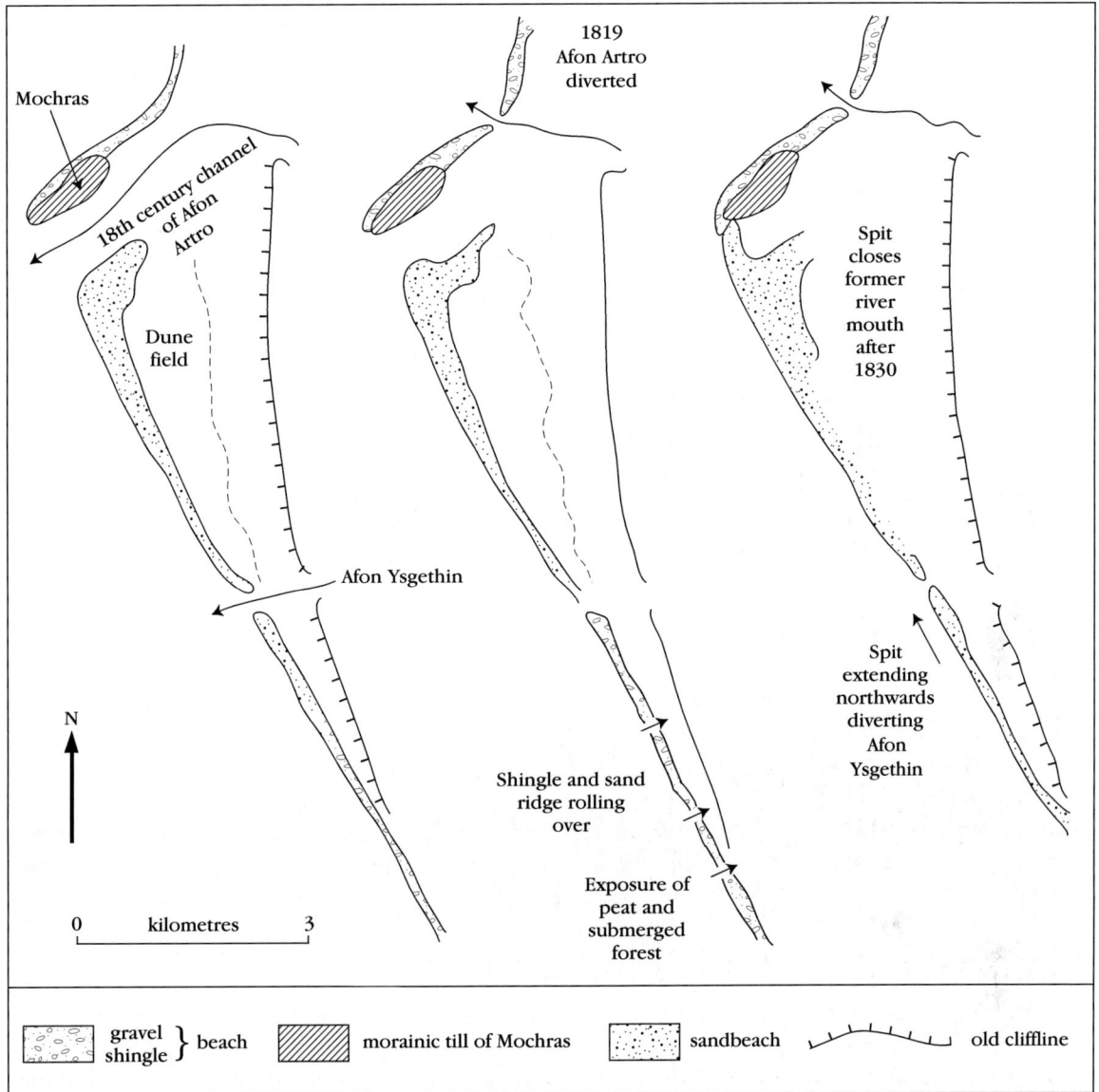

Figure 8.20 The historical development of Morfa Dyffryn. During the eighteenth century the sand beach was separated from the morainic hill of Mochras by the channel of the Afon Artro and formed a spit with recurves at its northern end. In 1829, the Afon Artro was diverted to the east of Mochras. About the same time, the southern beach was transgressing inland. By 1830, the spit had closed the former river mouth and had joined Mochras. In the south, a new spit was developing northwards across the mouth of the Afon Ysgethin.

been effective in providing a promontory against which the low-water beach has been aligned.

Despite its description in textbooks, Morfa Dyffryn cannot be regarded as a good example of a cuspate foreland, for the ness form is a cliffed headland rather than a coastal depositional structure. The presence of this relatively resistant headland has produced a situation in which the beach has tied the headland to the mainland and the beach has attained its present-day alignment as a result. The dune and beach system is better described as a tombolo, which makes Morfa Dyffryn a particularly large example. To some extent its size and alignment have been affected by the shallow area upon which it is built. There are many beaches whose low-tide alignment appears to be particularly influential in the long-term development of the position of the shoreline. It is apparent both at Holy Island,

Sand spits and tombolos

Figure 8.21 Aerial photograph of part of the northern sector of Morfa Dyffryn with sand transfers and the main geomorphological features numbered. 1 = till boulder and cobble beach derived from erosion of Mochras; 2 = main active zone of dunes and spit distal link with former island; 3 = major blowthrough; 4 = bar merging with beach – maintains sand supply to dunes; 5 = intertidal ridge and runnel; 6 = prevailing and dominant wave direction. (Photo: courtesy Cambridge University Collection of Aerial Photographs, Crown Copyright, Great Scotland Yard.)

Morfa Dyffryn

Figure 8.22 Active dunes of Morfa Dyffryn migrating eastwards (in the foreground) are affected by a large blowthrough to the centre right. (Photo: V.J. May)

Northumberland, and at Morfa Dyffryn, for example, that there has been much less change during the last 150 years in the alignment of the low-tide shoreline than of the high-tide shoreline. Because waves approaching the high-tide shoreline are refracted by the intertidal features especially on the low angles associated with sandy foreshores, the low tide shoreline plays an important role in the long-term development of the beach itself. At Morfa Dyffryn the low-tide alignment of the shoreline is strongly controlled by swell and the presence of Sarn Badrig, Mochras and the shoreline at Llanbedr, features that have not changed their positions significantly during the last 150 years. As a result, Morfa Dyffryn is not only a fine example of the process by which beaches align normal to the dominant waves (first outlined by Lewis, 1931, 1938), but also demonstrates the importance of the low tide coastal outline in controlling the alignment of the shoreline.

It is also a good example of sediment supply from the seabed. In this respect it is a comparatively rare feature in global terms for most sandy beaches are in deficit having passed the stage of sediment storage that characterized the Holocene transgression (Bird, 1985). Only about 20% of the world's sandy beaches are prograding, but Morfa Dyffryn has a shoreline that is maintained naturally, despite considerable transport of sand by wind into the dunes (Figure 8.22). Morfa Dyffryn differs from other features with which it has been compared such as Dungeness, Kent, and Benacre Ness, Suffolk, in being (a) dominated by extensive dunes, (b) tied to a headland, and (c) comparatively stable in position, although its dunes individually migrate inland at rates in excess of 6 m a^{-1} (Ranwell, 1972).

Conclusions

Morfa Dyffryn is distinguished by a beach and dunes whose alignment towards the dominant south-westerly waves is controlled by a till headland and the alignment of the low-tide shoreline. Its interest lies in its association with the subtidal and intertidal ridge of Sarn Badrig, its dunes and its comparative stability. The main present-day source of sediment appears to be the seabed. Morfa Dyffryn has been wrongly described in the past as a cuspate foreland because although its form is cuspate, this results from the presence of the headland at Mochras rather than from the realignment of beach sediments as occurs in true cuspate forelands. However, it is a fine example of a large sand tombolo and so an important and unusual feature of the British coastline.

ST NINIAN'S TOMBOLO, SHETLAND (HU 371 208)

J.D. Hansom

Introduction

St Ninian's tombolo, the largest geomorphologically active sand tombolo in Britain, is a classic geomorphological feature of national importance. The tombolo links the south-west Shetland Mainland to the small off-lying island of St Ninian's Isle (Figure 8.23). Although tombolos are by no means rare in an archipelago environment, they are numerically scarce relative to other classic forms of marine deposition (spits and bars). In the Northern Isles, tombolos (ayres) are generally formed of gravel, cobbles or occasionally boulders. St Ninian's tombolo is distinctive among its fellows in being composed mainly of sand. In addition, unlike many other ayres that are relict in terms of their evolution and relationship to contemporary sea level, St Ninian's tombolo is a geomorphologically active feature linked to a nearshore sediment circulatory system (Nature Conservancy Council, 1976; Smith, 1993; Bentley, 1996d). The tombolo is flanked on either end by areas of dunes and hill machair, enhancing the geomorphological interest. This almost perfectly formed feature set in a highly scenic part of the Shetland Isles must be one of the most outstanding tombolos in the world.

Description

St Ninian's tombolo is a large (*c.* 500 m in length along its central axis) sand tombolo linking St Ninian's Isle to the Shetland Mainland (Figures 8.23 and 8.24). By its very existence as a tombolo and its location, the beach is subject to wave activity from two completely opposing directions and is thus more liable to natural fluctuations of profile and beach area than a conventional arcuate beach. The beaches facing to the north and the south form long sweeping arcs stretching between the cliffs on either side of Bigton Wick to the north and St Ninian's Bay to the south (Figure 8.24). Waves approaching these beaches tend to break simultaneously along their entire length, suggesting that the planforms of each beach are in equilibrium with the approaching waves (Bentley, 1996d). The tombolo is strikingly symmetrical in plan (Smith, 1993; Bentley, 1996d) although it experiences changes in intertidal width and profile characteristics as a result of tidal and weather events. During low spring tides the tombolo is 60–70 m wide, while during the highest spring tides the central part may be completely covered in water. Typically the central part of the tombolo is 20–30 m wide at high tide (Figure 8.23).

St Ninian's tombolo is composed almost entirely of medium-grained sand (D_{50} = 0.24 mm) with a carbonate content of around 50%. However, there is evidence that the beach sand overlies a gravel base (Flinn, 1974; Smith, 1993). Flinn (1974) identified 'a rock base presumably of pebbles' at a depth of *c.* 2 m in two distinct locations using a probe and in several areas of the beach a scattering of flat pebbles lie on the surface. The tombolo appears to be nourished by nearshore sand sediment banks (Smith, 1993; Bentley, 1996d), although the exact mechanism of the sediment circulatory system at this site is imperfectly understood. The tombolo is a dynamic landform, repeatedly adjusting its form in response to tides and waves. During periods of constructive wave action sediment is transported from nearshore sources to the beach, raising the beach profile. The direction of sediment movement is reversed during periods of destructive wave action and the beach profile lowered. Changes in the height (and hence width) of the beach therefore occur with changing weather and marine conditions. In general, the tombolo is lower and narrower in winter, and is best-developed in summer (Mather and Smith, 1974). However short-term fluctuations, as a result of storms or periods of fine weather, can be imposed on this seasonal trend. For example, during the storms in which the oil tanker *Braer* ran aground (January 1993) the centre of the tombolo was totally underwater for several days. After such storms a temporary channel may form through the centre of the tombolo, with water typically flowing from north to south (Bentley, 1996d). There is also some evidence to suggest that the tombolo shifted slightly to the north following the January 1993 storms (Bentley, 1996d) although this may have been a temporary feature caused by predominantly southerly winds at the time.

Vegetated blown sand deposits veneer the slopes on either end of the isthmus resting on bedrock and a thin layer of till. Dunes (vegetated by marram *Ammophila arenaria*) adjacent to the beach at either end are backed by more

St Ninian's Tombolo

Figure 8.23 St Ninian's tombolo, looking south-west towards St Ninian's Isle. Dunes flank either extremity of the sandy tombolo. During the highest spring tides the central part may be completely covered in water. (Photo: G. Satterley/SNH.)

extensive areas of hill machair blown sand that encroaches on pasture. The dunes and blown sand forms the sink for most of the sediment circulation system at St Ninian's tombolo. Once sand is incorporated into the sand plain it is unlikely to be re-incorporated into the system unless there is substantial erosion of the blown-sand areas (Bentley, 1996d). The dunes and machair have a relatively subdued topography, as a result of the influence of the underlying rock surfaces.

The volume of blown sand at the eastern end of the tombolo is much greater than that at the western end; a reflection of the dominance of westerly winds in Shetland (Smith, 1993). However, the morphology of this area has been altered dramatically by commercial sand extraction during the 1970s. A former extraction pit has since been naturally infilled by blown sand (Bentley, 1996d) and at present the dunes form low (<5 m) rolling mounds and hollows with small *Ammophila*-clad embryo dunes forming in a number of places. However, the coastal edge is distinctly erosional in character, particularly on the northern flank and there are a number of small blowthroughs within the dune system. The edge of the higher blown-sand area forms a prominent triangular-shaped bench landward of the dunes, backed by a thin cover of sand that is cut by a number of arcuate erosion scars. At the west end of the tombolo the dunes adjacent to the beach are heavily eroded. There are a number of active blowthroughs developing in the steep eroded and undercut dune face. However, despite this evidence of frontal erosion there was substantial sand accretion at the foot of the dune face in March 1996 (Bentley, 1996d). The hill dunes to landward have undergone advanced deflation (Smith, 1993, Bentley, 1996d), with low erosional scarps extending for several tens of metres often parallel to the contours of the slope.

Interpretation

St Ninian's tombolo most likely formed during a period of rising relative sea level (Smith, 1993; Bentley, 1996d). The relative sea level history of Shetland is one of progressive submergence since the decay of the late Devensian ice-sheet

Sand spits and tombolos

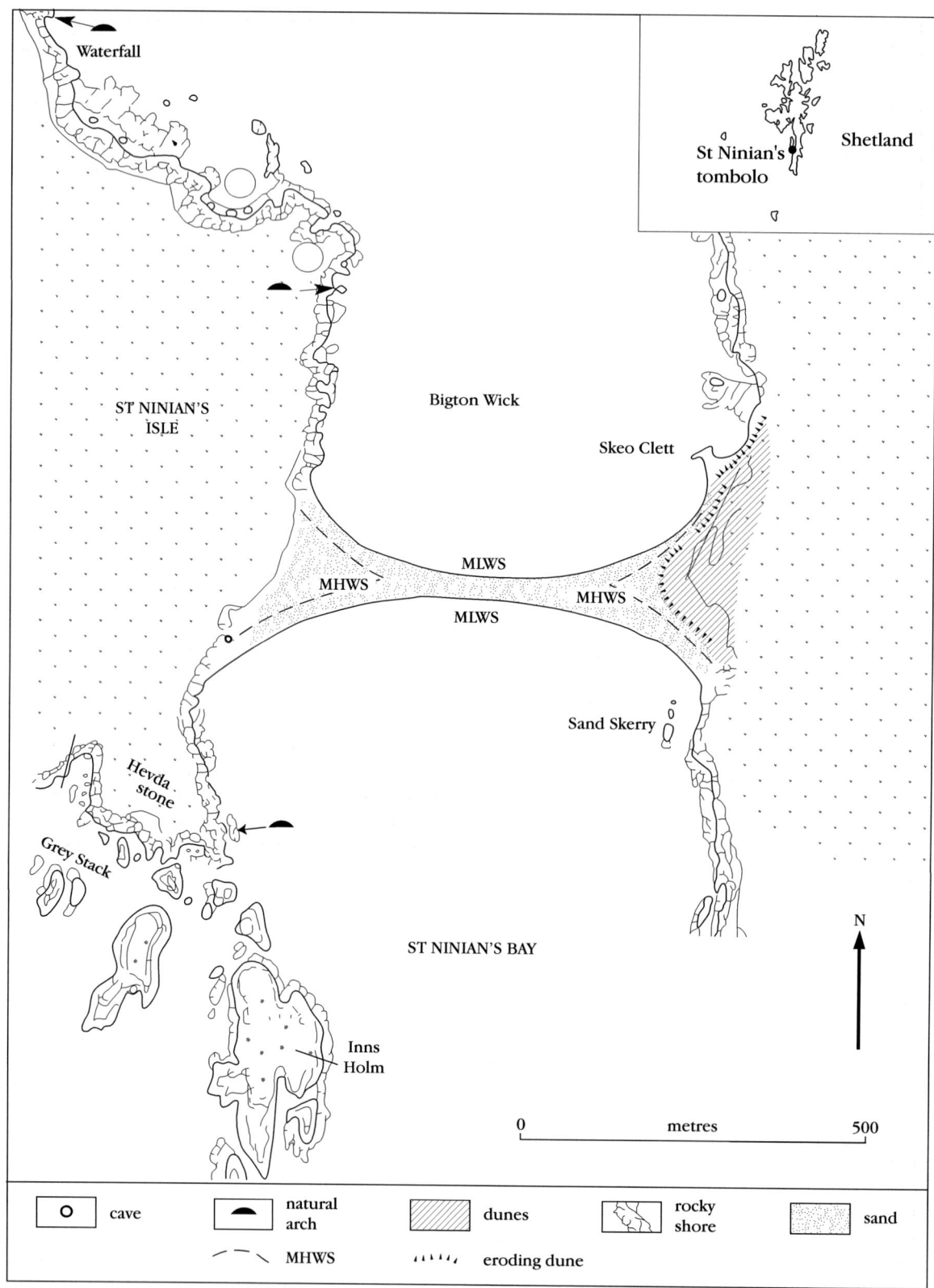

Figure 8.24 St Ninian's tombolo connects the Shetland mainland to an offshore island, and represents deposition from waves travelling south onto the north side, and vice versa in the south, in a very sheltered environment.

St Ninian's Tombolo

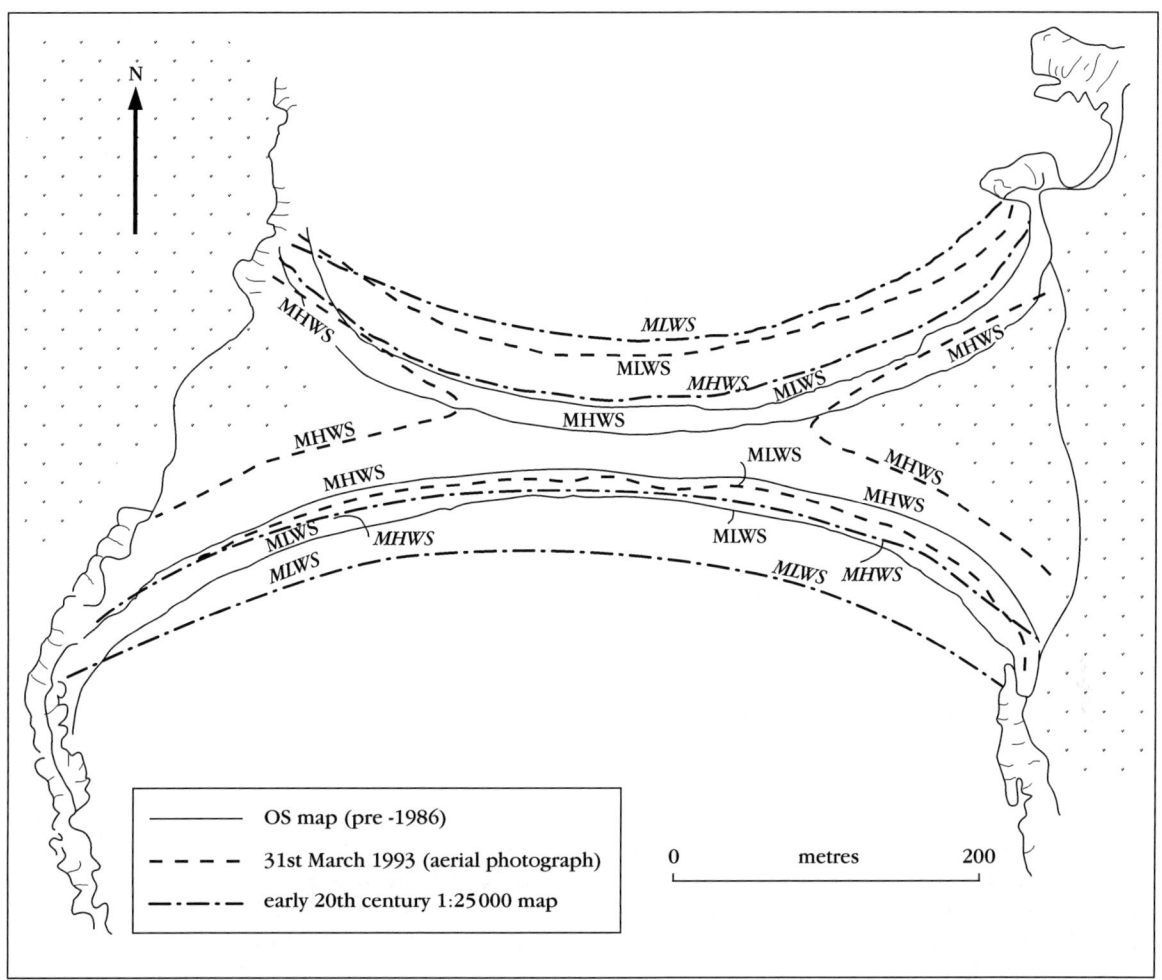

Figure 8.25 Change in tombolo position. In the early 20th century the tombolo was wider than at present, but with its axis in a similar position. Aerial photographs taken in March 1993 showed that the tombolo had migrated northwards by about 30 to 40 m. Subsequent topographical surveys later in the same year showed that the tombolo was migrating back southwards, suggesting that the northward shift was a temporary feature caused by southerly winds. MLWS, MHWS represent the position of mean low- and high-water springs, respectively. (Source: J. Swale, SNH.)

(Mykura, 1976). Dating of submerged peats in many of the sheltered voes and sounds indicate that c. 5500 years BP relative sea level stood around 9 m lower than at present (Hoppe, 1965). Depositional features such as tombolos, spits and bars are relatively common along recently submerged coasts (Johnson, 1919), although in the Northern Isles tombolos (ayres) are typically formed of shingle (e.g. the Ayres of Swinister, see GCR site report, this volume). The extensive sand tombolo of St Ninian's Isle, which is linked to the contemporary nearshore sediment circulation system, is distinctive in terms of its scale, composition and dynamism, and thus forms an unique component in the sand spit/tombolo GCR network.

There remains much scope for scientific research at the St Ninian's site, particularly to determine the evolution of this outstanding tombolo and investigate the complex relationship between sediment dynamics and relative sea-level change. At present, little is known concerning the exact evolution of the tombolo, except that it formed some time during the period of Holocene submergence of the Shetland coastline. The tombolo of St Ninian's Isle may have formed in a similar way to the South Ayre of Swinister. In this model, waves approaching from the west are diffracted and refracted around St Ninian's Isle, and eventually meet in the lee of the island. As the waves lose energy and deposit their load, sediment gradually builds up and eventually forms a connecting isthmus between St Ninian's Isle and the Shetland Mainland. Flinn (1997) also argues that the position of the tombolo is mainly related to diffraction around the island. It has been suggested that the tombolo has been in existence since at least the medieval times (Smith, 1993) when St Ninian's Isle was the site of a church. Smith (1993) attributes the quasi-permanency of this sand tombolo to the likelihood that the beach sand overlies a gravel base. If indeed it overlies a gravel core the implication is that at some time in the past, gravel was a relatively more important sediment source than at present. Although there is some evidence of a gravel core (Flinn, 1974; Smith, 1993; Bentley, 1996d), there is potential for further research using coring techniques and shallow seismic survey to determine the extent of the gravel base.

The earliest definitive evidence of the tombolo's existence is its depiction on an old chart drawn around 1700. The tombolo is clearly shown on a later chart drawn in 1743–1744, and George Thomas's 1830 map shows a feature similar to that of the present day. A comparison of old maps and aerial photographs suggests that although overtopping has occurred since 1822 (Flinn, 1997), the tombolo narrowed during the 20th century (Bentley, 1996d; Figure 8.25) and now has an increased frequency of overtopping of its centre. Beach profiles have been surveyed bi-annually since 1993 in order to determine any trends in the tombolo evolution. Although the record is too short to identify any long- or medium-term trends, some seasonal and inter-annual changes have been identified (Bentley, 1996d). Each winter there tends to be a loss of sand from the centre of the tombolo and a slight gain at the ends. This appears to be largely reversed during the summer when sand appears to return to the centre. However, in the interval between April 1993 and April 1994 there was a net loss of material from the centre of the tombolo, but at each end there were both localized gains and losses.

St Ninian's tombolo is linked to a nearshore sediment circulation system (Smith, 1993; Bentley, 1996d) where beach sediment is supplied from accumulations of sediment in the bays to the north and south. Smith (1993) expresses concern that the nearshore sediment bank has minimal possibilities of replenishment from nearby land sources, other than by wave erosion of blown-sand deposits at both the eastern and western ends. The seasonally wave-trimmed nature of the dunes at the western end and the sand extraction that substantially reduced the sand stocks at the eastern end in the 1970s may imply that the whole system is operating on a finite, and possibly decreasing, volume of sediment (Smith, 1993). In the context of continuing submergence, the supplies of sand via the nearshore and longshore sediment circulatory system that feed the tombolo are critical to its continued existence. In this respect any form of sand-extraction from the beach and surrounding windblown deposits is detrimental to the long-term existence of this outstanding tombolo.

Conclusions

St Ninian's tombolo, which connects St Ninian's Isle with the south-west Shetland Mainland, is the largest geomorphologically active sand tombolo in Britain. The size (*c.* 500 m long) and almost perfect symmetry of the tombolo is unique. The tombolo, composed of a shelly sand overlying a shingle base, is part of a dynamic and complex nearshore sediment circulation system. Although tombolos are relatively common along submerged coasts such as the Shetland Isles, it is the exceptional scale, composition and dynamism of St Ninian's tombolo that are of particular scientific interest. This interest is enhanced by the flanking windblown deposits of dunes and dune grassland. Conservation of this key site for coastal geomorphology is of the utmost importance; any disturbance of the sediment dynamics of the system may be critical to the tombolo's long-term existence.

COAST OF THE ISLES OF SCILLY (SV 910 165)

V.J. May

Introduction

The Isles of Scilly comprise five main islands, St Mary's, Tresco, St Martin's, Bryher and St Agnes, and over 100 smaller islands and islets (see Figure 8.2 for general location and Figure 8.26). Many of the larger islands have been formed by the linking together of smaller islands by sand tombolos, but some are linked by low terraces underlain by head and/or till (Barrow and Flett,

Isles of Scilly

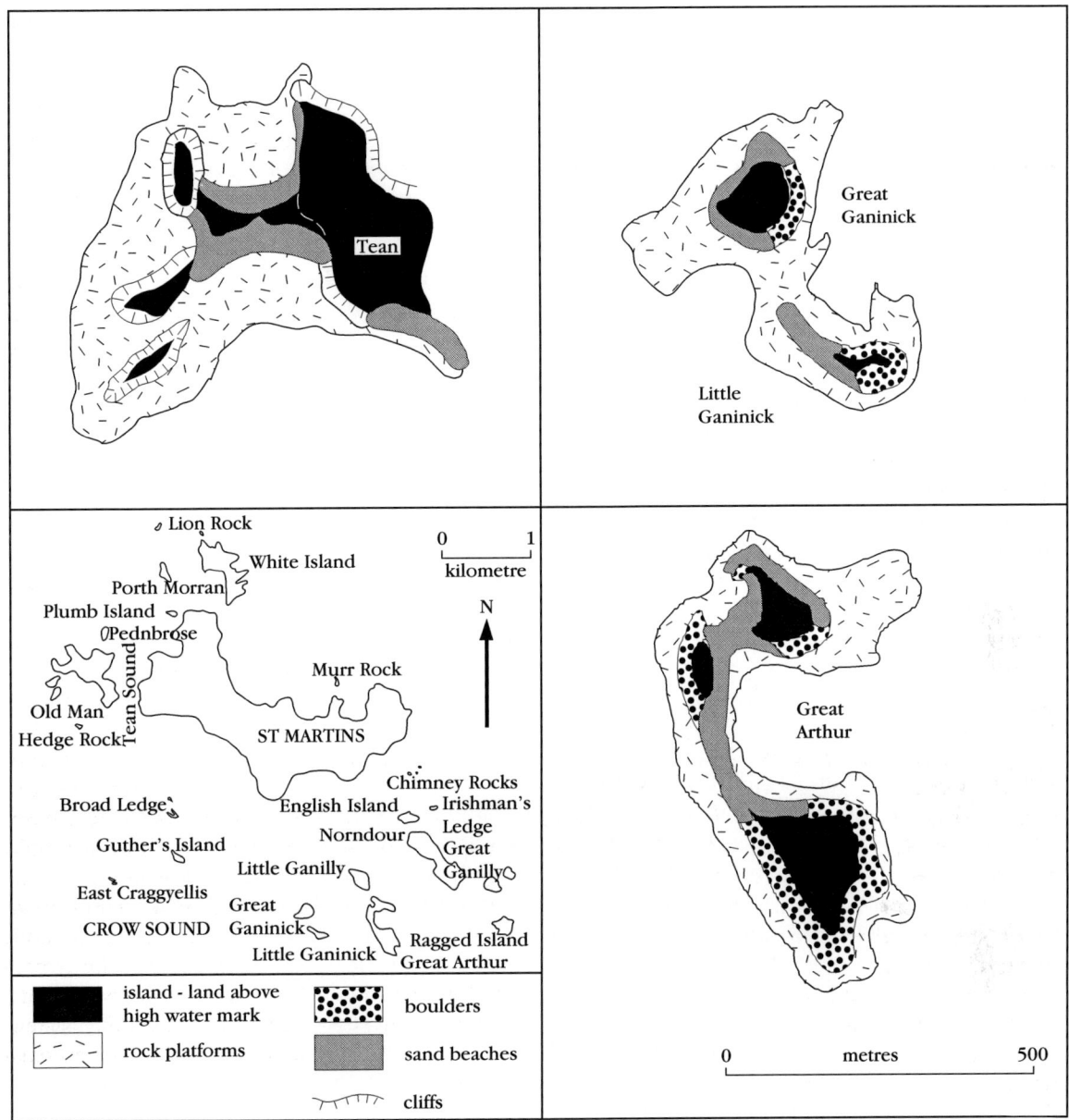

Figure 8.26 Key features of the coast of the Isles of Scilly.

1906; Mitchell and Orme, 1967). For example, the largest settlement on St Mary's, Hugh Town, is built mainly on a low isthmus fringed by sandy beaches. Apart from the islands of Scotland (mainly Orkney and Shetland) where Steers (1973) described over 40 such features, the Isles of Scilly contain the largest assemblage of tied islands in Great Britain. The beaches are predominantly sandy, derived from the weathering and erosion of head and/or till as well as the mainly granitic rocks that form the islands. Many of the linking beaches have been built upon, but the four islands that comprise this GCR site represent best the different stages of island linking.

The islands of Great and Little Ganinick represent early stages in the process of island linkage with a beach ridge extending from Little Ganinick towards Great Ganinick. On Great Arthur, the processes of beach development have linked two former islands and a third is gradually being joined to them. On Tean, not only are three small islands with links at various stages of the linking sequence, but there is evi-

463

dence of older beaches joining these islands and others. The changes in these developing features are affected both by the availability of sand derived from erosion on cliffs and platforms, and by the variation in wave direction and climate brought about by shelter and refraction around the larger islands of Scilly. There are no other sites in England and Wales in which these forms are common, let alone as well developed. Tied islands have received only limited attention in the literature (Gulliver, 1898–1899; Guilcher, 1954; Farquhar, 1967; Carter and Orford, 1988). The only comprehensive regional descriptions remain those of Steers (1973) and Mather and Smith (1974) for Orkney and Shetland. Although Steers (1981) drew attention to their unusual characteristics, there has been no detailed examination of them and their origins may need to be reconsidered in the light of Scourse's (1987, 1991) re-evaluation of the Pleistocene stratigraphy and Foster et al.'s (1991) evidence for surges and tsunamis in south-west England.

Description

The site comprises four islands, Tean, which lies at the northern end of Crow Sound between Tresco and St Martin's, and a group of three smaller islands (Great Arthur, Great and Little Ganinick) which lie at the south-eastern end of Crow Sound about 1.5 km north-east of St Mary's (Figure 8.26). Apart from waves that approach Tean from the north and the group of three islands from the ESE and south-east, all waves approaching the site are strongly refracted by the shallow seabed around the other islands. Each island is surrounded by platforms cut into granite. Sand, derived from the erosion both of these platforms and the cliffs and cliff-top sediments, supplies the beaches that rest upon the platforms. Their exact alignment depends upon wave direction and refraction between the major islands as well as over the platforms themselves.

At Great Ganinick, there are two beaches separated by zones of boulders. On the north side of the island (Figure 8.26), the beach forms a small cuspate foreland, whereas on the south side of the island the beach fringes the island. Little Ganinick has a single beach that trends north-westwards towards Great Ganinick. Although the sand of this beach spreads across the platform towards Great Ganinick, the sediment supply appears to be insufficient to link the two islands completely. In contrast at Great Arthur, the process is complete. Great Arthur comprises three rock islands joined by two sandy beaches. The larger of these is aligned with the rocky shore of the two southernmost former islands. Its eastern sheltered side is aligned towards small waves that refract into the bay from the east. The northern beach appears related to waves that pass between Great Ganinick and Little Ganilly before local refraction on the platform and headlands of Great Arthur.

On Tean (Figure 8.26), the process is demonstrated well by a set of beaches that not only tie small islands, but also show how these beaches may develop in the lee of islands without tying them completely. On Tean the balance between sediment supply and wave energy is such that double beaches have developed with a sandy flat between them. One beach has formed a cuspate form in the lee of a small island as the beach itself has been supplied with insufficient sediment to complete the link (Figure 8.26).

Interpretation

Tombolos have a limited research literature. Gulliver (1898–1899) suggested that the term 'tombolo' should be adopted from its Italian usage to include all beaches joining islands to the mainland. Where links between islands exist, the term is also used. Farquhar (1967) described a number of examples of tied islands, including the Isle of Portland and Holy Island. However, it is uncommon to find an assemblage of comparable forms at different stages of development within one area. Within Great Britain, there are two main areas where this occurs, the Isles of Scilly and Orkney and Shetland. Carter and Orford (1988) have described similar features on the drumlin coast of Clew Bay, County Mayo in the Republic of Ireland. Elsewhere within Europe, many of the best examples occur along the Italian coast. This site is, therefore, an important element in the assemblage of coastal landforms in southern Britain. It differs, both in its scale and the variety of forms within it, from Farquhar's other tied island site (Holy Island, Northumberland). Whereas Holy Island is affected by refracted North Sea swell, the features in Scilly are related to both refracted Atlantic swell and local wave systems within the island group.

In their study of linked islands in Clew Bay,

Carter and Orford (1988) emphasized that many of the links were established by solitary gravel ridges founded on coarse boulder frameworks. These links have been shown to facilitate sediment mobility and are sometimes marked by small crestal washovers, i.e. they have some slight tendency to transgressive behaviour. The principal factor in maintaining the beaches is the sediment supply from the continuing erosion of the cliffed drumlins. On the Isles of Scilly, erosion of the cliffs is slow, except in Pleistocene sediments, but there are some similarities with the Clew Bay features. First, many of the linking sandy beaches are commonly based on a more resistant foundation, in this case rock platforms strewn with boulders or possibly what is left or eroded ridges of till or head. Second, the beaches on Scilly are fed by erosion of low cliffs often cut into Pleistocene sediments (Mitchell and Orme, 1967; Steers, 1981). Erosion frequently exposes artefacts of archaeological importance.

The emplacement of some of the beaches may have resulted from the effects of surges and tsunamis (Foster et al., 1991). Although single ridges occur on certain of the islands (e.g. Great Arthur), there are several complex links that comprise beaches at several levels (e.g. Tean). As a result it may be necessary to rethink the linking process that has taken place in the Scillies. The linking forms of the islands thus offer a contrasting assemblage to those at Clew Bay and provide evidence of both similar and contrasting processes in different materials and on different timescales. The tied islands of the Scillies should be seen as important members of a group of contrasting and as yet poorly described features of the Atlantic coast of Europe.

Conclusions

Tied islands are rarely observed in England and Wales, but they are more common in the islands of Scotland and in Ireland. The Isles of Scilly include the largest British group of tied islands at various stages of development outside Orkney and Shetland. Their small size and variety makes them a very important location for further research into the relationships between sediment supply, sea level, wave patterns and beach development that bring about tied island formation; the site will be important for the study of the effects of sea-level rise on the completion of island tying.

CENTRAL SANDAY, ORKNEY (HY 6739 7242)

J.D. Hansom

Introduction

The south coast of the island of Sanday, northeast of the Orkney Mainland (see Figure 8.2 for general location), contains a unique assemblage of coastal depositional features, including tombolos, spits, sandflats, dunes and machair, most of which are relatively undisturbed by human activity (Figure 8.27). The most spectacular component of the assemblage of coastal landforms is the 2 km-long ayre, a gravel-cored sandy tombolo that connects the island of Tres Ness to the shore and encloses a large area of intertidal sandflats (Cata Sand), backed by the Plain of Fidge, a broad machair plain. Farther west, a second tombolo, Quoy Ayre, links the island of Els Ness to the mainland. While individually these features are of great geomorphological interest, collectively the complex and dynamic inter-relationships between the landforms of Central Sanday are unique in Scotland and are of national importance (Nature Conservancy Council, 1978). Although, this importance has been recognized for some time and the research potential of the site repeatedly emphasized (Steers, 1973; Mather et al., 1974; Nature Conservancy Council, 1978; Keast, 1994), it has failed to attract any detailed geomorphological study and interpretations of the complex evolution of this magnificent site remain speculative. However, research is now underway to establish the Late-Holocene shoreline response of the site in relation to changes in sea level and sediment supply (Rennie and Hansom, 2001).

Description

The extensive GCR site of Central Sanday, covering an area of c. 660 ha, consists of a complex series of depositional features. Two former islands (Tres Ness and Els Ness) are connected to the main island by dune-capped sand tombolos that partially enclose two embayments containing wide tidal sandflats (Little Sea and Cata Sand) (Figures 8.27 and 8.28). Short gravel-spits extend across the mouths of the embayments. Extensive areas of machair have formed landwards of the beaches.

Sand spits and tombolos

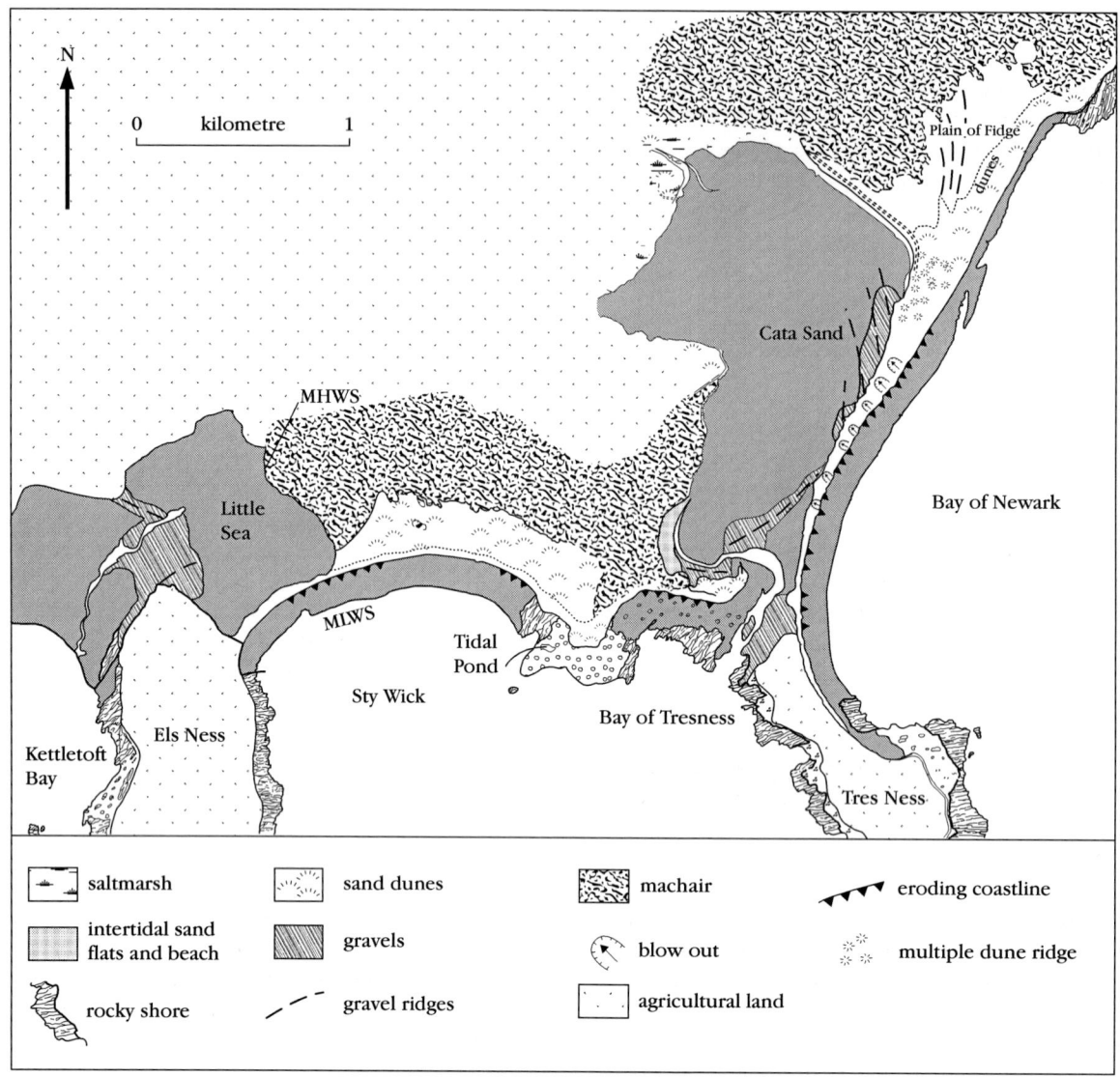

Figure 8.27 Geomorphological map of central Sanday showing the two tombolos that enclose Cata Sand and Little Sea. Note the orientation of the gravel ridges in Cata Sand. MHWS: Mean High-Water Springs; MLWS: Mean Low-Water Springs. (After Rennie and Hansom, 2001.)

The eastern part of this extensive site, the Bay of Newark is the largest and most complex beach unit in Orkney (Mather *et al.*, 1974; Nature Conservancy Council, 1978). The physiography is complex, consisting of sandflats, a dune-capped tombolo and the remnants of gravel ridges that underlie the site marking several stages in coastal evolution. The eastern end of the bay close to the Plain of Fidge consists of a complex of geomorphologically inactive sand dunes and parabolic blowthroughs. These steep, 7–10 m-high, longitudinal sand dune ridges trend almost at right-angles to the present-day coastline and although a few moribund stands of marram *Ammophila arenaria* survive, the Plain of Fidge contains the largest area of machair outwith the Western Isles (Mather *et al.*, 1974; Nature Conservancy Council, 1978). Two separate levels separated by an erosional scarp occur, the lower of these representing a deflation surface close to the water table. In places the scarp is undergoing erosion with a distinctive series of finger-like blowthroughs at the scarp edge. In the western part of the bay a 2 km-long dune-capped tombolo connects the island of Tres Ness to mainland Sanday, enclosing the tidal sandflat of Cata Sand (Figure 8.28). In plan, the tombolo is long, straight and narrows towards its southern end to only 30 m wide. The present-day beach consists almost exclusively of shell-rich, medium-grained sand (D_{50} = 0.29 mm) although gravel occurs at the extreme southern end of the tombolo where it hinges onto a low sandstone platform. The tombolo is capped by a single linear dune ridge, rising to *c*. 13 m in height, composed mainly of fixed dunes with local areas of mobile dunes (Keast, 1994). The most dynamic section of the dune ridge is at the narrow southern end of the tombolo, where unfixed dunes have been dissected by several blowthroughs, the largest of which is up to 40 m in wide. Mather *et al.* (1974) report no gravel at the base of these blowthroughs, however Keast (1994) found gravel at the base of the dunes, and substantial amounts were recorded during bi-annual field visits made by the author between 2000 and 2002. Unconfirmed reports suggest that the tombolo was breached at its southern end during storm conditions in the 1980s.

A system of gravel ridges underlie the machair of the Plain of Fidge at the mainland root of the sand tombolo and are also visible on Cata Sand where north-westerly relict gravel ridges diverge from the northward-trending dune-capped tombolo (Figure 8.28). The low gravel ridges form broad arcs, trending north and north-west from the outlet of Cata Sand. The linear dunes capping the tombolo rest on these gravel ridges, many of which are exposed at low tide (Nature Conservancy Council, 1978; Keast, 1994). The differences, both in composition and orientation, between these relict gravel features and contemporary sand landforms suggest a very different depositional environment in the past.

Farther west, a second dune-capped tombolo links the former island of Els Ness to the mainland of Sanday, enclosing the wide tidal sandflat of Little Sea. The tombolo (Quoy Ayre) forms the western part of the wide south-facing embayment of Sty Wick and is symmetrical in plan (Figure 8.27). The linear dune ridge capping the tombolo reaches a maximum height of 9 m towards the centre of the tombolo and consists almost entirely of highly stable and well-vegetated fixed dunes. Gravel is well-exposed at Quoy Ayre and appears to underlie the dune ridge.

The two sandflats (Little Sea and Cata Sand) are completely closed on their south and south-eastern sides by the dune-capped tombolos. Their outlets are towards the south-west, both of which are partly enclosed by spits. Short gravel spits project outwards from each flank of the outlet of Little Sea, but appear to be relatively inactive features in spite of lacking a cap of blown sand. Cata Sand is partly enclosed by a low, dune-capped, rounded spit, which extends 0.5 km eastwards across the outlet. This spit is underlain by low belts of gravel that are a south-west continuation of the relict ridge system visible in Cata Sand. Much of this short stubby spit is capped by stable sand dunes grading landwards into machair, although the tip of the spit supports low embryo dunes that are still developing. Comparison with aerial photographs and field evidence suggest that the spit tip is highly dynamic, alternating between short periods of erosion and accretion (Mather *et al.*, 1974; Keast, 1994). A smaller gravel spit projects northwards from Tres Ness on the other side of the outlet.

Interpretation

In spite of the wealth of the landform assemblage at Central Sanday, until recently there has been limited detailed geomorphological research carried out (Rennie and Hansom, 2001). Although the interpretation of the Central Sanday site is necessarily speculative, the inter-relationships of the landform assemblage within this dynamic system are of national importance (Steers, 1973; Mather *et al.*, 1974; Nature Conservancy Council, 1978; Keast, 1994), particularly as there has been almost no anthropogenic modification to the natural system. The two sand-capped tombolos are spectacular landforms yet the underlying gravel ridge orientations are at odds with the present-day coastal trend. The site is a key area for the study of constructive shoreline processes in an area of relative subsidence and so has great research potential.

Throughout much of the Holocene Epoch, the coastline of Orkney has undergone approximately similar amounts of submergence to Shetland (Lambeck, 1993) (see Figure 6.28). As a result, emerged shoreline features are absent in Orkney and are replaced by features of submergence so that the low gradient coast of Sanday has undergone significant alteration in planform and as bays became flooded, peninsulas became islands and beaches changed orien-

Sand spits and tombolos

Figure 8.28 Looking north-east along the dune-capped tombolo in the Bay of Newark. Older intertidal gravel ridges can be seen extending inland towards the north in Cata Sand. (Photo: J.D. Hansom.)

tation in response. On Sanday, the pattern of sea-level rise flooding embayments and isolated islands has been in part reversed by a healthy sediment supply that has connected or reconnected islands to the mainland. Historical map evidence exists in support of these changes. On John Thomson's map of 1832 (Figure 8.29), the low-lying former islands of Tres Ness and Els Ness that lie to the south of the mainland of Sanday, together with the offshore island of Start Point to the north, are depicted as long narrow peninsulas. The nature of these peninsulas is not known, but the form of Els Ness and Tres Ness suggests that they were, at least in part, complexes of gravel ridges, substantially wider than those at present, enclosing low-lying or flooded areas behind. The same map shows the area of Little Sea as a freshwater loch and the area that is now Cata Sand as a low area, possibly of seasonally-flooded machair or 'winter loch'. However, a map of 1847 shows both the loch at Little Sea and the low land at Cata Sand to be arms of the sea as they are today and, if both maps are accurate, marine inundation may have occurred between 1822 and 1847 (Black, 1847).

Central Sanday is a good example of a feature common to the sandy and dune-backed beaches of Scotland where a backbone of gravel provides the base on which wave-deposited or blown sand later accumulates (Mather *et al.*, 1974). The gravels then play an important shaping role in the evolution of what are now mainly sandy beach complexes. For example, the gravel ridges that originally began the process of enclosure of the Cata Sand basin are visible under the machair of the Plain of Fidge (Mather *et al.*, 1974), but their mainly north-west orientations differ from the contemporary north-trending dune-capped tombolo of the present-day coast. This suggests that at some time in the past the embayment was partly enclosed by gravel ridges deposited at different orientations to the contemporary constructional sand features. In addition, at some period prior to the tying of Tres Ness to the mainland of Sanday, gravel was a relatively more important beach material than at present, highlighting a change in sediment supply over time, and possible relationships to an altered offshore sediment supply.

The orientation of the gravel ridges led Mather *et al.*, 1974 to suggest that they were hinged on a point under the Plain of Fidge and so had extended southwards as a result of long-

Central Sanday

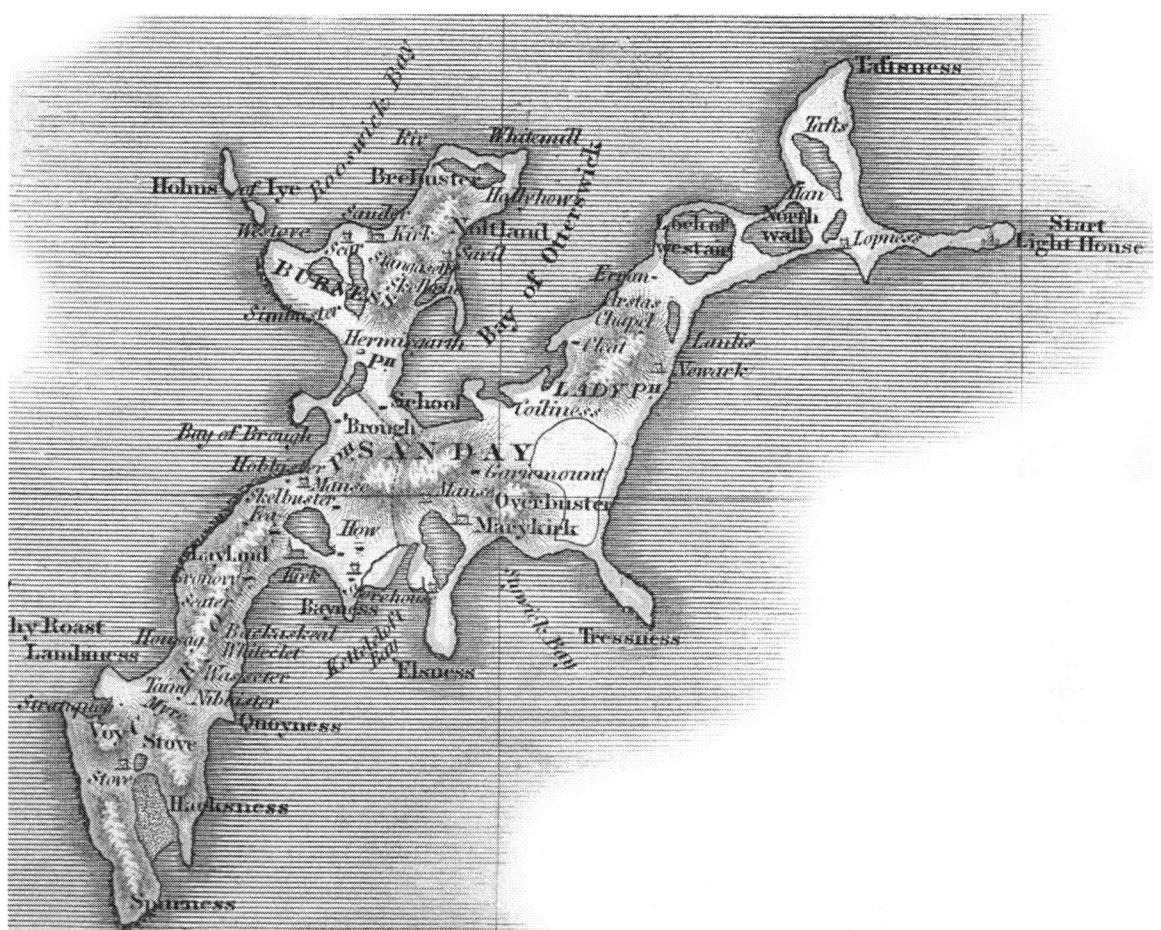

Figure 8.29 Coastline of Sanday in 1822 (from Thomson, 1832). Note the modern marine inlet at Cata Sand is mapped as a low area of land, possibly machair; the modern marine inlet of Little Sea is mapped as a freshwater lake and Start Island is mapped as a promontory with a lighthouse at the end.

shore drift from the north. Recent detailed mapping and Ground Penetrating Radar survey shows that the ridges recurve and splay northwards, suggesting drift from the south. However, not all the ridges recurve to the north, particularly those close to the outlet of Cata Sand (Rennie and Hansom, 2001). Although these relationships are not yet fully established it seems likely that gravel spit extension from the south resulted in partial closure of a wide and open bay at Cata Sand. Such spit extension requires a plentiful supply of coarse sediment, and the sequential drowning of areas of low-lying till-covered bedrock at Tres Ness, along with sediment driven onshore from glacial gravel banks offshore, may well be the source of much of the spit gravel. As sea level rose to its present-day level over the Holocene Epoch, the gravel was driven onshore from its source areas on the shallow Sanday shelf by storm conditions. Such a scenario requires an ongoing supply of gravel to allow the moving spits and barriers to keep pace with sea-level rise and extend along the coast. If the supply was insufficient then erosion of the updrift gravels would have fuelled distal extension (Hansom, 1999). Such reductions in gravel supply occur elsewhere in Scotland and co-incide with the increasing importance of sand as a beach material about 6500 years BP (Carter, 1988).

The variety of dune types, morphologies and processes at Central Sanday are unusual and are of interest in their own right, forming the most complex and complete beach–dune–machair system outside of the Western Isles (Nature Conservancy Council, 1978). The relationship of the dunes and machair with the tombolos, spits and relict gravel ridges on which they rest, adds to their scientific interest. Of particular interest are the north-west-trending longitudinal dune ridges near the south end of the Plain of Fidge. The orientations of this suite of dune ridges, which trend at a high angle to the present-day coastal edge, suggest that they are related to a period prior to the tying of Tres Ness and the mainland of Sanday. No dates exist for the onset of sand deposition in the Plain of Fidge or Sty Wick but, since the sand has a high shell content, it is likely to be sourced from offshore and thus may date to the period after 6500 years BP when coastal sand became a more important sediment source for beaches.

Conclusions

Central Sanday contains a wealth of undisturbed coastal landforms on a scale unique in Britain. The diversity and complexity of the assemblage of tombolos, spits, sandflats, dunes and machair is unsurpassed in the UK. Gravel underlies important elements of the now sand-dominated geomorphology, indicating a complex evolutionary history of the site that has yet to be fully investigated. The most spectacular components of the assemblage of coastal landforms are the two dune-capped tombolos, which connect the former islands of Tres Ness and Els Ness to the mainland. The tombolo connecting Tres Ness is over 2 km long, enclosing the tidal sandflat of Cata Sand, while the shorter, but no less spectacular, tombolo, Quoy Ayre, encloses the embayment of Little Sea. Dunes are well-developed on both tombolos, grading into machair on their landward sides. The variety and diversity of the dune and machair of Central Sanday, the largest area of machair outwith the Western Isles, is of great geomorphological interest in its own right. Collectively, the complex and dynamic inter-relationships between both the wind-blown and wave-constructed landforms of Central Sanday are of national importance (Nature Conservancy Council, 1978). Central Sanday is a rare example of where a healthy sediment supply has led to island tying and tombolo building even though relative sea level is rising. It provides an excellent comparison with the Isles of Scilly (see GCR site report in the present volume) where sediment supply reduction may now preclude island tying as sea level rises. There is great research potential at this site and it is a key area for the study of coastal evolution and development in an area of relative sea-level rise.

Chapter 9

Machair

J.D. Hansom

Introduction

INTRODUCTION

The machair lands of the north-western seaboard of both Scotland and Ireland represent a distinctive form of dune grassland system, unique to these areas. The machair system can be described as a flat or gently sloping, coastal dune-plain formed by windblown calcareous shell-sand, sometimes incorporating a mosaic of dunes to the seaward side and a species-rich grassland (managed by traditional low-intensity agriculture), wetland, loch and 'blackland' (mixtures of sand and peat) to the landward side. Often the dune cordon may be missing owing to frontal erosion, but a characteristic of machair surfaces is that they are lime-rich, subject to strong, moist, oceanic winds and show detectable current or historic biotic interference from grazing, cultivation, addition of natural fertiliser such as seaweed and, sometimes, artificial drainage (Ritchie, 1976; Hansom and Angus, 2001). As a result, the term 'machair' has meaning not only in a botanical and geomorphological sense but also in its strong cultural overtones, the spatial extent of machair overlapping closely with the current core areas of the Gaelic language in Ireland and, with the exception of Orkney and Shetland, in Scotland (Hansom and

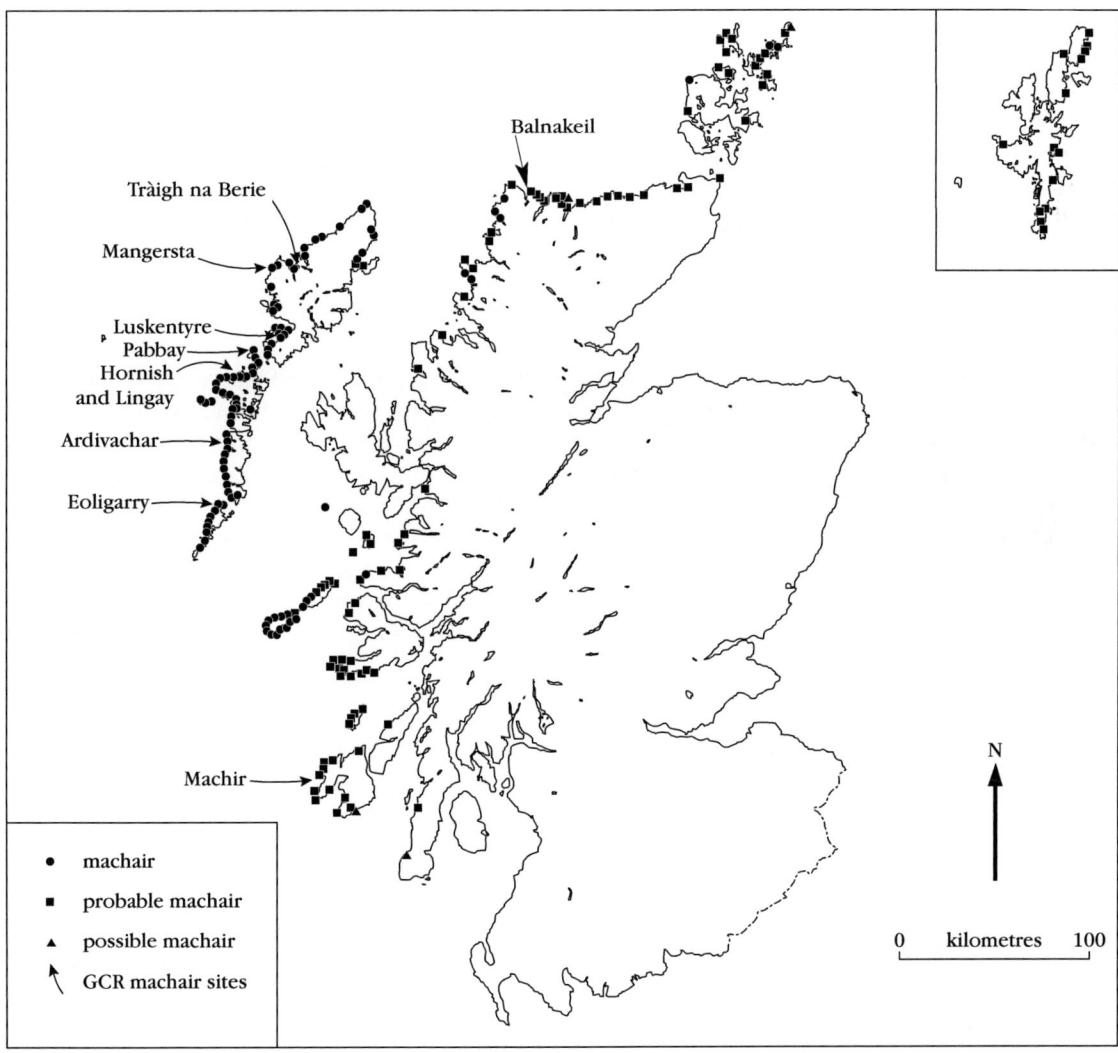

Figure 9.1 Distribution of machair in Scotland. Other than Sandwood, Torrisdale and Balta (see Chapter 7), all the sites included in the GCR fulfil both the geomorphological and vegetational definition of machair. Small vegetational differences in the above sites have resulted in the label 'probable machair'. Ongoing work that interprets the geomorphology and botany of machair aims to provide a definitive machair diagnostic test in the future and so the above classification will be subject to slight modification (Angus, 2003, pers. comm.). (After Hansom and Angus, 2001.)

Angus, 2001) (Figure 9.1). Since all of the machair lands have similarities in their land-use histories, it is likely that the present distribution and nature of machair systems owes as much to cultural factors as it does to biotic and abiotic influences.

The world extent of machair is about 30 000–40 000 ha of which 67% is found in Scotland and 33% in Ireland, although those figures are under review (Angus, 1994). Of the Scottish machair, nearly 40% has been selected for – and is protected internationally as – Special Areas of Conservation (SACs); about 80% of the Scottish resource is protected within Sites of Special Scientific Interest (SSSIs).

The following definition of machair is largely based on Ritchie (1976) and Angus (2001, 2002):

- a base of blown sand with a significant percentage of shell-derived materials lime-rich soils of pH normally greater than 7.0;
- a level or low-angled and smooth surface at a mature stage of geomorphological evolution;
- a sandy grassland-type vegetation devoid of long dune grasses and other key dune species;
- a detectable current or historic biotic interference resulting from grazing, cultivation, trampling and, sometimes, artificial drainage;
- an oceanic location with a moist, cool and windy climatic regime.

Recent work by Angus (2002) has stressed the importance of an integrated definition of machair that includes geomorphology and botany, but neither in isolation. As a result, several natural systems that in the past were identified as machair on, say, geomorphological grounds, do not fulfil the wider criteria. Therefore, such sites are deemed here not to be machair and have been included in Chapter 7 of the present volume. These sites are Sandwood and Torrisdale, both in Sutherland, and Balta in Shetland.

Hansom and Angus (2001) summarize the main features of the evolution of machair using data from a wide range of sources including Ritchie (1976), Mate (1991) and Gilbertson *et al.* (1999). Machair is in essence a sand plain produced from the normal cycle of deposition and erosion of sand dunes where a positive sand budget on the fronting beach results in coastal accretion, seaward movement of the coastal edge and landward wind-blow of surplus sand to be fixed by dune grasses. However, this developmental sequence conceals an erosional sub-cycle as the vegetation cover of the higher and older landward dunes become subject to processes (such as grazing) that may disrupt the vegetation cover and allow subsurface sand to be blown landwards (Figure 9.2). The resulting dunes lose their sand cores by deflation, the sand being removed landwards over an amorphous sand plain or machair and up cliff faces as climbing dunes. Since strong winds continue to modify these surfaces even at distance from the shore, they continue to suffer deflation, a process that halts only when the water table or substrate is reached. As a result, machair surfaces are often characterized by

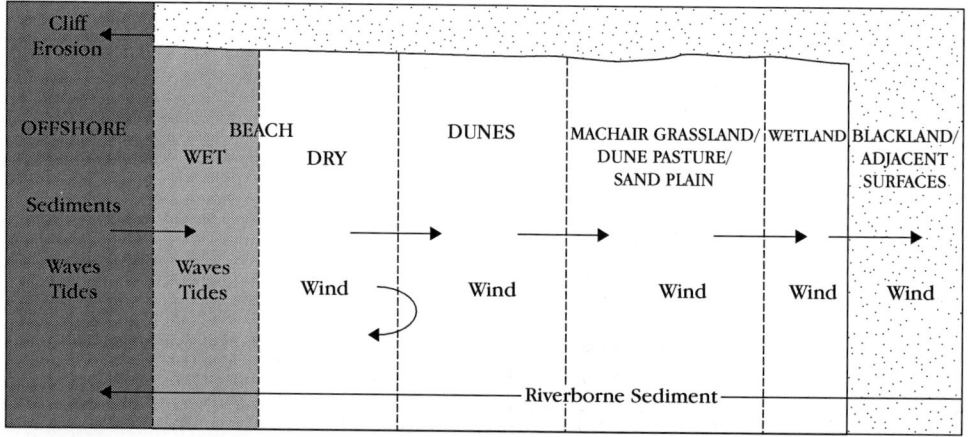

Figure 9.2 A diagrammatic representation of the beach–dune–machair system showing the general landward transport of sand broken by seaward returns via wind and streams. (After Mather and Ritchie, 1977.)

Introduction

steep eroded escarpments located between low-lying landward-dipping deflation surfaces and higher dune surfaces.

The conditions that favoured a positive sand budget were widespread on the Scottish coast as the Holocene rise in sea level began to slow markedly before about 6500 years BP (see Figure 6.28). For the first time waves were able to accomplish substantial shoreface modification and bring large amounts of sediment from the nearshore shelf onto beaches and then into sand dunes (Firth *et al.*, 1995) (Figure 9.3a). However, the finite nature of this sediment seems to have resulted in a progressive reduction in sediment availability and a switch to deficit sometime after 6500 years BP (Carter, 1988, 1992; Hansom, 1999) and the replacement of an accreting system by an erosional system (Figure 9.3b,c). It is likely that such a degradational cycle would be most advanced in those areas subject to high-energy wave conditions and isostatic submergence, conditions that are both met on the north-western seaboard of Scotland. As a result, erosion of the dune cordon in these western machair areas is now well advanced and in places the machair grassland itself is now suffering erosion. The positive beach sediment budgets that fed the embryo and foredunes once sited seawards of the fixed dunes and machair have long since been reversed by negative sediment budgets and the frontal dunes cannibalized (Figure 9.3c) (Hansom and Angus, 2001).

Support for the above geomorphological evidence of an initial surplus followed by declining sand supply comes from dating the sand layers in association with archaeological sites located originally within accreting dunes and now found on coasts undergoing erosion. Optically stimulated luminescence (OSL) dating of aeolian sands by Gilbertson *et al.* (1999) indicates that the carbonate sand of the Benbecula and North Uist machairs began to arrive from offshore about 8700 radiocarbon years BP and in Barra

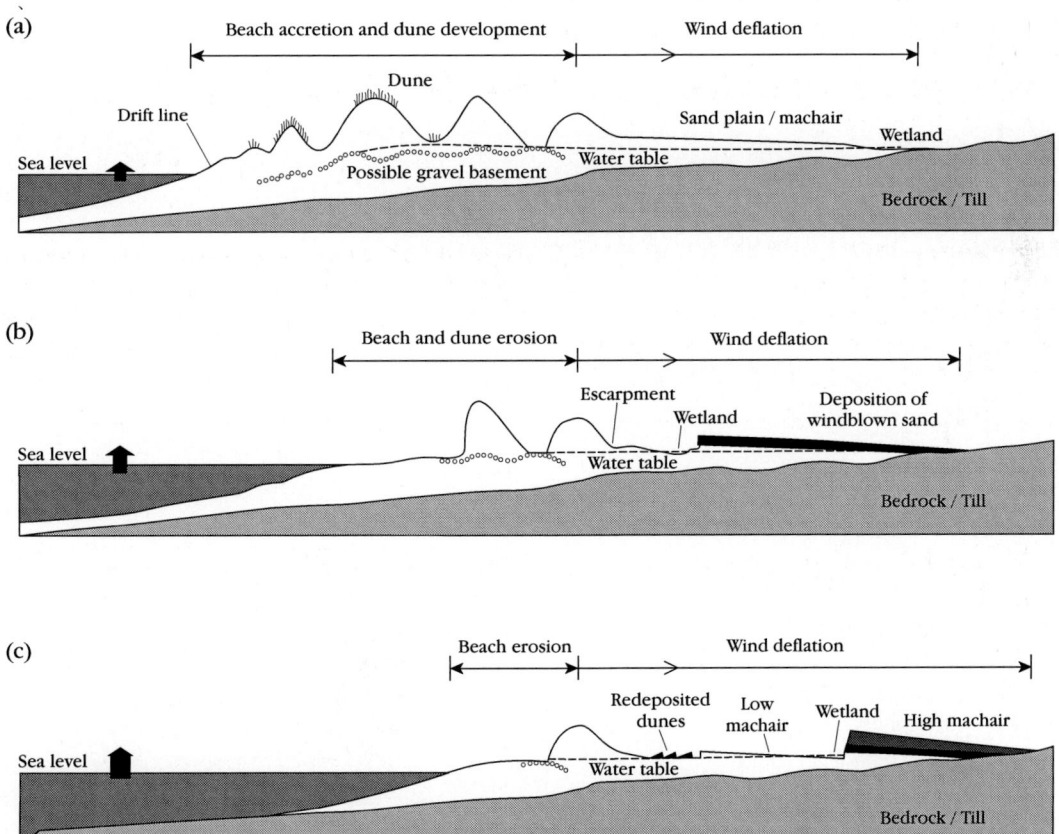

Figure 9.3 The Holocene development of machair from approximately 6500 thousand years ago to present, showing the switch from conditions of accretion of the dunes to erosion and recycling of dune sands into machair. (a) early–mid Holocene; (b) late Holocene; (c) present day. (After Hansom and Angus, 2001.)

about 6800 radiocarbon years BP. At Northton in South Harris, the onset of sand deposition that buried Neolithic remains occurred from 4500 radiocarbon years BP (Ritchie, 1979a). The discovery of exhumed archaeological sites within dunes is largely due to frontal erosion, although multiple cycles of erosion, deposition and deflation are concealed within this general erosional trend. However, in spite of the above, dating of the initiation of dune and machair formation in the Outer Hebrides remains problematic. Ritchie and Whittington (1994) report intertidal peats overlain by aeolian sands that date from 7800 radiocarbon years BP at Cladach Mór in North Uist and from 7700 radiocarbon years BP at the Landing Jetty in Pabbay. Yet other sites on these islands, for example at Quinish in Pabbay dating from 4300 radiocarbon years BP and at Borve in Benbecula dating from 5600 radiocarbon years BP, suggest that the arrival of aeolian sand and the initiation of machair development in the Hebrides was non-synchronous (Ritchie and Whittington, 1994). However, in the low and undulating coastal landscape of the Western and Northern Isles, rock basins close to sea level are likely to be affected by rising sea levels at different times and so the influence of local bathymetry and site factors represent important site-specific controls on the date of machair initiation (Figure 9.4).

Once established, the development of the machair plain is essentially erosion-driven, with new surfaces produced as old ones are consumed. However, there can be marked seasonal differences. For example, where deflation has exposed the water table, winter flooding may result in sand blowing onto wet surfaces and this results in a depositional flat surface rather than an erosional one. Archaeological studies provide evidence of fertile and stabilized sand surfaces around 2000 radiocarbon years BP (Ritchie, 1966). Several sites ascribed to the Iron Age and later are located on the low flat surfaces of machair that have been produced following deflation of earlier machair surfaces (Ritchie, 1979a). Gilbertson et al. (1996) document layers of thick organic palaeosols within the dunes and machair dating from Bronze Age to medieval times, together with periods of instability (particularly between the 9th and 13th centuries AD) as indicated from Viking settlements now buried below aeolian sand deposits (MacLaren, 1974).

Figure 9.4 A typical machair landscape of partly-drowned rock basins connected on the seaward side by wide sandy beaches and on the landward side by dune cordons backed by expanses of windblown machair sand. Looking north-east from North Uist over Vallay Strand in the foreground to Hornish and Lingay in the distance. (Photo: P. and A. Macdonald/SNH.)

Introduction

Historical evidence extends the above pattern of phased instability and stability of the machair into modern times. During the 16th century machair surfaces were stable with well-established agriculture, but the 17th century brought widespread sand-blow on much of the Scottish coast and burial of machair surfaces and buildings in the Outer Hebrides (Ritchie, 1966, 1979a; Lamb, 1991; Angus, 1997). Although probably more stable than it has been in the past, Hebridean machair is still actively forming and the present-day machair surface has probably formed over the same timescale as it has in the past, that is over periods of less than 100 years (Gilbertson *et al.*, 1999). Nevertheless, the present machair system as a whole represents the latest manifestation of a continuum of essentially similar processes operating since at least middle Holocene times.

The conservation value of machair

The geomorphological significance, and hence the Earth science conservation value, of machair arises from its importance to our understanding of:

1. the processes of machair erosion and accumulation;
2. the interaction of sediment supply and sea-level change;
3. the interaction of sediment, vegetation and land use.

As described above, it is believed that machair grassland has been modified by humans throughout its development. Traditionally, machair supports extensive grazing regimes and unique forms of cultivation that rely on cattle-grazing and low-intensity systems of rotational cropping. This traditional agriculture sustains a rich and varied dune and arable weed flora. Some of the arable weed species are now largely restricted in the UK to these traditionally managed areas. The habitat type also supports large breeding bird populations and is particularly important for waders and corncrake *Crex crex*.

The GCR site selection rationale for machair has been to represent the range and diversity of the geomorphological features (Table 9.1). In the present chapter, sites are arranged in a clockwise order around the coast, starting with the southernmost.

Machairs as biological SSSIs and Special Areas of Conservation (SACs)

In Chapter 1, it was emphasised that the SSSI site series is constructed both from areas nation-

Table 9.1 Machair GCR sites

Machair site	Main features	Other features	Tidal range (m)
Machir bay	Beach–dune–machair, high-level machair terraces, emerged beaches	Climbing dunes	3.0
Eoligarry	Vigorous erosional machair forms large blowouts, tombolo structure	Storm beach, wide intertidal, sheltered beach, archaeological dating	4.0
Ardivachar–Stoneybridge	Machair type site, high and low machair deflation corridors	Archaeological dating gravel barrier, palaeosols	3.6
Hornish and Lingay Strands	Flat, low-lying machair, water-table effects	Superimposed small dunes, artificial drainage	3.9
Pabbay	Climbing machair, conical dunes, wet machair	No rabbits	3.0
Luskentyre–Seilebost	Large beach–dune machair remnant of former larger system, 35m high dunes; growth/decay model site	Spits, blowouts	3.8
Mangersta	Eroded and deflated formerly extensive machair, advanced stage of erosion	Water table	3.8
Tràigh na Berie	Large dynamic beach–dune–machair dune cordon intact and well-nourished	Infill of valleys and lochs, no chronic erosion	3.8
Balnakeil	Dynamic climbing machair and dune blowouts, headland by-passing of sediment	Erosion of frontal edge, sand-fall over cliff	4.0

Machair

Table 9.2 Candidate Special Areas of Conservation supporting Habitats Directive Annex I habitat 'Machair' as a qualifying European feature. (Source: JNCC International Designations Database, July 2002.)

SAC name	Local authority	Machair extent (ha)
Coll Machair	Argyll and Bute	681
Monach Islands	Western Isles / Na h-Eileanan an Iar	292
North Uist Machair	Western Isles / Na h-Eileanan an Iar	1707
Sheigra–Oldshoremore	Highland	222
South Uist Machair	Western Isles / Na h-Eileanan an Iar	1785
Tiree Machair	Argyll and Bute	510

Bold type indicates a coastal GCR interest within the site

ally important for wildlife and GCR sites. An SSSI may be established solely for its geology/geomorphology, or its wildlife/habitat, or it may comprise a 'mosaic' of wildlife and GCR sites that may be adjacent, partially overlap, or be co-incident. Therefore, there are some areas of machair that are crucially important to the natural heritage of Britain that have been designated as SSSIs primarily for their wildlife conservation value, but implicitly will contain interesting coastal geomorphology features that are not included independently in the GCR because of the 'minimum number' criterion of the GCR rationale (see Chapter 1). These sites are are not described in the present geomorphologically focused volume.

In addition to being protected through the SSSI system for its national importance, machair is a 'Habitats Directive' Annex 1 habitat, eligible for selection as Special Areas of Conservation, (see Chapter 1). Furthermore, many machairs are of international ornithological importance, primarily for breeding waders, and for this reason may be designated Special Protection Areas under the 'Birds Directive'.

Because machair is a habitat unique to the north and west of Scotland and western Ireland, the UK has a special responsibility for machair, and has recently established a UK Machair Habitat Action Plan (Angus and Dargie, 2002)

Machair SAC site selection rationale

Site selection has taken account of the wide range of variation in physical type shown by Scottish machairs and has also been influenced by the UK's special responsibility for machair conservation. The largest sites have been selected, as these demonstrate the best structure and function and include the most diverse examples of transitions to other habitats. Sites have been selected from across the range of machair in the Outer and Inner Hebrides and on the Scottish mainland.

Table 9.2 lists machair SACs, and indicates which of these sites are also important (at least in part) as part of the GCR and are described in the present chapter.

MACHIR BAY, ISLAY, ARGYLL AND BUTE (NR 210 630)

Introduction

Machir Bay is a highly dynamic beach–dune–machair assemblage located on the exposed Atlantic coast of Islay (see Figure 9.1 for general location and Figure 9.5). The wide, high-energy beach is backed by a complex sequence of dune forms including low embryo dunes, an active foredune ridge, multi-ridged mature dunes, re-depositional sandhills and an extensive machair surface. The machair plain is of exceptional geomorphological interest as it drapes a number of topographical features including a series of high-level marine terraces, glacial deposits, talus slopes, and rock plateaux. Many streams drain through the dune and machair providing a strong hydrological control on morphology. Although several descriptions exist of the beach–dune–machair morphology of Machir Bay (Ritchie and Crofts, 1974; MacTaggart, 1996), greater interest has been shown in the emerged beaches, glacial terraces and relict clifflines that the machair partially obscures (Dawson, 1983; Dawson *et al.*, 1997).

Figure 9.5 Geomorphology of Machir Bay, Islay, showing a mix of machair types including substantial terraces at the rear of the system covered by high machair. (After Ritchie and Crofts, 1974.)

Machir Bay

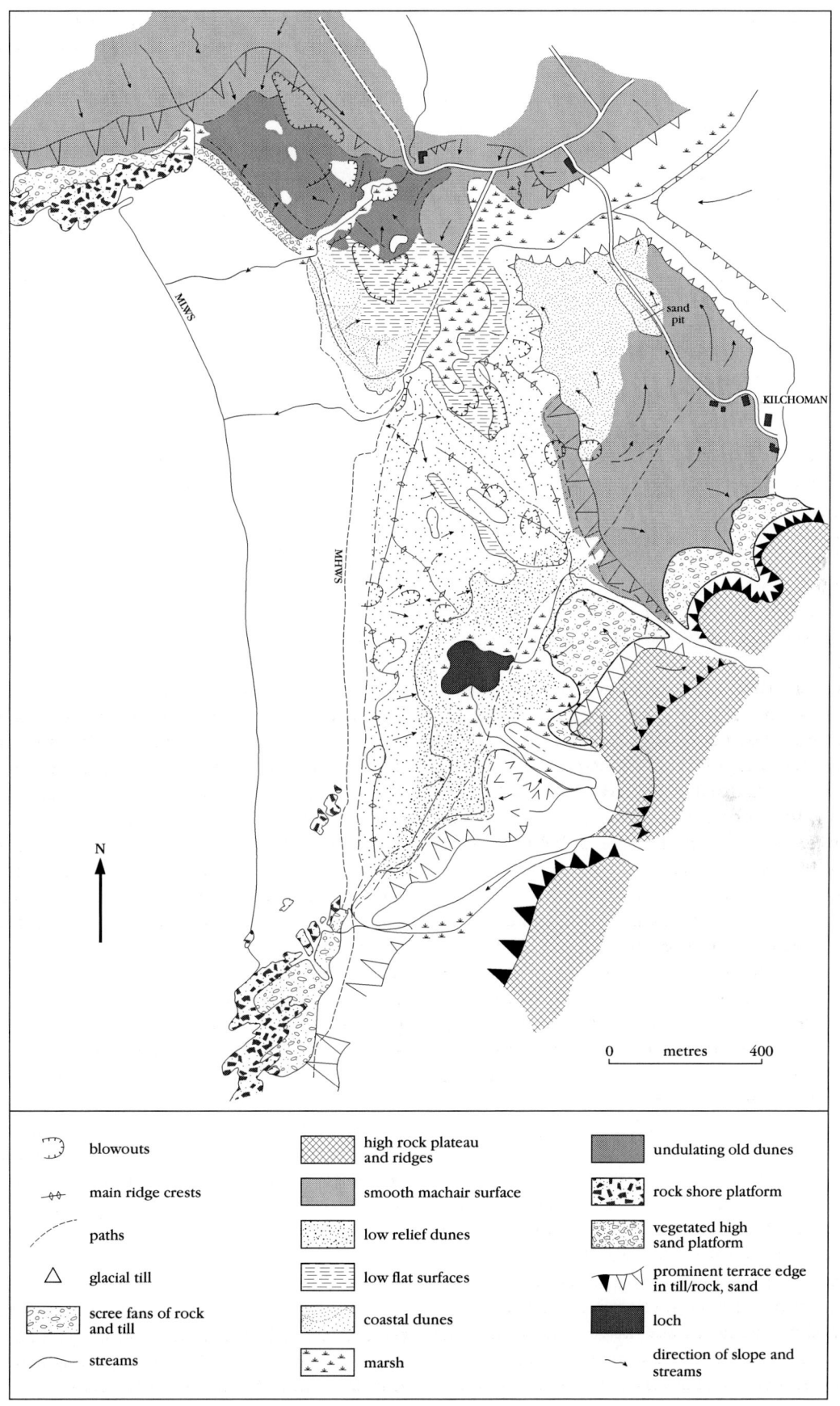

Machair

Description

The 2.1 km-long beach at Machir Bay on the exposed Atlantic coast of western Islay has an open south-west fetch and lies within a SW–NE-trending structural basin of Torridonian sandstone that represents an extension of the Loch Gorm depression. There is also widespread evidence of glaciation and the deposition of substantial quantities of glaciogenic materials both onshore and offshore. For example, till, moraine and various glaciofluvial and glaciomarine deposits are common on Islay (Dawson *et al.*, 1997). At Machir Bay, a relict cliff cut in Torridonian sandstone forms the southern margin of the bay, whereas the northern part of the bay is generally lower and merges into the flat plateau of the interior. A relict cliff cut in terraced glaciomarine gravels (Dawson *et al.*, 1997) at 23 m OD extends towards the north-west end of the Bay and lowers westwards with an average gradient of 9.8 m km^{-1} (Dawson, 1983). Although a conspicuous feature, it is partially obscured by a veneer of windblown sand that comprises part of the machair. The terrace is succeeded farther south by a smaller terrace fragment, also interpreted as outwash, which declines in altitude and terminates at an emerged ('raised') shoreline at 21.4 m OD (Dawson *et al.*, 1988). As a result of the occurrence of pre-existing terrace and cliff topography, the dune and machair landform assemblage of Machir Bay is asymmetrical in form being best developed in the south and east of the site.

The wide intertidal beach has a low gradient of 1–2° and is composed of medium-grade sand (0.23 mm mean diameter) with a calcium carbonate content of 34%. The beach exhibits considerable variation in profile and plan (Ritchie and Crofts, 1974) but the upper 20 m is rarely covered by seawater. The 3 m tidal range results in a 0.32 km-wide expanse of bare sand, broken only by the rocky intertidal outcrop of Carrig Chomain in the south. Extensive areas of aeolian sand ripples were noted on the beach face in July 1996 providing evidence of a sand supply to the dune system behind (MacTaggart, 1996). Two streams, the Allt Gleann na Ceardaich and the Allt na Criche, cross the beach in the centre and north of the bay. At the rear of the beach, the foredune ridge shows signs of periodic undercutting by storm waves at high tide, although subsequent deposition of embryo dunes partly obscures the erosional faces.

The foredune ridge landward of the beach is best developed in the south, where it reaches 15 m high and displays steep seaward slopes of around 20°. The dune face and crest are extensively covered in vigorous marram *Ammophila arenaria* growth, and this broad coastal dune continues, curving slightly inland, to the stream outlet in the centre of the bay. Ritchie and Crofts (1974) report that the frontal dunes are characteristically devoid of breaches or erosional hollows. However, a more recent report (MacTaggart, 1996) identifies a large, but relatively shallow, blowthrough in the foredune ridge close to the centre of the bay and several healed blowthrough forms farther south, indicating that foredune erosion may now be a more significant process than before. There is extensive evidence of local wave-erosion of the dune face, particularly in the south of the site (MacTaggart, 1996), and during the winter storms of 1989, the dune face receded by an estimated 5–10 m (MacTaggart, 1996). Fresh sand accumulations and low embryo dunes masked the lower slopes of the foredunes in July 1996 (MacTaggart, 1996).

Between the stream outlets the coastal edge consists of a 3 m-high dune ridge developed seawards of an older dune ridge (Ritchie and Crofts, 1974). Marram *Ammophila arenaria* colonization is patchy, and localized areas of erosion exist north of the outlet of the Allt Gleann na Ceardaich. No dunes occur north of the marshy outlet of the northern stream but instead a low altitude sand platform with a maximum altitude of *c.* 2 m slopes gently seawards onto the beach face and is vegetated for some distance down the beach face.

Older dunes and re-depositional sandhills are present landwards of the active dune ridge and again are best developed towards the south and south-east of the bay. A linear dune ridge seaward of the active foredune ridge is well defined in the centre of the bay (MacTaggart, 1996), while farther inland several other relict dune ridges trend north-west to south-east (Ritchie and Crofts, 1974). Smaller dune forms and ridges, blowthroughs, erosional scars and re-depositional forms add topographical diversity to this dune complex, which has an average relief amplitude of 5–10 m, although altitudes exceed 20 m OD in places (Ritchie and Crofts, 1974). At the base of an active blowthrough in this area, and in other blowthroughs within the dunes, MacTaggart (1996) identified outcrops of indurated aeolian calcarenite, blown sand cemented by the precipitation of calcium carbonate from subsurface water. In the south a distinctive amphitheatre-like depression supports a variety of machair and relict dunes on its sides and is floored by a semi-permanent, marshy loch (Figure 9.5).

The extensive machair at Machir Bay can be classified according to topographical situation. In the south, a series of fan-like deposits cover the face of the relict rock cliff as well as the screes beneath. Sand deposits banked against the relict cliff have been eroded to form a terrace feature. An extensive machair plain some 50 m above sea level has developed on the plateau surface of the glaciomarine gravel terrace cut into till and extending as far inland as the old church of Kilchoman. This plateau machair is predominantly stable, although characterized by several areas of bare sand (Ritchie and Crofts, 1974; MacTaggart, 1996). Higher areas of machair at up to 60 m OD have developed in ledges or depressions in the rocks to the south of the site. In the northern part of the site the machair surfaces are generally more subdued and high relief forms are rare. Numerous streams drain into Machir Bay, crossing both dunes and the cliffs in the south, resulting in an elevated and fluctuating water table that adds to the geomorphic diversity of the machair and dune forms. The two streams that cut across the beach in the centre and north of bay are responsible for locally high water-tables that form the base level for deflation of the dunes and machair.

Interpretation

Machir Bay contains a great variety of dune and machair forms, probably the result of the strong control of structure, subsurface morphology and hydrology and the dominance of winds from the north-west. It is this variety of dune and machair landforms and their relationships to a variety of geographical controls that is of outstanding geomorphological significance. The relatively undisturbed nature of Machir Bay provides a excellent opportunity to study the evolution of a variety of dune forms and the effects of water table and drainage controls on morphology, but as yet no detailed geomorphological research has been undertaken.

The emerged marine terraces and glacial deposits of Machir Bay have attracted greater scientific interest in the context of the Quaternary evolution of Western Isles of Scotland (Dawson, 1983). The coarse gravel terrace that underlies the extensive machair plain has been interpreted as a Lateglacial glaciomarine deposit that graded to a sea level lower than $c.$ 23 m OD during the decay of the last ice sheet (Dawson, 1983). The terrace declines in altitude westwards at an average gradient of 9.8 m km^{-1} towards Machir Bay where it passes beneath accumulations of blown sand. The terrace is succeeded farther south by a smaller Lateglacial terrace fragment, also interpreted as glaciomarine, which declines in altitude and terminates at an emerged shoreline at 21.4 m OD (Dawson, 1983).

In the period following deglaciation, rapid sea-level rise inundated many of the coastal bays on Islay and at Machir the lower parts of the southern rock cliffs and central glaciogenic deposits were re-occupied by the Holocene sea. Glaciogenic sediments available on the seabed were likely driven onshore to accumulate in sandy beaches such as those at Machir Bay. The subsequent process of sand delivery to the dunes and machair at Machir Bay is similar to that outlined for the Outer Hebrides (Ritchie, 1979a), but in Islay sea levels were falling in Late Holocene times. The late Holocene decline in sediment supply noted elsewhere in Scotland was delayed in Islay as a result of ongoing isostatic uplift that allowed waves to access new areas of sea-floor sediment while progressively elevating the rearmost coastal features by $c.$ 8 m over 6500 years. This resulted in the dune and machair landforms of Machir Bay being draped over pre-existing structures and deposits up to 60 m OD. The dominance of strong winds from the north-west also contributed to the apparent asymmetry in the distribution of windblown sand and imparted a strong southerly bias to this distribution. The machair and dune forms also reflect the strong hydrological and water-table controls of the Machir Bay basin, in particular around the depression in the south and close to the streams that cross the beach.

At present the coastal dunes at Machir Bay are undergoing a period of erosion. Frontal dune erosion appears to be caused by storm wave action undercutting the dune face, causing slumping of the vegetation (MacTaggart, 1996). However, it is unclear whether this is the seasonal effect of winter storms. Ritchie and Crofts (1974) suggested that Machir Bay was stable or accreting but MacTaggart (1996) observed that the foredune ridge was almost everywhere undergoing frontal erosion and that this appeared to contribute to blowthrough initiation and dune instability. However in the north, the continued extension of vegetation onto the low area at the back of the beach indicates that, in this sheltered part of the bay, accretion still occurs.

Conclusions

Machir Bay on the exposed Atlantic coast of western Islay is an excellent example of a well-developed and topographically diverse beach–dune–machair system. The site contains a great variety of dune and machair forms, as a result of the control of subsurface morphology and hydrology, and the dominance of winds from the north-west. The machair, which reaches altitudes of *c.* 60 m OD, is of exceptional geomorphological interest as it drapes a number of topographical features including a series of high-level glaciomarine terraces, talus slopes and rock plateaux.

EOLIGARRY, BARRA, WESTERN ISLES (NF 700 060)

Introduction

The sand isthmus of Eoligarry connects the rocky northern part of Barra, in the southern Outer Hebrides to the rocks of Ben Eoligarry Mór (see Figure 9.1 for general location, and Figure 9.6). Eoligarry is a wedge-shaped complex of sand dune and machair flanked to the east by an extensive intertidal beach named Tràigh Mhór, and to the west by Tràigh Eais, which faces the Atlantic Ocean. Tràigh Mhór, a vast, creamy-white beach of shell-sand, is well known for its locally harvested cockles, source of shells for building work (although the factory has now closed) and spectacular intertidal landing-strip for the air link between Barra and the Scottish mainland. The geomorphological interest of this beach lies in its size and gradient together with a well-defined series of intertidal bars and mega-ripples. Tràigh Eais, is a narrow, steep, high-energy beach composed of both gravel and sand with a high shell content. Between these beaches, the peninsula of Eoligarry is of scientific interest on account of the vigorous erosional processes at work. Some 20% of the area of Eoligarry above mean high-water springs consists of windblown bare sand, and most of the typical erosional and non-erosional landform features of Hebridean dune and machair landforms occur (Ritchie, 1971, 1979a; Hansom and Comber, 1996). The site has considerable geomorphological, archaeological and botanical interest (Farrow, 1974; Ritchie, 1979; Hansom and Comber, 1996; Gilbertson *et al.*, 1996, 1999)

Description

The GCR site of Eoligarry extends from MLWS on Tràigh Eais to MLWS on Tràigh Mhór, including all of the machair and dune area between and the hilltop machair of Ben Eoligarry Mór (Figure 9.6). It is noted for its range of classic machair erosional landforms within such a small area. The beach of Tràigh Eais on the exposed Atlantic coast of Eoligarry has a narrow, concave-upwards profile and is composed mainly of sand resting on a basement of gravel. Gravel up to *c.* 10 cm b-axis is evident on the upper profile, particularly in the north and also floors some of the depressions within the backing dunes. The break in slope between the upper and lower beach is indistinct in the south but in the north well-developed cusps occur in the prominent gravel storm ridge. The mean sand size on Tràigh Eais is *c.* 0.8 mm. There is evidence of an offshore bar lying about 80 m offshore from MLWS upon which waves break. Tràigh Eais is backed by a steep and unvegetated scarp eroded into the mature sand dunes. The extensive intertidal beach of Tràigh Mhór on the east coast of the Eoligarry isthmus, is characterized by a 1.3 km-wide platform of sand at low tide, although this narrows to 10 m at high tide (Figure 9.7). On Tràigh Mhór sand of 0.2–0.3 mm is found with a calcium carbonate shell component of 80% (Hansom and Comber, 1996). The interface between the low-tide platform and the narrow (*c.* 10 m wide) upper beach is marked by a distinct break in slope, with an associated change in colour, reflecting an increase in calcium carbonate content with distance up-beach. Distinctive bars and cusps composed of cockle shells occur on Tràigh Mhór, their genesis having been discussed by Farrow (1974). The mean diameter of the sand grains is 0.3 mm on the upper beach and 0.2 mm on the lower. The beach displays varied topography with an abundance of intertidal mega-ripples

Eoligarry

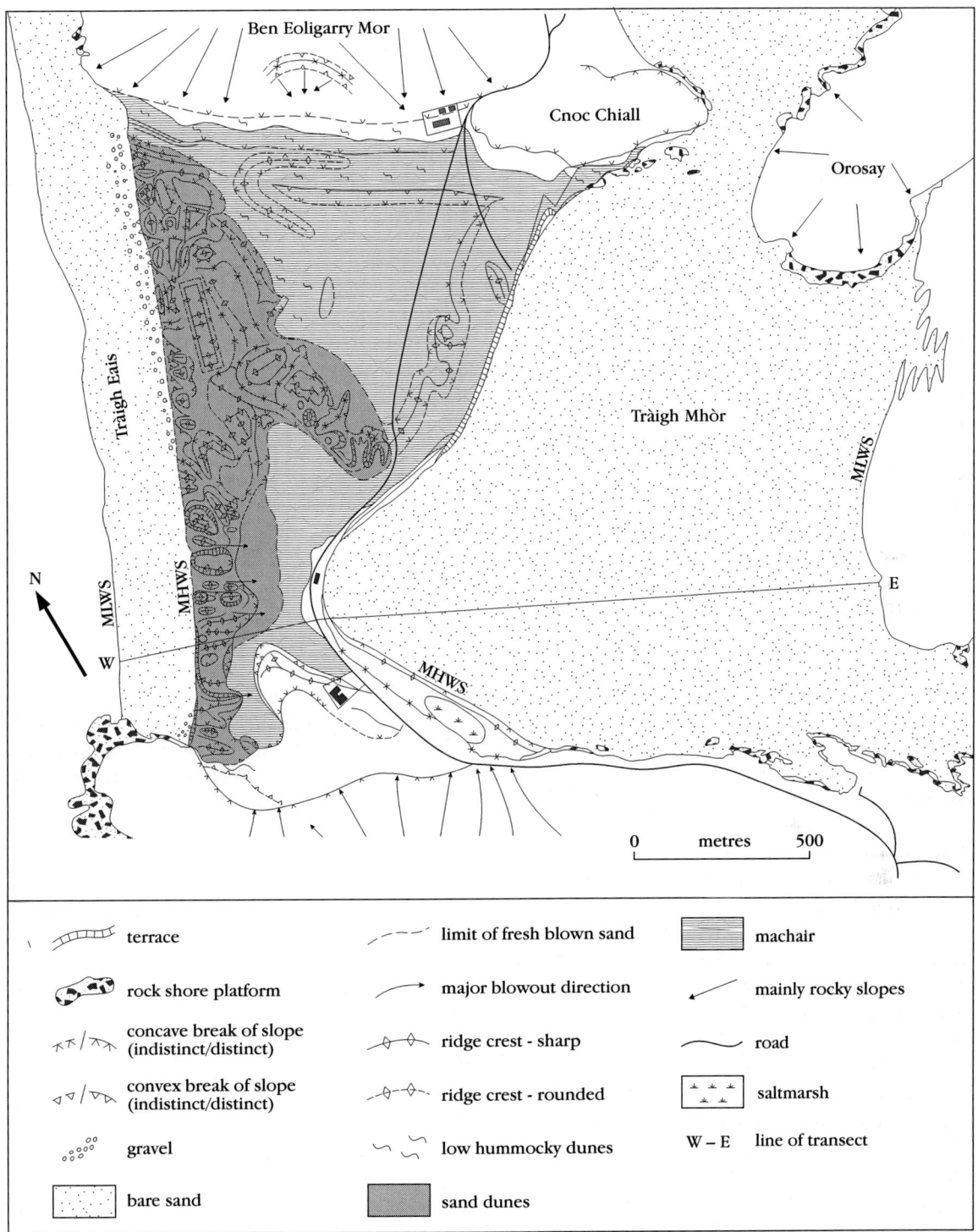

Figure 9.6 Geomorphology of the Eoligarry isthmus. Note the narrow Atlantic beach of Tràigh Eais and the extensive flat beach of Tràigh Mhór between which lie a cordon of high dunes punctuated deeply by blowthroughs. The otherwise extensive machair surfaces are extremely narrow at the southern end of the isthmus. The position of the west–east cross-section of Figure 9.7 is indicated. (After Hansom and Comber, 1996.)

and bar-forms, particularly to the north towards the Sound of Orosay. Erosion of the machair edge occurs in the north but the centre and south of the bay supports low embryo dunes at the rear of the beach.

The Eoligarry dunes and machair form a trianglar peninsula between Ben Eoligarry in the north (where it is over 1 km wide) and the mainland of Barra to the south (where it narrows to just over 300 m). The dune and machair landforms of Eoligarry can be divided into three main units: the high fringing dunes of Tràigh Eais on the western seaboard; the climbing dunes of Ben Eoligarry; and the various high and low undulating machair surfaces of the main isthmus. The dunes backing Tràigh Eais on the Atlantic coast represent an exceptional series of erosional aeolian landforms with a wave-eroded high dune ridge forming the frontal edge of the system. The dune ridge is lower and more continuous in the north, but reaches up to 20 m in the south where it is spectacularly dissected by several large and 15 m-deep V-shaped blowthroughs. Embryo dune forms are completely absent apart from minor re-deposition within the main blowthroughs. The blowthrough forms of the south represent spectacular examples of the erosional capability of winds crossing the isthmus from both the prevailing (south-westerly) direction, and also less frequently from the north-western sector (Hansom and Comber, 1996). They form simple SW–NE-trending linear blowthrough chutes through which large quantities of sand pass through the massive, knife-edged dune ridge that backs Tràigh Eais. The blowthroughs are highly dynamic, the volume of sand in transit making natural re-colonization by vegetation difficult and undercutting of the lateral flanks serving to widen the corridors and reduce the intervening dune segment. In several places, most of the original dunes have been removed, leaving vegetation-capped residuals standing alone in the centre of a deflational sand surface (Figure 9.8). To the rear of the seaward dune ridge lies an area of undulating fixed dunes that is widest in the north. Some of these dunes show ridges and depressions that are orientated both west–east and north–south, at odds with the preferred orientation of the large active blowthroughs in the south. These are ridges that were artificially constructed as part of dune rehabilitation works in the 1970s.

The south-facing flanks of Ben Eoligarry Mór are blanketed by sand blown from the beach and dunes below. The presence of such climbing dunes demonstrates the frequency and strength of the prevailing south-westerly winds that have forced the sand up to altitudes of 103 m. Surface instability, due to both high winds and the large rabbit population, has led to the formation of a large, bowl-shaped blowthrough which has recently been re-activated, leaving a scar of bare sand on the hillside (Hansom and Comber, 1996).

Two types of machair surface exist at Eoligarry, both characterized by high calcium carbonate contents and a distinctive calcareous grassland devoid of long dune grasses. The first is a low, almost horizontal, closely vegetated surface dominating the central–southern areas of the Eoligarry isthmus. The second is a higher, more undulating form, distinct from, but frequently grading into, the fixed dune systems to the west. It is also found in isolated zones particularly in the north of the area. Although both types are the result of the deposition of sand at volumes below those of the dunes to the west, they are nevertheless subject to secondary erosion themselves. Numerous large and small blowthroughs, with both westerly and south-easterly orientations, are eroded into the machair surface and serve to distribute sand over adjacent surfaces.

Interpretation

Eoligarry has been the subject of geomorphological and environmental archaeological research (Ritchie, 1971, 1979a; Hansom and Comber, 1996; Gilbertson *et al.*, 1996, 1999). The interpretation that follows is based on the above work and whereas some details of the evolution and development of the system are necessarily speculative, there is no doubt that the inter-relationships of the landform assemblage within this dynamic system are of national importance.

The key to the depositional history of Eoligarry, and its current erosion on the west and deposition on the east, lies in the altitude and geometry of the low-lying former peninsula of Ben Eoligarry and Orosay that once jutted northwards from the Barra mainland. Such low-lying peninsulas have become subject to increasing amounts of marine influence as a result of the rise in sea level that has affected the Hebrides throughout the Holocene Epoch. The

Eoligarry

depth of the former rock connection between Ben Eoligarry and the mainland is unknown but it must have been shallow enough to allow gravels, sourced from the nearshore zone, to be driven onshore to build an arcuate, west-facing barrier along the length of Tràigh Eais. In contrast, the connection between Orosay and Ben Eoligarry has become flooded by sea-level rise. The emplacement of one or more gravel ridges as part of a gravel barrier has played an important shaping role in the subsequent evolution of what are now largely sandy beach complexes. Once complete, the gravel barrier adjusted landwards and upwards in the face of ongoing sea-level rise. This is a common feature of the sandy and dune-backed beaches of Scotland where gravel is present on which wave-deposited beach and windblown sands have later accumulated (Hansom, 1988; Hansom and Angus, 2001). Where the source sediment is mixed, gravels are usually the first to arrive and are thrown up at the limit of storm waves, whereas sand arrives in quantity later.

The large influx of these coastal sediments is thought to have occurred around 6500 years BP (Hansom and Angus, 2001). It produced beaches with sufficient excess sand available on the upper profile to be blown into the extensive dune and machair systems that remain active today (Ritchie, 1971; Hansom and Comber, 1996). However, ongoing sea-level rise coupled with a reduction in offshore sand supply has also resulted in chronic erosion of many Hebridean beaches and the frontal undercutting of the sand dune systems that they support, such as occurs at Tràigh Eais. Some areas remained sheltered from severe waves and/or by a locally enhanced sediment supply and the effects of chronic frontal erosion have not yet occurred. Tràigh Mhór is such a beach, protected on all sides except the east and sheltered from Atlantic waves. It is a sediment trap within which a wide accretional beach has developed. Farrow (1974) identified onshore-moving bars composed of cockle shells and sand as a result of both tide and wave-induced onshore transport. Time-series maps and photographs allowed Farrow (1974) to demonstrate onshore movement of the cuspate bars over the period 1948–1965 and 1965–1973.

Optically stimulated luminescence (OSL) dating of aeolian sands (Gilbertson *et al.*, 1999) indicate that the carbonate sand of the Barra machair began to arrive about 6800 years BP. However, the arrival of aeolian sand and the initiation of machair development in the Hebrides was almost certainly non-synchronous (Ritchie and Whittington, 1994) and related to local bathymetry and sand supply as the sea level rose over the low-lying and undulating landscape. Thereafter within the Barra dunes and machair various palaeosols dating from Bronze Age to medieval occur together with evidence of periods of instability

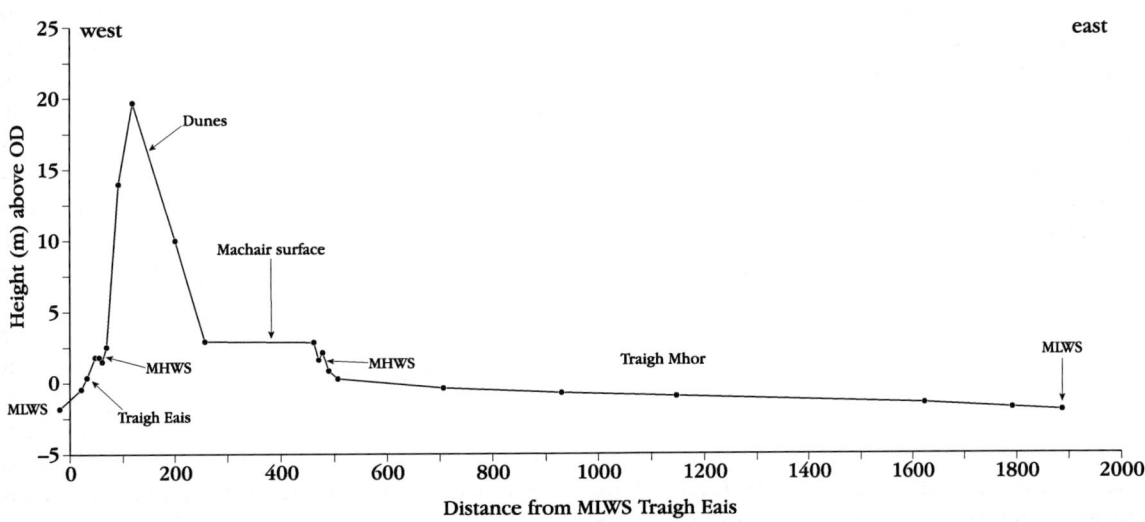

Figure 9.7 Representative cross-section levelled west to east over Eoligarry (see Figure 9.6 for line of section). Note the expanse of Tràigh Mhór and the relatively narrow cordon of coastal dunes that are currently undergoing severe wind erosion. (After Hansom and Comber, 1996.)

Figure 9.8 Looking south over the large blowthroughs on the west side of Eoligarry. Some remedial work has been undertaken but deflation is now so extensive that several dunes have been reduced to isolated 'buttes'. Tràigh Eais is to the right. (Photo: J.D. Hansom.)

and sand-blow that extend into modern times. Gilbertson *et al.* (1999) show that the upper surfaces of the Eoligarry machair are only 100 years old. Hansom and Comber (1996) and Gilbertson *et al.* (1999) emphasize that the machair landforms of Eoligarry are continually developing but are probably more stable now than in the past.

The main factors controlling the geomorphology of the dunes and machair features of Eoligarry are the degree of exposure and the availability of suitably sized sand for aeolian transport. Ritchie (1971) suggested that the high dunes and machair surfaces of Eoligarry are created and nourished by windblown sand from the beaches of Tràigh Eais. The vast extent of Tràigh Mhór derives some of its infill from sand blown across the narrow neck of land from the blowthroughs of the west. Sediment analysis of Tràigh Mhór beach shows that sands of 0.2 mm and 0.3 mm diameter are found on the lower and upper beach respectively whereas the dune sand to the west and rear of the beach is 0.37 mm (Hansom and Comber, 1996). Since the grain size of the dune would be expected to be finer than that of the feeder beach, the present, and probably the past, feeder zones for the dunes at Tràigh Mhór lie to the west. Nevertheless, Hansom and Comber (1996) identify a two-way flow of sand at Eoligarry as a result of easterly or north-easterly winds since, at the western extremity of the large blowthroughs, depositional 'tails' of aeolian sand up to 2 m high are streamlined to the south-west. In addition, along the eastern coastal edge, minor blowthrough features aligned towards the north-west have produced fans of bare sand on the surface inland. Once a blowthrough is initiated, local topography appears to exert an important control over its subsequent evolution.

Conclusions

The scientific importance of Eoligarry rests

largely in the outstanding range of well-developed active erosional features and processes that are unrivalled in any beach–dune–machair system of comparable size in the Hebrides. Most of the typical erosion forms of Hebridean dune and machair morphology are found at Eoligarry within an archaeological context that allows an unrivalled degree of dating precision of phased machair development. Scientific interest is further enhanced by the high- and low-energy flanking beaches of Tràigh Eais and Tràigh Mhór, and the resultant complex sediment interaction which occurs between beaches, dunes and machair plain.

ARDIVACHAR TO STONEYBRIDGE, SOUTH UIST, WESTERN ISLES (NF 740 464–NF 730 333)

Introduction

The 30 km-long stretch of South Uist coastline between Ardivachar Point in the north to Stoneybridge in the south (see Figure 9.1 for general location) includes excellent examples of almost every type of beach, dune and machair surface in the Western Isles. These landforms have developed in the context of a high-energy and open Atlantic coastline that is subject to ongoing submergence (Figure 9.9). In the north, the island of Gualan represents a remnant dune and machair system and in the south, the machair is replaced by a gravel beach backed by marsh and loch. Between these lies a beach–dune–machair system that demonstrates a close inter-relationship between water table and landform with well-developed, low, wet surfaces and a prominent high machair terrace fronted by a conspicuous escarpment that is actively subject to wind-blow. The extensive and well-developed nature of the system and the range and variety of erosional and depositional landforms is of outstanding geomorphological importance. Most of the research so far conducted on machair development has focused on this area, its scientific interest being further enhanced by the presence of numerous archaeological sites that provide not only a cultural context for the development of the landforms but also a dating control. It is a type area for machair development and geomorphology.

Description

The beach that extends between two outcrops of highly resistant Lewisian gneiss at Ardivachar and at Stoneybridge is effectively a single system broken only by the rocky outcrops of Sgeir Dremisdale and by the exit of the Howmore River (Figure 9.9). The river exit is characterized by several sand-bars and by localized backshore accretion with embryo dune formation. North of this point, the low-angled beach as far as Drimore is up to 160 m wide and composed of sand with a calcium carbonate content of 42%, although the dunes and machair behind can reach up to 84% (Ritchie, 1971). Seepage of ground water from the landward surfaces seasonally affects the beach so that the higher water-tables of winter intersect the gravels of the backshore storm ridge. Such impounding of the winter water table has profound geomorphological and ecological consequences, since the low-lying parts of the machair and dunes landward remain flooded to depths of up to 0.5 m for up to five months of the year (Ritchie, 1971). The backshore ridge is affected by storm wave activity and wave-transported gravels are found up to 50 m inland flooring the erosion and deflation hollows in the dunes behind.

Two areas of the beach depart from the above pattern of sand beach, gravel ridge and backing suites of dunes and machair. To the south, near Stoneybridge, the backing dune and machair zone disappears and is replaced by a superb broad-crested coarse gravel ridge that reaches up to 10 m above mean sea level and up to 50 m wide (Ritchie, 1971). Under storm conditions the ridge is subject to roll-over of the constituent gravels that encroach into the area of marsh and lochs on the landward side. South of Ardivacher Point, at West Gerinish, about 200 m of beach is now backed by a sloping gabion wall built to protect military installations. To the north of Ardivachar Point, the island of Gualan is an arcuate but narrow ridge of low dunes that is overwashed by gravels from the upper beach over all except the northernmost part of its 2.2 km length (Figure 9.10). The lower beach on the seaward side is sandy, low-angled and only a few hundred metres wide at low tide. At the northern end of Gualan, and to a lesser extent in the south, the intertidal sands widen to about 1 km, behind which occur embryo dunes backed by well-developed dunes of up to 4 m high (Keast, 1994). The eastern beach of Gualan represents

Figure 9.9 The extensive machair lands of South Uist, looking north, with the exit of the Howmore River in the foreground, Ardivachar, Loch Bee, Gualan and North Uist in the distance at the top of the photo. (Photo: P. and A. Macdonald/SNH.)

the westernmost limit of the intertidal sands of the narrows that separate South Uist from Benbecula (the South Ford). Much of the western coastal edge of Gualan is subject to erosion and this is manifest either in low cliffs cut into the underlying machair or by gravel washover in small lobes that extend eastwards considerable distances down the dune and machair backslope.

Elsewhere, the coastal edge is highly variable in elevation and morphology, although in general there is a tendency for erosion and retreat. In most places there is a steep seaward dune face, capped by a sharp crest at between 4 and 12 m OD. More complex areas of dunes occur where higher transverse ridges intersect the coast but in general the profile is relatively constant with a gentle concave backslope leading landwards to a low-lying and generally wet machair (a representative section of the northern part of the coast is shown as Figure 9.11). In some areas, for example at Drimore and Eochar, there are excellent examples of linear blowthroughs some of which have developed into fully formed deflation corridors and peripheral re-depositional hummocks.

Immediately inland of the coastal edge the machair surfaces are generally low features and rarely reach over a few metres above mean sea level. Where low ridges occur they can be shown to be composed of machair draped on top of underlying glaciogenic ridges (Ritchie, 1971). Nevertheless two distinctive machair landforms occur within the otherwise subdued topography of this zone. Extensive fields of hummocky dunes exist superimposed on top of the low flat machair plains. Between 1 and 3 m high and up to 8 m across, these hummocky dunes extend for considerable distances inland from the main coastal dune ridge. Dominated by marram *Ammophila* vegetation, they have the appearance of recently deposited features formed from sand recycled from frontal or blowthrough erosion. Inland, the most prominent of the machair landforms is a curved seaward-facing escarpment that stretches almost the entire length of the coastline. Well-developed at West Gerinish, Dremisdale and Drimore, the 16–20° scarp slope varies in height from 1 to 7 m. Several sections are under active wind erosion that has revealed a stratigraphy that includes several buried palaeosols. The upper and landward slope of the escarpment has a more gentle dip and shows signs of ongoing sand accretion on its surface. The upper surface slopes landwards to the margin of blown sand deposition, a boundary usually marked by either a coast-parallel series of shallow lochans and marshes (such as Loch Bee (Figures 9.9 and 9.10)), themselves subject to inundation of windblown sand, or by the rising surface of the hill land beyond.

Interpretation

The Ardivachar to Stoneybridge beach and machair system likely responded to Holocene sea level and sediment supply constraints in the same way as other machair systems in the Western Isles and were first developed sometime in the early to mid-Holocene before about 6500 years BP (Hansom and Angus, 2001). The relative sea-level curve for the Outer Hebrides has a form broadly similar to that of Orkney (see Figure 6.28) and is characterized by a slowing, or inflexion, in the rate of rapid rise in sea level at about 6500 years BP. This argument is supported by the existence of numerous sites in the Outer Hebrides where Holocene freshwater peat is now found in the intertidal or subtidal zones. In the Uists, submergence of up to 5 m is thought to have occurred since 8800 years BP (Ritchie, 1985). Such a rate of sea-level rise, across a pre-existing sediment covered shelf, is likely to be the main driver behind the existence of large areas of beach, dune and machair. However, at the local level, such as between Ardivachar and Stoneybridge, the main controls on the dune and machair geomorphology are the degree of local exposure to wind and wave, the availability of suitably sized sand for aeolian transport and, crucially for the timing of sand incursion, the local coastal configuration. The coast of the Uists and Benbecula is essentially composed of a series of low but varying altitude rock basins (Figure 9.9), the flooding of each of which was asynchronous. For example, at Peninerine on South Uist and at nearby Borve on Benbecula the first sand inundation occurred at about 6600 radiocarbon years BP, indicating the

Figure 9.10 (overleaf) The geomorphology of the area around Ardivachar, South Uist. A narrow cordon of active dunes separates the intertidal sandy beach from machair surfaces that are punctuated by erosional terraces. The west side of Loch Bee is subject to gradual infilling by machair sands. (After Ritchie, 1971.)

Machair

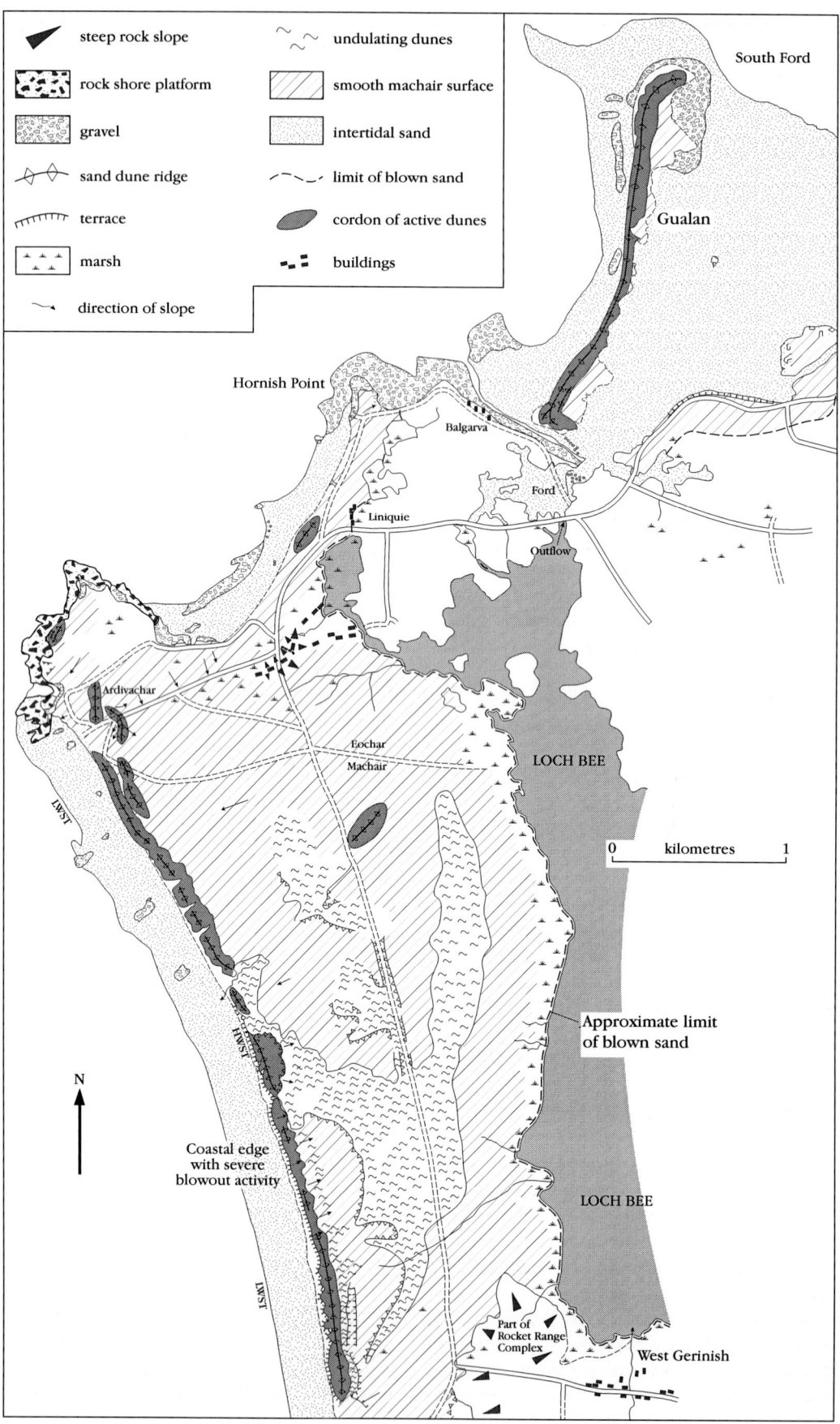

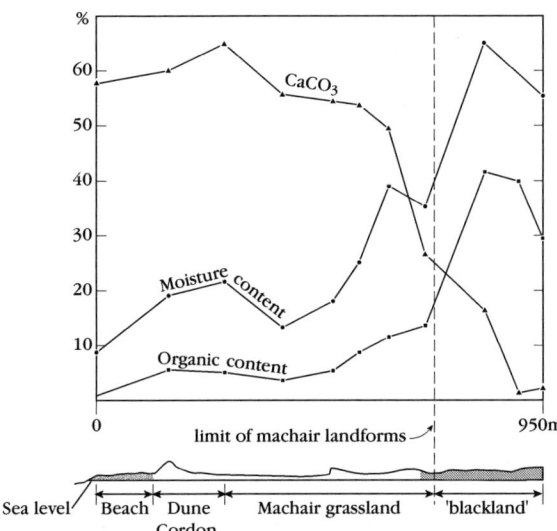

Figure 9.11 A representative cross-section over Dremisdale Machair, South Uist, showing the relationship between landforms, vegetation and soil characteristics. Note the landward decline of calcium carbonate content of the sand from high values close to the beach. Dremisdale is sited just north of the Howmore River exit (see Figure 9.9). (After Mather and Ritchie, 1977.)

proximity of the shore (Ritchie et al., 2001). On the other hand, on Pabbay to the north, intertidal peat at three locations suggests that whereas one part of the island was first affected by a major sandblow event c. 7700 radiocarbon years BP, another part was first affected c. 4300 radiocarbon years BP (Ritchie and Whittington, 1994). It appears most likely that progressive inundation by sand occurred as the coastline, forced by a rising relative sea level over much of the Holocene Epoch, moved eastwards. The result was the development of beach, dune and machair complexes which themselves became subject to erosion or flooding as the coastline progressed eastwards.

For example, between South Uist and Benbecula, the low basin that now contains the South Ford became flooded during the mid-Holocene and most of the central low-lying machair cover was lost. The narrow dune ridge of Gualan, north of Ardivachar, and small lateral fragments to the north and south of the South Ford, are probably all that remains of this formerly extensive machair. Tidal access to the Gualan machair was probably first gained as the dune cordon was breached in both the north and south ends. The breaches remain, but the process has advanced to the extent that the Gualan dune cordon is now thin and low with a coastal edge that is being eroded (Figure 9.10). Almost everywhere it is threatened by washover of gravels in storm conditions that will eventually lead to loss of the remaining dune and machair and further migration eastwards into the narrows.

Ongoing relative sea-level rise coupled with reductions in the offshore sand supply has subsequently resulted in erosion of many Hebridean beaches and the frontal undercutting of the sand dune and machair systems that they support (Figure 9.3; Hansom and Angus, 2001). The mainly sandy linear beach between Ardivachar and Stoneybridge reflects this and is mostly characterized by landward retreat of the coastal edge. Nevertheless, the sloping gabions at West Gerinish appear to be stable, with limited sign of either undercutting or outflanking. It may be that the sediment supply along this part of the coast has been locally healthy in recent years. Frontal erosion of the foot of the prominent, bare dune-slope by waves and deflation of the face by wind, has resulted in progressive landward migration of the seaward escarpment into the dune cordon and machair behind. Such progressive coast-parallel frontal removal of the dune cordon and the development of bare sand slopes exposed to the west, also provides conditions conducive to the development of linear blowthroughs and deflation corridors at right angles to the coastline. Good examples of this occur at Drimore and Eochar, together with the re-depositional dune landforms that form downwind of the erosional sites. Much of the remaining area landward of the seaward dune cordon is characterized either by low-lying and wet machair surfaces or high landward-sloping machair surfaces separated from the low-lying ground seawards by a steep, wind-eroded escarpment.

The low-lying machair demonstrates a sensitive inter-relationship between landform and water table. Although the higher and drier parts both to seaward and landward of the low-lying machair are subject to deflation for most of the year, the lower and wetter surfaces do not deflate when wet yet continue to accept deposition of windblown sand. The resultant landforms are thus virtually flat and with a gentle landward gradient away from the main sand source. No detailed studies exist of the geomorphological processes involved in this windblown re-depositional system, but it seems likely that ongoing deposition onto the usually wet and gently sloping surfaces results in their progressive elevation, particularly at the seaward margin. Low-lying, flat and wet ground may be progressively replaced by higher sloping sand surfaces.

Similar processes are likely to have been responsible in the past for the initial development of the escarpment that often lies landwards of the low-lying machair, a feature prominent at West Gerinish, Dremisdale and Drimore. It is well-known that the landward slope of the high machair and wetland behind is continually elevated by the receipt of substantial quantities of blown sand deflated from both the beach and dune cordon as well as directly blown from the front of the escarpment itself. Accretion of the top surfaces of high machair appears to be both episodic and of long standing since palaeosols and archaeological remains ranging in age from the Neolithic to recent historic are exposed at several places in the deflated edges of the high machair. For example, on the South Uist machair at Kildonan (NF 725 284), south of Stoneybridge, Iron Age structures and associated palaeosols dating from 2500 to 1500 radiocarbon years BP have been exposed by subsequent deflation (Gilbertson *et al.*, 1996) and at nearby Cladh Hallan (NF 734 219), Bronze Age middens and palaeosols from 4000 radiocarbon years BP are exposed in the eroding escarpment edge of the high machair (Gilbertson *et al.*, 1999). Such palaeosols often form a layer resistant to further deflation (Ritchie, 1979a,b) although there is also evidence of periods of instability and sand-blow that extend into modern times (Gilbertson *et al.*, 1996, 1999). Hansom and Comber (1996) and Gilbertson *et al.* (1999) emphasize that although machair landforms are continually developing, they are probably more stable now than in the past. As a result it is possible that the very marked relationships that exist between the supply of fresh shell-sand, soil characteristics, vegetation and landform noted by Ritchie (1976) (Figure 9.11) may change in the future as the supply of fresh sand alters.

Conclusions

The South Uist coastline between Ardivachar Point and Stoneybridge is an exposed Atlantic coastline subject to ongoing submergence and includes excellent examples of almost every type of beach, dune and machair surface in the Western Isles. The narrow overwashed island of Gualan represents all that remains of a formerly extensive dune and machair system. South of this lies a beach, dune and machair system that is undergoing frontal erosion by wind and wave but still demonstrates a close inter-relationship between water table and landform with well-developed, low, wet surfaces together with a prominent high-machair terrace surface fronted by a conspicuous escarpment that is actively subject to wind-blow. In the south the sand beach and machair is replaced by a high gravel beach backed by marsh and loch. The suite of machair landforms is extensive and the range and variety of erosional and depositional landforms is of international geomorphological importance. Interest is further enhanced by the occurrence of several archaeological sites that provide a cultural and dating context for machair development.

It is the type site on which much of the early work on machair development was based.

HORNISH AND LINGAY STRANDS (MACHAIRS ROBACH AND NEWTON), NORTH UIST, WESTERN ISLES (NF 860 750–NF 890 777)

Introduction

The beach–dune–machair systems of Hornish and Lingay Strands include Machairs Robach and Newton, Clachan Sands, and part of Vallaquie Strand (see Figure 9.1 for general location). The area provides excellent examples of most of the machair landform and vegetation types found in the Uists. Machairs Robach and Newton provide the best sites in the Western Isles for flat and low-lying machair landforms that have been influenced by water-table effects and modified by centuries of traditional grazing and cultivation practices. They also provide the finest examples in the Western Isles of old and high machair plateau forms that have either been dissected down to the water table by deflation to produce a distinctive scarp and table land appearance or undercut by wave activity to produce a truncated sequence. Recent depositional activity has enlarged the intertidal strands, and together with saltmarsh development adds to the complexity and scientific interest of the site.

Hornish and Lingay Strands

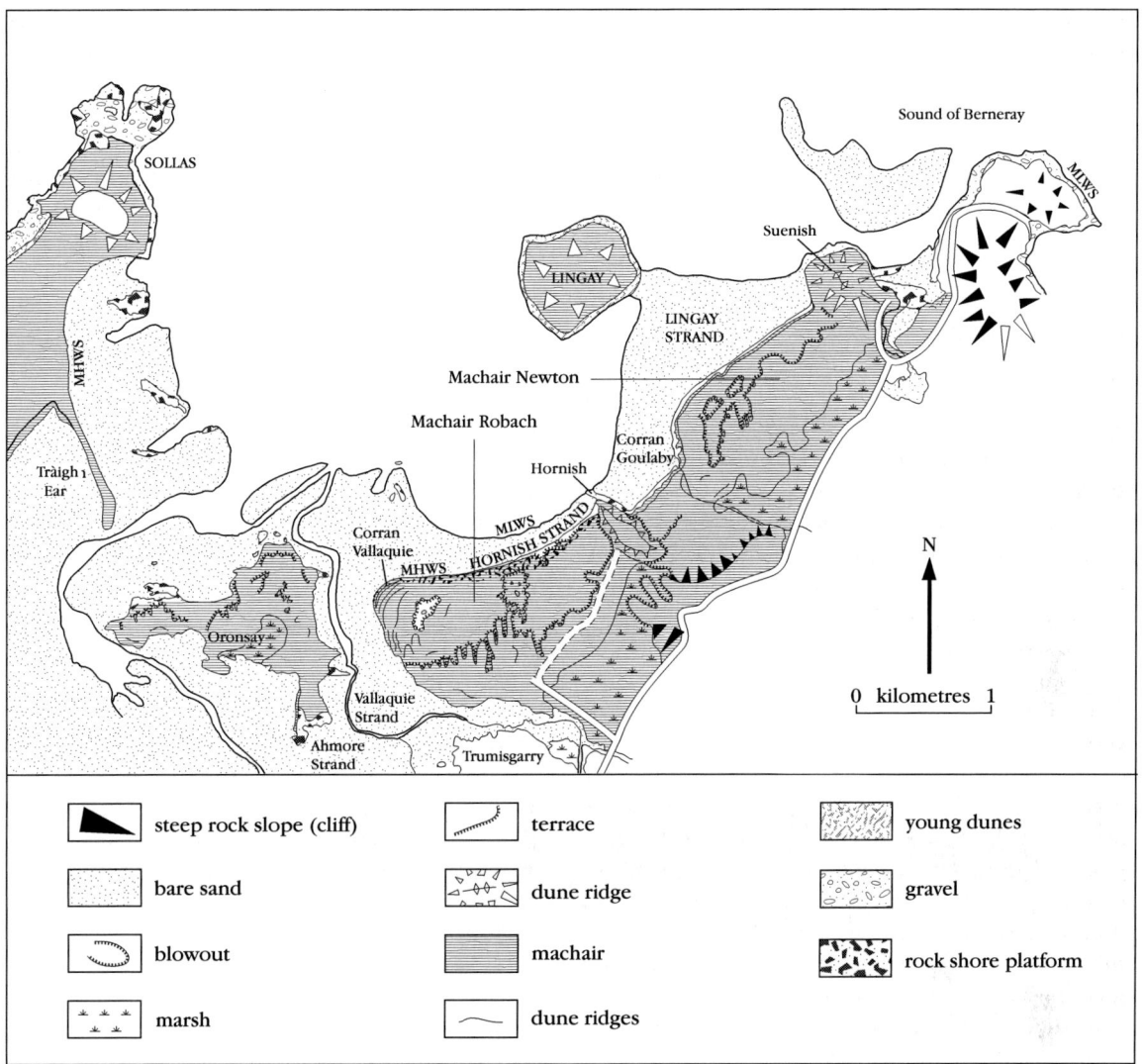

Figure 9.12 The geomorphology of Hornish and Lingay Strands including Machairs Robach and Newton. Well-developed beaches, dunes and machair have benefited from the relative protection from westerly waves offered by the headland at Sollas and the island of Lingay (see Figure 9.4). (After Ritchie, 1971.)

Description

Hornish and Lingay Strands face north-west on the north coast of North Uist. On their landward side, the beaches are backed by extensive machair surfaces that rise to the east and south and grade into hill land beyond (Figure 9.12). The northernmost part of the strand is hinged against a low rocky knoll at Suenish and the southern limit is the low island of Oronsay, itself fringed with fragments of rocky shore platform. The intervening sands stretch for 4.5 km in a gently curving bay broken only by a small rocky ridge of gneiss at Hornish where the beach narrows to only 100 m. At both extremities the beach widens to over 1 km in the north at the intertidal sand tombolo that extends to the island of Lingay and in the south at Corran Vallaquie. The beach gradient is shallow and low sand-bars are common in the fine sand. The shell content of sand at nearby Ahmore and Trumisgarry is 52% and 44% respectively. In the north of the site the coastal edge is mainly characterized by re-depositional young foredunes masking a retreating older machair edge and a narrow band of young dunes colonized by vigorous growths of marram *Ammophila arenaria* everywhere skirts the backshore. At Lingay

Strand, the protecting influence of Lingay Island is reflected in fresh deposition along the dune ridge and annual plants are common in the embryo dunes along the strandline. Along the entire length of this section northwards from Corran Goulaby, the foredune ridge is backed by a higher dune ridge, parts of which have sealed the seaward entrances of blowthroughs. South of Corran Goulaby and the exit of the Goulaby Burn, the coastal edge is undercut with low (1–2 m high) but active sand faces cut into the mature machair plain behind. South of the rocky outcrop at Hornish a relatively healthy 8 m-high foredune ridge extends to Corran Vallaquie and although fronted by seasonal accumulations of sand, the partly obscured underlying faces are undercut. At Corran Vallaquie a healthy foredune ridge is backed by several older dune ridges, suggesting progradation towards the island of Oronsay. Both Hornish and Lingay Strands are sheltered from westerly waves by the Sollas peninsula (Figure 9.12).

Landward of the fronting dunes a low machair complex slopes landwards towards an inner escarpment cut into higher machair deposits. Although the sand plain undulates on account of the deposition of small dunes on its surface, it lies at 1–2 m OD and so is close to the water table. It is regularly flooded, especially in winter, and in places is artificially drained by ditches. At 100–700 m inland from the coastal edge and marking the landward limit of the low, wet, machair plain is a prominent but irregular escarpment cut into a high machair plain. In the north the high plain lies some 3–8 m above the level of the lower surface and slopes landwards for 1.25 km to a poorly drained zone of marsh with small lochans where it meets the rising hill land beyond. Although the high machair plain is currently cultivated, it remains subject to windblown sand deposition that slowly infills the marsh zone behind. Just north of Hornish, the inner escarpment approaches the coastal edge and what remains of the low machair plain is actively being undercut by wave erosion. South of Hornish, Machair Robach extends up to 2.5 km inland and, although the landform sequence mirrors that of Machair Newton, the degree of deflation, especially of the higher machair surface, is more impressive and the zone that is locally eroded down to the water table widens towards the south-west. Flanking the deflated areas are eroded sandy faces cut into the surrounding machair surface and areas of marram-clad re-depositional dunes that are themselves subject to secondary deflation down to the water table. The escarpment of the high machair is punctuated in most places by linear blowthroughs that extend south or south-east leaving arms of high machair that may become detached and are then subject to enhanced blowthrough activity. The series of large dunes landward of Corran Vallaquie may have originated in this way.

Within the sheltered tidal inlet at Trumisgarry and protected by the expanse of Tràigh Vallaquie a small area of sandy saltmarsh is dissected by well-developed tidal creeks and numerous saltpans occur.

Interpretation

The Hornish–Lingay beach–dune–machair system probably responded to Holocene sea level and sediment supply constraints in the same way as other machair systems in the Western Isles. It probably developed sometime in early to mid-Holocene times before about 6500 years BP, the approximate date when sea-level rise and transgression slowed (Hansom and Angus, 2001). Although the start dates of the influx of sediments to beaches varied, the general trend is that the mid-Holocene was a period associated with extensive beach and dune development. However, ongoing sea-level rise (progressively exacerbated by land subsidence in the Western Isles) coupled with reductions in the offshore sand supply subsequently resulted in erosion of many Hebridean beaches and the frontal undercutting of the sand dune and machair systems that they support. Gilbertson *et al.* (1999) identified several periods of sand drift in the Uists dating from 9000–8300, 7500–7000, 6900–6400 and 5800–4200 radiocarbon years BP. In North Uist, Ritchie and Whittington (1994) show that organic deposits now exposed in the intertidal zone at Cladach Mór were first subject to sand incursion at 7600 radiocarbon years BP, one of the earliest records of offshore-sourced carbonate sand incursion in the Uists (Gilbertson *et al.*, 1999). At Camas Mór on the island of Vallay, 10 km west of Hornish Strand (Figure 9.4), the first sand incursion occurred at 6925 years BP. However, Ritchie *et al.* (2001) identify the major period of sand incursion at many sites in North Uist to be after 5200 radiocarbon years BP, this date agreeing with the well-known period of strong sand drift on the coasts of north-west

Europe. Within the constraints of the local topography, there is no evidence yet to assume that events at Hornish and Lingay departed substantially from the above general pattern, the development of machair surfaces most probably taking place sometime after the first arrival of large amounts of sand at the coast at 5200 radiocarbon years BP.

The ongoing dissection and deflation of the high machairs at Robach and Newton probably began as soon as they were formed and represent excellent examples of a constant cycle of deposition, erosion and re-deposition. Deflation continued until the water table was exposed, although in other places in the Western Isles this could equally be exposure of an underlying gravel basement. On Machair Robach this process is well advanced with an impressive summit accordance of remnants of high machair that allow the reconstruction of an original tableland that is now characterized by steep sandy windward scarps and gentle and stable backslopes. Re-deposition of eroded sand on top of the deflated surface produces a secondary spread of small superimposed dunes. In places, for example at Corran Goulaby just north of the rocky outcrop at Hornish, frontal erosion by both wind and wave has been so severe as to have removed both the fronting dune cordon and the high edge of the machair surface so that the low backslope now forms the coastal edge undergoing erosion. The control of sediment supply, water table and general dynamism of aeolian processes displayed at Machairs Robach and Newton, conforms well to the model of machair development suggested by Ritchie (1979a,b) and supported recently by Gilbertson *et al.* (1999), and Machair Robach is probably one of the best examples of machair erosion and development in the Western Isles. Viewed in the context of a generally submerging coastline, recycling of beach and sand dune sediments might be expected as wave erosion progressively enhances coastal instability and produces sand surfaces susceptible to wind-blow. So too might be the progressive flooding of low-lying basins to form intertidal strands, such as at Trumisgarry and Tràigh Ear, and the associated flooding of terrestrial deposits, some of which now appear as intertidal peats.

Although the coastal edge is often obscured by newly deposited windblown sand, the general underlying status of the beaches of the Uists appears to be characterized by erosion and sediment deficiency (Ramsay and Brampton, 2000e). However, this may be reversed where sediment sources are locally enhanced such as occurs at estuary mouths or downdrift of a longshore supply. For example, at Corran Vallaquie not only does there occur a healthy embryo-dune sequence but also several ridges of young dunes have developed as the shore has prograded towards Oronsay. The wider beach and its associated dunes have developed as a result of refraction-driven longshore drift from the east along the wide intertidal expanse of Hornish Strand. Sand may also be delivered from the south-west as a result of low-energy deposition within the Trumisgarry inlet and wind-blow from the upper beach on Vallaquie Strand.

Conclusions

Hornish and Lingay Strands provide excellent examples of most kinds of machair surfaces found in the Western Isles, together with wide tidal strands and inlets. The extensive beach–dune–machair systems have developed in the relative but variable shelter provided by offshore islands and skerries such as Lingay, Boreray and Berneray and this has resulted in varying degrees of erosion and deposition on the foreshore. Machair development also reflects these topographical effects, being in places protected from wave attack but still subject to ongoing deflation whereas in others it is subject to substantial wave erosion and removal of the original landforms, all set within a context of submergence. Machairs Robach and Newton are one of the sites of the highest geomorphological interest in the Highlands and Islands. Since 66% of the world resource of machair is found in the western seaboard of Scotland, there can be little doubt about the international scientific importance of Hornish and Lingay Strands and Machairs Robach and Newton.

PABBAY, HARRIS, WESTERN ISLES (NF 900 870)

Introduction

The island of Pabbay lies in the Sound of Harris some 8 km north of North Uist (see Figure 9.1 for general location). About 30% of the island area is covered by blown sand, most of which is located in the south-east, and the site is probably

Machair

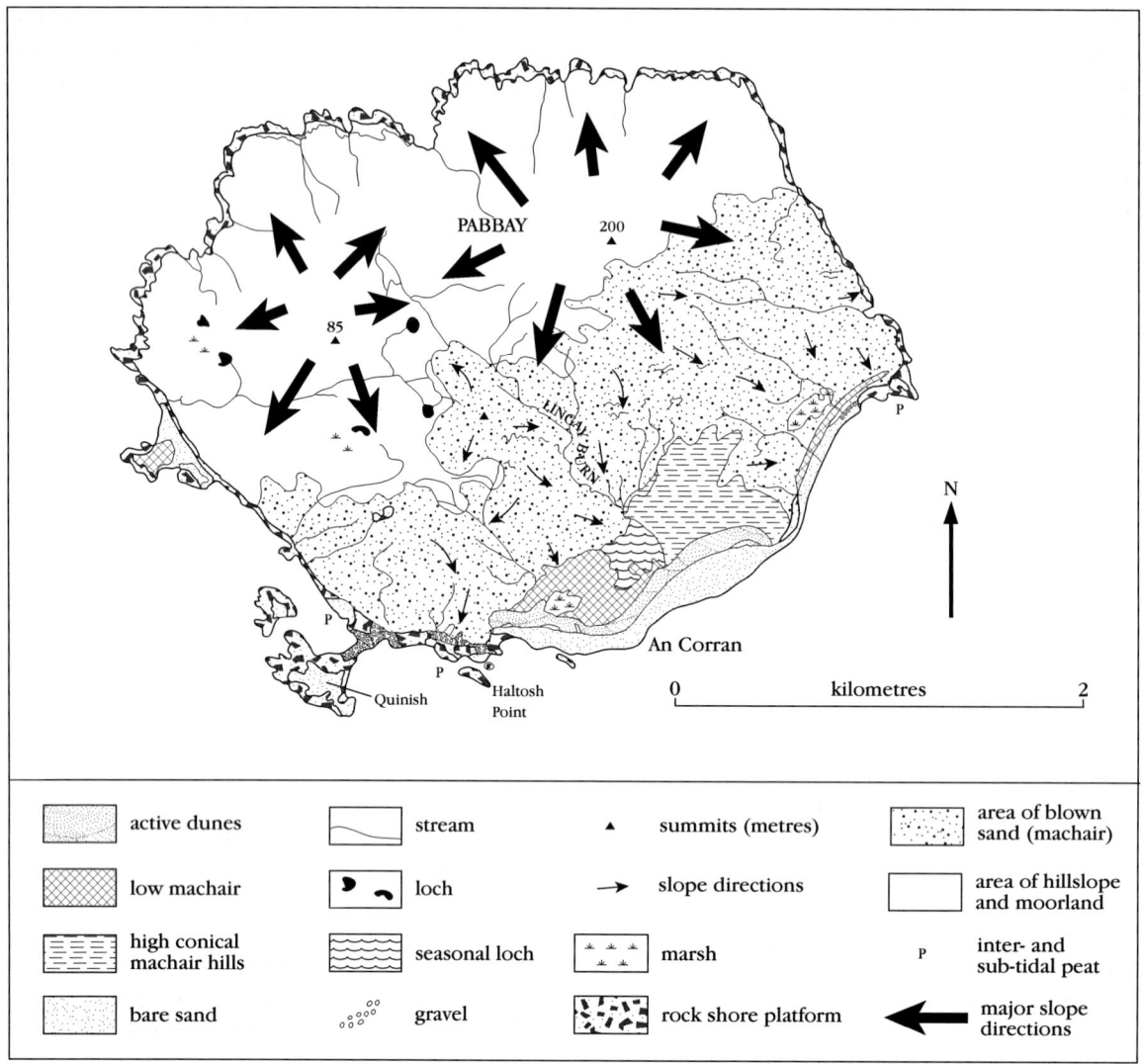

Figure 9.13 The geomorphology of Pabbay, Sound of Harris, showing the extensive area of climbing machair and the low-lying area east of Haltosh Point that has been infilled by beach and dune accretion since 1857, when it was a marine inlet. (After Ritchie, 1980.)

the best example of climbing dune habitat in the Western Isles. The machair on Pabbay is of national importance because it provides good examples of all the types of machair and dune surfaces found in the Outer Hebrides in addition to areas of unusual large conical dunes. Unusually, the machair areas face south-east rather than to the west as is the case in most of the Hebridean machair. There is no record of rabbits having reached the island, and since rabbits are thought to have had a major effect on dune and machair geomorphology, Pabbay provides a good comparative site. Several areas of intertidal and subtidal peats occur on Pabbay that may help elucidate the sequence of aeolian and sea-level events that led to machair development both on Pabbay and elsewhere in the Outer Hebrides.

Description

Pabbay is a distinctive, conical island, which has been uninhabited since the 1930s. The coastline is also distinctive being characterized by north and east coasts that are rocky, cliffed and indented with geos and inlets and south and west coasts that are low-lying and predominantly sandy (Figure 9.13). The coastal edge of the south and west coast comprises three crescentic sand embayments that have developed between low rocky skerries and platform remnants. Landwards of these embayments a beach–dune–machair complex has accumulated that has extended north to cover about 30% of the area of the island. Ritchie (1980) described the island as consisting of three main landform surfaces. The northern half of the island is characterized by ice-moulded bare rock and boulder surfaces, stripped of the original peat cover. The central–west plateau is covered by boggy moorland in the west and north and by windblown shell-sand elsewhere. The south is dominated by shallow basins filled with various types of machair and dune landforms.

The beaches along the south-west coast are mainly gravel storm ridges connecting low fragments of rock shore platform that are locally covered by a thin veneer of blown sand (MacTaggart, 1998c). Small pockets of intertidal organic deposits are occasionally visible near the offshore skerry of Quinish. Quinish is connected to Pabbay by a gravel tombolo that steepens at its landward end to obscure the machair edge. The gradient and height of the gravel ridge decline towards Haltosh Point where it is replaced by a predominantly sand beach. To the east, between Haltosh Point and the rocky cliffs of the north-east coast, lie two promontories that define three crescentic sandy embayments each composed mainly of shell-rich white sand. The first of these promontories, An Corran, is a triangular sandy foreland backed by low machair. The beach to the west is often partly covered by bars of small rounded gravels but the 1.5 km-long and 200 m-wide beach to the east is characterized by thick accumulations of sand that is subject to wind-blow (MacTaggart, 1998c). From comparison of aerial photographs the coastal edge of this promontory has retreated over the period from 1965 to present to form a steep scarp face that undercuts the seaward edge of the dune ridge, while progradation of several tens of metres has characterized the area to the east (MacTaggart, 1998c). The second promontory east of Haltosh Point (Figure 9.13) is a low rocky outcrop veneered by blown sand, beyond which is another sandy beach both shorter and narrower than the one to the west. A narrow gravel bank occurs along the coastal edge at the eastern end of this beach. Intertidal and subtidal peat deposits have been reported at the eastern end of the beach by Ritchie (1980).

Much of the coastal edge to the west of Haltosh Point is obscured by gravel beach deposits banked up against a steep and undercut machair face. Machair stratification is visible in places and displays alternating sequences of organic and sandy horizons. The coastal edge to the east of Haltosh Point is characterized mainly by localized areas of accreting and erosion-affected dune scarps that average 2–3 m high but a 15 m-high section undergoing erosion occurs in the face of a well-vegetated large dune ridge that runs inland and declines in height to the west of the first promotory. The ridge is punctuated by three shallow blowthroughs. Seawards of this large ridge lies a series of low accretional dunes that are vegetated to different degrees by primary colonizing species and marram *Ammophila arenaria*. At least two sequences of dune ridges are present, each running at a different angle from the present coast, the seaward sequence cross-cutting the landward sequence. The alignment of the large dune ridge defines a wider and more deeply indented bay centred on the outlet of Lingay Burn. Subsequent accretion has resulted in low dunes developing seawards of the large dune. The easternmost bay is characterized by a low machair edge undercut into several low hillocks, sections within which show thick accumulations of sand capped by an organic horizon that corresponds to the surface of the low machair plain behind. The organic layer is itself capped by windblown sands that have now been sectioned by coastal erosion. Ritchie (1980) reports several exposures of dune-foot organic horizons in the machair edge.

Behind the beach and dunes lies an extensive machair surface that displays three machair types: level or undulating machair, wet machair and hill-slope machair. Two large areas of wet machair occur landwards of the dunes backing the two eastern beaches. The largest of the wet machair areas was mapped by Ritchie (1980), and still remains, as a seasonal loch displaying distinctive vegetation zonation patterns dominated by bryophytes. The area of wet machair matches fairly closely with the extent of a marine inlet shown on the Admiralty Chart of 1857. To the south-west of the central wet machair lies low hummocky machair but to the north-east on the rising hillside lies a spectacular series of near-symmetrical conical sand dunes up to 8 m high and rising to 30 m OD. The dunes support machair vegetation with only small patches of marram *Ammophila arenaria*, although the intervening low-angled surfaces between the dunes contain damper habitats (Dargie, 1998). A small area of calcarenite occurs at the north end of the machair on Pabbay.

Interpretation

In common with much of Harris and the Uists, coastal development on Pabbay has been dominated by sea-level change and the availability of sediment. Throughout most of the Holocene Epoch sea-level rise has resulted in transgression of the seaboard of the Western Isles. However, since much of the coastline is characterized by undulating rocky surfaces at various altitudes and often with a variably thick veneer of glaciogenic deposits, the timing of inundation and subsequent beach and sand-dune deposition varies from site to site. Submergence of at least 2.8 m since 7700 radiocarbon years BP is indicated from the radiocarbon dating of freshwater peat from Pabbay (Ritchie, 1985), although dates elsewhere in the Uists suggest that the amount of submergence since 5100 radiocarbon years BP was of the order of 5 m. Certainly, the coastline of Pabbay was likely to have been characterized by variably drained low rocky basins at different altitudes and distances from the coast that began to fill with terrestrial peat at different times and may have been subject to the first wind-blow events at different times. Sea-level rise began to slow about 6500 years BP (see the introduction to the present chapter) but in the Outer Hebrides large amounts of sand from offshore sources had probably begun to arrive on shorelines prior to this date, resulting in wide beaches and the potential for substantial windblow and dune and machair development. It is thus likely that the lower levels of the first machair surfaces were formed early in the mid Holocene times, although the recycling of machair surfaces ensured that 'first wind-blow' events occurred over a wide range of dates. The abrupt change from a peat surface to a sandy upper section occurred at about 4400 radiocarbon years BP at Quinish on Pabbay (Ritchie, 1980). The former terrestrial peat deposits are now located 80 m seawards of MLWS, so it may be that much of the former low machair surface has been subsequently eroded and recycled landwards to form the higher machair, a process that continues today.

Historical records support the above reconstruction and indicate that the south-east part of Pabbay was formerly a large plain consisting of a sandy soil mixed with earth so fertile that Pabbay was known as the 'Granary of Harris' (Angus, 1997). However, in a sand storm in 1697 about 300 acres of arable land were overwhelmed by sand as well as land lost on the south-west side of the island 'where many people still alive have reaped crops of grain' (Walker (1764), cited in Angus, 1997). It is tempting to attribute the development of the climbing machair and conical dunes in the east of the island to the windblow events of 1697 particularly since the walls of ruined buildings have not yet been comlpetely buried by blown sand. Elton (1938) attributes this to stabilization of the higher surfaces into a closed turf resulting from a shift from predominantly arable farming before the end of the 18th century to pastoral farming thereafter. In spite of this possible stability at altitude, substantial changes at lower levels occurred as a result of sea-level rise. An early map of Pabbay in 1805 shows the coastline much as it is today yet the Admiralty Chart of 1857 shows a large tidal inlet behind the dunes. The first OS maps in 1881 show this loch sealed from the sea, presumably resulting from sand deposition (Angus, 1997). Since then, substantial accretion of dune and sand has resulted in the consolidation of the two forelands, with the high dune ridge probably marking the former position of the mean high-water springs in 1881. However, recent erosion of the west coast and accretion to the east (Angus, 2001) suggests that cycles of erosion and deposition, depending on storm approach direction, continue to influence the sand feed to the beaches and dunes of Pabbay.

Conclusions

Pabbay displays excellent examples of a range of beach–dune–machair landforms and is probably the best example of climbing dune habitat in the Western Isles (Dargie, 1998) with high conical dunes that are unique in terms of their scale and symmetry. The machair assemblage is also unusual in that it faces south-east whereas most other machair faces west. The island is also important for comparative study since there is no record of rabbits on Pabbay. The presence of several intertidal peat deposits together with an historical record of change allows the development of beach and machair to be placed in a temporal context.

LUSKENTYRE AND CORRAN SEILEBOST, HARRIS, WESTERN ISLES
(NG 057 973–NB 068 004)

Introduction

Luskentyre Banks and Corran Seilebost are twin peninsulas that enclose the vast intertidal sand beach of Tràigh Luskentyre on the west coast of Harris (see Figure 9.1 for general location). They are both unusual settings for machair development but have several distinctive features and may represent the remnants of a once larger system. Luskentyre is of interest as a dynamic beach–dune–machair system with substantial dune development, whereas Seilebost incorporates a beach–dune–machair system backed by saltmarsh and extensive areas of intertidal sand. Only the seaward tip of Seilebost is a true sand-spit with identifiable growth stages and a landward curve. Together the Luskentyre and Corran Seilebost system has the appearance of being set within a structural trough that has been subject to submergence and represents a single dynamic depositional complex that is of great geomorphological interest.

Description

The Luskentyre–Corran Seilebost site extends for 3 km from the northern tip of the Luskentyre Banks to the southern extremity of the Corran Seilebost spit. At its widest it reaches a little over 1 km from mean low-water spings to the south of Tràigh Rosamol to the southernmost part of Luskentyre Dunes (east of area (1) Figure 9.14). The intertidal sandflats and saltings of Tràigh Luskentyre are an integral part of the coastal geomorphological system; they are also important for ornithological reasons.

Tràigh Luskentyre occupies an inlet that extends inland 4 km to the south-east and is characterized by intertidal sandflats crossed by a single tidal channel that is subject to lateral migration. In 1996 the channel approached close to the tip of the spit at Corran Seilebost (see Figure 9.15; MacTaggart, 1997c). Well-developed mega-ripples on the intertidal sandflats indicate that ebb velocities are an important factor in the redistribution of sediments within the embayment. Sandy saltmarsh communities have developed in the more sheltered places within the inlet, such as in the lee of the promontory of Corran Seilebost and at the mouths of several freshwater streams that drain onto the sandflat. Enclosing the seaward end of the sandflats of Tràigh Luskentyre are the twin promontories of Luskentyre Banks in the north and Corran Seilebost in the south.

Luskentyre Banks forms a bulbous triangular-shaped foreland jutting out from the steep rocky slopes of mainland Harris towards the island of Taransay (Figure 9.15). The beach of Tràigh Rosamol in the north is hinged onto the southern part of the rocky promontory of Aird Groadnish and curves west to reach 300 m wide at the point of Luskentyre Banks. The beach continues to the south-east and narrows to 50 m at a second rocky promontory close to Luskentyre settlement. The beach is composed of fine-grained sand of 0.22 mm diameter, of which some 54% is shell (Ritchie and Mather, 1970b). Outcrops of calcarenite occur at mean low-water springs along this part of the foreland (Ritchie and Mather, 1970b). The southern end of Tràigh Rosamol is backed by a single ridge of 2–3 m-high dunes. The seaward edge of the dunes increases in height to the north where several marram-clad ridges have developed, probably reflecting a sand supply from the south. The northernmost extremity of Tràigh Rosamol is marked by a locally undercut dune edge and a deep sand-floored erosion corridor across which drains a small stream (Figure 9.14). To the east of the point, the coastal edge is accretionary with low dunes developed in front of the higher, more stable dunes behind. However, towards the south-east, the low accretionary ridge narrows and the backing dunes become progressively undercut to produce dune faces of up to 20 m high. In the extreme south-east of the site, the coastal edge is rocky with a thin veneer of machair.

Ritchie and Mather (1970b) and Harris and Ritchie (1989) described the landforms of the foreland in terms of areas approximating those numbered on Figure 9.14. In the east of the site in area 1, an area of dunes is dissected by a partly stabilized blowthrough system that is now being undercut at its seaward edge. To the north-west of this in area 2, an old and stable machair surface is dissected by a series of blowthroughs of varying sizes to produce a chaotic surface surrounding a long and stable blowthrough corridor that has extended to the north. Area 3 to the north-west of area 2 consists of a series of U-shaped blowthroughs, dune

Machair

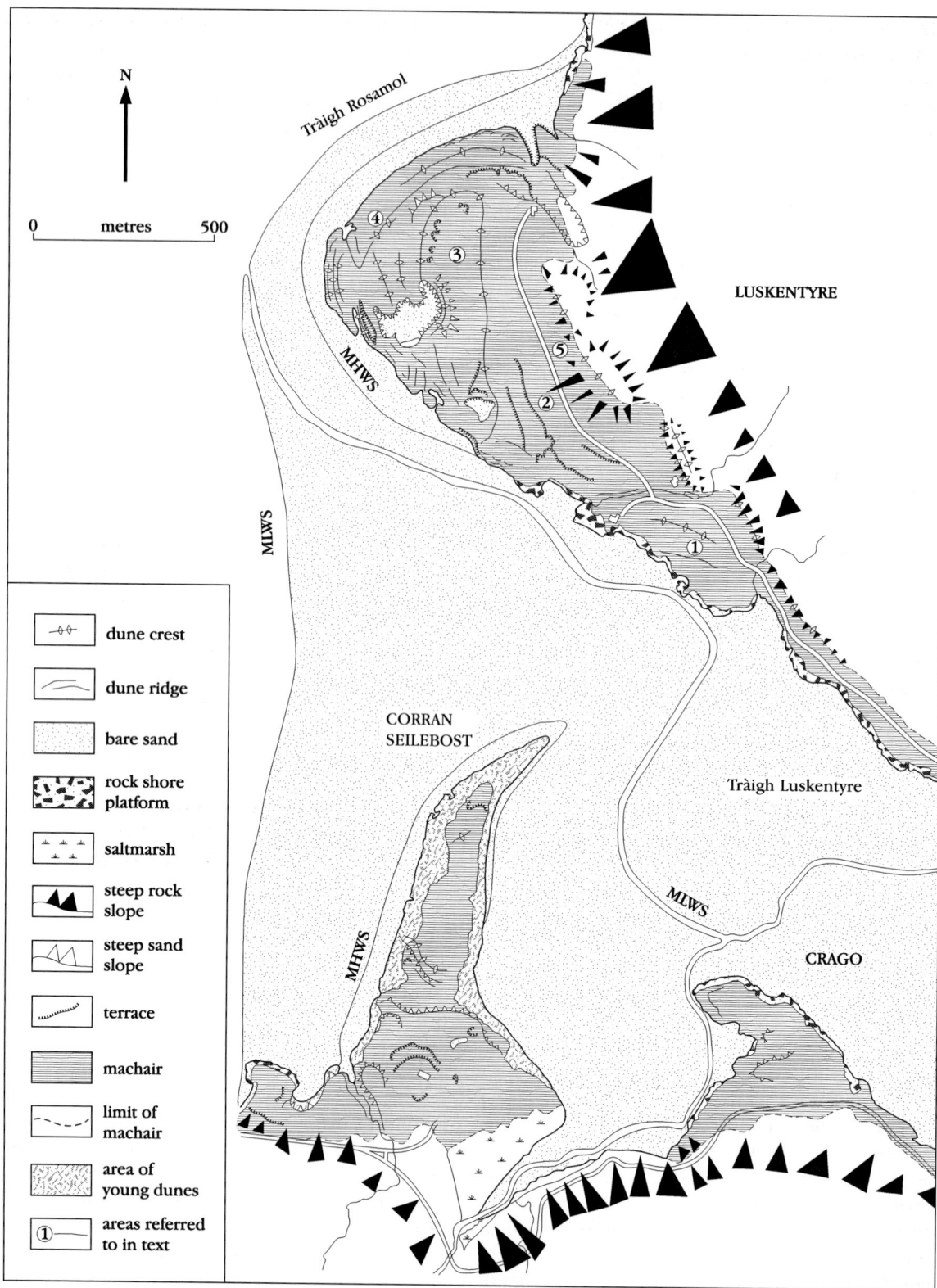

Figure 9.14 The geomorphology of Luskentyre and Corran Seilebost. Extensive intertidal sands front the dune and machair landforms of Luskentyre where several large blowthroughs occur. The dunes reach 35 m OD and are the highest free-standing dunes in Harris. Numbered locations on Luskentyre are referred to in the text. The complex of beach–dune–machair features may be the remnant of a once-larger machair that has been fragmented by submergence. (After Ritchie and Mather, 1970b.)

Luskentyre–Corran Seilebost

Figure 9.15 The rocky peninsula of Crago lies in the foreground of this aerial oblique looking north-west over Tràigh Luskentyre to the dunes and machair beyond. The western tip is highly active with clear evidence of extensive wind-blow. The free-standing spit of the tip of Corran Seilebost is also in view in the centre left. The island of Taransay (at the top) provides shelter from westerly waves. (Photo: P. and A. Macdonald/SNH.)

ridges and conical sand dunes that reach over 35 m OD, the highest free-standing dunes in Harris. The conical dunes appear to be recent re-depositional features superimposed on the top of an earlier central high ridge system. This high surface is impressively dissected in the south by both elongate and cauldron-form blowthroughs, the largest of which is a long-established feature surrounded on all sides by exposed sand escarpments of up to 35 m OD and which tower above the blowthrough floor lying at a few metres above OD (Figure 9.15; MacTaggart, 1997c). Outcrops of calcarenite occur on the floor of the blowthrough. Since Mather and Ritchie (1970) described the feature, the seaward end of the main blowthrough has been sealed by the deposition of a ridge of new dunes, although north-trending blowthroughs have subsequently developed along the advancing northern apex of the main blowthrough (MacTaggart, 1997c). In the northern part of area 3, the high sand ridge is dissected by a number of blowthroughs that have also extended northwards as a series of unvegetated sand waves with steeply sloping (30°) leading edges that inundate area 4 to the north. Area 3 is separated from the active coastal edge by area 4, an extensive, high, mature dune-plateau that reaches 20 m OD. Stable sand dunes have been superimposed on top of the older surface. The northernmost parts of areas 3 and 5 are truncated by the landward extension of the sand-floored erosion corridor at the north end of Tràigh Rosamol. The sides of this gorge-like form are cut into machair sand in the north and bedrock in the south and probably represent a fault-controlled stream incision that has become partly machair-filled. Area 5 represents a high, but gently sloping, surface of extensive hill machair that is used intensively for cultivation and grazing and that masks the underlying glacially moulded surfaces of the hill-slopes above.

Corran Seilebost is a peninsula of about 700 m wide at its root against the rocky knoll of Aird Horgabost in the south. From here it progressively narrows for 1.3 km northwards to a narrow neck of sand. On the west side, Tràigh Seilebost is a straight and gently sloping beach that extends seawards for 300 m and includes wave-built swash bars. On the east, the extremely gentle slope of the intertidal sandflats of Tràigh Luskentyre extends inland. The western coastal edge is backed by a nearly continuous foredune ridge of up to 10 m high, except in the south where the foreshore is backed by an undercut old and stable machair surface. Towards the north of the peninsula a healthy and prograding foredune ridge with extensive marram *Ammophila arenaria* cover has developed, although this becomes subject to wave undercutting towards a low sand-spit that extends north from the tip of the peninsula. The spit is highly dynamic and can lose much of the tip in a single storm, followed by a period of slow rebuilding. Where the spit begins to curve to the north-east, a few small blowthroughs have now been sealed by the development of foredunes across their seaward entrances (MacTaggart, 1997c). The dunes of the east face of the peninisula are lower than in the west but are subject to wave undercutting at high tide so that bare sand slopes characterize much of the Tràigh Luskentyre shore. Most of the central core of the peninsula is composed of dune ridges of varying sizes, some of these well-vegetated features recurving north and east where the spit has extended. In the south of the peninsula, the dunes extend to cover most of the edge of the higher machair surfaces behind. These older machair surfaces support several dune ridges and escarpments that trend west–east across the peninsula but they also occur as flat-topped features or form gentle aprons over the surrounding slopes. To the east of the machair area a small area of saltmarsh has developed with mature features such as tidal creeks and salt-pans. The seaward edge is undercut but the landward edge grades into sloping machair surfaces.

Interpretation

There seems no reason to suppose that the development of the Luskentyre–Corran Seilebost system should be substantially different in its response to Holocene sea level and sediment supply constraints than other machair systems in the Western Isles such as, for example at Mangersta or Hornish. Holocene sea-level rise slowed to a much reduced rate post-6500 years BP (Hansom and Angus, 2001). Although the start dates varied, the general trend is that the mid-Holocene was a time associated with an influx of sediments and extensive beach and dune development. However, ongoing sea-level rise (progressively exacerbated by land subsidence in the Western Isles) coupled with reductions in the offshore sand supply has subsequently resulted in erosion of many Hebridean beaches and the frontal undercutting of the sand dune and machair systems that they support.

Events at Luskentyre–Corran Seilebost appear to have followed this general trend, albeit exacerbated by two local effects. The low-gradient offshore zone resulted in a substantial and easily accessible source of beach and dune sediments, and the shelter offered by Taransay resulted in an essentially benign wave environment conducive to beach development within what had become a drowned tidal inlet. As a result the inlet became the locus of deposition with beach development at its entrance. Features of coastal deposition commonly occur where wave refraction reduces the capacity of waves to convey sediment. The swash-aligned orientation and location of Corran Seilebost can be explained in this manner. Similarly waves from both north and west influencing Luskentyre might also be expected to produce the triangular foreland of Corran Raah, on the lee shore of Taransay. In spite of the shelter afforded by Taransay, the extent and height of the dune and machair surfaces on both sides of Tràigh Luskentyre is impressive and, in the context of the inlet as a whole, Ritchie and Mather (1970b) suggested that the unusually shaped and sited twin promontories might represent fragments of a much more extensive dune and machair system. This hypothesis suggests that the original beach and dune probably developed only slightly seawards of the present location but that the machair surface may have extended over much of what is now Tràigh Luskentyre. Remnants of this machair surface fringe parts of Tràigh Luskentyre some distance inland, for example at Crago and farther east (Figure 9.14). Submergence over the later Holocene subsequently resulted in erosion of the fronting beach and dune, fragmentation of the fronting system and the submergence of Tràigh Luskentyre. The original direction of wind-blow within the enlarged system may have had a substantial westerly bias but once fragmentation occurred, the northern part of the system may have become more influenced by southerlies that drove blowthrough corridors northwards, a process that continues today. In addition, the seaward edges of the Seilebost saltmarsh are undercut and have a very similar appearance to wet machair. Submergence may have resulted in the conversion of a low-lying machair to what was reported by Dargie (1998) as a saltmarsh community with excellent examples of transitional saltmarsh–machair communities at its landward margin. Coring of the intertidal flats and saltmarshes may help resolve some of the above uncertainties. The outcrop of calcarenite below low water at Luskentyre may suggest submergence and/or retreat, since the formation of such crusts may be related to deposition or re-deposition of calcium carbonate under fluctuating water tables.

In the above scenario, the 'spit' of Corran Seilebost has not been formed by longshore drift from south–south-west, but by submergence of a more extensive dune-machair system and is not a true spit. In addition, the location of the estuarine channel will have influence on the extent and timing of erosion of both Corran Seilebost and Luskentyre.

Conclusions

Together the twin peninsulas of Luskentyre and Corran Seilebost represent a highly dynamic beach–dune–machair system that is one of the most scientifically interesting areas in the Western Isles. Not only do the sites contain the best examples in Harris of most of the features of machair landforms, but they have also formed in an unusual setting. It is possible that the present sites represent the remnants of a once more extensive beach–dune–machair system that has become fragmented by submergence. If this is the case, then the sites have added significance since they may form part of a suite of machair sites in the Western Isles that record, in the various stages of development of their landforms, a cycle of growth and decay that has affected machair over the Holocene Epoch.

MANGERSTA, LEWIS, WESTERN ISLES (NB 008 308)

Introduction

Mangersta Sands occupy a small embayment situated on the exposed west coast of Lewis, near Uig (see Figure 9.1 for general location). A 200 m-wide beach fronts a long, narrow machair that occupies a depression running inland. Mangersta is an excellent example of a beach–dune–machair complex in which much of the dune has been eroded and the machair surface has been deflated down to the water table. Little of the stripped surface has been re-colonized by vegetation and it seems likely that Mangersta represents an advanced stage in the cycle of growth and decay of beach–dune–machair complexes in exposed areas.

Description

The beach at Mangersta extends to 200 m in length and at low water is 200 m wide. It is contained within rock headlands cut into Lewisian gneiss and sits within a valley that is probably part of a pre-glacial drainage system (Ritchie and Mather, 1970b) (Figure 9.16). The low-gradient intertidal beach is composed of well-sorted, medium-grained sands (median diameter 0.39 mm), 38% of which is shell. The upper beach is characterized by a steeper–gradient arcuate storm ridge composed of subrounded gravel, through which water seeps from the surfaces behind. The gravel ridge is 100 m long and 20 m wide. A continuous cordon of frontal dunes behind the beach is lacking at Mangersta and

Machair

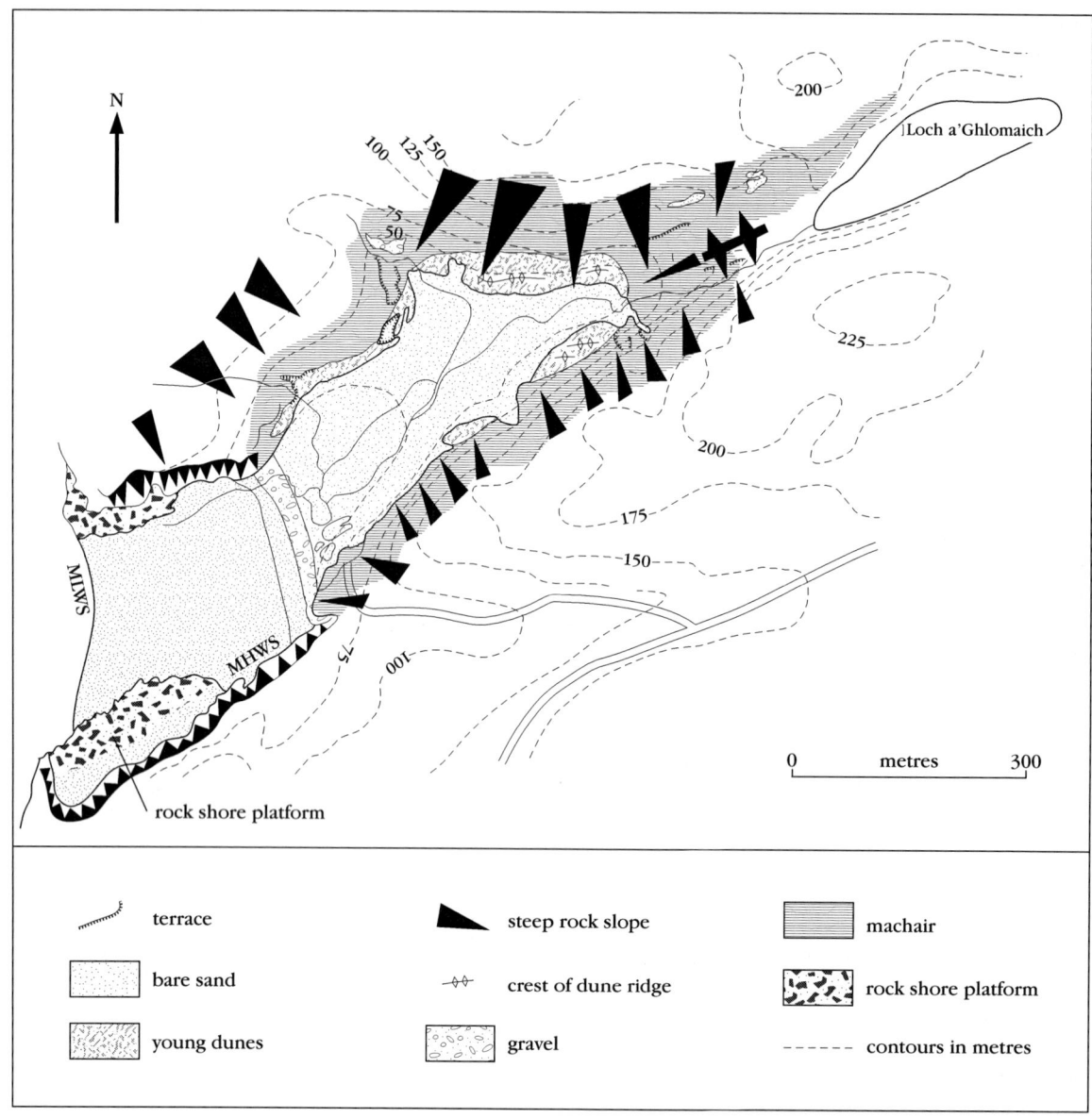

Figure 9.16 The geomorphology of the small embayment of Mangersta, Lewis. A narrow beach separates an area of low-lying sand from the sea. Mangersta is an example of a machair complex that has been eroded and deflated down to the water table. (After Ritchie and Mather, 1970b.)

only in the southern part are poorly developed and discontinuous young dunes found. Extending for at least 500 m inland of the gravel storm ridge at Mangersta, is a flat and bare sand surface that is deflated to the water table. The seaward part of the surface is littered with driftwood and a series of small braided streams cross the low-gradient (1°) slope (Figure 9.17). The flat-floored deflation surface abuts landwards against the steep rocky slopes surrounding the depression, the abrupt change of gradient being masked by small marram-clad sand dunes. In spite of new dune development and re-vegetation around the perimeter of the depression and to a limited extent around storm debris just above mean high-water springs, much of the central sand area was unvegetated in 1970 (Ritchie and Mather, 1970b) and remained so in 2001 when the photograph Figure 9.17 was taken. Most of the flanking dune orientations reflect deposition by winds from the south-west. A thin veneer of machair occurs on all of the

Mangersta

Figure 9.17 The central deflated surface of Mangersta viewed from the north-east. Climbing machair veneers the flanking slopes and occasional washover of the fronting beach occurs during storms. (Photo: J.D. Hansom.)

surrounding hill-slopes above the central depression. The lower parts of the machair are terrraced in places indicating that they have undergone erosion in the past, although much of the contact is obscured beneath the younger flanking dunes. The thin layer of hill machair is undergoing active erosion by runoff, and in places the underlying glacial till cover is visible in stream cuttings (Figure 9.17).

Interpretation

In common with much of the Western Isles, coastal development at Mangersta has been dominated by sea-level change and the availability of sediment. Throughout most of the Holocene Epoch, sea-level rise has resulted in transgression of the seaboard of the Western Isles. Although dates vary between locations in Scotland, it is generally thought that sea-level rise slowed after 6500 years BP (Hansom and Angus, 2001). A large influx of coastal sediments is thought to have occurred in response to the slowing of sea-level rise and beaches developed with sufficient excess sand available on the upper profile to be blown into the extensive dune and machair systems that remain active today (Ritchie, 1971; Hansom and Comber, 1996). However, ongoing sea-level rise, coupled with reductions in the offshore sand supply, subsequently resulted in chronic erosion of many Hebridean beaches and the frontal undercutting of the sand dune and machair systems that they support. The events at Mangersta appear to have followed this general trend, albeit exacerbated by local effects. The wider Mangersta coastline is characterized by cliffs cut into resistant rocks and so the potential sources of beach sediment are restricted to adjacent glaciogenic deposits. Since both the onshore and offshore extent of the Mangersta valley is limited (deep water exists close inshore at Mangersta), the amount of glaciogenic sediment available from both sources was probably also very restricted. It seems likely that given an initially limited supply of sediment, any beach development would eventually suffer sediment supply problems as the supply became exhausted. In the exposed wave and wind environment of Mangersta this situation may have been achieved earlier in the cycle of machair development and erosion than has occurred elsewhere.

The extent of hill machair at Mangersta, and the occurrence of erosional terraces at lower levels of the hill machair suggests that the central area once supported a machair surface probably sited landwards of a dune cordon behind the beach. The deflation of the central depression behind the beach is likely to be attributable to a reduction in the rate of sediment supply from an already restricted offshore contributing area. Reduction of beach sand leads to starvation and eventual removal by wind erosion of the landward dune area. The flat machair surfaces behind these dunes then underwent deflation down to either an underlying gravel basement or to the water table. Since the machair at Mangersta is crossed by streams and is often wet, it appears to have deflated down to the water table. Severely deflated machair occurs at several other sites in both the Western Isles and in the north-west Highlands, but nowhere is the amount of stripping so complete as at Mangersta (Ritchie and Mather, 1970b). The emplacement of the arcuate storm ridge on the upper beach at Mangersta may be a relatively recent development, since the ridge obscures any eroded frontal edge of the central machair depression. The gravels remain relatively unrounded and so have been relatively unaffected by wave abrasion. However, it may be that the gravels are part of a poorly-developed storm beach thrown up before the build-up of sand and subsequently exhumed from beneath the dune cover. Field visits in 2001 indicated that small amounts of gravel had been artificially added to the ridge in an attempt to reduce the likelihood of washover.

Conclusions

The causes of the severe deflation at Mangersta are not known with certainty, nor are the start and end dates of stripping episodes. For example, the effect of land-use practices or of rabbit-induced erosion cannot be discounted. Whatever the reason, Mangersta is the best example of stripped machair in Scotland, and so probably in the world. It seems likely that the fundamental and underlying cause for deflation and stripping may be related to the progressive exhaustion of a locally limited sediment supply.

Mangersta may thus represent an advanced stage in the cycle of growth and decay that, to varying extents depending on the local conditions, may characterize all beach–dune–machair complexes.

TRÀIGH NA BERIE, LEWIS, WESTERN ISLES (NB 103 360)

Introduction

Tràigh na Berie is one of the largest beach, dune and machair complexes on the Isle of Lewis in the Outer Hebrides of Scotland (Ritchie and Mather, 1970b; see Figure 9.1 for general location). Set in the rugged terrain of the ice-scoured Lewisian gneiss, Tràigh na Berie is relatively sheltered and contains a spectacular assemblage of soft coastal landforms including a wide beach, dunes, machair plain and hill machair. The coastal edge in the west and extreme east is now marked by erosion whereas the central part of the beach is experiencing accretion and embryo dunes characterize the coastal edge. Actively accreting dunes are unusual in Scotland and thus are of considerable geomorphological interest. The key geomorphological interest of Tràigh na Berie lies in the dynamism of the inter-relationships between the individual landform components of beach, dune, and machair.

Description

Tràigh na Berie, on the north-east side of the Valtos peninsula, is a wide 1.5 km-long sand beach extending in a long smooth curve between the rocky headlands of Sròn a'Chnip to the west and Stung to the east. Both headlands and the surrounding terrain are composed of resistant Lewisian gneiss. Behind the beach a wide dune and machair system extends inland onto steep ice-scoured slopes of gneiss. A linear depression cutting across the Valtos peninsula has been infilled with sand (Tràigh Teinish) which extends to a small beach to the south. A number of small lochs, located between the bedrock to the south and the advancing machair, are gradually being infilled by blown sand (Ritchie and Mather, 1970b; Figure 9.18). The main beach has a north-east aspect, facing into the relative shelter of Loch Roag. The adjacent coastline is highly crenulate and additional shelter is afforded to Tràigh na Berie by many small

Tràigh na Berie

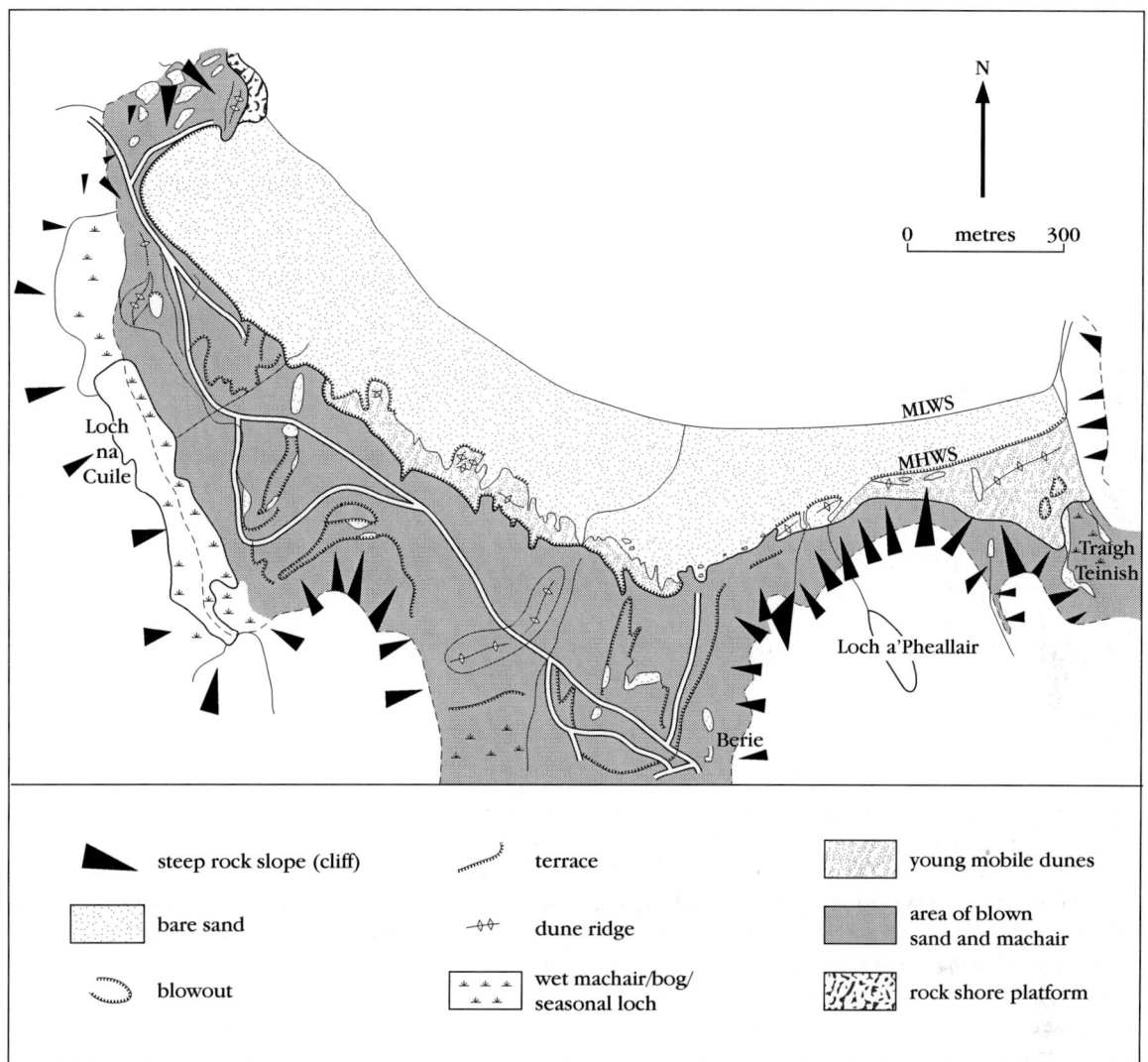

Figure 9.18 The geomorphology of Tràigh na Berie showing a mainly eroding coastal edge (see Figure 9.20). Several small lochs that lie behind the machair surfaces are subject to infill by blown sand. (Modified from Ritchie and Mather, 1970b.)

islets, the largest of which lie to the north-west and north-east (Figure 9.19). The sand at Tràigh na Berie beach is composed of fine, shell-rich sand of median diameter 0.2 mm and 47% calcium carbonate content. The upper beach, which forms the nourishment zone for the dune system, is over 50 m wide in the west but narrows to only 15 m in the east (Ritchie and Mather, 1970b). In spite of these intertidal widths, in 1970 the coastal edge in the extreme west was marked by a low sand-cliff, between 1–3 m high, eroded into machair and with no foredunes to the seaward side (Ritchie and Mather, 1970b). In contrast, the eastern end in 1970 was marked by rapid accretion with large quantities of sand accumulating in the form of broad dune ridges up to 12 m high across the mouth of the Tràigh Teinish depression. In view of this, Ritchie and Mather (1970b) suggested an easterly drift of sediment and an anticlockwise rotation of the beach arc, a view echoed by Ramsay and Brampton (2000e). However, by July 1994 most of the eastern coastal edge was undercut and showing signs of erosion. The development of embryo dunes was localized to only one central section of the beach, with little

Machair

Figure 9.19 An aerial view of Tràigh na Berie from the south-west taken in the mid-1980s shows a dynamic coastal edge subject to pressure from tourist caravans and resulting in substantial wind-blow and destabilization. Caravans are now restricted to the central section of the dune and machair area. (Photo: S. Angus.)

sign of coastal accretion at the east end (Keast, 1995). Figure 9.20 shows the undercut erosional edge in October 2001; it appears that the proportion of coastal edge experiencing erosion has increased greatly since it was first described in the 1970s.

Dune forms are best developed in the central–eastern part of the site (Figures 9.18 and 9.20) where the coastal edge is characterized by low mobile dunes, up to 5 m high but which increase in height towards the east (Keast, 1995). Along parts of this foredune ridge, wave undercutting has exposed parts of the dune face that are now subject to wind erosion and deflation. In 1970 the dune face in the central part of the beach was accreting with embryo dunes developing into short north–south-trending dune ridges separated by stretches of bare sand that coalesce in places to form continuous, coast-parallel foredune ridges (Ritchie and Mather, 1970b). In 1970 these embryo and foredunes abutted discontinuously against an abandoned machair edge, suggesting that the period of rapid foredune accretion observed in 1970 post-dates a previous period of machair erosion (Ritchie and Mather, 1970b). However, by 1995 the ridge had eroded in several places and had been completely breached around NB 110 357 (Keast, 1995) as a result of frontal wave erosion followed by wind deflation. In the east, within Tràigh Teinish, large broad dune ridges are orientated in a north–south direction and reflect the topographical channelling of winds.

Seawards of the beach and dune zone a wide machair plain, characterized by numerous erosional edges most of which are now healed, extends inland to the steep, ice-scoured gneiss slopes. The machair impounds several small lochs, which are gradually being infilled by blown sand to form marshy depressions. A major erosional terrace lies over 400 m landward of the high-water mark and separates the machair into two major units (Figure 9.18). The larger area of machair lying to the seaward side of the scarp is flat and scarred by numerous small and low erosional edges. The smaller machair unit to landward of the edge slopes gently landwards and grades into the wet machair, marshes and impounded lochs. Farther inland considerable areas of hill machair have developed on the steep gneiss slopes. In places the hill machair is eroded and blowthroughs and

Tràigh na Berie

Figure 9.20 This view of Tràigh na Berie from the east (October, 2001) shows a wide intertidal zone backed by an undercut dune cordon and machair to landward. (Photo: J.D. Hansom.)

exposed sand cliffs are present. At the extreme east end of the site, behind the coastal dune cordon, a level area of machair lies close to the water table.

Interpretation

Tràigh na Berie is a small basin that has become infilled with sand driven onshore by a rising sea level during the late Holocene. Since there is no evidence that gravel ridges underlie the beach, it can be assumed either that gravels were not a major constituent of the source materials or that gravel ridge development failed to occur, probably on account of the low levels of wave energy within the bay. It may also be the case that if the sand budget at Tràigh na Berie has been positive until recently, any gravel ridges still remain to be sectioned and exposed. Ritchie and Mather (1970b) described Tràigh na Berie as a relatively well-nourished and stable beach dominated in places by rapid accretion. The main source of sediment to the beach and dune system was, and still is, derived from the sand-covered seabed of West Loch Roag (Ritchie and Mather, 1970b). The high proportion of shell-derived beach sand suggests a significant offshore sand source but it is unknown whether the supply of sand at Tràigh na Berie has diminished recently. There is morphological evidence to suggest that the beach undergoes temporary changes from accretion to erosion and that the beach is currently mainly erosional (Figure 9.20).

The erosion of the machair edge in the west and rapid dune accretion in the east observed by Ritchie and Mather (1970b) led them to suggest that the beach appeared to be rotating in an anticlockwise direction. However, when re-examined in 1995 and 2001, this tendency was not obvious and erosion of the east and west ends was reported, a graphic example of the dynamism of both beach and dune that has resulted in the inclusion of Tràigh na Berie in the GCR network. The beaches and dunes appear to undergo alternating phases of erosion and accretion, an unusual occurrence for a site that seems well-served by a sand source and so relatively sheltered from both wind and wave.

Similarly, although on a longer timescale, the numerous erosional edges that characterize much of the machair plain indicate that the seaward part of the machair plain has experienced more than one episode of wind erosion (Ritchie and Mather, 1970b). The eroded machair scarp in the centre of the beach, which was fronted by actively accreting foredunes in 1970, may be the most recent expression of an erosional system that involves not only short cycles of alternate wind-generated scarp erosion and machair deflation down to the water table, but also of wave-generated frontal erosion of the seaward edge. The relative shelter of the site, plus its apparently healthy sand source may be the main reason why this beach–dune–machair system has not yet undergone the chronic frontal erosion found elsewhere on the more exposed shores of the west coast of the Hebrides. In this respect, Tràigh na Berie conforms with the model proposed in Figure 9.3 of the introduction to this chapter, with its dune cordon undergoing erosion but still largely intact. The sand removed from the dunes and machair edges is deposited elsewhere within the system. Although it remains to be fully investigated, the evidence of past erosional episodes within the machair is of great importance in the understanding of the way in which machair systems evolve over time.

It is possible that a contributory cause of the present instability in the central section of Tràigh na Berie is related to the impact of tourists using the beach (Angus and Elliott, 1992; Ramsay and Brampton, 2000e). Access through the dunes and machair to the beach certainly exacerbates erosion of the low machair faces at the western end of the beach. Wind erosion and re-deposition in the central section of the dune is also exacerbated by the presence of caravans within the dunes and the high density of pedestrian tracks over the dunes. However, the main cause of instability appears to be frontal erosion by waves and its subsequent effect on blowthrough development. The main drivers behind this situation are more likely to be sea-level rise and sediment supply than human impact.

Conclusions

Tràigh na Berie is a large spectacular beach–dune–machair unit set in an area of rugged gneiss upland, a very different setting from the open machair plains typical of the Uists. The geomorphology is controlled by a complex of inter-related marine and aeolian processes resulting in an extremely dynamic and variable system. The complexity of inter-relationships and vigour of processes is of very high geomorphological importance. Tràigh na Berie is of particular interest because it contains evidence of several stages of dune development as well as evidence of past erosional episodes in the machair plain. Tràigh na Berie is of outstanding importance in the context of elucidating machair evolution.

BALNAKEIL, SUTHERLAND (NC 390 700)

Introduction

The *c.* 3 km-long peninsula of An Fharaid encloses Balnakeil Bay and lies at the entrance to the Kyle of Durness, 15 km east of Cape Wrath on the extreme north-west tip of the Scottish mainland (see Figure 9.1 for general location). It contains a magnificent and spectacular array of beach and dune erosional and depositional landforms ranging from undercut dune faces to large and active dune blowthroughs. Large blowthrough corridors and transgressive sand waves carry windblown sand across the An Fharaid peninsula from the beaches on the west coast to cascade over a prominent cliff on the east, a landform assemblage unique in Britain if not in Europe. The wide, steep-sided, blowthrough corridors are up to *c.* 25 m deep and up to *c.* 400 m long (Figure 9.21). Mature machair vegetation has stabilized much of the higher blown sand deposits in the north and east of the peninsula leading to a marked contrast between these stable machair slopes and the lower altitude, highly active dunes and blowthrough corridors. The dunes of Balnakeil are some of the most active in Scotland and display important interactions between wave erosion and undercutting of the dune-toe and the effect of high velocity winds in driving the eroded dune sand inland and uphill. However, in spite of the clear geomorphological interest and importance of this site, it has attracted only limited scientific research, and much of the following description and interpretation is drawn from Ritchie and Mather (1969) and Hansom (1998).

Description

Balnakeil beach occupies the west side of the isthmus that connects the An Fharaid headland with the mainland to the north of Durness (Figure 9.21). The *c.* 3 km-long peninsula of An Fharaid consists of a block of gneiss tilted towards the south-west and connected at its southern margin to a similarly tilted block composed of schist beyond which the Durness limestones crop out (Ritchie and Mather, 1969). The blocks provide an inclined plane up which large quantities of sand are blown in a north-easterly direction from the beach and from frontal dunes undergoing erosion. The peninsula is up to 1.5 km wide close to the headland in the north, but narrows to *c.* 400 m in its central section at NC 390 700. The west coast is cliffed from NC 385 708 north to An Fharaid headland while the east coast is entirely cliffed with only a few small pocket beaches. The sand has a mean diameter of 0.292 mm with 52% of this being composed of calcium carbonate derived from shells (Ritchie and Mather, 1969). The spectacular assemblage of cliff, beach, dune and machair forms of the peninsula can best be described in two sections.

Balnakeil

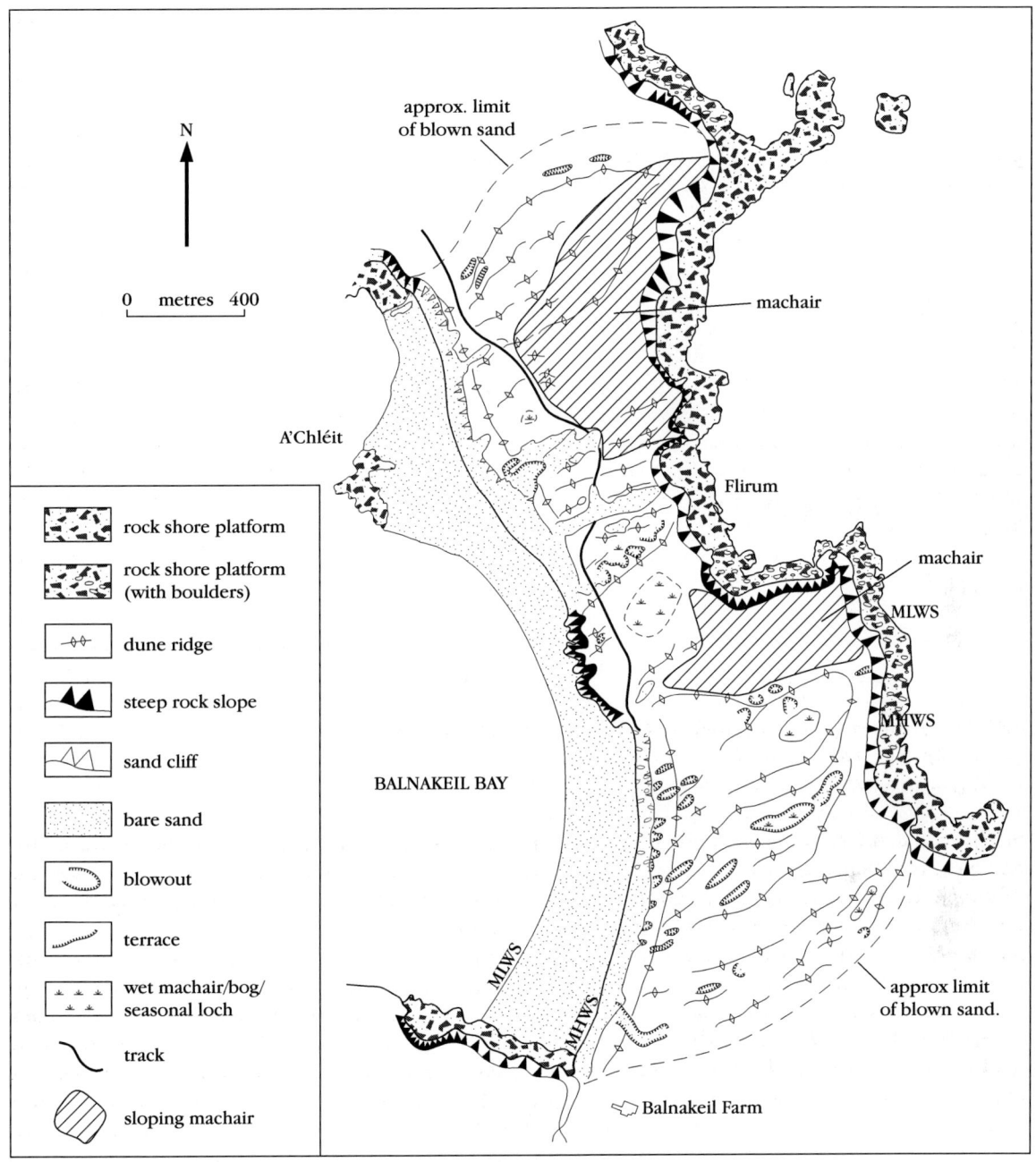

Figure 9.21 The geomorphology of Balnakeil–An Fharaid is dominated by the easterly transport of blown sand from the beaches of Balnakeil Bay into dune and machair surfaces that climb the slopes of the An Fharaid peninsula. Some of the larger linear blowthroughs channel sand to cascade over the cliff on the eastern edge of the peninsula. (After Hansom, 1998.)

The southern section includes the wide intertidal sand beach of south Balnakeil Bay and the mature dune system that extends across the peninsula to cliffs on the east coast. A broad, c. 10–15 m-high, coast-parallel dune ridge backs the 220 m-wide sandy beach and is erosional over much of its length. Although there is evidence of past wave-undercutting and basal slumping of the dune toe, the erosional contact is often covered by a small ramp of windblown sand to give the appearance of relative stability (Hansom, 1998). Higher up, to the rear of the

Figure 9.22 Looking north over the large linear blowthrough at Balnakeil, a highly dynamic feature that channels wind flow to the east to allow sand to be blown onto the adjacent climbing machair surfaces as well as to cascade over the cliff edge at Flirum, at the right of the scene. (Photo: J.D. Hansom.)

beach in the southern section, the dune face is well-vegetated with vigorous marram *Ammophila arenaria* growth, testifying to a healthy supply of fresh sand from the upper beach below. Several small blowthroughs occur in the coastal dune ridge, particularly around the stream outlet in the south close to Balnakeil Farm. Partial protection is afforded to the dune toe by the remnants of a tarmac roadway built in the 1950s along the approximate line of mean high-water springs, at the back of the beach.

To landward of the coastal dune ridge in the southern section, the undulating mature dunes are low in the west but increase in height as the system extends eastwards across the peninsula to climb An Fharaid headland as a thin machair cover. Damp dune slacks are common within the lower parts of the dune system and support wet machair vegetation communities. The bedrock surface is low in the west where it is veneered by sand but rises in the east to substantial cliffs of *c.* 30 m OD. The dune ridges trend in a north-easterly direction and reach heights of up to *c.* 20 m in places. They are mainly stable at present and support a mature vegetation cover, although there are several small blowthroughs and erosional scrapes probably initiated by sheep-rubbing and subsequently enlarged by wind. The highest dune in the southern section has a large circular blowthrough eroded into its north-west face. The windward face of the blowthrough is cut into bare sand and is highly active with wind eddy and scour of the lee face. This has led to undercutting and an impressive series of individual turf-block terraces in the process of sliding down the steep sand face of the blowthrough.

The northern section is the narrowest part of the peninsula and extends from Flirum on the east and the rock protuberance on the west coast, which separates Balnakeil Bay beach into a southern and northern section, to the rocky slopes north-east of the intertidal skerry of A'Chléit (Figure 9.21). The geological structure of this section consists of a low rocky slope rising from sea level in the west to up to 50 m-high cliffs near Flirum in the east. The skerry of A'Chléit is connected to the mainland by an intertidal sand tombolo, and acts as an intertidal pinning point for the 200 m-wide beach. To the north and south the beach width is restricted to less than 50 m. In spite of the volume of sand that has accumulated in the shelter of A'Chléit, the upper beach is erosional in this central section, with only a narrow beach at high tide. Exposure to north-westerly waves has resulted in undercutting and destabilization of the *c.* 15 m-high mature dunes that back the narrow beach, linear erosion of the dune toe–upper beach interface and the development of steep erosional faces in the bare sand of the dunes. The associated removal of dune-face vegetation has resulted in blowthrough activity in two main areas. One of these, centred on NC 390 702, continues through the peninsula east to Flirum and comprises large, steep-sided, linear blowthrough corridors that rise from sea level on the west shore to *c.* 30 m OD on the east shore (Figure 9.22). The presently unvegetated blowthrough feeds sands directly from Balnakeil Bay into a large transgressive sand wave that in the north inundates machair pasture and a small sheep enclosure and in the east cascades over the cliff. Dunes occur on the cliff edge in the east but sand is removed by waves at the foot. The active blowthrough corridor on the west is also flanked by a series of linear north-easterly orientated ridges capped by vigorous marram growth. The second blowthrough zone is centred on NC 387 704 where a vehicle access track is under threat of undermining by frontal erosion of the dunes on the west side. Emergency maintenance work has resulted in assorted rubble being tipped to protect the track, and sand repeatedly being cleared and bulldozed into the blowthrough. To the south of this point a large, wide blowthrough extends from mean high-water springs (at the wrecked hull of a boat) eastwards to *c.* 30 m OD. This large wide area of bare sand is punctuated at the western end by the upstanding remnants of old fixed dunes, complete with vegetation cap that are now eroded on all sides to produce pinnacles ranging in height from 3 m to 15 m. Landwards of the blowthrough vigorous marram growth has led to stabilization of the dune surfaces. A distinct boundary exists between the lower surfaces stabilized by marram *Ammophila arenaria* and the higher surfaces to the east charcterized by machair grassland. This boundary is likely to be mobile with rapid sand inundation favouring marram encroachment of machair.

Interpretation

In spite of the wealth of active dune processes at Balnakeil, with large blowthroughs feeding transgressive sand waves that allow sand to travel from the beaches on the west coast across the peninsula to cascade down the high cliffs on the east coast, the site has attracted surprisingly limited geomorphological research. Ritchie and Mather (1969) and Hansom (1998) described the geomorphology of the site and highlighted the importance of aeolian processes in both the past and present geomorphological development and evolution of the site, particularly in the central section. However, further geomorphological research at this spectacular and highly active site is clearly warranted.

Based on both theoretical considerations and field observations, Ritchie and Mather (1969) suggested a clockwise rotation of the beach over time with the erosion in the north being balanced by accretion in the south. In common with many Scottish beaches and particularly those in the north and west of Scotland, it is likely that Balnakeil Bay now has a much-reduced supply of sand from the offshore than previously (Mather and Ritchie, 1977; Hansom, 1999). This is associated with the reductions in offshore glaciogenic supply following the slowdown in the rate of Holocene sea-level rise in mid-Holocene times (Hansom, 1999). This theoretical argument is supported at Balnakeil by data that show the offshore seabed to be characterized by bedrock rather than sediment (BGS, 1991), the sediment that once rested within this zone having been transported onshore earlier in the Holocene Epoch. With limited sand supply from the offshore and no sources of river-borne sand nearby, the quantity of sediment within Balnakeil Bay is more or less finite and its distribution mainly the result of wave- and current-transport processes within the bay. The field observations that support the above view relate to the predominantly unidirectional waves that impinge on the beaches from the north-west and produce a southwards transport of sand. Some 20 years after the beach rotation suggestion of Mather and Ritchie (1977), similar landforms and processes were recorded by Hansom (1998) and so are clearly long-lived.

The northern beach remains narrow and waves access the base of the dunes regularly giving rise to chronic frontal erosion and slumping of the dunes above. Given the wave refraction patterns within Balnakeil Bay, much of the sand-loss is to the south but significant amounts of sand are also lost by wind-blow to the backing dunes via blowthrough corridors. The northern part of Balnakeil Bay is likely to operate at a budgetary loss in terms of sand input and output. However, in spite of an apparently rapid rate of erosion and dune-toe recession in the north, there is also some evidence suggesting that wave attack of the dune face may not be as rapid as supposed. In-situ timbers from the hull of a boat wrecked on the northern beach over 100 years ago are still exposed at MHWS, although the dune toe has receded a few metres beyond this (K. MacRea, pers. comm., 1996.).

In contrast, Mather and Ritchie (1977) noted accretion in the southern part of Balnakeil Bay, where both the intertidal zone and the upper beach are wider. Refracted waves break on the shore more or less simultaneously, and although longshore transport out of Balnakeil Bay into the Kyle of Durness to the south occurs, it is probably limited. This may not have been the case earlier in the Holocene Epoch, and there is evidence that the sand budget in the south of the Kyle of Durness was sufficient to feed wind-blown sand to South Balnakeil via the now-stabilized sand sheets on top of the intervening peninsula to the west of Balnakeil Farm. As a result, it is likely that the southern part of Balnakeil Bay is a partial sediment-trap, although the amount of sand transported to and from the southern part of the bay is unknown. Storm wave activity on the south beach probably results in onshore–offshore sediment exchanges, and this is the likely reason why there is evidence of not only accretion but also erosion on the south beach. The eroded remnants of a tarmac access road built by the Ministry of Defence in the 1950s runs along the back of the beach. Within a year of construction, the roadway was undermined by wave erosion and abandoned (K. MacRea, pers. comm., 1996). Since then, rather than being buried by sand accretion, its foundations remain exposed and in places subject to minor undermining by waves. The presence of the foundations now provide protection to the dune toe from wave undercutting, although vigorous marram growth on the seaward faces of the well-vegetated dunes above indicate a healthy blown-sand supply from the beach below (Hansom, 1998).

Wave-induced frontal erosion of the dune toe creates the conditions for wind-induced point erosion and blowthrough activity, since wave undercutting of the mature dunes backing the northern beach removes vegetation and allows unrestricted wind access to the bare sand surfaces. Wind deflation and sand transport has resulted in large areas of bare sand and massive blowthrough corridors that extend over $c.$ 400 m from the west coast of the peninsula to cascade in sandfalls over the eastern cliffs. The range of large- and small-scale wind erosional and re-depositional features within this system is spectacular and of immense scientific interest. The stabilized linear dune-ridges that flank the active, steep-sided blowthroughs provide evidence that until recently the ridges were the sides of large active blowthrough corridors which are now naturally stabilized by vegetation. This suggests a long-lived process that undergoes small shifts in location leading to time-transgressive zones of alternating erosional activity and vegetational stability.

Conclusions

Balnakeil Bay contains some of the most dynamic dune forms in Scotland. The rock-floored peninsula tilts towards the south-west, providing an inclined plane up which large quantities of blown sand progress in a north-easterly direction. Large, steep-sided and highly active blowthrough corridors feed transgressive sand waves that transport sand across the peninsula from the beaches on the west to cliffs on the east. It is possible that an earlier transgressive sand-wave system operated on the rocky peninsula between the Kyle of Durness and southern Balnakeil Bay. Sand cascades down the eastern cliffs and is effectively lost from the Balnakeil system. Mature machair vegetation has stabilized the blown sand veneer at higher altitudes, and the juxtaposition between this and the marram-dominated highly active ridges associated with the blowthrough corridors is striking and of great geomorphological interest. Balnakeil Bay is characterized by the juxtaposition of a range of dune erosional processes and landforms within a setting that is unparalleled in Britain, if not in Europe.

Chapter 10

Saltmarshes

Introduction

INTRODUCTION

E.C.F. Bird

Saltmarshes are vegetated areas in the upper part of the intertidal zone found on the shores of inlets, estuaries and embayments that are sheltered from strong wave action. The vegetation consists of halophytic (salt-tolerant) grasses, herbs and shrubs that can grow in the upper part of the intertidal zone, and are subjected to regular inundation by the sea. Their ecology has been described by Ranwell (1972), Adam (1990) and Packham and Willis (1997). Saltmarshes extend down to about mid-tide level, and have muddy, or sometimes sandy, substrates. They are generally bordered seawards by intertidal mudflats or sandflats, bare of vegetation or with carpets of algae, such as *Enteromorpha* spp., or seagrasses such as *Zostera* spp..

The distribution of active saltmarshes in Britain is shown in Figure 10.1 British saltmarshes are extensive where the tide range is large and the intertidal zone wide, with a very gentle transverse gradient, as on the shores of the Severn and Dee estuaries, Solway Firth (see GCR site report in the present chapter), and Bridgwater Bay in the Bristol Channel. A distinction has been made between these 'open marshes', which have spread seawards, and 'closed marshes', which occupy areas between landward recurves in the lee of spits such as Blakeney Point, Norfolk, and Culbin, Moray, and barrier islands such as Scolt Head Island, on the North Norfolk coast (see GCR site reports in Chapter 11) and Morrich More, Ross and Cromarty. Closed marshes become lagoons at high tide, then drain out through a system of converging tidal creeks as the tide falls (Steers, 1977).

Although small in areal terms, the west coast of Scotland and the Western Isles support many small and fringing saltmarshes, particularly where relative sea level has risen to create sheltered conditions and a complex shoreline (e.g. Loch Maddy, North Uist). As a whole, the 6567 ha of Scottish saltmarshes comprise some 15% of the British resource. Unlike the saltmarshes of southern and eastern Britain, they generally tend to be grazed, lack high sediment inputs and have a complete transition from halophytic to terrestrial vegetation. They are also characterized by mainly sandy substrates rather than the muds of the English saltmarshes.

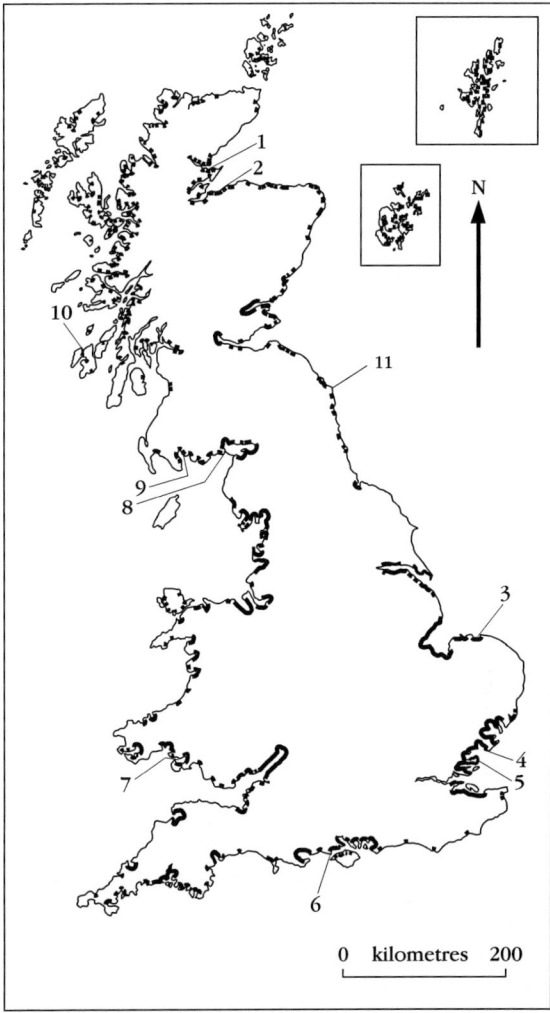

Figure 10.1 The generalized distribution of active saltmarshes in Great Britain. Key to GCR sites described in the present chapter or Chapter 11 (coastal assemblage GCR sites):
1. Morrich More; 2. Culbin; 3. North Norfolk Coast; 4. St Osyth Marsh; 5. Dengie Marsh; 6. Keyhaven Marsh, Hurst Castle; 7. Burry Inlet, Carmarthen Bay; 8. Solway Firth, North and South shores; 9. Solway Firth, Cree Estuary; 10. Loch Gruinart, Islay, 11. Holy Island. (After Pye and French, 1993.)

Extensive saltmarshes bordered by intertidal mudflats are seen in Poole Harbour, Dorset (Figure 10.2), Southampton Water, Portsmouth Harbour, Langstone Harbour and Chichester Harbour on the south coast of England. These large inlets formed during the Holocene marine transgression, and have persisted because the inflowing rivers are too small to supply much sediment, and because this is a subsiding coast.

Saltmarshes

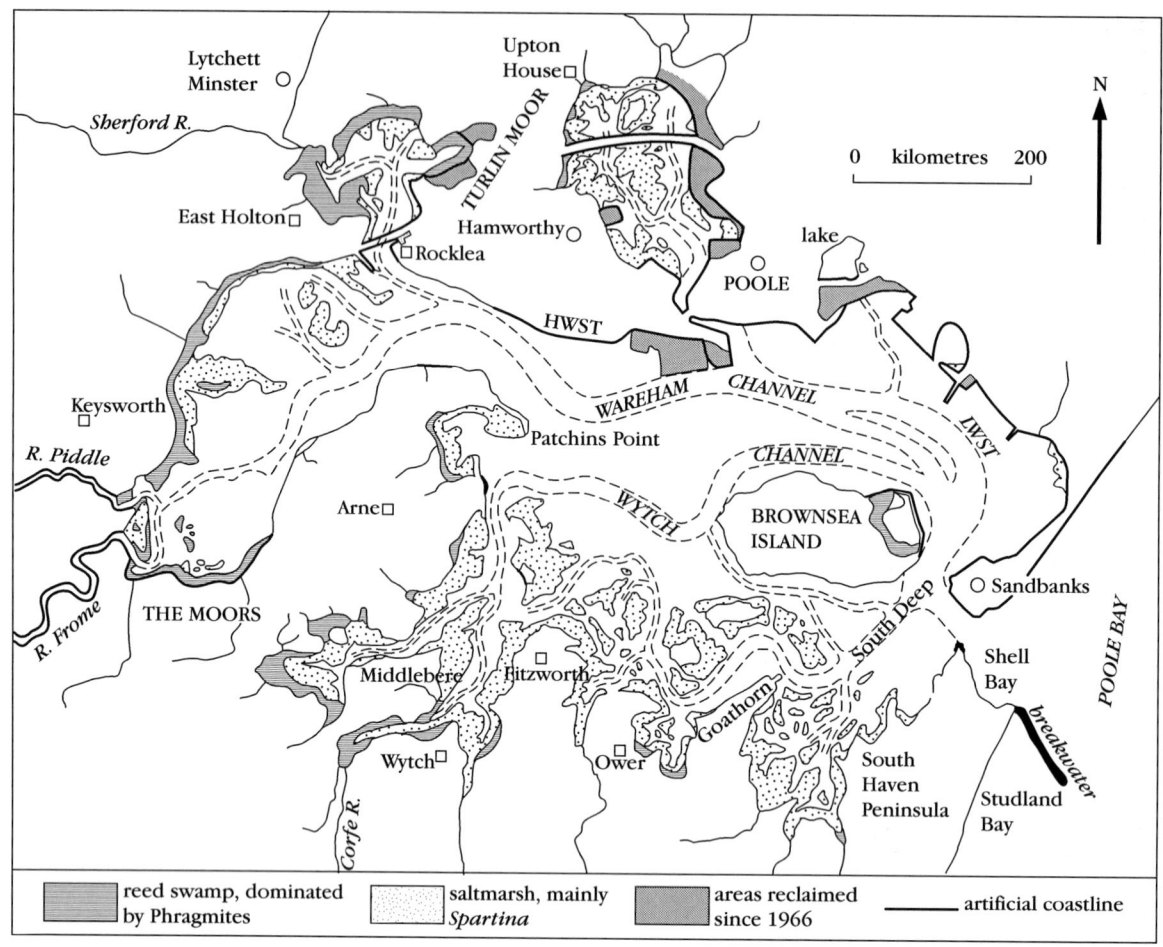

Figure 10.2 Marshes in Poole Harbour, Dorset. Common cord-grass *Spartina anglica* saltmarsh has developed here since 1899, and this is backed in the upper reaches by *Phragmites australis* reedswamp, where salinity is reduced by freshwater inflow. Saltmarsh has been reclaimed by embanking, especially near the northern urbanized fringes. (After Bird, 1984, p. 214; based on original map by V.J. May, updated to 2000)

On the west coast of Britain, notably in the Welsh and Scottish estuaries and in the Solway Firth (see GCR site reports in the present chapter), saltmarshes are generally firmer than those on the east coast because there are higher proportions of sand in the muddy sediment. Samphire *Salicornia* spp. are again the pioneers, but later colonization is dominated by grasses such as *Puccinellia*, which form a sward on marshland dissected by winding tidal creeks.

In Scotland, there are very few truly muddy saltmarshes; most of the marshes are sandy in character. Scottish saltmarshes tend to have little pioneer vegetation in comparison to those of England, although this rapidly gives way in the main to common saltmarsh-grass *Puccinellia maritima* with plantain *Plantago*, thrift *Armeria* and sea milkwort *Glaux* also common in the grazed swards. In sheltered shores of the Highland area, especially along sea lochs, patchy saltmarsh can develop on stony or rocky substrates. Saltmarsh vegetation has also been recorded on the cliff tops of St Kilda, resulting from wave spray in this exposed setting.

Evolution of a saltmarsh

Saltmarshes begin to form when vegetation colonizes the upper part of the intertidal zone (Pethick, 1984; Frey and Basan, 1985). Saltmarshes have been forming in sheltered sites around the coasts of Britain since the sea approached its present level about 5000–6000 years BP as a result of the Holocene marine transgression. For example, at St Osyth Marsh,

Introduction

Essex, (see GCR site report in the present chapter) shows by radiocarbon dating that saltmarsh development began at about 4200 years BP (Pethick, 1981).

In accreting intertidal zones, especially in areas where shelter from strong wave action is enhanced because of the growth of protective spits, barriers or shoals, the early stages in the development of a saltmarsh can be studied. Sandy saltmarshes are developing in the lee of The Bar, and its associated spits, within the barrier island complex at Culbin Sands, Moray (see GCR site report in the present chapter and in Chapter 11; Comber *et al.*, 1994), and behind Crow Point at Braunton Burrows, Devon (see GCR site report in Chapter 7).

On the North Norfolk coast (see GCR site report in Chapter 11), the evolution of saltmarshes – where stages in saltmarsh evolution can be traced from east to west between successively-formed recurves on spits at Blakeney Point and Scolt Head Island – has been documented, (Steers, 1960; Pethick, 1980a). There is initial colonization of muddy or sandy areas in the upper intertidal zone by individual halophytic plants (e.g. samphire *Salicornia* spp.), which expand vegetatively and eventually coalesce to form marshland dominated by single species such as common cord-grass *Spartina* spp.. Other species colonize, and the saltmarsh begins to trap sediment (mainly clay, silt and organic matter, sometimes with some fine-grained sand and shells) washed into the vegetated area by waves and currents as the tide rises, and retained by the filtering network of stems and leaves as it falls. Although often characterized as tide-dominated morphology, most saltmarshes are also influenced by wave action as the tide rises and falls. The vegetation diminishes wave action; swards of saltmarsh grass can reduce wave heights by up to 70% and wave energy by over 90%, and, as the water velocity diminishes, fine-grained sediment is deposited (Pethick, 1980b; Möller, 1998).

Sedimentation is also promoted by the growth of a subsurface root network, which binds the accreting sediment, by the presence of adhesive algal mats on the muddy surface, and by flocculation and precipitation of clay by the salt exuded from marsh plants (Pethick, 1984). In addition, mud that adheres to leaves and stems dries off and falls to the substrate. The processes of seaward spread and upward growth of a saltmarsh are aided by an abundant supply of sediment, but even in the absence of sediment input, saltmarshes can still aggrade by the accumulation of peat derived from the decaying vegetation.

In due course, a saltmarsh is built up to high spring tide level as a depositional terrace, only rarely submerged by the sea, that slopes gently from approximately the high spring tide line (HWOST) to the high neap tide line (HWONT), then more steeply (sometimes as a micro-cliff) to the mid-tide line (Figure 10.3). In the absence of saltmarsh vegetation the mudflats and sandflats remain as a more variable intertidal slope, and if the plant cover dies, or is cleared away, the saltmarsh terrace becomes dissected and degraded by erosion.

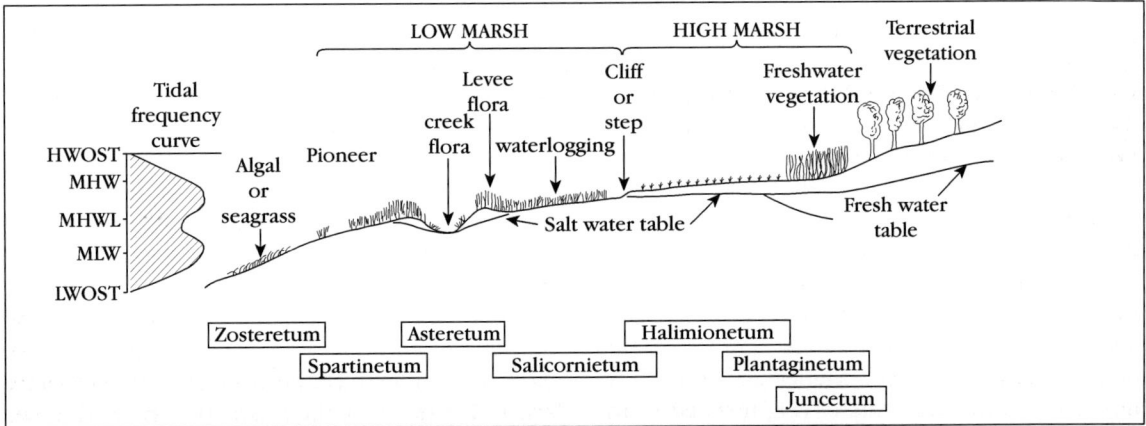

Figure 10.3 Typical saltmarsh vegetation zonation: the dominant species found in England and Wales at each level are named in the boxes. In Scotland, the sandy saltmarshes are dominated by common saltmarsh-grass *Puccinellia*. (After Carter, 1988, p. 344.)

Saltmarshes

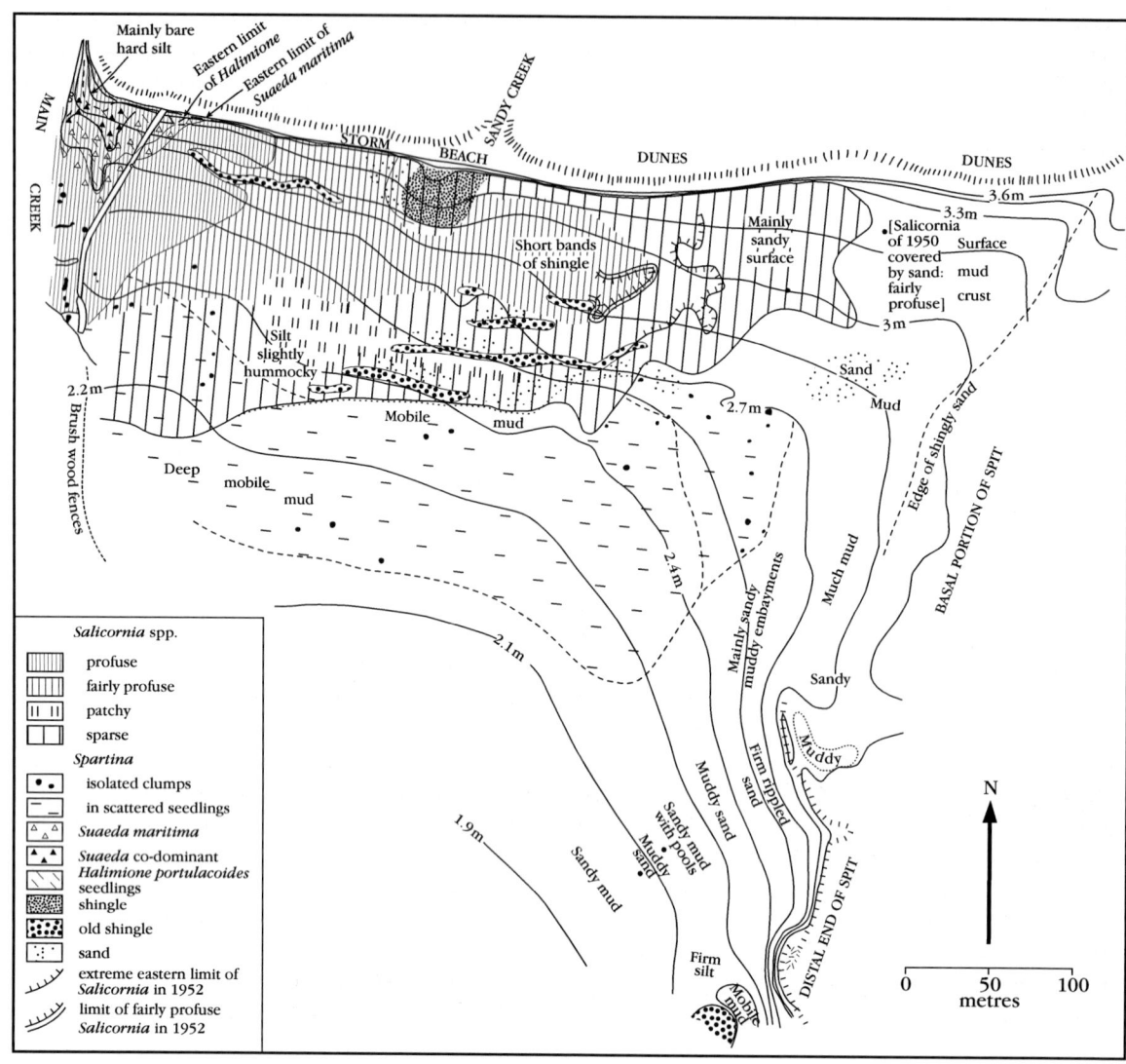

Figure 10.4 Map of the saltmarsh at Gibraltar Point, Lincolnshire, recording the position in 1951. The marsh was growing on the landward side of the spit; the area was re-surveyed in 1959, by which time 15–30 cm of sediment had built up the marsh surface over most of the area, and the low-lying mud and sand of 1951 had been colonized by common cord-grass *Spartina*. (After King, 1972a, p. 428.)

Saltmarsh species tolerate varying depths and durations of tidal submergence, and so spread forward to an intertidal contour that corresponds with these limits. As a result there is often a well-defined zonation of species parallel to the coastline, with plants such as samphire *Salicornia* spp. dominating the lowermost zone that is most frequently submerged by the tide, and sea-rush *Juncus maritimus* and other saltmarsh plants occupying higher zones that are less frequently submerged (Figures 10.3 and 10.4). These zones could simply represent the occupation by each species of a suitable habitat that moves seawards as accretion continues, but where the vegetation traps sediment that builds up and progrades the substrate, it also prepares the way for the seaward advance of the plant zones (a vegetation succession) on the developing saltmarsh terrace. The vegetation succession then continues with invasion of the landward edge of the saltmarsh by less halophytic species, such as common reed *Phragmites australis* and rushes (e.g. *Juncus* spp., *Scirpus* spp.) together with freshwater wetland species such as swamp scrub including willow *Salix* spp. and alder *Alnus* spp.. Eventually the transition is made to dry-land vegetation (Ranwell, 1972; Packham and Willis, 1997). These later stages in vegeta-

Introduction

tion succession can be seen locally at the back of saltmarsh terraces around Poole Harbour (Bird and Ranwell, 1964; Figure 10.2), but they have often been destroyed by hinterland drainage and land-claim. Succession from saltmarsh to freshwater swamp and land vegetation is likely to be accelerated by coastal emergence, and evidence of this may be found in Scottish sites where postglacial isostatic rebound is in progress, such as at Morrich More, in the Dornoch Firth (Hansom and Leafe, 1990). One of the few saltmarsh successions to woodland in the UK is in Sutherland, where the completion of an embankment in 1816 enabled alder *Alnus* to colonize the saltmarsh at the head of Loch Fleet. The resulting alder woods are so well established that they are now a National Nature Reserve.

Conventionally, algal mats (such as *Enteromorpha* spp.) and seagrasses (notably eelgrasses *Zostera* spp.), which grow on mudflats in the intertidal and subtidal zones, are not regarded as saltmarshes, even though they often prepare the way for saltmarsh encroachment. Seagrasses trap sediment because the plants are erect when the tide is high, and can form a sediment-filtering meadow in which wave heights are reduced by up to 40% and wave energy by 60% (Fonseca, 1996). The substrate is thus raised to form a seagrass bank or terrace that often grows upward and outward as the vegetation spreads. Nevertheless, some seagrass terraces are sharp-edged owing to wave activity or current scour.

Rates and patterns of accretion

Rates of vertical accretion have been measured on north Norfolk saltmarshes, notably at Scolt Head Island (Steers, 1960), where the progressive burial of artificial markers inserted in a saltmarsh showed vertical accretion of up to 8 mm a^{-1}, with variations related to marsh elevation and inundation frequency and the retention of sediment from turbid water overflowing from tidal channel margins. These saltmarshes grow vertically by accretion during ordinary high tides, but the higher parts receive sediment only in storm events (French and Spencer, 1993). In general, vertical accretion is relatively slow at the upper and lower limits of the saltmarsh, and more rapid in the intervening zone, where saltmarsh vegetation forms a relatively dense sediment-trapping cover and is regularly invaded by sediment-laden tidal water. Pethick (1981) found the upper limit of accretion on saltmarshes to lie below the level of the highest spring tides. Marshes, therefore, never quite attain this altitude.

Vertical accretion of up to 15 mm a^{-1} has been measured on saltmarshes in Essex, where coastal subsidence is in progress (Ranwell, 1972). In saltmarshes bordering the Severn estuary, French (1996) used evidence from heavy metal profiles and lead (^{210}Pb) dating to define distinct sedimentary units (between planes dating from 1840–1850, 1936±7, 1971±4 and 1958±4), and shows that vertical accretion (3–4 mm a^{-1}) has been proceeding at about the same rate as sea-level rise in the area. There are at least three saltmarsh terraces bordering the Severn estuary, representing cycles of marsh erosion and accretion. Accretion is most rapid (12.1 mm a^{-1}) on the lower terrace, submerged by every high tide, slower (6.4 mm a^{-1}) on the middle terrace, and slowest (2.3 mm a^{-1}) on the higher terrace, which is inundated only by high spring tides (French, 1996). However, where supply rates are high, accretion responds accordingly. For example, in the sandy saltmarshes of the inner Solway, Harvey (2000) has measured rates of vertical accretion of up to 51 mm a^{-1} in some areas and in excess of 20 mm a^{-1} over wider areas on account of a very healthy offshore sand supply.

The fine-grained sediment deposited in saltmarshes is derived largely from bordering intertidal mudflats and sandflats, which in turn have been supplied with clay, silt and sand by rivers, or similar sediment eroded from cliff and rocky shore outcrops. Fine-grained sediment has also been carried in from the sea floor, especially where there are mudrock outcrops or glacial or periglacial deposits in nearshore shallows, and organic matter derived from seaweeds and marine fauna, especially shelled organisms, has been swept on to marshes. Radionuclides contained in sea water and seabed sediments are also transported onto saltmarshes. Coring of the Solway saltmarshes has revealed ^{137}Cs and ^{241}Am peaks in the subsurface layer that relate to past high levels of emission from the Sellafield Nuclear Fuel Reprocessing Plant in Cumbria (Harvey and Allan, 1998; Harvey, 2000).

Providing there is a supply of fine-grained sediment, and wave action is gentle, saltmarshes can spread rapidly (Figure 10.4). The supply of mud to a saltmarsh increases where fluvial sediment yields are augmented by catchment soil erosion, where the dredging of channels increases muddy sediment in suspension, or where dredged material is dumped on or near marshes, accelerating vertical accretion and progradation. An excessive rate of mud deposition may however blanket and kill saltmarsh vegetation.

Sections through saltmarsh terraces, exposed in the banks of tidal creeks or in cliffs at the seaward edge, as on the saltmarsh at Morrich More, Ross and Cromarty (see GCR site report in the present chapter), generally comprise stratified deposits, with layers of coarser sands within the host sands or muds. These variations are related to wave conditions, storm waves washing sand into the saltmarsh, and mud accumulating as the tides rise and fall in calmer weather. In the Severn estuary the grain size of saltmarsh sediments diminishes from fine-grained sand to silt and clay landwards from the edge of the marsh as the result of sorting of sediment washed in from the seaward side, and there is a similar diminution vertically through the aggraded saltmarsh terrace because of progradation (Allen, 1996a). However, there are often storm-carried sediments, including sand and organic litter, on the upper saltmarsh (Stumpf, 1983), some of which may have been eroded from the seaward edge of the saltmarsh as occurs at Morrich More, Ross and Cromarty and Caerlaverock, Dumfries and Galloway.

Upward and outward growth of saltmarshes can be accelerated by an increase in the rate of sedimentation of the kind that occurred in Cornish estuaries in the 18th and 19th centuries when river sediment yields were augmented by mining waste. A grassy saltmarsh formed as the Fal delta grew rapidly between 1878 and 1973, when the river was carrying large quantities of kaolinite from the china clay workings on Hensbarrow Down (Ranwell, 1974). On the south coast of England rates of accretion have been very slow in areas where excavations made in saltmarshes (e.g. for salt manufacture) have persisted for many decades, as at Budleigh Salterton in Devon (see GCR site report in Chapter 6). In the Medway estuary in Kent large quantities of clay were cut for brick-making and cement production, leaving numerous pits and access canals; although this clay extraction ceased in the 1960s there has not yet been sufficient sediment deposition to obliterate them (French, 1997).

Micro-cliffs

At the seaward margins of many saltmarsh terraces there is a micro-cliff that may be up to 1.5 m high. Examples are seen on the Burry Inlet marshes in south Wales, where the marginal cliff forms a sharp drop to Llandridian Sands (see GCR site report for Carmarthen Bay, Chapter 11), and on the Dengie Peninsula in Essex (see GCR site report, this chapter), where the micro-cliff has been retreating at up to 10 m a^{-1}. Allen (1989) and Harvey (2000) found that saltmarsh micro-cliffs on sandy mud were bolder, often vertical, as in the Solway Firth, in comparison with the more subdued forms on soft mud in the Severn estuary. Where the top of the micro-cliff was bound by plant roots, recession was by way of calving, toppling and rotational slides of individual blocks of sediment together with stripping of surface vegetation (Harvey, 2000). In some sites cliffing is accompanied by continuing vertical accretion of muddy sediment in the saltmarsh, building up the saltmarsh terrace even though seaward advance has come to an end.

A saltmarsh micro-cliff may form in various ways. In some places it results from lateral movement of a tidal channel, undercutting the edge of a saltmarsh, but as this is a widespread phenomenon (there are now only a few sites where saltmarshes are spreading seawards), some more general explanation is required. It may be that, as on the sides of developing tidal creeks, seaward margins become oversteepened and cliffed, particularly during occasional storm wave episodes. In navigable estuaries, swash from boats will also tend to cut a cliff at the edge of the saltmarsh, while dredging, by steepening the submerged offshore slope, will also encourage the retreat of the marsh edge. Cliffing of this kind is repaired if there is an abundant supply of sediment to restore the profile and permit vegetation to spread again, but if there is a sediment deficit a saltmarsh cliff will persist. Alternatively, the cliffing of seaward margins of saltmarsh terraces could be a response to a rising sea level, deepening the adjacent water and allowing larger waves to attack the shore, and probably increasing tidal penetration in estuaries. This would also explain the widening and shallowing of tidal creeks that is occurring in saltmarshes in southern England, notably in Poole Harbour, Dorset.

Where the tidal range is large there is some-

Introduction

times a micro-cliff separating an upper (mature) saltmarsh of firm (often sandy) clay from a lower (pioneer) saltmarsh terrace on soft accreting mud (Pethick, 1992). A double terrace of this kind borders the Solway Firth, where an upper saltmarsh occurs landwards of the high-water line, and a lower saltmarsh seawards, as in the Nith estuary near Dumfries. Similar features are seen on the northern shores of Walney Island, in Cumbria (see GCR site report, Chapter 8), and in Loch Gruinart on Islay (see GCR site report in the present chapter). It is possible that the upper terrace has been cliffed and cut back during a stormy phase, and that the lower terrace represents a stage in rebuilding.

The effects of common cord-grass *Spartina anglica*

Many British saltmarshes have been modified by swards of common cord-grass *Spartina anglica*. This is a fertile hybrid that arose by the crossing of the native British species *S. maritima* with *S. alterniflora*, a non-native species that was accidentally introduced in the 1820s from the eastern USA. After the hybrid *S. anglica* originated in Southampton Water in about 1870 (Carey and Oliver, 1918) it was introduced to many estuaries, subsequently advancing across intertidal mudflats and rapidly building up marshland, and spreading to new areas (Figure 10.5). It has been used in the past to stabilize and land-claim tidal flats in estuaries in various parts of the world.

Early stages of *Spartina* invasion can be seen in the lee of Holy Island (see GCR site report in Chapter 11), and on the Humber mudflats behind the spit at Spurn Head (see GCR site report in Chapter 8) where clones spread on to sandy intertidal areas. In Poole Harbour, the arrival of *S. anglica* in 1899 was followed by the rapid expansion of saltmarshes into broader and higher terraces covered entirely by this plant (Figure 10.2). At the same time, intervening creeks and channels became narrower and deeper, indicating that there had been a transference of muddy sediment from these into the areas of spreading *Spartina*. On the north Norfolk coast and in the Dee estuary, experimental introduction of *S. anglica* modified natural saltmarshes and led to the evolution of broad depositional terraces in the intertidal zone (Oliver and Salisbury, 1913; Bird, 1963). Marker (1967) recorded the rapid spread of *S. anglica* intro-

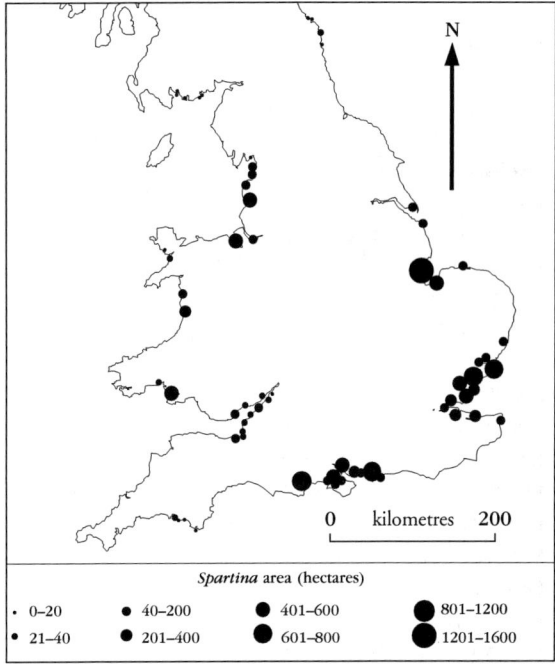

Figure 10.5 The distribution of *Spartina anglica* in England and Wales. (After Hubbard and Stebbings, 1967.)

duced in 1922, noting that it had become the pioneer colonist on accreting mudflats.

In Britain some of the older *Spartina* marshes now show evidence of die-back, especially along the seaward margins and in enclaves that become saltpans (Doody, 1984, 1990, 1992). The ecological reasons for die-back are not fully understood, but it is often associated with nitrogen deficiency, sulphide accumulation and waterlogging. At the seaward margins where the sward dies, sediment previously trapped is released, and there is a receding micro-cliff. Die-back of *Spartina* along creek margins has led to erosion of marsh edges and resulted in the widening and shallowing of tidal creeks and channels. The process may be cyclic in the sense that released mud is deposited in new or reviving *Spartina* marshes elsewhere. Bird and Ranwell (1964) reported that in some sectors in Poole Harbour *S. anglica* was still advancing, mainly in the upper estuary, whereas in others there was die-back and erosion, notably in Brand's Bay near the marine entrance. These trends have continued, although advance was very localized by the end of the 20th century. Recent *Spartina* die-back has been noted in the Solway marshes (Harvey, 2000).

In Scotland, north of the Solway, *Spartina* is not common. It occurs on the west coast at only

Saltmarshes

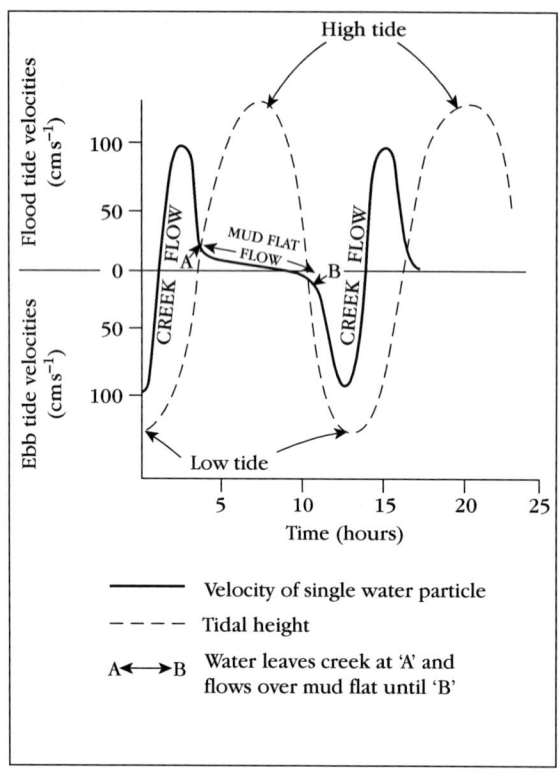

Figure 10.6 The velocities of a single water particle during a tidal cycle as it moves from a creek channel onto a mudflat surface. (After Pethick, 1984.)

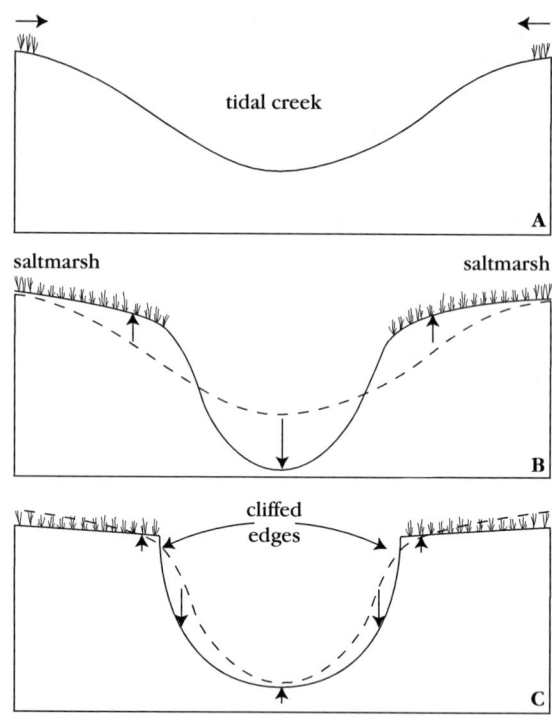

Figure 10.7 Stages in the evolution of a tidal creek as a saltmarsh encroachment takes place, forming terraces on either side of a deepening channel (B), the sides of which eventually become unstable (C). (After Bird, 1984, p. 213.)

a few isolated sites, such as Luskentyre, Western Isles (see GCR site report, Chapter 9) whereas at the east coast it is only a minor component reaching its northern limit in the Cromarty Firth (Hill, 1996, 1997).

Saltmarsh creeks

Studies of saltmarshes, particularly on the north Norfolk coast, have shown that as saltmarsh terraces are built up, the ebb and flow of the tide maintains a system of tidal creeks, the dimensions of which are related to the volume of water flowing up and down them as the tide rises and falls (Pethick, 1984, 1992; Figure 10.6). Typically dendritic and intricately meandering, they are channels within which the tide rises until the water floods the marsh surface. They are also drainage channels into which some of the ebbing water flows from the saltmarsh. They are thus like minor estuaries, particularly where they receive freshwater from hinterland runoff, or seepage from bordering beaches and dunes.

In the early stages, tidal creeks are relatively wide and shallow in cross-section, but as saltmarsh terraces rise and expand the creeks become narrower and deeper, and their banks higher and steeper, with frequent local slumping (Figure 10.7). Blocks of compacted mud, often with clumps of saltmarsh vegetation, collapse into the creek, especially where the banks are burrowed by crabs. Some tidal creeks are fringed by natural levees formed by deposition of sediment as the rising tide overflows, especially where such plants as orache *Atriplex* spp. have colonized the bordering banks. This pattern is more often found on the lower (and younger) seaward fringes of saltmarshes, the inter-creek areas becoming flatter as sedimentation proceeds.

Dendritic tidal creek systems give way to more rectilinear tidal creeks on some saltmarshes, as on the shores of Loch Gruinart on Islay and at Morrich More in the Dornoch Firth, north-east Scotland (see GCR site reports in the present chapter). Straight sub-parallel creeks across saltmarshes are more often found where the tide range is large and the transverse gradient small,

524

Introduction

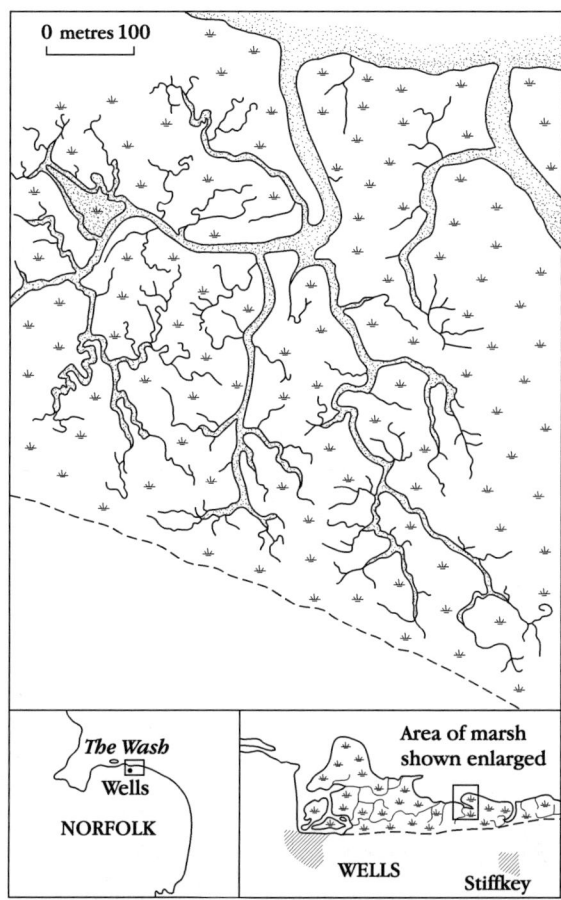

Figure 10.8 The intricate, dendritic creek network of a mature saltmarsh surface, Stiffkey, north Norfolk. (After Pethick, 1984, p. 159.)

accretion is in progress, to show asymmetry on meanders where erosion balances accretion, and to be rectilinear where erosion is dominant.

Studies of creek systems on the saltmarshes of Scolt Head Island in Norfolk (see GCR site report in Chapter 11) have shown that the exchange of water and sediment with the bordering marshes varied with current velocity as the creek water rose to over-bank levels. Vertical growth of the saltmarsh led to increasingly intermittent sediment transport in the creeks, fewer tides reaching the velocity required for the entrainment of channel sediments. However, some of the alternating submergence and drainage of a saltmarsh results from inflow and outflow across the seaward fringe rather than through creek systems (French and Stoddart, 1992).

Saltpans and pond holes

Saltpans (also known as 'pond holes') are small shallow depressions that may form as the result of the blocking of part of a tidal creek by slumping, as residual unvegetated areas within a developing saltmarsh, or as the result of local die-back of saltmarsh vegetation (Pethick, 1974; Steers, 1977). Some originated as the result of the collapse of subsurface cavities. They are flooded at high tide, and remain bare of plants because evaporation makes the trapped water hypersaline. Many become rounded as the result of scour by small waves and circulating currents generated by winds blowing across them. Numerous saltpans occur on saltmarshes between shingle recurves on Blakeney Point on the North Norfolk Coast (Figure 10.9; see also GCR site report in Chapter 11) and they are also extensive on saltmarsh terraces, such as those bordering the Cree estuary in south-west Scotland (see GCR site report in the present chapter). On many Scottish saltmarshes, creeks, abandoned owing to sea-level changes, form the basis of several series of linear saltpans that mirror the old creek pattern, and often drain via subsurface pipes, for example, Morrich More, Ross and Cromarty, has good examples of such abandoned networks (Smith, 1978; Hansom and Leafe, 1990.)

Freshwater swamps

Reference has been made to freshwater swamps on the landward margins of saltmarshes, where

or where the rate of seaward spread of saltmarsh has been rapid. In Bridgwater Bay on the southern shore of the Bristol Channel, where the tidal range is about 10 m, saltmarsh creeks run parallel and orthogonal to the coastline, whereas in Poole Harbour (Dorset), a microtidal estuarine embayment, creek patterns in bordering saltmarshes are mainly dendritic (Ranwell, 1972) (Figure 10.8).

The morphology of tidal creeks is related to sediment type, plant cover and tidal range. There are clearly defined steep-edged tidal creeks in saltmarsh terraces built largely of cohesive clay, as in Poole Harbour, but they become shallower and wider where the saltmarsh is sandier, as in the estuaries opening into Cardigan Bay. Saltmarsh creek patterns are trellised where linear cheniers of shelly sand have been deposited by storm surges, and channels have been cut through these. In cross-section, tidal creeks tend to be rounded furrows where

Saltmarshes

Figure 10.9 The distribution of saltpans on a saltmarsh at Blakeney Point, north Norfolk. (After Pethick, 1984, p. 164.)

they represent a late stage in vegetation succession to land vegetation, but freshwater swamps also fringe the shore in coastal lagoons, estuaries and sheltered embayments. They are dominated by common reed *Phragmites australis* often with rushes (e.g. *Juncus* spp.) and sedges (e.g. *Bolboschoenus maritimus* = *Scirpus maritimus*), which can grow out into water about 1 m deep, such as at Luce Sands, south-west Scotland (see GCR site report in Chapter 7). As in saltmarshes, freshwater swamps of this kind reduce wave action and current flow, and promote accretion of sediment, particularly silt and clay, in such a way as to build up a depositional terrace. Seasonal decay of freshwater swamp vegetation produces organic matter, which is deposited with the trapped sediment, and where the sediment supply is meagre this organic matter may accumulate on the depositional terrace as fibrous peat deposits. In due course the terrace is built up to high-water level, and land vegetation (scrub and woodland) then moves in. The outcome is progradation of the coastline by swamp encroachment, a process that is demonstrable on the shores of coastal lagoons where salinity is relatively low, as in Loe Pool in Cornwall, Slapton Ley in Devon (see GCR site reports in Chapter 6) and Abbotsbury Swannery

Introduction

at the western end of The Fleet in Dorset (see GCR site report for Chesil Beach in Chapter 6).

Impacts of human activity

Saltmarshes in Britain have been reduced in area by widespread coastal land-claim, generally by embanking upper marshes for conversion to pasture or cultivation. Extensive areas south of the Wash have been reclaimed in this way over the past few centuries (Kestner, 1962; Doody, 1987; Robinson, 1987; Halcrow, 1988, Hill, 1988; Mortimer, 2002), and where associated peat deposits have been compressed, dried out, or destroyed by burning or oxidation, the land surface has subsided. Over the same period, saltmarshes seaward of embankments and sea-walls have narrowed as the result of erosion ('coastal squeeze'), particularly on subsiding parts of the coast in east and south-east England (Doody, 1996). In Scotland, some 50% of the former intertidal area of the Forth estuary has been subject to claim over the last 400 years, resulting in the survival only of remnants of saltmarsh (Hansom *et al.*, 2001).

Saltmarshes have also been modified by pollution, and by the dumping of waste material of various kinds. Some saltmarshes have also been modified locally where channels have been excavated or tidal creeks widened and deepened to permit boat access and provide anchorages, and in some places excavated material has been dumped alongside to form hard ground for harbour facilities. Mention has been made of excavation of clay from the Medway marshes and elsewhere.

In recent years the costs of maintaining low-lying reclaimed areas by maintaining the flood banks have led to suggestions that such areas should be abandoned and allowed to revert to saltmarsh, which has plants, animals and birds with important nature conservation value (Doody, 1996). Such 'managed re-alignment' has been implemented at a few small sites, notably on Northey Island in the Blackwater estuary, Essex. A number of other land-claimed saltmarsh sites are under review, including other Essex sites, Slimbridge beside the Severn estuary, Porlock Marsh in north Devon and Skinflats in the Firth of Forth (Hansom *et al.*, 2001). A small RSPB site at Nigg Bay in Ross and Cromarty has recently undergone managed re-alignment, the first in Scotland to do so. The further advantage of such managed re-alignment is that given ideal conditions the re-established saltmarshes should build up to, and in future keep pace with, rising sea levels (French, 1997).

The GCR saltmarsh sites

The sites described in this chapter and those site described in Chapter 11 that contain saltmarsh in the assemblage (see Figure 10.1) represent the five saltmarsh types in Allen and Pye's (1992) classification that divides saltmarshes on the basis of their physical site-type and situation.

(1) Open coast marshes – Dengie, Solway Firth (North coast), North Norfolk, Morrich More,
(2) Estuarine back-barrier marshes – Culbin, Morrich More, North Norfolk, Keyhaven, Burry Inlet, Dengie (small area), St Osyth,
(3) Estuarine-fringing marshes – Cree, Solway Firth (south and north), St Osyth, Morrich More, Burry Inlet,
(4) Embayment marshes – Cree, Solway Firth (south),
(5) Loch- or fjord-head marshes – Loch Gruinart.

In this chapter, the site descriptions are arranged so that the northernmost on the North Sea coast is provided first, followed by the remainder in a clockwise order.

Further GCR sites that contain important saltmarsh localities are described in Chapter 11 of the present volume, where the saltmarsh forms part of a geomorphologically important coastal assemblage.

Since these GCR sites were first selected, there have been two national surveys of saltmarshes, Burd (1989) for the Nature Conservancy Council, and Pye and French (1993) for the former Ministry of Agriculture, Fisheries and Food. Readers are referred to these sources for a more comprehensive view of saltmarshes in Britain.

Saltmarshes as biological SSSIs and Special Areas of Conservation (SACs)

In Chapter 1, it was emphasized that the SSSI site series is constructed both from areas nationally important for wildlife and GCR sites. An SSSI may be established solely for its geology/

Introduction

Table 10.1 Candidate and possible Special Areas of Conservation in Great Britain supporting Habitats Directive Annex I coastal saltmarsh habitat(s) as qualifying European features. Non-significant occurrences of these habitats on SACs selected for other features are not included. (Source: JNCC International Designations Database, July 2002.)

SAC name	Local authority	Saltmarsh extent (ha)
Alde, Ore and Butley Estuaries	Suffolk	390
Carmarthen Bay and Estuaries/ Bae Caerfyrddin ac Aberoedd	Abertawe/ Swansea; Caerfyrddin/ Carmarthenshire; Penfro/ Pembrokeshire	2764
Chesil and the Fleet	Dorset	21
Culbin Bar	Highland; Moray	203
Dee Estuary/ Aber Dyfrdwy*	Cheshire; Fflint/ Flintshire; Wirral	2431
Dornoch Firth and Morrich More	Highland	539
Drigg Coast	Cumbria	162
Essex Estuaries	Essex	3770
Fal and Helford	Cornwall	70
Glannau Môn (Cors heli)/ Anglesey Coast (Saltmarsh)	Ynys Môn/ Isle of Anglesey	191
Humber Estuary*	City of Kingston upon Hull; East Riding of Yorkshire; Lincolnshire; North East Lincolnshire; North Lincolnshire	840
Kenfig/ Cynffig	Pen-y-bont ar Ogwr/ Bridgend	20
Mòine Mhór	Argyll and Bute	94
Morecambe Bay	Cumbria; Lancashire	1897
North Norfolk Coast	Norfolk	19
North Uist Machair	Western Isles / Na h-Eileanan an Iar	82
Pembrokeshire Marine/ Sir Benfro Forol	Penfro/ Pembrokeshire	274
Pen Llŷn a'r Sarnau/ Lleyn Peninsula and the Sarnau	Ceredigion; Gwynedd; Powys	748
Plymouth Sound and Estuaries	Cornwall; Devon; Plymouth	192
Severn Estuary/ Môr Hafren*	Bro Morgannwg/ Vale of Glamorgan; Caerdydd/ Cardiff; Casnewydd/ Newport; City of Bristol; Fynwy/ Monmouthshire; Gloucestershire; North Somerset; Somerset; South Gloucestershire	656
Solent Maritime	City of Portsmouth; City of Southampton; Hampshire; Isle of Wight; West Sussex	2276
Solway Firth	Cumbria; Dumfries and Galloway	4171
The Wash and North Norfolk Coast	Lincolnshire; Norfolk	3341

* Possible SAC not yet submitted to EC.
Bold type indicates a coastal GCR interest within the site

geomorphology, or its wildlife/habitat, or it may comprise a 'mosaic' of biological and GCR sites that may be adjacent, partially overlap, or be co-incident. Therefore there are a number of coastal SSSIs that are primarily selected for their wildlife conservation value, but implicitly will contain interesting coastal geomorphology features that are not included independently in the GCR because of the 'minimum number' criterion of the GCR rationale (see Chapter 1; Sherwood *et al.*, 2000). Therefore there are some areas of saltmarsh that are crucially important to the natural heritage of Britain are not described in the present geomorphologically focused volume, but are conserved for their habitat value as SSSIs.

In addition to being protected through the SSSI system for their national importance, certain types of saltmarsh are 'Habitats Directive'-Annex I habitats eligible for selection as SACs (see Chapter 1). As well as being eligible for SAC selection in their own right, saltmarshes can be an important component of some SACs selected for the Annex I type 'Estuaries'. Furthermore, many saltmarshes are of international ornithological importance, primarily for breeding waders and wintering wildfowl, and for this reason may be designated Special Protection Areas under the Birds Directive, and/or as Ramsar sites.

Saltmarsh SAC site selection rationale

For the two relatively widespread Annex I coastal saltmarsh types occurring in the UK, 'Salicornia and other annuals colonising mud and sand' and Atlantic salt meadows, sites have been selected to represent their geographical range and ecological variation. Generally, the largest areas of the habitat type have been selected. Preference has been given to sites where saltmarsh forms part of well-developed successional sequences, and there are transitions to other high-quality habitat assemblages at many of the selected sites. For the two rare saltmarsh types, 'Mediterranean and thermo-Atlantic halophilous scrubs' and 'Spartina' swards', all sites known to support significant examples have been selected as SACs. Only sites that are dominated by the native cord-grasses *Spartina maritima* and *S. alterniflora*, or the rare and local hybrid *S.* x *townsendii* have been considered for selection as *Spartina* swards, not stands of the widely introduced invasive common cord-grass *Spartina anglica*. Although a prominent feature of many estuaries, monoculture swards of the latter species are of little intrinsic value to wildlife, and in many areas *S. anglica* is considered a threat to the intertidal feeding-grounds used by large populations of wading birds and wildfowl. Attempts have been made to control *S. anglica* at several sites over many years.

Table 10.1 lists saltmarsh SACs, and indicates which of the sites are also important as part of the GCR and are described in this chapter.

CULBIN, MORAY (NH 980 615)

J.D. Hansom

Introduction

The assemblage of coastal landforms along the southern shore of the Moray Firth (see Chapter 11) is comparable to the barrier beach assemblage of the north Norfolk coast. The saltmarshes that have developed behind The Bar at Culbin represent the most recent features in the sequence of landform development. The area is therefore important for studying the evolution and development of saltmarshes in a national context. The marshes at Culbin (see Figure 10.1 for general location) are also distinctive in demonstrating a well-developed network of saltpans, but unusually few creeks (Burd, 1989; Comber *et al.*, 1994).

Description

The active back barrier (Allen and Pye, 1992) saltmarsh at Culbin is 203 ha in area (Pye and French, 1993; see Figure 11.5 in the present volume). The marsh edge varies from cliffed to ramped and the limited creek system is linear (Pye and French, 1993). Saltpans are common. Although the area includes both lateral erosion and accretion, Pye and French describe the site as accreting vertically. Present relative sea-level change is estimated to be close to 0 mm a^{-1}, but the area has undergone about 9 m of isostatic rise over the last 6500 years or so, over which time sea level has fallen to present level.

As part of an assemblage of coastal features, further description of the site is given in Chapter 11 of the present volume.

Interpretation

This benign environment has allowed saltmarsh to accrete rapidly on account of the western extension of the sheltering spit and bar features (Comber *et al.*, 1994). Intertidal sandflats and saltmarshes occur on the landward side of the Buckie Loch spit and The Bar in the shelter afforded by these large linear features. The saltmarsh areas are developing on an extensive sand-based intertidal zone. The seaward edge of the marsh may have a small undercut edge of *c.* 0.2 m (Ritchie *et al.*, 1978) or grade smoothly from sandflat to saltmarsh. The saltmarshes range from low developmental marsh surfaces characterized by intermittent stands of common saltmarsh-grass *Puccinellia* spp. and samphire *Salicornia* spp. to substantial areas of high marsh supporting a full vegetation cover merging to freshwater marsh species to the landward side. The two largest areas are identified as the marsh surface landward of the central section of The Bar and the area landward of Buckie Loch spit. The area of saltmarsh landward of Buckie Loch spit is expanding rapidly as distal extension of the spit continues to provide an increasingly lower-energy environment in which progradation can occur. Since the longshore extension of the protective spits is known (see Figure 11.2), then an approximate age can be placed on the initiation of saltmarsh: the marsh behind The Bar began to develop after 1858, whereas behind the Buckie Loch spit, saltmarsh was developing by 1730.

The extensive saltmarshes that have accreted in the shelter afforded by the two major spits may be susceptible to erosion by the landward migration of the protecting beach forms and the narrowing of their updrift proximal ends (Comber, 1993). Such activity is presently most severe at the neck of The Bar, where saltmarsh peat is exposed and is now being eroded on the foreshore. It is inevitable that since the sandflats and saltmarshes depend on the shelter provided by the Buckie Loch spit and The Bar, any erosion and movement in these protective features will force commensurate change in the sheltered areas behind.

Conclusions

The saltmarshes of Culbin are youthful, having largely developed in the last few hundred years. They display an intimate relationship to the shelter provided by the westward-moving gravel features of the outer coast. Westwards accretion is rapid, but proximal, erosion has led to foreshore exposures of immature saltmarsh peat.

MORRICH MORE, ROSS AND CROMARTY (NH 803 835–NH 892 830)

J.D. Hansom

Introduction

The saltmarshes at Morrich More (see Figure 10.1 for general location) have developed mainly within Inver Bay, protected by accretion of beaches to the north and west, but marshes also occur behind the two tidally connected islands in the north (see Chapter 11 for a description of the other coastal geomorphology features of interest). The marshes are the most extensive area of saltmarsh in the Highlands, and they form an integral part of a landform assemblage that has developed over the last 6500 years. The marsh sediments have an unusually high component of sand, and the marsh stratigraphy is layered – this feature is well displayed in the marsh edge along Inver Bay. The creek pattern over the saltmarsh is strongly linear and parallel. This distinctive pattern is largely confined to a few saltmarshes in Britain that are developing rapidly or are affected by isostatic uplift, as is particularly well demonstrated at Morrich More. The drainage dynamics of the saltmarsh at Morrich More exhibit an unusual drainage lag on the ebb tide, probably attributable to subsurface pipe networks (Leafe and Hansom, 1990). The saltmarsh has some fine examples of channel pans and primary pans. Together these attributes make Morrich More a key site for studies of saltmarsh geomorphology on an emerged coast.

Description

Pye and French (1993) describe Morrich More as including open coast, back-barrier and estuarine fringing marshes. The marsh edge morphology is variable from cliffed to ramped. The creek system is strongly linear, especially the younger parts, and saltpans are common (Hansom and Leafe, 1990; Leafe and Hansom, 1990). Extensive areas of intertidal sandflat have developed in the shelter provided by the barrier beaches of Innis Mhór and Patterson Island. Similar areas exist on the western flank of Morrich More and within Inver Bay. On the more elevated sections of sandflat, tidal inundation is of lower frequency and duration, thus allowing saltmarsh vegetation to colonize. The 260 ha of saltmarsh on Morrich More represents 5% of the remaining semi-natural saltmarsh in Scotland and 17% of that in the Highland Region, yet it is a distinctive system in its own right on account of the sandy nature of the substrate and its context of rapid isostatic uplift (Hansom and Leafe, 1990).

The saltmarshes are drained by creek systems

that extend into the Morrich along the axes of the inter-ridge swales or hollows. Thus, saltmarsh and emerged beach ridges interdigitate, and as the saltmarsh grows and accretes through time, the extremities of the beach ridges become progressively buried. Owing to isostatic uplift over 6500 years, the altitude over the Morrich More falls seawards to the west, north and east, and so saltmarshes have developed at the edges of a domed structure, with the oldest marshes close to the centre and the youngest in the west, north and east. The complex vegetation pattern of the Morrich More is dominated by this pattern of interdigitated ridges and slacks (Smith and Mather, 1973). The strandplain carries a rich flora of over 200 flowering species, ranging from intertidal sandflat species to *Juniperus-Calluna* heath on the oldest landward ridges. The vegetation succession of the Morrich More is discussed in further detail in Smith and Mather (1973) and Dargie (1989).

As part of an assemblage of coastal features, further description of the site is given in Chapter 11 of the present volume.

Interpretation

Relative sea-level change in the Dornoch Firth area is close to 0 mm a^{-1} (Pye and French, 1993) and the marsh is characterized by vertical accretion, but variable lateral erosion and accretion; accretion is relatively rapid in the back barrier area between the islands and the main Morrich More coastline. The size and relatively undisturbed history of 6500 years of continuous sedimentation of Morrich More has led to the presence of a full range of successional stages of embryo dune through to sand plain in association with dune slacks and interfingered saltmarsh. Dargie (1989) demonstrates the importance of the vegetational transitions at Morrich More, with those between the saltmarsh and dune systems being of particular complexity and therefore of high conservation value in view of the clear relationship with geomorphology. Vegetational transitions from saltmarsh to sand dune are extremely rare in Britain, and the transition on Morrich More from saltmarsh to calcareous dune, wet acid dune or dry dune grassland makes the upper saltmarsh vegetation, and its interaction with the domed geomorphology of the strandplain, uniquely important.

Conclusions

The scientific interest of Morrich More is outstanding both in terms of variety and scale of its coastal landforms including the strandplain, parabolic dunes, stabilized dunes, foredune succession, saltmarshes and sandflats. The saltmarshes are uniquely youthful to the west, north and east, away from the emerged surfaces of central Morrich. The site shows a well-developed vegetational transition from intertidal sandflat through saltmarsh to freshwater wetland and calcareous dome.

ST OSYTH MARSH, ESSEX (TM 090 144–TM 130 126)

V.J. May

Introduction

St Osyth Marsh (see Figure 10.1 for general location) is an important site for studies of saltmarsh morphology, and is one of the few marsh areas in Britain to have been dated, the maximum age being 4280 ± 45 years BP, by analysis of a peat seam preserved in grey-black clay at the site. The characteristic assemblage of saltmarsh features – creeks, saltpans and saltmarsh cliff – are all present at St Osyth Marsh, and reflect the maturity of the marsh systems (Hussey and Long, 1982). The saltpans have been intensively researched by geomorphologists (Pethick, 1970, 1984; Leeks, 1979), and provide much information relating to the formation and development of this unique coastal landform. This is one of the few sites in Britain where chenier development has been described fully (Greensmith and Tucker, 1975). One of the main interests is the process of breaching and secondary spit genesis brought about by landward roll-over across the marsh surface. This process is well displayed in the upper levels of the system.

Description

The site comprises two main areas: a narrow beach and saltmarsh that extends some 3 km westwards from St Osyth (TM 130 127) to Colne Point, where the shoreline turns towards NNW for a further 2.3 km. The area between St Osyth and Colne Point rarely exceeds 400 m in width and is limited landwards by a low sea defence embankment (see Figure 10.10a). At its widest, the area beyond Colne Point exceeds 1.4 km and is dominated by a well-developed saltmarsh system. Longshore sediment movement is from St Osyth towards Colne Point and into the Colne estuary. The spring tidal range is 3.8 m. The beach is a narrow ridge formed predominantly of sand and pebbles and resting on the seaward-facing edge of the saltmarsh. It is a thin deposit underlain by the saltmarsh clay. Such a description follows closely that given by Price (1955) to the features known as cheniers (or marsh beach ridges). These are much shallower sedimentary features than the barrier ridges of sand and shingle that front saltmarshes in such locations as Blakeney Point and Orfordness. Whereas the latter form independently of the development of saltmarsh, cheniers depend upon the presence of the marsh deposits for their foundation. It is common for the distal end of such features to form a small, narrow barrier spit. At Colne Point this spit shows a historical pattern of extension and shortening (Greensmith and Tucker, 1975; see Figure 10.10a) as well as destruction and reworking of the landward end of two older cheniers. The modern chenier is undergoing changes at present, which are probably related to gravel extraction between 1947 and 1962 at Colne Point (Robinson, 1953a) and to the recharging of the sediment supply by reworking of older chenier and tidal flat-deposits. Steers (1960, Plate 165) shows active excavation and the beginnings of a phase of breaching of the beach ridge to the north-west of Colne Point. Greensmith and Tucker (1975) describe a possible older chenier exposed in a low cliff at Colne Point.

Burd (1992) estimated that between 1973 and 1988 the Colne estuary marshes decreased in area by just under 12%. Although about 50 ha was gained by accretion, some 130 ha was lost by erosion and land-claim. The largest single loss was on either side of Colne Point. Much of this loss occurred within creeks, but Burd (1992) suggested that the methodology used to compare aerial photographs of the area may overestimate this apparently high erosion of creeks. The causes of these changes are discussed in the Dengie GCR site report below.

The saltmarsh east of Colne Point is drained by a main creek, which parallels the beach throughout its length. Creeks and saltpans form less than 10% of the surface area of the marsh. In contrast, to the west of Colne Point, there is a more complex pattern of creeks. Hussey and Long (1982) estimated that creeks and saltpans occupied over 26% of one hectare of emergent marsh; 68% was occupied by a common saltmarsh-grass–sea purslane (*Puccinellia maritima–Halimione* (=*Atriplex*) *portulacoides*) community. Although they cover less than 1% of the surface area, saltpans are an important morphological feature of much of this saltmarsh. Their shape and size vary greatly, ranging from 1–15 m² in area and 5–40 cm in depth (Leeks, 1979). Although some appear to be roughly circular ('sub-circular'), many others are linear features with similar shapes to creeks. Saltmarsh morphology is affected by the evolution of the beach at Colne Point (Butler, 1978), for when the beach is breached, the hydrodynamics of the creeks alter. For example, when the beach is intact, most creek drainage from the western marsh is towards the estuary at Sandy Point, some 2.5 km away. When the beach is breached, much of the upper marshland drainage reaches the sea via channels just west of Colne Point.

The marsh is underlain by a clearly defined seam of peat overlain by grey-black clay that contains root remains. This has been dated at 4280 ± 45 years BP (Butler, 1978; Hussey and Long, 1982). The surface sediment above the clay is mainly clay (52%) and silt (43%), with 5% sand. Much of the marsh is described by Hussey and Long (1982) as emergent, except near the mouth of the tidal creek where it is degrading. The surface of the emergent marsh lies at 2.30 m OD ± 0.15 m and is covered by about 99 tides per annum (Hussey and Long, 1982).

Interpretation

Two features of this site are of especially noteworthy: the presence of both modern and older cheniers, and the large number of saltpans. The modern chenier is poorly developed at the eastern proximal end of the beach, and the saltmarsh is undergoing erosion to the extent that there is a risk of breaching and flooding of the upper saltmarsh. There are two old cheniers: the older was speculatively dated by Greensmith and Tucker (1975) as having formed between about 1550 and 1200 years BP and the more recent, (much of which is recycled by the retreat of the modern beach) might have formed between 1200 and 250 years BP. This would be consistent with the date attributed to the marsh area of 4280 ± 45 years BP (Butler, 1978). On the Essex coast, these features appear to result from longshore transport that produces spits or fringing beaches, which during periods of higher wave-energy are carried on to the marsh edge. As the beach and ridge migrate inland, the saltmarsh is first buried and then exhumed. Erosion of the exhumed saltmarsh often produces a distinct marsh cliff. The presence of former saltmarsh standing at a higher altitude than the surrounding beach may cause local refraction of waves and so affect the alignment of the beach. The cheniers at Colne Point are composed mainly of sand and gravel, derived from cliff erosion and reworking of earlier beaches. The absence from a part of an eroded chenier at Colne Point of shells of the slipper limpet *Crepidula fornicata* (which was introduced into the Essex area between 1870 and 1880), provides an indicator of the minimum age of part of this feature (Greensmith and Tucker, 1975). Although cheniers occur elsewhere in Britain, they have rarely been described in detail and the Colne Point cheniers together with those to the south at Dengie Marsh (see GCR site report below) are the best examples of this unusual form.

The saltpans are also a feature that is particularly well represented here. The earliest descriptions in the Dovey estuary (Yapp *et al.*, 1917; Richards, 1934) identified two types of saltpan, the primary pan and the channel pan, which were described in more detail in north Norfok (Steers, 1946a; Pethick, 1974; Steers, 1977). The former are thought to have developed on the initial marsh surface as vegetation began to spread. Within small areas that were not vegetated evaporation of seawater produced highly saline conditions in which little plant life could survive or colonize. As a result, these hollows survive within the marsh topography. They often display circular forms that are attributed to the erosional effects of wavelets (Pethick, 1984), although this is rare here. There is, however, the possibility that some pans come into existence as a result of smothering of the surface by algal mats (Pethick, 1970) or, where cattle graze, by dung. Furthermore, the collapse of creek-banks may cut off sections of unvegetated mud that develop a partially rounded form under the influence of wavelets within the enclosed area. They are undoubtedly a feature of the marshes in north Norfolk, St Osyth, and elsewhere that need further investigation, especially in the light of improved understanding of creek sediment dynamics.

In contrast, the channel pans appear to originate when creeks are abandoned. Sedimentation at the mouth of the former creek blocks the exchange of water with active creeks and higher salinities maintain an absence of vegetation. Pethick (1984) suggests that possible causes may include changes in sea level that led to an abandonment of large numbers of creeks. Alternatively, as creeks are deepened because of saltmarsh accretion, the total volume of tidal flood water may require fewer channels. Sinuous channel pans may thus represent creek obsolescence. Although there is no evidence that subsurface piping systems occur in the marshes at St Osyth, their collapse elsewhere may also provide a mechanism for the development of elongated saltpans.

This site is of considerable importance to the understanding of saltmarsh morphology not only because it demonstrates the comparative longevity of such features in eastern England, but also because of its cheniers and saltpans. It has, however, another role as part of the coast protection of the Essex coast. The level of the saltmarsh is higher than the land that lies landwards of the site behind artificial sea defences. The continuing efficacy of the sea defences depends upon the continued presence of the beach and saltmarsh. Unfortunately, the construction of groynes at the northern end of the site has substantially reduced the supply of sediment to the beach, which is now seriously affected by erosion. As a result, not only the natural importance of the site, but also its coast protection role, are threatened.

If the sediment supply to the beach is not maintained, there is likely to be a deterioration of the proximal end of the beach, partial destruction of the saltmarshes and the sea-wall would become exposed. In these circumstances, it may prove prudent in the interest of maintaining the scientific interest to allow artificial beach-feeding by materials comparable to those that fed the beach in the past. The volume would need to be controlled so that it simulated the historical sediment transport patterns in magnitude and frequency. The coast protection needs would be furthered by such action. Like many sites on the English coast, the marsh and beach at St Osyth now depend upon human intervention for their future maintenance. The landward boundary is an artificial one (the sea-wall), without which the saltmarsh would by now have migrated well inland. The site is important, not least because it offers an opportunity to manipulate the coastal system in order to conserve features of national significance at the same time as providing insights into the links between sea defences, rising sea levels and saltmarsh development on sites restricted landwards by artificial structures.

Conclusions

One of the few dated saltmarsh sites in England and Wales, St Osyth Marsh is also important because of its cheniers, creeks and saltpans. Parts of the site are over 4000 years old and owe their preservation to the protective effects of both emergent saltmarsh and marsh-edge beaches. The cheniers at Colne Point are mainly in sand and gravel, unlike those farther south at Dengie, which are much more shelly. Understanding of the way in which saltmarshes and their protective cheniers develop has considerable importance for the protection of the low-lying Essex coastlands.

DENGIE MARSH, ESSEX (TR 030 089–TR 025 963)

V.J. May

Introduction

The Dengie peninsula (see Figure 10.1 for general location), like many parts of the Essex coast, has been progressively land-claimed by a series of embankments constructed on the upper marshes. These walls are now fronted by saltmarsh and intertidal flats in excess of 2 km wide. The tidal range is 3.8 m. The northern part of the saltmarsh is fronted by shell and sand ridges (Greensmith and Tucker, 1965, 1967, 1968, 1969, 1973a,b, 1975), and parts of the intertidal area is marked by the development of mud mounds (Greensmith and Tucker, 1965, 1967). Farther south, the main area of saltmarsh at Dengie has been the subject of a number of studies of sediment transport and creek development (Bayliss-Smith *et al.*, 1979; Reed, 1986, 1987, 1988; Reed *et al.*, 1985; Stoddart and Bayliss-Smith, 1985; Pye and French, 1993). Changes in the coastline were described by Robinson (1953a), but more recent changes in the marshland area have been estimated by Harmsworth and Long (1986) and Burd (1992).

Pethick (1989, 1991) has described the effects of, and recovery from, a storm event in January 1989.

Description

The Dengie peninsula is fronted by an extensive area of saltmarsh that is narrow at its northern and southern extremities and up to 600 m wide at Bridgewick. The marsh covers about 480 ha, of which about 16% (about 80 ha) was lost by erosion between 1970 and 1981 (Harmsworth and Long, 1986). Burd (1992) estimates that between 1973 and 1988 changes within six blocks of the saltmarsh, based upon measurements of the position of the marsh front, ranged from accretion of 1.1 m a^{-1} to erosion of 7.6 m a^{-1} (Figure 10.10b). Overall the marsh front lost on average 2.6 m a^{-1}. Mean low-water mark moved landwards at an average of 28.4 m a^{-1} at Sales Point, 8.7 m a^{-1} at St Peter's channel and 13.3 m a^{-1} at Watch House, thus steepening the foreshore (Pye and French, 1993). A total area of 473.8 ha in 1973 was reduced by 10% by 1988 (Burd, 1992), and this was the smallest net loss on all the Essex and north Kent saltmarshes.

On the landward side, the marsh is bounded by a 19th century sea-wall, the latest in a series of land-claims that started in the 16th century. The saltmarsh morphology at Dengie is particularly interesting. The marsh surface stands mainly at 2.5 m OD and is essentially planar with, unusually, no differentiation into upper and lower marsh. The marsh is dissected by numerous creeks and saltpan systems. Dengie also possesses many subsurface 'pipes'. The role of these subsurface features in bringing about saltpan formation was reported initially in Nigg Bay, Ross and Cromarty, Scotland (Kesel and Smith, 1978). It was subsequently recognized as occurring more widely, in the Ythan marsh, the marshes of south Wales, and in Shetland and Lewis (Smith, 1978).

Saltmarsh erosion was dominant along the Dengie peninsula (Harmsworth and Long, 1986) during the period 1960 to 1981, a trend that has been confirmed by Burd (1992). Net loss between 1973 and 1988 was 46.7 ha, much being accounted for by the January 1978 storm (Pye and French, 1993). Greensmith and Tucker (1965) showed that erosion was not uniform with a range of saltmarsh edge retreat from zero to 270 m. Boorman and Ranwell (1977), in

Dengie Marsh

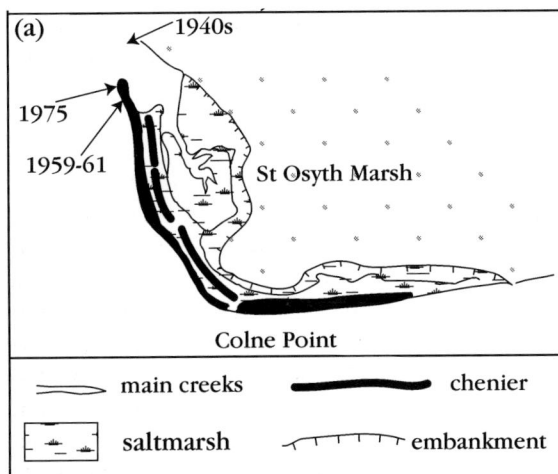

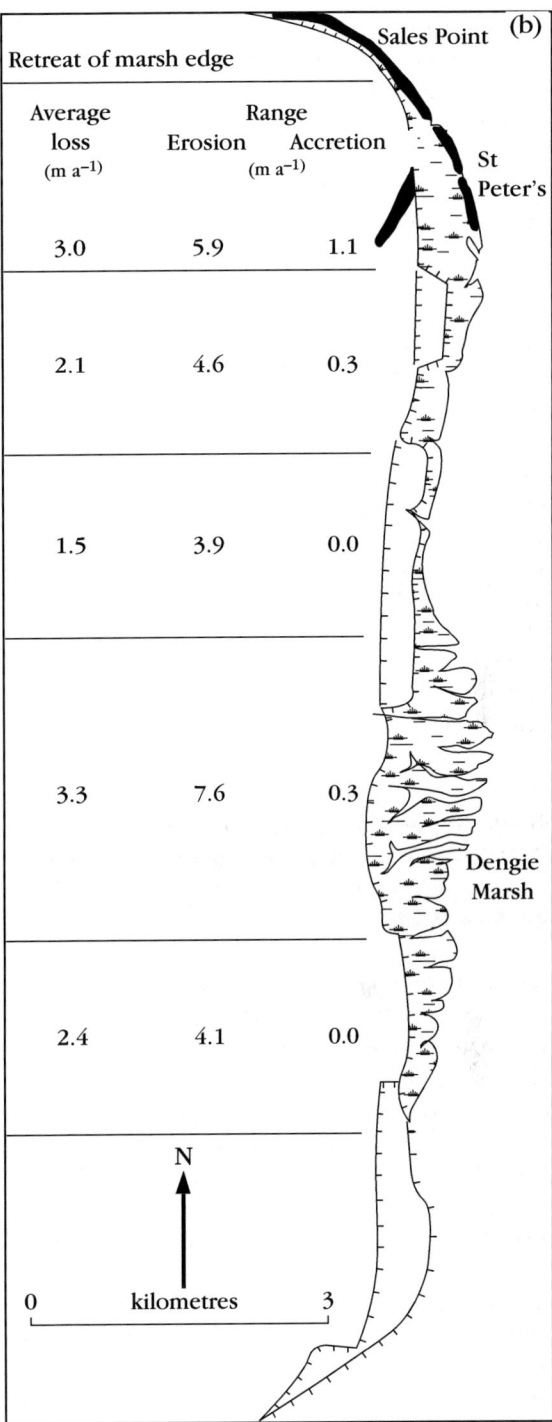

Figure 10.10 Cheniers and rates of change at (a) St Osyth Marsh (the arrows show the position of the northern end of the spit at different times) and (b) Dengie Marsh. Older cheniers occur landwards of the embankments at Dengie Marsh. Both the patterns of change in the marsh edge at Dengie Marsh and the spit at St Osyth show the tendency for these features to fluctuate in position, with erosion alternating with accretion. (Based in part on Greensmith and Tucker, 1975 and Burd, 1992.)

contrast, regarded Dengie as one of the few saltmarshes of south-east England that was accreting. Reed (1988) recorded short-term vertical accretion of 5–14 mm a^{-1}. However, the Anglian Water Authority (1980) reported the marsh as being dominated by erosion. Harmsworth and Long (1986) suggest that the characteristic pattern may be one of successive construction and erosion, depending upon rates of sea-level change and the degree of exposure to wave action. The outer edge of the marsh is marked by a distinct saltmarsh cliff (Harmsworth and Long, 1986), but much of the edge of the marsh, particularly at its northern end at Bradwell, is eroded into a zone of mud mounds (Greensmith and Tucker, 1965). The seaward rim of the marsh surface is intermittently overridden by transgressive shell ridges (Greensmith and Tucker, 1965, 1967, 1969, 1973a,b). As these cheniers transgress the saltmarsh, it is buried for a period of time before re-appearing seawards of the beach. Farther inland, the marsh lies over a series of chenier features especially at Sales Point. The coastline appears to have been affected by erosion after the Roman period for there is some evidence to suggest that the Roman fort of Othona, the present site of St Peter's Chapel, extended at least 120 m farther seawards than the present cliffline. Some fragments of submerged masonry within the marshland have been described as a Roman harbour (Johnson, 1976).

The chenier at Sales Point is predominantly

formed by shells (51%) and sand (47%), unlike the cheniers at Colne Point (St Osyth Marsh GCR site). There is no tendency to overall coarsening upwards, but the backslope is usually dominated by coarse-grained sediments, mainly large shell fragments (Greensmith and Tucker, 1965, 1969). The chenier cross-section was described as asymmetric, steepest on the landward side (Greensmith and Tucker, 1975). However, in January 1992, much of the northern part of the chenier had been eroded to form a cliff about 0.35 m in height and expose layering of the structure. Farther south, the form described by Greensmith and Tucker remained characteristic. Since 1947, the beach has migrated landwards at annual rates of up to 8 m, but it has also extended southwards as well as suffering occasional breaches. As a result, parts of the chenier and the saltmarsh undergo erosion, as well as providing, in the case of the chenier, reworked supplies of sediment for chenier building. The exposure of former saltmarsh on the foreshore and the growth of algal mats both affect the development of the chenier. The first refracts waves approaching the beach, the latter prevents re-distribution of sand and shells from the beach.

Interpretation

This is a rare example in Britain of an open coast marsh. Three issues have been examined separately in this site:

(1) the growth of cheniers and their relationship to changes of sea level,
(2) the reasons for the erosion of the saltmarsh, and
(3) the links between sedimentation and creek development.

Despite the separate approaches, there is a common theme – the link between the forms of the site and the effects of changes in the relationship between the land surface and sea level. Regional sea-level rise on the Essex coast has three components, summarized by Burd (1992) as follows:

(1) the general eustatic rise in sea level: estimated to have been 15 cm ± 10 to 20 cm over the past century (Rossiter, 1967; Robin, 1986; Woodworth, 1987, 1990a,b; Misdorp et al., 1990) in the North Sea,
(2) increased tidal range: for example 45 cm at Harwich since 1870, and
(3) isostatic changes in sea level: based on Shennan (1989), this is estimated at a rise in relative sea level of 1.5 mm a^{-1}.

The fringing beaches are of particular interest. Sales Point chenier is described by Greensmith and Tucker (1975) as a 'type-example'. It contrasts strongly with the cheniers of the Louisiana coast (Byrne et al., 1959), as well as others in New Zealand and Queensland, Australia, examined by Greensmith and Tucker (1975) in at least three respects. It lacks the upwards-coarsening of sediment of the Louisiana cheniers. It contrasts with them in the coarseness of the backslope, and in having its steepest slope on the landward side, unlike others where the seaward slope is steepest.

Greensmith and Tucker (1975) argued that the Essex cheniers resulted from periods of coastal instability and widespread erosion, in contrast to the type region of Louisiana where coastal stability predominated. They believe that the fundamental tectonic instability of the southern North Sea explains the contrast, and outline an evolutionary sequence for the development of the Essex chenier plain of which Dengie is part. Marine transgression, probably during the period 1434 ± 110 years BP to 1265 ± 100 years BP, brought about erosion of the upper tidal flats and saltmarsh, and initiated mud mounds (Greensmith and Tucker, 1967). Shells, sand and pebbles were released from the flats and provided a source for the initiation of cheniers. Extension by merging and elongation produced spits extending onto the tidal flats. Creeks both interrupted chenier development and were affected by it as tidal flows were diverted. In the later stages of chenier development, expansion of saltmarshes isolated some ridges inland. Several can be traced within the land-claim area between Sales Point and Dengie village. At the outer edge of the saltmarsh, new cheniers developed such as that at Sales Point, but they were subject to growth by elongation and also by erosion, particularly at their landward extremities. Nevertheless, they represent the later stages of a chenier plain thas has prograded over the last 2000 years (Greensmith and Tucker, 1975).

An increase in mean high-water level could be expected to give rise to a change in community composition as a result of increased frequency and duration of submergence. Increased erosion may result simply from changes in the wave

Dengie Marsh

climate (Boorman *et al.*, 1989). Despite the retreat of the marsh by some 40 m between 1955 and 1988 (1.2 m a^{-1}), Reed (1988) observed no signs that vegetation at Dengie was stressed. Vertical accretion is taking place at the same time as lateral erosion of the marsh front (Pethick, 1991) and so the marsh surface is able to keep pace with the regional rise in sea level. The adjacent mudflats adjust to sea-level rise by flattening their profile from the land towards the sea. In these circumstances, the saltmarsh is expendable as it provides a supply of sediment to the mudflats (Pethick, 1991). Pethick (1989, 1991) records that major storms in January 1989 brought about both retreat of the marsh front and flattening of the profile of the mudflats. A storm event in October 1989 (Pethick, 1992) with a calculated return period of 1 in 33 years and waves of significant height of 3.4 m produced marsh edge retreat of 5 m. The horizontal surface was also lowered. Between October 1989 and March 1990 the marsh surface recovered its pre-storm altitude and by October 1990 was slightly higher (Pethick, 1992). However, as the frequency and magnitude of extreme events may be increasing (Carter and Draper, 1988), the role of storms in this process of mudflat flattening may become increasingly significant if their frequency exceeds the recovery period.

Harmsworth and Long (1986) argued that the erosion of the saltmarsh edge may result from a change of sea level (or inundation) produced by sinking of the saltmarsh area, a pattern that contrasts with upwards growth during periods of sea-level still-stand. An alternative explanation, specific to Dengie, could be the increased exposure of the saltmarsh that resulted from the loss from the intertidal mudflats of beds of eelgrass *Zostera marina*, which was wiped out by disease in the 1930s (Harmsworth and Long, 1986).

The saltmarshes here are an important area for an examination of saltmarsh sedimentation and erosion, and the linkage between tidal dynamics and sediment transport from intertidal mudflats. Parts of the saltmarshes include subsurface pipes that affect drainage as well as the development of collapse features on the marshes and may account for some development of saltpans (Leeks, 1979). Reed (1988) noted that the Dengie peninsula is affected by a sea-level rise of about 3 mm a^{-1}. However, saltmarsh vegetation shows no sign of stress due to increased submergence incidence. Reed argues that this shows that continued accretion of the marsh surface is taking place. This accretion depends, however, upon the direct supply of sediment, during over-marsh tides, from the erosion of the marsh edge (Reed, 1988). The sediment pathways depend, however, upon the velocities that occur within creeks, for example in Bridge Creek on Dengie Marsh, velocity pulses occur on flood tides (Reed, 1987). More critically, she demonstrated that it is essential for the variability of velocity within creek systems to be observed before the time interval for sampling of velocity and discharge in tidal creeks is determined. It follows that calculation of sediment fluxes within creeks may be significantly affected by the temporal sampling as well as its spatial extent. Asymmetry of discharge in creeks plays a significant role in the distribution of sediment within the saltmarsh (Reed, 1987). Stoddart and Bayliss-Smith (1985) have shown, as a result of examining some 700 tidal cycles in these saltmarsh systems, that there is a strong positive sediment flux.

The relationship between the saltmarsh and the sea-walls is a particularly important one. First, the presence of the wall prevents migration of the saltmarsh inland under conditions of rising sea-level. Second, however, the saltmarsh acts as a very important element of coast protection (Pye and French, 1993). Frey and Basan (1985) have suggested that the height of waves crossing saltmarsh can be reduced by as much as 71% and wave energy by 92%. Maximum wave height in storms is reported as being 1 m lower where sea-walls are fronted by saltmarsh (Anglian Water Authority, 1980).

Conclusions

Many GCR sites include saltmarsh features, but Dengie Marsh is one of the few where the saltmarsh is the predominant reason for its inclusion in the GCR and where there have been detailed investigations over several years of the dynamics of the saltmarsh creeks and their role in the overall development of the saltmarsh.

Dengie Marsh is important, firstly because of the development of a substantial chenier plain of which the modern part remains intact and within the site. In particular, the Essex cheniers contrast with the type-form described in Louisiana. In Britain, such features have not been reported often, and where they do occur, for example in Poole Harbour or the Solent, they have not been described in detail. Secondly, Dengie Marsh is important as a research site for monitoring the relationship between saltmarsh development, sedimentation and creek hydrology. Although these processes have been examined elsewhere, this site is the only one in which the saltmarsh is not protected by artificial beach structures and so provides an important contrast with the sites in north Norfolk. Thirdly, the relationships between erosion of saltmarsh, changing sea level and coast protection have also been investigated at Dengie Marsh. The importance of the site in providing data upon which judgements about coast protection can be based is considerable and is a particularly important application of coastal saltmarsh research.

KEYHAVEN MARSH, HURST CASTLE, HAMPSHIRE (SZ 315 905)

V.J. May

Introduction

The Keyhaven saltmarshes (see Figure 10.1 for general location) are important for the range of geomorphological features they display, particularly the intricate pattern of saltmarsh creeks. The site is an important research area for examining the relationship between creek dynamics, tidal processes and sedimentation. The western part of the saltmarshes forms an integral part of the Hurst Castle Spit system (see GCR site report in Chapter 6), a classic site for the study of coastal geomorphology.

Description

Hurst Castle Spit protects a large area of saltmarshes, known as 'Keyhaven Marshes' (Figure 10.11). They are drained by an intricate pattern of creeks dominated by three major creeks – Mount Lake, alongside the spit, Keyhaven Lake and Hawker's Lake. The first two merge and drain into the Solent after being diverted by the modern recurves of the spit. Marsh-edge beaches ('cheniers' – see GCR site reports for St Osyth and Dengie above) are formed of shells and shingle. Their sand content is very low. Low-relief cheniers have developed along the marsh edge and provide some protection against erosion. Much of the saltmarsh edge is being eroded rapidly (6 m a^{-1} over the past 50 years: Pye and French, 1993), resulting in some patches of mud mounds. The upper intertidal zone is characterized by steep microcliffs and a strong concave upward profile within the upper part of the intertidal zone. The upper marsh lies at about 2.4 m OD with a seaward marsh edge at about 2.0 m OD. The elevation of the upper tidal flats is typically about 1.0 to 1.5 m OD (Pye and French, 1993). The seaward cliffs vary in height but are typically 0.7–1.5 m. The marsh surface varies in level by about 0.4 m. The surface of the marshes is characterized by a high proportion of eroded marsh, saltpans, and broad channels. There are only small areas of higher-level, species-rich saltmarsh, located mainly close to the spit and on its older recurves. Sea purslane *Atriplex portulacoides*, common sea-lavender *Limonium vulgare*, sea plantain *Plantago maritima*, sea meadow-grass *Puccinellia maritima*, common seablite *Suaeda maritima*, glasswort *Salicornia*

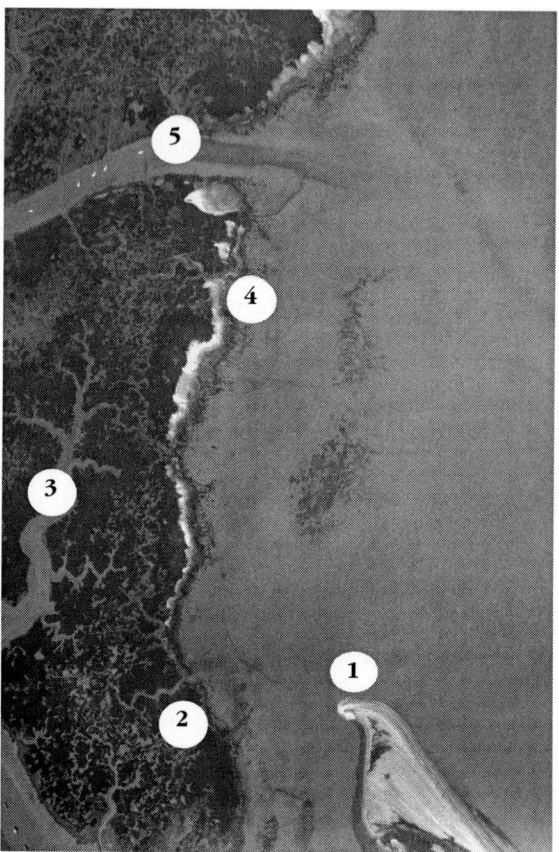

Figure 10.11 Keyhaven Marshes. (1) Distal point of Hurst Castle spit; (2) saltpans; (3) major creek; (4) retreating saltmarsh edge and chenier formation; (5) dominant channel draining upper marsh. (Photo: courtesy Cambridge University Collection of Aerial Photographs, Crown Copyright, Great Scotland Yard.)

spp., and sea aster *Aster tripolium* are common throughout these higher marshes. In contrast, the more extensive lower marshes are species-poor and dominated by common cord-grass *Spartina anglica*. The intertidal area close to the spit is often a stony mud. Before the late 19th century, much of this marsh stood as much as 1 m lower and was dominated by eelgrass *Zostera*. Colonization by *Spartina anglica* following its hybridization from the native *Spartina maritima* and the introduced *Spartina alterniflora* in Southampton Water led to a rapid build-up of the saltmarsh surface. The area of *Spartina*-dominated saltmarsh reached a maximum about 1930, after which the area declined (Bradbury, 1996). As the recurves of the modern spit have extended into the westernmost creek, they have increased local accretion of mudflats.

Bradbury (1996) describes the rapid short-term morphological and ecological evolution of the western Solent saltmarshes that include this site. There have been substantial losses of intertidal flat. Ke and Collins (1993) estimated the average annual loss of saltmarsh in the western Solent as 3.6×10^4 m^2 a^{-1}, at the same time as the saltmarsh surface is accumulating sediment at between 2 and 5 mm a^{-1}. Average erosion of the marsh edge was 3 m a^{-1} between 1992 and 1994, less than the open coast retreat but more than the fringing edge retreat of 1 m a^{-1} since 1950. Dyer (1980) showed that between 1950 and 1973 reduction in intertidal width varied between 180 and 360 m (7.8 m a^{-1}). There was a strong correlation between wind-generated wave-attack and the rate of erosion. Tidal range is 2.5 m on spring tides, but meteorological surges may raise waters levels by up to 50%. The upper marshes at Keyhaven are typically formed in sandy silts, becoming silty sand on the upper tidal flats.

Interpretation

These saltmarshes are remarkable for their rapid vertical accretion and areal extension with the arrival of *Spartina anglica* in the late 19th century. Their subsequent reduction in altitude and area was almost as rapid during the mid-20th century and is related to die-back of *Spartina* described in a series of papers (Braybrooks, 1957; Goodman, 1957, 1960; Goodman and Williams, 1961; Goodman *et al.*, 1959), which showed that it was associated with exceptionally poorly drained saltmarsh soils. Die-back occurred, however, both along channels and within the central parts of the marshes. In the latter, 'pan die-back' may have been associated with the restriction of drainage by rapid accretion around the edges of marshes. In the former, however, other factors, including algal mats, possibly resulting from local eutrophication and cloaking the surface, may have led to more extensive die-back. As channels widened, erosion of the marsh edges appears to have accelerated, although in many parts of the saltmarsh, die-back resulted in a lowering of the marsh surface rather than wholesale retreat of the marsh cliff. The saltmarshes that shelter behind the beach are also liable to damage from recreational use, as well as local pollution.

Conclusions

The development of saltmarsh in the lee of Hurst Castle Spit was limited until the arrival of common cord-grass *Spartina anglica* at the end of the 19th century. The geomorphological interest of this site lies in the rapid sedimentation and saltmarsh development associated with *Spartina* followed by an equally rapid decline and loss of saltmarsh area. Unlike the saltmarshes and cheniers of the Essex coast, those of the Keyhaven marsh are very recent in origin.

SOLWAY FIRTH SALTMARSHES (NX 829 492–NY 125 560)

J.D. Hansom

The saltmarshes ('merses') of the Solway Firth are an extensive group, comprising all those located to the east and upstream of Balcary Point (NX 829 492) and Skinburness (NY 125 560), on both the north (Scottish) and south (English) shores of the inner Solway Firth (see Figure 10.1 for general location and Figure 10.12). Defined in this way, the Solway Firth supports 3618 ha of saltmarsh (Pye and French, 1993), of which the GCR sites, Upper Solway Flats and Marshes on the south shore and the Solway Firth (North Shore), account for 2842 ha (76%). In addition, the saltmarshes at the Cree estuary in Wigtown Bay in the outer Solway Firth cover a further 553 ha. The Solway saltmarshes together account for almost 8% of British saltmarshes and although they display some different characteristics (Table 10.2), their common location, together with similarities, warrant their treatment within a combined section. The saltmarshes are, in the main, of the estuarine fringing type, being developed along the shores of the main Firth and its tributaries, although showing varying degrees of transition into open coast marsh at Caerlaverock on the Scottish shore. In addition, the saltmarsh at Moricambe Bay on the English shore shows many of the characteristics of a more enclosed embayment marsh (Table 10.2). The following text therefore describes the general topographic and hydrodynamic situation of the sites, and then seeks to describe and interpret the south shore group, the north shore group, and the Cree saltmarshes in turn.

The Solway Firth reaches almost 60 km wide between Burrow Head on the Scottish coast and St Bees Head on the English coast and extends over 130 km eastwards to the exits of the rivers Esk and Eden. With the exception of the Cree saltmarshes, the Solway saltmarshes are all located within the inner (eastern) Firth (Figure 10.12). The Firth is macrotidal; mean tidal range at Silloth on the Cumbrian coast reaches 8.4 m at springs and 4.8 m at neaps. On the northern coast the mean tidal range at Heston Islet in Auchencairn Bay is 7.4 m at springs and 3.9 m at neaps (Pye and French, 1993). The tidal streams generated can be significant especially at the mouths of tributary streams, at headlands and promontories and within channels between sandbanks. For example, the tidal stream in and out of the River Cree reaches 2.5 m s^{-1} at springs and similar velocities occur offshore of Southerness Point (Ramsay and Brampton, 2000f). The general situation is that the ebb tide runs for longer and flows at lower velocities than the flood tide. The extensive area of sandbanks retards the flood peak at successive locations upstream and contributes to a marked tidal asymmetry. This differential tidal flow accentuates the net deposition of sediment within the estuary as slower ebb currents are less able to transport sediment than the stronger flood (Comber et al., 1994).

The Solway Firth is exposed to waves from the south-west, although fetch lengths are rarely more than 250 km. As a result, most waves reach the shore as wind-waves generated in the Irish Sea or the Firth itself, or as refracted Atlantic swell (Ramsay and Brampton, 2000f). The net effect of what amounts to a unidirectional wave climate is that the Solway Firth, and in particular the inner Firth, is a sediment trap with sediment accreting on the extensive intertidal sandbanks. Thus there is a net build-up of sediment within the Solway, with little sediment escaping seawards (Perkins and Williams, 1966). One result of the predominantly eastward movement of sediment is that the Solway Firth saltmarshes are dominated by sandy sediments that are mainly marine in provenance.

Table 10.2 Characteristic geomorphological features of some of the main Solway Firth saltmarshes.

	Rockcliffe	Burgh	Moricambe Bay	Caerlaverock	Cree
Type	Fringing estuary	Fringing estuary	Fringing estuary, bay	Fringing estuary, transitional	Fringing estuary, bay
Marsh-edge morphology	Low cliffs and terraces	Low cliffs and terraces, locally ramped	Low cliffs and terraces, locally ramped	Low cliffs and terraces, rarely ramped	Ramped, locally cliffs and terraces
Creek system	Dendritic	Modified dendritic	Dendritic	Dendritic	Dendritic
Saltpans	Common	Common	Common	Infrequent	Common
Age of active marsh	>200 years	Unknown	Unknown	Pre-mid 19th century	Unknown
Mean sediment type					
Upper marsh	Sandy silt	Sand:fine sand /silt: clay	Sand:fine sand /silt: clay	Sand:silt:clay	Fine sand
Marsh edge	Sandy silt	Sandy silt	Sandy silt	Fine sand	Fine sand
Upper tidal flat	Sand to sandy silt	Sand to silty sand	Silty sand	Fine sand	Sand and gravel

Marshall (1962) showed that the saltmarshes of the Solway are usually composed of more than 90% fine-grained sand, with clay accounting for less than 4%. Since the average clay content of most British saltmarshes commonly exceeds 30%, and often is greater than 65%, the sand content of the Solway marshes is unusually high.

The combination of tidal regime, nature of the substrate and exposure to wave action influences the elevation at which pioneer marsh can become established. For example, in the south and west of Great Britain the lower limit of the pioneer *Spartina* tends to occur lower in the tidal frame in areas of restricted tidal range and on marshes sheltered from waves (Gray *et al.*, 1990). On sandy coasts, the lower level of all pioneer marsh vegetation lies higher in the tidal frame than on muddy coasts (Gray, 1992), and this may be because sandy substrates tend to occur in areas more exposed to wave action (Pye and French, 1993). Such sandy sediment is more prone to mobilization than muddy sediment and so mechanical removal of seedlings may occur before an adequate root structure has developed (Chapman, 1977). This may be one of the reasons why most Scottish saltmarshes, including those in the Solway Firth, tend to have little pioneer vegetation in comparison with those farther south and tend to be dominated by the communities found at higher tidal levels. Much of the Solway saltmarsh is characterized by a lawn-like sward dominated by closely cropped graminoid vegetation that has been grazed and/or traditionally stripped for Cumberland turf. These activities may have contributed to the extensive development of saltpans and creeks on the Solway marshes.

The Solway Firth is also characterized by extensive emerged flat surfaces that fringe the Firth, particularly in the north and east (Marshall, 1962). Many of the emerged flats also have a high sand content and display relict dendritic creek systems with numerous relict saltpans. The flat surfaces, locally known as 'carse', are emerged estuarine sandflats and saltmarshes and although fairly common in Scottish estuaries, they are nationally rare in the British context. They also provide evidence of past changes in sea level within the Solway Firth. For example, peat beds that now lie below sea level indicate times when sea level was lower than present, whereas the emerged carse, indicate times when sea level was higher than present. Haggart (1989) suggests that, over the early part of the Holocene Epoch, sea level rose from about −5 m OD at 10 000 years BP to reach a maximum of about +8.5 m OD at the peak of the transgression at about 6500 years BP. Haggart's (1989) Solway Firth Holocene sea-level curve is convincing and broadly matches the direction and timing of changes in relative sea-level elsewhere in Scotland such as in lower Strathearn (Perthshire) and the inner Moray Firth during mid-Holocene times. It is likely that the curve will gain support from palynological and other micro-palaeontological work under way at present in the Cree estuary, where it seems that the peak of the Holocene Transgression occurred at about 6500 years BP and reached 7–10 m OD (Firth *et al.*, 2000). This date compares well with the culmination of marine conditions at Crosscanonby in Cumbria, where Tooley (1985a) places the change in relative sea level sense from rising to falling at about 6800 years BP. Since then the overall trend has been mainly of a sea level falling towards the present day.

Based on sea-level curves and historical tide gauge records, Firth *et al.* (2000) estimate that present maximum rates of isostatic uplift are 1.8–1.95 mm a^{-1}, with the minimum rates in the range 0.4–0.56 mm a^{-1}. Since the lower estimates closely compare with uplift rates from recent geological evidence, they are probably a better estimate of actual rates. Since present-day global sea-level rise is estimated at about 1–2.5 mm a^{-1} (Houghton, 1994), the present status of the Solway coast is that it is subject to a slow rise in relative sea level.

SOLWAY FIRTH (NORTH SHORE), DUMFRIES AND GALLOWAY (NY 003 668–NY 118 652)

J.D. Hansom

Introduction

In spite of work by Marshall (1960, 1962), Steers (1973), Mather (1979) and Bridson (1980), the only recent detailed geomorphological research on the Solway (north shore) saltmarshes has concerned the inter-relationship between forms, sediments and radionuclides by Allan (1993), Harvey and Allan (1998) and Harvey (2000). Extensive saltmarshes occur adjacent to the River Nith that warrant detailed research, particularly concerning the recycling of eroded

Saltmarshes

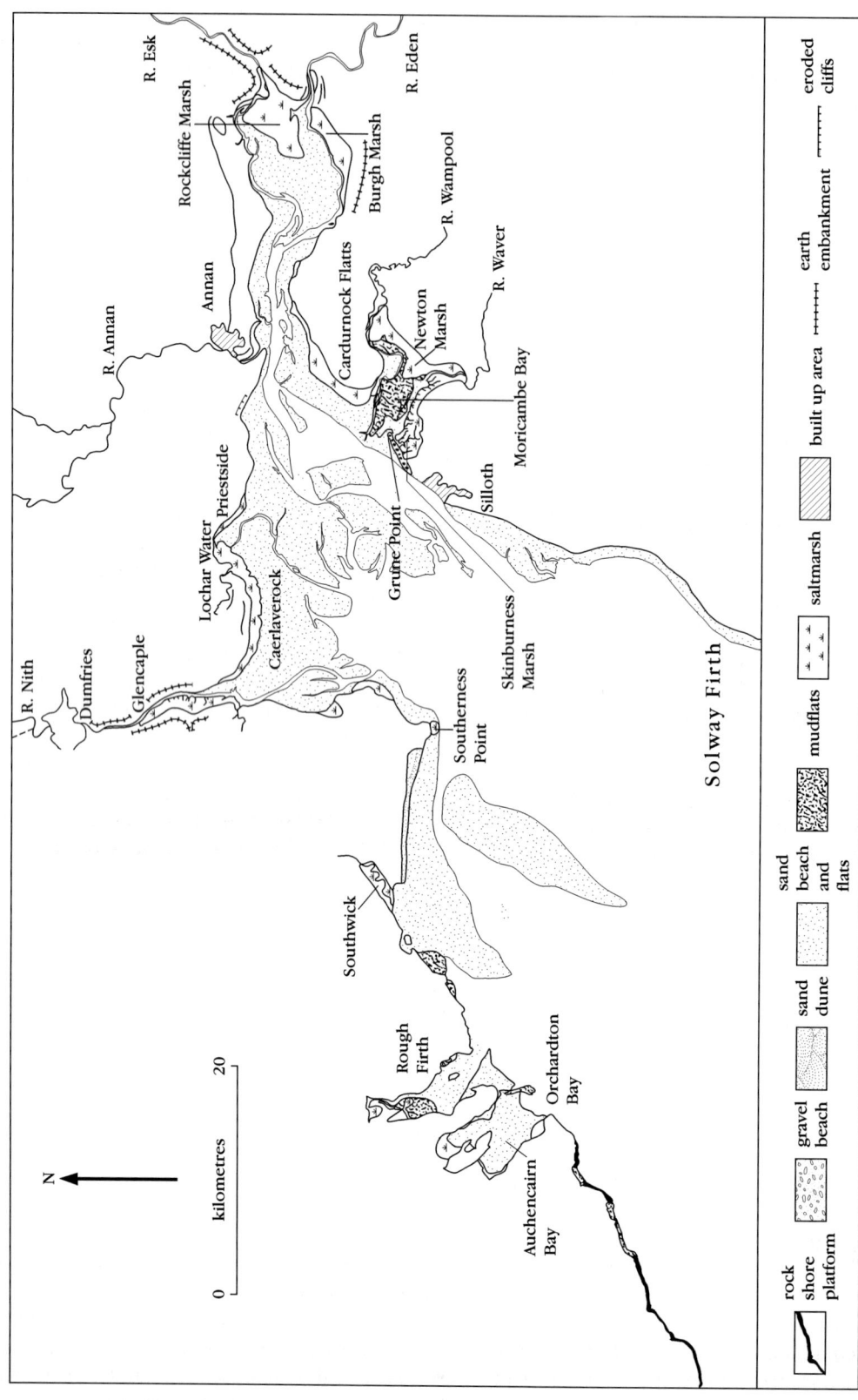

Figure 10.12 Location of the saltmarshes of the inner Solway Firth including the Upper Solway Flats and marshes on the south shore and the saltmarshes of the Solway Firth (north shore). The 2842 ha of saltmarsh found at these sites comprises 79% of all the saltmarsh in the Solway and 8% of all British saltmarshes. (After Pye and French, 1993.)

sediment in the developing marsh. Radionuclide-polluted sediment provides a useful method of quantifying such erosion and deposition.

Caerlaverock saltmarsh has been designated a Special Area of Conservation (SAC), a Special Protection Area (SPA), a National Nature Reserve (NNR) and it is part of the Nith Estuary National Scenic Area.

Description

The north Solway saltmarshes extend from the Nith estuary (NY 993 668) eastwards through the Caerlaverock National Nature Reserve across the mouth of Lochar Water to its exit at Priestside Merse (Figure 10.12). The intertidal sandflat in this area exceeds 10 km in width. Within the Nith estuary and the inner Solway, the active marshes are terraced with as many as four levels separated by unvegetated terrace edges. Flanking the west shore of the Nith, Kirkconnell Merse extends to 208 ha, much of this a *Puccinellia-Festuca* (common saltmarsh-grass –fescue) sward (Burd, 1989) (Figure 10.13). At its landward edge, the merse is entirely enclosed by earth banks, and the river edge is marked by a prominent terrace, which runs almost exactly north–south. The merse reaches 1 km at its widest to the west of Glencaple on the east bank of the Nith, but narrows southwards to disappear at Airds Point (Figure 10.13). Northwards, the merse narrows before widening and merging with Green Merse near Kelton. The marsh surface is traversed by a dense dendritic network of shallow creeks, with saltpans virtually absent other than in a narrow zone along the middle and rear of the marsh. Other than north of Kelton, where it reaches 200 m wide and is traversed by deep creeks, the saltmarsh on the east side of the Nith is narrow, intermittent and limited by the proximity of the channel of the Nith and by rising ground behind.

Caerlaverock Merse, including the 77 ha of Priestside Bank at its eastern end, extends to a total of 560 ha and is dominated by a mainly *Puccinellia-Festuca-Glaux* sward with small stands of reeds *Phragmites* (Burd, 1989; Figure 10.14). Common saltmarsh-grass *Puccinellia* and samphire *Salicornia* occur in the creeks. Caerlaverock is about 8 km long and widens from less than 100 m wide at the Nith mouth in the west, to almost 1 km wide at the Lochar Water in the east. The marsh sediments are formed almost entirely of fine-grained sand (0.2–0.002 mm in diameter) with some fine-grained silt and clay (Harvey, 2000). The landward boundary of Caerlaverock is marked by an earth bank, which extends for some distance inland along the Lochar Water. Landwards of the earth bank, a well-defined emerged beach occurs that shows two distinct surface levels at +8 m and +6.5 m OD (Steers, 1973). Using aerial photographs, Marshall (1962) indicated that these emerged beach surfaces were traversed by drainage channels that resembled old creeks and suggested that the surfaces represented emerged former saltmarshes.

The surface of Caerlaverock Merse is cut by several deeply incised creeks that run southwards to the marsh edge. However, the creeks are infrequent except in the vicinity of the Lochar Water, where a well-developed creek system exists draining eastwards. Elsewhere, the creeks are short and fed by small and infrequent surface streams. Saltpans are infrequent and tend to be located towards the rear of the saltmarsh surface. In the west of the saltmarsh, substantial lengths of the seaward edge are subject to erosion and are marked by a *c.* 1.5-m high, vertical, or in places stepped, terrace edge (see Figure 10.15). In places, the top surface of this terrace also shows extensive damage to the vegetation and surface sediment with vegetation having been stripped off for distances of up to 10 m inland. Lying below these stripped and eroded edges, and often masking the junction between terrace and sandflat below, lie numerous blocks of eroded saltmarsh sediment. The prominence and height of the terrace edge reduces to the east and is replaced by a low-angled and largely accreting foreshore. The amount of accretion increases towards the Lochar Water and this extremity of Caerlaverock is actively extending eastward, an extension favoured by the migration of the channel of the Lochar Water towards its eastern bank at Priestside.

Priestside Merse is situated on the east bank of the Lochar Water and comprises an area of grazed saltmarsh fronted by sandflat. Perkins (1973) reported that a wide saltmarsh with an erosional terrace at its front edge was present in 1964 but that by 1968 accretion was taking place at the front edge. Firth *et al.* (2000) observed a low 20–30 cm-high erosional bluff along this edge, fronted by a 5–10 m-wide zone of slumped material and turf blocks. Eastwards, the marsh

Saltmarshes

Figure 10.13 A view looking south over Kirkconnell Merse on the west side of the River Nith towards Airds Point in the middle distance and Southerness beyond. The saltmarsh is grazed and is crossed by many well-developed creeks that drain to a prominent terrace along the Nith. Part of the saltmarsh of Caerlaverock can be seen on the east side of the river and to the south lie extensive sandflats. (Photo: P. and A. Macdonald/SNH).

widens to 400 m before narrowing to a thin strip of *Puccinellia*-dominated sward at Powfoot (Mather, 1979). At Powfoot the saltmarsh is reduced to an area of 30 cm-high small hummocks capped by *Puccinellia*, locally known as 'dabs', the seaward side of which is marked by an abrupt boundary with the intertidal sandflat.

Although lying in the west, between Balcary Point and Southerness, and thus outwith the Solway Firth (North Shore) GCR site boundary, substantial saltmarshes have developed at the head of the major embayments of Auchencairn Bay, Orchardton Bay and Rough Firth and along the tidal channel of Southwick Water (Figure 10.12) as well at the head of some of the more sheltered small bays (e.g. Balcary Bay). These saltmarshes have developed in sheltered locations where extensive sandflats occur. A relatively extensive, but in the Solway context, rare, area of mudflat also occurs in Rough Firth. Many of the more extensive and mature saltmarshes, such as at Southwick, are dissected by a complex series of deep creek channels and tend to change abruptly from low marsh and pioneer communities such as samphire *Salicornia*, along the channel edges into areas of mature high marsh immediately above. At Orchardton, a substantial area of *Spartina*-dominated marsh has developed since the 1960s following the introduction of *S. anglica*. However, there are indications that this rapid accretion has been recently replaced by erosion, possibly associated with *Spartina* die-back (Harvey, 2000).

Interpretation

It is clear that the development of saltmarshes on the North Shore of the Solway is not a recent phenomenon and substantial evidence exists to indicate long-lived deposition related to shoreline emergence over the Holocene Epoch. For example, extensive level areas on either side of the Nith, and along the valley floor of its west bank tributaries, represent the emerged remnants of the Main Postglacial Shoreline and its intertidal estuarine flats (Firth *et al.*, 2000). At Barnkirk Point, the flat nature of the land immediately inland of the coastal edge is typical of emerged beaches and has been shown to include sedimentary evidence for Holocene sea-level change (Jardine, 1975). Along the Southwick section of the coast, modern salt-

Solway Firth (north shore)

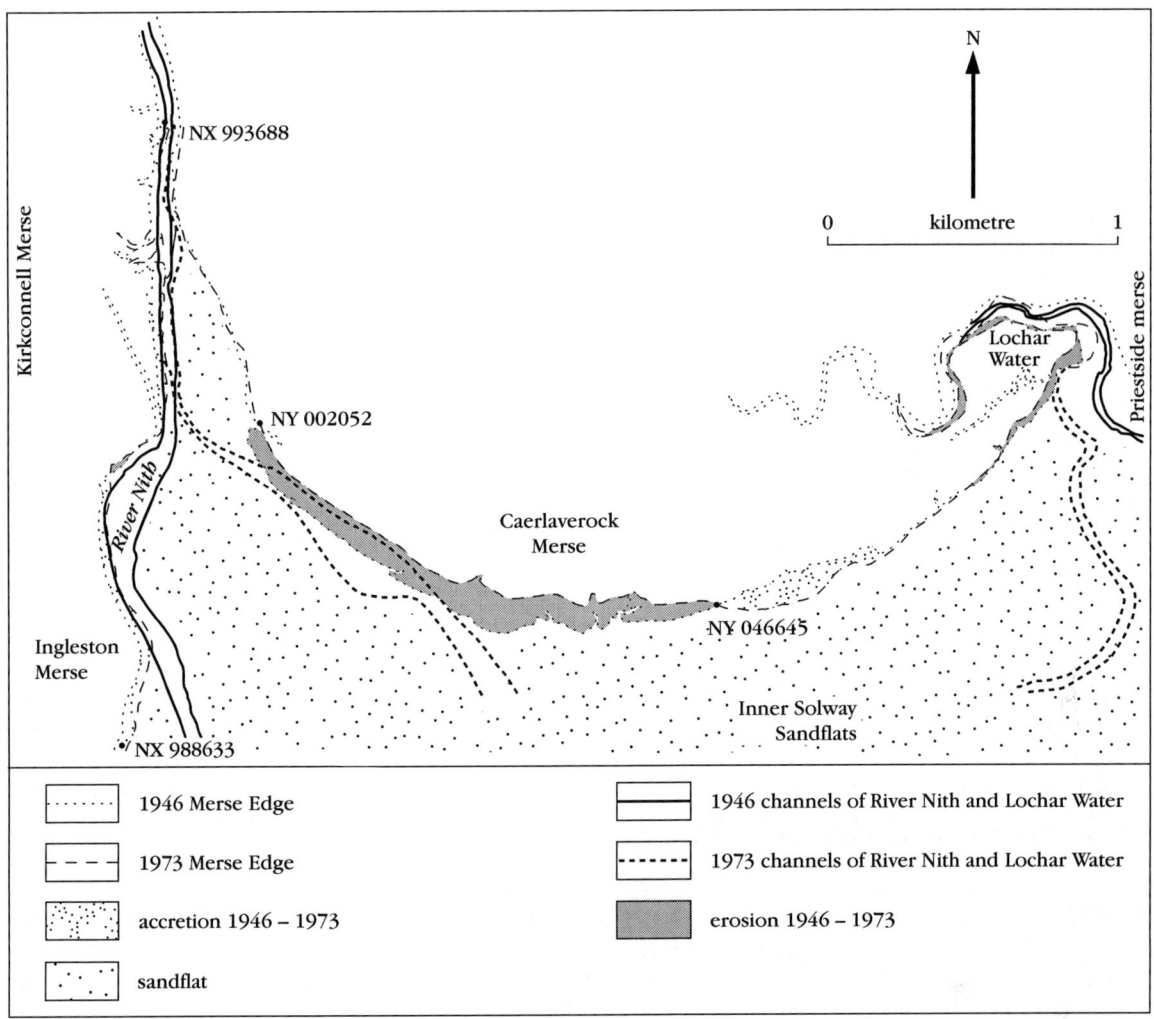

Figure 10.14 Erosion and accretion of the Caerlaverock saltmarsh edge between 1946 and 1973. Eastward migration of the main channel of the Nith resulted in erosion of the western side of Caerlaverock. Between 1973 and 1999, the channel had migrated back to the west of the bay and approximately occupied its 1946 route. In spite of this, Harvey (2000) has shown that erosion continues to dominate the west side of the saltmarsh (see Figure 10.15). In the east, close to the exit of the Lochar Water, accretion is the long-term trend. (After Rowe, 1978.)

marsh and sandflat are backed by relict cliffs 45–50 m high whose foot is adorned with emerged natural arches and stacks (e.g. at Needle's Eye and Lot's Wife). At the head of Auchencairn Bay, emerged estuarine flats up to 1 km wide lie at 8–9 m OD, and similar features lie at similar altitudes along the banks of the Urr Water as far north as Dalbeattie and make up much of the peat-covered surface behind the Mersehead Sands dune ridge. The uppermost of these surfaces at about 9 m OD and most likely date from the rise in relative sea level that occurred up to the peak of the Holocene transgression at about 6500 years BP. The lower surfaces relate to deposition that occurred on its subsequent fall (Haggart, 1989). The detailed position of terrace extent and edge location relates to the former routes of the main river channels and saltmarsh creek positions.

Such a process of adjustment to sea level and to the shifting positions of streams continues today and controls the way in which the saltmarsh and its feeder sandflats react to changing conditions. Within the Nith estuary and the inner Solway, the active marshes are terraced with as many as four levels, which Marshall (1962) suggests reflect periodic shifts and meandering of channels in the marshland or periods

Saltmarshes

Figure 10.15 Erosion on the west side of Caerlaverock saltmarsh. 1 m-high erosional terraces cut into the *Puccinellia*-dominated high-marsh surface are common on this shore but in places exposed to south-westerly wave activity, they are also subject to further surface-stripping of vegetation. (Photo: J.D. Hansom).

of increased tidal scour when high tides coincide with severe storms during periods of strong winds. However, it is also known that the moat of Caerlaverock Castle was filled by seawater in the 13th and 14th centuries, and this led Steers (1973) to suggest that there was very little marsh around it at that time. As a result, not all of the terraces may be related to sea-level changes and it is important to distinguish between the significance of the variations in altitude of the marshes and the presence of the terrace edges. The upper marshes appear likely to have been constructed at a higher relative sea level than today, but the terrace edges that separate them may not themselves represent a distinct change of sea level.

The development of most of the Solway marshes indicate that they undergo phases of erosion and deposition. For example at Caerlaverock the entire marsh seaward of the lower emerged beach seems to have developed in three phases between 1820 and 1962 (Steers, 1973). Prior to 1856 accretion had occurred in the lee of Saltcot Hill, although the site of accretion shifted to Bowhouse Scar between 1856 and 1898. Between 1898 and the 1920s, most accretion occurred west of Bowhouse Scar (Bridson, 1980). At present the oldest marsh occurs at the eastern end and is higher in altitude than the other parts. Accretion since the early 19th century was replaced by erosion during the 1920s, possibly because the piers of the Bowness to Annan viaduct, built in 1864, acted as groynes and accelerated accretion on their western side (Marshall, 1962). The rapid extension of Caerlaverock Merse and the eastern part of Kirckonnell Merse was probably related to attempts to improve the Nith river channel for navigation (Steers, 1973).

Between 1946 and 1955 erosion at Bowhouse, south-west of Caerlaverock Castle, totalled 38.1 m a^{-1} and the annual rate between 1955 and 1976 was *c.* 7.6 m (Bridson, 1980). Over similar timescales, the annual vertical accretion rate declined with altitude from 30 mm at +5 m OD, to 10 mm at +5.1 m and to very small amounts at +5.2 m (Steers, 1973). Between July 1959 and March 1961, common saltmarsh-grass *Puccinellia* marsh extended seawards by 4.9 m. However, individual events are also important, and the saltmarsh edge near Caerlaverock Castle was cut back by 3.3 m during a single storm between 30 October and 11 November 1960. Marshall concluded that generally erosion was exceeding accretion and this is borne out by saltmarsh-edge mapping conducted by Rowe (1978) (Figure 10.14) that shows the marsh edge at Caerlaverock to have retreated substantially over most of its western part while accreting in the east in the period 1946–1973. More recent work by Pye and French (1993), Hawker (1999) and Harvey (2000) show the process of erosion in the west and accretion in the east at the Lochar Water to be continuing.

The switch from deposition to erosion that has characterized the western part of Caerlaverock may be attributed to a number of factors.

(1) Changes in the position of the main channel of the Nith indicate a progressive easterly migration, accelerated by dredging related to navigational improvements and the dumping of the spoil on the west bank.
(2) The western side of the estuary is also more sheltered than the east side and Kirkconnell Merse showed a net increase of 51 ha between 1946–1973 (Rowe, 1978), a rate of about 2 ha per year. Over the same period, the west of Caerlaverock has undergone edge erosion via toppling failure of large blocks of saltmarsh sediment by undercutting at high tide during storm conditions. Even in this relatively sheltered part of the Solway, the short and steep waves that impinge on the saltmarsh edge are capable of significant erosion of unconsolidated sediments, and the stripping of layers of sediment and vegetation from the top surface of the terrace.
(3) The front edge of the saltmarsh at Caerlaverock is penetrated by a number of large creeks and within-creek erosion occurs through headward extension of the creek inland and vertical incision, forming a narrow channel. Loading pressures from overlying sediments may lead to the deformation of the lowermost saturated horizons of silty sand into the creek and Firth *et al.* (2000) suggest that the process may lead to localized lowering of the surface along the creek margins (and presumably the seaward edge) and the development of a noticeable tilt, or camber, towards the creek.
(4) It is also possible that the annual heavy grazing and trampling of the site by large numbers of wintering wildfowl (e.g. 13 000 barnacle geese), which locally damages the vegetation cover, may lower the resistance of the marsh surface to wave erosion. In this context much of the Solway saltmarsh is characterized by a lawn-like sward dominated by closely cropped graminoid vegetation. The grazing levels by cattle were recorded by Marshall (1962) as the highest on any British saltmarshes with 1 stock unit to 0.8–1.0 ha.
(5) The apparent present trend towards erosion of the saltmarsh edge may be controlled by the present rise in global sea level of about 1–2.5 mm a^{-1} (Houghton, 1994). Since the best estimates of actual rates of isostatic uplift in the Upper Solway lie in the range 0.4–0.56 mm a^{-1} (Firth *et al.*, 2000), the Upper Solway may be presently subject to a slow rise in sea level; the pattern of more widespread marsh-edge erosion, and more limited and local areas of saltmarsh accretion, is set to continue.

Work on the Solway by Harvey and Allan (1998) and Harvey (2000) relates to the way in which saltmarsh sedimentation can be informed by the inter-relationship between forms, sediments and radionuclides. The onshore movement of fine-grained sediments brings with it significant levels of radionuclides attached to the particles. These radionuclides are almost exclusively derived from the Sellafield Nuclear Fuel Reprocessing Plant in Cumbria and result in varying concentrations with depth in the saltmarsh (Figure 10.16). Coring of the saltmarsh shows that the depths of different peaks in radionuclide concentration varies spatially over the saltmarsh and so provide marker horizons from which sedimentation rates can be calculated over the time-span represented by the core length. Since erosion of older parts of the saltmarsh is ongoing then re-release of radionuclide-contaminated sediment may ultimately provide a method to estimate the relative contri-

Saltmarshes

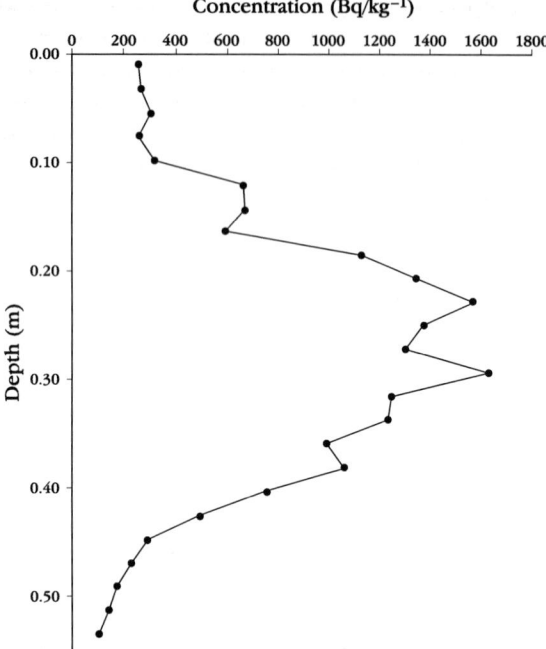

Figure 10.16 Elevated concentrations of the radionuclide ^{137}Cs occur at varying depths beneath the saltmarsh surfaces of all of the Scottish Solway saltmarshes, including at Caerlaverock. The time-integrated profile of Sellafield discharges shown here, comes from nearby Southwick saltmarsh, and shows peak concentrations at 0.30 m depth below the high-marsh surface, which represent input of sediments from the outer firth that peaked in 1978 and have declined since. These data can be used to calculate a sedimentation rate over time that can be compared with direct measurements using sedimentation plates or pins. (After Harvey, 2000.)

butions of old versus newly arrived sediment to accreting areas of the saltmarsh.

Conclusions

The saltmarshes of the Solway Firth north shore are nationally important because both their geomorphology, and that of their emerged counterparts, show evidence of past and present responses to isostatic uplift, changing sea levels and channel migration patterns. In addition, the top surface of the prominent marsh cliff that occurs along the seaward edge of the marsh has in exposed sites been subject to distinctive stripping of the vegetation cover during storm conditions. The old creek patterns that can be traced on the emerged marine surfaces landward of the present marshes represent evidence for the existence of higher and more extensive saltmarshes in the past, and radionuclides provide an additional means to estimate rates of sedimentation and thus age. Nationally, the marshes of the Solway north shore are key sites for the study of saltmarsh processes, morphology and evolution.

UPPER SOLWAY FLATS AND MARSHES (SOUTH SHORE), CUMBRIA (NY 143 569–NY 353 648)

J.D. Hansom

Introduction

The Upper Solway saltmarshes are classic estuarine marshes, which exhibit outstanding geomorphological features. A prominent marsh cliff occurs along most of the seaward edge of the Upper Solway marshes and pinpoints those parts of the marshes that are undergoing erosion. Creek systems in various stages of development are found on all of the saltmarshes and on Burgh and Rockcliffe Marshes exhibit a widely spaced dendritic pattern. Several types of saltpans are also found on the marshes. The saltmarshes exhibit some of the finest examples in Great Britain of marsh terraces that are believed to be formed by creek migration and isostatic uplift.

Description

The saltmarshes on the southern shore of the Solway Firth extend from Grune Point near Skinburness in the west, to Rockcliffe Marsh in the east at the mouth of the Esk (see Figure 10.1 for general location and Figure 10.12). Within Moricambe Bay, *c.* 1190 ha of saltmarsh extends from Skinburness Marsh south of Grune Point towards the south-east, and includes Newton Marsh, which lies between the estuaries of the rivers Waver and Wampool. Marsh is absent from the north-east shore of Moricambe Bay but occurs along a narrow fringe at Cardurnock Flatts. At Burgh Marsh, to the south of the River Eden estuary, the *c.* 524 ha (Burd, 1989) of saltmarsh is up to 1.25 km wide and is unbroken by large creeks. Rockcliffe Marsh extends to about 565 ha (Burd, 1989) and reaches over 3 km wide although it is punctuated by wide creek mouths. As with most of the Solway marshes, the Upper Solway marshes are often backed by low terraces that separate the present saltmarsh from the emerged carse surfaces, which lie to the landward side. All of these saltmarshes are based on a fine-grained sandy substrate that supports essentially similar plant communities, namely common saltmarsh-grass–fescue (*Puccinellia–Festuca*) and the dominant species, saltmarsh rush *Juncus gerardii*, which covers up to 64% of the site area (Burd, 1989). Rockcliffe is particularly important in terms of its size and its retention of an unusually wide transition between non-saline and saline habitats. Transitional grassland plants such as dog's-tail grass *Cynosurus*, buttercup *Ranunculus*, clover *Trifolium* and hawkbit *Leontodon* are found in abundance on these marshes but are not found at all in estuaries to the south of Ravenglass (Fahy *et al.*, 1993). *Spartina* is not yet widespread in the Solway but is found at Southwick in small amounts and at Orchardton, on the northern side of the Solway.

The northern shore of Grune Point is composed of a gravel spit upper beach fronted by 500 m of sandflats extending to low water. Active longshore drift of gravels occurs northwestward along the northern shore (Fahy *et al.*, 1993) and provides shelter for the development of saltmarshes within Moricambe Bay. The highest parts of Grune Point, particularly along the north-western side, are locally capped by dune sand. Skinburness Marsh lies south of Grune Point and is well terraced with extensive creek and saltpan development terminating at a rapidly accreting frontal margin close to Grune Point (Perkins, 1973) that is dominated by samphire *Salicornia* (Marshall, 1962). Elsewhere the marsh edge is cliffed and subject to intermittent erosion. The marshes of Moricambe Bay were noted by Marshall (1962) to be composed of over 90% fine-grained sand in both the saltmarshes and the bare sandflats, a characteristic shared throughout the Upper Solway. Most of the fine-grained sand was regarded by Marshall to be of marine origin, although some may have been reworked from the fine-grained sands of the emerged carse deposits. The modern marshes are terraced with small steps of between 0.3 m and 0.6 m in height occurring between the terraces. The higher of these terraces often separates the present marsh from the emerged carse behind. The emerged carse surface also displays terraces, the lower of which reaches up to 6.4 m OD, with the upper terrace attaining +7.3 m OD at Moricambe Bay and Burgh Marsh.

A major area of saltmarsh occurs along the southern shore of the Eden estuary at Burgh Marsh (Figure 10.12). The channel of the River Eden flows along the front edge of the marsh and appears to be migrating southwards in this area at the present time so that there is an absence of significant sandflat in front of the saltmarsh. However, extensive sandflat occurs on the northern side of the Eden channel and on the south side of the Esk channel. The highest parts of these sandflats have developed into saltmarshes dominated by saltmarsh rush *Juncus gerardii* with common saltmarsh-grass *Puccinellia* and fescue *Festuca* (Firth *et al.*, 2000). Allen (1989) reports relict, partially buried, erosional bluffs on the surface of Burgh Marsh, which are interpreted to indicate that oscillations in the proximity of the channel of the Eden over time, that have resulted in cycles of erosion and accretion.

The largest expanse of saltmarsh in the Upper Solway is Rockcliffe Marsh (Figure 10.17). Much of the upper central saltmarsh is relatively flat but around the margins and in the western part the marsh surface is incised by numerous creeks with extensive intervening areas of saltpans. Many of the creek mouths on the north-eastern margin adjacent to the Esk channel are narrow and deeply incised, although elsewhere on the north-west and southern margin the creeks are wide and characterized by areas of accretion with pioneer vegetation. Both the northern and southern margins have been expanding rapidly in response to the migration of the Esk and Eden channels towards the north and south respectively, leaving between a large low-energy zone that is occupied by the present marsh. Maps dating back to 1776 (ABP, 1991, in Black *et al.*, 1994) suggest that this is part of a long-term trend initiated by the southwards migration of the channel of the River Eden in the 19th century, but continued land uplift almost certainly contributes to the creation of a more emergent surface (Firth *et al.*, 2000). Between 1946 and 1973, Rockcliffe Marsh experienced a net expansion of 414 ha (15 ha a^{-1}) (Rowe, 1978).

Interpretation

The vegetation quality and degree of development on the Upper Solway marshes is thought to indicate that they comprise a relatively old, stabilized marsh system (Burd, 1989). The presence of eroded seaward edges up to 2 m high on

Saltmarshes

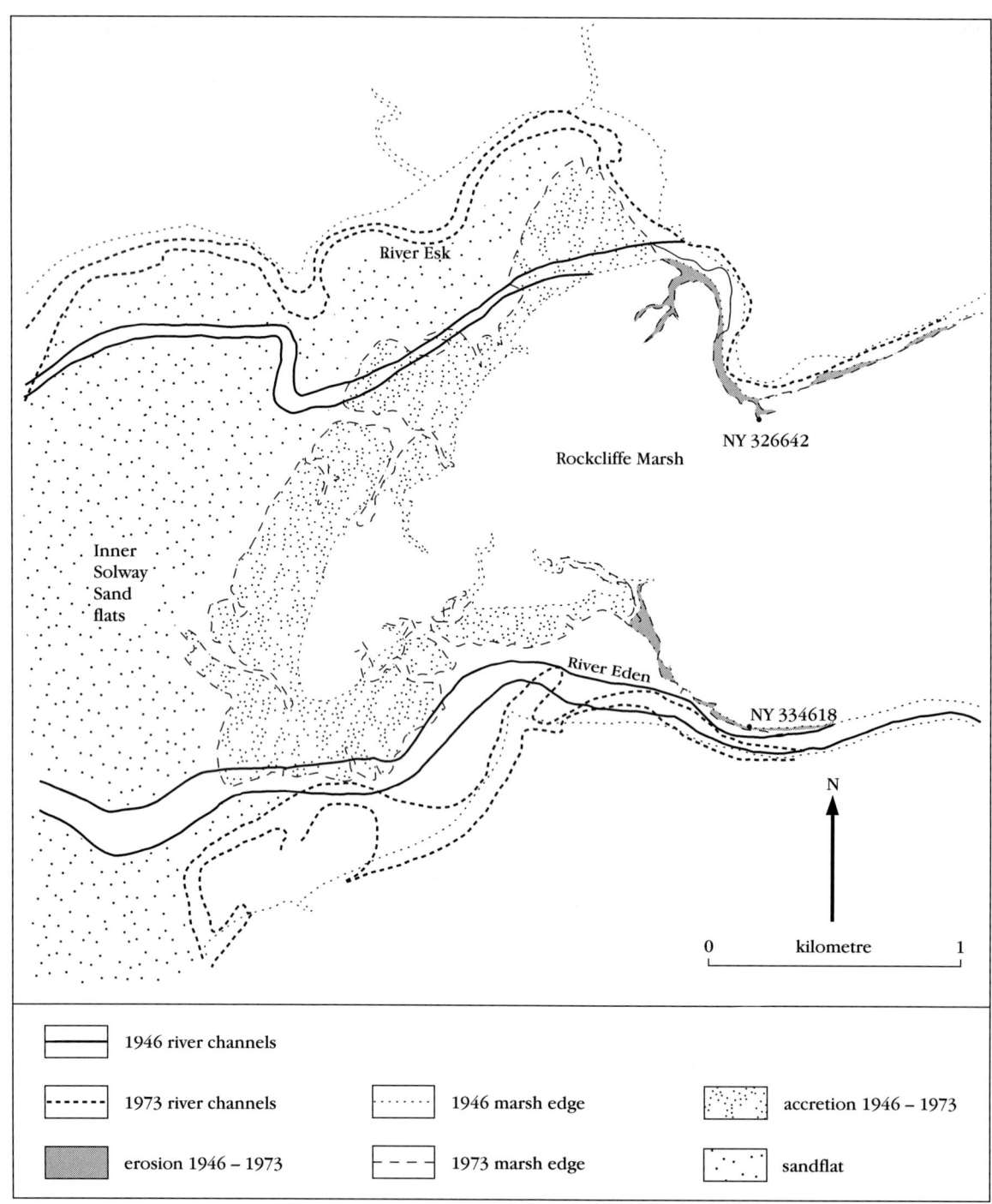

Figure 10.17 Migration of the Rivers Esk and Eden over the period 1846–1973 has contributed to rapid accretion and westward migration of Rockcliffe Marsh that continues today. A healthy supply of sediment comes from the extensive nearshore and intertidal sandflats, augmented by fluvial sediment from the two rivers. (After Rowe, 1978.)

many of the marshes lends support to this hypothesis and suggests that the developing marshes have been subject to cycles of erosion and deposition depending upon the relative proximity of river channels and the rate of sea-level change. It is likely that the broad transitions to mature upper marsh and freshwater communities that are so well displayed in the Upper Solway marshes are also related to the history of sea-level change experienced by the area. The transitions away from salt-affected vegetation so well-represented on the Upper Solway Marshes are of considerable importance because such zonations have been largely destroyed by land-claim in many other British saltmarsh systems. Although artificial embankments and walls are present along many of the intertidal reaches of rivers draining into the inner Solway and on some low-lying areas inland of Rockcliffe Marsh and Moricambe Bay, direct physical human impact on most of the Upper Solway saltmarshes remains minimal, although some of the saltmarshes have a history of turf-cutting and most are still grazed.

The Upper Solway Marshes also provide the finest examples in Britain of marsh terraces formed by the combined action of creek migration and land uplift. The terraces were first regarded by Dixon *et al.* (1926) as strong evidence for recent changes of sea level. They regarded the combination of gradual seawards decrease in altitude of the emerged 'carse' surfaces and the continued growth of Grune Point and small terraced flats on the modern saltmarsh as evidence for continuous uplift. However, Marshall (1962) interpreted the stepped nature of the marshes to be mainly erosional, since where they were present the terraces never graded into each other and the step was at an approximately constant height. This was thought to demonstrate alternation between erosion and accretion, probably the result of erosion by shifting river channels. The most likely scenario probably involves both of the above processes.

All of the marshes have eroded and accreted large areas during the 20th century. In Moricambe Bay, a loss of 39 ha of saltmarsh at Skinburness Marsh between 1860 and 1900 was balanced by accretion of 105 ha (Steers, 1946a). The *Salicornia*-dominated part of the marsh at Skinburness extended laterally by over 50 m between July 1959 and March 1961 (Marshall, 1962). At this time most of the edge of Burgh Marsh and the south-east edge of Rockcliffe Marsh the edge was characterized by high (2.0 m) cliffs, although elsewhere the marsh edge undergoing erosion was between 0.3 and 0.6 m above the adjacent sandflat (Marshall, 1962). Such erosion in this low wave-energy environment was attributed by Marshall to result largely from shifts in river channels rather than to wave activity. Indeed, with the possible exception of Cardurnock Flatts, all of the marshes are sheltered from substantial wave activity. The Moricambe Bay marshes are protected by Grune Point and a north-west-facing bay entrance that restricts the fetch of the dominant south-westerly waves. Rockcliffe Marsh lies at the head of a meandering estuary that reduces the access of westerly waves to only 1 km and is fronted by many kilometres of intertidal sandflats. As a result, patterns of erosion and accretion on the marshes are largely dictated by changes in river channels and by the long-term emergence of the coast.

Long-term estimates of erosion and accretion are possible by using areal comparisons of maps by Marshall (1962) and Pye and French (1993) over the period 1864 to 1993 (Table 10.3). These indicate that Rockcliffe and Skinburness Marshes gained area from 1864 to 1973, but

Table 10.3 Estimated areal accretion in hectares between 1864 and 1946, 1946 and 1973, 1973 and 1993 for selected inner Solway saltmarshes. (Based on data from Marshall, 1962; Rowe, 1978 and Pye and French, 1993.) All areas in ha. Caerlaverock Marsh is in the Solway Firth (north shore) GCR site.

Marsh	1864	1946	1993	1894–1964	1946–1973[1]	1946–1993[2]
Rockcliffe	664	709	565	+45	+414	−144
Burgh	688	534	524	−154	−82	−10
Skinburness	445	506	n/a	+61	+100	n/a
Caerlaverock	194	607	563	+413	−93	−44

1 Rowe (1978)
2 Pye and French (1993)

then reduced in area over the period 1946 to 1993. In the case of Rockcliffe, Rowe (1978) shows that rapid gains were made between 1966 and 1973 (430 ha) but the data of Pye and French (1993) indicate a subsequent loss in area by 1993. Such substantial and recent erosion seems unlikely given the location of Rockcliffe Marsh, and it may be that methodological differences in estimating accretion areas is responsible for this apparent anomaly. Where marsh edges undergoing erosion occur, mapping the boundaries is more secure than where accretion dominates and the actual edge is uncertain and seasonally mobile. Figure 10.17 shows the marsh edge at Rockcliffe between 1946 and 1978 as mapped by Rowe (1978). It is possible that the apparent present trend towards erosion of the saltmarsh edge suggested by Table 10.3 is related to the present rise in global sea level of about 1–2.5 mm a^{-1} (Houghton, 1994). The best estimates of actual rates of isostatic uplift in the Upper Solway lie in the range 0.4–0.56 mm a^{-1}, since these compare with uplift rates from recent geological evidence (Firth et al., 2000). It thus appears possible that, after a long period of emergence, the Upper Solway is presently subject to a slow rise in relative sea level and that this may produce a future trend of more-widespread marsh-edge erosion and more-limited and localized areas of saltmarsh accretion.

Conclusions

The Upper Solway Marshes together represent an area characterized by outstanding examples of emerged saltmarsh on which the geomorphological and vegetational effects of accretion in an inner estuary location have been accentuated in the past by isostatic uplift. In spite of this, some edges undergoing erosion, and distinct terraces on both the present and emerged marsh surfaces, indicate that changing locations of river and estuary channels are also responsible for cycles of erosion and accretion. At some places and times, such local effects may be more significant to the local development of the marsh than the longer-term effects of isostasy. Although little work has been done on development of these marshes, it is likely that the creek and saltpan networks relate closely to the interaction of erosion and accretion resulting from both local river regimes and general isostatic effects. After a long period of emergence, the more recent trend towards erosion of the saltmarsh edge may be a function of a slow rise in relative sea level.

CREE ESTUARY, OUTER SOLWAY FIRTH, DUMFRIES AND GALLOWAY (NX 465 545)

J.D. Hansom

Introduction

The saltmarshes of the Cree estuary (see Figure 10.1 for general location) demonstrate well the geomorphological features of fringing estuarine saltmarsh. The creek system is dendritic, especially in the north close to the Cree exit, although the creeks in the south have been artificially straightened. Saltpans are distributed over all marsh levels, particularly on the marshes adjacent to the Cree exit. Parts of the saltmarsh have developed very recently and independently of the complex of saltmarshes in the inner Solway Firth. For example, the marsh at the Baldoon Sands on the west side of the estuary has mainly developed since 1847 (Figure 10.18). In spite of this recent development, estuarine sedimentation has prevailed in the Cree estuary over most of the Holocene Epoch because all of the present saltmarshes are backed by extensive areas of emerged estuarine flats known in Scotland as 'carse'.

Description

The total extent of saltmarsh in Wigtown Bay amounts to 553 ha (Burd, 1989), 108 ha of which lie on the western shore of the Cree estuary (Rowe, 1978), extending for over 10 km northwards from near Jultock Point to the tidal limit of the River Cree itself to the south of the town of Newton Stewart. On the eastern shore of the Cree, north of Ravenshall Point, saltmarsh occurs only within a small area north of Wigtown Sands (Figure 10.18) and small, partly enclosed, areas farther upstream. The saltmarsh communities are mostly dominated by a common saltmarsh-grass–fescue (*Puccinellia-Festuca*) sward (Burd, 1989). Much of the shoreline within Wigtown Bay is characterized by extensive sandflats that are succeeded landwards by saltmarshes, brackish reedswamps and emerged Holocene carse deposits (Figure 10.19). Within the smaller bays, more-restricted areas of saltmarsh are present. However, in spite of the wealth of information gained on Holocene sea-level changes in this area and the wide variety of features present, few studies have been undertaken to assess the evolution and processes of the modern sandflats and saltmarshes.

North of Jultock Point (Figure 10.18) the flats of Baldoon Sands extend to 3.75 km wide and are backed by saltmarshes that widen from 50 m in the south to 1 km in the north adjacent to Wigtown Sands and the mouth of the River Bladnoch. The marshes are dissected by large and generally linear channels, and extensive saltpan systems occur in places. For example, at Baldoon the saltmarsh reaches 300 m wide and is traversed by several linear creeks that have been artificially deepened and straightened to facilitate more rapid drainage from the grazed marsh surface (Figure 10.20). There is only a very restricted amount of low marsh supporting primary colonizing vegetation. In places the marsh edge is marked by a small terrace but more often there occurs a low-angled ramp of partly-vegetated sand, which merges imperceptibly to the sandflat surface. The flora of the marsh surface is typical of a grazed Solway saltmarsh, being dominated by a fescue–common saltmarsh-grass *(Festuca–Puccinellia)* sward with abundant thrift *Armeria maritima* and sea milkwort *Glaux maritima*. At Crook of Baldoon, the marsh is backed by a rubble embankment that is used for flood protection and access and has been extended southwards recently, parallel to the fence-line, to enclose the rear of the high saltmarsh surface.

North of Wigtown, the sandflats give way to muddier sediments and marshes up to 1 km wide are located on the eastern side of the River Cree as well as on the west. North of Wigtown Sands towards Newton Stewart, the river channel is deeply incised and its muddy banks are characterized by failure and toppling of saltmarsh sediment and vegetation into the channel below. The estuary becomes increasingly restricted between either artificial earth embankments, boulder groynes or low cliffs cut into emerged Holocene estuarine deposits, but a range of high-marsh vegetation with occasional reedbed occurs (Figure 10.19). Minor depositional areas occurring on the inside loops of the meandering tidal channel are characterized by narrow and steeply sloping banks of silt and clay but strong river and tidal streams along much of

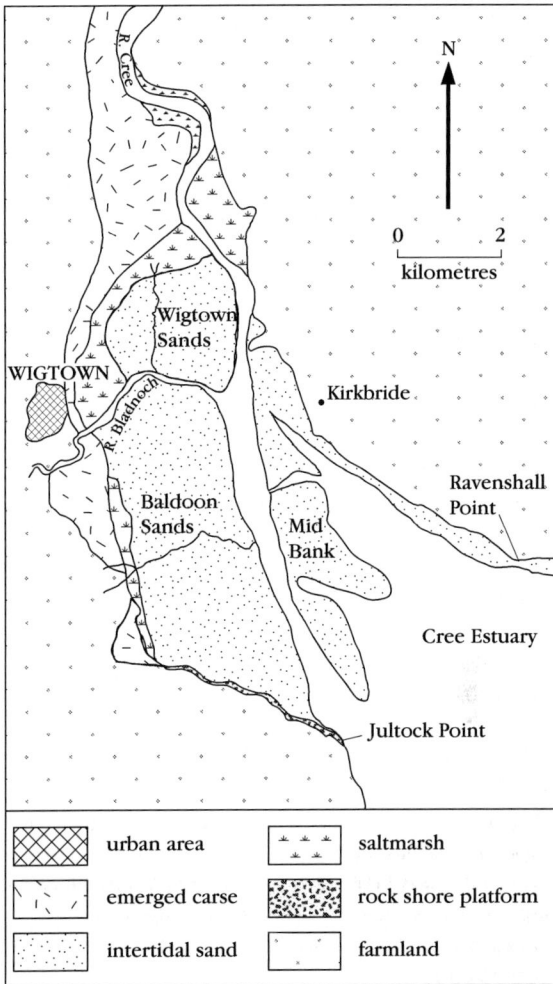

Figure 10.18 Saltmarshes and emerged carse surfaces in the Cree estuary, Wigtown Bay. The estuary is shallow and extensive sandflats are exposed at MLWS. Muddier sediments are restricted to the tidal reaches of the Cree River itself in the north where small saltmarshes fringe its course, particularly on the inside of meander loops.

this section have resulted in an incised channel and limited deposition (Figure 10.19).

South of the Cree exit, the foreshore is characterized by sand and gravels and no saltmarsh occurs until inside Fleet Bay, on the south-eastern side of Wigtown Bay, but outside the GCR site boundary. Here 1 km-wide mudflats with flanking saltmarshes occur between Skyreburn Bridge and The Canal. Low-marsh vegetation grades gradually landwards into high-marsh species and creek and saltpan development is limited, with little evidence of erosion. The

Saltmarshes

Figure 10.19 A view looking south over the meanders of the Cree estuary towards Jultock Point in the distance on the right (west) side. Extensive sandflats are visible to the south of the forest in the middle distance. Emerged carse surfaces (old saltmarsh deposits) dominate the foreground, separated from the present saltmarsh by small cliffs that can be seen in places along the main river channel. (Photo: P. and A. Macdonald/SNH.)

Canal is an artificial and embanked boulder-lined channel that is deeply incised at its landward end into emerged Holocene estuarine deposits.

Throughout the area of the Solway, there are large sections of the present shore backed by emerged Holocene marine features, good examples of which occur at the heads of the Cree and Fleet estuaries. In these areas the emerged estuarine silts and clays of the carselands attain altitudes of between 7–10 m OD. Within the Cree estuary carseland, sediments extend as far inland to the north as Newton Stewart and to within 2 km of Jultock Point in the south on the western shore. The carse deposits reach 3.5 km wide north of Wigtown, where they partially overlie and partially abut large peat beds at Moss of Cree, Carsegown Moss and Borrow Moss. The carselands in the Fleet estuary are more restricted, forming terraces up to 500 m wide.

Interpretation

The abundance of well-developed emerged landforms on the north coast of the Solway, and the Cree area in particular, has stimulated interest in the Holocene sea-level history, mainly by Jardine (1975, 1977, 1978) but also by Bishop and Coope (1977) and Haggart (1989). More recent work in the late 1990s by Wells, and reviewed by Firth *et al.* (2000), forms the basis of the following account. Recent stratigraphical evidence from cores taken from the emerged carselands flanking the Cree estuary has established the existence of a buried layer of at least 9 m of fine-grained sediments capped, at between −2 m and 0 m OD, by a thin layer of buried peat. At Carsewalloch Flow, the buried peat is itself buried by up to 9 m of emerged estuarine sediments deposited by the Main Postglacial Transgression at about 6500 years BP (Firth *et al.*, 2000), and that now form the carse surface. Estuarine microfossils from the basal

Cree Estuary, Outer Solway Firth

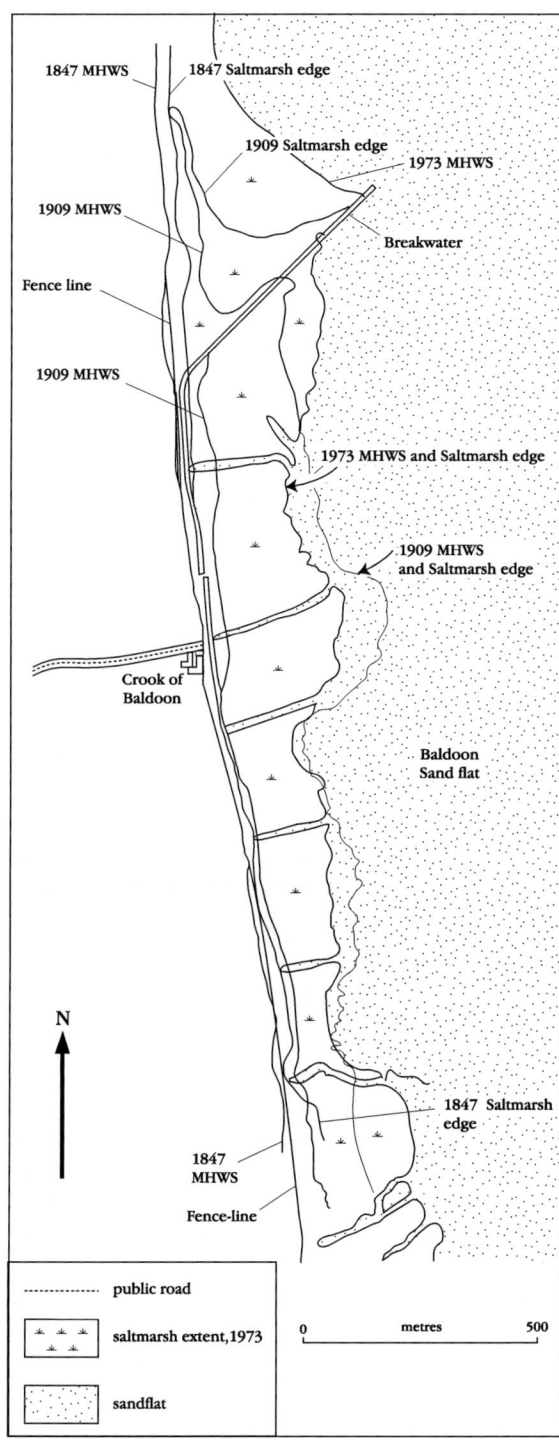

Figure 10.20 The linear saltmarshes at the Crook of Baldoon, on the western side of the Cree estuary appear to have undergone rapid accretion since the edge was mapped in 1847 but have mainly undergone edge retreat since 1973. Landward embankments along much of the Cree testify to land-claim for agricultural purposes in the past. Linear creeks at Baldoon have been artificially straightened and deepened.

sediments indicate that marine influences were dominant prior to 9600 years BP, but radiocarbon dating of the peat that caps these sediments shows sea level had fallen and had abandoned the area by at least 8300 years BP. However, radiocarbon dating of the uppermost part of the thin layer of peat indicates a transgression of the sea by 8000 years BP, followed by a rise to the Main Postglacial Shoreline at between 7–10 m OD by 6500 years BP. The subsequent fall of relative sea level from the uppermost carse surface to its present level was not uniform, because small areas of younger peats are found in the carse sediments at 5700 years BP, and evidence of a shoreline produced by a still-stand or transgression at 2000 years BP is also found. Over the past 2000 years the relative sea level in the Cree estuary and in the rest of the northern Solway Firth fell more or less smoothly to present levels. Such dating evidence produces a sea-level curve for the Cree that is broadly in agreement with elsewhere in Scotland (Haggart, 1989).

On the basis of sea-level curves, it is possible to determine the rate of crustal uplift in Scotland (Shennan, 1989; Firth *et al.*, 2000). Over the last few thousand years, the evidence from the Solway suggests a mean rate of crustal uplift of 1.0 mm a^{-1}. Since recent eustatic changes in sea level produce a rise of 1.0–2.4 mm a^{-1} (Trupin and Wahr, 1990), then the present relative sea-level trend on the north coast of the Solway is either stable or, more probably, slightly rising.

Set within this sea-level context, the saltmarshes of the Cree have developed on the fringes of an estuary where they, and their emerged couterparts, have benefited from the shelter provided by Burrow Head to the south and by a north-west–south-east orientation along the length of the inlet. The eastern shore of the Cree estuary is also indented by Fleet Bay, a major NNE–SSW-aligned embayment. Shelter within the Cree has been unchanged over thousands of years and has produced a largely unidirectional wave climate producing a sediment sink that has encouraged sediment influx but little removal. It appears that this system still operates, since many of the sandflat and saltmarsh systems of the Cree are accreting, although the relative contribution of the sediments of different provenance is more difficult to determine. The extensive sandbanks in the Cree and Water of Fleet estuaries suggest that most of the material passes landwards up the tidal channels and the presence of Sellafield-derived radionuclides attached to the sediment confirms the outer Solway as a major source (Harvey and Allan, 1998). Within the Cree mouth itself, increasing amounts of mud suggest a fluvial source augmented by active reworking of emerged Holocene sediments from the carse deposits.

The position of the main channel of the Cree is likely to have a major influence on the local erosion or accretion regime of the sandflat and saltmarsh. For much of its route south from Newton Stewart, the present channel is now guided by embankments designed to allow land-claim of the saltmarsh behind, but which also serve to contain and train the river along a relatively inflexible course, especially on the outer bends of meander loops. At the exit of the Cree to the north of Wigtown Sands, the river has been further deflected to the east by a series of boulder groynes, which, by protecting the northern extent of Wigtown Sands, its saltmarsh and claimed land, serves to direct the main channel of the river onto the eastern shore. As a result the eastern shore to the south of Creetown is scoured by stronger currents that limit the deposition of fine-grained sediments and so is mainly developed in sands and gravels with little or no saltmarsh. The land-claim embankments and the training breakwaters of the River Bladnoch at Wigtown on the west shore also serve to encourage depositional conditions on the west shore by deflecting the main channel of the Cree towards the east.

It is evident that there has been change to saltmarsh extents over the historical period. For example, Figure 10.20 shows the movement of the MHWS and saltmarsh extent at Baldoon, on the west side of the estuary. Although embankments enclose a substantial area of former saltmarsh, the relative positions of the 1847 MHWS and 1847 saltmarsh edge show a very narrow and linear saltmarsh, which suggests that the date of enclosure was just prior to 1847. Subsequent to this, accretion, and the construction of a 600 m-long breakwater, resulted in the eastward migration of the MHWS by up to 100 m and eastward migration of the 1909 saltmarsh edge by up to 400 m, presumably at a lower height and characterized by an extensive area of low marsh colonized by pioneer species. By 1973, the edge of the saltmarsh coincided with mean high-water springs and lay on average some 60 m west of the 1909 position and was characterized by an elevated marsh surface at the same altitude as MWHS. Only small patches of low marsh remain, and most of the saltmarsh surface is now regarded as high marsh in vegetational and geomorphological terms.

This process of erosion and steepening of the upper gradient of the saltmarsh continues today, but the main channel of the Cree lies well to the east and is unlikely to be the cause of erosion at Baldoon. More likely is a causal relationship with relative sea-level rise and sediment supply.

Conclusions

Fringing estuarine saltmarshes are greatly influenced by the location of the main channel of the estuary and tend to develop along lines flanking those areas that allow fine-grained sediment to settle. The saltmarshes of the Cree estuary demonstrate this attribute well. Where they have undergone lateral widening, the geomorphological features of dendritic creek systems and numerous saltpans are well developed, especially in the north close to the Cree exit. Parts of the saltmarsh have developed very recently and independently of the complex of saltmarshes in the inner Solway Firth, and substantial areas are less than 200 years old. The Cree estuary saltmarshes are also important because they are associated with extensive areas of emerged estuarine carseland that offers a well-constrained sea-level history and allows more recent sedimentation to be set within a detailed long-term context. Although many of the creeks in the south are artificially straightened and the saltmarshes and rivers of the Cree have not escaped artificial embanking and training to assist agricultural land-claim, enough of the natural features of fringing estuarine saltmarshes remain to make this large area of sandflat and saltmarsh of major conservation importance.

LOCH GRUINART, ISLAY, ARGYLL AND BUTE (NR 285 665)

J.D. Hansom

Introduction

The saltmarshes within Loch Gruinart, Islay (see Figure 10.1 for general location), demonstrate particularly well the geomorphological attributes of the type of saltmarsh found at the head and fringing the sides of long inlets such as the Scottish sea lochs. Such loch-head and fringing saltmarshes appear to be confined mainly to Scotland and Norway and develop in response to the exceptional shelter offered by long inlets and the constricted tidal dynamics found in such sites. They also differ from many saltmarshes elsewhere in Great Britain since they occur on an emerging coast. The saltmarshes of Loch Gruinart display distinct zoned drainage patterns with linear and narrow creeks and saltpans that are largely confined to the upper marsh. The saltmarshes not only form an integral part of the assemblage of coastal forms on Islay, but are nationally important for studies of saltmarsh geomorphology. The saltmarshes are the only GCR site to have developed in this setting. They are also distinctive in the coarseness of much of the substrate, a mixed gravel and sand, quite unlike its muddy counterparts in England and Wales. Other than recent work related to sea-level change (Dawson *et al.*, 1997), the coastal geomorphology of Loch Gruinart has not attracted any detailed attention.

Description

Loch Gruinart, sited on the north coast of Islay, is 2 km wide and 7 km long and is a mesotidal sea loch with a mean spring tidal range of 3.1 m (MacTaggart, 1998d). It faces due north towards Colonsay and Mull and so, in spite of the generally stormy nature of the Minch, the entrance to the loch benefits from the sheltering influence of these islands. As a result, Loch Gruinart contains the largest area of sand deposition on Islay (about 77% of all sand area; Ritchie and Crofts, 1974) (Figure 10.21). The low-lying area of Loch Gruinart is backed on all sides by a prominent emerged cliff whose foot lies at about 8 m OD. The topographic depression occupied by the loch continues inland as a low marshy depression and extends south where it is occupied by Loch Indaal, a sea loch on the south coast of Islay.

The western and southern side of the loch is composed of Torridonian Sandstone whereas the east is mainly Dalradian quartzite, grit and schist. The Loch Gruinart fault runs along the western shore and is paralleled by a fault that runs along the eastern shore of the loch a few kilometres to the east (Ritchie and Crofts, 1974). The Holocene sea-level history of Islay is relatively well known, and was characterized by submergence until between 6500–5000 years BP as a result of the main Holocene transgression. This was followed by emergence to the present time as a result of isostatic uplift (Dawson *et al.*, 1997).

The extensive intertidal sandflats and saltmarshes of Loch Gruinart lie in the shelter provided by a low rocky headland capped by low sand dunes at Ardnave Point on the west side and a large beach and dune complex that has developed on the east side at Killinallan (Figure 10.21). The beach of Tràigh Baile Aonghais, which fronts Killinallan, is wide and low-angled and fed by large amounts of sand from nearshore adjacent sandbanks. As a result the beach shows signs of recent accretion and embryo dune development. Ardnave Point and Killinallan Point, together with a relatively wide and shallow intertidal zone, have provided a sheltered environment within which sedimentation has resulted in the development of saltmarshes.

Extensive intertidal sandflats lie within Loch Gruinart, although there are also areas of gravels in the centre of the loch and along its margins (Figure 10.21). The surface of the intertidal flats are marked by mega-ripples at the mouth of the loch as a result of strong tidal streams. Along the margins of the loch, extensive areas of sandflats are colonized by algal mats and other primary colonizers. The occurrence of pioneer species suggests ongoing accretion, particularly to the north of the current areas of fringing saltmarsh. Landward of these sandflats are areas of fringing saltmarsh. On the western side, the saltmarsh surface is broken by low terraces that separate the lower marsh from the upper marsh surface, although in places the terrace edge is buried by later deposition. The saltmarsh surface is also punctuated by a range of different shapes of saltpan. Some of these are circular in form, others are linear and several are littered with stranded debris. In some areas in the mid-intertidal, small mounds of gravel and boulders have provided the nucleus for colonization by saltmarsh communities and mid-estuarine saltmarsh islands (a sub-type of fringing marshes) have developed.

Loch-head saltmarsh has developed across a 0.5 km-wide stretch of the southernmost part of the loch (Ritchie and Crofts, 1974; Figure 10.21). MacTaggart (1998d), using recent aerial photography together with the presence

Saltmarshes

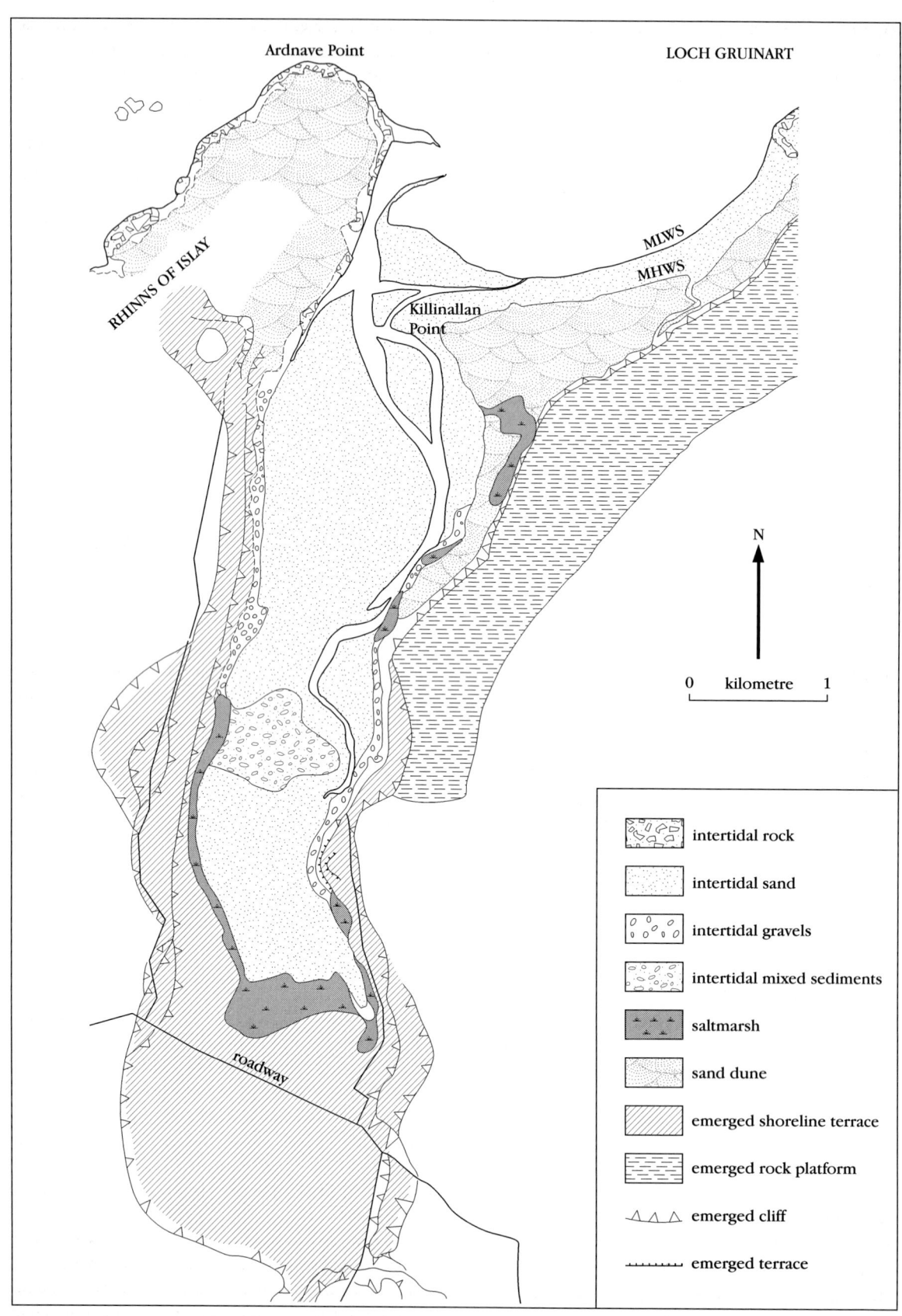

of pioneer species along the seaward edge of this stretch, suggested that progradation was ongoing. Distinct drainage patterns have developed over the saltmarsh surface. The narrow and linear creeks carry tidal flows over the upper and lower marsh surface and some of these join with artificial drainage ditches carrying freshwater from the adjacent hillsides across the upper marsh. Several of the creek sides show erosional undercutting and bank collapse. Saltpans are largely confined to the upper marsh where examples of circular and debris pans are common. The landward extent of the saltmarsh is constrained by an artificial embankment behind which are areas of reclaimed saltmarsh. A roadway crosses the southern part of the saltmarsh and emerged shoreline terrace (carse).

Interpretation

From the viewpoint of shelter, the loch-head and fringing saltmarshes of Loch Gruinart are quite normal in that they have developed in the benign wave environment offered by the presence of the rocky headland of Ardnave Point and the beach and dune complex at Killinallan. However, the saltmarshes are unusual in the British context in that they have developed, and continue to develop, on an emerging coast that is now characterized by a regional lack of coastal sediment supply. Nevertheless, the low-lying structural depression now occupied by lochs Gruinard and Indaal has been the focus for local deposition over much of the Holocene Epoch as a result of a combination of a supply of glaciogenic material from the adjacent low-gradient slopes and sea-level changes that have resulted in the inundation of the area at least twice in the last 10 000 years. Although the initial cutting of the prominent cliff probably took place soon after deglaciation, re-occupation and re-trimming occurred most probably at several times over this period before its final abandonment and the accretion of beaches and terraces at its foot later in the Holocene Epoch.

Detailed coring, biostratigraphy and dating indicates that marine–brackish–freshwater and marine–freshwater transitions occur in the diatoms that occur within the Gruinart subsurface sediments. Based on this evidence, Dawson *et al.* (1997) argue that the Rhinns of Islay were separated from the mainland of Islay by a marine inlet that formed following deglaciation about 13 000 years BP (Figure 10.22). Subsequent sea-level fall in the Lateglacial resulted in emergence for a period between 11 000–9000 years BP. However, the culmination of the Holocene transgression then resulted in the subsequent inundation of the Loch Gruinart area. Although Holocene sea-level rise in much of Scotland is generally thought to have been completed by *c.* 6000 years BP, the age of the sediments within Loch Gruinart appear to be young, dating from 2000 years BP, and suggest that a tidal strait may have existed between the Rhinns of Islay and the mainland of Islay between 8000–2000 years BP (Dawson *et al.*, 1997). Whatever the date of separation, ongoing isostatic uplift resulted in an increasingly shallow marine environment within Loch Gruinart that has been conducive to sedimentation and the resulting infill of the margins and heads of both lochs Gruinart in the north and Indaal in the south.

Set within the context of its Holocene sea-level history, the depositional regime of Loch Gruinart locally reflects progressive shallowing, shelter from waves at its entrance and the availability of locally derived sediments. As a direct result of its provenance the intertidal flats and saltmarshes are dominated by sandy sediments with little silt and mud, unlike the saltmarshes elsewhere in Britain. The underlying sediments include locally derived gravels that have become lag deposits that are undergoing encroachment by saltmarsh vegetation. It is also possible that ongoing uplift is reflected in the marked terracing that occurs on the fringing saltmarsh. Elsewhere, Pye and French (1993) describe seasonal erosion and accretion in saltmarshes resulting from storm waves or channel migration leading to terracing that may become obscured where conditions favour further accretion. At Loch Gruinart, channel migration may well be a function of ongoing shallowing, leading to abandonment of some upper saltmarsh surfaces and the relocation of accretion to lower surfaces at the rear of the intertidal sandflat. Such down-

◄Figure 10.21 The coastal geomorphology of Loch Gruinart, Islay is dominated by a history of sea-level changes with emerged erosional and depositional features flanking the north-south axis of the loch. In the shelter provided by Ardnave and Killinallan Points, extensive linear and loch-head saltmarshes have developed, some of which are extending onto intertidal gravels. (After Ritchie and Crofts, 1974.)

Saltmarshes

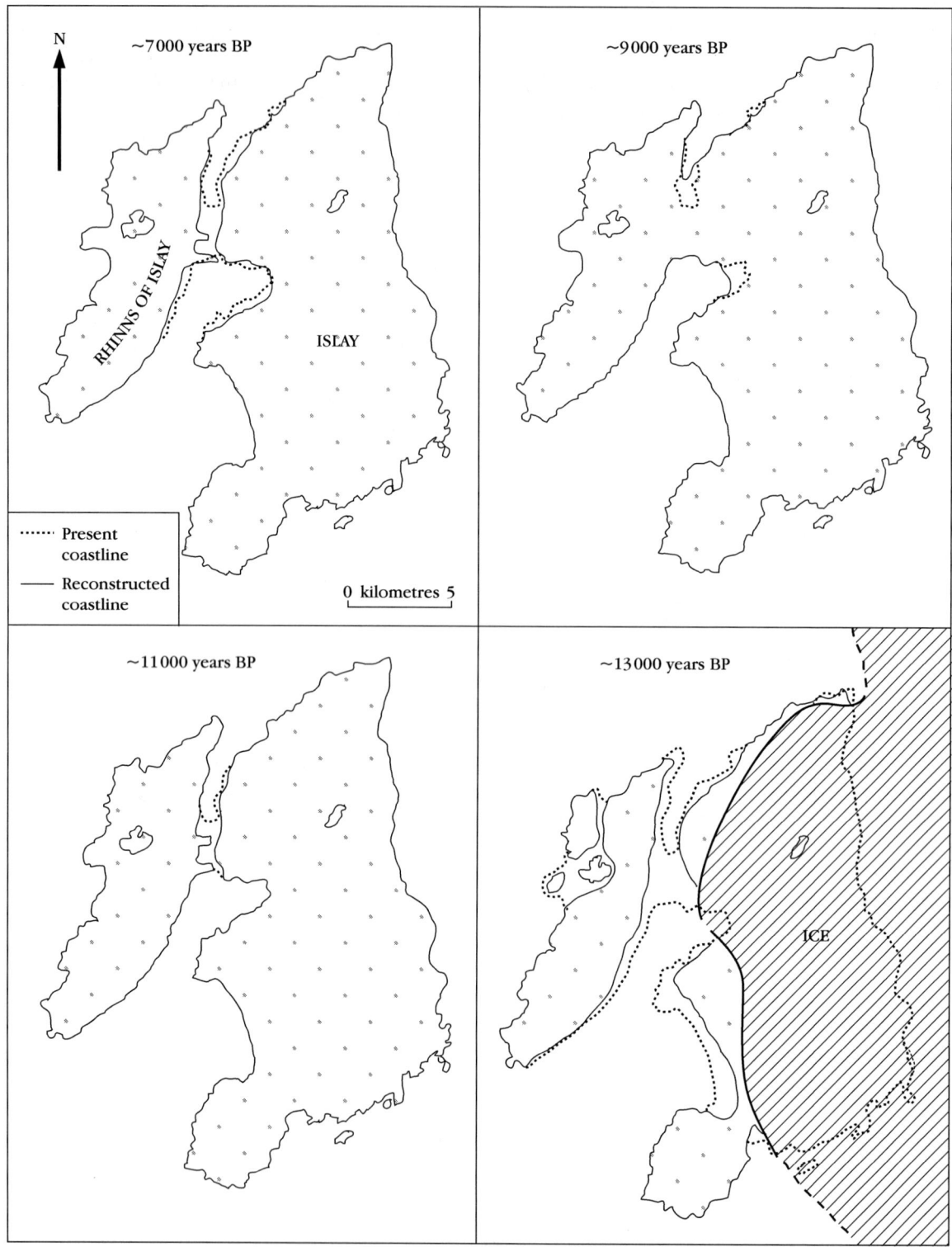

Figure 10.22 The changing coastline of the Loch Gruinart–Loch Indaal area, Islay, at 7000, 9000, 11 000 and 13 000 years BP, showing phases of marine inundation and land emergence. Since 7000 years ago the relative sea level has shown a more or less constant falling trend towards the position of the present coastline. (After Dawson and Dawson, 1997.)

marsh migration of sedimentation has almost certainly been exacerbated by the rapid uplift experienced by this part of the Islay coast and indicates that this would be an ideal site in which to study the effects of emergence on accreting saltmarshes.

It is possible that the distribution of saltpans solely on the upper levels of the marsh may be related to recent and rapid uplift of the upper marsh. However, it is equally likely that the development of saltpans requires a fairly dense and continuous cover of vegetation and this is found only on the upper marsh at Loch Gruinart. The occurrence of several collapsed pans may be related to the failure of subterranean pipe networks similar to those that exist elsewhere on Scottish saltmarshes (Leafe and Hansom, 1990).

Conclusions

The saltmarshes of Loch Gruinart are typical of the type of saltmarsh found at the head and fringing the sides of the Scottish sea lochs. Found in Scotland and Norway, they have been influenced by the ongoing emergence of the host coastline and so are nationally important for studies of saltmarsh geomorphology on emerging shores. The saltmarshes of Loch Gruinart display drainage patterns with linear and narrow creeks and saltpans that are largely confined to the upper marsh. In spite of being suited to the study of the effects of emergence on saltmarshes, the coastal geomorphology of Loch Gruinart has not yet attracted any detailed attention other than work related to sea-level change.

Chapter 11

Coastal assemblage GCR sites

Introduction

INTRODUCTION

V.J. May

There are several lengths of the British coast in which, in addition to outstanding specific features such as well-developed saltmarshes or gravel beaches, the total assemblage of individual features is also outstanding. There are seven sections of coast in Britain selected for the GCR that each contain a wide diversity of individual coastal forms that together form an integrated coastal system or 'coastal assemblage' (see Figure 1.2 for locations and Table 11.1, below, for an outline of the principal features). The sites are Morrich More in the Dornoch Firth, Ross and Cromarty, and Culbin in the Moray Firth in Scotland; Holy Island, Northumberland, the North Norfolk Coast, and The Dorset Coast in England; and Carmarthen Bay and Newborough Warren and Morfa Dinlle at the western end of the Menai Strait in Wales. The origins and dynamics of each site have been the subject of considerable debate. Each of the sites falls within a different part of the British coast and is affected by different tidal and wave conditions, sediment supply and sea-level histories. Carmarthen Bay is the only member of this group of sites that is predominantly macrotidal and faces the high-energy Atlantic wave environment; there are few other sites on the European coast that combine these features with a distinctive record of sea-level change. In contrast, the north Norfolk coast is dominated by large depositional structures mainly in sand and shingle but also sheltering important saltmarshes. The links between the longshore transport regime and the development of the structures has been a focus of debate. Both Carmarthen Bay and the north Norfolk coast include a wide range of predominantly depositional features in which cliff erosion plays a limited role in the sediment budget, and reworking of the existing beaches and shallow-water sediments is more important. Both lie in situations where glaciation has played a role in the development of the coast, either in providing sources of sediments or in producing a cliffed coastline within which the sediments have been deposited and reworked.

In contrast, the coast of south-eastern Dorset

Table 11.1 Main geomorphological features of the 'Coastal Assemblage' GCR sites.

Site	Main geomorphological features	Tidal range (m)
Culbin	Extensive dune system with dunes up to 30m high; parabolic dunes; emerged gravel strandplain and spits; sandy spits; gravel spits; extensive intertidal sandflats and saltmarshes; westerly shift.	3.6
Morrich More	Emerged sandy coastal strandplain with interdigitated saltmarsh and sandy beaches on either flank; offshore sandy islands and spit; large parabolic dune system; 1 km width intertidal sandflats in Dornoch Firth.	3.4
Carmarthen Bay	Major dunes; sand-spits and barrier beaches; hard-rock and easily eroded cliffs; rias; emerged beaches; extensive intertidal sandflats; and saltmarshes.	8.0
Newborough Warren and Morfa Dinlle	Major dunes (linear and parabolic); Holocene dunes; gravel spits; hard-rock and easily eroded cliffs; extensive intertidal sandflats; estuary; saltmarshes.	4.2
Holy Island	Barrier beaches; spits; emerged beach; longshore and offshore sediment sources (Huddart and Glasser, 2002)	4.1
North Norfolk Coast	Scolt Head Island, a major barrier island; Blakeney Point, a large shingle spit; intertidal flats; beaches; dunes; saltmarshes; cliffs. One of the few areas on the coastline of England and Wales where saltmarsh morphology, including saltpans, has been examined in detail.	6.4 (west) to 4.7 (east)
The Dorset Coast: Peveril Point to Furzy Cliff	Differential erosion to a longitudinal coastline; includes such classic landforms as Lulworth Cove. Hard-rock and soft-rock cliffs; platforms; landslides; pocket beaches; chines; submerged rock barriers.	1.7 (east) to 2.0 (west)

is cliffed and affected by sea-level change, but one where coastal alignment and forms owe much to geological structure and lithology. Erosion has produced an unrivalled variety of cliffs, bays and beaches. Beaches are formed mainly in flint and chert but, even though the chalk cliffs are undergoing erosion, many of the beaches are not supplied with significant quantities from such sources today. Changes in sea level and in the position of the coastline have left a legacy of hanging and deeply incised valleys, in contrast to Carmarthen Bay where sediment-rich, drowned estuaries and rias feature strongly.

The origins of the Purbeck coast are not well understood, even though parts have been very well described (e.g. Brunsden and Goudie, 1981), especially the geology (Damon, 1884; Strahan, 1898; Arkell, 1947; House, 1993). The sole evidence on this coast of higher sea levels is at Portland Bill, and although the coast east of St Alban's Head may preserve relict features, there is no other direct evidence of higher sea levels here. The effects of differential erosion are well known here. Unlike the other sites, this coast has increasingly been investigated underwater and so the nature of rocky seabed geomorphology can be used to further the interpretation of the features.

Although individual features such as Lulworth Cove and Stair Hole, Dorset, or Scolt Head Island, Norfolk, are outstanding in their own right, their importance is significantly increased by their association with other features of the adjacent coast. Such localities could be included within previous chapters of the present volume, but despite their individual importance, these features are best described within the wider regional context and in association with each other. Three of the sites (Carmarthen Bay, North Norfolk Coast and Dorset Coast) are highly segmented in terms of their morpho-sedimentology, with between 31 and 35 segments each, and averaging 1.7 km in length, based on the form and dynamics of the shoreline (Table 11.2). The coast of Caernarfon Bay includes seven of the CORINE categories (see p. 21, Chapter 1), a smaller number of segments and a similar mean segment length to Carmarthen Bay. This reflects the higher proportion of long sandy beaches. This variety reflects the impact of changing relative sea levels, the resistance of materials, and large-scale deposition.

Large-scale deposition is also a strong theme at Culbin, Morrich More and Holy Island where plentiful sediment has been available for beach building during much of the Holocene Epoch, aided by a falling relative sea level. All three sites combine internationally important features within complexes of gravel features, sand beaches, spits, dunes and saltmarshes. At Culbin, in the Moray Firth, a large gravel strandplain composed of gravel ridges and spits has become elevated over the Holocene Epoch and subsequently buried by large quantities of wind-blown sand. Much of the present-day coast is dominated by

Table 11.2 CORINE categories, data for the Carmarthen Bay, North Norfolk Coast, Purbeck (Dorset Coast) and Newborough Warren/Morfa Dinlle GCR sites; measurements are in km.

CORINE categories		Carmarthen Bay	North Norfolk	Purbeck	Newborough Warren and Morfa Dinlle
(A)	Hard-rock cliffs (with fringing beaches)	10	0	7	4
(B)	Soft rock cliffs (with fringing beaches)	1(1)	2(1)	21(4)	1
(C)	Pocket beaches	1	0	3	0
(D)	Coarse clastic beaches	2	3	0	1
(E)	Sandy beaches	9	13	0	5
(G)	Foreshores: fine sediments	4	11	11	1
(H)	Estuary	2	1	1	1
(J)	Port/harbour zone	3	0	0	0
(L)	Embankment	0	1	1	1
(X)	Mixed beaches	0	2	2	0
Mean segment length (km)		2.25	1.47	1.42	2.30
Total segments		32	35	31	14

sandy beaches and spits, which move westwards, some of the source sand eroded from sand dunes that have been blown eastwards, opposite to the direction of longshore drift. Most of the past and present gravels have migrated west and downdrift at rates of 15 m a^{-1} to form an impressive spit complex composed of largely unvegetated recurved ridges that now enclose extensive areas of sandy saltmarshes behind. In contrast to the longshore dominated system at Culbin, the Morrich More in the Dornoch Firth is essentially an emerged sandy strandplain composed entirely of a staircase of dune-capped sandy beach ridges whose orientation matches that of the approaching wave crests from the north-east (i.e. swash-aligned). Flanked by inlets on either side, the inter-ridge hollows have allowed tidal access and the development of saltmarsh so that the sandy beach ridges and saltmarsh hollows interdigitate. Since the coast is emerging, there is a close age association between the beaches and the saltmarsh with the youngest saltmarshes occurring towards the flanking inlets and the outer coast.

The south-western end of the Menai Strait brings together two sites of international importance in their own right, Newborough Warren and Morfa Dinlle. In both sites, the development of dunes plays a role, although this is the dominant interest at Newborough Warren. The spits and gravel ridges of the Abermenai spit and at Morfa Dinlle combine to provide evidence of the Holocene development of the shoreline of a major tidal estuary in which there has been little anthropogenic interference.

The present chapter describes sections of the British coastline that have been least affected by human interference. They show extraordinarily well the ways in which coasts of different geological materials and structures respond to marine and subaerial processes over a wide range of time and spatial scales. The Scottish site descriptions are followed by those for Wales and lastly for England.

CULBIN, MORAY (NH 980 615)

J.D. Hansom

Introduction

Culbin, located on the southern shore of the Moray Firth (see Figure 10.1 for general location), is a site of exceptional international interest for the scale, complexity and diversity of its coastal geomorphology. The site comprises an emerged sand and gravel strandplain covering over 30 km^2 containing numerous westward-trending spits and ridges, and is backed to landward by a prominent emerged cliff. An extensive and formerly mobile sand-dune system has developed on top of the gravel basement (Mackie, 1897; Steers, 1937; Ovington, 1950; Comber 1993). At one time it was the largest area of open sand dune in Britain, but most of the area was stabilized by conifer afforestation between 1922 and 1963. The active-process environments and landforms of Culbin are no less impressive than their emerged predecessors. A wide intertidal zone extends from Findhorn in the east to Nairn in the west, and displays several well-developed spit and bar features. Overall the Culbin foreland in the east is erosional whereas accretion occurs towards the west at the spit at Buckie Loch. Farther west still, the proximal (eastern) part of The Bar is erosional whereas its distal (western) end is extending (Comber *et al.*, 1994). Landwards of these constructive spit and barrier features, and also within Findhorn Bay, a series of extensive sandflats and sandy saltmarsh has developed in the low-energy, sheltered environment (Comber *et al.*, 1994).

Description

The Culbin coastline extends over *c.* 12 km from the mouth of the River Findhorn in the east to Nairn Links in the west (Figure 11.1). Culbin plays an integral part in understanding the physiographic evolution of the coast of the Moray Firth and should be viewed not in isolation but as part of a similar beach and dune coast that stretches from Spey Bay in the east to Whiteness Head Spit in the west (see site report in Chapter 6). The dominant westerly drift along the southern shore of the Moray Firth (Ramsay and Brampton, 2000c) is integral to the understanding of both contemporary and Holocene landform evolution in this area.

The Holocene gravel ridges that lie beneath the dunes of Culbin Forest provide an exceptional assemblage of emerged features related to higher relative sea levels over the last 6500 years. Radiocarbon dates from peat deposits found on top of the gravels but beneath the dunes show

Coastal assemblage GCR sites

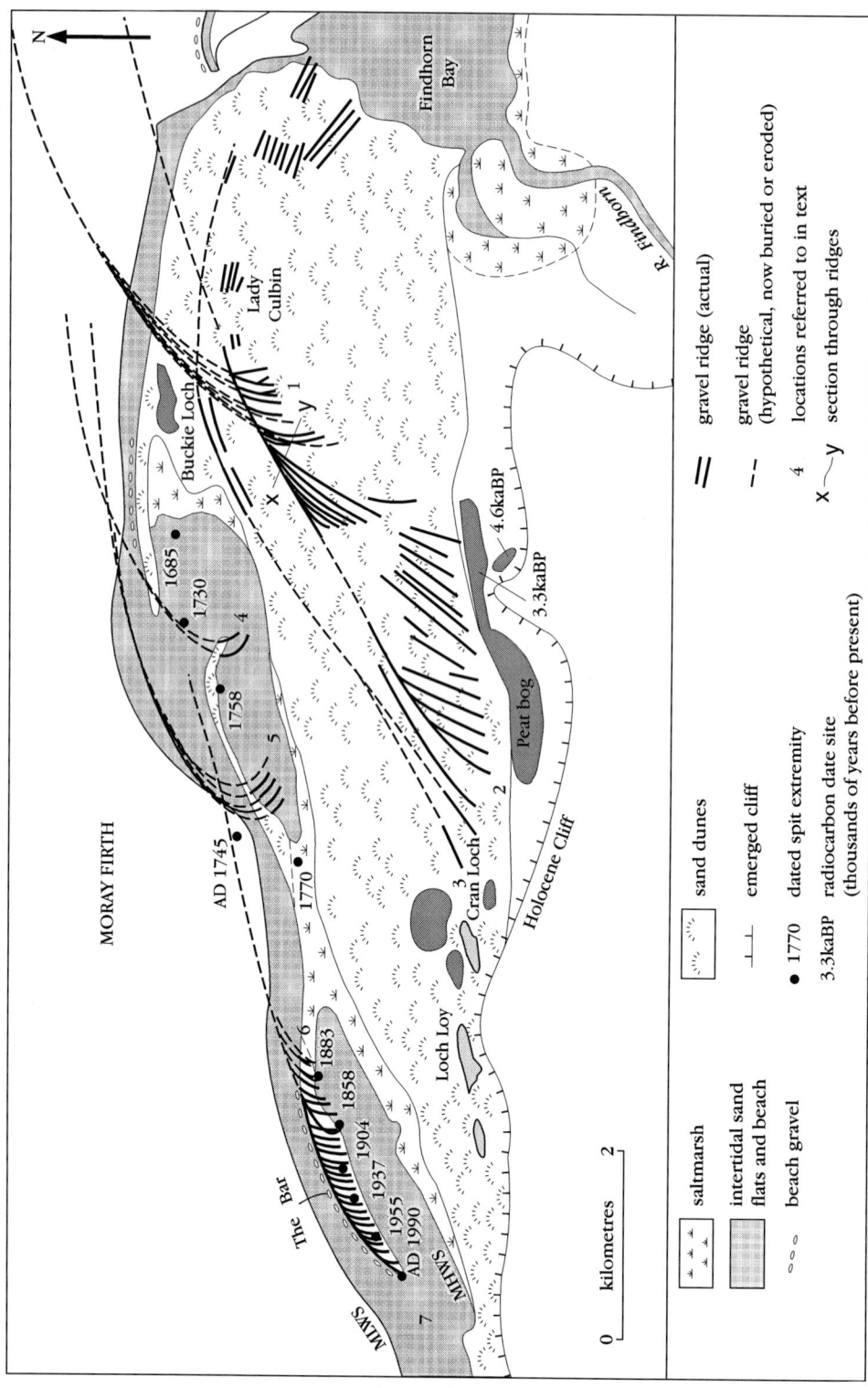

Figure 11.1 The GCR site of Culbin is a large and complex gravel strandplain composed of suites of partially visible emerged ridges capped by large sand dunes. The pattern of the underlying gravels can be reconstructed into a series of westward-extending gravel spits; the updrift erosion of the earlier spits fed the downdrift extension of the more recent ones that have been dated using historical maps and aerial photographs. Extensive sandflats and saltmarshes have developed in the shelter of the westward-extending spits. See Figure 11.2 for the section X–Y. (After Hansom, 1999.)

that sea level has fallen in this area during the past 6500 years (Comber, 1993). This isostatically driven fall in relative sea level in the Culbin area is reflected everywhere on the coast of the Moray Firth, and followed an earlier period of rapid sea-level rise (Smith, 1997). Given the altitudes and spatial locations of the emerged gravel ridges of Culbin, it is clear that they were emplaced during this phase of falling sea level (Comber, 1993, 1995).

Superimposed on these emerged gravels a major inland dune-system has developed, creating what was formerly the largest area of bare sand in Britain, prior to afforestation. The emerged marine beach deposits, which underlie the entire area of Culbin, are backed by an extensive abandoned cliff whose base lies at c. 9 m OD. The cliff can be traced around much of the Moray Firth (Hansom, 1988) and is the counterpart of the cliff at Spey Bay to the east (see GCR site report in Chapter 6). In the Culbin area the 5–7 m-high cliff is cut mainly into Late Devensian glaciogenic and glaciofluvial deposits and forms a divide between older (late Devensian) deposits to landward, and younger (Holocene) deposits to seaward (Firth, 1989).

The most striking landforms preserved within the Culbin dunes are the emerged gravel storm-ridges, found at altitudes of up to 10.9 m OD (Comber, 1993). Owing to the cover of dune sand, the gravel ridges are discontinuously exposed in the field, but can be traced on the ground and in aerial photography in an arcuate form, spanning approximately 5 km of sporadic exposure. These ridges represent abandoned upper beach deposits thrown up under high-energy storm events, and are composed of gravel clasts 30–50 mm in diameter. A narrow belt of ridges extends westwards across the north-eastern flank of Culbin, before splaying out southwards into a 'fan' at NH 997 630 (Figure 11.1, location 1). Landwards of this point, the approximately parallel ridges begin to splay out markedly into at least two distinct groups, which extend towards the south-west (Figure 11.1, locations 2 and 3). At the main apex of this 'fan' structure the landward ridges are truncated by the ridges to seaward, indicating erosion of the earlier features (Comber, 1995). Transects levelled across the entire sequence of ridges from the emerged cliff to the present-day beach display declining altitudes to the seaward from a maximum of 10.9 m OD to a minimum of 3.7 m OD. However, between some groups of ridges there is a marked ridge-crest fall of almost 3 m (Figure 11.2).

The Culbin dune system covers an area of approximately 13 km² and displays a range of forms unparalleled in any UK dune system. The orientation of the axes of most of the dunes is south-west to north-east, with blowthrough patterns preserved on account of their artificial fix-

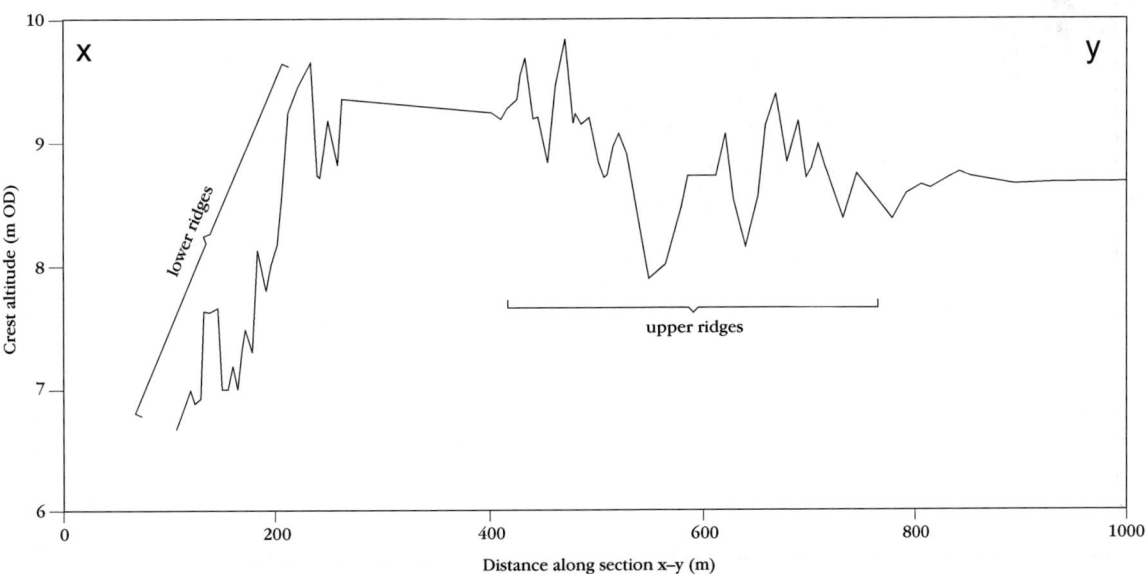

Figure 11.2 The gravel ridges over a 1000 m transect from x–y (see Figure 11.1) show two groups of emerged ridges, the most seaward of which decline rapidly in height towards the north-west. (After Comber, 1995.)

ing by afforestation. Three main dune types occur at Culbin: parabolic dunes; formerly transgressive dunes; and butte dunes. The parabolic dunes of Maviston, near Loch Loy in the west of the site (Figure 11.2), attain a maximum height of 15 m, with flanks up to 400 m long and a maximum width of *c.* 400 m, making these among the largest of their kind in Europe (Comber *et al.*, 1994). Prior to stabilization by planting, Ogilvie (1923) recorded the maximum height of the active Maviston dunes as 16.5 m, but the crests have since settled. Formation of the dunes progressed as a classic parabolic blowthrough sequence from an initial straight dune crest trending parallel to the coast. Destabilization of the central zone of the crest through the destruction of the vegetated surface produced an area of crestal instability, allowing sand to be blown downwind and allowing the dune crest to migrate and inundate the surfaces behind. Erosion of the lateral slopes of the widening blowthroughs accelerated the process, as did funnelling of the incident wind towards the unstable zone. While the central section of the dune continued to extend downwind, the flanks of the dune remained vegetated and thus stable, creating the distinctive parabolic dune forms found today. The exceptional size of the dunes at Maviston has meant that rates of movement were relatively low, retaining the essential form of the features since at least 1923 until stabilization by forestry (Ogilvie, 1923; Steers, 1937; Edlin, 1976).

Not all of the Culbin dunes display the effects of blowthrough activity, and examples of high dunes, reaching altitudes of up to 30 m, with the smoothed and rounded forms of previously unvegetated transgressive dunes are found, particularly in the west central area of the forest in the vicinity of the underlying gravel 'fan'. A good example of such a dune is Lady Culbin, located at NJ 013 640 (Figure 11.1). Ovington (1950) noted that the Lady Culbin dune moved at an average rate of 6.5 cm per day over a six-

Figure 11.3 Parabolic sand dunes undergoing erosion at the exit of the River Findhorn, Culbin. Erosion of this section of the Culbin foreland feeds sand to fuel accretion at the downdrift Buckie Loch spit. Harvesting of timber over a 20–30 m-wide dune edge zone is part of a management regime designed to reduce erosion caused by disruption of the dune surface by toppling, and to allow mechanical harvesting to be carried out in safety. (Photo: J.D. Hansom.)

Culbin

week period. Butte dunes are the eroded remnants of formerly vegetated and stabilized surfaces that have subsequently suffered erosion on all sides to leave a residual flat-topped stump flanked by steep, often unvegetated, slopes of sand. Good examples of butte dunes occur in the north-east of Culbin.

The contemporary coastal geomorphology of Culbin can be considered in terms of five landform assemblages: the Culbin foreland to the east of Buckie Loch; the Buckie Loch spit; The Bar (locations 6 and 7, Figure 11.1), and the extensive intertidal northern sandflats and saltmarshes in the shelter of the spits and barriers (Figure 11.1).

The Culbin foreland extends west from the Findhorn estuary to the Buckie Loch. The foreshore beach is composed mainly of sands, although gravel occurs on the foreshore at the mouth of the Findhorn. Much of this foreland coast is subject to severe erosion, which has resulted in the cutting of the backing sand dunes into prominent bare-sand cliffs up to 8 m high (Figure 11.3). Frontal erosion of these old dunes is recorded to be occurring at rates of up to 1 m a^{-1} (Ritchie *et al.*, 1978; Comber, 1993; Comber *et al.*, 1994). Dune erosion occurs despite the occurrence of relatively wide sandy beaches (up to *c.* 200 m) along the foreland. At the westward extremity of the former Buckie Loch a spectacular recurved spit (the Buckie Loch spit) extends westwards. The spit is presently *c.* 3 km long and up to 130 m wide in its central section, becoming narrower westwards. The spit foreshore is predominantly sandy but the upper foreshore sediment coarsens westwards so that a small gravel ridge has developed on its distal end. The rear of the spit is dominated by several suites of stabilized dune ridges, with actively accreting dunes occurring at both proximal and distal ends of the spit. Erosion of the updrift part of Buckie Loch Spit is resulting in occasional washover of low parts of the frontal dune and gradual infill of the Buckie

Figure 11.4 The magnificent gravel spit of The Bar at Culbin is extending westwards towards the town of Nairn at approximately 15 m a^{-1}. The sandy Buckie Loch spit can be seen in the upper middle distance and the narrow erosional neck at the eastern end of The Bar. This updrift erosion has truncated earlier ridges and is now encroaching into the area of saltmarsh that has developed behind the bar and may ultimately threaten the larger area of saltmarsh that lies in the right foreground. (Photo: P. and A. Macdonald/SNH.)

Coastal assemblage GCR sites

Figure 11.5 Spectacular recurves extend from the active outer beach ridges at The Bar at Culbin into the sheltered area behind. The inner recurved parts of the gravel ridges support small areas of heather, gorse, broom and pine whereas the intertidal flats between the gravel ridges support small areas of saltmarsh (see Figure 11.4). (Photo: P. and A. Macdonald/SNH.)

Figure 11.6 An extensive area of saltpans and linear creeks characterize the area of saltmarsh that has developed at the heads of the two intertidal lagoons that lie either side of the area where The Bar at Culbin is attached to the mainland (see Figure 11.4). (Photo: J.D. Hansom.)

Loch, which is now infilled.

The Bar at Culbin forms the most distinctive coastal feature on the southern shore of the Moray Firth and is a fine example of a 'flying barrier', with tidal lagoons and saltmarsh behind (Figure 11.4). The feature represents an attached gravel barrier orientated north-east–south-west and extending over a distance of 7.3 km. It is now attached to the mainland by a low neck of saltmarsh fronted by a gravel beach. At its narrowest the width of the saltmarsh behind the beach is only 250 m, the seaward edge being subject to burial by washover of gravels during storms. Recession of the beach over the saltmarsh has exposed peat on the foreshore. The eastern part of the Bar is mainly sandy, and is adorned with prominent sand dune ridges that sit on top of southward-trending gravel recurves that extend into the intertidal lagoon. The western end comprises the main part of the 'flying barrier' and displays a series of multiple gravel storm ridges backing an active gravel beach (Figure 11.5). Up to 13 major recurving gravel ridges occur landwards of the active ridge and represent recently abandoned shoreline features that have been dated (Figure 11.1, locations 4–7).

An extensive intertidal sandflat with multiple sand-bars occurs seawards of both the Buckie Loch spit and The Bar. These sandflats also extend into the channels and the intertidal zone on the landward side behind the Buckie Loch spit and The Bar, where the shelter afforded has allowed saltmarshes to develop (Figures 11.4 and 11.5). The marshes commonly have a small undercut edge of *c.* 0.2 m in height (Ritchie *et al.*, 1978); although some edges grade smoothly from sandflat to saltmarsh. The saltmarshes range from low developmental marsh surfaces characterized by intermittent stands of common saltmarsh-grass *Puccinellia* and samphire *Salicornia* (Figure 11.6) to substantial areas of high marsh supporting a full vegetation cover merging landwards to freshwater marsh species. The two largest areas are identified as the marsh surface landward of the central section of The Bar and the area landward of Buckie Loch spit. The marsh landward of The Bar is adorned with numerous saltpans and linear creeks (Figure 11.6). The area of saltmarsh landward of Buckie Loch spit is expanding rapidly as distal extension of the spit continues to provide a lower-energy environment in which progradation can occur. Buckie Loch itself was probably a former saltmarsh, abandoned by westward migration of the active marsh, and subsequently dominated by freshwater species before infilling occurred through washover, sand blow and colonization by trees.

Interpretation

The geomorphology of Culbin was central to Ogilvie's (1923) interpretation of the Holocene development of the Moray Firth. Ogilvie (1923) first described and mapped the emerged gravel features and relict cliff along the southern Moray Firth coast, linking the development of these features to a higher relative sea level and the reworking of vast quantities of glaciofluvial and glacial deposits from the Moray Firth coastal plain and the inner continental shelf as sea level adjusted following deglaciation. Ogilvie's account together with Steer's (1937) work on Culbin emphasized the importance of the westward direction of longshore drift along the southern shore of the Moray Firth to landform development throughout the Holocene Epoch. The excellent groundwork and elegant theories of coastal development provided by these early workers were pursued by Comber (1993, 1995), Hansom and Comber (1994), Comber *et al.* (1994) and Hansom (1999), who provide the most recent interpretation of the evolution of Culbin, and indeed the southern shore of the Moray Firth.

At the peak of the Holocene transgression (*c.* 6500 years BP) the high-stand of relative sea level at *c.* 9 m OD in the Culbin area impinged upon and re-trimmed a pre-existing cliff that probably had been cut initially during the Lateglacial period. During this time of higher relative sea level a marine corridor existed to the east of Culbin, south of the high ground of Burghead–Lossiemouth, which was then an offshore island (Ogilvie, 1923; Comber, 1993). Under conditions of net westerly drift, sediment from the River Spey is thought to have moved freely through this corridor and into a proto-Burghead Bay to be augmented by sediment from the River Findhorn. Combined with the net onshore movement of sediment under a rising sea level, a strongly positive sediment budget was created at the present-day location of Culbin (Hansom, 1999). The shoreline response to such rapid sediment input was to prograde seawards, and progradation at Culbin occurred in a similar fashion to many other gravel-dominated foreshore systems (e.g. Carter *et al.*, 1987) by developing a suite of multiple sub-parallel ridges. Transects across the gravel ridge 'fan' (Figure 11.1) demonstrate the trend in falling sea level that occurred after the peak of the Holocene transgression (Figure 11.2). The highest set of ridges in the 'fan' were deposited at the peak of the transgression, while most of the ridges landwards and eastwards occur at lower altitudes and were deposited as sea level fell and so are younger (Comber, 1993). The falling sea level reduced water depths in the marine corridor to the east of Culbin and it became blocked with westward-drifting sediment from the Spey. At Culbin, the loss of this sediment feed from the east was to dramatically reduce the amount of sediment available for storm ridge sedimentation. The sedimentary record of this decline in sediment supply at Culbin is represented by fewer gravel ridges deposited over time, creating a net steepening of the ridge suite to seaward and the beginning of erosion of the updrift gravels in the eastern part of Culbin itself.

The above interpretation suggests that the locus of gravel accumulation has shifted through time from the east of Culbin to the west, where it is now represented by The Bar. The attached gravel barrier has been migrating alongshore in a westerly direction towards the town of Nairn since least 1685 AD as documented by Ross (1992; Figure 11.1). This process of updrift erosion fuelling downdrift accretion has been a feature of the coastal development of the southern Moray Firth throughout the Holocene Epoch (Comber, 1995). Figure 11.1 shows a hypothesized development sequence, running numerically from 1–7, of Culbin Sands and The Bar proposed by Hansom and Comber (1994), Comber (1995) and Hansom (1999). The reworking of sediment from updrift to fuel downdrift accumulation can be seen to account for the truncated emerged gravel ridge sequences in the 'fan' (Figure 11.1, locations 1, 2 and 3). Radiocarbon dating of peat taken from depressions at the foot of the emerged cliff suggests that abandonment of gravel spit 3 occurred between 4600 and 3300 years BP when the River Findhorn abandoned a westerly course at Cran Loch to breach northwards through the gravel beach close to the present exit of the river (Comber, 1993). The process of spit destruction is suggested to have been repeated several times in the similar, though much younger, features that form The Bar and its predecessors. Hansom (1999) indicates the mode of emplacement of the gravel storm-ridges has remained the same since at least the mid-Holocene and represents a predictable response of sediment recycling within conditions of restricted sediment supply. In conditions of deficit, sediment within specific coastal cells is re-organized by erosion of some parts and re-deposition in others (Carter, 1988). Figure 11.1 also demonstrates the tendency over time of the Culbin foreshore to rotate clockwise to face north-east in an attempt to align itself normal to incident wave approach, evolving from a drift towards a swash alignment of the coast (Davies, 1980).

The entire length of The Bar at Culbin is subject to reworking as proximal erosion in the east fuels distal accretion in the west (Comber *et al.*, 1994). Where both contemporary and recently abandoned ridges have been truncated in the deeper water of the distal end, wave refraction has carried gravel around the tip (Figure 11.5). This results in recurves forming behind the sequential positions of the former distal ends of The Bar, some of the truncated remnants of which are now found in seemingly anomalous locations (for example Figure 11.1, locations 4 and 5). Between 1976 and 1989 The Bar continued to extend westwards at a mean rate of 14.6 m a^{-1} (Comber, 1993), demonstrating that the processes that created this section of the Scottish coastline over the Holocene Epoch still operate. Today's Bar is a direct descendant of many previous barriers and spits on this coast.

The Culbin dune system has been described extensively (e.g. Ogilvie, 1923; Steers, 1937; Ovington, 1950; Edlin, 1976). Several well-developed palaeosols are found at various sites throughout Culbin Forest and contain important information concerning the development of the dunes (Comber *et al.*, 1994). The palaeosol profiles are particularly mature, a feature unusual in dune systems of this size given their propensity to become remobilized under combined natural and anthropogenic pressure. The earliest documented reference to the mobile dune belt at Culbin was by Boethius in 1097 AD (in Craig, 1888), who referred to the inundation of parts of Moray by sand thrown up during storms in the North Sea, but several earlier periods are known including a major period of sand dune instability at c. 4500 years BP (Hickey, 1991). The most recent period of dune activity at Culbin coincides with the end of the most recent phase of wide-scale dune re-activation. This phase began in the 13th century and ended during the mid–late 17th century with the stormiest period of the 'Little Ice Age' (Hickey, 1991). The maturity of the Culbin palaeosol profiles suggests that some of the dunes remained stable for the early part of this dune mobilization phase (Comber *et al.*, 1994), supporting a full vegetation cover that prevented sand-blow. The dunes seem to have become re-activated relatively late in the sequence of dune activity. The documented story of the destruction of the Culbin estate by blown sand reports that the estate was overwhelmed over the course of a single storm in 1694 AD (Steers, 1937). However, it is more likely that the dunes were subject to an extended period of destabilization, with the final inundation occurring during the 1694 event. Destabilization was probably aided by the removal of the closed vegetation cover and, in particular, the removal of marram *Ammophila arenaria* for thatch (Comber *et al.*, 1994). In response to the loss of the important agricultural estate of Culbin, an Act of the Scottish Parliament was passed in 1695 to prevent pulling of 'bent' (marram) from sand dunes (Ross, 1992).

As demonstrated by Comber (1993) the contemporary coastal development of the Culbin foreshore can be directly linked to the Holocene evolution of the entire landform assemblage. The diverse process environment of the Culbin foreshore provides an excellent site for the study of a wide range of coastal processes and landforms. Erosion of the dunes west of the River Findhorn and on the updrift section of the Buckie Loch spit fuels downdrift accretion of the spit, which has been extending in a westerly direction at a mean rate of 22.3 m a^{-1} over the period 1870–1988 (Comber *et al.*, 1994). As erosion proceeds at the eastern extremity of the Buckie Loch spit, storm washover and marine incursion into the Buckie Loch occurs. A shallow lake in the late 19th century, the Buckie Loch is now a low, intermittently flooded area of grassland and deciduous trees fronted by a low dune-ridge. It is likely that westerly accretion will progressively seal the upper part of the Buckie Loch spit marsh, creating a new Buckie Loch farther to the west of the original (Comber *et al.*, 1994). By that time the present-day site of the Buckie Loch will have been all but removed, as erosion proceeds at the eastern end. Such change in both the Buckie Loch and The Bar has implications for the extensive saltmarshes that have accreted in the shelter afforded by the two major spits. Migration of the protecting structures and erosion of their updrift ends forces commensurate change in the sheltered environments behind and exposes the backing saltmarsh to erosion. Such activity is presently most severe at the neck, midway along The Bar, where saltmarsh peat is exposed on the intertidal zone as the foreshore transgresses landwards (Figure 11.4).

Conclusions

Culbin is an exceptional site for coastal geomorphology. Within Europe, no comparable suite of emerged gravel ridges and spits with capping sand dunes matches the scale, complexity and preservation of the features at Culbin. The gravel ridges record the fall of sea level from its mid-Holocene high at about 6500 years BP to its present-day level. In addition, a reduction in sediment supply forced a switch from widespread gravel accretion to a period of reworking of pre-existing gravel spit structures. Such internal reorganization of sediment has resulted in the sequential development of migrating spits, the most recent of which can be seen in the present-day Bar (Hansom, 1999). Resting on top of the ridges, the Culbin dunes once formed one of the largest areas of blown sand in Britain and although subsequently, and very successfully, stabilized by forestry, they are rated internationally as a geomorphologically important site for sand dune development. Culbin is also a key regional site for the interpretation of the history of Holocene landform evolution in the Moray Firth.

No less impressive are the active process environments of Culbin, with a dynamic migrating sand spit whose extension has led to the infill of a small lake and its imminent erosion. Culbin also has a spectacular example of a 'flying barrier', a spit whose eastern section is a dune-adorned sandy feature and whose western end is a superb rapidly extending gravel spit backed by numerous recurved gravel ridges (Figure 11.5). Both of these features are associated with the development of wide intertidal sandflats on both seaward and landward sides, the latter providing a sheltered environment for the development of saltmarshes. Culbin is a unique mix of spectacular Holocene emerged features together with equally impressive contemporary coastal processes and forms.

MORRICH MORE, ROSS AND CROMARTY (NH 803 835–NH 892 830)

J.D. Hansom

Introduction

Morrich More is a large coastal strandplain on the southern shore of the Dornoch Firth (see Figure 10.1 for general location) between Tain and Inver. Its development is related to a shallow offshore zone and the presence of abundant sandy material, which has been brought onshore and deposited in a series of sequential beach ridges under conditions of a falling relative sea level. The stratigraphical and morphological record contained within Morrich More is central to the understanding and reconstruction of the Holocene coastal development of the Dornoch Firth and wider Moray Firth (Hansom and Leafe, 1990; Hansom, 1991, 1999, 2001; Smith *et al.*, 1992; Firth *et al.*, 1995). Access to Morrich More is restricted on account of its use as a Royal Air Force weapons range.

Morrich More contains a diverse variety of constructional coastal landforms including an emerged strandplain, attached sandy barriers and spits, stabilized dunes including parabolic dunes, embryo and foredune succession, saltmarshes and sandflats. The importance of Morrich More, both within the British Isles and internationally, lies in the extent, scale and diversity of its geomorphology together with the fact that the transitional zones between accretionary landforms are well developed and preserved.

In view of the quantities of sediment involved, it is likely that the mid-Late Holocene seaward growth of Morrich More was the most rapid of any coastal feature in Great Britain.

The continuity between the Holocene and contemporary landforms of Morrich More make it an invaluable site for the reconstruction of past process environments as well as for study of the interaction of modern process-form relationships.

Description

The Morrich More strandplain consists of an alternating sequence of low dune-capped sand ridges separated by lower and wetter areas and saltmarsh. The entire landform covers an area of *c.* 34 km², of which *c.* 29 km² lies within the GCR site. The southern limit of the low-lying area of emerged sands is the *c.* 8–10 m OD base of a prominent slope that marks the line of an emerged cliff. Seawards of the cliff, a strandplain with 50 or so emerged sandy ridges extend *c.* 8 km to the north-east into the Dornoch Firth. The main ridges are marked on Figure 11.7. The majority of these features are aligned north-west–south-east and decline in altitude from 8.6 m OD close to the base of the cliff to 1.4 m OD at the present-day coastline. The emerged marine ridges of Morrich More are composed entirely of medium- to fine-grained sand, capped with dune sand and can be split into four distinct altitudinal groupings (Hansom, 1991). In the south of the site, the highest group of ridges are typically 1–1.5 m high, occur at 6.0 to 8.6 m OD and are spaced about 100 m apart. At 6.4 m OD beneath one of these higher ridges, a layer of woody peat yielded a radiocarbon age of 6445 years BP (Hansom and Leafe, 1990). A second set of ridges at heights of between 4.4 and 5.5 m OD occur to the seawards of the higher ridges and occupy the central section of Morrich More. Farther north and seawards, the lowest sets of sand ridges occur at altitudes of 2.5 to 4.0 m OD and 1.4 to 2.4 m OD. The latter group occur at the same altitude as the modern beach ridge. These lower ridges are more widely spaced than those to landward (Figure 11.8) and are adorned with windblown sand, taking their height to 4–5 m OD. As a result, the most recently deposited ridges on the outer coast, those of Patterson Island and Innis Mhór, are the largest and most prominent of any of the Morrich ridges.

In summary, as the emerged strandplain falls in altitude from over 8 m OD at the base of the cliff to 1.4 m OD at its seaward margin, there is a corresponding fall in the number of beach ridges, an increase in their spacing and a general increase in their prominence towards the open coast (Figure 11.7). Although many of the emerged ridges are covered in a thin veneer of blown sand, their form is obscured in three areas by extensive dune development, in the west, in the east and on the outer coast at Patterson Island and Innis Mhór. A feature of many of the emerged sand ridges is that they display truncated westward trending recurves at their northern ends, comparable to the modern features of the contemporary coast.

In the nearshore and intertidal zone, active sand-bars move onshore to produce a

Morrich More

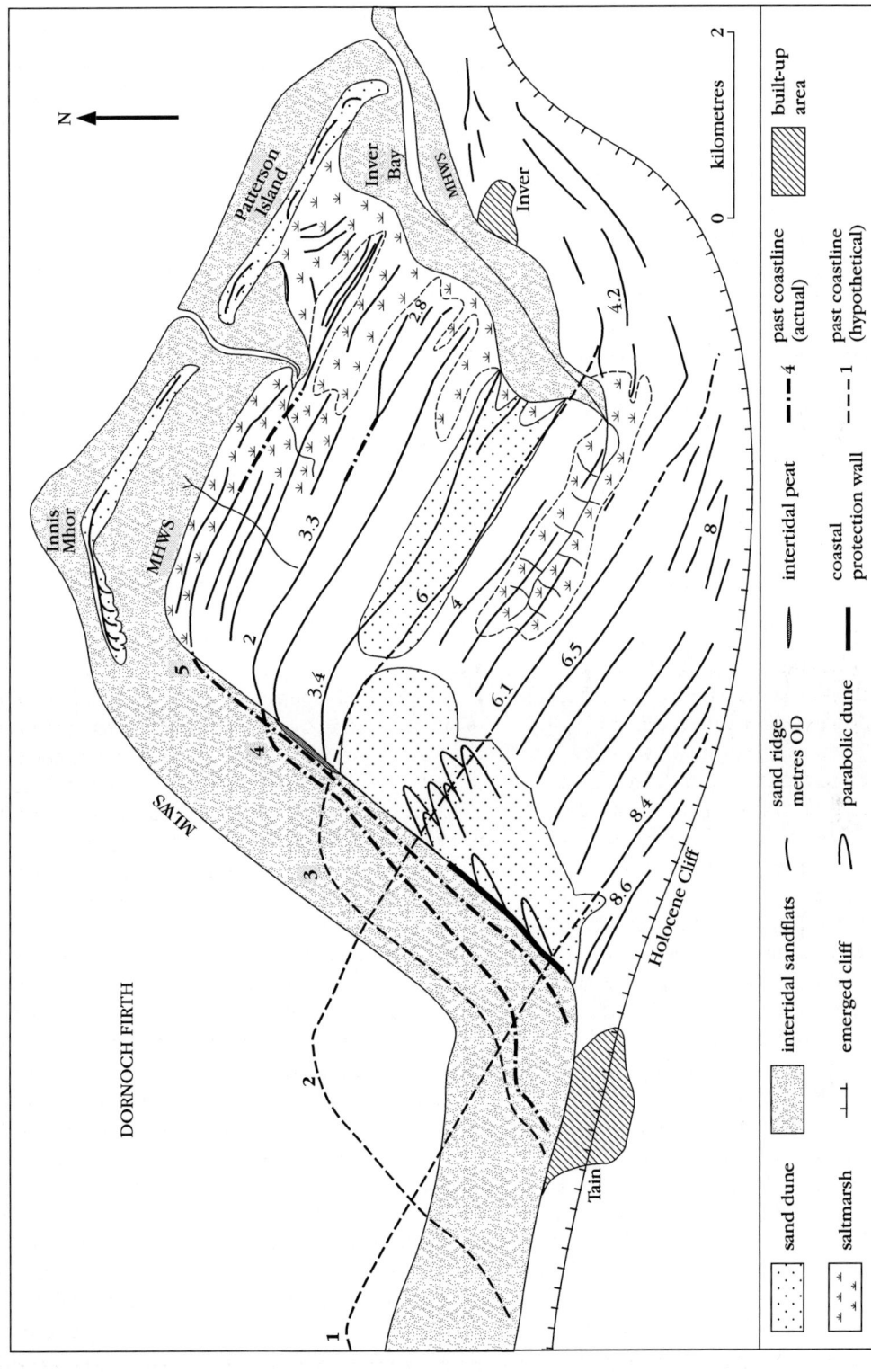

Figure 11.7 Morrich More is a series of emerged sand beach ridges adorned with sand dunes, which together form a progradational strandplain jutting out into the Dornoch Firth. Saltmarsh interdigitates between the sand ridges in the north and east of the structure. The numbered coastlines correspond to the approximate locations of reconstructed and actual coastal positions. Heights of the sand ridges are given in metres above OD. (After Hansom, 1999.)

Figure 11.8 Aerial view of Morrich More from the north-west. The islands of Patterson and Innis Mhór (the eastern end of which is seen to the right) are separated by a tidal channel, which connects with a large area of sand accretion in the shelter provided by Innis Mhór. The inlet of Inver Bay in the centre middle distance supports large areas of saltmarsh, which interdigitates between sand-beach ridges capped with blown sand. (Photo: P. and A. Macdonald/SNH.)

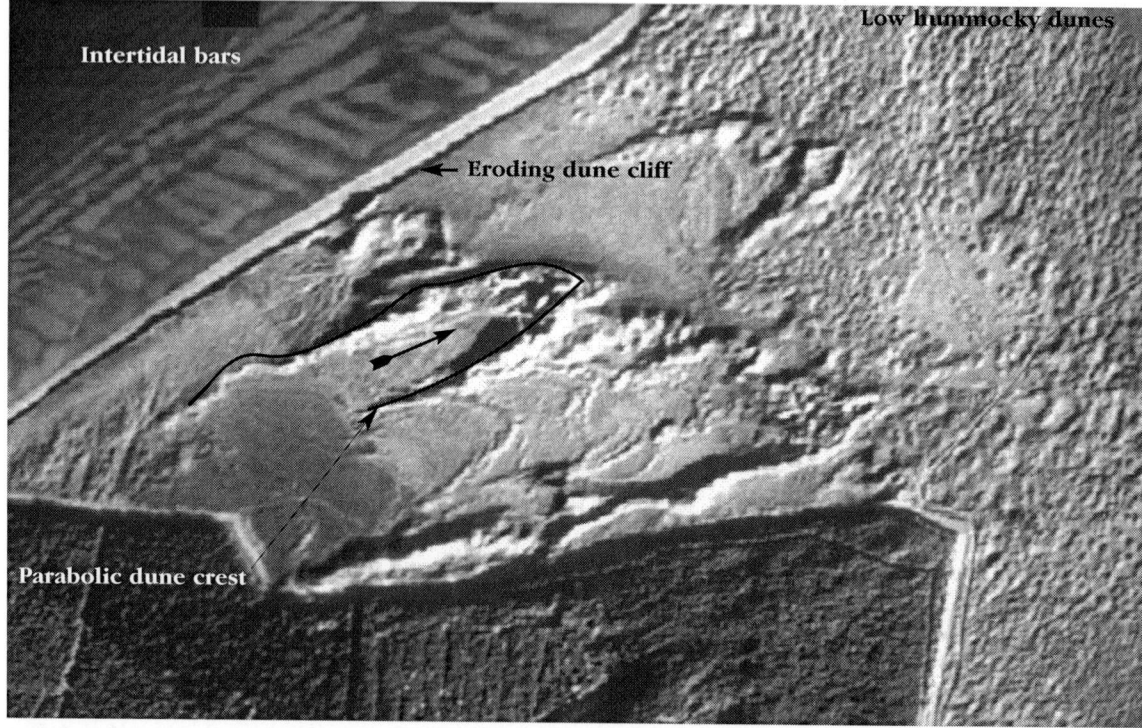

Figure 11.9 A remotely sensed image of the narrow upper beach and eroding edge on the western flank of Morrich More showing the large fixed parabolic dunes and the area of low dunes downwind. Intertidal bars are well-developed on the 1 km-wide intertidal flats on the Dornoch Firth side of Morrich More, and may indicate the direction of sediment movement under the flood tide (onshore) and ebb tide (alongshore to the northeast). North is at the top of the image. The arrow indicates direction of migration of the parabolic dunes. (After Hansom and Leafe, 1990.)

Morrich More

prominent sandy barrier (Hansom and Leafe, 1990). This barrier comprises two dune-capped high-tide 'islands' of Innis Mhór in the west and Patterson Island in the east, punctuated by a tidal channel in the middle (Figure 11.8). The fronting beach is shallow and sandy and is mainly backed by accreting sand dunes. Both extremities of the barrier islands are characterized by recurving sand spits. At the entrance to Inver Bay in the east, a simple low sand spit is deflected southwards into the bay. At the entrance to the Dornoch Firth in the west, a section of Innis Mhór characterized by erosion is connected to a *c.* 2 km-long westerly trending sandy spit that extends into the Firth. This spit is weakly recurved and is composed of several ridges that are now capped with dunes. The two dune-clad islands of Innis Mhór and Patterson Island are now connected to the main body of the strandplain by sandflat and saltmarsh, which is inundated to a depth of 1 m at high spring tides.

In contrast to the accretion and new dune development displayed on the north coast of Morrich More, the side that faces west to the Dornoch Firth is backed by an eroding dune edge whose base lies at MHWS and is fronted by a steep and narrow sandy beach (Figure 11.9). The height of the cliff undergoing erosion varies depending on the height of the sand dune behind, being usually about 2 m but reaching 14 m in the afforested dunes in the south-west, where mature trees topple onto the foreshore. In places along this shore, outcrops of peat are exposed on the intertidal beach. Extending from the foot of the beach, a prominent low-gradient, intertidal sandflat reaches 1 km in width before the low-tide channel of the Dornoch Firth is reached to the north-west. Low sand-bars exist on the sandflat surface (Figure 11.9). In response to erosion of the western flank of Morrich More, a series of low boulder revetments have been constructed over a 2 km stretch north-east from Tain. These have been effective in reducing erosion locally, but have contributed to accelerated erosion of the dune coast down-drift (Figure 11.7).

In addition to the extensive areas of intertidal

Figure 11.10 The large parabolic dunes on the western flank of Morrich More have migrated north-east (towards the camera) but have since stabilized, mainly by marram colonization. The low hummocky dunes in the foreground have been influenced in the past by sand blown from the parabolic dune field but are now also stable and covered mainly with marram and smaller areas of heather. (Photo: J.D. Hansom.)

sandflat on the western flank of Morrich More, similar areas also exist in the shelter provided by the barriers of Innis Mhór and Patterson Island and within Inver Bay. On the more-elevated sections of sandflat, tidal inundation is of lower frequency and duration, thus allowing saltmarsh vegetation to colonize. The 260 ha of saltmarsh on Morrich More represents 5% of the remaining natural saltmarsh in Scotland and it is a distinctive system in its own right on account of its relationship with the beach ridges (Hansom and Leafe, 1990). The saltmarshes are drained by linear creek systems, which extend into Morrich More along the axes of the inter-ridge swales or hollows (Figure 11.7). Thus, saltmarsh and emerged beach ridges interdigitate, and the extremities of the beach ridges become progressively buried. The complex vegetation pattern of Morrich More is dominated by this pattern of interdigititated ridges and slacks (Smith and Mather, 1973) and the strandplain carries a rich flora of over 200 flowering species, ranging from intertidal sandflat species to Juniper–*Calluna* heath on the oldest landward ridges (Stapleton and Pethick, 1996). The vegetation succession of Morrich More is discussed in further detail in Smith and Mather (1973) and Dargie (2000).

The most striking feature of the western sand dune margin of Morrich More are the exceptionally high parabolic dunes, which extend for up to 1 km inland and have a relief amplitude in excess of 14 m (Figure 11.9). The dunes are fixed by vegetation, mainly marram *Ammophila arenaria* and heather *Calluna vulgaris*, although the southern part is now artificially stabilized by afforestation. The highest dunes occur at the northern tip of the forested area and reach 20 m OD in a series of large parabolic dunes. These dunes exhibit exceptionally steep and often knife-edged slopes in excess of 30°. The alignment of the ridges and deflation hollows (250°) indicates a north-eastwards migration, sub-parallel to the western edge of Morrich More (Figure 11.10). Within the deflation hollows the water table is visible, and the sides reveal buried palaeosols. Only a small part of the most easterly of these dunes remain mobile with sand faces spilling forward, and the majority of the system has been stabilized by vegetation. The dunes of the outer coast islands of Innis Mhór and Patterson Island reach 7 m OD and are dominated by a mixture of marram and lyme-grass *Leymus arenarius*, both of which grow vigorously in the active aeolian depositional conditions on the outer coast.

Grass-covered low dune surfaces are confined to the western shore and its landward environments. A low undulating dune plain is truncated along the western edge by backshore erosion producing a scarp of about 1 m high and enabling the limited beach sand available to be blown up onto the dune surface behind. The dune surface carries browntop bent–sheep's fescue (*Agrostis tenuis*–*Festuca ovina*) and white clover *Trifolium repens* and locally is being stripped by wind-scour. A number of erosional scarps dissect the surface, taking the form of small linear cuts orientated east and north-east. The sand scarps often have small areas of sand accumulation at their downwind ends, which are progressively being colonized by marram.

Interpretation

Morrich More forms part of a network of sites used to infer the complex interaction between sea level, sediment supply and coastal evolution of the Dornoch Firth over the Holocene Epoch (Hansom, 1991, 2001; Firth *et al.*, 1995). The scientific importance of the immense Morrich More coastal strandplain has been recognized for many years. For example, Ogilvie (1923) described Morrich More as 'a wave-built sandy strandplain ... built out gradually throughout the uprising of the coast'. Further work (e.g. Smith and Mather, 1973; Smith, 1983) described Morrich More as an emerged strandplain built up during a succession of changing land–sea relationships during the Holocene Epoch. Smith (1986) goes further, suggesting that the entire Morrich More system is genetically a coastal strandplain created by *c.* 6500 years of shoreline accretion in the form of swash-bars thrown up by wave activity. However, perhaps as a result of the size and complexity of Morrich More, there was a lack of any detailed geomorphological studies until the work of Hansom and Leafe (1990). This work led to the reconstruction of the Holocene evolution of the Dornoch Firth (Hansom, 1991; Firth *et al.*, 1995) and emphasized the role that contemporary processes play in the continued development of Morrich More (Stapleton and Pethick, 1996; Hansom, 1999, 2001). The interpretation below is drawn mainly from this more recent research.

Hansom and Leafe (1990) suggest that at the peak of the Holocene transgression, in a situation of plentiful sediment supply, large amounts of sand were transported onshore and beach ridges began to develop rapidly close to the Holocene cliff. A radiocarbon date of 6450 years BP from peat found in a section at 6.4 m OD beneath one of these ridges (Hansom and Leafe, 1990) indicates that the ridge formed soon after this date. The age and altitude of the suite of sand ridges, at altitudes of 6.0 to 8.6 m OD, suggests development at the culmination of the Main Postglacial Transgression (Firth *et al.*, 1995). The progressive eastward progradation of the Morrich More shoreline was probably produced by onshore movement and vertical accretion of nearshore bars fed from offshore sediments, the narrow spacing and number of the landward-most ridges indicating that sediment supply was relatively abundant and that accretion was rapid at this time (Hansom and Leafe, 1990). Subsequently, as the rate of eustatic sea-level rise fell below the rate of isostatic uplift of the land, relative sea level fell and a second suite of ridges were formed between 4.4 and 5.5 m OD (Hansom, 1991; Firth *et al.*, 1995). Between 6400 years BP and *c.* 5000 years BP, the limited sea-level fall seems to have been conducive for large amounts of sands to move onshore to produce rapid shoreline regression

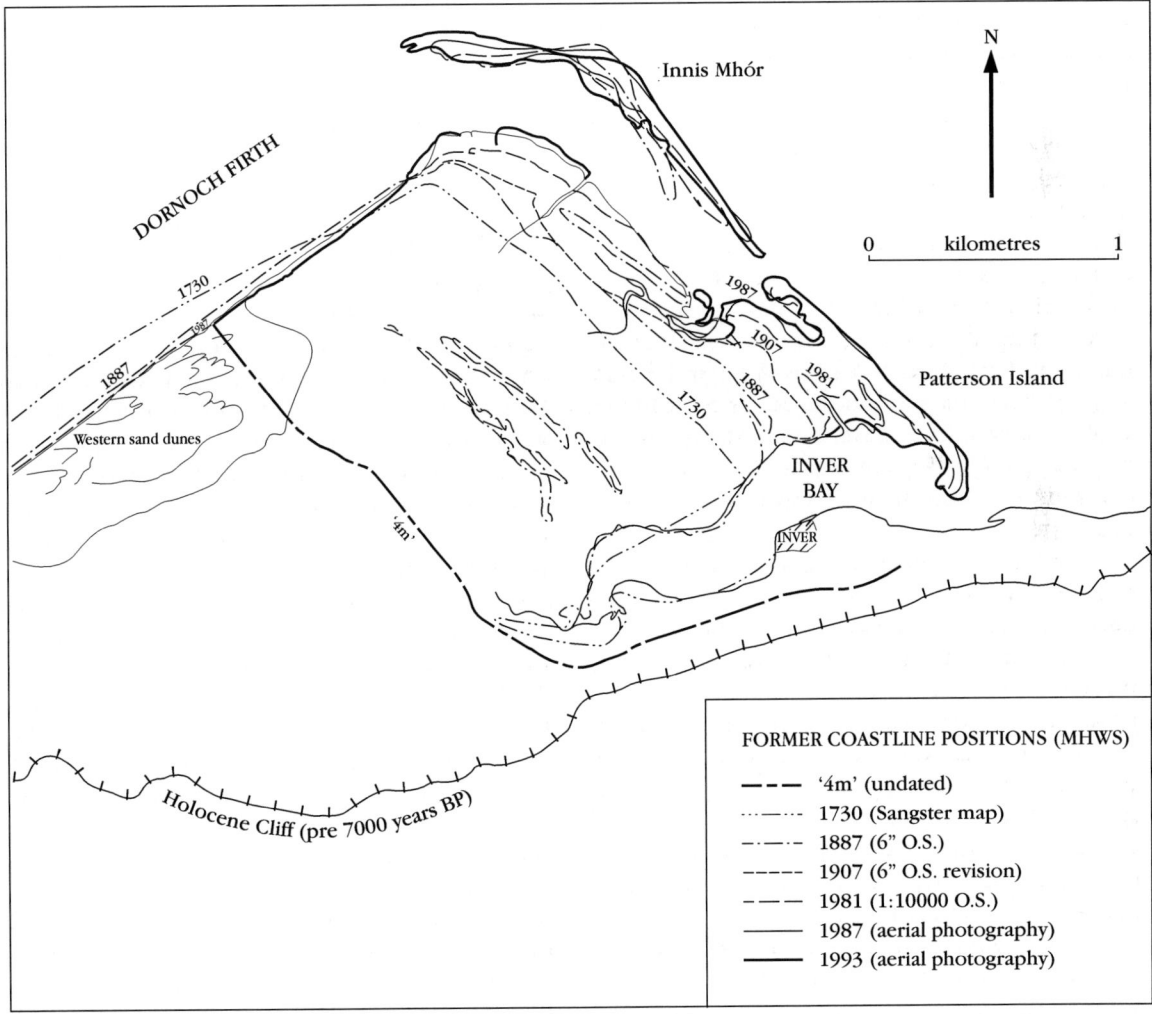

Figure 11.11 Former coastal positions of Morrich More, based mainly on historical sources and Ordnance Survey maps and aerial photography. The north-east coast has extended by about 1 km since 1730, but the western flank has eroded by varying amounts over the same period. Sediment eroded from the west is moved north-east by tidal streams and waves at high tide to be deposited in the area behind Innis Mhór and Patterson Island. (After Hansom and Leafe, 1990.)

and the addition of beach ridges. Although undated (except by regional sea-level curves) Hansom and Leafe (1990) suggest that the bulk of the Morrich More sands higher than 4 m OD, may have been in place by c. 5000 years BP. Seawards of the beach ridges at about 4 m OD, the true ridge altitude drops (although covered by sand dunes) and the spacing between beach ridges increases. It is hypothesized that the wider spacing between the most seaward, and thus most-recently deposited, ridges, indicates that rates of accretion and sediment supply have progressively reduced since 6500 years BP and certainly since deposition of the second group of higher ridges (Hansom and Leafe, 1990; Hansom, 1991; Firth et al., 1995). The postulated reduction in offshore sediment supply is supported in part by the morphology of the current beach ridge of Innis Mhór and Patterson Island, which, in relative terms, is the highest of the Morrich More ridges. It supports the assertion of Davies (1980) that during conditions of reduced sediment supply, beach ridges simply build higher, rather than constructing additional, ridges on the seaward beach face.

The general trend of north-eastward accretion of Morrich More over the Holocene Epoch is mirrored by the more recent changes that are known to have occurred since the first accurate map of the area was produced by Sangster in 1730 (Hansom and Leafe, 1990; Stapleton and Pethick, 1996). In 1730, MHWS of the outer coastline lay north-west of Inver and lay along the line of a prominent emerged ridge (Figure 11.11). Since then, Ordnance Survey maps and aerial photography of numerous dates have recorded the north-eastward migration of MHWS and MLWS as accretion has progressed on this shore. Such a migration has resulted in progressive narrowing of the entrance to Inver Bay and the shallowing of the bay itself. As the outer beaches moved north-eastward, the low-lying swales behind were flooded by tidal waters, promoting the development of sandflat and saltmarsh. The development of saltmarsh seems to be a rapid process, since the 1887 Ordnance Survey does not show Patterson Island at all and displays Innis Mhór as a small unvegetated sandbank. Patterson Island had emerged above MHWS by 1946 and although aerial photography shows a small dune area present, the intertidal area behind remained entirely intertidal sandflat until the late 1960s. Infilling is ongoing and comparisons between the 1981 map and 1987 aerial photography shows saltmarsh edge migration of 100 m over the six-year period as a result of accretion and colonization by saltmarsh plants (Hansom and Leafe, 1990).

Accretion on the outer north-eastern coast is matched by erosion on the inner Dornoch Firth coast of Morrich More (Figure 11.11). Erosion along the western flank of Morrich More has occurred at a mean rate of 0.47 m a^{-1}, with some 117 m lost between 1730 and 1990 (Hansom and Leafe, 1990). This has resulted in peat (radiocarbon-dated at 325 years BP), which probably developed in dune-slacks, being buried by dune migration eastwards and now being exposed on the foreshore by erosion. The narrow fringing beach of this north-western flank is now clearly erosional and is backed by a vertical erosional scarp that truncates the orientation of both dune ridges and emerged beach ridges. Although there is a lack of evidence concerning the length of time that erosion has predominated on the western shore of Morrich More, it is highly likely to have been ongoing for as long as deposition has built ridges on the north-eastern flank (Hansom and Leafe, 1990). Sand transport occurs northwards along the upper beach and so erosional activity on the north-western flank contributes directly to infill of the sandflat areas behind Innis Mhór and Patterson Island and indirectly contributes to deposition on the outer beach. Based on this evidence, Hansom and Leafe (1990) suggested that Morrich More has been migrating from west to east for some 7000 years and Hansom (1999) presents a five-stage schematic reconstruction of this migration based on breaks in the size and spacing of the emerged ridges (stages 1–5 on Figure 11.7).

The size and relatively undisturbed history of almost 7000 years of continuous sedimentation on Morrich More has led to the development of a range of successional stages of dune development from embryo dune on the outer shore, through foredune and mature dune forms, parabolic dune forms and low undulating dune plain, some of these in association with interfingered saltmarsh that grades into fresh-water marsh and dune slacks. Ratcliffe (1977) describes sand dune vegetation of Morrich More as 'one of the most important and distinctive dune systems in Europe', and Doody (1986) echoes this by regarding it as one of the finest sequences of natural vegetation in Great Britain. Dargie (1989) demonstrates the importance of the vegetational transitions at Morrich More, with those between the saltmarsh and dune systems being of particular complexity and therefore of high conservation value in view of a clear relationship with geomorphological conditions. Vegetational transitions from saltmarsh to sand dune are extremely rare in Britain and the transition on Morrich More from saltmarsh to calcareous dune, wet acid dune or dry dune grassland makes the upper saltmarsh vegetation, and its interaction with geomorphology, uniquely important.

The relationship between the mainly stabilized parabolic dunes in the west and the low undulating dune plain downwind to the northeast is unusual, since neither has an obvious upwind nourishment zone of foredune and young dunes (Figure 11.9). The orientation of the axes of both the parabolic dunes (250°) and the low dunes behind indicate dune forms produced by sand movement driven by south-westerly winds. Although some active sand movement still occurs over the crest of the northernmost of the parabolic dunes, and there remains active deposition from the beach to the southwest, the volumes are insufficient to sustain large-scale changes to dune forms. If the backshore erosion experienced by this side of Morrich More is a long-lived phenomenon, then it seems reasonable to assume that erosion has removed the feeder dune cordon that once fed the downwind dunes. Support for this assumption comes from the 1730 Sangster map that shows a coastal position c. 350 m to the west of its current location. The intertidal peat exposed on the beach and beneath the dunes also indicates erosion since its deposition, probably within a damp dune-slack some 325 years BP. It may also be likely that coastal recession itself contributed to dune instability and further movement of the parabolic dunes. Ogilvie (1923) noted that parabolic dunes on the western flank were then mobile and appeared to have moved some 183 m between 1873 and 1913. It seems probable that the earliest parabolic dunes of western Morrich More were fed from a beach and dune system in the west beyond Tain that now no longer exists.

Conclusions

The scientific interest of Morrich More is outstanding both in terms of the variety and scale of its coastal landforms, many of which have well-developed transitional zones between accretionary landforms, and because of a well-preserved morphological and stratigraphical record that records shoreline change and coastal development over the last 7000 years. The development of this large coastal strandplain is related to a shallow offshore zone and abundant sandy material, resulting in a series of sequential beach ridges deposited under conditions of a falling relative sea level. From a stratigraphical and morphological perspective, the extensive emerged strandplain of Morrich More is central to an understanding of the Holocene coastal development of both the Dornoch Firth and the wider Moray Firth. From a contemporary process and form viewpoint, there exists a diverse variety of forms including attached sandy barriers and spits, stabilized dunes including parabolic dunes, embryo and foredune succession, saltmarshes (some of which are interdigitated between beach and dune ridges) and sandflats. The importance of Morrich More, both within Great Britain and internationally, lies in the extent, scale and diversity of its Holocene and contemporary geomorphology and in the continuity that exists between them.

CARMARTHEN BAY, CARMARTHENSHIRE (SN 220 070–SN 421 868)

V.J. May

Introduction

Carmarthen Bay is formed by the estuaries of the rivers Taf, Twyi and Gwendraeth where they enter the sea between the Carboniferous Limestone headland of Worms Head on the Gower Peninsula and Caldey Island (see Figure 1.2 for general location). The bay includes several sub-units that would warrant selection as coastal geomorphology GCR sites in their own right. It is unusual, however, in containing features that are relatively uncommon in England and Wales and in being very little disturbed by human activity. At low tide, three interconnected units (Figures 11.12 and 11.13) are exposed:

1. Whiteford Burrows and Rhossili Bay (see Figures 11.17 and 11.18)
2. Cefn Sidan Sands, Tywyn and Pembrey burrows (see Figure 11.15), and

Coastal assemblage GCR sites

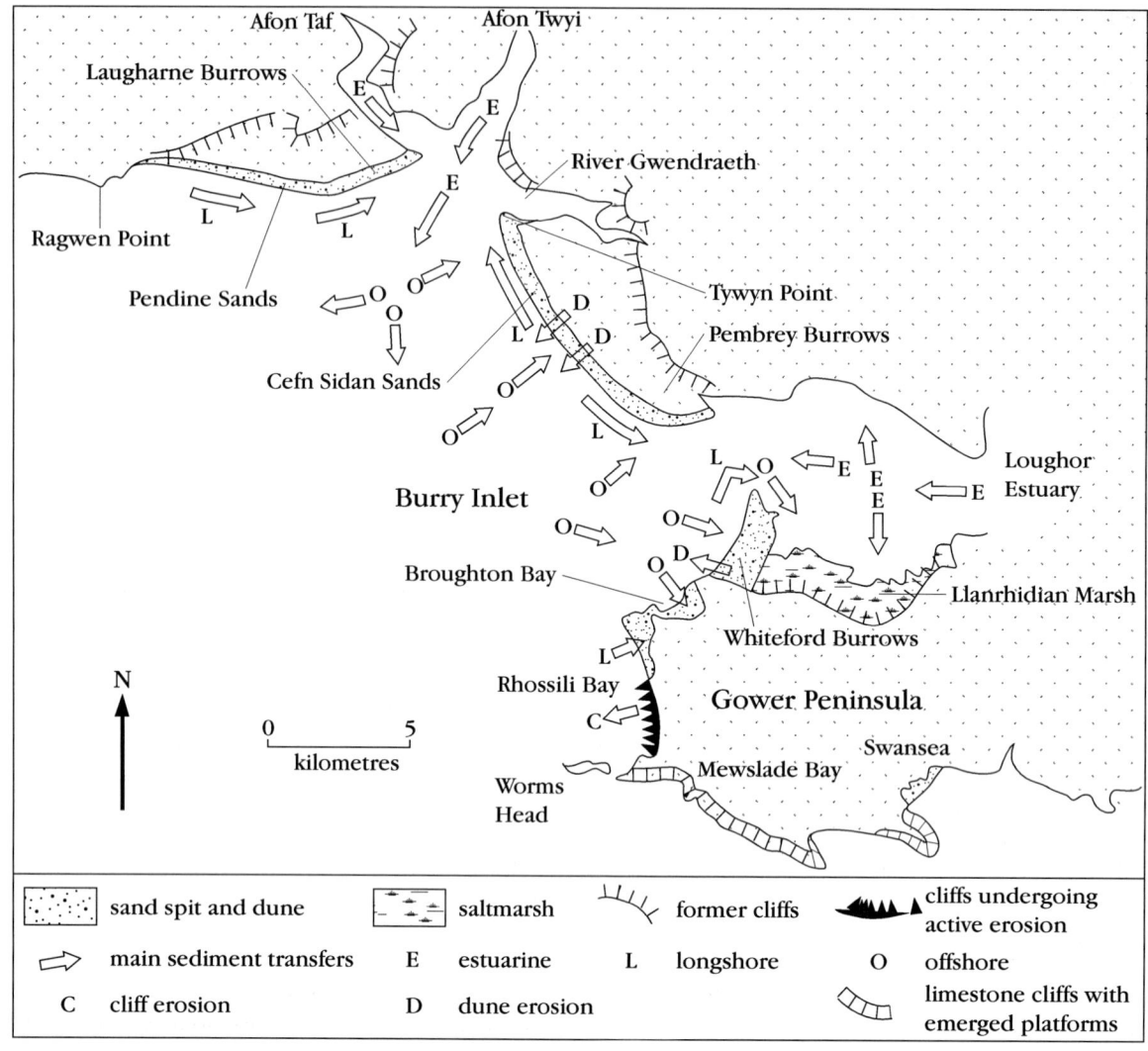

Figure 11.12 Sketch map of the key geomorphological features and sediment transfers of Carmarthen Bay. See also Figure 11.17 for details of the Rhossili Bay area. (Offshore transfers derived in part from Barber and Thomas, 1989.)

3. Pendine Sands and Laugharne Burrows (see Figure 11.14).

These three units can be further subdivided into 8 subsites (see 'Description' below).

The estuaries of the rivers Taf, Twyi and Gwendraeth separate units 2 and 3, whereas units 1 and 2 lie to the south and north of the Loughor estuary respectively. Parts of these estuaries are included in the site.

Both Pendine–Laugharne and Whiteford beaches form spits that trend away from a predominantly rocky cliffed coastline, whereas the barrier beach from Cefn Sidan to Pembrey links to the mainland only by former dunes and reclaimed marshland. This barrier feature is only one of three large barrier systems in England and Wales, the others being Scolt Head Island (North Norfolk Coast) and parts of the Holy Island, Northumberland, site. During the 20th century, the general trend has been for erosion at the proximal end of the spits, accretion at their distal ends and extensions of the forms into the estuaries (Figure 11.13). Cefn Sidan–Pembrey has been an area of general progradation, with extension of the beach as spits into estuaries of the rivers Loughor and Gwendraeth.

Apart from the coasts of north Norfolk and Holy Island, Carmarthen Bay contains the largest

Carmarthen Bay

assemblage of unmodified sandy beaches in England and Wales, but it has received very limited attention in the literature. North (1929) and Steers (1946a) concentrated mainly upon the hard-rock coastline and the evidence of changes in sea level. Savigear (1952) saw the eastward growth of the Pendine beach as controlling slope development on the abandoned cliffs behind the beach and marsh. Kahn (1968), Potts (1968), Jago (1980) and Jago and Hardisty (1984) described the sediments and the geomorphological processes that act upon individual parts of the site. Barber and Thomas (1989) have considered the whole bay in terms of sediment transport and its effects on the beaches. Emerged ('raised') beaches and periglacial and fluvioglacial deposits occur at several points around the bay and many writers have considered the Quaternary history of the area and Campbell and Bowen (1989) provide a comprehensive summary.

Description

There are eight major subsites, each of which is described in turn from west to east. These subsites form parts of an integrated whole that is bounded in the south-east by Worms Head and in the north-west by Ragwen Point. The eight units (for locations see Figure 11.12) are:

1. Pendine Sands and Laugharne Burrows
2. The estuaries of the Taf and Tywi
3. Cefn Sidan, Tywyn and Pembrey Burrows (including the Gwendraeth estuary)
4. The Loughor estuary (uncluding Llanrhidian saltmarsh)
5. Whiteford Burrows
6. Broughton Bay burrows and cliffs
7. Rhossili Bay
8. Worms Head and Mewslade Bay.

Consideration of processes at other GCR sites led to the decision to treat the whole of Carmarthen Bay as a single unit in the GCR, defined by the low-water limits of the intertidal zone at Ragwen Point and Rhossili. In addition, the cliffed coast around Worms Head, which is an important site in its own right, was integrated with the larger Carmarthen Bay GCR site. The site includes all the intertidal sand banks of the bay and the channels between them. The seaward boundary crosses the channels at their most seaward extent. Because the processes

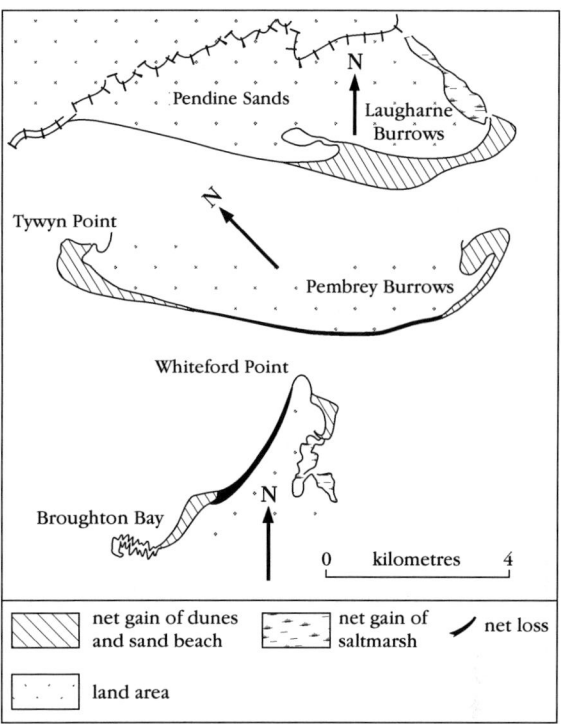

Figure 11.13 Variations in accretion and erosion since 1950, Carmarthen Bay.

that affect the site extend into deeper water, an appropriate geomorphological boundary lies along a straight line from Ragwen Point to Worms Head.

The site comprises 32 morpho-sedimentological units of which ten are hard-rock cliffs, seven are saltmarsh, six are stable or prograding beaches and three are sandy beaches affected by erosion. The remainder are till cliffs undergoing erosion (one unit), estuary/ria (two units), shingle beach (two units), and pocket beach (one unit). Depending upon tidal and wave conditions, sediments can move between most parts of the site, with the possible exception of the area south of Worms Head (Figure 11.12). Very large areas of sand are exposed at low tide, especially at the mouth of the rivers Taf and Twyi. At high tide, the behaviour of waves crossing these intertidal flats is affected by changes in channel position and bathymetry. The site is macrotidal, with a tidal range in excess of 8 m and currents that exceed 1 m s^{-1}. The maximum fetch in the direction of prevailing and dominant winds from the south-west exceeds 4000 km. Fetch to the SSE, i.e. across the Bristol Channel, is 70 km.

The north-western extremity of the site at Ragwen Point (SN 220 070) forms the low-water

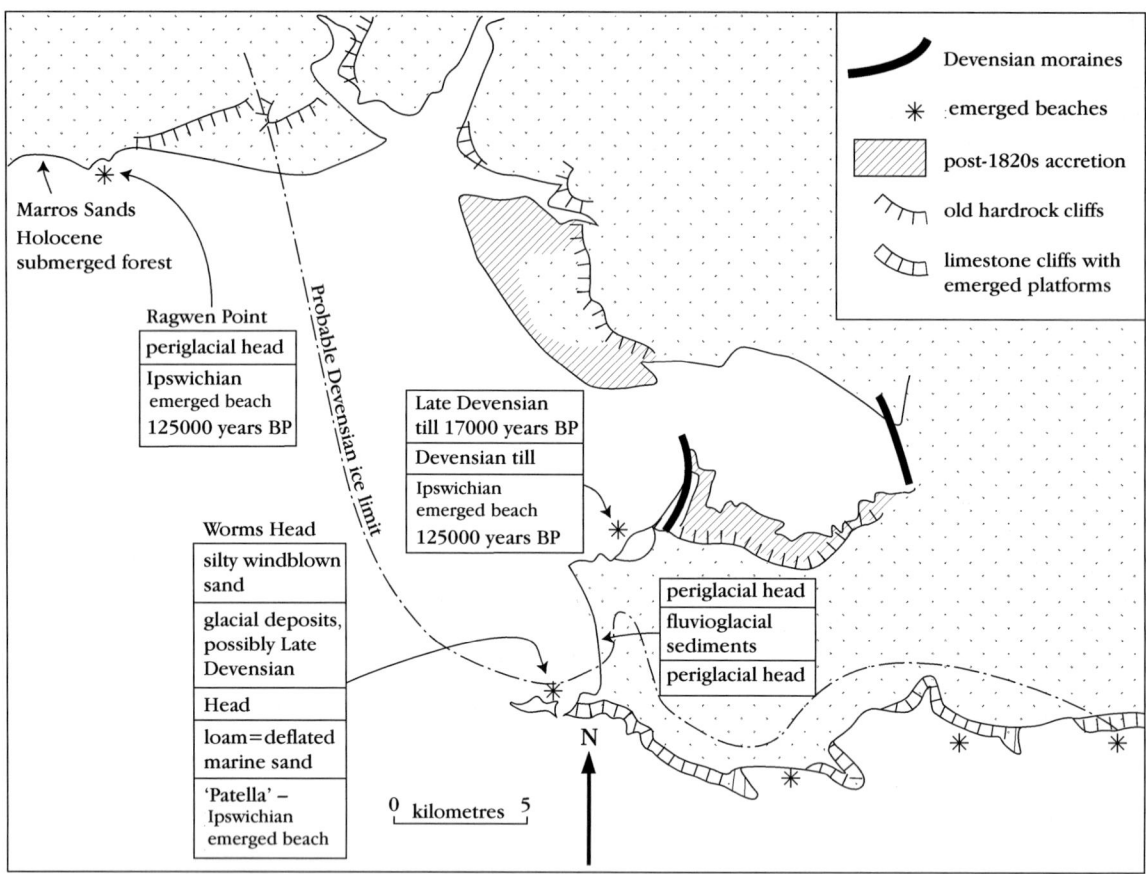

Figure 11.14 The Devensian geomorphology of Carmarthen Bay, with stylized sections through selected emerged beach sites. (After Campbell and Bowen, 1989.)

boundary of the sandy intertidal zone. A small pocket beach separates Ragwen Point from Gilman Point (SN 228 074) whence the cliffs that reach up to 30 m in height trend northwards to Dolwen Point (SN 233 079). The Pleistocene sequence (including emerged beaches) at Ragwen Point (Bowen, 1970, 1974; John, 1971, 1973; Campbell and Bowen, 1989; Figure 11.14) indicates the long time period over which the rock cliffs of this site have developed. As elsewhere in the site, several caves are cut into the cliffs.

Pendine Sands and Laugharne Burrows are dominated by dunes up to 700 m wide that extend some 9.5 km to Ginst Point (SN 331 078). The dunes attain heights in excess of 15 m, except in the vicinity of Wickett Pill where a stream draining marshland and a shallow lake existed up to the 1950s. Since then, there has been considerable accretion of Laugharne Burrows, but the shoreline of Pendine has remained virtually static. The intertidal zone is up to 1 km in width, notably opposite Wickett Pill. As the dunes grew eastwards they protected the rocky cliffs (Savigear, 1952) and the intervening marshland was reclaimed. North of Ginst Point, saltmarsh has developed in the lee of the dunes and in front of artificial embankments. The eastern part of the reclaimed marshland drains via Railsgate Pill into the Taf estuary at SN 306 099.

The 0.5 km-wide Taf estuary is bounded by rock cliffs on its western side, but on its eastern side a cobble spit (Black Scar) narrows the channel, and there has been some development of saltmarsh. Vertical cliffs about 30 m high form the shoreline around Wharley Point (SN 340 093) to the Twyi estuary at SN 352 100. Both sides of this estuary are formed by low vertical cliffs, but the eastern side is much lower than the western side. On the eastern side, the low cliff (SN 361 082) is replaced by a cobble

Carmarthen Bay

Figure 11.15 Pembrey: older dunes can be seen to the right of the photograph – these are now conifer plantations. Post-19th century accretion is evident in the middle and left background. A blowthrough is present in the foreground. (Photo: V.J. May.)

and gravel beach that widens southwards extending seawards into Salmon Point Scar (SN 355 070) and narrowing the estuary significantly at low tide.

The Gwendraeth estuary is dominated by sandflats and mudflats that merge into a growing saltmarsh in the lee of Tywyn Point (SN 357 065). The sand beach extends about 12 km from Tywyn Point through Tywyn and Pembrey Burrows to just south-west of Burry Port (SN 437 994). Intertidal sands extend seawards over 3 km at Cefn Sidan at the mouth of the rivers Taf and Twyi. Much of the area landward of the beaches of Cefn Sidan–Tywyn–Pembrey is dominated by former dunes in Tywyn Burrows, over 2 km-wide and reaching over 20 m in height (Figure 11.15). Much of the duneland is afforested. Seawards of the main zone of grey dunes, there is a zone of low-lying sandy hummocks, rarely higher than 2 m, fringed by a narrow ridge of younger, mainly active, dunes over 5 m in height, which extend into both the Gwendraeth and Loughor estuaries. Although the central part of the beach has suffered some retreat in recent years, the distal areas have extended several hundred metres during the last 150 years. This is a fine example of a progradational beach to which the supply of sediment is sufficient not only to allow the beach to grow in overall volume but also to be able to sustain the lateral growth of the spits.

The saltmarshes of the Burry Inlet comprise the most extensive area of saltmarsh in Wales: 2121 ha out of a total of 6712 ha (32%), and represent almost 5% of all British saltmarsh. Those of the south shore of the estuary from Whiteford Point to Loughor are of interest for the range of geomorphological features they display, particularly saltmarsh creeks, saltpans, erosion cliffs and range of sediments (Figure 11.16; Gillham, 1977). Berthlŵyd, Llanrhidian and Landimore marshes have developed in a sequence from east to west. The mature marshes at Berthlŵyd display well-developed terraces and a marsh cliff undergoing erosion: at Llanrhidian pans and creeks are present and display much dissection. At Landimore, an intricate and deep creek network is present. The sequence of marshes forms a key area for an understanding of saltmarsh dynamics, sediment transport and sea-level changes.

The marshes extend for about 15 km along

Coastal assemblage GCR sites

Figure 11.16 A deeply incised saltmarsh channel in the muds of Llanrhidian saltmarsh, Loughor Estuary. (Photo: J.D. Hansom.)

the northern shore of the Gower Peninsula and are up to 1.5 km wide. Landimore Marsh in the west lies in the shelter of Whiteford Burrows and is the youngest of the marshes. Llanrhidian Marsh is more exposed to waves entering Burry Inlet from the west, but Berthlŵyd Marsh is in the more sheltered upper part of the estuary. The marsh sediments have not been dated. The tidal range at springs is 6.6 m and at neaps 3.7 m (Pye and French, 1993). Fine-grained sediment deposition is restricted to the more sheltered upper intertidal zone and upper reaches of the estuaries. According to Carling (1981), grain size typically becomes finer with increasing elevation, and Pye and French (1993) record the mean grain size on the upper tidal flat as sands, on the marsh edge as sandy silts, and on the upper marsh as clayey silts. The marsh-edge is widely marked by a low cliff formed during periodic storm activity. Gently sloping ramped margins occur in areas of pioneer marsh progradation. There are some weakly developed terraces, the transition being marked by low cliff, ramp or residual mud-mound topography. Many creeks on the upper marshes show infilling in response to a reduction in tidal capacity while the marshes grow both vertically and laterally (Pye and French, 1993), but common cord-grass *Spartina anglica* was introduced to Loughor in 1931 and colonized rapidly in the 1950s and 1960s, but appears to have declined since (Hubbard and Stebbings, 1967; Burd, 1989). Smith (1978) identified subsurface piping patterns in this area, and attributes the development of saltpans (Packham and Willis, 1997) to their presence. At Landimore, an intricate and deep creek network is present. Small-scale mass-movements in rills and creeks in the muddy intertidal zone play an important role in the changes in creek morphology, the supply of sediment into the creeks, and in intertidal drainage patterns (Allen, 1989).

Whiteford Burrows extends some 3 km northwards from the northern side of the Gower Peninsula to Whiteford Point (SN 450 968; Figures 11.12 and 11.17), where its distal end is associated with a cobble bank to seaward. The burrows are formed of several lines of dunes reaching a maximum height of 24 m in the north-east. The main dune ridge is generally between 10 and 16 m in height, widening towards its landward end. To the west, a line of dune slacks separate more recently accumulated dunes from the main ridge. On the seaward side of the main ridge, a line of slacks rest on partly

Carmarthen Bay

rounded cobbles and shingle derived from Devensian till probably deposited by ice in the Loughor valley (Bridges, 1987). There has been some erosion of the proximal end in the recent past, but sand ridges have also extended into this area from Broughton Bay to the west.

The coast westwards to Burry Holms (SN 398 925; Figure 11.17) is dominated by vertical cliffs fronted by dunes up to altitudes of 50 m OD. At Broughton Burrows the dunes extend to sea level between Prissen's Tor (SN 425 937) and Twlc Point (SN 415 931) where the outlines of the bay appear to be fault-controlled. The cliffs are penetrated by several caves that have developed along joints in the Carboniferous Limestone. Since the late 1980s, marine erosion in Broughton Bay has exposed a very important Devensian multiple till sequence (Campbell and Bowen, 1989). Near Twlc Point, the base of this exposure is formed by an emerged beach conglomerate with fragments of marine shells, mainly flat periwinkle *Littorina littoralis,* resting on a Carboniferous Limestone platform (also prob-

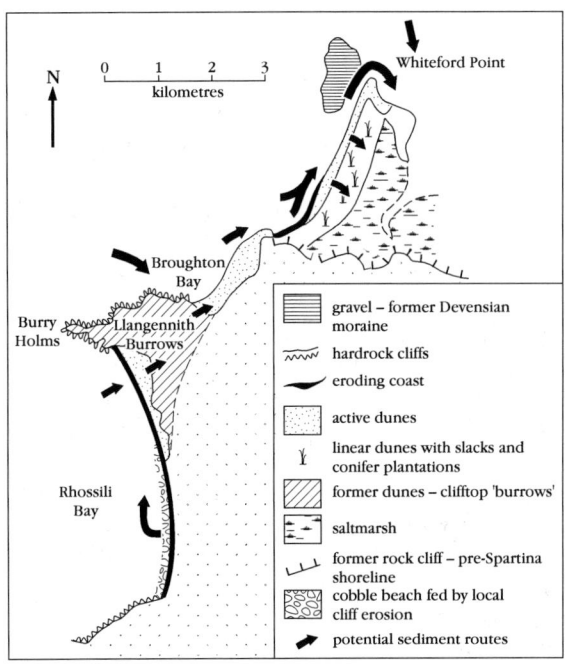

Figure 11.17 Geomorphological features of Rhossili Bay and Whiteford Burrows.

Figure 11.18 Rhossili Bay seen from the Carboniferous Limestone headland looking north towards Burry Holms and Llangennith Burrows. Rhosilli Down to the right of the photograph is fronted by low cliffs formed in periglacial head and fluvioglacial material. Erosion of these cliffs feeds the narrow fringing cobble beach. The wide intertidal beach is mainly sand. (Photo: J.D. Hansom)

Figure 11.19 Limestone intertidal solution features near Worms Head, typically up to 0.3 m in height. (Photo: V.J. May.).

ably marine in origin; Campbell, 1984). The emerged beach is overlain by head and tills that include some fragments of wood (willow/poplar Salicaceae). Shelly and stony tills in the upper part of the section show glaciotectonically-induced folds that are also exposed occasionally in plan form on the foreshore (Campbell and Bowen, 1989). The dunes that overlie this sequence are thought to have been in existence by Roman times (c. 2000 years BP) and to have been affected by renewed activity during intense late medieval storms (Lees, 1982, 1983).

Burry Holms, an island at high water, forms the northern end of Rhossili Bay, a c. 450 m-wide smoothly curving sandy beach directly aligned to the prevailing and dominant south-westerly waves (Figures 11.17 and 11.18). Just over 4 km in length, the bay has a northern shoreline formed by dunes. These rest on bedrock and reach altitudes of about 50 m OD, and are fed by sand blowing from the drying intertidal zone. The southern part of the bay is formed by low active cliffs in till and periglacial sediments (Campbell and Bowen, 1989). The shoreline is cut across the lithological change and is strongly wave-dominated. Shallow slides feed boulders and cobbles into a narrow fringing beach but the contribution of these cliffs to the beach sediment budget is slight. The beach appears to be in deficit, especially as there is no obvious sediment supply from either offshore or alongshore. Sand leaks from the beach via Llangennith and Broughton Burrows towards Whiteford Burrows.

The southern boundary of Rhossili Bay is formed by high limestone cliffs that are continued beyond a gap of 500 m to Worms Head (SN 384 878). Cliffs drop almost vertically into the sea on the northern side of the headland, but a shore platform is well developed around Worms Head and along the southern shoreline, where emerged beach, glaciogenic and periglacial head and colluvial sediments occur widely (Campbell and Bowen, 1989). Cemented *'Patella'* emerged beach deposits rest on the Carboniferous Limestone shore platform (Figure 11.15). Mewslade Bay, bounded to the east by the headland of Thurba (SN 421 868), has similar examples of a platform at present-day sea level and a higher platform planed across dipping strata with emerged beach sediments. The present-day platform is pitted by potholes and solution forms (Figure 11.19). The upper slopes have been severely weathered and screes cloak the lower part of the slopes resting, in places, on the emerged beach materials.

Interpretation

This old cliffed coastline preserves coastal forms of considerable age that have been retrimmed recently to provide the framework within which the extensive recent sedimentation has formed intertidal banks, beaches and saltmarshes. The interpretation of the coastline of Carmarthen Bay involves both explanation of the form and development of the coast during the Quaternary Period and the forms and processes that are active today.

Carmarthen Bay is an excellent example of a coastline whose outline was moulded by marine and subaerial processes throughout the Quaternary Period, but where the shoreline and its detail is much more recent in origin. Each subunit of the site is distinguished by features of major geomorphological interest. For example, the Loughor estuary saltmarshes form a key area for the understanding of saltmarsh dynamics, sediment transport, and sea-level changes, and the platforms and cliffs in Rhossili Bay are excellent examples of hard-rock erosional forms including features linked to former sea levels.

The Quaternary history of Carmarthen Bay indicates that it was affected by several phases of lower sea levels, glacial and glaciofluvial action, and periglaciation superimposed on a framework of river valleys. The interpretation of a number of cave and other sediments containing emerged beach deposits point to a sea level close to the present-day level about 200 000 years BP during Oxygen Isotope Stage 7 (Campbell and Bowen, 1989), and another period of emerged beach formation (sometimes referred to as the '*Patella*' beach) about 125 000 years BP during the Ipswichian (Oxygen Isotope Substage 5e) in which two emerged beach facies have been identified at the western end of Carmarthen Bay at Gilman Point (Bowen, 1970). Extensive periglacial activity led to head deposits forming during the onset of the Devensian and point to considerable cliff instability on parts of the coast. For example, John (1971, 1973) regards the presence of very large quartzite boulders in the head around Ragwen Point as an indicator of an intensive period of slope instability (Figure 11.14). At the time, there was apparently no mechanism removing this material beyond the cliff foot. The limit of Devensian ice probably crossed Carmarthen Bay from just south of Worms Head towards Pendine Point (Bowen, 1981a,b), since in-situ Devensian tills overlie the Ipswichian emerged beach at Broughton Bay (Bowen, 1984) but are absent farther west at Ragwen Point and Marros Sands. At Worms Head, glacial age deposits may be remnants of solifluxed pre-Ipswichian glacial materials or Late Devensian outwash (Bowen, 1970; Bowen *et al.*, 1985; Bowen and Sykes, 1988; Figure 11.14). At Rhossili, two phases of periglacial head are separated by fluvioglacial material. There is some doubt about the precise limit of the Devensian ice, although Stephens and Shakesby (1982) suggested that the bay might have been occupied by a large piedmont lobe. Its effect however was to leave substantial deposits of sand and gravel that were re-worked by a rising postglacial sea, and from which the sand of the coastal dunes and spits was blown. A loam lying above the Ipswichian emerged beach at Worms Head has been interpreted as deflated marine sand. Although there have undoubtedly been periods of localized erosion and trimming of the hard-rock slopes, the distribution of the Ipswichian emerged beach suggests that the broad outline of the bay was in place during the Ipswichian and that the Devensian saw only limited change in the hard-rock coasts, largely as a result of the protection offered by deposition of the Holocene sediments.

Modern processes have served to erode or protect parts of the older shoreline, although there is some doubt as to the extent that protection has lead to significant recent change in the older hard rock slopes. Savigear (1952), for example, argued that as the Pendine beach grew eastwards it protected former rocky cliffs and that the slope profiles demonstrate the effects of progressive protection of the cliff foot on slope profile. Attractive though this interpretation is, there are some difficulties: the slope profiles are not a simple progression from steep at the distal end to gentle at the proximal end of the spit; they have probably been affected by subaerial processes over a much longer time than the period of the extension of the spit, and more recent interpretations of the Pleistocene history of the area indicate that most change in these slopes occurred during intense periglacial conditions during Devensian times. Basal removal would have begun again with sea levels attaining present-day levels about 6500 years BP and would have then ceased sequentially from west to east. This may have done little more than exhume older cliffs.

The modern shoreline is a very dynamic one, as a result of the growth of spits, dune and saltmarsh development, changes in intertidal and deeper water bathymetry and erosion of both beaches and cliffs.

There are few sites in Britain that have been so little affected by anthropogenic interference and which contain such a fine assemblage of coastal forms. Carmarthen Bay is the only coastal assemblage GCR site that is directly affected by the Atlantic wave systems; it also contains four estuaries. Like the other coastal assemblage GCR sites, it gains its greatest importance from its completeness and the fact that it can be defined as a single unit in terms of its overall sediment dynamics.

It is apparent from the field studies that the individual parts of the site are interconnected with sediment moving between intertidal sands and the beaches, and therefore able to move from one subunit to another. Similarly there is evidence of sand movement from Rhossili Bay to the Whiteford Burrows spit and beyond.

In recent years concern about erosion of parts of the beaches, particularly at Pembrey, led the Coast Protection Authorities responsible for to commission studies of the bay treating it as a single coastal 'cell'. The Carmarthen Bay Study (Barber and Thomas, 1989) shows that the shape of the modern coastline is dominated by the high-energy, south-westerly wave regime interrupted by the discharges of the two major estuaries. Tidal streams in the bay have three dominant effects:

1. On the ebb tide they produce significant refraction of waves,
2. direct scour by currents occurs at the distal end of spits, particularly at Ginst and Tywyn Points, and
3. currents are also important mechanisms for moving sediments stirred up by wave action on the intertidal banks.

Waves are affected by refraction especially over the shallow areas of the estuaries, and by currents. Over the shallow inlets there is some partial breaking of waves, and the effects of bottom friction and shoaling on waves are particularly strong over the wide intertidal areas. As a result there is considerable stirring of sediments that can then be moved by tidal streams. Wind action also plays a major role in the movement of sand between intertidal drying areas and the dunes. This is especially important with dominant and prevailing westerly winds. There are distinct differences in the recent patterns of accretion on the three main beaches (Figure 11.13). Pendine Sands and Laugharne Burrows show no change along its proximal half where it is not protected by offshore intertidal banks, but substantial growth along the distal half at Laugharne where the effects of the offshore banks become significant. Pembrey Burrows shows a tendency to periodic retreat and advance along its main central section that faces directly into the main direction of wave approach, but persistent gain at the distal ends where sand appears to be transported from the intertidal banks. The accretion here is more than can be accounted for by longshore transfers from the erosion of the central section. Whiteford spit has shown most gain at its distal end within the Burry Inlet, probably by transfers from the offshore banks and by redistribution alongshore. Rhossili beach is a fine example of an Atlantic swell-dominated beach of considerable width (Figure 11.18), but for which there are very limited local sources of sediment, other than coarse shingle and cobbles from the till cliffs and sand from offshore.

Modelling of wave conditions in Carmarthen Bay (Barber and Thomas, 1989) showed that changes in bathymetry between 1977 and 1988 had significantly changed both the direction and the intensity of wave attack within the Loughor estuary and the joint estuary of the rivers Taf and Tywi. Sediment transport was also affected. According to the model, nearshore wave conditions were critically affected by the direction of approaching waves and the magnitude and direction of tidal streams. For example, during a model SSE storm with significant wave height (H_S) = 2.7 m and wave period (T_S) = 6.3 s), wave attack was concentrated on the central part of Cefn Sidan–Tywyn–Pembrey area two hours before high water. This could account for the observed erosion on this beach and the tendency for transport towards both ends of the beach. However, the most significant feature of both the model and the empirical studies is their definition of the bay as a single sedimentary unit.

Barber and Thomas (1989) argued that the approach of the Carmarthen Bay Study provides a 'modern' tool to aid decision-making regarding development and conservation proposals affecting a specific shoreline and its neighbours within a recognizable coastal 'cell' (Figure 11.12). From the point of view of coastal geomorphology, this recognition of the integrated characteristics of coastal processes within a large site provides considerable support for the designation of this area as a single site.

Further interest is added by the saltmarsh morphology. The saltmarshes of the Burry Inlet comprise the most extensive area of saltmarsh in Wales, and those of the south shore of the estuary from Whiteford Point to Loughor are of particular interest for the range of geomorphological features that they display, particularly saltmarsh creeks, saltpans, marsh terraces, erosion cliffs and the range of sediments. Understanding of the long-term dynamics of saltmarshes in the Bristol Channel has been focused mainly in the Severn estuary and there has been comparatively less attention given to those of the Burry Inlet. Within the Severn marshes, geomorphological and sedimentological techniques were used by Allen and Rae (1987, 1988) to elucidate the oscillations of late Holocene shorelines and to describe vertical saltmarsh accretion since the Roman period (c. 2000 years BP). Allen (1990a) has, however, voiced caution about relying upon saltmarsh sedimentation rates as a source of estimates of sea-level change especially because reworking of muddy intertidal sediments is an important process (Allen, 1987). Compaction, the intricate patterns of sedimentation, and local changes of elevation can each lead to errors. Postglacial stratigraphy of the Severn estuary indicates a long-term sea-level rise (about 3–4 m during the last 2000 years (Allen and Rae, 1988; Heyworth and Kidson, 1982) upon which are superimposed several periods of still-stand (Kidson and Heyworth, 1976; Allen, 1990a–c, 1991b, 1992). These interpretations probably also apply to the Burry Inlet, but there has been very limited investigation of them and the magnitude of change is thought to have been less in the Burry Inlet. Relative sea level continues to rise in the region: mean sea level at Avonmouth increased by 1.12 ± 0.62 mm a^{-1} between 1925 and 1980 (Woodworth *et al.*, 1991) and relative sea-level rise in Carmarthen Bay is estimated at between 1 and 2 mm a^{-1} (Pye and French, 1993).

Conclusions

Carmarthen Bay has perhaps the most varied assemblage of coastal features in the British Isles and is one of the few sites with limited anthropogenic disturbance. There are major dunes, sand spits and barrier beaches, both hard-rock and easily eroded cliffs, rias, emerged ('raised') beaches, extensive intertidal sand-flats and some of the most important saltmarshes in England and Wales. Ages of features in the site range from about 245 000 years old to modern, and the site is crossed by the probable limit of the Devensian ice in Britain.

There is no other site in England and Wales that has such well-developed spit and barrier beaches in a macrotidal regime dominated by south-westerly Atlantic wave conditions, and the site can be regarded as a single sedimentary unit. The interest of the site is heightened by the well-documented Quaternary sequences that occur within it at Ragwen Point, Broughton Burrows, Rhossili Bay and Worms Head, since it is possible to relate the modern features to earlier ones.

The comprehensive dating that confirms the longevity of many of the features, notably the cliff–platform sequences and some of the cliff-top dunes at the northern end of Rhossili Bay, is rare and of considerable significance to the understanding of glaciomarine margins and their subsequent development.

The saltmarshes of the Burry Inlet are of major regional and national interest for the range of geomorphological features they display, particularly saltmarsh creeks, saltpans, erosion cliffs and range of sediments, in a macrotidal environment.

THE COAST OF CAERNARFON BAY (NEWBOROUGH WARREN AND MORFA DINLLE)

V.J. May

Introduction

Newborough Warren and Morfa Dinlle are major dune areas on opposite sides of the western mouth of the Menai Strait (see Figure 1.2 for general location). Newborough Warren lies on the south-west facing shoreline of Anglesey (Ynys Môn) and is one of the three largest west coast dune systems in England and Wales, the others being Ainsdale, Lancashire and Braunton Burrows, Devon (see GCR site reports in Chapter 7). Morfa Dinlle forms the south-eastern coast of Caernarfon Bay, comprising a complex area of coast undergoing erosion, together with shingle ridges and dunes (Figure 11.20). Its geomorphology is linked both to the progressive erosion of the coast and to the long-term dynamics of the Menai Strait. The separation of Anglesey from the mainland of Wales by the Menai Strait is of major geomorphological significance. The gradual evolution of Caernarfon Bay owes much to the influence of the Menai Strait in providing a mechanism for moving and transporting large volumes of sediment. The action of the tidal flows has been to partition and transport various components of the sediment population, especially the gravels and the sands. Individually, Newborough Warren and Morfa Dinlle are of national significance, but taken together they are of major importance for understanding of the evolution of the coastline of north Wales and for the effects of sea-level change on coastal form and processes.

The Menai Strait

The double estuary that formed in the Menai Strait in the early postglacial period (see, for example, Greenly, 1919; Embleton, 1964) changed as sea level rose around 6500 years BP into the present-day complex tidal system that floods from both the east and the west. There is a residual flow towards the south-west (Harvey, 1967, 1968) owing to a combination of a higher tidal range at the north-eastern end and the relative phase of the tides. High water at Fort Belan (opposite Abermenai Point) is 1 hour prior to high water at Beaumaris (at the eastern end of the Menai Strait) and is 3 m lower. The resulting hydraulic gradient means that tidal flow in the western entrance to the Menai Strait is markedly ebb-dominated, leading to a pronounced ebb-tide delta whose outer ramparts are known as the 'Caernarfon Bar'. The flood tide in the Irish Sea sets from south to north and a flood-tide rampart has developed to the south of the entrance to the Strait, i.e. immediately west of Morfa Dinlle. Within the Strait, east of Fort Belan and Abermenai Point, flood-tide deltaic deposits have also developed, less extensive than the ebb-tide delta but nevertheless extremely important in their control of the estuarine dynamics. As the tidal delta deposits within the Strait developed during the Holocene Epoch, so the tidal prism in the estuary decreased. In response, the tidal flow velocities decreased in the entrance to the Strait and this allowed sedimentation to proceed, so reducing the entrance cross-sectional area (Pethick, 1997). Holocene infill of the Strait reduced the width from more than 3 km to less than 2 km and mean depth from 20 m to the present-day 10 m, with extensive sandflats across both east and west entrances. It is suggested by Pethick (1997) that the gravel ridges of Morfa Dinlle and the Abermenai spit are direct results of this gradual decrease in the tidal prism of the Menai Strait during the later part of the Holocene Epoch. As deposition proceeded in the Strait, so the cross-sectional area of the mouth decreased,

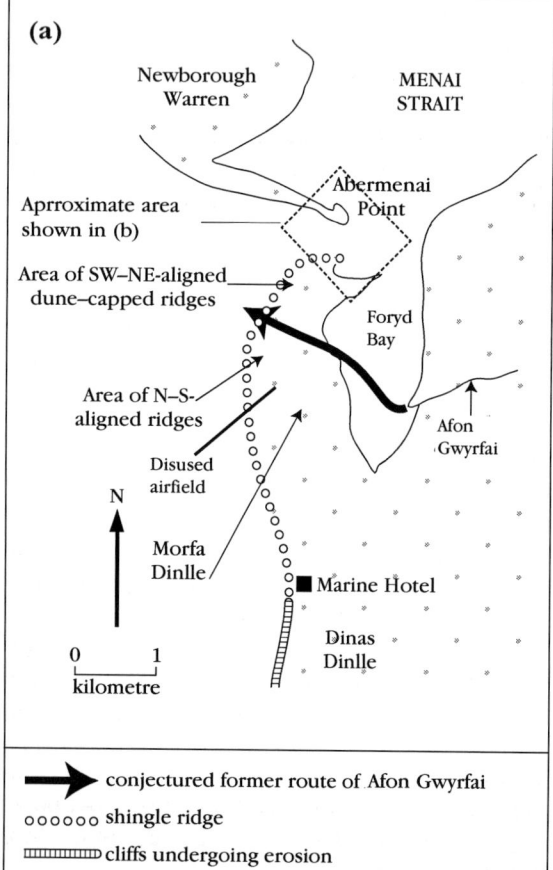

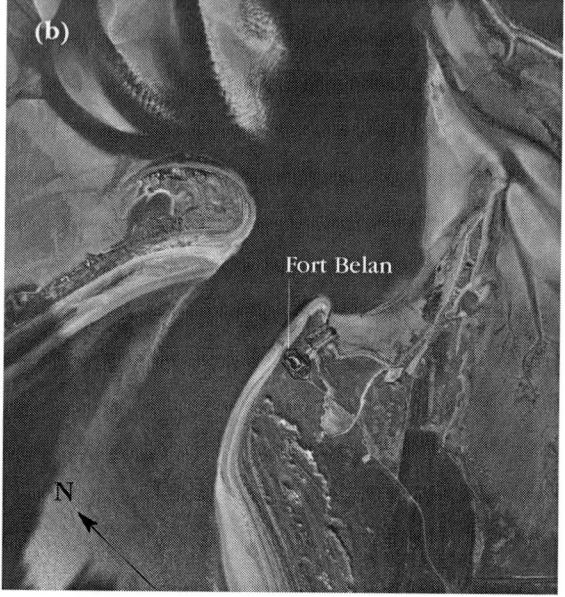

Figure 11.20a and b Sketch map of the key features of the area of Newborough Warren and Morfa Dinlle. (b) Detail of the entrance to the Menai Strait (Photo: courtesy Cambridge University Collection of Aerial Photographs © Countryside Council for Wales.)

and the Morfa Dinlle gravel ridges extended northwards to define an increasingly narrow mouth (now less than 400 m) between Abermenai Point and Fort Belan.

NEWBOROUGH WARREN, ISLE OF ANGLESEY (YNYS MÔN) (SH 367 648–SH 444 615)

V.J. May

Introduction

The shoreline of the major coastal dune system of Newborough Warren is controlled by the Menai Strait to the east, Afon Cefni to the west and Llanddwyn Island (Ynys Llanddwyn), which divides the shoreline between Malltraeth Bay and Llanddwyn Bay. There are large expanses of actively evolving and stabilized dunes, although much of the latter is afforested. East of Llanddwyn Island, parts of the dunes are cliffed and these are in a state of net sediment deficit, as sand is transported eastwards towards a spit that extends (in association with an artificial breakwater) to Abermenai Point. In Malltraeth Bay, the dunes exceed 30 m OD and rest upon and mask the rock outcrop landward of Llanddwyn Island. North-westwards, the beach extends into extensive intertidal sandy flats in the Malltraeth estuary. Like many coastal dunes, Newborough Warren was described by Steers (1946a), but Ranwell (1955, 1958, 1959, 1960) provided the most complete study of dune formation. The orientation of dunes to wind direction is demonstrated well here (Landsberg, 1956). More recently, Robinson (1980b) examined the links between the development of the dunes, the spit and the sandbanks at the mouth of the Menai Strait.

Description

The sand dunes at Newborough Warren are very mobile and it is one of the most-exposed west-coast dune areas; in the past sand travelled inland to cover agricultural land several kilometres from the shoreline (Wortham, 1913). Much of the western area of dunes has been planted with conifers masking the underlying geomorphological interest. This area is excluded from the GCR site. The site comprises six main morphological units.

1. The lower estuary of the Afon Cefni, much of which is formed by sandy, intertidal banks. The western side is formed by cliffs in Precambrian rocks.
2. Malltraeth Bay, where a large area of dunes extends about 2 km into, and about 1.5 km across, the Malltraeth estuary.
3. Llanddwyn Island (Ynys Llanddwyn), which projects about 1.75 km seawards of the beaches and divides the site into two separate beaches.
4. The beach and frontal dunes in an area stretching for about 2 km east of Llanddwyn Island; the area behind this is afforested and now has little geomorphological interest.
5. The eastern dunes of Newborough Warren.
6. The Abermenai spit, which is about 2.5 km in length.

Except for part of the afforested dune area within Malltraeth Bay, the GCR site area falls within the boundary of the Newborough Warren–Llanddwyn Island (Ynys Llanddwyn) National Nature Reserve.

Malltraeth Bay occupies the largest of the low valleys (traeths) that cross Anglesey (Ynys Môn) from north-east to south-west. Precambrian rocks of the Gwna and Fydlyn groups crop out along the western side of the bay and are overlain by dunes on its eastern side (Figure 11.21). The crests of the dunes are generally about 10 m OD, but there are three zones of higher dunes. The first (from SH 398 649 to SH 390 653) lies at the northern end of the dune area and is dominated by an undulating ESE–WNW ridge that exceeds 20 m in height. About 600 m seawards of this ridge, there is a second, more linear, feature (from SH 395 645 to SH 391 649), again exceeding 20 m in height. The present-day shoreline is backed by a wide zone of dunes generally higher than 15 m (between SH 392 639 and SH 388 647) but declining to about 10 m as the beach swings northwards into the Cefni estuary where it cuts across the ends of the two inner ridges. Each of these ridges becomes progressively aligned to SSE–NNW towards the alignment of the present-day beach. Whereas the earliest of the three ridges appears to have faced the dominant direction of wave approach, the later ones were less well adjusted to it. This may be a function of changing water depth and reduced sand supply. Robinson (1980b) suggests that it is probably a result of the sediment circulation within the estuary.

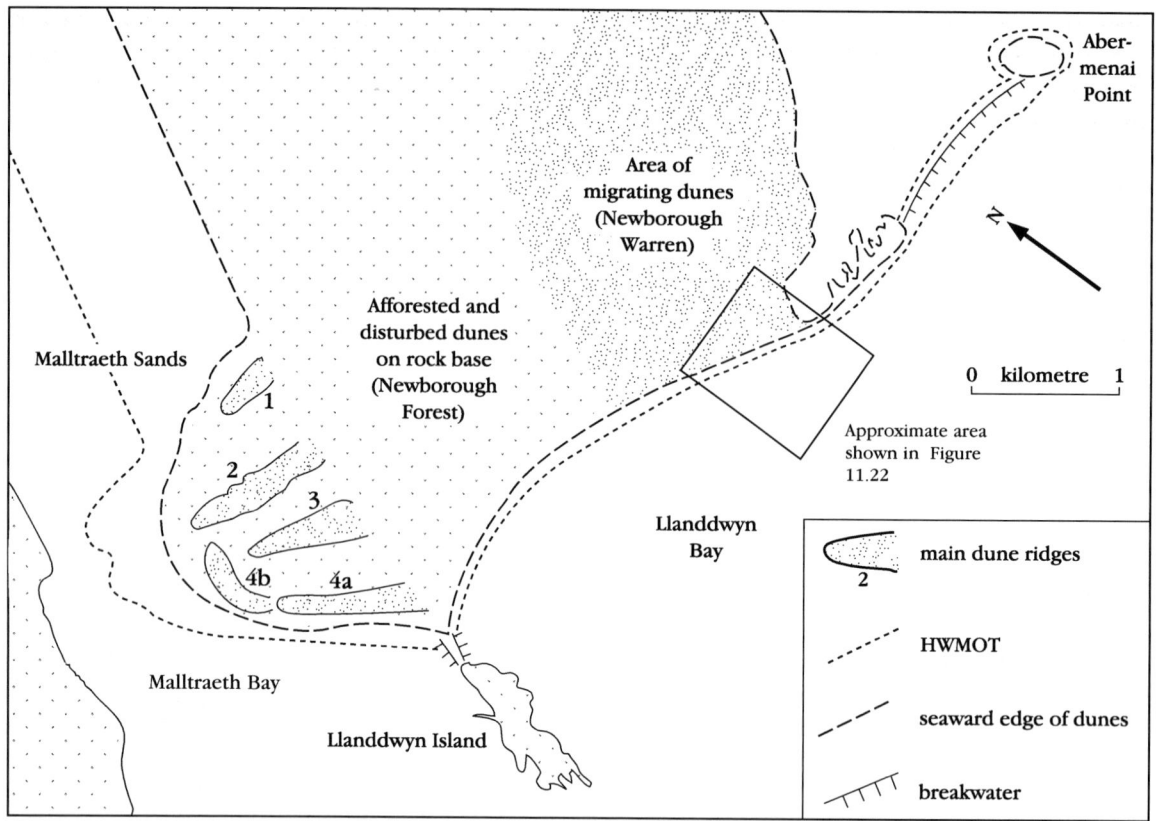

Figure 11.21 Key features of Newborough Warren. The western part of the site has a series of sand spits and dunes extending from a former rocky shore, of which Llanddwyn Island (Ynys Llanddwyn) is a seaward extension. The eastern shore changes from low, climbing dunes on a rock base to an extensive area of migrating dunes on a sandplain. The Abermenai spit owes its form in part to strengthening by an artifical breakwater.

Parts of the shore west of Llanddwyn Island have been artificially nourished in the past to reduce erosion and prevent loss of the forest. Of the three ridges constructed, the outermost was removed by storms in the 1970s, but the other two ridges remain fairly stable (Hansom, 1988).

Llanddwyn Island reaches about 12 m OD and is tied to the mainland by sand beaches that are banked on its landward connection with the rocks that underlie the rocky hill at Cerrig-Mawr. It not only divides the beach into the two bays (Figure 11.21), but also provides some protection to the western end of the beach in Llanddwyn Bay. As a result, the beach shows some of the characteristics of zeta-curve beaches in which wave energy is reduced by refraction and the planform of the beach is controlled by waves approaching from an easterly direction (Figure 11.21). But for the underlying rock ridge, this would be an excellent example of a sand tombolo.

Eastwards from Llanddwyn Island (SH 392 635 to SH 410 632), the GCR area comprises the beach and the fringing dune ridge only; the old dunes behind the line of younger 'yellow' dunes have been afforested. The fringing dunes are affected by erosion both from wave action and from recreational trampling, which is most concentrated in this area. The retreat of the beach is geomorphologically important to the beaches to the east, since it releases sand into the intertidal zone from where it is transported alongshore to feed both the active dunes of the eastern block of Newborough Warren and the Abermenai spit. Some parts of the intertidal area of Llanddwyn Bay are characertised by gravels underneath the sand. These are lag gravels produced by erosion of till deposits.

The eastern part of Newborough Warren is characterized by fine examples of migrating dune systems (Ranwell, 1955, 1958) that extend about 2 km inland. As on the western side of the site in Malltraeth Bay, the dunes tend to form linear sub-parallel ridges, although they are

Newborough Warren

Figure 11.22 Aerial photograph of part of Newborough Warren (see Figure 11.21 for location), which shows the contrasts between the coastal active and migrating dunes, the linear sub-parallel ridges that extend inland and area of mostly stabilized parabolic dunes. Intertidal sand ridges may provide a pathway for sand transport feeding the dunes in the east, but gravels underlie parts of the intertidal area in the west. (Photo: courtesy Cambridge University Collection of Aerial Photographs © Countryside Council for Wales.)

broken by areas of erosion and separated by well-developed slacks. As a result of the differential growth towards maximum height, and irregular erosion, much of the area is composed of parabolic dunes (Figure 11.22), the more landward of which stand above a sand plain.

The Abermenai spit is formed by two areas of low, recurved sand ridges on a shingle and gravel base. The spit was breached in 1868 and the distal part of the spit was separated from the mainland. After another breach in 1889, the Caernarfon Harbour Authority constructed protective works, now represented only by a line of broken stakes (Robinson, 1980b), which link the two areas of sand ridges. The spit, which exceeds 2.5 km in length, has an inner section about 1 km in length comprising several recurves almost at right angles to the main beach that are up to 450 m in length. Eastwards, for a further 1 km, the spit is dominated by washover forms and many low recurving ridges that represent steady extension eastwards along the line of the artificial breakwater. They are, in effect, a new distal part of the spit that is growing eastwards as sand is carried alongshore from the west.

Abermenai Point is the pre-breaching distal end of an older spit. The feature appears on all Admiralty charts and Ordnance Survey large-scale plans since the early 19th century (Robinson, 1980b), but is much older. The recurved end of the spit was named 'South Crook' during the late 13th century (Davies, 1942 quoted by Robinson, 1980b), but Robinson believed that other historical evidence, notably visits to the harbour of Abermenai by Grffydd ap Cynan in the late 11th century, suggests the possibility of an even older feature at this point. Although the breakwater has provided some armouring to the gap that existed after breaching, the growth of the western part of the spit demonstrates that sufficient sand has been available for continued growth to occur.

The presence of two relatively erosion-resistant points in the shoreline of Llanddwyn Bay, at Llanddwyn Island and an intertidal area of lag gravels produced from till erosion fronting the western set of recurves on Abermenai spit, has maintained the shoreline in its alignment towards the prevailing and dominant waves and there has been some growth of the dunes eastwards in the lee of the spit. Substantial amounts of sand also travel over the spit into Traeth Melynog, but there is at present no direct evidence of the volume or location of the seabed sources that are still capable of feeding the beaches.

Interpretation

Newborough Warren is of particular geomorphological interest because (i) the processes of dune-building and migration have been examined in detail (Ranwell, 1955, 1958, 1959, 1960, 1972), (ii) there is a marked contrast in the development of the beaches east and west of Llanddwyn Island, and (iii) the Abermenai spit is a good example of a major feature at the mouth of an estuary that has grown, been breached and rebuilt. In this respect it provides a good contrast with Spurn Head, Yorkshire, both in its relationship to wave conditions and in the pattern of breaching and rebuilding. First, it faces directly into the dominant and prevailing waves. Second, the tidal range is less (by about 2 m) than that of the River Humber, and third, the sediment supply to the Abermenai spit is substantially less, not least because much of the sand is blown inland. Steers (1946a) commented that there was no physiographical description of the site. There is still no interpretation of the ridges in Malltraeth Bay, but this is essential if the relationship between events on both sides of the Menai Strait are to be interpreted. Despite Ranwell's (1955, 1958) description of the key physiographical features of the dunes, there is no detailed analysis of the dynamics of the spit, although Robinson (1980b) describes the historical changes in the spit and the seabed in the approaches to the Menai Strait.

The migration of the dunes was examined by Ranwell (1955, 1958) who argued that the mechanics of dune development could be understood by studying a location where maximum erosion could be expected. Such a situation occurs at Newborough Warren where the prevailing and dominant south-westerly winds flow across a coastline oriented at right angles to its direction. Landsberg (1956) found a perfect correlation at this site between a calculated wind resultant and the orientation of parabola- or U-shaped dunes. Subsequently the rate of dune-building and dune travel in a region where entire dune ridges are successively moving landwards were measured by means of time-series levelled transects (Ranwell, 1958). The theoretical point of maximum erosion was shown to be at 18 m to windward of the crest of 15 m-high dunes. Zones of maximum accretion varied from 0 m to 18 m behind the crest in low, stable dune sections, to as much as 164 m to 183 m to leeward of the crest in high, unstable sections. Ranwell estimated that the dune nearest to the shoreline would need at least 50 years to grow to maximum height. Its mean rate of travel inland near the coast was estimated at 6.7 m a^{-1}. At least another 20 years or so would elapse while the dune travelled sufficiently far inland for a new embryo dune to develop. As a result, the cycle between the start of successive episodes of dune-building would take some 70 to 80 years to complete. Linear ridges, such as those found at Newborough Warren, reflect its ideal position for maximum uniform erosion, but it is rare for whole ridges to migrate uniformly (Ranwell, 1972). More commonly, blowthroughs occur in parts of the ridge (as occurs, for example, at Braunton Burrows,

Newborough Warren

Morfa Harlech, Morfa Dyffryn and Ainsdale – see GCR site reports in Chapter 7). Even in this optimally located system, parts of the coastal dunes reach maximum height more rapidly than others. As a result, irregular erosion of ridges produces parabolic or U-shaped dunes. Ranwell (1960) suggested that the cycle between dune-building and slack formation is about 80 years, i.e. very similar to the coastal dune-building cycle, but did not examine the implications of this similarity.

The volume of sand supplied to the dunes has been substantial. There is evidence of considerable deepening of Caernarfon Bay since the mid 18th century (Ranwell, 1955). During this period, the shallow-water seabed profile moved landwards; Ranwell argued that this probably accounts for much of the sand transported into the dunes. Robinson (1980b) argued that the continuing stability of the recurved end of Abermenai Point indicates that most of its sediment is derived from the tidal streams at the mouth of Menai Strait rather than from longshore sediment transport.

The beach systems of Newborough Warren also warrant considerable further investigation both with regard to their geomorphological history and their present-day dynamics. In particular, the effects of further retreat of the shoreline around Llanddwyn Island, perhaps isolating it from the mainland, should be investigated in order to assess the effects upon sand transport between the two bays. Robinson (1980b) argued that the varying curvature of the shoreline of Newborough Warren reflects the differing degree of shelter from dominant south-westerly waves, the full impact of the dominant waves in the central part of the spit and finally the influence of the transportational efficiency of tidal streams in the entrance to the Menai Strait. He regarded the complex system of sandbanks and channels as demonstrating the influence of ebb and flow channels on shallow-water sediment transport. The flood channels cross North Sands and run eastwards along the Abermenai spit and so may carry sand to the stable end of the spit.

Newborough Warren differs from the other large dune systems at Ainsdale and Braunton Burrows (see GCR site reports in Chapter 7) in the contrasts between its east and west parts. The western part of Newborough Warren is partially underlain by bedrock and extends into a small estuary, in contrast to the eastern part, which includes a recurved spit that extends into a large, deep channel. The dune cycle is better understood than in many other dune systems. The western part of the site has been marked by several separate phases of ridge building, a feature found in other dune systems such as South Haven Peninsula (see GCR site report in Chapter 7). The site is most important because it is possible to relate the dune succession to the geomorphological processes associated with dune-building and erosion (Ranwell, 1972).

Conclusions

One of the largest west-facing dune systems in England and Wales, Newborough Warren includes a recurved spit at the mouth of the Menai Strait, which has been breached periodically but has a distal end that is probably about 700 years old. Newborough Warren includes both very dynamic and very stable and long-lived features. The effects of two estuaries on the dynamics of the beach, especially in providing major nearshore sources of sediment, makes this site particularly important for geomorphological studies. The breaching and rebuilding of the Abermenai spit indicates the availability of large sediment inputs, as do the continuing growth and migration of the dunes where they have not been stabilized by afforestation. The site combines in one location features seen at Spurn Head (breaching and rebuilding, see GCR site report, Chapter 8) and Hurst Castle Spit (the effects of different wave directions at the mouth of an estuary on the development of recurves, see GCR site report, Chapter 6) with the major dune-building processes of large west coast dunes such as Braunton Burrows (Chapter 7) and those in Carmarthen Bay (Chapter 11) and Cardigan Bay. Its particular interest comes, then, from its hybrid form rather than from any single feature. Within the coastline of England and Wales, it forms an important member of the network of dune–beach–spit structures that range from the simple (e.g. East Head, see GCR site report, Chapter 8) to this complex site.

Its ecological importance (it is a National Nature Reserve and part of a Special Area of Conservation) depends on the relationships between sand supply and colonization and stabilization by vegetation. With a steepening nearshore slope, there is the likelihood of increased change and retreat associated with migration inland of the dune system, and this is already occurring in both the south-east and west parts

MORFA DINLLE, GWYNEDD (SH 435 557–SH 450 612)

V.J. May

Introduction

Morfa Dinlle, on the southern side of the western mouth of the Menai Strait (see Figure 11.20), comprises a complex coast undergoing erosion together with shingle ridges and dunes. At Dinas Dinlle, low cliffs about 25 m in height expose folded and faulted Devensian glaciogenic sediments that provide evidence for a possible advance of the Late Devensian ice-sheet (Campbell and Bowen, 1989). The sediments of the cliffs are also important in providing evidence that help elucidate the development of the western end of the Menai Strait (Bedlington, 1995; Harris *et al.*, 1996). Marine erosion of glacial deposits south of Dinas Dinlle has supplied a heterogeneous mix of sediments to the mainly north-eastwards moving drift system along Pen Llŷn (the Lleyn Peninsula) (Carter, 1990; Pethick, 1997). At the northern end of the cliffs, the coast has been re-inforced to protect the Marine Hotel (Figure 11.20a,b). A single shingle ridge extends northwards from Dinas Dinlle for about 2.5 km and has been protected, since 1976, by gabion mattresses along the ridge crest between Dinas Dinlle and an airfield. In places the lower seaward face of the ridge has been undermined leading to collapse of the gabions. The low-lying area between Dinas Dinlle and Morfa Dinlle village is believed to have been formed by deposition of gravel ridges, but these have been obliterated by the construction of the airfield. Morfa Dinlle itself comprises a series of shingle ridges capped in parts by low sand dunes. This area forms the GCR site.

There has been very little research into the character and dynamics of Morfa Dinlle. Steers (1946a) recorded that there was no physiographical description of the spit and noted that Morfa Dinlle is 'bordered by shingle which fans out in normal fashion at the distal end' (p. 120). He also regarded it as despoiled compared with the unspoilt Newborough Warren. Pethick (1997) also described and interpreted the site, upon which the following account is largely based. Both Carter (1990) and Pethick (1997) believe that the shingle features have grown northwards from Dinas Dinlle probably since about 4000–5000 years BP.

Description

Although the cliffs and shingle ridge to the south of Morfa Dinlle are not included in the present GCR site boundary, the geomorphological development of features within the boundary has depended upon their dynamics. The cliffs south of Dinas Dinlle are retreating irregularly, with occasional slumps and slippages across the face. Since 1875, the cliff at Dinas Dinlle has retreated by about 20 m, giving an annual rate of recession of less than 0.2 m a^{-1}. However, the rate of production of shingle-sized sediment from the erosion of the Dinas Dinlle cliff is considerably less than the rate of shingle accumulation on the north shore of Morfa Dinlle. Carter (1990) provided a first approximation for the rate of shingle input from the erosion of the Dinas Dinlle cliffs, suggesting that 1000 m^3 a^{-1} is released by erosion, of which only 15% is gravel (shingle). This total gravel input of 150 m^3 a^{-1} from the cliff would be insufficient to allow the observed rate of growth of the ridges, and it is concluded that erosion represents only a small proportion of the total input of gravel to the modern ridge system.

The modern shingle ridge system, in the north of Morfa Dinlle, is connected to the Dinas Dinlle cliffs by a shingle beach ridge running approximately north–south for approximately 2.5 km before the ridges curve along a SSW–NNE line. Although this beach ridge has been protected by coastal defences and received artificial sediment nourishment, several washover fans suggest a potential for landward movement. It is likely that the gabions (in place since 1976) have restricted the natural movement of the ridge leading to their ultimate destruction. The hard point at the Marine Hotel may now be beginning to impede down-drift sediment transport and so starve the ridge, but the field evidence for this is hard to assess without further study (Pethick, 1997).

Morfa Dinlle is characterized by a well-developed series of sub-parallel shingle ridges, partly obscured by wind-blown sand. The ridges in the northern region of Morfa Dinlle appear to be grouped into three sets.

1. Along the modern shoreline the shingle ridges run shore-parallel and are of recent origin. Nine distinct gravel ridges may be identified west of the relict dune field on Morfa Dinlle.
2. North of Warren Farm the shingle ridges run south-west–north-east and merge with the deposits that form the peninsula south of Fort Belan.
3. South of Warren Farm, the ridges run almost north–south and may indicate a period when the tidal mouth of the Afon Gwyrfai was at the location of Warren Farm. One of the most distinctive features of Morfa Dinlle is its separation from the mainland by the tidal inlet Foryd Bay, the estuary of the Afon Gwyrfai. South of the line of high sand dunes, traces of shingle ridges extend up to and in some cases into Foryd Bay, two of these ridges appearing to merge with the unnamed peninsula south of Fort Belan.

Between 1980 and 1990 approximately 250 m of new ridge formed along the northern shoreline of Morfa Dinlle (Carter, 1990). Although much of this new ridge was formed during a single storm event, the average rate of its development was 25 m a^{-1} (Pethick, 1997). The average volume of shingle contained in 1 m length of the ridge is 60 m^3 (Carter, 1990) so that the modern rate of accumulation of shingle is 1500 m^3 a^{-1} (not 900 m^3 a^{-1}, as reported by Carter (1990) and Pethick (1997)), albeit probably deposited during a single event. Detailed surveys of the extreme landward and seaward ridges show that their crest elevation increases from east to west by 0.7 m (Pethick, 1997). They are partly obscured by sand dunes. High dunes form a single line some 300 m landward of and parallel to the present-day shoreline (trending approximately SSW–NNE). The maximum elevation of the dune crest is 14 m OD and average crest elevation is 10 m OD (Pethick, 1997). The dunes are formed over the shingle ridge basement, providing a highly permeable substrate so that the slacks are dry and deflation down to the underlying shingle is possible. As well as this line of high dunes, the area is characterized by extensive, low, sand dunes whose structure is again related to the underlying shingle ridges. These low dunes continue to form on the present-day shoreline as sand from the nearshore ebb delta ramparts is blown onshore. The wind carries sand over the unvegetated seaward shingle ridge to be deposited as new embryo dunes on the vegetated second dune ridge.

Interpretation

The interpretation of the features at Morfa Dinlle depends on evidence from present-day rates of change, the evidence of the shingle ridge patterns and the Holocene history of the wider area.

The volume of material entering the system from cliff erosion has not been determined accurately, but Carter's approximation (1990) indicates it could be around 800–1200 m^3 a^{-1}, of which about 15% is probably gravel. The receding cliff exercises an important control over the recession and planform of the gravel spit (particularly at its proximal end near Dinas Dinlle). Carter (1990) and Pethick (1997) estimate that the solitary barrier is retreating landwards at a long-term (over a timescale of several centuries) rate of about 0.2 ma^{-1}, probably by phased storm overwashing. Pethick suggests that this landward movement continued during most of the Holocene Epoch. As the beach ridge transgressed the western extremities of the ridge systems, they would have been exposed on the shore and their sediments reworked and incorporated into the beach ridge. Longshore movement would then carry this reworked sediment to the north to form new ridges. A small proportion of material entering the system may also come from the seabed adjacent to the beach.

Harris *et al.* (1996) propose that the Dinas Dinlle hills, south of Morfa Dinlle, formed part of a more extensive push-moraine complex that extended westwards into the nearshore. The Dinas Dinlle moraine is one of a number of morainic ridges, possibly four in total, cut by the present-day coastline. They are composed of till units lying below an upper sand and gravel facies that would act as an easily eroded sediment source as Holocene sea level rose. This source was, and to some extent still is, responsible for the sediments that constitute the Morfa Dinlle complex (Pethick, 1997). The large quantities of sand and gravels produced by erosion of the morainic ridges during Holocene sea-level rise (perhaps 7000–6000 years BP) were moved northwards by prevailing longshore drift to form a series of spits connected to the seaward end of each of the morainic ridges. The rock-head immediately seaward of the present-day Dinas Dinlle coastline lies at −35 m (Harris *et al.*,

1996), suggesting a considerable depth of glaciogenic and Holocene coastal deposition.

It is also possible that a further moraine extended across what is now the mouth of the Afon Gwyrfai and formed the peninsula immediately south of Fort Belan. Pethick (1997) suggests (based upon preliminary study of surficial deposits and morphology) that such a morainic ridge would explain the complex topography of both the north-eastern area of Morfa Dinlle and the tidal section of the Afon Gwyrfai.

Pethick (1997) argued that the gravel ridges and sand dunes of Morfa Dinlle are a late Holocene phenomenon, certainly dating from post-4000 BP and probably much later than this. Dating the ridges themselves has, however, not been possible. As sea level continued to rise in the period 6000–4000 years BP, the coastline was forced eastwards and the continued erosion of the morainic ridges provided abundant sediment for the northward extension of the spits that consequently merged to form a single gravel beach between each of the morainic ridges to the south of Dinas Dinlle and extending to the north, perhaps as far as the present-day airport.

Peat deposits found in the intertidal area immediately west of Dinas Dinlle (Carter, 1990) are thought to date from 4000 years BP and confirm that a brackish-freshwater deltaic environment existed here at that time. This evidence, together with estimates of long-term cliff retreat, suggested to Carter that the coastline was then over 1 km west of its present-day position and that the gravel beach had already limited marine incursions to the east, although the Afon Gwyrfai would still have reached the open sea through a tidal inlet north of Dinas Dinlle (Figure 11.20a).

Successive shingle ridges extend north and east from the cliffs at Dinas Dinlle across the low marshlands towards this tidal inlet. Pethick (1997) conjectured that between 6000 and 4000 years BP the Afon Gwyrfai tidal inlet was pushed gradually northwards as the gravel beach continued to extend from Dinas Dinlle. However, the Gwyrfai was prevented from flowing north on its present-day course by the presence of the moraine that extended across the mouth of the present-day tidal mouth of the Gwyrfai from the eastern shore of Foryd Bay to just south of Fort Belan.

As the gravel spit extended northwards and eastwards, so the tidal mouth of the Gwyrfai was increasingly confined between the distal ends of the gravel spit and the moraine. At some stage it appears from the topographical evidence that the northern end of the gravel spit joined the western end of the moraine and blocked the mouth of the Gwyrfai. An extensive brackish lagoon was initially formed in Foryd Bay, but the waters of the Afon Gwyrfai eventually breached the moraine and tidal flow into the bay was re-established. Further detailed research is essential to test the validity or otherwise of the hypothesis. More recently, extensive land-claim of intertidal areas within Foryd Bay has reduced the tidal prism. The impact of these changes on the tidal entrance to the Bay may have been to reduce the overall dimensions of the tidal opening by northward extension of the shingle ridges. The impact of the changing tidal prism of Foryd Bay on the morphology of the Menai Strait is less obvious, owing to the relative discharges involved.

As the mouth of the Menai Strait narrowed, the coastal gravel beaches of Morfa Dinlle steadily advanced northwards and eastwards. The eastward movement of the coastline, which also resulted in the continued erosion of the morainic cliffs such as those at Dinas Dinlle, caused reworking of the gravel beach ridges as they were rolled landwards. Fresh sediments, eroded from the cliffs from Maen Dylan and Dinas Dinlle, were added to this reworked material. Today, however, the supply of new sediment from these sources is considerably less than the sediment inputs that were available from glacial debris present in early Holocene times.

As sediment moves north it falls more and more under the influence of the sediment circulation patterns of the Menai Strait. There is almost certainly a long-term exchange of material between the shoreline and the offshore area which, when understood, should explain the observed shoreline changes, including the supply of sand for the development of dunes above the gravel ridges. The evidence suggests that Morfa Dinlle is an active gravel-beach system, albeit with a relatively low rate of sediment input. This type of situation is increasingly unusual in England and Wales (especially on the west coast), since, over the past two centuries, human activities (notably shore protection) have acted to restrict sediment sources. Measurement of shingle characteristics over the northern sequence of shingle ridges shows a weak relationship between elevation and shingle mean diameter, and grain size increases towards the modern coastline. In general however, the shingle grain-size distribution seems to indicate a lack of pronounced structure suggesting that in-situ reworking of ridges has taken place, and lending support to the offshore seabed source hypothesis outlined above (Pethick, 1997).

The outgrowth of the gravel ridges supports an extensive 'dry-core' dune system in which the water table is usually below the deflation level, so that standing water is rarely, if ever, found in the system. The dunes have a degree of natural instability associated with geomorphological changes, themselves associated with grazing, pedogenesis and impact of human activities (Carter, 1990; Pethick, 1997). The main line of dunes is a relict formation (Carter, 1990) which has no direct sand supply from the beaches at the present time. However, the occurrence of blowthroughs suggests that some redistribution of sand is occurring and an extensive marram *Ammophila* cover exists. The crestline of the relict dunes follows a distinctive rectilinear line (Figure 11.20b) that seems to be caused by the interaction between the dunes and the underlying shingle ridge structure. The dune crestline appears to be held in position by the underlying shingle structures but, because the orientation of dune crestline and shingle ridges is slightly offset, at intervals the dune crests 'jump' from one underlying shingle ridge to another so forming the characteristic rectilinear pattern (Pethick, 1997). Reasons for the offset between dune crest and shingle ridge crests may be due to the difference in the prevailing wind direction, responsible for the dune orientation, and the orientation of the shoreline on which the shingle ridges formed. Wind-waves approach the shore at an oblique angle, so driving longshore currents towards the north. Further research on these dune systems is needed to interpret the sequence of coastal changes and related climatic variation.

The shingle ridges of Morfa Dinlle, with their superficial dune fields, represent the morphological response to two major postglacial events: the drowning of the Menai Strait to form a tidal estuary and the erosion of a series of glacial moraines to the south. The chronology of events is difficult to determine but it is suggested that the development of a number of shingle ridges along the open coast pre-dated the formation of the Menai Strait. As tidal flow was initiated into the Strait, so these shingle features extended and coalesced to form, in conjunction with the Abermenai spit, the mouth of the Strait. The subsequent decrease in the tidal prism of the Menai Strait, owing to sedimentary deposition, led to the progressive decrease in the width of its tidal entrance and consequently to the northerly movement of the Morfa Dinlle shingle ridges. The spatial pattern of shingle ridges displayed in the area consequently provides a record of the complex Holocene history of this region.

Conclusions

In spite of artificial protection at the southern end, Morfa Dinlle is now one of the last active drift-aligned gravel-ridge systems in the west of England and Wales. The dunes and the gravel ridges are of international geomorphological interest, because the shingle ridges complex of Morfa Dinlle together with the integral Newborough–Abermenai shingle system and their superficial sand dunes represent an extremely important, but relatively rare, geomorphological feature. Although single gravel ridges are widely distributed along the UK coast, few multiple ridge systems exist. Of these, the Dungeness (see GCR site report in Chapter 6), Culbin and Morrich More (see site reports in the present chapter) systems are the most extensive and best-known.

The geomorphological importance of the Morfa Dinlle site also rests in the topographical record of Holocene development of the shoreline of north Wales and in particular the Holocene development of the Menai Strait. The present-day pattern of shingle ridges provide an important record of the development of the Menai Strait during the Holocene Epoch, since their morphologies are directly related to tidal and sedimentary conditions in the Strait.

As a consequence of the relationship between the geomorphology of the Menai Strait and its western tidal entrance, the Morfa Dinlle and Newborough–Abermenai dunes and gravel ridges must be seen as integral components of a single system, defining the mouth of the Strait and responding to past changes. The relationship between the Morfa Dinlle ridges and the tidal dynamics of the Menai Strait, recorded in the topographical features of this site and the adjoining Abermenai–Newborough Warren area that together form the mouth of the Strait, is of international importance. The relationship between the mouth area of an estuary and its tidal dynamics is central to an understanding of estuarine management and, owing to the loss or destruction of comparable sites elsewhere, the Morfa Dinlle–Abermenai sites provide a unique opportunity for research into this complex interaction of open coast and tidal geomorphology.

HOLY ISLAND, NORTHUMBERLAND (NU 035 481–NU 171 362)

V.J. May

Introduction

The Holy Island GCR site (see Figure 10.1 for general location) includes one of the largest sandy beaches on the coastline of England and Wales. About 25% of the coastline between Edinburgh and Whitby is formed by predominantly sandy beaches (European Commission, 1998). To the north of Holy Island, the coast is predominantly cliffed, whereas to the south, hard-rock cliffs alternate with small sand beaches backed by narrow lines of dunes. The largest of these beaches is at Druridge Bay, but it lacks any substantial geomorphological interest, other than the effects of the removal of sand from its foreshore between 1960 and 1996.

Holy Island forms part of a suite of large sandy beach and dune systems along the British east coast that include the north Norfolk coast, Gibraltar Point, Lincolnshire, and the Sands of Forvie, and Rattray Head in Scotland. Similar to other English beaches, it is relatively narrow in comparison to larger Scottish dune systems such as the Sands of Forvie and the much more extensive beach–dune systems of the west coast. Unlike all except Rattray Head, its outline is controlled by the presence of major rocky outcrops that act as hinge points for sediment deposition and beach development. It lacks large amounts of gravel, although parts of the dunes lie upon a gravel base. It is dominated by progradation and there has been no interference with coastal processes by protection works and its relative remoteness has restricted pressures from recreation.

Geomorphological interest in the site has been comparatively limited, although Steers (1946a) drew attention to its considerable potential for research, and Carruthers *et al.* (1927) described the main features of the site. More recently, Robertson (1955) described the main ecological features of part of the site at Ross Links; Farquhar (1967) identified it as a key locality for tied island development; and King (1976) outlined its main geomorphological features. None of these later workers considered the site as a unit, though this is how Steers (1946a) described it. The description and discussion that follow regard the site as a single complex entity.

Description

A key area for coastal geomorphology, the Holy Island GCR site comprises three main units (Figure 11.23):

1. the dunes and the barrier beaches of Cheswick and Goswick Sands,
2. the dunes of the Snook and the cliff-top dunes and cliff–beach system on the north coast of Holy Island, and
3. the dunes and sandy beaches of Ross Links and Budle Bay.

In addition, there are hard-rock cliffs, an emerged ('raised') Holocene beach, saltmarsh and intertidal sandflats and mudflats. The site extends for about 20 km from Far Skerr in the north (NU 035 481) to Bamburgh in the south (NU 171 362). In the north, a predominantly sandy beach and dunes extend south-eastwards across Cheswick and Goswick Sands diverting eastwards the northern channel of the intertidal flats that lie between Holy Island and the mainland. The central part of the site is dominated by Holy Island whose eastern cliffs, cut into limestone and shales, provide the only erosion-resistant feature in the central part of this large site. Holy Island extends westwards in a large dune area known as 'the Snook'. South of Holy Island the main tidal inlet divides the rocky southern shore of the island from the northern sandy beaches of Ross Links. The plan form of Ross Links results from the wave refraction and energy distribution between Holy Island and Bamburgh. A prograding sandy shoreline extends southwards as a low sandy spit across the mouth of Budle Bay, a small tidal inlet floored mainly by sandy sediments. Its southern shore is formed by dunes banked against a hard-rock cliffline,

Holy Island

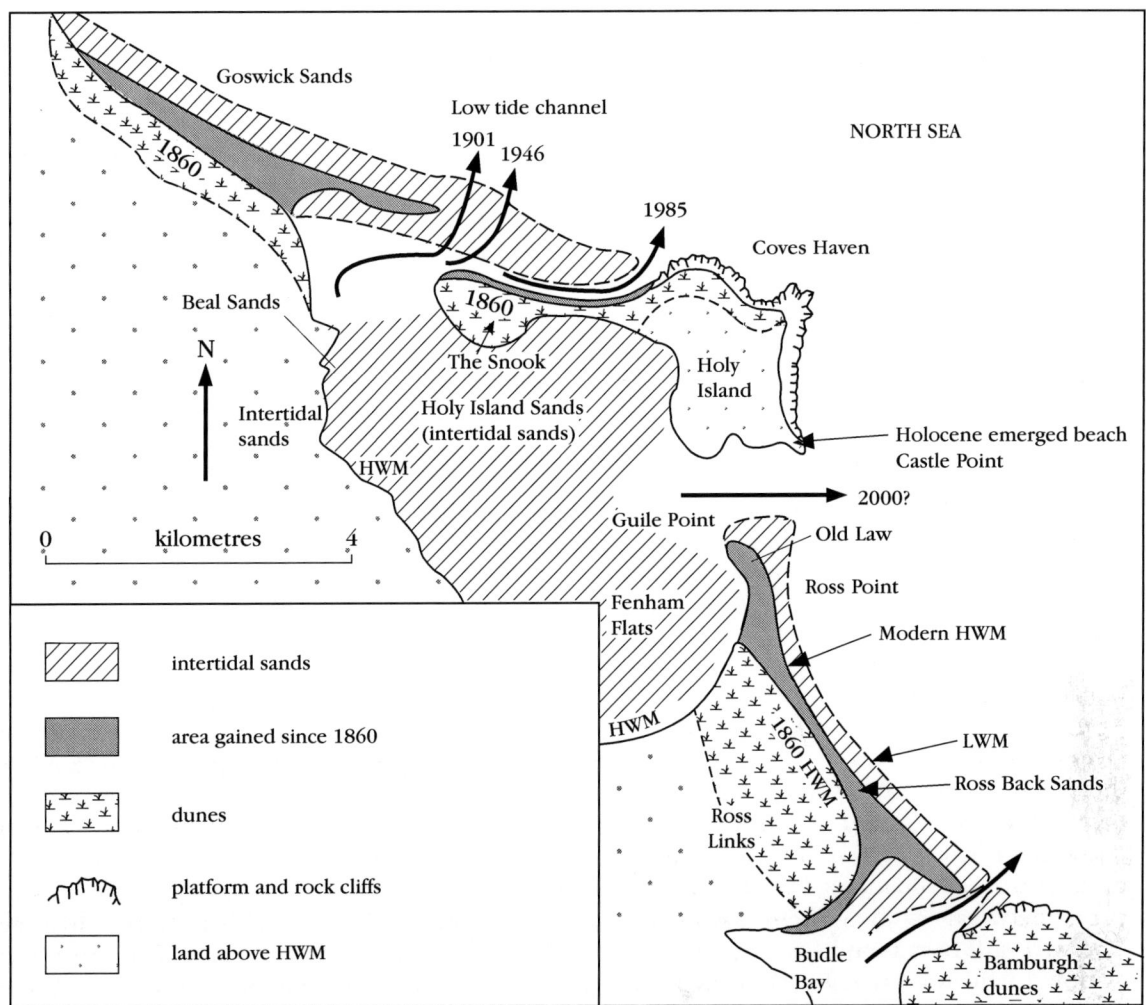

Figure 11.23 Sketch map of the key geomorphological and historical changes to Holy Island. Bold arrows show the dates of the main channels draining estuaries.

whose easternmost extremity is formed by the Whin Sill.

Despite the limitations of cartographic evidence, Ordnance Survey maps and plans of the area from the mid-19th century onwards show that there has been considerable accretion at both Goswick Sands and Ross Links (Figure 11.23). Steers commented (1946a, p. 452) that 'Unfortunately the physiography and ecology of the coast between Black Rocks and Budle Point have not been fully investigated. They should afford many interesting problems for research purposes.' There appears to have been no comprehensive examination of the coastal geomorphology of the whole site. Bird (1985) suggested that, similarly to the northern side of Rattray Head, the Sands of Forvie and the north Norfolk coast, the sand accumulation here may be derived from the seabed.

At its northern end at Cheswick and Goswick the site is bounded by low rock-cliffs formed in Lower Carboniferous shales. Rather more resistant limestone beds (Lowdean or Sandbanks Limestone) form small headlands and a number of reefs that extend seawards from the cliffs. About 250 m of reefs are exposed between high and low tides, acting as low groynes. No other exposures of bedrock occur for some 9 km to the south, until Holy Island itself.

Between Cheswick–Goswick and Holy Island, the shoreline is formed in sand. There are four main zones:

1. The landward side of the site is formed by dunes that were in place by the middle of the 19th century. Their maximum height reaches about 18 m at their northern end, where they rest on bedrock, but they decline in height towards the south-east. Over most of this area the dunes rarely exceed 8 m OD.
2. Seaward of this zone, there are several narrow lines of active dunes and an intermittent line of low vegetated embryo dunes.
3. This line is continued south-eastwards in Goswick Sands as a barrier beach extending towards Holy Island.
4. The final zone is an active intertidal beach that appears to be dominated by progradation. The sand that stands above high-water mark has gradually increased in area from the mid-19th century and extended towards Holy Island. It lacks any evidence of recurves and appears to have extended not as a result of longshore transport, but as a result of a gradual onshore accretion.

Over the same time period, the channel draining the northern part of the Beal Sands has been diverted eastwards and its dimensions have decreased (Figure 11.23). The shoreline alignment is strongly dominated by a swash-related curve between its two rocky extremities at Cheswick and Holy Island.

Holy Island itself was described by Farquhar (1967) as a situation where sand spits and sandbars were prevented from joining the island to the mainland by tidal streams. He also referred to 'the breached bars connecting Holy Island to the Northumberland coast' (p. 120). Steers (1946a) and King (1976) noted that shingle beaches have joined what were originally three or more separate islets to form the present-day Holy Island. Galliers (1970) suggested that the outline of Holy Island and its westward projection at the Snook have changed little between the publication of a map in 1610 and the present day.

The eastern part of Holy Island is formed mainly by Lower Carboniferous shales and thin, limestone strata, including the Lowdean or Sandbanks Limestone. Much of its surface is also covered by till and emerged beach sediments. On the northern side of the island, these are covered by dunes. In bays such as Coves Haven (Figure 11.26), where the dunes are aligned from north-west to south-east, and around Emmanuel Head, the shoreline is formed in part by vertical bedrock cliffs up to 18 m in height and by dunes banked against the underlying rock surfaces. The eastern shoreline is marked by several benches and rocky outcrops. The intertidal area is formed by boulder fields and some rocky platforms. At Castle Point, cobble ridges occur at both low-water mark and high-water mark, the higher ridge including several small recurves. King (1976) described a low gravel terrace of sub-rounded to rounded gravel at 3.6 m OD rising landwards to unweathered gravels at 5.5 m OD. This site has been identified as the only known emerged beach of Holocene age on the east coast of England (see GCR site report in Huddart and Glasser, 2002), and is an important reference site as an isostatic marker. King (1976) described the Snook as underlain by gravel ridges up to +3.6 m OD (a similar level to the Holocene beach) overlain by dunes that reach about 15 m OD. The dunes form both low ridges about 10 m in height roughly parallel to the shoreline (i.e. east–west) and crescentic-shaped mounds that vary in height between 10 m and 12 m. Absent from the earliest maps of the area, the Snook appears on 18th and early 19th century maps as a series of separated ridges. At Snipe Point, the dunes are aligned south-west–north-east. King (1976) suggested that the Snook dunes receive sand from the ridges to the north and north-west, that is Goswick Sands, and from the Holy Island Sands to the south. According to King, the symmetry of the dunes may reflect these two sources. The underlying gravels are exposed only in hollows, these forming most commonly where recreational access has produced local deflation.

South of Holy Island, the shallow lagoon between Holy Island and Ross Links drains the intertidal flats that are almost enclosed by the Snook and Goswick Sands to the north. Although there is some saltmarsh, much of this area is formed by extensive sandy and muddy areas crossed by very well-developed dendritic channel patterns. The more elevated sections of sandflat, where tidal inundation is of lower frequency and duration, are colonized by saltmarsh vegetation and in the west, dense stands of common cord-grass *Spartina* exist. Guile Point is the northernmost part of Ross Links, but the dunes that cover it are broken at Ross Point. There are several small rocky outcrops close to low-water mark seaward of Guile Point. Most of Ross Links is underlain by till, generally at about +3.8 m

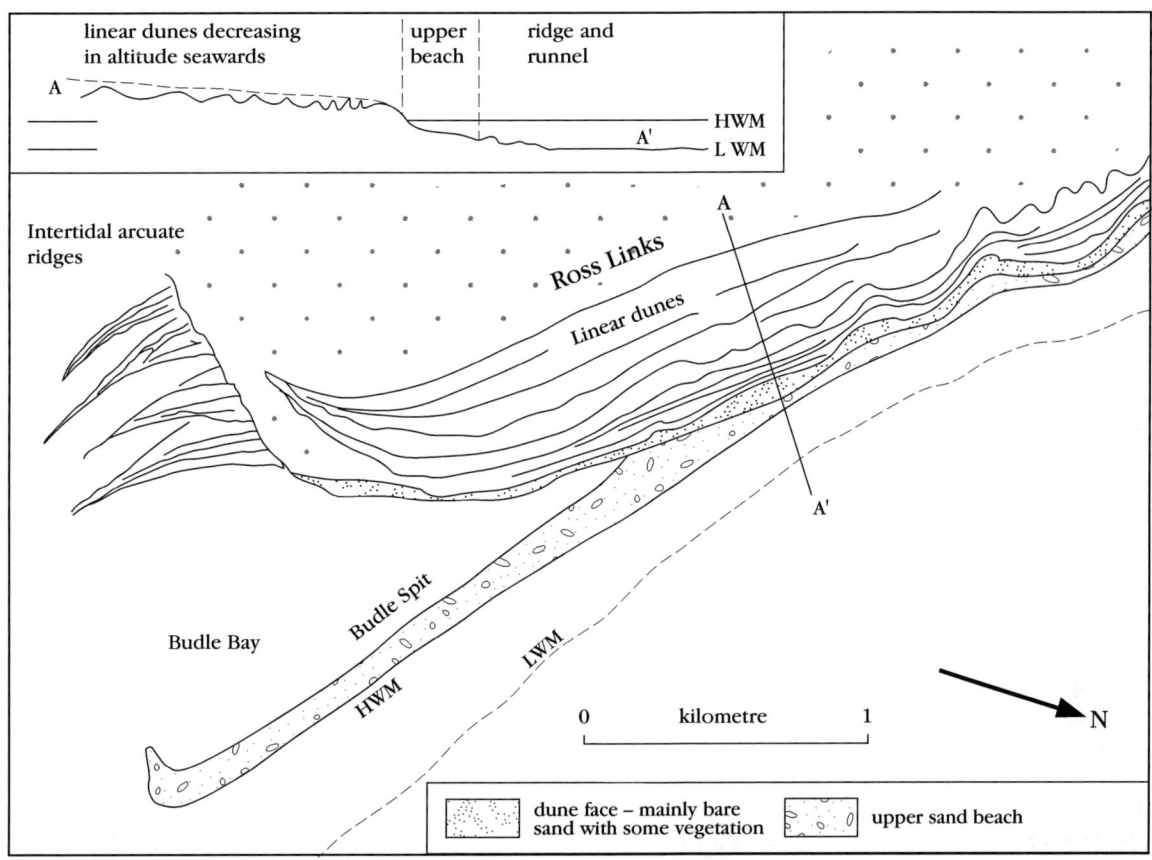

Figure 11.24 Cross-section and profile of the main geomorphological features of Budle Bay and Ross Links. (based in part on interpretation of aerial photographs, see Figure 11.25.)

OD, which is exposed between Guile Point and Ross Point. The relative resistance to erosion of this material has played a part in the recent development of Ross Links since it has fixed the northern end of the sand shoreline curve. Robertson (1955) divided the dunes of Ross Links into four zones from west to east:

1. The oldest part of Ross Links, wind-blown sand is underlain by glacial sand, regarded by Robertson as late-Glacial. The $CaCO_3$ content of this sand is very low. Blowthroughs expose buried podzols throughout this zone, the upper one having Bronze Age pottery in its A1 horizon. Brewis and Buckley (1928) suggested this surface could as a result be dated at about 3600 years BP.
2. Robertson's (1955) 'ancient beach' in which the sand is 'distinctly calcareous'. This has similar blowthroughs to the previous zone and is separated from the next zone by a single almost continuous dune ridge. Long Bog represented this feature best in Robertson's view because farther north it had been covered by dunes and was directly observable only in the blowthroughs.
3. The main area of Ross Links formed by linear dunes that rise from about +7 m OD at the landward boundary of the site to over +18 m OD. From a single ridge about 18 m high at Ross Point this zone widens from about 30 m to over 600 m in the south. There are between 7 and 16 sub-parallel sand dune ridges, each of which marks a period of dune-building (Figures 11.24 and 11.25). At the time of Robertson's 1955 survey the southern part of this zone was about 550 m wide and there were only 14 ridges.
4. The beach, Ross Back Sands and the sand-bar across Budle Bay. Robertson believed that this bar had built up following the 1953 storm surge, but it is an identifiable feature on 19th

Coastal assemblage GCR sites

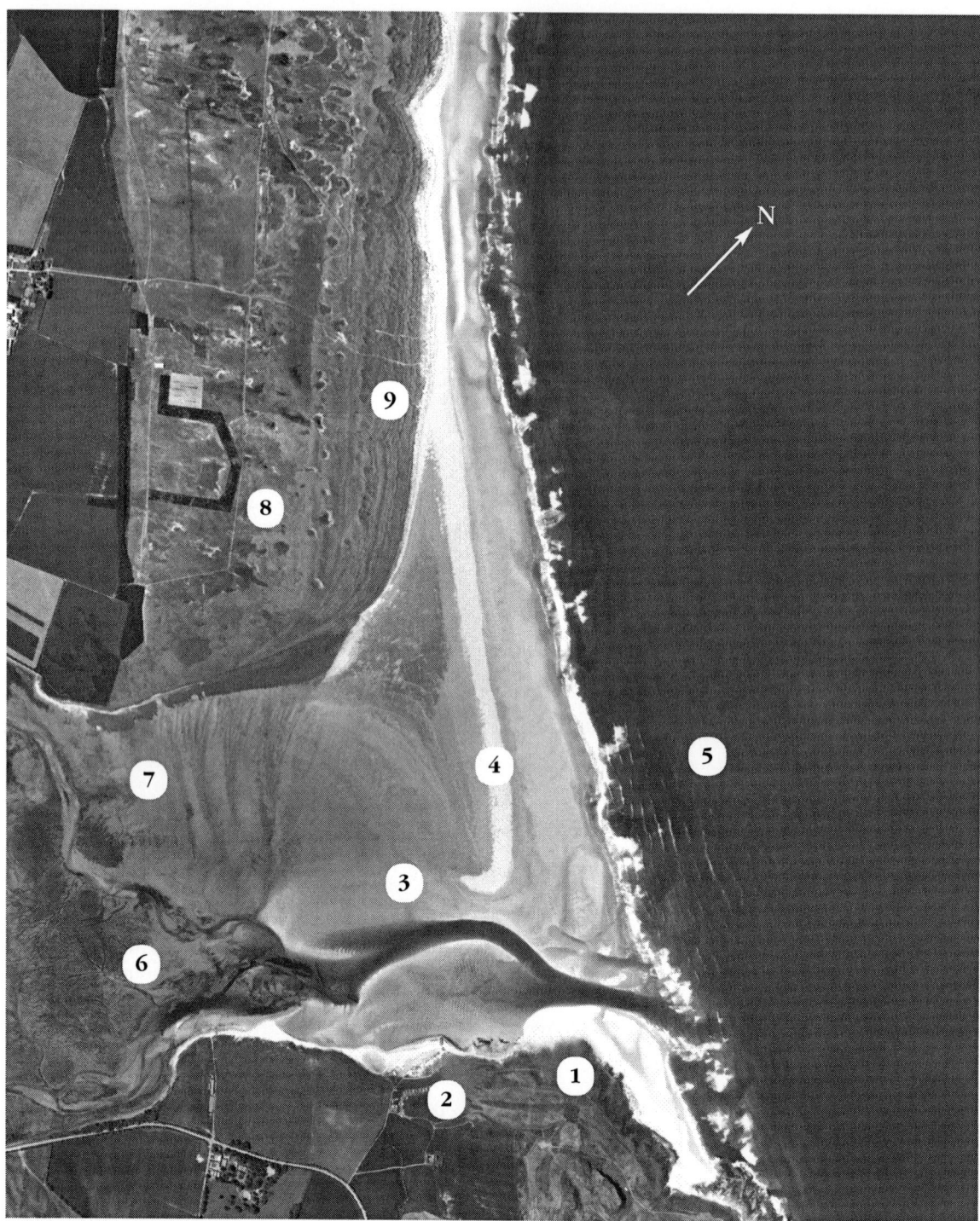

Figure 11.25 Aerial photograph mosaic showing the main features of Ross Links and Budle Bay. 1, cliff-foot dunes; 2, sand-waves in Budle Bay; 3, intertidal sandflats and mudflats; 4, Budle spit; 5, prevailing wave direction; 6, saltmarsh and intertidal mudflats; 7, possible former beach ridges; 8, dunes of Ross Links; 9, linear shore-parallel dunes decreasing in altitude towards shoreline. (Photo courtesy Cambridge University Collection of Aerial Photographs, Crown Copyright, Great Scotland Yard.)

Holy Island

Figure 11.26 Coves Haven, on the northern coast of Holy Island. The underlying Carboniferous Limestone is covered by till and emerged beach sediments, which are covered by dunes aligned from north-west to south-east. (Photo copyright English Nature.)

century maps.

King (1976) suggested that the curve of Ross Links shore results from the pattern of refracted swell. The refraction patterns would tend to move sand to north and south, and this accounts for the development of spits in both directions. The shoreline appears to have advanced as much as 250 m during the last 100 years (2.5 m a^{-1} is among the more rapid rates nationally). There has been much greater accretion across Budle Bay where a sand ridge narrows the mouth of this estuary to under 300 m at high-water mark. Evidence that this is the result of longshore transport is sketchy, and the detailed surface forms suggest that this beach may be the result of gradual seaward building of the shoreline. The low-water mark has always had an outline that has been tied strongly to the more resistant features of Guile Point and Bamburgh. It appears that during the last few decades sufficient sand has accumulated to sustain this beach and build it across the estuary.

Budle Bay and its southern shoreline show considerable evidence of sand movement by currents on parts of the intertidal estuary floor, and wave and wind action where sand has been banked up against the rocky outcrops (Figure 11.25). Here, as elsewhere on the site, the sand appears to have been derived from offshore, since there are no large inland or longshore sources.

Interpretation

Holy Island has been given surprisingly little attention by coastal geomorphologists, yet it is one of a small number of sites where accretion appears to dominate. On the coastline of Great Britain, fewer than 25% of all beaches are accreting. Sites in which progradation occurs throughout the site are thus very rare, for many are characterized by both erosion and accretion (e.g. see site reports for North Norfolk Coast, Dungeness, South Haven and Morfa Harlech). Growth in the distal parts of these beaches usually takes place at the expense of their landward parts. This is not the case in the Holy Island site (although sediment arriving here has been eroded from elsewhere). Barrier beaches are also rare on the British coast. In this site, the beaches at Cheswick and Goswick have many of the characteristics of such barrier beaches, i.e.

narrow strips of low-lying land formed entirely of beach sediments and frequently overwashed by waves. However, the Holy Island barriers are more properly described as bay barriers since they enclose embayments north and south of the island, unlike true barrier islands that lie separately from the land mass. The most important feature of the Holy Island barriers is the lack of significant longshore sediment feed to them. They appear to have grown primarily as a result of the addition of sand to the seaward face. Lengthening alongshore, which characterizes the Goswick Sands, is a function of beach growth in gradually deepening water rather than of spit extension. However, Hansom (pers. comm.) suggests that there is a southerly feed into these beaches.

The origins of the plentiful sand both offshore and nearshore have not been investigated in detail. One possible source is the reworking of glacial sediments filling depressions in the Carboniferous seabed rock surface, such sediments having a high ratio of glacial sand and gravel to till (Clayton, pers. comm.). There are only very limited fluvial sources. Robertson (1955) drew a distinction between the non-calcareous shell-free sand of the inner part of Ross Links and the calcareous sand containing marine shells, which formed his 'ancient beach' that developed in the British post-Mesolithic era (c. 8000 years BP). Since Mesolithic times, relative sea-level change along the Northumberland coast has been only about 2.6 m (Plater and Shennan, 1992). Plater and Shennan identified a transgressive phase up to 7630–7970 years BP, but consider that low rates of relative sea-level change (<1 mm a^{-1}) combined with local variations in sediment supply have been the most important processes here.

Holy Island differs from other large progradational sites in lacking extensive development of saltmarsh behind the beaches. Human activity along the beaches has been minimal. There has been no coast protection and there is little evidence that land-claim has been a significant process. There is saltmarsh in both the ineer Budle Bay and along the western shore of the National Nature Reserve (NNR). Parts of the dunes at Ross Links bear the scars of past use as a bombing range. Nonetheless, the site has some of the most pristine features anywhere on the English coast. The site's similarities to the features of Rattray Head on the Scottish coastline make for interesting comparison because Holy Island appears to represent along its northern side conditions comparable to an earlier stage of the development of Rattray Head and Strathbeg (see Chapter 8).

The southern extension of the sand ridge at Budle Bay poses a question as to its origins. It appears to result from gradual accumulation of sand across the bay as an extension of the shoreline curve to the north. There is, however, one piece of evidence that conflicts with this hypothesis. The Geological Survey Memoir (Carruthers et al., 1927) describes an area in a similar location to the sand ridge as 'raised beach', although Steers (1946a) was unconvinced by the account in the Memoir. If such materials were exposed in the past there is now no surface evidence of them. However, they would provide a base for the development of the present-day sand ridge. Further investigation of this location is needed to elucidate its history.

Both the cartographic evidence and the progradation of the dunes and beach throughout their length argue against any significant redistribution of sand alongshore. Although the development of the sand ridges, at both Ross Links and Goswick Sands, could be seen as resulting from longshore transport, both sediment cells are dominated by overall accretion. If longshore transport is occurring, there must be substantial inputs of sand to the beaches at Cheswick in the case of Goswick Sands and at Old Law in the case of Ross Links. There is some erosion from Far Skerr northwards towards Berwick, but it appears too limited to provide the volumes needed for the growth of the Goswick system. At Ross Links, the older dune ridges all have a predominantly linear sub-parallel form and there is no evidence of old recurves within the dunes. Robertson (1955) argued that on the basis of cartographic, soil and archaeological evidence that these ridges had formed between the beginning of the 16th century and the middle of the 18th century. The whole system is dominated by progressive movement seawards. Even considerable surface damage to the dunes as a result of bombing has not initiated shoreline retreat. There are, however, blowthroughs throughout the dunes mostly with an alignment SSE–NNW. The dune and beach system appears to have a strongly positive sediment balance.

Farquhar's (1967) suggestion that this site has been affected by breaching of beaches thus separating the islands from the mainland is not supported by either the cartographic or the field evidence. The only point at which there appears to have been an erosional break in an otherwise symmetrical shoreline is at Ross Point. Here the mid-19th century high tides appear to have passed between Ross Point and Old Law. Such a cut in the coast occurred in the underlying till and not in the beach. Robertson (1955) used cartographic evidence to suggest that although Old Law first appeared on Armstrong's map of 1769, it had been separated from the mainland by the time of Fryer's 1820 map. There is now sufficient accumulation of sand to ensure that the shoreline is a continuous one. There is no other evidence of breaching.

There is evidence in the cobble beaches of Holy Island itself of higher relative sea levels, but the general reduction in altitude of the dune ridges on Ross Links may be indicative of a falling relative sea level. However, such a hypothesis requires a fuller investigation of the site.

In summary, four issues needing further research are raised by the features of the Holy Island site, i.e.

1. the relationship of the dunes and beaches to any underlying till,
2. the development of the tied islands by beach growth or breaching,
3. the sources of sand for the substantial progradation at this site, and
4. their relationship to changes in sea level and wave climate.

Conclusions

Sand spits and barrier-type beaches characterize this predominantly prograding site. Holy Island is unusual in Britain in combining tied islands with barrier-beach development. The positive sediment budget for the site cannot be explained by longshore sediment transport alone and so an offshore source has to be postulated. The early development of Rattray Head in Scotland appears to have followed a similar pattern.

The significance of Holy Island lies, first, in the extensive progradation of sandy beaches, a rarity not only worldwide, but also on the coastline of Britain. Second, it illustrates well the role of different wave energy distributions through its contrasting beach forms and processes to the north and south of Holy Island. Third, the total assemblage and variety of contemporary and older coastal features makes it unusual. Fourth, it is a rare example of tied islands, in which several rocky islands have been joined by beaches. This is a very unusual form in England and Wales, although it is more common in Scotland. Finally, this site is one of only three locations in England and Wales where barrier-type beaches occur and is the only one that co-incides with conditions of coastal emergence.

The site is also of national and international importance as a National Nature Reserve, Special Area of Conservation (SAC) and Special Protection Area (SPA) under the Habitats and Birds Directives, a Ramsar site and a site of great archaeological and historical importance.

NORTH NORFOLK COAST, NORFOLK (TF 673 413–TG 153 437)

V.J. May

Introduction

The North Norfolk Coast GCR site extends from Hunstanton to Sheringham (see Figure 5.13). It includes not only internationally renowned locations such as Blakeney Point (see Figure 10.9) and Scolt Head Island, but also many smaller, but no less significant, beaches that form an integral part of the coastal system. Much of the site is characterized by a low upland fronted by gently sloping abandoned cliffs separated from sand and shingle beaches by extensive saltmarshes and intertidal flats. The saltmarshes of north Norfolk have been described as the finest coastal marshes in Great Britain (Steers, 1946b) and are among the best-documented and researched in the world. The marshes exhibit a progression of age and development from east to west, manifested through changes in marsh height and assemblage of geomorphological features. Creeks, saltpans and marsh stratigraphy are well exhibited on the north Norfolk marshes. The marshes have been a prime research site for the investigation of rates of saltmarsh accretion and tidal creek processes. At both the east and west ends of the site the beaches rest against retreating Chalk cliffs. Together with the intertidal flats and saltmarshes, the beaches of north Norfolk form one of the outstanding assemblages of coastal forms in Britain. Each of the

Coastal assemblage GCR sites

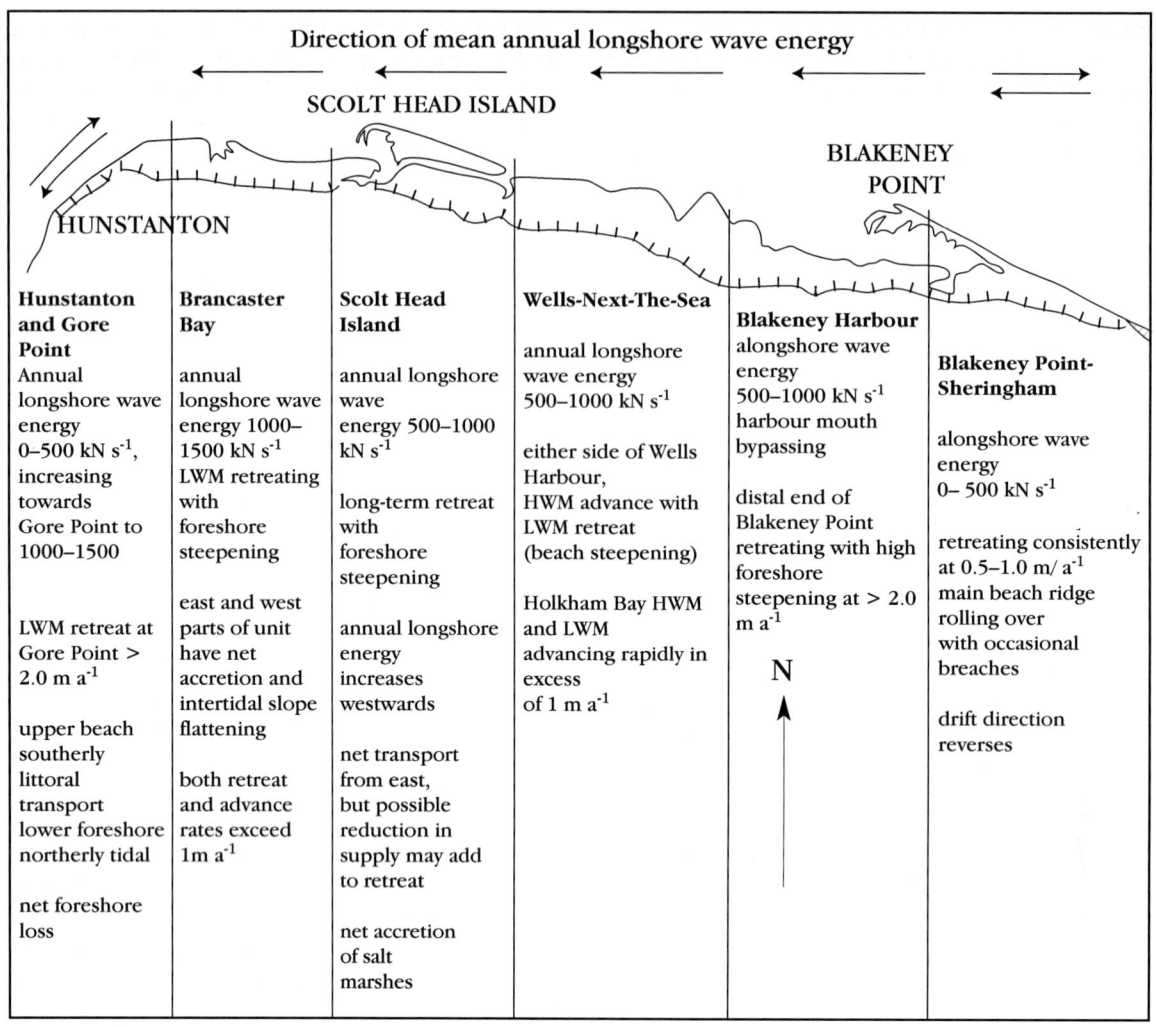

Figure 11.27 Summary features and recent dynamics of the North Norfolk Coast GCR site from Hunstanton to Sheringham, and east to west distance of about 40 km. (After Halcrow, 1988.)

major features is important in its own right: together they are of the highest importance. They have been extensively researched and are internationally famous (Redman, 1864; Wheeler, 1902; Oliver, 1913, 1929; Oliver and Salisbury, 1913; Hill and Hanley, 1914; Kendall, 1926; Steers, 1926a–c, 1927, 1929, 1934a–c, 1935a,b, 1936a,b, 1938a,b, 1939a, 1940, 1942, 1946a,b, 1948a,b, 1951a, 1952, 1953b, 1954, 1960, 1964a,b, 1971b, 1977, 1981; Steers and Kendall, 1928; Steers and Thomas, 1929a,b; Steers and Slater, 1932; Chapman, 1939, 1959; Burnaby, 1950; Grove, 1953; Steers and Grove, 1954; King, 1959, 1972b; Peake, 1960; Williams, 1960; Kidson, 1961; Hardy, 1964; 1966; Ranwell, 1964, 1968, 1972; Steers and Haas, 1964; Evans, 1965; Battarchaya, 1967; Roy, 1967; Zenkovich, 1967; Cambers, 1973, 1975; Pethick, 1974, 1980a,b, 1981, 1984, 1992; Barnes, 1977; Banham, 1979; Bayliss-Smith et al., 1979; McCave, 1978a–c; Murphy and Funnell, 1979; Straw and Clayton, 1979; White, 1979; Barfoot and Tucker, 1980; Bird, 1984, 1985; Bird and Schwartz, 1985; Goudie and Gardner, 1985; Carter, 1988; Funnell and Pearson, 1989; Stoddart et al., 1989; French et al., 1990; Pye et al., 1990; Bridges, 1991; Allen and Pye, 1992; French and Stoddart, 1992; Pye, 1992; French, 1993; Allison and Pye, 1994). The site is amongst those most quoted by textbooks concerned with physical geography, physical geology and geomorphology. Within the last decade, a major interdisciplinary collaborative study has gathered information allowing a more detailed and better-dated understanding of the Holocene evolution of this coastline (Chroston et al., 1999; Andrews *et al.*, 2000; and Andrews and Chroston, 2000).

Extending for some 50 km from Hunstanton in the west to Sheringham in the east, the features owe their origins in large part to the efficacy of longshore sediment transport both in the past and at present. The site comprises many separate morphological units that form six sediment cells (Cambers, 1975). Although Sir William Halcrow and Partners (Halcrow, 1988) also divide the coastline into six units, they identified slightly different boundaries, the units being lengths of shoreline that have 'coherent characteristics' but are not necessarily independent of adjacent cells (Figure 11.27). Here Cambers' cells are used as follows:

1. Hunstanton to Holme-next-the-Sea: Chalk and Carstone cliffs that are undergoing erosion are fronted by a wide sand and shingle beach that extends northwards beyond the cliffs to Holme-next-the-Sea where the fringing dunes reach their widest extent.
2. Holme-next-the-Sea to Brancaster: an area of dunes and beach ridges behind which lie both claimed marshland and natural saltmarsh.
3. Scolt Head Island: the best example of a barrier island on the British coast (Steers, 1981). Regular surveys since the early part of the 20th century make this one of the best-documented coastal sites anywhere in the world.
4. Gun Hill to Wells-next-the-Sea: dominated by a line of dunes known as 'Holkham Meals'.
5. Wells Channel to Blakeney Spit: a large number of small bars of sand, shingle and shells, and an unusual, recurved cuspate beach.
6. Blakeney Point to Sheringham: an excellent example of a recurved spit formed mainly of a single shingle ridge (over 9 km in length) extending from a shingle beach at the foot of retreating till cliffs between Weybourne and Sheringham. Generally, but not exclusively, the inland boundary is marked by a low bluff (an earlier now degraded cliffline) or land-claim embankments.

Description

Taken as a whole, this is a region of wide sandflats, a barrier island and a spit backed by tidal flats, saltmarshes or dunes. The seabed off the western part of the site is very shallow. Burnham Flats has depths of only 6 m as far as 10 km offshore. Tidal streams reach 0.77 m s^{-1}. East of Wells-next-the-Sea, a bank 7 km offshore has a water depth of only 3 m, but is separated from the coast by water of about 9 m depth between 1 and 2.5 km offshore. The tidal stream here reaches 1.08 m s^{-1}.

McCave (1978b) described the sediment characteristics of the area in detail, the main features being a long shingle barrier ending in Blakeney Point and tidal flats and saltmarshes dominated by muds, and dunes whose sand is better sorted than the beaches that feed them. The key trends along the shore are an increase in mean sand size towards the west and an increase in shingle on the beach eastwards from Blakeney Point. The size of this shingle increases eastwards from Blakeney to Sheringham (Hardy, 1964). Cambers (1975) estimated a potential west to east movement of sand along this coast of about 300 000 m^3 a^{-1}. Sir William Halcrow and Partners (Halcrow, 1988) show that the mean annual alongshore wave energy increases from between 0 and 500 kN s^{-1} at Sheringham to between 1000 and 1500 kN s^{-1} at Gore Point. However, the standard deviation is of a similar magnitude to the mean values, suggesting that the direction of alongshore energy could change from year to year.

Cambers (1975) measured coastal change by comparing the 1:10 560 Ordnance Survey maps for the 1880s with those of the 1950s, and showed that for East Anglia as a whole the total area gained from the sea was 58 370 m^2 compared to a loss of 134 817 m^2. The north Norfolk coast was, however, mainly characterized by accretion, the only areas of erosion being at Brancaster Spit, the central part of Scolt Head Island and at the eastern end of Blakeney spit. Between Burnham Harbour and Wells, accretion was greater than 8 m a^{-1}, and between Holkham Gap and Wells the dunes advanced seawards over 100 m between the 1880s and the 1950s. The *Anglian Coastal Management Atlas* (Halcrow, 1988) indicates that erosion has become more widespread in the 1980s. In particular, although annual rates of retreat of high-water mark have been lower than 1.5 m over the last 100 years (Figure 11.27), low- water mark has retreated by up to 4 m, so that the foreshore has generally been becoming steeper. There are in contrast many points along this coast where progradation has occurred, and the high-water mark has shifted seawards as the foreshore has steepened. In places the high-water mark and the low-water mark have both moved seaward and the foreshore slope has been maintained or even become shallower.

Open-coast and back-barrier saltmarshes, both active (2127 ha: Burd, 1989) and land-claimed (1500 ha), extend for about 35 km from Holme-next-the-Sea in the west to Cley next the Sea east of Blakeney Point. Much of the saltmarsh lies behind coastal barriers of sand (for example at Brancaster and Titchwell (Steers, 1934c, 1936a; Pye, 1992), shingle (Blakeney Point) or mixed sand and shingle (Scolt Head Island). Open-coast marshes landward of wide intertidal sandflats occur mainly at Thornham and Warham. Land-claimed marshes, which are not included in this GCR site, occur at Thornham, between Burnham Deepdale and Wells-next-the-Sea and landward of Blakeney Point. The marshes at Holme-next-the-Sea have been almost entirely reclaimed. Between Thornham and Titchwell the active marshes are mainly back-barrier marshes on which there has been some embanking. At Thornham and Gore Point, new back-barrier marshes have formed since the 1950s (Pye and French, 1993). Between Brancaster and Overy Staithe, parts of the predominantly back-barrier marshes have been reclaimed, but there are extensive active marshes in Brancaster Marsh, on Scolt Head Island and in Overy Marsh. At Brancaster, migration of the dune ridge landwards has covered parts of the back-barrier marsh (Pye and French, 1993).

The westernmost division lies between Hunstanton (TF 673 414) and Holme-next-the-Sea (TF 727 450). At Hunstanton, the coastline is dominated by near-vertical cliffs about 25 m in height cut in Carstone, Red Chalk and Lower Chalk (Figure 11.28). The Carstone forms a shore platform in which clearly visible rectangular jointing patterns have been only slightly eroded. The Lower Chalk collapses as the cliff is undermined and topples as large, tabular blocks. The cliffs are being eroded at about 0.3 m a^{-1}. Steers (1971b) suggested that their very steep nature results from the combined effects of the rate of marine erosion, the nature of the bedding and the strength of the rocks. In particular, the strength of the tabular Chalk forming the upper cliffs sometimes produces an upper overhang. A beach of sand and shingle extends northwards to Holme-next-the-Sea, where a line of fringing dunes that extend from the northern end of the Chalk cliffs reach their widest. Although the dune and beach ridges have been breached occasionally in the past, the general pattern is of gradual progradation fed by sediment moving north from the vicinity of Hunstanton. Ridges such as Gore Point, which extends westwards, are not permanent features, their presence and alignment appearing to depend upon the predominance of growth from the south or sediment supply from the north. East of Holme, the sand dunes are partially embanked and have built up over a former seawall (Steers, 1946b).

Steers (1981) described the role of shingle and shells in forming ridges upon which sand dunes subsequently form by reference to an example at Thornham. In 1914, a crescent-shaped sand and shell island developed in which dune plants colonized the ridge and played a significant role in raising its level by trapping wind-blown sand. Small dunes grew at each end of the ridge, behind which there are small recurves. Bridges (1991) cites a similar example that formed between 1930 and 1935. Here, as elsewhere in the site, saltmarshes have developed between the beaches and the former sea cliff (Peake, 1960).

For much of the distance between Holme-next-the-Sea (TF 728 450) and Brancaster (TF 797 452), the shoreline is formed by dune ridges up to 400 m in width (Figure 11.29). Behind the dunes at East Sands and at Brancaster Golf Course there are extensive saltmarshes that have not been subject to land-claim (Murphy and Funnell, 1979), whereas the central 2 km of this beach has no dune belt and is backed by embanked and land-claimed marshland. Accretion is dominant at both ends of the beach but the central part is affected by erosion, notably along the frontage of the Golf Course. A short length of armoured embankment has been constructed to control shoreline retreat. Erosion at Brancaster revealed two peats, one at between −0.08 m OD and −0.15 m OD that included forest remnants and beech *Fagus* sp. seeds, the other higher at between 2.5 and 3.5 m OD (Bridges, 1991). Funnell and Pearson (1989; see also Andrews *et al.*, 2000; Chroston *et al.*, 1999) showed that there were over 8 m of Holocene sediment with a broad channel running parallel to the main dune ridge. The modern ridges are dynamic, the eastern ridges have grown eastwards, but were farther seawards in 1937 than in 1951, a trend that has continued. The spit at East Sands has grown considerably since Steers' 1935 survey, although the reason is

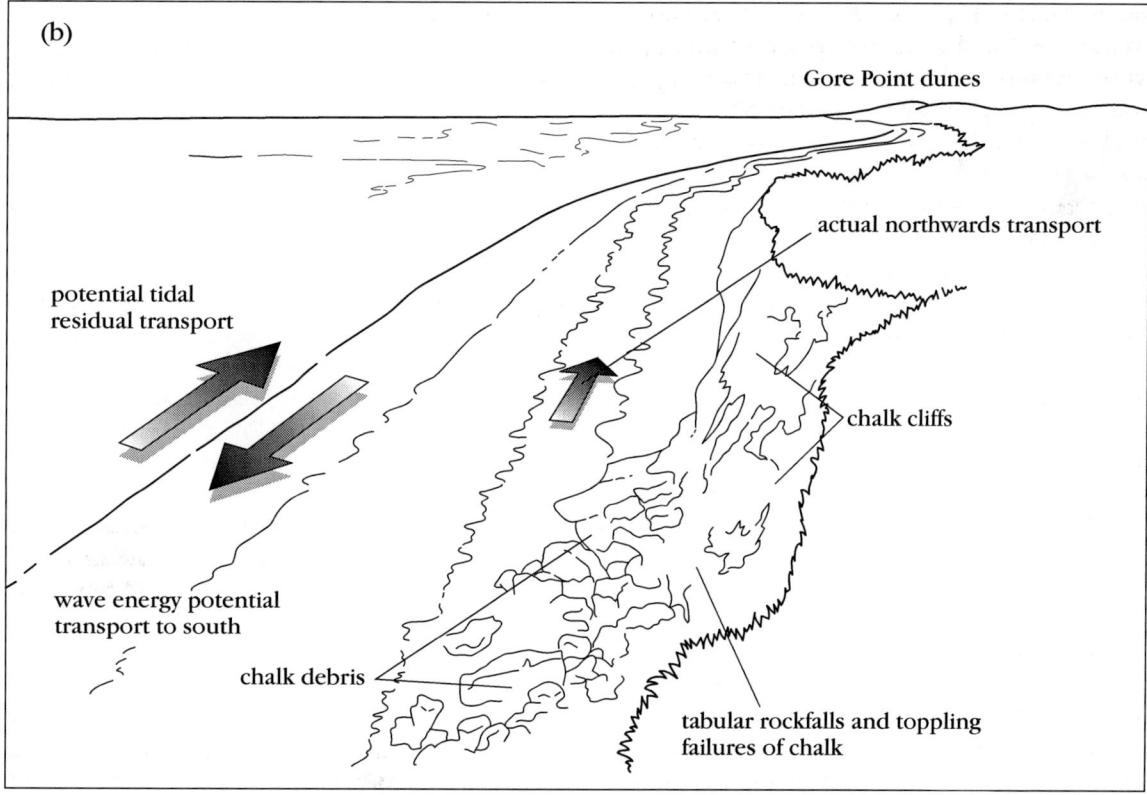

Figure 11.28 (a) The distinctive tabular chalk cliffs of Hunstanton, looking north. (b) Cartoon of potential transport mechanisms of rockfall debris from the failure of the chalk cliffs. (Photo: V.J. May.)

not clear. Cambers (1975) suggested that it may be associated with material deposited via the Harbour Channel, but it is difficult to conceive how this channel, draining a mainly muddy marshland area, could provide sufficient sand for the growth that has occurred. The channel has moved away from the beach and so it seems more likely that sand is moving within the intertidal area between Holme and East Sands.

Scolt Head Island (TF 793 467–TF 847 460; Figure 11.30) has been the subject of regular surveys since the early part of the 20th century, most of the early work being described in Steers (1960). The island is about 7 km in length with a predominantly sand beach about 900 m wide at low spring tides. Shingle patches occur, but most accumulates towards the top of the beach. More than 20 lateral shingle ridges run inland from the main beach: these trend south-west then turn towards the south. All the ridges are dominated by flint, most of which is well rounded (Roy, 1967). The western end of the island known as 'Far Point' is the youngest recurve, suggesting progressive growth of the island towards the west. Dunes have accumulated on most of the recurves and on the main ridge. The highest dunes are at the Headland, in the middle of the island on House Hills and on Norton Hills at the eastern end of the island. The Long Hills form an important line of high dunes associated with lateral ridges running towards Brancaster Harbour. A lower continuous line of dunes joins the Headland to Far Point. The dunes on the seaward face of the island are subject to erosion and have been breached in the past during storm surges (e.g. at Smuggler's Gap in 1938 and at The Breakthrough (TF 833 463) in 1953 and 1978). Sand from the intertidal beaches naturally replenished the breaches (Steers, 1960). However, despite this local supply of sand and the measures taken to heal the breach in 1953, the general retreat at Smuggler's Gap was over 20 m between 1953 and 1980 (Steers, 1981). At the breach site there is a very large washover fan of shingle. The dune crest is now low and a future breach is likely.

The erosion of the central part of the island contrasts with the accretion that is taking place at both ends, although it is much more substantial at the western end at Far Point. Erosion of the main beach is consistent with a gradual retreat of the shoreline as the beach extends westwards. Each lateral lobe represents an earlier beach at a more seaward position, as the island has moved westwards. Steers commented that the laterals form sharp angles with the main beach and also develop a further sharp bend along their length. He suggested that there is no completely satisfactory explanation of the formation of lateral ridges 'here or elsewhere' (1981, p. 360), but the Shoreline Management Plan (Halcrow, 1988) discusses their formation.

The marshes at Scolt Head Island have been a focus of attention since the 1920s (Steers, 1946b). They are separated by shingle recurves and appear to become younger to the west, i.e. from Plantago Marsh through Plover and Hut Marshes to Missel Marsh. The lowest parts of the saltmarshes at Scolt Head Island generally have a cover of glasswort *Salicornia* spp., cord-grass *Spartina anglica* and sea aster *Aster tripolium*. Sea purslane *Atriplex portulacoides*, sea lavender *Limonium vulgare* and sea meadow-grass *Puccinellia maritima* are more characteristic of the more mature marshes. The oldest marshes commonly have a cover of sea thrift *Armeria maritima*, long-leaved scurvy grass *Cochlearia anglica*, red fescue *Festuca rubra* and sea plantain *Plantago maritima* (Pethick, 1981). The easternmost marsh on Scolt Head Island, The Sloughs, has two distinct levels, the lower of which was colonized by vegetation towards the end of the 19th century.

Plover Marsh, in contrast, lies between two of the large recurved laterals of Scolt Head Island and has been gradually covered by the island's migrating dune ridge. At this site there is a washover feature where the shingle ridge has been overtopped. Continuing retreat on the eastern end of Scolt Island (c. 0.5 m a^{-1}) continues to reveal saltmarsh deposits. Exposed marsh deposits on the seaward side of the dunes were reported by Grove (1960) and dated at 441 ± 120 years BP (Joysey, 1967). Plover Marsh thus appears to have originated during the early 16th century.

Hut Marsh covers about 55 ha. The surface of its mature marsh lies at between 2.5 and 2.9 m OD (Stoddart *et al.*, 1989). Steep-sided deep creeks form a dendritic pattern, draining mostly into Hut Creek, which attains a width of over 15 m. In contrast, many of the small first-order creeks draining the upper marsh are usually less than 1 m wide and 0.5 m deep. The marsh had no vegetation in 1818 but had become vegetated and developed a creek system by 1880. The upper part of Hut Marsh lies about 7 cm higher than its eastern part, suggesting to Pethick

North Norfolk Coast

Figure 11.29 (a) Brancaster beach-dunes, sand and shingle beach with regular shore-normal cusps. (b) Eroded dune-face remnants of World War II defence structures and associated retreating foredune scarp. (Photos: V.J. May.)

Coastal assemblage GCR sites

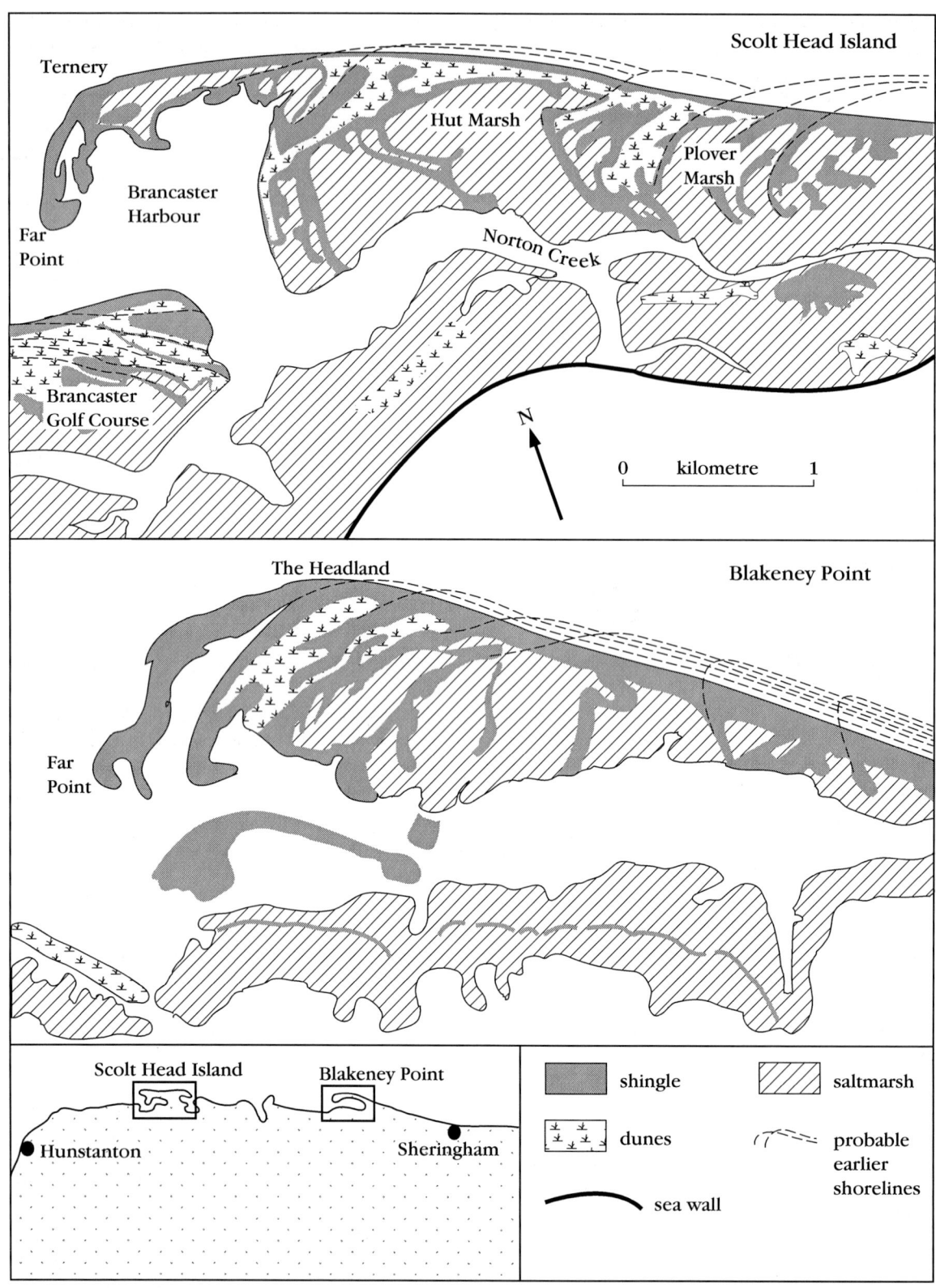

Figure 11.30 (a) Scolt Head Island geomorphological features. (Based mainly on Steers, 1946b, 1960; Bird, 1984, 1985; Halcrow 1988.) (b) Blakeney Point geomorphological features. (Based mainly on Steers, 1946b; Bird, 1984, 1985; Halcrow, 1988.)

(1980b) that it is about 50 years older. The highest rates of accretion occur in the central part of the marsh between the two major draining creeks, whereas the lowest rates occur on the highest parts which are least frequently flooded (Stoddart *et al.*, 1989). Stoddart *et al.* (1989) show that in the middle part of Missel Marsh, shrubby seablite *Suaeda vera* occurs on slightly higher areas. Samphire *Salicornia* spp. and green seaweed *Enteromorpha* spp. are seasonally abundant below the rims of the creeks. Pethick (1980a, 1981) has shown that the inception of the saltmarsh at Missel Marsh at the western end of Scolt Head Island occurred between 1880 and 1907. Steers (1960) estimated the rate of vertical accretion at 8.4 mm a^{-1} over a 22-year period from 1935 until 1957.

Between Gun Hill (TF 847 457) where the Burnham Overy channel drains the marshes at the eastern end of Scolt Head Island and Wells (TF 915 456), the coastline is dominated by a line of dunes known as 'Holkham Meals'. Accretion is predominant with dune crests reaching over 16 m to form the highest point within this site. With winds from the north-west, north or north-east, sand is blown off the beach surfaces very soon after they are exposed. Except at Overy Marsh, the former saltmarshes have mostly been landclaimed. The first enclosure took place in 1660 (Dutt, 1909), but Pethick (1980a) uses archaeological evidence (Clark, 1936, 1939) to show that the marshes are over 2000 years old and lower in altitude (Table 11.3). It appears from documentary evidence discussed by Steers that a channel flowed through Holkham Gap before the land-claim of the saltmarsh. The sandflats are at their widest either side of the Wells Channel, but it is not clear why this is the case. Holkham Bay is marked by slow progradation; dune barriers have been growing in the bay since the 1950s and the area behind the landward barrier is now muddy and colonized by samphire *Salicornia* and other saltmarsh plants (Clayton, pers. comm.). The dune front between Gun Hill and Holkham Gap (TF 890 450), which rises to 15 m, continues the alignment of the main beach at Scolt Head Island, but the dunes east of Holkham Gap have an arcuate form. Steers (1946b) regarded them as an offshore bar of shingle that became stabilized by dune-building. They were further fixed by afforestation during the mid-19th century.

From Wells to Blakeney Point (TF 991 444), there is less development of both dunes and beach ridges, but there are a large number of small offshore bars of sand, shingle and shells. Small beach ridges with limited dune growth fringe the marshes at the Stiffkey Meals, while on the eastern side of Wells Channel, a larger cuspate feature has developed. The growth of recurves at its eastern end suggest sediment transport towards the east, whereas its western tip has grown south-westwards. The role of the wide intertidal flats in modifying wave-energy distributions may also be important here.

Between Wells and Blakeney Point (TF 991 444), the marshes are open to the North Sea and include Wells, Warham, Lodge and Stiffkey marshes. This part of the North Norfolk Coast site is unusual in being the only lengthy stretch where saltmarsh, albeit with a narrow shingle fringe fronted by sand, forms the main feature of the coastline. Some of this marsh originated during the 1950s, with parts being colonized by vegetation only since the 1980s (Pethick, 1980a). The marshes are exposed to the north-east but locally sheltered by a 1.5–2.0 km-wide belt of intertidal sandflat with low onshore-migratory bars. The marshes are 800–1000 m wide and divided by a low shingle ridge. The upper marshes reach 2.8 m OD and are characterized by incised creeks and a floristically rich 'General Saltmarsh Community' (Spencer *et al.*, 1998b). The low marsh varies in height between 2.8 m OD just seaward of the ridge to 2.5 m OD at its seaward edge and is dominated by a pioneer community of common cord-grass *Spartina anglica* and sea-aster *Aster tripolium* and clumps of sea-purslane *Atriplex portulacoides*. Lateral growth of new marshes has taken place at Warham in the last 50 years (Pye and French, 1993). For example, *S. anglica*, first planted in 1907, covered 81 ha by the mid-1960s (Hubbard and Stebbings, 1967) and 149 ha by the late 1980s (Burd, 1989). The organic content of the marsh sediments is less than 15% by weight (French and Spencer, 1993). Aerial photographs show that the present-day low marsh at Stiffkey developed in the 1950s and 1960s (Spencer *et al.*, 1998b; Pethick, 1980a), but has been undergoing erosion since the late 1970s. The seaward margin has degraded into a hummocky topography drained by poorly defined anastomosing channels (see Figure 10.9; Pye and French, 1993). According to Cambers (1975), there is little change in the coastline here.

Blakeney Point (Figure 11.30) is a large shin-

Table 11.3 Summary of saltmarsh development in north Norfolk.

Time	Development
7500 years ago	First signs of marine incursion at $c.$ -7 m OD
Until 5500 years ago	Sediments accumulate as sea level rises
Between 5500 and 4500 years ago	Peats within saltmarsh muds and silts imply stability or perhaps fall in sea level
About 4000 years ago	Barrier features at Scolt Head and Blakeney probably in place (Allison, 1989)
About 3000 years ago	Coastline at Holkham is 3km north of its present position
About 2000 years ago	Romano-British remains indicate inner marshes at Brancaster and Burnham
Last few hundred years	Outer marshes develop at Scolt Head Island, Blakeney and at Warham
Since 1900	Open coast marshes grow rapidly with *Spartina* colonization between Wells and Stiffkey
Since 1950	New marshes at western Scolt, Thornham, Morston, western Blakeney. Dune ridges transgressing onto marsh at Brancaster

gle spit, comparable in size to Spurn Head. The shingle beach extends from Sheringham westwards for over 17 km, the first 5.5 km fringing low (up to about 30 m) till cliffs (Burnaby, 1950), and the central section forming a ridge fronting Salthouse Marsh and Fresh Marsh. The ridge is about 200 m wide and between 9 and 10 m in height. Hardy (1964) estimated that the whole structure contained about 2.3×10^6 m³ of shingle. The western part continues as a single ridge for a further 3 km before developing a series of long recurves trending southwards that are the most recent members of a set of over 20 shingle laterals of varying length. Blakeney Point has extended and shortened several times during the last 150 years. The morphological and cartographic evidence demonstrates that the spit has grown westwards. Steers (1927) estimated that the spit lengthened by 86.4 m a⁻¹ between 1886 and 1904 and by 45.7 m a⁻¹ between 1904 and 1925. Between 1649 and 1924 the ridge moved inland by an average of about 1 m a⁻¹ (Hardy, 1964). There is some debate about the extent to which longshore transport is consistently in this direction and it is possible that shingle moves one way and sand in the other (Battarchaya, 1967; Hardy, 1964; Steers, 1964b; Cambers, 1975). In recent years the ridge has been eroded by storms and then re-shaped by bulldozing material back into the increasingly narrow profile, similar to Hurst Castle Spit (see GCR site report in Chapter 6). This has led to a reduction in the shingle volume of the beach and in February 1995 a 200 m-wide breach through the ridge occurred. If the loss of shingle continues it is likely that Blakeney Point will become an isolated island such as Scolt Head, unless coast protection works are carried out to provide protection for low-lying settlements such as Salthouse. The geomorphological interest lies in allowing natural processes to continue unimpeded, though with the lost shingle restored by beach nourishment.

There are active marshes either side of the Blakeney Channel, but east of Blakeney, they have mostly been land-claimed. The marshes behind the shingle ridge from Salthouse to Blakeney Point increase in age eastwards, with the oldest probably developing first during the 15th century (Pethick, 1980a). Most recently, lateral growth of new marsh has taken place at the western end of Blakeney spit since the 1950s (Pye and French, 1993). Carey and Oliver (1918) reported thin coverings of samphire *Salicornia* spp. in the central marshes, whereas the marsh closer to Blakeney Point itself appears to be older (between 1818 and 1880: Pethick, 1980a).

Interpretation

Despite the long and detailed documentation of the north Norfolk coastline, the sources of the sediments forming the beaches and the direction of sediment transport is still open to debate. The direction of longshore transport has generally been described as eastwards and southwards along the Norfolk coast east of Sheringham, whereas the shingle features on the North Norfolk Coast site have been shown to develop towards the west (Redman, 1864; Wheeler, 1902; Steers, 1927, 1946b). This would suggest a division in the drift direction in the vicinity of Sheringham. Work by Sir William Halcrow and Partners (Halcrow, 1988) demonstrates that the direction of mean, annual, alongshore wave-energy west of Sheringham is from east to west.

Earlier, however, Hardy (1964) used marked shingle in a series of experiments, concluding that shingle moved eastwards except when winds were between north-east and south-east. With the prevailing westerly conditions at the time of his experiment, he found no evidence of a divergence of drift. A consideration of the distribution of shingle suggested that the only sources were the cliffs between Sheringham and Weybourne or small former islands landward of the spit. He also believed, on the basis of estimates of the volume of drift, that the spit was losing material by a slow net-transfer eastwards and that the grading of the shingle supported this view. Steers (1964b) believed that changes in wind and wave direction would explain the apparent paradox raised by Hardy's thesis. Variability of direction of alongshore wave energy and transport certainly occurs and the interpretation of the alongshore transport of sediment on this coast needs to take account of the higher-energy events that can cause transport in a direction different from the prevailing conditions.

The detailed examination of the sediment budget of the east Anglian coast reported by Cambers (1975) also addressed longshore drift in some detail. Cambers' general conclusion was that sand on the beaches of north Norfolk becomes finer-grained towards Sheringham to the east and that the gravel between Sheringham and Blakeney similarly becomes coarser towards the east. Cambers argued that present-day transport rates at Sheringham show an overall drift direction from west to east, but that a change in the orientation of the beach by between 4° and 5° could result in a pattern of no overall transport. A change greater than 5° would result in a reversal of the direction of drift, as would a similar change in the direction of wave approach. Although the spit formation at Blakeney was a response to wave energy and orientation in the past, the present-day changes in the orientation of the spit, with erosion at its eastern end and accretion at its western end, demonstrate that it has not reached equilibrium with the present-day energy conditions. Straw (1979) has described the 'eyes' (small ridges) landward of Blakeney Point as vestiges of a spit of Ipswichian age. This would suggest that wave and sediment transport conditions during the last interglacial were generally similar to those that produced the present-day spit.

The trend for the sand grain-size on the North Norfolk Coast site to become coarser towards the west was explained by Cambers as resulting from longshore transport during which the fine-grained fraction is preferentially moved (winnowed) offshore. The further the sand is transported, the longer the time period that it is exposed to processes that tend to winnow out the fines (McCave, 1978a–c). However, gravel is more likely to be sorted by selective transport and there is no reason why the dominant direction of transport for shingle should be the same as that for sand. More recently, Pethick (pers. comm.) has suggested that sand moves onshore from offshore banks (themselves probably supplied from the erosion of the Holderness (East Yorkshire) cliffs) and is then moved eastwards, crossing the major tidal inlets from time to time.

The age of the major spits and barriers was discussed by Steers, who concluded that about 500 to 600 years BP was 'at best a reasonable guess'. The cartographic and documentary record offers few clues to an earlier origin. Most land-claim has taken place since the mid-18th century. Steers (1946b) noted that there is evidence of occupation of the coastline as early as the Roman period, that the medieval ports of the north Norfolk coast were prosperous, and the earliest maps, from the late 16th century, include features that might have been forerunners of the present-day spits and barriers. The subsequent silting of the ports may be attributed to the more sheltered environment offered by the growth of Scolt Head Island and Blakeney Point spits, but it is not possible to date the shingle ridges as a result. Certainly sedimentation at the heads of the major tidal inlets and the consequent abandonment of such ports as Cley next the Sea long pre-dates the decline in the tidal prism resulting from the embanking of the former saltmarshes.

The well-developed saltmarshes that lie landwards of the ridges and, in particular, the saltmarshes between the laterals offer both further opportunities for dating of the origins of the spits and barriers (Pethick, 1980a, 1981) and for examination of the development of saltmarsh creeks and pans in marshes of different age. Pethick's (1980a) view that parts of the marshes are of pre-Roman date (before c. 2000 years BP) provides a potential earliest date for the initiation of the major beach structures has been superseded by later work; first by Funnell and Pearson (1989) and most recently by Andrews et al. (2000).

Under the 'Land–Ocean Interaction Study' (LOIS), it has been shown that a west–east channel cut in Chalk is now filled with Holocene sediments. It lies close to the present-day coastline; to seaward from Holme Point to Brancaster, and to landward from Scolt Head Island to Salthouse (Chroston *et al.* 1999). The channel has a very low eastward slope and probably carried water from the River Trent and the Wash rivers along the ice front of the Devensian glaciation some 18 000 years ago. West of Brancaster the coastline has moved landwards to lie south of the buried channel by at least 2 km during the last 6000 years, but east of Scolt Head Island the outer barrier lies on the northern end of the buried channel. Boreholes have penetrated the Holocene sediments to the Chalk floor, showing that the basal sediments lie between –7 and –11 m OD. The Holocene record includes terrestrial peats dated from >9000 to 7000 years BP, after which a steady rise of sea level deposited mud and silt (with rare saltmarsh peats) after 6000 years BP. Seven lithofacies are described, peat, back barrier muds and silts, muddy sand (marking tidal channels), pebbly sand (including washovers), rooted sand (vegetated dunes), interbedded sand (tidal flats) and gravel (beaches and barriers). Sedimentation behind the coastal barrier was fairly continuous, though datable peats suggest a period of stable or falling sea level about 5000 ± 500 a BP. The sand-gravel ridges known as 'meols' or 'meals' on Stiffkey Marsh are similar to cheniers and probably result from severe storms in the Little Ice Age, 300–500 years ago (Boomer and Woodcock, 1999; Knight *et al.*, 1998).

The LOIS investigations have yielded new data on the long-term, landward barrier movement along this coast. The sandy barrier facies have moved south and aggraded more or less in pace with rising sea level (Andrews *et al.*, 2000; Orford *et al.*, 1995). It is likely that the present landward movement of *c*. 1 m a^{-1} has been typical of the period of steady sea-level rise from 7400 years ago to the present, and this suggests that the present-day Holocene coastal prism, currently 3 km wide at Holkham, was 6 km wide 3000 years ago (Andrews and Chroston, 2000). This rate of retreat (which has destroyed virtually all of the Holocene sediments to seaward of the present barrier) matches that of the cliffs farther east (Clayton, 1989b), and it is likely that both are controlled by the rate at which water depths off East Anglia have increased as sea level has risen (and perhaps the sea floor been eroded) over several millennia.

Many of the barriers are backed by belts of sand dunes, and they form a further important feature, dated in places, but often of undetermined age (Knight *et al.*, 1998; Orford *et al.*, 2000). In their natural state they are stabilized by marram *Ammophila arenaria*, but around Holkham Bay they have been planted with Scots Pine trees as part of the land-claim carried out by the Holkham Estate over the last two centuries. Along the more exposed parts of the barriers (e.g. Scolt Head Island) the dunes are cliffed during storm surges and recover with the growth of a foredune in the years between. A few active blowthroughs survive, despite intermittent attempts to revegetate them.

Pethick (1980a) argues that the present-day north Norfolk marshes fall into three broad age groups that co-incide with the three periods of rising sea level of the last 2000 years (Table 11.3). They are the pre-Romano-British marshes about 2000 years old and associated with the Romano-British transgression (Godwin, 1940), medieval marshes about 400 years old developed during a 12th–14th century transgression (Tooley, 1974; Green and Hutchinson, 1961), and recent (post-1850), present-day sea-level rise. The marsh surfaces approach an asymptote at about 0.8 m below the highest tide level (Pethick, 1981). Kestner (1975) found similarly that levels of saltmarshes in the Wash always remains between 0.6 to 0.9 m below high-water ark of ordinary spring tides (HWMOST) Pethick identified a clear fall in the sedimentation rate with elevation and 'a very striking' co-incidence between the modal tide at 2.4 m OD and the marsh surface asymptote at 2.385 m OD (Figure 11.31b). This could probably be attributed to the infrequent flooding, and subsequent minimal accretion, of any marshes that attained 2.4 m OD.

On Hut Marsh, a tide-dominated back-barrier marsh on Scolt Head Island, 95% of total deposition is by direct settling (French and Spencer, 1993). Annual accretion varied between 8 mm a^{-1} adjacent to the larger channels and less than 1 mm a^{-1} on the saltmarsh surfaces farthest from the channels, i.e. on the highest parts of the marsh. Along the longest transport pathways, there was a reduction and then 'exhaustion' of suspended sediment. Storm events by causing surges, and thus higher water levels, accounted for a significant proportion of long-

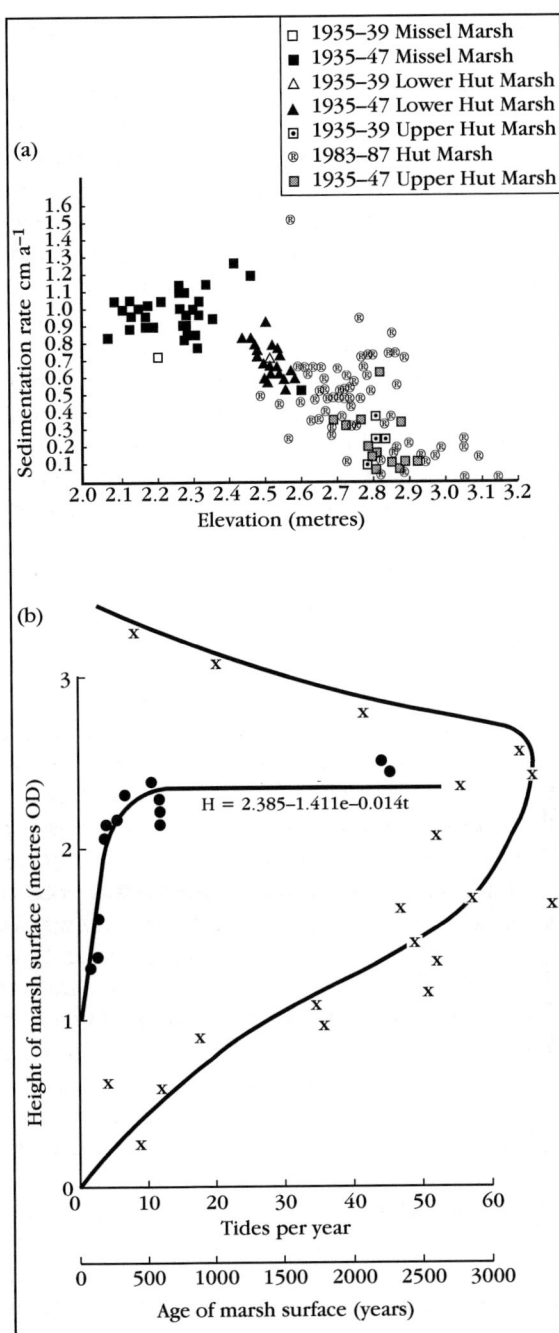

Figure 11.31 Relationship of saltmarsh elevation to tides and age of the marsh surface at Missel Marsh, Lower Hut Marsh, upper Hut Marsh and Hut Marsh. H = 2.385 − 1.411 e − 0.014t is a best-fit line based on the relationship between marsh height (H), tides per year (e) and age of marsh surface (t). (After Pethick, 1980b; 1981.)

term accretion on the highest surfaces. Although there is a general relationship between predicted tidal height and sediment deposition (Figure 11.31b), French and Spencer (1993) showed that this is disrupted by meteorological conditions and resuspension of muddy sediments within the creek system. The arithmetic mean vertical accretion rate for the marsh was 3.9 mm a^{-1}, which is higher than the present-day local rate of sea-level rise (1.5–2.0 mm a^{-1}: French and Spencer, 1993), a significant finding.

Although local rates of sedimentation within the marshes appear to have been fairly constant over recent decades, there are considerable variations across the marshes (Stoddart et al., 1989). Sedimentation is determined not only by the density of the drainage network, but also by channel size and velocity regime. Sediment is deposited in creeks during neap tides, but is mobilized by higher velocity pulses during spring tides (Bayliss-Smith et al., 1979; French, 1985; Green et al., 1986; Healey et al., 1981). They also carry sediment on to the marsh surface. The presence of sea-purslane *Atriplex portulacoides* along creek banks enhances deposition. Stoddart et al. (1989) recommended, on the basis of their studies on north Norfolk marshes, that future studies of sedimentation in macrotidal marshes should concentrate particularly upon the interaction between creeks and the vegetated surfaces and the transport pathways for sediment within marshes. More recent studies have shown that unsteady flows in creeks exhibit well-developed velocity and stress transients (French and Stoddart, 1992), which result from a discontinuous tidal prism and the interaction of shallow water tidal inputs with the hydraulically rough vegetated surfaces of the saltmarshes.

Investigation of fossiliferous concretions in the marshes and sandflats shows that they are both abundant and consist mainly of siderite, calcite and iron monosulfides (Allison and Pye, 1994). Siderite–magnesium–calcite–iron sulphide concretions form in oxygen-reduced sediments with active sulphate-reducing bacteria (Pye et al., 1990). Whole concretions form within tens of years, with mineralization becoming visible within months. Iron diagenesis is described (Allison and Pye, 1994) as 'exceedingly active', although the concentration of dissolved iron in pore water rarely exceeds 1 ppm here. Tidal pumping produces both horizontal and vertical movements of up to 60 cm per day.

The north Norfolk marshes have an important role internationally in providing both long-term, and shorter-term – but more detailed – bases for future comparative studies. The LOIS work has greatly increased our knowledge of the older Holocene sediments and the changes that occurred as sea level reached the present line of barriers some 7500 years ago. The saltpans that characterize many of the marshes also form an important element that has been examined both here and in Essex (St Osyth and Dengie marshes, see GCR site reports). Saltmarshes form an integral part of this site and are particularly important as one of 11 GCR sites identified as being geomorphologically characteristic for their saltmarsh morphology (see Chapter 10). Much of the work carried out by Steers and others on this coastline concentrated upon marsh sedimentation within the sheltered environment landward of the beaches. Because of the length of record and the opportunities to date the marshes, these marshes also have considerable significance internationally as areas to be compared with other detailed marsh surveys (for example, Richards, 1934; Chapman, 1938; Guilcher and Berthois, 1957; Ranwell, 1964; Pestrong, 1965; Harrison and Bloom, 1977).

Steers (1981) described the Scolt Head Island complex of beaches, recurves and saltmarshes as the best in Britain and probably also in Europe, on the basis of the assemblage of such features in a relatively small area. The inclusion of other major features such as Blakeney Point adds to the significance of the site. The North Norfolk Coast GCR site is especially important because the 70-year record of regular surveys provides an unrivalled baseline against which assessment of the changes in coastal dynamics associated with present-day sea-level change can be judged.

Conclusions

The coastal features on the North Norfolk Coast GCR site are of outstanding geomorphological importance. It is an extensive site, over 50 km in length, which includes such internationally renowned coastal features as Scolt Head Island, a major barrier island, and Blakeney Point, a large shingle spit. Both have been studied for many decades, the former regularly for over 80 years. Smaller, less well-known parts of the site, intertidal flats, beaches, dunes, saltmarshes and cliffs, are integral to its patterns of sediment transport.

The north Norfolk marshes have an important international role in providing both long-term and short-term bases for future comparative studies. Because of the length of record and the opportunities to date the marshes, these marshes also have considerable significance internationally as areas to be compared with other detailed marsh surveys. Recent research during the 1990s has demonstrated in greater detail than previously the close links between salt-marsh morphology, sedimentation processes and vegetation. As a result, the saltmarshes are of European and international significance.

The assemblage of largely unspoilt coastal features of this site and its 80-year record of research make this site an internationally important location for the understanding of the geomorphology of beaches and saltmarshes. Sites elsewhere are of equal importance in magnitude and naturalness as the spits, barrier beaches and marshes here, but few have been described in detail and none can rival the detailed record of this site.

The saltmarshes are one of the few areas on the coastline of England and Wales where saltmarsh morphology, including saltpans, has been examined in detail. The North Norfolk Coast GCR site is also a Special Area of Conservation under the Habitats Directive, a Special Protection Area under the Birds Directive, and a Ramsar site; virtually the whole coastline is owned and/or managed by conservation organizations.

THE DORSET COAST: PEVERIL POINT TO FURZY CLIFF, DORSET
(SY 697 816–SZ 041 786)

V.J. May

Introduction

The southern flank of the Isle of Purbeck and the coast west from Lulworth Cove is well known for its geological structures (Strahan, 1898; Arkell, 1947; House, 1993), and this predominantly cliffed coast is one of the most important locations in Britain for demonstrating relationships between rock structure, rock strength and coastal landforms (Horsfall, 1993; see Figure 1.2 for general location). The whole site forms one of the most frequently described British examples of a longitudinal coastline (e.g. Holmes, 1965; Bird, 1984; King, 1959, 1972). Within it, there are several classic coastal localities, of which Lulworth Cove is probably the most well known. This scale and range of features has attracted most attention (for example, Allison, in press; Brunsden and Goudie, 1981; Burton, 1937; Komar, 1976; Sparks, 1971; Small, 1970, 1978). A series of small bays containing beaches distinguished by local grading of sediment fed from distinct, identifiable sources (Arkell, 1947; Heeps, 1986) provides unrivalled opportunities for the study of beach development. Overall, the range of features developed on different rock-types and at a variety of scales makes this coast of paramount importance for understanding relationships between coastal form, processes and materials. This coastline is extensively used for educational purposes and tourism, and is attracting increasing research interest. In recognition of the site's importance for coastal geology and geomorphology, it is one of around 70 GCR sites that form the Dorset and East Devon Coast World Heritage site (see also Ballard Down, Budleigh Salterton, Chesil Beach, Ladram Bay Lyme Regis to Golden Cap and South Haven Peninsula GCR site reports, this volume).

Description

The main geological strata and geomorphological units of the area are shown in figures 11.32 and 11.33. The geology includes strata from the Upper Chalk to the Oxford Clay. The longitudinal coastline of south-east Purbeck between Durlston Head and Worbarrow Tout contrasts with transverse sections from Portland Stone to Chalk in Worbarrow Bay and across the Purbeckian strata in Durlston Bay. Cliffs about 30 m in height are nearly vertical and truncate several small valleys, such as at Seacombe and Winspit Bottom, which show varying levels of adjustment to changes in base level. At St Aldhelm's Head (sometimes referred to as 'St Alban's Head'), these simple cliff-forms are replaced by large landslips that become increasingly active as clays and shales of the Portland Sand and Kimmeridge Clay are exposed around Chapman's Pool. The role of landslide debris in cliff protection is well exemplified here. The cliffs exceed 120 m but fall to 60 m or less west of Hounstout where Kimmeridgian rocks form both near-vertical cliffs and extensive platforms that extend several hundred metres offshore. To the west of Kimmeridge, the cliffs are dominated by resistant outcrops of Portland and Purbeck beds that are seen in alternating high, often vertical, cliffs and narrow submarine ridges that occasionally appear at low tide, as at Man o' War Rocks, immediately east of Durdle Door. The overall development of the bays, which include Lulworth Cove, has been seen as exemplifying different stages of coastal evolution. The relationships between structure, rock material and coastal form are well exemplified between Lulworth Cove and Bat's Head as caves, arches, rock offshore reefs, headlands and bays are developed to a greater or lesser degree.

The complex structural patterns of the cliffs and platforms between Ringstead and Black Head give rise to a great variety of forms providing an excellent location for examination of differential erosion processes. From Black Head westwards to Furzy Cliff, the cliffs are dominated by several different mass-movement systems that feed the beaches with a variety of sediment ranging in size from clay to boulders.

The area is microtidal, with a range less than 2.0 m, and tidal streams are generally weak. The prevailing and dominant wave trains are south-westerly, at times with origins in the southern Atlantic Ocean, and periods of 10 seconds or more are common. Waves from the south-east are characterized by a period of 5 seconds and are restricted in fetch. They tend to degenerate with the onset of south-westerly or northerly winds. Nevertheless they can be important locally in moving beach material within the semi-enclosed bays and exposing bedrock to erosion (Heeps, 1986).

The western end of the site lies at the north-eastern extremity of Weymouth beach. The beach is in deficit, and erosion of the Oxford Clay has contributed to mass-movements in Furzy Cliff. Cliff-top retreat has averaged about 1 m a^{-1} in recent years (Figure 11.34); most change occurs in relatively infrequent landslides. May (1964) described a rotational slip at the western end in January 1964, and more recently the whole of the cliff has become affected. Within a matter of weeks much of the material brought to the cliff foot by the slides is removed by wave activity. Slides occur frequently, with spatially separated larger events taking place about every eight years. In contrast, the cliff foot east of Bowleaze Cove (cut in Osmington Oolite and Bencliff Grit overlying Nothe Clay and Nothe Grit) is naturally armoured by boulders derived from rockfalls and reveals very little retreat of the lower cliff face. However, there have been two major failures on the upper slopes, separated by some 70 years (1900 and

1971). The first, described by Richardson (1900) affected the whole cliff, but the area remained largely unchanged apart from progressive degradation and establishment of a mature vegetation cover of brambles and grasses, interspersed with waterlogged hollows dominated by rushes *Juncus* behind the slip elements. During the 1970s this cliff once again became very active with a series of shear planes dividing a staircase of slide blocks reaching some 80 m inland of the cliff edge of the early 1970s. This part of the site thus demonstrates very vividly the effects and interplay of marine and subaerial processes on this coast.

Between Redcliff Point and Bran Point, the nature of the cliffs varies greatly with lithology and structure, and also upon the form of the cliff foot and intertidal zone. At the western end there is a small shingle beach that normally protects the cliff foot from direct erosion, whereas farther to the east the Nothe Grit forms a resistant foot to the cliff and a beach is absent. Eastwards from Osmington Mills, outcrops of the Corallian strata marked by a series of struc-

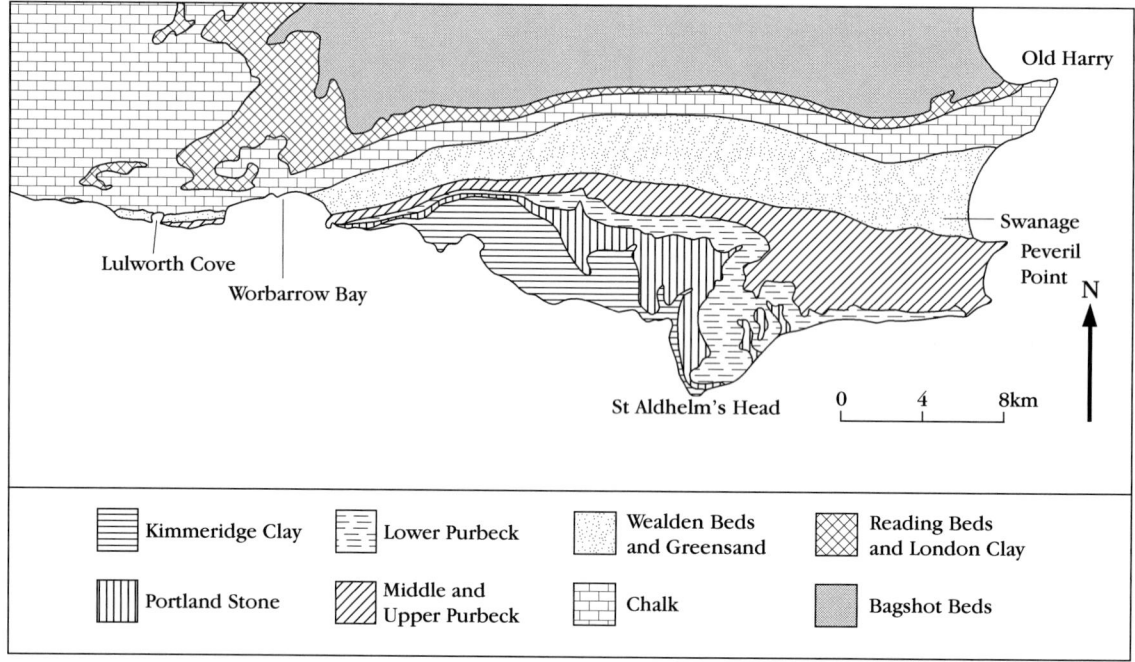

Figure 11.32 Geological map of the Dorset coast from Lulworth Cove to Studland Bay.

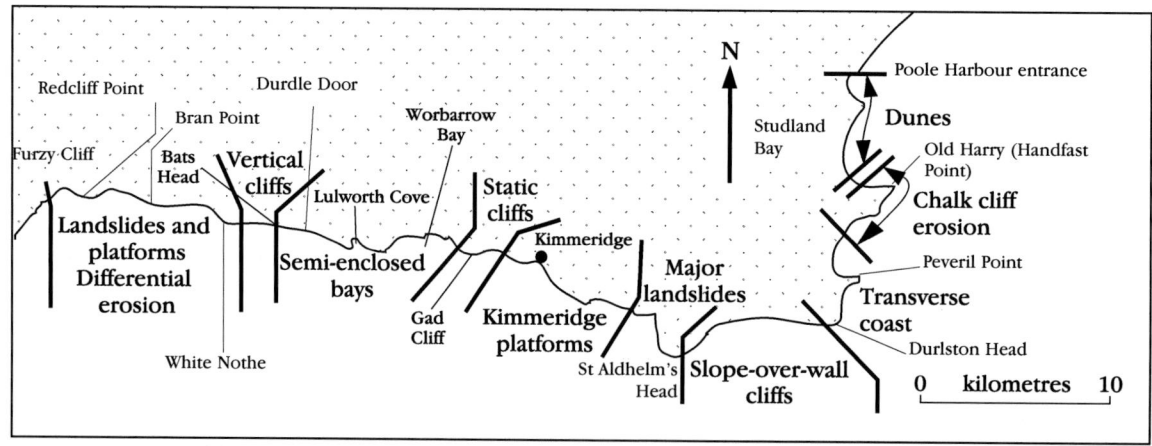

Figure 11.33 Summary geomorphological character of the coast between Furzy Cliff and Poole Harbour.

turally controlled stepped platforms. Some erosion of the detail of the platforms depends upon cobbles that are rolled along the weaker junctions. Arkell (1947, 1951a, 1955) suggested that rare events (such as the Martinstown storm of 18 July 1955, when over 280 mm rain fell, over 180 mm of which fell in 4.5 hours) may have played a significant role in re-shaping much of the coastal slope east and west of Osmington where it is dominated by clays.

Ringstead Bay, cut into the Kimmeridgian strata, lies between Bran Point and White Nothe. From cliffs about 30 m high at Bran Point it falls to a series of slumped and heavily vegetated slopes that are only 5 m high at Ringstead Bay (Figure 11.35). At its eastern end there is an active cliff between 2 and 35 m in height that retreated more than 3 m between 1996 and 1998 into the foot of the White Nothe landslide complex. The cliff top behind the landslides, however, attains an altitude of 150 m.

The beach at Ringstead is formed almost entirely of rounded oxidized flint, ranging in size from coarse sand to cobble, the latter mainly where Chalk enters the beach from the White Nothe cliffs. Heeps (1986) showed that this beach has a balanced sediment budget although considerable movements of sediment occur within Ringstead Bay. This beach, like most others to the east, has a very abrupt seaward boundary about 20–30 m offshore, where it rests on a rock platform. The beach moves between the ends and centre of the bay and between the upper and the lower beach. Thus over the period of about 15 months (1983–1984) when profiles were surveyed, a loss of about 440 m³ was balanced by deposition of almost exactly the same amount. There are extensive submerged and intertidal platforms formed mainly in Corallian strata, and these filter reduce the wave energy approaching this beach. Seaweed growth has been observed on much of the platform and Heeps (1986) recorded weed-rafting of material from platform to the beach, but suggested that it plays only a small part in augmenting the sediment within Ringstead Bay.

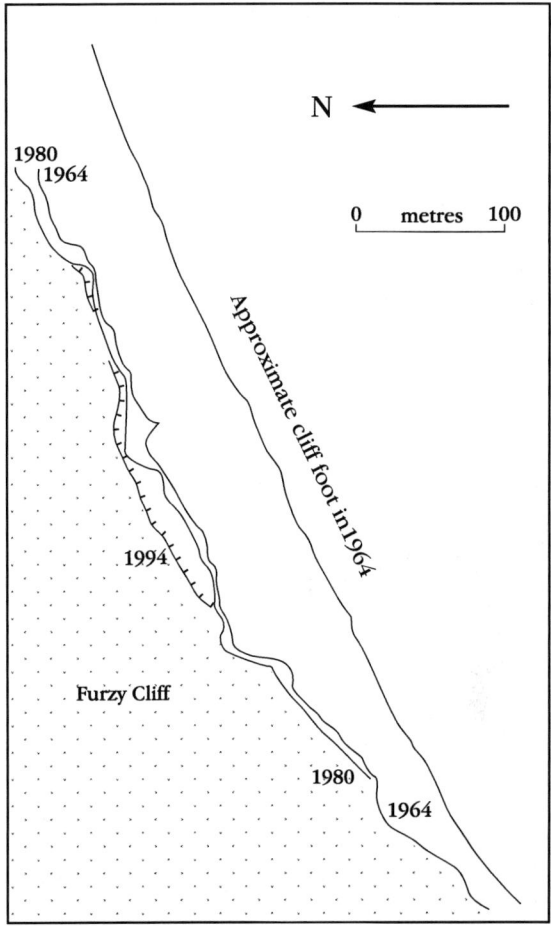

Figure 11.34 Cliff retreat at Furzy Cliff.

Table 11.4 Rates of cliff-top retreat since c. 1900 on the Dorset Coast.

Mean annual rate (m a⁻¹)	Rock type	Location of retreat
0.01	Portland Stone	Durlston Head to Winspit
0.18	Chalk	Hambury Tout to White Nothe
0.22	Chalk	Worbarrow Bay
0.25	Purbeck Beds	Durlston Bay
0.37	Jurassic clays	Furzy Cliff to Shortlake
0.38	Wealden	Worbarrow Bay
0.39	Kimmeridge clays and shales	Kimmeridge
0.41	Kimmeridge clays	Ringstead
0.43	Kimmeridge clays	Chapman's Pool
0.50	Wealden	Lulworth Cove

Since 1986, there has been considerable erosion of the low cliffs fronting the caravan park and boulders have been placed at the foot of the cliffs to reduce erosion, together with a small rubble groyne that has affected sediment movement on this section of the beach. By 1995 the central part of the beach had been lowered by over 1.5 m and only a thin veneer of shingle rested on the underlying clay and shale. The rapid depletion of the beach is attributed to a series of five southerly storms in 1993 that removed virtually all of the shingle. Unlike the beaches to the east and west, which have also been lowered from time to time, the central beach has not recovered. A coast protection scheme prepared by West Dorset District Council in late 1994 aimed to replenish the beach and to provide some stability. By October 1995, the eastern beach was severely depleted at its western end. During spring 1996, beach replenishment and further rock armouring of the cliff foot were carried out by West Dorset District Council. By June of the same year, 40% of the surface of the beach to the west comprised angular flint clasts derived from the replenishment site, and by the autumn of 1996 this material had been integrated into the top 0.5 m of the beach towards Bran Point. The replenished beach had been lowered by over 1 m but neither the former cliffs nor the underlying clay had been re-exposed. Between spring 1996 and mid-February 1998, parts of the clay cliff at the western end of the replenishment site had retreated by up to 2.8 m.

At the eastern end of Ringstead Bay, the coastal landscape is dominated by the high complex cliffs between Burning Cliff and White Nothe (Figure 11.35). The profile from the cliff top to the beach is distinguished by a near-vertical upper cliff that is interspersed with gentler grassed slopes almost reaching the cliff-top edge. Some of these slopes comprise angular flint screes cloaked by vegetation and a vestigial soil cover. In places, however, the screes themselves are exposed, forming distinct features at angles between 27° and 32° that fall to a hollow formed behind the large rotational-slip blocks that characterize the middle cliff (Figure 11.36). Below these the cliff has become active in recent years and large areas are affected by shear planes, slide scars and several zones of movement. Former-slip blocks are exposed within some of these areas.

The Kimmeridge Clay crops out at the eastern end of the bay, but is progressively cloaked by chalk landslide forms. Failures in the clays are partly responsible for a complex 'staircase' of rotational-slip blocks and active mass-movements that feed chalk to the eastern end of Ringstead Bay. The beach here, however, receives only small quantities of such material and is composed mainly of rounded flint clasts. Chalk boulders and fresh flint nodules are transported to the beach by a series of landslides. Many of the boulders remain in the intertidal zone and provide effective protection to the foot of the cliff. Both chalk and flint are broken down into shingle-sized fragments, but little reaches the main beach in Ringstead Bay because it is trapped behind the boulder ramparts on the middle and lower shoreline.

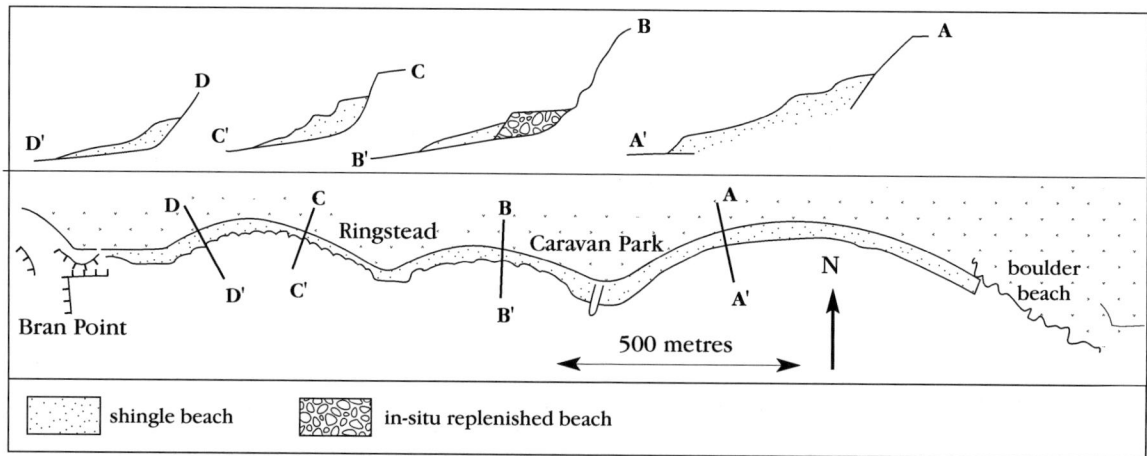

Figure 11.35 Cross-sections of the beaches of Ringstead Bay, and sketch map showing locations of sections.

At White Nothe itself, the cliffs are much less complex, the whole cliff is composed of Chalk here. From White Nothe to Bat's Head, the cliffs truncate a series of dry or 'combe' valleys. The beach is formed of newly deposited chalk and flint mixed with subsidiary amounts of rounded oxidized flint, the range of clasts showing a diversity in the degree of roundness. Although the beach rests on a Chalk platform, the platform is poorly developed. This is the only beach within the Chalk sector of the English coast where recently produced Chalk and flint clasts dominate the beach. The erosion of the eastern side of White Nothe, together with some landslide debris transported from the eastern end, provide the main source of clasts. There is little evidence of major falls from the cliffs behind the beach, although many small falls occur; much of the contemporary erosion is of the toes of relict talus slopes. Offshore, side-scan sonar surveys (Heeps, 1986, 1987) have shown that there are important extensions of the reef features that characterize the coast to the east and extend from off White Nothe to Worbarrow Tout. Geological structures are revealed particularly well in many of the seabed forms, despite some cloaking by thin veneers of rippled sand. The foresyncline of the Purbeck monocline that is cut by the cliffs west of Bat's Head forms a distinctive feature of the submerged coast (Figures 11.37– 11.39.)

The section of coast from Bat's Head to Man o'War Bay has been described in detail by Heeps (1986). It has three main morphological elements, high Chalk cliffs, lower cliffs or bay floors cut into the Lower Cretaceous and Upper Jurassic strata, and a discontinuous rock reef formed in Portland Beds. Its best-known feature is Durdle Door, an arch cut through the near-vertical strata of the Portland Stone. The reef continues across the bays both to the east and west of Durdle Door. Because it reduces much of the wave energy approaching the beaches and prevents the beach sediments from leaving the bays other than in suspension, the reef plays a significant part in making these bays closed sediment cells. Heeps (1986) showed that, whereas there is some supply of flint and Chalk into the beach, the overall volume of sediment held within the bay west of Durdle Door changes little although there can be substantial changes in form from time to time depending upon wave conditions. The mainly flint shingle beach rests upon a narrow chalk platform that is exposed

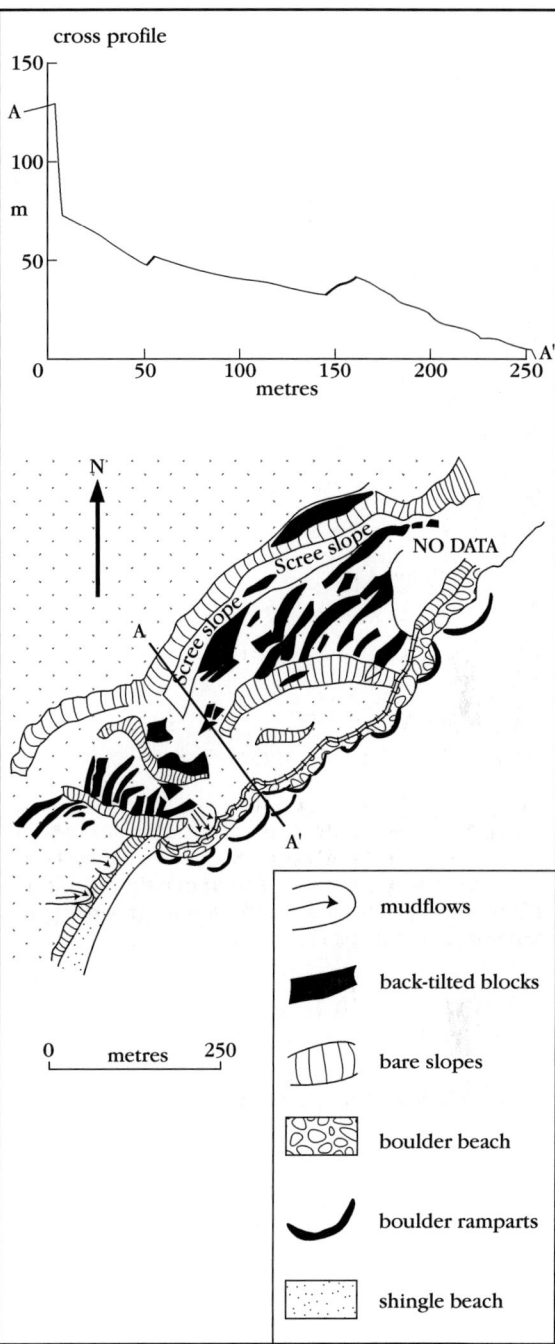

Figure 11.36 Geomorphology and cross profile of the White Nothe landslides in the Dorset Coast GCR site.

when shingle is re-distributed within the bay. The platform slopes at angles of up to 10°, is about 15 m in width, and is covered by fine shingle beach rarely thicker than 1 m, with little berm development. Beyond the beach edge,

Figure 11.37 Sonargraph of the seabed between White Nothe and Bat's Head, showing an area of tilted strata (dipping from north-east to south-west) and joint and block patterns. The upper right hand area shows blocks broken from the strata. (Image by permission of C. Heeps.)

there are scattered clasts on the platform. The seabed drops steeply outside the reef to depths of about 14 m. At the western end of the bay, Bat's Head is penetrated by a small arch along the vertically dipping bedding planes of the Chalk. A stack rises from the platform. Erosion is more rapid where dry valleys floored by solifluction materials that reach almost to highwater mark have been truncated by the retreat of the coastline.

Daily surveys of the Durdle Door beach (Heeps, 1986) for a month in the winter of 1983 show that prolonged exposure to easterly wave regimes resulted in changes that although small in magnitude were significant with respect to sediment circulation within a morphologically constrained unit (Figure 11.40). Storm action (associated with south-westerly waves) accounted for large changes in the beach volume, but the process was a simple onshore–offshore exchange, the material being returned by post-storm conditions. Easterly waves occurred less frequently, but their effect was much greater. In particular, swell from the east initiated long-shore drift that brought about exposure of the Chalk platform in the west. Heeps regarded such conditions as the 'extreme event' experienced by beaches along the south-east Dorset coast. During the longer 15-month period of Heeps' survey, the beach showed almost no variation in volume with 618 m³ removed against 605 m³ deposited. Her survey also suggested that within this bay there is a tendency for a sediment parting towards the western end of the Durdle Door bay.

Man o'War Bay has a small shingle beach, sheltered by Man o'War Rocks, which develops a partial cuspate form in the lee of the easternmost part of the barrier. Cusps are commonly present on the beach with a distinctive grading of their dimensions from large in the centre of the bay to smaller at both extremities. Sediment size is also graded from smaller (− 2 phi) material at the ends of the bay and larger in the centre (> −3.5 phi). Both the cusp and sediment grading are characteristic of these bays. Remnant upper berms as well as contemporary berms, both with cusps, are common.

The cliffs rise eastwards from Durdle Door to a height of 138 m at Hambury Tout. Stair Hole and Lulworth Cove form the best-known features of this coast. The rounded cove cut into the Chalk back wall contrasts with the more linear form of the adjacent Stair Hole. The mudslides in the Wealden strata of Stair Hole are gradually cutting back its landward side, and wave action has opened out a series of arches in the hard Jurassic limestones that form its outer side. Waves now access the toe of the mudslides, and during storm periods and after prolonged rainfall, they become very active. Lulworth Cove is known internationally as the classic form of a near-circular bay resulting from the differential erosion of weaker strata behind a resistant and protecting outer wall of harder rock. It is described in almost all texts on physical geography both within the British Isles and worldwide.

Both east and west of Lulworth Cove, the coast is formed by vertical cliffs in the Portland Stone. At Mupe Bay, the coastline is cut into the younger beds and a small bay with a shingle beach has formed. The Portland and Purbeck strata that dip steeply here have been eroded to close to sea level but extend seawards across

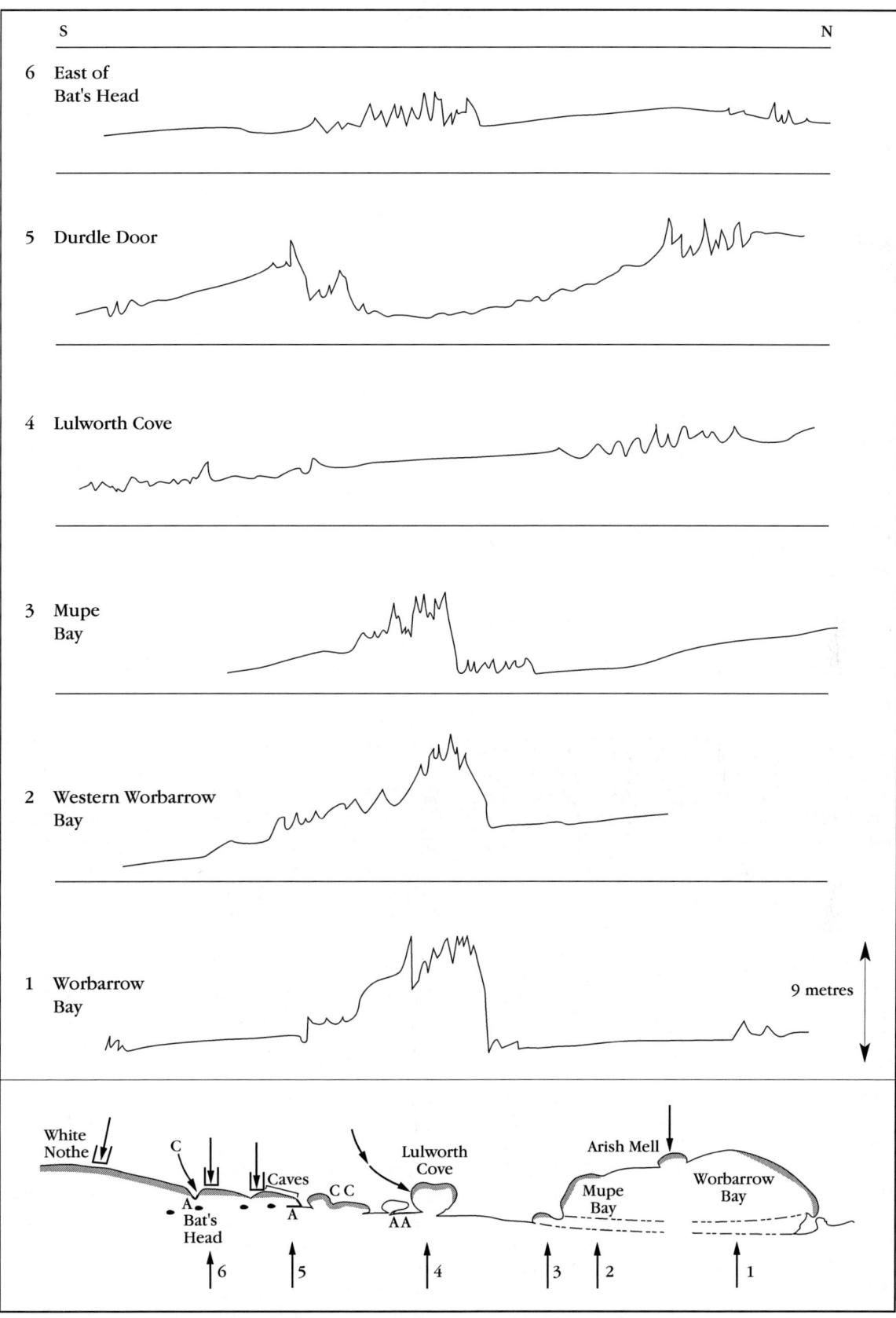

Figure 11.38 Profiles of the submerged rock ridges between Bat's Head and Worbarrow Tout. The ridges are formed by steeply dipping Purbeckian and Portlandian strata between the headlands along this coast. Typically the northern face of the ridges is formed by the structural surface of strata, which here dip northwards (inland). Opposite Lulworth Cove, the ridges are less than 1 m in height, whereas in Worbarrow Bay they attain 9 m. C = Cave; A = Arch; hanging dry valleys are shown with a bracketed arrowhead, whereas valleys draining to sea level are shown without a bracket. Locations of the profiles 1 to 6 are indicated in the plan view.

Coastal assemblage GCR sites

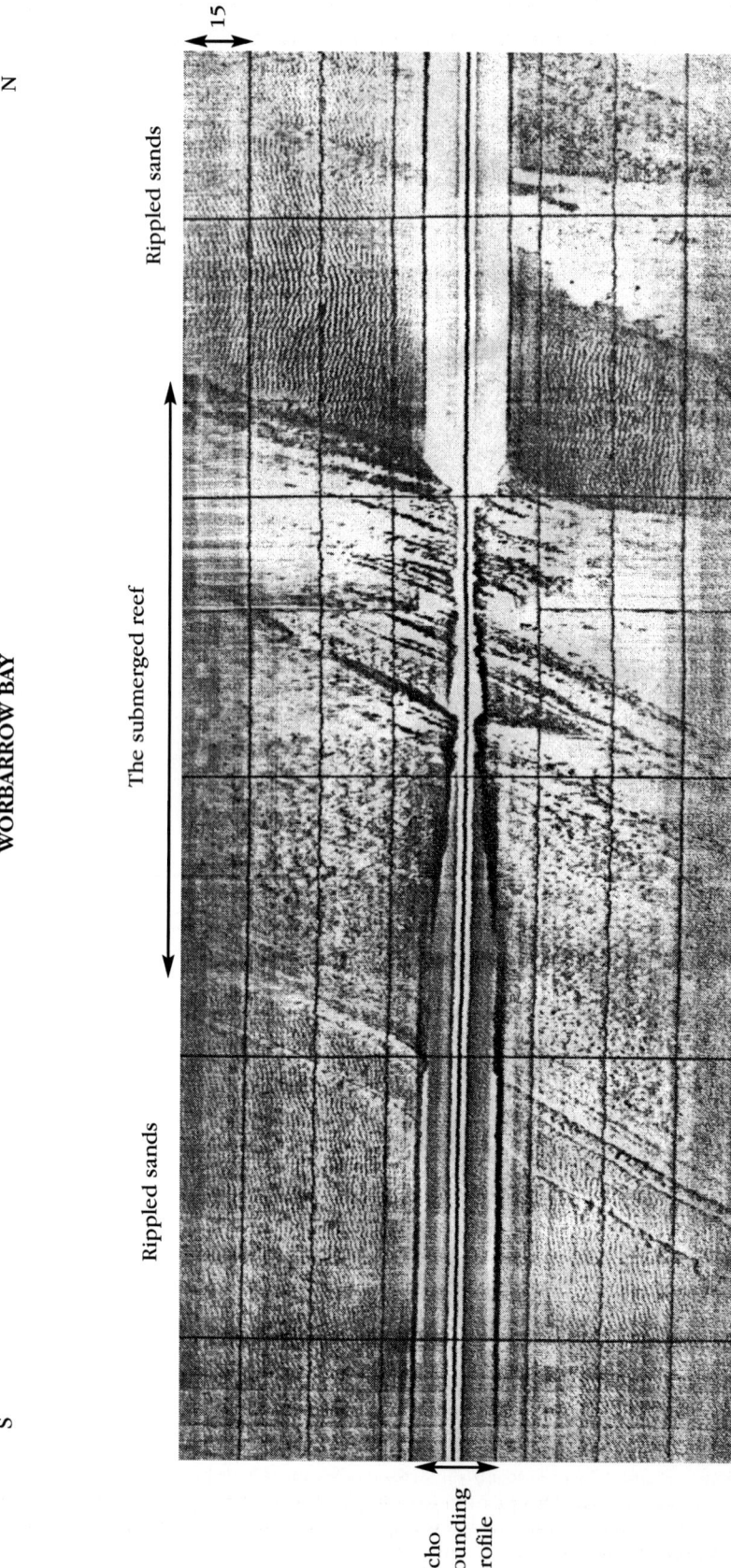

Figure 11.39 Sonargraph of the Worbarrow Bay submerged rock reef. The echo-sound profile shows the depth profile of the seabed and the distinct, steep, backwall of the ridge. Either side of the echo-sound trace, the sonar image shows the seabed patterns of the individual ridges formed by the eroded, dipping strata. The alignment of the sand ripples indicates that sand movement is alongshore. (Photo reproduced by permission of C. Heeps.)

The Dorset Coast

Worbarrow Bay to re-emerge at Worbarrow Tout. The detailed form of this offshore feature has been described by Heeps (1986) as have the nature and behaviour of the cliffs and beaches in Worbarrow Bay. The submerged reef appears to describe an arc from Mupe Rocks to Worbarrow Tout with its apex opposite Arish Mell. The reef stands as a series of prominent ridges and troughs separating a flat plain both to landward and seaward. A second reef trending southwest–north-east is aligned with Gad Cliff. It is about only 1.5 m in height and is made up of a series of very jagged, parallel ridges crossed by numerous small faults. Bifurcating and sinuous sand and shell ripples cover much of the intervening floor of the bay. The main Worbarrow Bay Portland Stone reef is a substantial feature, up to 9 m in height above seafloor (Figure 11.39). It is rugged, with a crest marked by a series of parallel ridges separated by troughs that are up to 3 m deep. On its landward side the reef has an almost-vertical wall. The seaward slope is gentler. The average depth of water seaward of this feature is 23 m, whereas to landward it is only 18 m. Opposite Arish Mell it stands at 6 m high (c. –12 m OD) and has a gap, probably the line of the former valley through the ridge. The presence inside the reef of unbroken shells in profusion suggests that this is a low-energy environment with little sediment movement induced by wave or current action. There are large boulder accumulations to seaward of Gad Cliff and opposite the main cliff falls at Cow Gap.

The shape of Worbarrow Bay is strongly controlled by the outcrop pattern of Chalk and Upper Greensand, Wealden and the Purbeck and Portland Beds (Figures 11.32 and 11.41). The Chalk cliffs drop directly into the sea and there is no intertidal platform except at the extreme western end of the bay. A large shingle beach rests against the slumped cliffs cut in Wealden strata, which are affected by many small landslides (Allison and Brunsden, 1990). Despite this inherent instability and the development of mud fans on the upper beach, especially after prolonged rainfall (Heeps, 1986), the cliffs supply little sediment to the predominantly flint beach. Changes in the beach are mainly onshore–offshore; there is some longshore movement within the bay, but there is no dominant direction of movement. There is a well-developed upper berm at about 5.45 m OD that is affected only during major storm events and

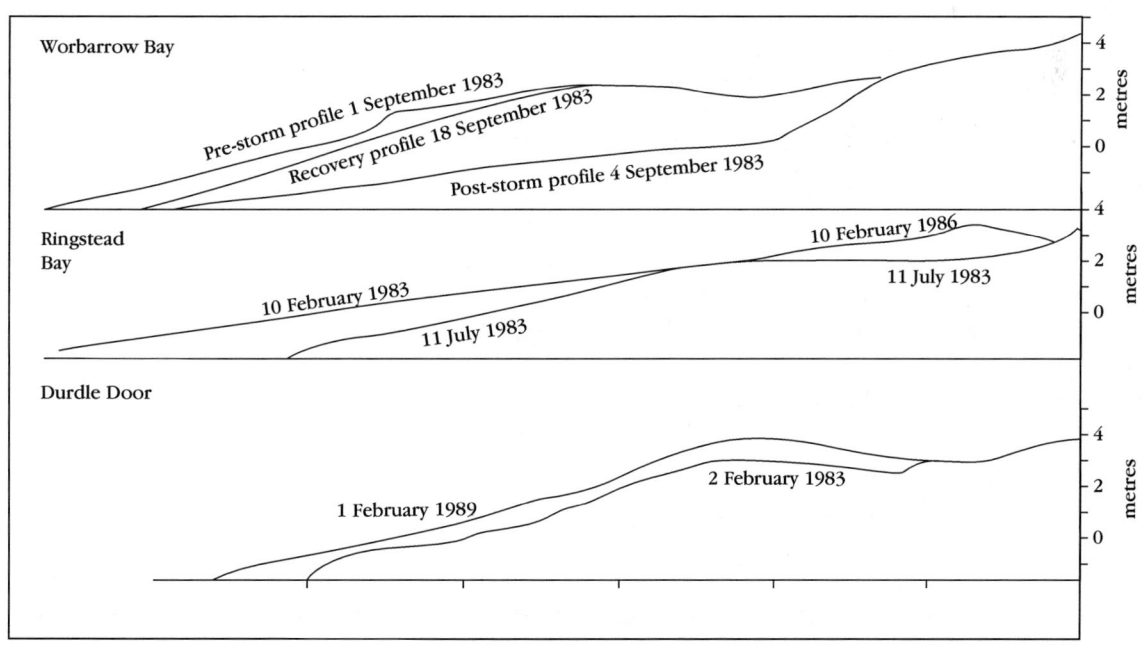

Figure 11.40 Changes in shingle beach profiles in Worbarrow Bay, Ringstead Bay and Durdle Door. Erosion and recovery of beaches can take place over very short timescales. These profiles are based upon monthly resurveys of sample transects. (After Heeps, 1986.)

can be as steep as 30°. These steep profile areas are the most exposed and are composed of the coarsest and best-sorted sediments (Arkell, 1947; Heeps, 1986).

The cliffs at Gad Cliff are very exposed (Figure 11.42), and are characterized by an upper cliff predominantly in Portland Stone and a vegetated lower slope on the Portland Sand and the Kimmeridge Clay. There are large boulder accumulations at the cliff foot that very effectively protect the cliff. The boulder fields continue offshore for several hundred metres. East of Gad Cliff, the cliffs decrease in height towards Kimmeridge Bay. Low (c. 15 m) steep cliffs in gently dipping clay and shale strata stand behind a very well-developed series of wide platforms of Kimmeridgian cementstone (Figure 11.43; Arkell, 1947). The form is mainly controlled by geological structure.

Between Hounstout and St Aldhelm's Head, the coast is dominated both by high (to 130 m) limestone-capped cliffs and large undercliffs. Areas of very rapid change (Figures 11.44–11.47) occur via landslides wherever the Kimmeridge Clay occurs at sea level. At Chapman's Pool, a small semi-circular cove has developed between headlands formed of large boulder accumulations resulting from the landslides. Small platforms are exposed in the intertidal zone in Chapman's Pool. The stepped profile described by Brunsden (1973) for west Dorset is found here in both active and passive landslide areas, although in the latter there has been widespread degradation of many of the clay slopes. Flat-topped ridges of Portlandian limestone and sandstone lying on Kimmeridge Clay are truncated by landslides of various (but as yet undetermined) ages. The strata dip southwards and seawards at about 2° and there is a fault to the east of St Aldhelm's Head. There are three major spatial units within this landslide region (May, 1997a).

1. An active landslide at Hounstout cliff distinguished by clay slopes, frequent rockfalls, mudslides and widespread gullying. Vegetation occurs mainly on the back-tilted, rotational slip blocks. There is much standing surface water and during rainfall, runoff is high. There are many small, frequent, mass-movements. These landslides have been particularly active since about 1970 (Jones, 1980).
2. At Emmet's Hill, the landslides are not particularly active today, although back-tilted blocks of limestone and sandstone indicate significant past movements. The hollows between these blocks are filled by debris from localized rockfalls and by downwashed sand and clay. Small trees (10 m high) grow in some of these

Figure 11.41 Looking eastwards across Worbarrow Bay, showing the outcrop of Purbeck and Portland beds at Worbarrow Tout, the Wealden cliffs undergoing erosion and the shingle storm-beach with cusp development. (Photo: V.J. May.)

The Dorset Coast

Figure 11.42 Gad Cliff. The upper cliff is in Portland Stone; the debris and boulder field, well-vegetated by scrub, lies on Portland Sand and upper Kimmeridge Clay. The boulder beach has alternating ramparts and baylets related to differential erosion (associated with bedding in the Kimmeridge Clay) and debris toes. (Photo: V.J. May.)

hollows. Most movements of the 20th century have been rockfalls from the high limestone and sandstone cliffs, but since 1990, the slopes in Kimmeridgian clays have been affected by shallow slides. At the cliff foot, there has been increased marine erosion.

3. St Aldhelm's Head is marked by a wide undercliff cloaked by boulders. There are a number of large debris fans produced by both natural processes and quarrying, the latter during the 19th and 20th centuries. Back-tilted blocks appear to be absent. Small (1970) suggested that these slopes may have formed under frost action during the Quaternary Period, but there is as yet no firm evidence to support this view.

East of St Aldhelm's Head, the undercliff narrows rapidly and the cliffs become lower in height, but steeper. Immediately east of St Aldhelm's Head, although there is no true undercliff, a steep grassed slope on the Portland Sand lies below the steeper upper cliff of Portland Stone. Boulders protect this from erosion in a similar way to the cliffs at Gad Cliff.

To the east towards Durlston Point, the cliffs are steep. Sometimes they plunge directly into the sea, in other places they have a small coastal platform. This is most characteristic of the mouths of the small hanging valleys that characterize this coastline at Winspit, Seacombe and Dancing Ledge. At the mouth of Seacombe, the cliff face truncates a former stream valley infilled by angular debris. The slope-over-wall form of much of this coastline has been modified in places by lynchets (man-made terraces). The lower part of the slope is often cloaked by angular debris, the thickness of which appears to deepen downslope. Some slope sections are well-exposed where quarrying has cut into the slope. The valleys are floored by angular debris. At Seacombe the valley is flat-floored almost to its mouth, whereas the Winspit valley is distinguished by a series of incised meanders. Although these features are not parts of the original GCR site, they are related to the origin of the coast and the extent to which it is a contemporary feature or a reworking of earlier forms.

At Durlston Head, the coastline changes direction to become a transverse one. Much of the cliff has been affected by landslides; they recur

Figure 11.43 Broad Bench, a Kimmeridge shale platform, showing block removal at platform edge. (Photo: V.J. May.)

Figure 11.44 The cliffs between Hounstout and St Aldhelm's Head are characterized by large landslides of as yet undetermined age. Large Portland Stone boulder fields provide protection against wave attack, but where they are absent shoreline retreat produces steep lower cliffs. (Photo: V.J. May)

The Dorset Coast

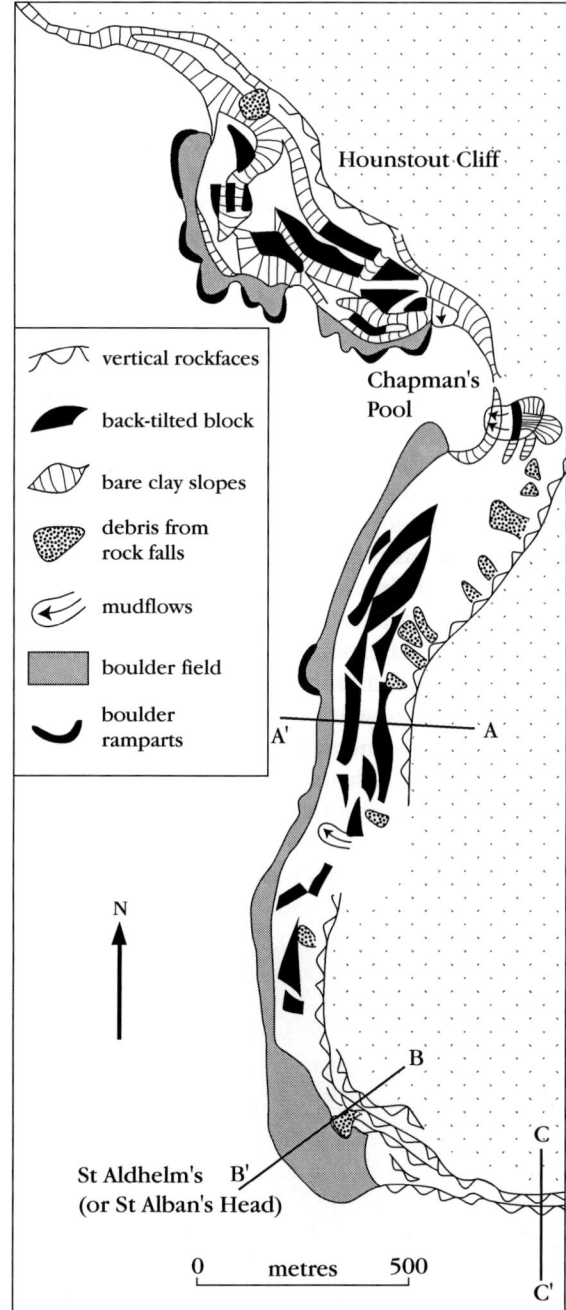

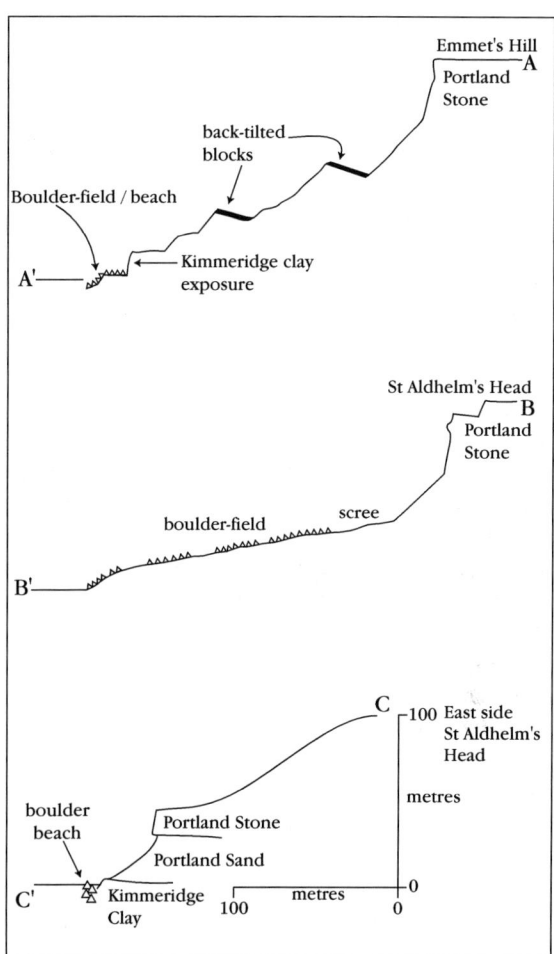

Figure 11.45 Emmet's Hill and St Aldhelm's Head cross-profiles. For locations of profiles A–C, see Figure 11.47.

Figure 11.46 Landslide features between Hounstout and St Aldhelm's Head. (After May, 1997a.)

sporadically. In late 1994, for example, a small slide affected the cliff to the north of Durlston Castle. Part of this cliff has now been modified by dumping of boulders in an attempt to protect cliff-top dwellings. The southern cliffs of Durlston Bay are more active than their well-vegetated characteristics imply. The northern extremity of the GCR site lies at Peveril Point, where a series of intertidal reefs reflect the synclinal tectonic structure here.

Interpretation

Five issues have been the focus of debate and research along this coastline:

1. The development of the headland–bay topography east and west of Lulworth Cove.

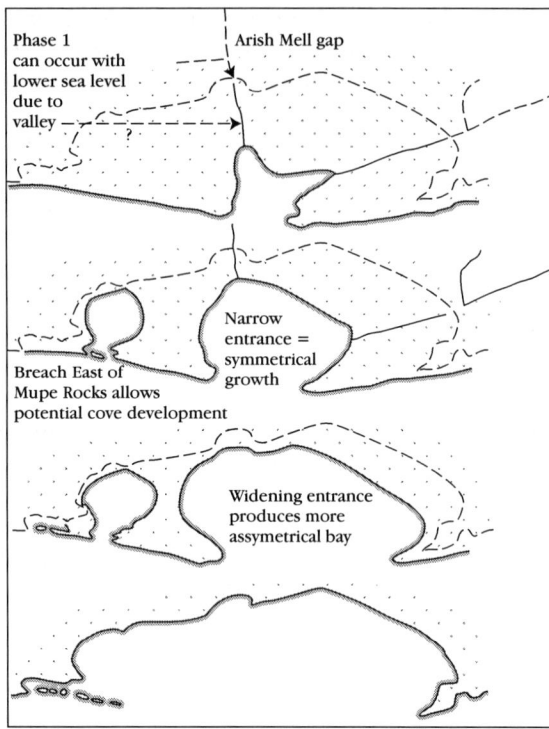

Figure 11.48 Development of Worbarrow Bay from initial flooding of former valley along line of Arish Mell gap. The breakthrough in the vicinity of Mupe Rocks develops in early stages in similar way to that at Stair Hole. Subsequently, both Worbarrow and Mupe bays merge with increasing asymmetry as the eastern shoreline of Worbarrow Bay becomes more exposed to waves from English Channel.

attracted attention. Between Durlston Point and Furzy Cliff, 23 valleys are cut by the coastline. Of these, 11 have been truncated by the sea, four having permanent or occasional waterfalls and six valleys reach the coast by a gorge or deeply incised valley. At Lulworth and Arish Mell, partially drowned valleys are cut in the Portland beds. Four occur in the Portland and Purbeck beds, seven in the Kimmeridge Clay, and eight in the Chalk. Three are in the clays of the Wealden or Oxford Clay. The rates of present-day cliff retreat in the clays are such that the lack of incision of streams could be attributed to their inability to keep pace in their downcutting with the retreating cliffline. In the Chalk, the dry valleys are truncated by the sea in a similar way to those in other Chalk sites at the Seven Sisters (Sussex) and between Kingsdown and Dover (Kent; see GCR site reports). In the Isle of Purbeck, the valleys at Winspit and Seacombe both have floors that are not graded to present-day sea level. Both have deeply incised rock valleys that are filled by angular debris. At Seacombe a cross-section of this infilled valley is exposed in the cliff face (Figure 11.47), whereas at Winspit it has mostly been excavated by a series of incised meanders. The coastal slopes around these valleys have a slope-over-wall form. The lower cliff cuts across the lower part of the 'slope', which is characteristically cloaked by angular debris.

There are several possible interpretations for this combination of forms. The first is that the truncated valleys are a result of the varying ability of coastal streams to keep pace with cliff retreat in different lithologies. A second possibility is that the present-day cliffs are being reworked, having been cloaked by angular slope deposits during the last glacial period. The last 6000 years have been a time during which debris has been removed, with some erosion of the cliffs and some removal of the valley infill, especially in the Winspit valley where the stream has a higher discharge. This is comparable to present-day retreat rates of about 0.03 m a^{-1}. Both to the west of Winspit and at Gad Cliff, there are high cliffs that have a vegetated slope below a vertical face. Both show very little sign of retreat and owe their present stability largely to the protection afforded by boulder accumulations at the foot of the slope. In the absence of specific dating of the slopes it is possible to only speculate on the period over which they have had this form beyond the last couple of centuries. Offshore from Gad Cliff, there are large accumulations of boulders that can be interpreted as resulting from cliff retreat during a lower or rising sea level and which would reduce the energy of waves approaching this coastline. The large landslide at St Aldhelm's Head appears to be covered in parts by angular debris, which Small (1978) suggested may have a periglacial origin. Most of the evidence for reworking of an earlier coastline is circumstantial (Mottram, 1972), in the absence of dated sediment. However, it is difficult to explain many of the forms of the hard-rock coastline without considering this possibility.

Heeps' (1986) study of the submarine topography and the additional interpretation of later side-scan sonar records reveal an intricate morphology in which boulder accumulations, submerged platforms, ridges and troughs are cloaked by a veneer of sand, shell and coarser sediments (May and Drayson, 2001). There is some evidence from repeated seabed surveys that the veneer is subject to slight changes, rather than major changes. Some of the larger boulder fields are associated with cliffs where collapse carries boulders of sufficient size to the shoreline. That these areas of boulders continue offshore argues that they have been active over much of the period of Holocene sea-level rise and of present-day sea level. Much more work needs to be done on this part of the site, but there are few other areas where the nature of a cliffed coastal zone has been investigated from the cliff top to the usual seaward limit of wave action.

Concerning the question of the grading of sediments within the bays, Arkell (1947) commented on the tendency towards coarser sediment in the centres of the bays. Heeps (1986) confirmed that this is generally the case and put forward reasons for the wave-sorting that occurs. She also demonstrated that most of these bays are closed sediment-cells, in that the enclosing headlands and the offshore barriers prevent outward sediment transfers of particles other than clays in suspension. Inward transfers also appear very limited because of the nature of the barriers and their height. The beaches, therefore, depend solely upon the addition of shingle from the erosion of the cliffs and the longevity of flint already within them. Measurement of the sedimentary product of rockfalls from the Chalk (May and Heeps, 1985; Heeps, 1986) shows that most chalk is broken down into pebbles, and then shelly fragments and fines within a few months of the original cliff collapse. Boulders of limestone and sandstone that are upwards in size of 0.5 m may survive for longer in the intertidal and shallow water zone on the seabed if they are large and resistant enough.

The relationship between subaerial processes and the development of the beaches has attracted attention in the literature. Although Bray *et al.* (1992) indicate that the net direction of sediment transport in Weymouth Bay is from east to west, this is true only of the offshore transport routes where the sediments become finer westwards. Along the shoreline itself, transport is largely confined within the bays, where reversals of drift as well as localized onshore–offshore movements occur regularly. Of the shingle beaches, only one, east of White Nothe, appears to be formed mainly of contemporary material. All the others contain substantial quantities of well-rounded, oxidized flint clasts together with varying volumes of angular and subangular, grey flint. Davies' (1972) view that flint nodules are readily quarried from the Chalk platforms and so provide an 'abundant source of pebbles' (p. 118) is not borne out in the short term by Heeps' (1986) examination of the beaches in this area.

In summary, strike-aligned coasts are comparatively unusual in Britain. Most wave attack here is from the south-west, obliquely to the coastline, with a potential fetch of several thousand kilometres at the eastern end of the site. However, because the location is microtidal, much wave energy is concentrated in a narrow band at the base of the cliffs. A submerged rock reef along part of the coastline reduces wave energy inputs to the beaches, and sediment is conserved within each bay.

Conclusions

The Dorset Coast GCR site contains some of the most visually appealing coastal landforms in Britain. It is a world-renowned example of a longitudinal coastline, includes such classic landforms as Worbarrow Bay and Lulworth Cove, and provides an excellent field observatory for studies of cliff, beach and nearshore geomorphology. Lulworth Cove and Durdle Door are also known worldwide for their scenic value. The scientific importance of this site comes from the variety of erosional features, the debate about the sequence of their development, the linked submarine forms, the closed sediment systems of the bays, and the very large number of truncated (hanging) valleys.

This coast is a complex one, whose interpretation ranges from the analysis of the present-day processes to the unravelling of the longer geomorphological history of the site. The classic landforms have been studied at scales that often ignore the detail of smaller features and localized processes but these combine with the larger forms to suggest a complex and lengthy history for this coastline.

It is an unusual coast in having been the focus of both traditional geological and geomorphological study and submarine survey that now allows the whole coastal system to be described and analysed.

This coast is also important in containing the largest set of closed sediment-cells to have been described anywhere in the British Isles. There remain a large number of sub-sites within this site that warrant more detailed examination. It is also an most important rocky coastline on account of the assemblage of forms that occur within it, from the scale of the longitudinal coast to the individual features of the seabed, beaches and cliffs. In this totality of interest, it is of international importance to coastal geomorphology.

The Purbeck coast is also the location of Britain's first Voluntary Marine Reserve at Kimmeridge, is designated as a Special Area of Conservation under the European Union Habitats Directive and includes Jurassic stratotypes. Indeed, so important is the geology and geomorphology of this coast, that this site forms part of the Dorset and East Devon Coast World Heritage Site, which was declared on account of its Earth science features of interest in December 2001.

References

In this reference list the arrangement is alphabetical by author surname for works by sole authors and dual authors. Where there are references that include the first-named author with others, the sole-author works are listed chronologically first, followed by the dual author references (alphabetically) followed by the references with three or more authors listed *chronologically*. Chronological order is used within each group of identical authors.

Acton, J. and Dyer, C. (1975) Mapping of tidal currents near the Skerries Bank. *Journal of the Geological Society of London*, **131**, 63–7.

Adam, P. (1990) *Saltmarsh Ecology*, Cambridge Studies in Ecology, Cambridge University Press, Cambridge, 461 pp.

Adlam, W.J. (1961) The origin and source of some of the features of Chesil Beach, Dorset. *Southern Geographer*, **2**, 1–8.

Agar, R. (1960) Postglacial erosion of the North Yorkshire coast from the Tees estuary to Ravenscar. *Proceedings of the Yorkshire Geological Society*, **32**, 409–28.

Al-Bakri, D. (1986) Provenance of the sediments in the Humber estuary and the adjacent coasts, eastern England. *Marine Geology*, **72**, 171–86.

Allan, R.L. (1993) Distribution, geochemistry and geochronology of Sellafield waste in contaminated Solway Firth floodplain deposits. Unpublished PhD thesis, University of Glasgow.

Allen, J.R.L. (1987) Dessication of mud in the temperate intertidal zone: studies from the Severn estuary and eastern England. *Philosophical Transactions of the Royal Society of London*, **B315**, 127–56.

Allen, J.R.L. (1989) Evolution of salt marsh cliffs in muddy and sandy systems: a qualitative comparison of British west-coast estuaries. *Earth Surface Processes and Landforms*, **14**, 85–92.

Allen, J.R.L. (1990a) Salt-marsh growth and stratification: a numerical model with special reference to the Severn estuary, southwest Britain. *Marine Geology*, **95**(2), 77–96.

Allen, J.R.L. (1990b) Constraints on measurement of sea level movement from saltmarsh accretion rates. *Journal of the Geological Society*, **147**, 5–7.

Allen, J.R.L. (1990c) The formation of coastal peat marshes under an upward tendency of relative sea level. *Journal of the Geological Society*, **147**, 743–5.

Allen, J.R.L. (1991a) Fine sediment and its sources, Severn estuary and inner Bristol Channel, southwest Britain. *Sedimentary Geology*, **75**, 57–66.

Allen, J.R.L. (1991b) Saltmarsh accretion and sea level movement in the inner Severn estuary, southwest Britain: the archaeological and historical contribution. *Journal of the Geological Society*, **148**, 485–94.

Allen, J.R.L. (1992) Tidally-influenced marshes in the Severn estuary, southwest Britain. In *Saltmarshes. Morphodynamics, Conservation and Engineering Significance* (eds J.R.L. Allen and K. Pye), Cambridge University Press,

References

Cambridge, pp. 123–47.

Allen, J.R.L. (1996a) Shoreline movement and vertical textural patterns in salt marsh deposits. *Proceedings of the Geologists' Association*, **107**, 15–23.

Allen, J.R.L. (1996b) The sequence of early land-claims on the Walland and Romney Marshes, southern Britain: a preliminary hypothesis and some implications. *Proceedings of the Geologists' Association*, **107**, 271–80.

Allen, J.R.L. and Pye, K. (eds) (1992) *Saltmarshes. Morphodynamics, Conservation and Engineering Significance*, Cambridge University Press, Cambridge, 184 pp.

Allen, J.R.L and Rae, J.E. (1987) Late Flandrian shoreline oscillations in the Severn estuary: a geomorphological and stratigraphical reconnaissance. *Philosophical Transactions of the Royal Society of London*, **B135**, 185–230.

Allen, J.R.L. and Rae, J.E. (1988) Vertical salt-marsh accretion since the Roman Period in the Severn estuary, southwest Britain. *Marine Geology*, **83**, 225–35.

Allen, J.R.L., Friend, P.F., Lloyd, A. and Wells, H. (1994) Morphodynamics of intertidal dunes: a year-long study at Lifeboat Station Bank, Wells-next-the-Sea, Eastern England. *Philosophical Transactions of the Royal Society of London*, **A347**(1682), 291–344.

Allison, H. (1989) The sedimentary history of Scolt Head Island. In *Blakeney Point and Scolt Head Island* (eds H. Allison and J.P. Morley), The National Trust, Blickling, pp. 28–32.

Allison, P.A. and Pye, K. (1994) Early diagenetic mineralization and fossil preservation in modern carbonate concretions. *Palaios*, **9**, 561–75.

Allison, R.J. (1986) Mass movement and coastal cliff development of the Isle of Purbeck, Dorset. Unpublished PhD thesis, University of London.

Allison, R.J. (1989) Rates and mechanisms of change in hard rock coastal cliffs. *Zeitschrift für Geomorphologie, Supplementband*, **73**, 125–38.

Allison, R.J. (1990) *Landslides of the Dorset Coast, British Geomorphological Research Group Field Guide*, British Geomorphological Research Group, London, 125 pp.

Allison, R.J. (ed.) (1992) *The Coastal Landforms of West Dorset*, Geologists' Association Guide, No. **47**, Geologists' Association, London, 134 pp.

Allison, R.J. (ed.) (1999) *Dorset Revisited: Position Papers and Research Statements, British Geomorphological Research Group Spring Field Meeting 1999*, West Dorset Coastal Research Group, 109 pp.

Allison, R.J. (in press) *The Coastal Landforms of East Dorset*, Geologists' Association Guide, Geologists' Association, London.

Allison, R.J. and Brunsden, D. (1990) Some mudslide movement patterns. *Earth Surface Processes and Landforms*, **15**(4), 297–311.

Allison, R.J. and Kimber, O.G. (1998) Modelling failure mechanisms to explain rock slope change along the Isle of Purbeck coast, UK. *Earth Surface Processes and Landforms*, **23**, 731–50.

Ameen, M.S. (1990) Macrofaulting in the Purbeck–Isle of Wight monocline. *Proceedings of the Geologists' Association*, **101**, 31–46.

Ameen, M.S. and Cosgrove, J.W. (1990a) Kinematic analysis of the Ballard Fault, Swanage, Dorset. *Proceedings of the Geologists' Association*, **101**, 119–29.

Ameen, M.S. and Cosgrove, J.W. (1990b) A kinematic analysis of meso-fractures from Studland Bay, Dorset. *Proceedings of the Geologists' Association*, **101**, 303–14.

Andrews, J. and Chroston, N. (2000) Holocene evolution of the North Norfolk barrier coastline in the Holkham–Blakeney–Cley area. In *Norfolk and Suffolk* (eds S.G. Lewis, C.A. Whiteman and R.C. Preece), Quaternary Research Association Field Guide, Quaternary Research Association, London, pp. 131–47.

Andrews, J.E., Boomer, I., Bailiff, I.K., Balson, P., Bristow, C., Chroston, P.N., Funnell, B.M., Harwood, G.M., Jones, R.W., Maher, B.A. and Shimmield, G. (2000) Sedimentary evolution of the north Norfolk barrier coastline in the context of Holocene sea-level change. In *Holocene Land-Ocean Interaction and Environmental Change around the North Sea* (eds I. Shennan and J.E. Andrews), *Geological Society Special Publication*, No. **166**, The Geological Society, London, pp. 219–51.

Anglian Water Authority (1980) Improvement of Seawalls of the Dengie Peninsula. Saltings Regeneration – Pilot Scheme. *Engineer's Report*, **ROI 1537**, Essex Rivers Division, Anglian Water Authority, Colchester.

Angus, S. (1994) The conservation importance of the machair systems of the Scottish islands, with particular reference to the Outer

References

Hebrides. In *The Islands of Scotland: a Living Marine Heritage* (eds J.M. Baxter and M.B. Usher), *Natural Heritage of Scotland Series*, No. **1**, HMSO, London, pp. 95–120.

Angus, S. (1997) *The Outer Hebrides, Volume 1: the Shaping of the Islands*, White Horse Press, Cambridge, 250 pp.

Angus, S. (2001) *The Outer Hebrides, Volume 2: Moor and Machair*, White Horse Press, Cambridge, 352 pp.

Angus, S. (2002) The definition of machair. Unpublished discussion paper, 2nd Septmber, 2002. Scottish Natural Heritage, Inverness.

Angus, S. and Dargie T. (2002) The UK Machair Habitat Action Plan: progress and problems *Botanical Journal of Scotland* **54** (1), 63–74.

Angus, S. and Elliott, M.M. (1992) Problems of erosion in Scottish machair with particular reference to the Outer Hebrides. In *Coastal Dunes: Geomorphology, Ecology and Management for Conservation: 3rd European Dune Congress* (eds R.W.G. Carter, T.G.F. Curtis and M.J. Sheehy-Skeffington), A.A. Balkema, Rotterdam, pp. 93–112.

Anon. (1887 [written 1799]) Letter dated 9 February 1799. In *A Popular Guide to the Geology of the Isle of Wight; with a note on its relation to that of the Isle of Purbeck* (ed. M.W. Norman), Knight's Library, Ventnor.

Anthony, E.J. and Dolique, F. (2001) Natural and human influences on the contemporary evolution of gravel shorelines in northern France between the Seine estuary and Belgium. In *Ecology and Geomorphology of Coastal Shingle* (eds J.R. Packham, R.E. Randall, R.S.K. Barnes and A.S. Neal), Westbury Academic and Scientific Publishing, Otley, pp. 132–48.

Anthony, J.C. (1998) An investigation into the ecological changes on Braunton Burrows Dune system. Unpublished BSc thesis, University of Nottingham.

Arber, E.A.N. (1911) *The Coast Scenery of North Devon, Being an Account of the Geological Features of the Coast-Line Extending from Porlock in Somerset to Boscastle in North Cornwall*, J.M. Dent, London, 261 pp.

Arber, M.A. (1940) Outline of south-west England in relation to wave attack. *Nature*, **146** (3688), 27–8.

Arber, M.A. (1941) The coastal landslips of west Dorset. *Proceedings of the Geologists' Association*, **52**, 273–83.

Arber, M.A. (1949) Cliff profiles of Devon and Cornwall. *Geographical Journal*, **114**, 191–7.

Arber, M.A. (1973) Landslips near Lyme Regis. *Proceedings of the Geologists' Association*, **84**, 121–33.

Arber, M.A. (1974) The cliffs of north Devon. *Proceedings of the Geologists' Association*, **85**, 147–57.

Arkell, W.J. (1943) The Pleistocene rocks at Trebetherick Point, Cornwall: their interpretation and correlation. *Proceedings of the Geologists' Association*, **54**, 141–70.

Arkell, W.J. (1947) *The Geology of the Country Around Weymouth, Swanage, Corfe and Lulworth*, Memoir of the Geological Survey of Great Britain, sheets 341–343 with parts of 327, 328 and 329 (England and Wales), HMSO, London, 386 pp.

Arkell, W.J. (1951a) Dorset Geology, 1940–1950. *Proceedings of the Dorset Natural History and Archaeological Society*, **72**, 176–94.

Arkell, W.J. (1951b) The structure of Spring Bottom Ridge and the origin of the mudslides, Osmington, Dorset. *Proceedings of the Geologists' Association*, **62**, 21–30.

Arkell, W.J. (1954) The effects of storms on Chesil Beach in November 1954. *Proceedings of the Dorset Natural History and Archaeological Society*, **76**, 141–5.

Arkell, W.J. (1955) Geological results of the cloudburst in the Weymouth district, 18th July 1955. *Proceedings of the Dorset Natural History and Archaeological Society*, **77**, 90–7.

Armstone, J.C. (1970) Recreation and conservation in a coastal dune system; a study of East Head, Chichester Harbour. Unpublished MSc thesis, University of Salford.

ASH Consulting Group (1994) *Coastal Erosion and Tourism in Scotland: a Review of Protection Measures to Combat Coastal Erosion Related to Tourist Activities and Facilities*, Scottish Natural Heritage Review, No. **12**, Scottish Natural Heritage, Edinburgh, 115 pp.

Ashall, J., Duckworth, J., Holder, C. and Smart, S. (1992) *Sand Dune Survey of Great Britain. Site Report No. 122: Newborough Warren and Forest, Ynys Mon, Wales, 1991*, Joint Nature Conservation Committee, Peterborough.

Ashall, J., Duckworth, J., Holder, C. and Smart, S. (1995) *Sand Dune Survey of Great Britain. Site Report No. 123: Aberffrawm Ynys Mon, Wales, 1991*, Joint Nature Conservation Committee, Peterborough.

Ashton, W. (1920) *The Evolution of a Coastline*, W. Ashton and Sons, Southport, 302 pp.

References

Atkinson, D. and Houston, J.A. (eds) (1993) *The Sand Dunes of the Sefton Coast*, National Museums and Galleries on Merseyside, Liverpool, 194 pp.

Austen, R.A.C. (1851) On the superficial accumulation of the coasts of the English Channel and the changes they indicate. *Quarterly Journal of the Geological Society of London*, 7, 118–36.

Austin, S., Clayton, K., Pitty, A. and Stone, S. (1994) Neotectonics of the British Isles: structural stability during the Pliocene/Quaternary and at the present day. Unpublished Nirex Report, 161 pp.

Baden-Powell, D.F.W. (1930) On the geological evolution of Chesil Bank. *Geological Magazine*, 67, 499–513.

Balaam, N.D., Levitan, B. and Strater, V. (eds) (1987) *Studies in Palaeoeconomy and Environment in South West England*, *British Archaeological Reports British Series*, No. 181, British Archaeological Reports, Oxford, 264 pp.

Balchin, W.G.V. (1946) The geomorphology of the north Cornish coast. *Transactions of the Royal Geological Society of Cornwall*, 17 (6), 317–44.

Balchin, W.G.V. (1954) *Cornwall*, Hodder and Stoughton, London, 128 pp.

Banham, P.R. (1979) Some effects of the sea-flood at Wells-next-the-Sea, 11th January 1978. *Transactions of the Norfolk and Norwich Naturalists' Society*, 25, 263–5.

Barber, P.C. and Thomas, R.P. (1989) Case study at Carmarthen Bay. In *Coastal Management: Proceedings of the Conference organised by the Institution of Civil Engineers and held at Bournemouth on 9–11 May 1989*, Thomas Telford, London, pp. 243–62.

Barfoot, P.J. and Tucker, J.J. (1980) Geomorphological changes at Blakeney Point, Norfolk. *Transactions of the Norfolk and Norwich Naturalists' Society*, 25, 49–60.

Barnes, F.A. and King, C.A.M. (1951) A preliminary survey at Gibraltar Point, Lincolnshire. *Bird Observatory and Field Research Station, Gibraltar Point, Lincolnshire*, Report for 1951, 41–59.

Barnes, F.A. and King, C.A.M. (1953) The Lincolnshire coastline and the 1953 storm flood. *Geography*, 38, 141–60.

Barnes, F.A. and King, C.A.M. (1955) Beach changes in Lincolnshire since the 1953 storm surge. *East Midland Geographer*, 1, 18–28.

Barnes, F.A. and King, C.A.M. (1957) The spit at Gibraltar Point, Lincolnshire. *East Midland Geographer*, 1, 22–31.

Barnes, F.A. and King, C.A.M. (1961) Salt marsh development at Gibraltar Point, Lincolnshire. *East Midland Geographer*, 2, 20–31.

Barnes, R.S.K. (1977) *The Coastline*, John Wiley and Sons Ltd, Chichester, 356 pp.

Barnes, R.S.K. (2001) The geomorphology and ecology of the shingle-impounded coastal lagoon systems of Britain. In *Ecology and Geomorphology of Coastal Shingle* (eds J.R. Packham, R.E. Randall, R.S.K. Barnes and A. Neal), Westbury Academic and Scientific Publishing, Otley, pp. 320–35.

Barrow, G. and Flett, J.S. (1906) *The Geology of the Isles of Scilly*, Memoir of the Geological Survey of Great Britain, sheets 357 and 360 (England and Wales), HMSO, London, 37 pp.

Barton, M.E. (1990) Stability and recession of the Chalk cliffs at Compton Down, Isle of Wight. In *Chalk: Proceedings of the International Chalk Symposium held at Brighton Polytechnic on 4–7 September 1989* (ed. G.P. Birch), Thomas Telford (for the Institution of Civil Engineers), London, pp. 541–4.

Barton, M.E. (1991) The natural evolution of the soft rock cliff at Shanklin, Isle of Wight, and its planning and engineering implications. In *Slope Stability Engineering: Developments and Applications. Proceedings of the International Conference on Slope Stability Organized by the Institution of Civil Engineers and held on the Isle of Wight on 15–18 April 1991* (ed. R.J. Chandler), Thomas Telford, London, pp. 181–8.

Bartrum, J.A. (1916) High water Rock Platforms: a phase of Shoreline Erosion. *Transactions of the New Zealand Institute*, 48, 132–4.

Basa, T., Greensmith, J.T. and Vita-Finzi, C. (1997) The sub-surface Holocene Middle sands of Dungeness. *Proceedings of the Geologists' Association*, 108(2), 105–12.

Bascom, W.N. (1954) The control of stream outlets by wave refraction. *Journal of Geology*, 62, 600–5.

Battarchaya, G. (1967) The recent sedimentation of Blakeney Point, north Norfolk. Unpublished diploma thesis, Imperial College, London.

Bayliss-Smith, T.P., Healey, R., Lailey, R., Spencer, T. and Stoddart, D.R. (1979) Tidal flows in salt-marsh creeks. *Estuarine, Coastal and*

References

Marine Science, 9, 235–55.

Bedlington, D.J. (1995) Holocene sea-level changes and crustal movements in north Wales and Wirral. Unpublished PhD thesis, University of Durham.

Belderson, R.H. and Stride, A.H. (1969) Tidal currents and sand wave profiles in the northeastern Irish Sea. *Nature*, 222, 74–5.

Bentley, M. (1995) *Whiteness Head*, Unpublished Earth Science Documentation Series, Scottish Natural Heritage, Perth.

Bentley, M. (1996a) *Ayres of Swinister*, Unpublished Earth Science Documentation Series, Scottish Natural Heritage, Perth.

Bentley, M. (1996b) *Invernaver SSSI*, Unpublished Earth Science Documentation Series, Scottish Natural Heritage, Perth.

Bentley, M. (1996c) *Dunnet Links SSSI*, Unpublished Earth Science Documentation Series, Scottish Natural Heritage, Perth.

Bentley, M. (1996d) *St. Ninian's Tombolo SSSI*, Unpublished Earth Science Documentation Series, Scottish Natural Heritage, Perth.

Benton, M.J., Cook, E. and Turner, P. (2002) *Permian and Triassic Red Beds and the Penarth Group of Great Britain*, Geological Conservation Review Series, No. 24, Joint Nature Conservation Committee, Peterborough, 337 pp.

Bickmore, D.P. and Shaw, M.A. (eds) (1963) *Atlas of Britain and Northern Ireland*, Clarendon Press, Oxford, 234 pp.

Binney, E.W. and Talbot, J.H. (1843) *On the petroleum found in the Downholland Moss, near Ormskirk*. Paper read at the Fifth Annual General Meeting of the Manchester Geological Society, 6 October, 1843.

Birch, G.P. (1990) Engineering geomorphological mapping for cliff stability. In *Chalk: Proceedings of the International Chalk Symposium held at Brighton Polytechnic on 4–7 September 1989* (ed. G.P. Birch), Thomas Telford (for the Institution of Civil Engineers), London, pp. 545–9.

Bird, E.C.F. (1963) Changes at Blakeney Point since 1953. *Transactions of the Norfolk and Norwich Naturalists' Society*, 9, 502–42

Bird, E.C.F. (1968) *Coasts*, Australian National University Press, Canberra, 320 pp.

Bird, E.C.F. (1972) The physiography of The Fleet. *Proceedings of the Dorset Natural History and Archaeological Society*, 93, 119–24.

Bird, J.B. (1977) Beach changes and recreation planning on the west coast of Barbados. *W.I. Geog. Polon*, 36, 31–41.

Bird, E.C.F. (1980) Seatown beach. *Dorset County Magazine*, 89, 1–5.

Bird, E.C.F. (1984) *Coasts. An Introduction to Coastal Geomorphology*, 3rd edn, Blackwell Scientific Publications, Oxford, 320 pp.

Bird, E.C.F. (1985) *Coastline Changes: a Global Review*, Wiley-Interscience, Chichester, 220 pp.

Bird, E.C.F. (1987) The effects of quarry waste disposal on beaches on the Lizard Peninsula, Cornwall. *Journal of the Trevithick Society*, 14, 83–92.

Bird, E.C.F. (1989) The beaches of Lyme Bay. *Proceedings of the Dorset Natural History and Archaeological Society*, 111, 91–7.

Bird, E.C.F. (1996) *Beach Management*, Wiley-Interscience, Chichester, 281 pp.

Bird, E.C.F. (1997) *The Shaping of the Isle of Wight; with an Excursion Guide*, Ex Libris Press, Bradford on Avon, 176 pp.

Bird, E.C.F. (1998) *The Coasts of Cornwall. Scenery and geology with an excursion guide*, Alexander Associates, Fowey, 237 pp.

Bird, E.C.F. (2000) *Coastal Geomorphology: an Introduction*, John Wiley and Sons Ltd, Chichester, 340 pp.

Bird, E.C.F. and May, V.J. (1976) *Shoreline Changes in the British Isles During the Past Century*, Division of Geography, Bournemouth College of Technology, 46 pp.

Bird, E.C.F. and Ranwell, D.S. (1964) Spartina salt marshes in southern England. IV The physiography of Poole Harbour, Dorset. *Journal of Ecology*, 52, 355–66.

Bird, E.C.F. and Rosengren N. J. 1986 *Coastal Cliff Management: An Example from Black Rock Point, Melbourne, Australia*, University of Melbourne, Parkville, Victoria

Bird, E.C.F. and Schwartz, M.L. (eds) (1985) *The World's Coastline*, Van Norstrand Rheinhold, New York, 1071 pp.

Birnie, J. (1981) Environmental changes in Shetland since the end of the last glaciation. Unpublished PhD thesis, University of Aberdeen.

Bishop, W.W. and Coope, R. (1977) Stratigraphical and faunal evidence for Lateglacial and early Flandrian environments in south-west Scotland. In *Studies in the Scottish Lateglacial Environment* (eds J.M. Gray and J.J. Lowe), Pergamon Press, Oxford, pp. 61–88.

References

Black, A. and Black, C. (1847) *Orkney Islands*, Map sheet published by A. and C. Black, Edinburgh.

Black, D.L., Hansom, J.D. and Comber, D.P.M. (1994) Estuaries Management Plans – Coastal Processes and Conservation: Solway Firth. Unpublished report to English Nature, Peterborough, from Coastal Research Group, Department of Geographical and Topographical Science, University of Glasgow, 91 pp + maps.

Black, W.J. (1879) Remarks on the Chesil Bank. *Transactions of the Manchester Geological Society*, 15, 43–50.

Blackbourn, G.A. (1985) *Geological Field Guide to Foula*, Britoil Corporate Exploration Stratigraphic Laboratory, Shetland.

Blackbourn, G.A. and Russell, K. (1981) *Foula: Field Trip Guide*, The British National Oil Corporation, Exploration Division.

Blackley, M.W.L. (1979) Beach changes between Aldeburgh and Southwold, March 1978–May 1979. *Sizewell-Dunwich Banks Field Study Topic*, Report 3, *Institute of Oceanic Studies Science Report*, 90, 18 pp.

Blackley, M.W.L., Carr, A.P. and Gleason, R. (1972) Tracer experiments in the Tor–Torridge estuary with particular reference to Braunton Burrows NNR. *Unit of Coastal Sedimentation*, Report No. 1972/22, Nature Conservancy Council, Taunton, 19 pp.

Bloom, A.L. (1965) Coastal isostatic downwarping by postglacial rise of sealevel. In *Abstracts for 1965: abstracts of papers submitted for six meetings with which the Society was associated* (Geological Society of America), *Geological Society of America Special Paper*, No. 82, Geological Society of America, New York, pp. 1–14.

Bloom, A.L. (1967) Pleistocene shorelines: a new test of isostasy. *Geological Society of America Bulletin*, 78, 1477–94.

Bloom, A.L. (1978) *Geomorphology: a Systematic Analysis of Late-Cenozoic Landforms*, Prentice-Hall, London, 510 pp.

Bluck, B.J. (1967) Sedimentation of beach gravels: examples from south Wales. *Journal of Sedimentary Petrology*, 37, 128–56.

Bond, W.R.G. (1951) Theories as to the origin of Chesil Beach. *Proceedings of the Dorset Natural History and Archaeological Society*, 73, 163–70.

Boomer, I. and Woodcock, L. (1999) The nature and origin of the Stiffkey Meals, North Norfolk coast. *Bulletin of the Geological Society of Norfolk*, 49, 3–13.

Boorman, L.A. (1993) Dry coastal ecosystems of Britain: dunes and shingle beaches. In *Dry Coastal Ecosystems: Polar Regions and Europe* (ed. E. van der Maarel), Elsevier, Amsterdam, pp. 197–228.

Boorman, L.A. and Fuller, R.M. (1977) Studies of the impact of paths on the dune vegetation at Winterton, Norfolk, England. *Biological Conservation*, 12, 203–16.

Boorman, L.A. and Ranwell, D.S. (1977) *Ecology of Maplin Sands and the Coastal Zones of Suffolk, Essex and North Kent*, Institute of Terrestrial Ecology, Cambridge, 56 pp.

Boorman, L.A., Goss-Custard, J.D. and McGroty, S. (1989) *Climate Change, Rising Sea Level and the British Coast*, Institute of Terrestrial Ecology Research Publication, No. 1, HMSO, London, 24 pp.

Borlase, W. (1755) Letter to the Rev. Charles Lyttleton. *Philosophical Transactions of the Royal Society of London*, 69, 373–8.

Bourne, W.R.P., Gimingham, C.H., Morgan, N.C. and Britton, R.H. (1973) The Loch of Strathbeg. *Nature*, 242, 93–5.

Bowden, K.F. (1960) Circulation and mixing in the Mersey Estuary. *International Association of Scientific Hydrology: Surface Water*, 51, 352–60.

Bowen, A.J. and Inman, D.L. (1969) Rip currents 2, Laboratory and field observation. *Journal of Geophysical Research*, 74, 5479–90.

Bowen, A.J. and Inman, D.L. (1971) Edge waves and crescentic bars. *Journal of Geophysical Research*, 76, 8662–71.

Bowen, D.Q. (1970) South-east and central south Wales. In *The Glaciations of Wales and Adjoining Regions* (ed. C.A. Lewis), Geographies for Advanced Study, Longman, London, pp. 197–228.

Bowen, D.Q. (1974) The Quaternary of Wales. In *The Upper Palaeozoic and post-Palaeozoic rocks of Wales* (ed. T.R. Owen), University of Wales Press, Cardiff, pp. 373–426.

Bowen, D.Q. (1981a) 'The South Wales end-moraine': fifty years after. In *The Quaternary of Britain* (eds J. Neale and J. Flenley), Pergamon Press, Oxford, pp. 60–7.

Bowen, D.Q. (1981b) Sheet 1.3. In *National Atlas of Wales* (eds H. Carter and H.M. Griffiths), University of Wales Press, Cardiff.

Bowen, D.Q. (1984) Introduction, Western

References

Slade and Eastern Slade, Langland Bay, Abermawr, Fishguard–Newport–Cardigan. In *Field Guide: Annual Field Meeting April 1984: Wales: Gower, Preseli, Fforest Fawr* (eds D.Q. Bowen and A. Henry), Quaternary Research Association, Cambridge, pp. 1–102.

Bowen, D.Q. and Sykes, G.A. (1988) Correlation of marine events and glaciations on the north-east Atlantic margin. *Philosophical Transactions of the Royal Society of London*, **B318**, 619–35.

Bowen, D.Q., Sykes, G.A., Reeves, A., Miller, G.H., Andrews, J.T., Brew, J.S. and Hare, P.E. (1985) Amino acid geochronology of raised beaches in south-west Britain. *Quaternary Science Reviews*, **4**, 279–318.

Brachi, R.A., Collins, K.J. and Roberts, C.D. (1979) A winter survey of semi-enclosed rocky coastline: survey strategy and field techniques, Kimmeridge, Dorset. *Progress in Underwater Science, New Series*, **4**, 37–47.

Bradbury, A. (1996) Western Solent saltmarsh study. In *Saltmarsh Management for Flood Defence* (eds Sir William Halcrow and Partners), *Research Seminar Proceedings November 1995*, National Rivers Authority, Bristol, pp. 152–68.

Bradbury, A.P. and Kidd, R. (1998) Hurst Spit stabilisation scheme – design and construction of beach recharge. In *Proceedings of the 33rd MAFF Conference of River and Coastal Engineers, Keele University, 1–3 July 1998*, Ministry of Agriculture, Fisheries and Food, London, pp. 1.1.1–1.1.13.

Bradbury, A.P. and Powell, K.A. (1992) Short-term profile response of shingle spits to storm wave action. In *Coastal Engineering, 1992: Proceedings of the 23rd International Conference, October 4–9 1992, Venice, Italy* (ed. B.L. Edge), American Society of Civil Engineers, New York, pp. 2694–707.

Bray, M.J. (1986) *A Geomorphological Investigation of the South-west Dorset Coast: Report to Dorset County Council and West Dorset District Council, Volume 1: Patterns of Sediment Supply*, London School of Economics, London, 144 pp.

Bray, M.J. (1990a) *A Geomorphological Investigation of the South-west Dorset Coast: Report to Dorset County Council and West Dorset District Council, Volume 2: Patterns of Sediment Transport*, London School of Economics, London, 798 pp.

Bray, M.J. (1990b) Landslide and littoral zone sediment transport. In *Landslides of the Dorset Coast* (ed. R.J. Allison), British Geomorphological Research Group Field Guide, British Geomorphological Research Group, London, pp. 107–17.

Bray, M.J. (1992) Coastal sediment supply and transport. In *The Coastal Landforms of West Dorset* (ed. R.J. Allison), Geologists' Association Guide, No. 47, Geologists' Association, London, pp. 94–105.

Bray, M.J. (1996) Beach budget analysis and shingle transport dynamics in west Dorset. Unpublished PhD thesis, University of London.

Bray, M.J. (1997) Episodic shingle supply and the modified development of Chesil Beach, England. *Journal of Coastal Research*, **13**, 1035–49.

Bray, M.J. (1999) The coastal system of west Dorset. In *Dorset Revisited: Position Papers and Research Statements* (ed. R.J. Allison), *British Geomorphological Research Group Spring Field Meeting 1999*, West Dorset Coastal Research Group, pp. 7–16.

Bray, M.J. and Duane, W.J. (2001) Porlock Bay: geomorphological investigation and monitoring. *Environment Agency Report*, **STCG024**, 113 pp.

Bray, M.J. and Hooke, J.M. (1997) Prediction of soft-cliff retreat with accelerating sea-level rise. *Journal of Coastal Research*, **13**, 453–67.

Bray, M.J., Carter, D.J. and Hooke, J.M. (1991) *Coastal Sediment Transport Study: Report to SCOPAC*, Portsmouth Polytechnic, Portsmouth, 5 volumes.

Bray, M.J., Carter, D.J. and Hooke, J.M. (1992) *Sea-level Rise and Global Warming: Scenarios, Physical impacts and Policies: Report to SCOPAC*, Portsmouth Polytechnic, Portsmouth.

Bray, M.J., Carter, D.J. and Hooke, J.M. (1995) Littoral cell definition and budgets for central south England. *Journal of Coastal Research*, **11**, 381–400.

Braybrooks, E.M. (1957) The general ecology of *Spartina townsendii* (sic *S. anglica*) with special reference to sward build-up and degradation. Unpublished MSc thesis, University of Southampton.

Breeds, J. and Rogers, D. (1998) Dune management without grazing – a cautionary tale. *Enact: Management for Wildlife*, **6**, 19–22.

Brewis, P., and Buckley, F., (1928) Notes on prehistoric pottery and a bronze pin from Ross

References

Links, Northumberland. *Archaeol Aeliana*, 4 ser, 5, 13–25

Bridges, M. (1987) *Classic Landforms of the Gower Coast*, Classic Landform Guides, No. 7, Geographical Association, Sheffield, 48 pp.

Bridges, E.M. (1991) *Classic Landforms of the North Norfolk Coast*, Classic Landform Guides, No. **12**, Geographical Association, Sheffield, 48 pp.

Bridson, R.H. (1980) Saltmarsh – its accretion and erosion at Caerlaverock National Nature Reserve, Dumfries. *Transactions of the Dumfriesshire and Galloway Natural History and Antiquarian Society*, **55**, 60–7.

Briquet, A. (1930) *Le littoral du Nord de la France*, Librairie Armand Colin, Paris, 439 pp.

Bristow, H.W. and Whitaker, W. (1869) On the formation of the Chesil Bank, Dorset. *Geological Magazine*, 6, 433–40.

Bristow, C.S. and Bailey, S.D. (2001) Non-invasive investigation of water table and structures in coastal dunes using Ground-Penetrating Radar (GPR): implications for dune management. In *Coastal Dune Management: Shared Experience of European Conservation Practice* (eds J.A. Houston, S.E. Edmondson and P.J. Rooney), Liverpool University Press, Liverpool, pp. 408–17.

Bristow, C.S., Chroston, N.P. and Bailey, S.D. (2000) The structure and development of coastal dunes: insights from ground-penetrating radar (GPR) surveys, Norfolk, England. *Sedimentology*, 47, 923–44.

Bristow, H.W. (1889) *The Geology of the Isle of Wight*, 2nd edn (revised and enlarged by C. Reid and A. Strahan), Memoir of the Geological Survey of Great Britain, HMSO, London, 138 pp.

BGS 1977 (Cited chap 2

British Geological Survey (1991*) Geology of the United Kingdom, Ireland and the adjacent continental shelf*, 1:1 000 000 scale map, British Geological Survey, Keyworth.

British Geological Survey (1996) Chapter 2.3: Wind and water. In *Coasts and Seas of the United Kingdom. Region 3. North-East Scotland: Cape Wrath to St Cyrus* (eds J.H. Barne, C.F. Robson, S.S. Kasnowska, J.P. Doody and N.C. Davidson), Coastal Directories Series, Joint Nature Conservation Committee, Peterborough, pp. 28–32.

British Geological Survey (1997a) Chapter 2.3: Wind and water. In *Coasts and Seas of the United Kingdom. Regions 15 and 16. North-West Scotland: the Western Isles and West Highland* (eds J.H. Barne, C.F. Robson, S.S. Kasnowska, J.P. Doody, N.C. Davidson and A.L. Buck), Coastal Directories Series, Joint Nature Conservation Committee, Peterborough, pp. 31–6.

British Geological Survey (1997b) Chapter 2.3: Wind and water. In *Coasts and Seas of the United Kingdom. Region 14. South-West Scotland: Ballantrae to Mull* (eds J.H. Barne, C.F. Robson, S.S. Kasnowska, J.P. Doody, N.C. Davidson and A.L. Buck), Coastal Directories Series, Joint Nature Conservation Committee, Peterborough, pp. 28–32.

British Geological Survey, and Scott Wilson Resource Consultants (1997a) Chapter 2.3: Wind and water. In *Coasts and Seas of the United Kingdom. Region 1. Shetland* (eds J.H. Barne, C.F. Robson, S.S. Kasnowska, J.P. Doody, N.C. Davidson and A.L. Buck), Coastal Directories Series, Joint Nature Conservation Committee, Peterborough, pp. 27–30.

British Geological Survey, and Scott Wilson Resource Consultants (1997b) Chapter 2.3: Wind and water. In *Coasts and Seas of the United Kingdom. Region 2. Orkney* (eds J.H. Barne, C.F. Robson, S.S. Kasnowska, J.P. Doody, N.C. Davidson and A.L. Buck), Coastal Directories Series, Joint Nature Conservation Committee, Peterborough, pp. 26–9.

British Maritime Technology Ltd (1986) *Area 11, North Sea*. In *Global wave statistics* (British Maritime Technology Ltd), Unwib, Woking.

Bromhead, E.N., Chandler, M.P. and Hutchinson, J.N. (1991) The recent history and geotechnics of landslides at Gore Cliff, Isle of Wight. In *Slope Stability Engineering: Developments and Applications. Proceedings of the International Conference on Slope Stability Organized by the Institution of Civil Engineers and held on the Isle of Wight on 15–18 April 1991* (ed. R.J. Chandler), Thomas Telford, London, pp. 189–96.

Brooks, N.P. (1988) Romney Marsh in the Early Middle Ages. In *Romney Marsh: Evolution, Occupation, Reclamation* (eds J. Eddison and C. Green), Oxford University Committee for Archaeology Monograph, No. **24**, OUCA, Oxford, pp. 90–104.

Browne, I. (1994) Seismic stratigraphy and relict coastal sediments off the east coast of Australia. *Marine Geology*, **122**, 81–107.

Browne, M.A.E. and Jarvis, J. (1983) Late Devensian marine erosion in St Andrews Bay,

References

east-central Scotland. *Quaternary Newsletter*, **41**, 11–17.

Brunsden, D. (1968) Moving cliffs of Black Ven. *Geographical Magazine*, **41**, 372–4.

Brunsden, D. (1973) The application of system theory to the study of mass-movement. *Geologia applicata e idrogeologia*, **8**, 185–207.

Brunsden, D. (1974) The degradation of a coastal slope, Dorset, England. In *Progress in Geomorphology* (eds E.H. Brown and R.S. Waters), *Institute of British Geographers Special Publication*, No. 7, Institute of British Geographers, London, pp. 79–98.

Brunsden, D. (1996) The landslides of the Dorset Coast: some unresolved questions. *Proceedings of the Ussher Society*, **9**, 1–7.

Brunsden, D. (1999) Chesil Beach – two ideas. In *Dorset Revisited: Position Papers and Research Statements* (ed. R.J. Allison), *British Geomorphological Research Group Spring Field Meeting 1999*, West Dorset Coastal Reasearch Group, pp. 19–21.

Brunsden, D. and Chandler, J.H. (1996) Development of an episodic landform change model based upon the Black Ven mudslide, 1946–1995. In *Advances in Hillslope Processes 2* (eds M.G. Anderson and S.M. Brooks), John Wiley and Sons Ltd, Chichester, pp. 869–96.

Brunsden, D. and Goudie, A. (1981) *Classic Coastal Landforms of Dorset*, Classic Landform Guides, No. 1, Geographical Association, Sheffield, 42 pp.

Brunsden, D. and Goudie, A. (1997a) *Classic Landforms of the West Dorset Coast*, Classic Landform Guides, Geographical Association, Sheffield, 56 pp.

Brunsden, D. and Goudie, A. (1997b) *Classic Landforms of the East Dorset Coast*, Classic Landform Guides, Geographical Association, Sheffield, 48 pp.

Brunsden, D. and Jones, D.K.C. (1972) The morphology of degraded landslide slopes in south-west Dorset. *Quarterly Journal of Engineering Geology*, **5**, 205–22.

Brunsden, D. and Jones, D.K.C. (1976) The evolution of landslide slopes in Dorset. *Philosophical Transactions of the Royal Society of London*, **A283**, 605–31.

Brunsden, D. and Jones, D.K.C. (1980) Relative time scales and formative events in coastal landslide systems. In *Coasts under Stress* (eds A.R. Orme, D.B. Prior, N.P. Psuty, and H.J. Walker), *Zeitschrift für Geomorphologie*, *Supplementband*, No. 34, Borntraeger, Berlin, pp. 1–19.

Brunsden, D. and Thornes, J.B. (1979) Landscape sensitivity and change. *Transactions of the Institute of British Geographers, New Series*, **4**(4), 463–84.

Brunsden, D., Ibsen, M.L., Lee, M. and Moore, R. (1995) The validity of temporal archive records for geomorphological processes. *Quaestiones Geographicae, Special Issue*, **4**, 79–92.

Brunsden, D., Coombe, K., Goudie, A.S. and Parker, A.G. (1996) The structural geomorphology of the Isle of Portland, southern England. *Proceedings of the Geologists' Association*, **107**, 209–30.

Buchan, G.M. (1976) Dynamics of a river-mouth spit-bar and related processes in Aberdeen Bay. Unpublished PhD thesis, University of Aberdeen, 2 volumes.

Buchan, S. (1931) On dykes in east Aberdeenshire. *Transactions of the Edinburgh Geological Society*, **12**, 323–8.

Burd, F. (1986) Salt marsh survey of Great Britain: Cornwall County Report.

Burd, F. (1989) *The Saltmarsh Survey of Great Britain: An Inventory of British Saltmarshes*, Research and Survey in Nature Conservation, No. 17, Nature Conservancy Council, Peterborough, 180 pp.

Burd, F. (1992) *Erosion and Vegetation Change in the Saltmarshes of Essex and North Kent, Between 1972 and 1988*, Research and Survey in Nature Conservation, No. 42, Nature Conservancy Council, Peterborough, 116 pp.

Burden, R.J. (1997) A hydrological investigation of three Devon sand dune systems: Braunton Burrows, Northam Burrows and Dawlish Warren. Unpublished PhD thesis, University of Plymouth.

Burnaby, T.P. (1950) The tubular Chalk stacks of Sheringham. *Proceedings of the Geologists' Association*, **61**, 226–41.

Burnett, J.H. (1964) *The Vegetation of Scotland*, Oliver and Boyd, Edinburgh.

Burnham, C.P. (1983) Soil formation on bare shingle at Dungeness, Kent. *Seesoil*, **1**, 42–56.

Burnham, C.P. and Cook, H.F. (2001) Hydrology and soils of coastal shingle with specific reference to Dungeness. In *Ecology and Geomorphology of Coastal Shingle* (eds J.R. Packham, R.E. Randall, R.S.K. Barnes and A.S. Neal), Westbury Academic and Scientific

References

Publishing, Otley, pp. 107–31.

Burrin, P. (1988) The Holocene floodplain and alluvial deposits of the Rother valley and their bearing on the evolution of Romney Marsh. In *Romney Marsh: Evolution, Occupation, Reclamation* (eds J. Eddison and C. Green), Oxford University Committee for Archaeology Monograph, No. 24, OUCA, Oxford, pp. 31–52.

Burrows, A.J. (1884–1885) Romney Marsh, past and present. *Transactions of the Survey Institute*, 17, 338.

Burton, H.St.J. (1937) The origin of Lulworth Cove. *Geological Magazine*, 74, 377–83.

Bury, H. (1920) The chines and cliffs of Bournemouth. *Geological Magazine*, 57, 71–6.

Butler, R.J. (1978) Salt marsh morphology and the evolution of Colne Point in Essex, England. Unpublished PhD thesis, Queen Mary College, University of London.

Butler, R.J., Greensmith, J.T. and Wright, L.W. (1981) *Shingle Spits and Salt Marshes in the Colne Point area of Essex, a Geomorphological Study*, Occasional Papers in Geography, No. 18, Department of Geography, Queen Mary College, London, 51 pp.

Byrne, J.V. (1964) An erosional classification for the north Oregon coast. *Annals of the Association of American Geographers*, 54, 329–35

Byrne, J.V., Leroy, D.O.S. and Riley, C.M. (1959) The chenier plain and its stratigraphy, southwestern Louisiana. *Transactions – Gulf Coast Association of Geological Societies*, 9, 237–60.

Calkin, J.B. (1968) *Ancient Purbeck. An Account of the Geology of the Isle of Purbeck and its Early Inhabitants*, The Friary Press, Dorchester, 61 pp.

Callander, J.G. (1911) Notice of the discovery of two vessels of clay on the Culbin Sands, the first containing wheat and the second from a kitchen midden, with a comparison of the Culbin Sands and Glen Luce Sands and of the relics found on them. *Proceedings of the Society of Antiquaries of Scotland*, 45, 158–81.

Callow, W.J., Baker, M.J. and Hassall, G.I. (1966) National Physical Laboratory Radiocarbon Measurements: IV. *Radiocarbon*, 8, 340–7.

Cambers, G. (1973) The retreat of unconsolidated Quaternary cliffs. Unpublished PhD thesis, University of East Anglia.

Cambers, G. (1975) Sediment Transport and Coastal Change. *East Anglian Coastal Research Programme*, Report 3, University of East Anglia, Norwich, 65 pp.

Cambers, G. (1976) Temporal scales in coastal erosion systems. *Transactions of the Institute of British Geographers, New Series*, 1 (2), 246–56.

Campbell, S. (1984) The nature and origin of the Pleistocene deposits around Cross Hands and on west Gower, South Wales. Unpublished PhD thesis, University of Wales.

Campbell, S. (1998) Westward Ho! In *Quaternary of South-West England* (S. Campbell, C.O. Hunt, J.D. Scourse and D.H. Keen), Geological Conservation Review Series, No. 14, Joint Nature Conservation Committee, Peterborough, pp. 224–33.

Campbell, S. and Bowen, D.Q. (1989) *Quaternary of Wales*, Geological Conservation Review Series, No. 2, Nature Conservancy Council, Peterborough, 237 pp.

Campbell, S. and Gilbert, A. (1998) The Croyde–Saunton Coast. In *Quaternary of South-West England* (S. Campbell, C.O. Hunt, J.D. Scourse and D.H. Keen), Geological Conservation Review Series, No. 14, Joint Nature Conservation Committee, Peterborough, pp. 214–24.

Campbell, S., Hunt, C.O., Scourse, J.D. and Keen, D.H. (1998) *Quaternary of South-West England*, Geological Conservation Review Series, No. 14, Joint Nature Conservation Committee, Peterborough, 439 pp.

Canning, A.D. and Maxted, K.R. (1979) *Coastal Studies in Purbeck*, Purbeck Press, Swanage, 86 pp.

Carey, A.E. and Oliver, F.W. (1918) *Tidal Lands. A Study of Shore Problems*, Blackie, London, 284 pp.

Carling, P.A. (1981) Sediment transport by tidal currents and waves: observations from a sandy intertidal zone (Burry Inlet, south Wales). In *Holocene Marine Sedimentation in the North Sea Basin* (eds S.D. Nio, R.T.E. Schuttenhelm and T.C.E. Van Weering), International Association of Sedimentologists Special Publication, No. 5, Blackwell Scientific Publications, Oxford, pp. 65–80.

Carling, P.A. (1982) Temporal and spatial variation in intertidal sedimentation rates. *Sedimentology*, 29, 17–23.

Carr, A.P. (1962) Cartographical error and

References

historical accuracy. *Geography*, **47**, 135–44.

Carr, A.P. (1965) Shingle spit and river mouth: short term dynamics. *Transactions of the Institute of British Geographers*, **36**, 117–29.

Carr, A.P. (1967) The London Clay surface in part of Suffolk. *Geological Magazine*, **104**, 574–84.

Carr, A.P. (1969a) Size grading along a pebble beach: Chesil Beach, England. *Journal of Sedimentary Petrology*, **39**, 297–311.

Carr, A.P. (1969b) The growth of Orford Spit: cartographic and historical evidence from the sixteenth century. *Geographical Journal*, **135**, 28–39.

Carr, A.P. (1970) The evolution of Orfordness, Suffolk, before 1600 AD: geomorphological evidence. *Zeitschrift für Geomorphologie, Neue Folge*, **14**, 289–300.

Carr, A.P. (1971a) Experiments on longshore transport and sorting of pebbles: Chesil Beach, England. *Journal of Sedimentary Petrology*, **41**, 1084–104.

Carr, A.P. (1971b) South Haven Peninsula: physiographic changes in the twentieth century. In *Captain Cyril Diver, a Memoir* (ed. P. Merrett), Nature Conservancy Council, Furzebrook Research Station, pp. 32–8.

Carr, A.P. (1971c) Orford, Suffolk: further data on the Quaternary evolution of the area. *Geological Magazine*, **108**, 311–16.

Carr, A.P. (1972) Aspects of spit development and decay: the estuary of the River Ore, Suffolk. *Field Studies*, **4**, 633–53.

Carr, A.P. (1973) Present-day beach process studies and the evolution of the coastline near Orford, Suffolk. Unpublished PhD thesis, University of London, 325 pp.

Carr, A.P. (1974) Differential movement of coarse sediment particles. In *Proceedings of the 14th International Conference on Coastal Engineering, Copenhagen, Denmark, June 1974*, American Society of Civil Engineers, New York, pp. 851–67.

Carr, A.P. (1979) Longterm changes in the coastline and offshore banks. *Sizewell-Dunwich Banks Field Study Topic*, Report **2**, *Institute of Oceanic Studies Science Report*, **89**.

Carr, A.P. (1980) Chesil beach and the adjacent area: outline of existing data and suggestions for future research. Institute of Oceanographic Sciences Report to Dorset County Council and Wessex Water Authority, 21 pp.

Carr, A.P. (1981) Evidence for the sediment circulation along the coast of East Anglia. *Marine Geology*, **40**, M9–M22.

Carr, A.P. (1983a) Chesil Beach: environmental, economic and sociological pressures. *Geographical Journal*, **149**, 53–62.

Carr, A.P. (1983b) Shingle beaches: aspects of their structure and stability. In *Shoreline Protection: Proceedings of a Conference Organized by the Instituion of Civil Engineers and held at the University of Southampton on 14-15 September 1982* (eds Institution of Civil Engineers), Thomas Telford, London, pp. 97–104.

Carr, A.P. (1986) The estuary of the River Ore, Suffolk: three decades of change in a longer-term context. *Field Studies*, **6**, 43–58.

Carr, A.P. (1999) Chesil Beach and the Fleet. In *Dorset Revisited: Position Papers and Research Statements* (ed. R.J. Allison), British Geomorphological Research Group Spring Field Meeting 1999, West Dorset Coastal Research Group, pp. 17–18.

Carr, A.P. and Baker, R.E. (1968) Orford, Suffolk: evidence for the evolution of the area during the Quaternary. *Transactions of the Institute of British Geographers*, **45**, 107–23.

Carr, A.P. and Blackley, M.W.L. (1969) Geological composition of the pebbles of Chesil Beach, Dorset. *Proceedings of the Dorset Natural History and Archaeological Society*, **90**, 133–40.

Carr, A.P. and Blackley, M.W.L. (1973) Investigations bearing on the age and development of Chesil Beach, Dorset, and the associated area. *Transactions of the Institute of British Geographers*, **58**, 99–111.

Carr, A.P. and Blackley, M.W.L. (1974a) A statistical analysis of the beach metaquartzite clasts from Budleigh Salterton, Devon. In *Proceedings of the 14th International Conference on Coastal Engineering, Copenhagen, Denmark, June 1974*, American Society of Civil Engineers, New York, pp. 302–13.

Carr, A.P. and Blackley, M.W.L. (1974b) Ideas on the origin and development of Chesil Beach, Dorset. *Proceedings of the Dorset Natural History and Archaeological Society*, **95**, 1–9.

Carr, A.P. and Blackley, M.W.L. (1975) A statistical analysis of the beach metaquartzite clasts from Budleigh Salterton, Devon. *Proceedings of the Ussher Society*, **3**, 302–15.

Carr, A.P. and Gleason, R. (1972) Chesil Beach, Dorset, and the cartographic evidence of Sir John Coode. *Proceedings of the Dorset*

References

Natural History and Archaeological Society, **93**, 125–31.

Carr, A.P. and Graff, J. (1982) The tidal immersion factor and shore platform development: discussion. *Transactions of the Institute of British Geographers, New Series*, **7** (2), 240–5.

Carr, A.P. and Seaward, D.R. (1990) Chesil Beach: changes in crest height 1969–1990. *Proceedings of the Dorset Natural History and Archaeological Society*, **112**, 109–12.

Carr, A.P. and Seaward, D.R. (1991) Chesil Beach: landward recession 1965–1990. *Proceedings of the Dorset Natural History and Archaeological Society*, **113**, 157–60.

Carr, A.P., Gleason, R. and King, A.C. (1970) Significance of pebble size and shape in sorting by waves. *Sedimentary Geology*, **4**, 89–101.

Carr, A.P., Blackley, M.W.L. and King, H.L. (1982) Spatial and seasonal aspects of beach stability. *Earth Surface Processes and Landforms*, **7**, 267–82.

Carruthers, R.G., Dinham, C.H., Burnett, G.A. and Maden, J. (1927) *The Geology of Belford, Holy Island, and the Farne Islands*, 2nd edn, Memoir of the Geological Survey of Great Britain, Sheet 4 (England and Wales), HMSO, London, 195 pp.

Carson, M.A. and Kirkby, M.J. (1972) *Hillslope Form and Process*, Cambridge Geographical Studies, No. 3, Cambridge University Press, Cambridge, 475 pp.

Carter, R.W.G. (1982) Recent variations in sea-level on the north and east coasts of Ireland and associated shoreline response. *Proceedings of the Royal Irish Academy, Section B*, **82**, 177–87.

Carter, R.W.G. (1985) Approaches to sand dune conservation in Ireland. In *Sand Dunes and their Management* (ed. J.P. Doody), Focus on Nature Conservation, No. **13**, Nature Conservancy Council, Peterborough, pp. 29–41.

Carter, R.W.G. (1988) *Coastal Environments: an Introduction to the Physical, Ecological and Cultural Systems of Coastlines*, Academic Press, London, 617 pp.

Carter, R.W.G. (1990) Morfa Dinlle. Unpublished report to the Countryside Council for Wales.

Carter, R.W.G. (1992) How the British coast works: inherited and acquired controls. In *Conserving Our Landscape: Proceedings of the Conference 'Conserving Our Landscape, Evolving Landforms and Ice-Age Heritage', Crewe, May 1992* (eds C. Stevens, J.E. Gordon, C.P. Green and M.G. Macklin), English Nature, Peterborough, pp. 63–8.

Carter, D.J.T. and Draper, L. (1988) Has the North-east Atlantic become rougher? *Nature*, **332**, 494.

Carter, R.W.G. and Orford, J.D. (1981) Overwash processes along a gravel beach in southeast Ireland. *Earth Surface Processes and Landforms*, **6**, 413–26.

Carter, R.W.G. and Orford, J.D. (1984) Coarse clastic barrier beaches: a discussion of the distinctive dynamic and morphosedimentary characteristics. *Marine Geology*, **60**, 377–89.

Carter, R.W.G. and Orford, J.D. (1988) Conceptual model of coarse clastic barrier formation from multiple sediment sources. *Geographical Review*, **78**, 221–39.

Carter, R.W.G. and Orford, J.D. (1993) The morphodynamics of coarse clastic beaches and barriers: a short- and long-term perspective. In *Beach and surf zone morphodynamics* (ed. A.D. Short), *Journal of Coastal Research Special Issue*, No. **15**, Coastal Education and Research Foundation (CERF), Fort Lauderdale, pp. 158–70.

Carter, R.W.G., Johnston, T.W. and Orford, J.D. (1984) Stream outlets through mixed sand and gravel coastal barriers; examples from southeast Ireland. *Zeitschrift für Geomorphologie, Neue Folge*, **28** (4) 427–42.

Carter, R.W.G., Johnston, T.W., McKenna, J. and Orford, J. (1987) Sea level, sediment supply and coastal changes: examples from the coast of Ireland. *Progress in Oceanography*, **18**, 79–101.

Carter, R.W.G., Forbes, D.L., Jennings, S.C., Orford, J.D., Shaw, J. and Taylor, R.B. (1989) Barrier and lagoon coast evolution under differing relative sea-level regimes: examples from Ireland and Nova Scotia. *Marine Geology*, **88**, 221–42.

Castleden, R. (1982) *Classic Landforms of the Sussex Coast*, Classic Landform Guides, No. **2**, Geographical Association, Sheffield, 39 pp.

Caton, P.G. (1976) *Maps of Hourly Wind Speed over the United Kingdom 1965–1973*, Climatological Memorandum, No. **79**, Meteorological Office, Bracknell, 4 pp.

Catt, J.A. (1977) *Yorkshire and Lincolnshire: Guidebook for Excursion, International Union for Quaternary Research*, **C7**, International Union for Quaternary Research, Birmingham, 56 pp.

References

Catt, J.A. and Penny, L.F. (1966) The Pleistocene deposits of Holderness, East Yorkshire. *Proceedings of the Yorkshire Geological Society*, **35**, 375–400.

Chandler, J.H. and Brunsden, D. (1995) Steady state behaviour of the Black Ven mudslide: the application of archival analytical photogrammetry to studies of landform change. *Earth Surface Processes and Landforms*, **20**, 255–75.

Chandler, R.J. (1991) (ed.) *Slope Stability Engineering: Developments and Applications. Proceedings of the International Conference on Slope Stability Organized by the Institution of Civil Engineers and held on the Isle of Wight on 15–18 April 1991*, Thomas Telford, London, 443 pp.

Chapman, V.J. (1938) Studies in salt marsh ecology. *Journal of Ecology*, **26**, 144–79.

Chapman, V.J. (1939) Scolt Head Part 2. Marsh development in Norfolk. *Transactions of the Norfolk and Norwich Naturalists' Society*, **14**, 395–7.

Chapman, V.J. (1959) Studies in salt marsh ecology IX. Changes in salt marsh vegetation at Scolt Head Island. *Journal of Ecology*, **47**, 619–39.

Chapman, V.J. (ed.) (1977) *Wet Coastal Ecosystems*, Ecosystems of the World, No. **1**, Elsevier, Amsterdam, 428 pp.

Chastain, C. (1976) Hydrodynamic process and sea cliff/platform erosion, Favourite Channel, Alaska. Unpublished PhD thesis, University of California, Santa Cruz.

Chisholm, N.W.T. (1996) Morphological changes at Braunton Burrows, north-west Devon. *Proceedings of the Ussher Society*, **9**, 25–30.

Chorley, R.J., Schumm, S.A. and Sugden, D.E. (1984) *Geomorphology*, Methuen, London and New York, 607 pp.

Chroston, P.N., Jones, R. and Makin, B. (1999) Geometry of Quaternary sediments along the north Norfolk coast, UK: a shallow seismic study. *Geological Magazine*, **136**, 465–74.

Churchill, D.M. (1965) The displacement of deposits formed at sea-level 6,500 years ago in southern Britain. *Quaternaria*, **7**, 239–57.

Churchill, D.M. and Wymer, J.J. (1985) The kitchen midden site at Westward Ho!, Devon, England: ecology, age and relation to changes in land and sea level. *Proceedings of the Prehistoric Society*, **31**, 74–84.

Clark, M.J. (1974) Conflict at the coast. *Geography*, **59**, 93–103.

Clark, M.J. and Small, R.J. (1967) An investigation into the feasibility of the seabed movement of shingle in Christchurch Bay. *Coastal Research Report*, No. **1**, Department of Geography, University of Southampton.

Clark, A.R., Palmer, J.S., Firth, T. and McIntyre, G. (1993) The management and stabilisation of weak sandstone cliffs at Shanklin, Isle of Wight. In *The Engineering Geology of Weak Rock: proceedings of the 26th Annual Conference of the Engineering Group of the Geological Society, Leeds, United Kingdom, 9–13 September 1990* (eds J.C. Cripps et al.), Engineering Geology Special Publication, No. **8**, A.A. Balkema, Rotterdam, pp. 375–84.

Clark, R. (1936) Holkham Camp. *Proceedings of the Prehistoric Society*, **2**, 231–3.

Clark, R. (1939) The Iron Age in Norfolk and Suffolk. *Archaeological Journal*, **96**, 1–113.

Clarke, R.H. (1970) Quaternary sediments off south-east Devon. *Quarterly Journal of the Geological Society of London*, **129**, 277–318.

Clayden, A.W. (1906) *History of Devonshire Scenery*, Chatto and Windus, London, 202 pp.

Clayton, K.M. (ed.) (1964) *A Bibliography of British Geomorphology*, George Philip, London, 211 pp.

Clayton, K.M. (1980) Beach sediment budgets and coastal modification. *Progress in Physical Geography*, **4**, 471–86.

Clayton, K.M. (1989a) Implications of climatic change. In *Coastal Management. Proceedings of the Conference organised by the Institution of Civil Engineers and held at Bournemouth on 9–11 May 1989*, Institution of Civil Engineers, Thomas Telford, London, pp. 165–76.

Clayton, K.M. (1989b) Sediment input from the Norfolk cliffs, eastern England – a century of coast protection and its effect. *Journal of Coastal Research*, **5**, 433–42.

Clayton, K.M. (1992) *Coastal Geomorphology*, 2nd edn, Nelson, Walton-on-Thames, 72 pp.

Clayton, K.M. (1998) Neotectonics of the British Isles: the Geomorphological Evidence for Tectonic Activity in the Last Five Million Years. *Nirex Safety Studies Report* (NSS/R Series), **NSS R236**, Nirex, Harwell, 133 pp.

Clayton, K. and Shamoon, N. (1998) A new approach to the relief of Great Britain – II. A classification of rocks based on relative resistance to denudation. *Geomorphology*, **25**, 155–71.

Clayton, K. and Shamoon, N. (1999) A new

approach to the relief of Great Britain – III. Derivation of the contribution of neotectonic movements and exceptional regional denudation to the present relief. *Geomorphology*, **27**, 173–89.

Clayton, K.M., McCave, I.N. and Vincent, C.E. (1983) The establishment of a sand budget for the East Anglian coast and its implications for coastal stability. In *Shoreline Protection: Proceedings of a Conference Organized by the Instituion of Civil Engineers and held at the University of Southampton on 14–15 September 1982* (eds Institution of Civil Engineers), Thomas Telford, London, pp. 91–6.

Clements, M. (1994) The Scarborough exerience – Holbeck landslide, 3/4 June 1993. *Proceedings of the Institution of Civil Engineers: Municipal Engineers*, **103**, 63–70.

Clough, C.T., Barrow, G., Crampton, C.B., Maufe, H.B., Bailey, E.B. and Anderson, E.M. (1910) *The Geology of East Lothian, Including Parts of the Counties of Edinburgh and Berwick*, Memoir of the Geological Survey of Great Britain, Sheet 33, with parts of 34 and 31 (Scotland), HMSO, Edinburgh, 226 pp.

Cobb, R.T. (1957) Shingle Street, Suffolk. *Report of the Field Studies Council*, 1956–57, 31–42.

Cockburn, A.M. (1935) The geology of St Kilda. *Transactions of the Royal Society of Edinburgh*, **58**, 511–47.

Codrington, T. (1870) Some remarks on the formation of Chesil Bank. *Geological Magazine*, **7**, 23–5.

Colenutt, G.C. (1928) The cliff-founder and landslide at Gore Cliff, Isle of Wight. *Proceedings of the Isle of Wight Natural History and Archaeological Society*, **1**, 561–70.

Colquhon, R.S. (1969) Dune erosion and protective works at Pendine, Carmarthenshire, 1961–68. In *Proceedings of the 11th Conference on Coastal Engineering, London, England, September 1968*, American Society of Civil Engineers, New York, pp. 708–18.

Comber, D.P.M. (1993) Shoreline response to relative sea level change: Culbin Sands, northeast Scotland. Unpublished PhD thesis, University of Glasgow.

Comber, D.P.M. (1995) Scottish Landform Example – 11. Culbin Sands and the Bar. *Scottish Geographical Magazine*, **111**(1), 54–7.

Comber, D.P.M., Hansom, J.D. and Fahy, F.M. (1993) Taw–Torridge Estuary: Coastal Processes and Conservation. *Report by the Coastal Research Group*, University of Glasgow for English Nature, Peterborough, 61 pp.

Comber, D.P.M., Hansom, J.D. and Fahy, F.M. (1994) Culbin Sands, Culbin Forest and Findhorn Bay SSSI: Documentation and Management Prescription. *Scottish Natural Heritage Research, Survey and Monitoring Report*, No. **14**, 98 pp.

Conway, B.W. (1974) The Black Ven landslip, Charmouth, Dorset. *Report of the Institute of Geological Sciences*, **74/3**, 16 pp.

Conway, B.W. (1979) The contribution made to cliff instability by head deposits in the west Dorset coastal area. *Quarterly Journal of the Geological Society of London*, **12**, 267–79.

Coode, J. (1853) Description of the Chesil Bank, with remarks upon its origin, the causes which have contributed to its formation and upon the movement of shingle generally (including appendices). *Minutes of the Proceedings of the Institution of Civil Engineers*, **12**, 520–46.

Cooke, W. (1808) *A New Picture of the Isle of Wight, to which are Prefixed, an Introductory Account of the Island and a Voyage round its Coast*, Vernor, Hood and Sharpe, London, 169 pp.

Coombe, E.D.K. (1996) Implications of an investigation by coring into the sediments of the Fleet Lagoon, Chesil Beach, Dorset, England. Unpublished Dphil thesis, Oxford University.

Coombe, E.D.K., Bruce, A.S., Goudie, A.S., Juggins, S., Parker, A.G. and Whittaker, J.E. (in press) The evolution of the Fleet Lagoon, Dorset, southern England, during the last c. 6000 years; preliminary evidence from a multidisciplinary study.

Cooper, L.H.N. (1948) A submerged ancient cliff near Plymouth. *Nature*, **161**, 280.

Cooper, R. (in press) *Mass Movements in Great Britain*, Geological Conservation Review Series, Joint Nature Conservation Committee, Peterborough.

Cope, F.W. (1939) Oil occurrences in south-west Lancashire (with a biological report by Kathleen B. Blackburn). *Bulletin of the Geological Survey of Great Britain*, **2**, 18–25.

Cornish, V. (1898a) On sea beaches and sandbanks. *Geographical Journal*, **11**, 528–43 and 628–51.

References

Cornish, V. (1898b) On the grading of the Chesil Beach shingle. *Proceedings of the Dorset Natural History and Antiquarian Field Club*, **19**, 113–21.

Cornish, V. (1912) Ocean waves, sea-beaches and sandbanks. *Journal of the Royal Society of Arts*, **60**, 1105–10 and 1121–6.

Cottingham, A.M. (1994) Water table and afforestation of the Newborough dune system. Unpublished BSc thesis, University of Wales, Bangor.

Cotton, C.A. (1941) *Landscape as Developed by the Processes of Normal Erosion*, Cambridge University Press, Cambridge, 301 pp.

Cotton, C.A. (1951) Atlantic gulfs, estuaries and cliffs. *Geological Magazine*, **88**, 113–28.

Cox, B.M. and Gallois, R.W. (1981) The stratigraphy of the Kimmeridge Clay of the Dorset type area and its correlation with some other Kimmeridgian sequences. *Report of the Institute of Geological Sciences*, **80/4**, 44 pp.

Craig, G. (1888) The Culbin sandhills. *Transactions of the Edinburgh Geological Society*, **5**, 524–31.

Craig-Smith, S.J. (1971a) Report on the behaviour of 14 selected beach sections during the 8 months January to June 1971. *East Anglian Coastal Study*, Report **1**, University of East Anglia, Norwich, 27 pp.

Craig-Smith, S.J. (1971b) The changing system. *East Anglian Coastal Study*, Report **3**, University of East Anglia, Norwich, 68 pp.

Craig-Smith, S.J. (1973) Sediment characteristics within the coastal-offshore system. *East Anglian Coastal Study*, Report 4, University of East Anglia, Norwich, 65 pp.

Crampton, C.B. and Carruthers, R.G. (1914) *The Geology of Caithness*, Memoir of the Geological Survey of Great Britain, sheets 110 and 116 with parts of 109, 115 and 117 (Scotland), HMSO, Edinburgh, 194 pp.

Crawford, R.M.M. and Wishart, D. (1966) A multivariate analysis of the development of dune slack vegetation in relation to coastal accretion at Tentsmuir, Fife. *Journal of Ecology*, **54**, 729–43.

Cunliffe, B.W. (1980) The evolution of Romney Marsh: a preliminary statement. In *Archaeology and Coastal Change* (ed. F.H. Thompson), Society of Antiquaries of London Occasional Papers, New Series, No. **1**, Society of Antiquaries of London, London, pp. 37–55.

Cunliffe, B.W. (1988) Romney Marsh in the Roman Period. In *Romney Marsh: Evolution, Occupation, Reclamation* (eds J. Eddison and C. Green), Oxford University Committee for Archaeology Monograph, No. **24**, OUCA, Oxford, pp. 83–7.

Currie, A. (1974) *The Vegetation of the Shingle Spit at Whiteness Head, Inverness*, Nature Conservancy Council, Peterborough.

Daley, B. and Balson, P. (1999) *British Tertiary Stratigraphy*, Geological Conservation Review Series, No. **15**, Joint Nature Conservation Committee, Peterborough, 388 pp.

Damon, R. (1884) *Geology of Weymouth, Portland, and the Coast of Devonshire, from Swanage to Bridport-on-the-sea: with Natural History and Archaeological Notes*, New and enlarged edition, E. Stanford, London, 250 pp.

Darbyshire, J. (1982) *An Investigation of Beach Cusps in Hell's Mouth Bay*. Paper read to the British Geomorphological Research Group Small-Scale Coastal Processes Conference, Polytechnic of Wales, Pontypridd.

Darbyshire, M. (1958) Waves in the Irish Sea. *The Dock and Harbour Authority*, **39**, 245–8.

Dargie, T.C.D. (1989) Morrich More S.S.S.I., Ross and Cromarty. Vegetation Survey 1988. *Nature Conservancy Council CSD Report*, No. **915**, 126 pp + maps.

Dargie, T.C.D. (1994) Sand dune vegetation of Scotland: Sandwood Bay (Southern Parphe SSSI), Sutherland. Unpublished Sand Dune Survey report to Scottish Natural Heritage, Edinburgh.

Dargie, T.C.D. (1998) Sand dune vegetation of Scotland: Western Isles. *Scottish Natural Heritage Research, Survey and Monitoring Report*, No. **96**, 3 volumes.

Dargie, T.C.D. (2000) *Sand Dune Vegetation Survey of Scotland*, Scottish Natural Heritage Commissioned Report No. F97AA401, SNH, Edinburgh, 193pp.

Darton, D.M., Dingwall, R.G. and McCann, D.M. (1981) Geological and geophysical investigations in Lyme Bay. *Report of the Institute of Geological Sciences*, **79/10**.

Darwin, C. (1842) *On the Structure and Distribution of Coral Reefs: Being the First Part of the Geology of the Voyage of the Beagle, Under the Command of Capt. Fitzroy, R.N. During the Years 1832 to 1836*, Smith, Elder and Co., London, 214 pp.

Davidson-Arnott, R.G.D. (1986) Rates of erosion of till in the nearshore zone. *Earth Surface Processes and Landforms*, **11**, 53–8.

References

Davies, C.M. (1983) Pleistocene rockhead surfaces underlying Barnstaple Bay (SW England). *Marine Geology*, **54**, M9–M16.

Davies. G.M. (1935) *The Dorset Coast: a Geological Guide*, Murby and Co., London, 126 pp.

Davies, G.M. (1956) *The Dorset Coast: a Geological Guide*, 2nd edn, A. and C. Black, London, 128 pp.

Davies, J. (1997) A sediment budget model of the southwest coast of the Isle of Wight. Unpublished BSc thesis, Bournemouth University.

Davies, J.L. (1958) Wave refraction and the evolution of shoreline curves. *Geographical Studies*, **5**, 1–14.

Davies, J.L. (1972) *Geographical Variation in Coastal Development*, Geomorphology Texts, No. 4, Oliver and Boyd, Edinburgh, 204 pp.

Davies, J.L. (1980) *Geographical Variation in Coastal Development*, 2nd edn, Geomorphology Texts, No. 4, Longman, London, 212 pp.

Davies, P. and Williams, A.T. (1986) Cave development in Lower Lias coastal cliffs, the Glamorgan Heritage Coast, Wales, UK. In *Proceedings of the Iceland Coastal and River Symposium* (ed. G. Sigbjarnarson), National Energy Authority, Reykjavik, pp. 75–92.

Davies, P., Williams, A.T. and Bomboe, P. (1991) Numerical modelling of Lower Lias rock failures in the coastal cliffs of south Wales, UK. In *Coastal Sediments '91: Proceedings of a Speciality Conference on Quantitative Approaches to Coastal Processes, Seattle, Washington, June 25–27 1991* (eds N.C. Kraus, K.J. Gingerich and D.L. Briebel), American Society of Civil Engineers, New York, pp. 1599–612.

Davies, W. (1963) Recent sediments of the Gibraltar Point area, Lincolnshire. Unpublished PhD thesis, Imperial College, London.

Davis, W.M. (1912) A geographical pilgrimage from Ireland to Italy. *Association of American Geographers Annals*, **2**, 73–100.

Dawson, A.G. (1979) Raised shorelines of Jura, Scarba and northeast Islay. Unpublished PhD thesis, University of Edinburgh.

Dawson, A.G. (1980a) The low rock platform in western Scotland. *Proceedings of the Geologists' Association*, **91**, 339–44.

Dawson, A.G. (1980b) Shore erosion by frost: an example from the Scottish Lateglacial. In *Studies in the Lateglacial of North-West Europe: Including Papers Presented at a Symposium of the Quaternary Research Association held in University College London, January 1979* (eds J.J. Lowe, J.M. Gray and J.E. Robinson), Pergamon Press, Oxford, pp. 45–53.

Dawson, A.G. (1982) Lateglacial sea-level changes and ice limits in Islay, Jura and Scarba, Scottish Inner Hebrides. *Scottish Journal of Geology*, **18** (4), 253–66.

Dawson A.G. (1983) *Islay and Jura, Scottish Hebrides*, Quaternary Research Association Field Guide, Quaternary Research Association, Cambridge, 29 pp.

Dawson, A.G. (1984) Quaternary sea-level changes in western Scotland. *Quaternary Science Reviews*, **3** (4), 345–68.

Dawson, A.G. (1991) Scottish landform examples – 3. The raised shorelines of northern Islay and western Jura. *Scottish Geographical Magazine*, **107** (3), 207–12.

Dawson, A.G. (1993) West coast of Jura, Inner Hebrides. In *Quaternary of Scotland* (eds D.G. Sutherland and J.E. Gordon), Geological Conservation Review Series, No. 4, Chapman and Hall, London, 382–8.

Dawson, A.G. and Dawson, S. (1997) Holocene relative sea-level changes, Gruinart, Isle of Islay. In *Isla and Jura* (eds A.G. Dawson and S. Dawson), Quaternary Research Association Field Guide, Quaternary Research Association, London, pp. 78–98.

Dawson, A.G., Mathews, J.A. and Shakesby, R.A. (1987) Rock platform erosion on periglacial shores: a modern analogue for Pleistocene rock platforms in Britain. In *Periglacial Processes and Landforms in Britain and Ireland* (ed. J. Boardman), Cambridge University Press, Cambridge, pp. 173–82.

Dawson, A., Benn, D.I. and Dawson, S. (1997) Late Quaternary glaciomarine sedimentation in the Rhinns of Islay, Scottish Inner Hebrides. In *Islay and Jura* (eds A.G. Dawson and S. Dawson), Quaternary Research Association Field Guide, Quaternary Research Association, London, pp. 66–77.

Dawson, A.G., Smith, D.E. and Dawson, S. (2001) Potential impacts of climate change on sea levels around Scotland. *Scottish Natural Heritage Research, Survey and Monitoring Report*, No. **178**.

de Boer, G. (1963) Spurn Point and its predecessors. *The Naturalist*, **1963**, 113–20.

References

de Boer, G. (1964) Spurn Head: its history and evolution. *Transactions of the Institute of British Geographers*, **34**, 71–89.

de Boer, G. (1967) Cycles of change at Spurn Head, Yorkshire, England. *Shore and Beach*, **35**, 13–20.

de Boer, G. (1968) A history of the Spurn lighthouses. *East Yorkshire Local History Series*, **24**, 72 pp.

de Boer, G. (1969) The historical variations of Spurn Head: the evidence of early maps. *Geographical Journal*, **135**, 17–27.

de Boer, G. (1973) The two earliest maps of Hull. *Post-Medieval Archaeology*, **7**, 79–87.

de Boer, G. (1981) Spurn Point: erosion and protection after 1849. In *The Quaternary in Britain* (eds J. Neale and J. Flenley), Pergamon Press, Oxford, pp. 206–15.

de Boer, G. and Carr, A.P. (1969) Early maps as evidence for coastal changes. *Geographical Journal*, **135**, 17–39.

de Boer, G. and Skelton, D.A. (1969) The earliest English chart with soundings. *Imago Mundi*, **23**, 9–16.

de la Beche, H.T. (1830) Notes on the formation of extensive conglomerate and gravel deposits. *Philosophical Magazine, Series 2*, **7**, 161–71.

de la Beche, H.T. (1839) *Report on the Geology of Cornwall, Devon and West Somerset*, Memoir of the Geological Survey of Great Britain (England and Wales), HMSO, London, 624 pp.

de Luc, J.A. (1811) *Geological Travels*, Volume 2, F.C. and J. Rivington, London.

de Martonne, E. (1913) *Traité de Géographie Physique*, Librairie Armand Colin, Paris, 922 pp.

de Moor, G. (1979) Recent beach evolution along the Belgian North Sea coast. *Bulletin Société Belge de Géologie*, **88**, 143–57.

de Rance, C.E. (1869) *The Geology of the Country between Liverpool and Southport*, Memoir of the Geological Survey of Great Britain, Sheet 90SE (England and Wales), HMSO, London, 9 pp.

de Rance, C.E. (1872) *The Geology of the Country around Southport, Lytham and South Shore*, Memoir of the Geological Survey of Great Britain, Sheet 90NE (England and Wales), HMSO, London, 15 pp.

de Rance, C.E. (1877) *The Superficial Geology of the Country adjoining the Coasts of Southwest Lancashire*, Memoir of the Geological Survey of Great Britain, sheets 90 and 91SW and parts of 89NW and SW, 79NE and 91SE, (England and Wales), HMSO, London, 139 pp.

de Rance, C.E. (1878) *Geology of the Country around Preston, Blackburn and Burnley*, Memoir of the Geological Survey of Great Britain, Sheet 89NW (England and Wales), HMSO, London, 15 pp.

Denness, B. (1972) The reservoir principle of mass-movement. *Report of the Institute of Geological Sciences*, **72/7**, 13 pp.

Denness, B., Conway, B.W., McCann, D.M. and Grainger, P. (1975) Investigation of a coastal landslip at Charmouth, Dorset. *Quarterly Journal of Engineering Geology*, **8**, 119–40.

Derbyshire, E. (1976) *Geomorphology and Climate*, John Wiley and Sons Ltd, London, 512 pp.

Deshmukh, I.J. (1974) Fixation, accumulation and release of energy by *Ammophila arenaria*, Tentsmuir. Unpublished PhD thesis, University of Dundee.

Devoy, R.J.N. (1982) Analysis of the geological evidence for Holocene sea-level movements in south-east England. *Proceedings of the Geologists' Association*, **93**, 65–90.

Dewey, H. (1909) On overthrusts at Tintagel (North Cornwall). *Quarterly Journal of the Geological Society of London*, **65**, 265.

Dewey, H. (1914) The geology of north Cornwall. *Proceedings of the Geologists' Association*, **25**, 154–79.

Diver, C. (1933) The physiography of the South Haven Peninsula, Studland Heath, Dorset. *Geographical Journal*, **81**, 404–27.

Dix, J., Long, A. and Cooke, R. (1998) The evolution of Rye Bay and Dungeness Foreland: the offshore seismic record. In *Romney Marsh: Environmental Change and Human Occupation in a Coastal Lowland* (eds J. Eddison, M. Gardiner and A. Long), *Oxford University Committee for Archaeology Monograph*, No. 46, OUCA, Oxford, pp. 1–12.

Dixon, E.E.L., Maden, J., Trotter, F.M., Hollingworth, S.E. and Tonks, L.H. (1926) *The Geology of the Carlisle, Longtown and Silloth District*, Memoir of the Geological Survey of Great Britain, sheets 11, 16 and 17 (England and Wales), HMSO, London, 118 pp.

Dixon, N. and Bromhead, N.E. (1991) The mechanics of first-time slides in the London Clay Cliffs at the Isle of Sheppey, England. In *Slope Stability Engineering: Developments*

References

and Applications. *Proceedings of the International Conference on Slope Stability Organized by the Institution of Civil Engineers and held on the Isle of Wight on 15–18 April 1991* (ed. R.J. Chandler), Thomas Telford, London, pp. 277–82.

Dobson, R.S. (1967) Some applications of a digital computer to hydraulic engineering problems. *Technical Report*, No. 80, Department of Civil Engineering, Stanford University, California.

Donovan, D.T. and Stride, A.H. (1961) An acoustic survey of the sea floor south of Dorset and its geological interpretation. *Philosophical Transactions of the Royal Society of London*, **B244**, 299–330.

Donovan, D.T. and Stride, A.H. (1975) Three drowned coastlines of probable Tertiary age around Devon and Cornwall. *Marine Geology*, **19**, 35–40.

Doody, J.P. (ed.) (1984) *Spartina anglica in Great Britain*, Focus on Nature Conservation, No. **5**, Nature Conservancy Council, Peterborough.

Doody, J.P. (ed.) (1985) *Sand Dunes and their Management*, Focus on Nature Conservation, No. **13**, Nature Conservancy Council, Peterborough, 263 pp.

Doody, J.P. (1987) The impact of 'reclamation' on the natural environment of the Wash. In *The Wash and its Environment* (eds J.P. Doody and B. Barnett), Focus on Nature Conservation, No. **5**, Nature Conservancy Council, Peterborough, pp. 165–72.

Doody, J.P. (1989) Management for nature conservation. In *Coastal Sand Dunes: Proceedings of the Symposium organised by the Royal Society of Edinburgh and held at 22, 24 George Street, Edinburgh, 17–19 May 1989* (eds C.H. Gimingham, W. Ritchie, B.B. Willetts and A.J. Willis), *Proceedings of the Royal Society of Edinburgh*, **96B**, pp. 247–65.

Doody, J.P. (1990) *Spartina* – friend or foe? A conservation viewpoint. In *Spartina anglica – a research review* (eds A.J. Gray and P.E.M. Benham), Institute of Terrestrial Ecology Research Publication, No. 2, HMSO, London, pp. 77–9.

Doody, J.P. (1992) The conservation of British saltmarshes. In *Saltmarshes, Morphodynamics, Conservation and Engineering Significance* (eds J.R.L. Allen and K. Pye), Cambridge University Press, Cambridge, pp. 80–114.

Doody, J.P. (1996) Management and use of dynamic estuarine shorelines. In *Estuarine Shores: Evolution, Environments and Human Alterations* (eds K.F. Nordstrom and C.T. Roman), John Wiley and Sons Ltd, Chichester, pp. 421–34.

Doody, P. (1986) Proof of Evidence. *Tain Public Inquiry, Land and Marine Engineering Ltd Development Proposal*.

Dossor, J. (1955) The coast of Holderness: the problem of erosion. *Proceedings of the Yorkshire Geological Society*, **30** (2), 133–45.

Dowker, C. (1897) On Romney Marsh. *Proceedings of the Geologists' Association*, **15**, 220–33.

Draper, L. (1966) The analysis and presentation of wave data – a plea for uniformity. In *Proceedings of the 10th International Conference on Coastal Engineering, Tokyo, Japan, September 1966*, American Society of Civil Engineers, New York, pp. 1–11.

Draper, L. (1991) Wave Climate Atlas of the British Isles. *Offshore Technology Report*, **OTH 89 303**, HMSO for Department of Energy, London.

Draper, L. and Blakey, A. (1969) Waves at the Mersey Bar light vessel. *NIO Internal Report*, **A37**, 1–4.

Draper, L. and Bownass, T.M. (1983) Wave devastation behind Chesil beach. *Weather*, **38**, 346–52.

Draper, L. and Squire, E.M. (1967) Waves at Ocean Weather Station India. *Transactions of the Royal Institute of Naval Architects*, **199**, 85–93.

Drew, F. (1864) *The Geology of the Country between Folkstone and Rye, including the whole of Romney Marsh*, Memoir of the Geological Survey of Great Britain, Sheet 4 (England and Wales), HMSO, London, 27 pp.

Dugdale, R.E. (1977) Sediment movement in the nearshore zone, Gibraltar Point, Lincolnshire. Unpublished PhD thesis, University of Nottingham.

Durrance, E.M. (1969) The structure of Dawlish Warren. *Proceedings of the Ussher Society*, **2**, 91–101.

Dutt, W. (1909) *Norfolk and Suffolk Coast*, T. Fisher Unwin, London, 413 pp.

Duvivier, J. (1961) The Selsey coast protection scheme. *Proceedings of the Institution of Civil Engineers*, **20**, 481–506.

Dyer, K.R. (1970) Sediment distribution in Christchurch Bay. *Journal of the Marine*

References

Biological Association, UK, **50**, 673–82.

Dyer, K.R. (1980) Sedimentation and sediment transport. In *The Solent Estuarine System: an appraisal of Current Knowledge*, Natural Environment Research Council Publications, Series C, No. 22, Natural Environment Research Council, Swindon, pp. 20–4.

Earll, R. and Pagett, R.M. (1984) *A Classification and Catalogue of the Sea Lochs of the Western Isles*, Nature Conservancy Council Information and Library Services, Peterborough, 29 pp.

Eddison, J. (1983a) Flandrian barrier beaches of the coast of Sussex and south-east Kent. *Quaternary Newsletter*, **39**, 25–9.

Eddison, J. (1983b) The evolution of the barrier beaches between Fairlight and Hythe. *Geographical Journal*, **149**, 39–53.

Eddison, J. (ed.) (1995) *Romney Marsh: the Debatable Ground*, Oxford University Committee for Archaeology Monograph, No. 41, OUCA, Oxford, 174 pp.

Eddison, J. (1998) Catastrophic changes: the evolution of the barrier beaches of Rye May. In *Romney Marsh: Environmental Change and Human Occupation in a Coastal Lowland* (eds J. Eddison, M. Gardiner and A. Long), Oxford University Committee for Archaeology Monograph, No. 46, OUCA, Oxford, pp. 65–88.

Eddison, J. and Green, C. (eds) (1988) *Romney Marsh: Evolution, Occupation and Reclamation*, Oxford University Committee for Archaeology Monograph, No. 24, OUCA, Oxford, 208 pp.

Eddison, J., Gardiner, M. and Long, A. (eds) (1998) *Romney Marsh: Environmental Change and Human Occupation in a Coastal Lowland*, Oxford University Committee for Archaeology Monograph, No. 46, OUCA, Oxford, 220 pp.

Edlin, H.L. (1976) The Culbin Sands. In *Environment and Man, Volume 4, Reclamation* (eds J. Lenihan and W.W. Fletcher), Blackie, London, pp. 1–31.

Elliott, J. (1847) Account of the Dymchurch wall, which forms the sea defences of Romney Marsh. *Minutes of the Proceedings of the Institution of Civil Engineers*, **6**, 466–84.

Ellis, N.V. (ed.), Bowen, D.Q., Campbell, S., Knill, J.L., McKirdy, A.P., Prosser, C.D., Vincent, M.A. and Wilson, R.C.L. (1996) *An Introduction to the Geological Conservation Review*, Geological Conservation Review Series, No. **1**, Joint Nature Conservation Committee, Peterborough, 131 pp.

Elton, C. (1938) Notes on the ecological and natural history of Pabbay and other islands in the Sound of Harris. *Journal of Ecology*, **26**, 275–97.

Embleton, C. (1964) The deglaciation of Arfon and southern Anglesey, and the origin of the Menia Strait. *Proceedings of the Geologists' Association*, **75**, 407–30.

Embleton, C. (1993) Seismicity, tectonics and geomorphology in Britain. *Proceedings of the National Science Council, A: Physical Sciences and Engineering*, **17**, 1–19.

Emery, K.O. and Kuhn, G.G. (1982) Sea cliffs: their processes, profiles, and classification. *Geological Society of America Bulletin*, **93**, 644–54.

Empsall, B. (1989) Workington ironworks reclamation. In *Proceedings of the Conference of the Institute of Engineers*, pp. 279–92.

Englefield, H.C. (1816) *A Description of the Principal Picturesque Beauties, Antiquities and Geological Phenomena of the Isle of Wight. With Additional Observations on the Strata of the Island, and their Continuation in the Adjacent Parts of Dorsetshire, by Thomas Webster, Esq.*, Payne and Foss, London, 238 pp.

Erdtman, G. (1926) Some micro-analyses of Moorlog from the Dogger Bank. *Essex Naturalist*, **21**, 107–12.

Esler, D. (1976) Coastal dune stability. Unpublished PhD thesis, University of Aberdeen.

Esler, D. (1983) Contemporary morphology of the face of the foredunes at Forvie. In *Northeast Scotland Coastal Field Guide and Geographical Essays* (ed. W. Ritchie), Department of Geography, University of Aberdeen, Aberdeen, pp. 52–60.

European Commission (1998) *CORINE –érosion cotière*, Commission of the European Community, Luxembourg, 170 pp.

Evans, G. (1960) Recent sedimentation in The Wash. Unpublished PhD thesis, Imperial College, London.

Evans, G. (1965) Intertidal flat sediments and their environments of deposition in the Wash. *Quarterly Journal of the Geological Society of London*, **121**, 209–45.

Evans, H.M. (1912) Sand formation against the Saunton Down Cliff, north Devon. *Report and Transactions of the Devonshire Association*

References

for the Advancement of Science, Literature and Art, **44**, 692–702.

Evans, J.R., Kirby, J.R. and Long, A.J. (2001) The litho- and biostratigraphy of a late Holocene tidal channel in Romney Marsh, southern England. *Proceedings of the Geologists' Association*, **112**, 111–30.

Everard, C.E. (1954) The Solent River: a geomorphological study. *Transactions of the Institute of British Geographers*, **20**, 41–58.

Everard, C.E. (1962) Mining and shoreline evolution near St. Austell, Cornwall. *Transactions of the Royal Geological Society of Cornwall*, **19**, 199–219.

Fahy, F.M., Hansom, J.D. and Comber, D.P.M. (1993) Estuaries Management Plans – Coastal Processes and Conservation: Duddon Estuary [and] recommendations. Unpublished report to English Nature, Peterborough, from Coastal Research Group, Department of Geographical and Topographical Science, University of Glasgow, 2 volumes.

Fairbridge, R.W. (1968) Beach. In *The Encyclopedia of Geomorphology* (ed. R.W. Fairbridge), Reinhold Book Corporation, New York, pp. 62–8.

Farquhar, O.C. (1967) Stages in island linking. *Oceanographic Marine Biology Annual Review*, **5**, 119–39.

Farquhar, O.C. (1973) Stages in island linking. In *Spits and Bars* (ed. M.L. Schwartz), Dowden, Hutchinson and Ross, Stroudsberg, Pennsylvania, pp. 308–30.

Farrow, C. (1974) The ecology and sedimentation of the *Cardium* shellsands and transgressive banks of Traigh Mhor, Isle of Barra, Outer Hebrides. *Transactions of the Royal Society of Edinburgh*, **69**, 203–21.

Ferentinos, G. and McManus, J. (1981) Nearshore processes and shoreline development in St Andrews Bay, Scotland, U.K. In *Holocene Marine Sedimentation in the North Sea Basin* (eds S.-D. Nio, R.T.E. Schuttenhelm and T.J.C.E. van Weering), *Special Publication of the International Association of Sedimentologists*, No. 5, Blackwell Scientific Publications, Oxford, pp. 161–71.

Ferry, B.W. and Waters, S.J.P. (1988) Natural wetlands on shingle at Dungeness, Kent, England. *Biological Conservation*, **43**, 27–41.

Ferry, B.W., Barlow, S.L. and Waters, S.J.P. (1989) The shingle ridge succession at Dungeness. *Botanical Journal of the Linnaean Society*, **101**, 19–30.

Finlay, T.M. (1930) The Old Red Sandstone of Shetland: Part II North-western area. *Transactions of the Royal Society of Edinburgh*, **56**, 671–94.

Firth, C.R. (1989) Late Devensian raised shorelines and ice-limits in the inner Moray Firth area, northern Scotland. *Boreas*, **18**, 5–21.

Firth, C.R. and Haggart, B.A. (1989) Loch Lomond Stadial and Flandrian shorelines in the inner Moray Firth area, Scotland. *Journal of Quaternary Science*, **4**, 37–50.

Firth, C.R. and Smith, D.E. (1993) Holocene sea level changes in Shetland. In *Shetland Isles* (eds J. Birnie, J. Gordon, K. Bennett and A. Hall), Quaternary Research Association Field Guide, Quaternary Research Association, Cambridge, p. 17.

Firth, C.R., Smith, D.E., Hansom, J.D. and Pearson, S.G. (1995) Holocene spit development on a regressive shoreline, Dornoch Firth, Scotland. In *Coastal Evolution in the Quaternary: IGCP Project 274* (eds O. van de Plassche, M.J. Chrzastowski, J.D. Orford, A.C. Hinton and A.J. Long), *Marine Geology Special Volume*, **124**, Elsevier, Amsterdam and London, pp. 203–14.

Firth, C.R., Collins, P.E.F. and Smith, D.E. (2000) *Focus on Firths: Coastal Landforms, Processes and Management Options. V. The Solway Firth*, Scottish Natural Heritage Review, No. **128**, Scottish Natural Heritage, Edinburgh, 120 pp.

Fisher, O. (1866) On the disintegration of a chalk cliff. *Geological Magazine*, **3**, 354–6.

Fisher, O. (1873) On the phosphatic nodules of the Cretaceous rocks of Cambridgeshire. *Quarterly Journal of the Geological Society of London*, **29**, 52–63.

Fitton, W.H. (1847) A stratigraphical account of the section from Atherfield to Rocken End, on the south-west coast of the Isle of Wight. *Quarterly Journal of the Geological Society of London*, **3**, 289–327.

Fitzroy, R. (1853) Discussion of Coode, J. (1853) Description of the Chesil Bank, with remarks upon its origin, the causes which have contributed to its formation, and upon the movement of shingle generally. *Minutes of the Proceedings of the Instituion of Civil Engineers*, **12**, 520–57.

Fleming, C.A. (1965) Two-storied cliffs at the Auckland Islands. *Transactions of the Royal Society of New Zealand (Geology)*, **3**, 171–4.

Flinn, D. (1964) Coastal and submarine features

References

around the Shetland Islands. *Proceedings of the Geologists' Association*, **75**, 321–40.

Flinn, D. (1969) On the development of coastal profiles in the north of Scotland, Orkney and Shetland. *Scottish Journal of Geology*, **5**, 393–9.

Flinn, D. (1974) The coastline in Shetland. In *The Natural Environment of Shetland: Proceedings of the Nature Conservancy Council Symposium held in Edinburgh, 29–30 January 1974* (ed. R. Goodier), Nature Conservancy Council, Edinburgh, pp. 13–25.

Flinn, D. (1978) The most recent glaciation of the Orkney–Shetland Channel and adjacent areas. *Scottish Journal of Geology*, **14**, 109–23.

Flinn, D. (1997) The role of wave diffraction in the formation of St. Ninian's Ayre (tombolo) in Shetland, Scotland. *Journal of Coastal Research*, **13**(1), 202–8.

Flint, K. (1980) Channel adjustment and basin morphology in relation to cliff retreat, with reference to chines on the Isle of Wight. *Horizon*, **28**, 54–66.

Flint, K.E. (1982) Chines on the Isle of Wight: channel adjustment and basin morphology in relation to cliff retreat. *Geographical Journal*, **148**, 225–36.

Fonseca, M.S. (1996) The role of seagrasses in nearshore sedimentary processes: a review. In *Estuarine Shores: Evolution, Environments and Human Alterations* (eds K.F. Nordstrom and C.T. Roman), John Wiley and Sons Ltd, Chichester, pp. 261–86.

Fontolan, G. and Simeoni, U. (1999) Holocene cuspate forelands in the Strait of Magellan, southern Chile. *Revista Geologica de Chile*, **26**, 175–86.

Foote, Y., Moses, C., Robinson, D., Saddleton, P. and Williams, R. (2002) European shore platform erosion dynamics. *Oceanology International 2000 Brighton Conference*.

Forbes, D.L. and Boyd, R. (1987) Gravel ripples on the Inner Scotian Shelf. *Journal of Sedimentary Petrology*, **57**, 46–54.

Forbes, D.L., Orford, J.D., Carter, R.W.G., Shaw, J. and Jennings, S.C. (1995) Morphodynamic evolution, self-organisation, and instability of coarse-clastic barriers on paraglacial coasts. *Marine Geology*, **126**, 63–85.

Forteath, G.N.R. (1977) Studies in macrovertebrate communities of the Loch of Strathbeg. Unpublished PhD thesis, University of Aberdeen.

Foster, H.D. (1970) Sarn Badrig, a submarine moraine in Cardigan Bay, north Wales. *Zeitschrift für Geomorphologie*, **14**, 475–86.

Foster, I.D., Albon, A.J., Bardell, K.M., Fletcher, J.L., Jardine, T.C., Mothers, R.J., Pritchard, M.A. and Turner, S.E. (1991) High energy coastal sedimentary deposits: an evaluation of depositional processes in southwest England. *Earth Surface Processes and Landforms*, **16**, 341–56.

Fox, H.R. (1978) Aspects of beach sand movement at Gibraltar Point, Lincolnshire. Unpublished PhD thesis, University of Nottingham.

Francis, E.H. (1975) Dunbar. In *The Geology of the Lothians and South East Scotland: an Excursion Guide*, 2nd edn (eds G.Y. Craig and P.McL.D. Duff), Scottish Academic Press, Edinburgh, pp. 93–106.

Fraser, E.J. (1979) Aspects of sediment movement on the foreshore of south Lincolnshire. Unpublished PhD thesis, University of Nottingham.

French, J.R. (1985) Modification of tidal flows on sedimentation by salt marsh vegetation. Unpublished BA thesis, University of Cambridge.

French, J.R. (1993) Numerical simulation of vertical marsh growth and adjustment to accelerated sea level rise, North Norfolk, UK. *Earth Surface Processes and Landforms*, **18**, 63–81.

French, J.R. (1996) Function and optimal design of saltmarsh channel networks. In *Saltmarsh Management for Flood Defence, Research Seminar Proceedings November 1995* (eds Sir William Halcrow and Partners), National Rivers Authority, Bristol, pp. 85–95.

French, J.R. and Spencer, T. (1993) Dynamics of sedimentation in a tide-dominated backbarrier salt marsh, Norfolk, U.K. *Marine Geology*, **110**, 315–31.

French, J.R. and Stoddart, D.R. (1992) Hydrodynamics of salt marsh creek systems: implications for marsh morphological development and material exchange. *Earth Surface Processes and Landforms*, **17**, 235–52.

French, J.R., Spencer, T. and Stoddart, D. (1990) *Backbarrier Salt Marshes of the North Norfolk Coast: Geomorphic Developments and Response to Rising Sea-levels*, Discussion Papers in Conservation, No. **54**, Ecology and Conservation Unit, University College

References

London, London, 35 pp.

French, P.W. (1996) Implications of a salt marsh chronology for the Severn estuary based on independent lines of dating evidence. *Marine Geology*, **135**, 115–25.

French, P.W. (1997) *Coastal and Estuarine Management*, Routledge Environmental Mangement Series, Routledge, London, 272 pp.

Frey, R.W. and Basan, P.B. (1978) Coastal salt marshes. In *Coastal Sedimentary Environments* (ed. R.A. Davis), Springer-Verlag, New York, pp. 101–69.

Frey R.W. and Basan, P.B. (1985) Coastal salt marshes. In *Coastal Sedimentary Environments*, 2nd edn (ed. R.A. Davis), Springer-Verlag, New York, pp. 225–301.

Fuller, R.M. (1985) An assessment of damage to the shingle beaches and their vegetation. In *Dungeness – Ecology and Conservation, Report of a meeting held at Botany Department, Royal Holloway and Bedford New College, on 16 April 1985* (eds B. Ferry and S. Waters), Focus on Nature Conservation, No. 12, Nature Conservancy Council, Peterborough, pp. 25–42.

Fuller, R.M. and Randall, R.E. (1988) The Orford shingle, Suffolk, UK – classic conflicts in coastline management. *Biological Conservation*, **46**, 95–114.

Funnell, B.M. and Pearson, I. (1989) Holocene sedimentation on the North Norfolk barrier coast in relation to relative sea-level change. *Journal of Quaternary Science*, **4**, 63–81.

Galliers, J.A. (1970) *The Geomorphology of Holy Island, Northumberland*, University of Newcastle-upon-Tyne Department of Geography Research Series, No. 6, University of Newcastle-upon-Tyne, Newcastle-upon-Tyne, 34 pp.

Garcia-Novo, F. (1976) Ecophysiological aspects of the distribution of *Elymus arenaria* and *Cakile maritima* on the dunes at Tentsmuir. *Oecologia Plant*, **11**, 13–25.

Gardner, J.S. (1879) *Geology of the Isle of Wight*, 2nd edn, Memoir of the Geological Survey of Great Britain, HMSO, London.

Gardner, J.S. (1889) Flora of Alum Bay. In *Geology of the Isle of Wight*, 2nd edn, Memoir of the Geological Society of Great Britain, HMSO, London.

Geikie, J. (1878) On the glacial phenomena of the Long Island or Outer Hebrides: Second paper. *Quarterly Journal of the Geological Society of London*, **34** (4), 819–70.

Gemmell, S.G.L. (2000) Sediment transfer from gravel-bed rivers to beaches. Unpublished PhD thesis, University of Glasgow.

Gemmell, S.G.L., Hansom, J.D. and Hoey, T.B. (2001a) The geomorphology, conservation and management of the River Spey and Spey Bay SSSIs, Moray. *Scottish Natural Heritage Research, Survey and Monitoring Report*, No. 57, 215 pp.

Gemmell, S., Hansom, J.D. and Hoey, T. (2001b) River-coast sediment exchanges: the Spey Bay sediment budget and management implications. In *Ecology and Geomorphology of Coastal Shingle* (eds J.R. Packham, R.E. Randall, R.S.K. Barnes and A. Neal), Westbury Academic and Scientific Publishing, Otley, pp. 159–67.

George, T.N. (1932) The Quaternary beaches of Gower. *Proceedings of the Geologists' Association*, **43**, 291–324.

George, T.N. (1933) The coast of Gower. *Proceedings of the Swansea Scientific and Field Naturalists Society*, **1**, 23–48.

Gibbs, P. (1980) Observation of short-term profile changes on Chesil Beach, Dorset. *Proceedings of the Dorset Natural History and Archaeological Society*, **102**, 77–82.

Gierloff-Emden, H.G. (1980) *Geographie des Meeres, Ozeane und Küsten*, De Gruyter, Berlin, 1455 pp.

Gilbert, A. (1996) The raised shoreline sequence at Saunton in North Devon. In *Devon and East Cornwall* (eds D.J. Charman, R.M. Newnham and D.G. Croot), Quaternary Research Association Field Guide, Quaternary Research Association, London, pp. 40–7.

Gilbert, C.J. (1930) Land oscillation during the closing stages of the Neolithic depression. *2nd Report of the Committee of Pliocene and Pleistocene Terrain*, p. 96.

Gilbert, C.J. (1933) The evolution of Romney Marsh. *Archaeologia Cantiana*, **45**, 246–72.

Gilbertson, D.D. and Mottershead, D.W. (1975) The Quaternary deposits at Doniford, west Somerset. *Field Studies*, **4**, 117–22.

Gilbertson, D.D., Grattan, J. and Schwenninger, J-L. (1996) A stratigraphic survey of the Holocene coastal dune and machair sequences. In *The Outer Hebrides: the Last 14,000 Years* (eds D.D. Gilbertson, M. Kent and J. Grattan), Sheffield Environmental and Archaeological Research Campaign in the Hebrides, No. 2, Sheffield Academic Press,

References

Sheffield, pp. 72–101.

Gilbertson, D.D., Schwenninger, J.-L., Kemp, R.A. and Rhodes, E.J. (1999) Sand-drift and soil formation along an exposed North Atlantic coastline: 14,000 years of diverse geomorphological climatic and human impacts. *Journal of Archaeological Science*, **26** (4), 439–469.

Gilham, M.E. (1977) *Sand Dunes*, Heritage Coast Joint Management and Advisory Committee.

Gleason, R. and Hardcastle, P.J. (1973) The significance of wave parameters in the sorting of beach pebbles. *Estuarine and Coastal Science*, **1**, 11–18.

Gleason, R., Blackley, M.W.L. and Carr, A.P. (1975) Beach stability and particle size distribution, Start Bay. *Quarterly Journal of the Geological Society of London*, **131**, 83–101.

Godard, A. (1965) *Recherches de Géomorphologie en Ecosse du Nord-Ouest*, Publications de la Faculté de l'Université de Strasbourg. Fondation Baulig, No. 1, les Belles Lettres, Paris, 703 pp.

Godwin, H. and Willis, E.H. (1961) Cambridge University Natural Radiocarbon Measurements III. *Radiocarbon*, **3**, 60–76.

Godwin, H. (1940) Studies in the post-glacial history of British vegetation, IV. Post-glacial changes of land and sea in the English Fenland. *Philosophical Transactions of the Royal Society of London*, **B230**, 239–303.

Goodman, P.J. (1960) Investigations into 'die-back' in *Spartina townsendii* agg. II. The morphological structure and composition of the Lymington sward. *Journal of Ecology*, **48**, 711–24.

Goodman, P.J. and Williams, W.T. (1961) Investigations into 'die-back' in *Spartina townsendii* agg. III. Physiological correlates of 'die-back'. *Journal of Ecology*, **49**, 391–8.

Goodman, P.J., Braybrooks, E.M. and Lambert, J.M. (1959) Investigations into 'die-back' in *Spartina townsendii* agg. I. The present status of *Spartina townsendii* in Britain. *Journal of Ecology*, **47**, 651–77.

Gordon, J.E. and Sutherland, D.G. (1993) (eds) *Quaternary of Scotland*, Geological Conservation Review Series, No. 6, Chapman and Hall, London, 695 pp.

Gordon, R. (1843) Description of the Sheriffdoms of Aberdeen and Banff. In *Collections for a History of the Shires of Aberdeen and Banff* (ed. J. Robertson), *Spalding Club Publications*, No. 9, Spalding Club, Aberdeen.

Goudie, A. (1972) Vaughan Cornish: geographer (with a bibliography of his published works). *Transactions of the Institute of British Geographers*, **55**, 1–16.

Goudie, A. (1981) *The Human Impact*, Blackwell Scientific Publications, Oxford, 326 pp.

Goudie, A. (1990) *The Landforms of England and Wales*, Blackwell Scientific Publications, Oxford, 394 pp.

Goudie, A.. and Brunsden, D. (1994) *The Environment of the British Isles – an Atlas*, Oxford University Press, Oxford, 184 pp.

Goudie, A. and Gardner, R. (1985) *Discovering Landscape in England and Wales*, Allen and Unwin, London, 177 pp.

Grainger, P. and Kalaugher, P.G. (1987) Cliff-top recession related to the development of coastal landsliding west of Budleigh Salterton, Devon. *Proceedings of the Ussher Society*, **6**, 1169–74.

Grainger, P. and Kalaugher, P.G. (1988) Hazard zonation of coastal landslides. In *Proceedings of the Fifth International Symposium on Landslides, 10-15 July 1988, Lausanne* (ed. C. Bonnard), A.A. Balkema, Rotterdam, pp. 1169–74.

Gray, A.J. (1980) Saltmarshes and reclaimed land. In *Wild Flowers: their Habitats in Britain and Northern Europe* (eds G. Halliday and A.W. Malloch), Peter Lowe, London, pp. 123–33.

Gray, A.J. (1992) Saltmarsh plant ecology: zonation and succession revisited. In *Saltmarshes: Morphodynamics, Conservation and Engineering Significance* (eds J.R.L. Allen and K. Pye), Cambridge University Press, Cambridge.

Gray, A.J., Banham, P.E.M. and Raybould, A.F. (1990) *Spartina anglica* – the evolutionary and ecological background. In *Spartina anglica – a Research Review* (eds A.J. Gray and P.E.M. Benham), Institute of Terrestrial Ecology Research Publication, No. 2, HMSO, London, pp. 5–10.

Gray, J.M. (1974) The main rock platform of the Firth of Lorn, western Scotland. *Transactions of the Institute of British Geographers*, **61**, 81–99.

Gray, J.M. (1978) Low-level shore platforms in the south-west Scottish highlands: altitude, age and correlation. *Transactions of the*

References

Institute of British Geographers, New Series, **3** (2), 151–64.

Green, C.P. (1988) Palaeogeography of marine inlets in the Romney Marsh area. In *Romney Marsh: Evolution, Occupation and Reclamation* (eds J. Eddison and C. Green), Oxford University Committee for Archaeology Monograph, No. **24**, OUCA, Oxford, pp. 167–74.

Green, C. and Hutchinson, J. (1960) Archaeological evidence. In *The Making of the Broads. A Reconsideration of their Origin in the Light of New Evidence*, (eds J.M. Lambert, J.N. Jennings, C.T. Smith, C. Green and J.N. Hutchinson), Royal Geographical Society Research Series, No. **3**, John Murray, London, pp. 113–46.

Green, C., Larwood, C.P. and Martin, A.J. (1953) The coastline of Blofeld and Flegg. *Transactions of the Norfolk and Norwich Naturalists' Society*, **17**, 27–42.

Green, C.D. (1973) The sediments of the south side of the entrance to the Tay estuary. Unpublished PhD thesis, University of Dundee.

Green, C.P. and McGregor, D.F.M. (1986) *Dungeness: a Geomorphological Assessment*, Nature Conservancy Council, London, 2 volumes.

Green, C.P. and McGregor, D.F.M. (1988) *Orfordness: a Geomorphological Assessment*, Nature Conservancy Council, London, 2 volumes.

Green, C.P. and McGregor, D.F.M. (1990) Orfordness: geomorphological conservation perspectives. *Transactions of the Institute of British Geographers*, **15**, 48–59.

Green, H.M., Stoddart, D.R., Reed, D.J. and Bayliss-Smith, T.P. (1986) Saltmarsh tidal creek dynamics, Scolt Head Island, Norfolk, England. In *Proceedings of the Iceland Coastal and River Symposium* (ed. G. Sigbjarnarson), National Energy Authority, Reykjavik, pp. 93–103.

Green, R.D. (1968) *Soils of Romney Marsh, Soil Survey of Great Britain Bulletin*, No. **4**. Agricultural Research Council, Harpenden.

Green, R.D. and Askew, G.P. (1958) Kent. *Report of the Soil Survey of Great Britain*, **9**, 27–30.

Greenly, E. (1919) *The Geology of Anglesey*, Memoir of the Geological Survey (England and Wales), HMSO, London, 2 volumes, 980 pp.

Greensmith, J.T. and Gutmanis, J.C. (1990) Aspects of the late Holocene depositional history of the Dungeness area, Kent. *Proceedings of the Geologists' Association*, **101** (3), 225–37.

Greensmith, J.T. and Tucker, E.V. (1965) Salt marsh erosion in Essex. *Nature*, **206**, 606–7.

Greensmith, J.T. and Tucker, E.V. (1967) Morphology and evolution of inshore shell ridges and mudmounds on modern intertidal flats, near Bradwell, Essex. *Proceedings of the Geologists' Association*, **77**, 329–46.

Greensmith, J.T. and Tucker, E.V. (1968) Imbricate structure in Essex offshore shellbanks. *Nature*, **220**, 1115–16.

Greensmith, J.T. and Tucker, E.V. (1969) The origin of Holocene shell deposits in the chenier plain facies of Essex (Great Britain). *Marine Geology*, **7**, 403–25.

Greensmith, J.T. and Tucker, E.V. (1973a) Holocene transgressions and regressions on the Essex coast, outer Thames estuary. *Geologie Mijnbouw*, **52**, 193–203.

Greensmith, J.T. and Tucker, E.V. (1973b) Peat balls in late-Holocene sediments of Essex, England. *Journal of Sedimentary Petrology*, **43**, 894–7.

Greensmith, J.T. and Tucker, E.V. (1975) Dynamic structures in the Holocene chenier plain setting of Essex, England. In *Nearshore Sediment Dynamics and Sedimentation* (eds J. Hails and A. Carr), John Wiley and Sons Ltd, Chichester, pp. 251–72.

Greenwood, B. (1969) Sediment parameters and environmental discrimination: an application of multivariate statistics. *Canadian Journal of Earth Science*, **6**, 1347–58.

Greenwood, B. (1978) Spatial variability of texture of a beach-dune complex, North Devon, England. *Sedimentary Geology*, **21**, 27–44.

Gregory, K.J. (ed.) (1997) *Fluvial Geomorphology of Great Britain*, Geological Conservation Review Series, No. **13**, Chapman and Hall, London, 347 pp.

Gresswell, R.K. (1937) The geomorphology of the south-west Lancashire coastline. *Geographical Journal*, **90**, 335–49.

Gresswell, R.K. (1953a) The coast. In *A Scientific Survey of Merseyside* (ed. W. Smith), British Association for the Advancement of Science, Liverpool, pp. 49–52.

Gresswell, R.K. (1953b) *Sandy Shores in South Lancashire: the Geomorphology of South-West Lancashire*, Liverpool Studies in Geography,

References

Liverpool University Press, Liverpool, 194 pp.

Gresswell, R.K. (1957) Hillhouse coastal deposits in south Lancashire. *Liverpool and Manchester Geological Journal*, **2**, 60–78.

Gresswell, R.K. (1964) The origin of the Mersey and Dee Estuaries. *Geological Journal*, **4**, 77–85.

Griggs, G.B. and Trenhaile, A. (1994) Coastal cliffs and platforms. In *Coastal Evolution: Late Quaternary Shoreline Morphodynamics* (eds R.W.G. Carter and C.D. Woodroffe), Cambridge University Press, Cambridge, pp. 425–50.

Grove, A.T. (1950) Tentsmuir–Fife. Soil blowing and coastal changes. Unpublished report to the Nature Conservancy, Edinburgh, 24 pp.

Grove, A.T. (1953) The sea flood on the coasts of Norfolk and Suffolk. *Geography*, **38**, 164–70.

Grove, A.T. (1955) The mouth of the Spey. *Scottish Geographical Magazine*, **71**, 104–7.

Grove, A.T. (1960) Beach profiles at Scolt Head Island. In *Scolt Head Island* (ed. J.A. Steers), Heffer, Cambridge, pp. 67–9.

Groves, T.B. (1875) The Chesil Bank. *Nature*, **11**, 506–7.

Gubbay, S. (1990) *Future for the Coast: Proposals for a UK Coastal Zone Management Plan*, Marine Conservation Society, Ross-on-Wye, 31 pp.

Guilcher, A. (1954) *Morphologie Littorale et Sous-Marine*, Presses Universitaires de France, Paris, 210 pp.

Guilcher, A. (1958) *Coastal and Submarine Morphology* (translated by B.W. Sparks and R.H.W. Kneese), Methuen, London, 274 pp.

Guilcher, A. (1965) Drumlin and spit structures in the Kenmare River, south-west Ireland. *Irish Geographer*, **5**, 7–19.

Guilcher, A. and Berthois, L. (1957) Cinq années d'observation sedimentologiques dans quatre estuaires – témoins de l'Ouest de la Bretagne. *Revue de Géomorphologie Dynamique*, **8**, 67–86.

Guilcher, A., Bodéré, J.C., Coudé, A., Hansom, J.D., Moign, A. and Peullvast, J.P. (1986) Le problème des strandflats en cinq pays de hautes latitudes. *Revue de Géologie Dynamique et de Géographie Physique* **27**(1), 47–79.

Gulliver, F.P. (1897) Dungeness foreland. *Geographical Journal*, **9**, 536–46.

Gulliver, F.P. (1898–1899) Shoreline topography. *Proceedings of the American Academy of Arts and Science*, **24**, 149–255.

Haggart, B.A. (1986) Relative sea level changes in the Beauly Firth, Scotland. *Boreas*, **15**, 191–207.

Haggart, B.A. (1987) Relative sea level changes in the Moray Firth area, Scotland. In *Sea-Level Changes* (eds M.J. Tooley and I. Shennan), Blackwell Scientific Publications, Oxford, pp. 67–108.

Haggart, B.A. (1989) Variations in the pattern and rate of isostatic uplift indicated by a comparison of sea level curves from Scotland. *Journal of Quaternary Science*, **4**, 67–76.

Hails, J.R. (1975a) Submarine geology, sediment distribution and Quaternary history of Start Bay, Devon: Introduction. *Journal of the Geological Society*, **131**, 1–5.

Hails, J.R. (1975b) Sediment distribution and Quaternary history of Start Bay. *Journal of the Geological Society*, **131**, 19–36.

Hails, J.R. (1975c) Some aspects of the Quaternary history of Start Bay, Devon. *Field Studies*, **4**, 207–22.

Hails, J.R. and Carr, A.P. (eds) (1975) *Nearshore Sediment Dynamics and Sedimentation: an Interdisciplinary Review*, John Wiley and Sons Ltd, London, 316 pp.

Halcrow, Sir W. and Partners (1980) Coast Protection at the Pebble Ridge, Westward Ho!. *Final report to Torridge District Council*.

Halcrow, Sir W. and Partners (1988) *Anglian Coastal Management Atlas*, Anglian Water, Peterborough, 97 pp.

Halcrow Maritime (1999) *Poole and Christchurch Bays. Shoreline Management Plan*, 3 volumes, Halcrow Maritime, Swindon.

Hall, A.M. (1986) Deep weathering patterns in north-east Scotland and their geomorphological significance. *Zeitschrift für Geomorphologie, Neue Folge*, **30** (4), 407–22.

Hall, A.M. (1989) Pre-Late Devensian coastal rock platforms around Dunbar. *Scottish Journal of Geology*, **25** (3), 361–5.

Hall, B.R. (1954–1955) Borehole records from the mosses of south-west Lancashire. *Soil Survey of England and Wales*, MS 65.

Hall, T.M. (1870) The raised beaches and submerged forests of Barnstaple Bay. *The Student and Intellectual Observer of Science, Literature and Art*, **4**, 338–49.

Hall, T.M. (1879) The submerged forest of Barnstable Bay. *Quarterly Journal of the Geological Society of London*, **35**, 106.

Hallegoüet, B. and Moign, A. (1979) Progradation et érosion d'un secteur littoral

References

sableux en Bretagne nord: mesures et bilan. In *Les Côtes Atlantiques de l'Europe: Évolution, Aménagement, Protection* (ed. A. Guilcher), *Actes de Colloques*, No. 9, Centre National pour l'Exploitation des Océans (CNEXO), Brest, pp. 45–54.

Halliwell, A.R. (1975) The dredging requirements for the entrance channel serving McDermotts Fabrication Yard, Ardersier, Moray Firth. *Report to Blyth and Blyth*, Heriot-Watt University, Edinburgh.

Halliwell, A.R. and O'Connor, B.A. (1966) Suspended sediment in a tidal estuary (Mersey). *Geophysical Journal of the Royal Astronomical Society*, 32, 439–58.

Halsey, S.D. (1979) Nexus: new model of barrier island development. In *Barrier Islands* (ed. S.P. Leatherman), Academic Press, New York, pp. 185–209.

Hamilton, H. (ed.) (1965) *The Third Statistical Account of Scotland: the Counties of Moray and Nairn*, Collins, Glasgow, 462 pp.

Hannah, G.W. (1986) The evolution of Bridport Harbour. *Proceedings of the Dorset Natural History and Archaeological Society*, 108, 27–31.

Hansom, J.D. (1983) Ice-formed intertidal boulder pavements in the sub-Antarctic. *Journal of Sedimentary Petrology*, 53 (1), 135–45.

Hansom, J.D. (1988) *Coasts*, Cambridge Topics in Geography, Cambridge University Press, Cambridge, 96 pp.

Hansom, J.D. (1991) Holocene coastal development in the Dornoch Firth. In *Late Quaternary Coastal Evolution in the Inner Moray Firth: Field Guide* (eds C.R. Firth and B.A. Haggart), West London Press, City of London Polytechnic, London, pp. 45–55.

Hansom, J.D. (1998) The geomorphology of Balnakeil/An Fharaid dune complex. Unpublished Survey and Monitoring Report, Scottish Natural Heritage, Edinburgh.

Hansom, J.D. (1999) The coastal geomorphology of Scotland: understanding sediment budgets for effective coastal management. In *Scotland's Living Coastline* (eds J. Baxter, K. Duncan, S. Atkins and G. Lees), Natural Heritage of Scotland Series, No. 7, The Stationery Office, London, pp. 34–44.

Hansom, J.D. (2001) Coastal sensitivity to environmental change: a view from the beach. *Catena*, 42, 291–305.

Hansom, J.D. and Angus, S. (2001) Tir a'Mhachair (Land of the Machair): sediment supply and climate change scenarios for the future of the Outer Hebrides machair. In *Earth Science and the Natural Heritage: Interactions and Integrated management* (eds J.E. Gordon and K.F. Lees), Natural Heritage of Scotland Series, No. 9, The Stationery Office, Edinburgh, pp. 68–81.

Hansom, J.D. and Black, D.L. (1996) *Coastal Processes and Management of Scottish Eestuaries II: Estuaries of the Outer Moray Firth*, Scottish Natural Heritage Review, No. 51, Scottish Natural Heritage, Edinburgh, 94 pp.

Hansom, J.D. and Comber, D.P.M. (1994) Culbin: coastline on the move. *Geological Conservation in the Moray Firth*. Scottish Natural Heritage Regional Conference Proceedings, Elgin, March 1994 (Unpublished).

Hansom, J.D. and Comber, D.P.M. (1996) Eoligarry SSSI documentation and management prescription. *Scottish Natural Heritage Research, Survey and Monitoring Report*, No. 49, 73 pp.

Hansom, J.D. and Evans, D.J.A. (1995) Scottish landform examples – 13. The Old Man of Hoy. *Scottish Geographical Magazine*, 111 (3), 172–4.

Hansom, J.D. and Kirk, R.M. (1989) Ice in the intertidal zone: examples from Antarctica. *Essener Geographische Arbeiten*, 18, 211–36.

Hansom, J.D. and Kirk, R.M. (1991) Change on the coast: the need for management. In *Aspects of Environmental Change* (eds T.R.R. Johnston and J.R. Flenley), Massey University Press, Palmerston North, New Zealand, pp. 49–62.

Hansom, J.D. and Leafe, R.N. (1990) The geomorphology of Morrich More: development of a scientific database and management prescription. *Nature Conservancy Council CSD Report*, No. 1161, 174 pp + maps.

Hansom, J.D. and Rennie, A. (2003) Assessment of rates and causes of change in Scotland's beaches and dunes. *Scottish Natural Heritage Report*, Battleby, Scotland.

Hanson, J.D., Comber, D.P.M. and Fahy, F.M. (1993) Ribble Estuary. Estuary management plans. *Coastal Processes and Conservation Report*, University of Glasgow for English Nature, Peterborough.

Hansom, J.D., Lees, R.G., Maslen, J., Tilbrook, C. and McManus, J. (2001) Coastal dynamics and

References

sustainable development: the potential for managed realignment in the Firth of Forth. In *Earth Science and the Natural Heritage: Interactions and Integrated Management* (eds J.E. Gordon and K.F. Lees), Natural Heritage of Scotland Series, No. 9, The Stationery Office, Edinburgh, pp. 148–60.

Hansom, J.D., Hall, A., Barltrop, N. and Jarvis, J. (in press) Extreme storm waves in the North Atlantic and their impact on cliff-top processes in the Shetland Islands. *Science*.

Hardcastle, P.J. and King, A.C. (1972) Sea wave records from Chesil Beach, Dorset. *Civil Engineering*, 67, 299–300.

Hardisty, J. (1982) Sediment dynamics in the inter-tidal profile. Unpublished PhD thesis, University of Hull.

Hardisty, J. and Laver, A.J. (1989) Breaking waves on a macrotidal barred beach: a test of McCowan's criteria. *Journal of Coastal Research*, 5, 79–82.

Hardy, J.R. (1964) The movement of beach material and wave action near Blakeney Point, Norfolk. *Transactions of the Institute of British Geographers*, 34, 53–69.

Hardy, J.R. (1966) An ebb-flood channel system and coastal changes near Winterton, Norfolk. *East Midland Geographer*, 4, 24–30.

Harlow, D.A. (1979) The littoral sediment budget between Selsey Bill and Gilkicker Point, and its relevance to coastal protection works on Hayling Island. *Quarterly Journal of Engineering Geology*, 12, 257–65.

Harlow, D.A. (1982) Sediment processes, Selsey Bill to Portsmouth. Unpublished PhD thesis, University of Southampton.

Harmsworth, G.C. and Long, S.P. (1986) An assessment of saltmarsh erosion in Essex, England, with reference to the Dengie Peninsula. *Biological Conservation*, 35, 377–87.

Harper, S.A. (1976) The dynamics of salt marsh development in south-east Lincolnshire. Unpublished PhD thesis, University of Nottingham.

Harper, S.A. (1979) Sedimentation on the New Marsh at Gibraltar Point, Lincolnshire. *East Midland Geographer*, 7, 153–67.

Harris, C., Williams, G., Brabham, P., Easton, G. and McCarroll, D. (1996) Glaciotectonized Quaternary sediments at Dinas Dinlle, Gwynedd, north Wales, and their bearing on the style of deglaciation in the eastern Irish Sea. *Quaternary Science Reviews*, 16, 109–37.

Harris, T. and Ritchie, W. (1989) *Dune and Machair Erosion in the Luskentyre area: a Preliminary Survey*, Nature Conservancy Council, Edinburgh.

Harrison, E.A. and Bloom, A.L. (1977) Sedimentary rates on tidal salt marshes in Connecticut. *Journal of Sedimentary Petrology*, 47, 1484–90.

Hart, B.S. (1991) A study of pebble shape from gravely shoreface deposits. *Sedimentary Geology*, 73, 185–9.

Hartnall, T.J. (1982) Vegetation and sediment supply in relation to morphological development on the salt-marsh at Gibraltar Point, Lincolnshire. Unpublished PhD thesis, University of Nottingham.

Harvey, J.G. (1967) Drifter studies in the Irish Sea. In *Liverpool Essays in Geography: a Jubileee Collection Publication* (eds R.W. Steel and R. Lawton), Longman, London, pp. 137–56.

Harvey, J.G. (1968) The flow of water through the Menai Strait. *Geophysical Journal of the Royal Astronomical Society*, 15, 517–28.

Harvey, M.M. (2000) Geomorphological controls on the sedimentation patterns of, and distribution of anthropogenic radionuclides in, coastal salt marshes, south-west Scotland. Unpublished PhD thesis, University of Glasgow.

Harvey, M.M. and Allan, R.L. (1998) The Solway Firth saltmarshes. *Scottish Geographical Magazine*, 114 (1), 42–5.

Haslett, S. (2000) *Coastal Systems*, Routledge, London, 218 pp.

Hawker, D. (1999) Saltmarsh in the Solway Firth European Marine Site. Unpublished report to Scottish Natural Heritage, Battleby, Perth.

Hawley, N. (1982) Intertidal sedimentary structures on macrotidal beaches. *Journal of Sedimentary Petrology*, 52, 785–95.

Headworth, M.G. (1983) The gravel aquifer at Dungeness, Kent. *Journal of the Geological Society*, 140, 334–43.

Healey, R.G., Pye, K., Stoddart, D.R. and Bayliss-Smith, T.P. (1981) Velocity variations in salt-marsh creeks, Norfolk, England. *Estuarine, Coastal and Shelf Science*, 13, 535–45.

Heathwaite, A.L. and O'Sullivan, P.E. (1991) Sequential inorganic chemical analysis of a core from Slapton Ley, Devon, UK. *Hydrobiologia*, 214, 125–35.

Heeps, C. (1986) Sediment circulation in mixed

References

gravel and shingle bayhead beaches on the south-east Dorset coast. Unpublished PhD thesis, Council for National Academic Awards (CNAA).

Heeps, C. (1987) Recognition and prediction of seabed sediment patterns. *Report to Ministry of Defence Procurement Executive Admiralty Research Establishment (Portland)*.

Heijne, I.S. and West, G.M. (1991) Chesil Sea Defence Scheme. Paper 2: design of interceptor drain. *Proceedings of the Institution of Civil Engineers*, **90**, 799–817.

Henderson, G. (1979) A study of wave climate and wave energy in Poole and Christchurch Bays. Unpublished PhD thesis, University of Southampton.

Henson, M.R. (1970) The Triassic rocks of south Devon. *Proceedings of the Ussher Society*, **2**, 172–7.

Heron-Allen, E. (1911) *Selsey Bill –Historic and Pre-Historic*, Duckworth and Co., London, 404 pp.

Hesp, P. (2002) Foredunes and blowouts: initiation, geomorphology and dynamics. *Geomorphology*, **48**, 245–68.

Hewett, D.G. (1970) The colonization of sand dunes after stabilization with Marram grass (*Ammophila arenaria*). *Journal of Ecology*, **58**, 653–68.

Hewett, D.G. (1971) The effects of the cold winter of 1962/3 on Juncus acutus at Braunton Burrows, Devon. *Report and Transactions of the Devon Association for the Advancement of Science, Literature and Art*, **102**, 193–201

Hey, R.W. (1967) Sections in the beach-plain deposits of Dungeness, Kent. *Geological Magazine*, **104**, 361–70.

Heyworth, A. and Kidson, C. (1982) Sea level changes in southwest England and Wales. *Proceedings of the Geologists' Association*, **93**, 91–111

Hickey, K. (1991) Documentary records of coastal storms in Scotland 1500–1991 AD. Unpublished PhD thesis, University of Coventry.

Hill, M.I. (1988) Saltmarsh vegetation of the Wash. An assessment of change from 1971 to 1985. *Research and Survey in Nature Conservation*, **13**, Nature Conservancy Council, Peterborough, 82 pp.

Hill, M.I. (1996) Saltmarsh. In *Coasts and Seas of the United Kingdom. Region 3. North-East Scotland: Cape Wrath to St Cyrus* (eds J.H. Barne, C.F. Robson, S.S. Kasnowska, J.P. Doody and N.C. Davidson) Coastal Directories Series, Joint Nature Conservation Committee, Peterborough, pp. 61–4.

Hill, M.I. (1997) Saltmarsh. In *Coasts and Seas of the United Kingdom. Regions 15 and 16. North-West Scotland: the Western Isles and West Highland* (eds J.H. Barne, C.F. Robson, S.S. Kasnowska, J.P. Doody, N.C. Davidson and A.L. Buck), Coastal Directories Series, Joint Nature Conservation Committee, Peterborough, pp. 72–6.

Hill, M.O. and Wallace, H.L. (1989) Vegetation and environment in afforestation of sand dunes at Newborough Warren, Anglesey. *Forestry*, **62**, 249–67.

Hill, T.G. and Hanley, J.A. (1914) The structure and water content of shingle beaches. *Journal of Ecology*, **2**, 21–38.

HMSO (1947) *Report of the Wildlife Conservation Special Committee*, **Cmd. 7122**, HMSO, London.

Holmes, A. (1944) *Principles of Physical Geology*, Nelson, London, 532 pp.

Holmes, A. (1965) *Principles of Physical Geology*, 2nd edn, Nelson, London, 1288 pp.

Hook, B.J. and Kemble, J.R. (1991) Chesil Sea Defence Scheme. Paper 1: concept, design and construction. *Proceedings of the Institution of Civil Engineers*, **90**, 783–98.

Hooke, J.M., Bray, M.J. and Carter, D.J. (1996) Sediment transport analysis as a component of coastal management – a UK example. *Environmental Geology*, **27**, 347–57.

Hope-Simpson, J.F. (1985) Monitoring by photographs on Braunton Burrows: relevance to nature conservation purposes. In *Sand Dunes and Their Management* (ed. J.P. Doody), Focus on Nature Conservation, No. 13, Nature Conservancy Council, Peterborough, pp. 175–85.

Hope-Simpson, J.F. (1997) Dynamic plant ecology of Braunton Burrows, southwestern England. In *Dry Coastal Ecosystems. General Aspects Ecosystems of the World 2C* (ed. E. van der Maarel), Elsevier, Amsterdam, pp. 437–52.

Hope-Simpson, J.F. and Jefferies, R.L. (1966) Observations relating to vigour and debility in marram grass (*Ammophila arenaria* (L.) Link). *Journal of Ecology*, **54**, 271–4.

Hope-Simpson, J.F. and Yemm, E.W. (1979) Braunton Burrows: developing vegetation in dune slacks 1948–1977. In *Ecological Processes in Coastal Environments* (eds R.L.

References

Jeffries and A.J. Davey), Blackwell Scientific Publications, Oxford, pp. 113–28.

Hoppe, G. (1965) *Submarine Peat in the Shetland Islands*, Institute of British Geographers Special Publication, No. 7, Institute of British Geographers, London, 197–210.

Horsfall, D. (1993) Geological controls on coastal morphology. *Geography Review*, 7, 16–22.

Hosking, K.F.G. and Ong, P.M. (1963) The distribution of tin and certain other 'heavy' metals in the superficial portions of the Gwithian/Hayle beach of west Cornwall. *Transactions of the Royal Geological Society of Cornwall*, 19, 351–92.

Houghton, J. (1994) *Global Warming: the Complete Briefing*, Lion, Oxford, 192 pp.

Houghton, J.T., Jenkins, G.J. and Ephraums, J.J. (1990) *Climate Change. The IPCC Scientific Assessment*, Cambridge University Press, Cambridge, 365 pp.

House, M.R. (1969) *The Dorset Coast from Poole to the Chesil Beach*, 2nd edn, Geologists' Association Guide, No. 22, Benham and Co., Colchester, 21 pp.

House, M.J. (1989) *Geology of the Dorset Coast*, Geologists' Association Guide, Geologists' Association, London, 162 pp.

House, M.R. (1993) *Geology of the Dorset Coast*, 2nd edn, Geologists' Association Guide, Geologists' Association, London, 164 pp.

Hubbard, J.C.E. and Stebbings, R.E. (1967) Distribution, date of origin and acreage of *Spartina townsendii* (*s.l.*) marshes in Great Britain. *Proceedings of the Botanical Society of the British Isles*, 7, 1–7.

Huddart, D. (1992) Coastal environmental changes and morphostratigraphy in southwest Lancashire, England. *Proceedings of the Geologists' Association*, 103, 217–36.

Huddart, D. and Carter, P.A. (1977) The coasts of north-west England. In *The Quaternary History of the Irish Sea* (eds C. Kidson and M.J. Tooley), Seel House Press, Liverpool, p. 345.

Huddart, D. and Glasser, N.F. (2002) *Quaternary of Northern England*, Geological Conservation Review Series, No. 25, Joint Nature Conservation Committee, Peterborough, 745 pp.

Hughes, T.Mck. (1887) On the ancient beach and boulders near Braunton and Croyde, in North Devon. *Quarterly Journal of the Geological Society of London*, 43, 657–70.

Hunsdale, R. and Sanderson, D.J. (1998) Fault size distribution analysis – an example from Kimmeridge Bay, Dorset, UK. In *Development, Evolution and Petroleum Geology of the Wessex Basin* (ed. J.R. Underhill), *Geological Society of London Special Publication*, No. 133, Geological Society of London, Bath, pp. 299–310.

Huntley, D. and Bowen, A. (1973) Field observations of edge waves. *Nature*, 24, 160–1.

Huntley, D. and Bowen, A. (1975a) Field observations of edge waves and a discussion of their effect on beach material. *Journal of the Geological Society of London*, 131, 69–81.

Huntley, D. and Bowen, A. (1975b) Comparison of the hydrodynamics of steep and shallow beaches. In *Nearshore Sediment Dynamics and Sedimentation* (eds J. Hails and A. Carr), John Wiley and Sons Ltd, London, pp. 69–109.

Hussey, A. and Long, S.P. (1982) Seasonal changes in weight of above- and below-ground vegetation and dead plant material in a salt marsh at Colne Point, Essex. *Journal of Ecology*, 70, 757–71.

Hutcheson, A. (1914) The archaeology of Tentsmuir. *Proceedings and Transactions of the Dundee Naturalists' Society*, 1 (1), 31–40.

Hutchinson, J.N. (1965) *A Reconnaissance of Coastal Landslides in the Isle of Wight*, Note En 11/65, Building Research Station, Watford.

Hutchinson, J.N. (1972) Field and laboratory studies of a fall in Upper Chalk cliffs at Joss Bay, Isle of Thanet. In *Stress-Strain Behaviour of Soils: Proceedings of the Roscoe Memorial Symposium, University of Cambridge, 29–31 March, 1971* (ed. R.H.G. Parry), G.T. Foulis, Henley-on-Thames, pp. 692–706.

Hutchinson, J.N. (1973) The response of London clay cliffs to differing rates of toe erosion. *Geologica Applicata e Idrologica*, 8, 221–39.

Hutchinson, J.N. (1976) Coastal slides in cliffs of Pleistocene deposits between Cromer and Overstrand, Norfolk, England. In *Lauritis Bjerrum Memorial Volume: Contributions to Soil Mechanics* (eds N. Janbu, F. Jørstad and B. Kjærnsli), Norges Geotekniske Institutt, Oslo, pp. 155–82.

Hutchinson, J.N. (1980) Various forms of cliff instability arising from coast erosion in the UK. *Fjellsprengningsteknikk Bergmekanikk/ Geoteknikk 1979*, 19.1–19.32.

References

Hutchinson, J.N. (1983) Engineering in a landscape. Inaugural lecture, 9th October 1979. Imperial College of Science and Technology, University of London.

Hutchinson, J.N. (1984) Landslides in Britain and their countermeasures. *Journal of Japan Landslide Society*, **21-1**, 1–24.

Hutchinson, J.N. (1987) Some coastal landslides of the southern Isle of Wight. In *Wessex and the Isle of Wight Field Guide Prepared to Accompany the Annual Field Meeting Held at Southampton and Cowes 21–25 April 1987* (ed. K.E. Barber), Quaternary Research Association, Cambridge, pp. 123–35.

Hutchinson, J.N. (1991) The landslides forming the South Wight Undercliff. In *Slope Stability Engineering: Developments and Applications. Proceedings of the International Conference on Slope Stability Organized by the Institution of Civil Engineers and held on the Isle of Wight on 15–18 April 1991* (ed. R.J. Chandler), Thomas Telford, London, pp. 157–68.

Hutchinson, J.N. and Gostelow, T.P. (1976) The development of an abandoned cliff in London clay at Hadleigh, Essex. *Philosophical Transactions of the Royal Society*, **283**, 557–604.

Hutchinson, J.N., Bromhead, E.N. and Lupini, J.F. (1980) Additional observations on the landslides at Folkestone Warren. *Quarterly Journal of Engineering Geology*, **13**, 1–31.

Hutchinson, J.N., Chandler, M.P. and Bromhead, E.N. (1981) Cliff recession on the Isle of Wight SW Coast. In *Proceedings of the Tenth International Conference on Soil Mechanics and Foundation Engineering, Stockholm 15–19 June 1981* (ed. N. Flodin), A.A. Balkema, Rotterdam, pp. 429–34.

Hutchinson, J.N., Brunsden, D. and Lee, E.M. (1991) The geomorphology of the landslide complex at Ventnor, Isle of Wight. In *Slope Stability Engineering: Developments and Applications. Proceedings of the International Conference on Slope Stability Organized by the Institution of Civil Engineers and held on the Isle of Wight on 15–18 April 1991* (ed. R.J. Chandler), Thomas Telford, London, pp. 213–18.

Huxley, T.H. (1884) *Physiography*, Macmillan, London, 384 pp.

Hydraulics Research (1979) West Bay, Bridport, Dorset: a sea defence and coast protection study. *Hydraulics Research Report*, No. **EX863**, 12 pp.

Hydraulics Research (1985) West Bay Harbour: a numerical study of beach changes east of the harbour entrance. *Hydraulics Research Report*, No. **EX1301**, 33 pp.

Hydraulics Research (1991a) West Bay Harbour: analysis of recent beach changes east of the harbour. *Hydraulics Research Report*, No. **EX2272**, 16 pp.

Hydraulics Research (1991b) West Bay Bridport: a random wave physical model investigation. *Hydraulics Research Report*, No. **EX2187**, 53 pp.

Hydraulics Research (1997) Coastal cells in Scotland. *Scottish Natural Heritage Research, Survey and Monitoring Report*, No. **56**.

Hydraulics Research Station (1969) *The southwest Lancashire coastline; a report of sea defences*, Report **EX 450**, Hydraulics Research Station, Wallingford, 43 pp.

Hydrographer of the Navy (1978) *Admiralty Chart 2825 Lochs on the east coast of Uist*, Hydrographer of the Navy, Taunton.

Hydrographer of the Navy (1997) *Admiralty Tide Tables 1998*, Hydrographer of the Navy, Taunton, 4 volumes.

Idle, E.T. and Martin, J. (1975) The vegetation and land use history of Torrs Warren, Wigtownshire. *Transactions of the Dumfries and Galloway Natural History Society*, **51**, 1–9.

Innes, J.B. and Long, A.J. (1992) A preliminary investigation of the 'Midley Sand' deposit, Romney Marsh. *Quaternary Newsletter*, **67**, 32–9.

Innes, J.B. and Tooley, M.J. (1993) The age and vegetational history of the Sefton coast dunes. In *The Sand Dunes of the Sefton Coast* (eds D. Atkinson and L. Houston), National Museums and Galleries on Merseyside, Liverpool, pp. 41–4.

Institute of Estuarine and Coastal Studies (IECS) (1992) *Spurn Heritage Coast Study Final Report*, Institute of Estuarine and Coastal Studies, Hull University

Isla, F.I. (1993) Overpassing and armouring phenomena on gravel beaches. *Marine Geology*, **110**, 369–76.

Jackson D.L. (2000) Guidance on the interpretation of the Biodiversity Broad Habitat Classification (terrestrial and freshwater types): Definitions and the relationship with other habitat classifications. *JNCC Report*, No. **307**.

References

Jago, C.F. (1980) Contemporary accumulation of marine sand in a macrotidal estuary, south-west Wales. In *Shallow Marine Processes and Products* (eds A.H. Bouma, D.S. Gorsline, C. Monty and G.P. Allen), *Sedimentary Geology Special Issue*, Elsevier, Amsterdam, No. 26, pp. 21–49.

Jago, C.F. and Hardisty, J. (1984) Sedimentology and morphodynamics of a macrotidal beach, Pendine Sands, SW Wales. *Marine Geology*, 60, 123–54.

James, P.A. and Wharf, A.J. (1989) Timescales of soil development in a coastal sand dune system, Ainsdale, North-west England. In *Perspectives in Coastal Dune Management* (eds F. van der Meulen, J.D. Jungerius and J. Visser), SPB Academic Publishing, The Hague, pp. 287–95.

Jardine, W.G. (1975) Chronology of Holocene marine transgression and regression in south-western Scotland. *Boreas*, 4, 173–96.

Jardine, W.G. (1977) The Quaternary marine record in southwestern Scotland and the Scottish Hebrides. In *The Quaternary History of the Irish Sea* (eds C. Kidson and M.J. Tooley), Seal House Press, Liverpool, pp. 99–118.

Jardine, W.G. (1978) Holocene coastal sediments and former shorelines of Dumfriesshire and Eastern Galloway. *Transactions of the Dumfriesshire and Galloway Natural History and Antiquarian Society, 3rd Series*, 55, 1–59.

Jay, H., McCue, J. and Hendry, M. (2001) The use of modelling in the management of shingle coastlines: a case study from East Sussex. In *Ecology and Geomorphology of Coastal Shingle* (eds J.R. Packham, R.E. Randall, R.S.K. Barnes and A. Neal), Westbury Academic and Scientific Publishing, Otley, pp. 402–8.

Jehu, T.J. (1918) Rock-boring organisms as agents in coast erosion. *Scottish Geographical Magazine*, 34, 1–11.

Jennings, S.C. and Orford, J.D. (1999) The Holocene inheritance embedded within contemporary coastal management problems. In *Proceedings of the 34th MAFF Conference of River and Coastal Engineers, Keele University, 30 June–2 July 1999*, Ministry of Agriculture, Fisheries and Food, London, pp. 9.2.1–9.2.15.

Jennings, S.C. and Smyth, C. (1990) Holocene evolution of the gravel coastline of East Sussex. *Proceedings of the Geologists' Association*, 101, 213–24.

Jennings, S.C. and Smyth, C. (1991) Holocene evolution of the gravel coastline of East Sussex: reply to correspondence. *Proceedings of the Geologists' Association*, 102, 306–8.

Jennings, S.C., Orford, J.D., Canti, M., Devoy, R.J.N. and Straker, V. (1998) The role of relative sea-level rise and changing sediment supply on Holocene gravel barrier development; the example of Porlock, Somerset, UK. *The Holocene*, 8, 165–81.

Jerwood, L.C., Robinson, D.A. and Williams, R.B.G. (1990a) Experimental frost and salt weathering of chalk: I. *Earth Surface Processes and Landforms*, 15, 611–24.

Jerwood, L.C., Robinson, D.A. and Williams, R.B.G. (1990b) Experimental frost and salt weathering of chalk: II. *Earth Surface Processes and Landforms*, 15, 699–708.

Job, D. (1989) Beach profiles and wave action. *Geography Review*, 2, 11–14.

Job, D. (1993) Coastal management: Start Bay, south Devon. *Geography Review*, 7, 13–17

John, B.S. (1971) Pembrokeshire. In *The Glaciations of Wales and Adjoining Regions* (ed. C.A. Lewis), Geographies for Advanced Study, Longman, London, pp. 229–65.

John, B.S. (1973) Vistulian periglacial phenomena in south-west Wales. *Biuletyn Peryglacjalny*, 22, 185–211.

John, B.S. (1978) Valiant cliffs of Pembrokeshire. *Geographical Magazine*, 50, 467–70.

Johnson, D.W. (1919) *Shore Processes and Shoreline Development*, John Wiley and Sons Ltd, New York, 584 pp.

Johnson, D.W. (1925) *The New England-Acadian Shoreline*, John Wiley and Sons Ltd, New York, 608 pp.

Johnson, D.W. and Reid, W.G. (1910) The form of Nantasket Beach. *Journal of Geology*, 10, 162–87.

Johnson, S. (1976) *The Roman Forts of the Saxon Shore*, Elek, London, 172 pp.

Jolliffe, I.P. (1961) The use of tracers to study beach movements and the measurement of longshore drift by a fluorescent technique. *Revue de Géomorphologie Dynamique*, 12, 81–95.

Jolliffe, I.P. (1964) An experiment designed to compare the relative rates of movement of beach pebbles. *Proceedings of the Geologists' Association*, 75, 67–86.

Jolliffe, I.P. (1979) West Bay and the Chesil bank, Dorset. Coastal regimen conditions, resource

References

use and the possible environmental impact of mining activities on coastal erosion and flooding. Report to West Dorset District Council and Dorset County Council, 87 pp.

Jolliffe, I.P. (1983) Coastal erosion and flood abatement; what are the options? *Geographical Journal*, **149**, 62–7.

Jolliffe, I.P. and Wallace, H. (1973) The role of seaweed in beach supply and in shingle transport below low tide level. In *Oceans 2000: Third World Congress of Underwater Activities* (ed. British Sub Aqua Club), British Sub Aqua Club, London, pp. 189–96.

Jones, M.E. (1980) Landslips: solutions to the paths ban at Chapman's Pool. *Dorset: the County Magazine*, **87**, 6–17.

Jones, M.E., Allison, R.J. and Gilligan, J. (1984) On the relationship between geology and coastal landform in central southern England. *Proceedings of the Dorset Natural History and Archaeological Society*, **105**, 107–18.

Jones, D.K.C. and Lee, E.M. (1994) *Landsliding in Great Britain*, HMSO, London, 361 pp.

Jope, E.M. and Jope, H.M. (1959) A horde of 15th century coins from Glen Luce sand dunes and their context. *Medieval Archaeology*, **3**, 259–79.

Joysey, K. (1967) Report on the symposium on Scolt Head Island. *Proceedings of the Linnean Society*, **178**, 76.

Kahn, A.S. (1968) The recent sediments of the area between the Loughor and Towy estuaries, Carmarthen Bay. Unpublished PhD thesis, Imperial College, University of London.

Kalaugher, P.G. and Grainger, P. (1981) A coastal landslide at West Down Beacon, Budleigh Salterton, Devon. *Proceedings of the Ussher Society*, **5**, 217–21.

Kalaugher, P.G. and Grainger, P. (1991) The influence of changes in sea level on coastal cliff instability in Devon. In *Quaternary Eengineering Geology: Proceedings of the 25th Annual Conference of the Engineering Group of the Geological Society* (eds A. Forster, M.G. Culshaw and J.A. Little), *Engineering Geology Special Publication*, No. 7, The Geological Society, London, pp. 361–7.

Kalaugher, P.G., Grainger, P. and Hodgson, R.L.P. (1995) Tidal influence on the intermittent surging movements of a coastal mudslide. *Proceedings of the Ussher Society*, **8**, 416–20.

Ke, X. and Collins, M.B. (1993) *Saltmarsh Protection and Stablisation West Solent*, Report No. **SUDO/93/6/C**, Southampton University Department of Oceanography, Southampton.

Kear, B.S. (1985) Soil development and soil patterns in north-west England. In *The Geomorphology of North-West England* (ed. R.H. Johnson), Manchester University Press, Manchester, pp. 80–93.

Keast, S. (1994) The Ardivachar to Stoneybridge Proposed SSSI. Unpublished Earth Science Documentation Series, Scottish Natural Heritage, Perth.

Keast, S. (1995) Traigh na Berie SSSI, Isle of Lewis. Unpublished Earth Science Documentation Series, Scottish Natural Heritage, Perth.

Keatch, D.R.A. (1965) Geomorphology of the coast from Cardiff to Porthcawl, with special reference to the transportation of beach material. Unpublished PhD thesis, University of Wales.

Keene, P. (1986) *Classic Landforms of the North Devon Coast*, Classic Landform Guides, No. 6, Geographical Association, Sheffield, 48 pp.

Keene, P. (1989) Classic landforms of the North Devon Coast. *Proceedings of the Ussher Society*, **7** (2), 192–3.

Keene, P. (1992) *Coastal Management and Coastal Erosion at Westward Ho!*, Thematic Trails, Oxford.

Keene, P. (1996) *Classic Landforms of the North Devon Coast*, New edition, Classic Landform Guides, Geographical Association in conjunction with the British Geomorphological Research Group, Sheffield, 48 pp.

Kelland, N.C. (1975) Submarine geology of Start Bay determined by continuous seismic profiling and core sampling. *Quarterly Journal of the Geological Society of London*, **131**, 7–17.

Kelland, N.C. and Hails, J.R. (1972) Bedrock morphology and structures within overlying sediments, Start Bay, south-west England, determined by continuous seismic profiling, side-scan sonar and core sampling. *Marine Geology*, **13**, M19–M26.

Kellock, E. (1969) Alkaline basic igneous rocks in the Orkneys. *Scottish Journal of Geology*, **5**, 140–53.

Kendall, O.D. (1926) Scolt Head Island Part 2. The mapping of Scolt Head Island. *Transactions of the Norfolk and Norwich Naturalists' Society*, **12**, 246–54.

Kendall, W.B. (1907) Waste of coastline, Furness and Walney in 1000 years. *Barrow*

References

Naturalists' Field Club Annual Report, **18**, 78.

Kenyon-Bell, C. (1948) Sea defences at Westward Ho!. *Report to the Urban District Council of Northam.*

Kesel, R.H. and Smith, J.S. (1978) Tidal creek and pan formation in intertidal salt marshes, Nigg Bay, Scotland. *Scottish Geographical Magazine*, **94**, 159–68.

Kestner, F.J.T. (1962) The old coastline of The Wash. *Geographical Journal*, **128**, 457–8.

Kestner, F.J.T. (1975) The loose boundary regime of The Wash. *Geographical Journal*, **141**, 389–414.

Kidson, C. (1950) Dawlish Warren: a study of the evolution of the sand spits across the mouth of the River Exe in Devon. *Transactions of the Institute of British Geographers*, **16**, 69–80.

Kidson, C. (1960) The shingle complexes of Bridgwater Bay. *Transactions of the Institute of British Geographers*, **28**, 75–87.

Kidson, C. (1961) Movement of beach materials on the east coast of England. *East Midland Geographer*, **16**, 3–16.

Kidson, C. (1963) The growth of sand and shingle spits across estuaries. *Zeitschrift für Geomorphologie*, **7**, 1–22.

Kidson, C. (1964a) The coasts of south and south-west England. In *Field Studies in the British Isles* (ed. J.A. Steers), Nelson, London, pp. 26–42.

Kidson, C. (1964b) Dawlish Warren, Devon: late stages in sand spit evolution. *Proceedings of the Geologists' Association*, **75**, 167–84.

Kidson, C. (1977) The coast of South West England. In *The Quaternary History of the Irish Sea* (eds C. Kidson and M.J. Tooley), Seel House Press, Liverpool, pp. 267–98.

Kidson, C. and Carr, A.P. (1959) The movement of shingle over the sea bed close inshore. *Geographical Journal*, **125**, 380–9.

Kidson, C. and Carr, A.P. (1960) Dune reclamation at Braunton Burrows, Devon. *Chartered Surveyor*, **93**, 298–303.

Kidson, C. and Heyworth, A. (1976) The Quaternary deposits of the Somerset Levels. *Quarterly Journal of Engineering Geology*, **M9**, 217–35.

Kidson, C. and Tooley, M.J. (1977) *The Quaternary History of the Irish Sea*, Seel House Press, Liverpool, 345 pp.

Kidson, C., Carr, A.P. and Smith, D.B. (1958) Further experiments using radio-active methods to detect the movement of shingle over the sea bed and alongshore. *Geographical Journal*, **124**, 210–18.

Kidson, C., Collin, R.L. and Chisholm, N.W.T. (1989) Surveying a major dune system – Braunton Burrows, north west Devon. *Geographical Journal*, **155**, 94–105.

King, C.A.M. (1951) Depth of disturbance of sand on beaches by waves. *Journal of Sedimentary Petrology*, **21**, 131–40.

King, C.A.M. (1953) The relationship between wave incidence, wind direction and beach changes at Marsden Bay, County Durham. *Transactions of the Institute of British Geographers*, **19**, 13–23.

King, C.A.M. (1959) *Beaches and Coasts*, Edward Arnold, London, 403 pp.

King, C.A.M. (1964) The character of the offshore zone and its relationship to the foreshore near Gibraltar Point, Lincolnshire. *East Midland Geographer*, **3**, 230–43.

King, C.A.M. (1968a) Beach measurements at Gibraltar Point, Lincolnshire. *East Midland Geographer*, **4**, 295–300.

King, C.A.M. (1968b) Spitsim. *British Geomorphological Research Group Occasional Paper*, **6**, 63–72.

King, C.A.M. (1970) Changes in the spit at Gibraltar Point, Lincolnshire. *East Midland Geographer*, **5**, 19–30.

King, C.A.M. (1971) The relationship between wave incidence, wind direction, and beach changes at Marsden Bay, Co. Durham. In *Introduction to Coastline Development* (ed. J.A. Steers), Macmillan, London, 229 pp.

King, C.A.M. (1972a) *Beaches and Coasts*, 2nd edn, Edward Arnold, London, 570 pp.

King, C.A.M. (1972b) Some spatial aspects of the analysis of coastal spits. In *Spatial Analysis in Geography* (ed. R.J. Chorley), Methuen, London, pp. 355–69.

King, C.A.M. (1973) Dynamics of beach accretion in south Lincolnshire, England. In *Coastal Geomorphology* (ed. D.R. Coates), Publications in Geomorphology, Binghampton, pp. 73–90.

King, C.A.M. (1976) *Northern England*, Geomorphology of the British Isles Series, Methuen, London, 213 pp.

King, C.A.M. (1978) Coastal geomorphology in the United Kingdom. In *Geomorphology: Present Problems and Future Prospects* (eds C. Embleton, D. Brunsden and D.K.C. Jones), Oxford University Press, Oxford, pp. 224–50.

King, C.A.M. (1982) Ridges and runnels. In *The*

References

Encyclopedia of Beaches and Coastal Environments (ed. M.L. Schwartz), Dowden, Hutchinson and Ross, Stroudsberg, Pennsylvania, 692 pp.

King, C.A.M. and Barnes, F.A. (1964) Changes in the configuration of the inter-tidal beach zone of part of the Lincolnshire coast since 1951. *Zeitschrift für Geomorphologie*, **8**, 105–26.

King, C.A.M. and McCullagh, M.J. (1971) A simulation model of a complex recurved spit. *Journal of Geology*, **79**, 22–37.

King, C.A.M. and Williams, W.W. (1949) The formation and movement of sand bars by wave action. *Geographical Journal*, **113**, 70–85.

Kirk, W. (1955) Prehistoric sites at the Sands of Forvie, Aberdeenshire. *Aberdeen University Review*, **35**, 150–71.

Knight, J., Orford, J.D., Wilson, P., Wintle, A.G. and Braley, S. (1998) Facies, age and controls on recent coastal sand dune evolution in north Norfolk, eastern England. In *Proceedings of the Palm Beach International Coastal Symposium, 19-23 May, 1998* (eds C.W. Finkl and P. Bruun), *Journal of Coastal Research Special Issue*, No. **26**, Coastal Education and Research Foundation, Florida, pp. 154–61.

Koh, A. (1992) Black Ven. In *The Coastal Landforms of West Dorset* (ed. R.J. Allison), Geologists' Association Guide, No. **47**, Geologists' Association, London, pp. 67–79.

Komar, P.D. (1976) *Beach Processes and Sedimentation*, Prentice-Hall, Englewood Cliffs, New Jersey, 429 pp

Komar, P.D. (1977) Computer simulation of turbidity current flow and the study of deep-sea channels and fan sedimentation. In *The Sea: Ideas and Observations on Progress in the Study of the Seas; Volume 6: Marine Modelling* (eds E.D. Goldberg, I.N. McCave, J.J. O'Brien and J.H. Steele), Wiley-Interscience, New York, pp. 603–21.

Lacey, S. (1987) Coastal sediment processes in Poole and Christchurch Bays and the effects of coast protection works. Unpublished PhD thesis, University of Southampton.

Ladle, M. (1981) *The Fleet and Chesil Beach: Structure and Biology of a Unique Coastal Feature*, Dorset County Council, Dorchester, 75 pp.

Lake, R.D. and Shepherd-Thorn, E.R. (1987) *Geology of the Country around Hastings and Dungeness*, Memoir of the British Geological Survey, sheets 320 and 321 (England and Wales), HMSO, London, 81 pp.

Lamb, H.H. (1982) *Climate, History and the Modern World*, Methuen, London, 387 pp.

Lamb, H.H. (1991) *Historic storms of the North Sea, British Isles and Northwest Europe*. Cambridge University Press, Cambridge, 250 pp.

Lambarde, W. (1576 [written 1570]) *A Perambulation of Kent: Conteining the Description, Hystorie, and Customes of that Shire*, W. Burrill, London.

Lambeck, K. (1992) Glacial rebound and sea-level change in the British Isles. *Terra Nova*, **3**, 379–89.

Lambeck, K. (1993) Glacial rebound of the British Isles – 1. Preliminary model results. *Geophysical Journal International*, **115** (3), 941–59.

Landsberg, S.Y. (1955) The morphology and vegetation of the Sands of Forvie. Unpublished PhD thesis, University of Aberdeen.

Landsberg, S.Y. (1956) The orientation of dunes in Britain and Denmark in relation to winds. *Geographical Journal*, **122**, 176–89.

Lang, W.D. (1914) The geology of Charmouth Cliffs, beach and foreshore. *Proceedings of the Geologists' Association*, **25**, 293–360.

Lang, W.D. (1928) Landslips in Dorset. *Natural History Magazine*, **1**, 201–9.

Lang, W.D. (1932) The geology of Golden Cap. *Proceedings of the Dorset Natural History and Archaeological Society*, **54**, 145–72.

Lang, W.D. (1942) Geological notes 1941–42. *Proceedings of the Dorset Natural History and Archaeological Society*, **64**, 129–30.

Lang, W.D. (1944) Geological notes 1943–44. *Proceedings of the Dorset Natural History and Archaeological Society*, **66**, 129.

Lang, W.D. (1955) Mudflows at Charmouth. *Proceedings of the Dorset Natural History and Archaeological Society*, **75**, 151–6.

Lang, W.D. (1959) Report on Dorset natural history: Geology. *Proceedings of the Dorset Natural History and Archaeological Society*, **80**, 22.

Latham, J.-P., Hoad, J.P. and Newton, M. (1998) Abrasion of a series of tracer materials on a gravel beach, Slapton Sands, Devon, UK. In *Advances in Aggregates and Armourstone Evaluation* (ed. J.-P. Latham), *Geological Society Engineering Geology Special Publication*, No. **13**, The Geological Society,

References

Bath, pp. 121–35.

Leach, A.L. (1933) The geology and scenery of Tenby and the south Pembrokeshire coast. *Proceedings of the Geologists' Association*, **44**, 187–216.

Leafe, R.N. and Hansom, J.D. (1990) Estimating the effects of groundwater variation on the spectral response of a tidal salt marsh. In *Proceedings of the NERC Workshop on Airborne Remote Sensing, 1989, I.F.E., Windermere*, Natural Environment Research Council, pp. 169–79.

Leddra, M.J. and Jones, M.E. (1990) Steady-state flow during undrained loading of chalk. In *Chalk: Proceedings of the International Chalk Symposium, held at Brighton Polytechnic on 4–7 September 1989* (ed. G.P. Birch), Thomas Telford (for the Institution of Civil Engineers), London, pp. 245–52.

Lee, E.M. (1992) Urban landslides: impact and management. In *The Coastal Landforms of West Dorset* (ed. R.J. Allison), Geologists Association Guide, No. **47**, Geologists' Association, London, pp. 80–93.

Leeks, G. (1979) Mudlarks in the Essex marshes. *Geographical Magazine*, **51**, 665–70.

Lees, B.J. (1979) Sediment transport measurements in the Sizewell–Dunwich Banks area, East Anglia, UK. In *Holocene Sedimentation in the North Sea Basin. Selected Papers from the International Association of Sedimentologists Meeting, Texel, Netherlands, September 1979* (eds S.D. Nio, R.T.E. Schuttenheim and T.C.E. van Veering), *International Association of Sedimentologists Special Publication*, No. **5**, Blackwell Scientific Publications, Oxford.

Lees, B.J. (1980) Introduction and geological background. *Sizewell–Dunwich Banks Field Study Topic*, Report **1**, *Institute of Oceanographic Sciences Science Report*, 88.

Lees, D.J. (1982) The evolution of Gower coasts. In *Geographical Excursions from Swansea. Volume 1, Physical Environment* (ed. G. Humphrys), Department of Geography, University College, Swansea, pp. 15–32.

Lees, D.J. (1983) Post-glacial sand-dune history and archaeology at Broughton Bay, Gower. In *British Geomorphological Research Group, Spring Meeting, May 13–15, 1983. Field Excursion Handbook*, Swansea, pp. 27–9.

Leland, J. (1906–10 [written 1535–1543]) *Itinerary of John Leland in or about the years 1535–1543* (ed. L.T. Smith), George Bell and Sons, London, 5 volumes.

Lennon, G.W. (1963) The identification of weather conditions associated with the generation of major storm surges along the west coast of the British Isles. *Quarterly Journal of the Royal Meteorological Society*, **89**, 381–94.

Lennon, G.W., Gumbel, E., Barricelli, N.A. and Jenkinson, A.F. (1963) A frequency investigation of abnormally high tidal levels at certain west coast ports. *Proceedings of the Institution of Civil Engineers*, **25**, 451–84.

Lewin, T. (1862) *The Invasion of Britain by Julius Caesar, with Replies to the Remarks of the Astronomer-Royal and of the late Camden Professor of Ancient History at Oxford*, 2nd edn, London.

Lewis, D. (1992) The sands of time: Cornwall's Hayle to Gwithian Towans. In *Coastal Dunes: Geomorphology, Ecology and Management: Proceedings of the 3rd European Dune Congress, Galway, Ireland, 17–21 June 1992* (eds R.W.G. Carter, T.G.F. Curtis and M.J. Sheehy-Skeffington), A.A. Balkema, Rotterdam, pp. 463–73.

Lewis and Duvivier (1976) Study of littoral movements: Selsey Bill to Pagham Harbour. *Consultants' Report to Chichester District Council and Southern Water Authority*.

Lewis, W.V. (1931) Effect of wave incidence on the configuration of a shingle beach. *Geographical Journal*, **78**, 131–48.

Lewis, W.V. (1932) The formation of Dungeness foreland. *Geographical Journal*, **80**, 309–24.

Lewis, W.V. (1937) The formation of Dungeness and Romney marsh. *Transactions of the Southeastern Union of Science Societies*, **42**, 65–70.

Lewis, W.V. (1938) The evolution of shoreline curves. *Proceedings of the Geologists' Association*, **49**, 107–27.

Lewis, W.V. and Balchin, W.G.V. (1940) Past sea levels at Dungeness. *Geographical Journal*, **96**, 258–85.

Long, A.J. and Fox, S. (1988) The geomorphology of Denge Beach. Unpublished report for the Nature Conservancy Council and the Romney Marsh Research Trust.

Long, A.J. and Hughes, P.D.M. (1995) Mid and late Holocene evolution of the Dungeness foreland, U.K. *Marine Geology*, **124** (1–4), 253–71.

Long, A.J. and Innes, J.B. (1993) Holocene sea-level changes and coastal sedimentation in

References

Romney Marsh, south-east England, UK. *Proceedings of the Geologists' Association*, **104**, 223–37.

Long, A.J. And Innes, J.B. (1995a) A palaeoenvironmental investigation of the 'Midley Sand' and associated deposits at the Midley Church Bank, Romney Marsh. In *Romney Marsh: the Debatable Ground* (ed. J. Eddison), Oxford University Committee for Archaeology Monograph, No. **41**, OUCA, Oxford, pp. 37–50.

Long, A.J. and Innes, J.B. (1995b) The back-barrier and barrier depositional history of Romney Marsh, Walland Marsh and Dungeness, Kent, England. *Journal of Quaternary Science*, **10** (3), 267–83.

Long, A.J., Plater, A.J., Waller, M.P. and Innes, J.B. (1996) Holocene coastal sedimentation in the eastern English Channel: new data from the Romney Marsh region, United Kingdom. *Marine Geology*, **136**, 97–120.

Long, A.J., Waller, M., Hughes, P. and Spencer, C. (1998) The Holocene depositional history of Romney Marsh Proper. In *Romney Marsh: Environmental Change and Human Occupation in a Coastal Lowland* (eds J. Eddison, M. Gardiner and A. Long), Oxford University Committee for Archaeology Monograph, No. **46**, OUCA, Oxford, pp. 45–64.

Long, D., Smith, D.E. and Dawson, A.G. (1989) A Holocene tsunami deposit in eastern Scotland. *Journal of Quaternary Science*, **4**, 61–6.

Lovegrove, H. (1953) Old shorelines near Camber Castle. *Geographical Journal*, **119**, 200–7.

Lumsden, G.I. and Davies, A. (1965) The buried channel of the River Nith and its marked change in level across the Southern Upland fault. *Scottish Journal of Geology*, **1**, 134–43.

Lyell, C. (1835) *Principles of Geology, Being an Attempt to Explain the Former Changes of the Earth's Surface, by Reference to Causes now in Operation*, 4th edn, Murray, London, 4 volumes.

Lyell, C. (1867) *Principles of Geology, Being an Attempt to Explain the Former Changes of the Earth's Surface, by Reference to Causes now in Operation*, 10th edn, Murray, London, 4 volumes.

Mackie, W. (1897) The sands and sandstones of eastern Moray. *Transactions of the Edinburgh Geological Society*, **7**, 148–72.

Mackintosh, D. (1868) On the mode and extent of encroachment of the sea on some parts of the shores of the Bristol Channel. *Quarterly Journal of the Geological Society of London*, **24**, 277–90.

MacLaren, A. (1974) A Norse house on Drimore machair, South Uist. *Glasgow Archaeological Journal*, **3**, 9–18.

MacTaggart, F. (1996) Rinns of Islay SSSI. Unpublished Earth Science Documentation Series, Scottish Natural Heritage, Perth.

MacTaggart, F. (1997a) Loch Maddy–Sound of Harris SSSI. Unpublished Earth Science Documentation Series, Scottish Natural Heritage, Perth.

MacTaggart, F. (1997b) Barry Links SSSI. Unpublished Earth Science Documentation Series, Scottish Natural Heritage, Perth.

MacTaggart, F. (1997c) Luskentyre Banks and Saltings (incorporating Luskentyre–Corran Seilebost GCR site). Unpublished Earth Science Documentation Series, Scottish Natural Heritage, Perth.

MacTaggart, F. (1998a) West Coast of Jura. Unpublished Earth Science Documentation Series, Scottish Natural Heritage, Perth.

MacTaggart, F. (1998b) Sands of Forvie and Ythan Estuary SSSI and Foveran Links SSSI. Unpublished Earth Science Documentation Series, Scottish Natural Heritage, Perth.

MacTaggart, F. (1998c) Pabbay SSSI. Unpublished Earth Science Documentation Series, Scottish Natural Heritage, Perth.

MacTaggart, F. (1998d) Gruinart Flats SSSI. Unpublished Earth Science Documentation Series, Scottish Natural Heritage, Perth.

MacTaggart, F. (1999) Balta SSSI. Earth Science Documentation Series, Scottish Natural Heritage, Battleby, Perth

MacTaggart, F. (1999) Southern Parphe SSSI. Unpublished Earth Science Documentation Series, Scottish Natural Heritage, Perth.

Maddock, H.E. (1875) Changes in the coast-line, especially between Beachy Head and Hastings. *Eastbourne Natural History Society*, **1**, 1–6.

Maddrell, R.J. (1996) Managed coastal retreat, reducing flood risks and protection costs, Dungeness nuclear power station, U.K. *Coastal Engineering*, **28**, 1–15.

Madgett, P.A. (1975) Re-interpretation of Devensian till stratigraphy in eastern England. *Nature*, **253**, 105–7.

Madgett, P.A. and Catt, J.A. (1978) Petrography,

stratigraphy and weathering of late Pleistocene tills in east Yorkshire, Lincolnshire, and north Norfolk. *Proceedings of the Yorkshire Geological Society*, **42**, 55–108.

Madgett, P.A. and Inglis, E.A. (1987) A re-appraisal of the erratic suite of the Saunton and Croyde areas, North Devon. *Report and Transactions of the Devonshire Association for the Advancement of Science, Literature and Art*, **119**, 135–44.

Marex (1975) Environmental conditions west of the Shetlands – Aug 1974. Report No. **158**, Marine Exploration Ltd, Cowes.

Marker, M.E. (1967) The Dee estuary: its progressive silting and salt marsh development. *Transactions of the Institute of British Geographers*, **41**, 65–71.

Marshall, J,R, (1960) The physiographic development of Caerlaverock Merse. *Transactions of the Dumfriesshire and Galloway Natural History and Antiquarian Society*, **39**, 102–23

Marshall, J.R. (1962) The morphology of the upper Solway salt marshes. *Scottish Geographical Magazine*, **78** (2), 81–99.

Martin, J.M. (1872) Exmouth Haven and its threatened destruction. *Report and Transactions of the Devon Association for the Advancement of Science, Literature and Art*, **5**, 84–9.

Martin, J.M. (1876) The changes at Exmouth Haven. *Report and Transactions of the Devon Association for the Advancement of Science, Literature and Art*, **8**, 453–60.

Martin, J.M. (1893) The changes at Exmouth Haven. *Report and Transactions of the Devon Association for the Advancement of Science, Literature and Art*, **25**, 406–15.

Mason, S.J. and Hansom, J.D. (1988) Cliff erosion and its contribution to a sediment budget for part of the Holderness coast, England. *Shore and Beach*, **56**, 30–8.

Mason, S.J. and Hansom, J.D. (1989) A Markov model for beach changes on the Holderness coast of England. *Earth Surface Processes and Landforms*, **14**, 731–43.

Masselink, G. and Short, A.D. (1993) The effect of tide range on beach morphodynamics and morphology: a conceptual beach model. *Journal of Coastal Research*, **9**, 785–800.

Mate, I.D. (1991) The theoretical development of machair in the Hebrides. *Scottish Geographical Magazine*, **108** (1), 35–8.

Mather, A.S. (1979) *Beaches of Southwest Scotland*, Department of Geography, University of Aberdeen, Aberdeen, 2 volumes.

Mather, A.S. and Ritchie, W. (1977) *The Beaches of the Highlands and Islands of Scotland*, Department of Geography, University of Aberdeen, Aberdeen, 201 pp.

Mather, A.S. and Smith, J.S. (1974) *Beaches of Shetland*, Department of Geography, University of Aberdeen, Aberdeen, 103 pp.

Mather, A.S., Smith, J.S. and Ritchie, W. (1974) *The Beaches of Orkney*, Department of Geography, University of Aberdeen, Aberdeen, 168 pp.

Mathieson, J. (1928) The antiquities of the St Kilda Group of islands. *Proceedings of the Society of Antiquaries of Scotland*, **62**, 123–32.

May, V.J. (1964) A study of recent coastal changes in south-east England. Unpublished MSc thesis, University of Southampton.

May, V.J. (1966) A preliminary study of recent coastal changes and sea defences in south-east England. *Southampton Research Series in Geography*, **3**, 3–24.

May, V.J. (1971a) The retreat of chalk cliffs. *Geographical Journal*, **137**, 203–6.

May, V.J. (1971b) Hengistbury Head. In *Field Studies in South Hampshire and the Surrounding Region* (ed. M.J. Clark), Southampton Branch of the Geographical Association, Southampton, pp. 111–16.

May, V.J. (1971c) South Haven Peninsula and Ballard Down, Hengistbury Head. In *Field Studies in South Hampshire and the Surrounding Region* (ed. M.J. Clark), Southampton Branch of the Geographical Association, Southampton, pp. 90–7.

May, V.J. (1975) Cliff erosion and beach development, Shipstal Point. *Proceedings of the Dorset Natural History and Archaeological Society*, **97**, 8–12.

May, V.J. (1985) Geomorphological aspects of Dungeness. In *Dungeness – Ecology and Conservation, Report of a meeting held at Botany Department, Royal Holloway and Bedford New College, on 16 April 1985* (eds B. Ferry and S. Waters), Focus on Nature Conservation, No. **12**, Nature Conservancy Council, Peterborough, pp. 2–12.

May, V.J. (1990) Unravelling the coastline at Lulworth: from muddy boots to submarine acoustics. *Journal of the National Association of Field Studies Officers*, **1**, 19–21.

References

May, V.J. (1992) Coastal tourism, geomorphology and geological conservation: the example of south central England. In *Tourism vs Environment: the Case for Coastal Areas* (ed. P.P. Wong), Kluwer Academic, Dordrecht, pp. 3–10.

May, V.J. (1997a) Physiography of coastal cliffs. In *Dry Coastal Ecosystems: General Aspects* (ed. E. van der Maarel), Elsevier, Amsterdam, pp. 29–41.

May, V.J. (1997b) Studland beach: changes in the beach and dunes and their implications for shoreline management between Poole Harbour and Old Harry. *Report to the National Trust*.

May, V.J. (1999) Where will our coast be in 2020? Strategies for managing retreat in comercially valuable and intensively used dunes and cliffed coasts of high nature conservation value. In *Vision 2020: the People, the Coast, the Ocean, Coastal Zone '99 Conference, San Diego* (ed. Urban Harbors Institute), pp. 212–14.

May, V.J. (in press) Chalk coasts. In *Encyclopaedia of Coastal Science* (ed. M. Schwartz), Kluwer Academic, Dordrecht.

May, V.J. and Heeps, C. (1985) The nature and rates of change on chalk coastlines. In *Geomorphology of Changing Coastlines* (ed. E.C.F. Bird), *Zeitschrift für Geomorphologie*, *Supplementband*, No. **57**, Gebrüder Borntraeger, Berlin, pp. 81–94.

May, V.J. and Heeps, C. (in press) The Studland peninsula. In *Coastal Landforms of East Dorset* (ed. R. Allison), Geologists' Association Guide, Geologists' Association, London.

May, V.J. and Schwartz, M.L. (1981) Worldwide coastal sites of special scientific interest. In *Coastal Dynamics and Scientific Sites* (eds E.C.F. Bird and K. Koike), Komazawa University, Tokyo, pp. 91–118.

McCann, S.B. (1961) The raised beaches of western Scotland. Unpublished PhD thesis, University of Cambridge.

McCann, S.B. (1964) The raised beaches of north-east Islay and western Jura, Argyll. *Transactions of the Institute of British Geographers*, **35**, 1–16.

McCann, S.B. (1968) Raised shore platforms in the Western Isles of Scotland. In *Geography of Aberystwyth. Essays Written on the Occasion of the Departmental Diamond Jubilee 1917/18–1967/68* (eds E.G. Bowen, H. Carter and J.A. Taylor), University of Wales Press, Cardiff, pp. 22–34.

McCave, I.N. (1978a) Grain size trends and transport along beaches: example from East Anglia. *Marine Geology*, **28**, M43–M57.

McCave, I.N. (1978b) Fine sediment sources and sinks around the East Anglian coast. *Journal of the Geological Society of London*, **144**, 149–52.

McCave, I.N. (1978c) Sediments of the East Anglian coast. *East Anglian Coastal Research Programme*, Report 6, University of East Anglia, Norwich, 94 pp.

McCullagh, M.J. and King, C.A.M. (1970) Spitsym: a Fortran IV computer program for spit simulation. *Kansas State Geological Survey/University of Kansas Contribution*, **50**, 1–20.

McFarlane, P.B. (1955) Survey of two drowned river valleys in Devon. *Geological Magazine*, **92**, 419–29.

McGregor, D.F.M. and Green, C.P. (1989) Geomorphological conservation assessment of coastal shingle systems. *Geoöko Plus*, **1**, 189–90.

McInnes, I.J. (1964) The Neolithic and Bronze Age pottery of Luce Sands, Wigtownshire. *Proceedings of the Society of Antiquities, Scotland*, **97**, 40–81.

McLeod, C.R., Yeo, M., Brown, A.E., Burn, A.J., Hopkins, J.J., and Way, S.F. (eds.) (2002) *The Habitats Directive: Selection of Special Areas of Conservation in the UK*. 2nd edn. Joint Nature Conservation Committee, Peterborough.

McManus, J. and Wal, A. (1996) Sediment accumulation mechanisms on the Tentsmuir coast. In *Fragile Enviroments: the Use and Management of Tentsmuir National Nature Reserve, Fife* (ed. G. Whittington), Scottish Cultural Press, Edinburgh, pp. 1–15.

Middlemiss, F.A. (1983) Instability of Chalk cliffs between the South Foreland and Kingsdown, Kent, in relation to geological structure. *Proceedings of the Geologists' Association*, **94**, 115–22.

Miller, J. (1979) The Physical Landscape. In *A St Kilda Handbook* (ed. A. Small), *University of Dundee Occasional Paper*, No. **5**, National Trust for Scotland, Edinburgh, pp. 11–16.

Miller, J.A. and Mohr, P.A. (1965) Potassium-argon age determinations on rocks from St Kilda and Rockall. *Scottish Journal of Geology*, **1** (1), 93–9.

References

Miller, R. (1976) *Orkney*, Batsford, London, 192 pp.

Misdorp, J., Dronkers, J. and Spradley, J.R. (eds) (1990) *Strategies for Adaptation to Sea Level Rise*, Intergovernmental Panel on Climate Change, Ministry of Transport and Public Works, The Hague.

Mitchell, G.F. (1960) The Pleistocene history of the Irish Sea. *Advancement of Science*, **17**, 313–25.

Mitchell, G.F. and Orme, A.R. (1967) The Pleistocene deposits of the Isles of Scilly. *Quarterly Journal of the Geological Society of London*, **123**, 59–92.

Möller, I. (1998) Wave attenuation over salt marsh surfaces. Unpublished PhD thesis, University of Cambridge, Cambridge.

Monkhouse, F.J. (1954) *Principles of Physical Geography*, University of London Press, London, 452 pp.

Monkhouse, F.J. (1965) *A Dictionary of Geography*, Edward Arnold, London, 344 pp.

Moore, J.R. (1968) Recent sedimentation in northern Cardigan Bay, Wales. *Bulletin of the British Museum (Natural History), Mineralogy*, **2** (2), 1–131.

Moore, R. (1991) The chemical and mineralogical controls upon the residual strength of pure and natural clays. *Geotechnique*, **41**, 35–47.

Moore, R. and Brunsden, D. (1996) A physiochemical mechanism of seasonal mudsliding. *Geotechnique*, **46**, 259–78.

Morey, C.R. (1976) The natural history of Slapton Ley Nature Reserve, IX. The morphology and history of the lake basins. *Field Studies*, **4**, 353–68.

Morey, C.R. (1980) The origin and development of a coastal lagoon system, Start Bay, south Devon. Unpublished MPhil thesis, Council for National Academic Awards (CNAA).

Morey, C.R. (1983) The evolution of a barrier-lagoon system – a case study from Start Bay. *Proceedings of the Ussher Society*, **5**, 454–9.

Mörner, N.-A. (1972) Isostasy, eustasy and crustal sensitivity. *Tellus*, **24**, 586–92.

Mörner, N.-A. (1973) Eustatic changes during the last 3000 years. *Palaeogeography, Palaeoclimatology, Palaeoecology*, **13**, 1–14.

Mortimer, D. (2002) *Wash and North Norfolk Coast European Site. Management Scheme*, Report for English Nature, Peterborough.

Mortimore, R.N., Wood, C.J. and Gallois, R.W. (2001) *British Upper Cretaceous Stratigraphy*, Geological Conservation Review Series, No. **23**, Joint Nature Conservation Committee, Peterborough, 558 pp.

Mottershead, D.N. (1967) The evolution of the Valley of the Rocks. *Exmoor Review*, **8**, 69–72.

Mottershead, D.N. (1981) The persistence of oil pollution on a rocky shore. *Applied Geography*, **1**, 297–304.

Mottershead, D.N. (1986) *Classic Landforms of the South Devon Coast*, Classic Landform Guides, No. **5**, Geographical Association, Sheffield, 48 pp.

Mottershead, D.N. (1989) Rates and patterns of bedrock denudation by coastal salt spray weathering: a seven year record. *Earth Surface Processes and Landforms*, **14** (5), 383–98.

Mottershead, D.N. (1998) Coastal weathering of greenschist in dated structures, south Devon, UK. *Quarterly Journal of Engineering Geology*, **31** (4), 343–46.

Mottram, B. (1972) Some aspects of the evolution of parts of the Dorset coast. *Proceedings of the Dorset Natural History and Archaeological Society*, **94**, 21–6.

Motyka, J.M. and Brampton, A.H. (1993) Coastal Management: Mapping of Littoral Cells. *HR Wallingford Ltd Report*, No. **SR328**.

Murphy, P. and Funnell, B.M. (1979) Preliminary Holocene stratigraphy of Brancaster Marshes. *Bulletin of the Geological Society of Norfolk*, **31**, 11–16.

Murthy, T.K.S. and Cook, J. (1962) Maximum wave heights in Liverpool Bay. *Vickers-Armstrong Department of Design*, Report **V3031/HYDRO/04**.

Mykura, W. (1976) *British Regional Geology. Orkney and Shetland*, Natural Environment Research Council, Institute of Geological Sciences, HMSO, Edinburgh, 149 pp.

Nature Conservancy Council (1987) *St Kilda: National Nature Reserve*, Nature Conservancy Council, North West Scotland Region, Inverness.

Nature Conservancy Council, Geology and Physiography Section (1976) Shetland: localities of geological and geomorphological importance. *NCC Report*, **NC 158K**, Geology and Physiography Section, Nature Conservancy Council, Newbury, 64 pp.

Nature Conservancy Council, Geology and Physiography Section (1978) *Orkney: Localities of Geological and Geomorphological Importance*, Geology and

References

Physiography Section, Nature Conservancy Council, Newbury, 47 pp.

Neate, D.J.M. (1967) Underwater pebble grading of Chesil Bank. *Proceedings of the Geologists' Association*, **78**, 419–26.

Needham, S. (1988) A group of Early Bronze Age axes from Lydd. In *Romney Marsh: Evolution, Occupation, Reclamation* (eds J. Eddison and C. Green), Oxford University Committee for Archaeology Monograph, No. **24**, OUCA, Oxford, pp. 77–82.

Nicholls, R.J. (1984) The formation and stability of shingle spits. *Quaternary Newsletter*, **44**, 14–21.

Nicholls, R.J. (1985) The stability of shingle beaches in the eastern half of Christchurch Bay. Unpublished PhD thesis, University of Southampton.

Nicholls, R.J. (1987) Evolution of the upper reaches of the Solent River and the formation of Poole and Christchurch Bays. In *Wessex and the Isle of Wight Field Guide Prepared to Accompany the Annual Field Meeting Held at Southampton and Cowes 21-25 April 1987* (ed. K.E. Barber), Quaternary Research Association, Cambridge, pp. 99–114.

Nicholls, R.J. and Clark, M.J. (1986) Flandrian peat deposits at Hurst Castle spit. *Proceedings of the Hampshire Field Club and Archaeological Society*, **42**, 15–21.

Nicholls, R.J. and Webber, N.B. (1987a) The past, present and future evolution of Hurst Castle spit, Hampshire. *Progress in Oceanography*, **18**, 119–37.

Nicholls, R.J. and Webber, N.B. (1987b) Coastal erosion in the eastern half of Christchurch Bay. In *Planning and Engineering Geology* (eds M.G. Culshaw, F.G. Bell, J.C. Cripps and M. O'Hara), *Engineering Special Publication*, No. **4**, Geological Society of London, London, pp. 549–54.

Nicholls, R.J. and Webber, N.B. (1987c) Aluminium pebble tracer experiments on Hurst Castle spit. In *Proceedings of Coastal Sedimentology '87'*, American Society of Civil Engineers, New York, pp. 1563–77.

Nicholls, R.J. and Webber, N.B. (1989) Characteristics of shingle beaches with reference to Christchurch Bay, S. England. In *Proceedings of the 21st Coastal Engineering Conference, Malaga, Spain* (ed. B.L. Edge), American Society of Civil Engineers, New York, pp. 1922–36.

Norman, M.W. (1887) *A Popular Guide to the Geology of the Isle of Wight; with a Note on its Relation to that of the Isle of Purbeck*, Knight's Library, Ventnor, 240 pp.

North Norfolk District Council, Great Yarmouth Borough Council, Waveney District Council and the National Rivers Authority (1996) *Sheringham to Lowestoft Shoreline Management Plan.*

North, F.J. (1929) *The Evolution of the Bristol Channel, with Special Reference to the Coast of South Wales*, National Museum of Wales, Cardiff, 103 pp.

Nowell, D.A.G. (1995) Faults in the Purbeck–Isle of Wight monocline. *Proceedings of the Geologists' Association*, **106**, 145–50.

Nowell, D.A.G. (1998) Structures affecting the coast around Lulworth Cove, Dorset, and syn-sedimentary Wealden faulting. *Proceedings of the Geologists' Association*, **108**, 257–68.

Nunny, R.S. (1995) Lyme Bay Environmental Study. Volume 1: Hydrography; Volume 2: Sediments. *Project Report by Ambios Environmental Consultants Ltd for Kerr McGee Oil (UK) plc*, 18 pp + appendices.

Oertel, G.F. (1979) Barrier island development during the Holocene recession, south-eastern USA. In *Barrier Islands* (ed. S.P. Leatherman), Academic Press, New York, pp. 273–90.

Ogilvie, A.G. (1923) The physiography of the Moray Firth coast. *Transactions of the Royal Society of Edinburgh*, **53**, 377–404.

Oldale, R.N. (1985) A drowned Holocene barrier spit off Cape Ann, Massachusetts. *Geology*, **13**, 375–7.

Oliver, F.W. (1913) Some remarks on Blakeney Point, Norfolk. *Journal of Ecology*, **1**, 4–15.

Oliver, F.W. (1929) Blakeney Point reports. *Transactions of the Norfolk and Norwich Naturalists' Society*, **12**, 630–53.

Oliver, F.W. and Salisbury, E.J. (1913) Topography and vegetation of Blakeney Point. *Transactions of the Norfolk and Norwich Naturalists' Society*, **9**, 502–42.

Onyett, D. and Simmons, A. (1983) *East Anglian Coastal Research Project Final Report*, Geobooks, Norwich, 125 pp.

Orford, J.D. (1977) Some aspects of beach ridge development on a fringing gravel beach, Dyfed, west Wales. In *Les Côes Atlantiques de l'Europe: Évolution, Aménagement, Protection* (ed. A. Guilcher), *Actes de Colloques*, No. **9**, Centre National pour l'Exploitation des Océans, Brest, pp. 35–44.

Orford, J.D. and Carter, R.W.G. (1982) Crestal

References

overtop and washover sedimentation on a fringing sandy gravel barrier coast, Carnstone Point, SE Ireland. *Journal of Sedimentary Petrology*, **52**, 265–78.

Orford, J.D. and Jennings, S. (1998) The importance of different time-scale controls on coastal management strategy: the problem of Porlock gravel barrier, Somerset, UK. In *Coastal Defence and Earth Science Conservation* (ed. J. Hooke), The Geological Society Publishing House, Bath, pp. 87–102.

Orford, J.D. and Wright, P. (1978) What's in a name? – descriptive or genetic implications of 'ridge and runnel' topography. *Marine Geology*, **28**, M1–M8.

Orford, J.D., Carter, R.W.G. and Jennings, S.C. (1991) Coarse clastic barrier environments: evolution and implications for Quaternary sea level interpretation. *Quaternary International*, **9**, 87–104.

Orford, J.D., Carter, R.W.G., McKenna, J. and Jennings, S.C. (1995) The relationship between the rate of mesoscale sea-level rise and the rate of retreat of swash-aligned gravel-dominated barriers. *Marine Geology*, **124**, 177–86.

Orford, J.D., Carter, R.W.G. and Jennings, S.C. (1996) Control domains and morphological phases in gravel-dominated coastal barriers. *Journal of Coastal Research*, **12**, 589–605.

Orford, J.D., Wilson, P., Wintle, A.G., Knight, J. and Braley, S. (2000) Coastal dune initiation and development in Northumberland and Norfolk: Holocene and Recent scenarios. In *Holocene Land-Ocean Interaction and Environmental Change around the North Sea* (eds I. Shennan and J.E. Andrews), *Geological Society of London Special Publication*, No. **166**, Geological Society of London, London, pp. 197–217.

Orford, J.D., Jennings, S.C. and Forbes, D.L. (2001) Origin, development, reworking and breakdown of gravel-dominated coastal barriers in Atlantic Canada: future scenarios for the British coast. In *Ecology and Geomorphology of Coastal Shingle* (eds J.R. Packham, R.E. Randall, R.S.K. Barnes, and A. Neale), Westbury Academic and Scientific Publishing, Otley, pp. 23–55.

Orme, A.R. and Orme, A.J. (1988) Ridge and runnel enigma. *Geographical Review*, **78**, 169–84.

Ovington, J.D. (1950) The afforestation of Culbin Sands. *Journal of Ecology*, **38**, 303–19.

Owen, D.E. (1934) The Carboniferous rocks of the north Cornish coast and their structures. *Proceedings of the Geologists' Association*, **45** (4), 451–71 + pls 40–2.

Owen, A.E.B. (1952) Coastal erosion in east Lincolnshire. *The Lincolnshire Historian*, **9**, 340–1.

Owen, A.E.B. (1974–1975) Hafdic: a Lindsey name and its implications. *The English Place Name Society Journal*, **7**, 45–56.

Owens, J.S. and Case, G.O. (1908) *Coast Erosion and Foreshore Protection*, St Bride's Press, London, 144 pp.

Packham, J.R. and Liddle, M.J. (1970) The Cefni salt marsh, Anglesey, and its recent development. *Field Studies*, **3**, 331–56.

Packham, J.R. and Willis, A.J. (1997) *Ecology of Dunes, Salt marsh and Shingle*, Chapman and Hall, London, 335 pp.

Packham, J.R. and Willis, A.J. (1997) Braunton Burrows in context: a comparative management study. In *Coastal Dune Management: Shared Experience of European Conservation Practice* (eds J.A. Houston, S.E. Houston and P.J. Rooney), Liverpool University Press, Liverpool, pp. 65–79.

Packham, J.R., Randall, R.E., Barnes, R.S.K. and Neal, A. (2001) *Ecology and Geomorphology of Coastal Shingle*, Westbury Academic and Scientific Publishing, Otley, 460 pp.

Palmer, H.R. (1834) Observations on the motions of shingle beaches. *Philosophical Transactions of the Royal Society of London*, **A124**, 567–76.

Parker, W.R. (1971) Aspects of the marine environment at Formby Point, Lancashire. Unpublished PhD thesis, University of Liverpool.

Parker, W.R. (1975) Sediment mobility and erosion on a multi-barred foreshore (southwest Lancashire, UK). In *Nearshore Sediment Dynamics and Sedimentation* (eds J. Hails and A. Carr), John Wiley and Sons Ltd, Chichester, pp. 151–80.

Paskoff, R. (1978) Sur l'evolution geomorphologique du grand escarpment cotier du desert Chilien. *Geographie, Physique et Quaternaire*, **32**, 351–60.

Paskoff, R. (1985) *Les Littoraux*, Masson, Paris, 190 pp.

Paskoff, R. and Sanlaville, P. (1978) Observations géomorphologiques sur les côtes de l'archipel Maltais. *Zeitschrift für Geomorphologie, Neue Folge*, **22**, 310–28.

References

Paterson, I.B. (1981) The Quaternary geology of the Buddon Ness area of Tayside, Scotland. *Report of the Institute of Geological Sciences*, **81/1**, 9 pp.

Peacock, D.C.P. and Sanderson, D.J. (1993) Estimating strain from fault slip using a line sample. *Journal of Structural Geology*, **15**, 1513–16.

Peacock, D.C.P. and Sanderson, D.J. (1994) Geometry and development of relay ramps in normal fault systems. *Bulletin of the American Association of Petroleum Geologists*, **78**, 147–65.

Peacock, G. (1869) On the encroachment of the sea on Exmouth Warren. *Advancement of Science*, **39**, 166.

Peacock, J.D., Austin, W.E.N., Selby, I., Graham, F.D.K., Harland, R. and Wilkinson, I.P. (1992) Late Devensian and Flandrian palaeoenvironmental changes in the Scottish continental shelf west of the Outer Hebrides. *Journal of Quaternary Science*, **7**, 145–61.

Peake, J.F. (1960) A salt marsh at Thornham in north-west Norfolk. *Transactions of the Norfolk and Norwich Naturalists' Society*, **19**, 56–62.

Peake, D.S. (1961) Glacial changes in the Alyn river system and their significance in the glaciology of the north Welsh border, *Quarterly Journal of the Geological Society of London*, **117**, 335–66.

Pengelly, W. (1870) Modern and ancient beaches of Portland. *Report and Transactions of the Devon Association for the Advancement of Science, Literature and Art*, **4**, 195–205.

Penny, L.F., Coope, G.R. and Catt, J.A. (1969) Age and insect fauna of the Dimlington silts, east Yorkshire. *Nature*, **224**, 65–7.

Perkins, E.J. (1973) *The Marine Flora and Fauna of the Solway Firth*, Dumfriesshire and Galloway Natural History and Antiquarian Society, Dumfries, 112 pp.

Perkins, E.J. and Williams, B.R.H. (1966) The Biology of the Solway Firth in Relationship to the Movement and Accumulation of Radioactive Materials. II. The Distribution of Sediments and Benthos. *UKAEA Production Report*, **587(cc)**, Chapelcross.

Perkins, J.W. (1980) *Cliff and Slope Stability, South Wales*, Department of Extra-Mural Studies, University College, Cardiff, 206 pp.

Pestrong, R. (1965) The development of drainage patterns on tidal marshes. *Stanford University Publication, Geological Sciences*, **10**, 1–87.

Pethick, J. (1970) Salt-marsh morphology. Unpublished PhD thesis, University of Cambridge.

Pethick, J. (1974) The distribution of salt pans on tidal salt marshes. *Journal of Biogeography*, **1**, 57–62.

Pethick, J. (1980a) Salt-marsh initiation during the Holocene transgression: the example of the north Norfolk marshes, England. *Journal of Biogeography*, **7**, 1–9.

Pethick, J. (1980b) Velocity surges and asymmetry in tidal channels. *Estuarine Coastal Marine Science*, **11**, 331–45.

Pethick, J. (1981) Long-term accretion rates on tidal salt-marshes. *Journal of Sedimentary Petrology*, **51**, 571–7.

Pethick, J. (1984) *An Introduction to Coastal Geomorphology*, Edward Arnold, London, 260 pp.

Pethick, J. (1989) Essex saltmarsh erosion. *University of Hull, Institute of Estuarine and Coastal Studies, National Rivers Authority Project, Quarterly Report*, **4**.

Pethick, J. (1991) Essex saltmarsh erosion. *University of Hull, Institute of Estuarine and Coastal Studies, National Rivers Authority Project, Quarterly Report*, **7**.

Pethick, J. (1992) Salt marsh geomorphology. In *Salt Marshes* (eds J.R.L. Allen and K. Pye), Cambridge University Press, Cambridge, pp. 41–62.

Pethick, J. (1996) Coastal slope development: temporal and spatial periodicity in the Holderness cliff recession. In *Advances in Hillslope Processes, Volume 2* (eds M.G. Anderson and S.M. Brooks) John Wiley and Sons Ltd, Chichester, pp. 897–917.

Pethick, J. (1997) The geomorphology of Morfa Dinnle. *Report to Countryside Council for Wales*, 13 pp.

Pethick, J. and Leggett, D. (1993) The morphology of the Anglian coast. In *Coastlines of the Southern North Sea* (eds R. Hillen and H.J. Verhagen), American Society of Civil Engineers, New York, pp. 52–64.

Pethick, J., Leggett, D. and Husain, L. (1990) Boundary layers under salt marsh vegetation developed in tidal currents. In *Vegetation and Erosion* (ed. J.B. Thornes), John Wiley and Sons Ltd, Chichester, pp. 113–24.

Phillips, A.W. (1962) Some aspects of the coastal geomorphology of Spurn Head, Yorkshire. Unpublished PhD thesis, University of Hull.

References

Phillips, A.W. (1963) Tracer experiments at Spurn Head, Yorkshire. *Shore and Beach*, **31**, 30–5.

Phillips, A.W. (1964) Some observations of coastal erosion: studies at south Holderness and Spurn Head. *The Dock and Harbour Authority*, **45**, 64–6.

Phillips, A.W. (1969) A seabed drifter investigation in Morecambe Bay. *The Dock and Harbour Authority*, **49**, 571.

Phillips, A.W. and Rollinson, W. (1971) *Coastal Changes on Walney Island, North Lancashire*. Department of Geography, University of Liverpool, Research Paper, **8**, 36 pp.

Pile, J. (1996) Defining the coastal zone using environmental science data: implications for cost-benefit analysis in areas prone to landslides. Unpublished MSc thesis, Bournemouth University.

Pingree, R.D. (1978) The formation of The Shambles and other banks by tidal stirring of the seas. *Journal of the Marine Biological Association, UK*, **58**, 211–26.

Pinot, J. (1963) Quelques accumulations de galets de la côte tregoroise. *Annales de Géographie*, **72**, 13–31.

Pirkis, D.H.B. (1963) The coastline of Foula. *Brathay Expedition Group Annual Report and Account of Expeditions in 1963*.

Pitty, A.F. (1971) *Introduction to Geomorphology*, Methuen, London, 526 pp.

Plater, A.J. (1992) The late Holocene evolution of Denge Marsh, southeast England: a stratigraphic, sedimentological and micropalaeontological approach. *The Holocene*, **2**, 63–70.

Plater, A.J. and Shennan, I. (1992) Evidence of Holocene sea level change from the Northumberland coast, eastern England. *Proceedings of the Geologists' Association*, **103**, 201–16.

Plater, A. and Long, A. (1995) The morphology and evolution of Denge Beach and Denge Marsh. In *Romney Marsh: the Debatable Ground* (ed. J. Eddison), *Oxford University Committee for Archaeology Monorgraph*, No. 41, OUCA, Oxford, pp. 8–36.

Plater, A., Huddart, D., Innes, J.B., Pye, K., Smith, A.J. and Tooley, M.J. (1993) Coastal and sea level changes. In *The Sand Dunes of the Sefton Coast* (eds D. Atkinson and J.A. Houston), National Museums and Galleries on Merseyside, pp. 23–34.

Pontee, N.P. (1995) The morphodynamics and sedimentary architecture of mixed sand and gravel beaches, Suffolk, UK. Unpublished PhD thesis, University of Reading.

Potter, J.A. and Hosie, C.A. (2001) Using behaviours to identify rabbit impacts on dune vegetation at Aberffraw, North Wales. In *Coastal Dune Management: Shared Experience of European Conservation Practice* (eds J.A. Houston, S.E. Edmonson and P.J. Rooney), Liverpool University Press, Liverpool, pp. 108–16.

Potts, E.A. (1968) The geomorphology of the sand dunes of south Wales, with special reference to Gower. Unpublished PhD thesis, University College, Swansea.

Precheur, P. (1960) *Le Littoral de la Manche de Ste Adresse àAult*, SFIL, Poitiers, 138 pp.

Prestwich, J. (1875) On the origin of Chesil Bank, and on the relation of the existing beaches to past geological changes independent of the present coast action. *Minutes of the Proceedings of the Institution of Civil Engineers*, **40**, 61–114.

Prestwich, J. (1892) The raised beaches and 'head' or rubble-drift of the south of England: their relation to the valley drifts and to the glacial period and on a late post-glacial submergence. *Quarterly Journal of the Geological Society of London*, **48**, 263–342.

Price, W.A. (1955) Environment and formation of the chenier plain. *Quaternaria*, **2**, 75–86.

Pringle, A.W. (1981) Beach development and coastal erosion in Holderness, north Humberside. In *The Quaternary in Britain: Essays, Reviews and Original Work on the Quaternary Published in Honour of Lewis Penny on his Retirement* (eds J. Neale and J. Flenley), Pergamon Press, Oxford, pp. 194–205.

Pringle, A.W. (1985) Holderness coast erosion and the significance of ords. *Earth Surface Processes and Landforms*, **10**, 107–24.

Prior, D.B. (1977) Coastal mudslide morphology and processes on Eocene clays in Denmark. *Geografisk Tidskrift*, **76**, 14–33.

Prior, E.S. (1919) The Bridport shingle. A discussion of pebbles. *Proceedings of the Dorset Natural History and Antiquarian Field Club*, **40**, 52–65.

Psilovikos, A.A. (1974) An examination of some methods of sedimentary analysis with reference to samples from the Gibraltar Point area, Lincolnshire. Unpublished MPhil thesis, University of Nottingham.

Psilovikos, A.A. (1979) Sediment analysis at

References

Gibraltar Point, Lincolnshire. *East Midland Geographer*, **7**, 128–33.

Psuty, N.P. and Moreira, M.E. (1990) Nourishment of a cliffed coastline, Praia da Rocha, The Algarve, Portugal. *Journal of Coastal Research, Special Issue*, 6, pp. 21–32.

Pye, K. (1990) Physical and human influences on coastal dune development between the Ribble and Mersey estuaries, northwest England. In *Coastal Dunes: Form and Process* (eds K.F. Nordstrom, N.P. Psuty and R.W.G. Carter), John Wiley and Sons Ltd, Chichester, pp. 339–59.

Pye, K. (1991) Beach deflation and backshore dune formation following erosion under storm surge condition: an example from Northwest England. In *Sand, Dust and Soil in their Relation to Aeolian and Littoral Processes* (eds O.E. Bardoff-Nielsen and B.B. Willetts), *Acta Mechanica*, Supplementum 2.

Pye, K. (1992) Saltmarshes on the barrier coastline of North Norfolk, eastern England. In *Saltmarshes. Morphodynamics, Conservation and Engineering Significance* (eds J.R.L. Allen and K. Pye), Cambridge University Press, Cambridge, pp. 148–78.

Pye, K. (2001) The nature and geomorphology of coastal shingle. In *Ecology and Geomorphology of Coastal Shingle* (eds J.R. Packham, R.E. Randall, R.S.K. Barnes and A. Neal), Westbury Academic and Scientific Publishing, Otley, pp. 2–22.

Pye, K. and French, P.W. (1993) Erosion and Accretion Processes on British Saltmarshes. *Final Report to the Ministry of Agriculture, Fisheries and Food*, Cambridge Environmental Research Consultants, Cambridge, 5 volumes.

Pye, K. and Neal, A. (1993) Late Holocene dune formation on the Sefton coast, northwest England. In *The Dynamics and Environmental Context of Aeolian Sedimentary Systems* (ed. K. Pye), *Geological Society of London Special Publication*, No. **72**, Geological Society, London, pp. 201–17.

Pye, K. and Neal, A. (1994) Coastal dune erosion at Formby Point, north Merseyside, England: causes and mechanisms. *Marine Geology*, **119**, 39–56.

Pye, K. and Smith, A.J. (1988) Beach and dune erosion and accretion on the Sefton coast, northwest England. In *Dune/beach Interaction* (ed. N.P. Psuty), *Journal of Coastal Research Special Issue*, No. 3, Coastal Education and Research Foundation, Charlottesville, pp. 33–6.

Pye, K., Stokes, S. and Neal, A. (1995) Optical dating of aeolian sediments from the Sefton coast, northwest England. *Proceedings of the Geologists' Association*, **106**, 281–92.

Pye, K., Dickson, J.A.D., Schiavon, N., Coleman, M.L. and Cox, M. (1990) Formation of siderite–Mg calcite–iron sulphide concretions in intertidal marsh and sandflat sediments, north Norfolk, England. *Sedimentology*, **37**, 325–43.

Quelennec, R.E. (1988) *Compte-rendu du Deuxiène Séninaire du Projet 'CORINE Érosion Cotière'*, Marseille, 3 Decembre 1988, Bureau de Recherches Géologiques et Minières, Marseille, 21 pp.

Ralston, I.B.M. (1983) Relationships between archaeological sites and geomorphology in the coastal zone of northeast Scotland. In *Northeast Scotland Coastal Field Guide and Geographical Essays* (ed. W. Ritchie), Department of Geography, University of Aberdeen, Aberdeen, pp. 111–25.

Ramsay, D.L. and Brampton, A.H. (2000a) Coastal Cells in Scotland: Cell 5 – Cape Wrath to the Mull of Kintyre. *Scottish Natural Heritage Research, Survey and Monitoring Report*, No. **147**, 92 pp.

Ramsay, D.L. and Brampton, A.H. (2000b) Coastal Cells in Scotland: Cell 1 – St Abbs Head to Fife Ness. *Scottish Natural Heritage Research, Survey and Monitoring Report*, No. **143**, 96 pp.

Ramsay, D.L. and Brampton, A.H. (2000c) Coastal Cells in Scotland: Cell 3 – Cairnbulg Point to Duncansby Head. *Scottish Natural Heritage Research, Survey and Monitoring Report*, No. **145**, 110 pp.

Ramsay, D.L. and Brampton, A.H. (2000d) Coastal cells in Scotland: Cell 2 – Fife Ness to Cairnbulg Point, *Scottish Natural Heritage Research, Survey and Monitoring Report*, No. **144**, 101 pp.

Ramsay, D.L. and Brampton, A.H. (2000e) Coastal Cells in Scotland: Cells 8 & 9 – The Western Isles. *Scottish Natural Heritage Research, Survey and Monitoring Report*, No. **150**, 111 pp.

Ramsay, D.L. and Brampton, A.H. (2000f) Coastal Cells in Scotland: Cell 7 – Mull of Galloway to the inner Solway Firth. *Scottish Natural Heritage Research, Survey and Monitoring Report*, No. **149**, 72 pp.

References

Ramsay, R.C., Crossley, L.F., Rukin, G., Chaplin, A.W. and Howlett, J.H.D. (1977) Coastal Erosion and Tidal Flooding Risk on the Holderness Coast. *Summary and Report of Informal Working Group of Technical Officers* (mimeographed), 8 pp.

Ramsey, L.F. (1934) West Wittering Harbour. *Sussex County Magazine*, 8.

Ramster, J.W. and Hill, H.W. (1969) Current system in the northern Irish Sea. *Nature*, 224, 59–61.

Randall, R.E. (1973) Shingle Street, Suffolk: an analysis of a geomorphic cycle. *Bulletin of the Geological Society of Norfolk*, 24, 15–35.

Randall, R.E. (1977) Shingle Street and the sea. *Geographical Magazine*, 49, 569–73.

Randall, R.E. and Fuller, R.M. (2001) The Orford Shingles, Suffolk, UK: evolving solutions in coastline management. In *Ecology and Geomorphology of Coastal Shingle* (eds J.R. Packham, R.E. Randall, R.S.K. Barnes and A. Neal), Westbury Academic and Scientific Publishing, Otley, pp. 242–60.

Ranwell, D.S. (1955) Slack vegetation, dune system development and cyclical change at Newborough Warren, Anglesey. Unpublished PhD thesis, University of London.

Ranwell, D.S. (1958) Movement of vegetated sand dunes at Newborough Warren, Anglesey. *Journal of Ecology*, 46, 83–100.

Ranwell, D.S. (1959) Newborough Warren, Anglesey. I. The dune system and dune slack habitat. *Journal of Ecology*, 47, 571–601.

Ranwell, D.S. (1960) Newborough Warren, Anglesey. II. Plant associes and succession cycles of the sand dune and dune slack vegetation. *Journal of Ecology*, 48, 117–41.

Ranwell, D.S. (1964) *Spartina* salt marshes in southern England. II. Rate and seasonal pattern of sediment accretion. *Journal of Ecology*, 52, 79–94.

Ranwell, D.S. (1968) Coastal marshes in perspective. *Regional Studies Group Bulletin, Strathclyde*, 9, 1–26.

Ranwell, D.S. (1972) *Ecology of Salt Marshes and Sand Dunes*, Chapman and Hall, London, 258 pp.

Ranwell, D.S. (1974) The salt marsh to tidal woodland transition. *Hydrobiological Bulletin*, 8, 139–51.

Ratcliffe, D.A. (ed.) (1977) *A Nature Conservation Review. The Selection of Biological Sites of National Importance to Nature Conservation in Britain. Volume 1*, Cambridge University Press, Cambridge, for the Natural Environment Research Council and the Nature Conservancy Council, 401 pp.

Reade, T.M. (1872) The post-glacial geology and physiography of west Lancashire and the Mersey estuary. *Geological Magazine*, 9, 111–19.

Reade, T.M. (1881) On a section of the Formby and Leasowe Marine Beds and Superior Peat Bed, disclosed by cuttings for the outlet sewer at Hightown. *Proceedings of the Liverpool Geological Society*, 4, 269–77.

Reade, T.M. (1902) Glacial and post-glacial features of the Lower valley of the River Lune and its estuary. *Proceedings of the Liverpool Geological Society*, 9, 163–93.

Reade, T.M. (1908) Post-glacial beds at Great Crosby as disclosed by the new outfall sewer. *Proceedings of the Liverpool Geological Society*, 10, 249–61.

Redman, J.B. (1852) On the alluvial formations and the local changes of the south coast of England (including appendix). *Minutes of the Proceedings of the Institution of Civil Engineers*, 11, 162–223.

Redman, J.B. (1864) The east coast between the Thames and the Wash estuaries. *Minutes of the Proceedings of the Institution of Civil Engineers*, 23, 186–256.

Redstone, V.B. (1908) The Suffolk shore. In *Memorials of Old Suffolk* (ed. V.B. Redstone), Bemrose and Sons, London, pp. 221–43.

Reed, D.J. (1986) Suspended sediment transport in salt marsh creeks. Unpublished PhD thesis, University of Cambridge.

Reed, D.J. (1987) Temporal sampling and discharge asymmetry in salt marsh creeks. *Estuarine, Coastal and Shelf Science*, 25, 459–66.

Reed, D.J. (1988) Sediment dynamics and deposition in a retreating coastal salt marsh. *Estuarine, Coastal and Shelf Science*, 26, 67–79.

Reed, D.J. (1990) The impact of sea-level rise on coastal salt marshes. *Progress in Physical Geography*, 14, 465–81.

Reed, D.J., Stoddart, D.R. and Bayliss-Smith, T.P. (1985) Tidal flows and sediment budgets for a salt-marsh system, Essex, England. *Vegetatio*, 62, 375–80.

Reid, C. (1885) *The Geology of Holderness, and the Adjoining Parts of Yorkshire and Lincolnshire*, Memoir of the Geological Survey of Great Britain (England and Wales),

References

HMSO, London, 177 pp.

Reid, C. (1898) The Eocene deposits of east Devon. *Quarterly Journal of the Geological Society of London*, **54**, 234–8.

Reid, C. (1907) The Geology of the Country around Mevagissey, Memoir of the Geological Survey of Great Britain, Sheet 353 (England and Wales), HMSO, London, 73 pp.

Reid, C. and Flett, J.S. (1907) *The Geology of the Land's End District*, Memoir of the Geological Survey of Great Britain, sheets 351 and 358 (England and Wales), HMSO, London, 158 pp.

Rendel Geotechnics Consultants (1995) Coastal Planning and Management: Applied earth science mapping: Fraserburgh to Scotstown Head, Grampian Region. Coastal applied earth science mapping case study. Unpublished open file report to Department of the Environment, Transport and the Regions, London.

Rennie A. and Hansom, J.D. (2001) Shoreline response to changes in sediment supply and sea level on a submerging coast, Sanday, Orkney. Unpublished Annual Progress Report to Scottish Natural Heritage, Edinburgh.

Reynolds, D.H.B. (1986) Dungeness Foreland and its shingle and what lies under them. *Geographical Journal*, **152**, 81–7.

Rhind, D.W. (1965) Evidence of sea-level changes along the coast north of Berwick. *Proceedings of the Berwickshire Naturalists Club*, **37**, 10–15.

Rhind, P.M., Blackstock, T.H., Hardy, H.S., Jones, R.E. and Sandison, W. (2001) The evolution of Newborough Warren dunes system with particular reference to the past four decades. In *Coastal Dune Management: Shared Experience of European Conservation Practice* (eds J.A. Houston, S.E. Houston and P.J. Rooney), Liverpool University Press, Liverpool, pp. 345–79.

Rice, R.J. (1977) *Fundamentals of Geomorphology*, Longman, London, 387 pp.

Richards, F.J. (1934) The salt marshes of the Dovey estuary, IV. The rates of vertical accretion, horizontal extension and scarp erosion. *Annals of Botany*, **48**, 235–59.

Richards, J. and Pye, K. (2001) The cheniers of the Essex coast: Sedimentology and management for flood defence. In *Ecology and Geomorphology of Coastal Shingle* (eds J.R. Packham, R.E. Randall, R.S.K. Barnes and A. Neal), Westbury Academic and Scientific Publishing, Otley, pp. 167–71.

Richards, K.S. and Lorriman, N.R. (1987) Basal erosion and mass movement. In *Slope Stability* (eds M.G. Anderson and K.S. Richards), John Wiley and Sons Ltd, pp. 331–57.

Richardson, N.M. (1900) A recent landslip on Jordan Cliff, with a suggestion as to the causes of hill terraces. *Proceedings of the Dorset Natural History and Archaeological Society*, **21**, 91–100.

Richardson, N.M. (1902) An experiment on the movements of a load of brickbats deposited on the Chesil Beach. *Proceedings of the Dorset Natural History and Archaeological Society*, **23**, 123–33.

Riddell, K.J. and Fuller, T.W. (1995) The Spey Bay geomorphological study. *Earth Surface Processes and Landforms*, **20**, 671–86.

Ringrose, P.S. (1989) Recent fault movement and palaeoseismicity in western Scotland. *Tectonophysics*, **163**, 305–14.

Ringrose, P.S., Hancock, P., Fenton, C. and Davenport, C.A. (1991) Quaternary tectonic activity in Scotland. In *Quaternary Engineering Geology: Proceedings of the 25th Annual Conference of the Engineering Group of the Geological Society* (eds A. Forster, M.G. Culshaw and J.A. Little), Engineering Geology Special Publication, No. 7, The Geological Society, London, pp. 679–86.

Ritchie, W. (1966) The post-glacial rise in sea-level and coastal changes in the Uists. *Transactions of the Institute of British Geographers*, **39**, 79–86.

Ritchie, W. (1967) The machair of South Uist. *Scottish Geographical Magazine*, **83**, 161–73.

Ritchie, W. (1968) *The Coastal Geomorphology of North Uist*, O'Dell Memorial Monograph, No. 1, University of Aberdeen, Department of Geography, Aberdeen, 32 pp.

Ritchie, W. (1971) *The Beaches of Barra and the Uists: a Survey of the Beach, Dune and Machair Areas of Barra, South Uist, Benbecula, North Uist and Berneray*, Department of Geography, University of Aberdeen, Aberdeen, 83 pp.

Ritchie, W. (1976) The meaning and definition of machair. *Transactions and Proceedings of the Botanical Society of Edinburgh*, **42**, 431–40.

Ritchie, W. (1979a) Machair development and chronology of the Uists and adjacent islands. In *The Natural Environment of the Outer Hebrides* (ed. J.M. Boyd), *Proceedings of the Royal Society of Edinburgh*, **77B**, 107–22.

References

Ritchie, W. (1979b) *The Beaches of Fife*, Department of Geography, University of Aberdeen, Aberdeen, for the Countryside Commission for Scotland (Perth), 92 pp.

Ritchie, W. (1980) The beach, dunes and machair landforms of Pabbay, Sound of Harris. In *Sand Dune Machair 3: Report on Meeting in the Outer Hebrides, 14-16 July 1978* (ed. D.S. Ranwell), Institute of Terrestrial Ecology, Cambridge, pp. 13–19.

Ritchie, W. (ed) (1983) *Northeast Scotland Coastal Field Guide and Geographical Essays*, Department of Geography, University of Aberdeen.

Ritchie, W. (1983) The Sands of Forvie. In *Northeast Scotland Coastal Field Guide and Geographical Essays* (ed. W. Ritchie), Department of Geography, University of Aberdeen, pp. 12–19.

Ritchie, W. (1984) *A Preliminary Study of the West Coast of Orkney*, Shell UK Exploration and Production, Aberdeen.

Ritchie, W. (1985) Intertidal and subtidal organic deposits and sea level changes in the Uists, Outer Hebrides. *Scottish Journal of Geology*, **21**, 161–76.

Ritchie, W. (1992) Scottish landform examples – 4 Coastal parabolic dunes of the Sands of Forvie. *Scottish Geographical Magazine*, **108** (1), 39–44.

Ritchie, W. (1997) The geomorphology of the sands of Forvie. In *The Ythan, Festschrift for Professor George M. Dunnet* (ed. M.L. Gorman), University of Aberdeen, Aberdeen.

Ritchie, W. and Crofts, R. (1974) *The Beaches of Islay, Colonsay and Jura*, Department of Geography, University of Aberdeen, Aberdeen, 195 pp.

Ritchie, W. and Mather, A.S. (1969) *The Beaches of Sutherland: a Survey of the Beach, Dune and Machair Areas of North and West Sutherland*, Department of Geography, University of Aberdeen, Aberdeen.

Ritchie, W. and Mather A.S. (1970a) *The Beaches of Caithness: a Survey of the Beach, Dune and Dune Pasture Areas of Caithness*, Department of Geography, University of Aberdeen, Aberdeen, 68 pp.

Ritchie, W. and Mather, A.S. (1970b) *The Beaches of Lewis and Harris*, Department of Geography, University of Aberdeen, Aberdeen, 113 pp.

Ritchie, W. and Mather, A.S. (1984) *The Beaches of Scotland*, Department of Geography, University of Aberdeen, Aberdeen, 130 pp.

Ritchie, W. and Whittington, G. (1994) Non-synchronous aeolian sand movements in the Uists: the evidence of the intertidal organic and sand deposits at Cladach Mór, North Uist. *Scottish Geographical Magazine*, **110** (1), 40–6.

Ritchie, W., Rose, N. and Smith, J.S. (1978) *The Beaches of Northeast Scotland*, Department of Geography, University of Aberdeen, Aberdeen, 278 pp.

Ritchie, W., Whittington, G. and Edwards, K.J. (2001) Holocene changes in the geomorphology and vegetational history of the Atlantic littoral of the Uists, Outer Hebrides, Scotland. *Proceedings of the Royal Society of Edinburgh*, **92**, 121–36.

Robertson, D.A. (1955) The ecology of the sand dune vegetation of Ross Links, Northumberland, with special reference to secondary succession in the blowouts. Unpublished PhD thesis, University of Durham.

Robertson, I. (1990) Erosion and stability of till cliffs on the Holdemess coast. Unpublished PhD thesis, University of Newcastle-upon-Tyne.

Robertson-Rintoul, M.J. (1985) The morphology and dynamics of parabolic dunes within the context of the coastal dune systems of mainland Scotland. Unpublished PhD thesis, University of Oxford.

Robertson-Rintoul, M.J. (1990) A quantitative analysis of the near-surface wind flow pattern over coastal parabolic dunes. In *Coastal Dunes. Form and Process* (eds K.F. Nordstrom, N. Psuty and R.W.G. Carter), Coastal Morphology and Research, John Wiley and Sons Ltd, Chichester, pp. 57–78.

Robin, G. de Q. (1986) Changing the sea level. In *The Greenhouse Effect, Climatic Change and Ecosystems* (eds B. Bolin, B.R. Doos, J. Jager and R.A. Warrick), SCOPE, No. **29**, John Wiley and Sons Ltd, Chichester.

Robinson, A.H.W. (1953a) The changing coastline of Essex. *Essex Naturalist*, **29**, 78–93.

Robinson, A.H.W. (1953b) The storm surge of 31 January–1 February 1953. *Geography*, **38**, 141–60.

Robinson, A.H.W. (1955) The harbour entrances of Poole, Christchurch and Pagham. *Geographical Journal*, **121**, 33–50.

Robinson, A.H.W. (1961) The hydrography of Start Bay and its relationship to beach

References

changes at Hallsands. *Geographical Journal*, **121**, 63–77.

Robinson, A.H.W. (1964) The inshore waters, sediment supply and coastal changes of part of Lincolnshire. *East Midland Geographer*, **3**, 307–21.

Robinson, A.H.W. (1966) Residual currents in relation to shoreline evolution of the East Anglian coast. *Marine Geology*, **4**, 57–84.

Robinson, A.H.W. (1968) The use of the sea bed drifter in coastal studies with particular reference to the Humber. In *Küstengeomorphologie/Coastal Geomorphology, Zeitschrift für Geomorphologie, Supplementband*, **7**, 1–23.

Robinson, A.H.W. (1980a) Erosion and accretion along part of the Suffolk coast of East Anglia, England. *Marine Geology*, **37**, 133–46.

Robinson, A.H.W. (1980b) The sandy coast of south-west Anglesey. *Transactions of the Anglesey Antiquarian and Field Club*, **(1980)**, 37–66.

Robinson, A.H.W. and Milward, J. (1983) *The Shell Book of the British Coast*, David and Charles, Newton Abbot, 560 pp.

Robinson, D. (1987) The Wash: geographical and historical perspectives. In *The Wash and its Environment* (eds J.P. Doody and B. Barnett), Focus on Nature Conservation, No. **5**, Nature Conservancy Council, Peterborough, pp. 23–33.

Robinson, D.A. and Jerwood, L.C. (1987a) Subaerial weathering of chalk shore platforms during harsh winters in southeast England. *Marine Geology*, **77**, 1–14.

Robinson, D.A. and Jerwood, L.C. (1987b) Frost and salt weathering of chalk shore platforms near Brighton, Sussex, U.K. *Transactions of the Institute of British Geographers, New Series*, **12** (2), 217–26.

Robinson, D.A. and Jerwood, L.C. (1987c) Subaerial weathering of chalk shore platforms during harsh winters in south-east England. *Marine Geology*, **77**, 1–14.

Robinson, I.S., Warren, L. and Longbottom, J.F. (1983) Sea level fluctuations in the Fleet, an English tidal lagoon. *Estuarine, Coastal and Shelf Science*, **16**, 651–68.

Robinson, L.A. (1974) Towards a process-response model for cliffed coasts. Unpublished PhD thesis, University of Leeds.

Robinson, L.A. (1976a) The micro-erosion meter technique in a littoral environment. *Marine Geology*, **22**, 51–8.

Robinson, L.A. (1976b) The morphology of the north-east Yorkshire coastline. *Zeitschrift für Geomorphologie, Neue Folge*, **20**, 331–49.

Robinson, L.A. (1977a) Marine erosion processes at the cliff foot. *Marine Geology*, **23**, 257–71.

Robinson, L.A. (1977b) Erosion processes on the shore platform of north-east Yorkshire. *Marine Geology*, **23**, 339–61.

Robinson, L.A. (1977c) The morphology and development of the north-east Yorkshire shore platforms. *Marine Geology*, **23**, 237–55.

Rodwell, J.S. (ed.) (2000) *British Plant Communities. Volume 5. Maritime Communities and Vegetation of Open Habitats*, Cambridge University Press, Cambridge, 512 pp.

Rogers, E.H. (1946) The raised beach, submerged forest and kitchen midden of Westward Ho!, and the submerged stone row of Yelland. *Proceedings of the Devon Archaeological Exploration Society*, **3**, 109–35.

Rogers, I. (1908) On the submerged forest at Westward Ho!. *Report and Transactions of the Devon Association for the Advancement of Science, Literature and Art*, **40**, 249–59.

Rogers, J.J. (1859) Strata of the Cober Valley, Loe Pool, near Helston. *Transactions of the Royal Geological Society of Cornwall*, **7**, 352–4.

Ross, S. (1992) *The Culbin Sands – Fact and Fiction*. University of Aberdeen, Centre for Scottish Studies, Aberdeen, 196 pp.

Rossiter, J.R. (1967) An analysis of sea level variations in European Waters. *The Geophysical Journal of the Royal Astronomical Society*, **A246**, 371–400.

Rossiter, J.R. (1972) Sea-level observations and their secular variation. *Philosophical Transactions of the Royal Society of London*, **A272**, 131–9.

Rowe, S.M. (1978) *An Investigation of the Erosion and Accretion Regime on the Saltmarshes of the Upper Solway Firth from 1946–1975*, Nature Conservancy Council, Peterborough.

Roy, P.S., Cowell, P.J., Ferland, M.A. and Thom, B.G. (1994) Wave-dominated coasts. In *Coastal Evolution: Late Quaternary Shoreline Morphodynamics* (eds R.W.G. Carter and C.D. Woodroffe), Cambridge University Press, Cambridge, pp. 121–86.

Roy, P.S. (1967) The recent sedimentology of Scolt Head Island, Norfolk. Unpublished PhD

References

thesis, Imperial College, London.

Roy, W. (1747) *Military Survey of Scotland*, sheet 31.

Royal Commission on Coastal Erosion and Afforestation (1907–11) *Report and Minutes of Evidence 1907; Report and Minutes of Evidence 1909; Report 1911*, HMSO, London.

Russell, J.K. (1978) Tidal currents and sediment movement in the nearshore zone at Gibraltar Point. Unpublished PhD thesis, University of Nottingham.

Russell, R.C.H. (1956) Discussion following Williams, W.W. An east coast survey: some recent changes in the coast of East Anglia. *Geographical Journal*, **122**, 317–34.

Russell, R.J. (1971) Water-table effects on sea coasts. *Geological Society of America Bulletin*, **82**, 2343–8.

Salisbury, E.J. (1952) *Downs and Dunes: their Plant Life and its Environment*, Bell, London, 328 pp.

Sarre, R.D. (1989) Aeolian sand drift from the intertidal zone on a temperate beach: potential and actual rates. *Earth Surface Processes and Landforms*, **14** (3), 247–58.

Sarrikostis, E. and McManus, J. (1987) Potential longshore transports on the coasts north and south of the Tay estuary. *Proceedings of the Royal Society of Edinburgh*, **92B**, 335–44.

Savigear, R.A.G. (1952) Some observations on slope development in south Wales. *Transactions of the Institute of British Geographers*, **18**, 31–51.

Schumm, S.A. and Lichty, R.W. (1965) Time, space and causality in geomorphology. *American Journal of Science*, **263**, 110–9.

Scott, P.A. (1976) Beach development along the Holderness coast, north Humberside, with special reference to ords. Unpublished PhD thesis, University of Lancaster.

Scourse, J.D. (1987) Periglacial sediments and landforms in the Isles of Scilly and west Cornwall. In *Periglacial Processes and Landforms in Britain and Ireland* (ed. J. Boardman), Cambridge University Press, Cambridge, pp. 225–36.

Scourse, J.D. (1991) Glacial deposits of the Isles of Scilly. In *Glacial deposits in Great Britain and Ireland* (eds J. Ehlers, P.L. Gibbard and J. Rose), A.A. Balkema, Rotterdam, pp. 291–300.

Searle, S.A. (1975) *The Tidal Threat*, The Dunes Group, Chichester, 26 pp.

Sedgwick, A. and Murchison, R.I. (1840) On the Physical Structure of Devonshire: and on the Subdivisions and Geological Relations of its Older Stratified Deposits. *Transactions of the Geological Society of London, Series 2*, **5**, 633–703.

Shackley, M. (1981) *Environmental Archaeology*, Allen and Unwin, London, 213 pp.

Shennan, I. (1989) Holocene crustal movements and sea-level changes in Great Britain. *Journal of Quaternary Science*, **4**, 77–89.

Shepard, F.P. (1952) Revised nomenclature for depositional coastal features. *Bulletin of the American Association of Petroleum Geologists*, **36** (10), 1902–12.

Shepard, F.P. (1963) *Submarine Geology*, 2nd edn, Harper and Row, New York, 557 pp.

Sheppard, T. (1906) List of papers, maps, etc., relating to the erosion of the Holderness coast, and to changes in the Humber estuary. *Transactions of the Hull Geological Society*, **6** (1), 43–57.

Sheppard, T. (1912) *The Lost Towns of the Yorkshire Coast and Other Chapters Bearing Upon the Geography of the District*, A. Brown and Sons, London, 329 + xviii pp.

Sherwood, B.R. Gardiner, B.G. and Harris, T. (2000) *British Saltmarshes*. Forest Text for the Linnean Society of London, 418pp.

Shi, Z. (1992) Application of the 'Pjerup approach' for the classification of sediments in the microtidal Dyfi estuary, west Wales, U.K. *Journal of Coastal Research*, **8**, 482–91.

Shi, Z. and Lamb, H. (1991) Post-glacial sedimentary evolution of a microtidal estuary, Dyfi estuary, west Wales, U.K. *Sedimentary Geology*, **73**, 227–46.

Silvester, R. (1960) Stabilization of sedimentary coastlines. *Nature*, **188**, 467–9.

Simons, D.B. and Richardson, E.V. (1961) Forms of bed roughness in alluvial channels. *American Society of Civil Engineers Journal of Hydrologic Engineering*, **87**, 81–105.

Single, M.B. and Hansom, J.D. (1994) Torrs Warren–Luce Sands SSSI: Documentation and Management Prescription. *Scottish Natural Heritage Research, Survey and Monitoring Report*, No **13**, 85 pp.

Sissons, J.B. (1967) *The Evolution of Scotland's Scenery*, Oliver and Boyd, Edinburgh, 259 pp.

Sissons, J.B. (1972) Dislocation and non-uniform uplift of raised shorelines in the western part of the Forth valley. *Transactions of the Institute of British Geographers*, **55**, 145–59.

References

Sissons, J.B. (1974) Lateglacial marine erosion in Scotland. *Boreas*, **3**, 41–8.

Sissons, J.B. (1976) *Scotland*, The geomorphology of the British Isles Series, Methuen, London, 150 pp.

Sissons, J.B. (1981) British shore platforms and ice sheets. *Nature*, **291**, 473–5

Sissons, J.B. (1982) The so-called high 'interglacial' rock shoreline of western Scotland. *Transactions of the Institute of British Geographers, New Series*, **7** (2), 205–16 + map.

Sissons, J.B. and Cornish, R. (1982) Differential glacio-isostatic uplift of crustal blocks at Glen Roy, Scotland. *Quaternary Research*, **18**, 268–88.

Slade, G.O. (1962) Westward Ho! pebble ridge. *The Surveyor and Municipal and County Engineer*, **23** (June), 812–14.

Sly, P.G. (1966) Marine geological studies in Liverpool Bay and adjacent areas. Unpublished PhD thesis, University of Liverpool.

Small, R.J. (1970) *A Study of Landforms*, Cambridge University Press, Cambridge, 486 pp.

Small, R.J. (1978) *A Study of Landforms*, 2nd edn, Cambridge University Press, Cambridge, 502 pp.

Smart, J.G.O. (1966) *Geology of the Country Around Canterbury and Folkestone*, Memoir of the Geological Survey of Great Britain, sheets 289, 305 and 406 (England and Wales), HMSO, London, 337 pp.

Smeaton, J. (1791) *A Narrative of the Building and a Description of the Construction of the Edystone Lighthouse with Stone: to which is Adjoined, an Appendix, Giving Some Account of the Lighthouse on Spurn Point, Built Upon a Sand*, J. Smeaton, London, 198 pp.

Smith, D.E. (1997) Sea-level change in Scotland during the Devensian and Holocene. In *Reflections on the Ice Age in Scotland* (ed. J.E. Gordon), Scottish Association of Geography Teachers/Scottish Natural Heritage, Glasgow, pp. 136–51.

Smith, D.E., Firth, C.R., Turbayne, S.C. and Broooks, C.L. (1992) Holocene relative sea-level changes and shoreline displacement in the Dornoch Firth area, Scotland. *Proceedings of the Geological Association*, **103**, 237–57.

Smith, J. (1903) Torrs Warren. *Annals of the Andersonian Society, Glasgow*, **111**, 34–52.

Smith, J. (1993) The Houb, Dales Voe: coastal processes. In *Shetland Isles* (eds J. Birnie, J. Gordon, K. Bennett and A. Hall), Quaternary Research Association Field Guide, Quaternary Research Association, Cambridge, p. 60.

Smith, J.S. (1968) Shoreline evolution in the Moray Firth. Unpublished PhD thesis, University of Aberdeen.

Smith, J.S. (1974) A report on the current situation at Whiteness Head, Inverness-shire, with particular reference to the environmental effects of the McDermott fabrication yard. *Report for Inverness County Council and the Highlands and Islands Development Board*, Department of Geography, University of Aberdeen.

Smith, J.S. (1978) A new method of salt-pan formation in intertidal marshes. *Transactions of the Botanical Society of Edinburgh*, **43**, 127–30.

Smith, J.S. (1983) The Morrich More. In *North-East Scotland Coastal Field Guide and Geographical Essays* (ed. W. Ritchie), International Geographical Union Coastal Commission, Department of Geography, University of Aberdeen, pp. 43–5.

Smith, J.S. (1986) The coastal topography of the Moray Firth. In *The Marine Environment of the Moray Firth* (ed. R. Ralph), *Proceedings of the Royal Society of Edinburgh*, **91B**, pp. 1–12.

Smith, J.S. (1993) St. Ninian's Tombolo GCR Site. Unpublished GCR report, Scottish Natural Heritage.

Smith, J.S. and Mather, A.S. (1973) *The Beaches of East Sutherland and Easter Ross*, Department of Geography, University of Aberdeen, Aberdeen, 97 pp.

Smith, P.H. (1999) *The Sands of Time. An Introduction to the Sand Dunes of the Sefton Coast*, National Museums and Galleries on Merseyside, Liverpool.

Sneddon, P. and Randall, R.E. (1993a) *Coastal Vegetated Shingle Structures of Great Britain: Main Report*, Joint Nature Conservation Committee, Peterborough.

Sneddon, P. and Randall, R.E. (1993b) *Coastal Vegetated Shingle Structures of Great Britain. Appendix 1: Shingle Sites in Wales*, Joint Nature Conservation Committee, Peterborough.

Sneddon, P and Randall, R.E. (1994a) *Coastal Vegetated Shingle Structures of Great Britain. Appendix 2: Shingle Sites in Scotland*, Joint Nature Conservation

References

Committee, Peterborough.

Sneddon, P. and Randall, R.E. (1994b) *Coastal Vegetated Shingle Structures of Great Britain. Appendix 3: Shingle Sites in England*, Joint Nature Conservation Committee, Peterborough.

So, C.L. (1965) Coastal platforms of the Isle of Thanet, Kent. *Transactions of the Institute of British Geographers*, **37**, 147–56.

Somerville, A.A., Hansom, J.D., Sanderson, D.C.W. and Housley, R.A. (2003) Optically stimulated luminescence dating of large storm events in northern Scotland. *Quaternary Science Reviews*, **22**, 1085–92.

Sparks, B.W. (1960) *Geomorphology*, Longman, London, 371 pp.

Sparks, B.W. (1971) *Rocks and Relief*, Longman, London, 404 pp.

Spearing, H.G. (1884) On the recent encroachment of the sea at Westward Ho!, north Devon. *Quarterly Journal of the Geological Society of London*, **40**, 474.

Spencer, C.D., Plater, A.J. and Long, A.J. (1998a) Holocene barrier beach evolution: the sedimentary record of Walland Marsh. In *Romney Marsh: Environmental Change and Human Occupation in a Coastal Lowland* (eds J. Eddison, M. Gardiner and A. Long), *Oxford University Committee for Archaeology Monograph*, No. **46**, OUCA, Oxford, pp. 13–30.

Spencer, C.D., Plater, A.J. and Long, A.J. (1998b) Rapid coastal change during the mid- to late-Holocene: the record of barrier estuary sedimentation in the Romney Marsh region, southeast England. *The Holocene*, **8**, 143–63.

Stamp, D.L. (1947) *Britain's Structure and Scenery*, 2nd edn, The New Naturalist, No. 4, Collins, London.

Stapleton, C. and Pethick, J. (1996) *Coastal Processes and Management of Scottish estuaries I: The Dornoch, Cromarty and Beauly/Inverness Firths*, Scottish Natural Heritage Review, No. **50**, Scottish Natural Heritage, Edinburgh, 99 pp + maps.

Steers, J.A. (1925) The Suffolk shore. *Proceedings of the Suffolk Institute of Archaeology and Natural History*, **19**, 1–14.

Steers, J.A. (1926a) Orford Ness; a study in coastal physiography. *Proceedings of the Geologists' Association*, **37**, 306–25.

Steers, J.A. (1926b) Scolt Head. *Transactions of the Norfolk and Norwich Naturalists' Society*, **12**, 84–8.

Steers, J.A. (1926c) Scolt Head Island part 1. The physiographical evolution of Scolt Head Island. *Transactions of the Norfolk and Norwich Naturalists' Society*, **12**, 229–45.

Steers, J.A. (1927) The East Anglian coast. *Geographical Journal*, **69**, 24–48.

Steers, J.A. (1929) Geographical work on Scolt Head Island and adjacent areas. *Transactions of the Norfolk and Norwich Naturalists' Society*, **12**, 664–7.

Steers, J.A. (1933) Scolt Head Island report for 1931–32. *Transactions of the Norfolk and Norwich Naturalists' Society*, **13**, 293–5.

Steers, J.A. (1934a) Scolt Head Island report for 1933. *Transactions of the Norfolk and Norwich Naturalists' Society*, **13**, 324–32.

Steers, J.A. (1934b) Scolt Head Island. *Geographical Journal*, **83**, 479–94.

Steers, J.A. (1934c) *Scolt Head Island: the Story of its Origin: the Plant and Animal Life of the Dunes and Marshes*, Heffer, Cambridge, 234 pp.

Steers, J.A. (1935a) Scolt Head Island report for 1933–34. *Transactions of the Norfolk and Norwich Naturalists' Society*, **13**, 418–21.

Steers, J.A. (1935b) A note on the rate of sedimentation on a salt marsh on Scolt Head Island, Norfolk. *Geological Magazine*, **72**, 443–5.

Steers, J.A. (1936a) Scolt Head Island report for 1934–35. *Transactions of the Norfolk and Norwich Naturalists' Society*, **14**, 55–60.

Steers, J.A. (1936b) Some notes on the north Norfolk coast from Hunstanton to Brancaster: a supplement to the paper on Scolt Head Island, *Geographical Journal*, **87**, 35–46

Steers, J.A. (1937) The Culbin Sands and Burghhead Bay. *Geographical Journal*, **90**, 498–528.

Steers, J.A. (1938a) Scolt Head Island report for 1936–37 and some notes on the north Norfolk coast. *Transactions of the Norfolk and Norwich Naturalists' Society*, **14**, 210–16.

Steers, J.A. (1938b) The rate of sedimentation on salt marshes on Scolt Head Island, Norfolk. *Geological Magazine*, **75**, 26–39.

Steers, J.A. (1939a) Scolt Head Island Part 1. Report for 1938. *Transactions of the Norfolk and Norwich Naturalists' Society*, **14**, 391–4.

Steers, J.A. (1939b) Sand and shingle formations in Cardigan Bay. *Geographical Journal*, **94**, 209–27.

Steers, J.A. (1940) Scolt Head report for 1939.

References

Transactions of the Norfolk and Norwich Naturalists' Society, **15**, 41–6.

Steers, J.A. (1942) The physiography of East Anglia (Presidential address). *Transactions of the Norfolk and Norwich Naturalists' Society*, **15**, 231–58.

Steers, J.A. (1946a) *The Coastline of England and Wales*, Cambridge University Press, Cambridge, 644 pp.

Steers, J.A. (1946b) Twelve years' measurement of accretion on Norfolk salt marshes. *Geological Magazine*, **85**, 163–6.

Steers, J.A. (1948a) Accretion on Scolt Head Island marshes. *Transactions of the Norfolk and Norwich Naturalists' Society*, **16**, 280–2.

Steers, J.A. (1948b) Notes on the new map of Blakeney Point. *Transactions of the Norfolk and Norwich Naturalists' Society*, **16**, 283.

Steers, J.A. (1951a) Recent changes on the marshland coast of north Norfolk. *Transactions of the Norfolk and Norwich Naturalists' Society*, **17**, 206–13.

Steers, J.A. (1951b) Notes on the erosion along the coast of Suffolk. *Geological Magazine*, **88**, 435–9.

Steers, J.A. (ed.) (1952) *A Guide to Blakeney Point and Scolt Head Island*, The National Trust, London, 53 pp.

Steers, J.A. (1953a) *The Sea Coast*, New Naturalist Series, No. **25**, Collins, London, 276 pp.

Steers, J.A. (1953b) The east coast floods, January 31–February 1, 1953. *Geographical Journal*, **119**, 280–98.

Steers, J.A. (1954) Shoreline changes on the marshland coast of north Norfolk, 1951–53. *Transactions of the Norfolk and Norwich Naturalists' Society*, **17**, 322–6.

Steers, J.A. (ed.) (1960) *Scolt Head Island*, 2nd edn, Heffer, Cambridge, 269 pp.

Steers, J.A. (1962) Coastal cliffs: report of a symposium. *Geographical Journal*, **128**, 303–20.

Steers, J.A. (1964a) *The Coastline of England and Wales*, 2nd edn, Cambridge University Press, Cambridge, 762 pp.

Steers, J.A. (ed.) (1964b) *Blakeney Point and Scolt Head Island*, The National Trust, London, 76 pp.

Steers, J.A. (1969) The Sea Coast, 4th edn, New Naturalists Series, No. **25**, Collins, London, 276 pp.

Steers, J.A. (ed.) (1971a) *Introduction to Coastline Development*, Geographical Readings, Macmillan, London, 229 pp.

Steers, J.A. (1971b) *Blakeney Point and Scolt Head Island*, 2nd edn, The National Trust, London, 84 pp.

Steers, J.A. (1973) *The Coastline of Scotland*, Cambridge University Press, Cambridge, 335 + xvi pp.

Steers, J.A. (1977) Physiography. In *Wet Coastal Ecosystems* (ed. V.J. Chapman), Ecosystems of the World, No. **1**, Elsevier, Amsterdam, pp. 31–60.

Steers, J.A. (1981) *Coastal Features of England and Wales. Eight Essays*, Oleander Press, Cambridge, 206 pp.

Steers, J.A. and Grove, A.T. (1954) Shoreline changes on the marshland coast of north Norfolk, 1951–53. *Transactions of the Norfolk and Norwich Naturalists' Society*, **17**, 322–6.

Steers, J.A. and Haas, J.A. (1964) An aid to stabilization of sand dunes; experiments at Scolt Head Island. *Geographical Journal*, **130**, 265–7.

Steers, J.A. and Jensen, J.A.P. (1953) Winterton Ness. *Transactions of the Norfolk and Norwich Naturalists' Society*, **17**, 259.

Steers, J.A. and Kendall, O.D. (1928) Scolt Head Island. *Transactions of the Norfolk and Norwich Naturalists' Society*, **12**, 461–7.

Steers, J.A. and Slater, L. (1932) Scolt Head Island 1931. *Transactions of the Norfolk and Norwich Naturalists' Society*, **13**, 130–3.

Steers, J.A. and Thomas, H.D. (1929a) Vegetation and sedimentation as illustrated in the region of the Norfolk salt marshes. *Proceedings of the Geologists' Association*, **40**, 341–52.

Steers, J.A. and Thomas, H.D. (1929b) The visit to Blakeney Point and Hunstanton. *Proceedings of the Geologists' Association*, **40**, 353–6.

Steers, J.A., Stoddart, D.R., Bayliss-Smith, T.P., Spencer, T. and Durbridge, P.M. (1979) The storm surge of 11 January 1978 on the east coast of England. *Geographical Journal*, **145**, 192–205.

Stephens, N. (1966) Some Pleistocene deposits in North Devon. *Biuletyn Peryglacjalny*, **15**, 103–14.

Stephens, N. (1974) Some aspects of the Quaternary of South-West England; Westward Ho!; The Fremington area; North Devon; Hartland Quay and Damhole Point; Chard area and the Axe Valley section. In *Exeter Field Meeting, Easter 1974* (ed. A. Straw), Field Handbook, Quaternary Research

References

Association, Exeter, pp. 5–7, 25–7, 28–9, 35–42, 45, 46–51.

Stephens, N. and Shakesby, R.A. (1982) Quaternary evidence in South Gower. In *Geographical Excursions from Swansea. Volume 1: Physical Environment* (ed. G. Humphrys), Department of Geography, University College, Swansea, pp. 33–50.

Stephens, N. and Synge, F.M. (1966) Pleistocene shorelines. In *Essays in Geomorphology* (ed. G.H. Dury), Heinemann, London, pp. 1–52.

Stephenson, W.T. and Kirk, R.M. (2000a) Development of shore platforms on Kaikoura Peninsula, South Island, New Zealand. Part I: the role of waves. *Geomorphology*, **32**, 21–41.

Stephenson, W.T. and Kirk, R.M. (2000b) Development of shore platforms on Kaikoura Peninsula, South Island, New Zealand. Part 2: the role of subaerial weathering. *Geomorphology*, **32**, 43–56.

Stoddart, D.R. and Bayliss-Smith, T.P. (1985) Tidal flows and sediment budgets for a salt-marsh system, Essex, England. *Vegetatio*, 375–80.

Stoddart, D.R., Reed, D.J. and French, J.R. (1989) Understanding salt-marsh accretion, Scolt Head Island, Norfolk, England. *Estuaries*, **12**, 228–36.

Stone, J., Lambeck, K., Fifield, L.K., Evans, J.M. and Cresswell, R.G. (1996) A Lateglacial age for the Main Rock Platform, western Scotland. *Geology*, **24** (8), 707–10.

Stove, C.G. (1978) The hydrology, circulation and sediment movements of the Ythan estuary. Unpublished PhD thesis, University of Aberdeen.

Strahan, A. (1898) *The Geology of the Isle of Purbeck and Weymouth*, Memoir of the Geological Survey of Great Britain, Sheet 17, HMSO, London, 278 pp.

Strahan, A. (1907) Evidence given to Royal Commission on Coastal Erosion and Afforestation (1907–1911). Report and minutes of evidence, 149.

Straw, A. (1979) Eastern England. In *Eastern and Central England* (eds A. Straw and K.M. Clayton), Methuen, London, pp. 1–139.

Straw, A. and Clayton, K.M. (1979) *The Midlands and Eastern England*, Methuen, London, 247 pp.

Stuart, A. and Hookway, R.J.S. (1954) Coastal Erosion at Westward Ho!. *Report to the Coast Protection Committee (Special)*, Devon County Council, 13 pp.

Stuart, A. and Simpson, B. (1937) The shore sands of Devon and Cornwall. *Transactions of the Royal Geological Society of Cornwall*, **17**, 13–40.

Stumpf, R.P. (1983) The process of sedimentation on the surface of a salt marsh. *Estuarine, Coastal and Shelf Science*, **17**, 495–508.

Suess, E. (1906) *The Face of the Earth (Das Antlitz der Erde)* (translation by W.J. Sollas), Clarendon Press, Oxford.

Summers, I. (1985) Recyling at Dungeness. In *Shingle Beaches: Recycling and Nourishment* (ed. S.M. Bevan), Hydraulics Research, Wallingford, p. 23.

Sunamura, T. (1973) Coastal cliff erosion due to waves – field investigations and laboratory experiments. *Journal of the Faculty of Engineering, the University of Tokyo*, **32**, 1–86.

Sunamura, T. (1975) A laboratory study of wave-cut platform. *Journal of Geology*, **83**, 389–97

Sunamura, T. (1976) Feedback relationship in wave erosion of laboratory rocky coasts *Journal of Geology*, **84**, 427.

Sunamura, T. (1977) A relationship between wave-induced cliff erosion and erosive force of waves. *Journal of Geology*, **85**, 613–18.

Sunamura, T. (1981) Predictive models for rocky coast erosion. *Transactions of the Japanese Geomorphological Union*, **2** (1), 131–4.

Sunamura, T. (1983) Processes of sea cliff and platform erosion. In *CRC Handbook of Coastal Processes and Erosion* (ed. P.D. Komar), CRC Press, Boca Raton, Florida, pp. 233–65.

Sunamura, T. (1992) *Geomorphology of Rocky Coasts*, Coastal Morphology and Research Series, John Wiley and Sons Ltd, Chichester, 302 pp.

Sutherland, D.G. (1984) The submerged landforms of the St. Kilda archipelago, western Scotland. *Marine Geology*, **58**, 435–42.

Sutherland, D.G., Ballantyne, C.K. and Walker, M.J.C. (1984) Late Quaternary glaciation and environmental change on St Kilda, Scotland, and their palaeoclimatic sugnificance. *Boreas*, **13**, 261–72.

Suthons, C.T. (1963) Frequency of occurrence of abnormally high sea levels on the east and south coasts of England. *Proceedings of the Institution of Civil Engineers*, **25**, 433–49.

Suzuki, T., Takahashi, K., Sunamura, T. and Terada, M. (1970) Rock mechanics on the formation of washboard-like relief on wave-

References

cut benches at Arasaki, Miura Peninsula, Japan. *Geographical Review of Japan*, **43**, 211–22 [in Japanese; English abstract].

Swift, D.J.P. (1976) Coastal sedimentation. In *Marine Sediment Transport and Environmental Management* (eds D.J. Stanley and D.J.P Swift), John Wiley and Sons Ltd, New York.

Swinnerton, H.H. (1931) Post-glacial deposits of the Lincolnshire coast. *Quarterly Journal of the Geological Society of London*, **87**, 360–75.

Swinnerton, H.H. (1936) The physical history of east Lincolnshire. *Transactions of the Lincolnshire Naturalists' Union*.

Synge, F.M. and Stephens, N. (1966) Late- and post-glacial shorelines and ice-limits in Argyll and north-east Ulster. *Transactions of the Institute of British Geographers*, **39**, 101–25.

Tansley, A.G. (1939) *The British Islands and their Vegetation*, Cambridge University Press, Cambirdge, 2 volumes, 484 pp. and 545 pp.

Tansley, A.G. (1945) *Our Heritage of Wild Nature: a Plea for Organised Nature Conservation*, Cambridge University Press, Cambridge, 74 pp.

Thom, C. and Thom, F. (eds) (1983) *Domesday Book: 7 Dorset*, Phillimore, Chichester, 61 pp.

Thomas, C. (1953–1954) *Proceedings of the West Cornwall Field Club, New Series*, Appendix to Volume 1, **19**, 53–6 and 59.

Thomson, J. (1832) Maps of the Orkney Isles. In *The Atlas of Scotland*, Thomson, Edinburgh.

Thompson, C. (1923) The erosion of the Holderness coast. *Proceedings of the Yorkshire Geological Society*, **20** (1), 32–9.

Thompson, T. (1824) *The History of the Church and Priory of Swine in Holderness*, J. Nichols and Son, Hull, 326 pp.

Thorn, B. (1960) *The Design of Sea Defence Works*, Butterworths Scientific Publications, London, 106 pp.

Thorpe, K. (1998) *Marine Nature Conservation Review Sectors 1 and 2. Lagoons in Shetland and Orkney: Area Summaries*, Joint Nature Conservation Committee, Peterborough.

Tija, H.D. (1985) Notching by abrasion on a limestone coast. *Zeitschrift für Geomorphologie*, **29**, 367–72.

Ting, S. (1936) Beach ridges and other shore deposits in south-west Jura. *Scottish Geographical Magazine*, **52**, 182–7.

Ting, S. (1937) The coastal configuration of western Scotland. *Geografiska Annaler*, **19**, 62–83.

Tooley, M.J. (1969) Sea-level changes and the development of coastal plant communities during the Flandrian in Lancashire and adjacent areas. Unpublished PhD thesis, University of Lancaster.

Tooley, M.J. (1970) The peat beds of the south-west Lancashire coast. *Nature in Lancashire*, **1**, 19–26.

Tooley, M.J. (1971) Evolution of the Fylde coast. In *Aspects of Fylde Geography* (ed. A.R. Wilson), Geographical Association, Blackpool, pp. 1–7.

Tooley, M.J. (1973) Flandrian sea-level changes in north-west England and pan-North-west European correlations. *Abstracts, INQUA 9th Congress, Christchurch, New Zealand*, 373–4.

Tooley, M.J. (1974) Sea level changes during the last 9000 years in north-west England. *Geographical Journal*, **140**, 18–42.

Tooley, M.J. (1976) Flandrian sea level changes in west Lancashire and their implications for the 'Hillhouse Coastline'. *Geological Journal*, **11**, 137–52.

Tooley, M.J. (ed.) (1977a) *The Isle of Man, Lancashire Coast and Lake District, International Union for Quaternary Research Guidebook for Excursion*, **A4**, GeoAbstracts, Norwich, 60 pp.

Tooley, M.J. (1977b) The Quaternary of north-west England and the Isle of Man. In *The Isle of Man, Lancashire Coast and Lake District* (ed. M.J. Tooley), *International Union for Quaternary Research Guidebook for Excursion*, **A4**, GeoAbstracts, Norwich, pp. 5–7.

Tooley, M.J. (1978) *Sea-level Changes in North West England during the Flandrian Stage*, Clarendon Press, Oxford, 232 pp.

Tooley, M.J. (1982) Sea level changes in northern England. *Proceedings of the Geologists' Association*, **93**, 43–51.

Tooley, M.J. (1985a) Sea level changes and coastal morphology in northeast England. In *The Geomorphology of North-West England* (ed. R.H. Johnston), Manchester University Press, Manchester, pp. 94–121.

Tooley, M.J. (1985b) Climate, sea level and coastal changes. In *The Climatic Scene* (eds M.J. Tooley and G.M. Sheail), Allen and Unwin, London, pp. 206–34.

Tooley, M.J. (1992) Recent sea-level changes. In *Saltmarshes. Morphodynamics, Conservation and Engineering Significance* (eds J.R.L. Allen and K. Pye), Cambridge University Press,

References

Cambridge, pp. 19–40.

Tooley, M.J. (1993) Long term changes in eustatic sea level. In *Climate and Sea Level Change*, (eds R.A. Warwick, E.M. Barrow and T.M.L. Wigley), Cambridge University Press, Cambridge, pp. 81–107.

Tooley, M.J. (1995) Romney Marsh: the Debatable Ground. In *Romney Marsh: the Debatable Ground* (ed. J. Eddison), Oxford University Committee for Archaeology Monograph, No. 41, OUCA, Oxford, pp. 1–7.

Tooley, M.J. and Kear, R. (1977) Shirdley Hall Sand Formation. In *The Isle of Man, Lancashire Coast and Lake District* (ed. M.J. Tooley) International Union for Quaternary Research Guidebook for Excursion, A4, GeoAbstracts, Norwich, pp. 9–12.

Tooley, M.J. and Shennan, I. (eds) (1987) *Sea-level Changes*, Institute of British Geographers Special Publication, No. 20, Blackwell Scientific Publications, Oxford, 397 pp.

Tooley, M.J. and Switsur, R. (1988) Water level changes and sedimentation during the Flandrian Age in the Romney Marsh area. In *Romney Marsh: Evolution, Occupation, Reclamation* (eds J. Eddison and C. Green), Oxford University Committee for Archaeology Monograph, No. 24, OUCA, Oxford, pp. 53–71.

Torr, C. (1923) *Small Talk at Wreyland*, Series III, Cambridge University Press, Cambridge, 112 pp.

Toy, H.S. (1934) The Loe Bar near Helston, Cornwall. *Geographical Journal*, **83**, 40–8.

Travis, C.B. (1922) On peaty beds in the Wallasey Sand-hills. *Proceedings of the Liverpool Geological Society*, **13**, 207–14.

Travis, C.B. (1926) The peat and forest bed of the south-west Lancashire coast. *Proceedings of the Liverpool Geological Society*, **14**, 263–77.

Travis, C.B. (1929) The peat and forest beds of Leasowe, Cheshire. *Proceedings of the Liverpool Geological Society*, **15**, 157–78.

Travis, W.G. (1908) On plant remains in peat in the Shirdley Hill Sand at Aintree, South Lancashire. *Proceedings of the Liverpool Botanical Society*, **1**, 47–52.

Trenhaile, A.S. (1969) A geomorphological investigation of shore platforms and high-water rock ledges in the Vale of Glamorgan. Unpublished PhD thesis, University of Wales.

Trenhaile, A.S. (1971) Lithological control of high-water rock ledges in the Vale of Glamorgan, Wales. *Geografiska Annaler*, **53A**, 59–69.

Trenhaile, A.S. (1972) The shore platforms of the Vale of Glamorgan. *Transactions of the Institute of British Geographers*, **56**, 127–44.

Trenhaile, A.S. (1974a) The geometry of shore platforms in England and Wales. *Transactions of the Institute of British Geographers*, **62**, 129–42.

Trenhaile, A.S. (1974b) The morphology and classification of shore platforms in England and Wales. *Geografiska Annaler*, **56A**, 103–10.

Trenhaile, A.S. (1978) The shore platforms of Gaspé, Québec. *Annals of the Association of American Geographers*, **68**, 95–114.

Trenhaile, A.S. (1980) Shore platforms: a neglected coastal feature. *Progress in Physical Geography*, **4**, 1–23.

Trenhaile, A.S. (1982) A reply to A.P. Carr and J. Graff. *Transactions of the Institute of British Geographers, New Series*, **7**, 246–7.

Trenhaile, A.S. (1983) The width of shore platforms: a theoretical approach. *Geografiska Annaler*, **65A**, 147–58.

Trenhaile, A.S. (1987) *The Geomorphology of Rock Coasts*, Oxford Research Studies in Geography, Clarendon Press, Oxford, 384 pp.

Trenhaile, A.S. (1997) *Coastal Dynamics and Landforms*, Clarendon Press, Oxford, 366 pp.

Trenhaile, A.S. and Byrne, M.-L. (1986) A theoretical investigation of the Holocene development of rock coasts, with particular reference to shore platforms. *Geografiska Annaler*, **68A**, 1–14.

Trenhaile, A.S. and Layzell, M.G.J. (1981) Shore platform morphology and the tidal duration factor. *Transactions of the Institute of British Geographers, New Series*, **6** (1), 82–102.

Trenhaile, A.S., Pérez Alberti, A., Martínez Cortizas, A., Costa Casais, M. and Blanco Chao, R. (1999) Rock coast inheritance: an example from Galacia, northwestern Spain. *Earth Surface Processes and Landforms*, **24** (7), 605–21.

Trudgill, S.T. (1987) Bioerosion of intertidal limestone, County Clare, Eire. B. Zonation, process and form. *Marine Geology*, **74**, 111–21.

Trueman, A.E. (1922) Liassic rocks of Glamorgan. *Proceedings of the Geologists' Association*, **33**, 245–84.

Trueman, A.E. (1930) The Lower Lias of Nash

References

Point, Glamorgan. *Proceedings of the Geologists' Association*, **41**, 148–59.

Trupin, A. and Wahr, J. (1990) Spectroscopic analysis of global tide gauge sea level data. *Geophysical Journal International*, **100**, 441–53.

Tucker, M.J. (1963) Analysis of records of sea waves. *Proceedings of the Institution of Civil Engineers*, **26**, 304–16.

Twidale, C.R. (1968) *Geomorphology: With Special Reference to Australia*, Nelson, Melbourne, 406 pp.

UKDMAP (1998) *United Kingdom Digital Marine Atlas*, 3rd edn, British Oceanographic Data Centre, Proudman Oceanographic Laboratory, Bidston.

Ussher, W.A.E. (1908) *The Geology of the Quantock Hills and of Taunton and Bridgwater*, Memoir of the Geological Survey of Great Britain, Sheet 295 (England and Wales), HMSO, London, 109 pp.

Valentin, H. (1954) Der Landverlust in Holderness, Ostengland, von 1852 bis 1952. *Die Erde*, **6**, 296–315.

Valentin, H. (1961) The central west coast of Cape York peninsula. *Australian Geography*, **8**, 65–72.

Valentin, H. (1971) Land loss at Holderness. In *Applied Coastal Geomorphology* (ed. J.A. Steers), Macmillan, London, pp. 116–37.

van Straaten, L.M.J.U. (1959) Littoral and submarine morphology of the Rhone delta. In *Second Coastal Geography Conference, held on April 6–9, 1969 at the Coastal Studies Institute, Louisiana State University* (ed. R.J. Russell), Coastal Studies Institute, Louisiana State University, Washington D.C., pp. 233–64.

Victoria University of Manchester (1983) Holderness Coast Erosion: Pre-feasibility Study: Final Report. Unpublished report prepared for the Humberside Joint Advisory Committee on Coast Protection.

Viles, H.A. and Spencer, T. (1995) *Coastal problems: Geomorphology, Ecology and Society at the Coast*, Edward Arnold, London, 350 pp.

Vincent, C.E. (1979) Longshore sand transport rate – a simple model for the East Anglian coastline. *Coastal Engineering*, **3**, 113–36.

Vladimirov, A.T. (1961) The morphology and evolution of the lagoon coast of Sakhalin. *Trudy Institute Okeanology*, **48**, 145–71.

Vollans, E. (1995) Medieval salt-making and the inning of the tidal marshes at Belgar, Lydd. In *Romney Marsh: the Debatable Ground* (ed. J. Eddison), Oxford University Committee for Archaeology Monograph, No. **41**, OUCA, Oxford, pp. 118–26.

von Weymarn, J. (1974) Coastline development in Lewis and Harris, Outer Hebrides with particular reference to the effects of glaciation. Unpublished PhD thesis, University of Aberdeen.

Wal, A. (1992) Sedimentological effects of aeolian processes active in the Tentsmuir area, Fife, Scotland. Unpublished PhD thesis, University of St Andrews.

Wal, A. and McManus, J. (1993) Wind regime and sand transport on a coastal beach-dune complex, Tentsmuir, eastern Scotland. In *The Dynamics and Environmental Context of Aeolian Sedimentary Systems* (ed. K. Pye), Geological Society Special Publication, No. **72**, Geological Society, London, pp. 159–71.

Walker, M.J.C. (1984) A pollen diagram from St Kilda, Outer Hebrides, Scotland. *New Phytologist*, **97**, 369–404.

Waller, M. (1993) Flandrian vegetation history of south-eastern England. Pollen data from Pannel Bridge, East Sussex. *New Phytologist*, **124**, 345–69.

Waller, M. (1994) Flandrian vegetation history of south-eastern England. Stratigraphy of the Brede valley and pollen data from Brede Bridge. *New Phytologist*, **126**, 369–92.

Waller, M.P., Burrin, P.J and Marlow, A. (1988) Flandrian sedimentation and palaeoenvironments in Pett Level, the Brede and lower Rother valleys and Walland Marsh. In *Romney Marsh: Evolution, Occupation, Reclamation* (eds J. Eddison and C. Green), Oxford University Committee for Archaeology Monograph, No. **24**, OUCA, Oxford, pp. 3–30.

Walton, K. (1956) Rattray: a study of coastal evolution. *Scottish Geographical Magazine*, **72**, 85–96.

Walton, K. (1959) Ancient elements in the coastline of north-east Scotland. In *Essays in Memory of Alan G. Ogilvie*, University of Edinburgh, Edinburgh.

Walton, K. and Ritchie, W. (1972) The evolution of the Sands of Forvie and the Ythan estuary. In *Northeast Scotland: Geographical Essays* (ed. C.M. Clapperton), Department of Geography University of Aberdeen, Aberdeen.

Ward, E.M. (1922) *English Coastal Evolution*, Methuen, London, 262 pp.

References

Ward, G. (1931) Saxon Lydd. *Archaeologia Cantiana*, **43**, 29–37.

Warrington, G., Cope, J.C.W. and Ivimey-Cook, H.C. (1994) St Audrie's Bay, Somerset, England: a candidate Global Stratotype Section and Point for the base of the Jurassic System. *Geological Magazine*, **131**, 191–200.

Wass, M. (1995) The proposed northern course of the Rother: a sedimentological and microfaunal investigation. In *Romney Marsh: the Debatable Ground* (ed. J. Eddison), Oxford University Committee for Archaeology Monography, No. 41, OUCA, Oxford, pp. 51–77.

Waters, S. (1985) Vegetation of natural wetlands. In *Dungeness – Ecology and Conservation, Report of a meeting held at Botany Department, Royal Holloway and Bedford New College, on 16 April 1985* (eds B. Ferry and S. Waters), Focus on Nature Conservation, No. 12, Nature Conservancy Council, Peterborough.

Watkin, E.E. (1976) *Ynyslas Nature Reserve Handbook*, Nature Conservancy Council and University College of Wales, Aberystwyth.

Webster, T. (1816) Additional observations, chiefly geological, on the Isle of Wight. In *A Description of the Principal Picturesque Beauties, Antiquities and Geological Phenomena of the Isle of Wight* (ed. H.C. Englefield), Payne and Foss, London, p. 238.

West, I.M. (1972) The origin of the supposed raised beach at Porth Neigwl, north Wales. *Proceedings of the Geologists' Association*, **83**, 191–5.

West, I.M. (1980) Geology of the Solent estuarine system. In *The Solent Estuarine System: an Assessment of Present Knowledge* (ed. M. Burton), *Natural Environment Research Council Publications Series C*, No. 22, NERC, Swindon, pp. 6–18.

Wetherill, P.J. (1980) Hydrodynamic process, sediment movement and coastal inlet morphodynamics at the mouth of the River Ythan, Aberdeenshire. Unpublished PhD thesis, University of Aberdeen.

Whalley, R.W. (1977) Walney Island. Unpublished BA thesis, University of Southampton.

Wheeler, W.H. (1902) *The Sea Coast*, Longman, London, 361 pp.

Whittaker, J.E. (1980) The Fleet Dorset: a seasonal study of the watermass and its vegetation. *Proceedings of the Dorset Natural History and Archaeological Society*, **100**, 73–99.

White, D.J.B. (1979) The effects of the storm of 11th January 1978 at Blakeney Point. *Transactions of the Norfolk and Norwich Naturalists' Society*, **25**, 267–9.

White, H.J.O. (1921) *A Short Account of the Geology of the Isle of Wight*, Memoir of the Geological Survey of Great Britain, Sheet 10 (England and Wales), HMSO, London, 219 pp.

Whittington, G. (ed.) (1996) *Fragile Environments: The Use and Management of Tentsmuir National Nature Reserve, Fife*, Scottish Cultural Press, Dalkeith, 120 pp.

Whittow, J.B. (1957) The Lleyn Peninsula, north Wales: a geomorphological study. Unpublished PhD thesis, University of Reading.

Whittow, J.B. (1960) Some comments on the raised beach platform of south-west Caernarvonshire and on an unrecorded raised beach at Porth Neigwl, north Wales. *Proceedings of the Geologists' Association*, **71**, 31–9.

Whittow, J.B. (1965) The interglacial and postglacial strandlines of north Wales. In *Essays in Geography for Austin Miller* (eds J.B. Whittow and P.D. Wood), University of Reading, Reading, pp. 94–117

Wilks, P.J. (1977) Holocene sea-level change in the Cardigan Bay area. Unpublished PhD thesis, University of Wales (Aberystwyth).

Wilks, P.J. (1979) Mid-Holocene sea-level and sedimentation interactions in the Dovey estuary area, Wales. *Palaeogeography, Palaeoclimatology, Palaeoecology*, **26**, 17–36.

Williams, A.T. and Caldwell, N.E. (1981) Sediment disturbance by 'beach paddlers' along the Glamorgan Heritage Coast, United Kingdom. *Shore and Beach*, **49**, 30–3.

Williams, A.T. and Caldwell, N.E. (1988) Particle size and shape in pebble beach sedimentation. *Marine Geology*, **82**, 199–215.

Williams, A.T. and Davies, P. (1984) Cliff failure along the Glamorgan Heritage Coast, Wales, UK. In *Mouvements de Térrains* (ed. J.-C. Flageollet), *Série Documents du Bureau de Recherches Géologiques et Minières*, **83**, Bureau de Recherches Géologiques et Minières, pp. 109–19.

Williams, A.T. and Davies, P. (1987) Rates and mechanisms of coastal cliff erosion in Lower Lias rocks. In *Coastal Sediments '87: proceedings of a Speciality Conference on Advances*

References

in Understanding of Coastal Sediment Processes, New Orleans, Louisiana, May 12–14 1987 (ed. N.C. Kraus), American Society of Civil Engineers, New York, pp. 1855–70.

Williams, A.T., Caldwell, N.E. and Yule, A.P. (1981) Beach morphology changes at Ynyslas Spit, Dyfed, Wales. *Cambria*, **8**, 51–69.

Williams, A.T., Davies, P. and Bomboe, P. (1993) Geometrical simulation studies of coastal cliff failures in Liassic strata, south Wales, U.K. *Earth Surface Processes and Landforms*, **18**(8), 703–20.

Williams, W.W. (1947) The determination of gradients of enemy held beaches. *Geographical Journal*, **109**, 76–93.

Williams, W.W. (1956) An east coast survey: some recent changes in the coast of East Anglia. *Geographical Journal*, **122**, 317–34.

Williams, W.W. (1960) *Coastal Changes*, Routledge and Kegan Paul, London, 220 pp.

Williams, W.W. and Fryer, D.H. (1953) Benacre Ness: an east coast erosion problem. *Royal Institute of Chartered Surveyors Journal*, **32**, 772–81.

Willis, A.J. (1963) Braunton Burrows: the effects on vegetation of the addition of mineral nutrients to the dune soils. *Journal of Ecology*, **51**, 353–74.

Willis, A.J. (1965) The influence of mineral nutrients on the growth of *Ammophila arenaria*. *Journal of Ecology*, **53**, 735–45.

Willis, A.J. (1967) A new location for *Liparis loeselii*. *Proceedings of the Botanical Society of the British Isles*, **6**, 352–3.

Willis, A.J. (1985) Dune water and nutrient regimes – their ecological relevance. In *Sand Dunes and their Management* (ed. J.P. Doody), Focus on Nature Conservation, No. **13**, Nature Conservancy Council, Peterborough, pp. 159–74.

Willis, A.J. (1989) Coastal sand dunes as biological systems. *Proceedings of the Royal Society of Edinburgh*, **96B**, 17–36.

Willis, A.J., Folkes, B.F., Hope-Simpson, J.F. and Yemm, E.W. (1959a) Braunton Burrows: the dune system and its vegetation, Part I. *Journal of Ecology*, **47**, 1–24.

Willis, A.J., Folkes, B.F., Hope-Simpson, J.F. and Yemm, E.W. (1959b) Brauton Burrows: the dune system and its vegetation, Part II. *Journal of Ecology*, **47**, 249–88.

Willis, A.J. and Jefferies, R.L. (1963) Investigations on the water relations of sand-dune plants under natural conditions. In *The Water Relations of Plants* (eds A.J. Rutter and F.H. Whitehead), Blackwell Scientific Publications, Oxford, pp. 168–89.

Wilson, G. (1951) The tectonics of the Tintagel area, north Cornwall. *Quarterly Journal of the Geological Society of London*, **106**, 393–432.

Wilson, G. (1952) The influence of rock structures on coast-line and cliff development around Tintagel, north Cornwall. *Proceedings of the Geologists' Association*, **63**, 20–48.

Wilson, G. (1971) The influence of rock structures on coast-line and cliff development around Tintagel, north Cornwall. In *Introduction to Coastline Development* (ed. J.A. Steers), Macmillan, London, pp. 133–61.

Wilson, K. (1960) The time factor in the development of dune soils at South Haven Peninsula, Dorset. *Journal of Ecology*, **48**, 341–59.

Wilson, V., Welch, F.B.A., Robbie, J.A. and Green, G.W. (1958) *The Geology of the Country around Bridport and Yeovil*, Memoir of the Geological Survey of Great Britain, sheets 327 and 312 (England and Wales), HMSO, London, 239 pp.

Winkelmolen, A.M. (1978) Size, shape and density sorting of beach material along the Holderness coast, Yorkshire. *Proceedings of the Yorkshire Geological Society*, **42**, 109–41.

Wood, A. (1968) Beach platforms in the Chalk of Kent. *Zeitschrift für Geomorphologie*, Neue Folge, **12**, 107–13.

Woodworth, P.L. (1987) Trends in the U.K. mean sea level. *Marine Geodesy*, **11**, 57–87.

Woodworth, P.L. (1990a) Measuring and predicting long term sea level changes. *NERC News*, **1990** (October), 22–25.

Woodworth, P.L. (1990b) A search for accelerations in records of European mean sea level. *International Journal of Climatology*, **10**, 129–43.

Woodworth, P.L., Shaw, S.M. and Blackman, D.L. (1991) Secular trends in mean tidal range around the British Isles and along the adjacent European coastline. *Geophysical Journal International*, **104**, 593–609.

Woodworth, P.L. (1993) Sea level changes. In *Climate and Sea Level Change* (eds R.A. Warrick, E.M. Barrow and T.M.L. Wigley), Cambridge University Press, Cambridge, pp. 379–91.

Wooldridge, S.W. and Linton, D.L. (1955)

References

Structure, Surface and Drainage in South-east England, George Philip, London, 176 pp.

Wooldridge, S.W. and Morgan, R.S. (1937) *The Physical Basis of Geography. An Outline of Geomorphology*, University Geographical Series, Longman, London, 445 pp.

Worth, R.H. (1904) Hallsands and Start Bay. *Reports and Transactions of the Devonshire Association for the Advancement of Science*, 36, 302–46.

Worth, R.H. (1907) Minutes of evidence to the Royal Commission on Coast Erosion, 1907, 1 (2), Appendix 14, 177–84.

Worth, R.H. (1909) Hallsands and Start Bay, Part II. Hallsands. *Reports and Transactions of the Devonshire Association for the Advancement of Science*, 41, 301–8.

Worth, R.H. (1923) Hallsands and Start Bay, Part III. *Reports and Transactions of the Devonshire Association for the Advancement of Science*, 55, 131–42.

Wortham, W.H. (1913) *Some Features of the Sand Dunes in the s.w. Corner of Anglesey*. Report of the British Association for the Advancement of Science, 1913.

Wray, D.A. and Cope, F.W. (1948) *The Geology of Southport and Formby*, Memoir of the Geological Survey of Great Britain, sheets 75 and 83 (England and Wales), HMSO, London, 54 pp.

Wright, L.D. and Short, A.D. (1983) Morphodynamics of beaches and surf zones in Australia. In *CRC Handbook of Coastal Processes and Erosion* (ed. P.D. Komar), CRC Press, Boca Raton, pp. 25–64.

Wright, L.D. and Short, A.D. (1984) Morphodynamic variability of surf zones and beaches: a synthesis. *Marine Geology*, 56, 93–118.

Wright, L.W. (1967) Some characteristics of the shore platforms of the English Channel coast and the northern part of North Island, New Zealand. *Zeitschrift für Geomorphologie, Neue Folge*, 11, 36–46.

Wright, L.W. (1969) Shore platforms and mass movements: a note. *Earth Science Journal*, 3, 44–7.

Wright, L.W. (1970) Variation in level of the cliff/shore platform junction along the south coast of Great Britain. *Marine Geology*, 9, 347–53.

Wright, P. (1976) The morphology, sedimentary structures and processes of the foreshore at Ainsdale, Merseyside. Unpublished PhD thesis, University of Reading.

Wright, P. (1981) Aspects of the coastal dynamics of Poole and Christchurch Bays. Unpublished PhD thesis, University of Southampton.

Wright, P. (1984) Facies development on a barred (ridge and runnel) coastline – the case of south-west Lancashire (Merseyside). In *Coastal Research – UK Perspectives* (ed. M.W. Clark), Geobooks, Norwich, pp. 105–18.

Wright, R. (1981) *The Beaches of Tayside*, Department of Geography, University of Aberdeen, Aberdeen, 66 pp.

Wright, R. and Harris, T.A. (1988) Change detection in physically fragile coastal areas of Scotland by remote sensing. In *Proceedings of the International Symposium on Remote Sensing of the Coastal Zone, Gold Coast, Queensland, 7–9 September 1988*, Department of Geographic Information, Brisbane.

Wright, W.B. (1911) On a pre-glacial shoreline in the Western Isles of Scotland. *Geological Magazine, Decade 5*, 8, 97–109.

Yapp, R.H., Johns, D. and Jones, O.T. (1917) The salt marshes of the Dovey estuary, Part II. *Journal of Ecology*, 5, 65–103.

Yates, R.A. (1968) Surveying techniques in coastal geomorphology. In *Geography at Aberystwyth* (eds E.G. Bowen, H. Carter and J.A. Taylor), University of Wales Press, Cardiff, pp. 129–42.

Young, A. (1972) *Slopes*, Geomorphology Texts, No. 3, Oliver and Boyd, Edinburgh, 288 pp.

Zenkovich, V.P. (ed. J.A. Steers) (1967) *Processes of Coastal Development*, Oliver and Boyd, Edinburgh, 738 pp.

Glossary

This glossary provides brief explanations of the technical terms used in the introductions to the chapters and in the 'conclusions' sections of the site reports. These explanations are not rigorous scientific definitions but are intended to help the general reader. Words in **bold** type indicate an internal reference to another glossary entry.

Abrasion: the process of mechanical wearing away of parts of rocks or **fossils** by **sediment**-laden water, air or ice. The process produces an increasingly smoothed and rounded outline shape.

Absolute age: the actual age of formation of a natural feature, rock or **fossil**, determined by **absolute dating**. Usually given as 'years **BP**', actually meaning years before AD 1950 by international agreement.

Absolute dating: a method of determining the **absolute age** of formation of a rock, mineral or fossil, by techniques such as **radiocarbon dating**.

Accretion: build-up or accumulation of **sediment**.

Aeolian: descriptive of **sediment** transported and deposited by the wind.

Aggradation: the building upwards of a river valley or floodplain by accumulation of **fluvial** deposits; can also be applied to material deposited by other agencies, such as wind or waves.

Alluvial fan: a cone-shaped deposit made up of water-laid deposits, and also some material transported by mud flows.

Alluvial: a term applied to the environments, action, and products of rivers or streams. Alluvial deposits are composed of **clastic** material deposited in the river **floodplain**.

Alluviation: the process of the accumulation of material deposited by river water, usually located along the river valley and tending to be predominantly fine-grained **silt** or **sand**.

Alluvium: sediment transported and deposited by rivers.

Anastomosing: descriptive of a system that branches or contains a network; for example the channel pattern of a **braided** stream.

Anthropogenic: produced or induced by human activity.

Anticline: an arch-shaped upfold of rocks produced by tectonic activity (cf. **syncline**) with younger strata on the outermost part of the arch and older rock in its core.

Arch: a hollow cut by wave **erosion** through a rocky headland.

Archipelago: a group of islands.

Arcuate: arc-shaped.

Arenaceous: descriptive of **clastic sediments** made up of **sand**-sized particles, cf. **argillaceous**.

Argillaceous: 'clay', descriptive of fine-grained detrital **sediments** made of **silt**, or **clay**-sized particles, cf. **arenaceous**.

Ayre: a regional term used mainly in Shetland for a **tombolo** or a long narrow **spit** of **shingle** or **sand**, usually formed across a shallow bay or **voe**.

Glossary

Ball and low: *see* **ridge and runnel**.

Bar: a low, elongated body of **sediment**, such as **sand** or **gravel** laid down in shallow water, built up by wave action offshore, and lying more or less parallel to the general coastline and sometimes attached to it; cf. **spit**, **tombolo**. Sometimes known as a 'barrier beach', where it emerges above high tide level.

Barrier beach: *see* **bar**

Beach: the strip of land along the margin of a body of water that is washed by waves or tides sufficiently to prevent all or most terrestrial plant growth.

Beach feeding: coastal engineering involving the importation of **sediment**, usually obtained from the seabed. Also known as beach nourishment, beach replenishment.

Beach plain: a continuous level or undulating area formed by closely-spaced successive embankments of wave-deposited beach material added more or less uniformly to a prograding shoreline.

Beach ridge: a low, rectilinear or curvilinear mound of beach or beach and **dune** material (**sand**, **gravel**, **shingle**) heaped up by wave action on the backshore of a **beach** above the present limit of storm waves or the reach of ordinary tides, and running roughly parallel to the shore.

Bedding plane: a planar feature in **sedimentary** rocks representing an original surface of deposition. Conspicuous bedding planes may indicate a short interruption in, or change in character of, **sediment** deposition.

Bedrock: the rock, usually solid, underlying soil and other unconsolidated surficial material.

Berm: a gently inclined ridge of sediment at the top of the beach.

Bio-erosion: **weathering** and **erosion** resulting from the activities of animals.

Biogenic: descriptive of sediments that of biological origin, e.g. shellfish fragments

Bio-geochemical cycling: the movement of chemical elements from organism to physical environment to organism, in a more or less circular pathway. It is termed 'nutrient cycling' if the elements involved are essential to life. An element may be solid, liquid, or gaseous, or form different chemical compounds, in the various parts of the cycle.

Blow-hole: a hole in a clifftop leading to a sea **cave**, through which air is forced by the action of the sea.

Blowthrough (also **blowout**): A localized area of **deflation**, especially on a coastal **sand dune**. Deflation may have begun through the removal of vegetation by grazing animals or by trampling, or through the cutting off of sand supplies as new dunes develop near the shoreline.

Bog: ground that is waterlogged and spongy. It consists mainly of mosses and contains acidic decaying vegetation that may develop into **peat**.

Boulder clay: *see* **till**.

Boulder: an unattached rock particle with a diameter of more than 256 mm.

BP: before present, *see* **absolute age**.

Brackish: waters with salinities intermediate between fresh and marine.

Braided channel: a stream or river channel that branches frequently and rejoins after separation by **bars**.

Braided river (**braided stream**): a stream or river that divides into an interlacing network of several small branching and rejoining shallow channels, separated by **bars** or islands.

Breaker zone: the area between the outermost breakers (waves that have become too steep to remain stable, and are therefore breaking into foam near the shore) and the extent of wave uprush on the **beach**.

Breccia: a **sedimentary rock** consisting of angular pebbles (cf. **conglomerate**).

Cainozoic Era: the youngest era of geological time, spanning from approximately 65 million years ago to the present, and consisting of the **Tertiary** and **Quaternary periods**.

Calcareous: containing large quantities of calcium carbonate ($CaCO_3$).

Calcarenite: **limestone** formed mainly of **sand**-size calcium carbonate fragments.

Carboniferous: a geological period from about 345 to 280 million years ago; rocks formed in this period in the British Isles are characterized by a lower marine **carbonate** sequence and upper marginal marine and terrestrial sequences containing coal deposits.

Carr: a regional term (mainly used in north-east England and south-east Scotland) for a **reef**; also a term for a mire with scrub dominated by alder and willow.

Carse: a Scottish term for areas of emerged or land-claimed estuarine flats.

Glossary

Catchment: a term often synoymous with drainage basin; the area that collects the water flowing to a particular river; *see* **watershed**.

Cave: best defined as 'a natural cavity large enough to be entered by a person'. Most caves are formed by dissolution of **limestone** within **karst** landscapes, but others include sea, **glacier**, lava and **tectonic caves**. Sea caves can be defined as a hollow normally eroded in a **cliff**, with the penetration being greater than the width at the entrance.

Cement: the mineralogical 'glue' that holds particles together in **sedimentary** rocks.

Chalk: a poorly lithified, weak, friable and porous white **limestone**. Stratigraphically, *the Chalk* (a proper noun with a capital letter) is used synonymously in Britain with all of those rocks that formed during the Late **Cretaceous** Epoch.

Chenier: a ridge of **sand** and shells deposited at the **swash** limit on a **saltmarsh** or other coastal plain by a storm **surge**.

Chine: a sharply incised valley intersected by a sea **cliff**.

Chert: a fine-grained **silica**-rich rock that commonly occurs as nodules or bands within **limestones**.

Clast (adj. **clastic**): a sedimentary particle, usually larger than 4 mm diameter; a fragment of a pre-existing rock or **fossil** (*bioclast*).

Clay: very fine-grained **sediment**, <0.004 mm in size; a soft very fine-grained **sedimentary** rock composed primarily of clay-sized particles.

Cleft: a **fissure** or crevice.

Cleit: a small unmortared drystone turf-roofed building peculiar to the St Kilda island group, used by the original islanders to store food and fuel.

Cliff: any slope steeper than 45°.

Climate change: long-term trends or shifts in climate caused by natural mechanisms or by human activity.

Climbing dune: a **sand dune** formed by the piling-up of **sand** by wind against a **cliff** or hillslope.

Cnoch-and-lochan: a glacially scoured series of rock knolls and tarns

Coastal Management Plan: a plan that sets the management of a section of coast into its physical (erosion/deposition) and human contexts (planning, recreation) in order to minimize conflicts of interest between all coastal dunes. cf **shoreline management plan**.

Coastal squeeze: the process whereby, in the face of rising sea levels, an area of **intertidal** habitat, such as **saltmarsh**, mudflat or saline lagoon is prevented from migrating landwards owing to the presence of a hard boundary such as a sea defence or natural slope.

Coastal zone: the space in which terrestrial environments influence marine (or lacustrine) environments and vice versa. The coastal zone is of variable width, and may also change over time. Delimitation of zonal boundaries is not normally possible; more often such limits are marked by an environmental gradient or transition. At any one locality, the coastal zone may be characterized according to physical, biological or cultural criteria, which need not, and rarely do, co-incide.

Cobble: a rock particle with a diameter of between 64 and 256 mm with a generally rounded or subrounded shape.

Col: the highest point on a divide between two valleys.

Combe: A short valley, especially in chalk areas.

Conglomerate: a **sedimentary** rock consisting of rounded pebbles (cf. **breccia**).

Contemporaneous: formed or occurring at the same time.

CORINE: Co-ORdination of INformation on the Environment (a European Union biotopes classification initiative).

Corrasion: **erosion** by the mechanical pounding, scraping, and battering action of water or ice carrying pieces of rock, which wears away the land surface.

Corrosion: **weathering** by the wearing away of materials by chemical action. Corrosion usually involves the combined action of oxygen and water on a metal, and can be speeded by the presence of such substances as salt, acids, or bases, or air pollutants like sulphur dioxide.

Corrie: a Scottish term for a cirque, a deep, steep-walled, hollow in a mountain caused by glacial erosion; = cwm in Wales, = coire (Gaelic).

Cove: *see* **pocket beach**.

Creek: a small narrow inlet of the sea, longer than it is wide.

Creep: the slow mass-movement of material

Glossary

down relatively steep slopes, mainly under the force of gravity, but also influenced by saturation with water and alternate freezing and thawing; cf. **landslide**.

Crenulate: finely indented or notched.

Cretaceous: a period of geological time from about 142 to 65 million years ago.

Crevice: a narrow crack in a hard substratum <10 mm wide at its entrance, with the penetration being greater than the width at the entrance.

Crystallization: the formation of crystals or a crystalline structure.

Cuesta: an asymmetric ridge, with a steep slope on one side, and a shallow slope on the other.

Current: horizontal movement of water in response to meteorological, oceanographical and topographical factors; a steady flow in a particular direction. 'Current' refers to residual flow after any tidal element (**tidal streams**) has been removed.

Cusp (adj. **cuspate**): a complex depositional feature formed when **longshore drift** from two directions meets to produce a series of ridges at right angles to each other, forming a low-lying triangular foreland.

Debris flow: an avalanche-like break up and displacement of sediments down a slope, resulting in a chaotic jumble of fragments of different sizes in a muddy matrix.

Deflation: a process of surface-lowering (erosion) by wind action

Delta (adj. **deltaic**): a fan-shaped or irregular mass of **sediment** deposited where a river enters a lake or the sea.

Dendritic: a drainage pattern whose shape resembles the pattern made by the branches of a tree or veins of a leaf.

Denudation: the combined processes of **weathering** and **erosion** that wear down landscapes.

Desiccation: removal of moisture, which may produce desiccation cracks in the surface.

Devensian: the term for the last **glacial period** in Britain (maximum *c.* 18 000 years **BP**)

Diagenesis (adj. **diagenetic**): the alteration of the mineralogy and texture of **sediments** and **fossils** when they are close to the Earth's surface by chemical and physical processes; the term excludes **metamorphic** alteration.

Differential weathering: weathering that occurs at different rates, due to differences in composition and resistance of a rock and/or differences in the intensity of **weathering**.

Dip: the angle between a surface and a horizontal plane.

Dissolution: natural process of dissolving a solid; specifically in **karst** processes, the dissolving of **carbonate** rock to create a liquid solution of calcium and bicarbonate ions in water; also known as **solution**.

Distal: far from source.

Drift: a term used to characterize all unconsolidated rock debris transported from one place to another. *See also* **longshore drift**.

Drowned features: landscape features that are submerged as a result of changes in the relative levels of sea to land.

Drumlin: a low, rounded hill of **glacial till**, which was moulded into a streamlined shape by a **glacier** ice passing over it. Its long axis is parallel to the direction of flow of the **ice sheet** beneath which it formed.

Dry-core dunes: a **dune** system in which the **water table** is usually below the **deflation** level, so that standing water is rarely, if ever, found in the system

Dry valley: a fluvial valley cut by a **subaerial** stream or river then abandoned and left dry due to underground drainage, so that it now seldom, if ever, has water flowing along it in the form of a stream channel.

Dune: in the terrestrial environment, a mound or ridge of unconsolidated windblown **sediment**.

Dune slack: Flat-bottomed, hollow zone within a **sand dune** system that has developed over impervious **strata**. The slack may result from **erosion** or **blowthrough** of the **dune** system, and the flat base level is therefore close to or at the permanent **water table** level. Characteristically, dune slacks have rich, marshy flora, with willows (*Salix* species) as typical woody colonizers.

Dyke: a vertically or sub-vertically orientated band of rock. The term is generally applied to **igneous** rocks which have 'intruded' or 'cut through' pre-existing rocks, although **sedimentary** (Neptunian) forms occur.

Ebb-tide: outgoing or falling tide.

Edge waves: infragravity oscillations forced by resonance within within the surf zone. They interact with incoming waves to produce cir-

Glossary

culation cells whose currents transport sediment and may control net longshore movements.

Embayment: a type of marine inlet typically where the line of the coast follows a concave sweep between rocky headlands, sometimes with only a narrow entrance to the embayment.

Emerged beach: a former **beach** now situated above the level of the present shoreline as a result of earth movement, or changes in relative sea level. Also called a 'raised beach'.

English Channel: the arm of the Atlantic Ocean between southern England and northern France, linked with the **North Sea** by the Strait of Dover.

Ephemeral: short-lived, intermittent.

Erosion surface: a surface shaped by the processes of **erosion**.

Erosion: the wearing away of the land's surface by mechanical processes such as the flow of water, ice or wind; cf. **weathering**.

Erosional notch: a notch in a cliff resulting directly from erosinal processes.

Erratic: a large **clast** left behind by melting ice and composed of rock not found locally.

Esker: a sinuous ridge of sand and gravel deposited by a **meltwater** stream flowing within a tunnel under a **glacier** or **ice sheet**.

Estuary: an inlet of the sea reaching into a river valley as far as the upper limit of tidal rise. (There are many alternative definitions of the term, most based on the dilution of seawater by freshwater derived from land drainage.)

Eustatic: concerning worldwide changes in sea level (as distinct from changes when land locally sinks into, or rises from, the sea). Eustatic changes of sea level may be caused by **ice ages** or may reflect periods of major **tectonic** activity.

Eyes: a Norfolk term for small inter-tidal ridges.

Fan: a low-lying accumulation of **sediment** with a roughly triangular outline. See **alluvial fan**.

Fault: a fracture within a rock along which there has been displacement due to **tectonic** deformation (e.g. earthquakes).

Feeder bluff: a term for a **cliff** undergoing **erosion**, which serves as a source of **beach sediments**.

Fen: see **bog**.

Fetch: the distance across water over which the wind blows from a particular direction uninterrupted by land.

Fissure: a fracture surface, or crack within a rock along which a clear separation can be seen. Often filled with material, frequently mineral-bearing.

Fjard: a series of shallow basins connected to the sea via shallow and often intertidal sills. Fjards are found in areas of low-lying ground that have been subject to **glacial** roughening. They have a highly irregular outline, no main channel, and lack the high relief and 'U'-shaped cross-section of fjordic inlets.

Fjord: a long, narrow, steep-sided inlet of the sea having a shallow entrance sill. Fjords are **glacially** over-deepened and may have a series of sills and basins, often having deep water at the head. They are commonly surrounded by high ground, and, in cross-section, have a deep 'U'-shape. See also **glacial valley**.

Flagstone: a hard, fine-grained **sandstone** that slits uniformly along **bedding planes** into slabs.

Flandrian: see **Holocene**.

Flint: a variety of chert, a hard, glassy and non-crystalline mineral form of silicon dioxide (quartz), frequently found in **carbonate** sediments, where it has developed from dissolved silica derived from sponges.

Floodplain: the level surface next to a river that is water covered during times of flood.

Flood-tide (or **flow**): incoming or rising tide.

Fluvial: relating to a river or river activity.

Fluvio-glacial: See **glaciofluvial**.

Flying barrier: a looped **bar** or **spit** formed on the landward side of an island that is subsequently reduced below sea level by wave **erosion** before being destroyed.

Fold: a flexure in rocks.

Foliation: the planar arrangement of minerals, or other textural or structural features in rocks.

Foredune: A ridge of irregular **sand dunes**, typically found adjacent to **beaches** on low-lying coasts, and partially covered with vegetation.

Foreshore: the outer, or lower, seaward-sloping zone of a shore or **beach**. Also applied to the area of land between a body of water, and land that is occupied or cultivated.

Fossil: the preserved remains of animals and plants.

Full: a regional term for a **gravel beach ridge** in England and Wales.

Gabion: a wire basket, filled usually with stone,

Glossary

used for structural purposes such as retaining walls, **revetments**, slope protection, and similar applications.

Gault clay: a glutinous marine deposit of Early **Cretaceous** age found in south-eastern England and in France, containing abundant **fossils**.

GCR: *see* **Geological Conservation Review**.

Geo: (from Norse ' gja' meaning 'cleft') a northern Scottish term for a steep-sided narrow inlet of a cliffed coastline, often eroded along a major near-vertical **joint** or **fault**.

Geological Conservation Review (GCR): a review programme that assessed and selected nationally important geological and **geomorphological** sites in Great Britain with a view to their long-term conservation as **SSSIs**.

Geomorphology: the study of the landforms and the processes that formed them.

Geotechnology: the application of scientific methods and engineering techniques to the exploitation and use of natural resources.

Glacial: relating to the activity and presence of **glaciers** or ice.

Glacial advance: a time interval marked by an advance or expansion of a **glacier**.

Glacial age: a subdivision of a glacial epoch.

Glacial cycle: a major climatic oscillation of the order of 100 000 years, during which the ice sheets advanced and subsequently retreated and recurrent at fairly regular times.

Glacial deposit: a deposit or **drift** transported by **glaciers** or icebergs, and deposited directly on land or in the sea.

Glacial drainage: the system of **meltwater** streams flowing from a **glacier** or **ice sheet**.

Glacial drift: *see* **glacial deposit**.

Glacial epoch or period: any period of geological time during which the climate was cold in both the northern and southern hemispheres and **ice sheets** and **glaciers** covered a larger total area than those of the present day.

Glacial erosion: the **erosion**, by, for example grinding, gouging and scratching, by the movement of a **glacier** with rock fragments within it, and also by **meltwater** streams.

Glacial lake: a lake fed primarily by the **meltwater** of a **glacier**, and found beyond the margins of the **glacier**.

Glacial maximum: the time or position of the greatest advance of a **glacier** or **ice sheet**.

Glacial recession: a time marked by a decrease in the size and volume of a **glacier**.

Glacial stage: a major subdivision of a **glacial epoch**, for example one of the major cycles of growth and disappearance of the **Pleistocene ice sheets**.

Glacial valley: a deep, steep-sided U-shaped valley, influenced by a **glacier** that has widened and deepened a pre-existing river valley by **glacial erosion**. Sometimes called a 'glacial trough'.

Glaciation: a term to describe the formation, movement and recession of **glaciers** and **ice sheets**.

Glacier surge: a period of very rapid flow and growth of a **glacier**.

Glacier: a large body of ice formed in part on land by the compaction of snow, which moves slowly by **creep** downslope, or outwards in all directions under the influence of gravity.

Glacio-eustatic: relating to changes in sea level due to seawater being 'locked up' in **ice sheets** and vertical movements of the Earth's crust due to loading and unloading of the crust by the weight of the **ice sheets**. *See* **eustatic**.

Glaciofluvial: relating to the **meltwater** streams which flow from melting **glacier** ice, and to the deposits and landforms created by such streams, for example **outwash plains**.

Glaciogenic: of or relating to **glaciers** and **glaciations**.

Glacio-isostatic: relating to vertical crustal movements associated with the addition (causing crustal depression) and removal (leading to crustal uplift) of **glaciers**. *See* **isostasy**.

Glaciolacustrine: relating to **glacial lakes**.

Glaciomarine sediments: glacially eroded, terrestrially derived **sediments** (**clay**, **silt**, **sand**, and **gravel**) deposited in the marine environment. The sediments may accumulate by ice rafting, as an ice-contact deposit or by **aeolian** transport.

Glaciotectonic: the deformation of rocks or **sediments** caused by **glacial** movement.

Gneiss: a coarse-grained **metamorphic** rock, composed of alternating light and dark bands, formed at very high temperatures and pressures.

Granite: a pale-coloured, coarse-grained **plutonic igneous** rock, commonly occurring as large intrusions but also found in veins.

Granule: a sediment particle of very coarse **sand**, 2–4 mm diameter.

Glossary

Gravel: sediment particles, the term used to describe beach sediments in the **pebble** size range, which may be formed from rock or shell fragments.

Greensand: a greenish **sandstone** consisting mainly of quartz and glauconite, deposited during the **Cretaceous** Period.

Grit: coarse, angular **sediment** particles.

Groyne: a wall or jetty built out from a riverbank or seashore, intended to combat the effects of **longshore drift** and control local **erosion**.

Gully: A vertical space between two rock walls, at least 0.5 m wide and 0.5 m deep.

Hanging valley: a tributary valley whose floor is higher than the floor of the main valley. Usually the result of **glaciation**, but also produced by marine **erosion** leading to cliff retreat.

High Rock Platform: a **shore platform** formed during the **Quaternary** Period and now found in Scotland at an elevation of c. 33 m **OD**, owing to a fall in sea level relative to the land.

Hog's-back: narrow, symmetrical ridge, underlain and controlled by a resistant **bed** dipping at some 40° or more.

Holocene Epoch: a geological time division; the most recent global epoch, which began approximately 10 000 years **BP** and is characterized by sea-level rise in all those places in Britain not affected by isostatic uplift. It is roughly equivalent to the European **Flandrian Stage**.

Honeycomb weathering: a form of chemical **weathering** in which numerous pits occur on a rock exposure, causing the surface to look similar to a large honeycomb. It typically occurs in arid regions, affecting granular rocks such as **sandstones** and **tuffs**.

Hydrography: the scientific study of seas, lakes and rivers (cf. '**hydrology**').

Hydrology: the study of the distribution, conservation, use etc. of the water of the Earth and its atmosphere (cf. '**hydrography**').

Hydration: Chemical combination of water with another substance, e.g. the addition of water to a mineral (such as anhydrite) to produce a hydrous phase (in this case, gypsum). Hydration may be important in the **weathering** of rocks.

Ice Age: a name often applied to the **Pleistocene Epoch** during which large areas were repeatedly covered by **ice sheets** and **glaciers**.

Igneous: a rock that has formed from molten rock (magma), either by volcanic activity or intrusive processes. It consists of interlocking crystals, the size of which depends on the rate of cooling of the magma.

Interfluve: an area of higher ground separating two river valleys.

Interglacial: a period of relatively warm climate between two episodes of **glaciation** where ice is in retreat.

Interstadial: a relatively short period within a major phase of **glaciation** when ice was not advancing and climate conditions were comparatively warm.

Intertidal: the area of the shore between the highest and the lowest tides; cf. **littoral**.

Irish Sea: the area of sea between Great Britain and Ireland, from St George's Channel in the south to North Channel in the north, including all **estuaries** except the Firth of Clyde.

Isostasy: the condition of equilibrium, comparable to buoyancy, of the Earth's crust floating in the underlying layer (aesthenosphere) of the Earth. Crustal loading, for example by ice, water or volcanic flows, leads to isostatic depression, and the crust sinks deeper into the aesthenosphere. The removal of weight leads to isostatic uplift or rebound, and the crust rises. The depression and rebound occur over long timescales, of the order of thousands of years

Joint: a fracture in a rock that exhibits no displacement across it (unlike a **fault**). May be caused by shrinkage of **igneous** rocks as they cool in the solid state, or, in **sediments**, by regional extension or compression of **sediment** caused by earth movements.

Jurassic: a period of geological time, from 195 to 140 million years ago.

Kame: a low mound of stratified **sand** and **gravel** originally deposited on top of, or at the margin of, a **glacier** or **ice sheet** by **meltwaters**, and remaining as a topographical feature after the ice has melted.

Karst: a distinctive terrain created by **erosion** of a soluble rock, where the topography and landforms are a consequence of efficient

Glossary

underground drainage; characterized by **caves**, **sinkholes** and **dry valleys** and mainly developed on **limestone**.

Kettlehole: a depression in glacial or **glaciofluvial sediments**, resulting from the melting of a mass of **glacier** ice that was buried in **sediment**.

Lacustrine: relating to, formed within in, or produced by, lakes.

Lagoon: a shallow body of coastal saltwater (from brackish to hypersaline) partially separated from an adjacent sea by a barrier of sand or other sediment, or less frequently by rocks.

Lamina (pl. **laminae**): a thin layer within a **sedimentary** rock, typically less than 1 cm thick.

Land-claim: the process of creating usable land from flooded or intertidal land, usually involving impoundment and drainage.

Landform: a natural feature of the surface of the land.

Landslide: a large, rapid mass-movement of material down relatively steep slopes, mainly under the force of gravity, but also influenced by saturation with water and alternate freezing and thawing; cf. **creep**, **rockfall**.

Last Glacial Maximum: the time of the last great glacier advance, when **ice sheets** and **glaciers** reached their maximum thickness and extent. Dated to between 22 000 and 18 000 years **BP**.

Late-glacial: relating to the time of the end of the last glaciation of the **Pleistocene Epoch**, part of the **Devensian Stage**.

Lateral moraine: a low ridge-like **moraine**, built along on the side margin of a **glacier**.

Lava: molten rock extruded onto the Earth's surface, or the resultant solid rock.

Leach: to dissolve or remove from a soil or rock.

Levee: a broad ridge alongside a river or stream, deposited by floodwaters when they overtop the channel banks.

Lias: in Britain, a stratigraphical unit of rocks formed during the early part of the **Jurassic Period**.

Limestone: sedimentary rock composed largely of calcium carbonate ($CaCO_3$) in the form of the mineral **calcite**, often derived from the shells of organisms, and soluble in weak acids including rain and soil water; strong, well-lithified limestones may stand in high vertical **cliffs** and can span large **cave** passages formed within them by dissolutional enlargement of fractures.

Links: relatively flat land along a seashore that is typically sandy and turf covered.

Lithology: descriptive of the constitution of a **sediment** or a rock, including texture, composition and colour, and size, shape and mineral composition of constituent crystals or **clasts** in the rock.

Littoral: the area of the seashore that is occupied by marine organisms that are adapted to, or need, alternating exposure to air and wetting by submersion, splash or spray; cf. **intertidal**.

Loch Lomond Stadial: a relatively cold period during the late glacial between 11 000 and 10 000 years **BP**.

Lodgment till: a **glacial** deposit laid down underneath an **ice sheet** or valley **glacier**. It is usually **clay**-rich and contains **boulders**.

Loess: a fine-grained **sediment** of windblown **silt** and **clay**, largely derived from cold **periglacial** deserts.

Longshore drift: movement of **sand** and **shingle** along the shore.

Low: inter-ridge hollow in southern England. See **swale**.

Machair: a type of coastal **dune** pasture on lime-rich sand, which has developed on a level coastal plain in wet, windy Atlantic seaboard conditions. Machair is globally restricted to north and west Scotland and western Ireland.

Macro-tidal: an **estuary** or other inlet with an average **spring tidal range** greater than 4 m.

Managed re-alignment or '**retreat**': allowing the coastline to recede to a new line of defence (natural or man-made), usually accompanied by measures to encourage the development of environmentally beneficial mudflat or **saltmarsh** areas seaward of the new defence line.

Marine regression: the withdrawal of the sea from large areas of land, due to a fall in sea level relative to the land.

Marine transgression: the encroachment of the sea across large areas of land, due to a rise in sea level relative to the land.

Mass movement: the downslope movement of rock fragments and soil under the influence of gravity. The material concerned is not

Glossary

incorporated into water or ice, and moves of its own accord, but slides are often triggered by increase in water pressure on rocks and soil.

Mass wasting: the dislodging and transport of soil and sediment due to gravity. Processes include **solifluction** and rock-falls.

Matrix: the **sediment**, usually very fine-grained, which infills the spaces between larger grains.

Meltwater: water produced by the melting of snow and ice.

Metamorphism (adj. **metamorphic**): the process of alteration of **igneous** and **sedimentary** rocks by increases in pressure and/or temperature (but without melting) within the Earth's crust.

Moraine: a ridge of unsorted, unstratified **glacial till** deposited on top of or at the margins of a **glacier** or **ice sheet**.

Morphology: the form and structure of the landscape.

Mud: fine-grained **sediment** particles of **silt** and/or **clay** with a diameter of <0.0625 mm.

Mudflat: an expanse of **mud** or muddy **sediment** in the **intertidal** zone.

Mudflow: The **mass movement** of fine-grained material held in suspension by water.

Mudstone: a very fine-grained rock.

National Grid: a metric grid used in maps, based on the Tranverse Mercator Projection and developed by the Ordnance Survey for use in Great Britain.

National Nature Reserve (NNR): a site of national or international importance for nature conservation, declared under the National Parks and Access to the Countryside Act, 1949 and the Wildlife and Countryside Act, 1981.

Naze: a local term used in Essex for a headland.

Neap tide: the astronomical tide of minimum **tidal range**, occurring at the time of the first and third quarters of the moon.

Negative surge: the depression of sea level below Lowest Astronomical Tide by meteorological conditions; cf. **surge**.

Neotectonic: concerning tectonic (crustal) movements operating in the recent geological past (especially of Quaternary age) and those still occurring today.

Ness: a large low-lying **foreland** or promontory; also used in south-east England for a headland.

Nothe: a local term used in Dorset for a headland.

North Sea: the sea to the east of Great Britain, east of 4° W to the north of Scotland, north of 51° N at the Strait of Dover, and south of 61° N.

Oceanography: the branch of science dealing with the physical, chemical, geological and biological features of the oceans and ocean basins.

OD: *see* **Ordnance Datum**.

Ord: a term used in Yorkshire to describe areas of low, denuded beach that migrate along the direction of **longshore drift**, and are separated by fuller areas of beach.

Ordnance Datum: the fixed reference point for heights and contours shown on Ordnance Survey maps, which is based on mean sea level (MSL) as recorded at Newlyn (Cornwall) over a seven-year period from 1915 to 1921. This is not the same as chart datum, which is the set reference point on marine charts for water depth in relation to tides. On metric charts for which the UK Hydrographic Office is the charting authority, chart datum is a level as close as possible to Lowest Astronomical Tide (LAT), the lowest predictable tide under average meteorological conditions.

Overflow channel: an eroded trough cut by water spilling over from another channel or a standing water body like a lake.

Overwash fan: *see* **washover fan**.

Oxidation: the chemical process of removing electrons from an element or compound (e.g. the oxidation of iron compounds from ferrous to ferric); frequently together with the removal of hydrogen ions.

Oxygen isotope analysis: a method for estimating past ocean temperatures. The ratio of the stable oxygen isotopes, ^{18}O and ^{16}O, is temperature dependent in water, ^{18}O increasing as temperature falls. Oxygen incorporated in the calcium-carbonate shells of marine organisms will reflect the prevailing $^{18}O:^{16}O$ ratio.

Oyces: a Shetland term used to describe tidal lagoons enclosed by spits or barriers.

Palaeosol: an ancient or 'fossilized' soil.

Palynology: the study of microscopic plant fos-

Glossary

sils, such as spores, pollen, algal cysts and acritarchs and their distribution, which has proved to be of considerable use in correlating sedimentary deposits.

Parabolic dune: a **sand dune** with a long, scoop-shaped form, convex in the downwind direction so that its horns point upwind, fixed by vegetation while the convex surface migrates downwind. Typically found where strong onshore winds supply abundant sand.

Pea-shingle: a clean **gravel**, the individual particles of which are similar in size to peas.

Peat: an unconsolidated deposit of semi-carbonized plant remains formed in a water-saturated environment, such as a **bog** or **fen**.

Pebble: a rock particle with a diameter of between 4 and 64 mm, typically implying a degree of rounding.

Pedestal: a thin neck or column of rock topped by a wider mass, produced by undercutting due to wind **abrasion** or differential **weathering**.

Periglacial: zone or environment peripheral to **glaciers**, so that it is very cold but is not covered by ice sheets, characterized by the action of intense frost, often combined with largely permanently frozen ground known as 'permafrost'.

Piping: natural subsurface channels of different sizes, which can form an interconnecting network, often in soils with significant amounts of swelling **clays**.

Planation surface: a term used in Britain to describe a fairly level plain resulting from prolonged **erosion** by rivers, slope processes, marine **erosion**, or other types of erosional activity.

Planation: the process of **erosion** by which the surface undergoing erosion becomes flat or level.

Platform: *see* **shore platform**.

Pleistocene: the first epoch of the **Quaternary** Period, from about 1 800 000 to 10 000 years ago, composed of alternations of great cold with stages of relative warmth, and sometimes referred to as the **Ice Age**.

Plunging cliff: a **cliff** that descends directly in to deep water.

Pocket beach: a beach contained within bounding headlands; a cove.

Postglacial: referring to the time interval since the total disappearance of **glaciers** at middle latitudes.

Preglacial: referring to the time prior to a **glacial** period. Also said of material underlying glacial deposits.

Progradation: the seaward-migration of a shoreline.

Quartzite: a **metamorphic** rock formed from more or less pure quartz **sandstones**.

Quaternary Period: a geological time division ranging from about 1.8 million years **BP** to the present day, it is the latest period of geological time, and the second period of the **Cainozoic Era**. It is divided into two epochs, the **Pleistocene** and the **Holocene**. *See* 'Ice Age'.

Radiocarbon dating: a method of ascertaining the age of a sample by measuring amounts of carbon-14 (^{14}C) within organic material. The method is based on the assumption that upon removal from the Earth's carbon cycle (for example when an organism dies), carbon-14 content is 'frozen', and then the proportion reduces over time through radioactive decay; by measuring the relative abundance of stable and radioactive carbon in a sample and knowing the decay rate of carbon-14 the elasped time can be calculated.

Raised beach: *see* **emerged beach**.

Ramsar site: a site designated under the international Convention on Wetlands of International Importance especially as Waterfowl Habitat ('the Ramsar Convention').

Recurve: a landward-curving sand or gravel ridge produced by the successive extension of a **spit**.

Reef: (1) a ridge of rock or coarse material, the top of which lies close to the surface of the sea, and may be exposed at low tide; (2) an elevated structure on the seabed built by **calcareous** or other concretion-forming organisms, or by chemical precipitation; (3) an artificial structure deliberately constructed or placed on the seabed with the intention of influencing the local environment, for example to enhance fisheries or to absorb wave energy.

Refraction: the change in the approach angle of a **wave** as it moves towards the shore. As water becomes shallow, waves slow down. Refraction causes waves to converge on headlands and diverge in bays, concentrating

Glossary

wave energy on headlands rather than on beaches.

Regolith: unconsolidated, weathered, broken rock debris, mineral grains, and superficial deposits that overlie unaltered bedrock.

Regression: *see* **marine regression**

Relict: descriptive of a geological or geomorphological feature surviving in its primitive form, i.e. no longer actively forming or evolving.

Revetment: a facing of rocks, sandbags, etc., intended to protect a coastline from **erosion**.

Ria: a drowned river valley in an area of high relief; most have resulted from the post-glacial rise in sea level relative to the land.

Ridge and runnel: multiple broad intertidal **bars** and **swales** running parallel to the coastline, the ridge crests typically being spaced at intervals of about 100 m, with an amplitude of about 1 m.

Rill dissection: the **erosion** of soil by water running through little streamlets, or head-cuts, forming rills, or small **gullies**.

Rockfall: a type of mass-movement where coarse material moves rapidly downwards from one part of the slope to another.

SAC: *see* **Special Area of Conservation**.

Saltmarsh: an area of **alluvial** or **peat** deposits, colonized by herbaceous and small shrubby terrestrial vascular plants, almost permanently wet and frequently inundated with saline waters.

Saltpan: a small, shallow, undrained, natural depression containing a salt deposit produced by the accumulation and subsequent evaporation of water; or a shallow pool of somewhat salty water occupying such a depression. They are flooded at high tide, and remain bare of plants because evaporation makes the trapped water hypersaline.

Sand: particles in the size categories 0.062 mm diameter (very fine-grained sand) to 4 mm diameter (granules).

Sand dune: *see* **dune**.

Sandflat: an expanse of **sand** or sandy **sediment** in the **intertidal** zone.

Sandstone: a **sedimentary** rock made of **lithified sand**.

Sarn: a Welsh word for 'causeway', used in west Wales for a roughly linear boulder or cobble reef derived from glacial moraine, lying at shallow depth (maximum depth about 10 m below chart datum), and completely covered at low tide.

Scar: Lag deposits of boulder and cobble dominated areas in the **intertidal** zone produced by erosion of glacial till.

Scour: the effect of abrasion, usually by water- or wind-borne sand or gravel, on a surface.

Scree: *see* **talus**.

Seabed: the sea floor.

Sea loch: a marine inlet in Scotland that has fjordic or fjardic features, entered by the tide (on each cycle), and with a salinity generally greater than 30 ‰. Brackish conditions may be periodically established, particularly in the surface layers.

Sediment budget: the inputs of sediments from various sources minus the outputs to various sinks. Can be positive (implies accretion) or negative (undergoing erosion)

Sediment cell: a compartment of coastline, divided from neighbouring sections of coast in terms of **longshore drift**, **current** flow, and wave convergence and divergence. Also known as 'coastal cell' or 'coastal processes cell'.

Sediment: loose material derived from the **weathering** and **erosion** of pre-existing rocks, or from biological activity (e.g. shells and organic matter) or from chemical precipitation (e.g. **evaporites**).

Sedimentation: the process of settling of suspended solid particles from water by gravity.

Sedimentary rocks: rocks formed by the **lithification** (cementation) of sediment. Sedimentary rocks may be composed of mineral or rock particles (**clasts**) to form sandstones, claystones or sediments of biological origin to form limestone and peat, or of chemical precipitation to form **evaporites**.

Shale: a fine-grained **sedimentary** rock, composed of **clay** particles, that splits easily into thin layers.

Shell-sand: a **sediment** comprising predominantly shell fragments of **sand**-size.

Shingle: beach **pebbles**, normally well-rounded as a result of abrasion.

Shoal: a mound or other structure raised above the seabed in shallow water that is composed of, or covered by, unconsolidated material and may be exposed at low water.

Shore platform / intertidal platform: a surface forming a level rock platform in the **intertidal** zone. Sometimes referred to as a 'wave-cut platform', but others can be lev-

Glossary

elled by weathering or may be structurally controlled.

Shoreline management plan: a plan that focused on the management of coastal erosion and defence, usually within a coastal sediment cell.

Sidereal year: the time required for one complete revolution of the Earth about the Sun (c. 365.25 days).

Significant wave height: the mean height of the highest one third of all waves.

Silt: fine-grained **sediment** particles ranging from 0.004–0.0625 mm in size.

Site of Special Scientific Interest (SSSI): an area of land or water notified by the statutory nature conservation agencies under the Wildlife and Countryside Act 1981 (as amended) as being of special nature and/or geological conservation importance; the principal designation under which **GCR** sites are protected.

Skerry: (pl: **skerries**): low-lying rocky island or reef, often without terrestrial vegetation, and frequently swept by the sea.

Slope-over-wall: a cliff form characterized by a convex upper profile and a vertical or subvertical lower profile. Formed by periglacial processes followed by sea-level rise, which trims the lower cliff.

Slump: contorted structures produced by the **mass-movement** of unconsolidated sediment.

SPA: *see* **Special Protection Area**.

Special Area of Conservation (SAC): a site of European Community importance designated by an EU Member State under the 1992 Habitats Directive 92/43/EEC for the conservation of natural habitats or species listed in Annexes I and II of the Directive. These SACs, together with **Special Protection Areas** (SPAs) classified under the 1979 Birds Directive 79/409/EEC, collectively form the Natura 2000 Network. Most SACs above low-water mark are underpinned by **Sites of Special Scientific Interest**.

Special Protection Area (SPA): a site of environmental importance recognized at a European level, classified by an EU Member State under the 1979 Birds Directive 79/409/EEC on the conservation of wild birds. These SPAs, together with **Special Areas of Conservation** (SACs), designated under the 1992 Habitats Directive 92/43/EEC collectively form the Natura 2000 Network.

In Britain most SPAs above low-water mark are underpinned by **Sites of Special Scientific Interest**.

Spit: a low, elongated accumulation of **sediment** such as **sand** or **shingle**, projecting from the shore into a water body; cf. **bar**, **tombolo**.

Spring tide: the astronomical tide of maximum **tidal range**, occurring at or just after new moon and full moon. The most marked spring tides (equinoctal springs) occur at the spring and autumn equinoxes.

SSSI: *see* **Site of Special Scientific Interest**.

Stack: (from Gaelic 'stac') a pillar of rock formed by wave action, often when a sea **arch** collapses.

Strandplain: a prograded shoreline built seawards by waves and **tidal streams** or **currents**, and continuous for some distance along the coast.

Strata: (singular: **stratum**): layers within **sedimentary** rocks. The term is often used instead of 'beds'.

Strait: any deep (>5 m depth) tidal channel between two bodies of open coastal water. Strictly, a strait is the stretch of water between an island and its mainland (or adjacent islands).

Strath: a Scottish term for a broad level-floored river valley.

Stratigraphy: the study of rock successions and their distribution in space and time preserved from the geological past, in order to reveal the history of the succession of events and life of the past.

Stump: the rock mass that remains after a **stack** has been worn down.

Subaerial weathering: the normal processes of **weathering** that loosen rock fragments and transport debris downslope; in the intertidal zone, it may affect rocks down to the level of permanent saturation.

Subcell: a subdivision of a **sediment cell**.

Submerged coast: a former coastline that is preserved underwater, following submergence by rising sea levels relative to the land.

Surge: the elevation of sea level above Highest Astronomical Tide by low pressure meteorological conditions, also known as a 'storm surge'; cf. **negative surge**.

Swale: a long, narrow depression, approximately parallel to the shoreline, between two ridges on a **beach**, and draining through transverse channels as the tide falls. A com-

Glossary

ponent of **ridge and runnel** topography. *See* **low**.

Swell: sea waves that have left the area where they were generated by the wind, or that have remained after the generating wind has disappeared.

Talus: an accumulation of rock litter at the foot of a slope, generally with a wide size-range (up to several metres) and ungraded; commonly used to denote debris shed from the high part of a reef slope and transported basinward by gravity ('reef talus', 'talus apron'). Also called scree.

Tectonism (adjective **tectonic**): deformation of the Earth's **crust** and the consequent structural effects (e.g. **faulting, folding** etc.), often associated with crustal plate movements and mountain-building.

Tertiary: a period of geological time ranging from about 65 to 2 million years ago, preceding the **Quaternary** Period.

Tidal prism: the volume of water that passes in or out of an inlet during a tidal cycle.

Tidal range: the difference in water height between Extreme High Water of Spring Tides and Extreme Low Water of **Spring Tides**.

Tidal stream: the alternating horizontal movement of water associated with the rise and fall of the tide (cf. **current**).

Tied island: an island connected to the mainland, or to another island, by a **tombolo**.

Till: unsorted, non-stratified **sediment** deposited directly by **glacial** ice without the intervention of water; commonly known as 'boulder clay' or 'glacial till'.

Tombolo: a **spit** that links an island to the mainland or to another island, formed by deposition when waves are refracted round the island.

Tràigh: (pronounced 'try') a Gaelic term for a **beach**, shore, or strand.

Training: the building of walls or embankments within an **estuary** to direct **current** flow and stabilize a shifting channel.

Transgression: *see* **marine transgression**

Triassic: a period of geological time ranging from about 230 to 195 million years ago.

Tsunami: a wave train, or series of waves, generated in a body of water by an impulsive disturbance that vertically displaces the water column, including earthquakes, **landslides**, volcanic eruptions, explosions, or the impact of cosmic bodies such as meteorites.

Unconformity: the surface that separates two **sedimentary** sequences of different ages; it represents a gap in the geological record when there was **erosion**, and/or **tectonism** and/or no deposition. There is often an angular discordance between the two sequences.

Undercliff: a subordinate cliff comprising material fallen from the cliff above; the lower part of a cliff whose upper part has undergone landsliding.

Voe: a regional term used in Shetland for an inlet, usually glacially eroded.

Wave: a ridge of water between two depressions. As waves approach a shore, they curl into an arc and break. The energy of surface waves is responsible for most coastal **erosion**.

Wave exposure: the degree of wave action on an open shore, governed by the distance of open water over which the wind may blow to generate waves (the **fetch**) and the strength and incidence of the winds.

Washover fan: a fan-shaped body of **sediment** deposited by marine waters flowing landwards through or across a coastal barrier such as a **bar** or island. Such features are formed especially during storms when the barriers are likely to be overtopped.

Waterfall: a steep fall of water along the course of a river or stream. River **erosion** may produce a waterfall where the river crosses a band of hard rock; the waterfall will continually retreat upstream through **erosion**. Coastal **erosion** may produce a waterfall where cliff retreat, proceeding faster than the downcutting river erosion, forms a **hanging valley**.

Water layer weathering: **weathering** processes related to the wetting and drying of sea **cliffs** and **shore platforms** by waves, spray and tides.

Water table: the level within a rock mass below which all voids are filled with groundwater; above it the **vadose zone** is freely draining, and below it the **phreatic zone** is totally and permanently saturated.

Glossary

Watershed: the boundary delimiting a river drainage basin as the basic hydrological unit.

Wave quarrying: the prising or pulling away of pieces of rock by the shock of impact of breaking waves.

Weathering: the process by which rocks are broken down in place by physical, chemical and biological processes; the term does not infer any transportation of the weathered rock material (cf. **erosion**).

Wetland: an area of low-lying land where saturation with water is the dominant factor in determining the nature of soil development and the types of fauna and flora living in the soil and on its surface. Examples include **bogs**, **fens**, marshes and swamps.

World Heritage Site: a site designated under the international Convention Concerning the Protection of the World Cultural and Natural Heritage (the 'World Heritage Convention').

Zawn: Cornish word for a narrow gully or cleft in the rocks leading down to the sea.

Index

Note: Page numbers in **bold** and *italic* type refer to **tables** and *figures* respectively

abrasion 39, 93
  cliffs 39–40
  shore platforms 47, 83, 91, 108
  Villians of Hamnavoe 69, 71
  see also erosion
Aesha Head, Papa Stour 74
Ainsdale, Lancashire *330*, **331**, **332**, **333**, 359–64
  Downholland Moss 362–3
  dunes 221, 223, 360, 362, 363, 364
  foreshore
    bedforms 359
    ridge-and-runnel zone 360, *361*, 362, 364
  Formby Point 363
  sediment transport *362*, 364
  slacks 360, 363
Alde, Ore and Butley estuaries **528**
Anglesey (Ynys Môn) 595
  saltmarsh **528**
  see also Newborough Warren; Tywyn Aberffraw
anthropogenic influences 20, 23, 28, 527
  beach erosion 226–7
  saltmarshes 527
  soft-rock cliff retreat 137–8
  see also beach nourishment; coastal defences/ protection/engineering; coastal management; GCR sites; machair; Shoreline Management Plans

arches 43, *43*, 57, 111
  Aesha Head, Papa Stour 74
  Bullers of Buchan 104
  Castles of Yesnaby and Qui Ayre arch, Yesnaby 84, *85*
  Foula, dissecting Gaada Stack 79
  Green Bridge of Wales 119, *120*, 121
  St Kilda 62
  see also Ladram Bay; Old Man of Hoy; stacks
Ardivachar to Stoneybridge, South Uist, Western Isles **332**, *473*, **477**, 487–92
  beach-dune–machair system 212, 487, 489, 491
  Gualan Island 487, *488*, 489, *490*, 491, 492
  Howmore River 487
  machair 487, 489, 491–2
  West Gerinish 487, *490*, 491
attrition 181, 257
  see also abrasion; erosion; machair
Ayres of Swinister, Shetland 184, 213, 220, 281–5

Ballard Down, Dorset *132*, **133**, 176–81
  beaches 180, 181
  cliffs 177, *178*, 179, 179–81
  Handfast Point 176, 177, 179, *179*, 180
  Old Harry Rocks 176, 177, 181

  Studland Bay 176
  Swanage Bay 176, 177
  see also Dorset Coast
Balnakeil, Sutherland 208, 225, **332**, **477**, 510–14
  An Fharaid Peninsula 510, 512
  blowthrough corridors 510, *511*, 513, 514
  dunes 510, 514
  skerry of A'Chléit 512
Balta Island, Shetland *330*, **331**, **332**, 384–6, *420*
  dune (aeolian) calcarenite 225–6, *385*
  dune grassland system 386
  South Links 385, *384*, 385–6
barrier beaches/islands 10, **203**, 220, 231–2, *232*, 584, 609–10
  The Bar, Culbin *571*, *572*, 573, 575
  Budleigh Salterton 252
  Cefn Sidan to Pembrey 584
  Chesil Beach 219–20, 256
  Dungeness and Rye Harbour 311–12
  Morrich More 576–9
  Porlock, Somerset 266, 267–8, *268*, 269–71, *269*
  Sandwood Bay 220, *371*, 372, *372*
  Slapton Sands and Beesands 244
  see also North Norfolk Coast; Walney Island

Index

Barry Links, Angus 222, *330*, **331**, **333**, 400–7, *410*, **418**
  Buddon Ness 235, **332**, 402, 403
  Gaa Sands *401*, 402, *403*, *410*
  parabolic dunes 223, 400–1, 404, 405, 406
bars *232*, 241
  crescentic **203**
  and troughs 214–5
    see also ords, ridge-and-runnel beaches; Tentsmuir
bay-head beaches 417–18
  see also Ayres of Swinister; Dunnet Bay; Luce Sands; Tywyn Aberffraw
beach morphology **203**, *204*, 205
  see figures 5.1 and 5.2 for beach terminology synonyms
beach nourishment 20, 200, 227
  Dungeness 312
  Hurst Castle Spit 272–3, 277
  Newborough Warren 595–6
  Orfordness 304
  Ringstead Bay 628
  Rye Harbour 314
  Spey Bay 227, 294
  Westward Ho! cobble beach 238
  see also anthropogenic influences
beach ridges 217
  gravel and shingle 231–2, 235
  on saltmarshes 217
  see also ords; ridge-and-runnel beaches
beaches 9, 13, 36, 72, 203, *205* 212, 332
  Ainsdale 359–64
  Ayres of Swinister 281–5
  Balta Island 384–6
  Barry Links 400–7
  Benacre Ness 301–4
  Braunton Burrows 348–54
  Budleigh Salterton 251–4
  Central Sanday 464–9
  chalk pebble, Sewerby 185
  Chesil Beach 254–66

Dawlish Warren
drift-aligned **203**, 213
Dungeness 315–25
Dunnet Bay 380–4
East Head (Chichester Harbour) 426–30
Forvie 393–400
Gibraltar Point 439–43
Hallsands 248–50
Hurst Castle Spit 271–7
intertidal zone 22–3 *205*, 267
  Culbin 567, 571, 573
  Dunbar 108
Isles of Scilly 461–4
lateral grading 215–16
Loe Bar 241–4
Luce Sands 364–70
Marsden Bay 337–40
Morfa Dyffryn 452–7
Morfa Harlech 449–52
Orfordness and Shingle Street 304–10
Oxwich Bay 354–6
Pagham Harbour 278–81
Porlock 266–71
prograding 216–17, 334, 582
Pwll-ddu 422–4
reflective or dissipative 214, 231
Rye Harbour 313–4
St Ninian's Tombolo 457–61
Sandwood Bay 370–5
Slapton Sands 246–7
South Haven Peninsula 340–5
Spey Bay 290–7
Spurn Head 430–4
Strathbeg 387–93
swash-aligned **203**, 212, 232
Tentsmuir 407–13
Torrisdale Bay and Invernaver 375–80
Tywyn Aberffraw 356–9
Upton and Gwithian Towans 345–8
Walney Island 443–5
West Coast of Jura 297–301
Westward Ho! Cobble Beach 238–41
Whiteness Head 285–90
Winterton Ness 446–9
Ynyslas 424–6
zeta-curved 191, **203**, 212,

251, 252, 596
  see also Chapters 5–8; boulder beaches; flint beaches; fringing beaches; machair; sandy beaches; shingle beaches; spits; tombolos
Beachy Head to Seaford, East Sussex *132*, **133**, 170–6
  Birling Gap **136**, 173–4, 175
  cliff retreat 171, 175
  cliff–beach–platform 171, 176
  Cuckmere Haven 172, 174
  Seaford Head and Hope Gap 172
  Seven Sisters 44, 171, 174, 176
  shore platforms 172, 173, 174
  subaerial weathering 175–6
Benacre Ness, Suffolk *132*, **134**, *233*, **235**, 301–4, *304*, 447
  accretion and erosion 302–4, *302*
  Benacre Broad beach 302
  Covehithe cliffs, 302
  northward migration 219, 301–2, 302, *303*
berms *204*, 205
bevelled cliffs 48, 180
bio-erosion 40, 93, 167
Blackpool, ridge-and-runnel beach 12
Blakeney Point *see* North Norfolk Coast
blowholes 42, 57, 87
  Bullers of Buchan 103, 106, *106*
  Flamborough Head 183
  South Pembrokeshire Cliffs 117
  Villians of Hamnavoe 68
Blue Anchor–Watchet–Lilstock, Somerset *132*, **133**, 145–8
  cliffs 145, 148
  intertidal platforms 145
    development of cobble fields 147
    micro-cuestas *146*, 147–8
    'washboard relief' 148
  St Audrie's Bay 145
  tidal regime 147

Index

Boreray, St Kilda 62, 65
boulder beaches 9, 72, 75, 87, 107, 111, 340
   Bullers of Buchan 103, 104, *105*, 106
   Foula 77, *78*, 79
   Portland Limestone 634, *635*, 638, 639
   South Pembroke Cliffs 119
   Villians of Hamnavoe 70
Brancaster Bay **612**
Braunton Burrows, Devon 214, 226, *330*, **331**, **332**, **333**, 348–54, **418**, *420*
   Airy Point *349*
   dunes 224, 348, 349, 350, 351, 352, 353, 354
     blowthroughs 350
     military use 351–2, 352
     rabbit population 352
   sand and gravel extraction, Torridge estuary 353
   sandy saltmarsh 519
   sediment budget 352
   shore platform 351, *351*, 353
   spits 219
   submerged forest 353
Bridgwater Bay, saltmarsh creeks 524–5
Bridlington Bay, ridge-and-runnel beach 12
   see also Flamborough Head; Holderness
Buddon Sand **332**, 405, *405*
   see also Barry Links
Budleigh Salterton *132*, **133**, *233*, **234**, 251–4
   Otter estuary 253
   pebbles as a 'tracer' 253
   pebble source 251, 252
   sediment grading 253
   shore platforms 251
   wave refraction 251, 253
   West Down Beacon landsliding 252, 253
Bullers of Buchan 57, *58*, **59**, 103–7
   blowholes, The Pot 103, 106, *106*
   boulder beaches, 103, 104, *105*, 106
   caves, North and South Seals 104

conical stacks 104
   Dunbuy Island 106
   geos 103–4, *105*, *106*
   glacial legacy 104, 106
   shore platforms 104, 106
Caernafon Bay Coast 566, 593–5, 599
   Menai Strait 594–5, *594*
   Morfa Dinlle and Newborough–Abermenai dunes/gravel ridges 603–4
   Newborough Warren and Morfa Dinlle 593–4
   see also Morfa Dinlle; Newborough Warren
calcium carbonate **332**, 335
   beaches/dunes 224
Cardigan Bay 452
   Aberystwyth to Mochras 453
Carmarthen Bay *58*, **59**, *330*, **333**, **418**, *420*, **565**, 566, **566**, 583–93
   Broughton Bay 589
   Burry Holms 590
   Burry Inlet *517*, 522, 587
     saltmarshes 587, 592, 593
   caves 586, 589
   Cefn Sidan to Pembrey, barrier beach 584, 587
   cliffs 589, 590
   dunes
     Broughton Burrows 589
     Pembrey Burrows, 587
     Pendine Sands to Laugharne Burrows 586
     Tywyn Burrows 587
     Whiteford Burrows 587
   emerged beaches 590
     Gilman Point 591
     Twic Point 589
   Gwendraeth estuary 587
   Pendine Sands and Laugharne Burrows beach **332**
     saltmarsh 586
     spit 584
   Ragwen Point 585
   Rhossili Bay 12, 334, *589*, 509, 590, 592
   saltmarshes 592–3
     Berthlwyd, Llanrhidian and Landimore 587, *588*
     Burry Inlet 587, *589*, 592, 593

Loughor estuary 590
   Whiteford Point to Loughor 587
   sandy beaches
     Pembrey Beach **332**, 587, *587*, 592
     Pendine beach 591, 592
     Rhossili Bay 590
     Tywyn Point to Burry Port 587, *587*
   shore platforms
     Mewslade Bay 590
     Worms Head 590, 591
   Taf estuary 585–6
   Tywyn Burrows 587
   Whiteford Burrows 587
   Whiteford spit 584, 592
Caves see marine/sea caves
Central Sanday, Orkney 213, 235–6, **418**, *420*, 464–9
   Bay of Newark 464, *465*
   beach–dune–machair system 468
   Cata Sand 464–5, *465*, *466*, 467, 468
   gravel ridges 466, 468
   Little Sea 464, *465*, *466*, *467*, 468
   Plain of Fidge, machair plain 464, 465–6, 468
   proto-barriers 220
   Quoy Ayre 466
Chalk cliffs 629, 641
   and flint beaches 3, 6–7
   Ballard Down 176–81
   Beachy Head to Seaford Head 170–6
   Flamborough Head 181–7
   Joss Bay 187–91
   Kingsdown to Dover 165–70
   sand supply 331
   shore platforms 135–6
   south-west Isle of Wight 160, *160*, 163, *163*, 164
   White Nothe to Bat's Head 629
   Worbarrow Bay 633
cheniers 217, *232*, 525
   Dengie Marsh 535–6, *535*
   Keyhaven Marshes 273, 538, *538*
   St Osyth Marsh 531, 532–3
Chert see flint

Index

Chesil Beach, Dorset 203, 220, 227, 231, *233*, **234**, 254–66
  beach material
    longshore size grading 215, *216*, 254, 258–60, 266
    sources 207, 254, 258, 260–1, 261
    vertical sorting 259
  borehole information 256–7, *257*, 259
  The Fleet 25–8, 220, 256, 526–7, **528**, 262, 263, 265
  initiation and development 208, 263, 265, 266
  longshore sediment transport 259–61, 262, 265
Clew Bay, linked islands 463
cliff failure 34, 35–6, 37, 84, 103, 134
  rotational landslides 135, 136, 162
  toppling
    in Chalk 136, 174, 614
    Gore Cliff 161
    Nash Point 150–1
    Torridonian sandstone 36, 370
  Isle of Wight, *162*
  Lyme Regis to Golden Cap 151–8
  Nash Point 150–1
  see also mass movements
cliff profiles
  Ballard Down 180
  Beachy Head 172–3
  benched, St Abb's Head 111, 113
  bevelled 48, *50*, 115
    Bullers of Buchan 103
  convex, Sandwood Bay 370
  development of *49*
  geological control 5
  sandstone 90
  Joss Bay 189
  multi-storied, St Kilda 66
  slope-over-wall 5, 8, *626*, 640
    Loe Bar 241, *243*
    Oxwich Bay 355
    Pwll-ddu 422
    St Aldhelm's Head *637*
    Solfach 126
    Tintagel 115

  stepped, Villians of Hamnavoe 68, 71
  vertical 589
    Dover to Kingsdown 165–70
    Hoy, Old Red Sandstone 84
    Joss Bay 187, 188
    Nash Point 149
    Orkney west coast 83
    Porth Neigwl 193
    Sandwood Bay 370
    South Pembroke Cliffs 119, *119*
  see also bevelled cliffs; chapters 2–4
cliff retreat 125, 131
  Birling Gap 174
  Covehithe cliffs 302
  Dinas Dinlle 600, 601
  Dorset coast 627, **627**, 641
  Foula 76–80
  Furzy Cliff, Dorset 627, *627*
  Hallsands 250
  Isle of Wight 158, 161, 164
  Joss Bay 187, 188–9
  Kingsdown to Dover 165, 166–7, 169
  in the Lias 145, 147
  Nash Point 150–1
  Norfolk Coast 38, *614*
  North Yorkshire **143**
  Porth Neigwl 192–3, 194
  processes of 12, 14, 36, *37*
  Robin Hood's Bay 141
  St Kilda 60–8
  see also cliffs; landslides; mass movements
cliffs 3, 6–7, 40, 55–199
  Ballard Down 176
  Beachy Head to Seaford Head 170
  biological SSSIs and SACs 51–3
  Blue Anchor–Watchet–Lilstock 145
  Bullers of Buchan 103
  characteristic medium-scale features 41–3
  classifications of 40, **41**, 41, 42
  Cotton's two-cycle model 116

  Dunbar, East Lothian 107
  Duncansby to Skirza Head 86
  Flamborough Head 181
  Foula, Shetland 76
  geological controls on 40–1
    sandstone, shale and chalk 131
    stiff clays 131
  Hartland Quay, Devon 121
  Holderness bayhead, upland coasts 131
  influence of inland topography 44–5, 57
  Joss Bay 187
  Kingsdown to Dover 165
  Ladram Bay, Devon 138
  Loch Maddy–Sound of Harris coastline 94
  Lyme Regis to Golden Cap, Dorset 151
  multi-faceted, Ballard Down 179, *179*
  Nash Point 148
  Northern Islay 98
  Papa Stour 72
  parallel retreat 131
  Porth Neigwl 191
  Robin Hood's Bay 141
  St Abb's Head 110
  St Kilda 60
  and shore platforms 33–4, *34*, 49–1
  Solfach 126
  South Pembroke Cliffs 117
  South-west Isle of Wight 158
  submerged, St Kilda 65–6
  Tarbat Ness 90
  Tintagel 113
  Villians of Hamnavoe 68
  West Coast of Orkney 81
  see also Chapters 2–4; bevelled cliffs; cliff retreat; inland cliffs; plunging cliffs; soft-rock cliffs
climate change
  coastal squeeze 18
  rising sea-level 17
coastal barriers 219–20
  drift-aligned 219
  swash-aligned 219, 220, 241
    and drift-aligned, Porlock 266, *268*
    Sandwood Bay,

720

Index

Sutherland 220, 370, *371*, 372, *372*
see also barrier beaches/islands
coastal defences/ protection/ engineering 19–20, 23, 226, 238, 341, 406, 441, 533
  beach erosion 226
  East Head, West Sussex 429
  Formby–Ainsdale coast 364
  Hurst Castle Spit 272
  Joss Bay 188
  Ringstead Bay 628
  sediment supply to beaches 206, 208
  Slapton Sands and Hallsands 245
  soft-rock cliffs 137
  urbanization 18
  Westward Ho! 238
  see also anthropogenic influences
coastal dunes see dunes
coastal flooding
  surges 12
    North Sea (1953) 12
    Towyn, North Wales 12
  see also saltmarshes
coastal management 18–21
  narrowing of saltmarshes 16, 527
  sediment cells and sub-cells *13*, 14–15
  see also anthropogrenic influences
coastal slope processes 36–8
  see also landslides; mass movements; toppling
cobbles 231
  Blue Anchor–Watchet–Lilstock 147
  Devon 122
  Golden Cap 154
  Holy Island 606, 611
  Westward Ho! 238–40
  see also shingle; gravel
Conachair, St Kilda 60
Coombe Rock, Seven Sisters 174
Corran Seilebost see Luskentyre and Corran Seilebost, Harris, Western Isles
corrasion 144
  see also erosion
Covehithe Ness see Benacre Ness, Suffolk
Crackington Haven, Cornwall 131
Cree Estuary, Outer Solway Firth *517*, 525, 539
  Baldoon Sands *553*, *555*, 556
  carse deposits 552, *553*, 554
  Carsewalloch Flow, radiocarbon dating 554–5
  Cree River channel and estuary 553, 554, 556
  creek system, dendritic 552
  crustal uplift 555
  emerged Holocene features 554
  Fleet Bay 553–5
  micropalaeontology 541
  rise and fall in relative sea level 555
  saltmarshes 552, 553, *553*, 555
  saltpans 552
  sandflats 552–3, 554, 555–6
  wave climate 555
  Wigtown Bay 552–3, *554*
Culbin, Moray 235–6, *330*, **331**, **332**, **333**, **418**, *517*, 529–30, **565**, 567–75
  back barrier saltmarsh 529, *571*
  The Bar 218, 220, **528**, 567
    'flying barrier' *571*, *572*, 573, 575
    gravel ridges *572*, 573
    recurves *572*, 574
  Buckie Loch 571, 575
  Culbin foreland 567, *570*, 571
  gravel ridges 567, *568*, *569*, 574, 575
  Holocene development *568*, 573–4
  intertidal zone 567, 571, 573
  Lateglacial cliff retrimmed 49, 573
  longshore drift 573
    reduction of sediment supply 574, 575
  sand dune system 569–71, 574–5
    butte dunes 570–1
    conifer afforestation 567, 575
    Lady Culbin *568*, 570
    parabolic dunes 569–70, *570*
  sandy saltmarshes 519, 529, 530, 567, *568*, *572*, 573, 575
    freshwater transition 530
  sea-level changes 567, 568, 569
  see also Spey Bay; Whiteness Head
cuspate forelands **203**, 219, 235
  Dungeness 315–25
  see also Benacre Ness; Morfa Dyffryn; Morfa Harlech; Orfordness and Shingle Street

Dawlish Warren, Devon *330*, **333**, **418**, *420*, 435–9
  Bull Hill Sand 437
  Checkstone Reef 437
  Pole Sand 437
  spit *211*, 419, 435, 437, *438*
  Warren Point 437
Dengie Marsh, Essex **233**, **234**, *304*, *517*, 534–8
  cheniers 535–6
  Sales Point chenier 535–6
  saltmarshes 522, 534, *535*, 537
  sea-level rise 536, 537
  sediment pathways 537
  storm events 537
Devensian ice 121, 125, 433, *586*, 591
differential erosion
  Isle of Wight 158, 165
  Lulworth Cove 630
  Orkney 84, 86
  Shetland 76
  see also geological control; jointing; structural control
Dinas Dinlle see Morfa Dinnlle
Dornoch Firth and Morrich More SAC **528**
  see also Morrich More
Dorset Coast: Furzy Cliff to

Index

Peveril Point 37–8, *58*, **59**, *132*, **133**, *233*, **234**, **565**, 566, 624–42
  Bat's Head to Man O'War Bay 625, 629–30, *631*
  bays, closed sediment cells 629, 641, 642
  Bowleaze Cove 625
  Chapman's Pool 634, *637*, 639
  Durdle Door *626*, 629, *631*, *633*, 642
  Furzy Cliff to Poole Harbour 625, *626*, 627
  Gad Cliff 633, 635, *635*, 640, 641
  Hounstout to St Aldhelm's Head 634–5, *636*, *637*
    Emmet's Hill 634, *637*
  Kimmeridge Bay 625, *626*, 634, *636*, 639
  Lulworth Cove 566, 624, *626*, 630, 637–40, 642
  Man o'War Bay 629, 639
  mass movements 625, 627, 628, *629*, 634, 635, *636*, 637–8
  Mupe Bay 633, 639
  Purbeck coast 566, **566**, *626*, 639, 642
  Redcliff Point to Bran Point 626, 627
  Ringstead Bay 627–9, *628*
  Ringstead to Black Head 625
  St Aldhelm's Head to Durlston Point 625, 635, 637, 641
  Stair Hole 566, 630, 638–9
  submerged reef 629, 630, *631*, *632*, 633
  White Nothe to Bat's Head 629
  Worbarrow Bay *626*, 633, 638, 639
  see also Ballard Down; Chesil Beach; South Haven Peninsula
Dorset and East Devon Coast World Heritage site 29, 152, 177, 251, 256, 625
  see also Budleigh Salterton; Dorset Coast
Dunbar, East Lothian *58*, **59**, 107–10, **332**
  cliff forms 108
  emerged areas 108, *109*
  gravel and boulder beaches 107
  shore platforms 107, 108–10
  skerries 108
Duncansby to Skirza Head, Caithness *58*, **59**, 86–9, *89*
  Duncansby stacks 86, 87, 88, 89, *89*
  Skippie Geo to Fast Geo, Wife Geo 87, *88*
  Skirza Head to Skippie Geo 41, 87, *88*
dunes 203, 221–6, 334–5
  Ainsdale 359–64
  Balta Island 384–6
  Barry Links 400–7
  and beaches see beaches
  Braunton Burrows 348–54
  classification 222, **331**, 422
  cliff-top *223*, 225
  climbing dunes 225, 345, 483, 512
    Pabbay 495–6, 498
    Sandwood Bay 374, 375
    Torrisdale and Invernaver 376, 379
  dune topography
    accreting sequence 221–2, *222*, *223*
    blowthroughs 221, 222, 223, *223*
    foredune initiation 221–2
  Dunnet Bay 380–4
  Forvie 393–400
  Holy Island 604–6
  linear-dune systems 334–5
  Luce Sands 364–70
  Marsden Bay 337–40
  migration inland 334
  Oxwich Bay 354–6
  Sandwood Bay 370–5
  South Haven Peninsula 340–5
  Strathbeg 387–93
  Tentsmuir 407–13
  Torrisdale Bay and Invernaver 375–80
  transgressive dunes 224
  Tywyn Aberffraw 356–9
  Upton and Gwithian Towans 345–8
  vegetation 221–2, 224–5, **332**
  see also dune systems; machair; parabolic dunes; Chapter 7; Chapter 11
Dungeness, Kent 28, 203, 210, 226, 231, 310–12, *311*, 315–25
  Beach Bank soil series 311, 316, 317, 318
  Broomhill 320
  cuspate foreland 315, 319
  Camber 311, 312
  Denge Beach *316*, 319, 322
  Lydd soil series 316
  marshes 324, 325
  Denge Marsh 323, 324
  mounds in Newer Marsh 319
  natural pits 318, 320
  the ness 320, 321–2
  New Romney and Hythe 311, *311*
  nourishment 312
  power station 318–21
  Rye Bay 312, *313*, 319, 324–5
  sea-level rise 323
  shingle/beach ridges 217, 232, 315, 316, 318–20, 323, 324
    dating 215, 310–11, *311*, 325
  Jury's Gap to Dungeness 312, 316–17, *316*
  see also Rye Harbour
Dunnet Bay, Caithness 222, *330*, **331**, **332**, **333**, 380–4
  beach–dune–links system 332, 380, *381*,*382*, *382*
  blowthroughs 223, 224, 382, 383, 384
  Burn of Midsand 382
  dune ridge 380, *382*, 383
Dunwich 9, 135, *137*

East Head, West Sussex *330*, **333**, 418, *420*, 426–30
  beaches 426, *427*, 429
  coast protection 427
  Hayling Island 427
  saltmarsh 426, *428*
  shingle ridges 428
  spit 420, *427*, 427–8

Index

emerged coasts 203, 248, *248*, 297–301
  Braunton Burrows 353
  Dunbar 108, *109*
  south-west Britain 225
  see also glacio-isostasy; isostatic recovery
Eoligarry, Barra, Western Isles 225, **332**, **418**, *473*, **477**, 482–7
  Ben Eoligarry Mór 483, 485
  Ben Eoligarry-Orosay connection 485
  dunes, Traìgh Eais 483, *484*, 486
  machair 483, 485
  sea-level rise 485
  Traìgh Eais 482, 483
  Traìgh Mhór 482–3, *484*, 486
erosion
  erosional scour by glacial ice 95
  honeycomb weathering 40, 108
  mechanical wave erosion 38–40
  intertidal 46, *47*
  resistance to, British rock types **9**
  selective 41, 73
    Tarbat Ness 93
  solutional and bioerosional processes 40, 93
  weathering erosion 40
  see also abrasion; attrition; cliff retreat; corrosion; differential erosion; frost; geological control; marine erosion; tafoni; solution; structural control

faulting
  Dùn, Hirta, St Kilda 62
  Flamborough Head 182, 183
  Foula 77
  and landslides 122
  Papa Stour 73, *74*, 75, 76, **76**
  St Abb's Head 111, 113
  Tintagel 114, 115, *116*, *117*
  see also differential erosion; structural control
fetch 12, 13, 66, 358
fjards 95
Flamborough Head, Yorkshire 9, 13, 35, *132*, **133**, 181–7, 196
  beaches, Bempton Cliff 182
  Bempton to Long Ness (Bempton Cliffs) 182
    Staple Nook 183
  Cattlemere Hole to Sewerby 182, 183, 185
  caves, stacks, arches and blowholes 182, 183, *184*, *186*
  Chalk 181, 182, 183, 187
  Long Ness to Cattlemere Hole 182, 183, 185, 187
  North Sea wave climate 181–2, *182*
  shore platforms 182, 185, 187
Flandrian *see* Holocene marine transgression; sea-level history
flint beaches 207
  Benacre Ness 302
  brown flints 207, 268
  flint and chalk 3, *6–7*, *9*, 168, 176, 190
    Ballard Down 180
    Birling Gap 175
    Joss Bay 187
    White Nothe to Bat's Head 629
  flint and chert, Seatown 153
  Hurst Castle Spit 272
  Orfordness 306
  oxidized 174, 175, 627–8, 641
  Ringstead Bay 627–8, 629, 641
  south-west Isle of Wight 158, 159, 160, 164
  see also cobbles; shingle; gravel; Chapter 6
Foreness Point *see* Joss Bay
Forvie, Aberdeenshire *330*, **331**, **332**, **333**, 393–400, *396*, **418**
  deflation surfaces 397, 398
  Foveran 392, 395, *395*, 398
  North Forvie *394*, 395, 399, 400
    parabolic dunes 223, 395, *396*, 397, 399, 400
  Rockend *394*, 395, 397
  sand movement/sediment circulation 377, 393, 395, *396*, 399, *399*
  South Forvie *394*, *396*, 398
    bare sand dome 385, 395, 397, *397*, 400
  spits and bars, River Ythan estuary *395*, *396*, 397
  transgressive dunes 224
Foula, Shetland 58, **59**, 76–80
  The Brough stack 80
  Da Ness to Wurr Wick, boulder ridge 79
  Hellabrick's Wick to Smallie 78, 79
  Hiora Wick 80
  inherited cliff 79, 80
  The Kame 76, 78, 79
  The Noup 77, 78
  Scarf Geo 77, 78
  Shoabill to Hellabrick's Wick 77, 78, 79
  shore platform development 79, 80
  Smallie to Wester Hoevdi 79
  Soberlie to Da Ness 78, 79
  South Ness 77, 78, 80
  stacks and arches 79
  Strem Ness 77, 78, 80
  Wester Hoevdi to Soberlie Hill 78, 79
  Wurr Wick to Shoabill 77, 78
fringing beaches
  cliff-foot 231, *232*
  compartmentalization of 232
  Morfa Dyffryn 452
frost, erosion of Chalk platforms 47
Furzy Cliff to Peveril Point *see* Dorset Coast: Furzy Cliff to Peveril Point

Gaa Sands, Barry Links *401*, 402, *403*, **410**
Galti Stacks, Papa Stour 73, *74*
GCR Networks 21, 23
  CORINE classification of coasts 21, **21**, 329
GCR sites
  legal protection of 28–30
  international measures 29–30
  Sites of Special Scientific Interest (SSSIs) 28
  Special Areas of

723

Index

Conservation (SACs) 30
main features *24–7*
selection guidelines 20–3
geological control on coastline 5, *6–7*, 8–10
see also structural control; jointing
geos
development of 41
Duncansby to Skirza Head 41, 86–9
Foula, Kubbi a'Skeld and Sloag a Ruscar 77, *78*
Papa Stour 73, 74, 75, 76
South Pembroke Cliffs, 119
St Kilda 61, 62
Geo na h-Airde 41, 62
Villians of Hamnavoe 68
Gibraltar Point, Lincolnshire 28, 214, 329, **418**, *420*, 439–43
bars and troughs 215
foredunes 440
foreshore, ridges and runnels 440, 441
Inner Knock *440*, 441
intertidal sandbanks 440
longshore drift 210
New Marsh (saltmarsh) *440*, 441, 442
Old Marsh *440*, 442
saltmarsh *520*
Skegness Middle Bank 441
spit 440–1
Wainfleet Swatchway 440–1, *440*
West Dunes *440*
glacial till 10, 62, 77, 141, 607, 331
Devensian
Carmarthen Bay 589
Flamborough Head 183, 187
Porth Neigwl 191, 192
pre-Devensian, South Pembroke Cliffs 121
see also Skipsea Till; Holderness
glacio-isostasy, and sea level 49
see also Culbin; Dunbar; emerged coasts; isostatic recovery; Jura; Morrich More; relative sea level; Chapter 10
gravel beaches 107, 111, 204, 206, 236
Ayres of Swinister 281–5
conservation value of 235–6
emergent
northern Islay 98, 99, 100
Tarbat Ness 90, 206–7
gravel extraction 227, 235, 250, 353
Hallsands 227, 245, 248, 249–50
Seatown beach 153, 227
glaciogenic 206, 207
limestone gravel 207
provision of materials by landsliding 131
sea level and gravel supply
see Culbin; Chesil Beach; Sanday
Spey Bay 290–7
Taf estuary 586
see also cobbles; shingle beaches
groynes 19, 168, 175, 272–3, 432

Habitats Directive 30
Annex I habitat types **29**, 30, 51–3, 236, **237**, **336**
Hallsands, Devon *58*, **59**, 248–50
beach levels *248*, 250
cliffs 248, 249
gravel extraction, beach erosion following 227, 245, 248, 249, *249*, 250
see also Slapton Sands
Handfast Point see Ballard Down
hanging valleys 153
Hartland Quay 57, 122, 124–5
Porth Neigwl 193, 194
Tintagel 114
hard-rock cliffs 3, *6–7*, 23, 33, 36, 55–128
see also cliffs; Chapter 3
Hartland Quay, Devon 57, *58*, **59**, 121–6
beaches, shingle and cobble 122
Blagdon Cliff landslide 122
Blegberry Water and Abbey River 122, *123*, 125
cliffs 122
coastal waterfalls 122–6
hanging valleys 122, 124–5
Hartland Point 120, 121
Keivill's Wood, rotational slip scar 122
Milford Water *123*, 125
Speke's Mill Mouth 122, 124
shore platforms 122, 125, 126
Tichberry Water 122, *123*
Wargery Water 122–3, *123*
'Hell's Mouth' see Porth Neigwl
Hengistbury Head 134
High Rock Platform
Northern Islay 98–9, 99, 100, *101*, 102, *102*
west coast of Jura 297–8, 300
high-energy environments 149
Machir Bay 478
Nash Point 150
Papa Stour, Shetland 73
St Kilda 60, 66
south west Hoy, Orkney 81
Traigh Eais, Eoligarry, Barra 482
Villians of Hamnavoe 69, 71
Hirta, St Kilda
cliff failure 197–8
Dùn 62, *603*
small geos and caves 62
Holderness, Yorkshire 14, *132*, **134**, 135, 195–9, 331
cliff recession 195–6, **195**, 196, 197, 198
historic accounts 195
longshore transport of sediment 198
ord movement and erosion 197, 198, 207
relationship between cliff height and erosion *198*
forms spit to Spurn Head 195, 207, 434
ords 14, 195, 196, 197, 434
shore platform obscured 33
see also Spurn Head
Holkham Bay, Norfolk 208, 619
see also North Norfolk Coast
Holocene marine transgression 13, 15, *16*, 44
building of beaches 3, *6–7*, 207–8

724

Index

see also sea-level history
Holy Island, Northumberland
  58, **59**, 208, *330*, **331**, **333**,
  334, **418**, *420*, **565**, 604–11
  Budle Bay 605, *605*, *608*,
    609, 610
  Budle spit 607, *607*, *608*,
    609
  Castle Point, cobble ridges
    606
  Cheswick and Gosford
    Sands 604–6
    barrier beaches 609–10
    reefs 605
    sandy beach and dunes
      604
  cobble beaches 606, 611
  Coves Haven 606
  dissipative beaches 214
  dunes 604, *605*, 606
  eastern cliffs and shoreline
    604, 606
  Emmanuel Head 606
  emerged beach sediments
    605, 606
  Ross Back Sands 6–9
  Ross Links 606, *607*, *608*,
    609
  Ross Point 610–11
  saltmarsh 606–7, *608*, 610
  Spartina invasion 523
Hornish and Lingay Strands
  (Machairs Robach and
  Newton), North Uist *473*,
  477, 492–5, *493*
  blowthoughs
  Corran Goulaby 494, 495
  machair 492, 494
  Machair Newton
  Machair Robach 494, 495
  Orosay
  Sollas Peninsula *493*, 494
Humber Estuary **528**
Hunstanton, Norfolk 35
Hurst Castle Spit, Hampshire
  211, 218, *233*, **234**, 271–7
  active recurve 271, 272, 273,
    275, 277
  anthropogenic effects 272–3,
    277
  Hordle Cliff 272
  Hurst Beach 272, *274*, 277
  Keyhaven Marshes 271,
    273–4, *275*, 276, 277

sediment supply 209
Shingles Bank 272, 276–7,
  273
spit with recurves 217, *218*,
  272, 273, *275*, 276–7
see also Keyhaven Marshes,
  Hampshire

inheritance 15, 50, 76
  evidence of 48–51
  Holocene submergence,
    western Scotland 51, 57
  plunging cliffs 50–1
inlets and submerged coasts 23
  see also geos; Loch Maddy;
    rias; Sanday
Irish Sea
  sediment transport 14
  wave height 12
  see also Figure 1.2 for locations of Irish Sea GCR sites
Islay *see* Northern Islay
Isle of Wight, south-west 34–5,
  132, **133**, 137, 158–66, *233*,
  234
  Atherfield Point to The
    Undercliff, active cliffs 159
  beaches
    boulder and cobble 160
    flint 158, 159, 160, 164
  Blackgang Chine, base failures 161, *163*
  Blackgang Chine to
    Atherfield Point 159, *159*
  Brook Bay 159, *159*
  chines 158, 159–60
    origins 163–4
  cliff retreat 158, 161, *161*,
    164
    and cliff failure 163
  cliffs 131, 158, 162–3
    Compton Down 163, *163*,
      164
  differential erosion 158, 165
  Freshwater Bay *159*
  Gore Cliff 161
  Hanover Point to Freshwater
    Bay 159–60, *159*
  landslides 159, 162, 164
    'Gore Cliff landslip' 161
    Rocken End and
      Blackgang Chine 161
  The Needles 160, *160*
  platform development 158

St Catherine's Point to
  Blackgang Chine 159, *159*
Scratchell's Bay 160, 164
Tennyson Down (South)
  159, 160, 163
toppling failure,
  The Undercliff *159*, 161, 164
Isles of Scilly *see* Scilly, Isles of
isostatic recovery 15, 301
  emerged gravel ridges 235
  prograding beaches 216–17,
    334, 582
  Scotland 50, 100–1, 102
  submergence
  see also emerged coasts;
    glacio-isostasy; Orkney;
    Shetland

jointing 57
  Bullers of Buchan 103, *105*
  cliff failure/instability 166–7,
    168, 170, 171, 174
  development of geos, caves,
    arches and stacks 41
  Duncansby 86–9
  Foula 76–80
  Handfast Point 177, 179,
    180
  Ladram Bay 138, *139*
  Marsden Bay 339, 340
  Nash Point 150
  Papa Stour 72–6
  St Kilda 60–8
  South Haven Peninsula 344
  Tintagel 115
  West coast of Orkney 81
  see also structural control
Joss Bay, Kent *132*, **133**,
  187–91
  Botany Bay 187, 189, 190
  cave–arch–stack development 188, 189
  cliff profiles 189, 190
  cliff retreat 187–9
  dry valleys 188
  Foreness Point 188, 225
  intertidal platforms 187,
    188, 189–90
  jointing 188
  micro-cliffs 190–1
  south of White Ness 188,
    189
Jura, west coast, Argyll and
  Bute *233*, **235**, 297–301

Index

Colonsay Ridge *298*, 301
Corpach 300, 301
emerged beach ridges 297, 299–301, *299*
High Rock Platform (Shian Bay to Ruatallain) 297–8, 300
Low Rock Platform 298–300
Main Rock Platform 298–300, 301
sea levels 301

Keyhaven Marshes, Hampshire 271, 273–4, *275*, 277, *517*, 538–9
  coastal protection work 75
  creek pattern 538
  erosion of saltmarsh edge 538, *538*, 539
  marsh-edge cheniers 538, *538*
  Hurst Castle Spit 538
  species-rich high marsh 538–9
  vegetation 276, 539
  see also Hurst Castle Spit
Kingsdown to Dover, Kent *132*, **133**, 165–70, *233*, **234**
  beaches 165, 168, 226–7
  boulder ramparts 165, 168, *169*, 170
  cliff falls 166–7, 168, 170
  cliff retreat 165, 166–7, 168, 170
  harbour/groyne construction 168
  shore platforms 165, 167, 168, *169*, 170

Ladram Bay, Devon *132*, **133**, 138–41
  cliffs 138, *139*
  shingle and cobble beaches 138, *139*
  shore platforms 138, *139*, 140
  stacks 138–9, *139*, 140, *140*, 141
landslides 5, 34, 36, 37–8, 122, 137–8, 155
  Budleigh Salterton 252, 253
  in clay 131, 134, 141–2, *142*
  Dorset Coast 37–8, 633, 634, *636*, 637–8, *637*

Lyme Regis to Golden Cap 131, 134, 152, 154–5, 156, *156*, 157, 261
  Isle of Wight 5, 158, 159, 161, 162, 164
  rotational, and cliff failure 135, 136, 162
  see also mass movements
Lias 141–2, *142*, 152, 156
  biozones 150
  cliff retreat 145, 147
lithological control
  and cliff development 41, 57
  Duncansby to Skirza Head 90
  St Kilda 62, 66
  south of Aesha Head, Papa Stour 73–4
  Villians of Hamnavoe 71–2
  west coast of Orkney 81
  see also geological control; structural control
Loch Gruinart, Islay **332**, *517*, 523
  Ardnave Point 557, *558*
  emerged cliff 557, 559
  Holocene sea-level history 559, *560*
  intertidal sandflats 557, *558*
  Killinallan Point and Tràgh Baile Aonghais beach 557, *558*
  saltmarshes 557, 559
  saltpans 557, 559, 561
  tidal creeks 524, 559
Loch Maddy–Sound of Harris, North Uist, Western Isles *58*, **59**, 94–8
  geology 95, *96*
  inner Minch coastline 95
  Loch Aulasary 96, *97*
  Loch Blashaval 95, *96*, *97*
  Loch Maddy
    fjard 95
    intertidal areas 95
    links with inland lochs 95
    skerries and reefs 94
    submergence of inherited glaciated surfaces 51, 57
  Loch Mhic Phàil 97
    tidal ponds 95–6
  Loch Siginish 96, *97*
  Outer Hebridean Thrust Plane 95, *97*

Sound of Harris
  the Rangas 95
  Tahay, Vaccasay and Opsay 95
  submerged landscape 94, 96, *96*, 98
  wave climates 94, 98
Loe Bar, Cornwall *233*, **234**, 241–4
  bay–bar 241, 244
  ria 241, 244
  breaching 241, 244
  evolution of 242–3, 244
  sediment sink 244
  beach
    closed sediment cell 241
    flint, origin 208, 243–4
    swash-aligned 212, 220
  beach cusps 213, *243*
  slope-over-wall cliffs 241, *243*
  Loe Pool 241, 526
  Porthleven Sands 241
  washover fans 241
  see also Pagham Harbour; Slapton Sands and Hallsands
longshore drift 12, 36, 209–10
  Chesil Beach 259–61, 262, 265
  Culbin, Moray 573, 574, 575
  direction and rate of transport 13–14
  edge-wave activity 193, 209
  Gibraltar Point 210
  Hornish Strand 495
  Nash Point 150
  Orfordness 210
  prevention of 175
  sorting by and lateral grading 215
Lossiemouth **332**
Low Rock Platform
  age 99–100
  Northern Islay 99, 100, 102
  west coast of Jura 198–9, 299–300
Luce Sands, Dumfries and Galloway 224, *330*, **331**, **332**, **333**, 364–70
  emerged Holocene gravel ridges 365, 368
  Holocene development, emerged features 369

Index

intertidal beach 365
  bars and troughs 214
  bay-head beach 369
  ridges and runnels 365
prograding beach 216–17
sediment sources *366*, 369
Torrs Warren–Luce Sands
  dune system 365–6, *367*, 368, *368*
    anthropogenic influence 369–70
    blowthroughs 369
    dune slacks 365, 369
    foredune ridges 369
Luskentyre and Corran Seilebost, Harris, Western Isles **332**, *473*, **477**, 499–503
  calcarenite 499, 503
  Corran Seilebost
    beach–dune–machair system 499, 503
    dune ridges 502
    dynamic spit 499, 502
    late Holocene submergence 503
  Luskentyre 499, 501–2
    beach–dune–machair system 499
    dunes 499, 501–2
    machair 499, 502, 503
  Luskentyre Banks 499, 501–2
  swash-alignment 502–3
  Tràigh Luskentyre 221, 499, *500*
  Tràigh Rosamol 499, 502
  Tràigh Seilebost 221, 502
Lyme Regis to Golden Cap, Dorset 34–5, *132*, **133**, 151–8, *233*, **234**
  beaches
    sediment source 152, 153–4, 206
    Golden Cap to Charmouth 154
    lateral grading 215
  Black Ven
    landslides 152, *154*, 155, 156
  Charmouth
    cliff instability 156, 157
    unstable beach 157
  cliffs 151, 152, *153*
  Fairy Dell 154–6

Golden Cap 157, 261
Lyme Regis
  cliff-beach system 157–8, 226
  major landslides 131, 134, 152, 153, *154*, 155, 156, *156*
  pocket beaches 152
Seatown beach
  sediment transfer 152, *153*
  shingle extraction 153, 227
sediment movement 157
The Spittles, re-activated landslide 156, *156*
west of Golden Cap
  landslides 153
  shore platform 153

machair 3, *6–7*, 30, 203, 225
  anthropogenic modification 474, 476–7, 477–8
  archaeological sites 475, 476
  Ardivachar to Stoneybridge 487–92
  Balnakeil 510–14
  conservation value 477
  cultural overtones 473
  dating of initiation and formation 475–6
  definition of 473, 474
  deflation surfaces 476
  Eoligarry, Barra 482–7
  evolution 474–5, *474*, 475, *475*, 476, 477
  Hornish and Lingay Strands 492–5
  Luskentyre and Corran Seilebost 499–503
  Machir Bay, Islay 478–82
  Mangersta, Lewis 503–6
  North Uist **528**
  Pabbay, Harris 495–8
  sediment budgets 475
  Special Areas of Conservation 474
  Tràigh na Berie, Lewis 506–10
  see Chapter 9
Machir Bay, Islay, Argyll and Bute **332**, *473*, **477**, 478–82
  calcarenite 481

dunes 478, 480
dynamic
  beach–dune–machair assemblage 478, 481
  emerged features 478, 480, 481
  foredune ridge 480, 481–2
  glacial terraces 478, 480, 481
  Holocene transgression 481
  machair plain 478, 481
  relict clifflines 478, 480
  Tràigh Mhór, bars and cusps 482
Main Postglacial Shoreline
  Moray Firth 289
  Solway Firth 544, 555
Main Rock Platform
  age 99–100
  Northern Islay 99, 100–1, 102
  west coast of Jura 298, 299–300, 301
Mangersta, Lewis, Western Isles **332**, *473*, **477**, 503–6
  beach–dune–machair cycle 503
  climbing machair 504–5, *505*, 506
  deflation surface behind storm ridge 504, *504*, 506
  development 505
    sediment supply 505–6
  sandy beach 502, *504*
    gravel storm ridge 503, 506
  stripped machair 503, 506
marine/sea caves 41–3, 57
  arch and stack development 41, 43, *43*
  blowhole development 42
  Bullers of Buchan 104
  emerged 99
  and lithology 42
  Nash Point 151
Marsden Bay, County Durham 214, *233*, **234**, *330*, 337–40
  limestone cliffs 338, *338*, 339, 340
  sandy beach 212, 332, 334, 339–40
    sources of sand 338–9
mass movements
  on coastal slopes 36

727

Index

cliff failure 134
   Bullers of Buchan 103
   Holderness 197
  deep-seated *38*
  dip-slip slides 37, 57, 78, 80
  Dorset Coast 155, 625, 633, 634, 635, *637*
  Hartland Quay, localized 122
  St Kilda 66–7
  soft-rock cliffs 131
   Blackgang Chine 161–2, *162, 163*
   Ladram Bay 138, 140
   Robin Hood's Bay 141
   Warden Point, Isle of Sheppey 131
  see also landslides; soft/weak rock cliffs
merse *see* Solway Firth Saltmarshes
Mochras *see* Morfa Dyffrn
Moray Firth
  longshore drift 210, 212
  progradation on sandy coasts 215–16
  sea-level history *284*, 288–9
  see also Culbin; Spey Bay; Tarbat Ness; Whiteness Head
Morfa Dinlle **566**, 593, 600–4
  Afon Gwyrfai tidal inlet 602
  Dinas Dinlle 600, *601*, 602
  dunes 601, 602, 603
  Foryd Bay, land-claim 602
  gravel beach system 602–3
  gravel ridges 595, 600–1, 602, 603
  moraine across mouth of Afon Gwyrfai 601–2
  sediment movement 601, 602
  Warren Farm, ridges 601
  see also Newborough Warren; Newborough Warren and Morfa Dinlle
Morfa Dyffryn, Merioneth, Gwynedd 223, *330*, **332**, **333**, 335, **418**, *420*, 421, 452–7
  cuspate foreland 219, 452
  dunes 452–3
  fringing beach 452
  historical development 453, *454*

multiple recurved ridges 419
sandy beach 453
Sarn Badrig *450*, 452, 453
the Sarns
  and sediment movement in Cardigan Bay 452
  offshore cobble and boulder banks 452
sediment supply *454*, 456
southern spit, Afon Ysgethin 453, *454*
Morfa Harlech, Merioneth, Gwynedd 223, *330*, **331**, **332**, **333**, **418**, *420*, 449–52
  beach and dunes 449–50, *450*
  cuspate foreland 219
  intertidal flats 451
  processes and sediment transfers *451*
  prograding 449, 451
  recurved zones 449, *450*
  relict cliffs 449
  spit 449, 451–2
  ridge and runnels 450
  swash-alignment 449, 451
Morrich More, Ross and Cromarty *330*, **331**, **333**, *418*, *517*, 530–1, **565**, 576–83
  barrier islands
   with recurving spits 579
   fronting beach with accreting sand dunes *578*, 579
   Innis Mhór and Patterson Island 576, *577*, 579
  beach ridges 217, 567, 576, 580–1, *581*–2
  and Dornoch Firth **528**
  dune development 580, 582–3
  emerged cliff 576
  emerged strandplain *577*, 580
   dune-capped ridges and saltmarsh hollows 567, 576, 580, 582
   latest ridges (Innis Mhór and Patterson Island) 576, *577*
  eroding dunes, facing Dornoch Firth *578*, 579, 582

north-eastward progradation 581, *581*, 582
  Inver Bay 582
parabolic dunes *578*, *579*, 580, 583
postglacial isostatic recovery and progration 216–17, 582
saltmarshes 530, 531, 582
sand-bars, intertidal 576, *578*, 579–80, *581*
tidal creeks 524
Nash Point, Glamorgan *132*, **133**, 148–51, *233*, **234**
  cliff retreat 150–1
  cliffed coastline 149, *149*, 150
  intertidal platforms *146*, 149, 150
  jointing 149, 151
  limestone gravel beaches 207
  tidal range 150
neotectonic movements
  influence of 49–50
  local 15
nesses
  acute and oblique *232*
  relationship with offshore banks and shoals 419, 447
  see also cuspate forelands; Dungeness; Holderness; Orfordness and Shingle Street; Winterton Ness
Newborough Warren *330*, **331**, **332**, **333**, 335, **418**, *420*, **565**, **566**, 593
  Abermenai Point 594, *594*
   pre-breaching end of older spit 598
  Abermenai spit 594, *594*, 597–8
  breaching and rebuilding 597, 599
  recurves 597
  beach systems 599
  dunes 593
   afforestation 595, *596*
   blowthroughs 598–9
   dune-building 567, 598, 599
   Maltraeth Bay/Sands 595
   migration 222, 224, 595, *596*, *596*, 598

Index

sand supply 599
Llanddwyn Island 595–6
  tied to mainland 596
Maltraeth Bay/Sands 595, *596*
  dune ridges/zones 595, 598
sandbank and channel systems 599
see also Morfa Dinlle
North Norfolk Coast 38, *132*, **134**, *330*, 333, 334, *517*, **528**, **565**, **566**
  abandoned cliffs 611
  barriers *220*, 622
  Blakeney Harbour **612**
  Blakeney Point 10, 214, 231, *233*, **234**, *304*, **331**, **612**, 613
    geomorphological features *618*
  Blakeney Spit 217–18, 613, 619–1, 624
    saltmarshes 619
    saltpans 525, *526*
  Brancaster (East Sands) spit 613, 616
  Brancaster Marsh 614
  Burnham flats 613
  Burnham Overy channel 619
  glacial sediments 622
  Gun Hill to Wells-next-the-Sea 613, 619
  Holkham Bay 619, 622
  Holme-next-the-Sea to Brancaster 613, 614, *617*
  Hunstanton to Holme
    Chalk and Carstone cliffs 613, 614
    sand and shingle beach 613, 614, *615*
  Hut Marsh 616, 622, *623*
  intertidal flats 611, 619
  introduction of *Spartina anglica* 523
  LOIS study 622–3
  longshore sediment transport 612–13
  Missel Marsh 616, 619, *623*
  origins/age of major spits and barriers 622
  saltmarshes 611, **620**, 622–4
    back-barrier 613–14
    creeks, saltpans and marsh stratigraphy 611
    evolution 519
    land-claim 614
    rates of sedimentation 623
    saltpans 624
    vertical accretion, Scolt Head Island 521
  Scolt Head Island 217–18, 231, *233*, **234**, **331**, **418**, *420*, 566, **612**, 624
    accretion rates 616
    barrier island backed by saltmarsh 220, 613
    erosion 613, 616
    geomorphological features *618*
    Plover Marsh 616
    saltmarshes 521, 525, 614, 616, *618*
    sand dunes 616
    shingle ridges 616
    The Sloughs 616
  sediment sources and transport 613, 621, 622
  shore platform, Carstone 614
  Thornham and Gore Point 614
  Wells Channel to Blakeney Spit
    sediment cell 613
    Warham 619
  Wells to Blakeney Point 619–20
    Stiffkey Meals marshes 619, 622
  Wells-next-the-Sea 231, *233*, **418**, *420*, 566, **612**
North Uist Machair **528**
Northern Islay, Argyll and Bute *58*, **59**, 98–102
  Coir Odhar end moraine 99, *101*
  emerged features 98, 99, 100, 102
  High Rock Platform 98–9, 100, 102, *101*, *102*
  Low Rock Platform 99, 100
  Main Rock Platform 99, 100–1, 102
  Mala Bholsa to Rubha a'Mhàil 99, *101*, *102*
  rock platforms 100

notching 190
  and cliff falls in Chalk 189
  Duncansby to Skirza Head 87
  Hirta, St Kilda 62
  Orkney west coast 83

offshore banks 219
  Morfa Dyffryn, cobble and boulder 452
  sand and gravel 332
  sand movement from 622
Old Harry Rocks *see* Ballard Down; Dorset Coast
Old Man of Hoy 35, 51, 80, 81
  development of 43, *83*, 84
  see also cliffs; stacks; arches
ords
  Holderness 14, 196, 197, 434
  migration 13, 14, 195
  Spurn Head 197, 434
  see also ridge and runnel
Orfordness and Shingle Street 10, 14, *211*, *233*, **235**, 304–10, *304*, *307*
  anthropogenic influence 227, 305, 307
  developmental phases 220, *305*, 309–10
  displacement of River Ore mouth 307
  longshore drift 210
  ridge patterns 306, *307*
  spit 235
    deflected River Alde 217
    historical changes in 306, 307, *308*, 309, *309*, 310
  see also Shingle Street
Orkney and Shetland Islands *330*
  'ayres', small gravel barriers 220
  Holocene submergence 468
  Skara Brae, Neolithic settlement 224
  see also Ayres of Swinister; Balta; Central Sanday; Foula; Papa Stour; Villians of Hamnavoe
Orkney, west coast of *58*, **59**, 81–6
  Breckness to Hole O'Row cliffs *82*, 83–4

729

Index

cliffs inherited 51, 84
Kame of Hoy to The Pow 81, 82, 83
   Selwick, shore platform 83
lithology and structure 81, 84, 86
plunging cliffs 81, 83
Rora Head to Kame of Hoy 82
   intertidal platform 81
   Old Man of Hoy 81, *83*, 84
   St John's Head, cliffs, caves and stacks 81
   shore platforms 81
   see also Old Man of Hoy
shore platforms 81, 83
Yesnaby *82*, 83
   stacks and arches 84, *85*
Overwash see washover
Oxwich Bay, Glamorgan *330, 331, 332, 333*, 354–6, *420*
   beach 212, 356
   blowthroughs
   intertidal peat
   Nicholaston Burrows *354*, 355, 356
   Oxwich burrows, single dune ridge *354*, 355
   shore platform 356
   sand transfer 356

Pabbay, Harris, Western Isles *473*, **477**, 495–8
   beach–dune–machair complex 497, 498
   blown sand 495, 498
   climbing dunes 495–6, 498
   dunes, accretionary 497
   emergence and submergence 498
   intertidal and sub-tidal peats 496
   machair surface 496, 497–8
   Quinish *496*, 498
   gravel tombolo 497
   sand embayments, Haltosh Point to rocky cliffs *496*, 497
Pagham Harbour, West Sussex *233*, **234**, 278–81
   formation of entrance 218–19
   intertidal gravels 278
   'Pagham Delta' 278
   Selsey Bill 278, 279, *280*
   shingle spit 278, *279*, 280, 281
   spit–bay–bar–spit sequence 278
   weed-rafting of shingle 278, 280
Papa Stour, Shetland *58*, **59**, 72–6
   Aesha Head northwards 74–5
   Fogla and Lyra skerries 75
   Stourhund 75
   Christie's Hole 73, *74*
   coastal platforms 75–6
   Cribbie Peninsula *74*, 75
   Galti Stacks 73, *74*
   geos 73, *74*, 75, 76
   Lamba Ness, fault-controlled stacks 76
   submerged dissected plateau 73
   Wilma Skerry 73
parabolic dunes 221, 222, 223, *223*, 351, 395
   Barry Links 400–1, 404, 405, 406
   Forvie 223, 395, *396*, 397, 399
   geomorphological processes 400
   grown from blowthroughs 224
   internal structure, Tywyn Aberffraw 359
   Maviston, Culbin 569–71
   Morrich More *578, 579*, 580, 583
periglacial activity/processes 180, 591
   in dolomitized Chalk, Flamborough Head 183
   gravel from 207
   Joss Bay 188
   northern Islay 100
   Seven Sisters 174
   South Pembroke coast 121
pitting features 108
   Tarbat Ness 91, 93
   see also erosion
platforms see shore platforms
plunging cliffs 33, 45, 75, 77

Conachair, St Kilda 45
Foula 77, 80
legacy of inheritance 50–1
retreat 45
Stac Armin and Stac Lee, St Kilda 65
see also cliffs; chapters 2–3
Plymouth Sound and Estuaries **528**
pocket beaches 204, *232*, 300
   Balnakeil east coast 510
   Beachy Head to Seaford 173
   Ladram Bay 139
   Lyme Regis to Golden Cap 152
   Ragwen Point 585
   sources of materials 131
   swash-aligned **203**
   see also beaches; Chapter 7
Poole Harbour
   dendritic creek systems 525
   effects of *Spartina anglica* 523
   marshes 517, *518*, 522
   see also South Haven Peninsula; Dorset Coast
Porlock, Somerset 236, 266–71
   1996 breaching event 266, 268–9, *268, 269*, 270–1
   barrier reworking 270
   Exmoor Plateau 267, 270
   longshore segmentation 266
   relative sea-level rise 270–1
   saltmarsh development 266, 271
   sediment budget 267, 270
   swash-aligned and drift-aligned sections 266, 267–8, *268*, 270
Porth Neigwl, Lleyn Peninsula *132*, **134**, 191–4
   bars 191, 193
   beach cusps 191, 193
   beaches 191, 192, 193, 207, 214
   cementation 194
   emerged 191–2
   cliffs 191, 192–3, 194
   confined cliff–beach system 191
   hanging and incised valleys 193, 194
Portland see Chesil Beach; Dorset Coast

Index

Purbeck coast *see* Dorset Coast
Pwll-ddu, Glamorgan **418**, *420*, 422–4
   bay-head beach–dune system 422, 423
   beach ridges 422, 423, 424
   cliff retreat 422–3
   hindshore dunes 422
   ria 423
   slope-over-wall cliffs and shore platforms 422
   spit and beach systems 422

Quarrying
   see marine erosion
Quinish *496*, 498
Qui Ayre, Sanday 84

raised beaches
   see emerged beaches
Rattray *see* Strathbeg
reefs 43, 94, 104, 605
relative rock resistance and coastal form 8, *8*
rias
   Carmarthen Bay and Loe Bar GCR sites 128
   Pwll ddu 422–4
   south-west England 126, 207
   see also geos; inlets
ridge-and-runnel beaches 12, *204*, *205*, 214–15, *215*, 334, 364, 408
   Ainsdale 360, *361*, 362
   Braunton Burrows *349*
   See also ord
Robin Hood's Bay, Yorkshire *132*, **133**, 141–5
   beaches, from glacial drift deposits 207
   cliffs 141–2, *142*
   mass-movement and cliff retreat 141
   subhorizontal shore platforms 45, 141, 143, *144*
      curving ridges and troughs 142
      erosion 144
      inheritance 143–5
      role of debris 143–4
rockfalls 36
   ancient, protective 116
   and coastal subsidence 36
   small high-frequency 174

   see also cliffs; landslides; mass movements
rock resistance *see* erosion; geological control; structural control; *see also* Figure 1.3 and Table 1.4
Rye Harbour, East Sussex *233*, **234**, *330*
   beach ridges 311
   double spit *313*, 314
   drift-aligned beach 213
   shingle beach 313, *313*, 314, 315
   Winchelsea 314

SACs *see* Special Areas of Conservation
St Abb's Head, Berwickshire *58*, **59**, 110–13
   arch at Hope's Heugh 111
   benched cliff profiles 111, 113
   geological control 110, 111, *112*, 113
   Hope's Heugh to Pettico Wick 111
   Horsecastle Bay, gravel storm beach 111, *112*
   scree slopes and boulder fields 113
   shore platforms 111, 113
   Wuddy Heugh 111, *112*
St Kilda, Western Isles 29, 49, *58*, **59**, 60–8
   Boreray 62, 65
   Conachair
      Britain's highest sea cliff 60
   extreme wave and weather conditions 60, 66
   geos 61
   Hirta 61–2, 66
   Soay 62
   Stac an Armin 60, *61*, *64*, 65
      Britain's highest stack 60
   Stac Lee *61*, 65
   submerged landforms 65–6, *65*, 67
   volcanic history and geological structure 60, 66
St Ninian's Tombolo, Shetland 219, **332**, **418**, *420*, 457–61
   dunes 457, *458*

   evolutionary pattern 457, *460*, 461
   sediment circulatory system 457, 460, 461
St Osyth Marsh, Essex *233*, **234**, *304*, *517*, 531–4
   beach feeding 533
   chenier development 217, 531, 532–3
   Colne Point structure 531
   spit 532
   saltmarshes 531, 532
   saltpans 531, 532, 533, *535*
   sea defences 533
salt weathering
   and frost action, within spray zones 40
   west coast of Britain 518
   see also Dornoch Firth; erosion; pitting; Tafoni; Tarbat Ness
saltmarsh creeks 524–5
   dendritic systems 524, *525*
   morphology related 525
   saltpans (pond holes) 525
   stages in evolution of *524*
   straight parallel creeks 524
saltmarshes 3, 6–7, 10, 28
   beach ridges on 217
   bordering intertidal mudflats 517–18
   Culbin 529–30
   Dengie Marsh 534–8
   effects of *Spartina anglica* (common cord-grass) 523, *523*
   evolution of 518–21
   Keyhaven Marshes 271, 273–4, *275*, 276
   land-claim 19, 527
   Loch Gruinart, at estuary heads 10–11
   managed re-alignment 527
   micro-cliffs 131, 522–3
   modified by pollution, dumping and excavation 527
   Morrich More 530–1
   North Norfolk Coast 220, **220**
   open and closed marshes 517
   rates and patterns of accretion 521–2

Index

saltmarsh sites 527
saltpans 525, *526*
  St Osyth Marsh, Essex 532, 533, *535*
  Scotland and Western Isles 517
  St Osyth Marsh 531–4
  Solway Firth Saltmarshes 539–36
  Whiteness Head 288
Sanday *see* Central Sanday
sand spits 417, *417*, 418, 421–2
  evolution of 419
  across infilling estuaries, Morfa Harlech 449–52
  breaching of proximal end 419
  development of separate and distinct ridges 419
  double spits 343, 419, 420, 421
  sediment supply 419
  and tidal range **418**, 419
  see also spits
Sands of Forvie *see* Forvie
Sandwood Bay, Sutherland *330*, **331**, 370–5, *373*, *398*
  beach–dune complex 370, *371*, 372–4
  cliffs 370
  exposures of bedrock 370
  dune system, 372, 373, *373*, 375
    blowthroughs 375
    accretion and recycling 374, 375
  Sandwood Loch 221, 370, *371*, 372, *372*, 374
  sandy beach 371–2, *371*, *372*
sandy beaches 3, *6–7*, 329, 461
  Carmarthen Bay 585
  dunes behind 221, *222*, 335
    Atlantic coasts 221
    North Sea coast 221
  in embayments 332
  Joss Bay 187
  Newborough Warren 595–6
  prograding 216–17
  Rhossili Bay 590
  Sandwood Bay 371–2, *371*, *372*
  Tintagel 114
  wave energy 204

see also Ainsdale; Holy Island; machair; Strathbeg
Scilly, Isles of 461–4
  beaches 208, 213, 462
  granite platforms surround islands 462
  Great Arthur 462, *463*
  Great and Little Ganinick 62, *463*
  Tean 462, *463*
  tied islands 461, 462, *463*, *463*
  tombolo evolution 219

Scolt Head Island *see* North Norfolk Coast
sea-level history 15–18, 48
  evidence for trend 15
  Holocene rise in the Uists 96, 98
  Holocene transgression 15, *16*
  Late Devensian 67
  Loch Lomond Stadial, local evidence of 14
  local histories of 15–17, *16*
  neotectonic land movements 49 round British coast 48
  Tarbat Ness 93
  variations in, and cliff abandonment 48
  see also Holocene history; relative sea-level rise
sediment cells and sub-cells *13*, 14–15, 20, 203, 613
  and coastal management 14–15
  Dorset Coast, closed cells 629, 641, 642
sediment transport 406
  Ainsdale *362*, 364
  Benacre Ness 302, 303–4
  by wind, Tentsmuir 408
  Chesil Beach 259–61
  Irish Sea 14
  local studies of 14
  North Norfolk Coast 612–13, 621
  Whiteness Head 285–6
  within bays 641
  see also longshore drift; sediment budgets; sediment cells
Selsey Bill 278, 280, *280*

Seven Sisters, Sussex, cliff form 44, *44*, 171, 174, 176
  see also Beachy Head to Seaford
Severn Estuary
  saltmarshes
    micro-cliffs 522
    vertical accretion 521
Shell Bay *see* South Haven Peninsula
Shetland Islands *330*
  absence of modern shore platforms 75–6
  cliff-top boulder deposits **70**
  cliffs inherited 76
  Holocene submergence 69
  Ayres of Swinister 284–5
  see also Balta Island; St Ninian's Tombolo; Papa Stour; Villians of Hamnavoe
shingle beaches 3, *6–7*, 122, 141–2, 203, 206, 207
  characteristic zonations 236, 237
  classification 231, *232*
  with cobbles
    Budleigh Salterton 251, 253
    Ladram Bay 138, 139
  conservation value of 235–6
  flint
    Atherfield Point to Blackgang 158
    Freshwater Bay 160
  form of 235
  offshore gravels sources during Holocene transgression 3, *6–7*
  Pagham Harbour 278
  permeable 204, 231
  shingle, coarse sand and flint, Loe Bar 208, 241, 243
  Swanage Bay 177
  see also gravel beaches
shingle spits, Blakeney Point and Scolt Head Island, derivation of 217–18
Shingle Street
  beach 14
  disappearance of lagoons 306
shore platforms 12, 22, 33–4, 34, 36, 45–8, 49, 57, 67, 73,

732

Index

75, 81, 83, 95, 108, 111, 113, 131, 185
Blue Anchor–Watchet–Lilstock 145–8
Bullers of Buchan 106
Chalk 40, 47
Dunbar 107–8
Duncansby to Skirza Head 87, 90
Foula 77
Ladram Bay 138, 139, 140, *140*
North Yorkshire 47–8
northern Islay 98–9
ramp platforms 45, 79
Robin Hood's Bay 45, 141, 142–5
St Kilda 67
Selwick, Orkney 83
Sewerby 185, 187
subaerial weathering during severe winters 175–6
Shoreline Management Plans 20
Sites of Special Scientific Interest (SSSIs) 23, 28
   biological
      dunes and sandy beaches 336–7
      gravel and shingle structures 236–8
      machair 477–8
      saltmarshes 527–9
   composite 28
   skerries
      Bullers of Buchan 104
      Dunbar 108
      Loch Maddy–Sound of Harris 94, 95
      Papa Stour 73
Slapton Bar 232
Slapton Sands, Devon 246–7
   edge waves and incident waves 248
   near shore circulation cells 248
   shingle barrier beach 246
   sediment source 208
   and Slapton Ley 246, 247–8
      swamp encroachment 526
   Slapton Ley 208, 246, 247
   source of flints 9
   Torcross beach 246

Slapton Sands and Hallsands, Devon *233*, **234**
   beaches 244, 245
   Beesands 244, 245, *246*
   coastal protection works 245
   relict coastline submerged 245, *245*
   Skerries Bank *245*, 246
   Start Bay
      zeta-curved 244
   see also Hallsands, Devon
soft/weak-rock cliffs 3, *6–7*, 33, 45, 129–99
   anthropogenic influences 35, 137–8
   classification 131, 134
   conservation value 138
   cut in Chalk 35
   retreat in 134–7
   see also cliffs
Solfach 58, **59**, 126–8
   cliffs, slope-over-wall 126
   ria and infilled Gwada valley 126, 127–8, *127*
solution piping
   Redend Point 344
   Short Cliff 175
solutional weathering 40
   South Pembrokeshire platform 121
   Tarbat Ness 91, 93
   see also erosion
Solway Firth (North Shore) *517*, 539, 541–8
   Barnkirk Point 544–5
   Caerlaverock Merse/saltmarsh 540, *542*, 543, **545**, 546–7
   Kirkconell Merse 543, *544*, 545, 546
   Nith valley 544, 547
   Priestside Merse *542*, 543–4, 545
   saltmarshes 543, 544, 545–6, 547–8, *548*
Solway Firth Saltmarshes **528**, 539–41, *542*
   Auchencairn Bay 545
   characteristic features **540**
   elevation of pioneer marsh 541
   emerged mudflats and saltmarsh 541
   micro-cliffs 522

Moricambe Bay 540
phases of erosion and deposition 546
rates of vertical accretion 521
sediment trap 540–1
Southerness Point 540, *542*
Southwick section 545
see also Cree Estuary, Outer Solway Firth; Upper Solway Flats and Marshes (South Shore)
South Haven Peninsula, Dorset 217, *330*, **331**, **332**, **333**, 335, 340–5, *420*
   foredune–beach–shallow water system 344
   Little Sea *341*, 342
   Poole Harbour 343
   rate of soil evolution 223
   Redend Point *341*, 341, 344
      dune–rock interface 346
   extent of dunes 344
   sandy beach/dune ridges 217, 340, 341–3, 344, 345
   sediment sources 345
   Shell Bay *341*, 340, 343, 344
   small headlands, Poole Harbour 342
   see also Dorset Coast
South Pembroke Cliffs, Pembrokeshire 58, **59**, 117–21, *118*
   arch and stack development, Green Bridge of Wales 119, *120*, 121
   blowholes, Devil's Barn 119
   Bullslaughter Syncline 119
   caves and arches 119
   cliffs 118, 119, *119*
   Devensian periglaciation 121
   Elegug Stacks 118
   Flimston Castles, Devil's Cauldron 119
   Flimston 'coastal flats', marine erosion surface 118, *118*
   geos
      Huntman's Geo and Stemmis Ford *118*, 119
Special Areas of Conservation (SACs) 30
   candidate and possible sites **237**, **336**, **478**

733

Index

coastal gravel/shingle site
  selection rationale 236–8
dune site selection rationale 337
The Fleet 258
machair 474
  selection rationale 478
North Norfolk Coast 624
saltmarshes 529
  selection rationale 529
sea cliff sites, selection rationale 53
Spey Bay, Moray 203, 207, 290–7
  beach replenishment 227, 294
  Boar's Head Rock 292
  to Burghead Bay, gravel spits 218, 292–3, 295
  coastal recession 292, 293, 294
  contemporary gravel beach 291, 292, 293
  emerged gravel ridge strandplain 290, 291, *291*, *294*, 295
  gravels 293, 295–6
  sandy beach and dunes, Lossiemouth 292, 293, 296
  sediment budget for 295–7, *296*
  Spey Mouth 293–4, *295*
    fluvial–coastal interaction 290, 297
  spit extension 231
spits **203**, 217–19, 231
  in areas of low tidal range 3, *6–7*, 10
  breaching 419, 429–30, *431*
  Central Sanday 464–9
  development of recurves 232
  Dawlish Warren 435–9
  East Head (Chichester Harbour) 426–30
  Gibraltar Point 440–1
  Isles of Scilly 461–4
  Morfa Dyffryn 453
  Morfa Harlech 449, 451–2
  multiple *232*
  with ness *232*
  paired 219, *232*, 278, 280, 427
  Pwll-ddu 422
  with recurves *232*
  St Ninian's Tombolo 457–61
  Spurn Head 430–5
  and tombolos 3, *6–7*
  Walney Island 443–5
  Winterton Ness 446–9
  Ynyslas 424, 425
  see Chapter 8
Spurn Head, Yorkshire 28, *211*, 223, 329, **418**, *420*, 523
  Chalk Banks 431, *431*, 433
  cliffed area of Skipsea Till 430, 433
  foreshore (Humberside) 430
  Holderness cliffs 207, 331, 434, *434*
  Holocene sea level rise 433
  Kilnsea Warren 430
  Old Den 430, *431*, 433, 434
  ords 197, 434
  recent retreat 434
  spit
    1953 storm surge 433
    breaching 431, 433–4, 435
    changes in coast alignment 430
    cyclic evolution 419, 430, *431*, 433, 435
    morainic foundation for 421
  wave and tidal processes 434
  see also Holderness
SSSIs *see* Sites of Special Scientific Interest (SSSIs)
Stac an Armin *61*, *64*
  Britain's highest stack, St Kilda 60
  plunging cliffs 65
stacks 35, 57, 111
  and arches, Yesnaby 84, *85*
  controlled by faulting 76
  development of 43, *43*
    South Pembroke Cliffs 119, *120*
  dip-controlled, Gaada, Sheepie and The Brough, Foula *78*, *79*
  Duncansby Head 86, 87, *88*, 89, 90, *90*
  emerged, Tarbat Ness 90
  Handfast Point, role 176–80
  and jointing, Ladram Bay 138–9, *139*, 140, *140*
  Marsden Bay 339
  Old Man of Hoy 35, 43, 51, 80, 81, 83, 84
  Papa Stour 73, 74
  St Kilda 60, 61, 62, *64*, 66
  Stac an Armin 60, *61*, *64*, 65
  triangular, Logat Stacks, Foula *78*, 79
  Upton and Gwithian Townas 346–9
Stac Lee 61
storm beaches 13, 111, 142, *205*
  Villians of Hamnavoe 68, 69
storm events, effects on saltmarshes 323, 537, 624
storm surges 440
  1953, effects at Scolt Head Island and Blakeney Point 214
  Ainsdale 363–4
  Benacre Ness 302
  Spurn Head 431
  and beach ridges on saltmarsh 217
storm waves 12
  and beach profiles 214
  St Kilda 66
  Villians of Hamnavoe 68, 71, 72
Strathbeg, Aberdeenshire *330*, **331**, *332*, *333*, 387–93
  beach 388, *389–91*, 393
  dune ridges 367, 388
  dunes 387, 388
  emerged gravel ridges 388
  evolution of the area 391, *391–2*, *391*, *392*
  Loch of Strathbeg 388, 388–9
    closure of 387, *387*, 391, 392, 393
  Rattray, rise and decline of 392, 393
  Rattray Head, spit growth changing direction 219
  relict cliff 388, *389–91*
  sand cliff 388
  stream outlet 388
  'winter lochs' 388
structural control 12, 41–2, 47, 71–2, 73, 90, 103, 104, 116, 250
  on cliff form, Kingsdown to Dover 165

734

Index

on cliffs and shore platforms 57
and formation of geos and stacks, Duncansby to Skirza Head 89–90
on west Orkney cliffs 81
South Pembroke Cliffs, caves, arches and geos 119
see also cliffs; shore platfoms
Studland Bay see Ballard Down
subaerial erosion
  Bullers of Buchan 104, 106–7
  St Kilda 66–7
submergence
  of previously glaciated areas 51, 57
    see also glacioisostasy; isostatic recovery; Loch Maddy–Sound of Harris;
surges 150
  and negative surges 12
  coastal hazard 20
  see also storm surges
Swanage Bay 177
  see also Ballard Down

tafoni 43, 57, 91
  see also erosion
Tarbat Ness, Ross and Cromarty 58, 59, 90–4
  emergent gravel beaches and platforms 90
  general morphology 92, 93–4
  Old Red Sandstone 90, 91, 92
  relict landforms 91, 93
  shore platforms 91, 92, 93, 94
  south-east coast
    overhanging cliffs 91
    remnants of higher shore platforms, Craig Ruadh 91, 92
  strike-aligned 91
  Wilkhaven, relict cliff and stack 93–4
Tentsmuir, Fife *330*, **331**, *332*, **333**, 407–13, **418**
  Abertay Sands *401*, 407, 408, *409*, *410*
  afforestation 408
  coastal progradation
    mechanisms 408, 410
    rates and form of 410–11, *412*
  dune ridges and slacks 408
  evolution of parallel dunes 223–4
  long-term accretion 407, 408, 411
  morphology linked to Abertay Sands 408, *409*
  sand sources 208, 411, *412*
  Tentsmuir Point 408, 410, 412–13
  see also Barry Links
Thanet, coastal platforms 46–7
tidal streams 11, 276, 540
tied islands see Scilly, Isles of; tombolos
till cliffs
  see glacial till
Tintagel, Cornwall 58, **59**, 113–17, *116*, *117*, 207
  Bossiney Haven, Elephant Rock, 115, *115*
  Castle Fault and Caves Fault Zone 115
  cliffs
    development of by undercutting 116
    slope-over-wall features 114, *114*, 115, 116
  landforms controlled by structure and rock strength 114
  sandy beaches, Trebarwith Strand and Bossiney Haven 114
  Start Point to Dennis Point, Gull Rock stack 114
  Tintagel Island, to north a more complex coastline 114
  Trebarwith Strand 115
  West Cove–Tintagel–Bossiney Haven
    slope-over-wall forms, bevelled 115
    Willapark, hogback cliff 115
tombolos 203, *232*, **461**, 493
  Central Sanday 464–9
  gravel
    Ayres of Swinister *281*, 282–3
    Pabbay 497
  and spits 3, *6–7*
  stages of evolution seen in Scilly Isles 219, 462
  see also Central Sanday, Orkney; Chesil Beach; St Ninian's Tombolo
toppling 150–1, 161
  in Chalk 136, 174, 614
  in Torridonian sandstone cliffs 36, 370
  see also mass movements
Torrisdale Bay and Invernaver, Sutherland *330*, **331**, *332*, 375–80, **418**, *420*
  Borgie terrace 375, 377, 379
  Druim Chuibhe, glacially scoured bedrock ridge 375, 379
    climbing dunes 378, 379
    dune heathland 379
  main dune ridge 378
  Naver terraces *377*, 379
    archaeological importance 375, 377
  outwash terraces, Borgie and Naver valleys 379
  sand bars, across mouths of Borgie and Naver rivers 377
  Torrisdale Bay 377, *378*, 379
Tràigh Mhór, Barra, beach of inwashed cockle shells 208
Tràigh na Berie, Lewis, Western Isles *332*, *473*, *475*, **477**, 506–10
  coastal edge 506, 507, 508, *508*, 510
  dunes *507*, 508, *509*
  hill machair developed

inland 508
  machair plain *507*, 508, 509
  sandy beach 506, *507*, 509
  shelter 506, *508*
  Tràigh Teinish 506, *506*, 508
tsunamis 243
  long-distance effects of 12
Tywyn Aberffraw, Anglesey *330*, **331**, *332*, **333**, *335*, 356–9
  bay-head beach 358
  beach 356, 357, 358
  cliffs 356, 357
  dunes 222, 357, 358–9
  fetch 358
  key features and profile *357*
  sand plain and fixed dunes 357
  see also Morfa Dinnlle; Dinas Dinlle; Newborough Warren

Upper Solway Flats and Marshes (South Shore) *517*, 539, 548–52
  Burgh Marsh **540**, *542*, 548–9
  creek systems 548
  erosion and accretion 551–2, **551**
  Grune Point 542, 549
  marsh terraces 548, 549, 551
  Moricambe Bay 549, 551
    includes Newton Marsh **540**, *542*, 548
  Rockcliffe Marsh **540**, *542*, 549, *550*, 551, **551**
  saltpans 548, 549
  sea-level change 551, 552
  Skinburness Marsh *542*, 548, 549, 551, **551**
Upton and Gwithian Towns, Cornwall 58, **59**, *330*, **331**, **333**, 345–8
  beach erosion 226
  coastline processes 347–8
  dunes 225, 345, 346, 348
  erosion of rocky coast 346
  exhumed cliff 346, 348
  Peter's Point to northern boundary, coves, stacks and caves 345, *346*, *347*

Ventnor–Shanklin, Isle of Wight, landsliding 5
Villians of Hamnavoe, Shetland 58, **59**, 68–72
  Atlantic storm waves 68, 69, 71
  bedrock, wave-scoured with imbricate boulder clusters 68, 70, 72
  Burn of Tingon, staircase waterfall 70, 72
  Eshaness, cliff-top boulder deposits 70–1, *71*
  Grind of the Navir, cliff-top boulder ridges 70–1, *71*
  high altitude abrasion and scour features 69, 71
  Hole of Geuda (blowhole) 70, 71, 72
  South Gill *68*, 70
  South Head *68*, 70–2

Walney Island, Lancashire 221, **332**, **418**, *420*, 443–45, 523
  barrier island 445
  Haws Point spit, twentieth century changes 444
  Hillock Whins to Haws Hole 443
  Hillock Whins to South End 443
    shingle and boulder beach 443
    shoreline formed as series of spits and tombolos 444
  North End Haws, relationship to Sandscale Haws *444*, 444
  'scars' 444, 445
  offshore sediment sources 445
  spits 444, 445
washover fans 12, 241, 600
wave climates 22
wave energy 10, *10*, *11*, 12, 22, 204
wave erosion
  by breaking waves 39, *39*
  Flamborough Head 182
  Hornish and Lingay Strands, machair erosion 495
  mechanical abrasion 39–40
  erosion rate and rock structure 41
  quarrying and erosion 39–40
  wave quarrying 38–9
wave quarrying 38–9, 144
  and cave excavation 42
  Hole of Geuda 71, 72
  joints and fractures 39
  St Kilda 62, 66
  secondary processes 38
  Tarbat Ness 93
  see also differential erosion; erosion; structural control
wave spray processes 40
  Tarbat Ness 90, 93
Weak-rock cliffs *see* soft/weak-rock cliffs
weathering
  microforms
    Tarbat Ness 90, *91*, 93
  subaerial 40, 57, 175–6
  and mass movements 36
  see also erosion
Westward Ho! cobble beach, Devon 223, *233*, **234**, 238–41, **332**, *349*
  artificial beach feeding 238
  coastal protection works 238
  effects of Holocene transgression 239
  Grey Sand Hill 238
  intertidal area 238
  lateral grading of sediments 215
  Northam Burrows 238, 239
  peat and blue clays 238–9
  relict dunes 225
  similarities with Budleigh Salterton beach 240
  Taw–Torridge estuary alignment 240
Whiteness Head, Moray *233*, **235**, 285–90
  alignment of coastal features 288
    reduction in sediment supply 289–90
  dune areas 287
  gravel-ridged spit 286–7
  historical evolution/change 285, 287, *287*, *288*, 289, 290
  recurves 218, 285, *286*
  longshore sediment supply

Index

288, 289–90
saltmarsh 286, *286,* 288
land-claim 290
sediment transport, longshore extension 285–6
wind-stripping, of soil and turf
Bullers of Buchan 104
St Kilda 62, 66
Villians of Hamnavoe 70, 71–2
west Orkney 83
Winterton Ness, Norfolk *330,* **333, 418,** *420,* 446–9
beach, sand and shingle 447
dunes 223–4, 446, 449
effects of 1953 floods 447
northward migration *446,* 447

sediment transfers 447, *448,* 448
Shoreline Management Plan 447
southward migration 219, *446*
Worbarrow Bay *see* Dorset Coast
World Heritage Convention and Sites 29

Yorkshire 141
cliff retreat rates **143**
Jurassic cliffs and shore platforms 47–8
see also Flamborough Head; Holderness; Robin Hood's Bay; Spurn Head
Ynyslas, Ceredigion 203, 223, *233,* **234,** *330,* **332, 333,** *420,* 424–6
Borth Bog 425
intertidal banks 425
sand movement 425
sandflats 424
shingle ridge 425
South Banks 425
spit 424, 425
submerged forest and peat 425

zeta-curve bays, Start Bay 244
zeta-curve beaches 191, **203,** 212, 251, 252, 596